ADVANCES IN ASSISTED
REPRODUCTIVE TECHNOLOGIES

ADVANCES IN ASSISTED
REPRODUCTIVE TECHNOLOGIES

Edited by

Shlomo Mashiach, M.D., Zion Ben-Rafael, M.D., Neri Laufer, M.D.,
and
Joseph G. Schenker, M.D.

PLENUM PRESS • NEW YORK AND LONDON

Library of Congress Cataloging-in-Publication Data

World Congress on In Vitro Fertilization and Alternate Assisted
   Reproduction (6th : 1989 : Jerusalem)
   Advances in assisted reproductive technologies / edited by Shlomo
Mashiach ... [et al.].
        p.   cm.
   "Proceedings of the Sixth World Congress on In Vitro Fertilization
and Alternate Assisted Reproduction, held April 2-7, 1989, in
Jerusalem, Israel"--T.p. verso.
   Includes bibliographical references.
   Includes index.
   ISBN-13: 978-1-4612-7907-5   e-ISBN-13: 978-1-4613-0645-0
   DOI: 10.1007/978-1-4613-0645-0
   1. Fertilization in vitro, Human--Congresses. 2. Human
reproductive technology--Congresses.   I. Mashiach, Shlomo.
II. Title.
   [DNLM: 1. Fertilization in Vitro--congresses. 2. Reproduction
Technics--congresses.   WQ 205 W9275a 1989]
RG135.W68   1989
618.1'78059--dc20
DNLM/DLC
for Library of Congress   90-7852
                                                              CIP

Editorial Assistance:
Shaina Henigman, Tel Aviv, Israel

Congress organized by:
KENES, Organizers of Congresses and Special Events, Ltd.
Tel Aviv, Israel

Proceedings of the Sixth World Congress on In Vitro Fertilization and
Alternate Assisted Reproduction, held April 2-7, 1989, in Jerusalem, Israel

© 1990 Plenum Press, New York
Softcover reprint of the hardcover 1st edition 1990
A Division of Plenum Publishing Corporation
233 Spring Street, New York, N.Y. 10013·

PREFACE

The World Congress of In Vitro Fertilization and Alternate Assisted Reproduction, held in Jerusalem, Israel, 2-7 April, 1989, was the sixth in the sequence of these Congresses, but was the first to emphasize the major importance and the place of assisted reproductive technologies in the treatment of infertility.

The eternal City of Jerusalem witnessed the gathering of more than 1500 participants from all over the world who shared and exchanged knowledge and up-to-date experience in this ever-evolving field.

The high quality scientific contributions to the Congress culminated in the publication of this Proceedings. It embraces all-important aspects in the field of in vitro fertilization and alternate assisted reproduction.

Papers on controversies and diversities of methods to stimulate the ovaries, imaging techniques, basic research and state-of-the-art papers on ovarian physiology, the role of GnRH and its analog, clinical aspects of IVF treatment and cryopreservation, up-to-date techniques in assisted reproductive technologies that are quickly developing in conjunction with IVF, were included. When should IVF be preferable to surgery? What are the expected up-to-date world results and what are the psychological, moral, ethical and religious implications? These are all the concerns of the treating team and are addressed here. Male factor infertility remains a frustrating problem, but advances in the understanding of sperm-egg interaction, sperm evaluation and preparation are reported. Micromanipulation emerges as a possible alternative to bring some relief to this problem, but it also promises to be central in promoting the field of prenatal genetic analysis. Both these breakthrough methods are maturing as fields which require the coordination of many sub-specialities. In fact, IVF is unique in the field of medicine in merging the efforts and skills of the clinicians, biologists, scientists, geneticists, endocrinologists, embryologists, psychologists and paramedical groups.

Chapters on implantation, embryo assessment and improvements of laboratory techniques, subjects which are crucial to the success of the entire procedure, and chapters on gamete donation and surrogacy which broaden the treatment to previously sterile patients, are also included.

This book is of interest to the teams which are just starting IVF, as well as to established groups who want to diversify and broaden their capabilities and knowledge.

We are confident that this book will provide a useful, up-to-date scientific and clinical guide for future academic and clinical investigation.

S. Mashiach, M.D., Z. Ben-Rafael, M.D., N. Laufer, M.D., and J. G. Schenker, M.D.

ACKNOWLEDGMENT

The publication of this book, and the organization of the VIth World Congress of In Vitro Fertilization and Alternate Assisted Reproduction which was held in Jerusalem, Israel, 2-7 April, 1989, brought much satisfaction to all those involved.

We feel, however, that our special and grateful thanks should go to Zion Ben-Rafael, M.D., whom we all met as Secretary of the Congress, and who is a co-editor of this book. He was the guiding force of the book, and his energy and enthusiasm in the preparation hereof and in bringing it to its fruition is highly appreciated by all his colleagues.

The Editors

CONTENTS

Chapter 3. OVARIAN PHYSIOLOGY

Chapter 4. GnRH AND GnRH ANALOGS: BASIC ASPECTS

Chapter 5. ALTERNATE ASSISTED REPRODUCTION

Chapter 6. IVF AND MICROSURGERY IN CASES OF MECHANICAL INFERTILITY

Chapter 7. CLINICAL ASPECTS OF IVF TREATMENT:I

Chapter 8. MALE INFERTILITY

Chapter 9. NEW IMAGING TECHNIQUES AND REPRODUCTION

Chapter 10. GAMETE AND EMBRYO CRYOPRESERVATION

Chapter 11. GAMETE, EMBRYO DONATION AND SURROGACY

Chapter 12. IMPLANTATION

Chapter 13. LABORATORY TECHNIQUES

Chapter 14. ENHANCEMENT OF FERTILIZATION
BY MICROMANIPULATION

Chapter 15. GENETIC ANALYSIS OF THE PRE-EMBRYO

Chapter 16. PSYCHOSOCIAL ASPECTS OF IVF AND ALTERNATE ASSISTED REPRODUCTION

Chapter 17. WORLD COLLABORATIVE RESULTS AND OUTCOME OF IVF PREGNANCIES

Chapter 18. ETHICS, RELIGION AND LEGAL ISSUES

Chapter 19. EDITOR'S PERSPECTIVE

IS THERE A BEST STIMULATION PROTOCOL?

Howard W. Jones, Jr., Georgeanna S. Jones, Suheil J. Muasher, and Debbie Jones

Jones Institute for Reproductive Medicine, Department of Obstetrics and Gynecology
Eastern Virginia Medical School
Norfolk, Virginia, USA

INTRODUCTION

> *"And there sat Arthur on the dais throne.*
> *...'What is it?*
> *The phantom of a cup that comes and goes?'*
> *'Nay, what phantom?' answered Percival.*
> *The cup, the cup itself from which our Lord*
> *Drank at the last sad supper, and if a man*
> *Could touch or see it, he was healed at once.*
> *And now the Holy Thing is here again*
> *Among us; fast thou too and pray*
> *And tell thy brother knights to fast and pray*
> *That so perchance the vision may be seen*
> *By thee and those, and all the world be healed.' "*
>   *(From Idylls of the King)*

It is unfortunate that we do not have an Alfred Tennyson to describe our efforts, but we fancy ourselves modern knights as we too seek a Holy Grail in the form of a universal stimulation protocol which can apply to all patients and which can produce many fertilizable eggs and a pregnancy and term delivery rate better than natural fecundity. Does such a holy stimulation exist? The main purpose of this exercise is to examine that question in the context of stimulation with gonadotropins, the topic assigned to me by the organizing committee.

Advances in Assisted Reproductive Technologies, Edited by
S. Mashiach *et al.,* Plenum Press, New York, 1990

1

PATIENTS AND METHODS

From 1 January 1981-31 December 1988 (Norfolk series 1-33) in the Norfolk program, 3,689 attempts at stimulation were carried out with all types of gonadotropic stimulation in 1,637 patients. For the following analyses, various subsets of this patient material were used.

The method of stimulating and monitoring the patients has been reported.[1] The technique of follicular aspiration was first by laparoscopy, with a switch to transvaginal ultrasound-guided aspiration during the latter part of 1986. There was no demonstrable change in the number of mature eggs or in the pregnancy rates accompanying this switch. The method of culturing and diagnosing the maturity of the oocyte[2] and the method of transfer[3] have remained constant.

For this analysis no cryopreservation data are included. These data would increase the pregnancy rates but would be incomplete, since many cryopreserved pre-zygotes remain to be transferred.

For the purposes of this discussion, the word *attempt* designates the admission of a patient into a cycle of stimulation. The word *cycle* (harvest) applies to the procedure by which the eggs are obtained (laparoscopic aspiration or ultrasound-guided transvaginal aspiration). The word transfer refers to the *transfer* of at least one fertilized and cleaving egg which, on aspiration, was in either metaphase II or late metaphase I. Thus, the cancellation rate is the percentage of the number of patients who failed to reach harvest after an attempt at stimulation.

The Stimulation Regimes

Five stimulation regimes will be compared:

1. 2 hMG: Two ampules of hMG (Pergonal; Serono) were given beginning on day three of the menstrual cycle, and continued until a satisfactory response was obtained.
2. Combo I: Two ampules of hMG were given beginning on day three of the menstrual cycle, supplemented by two ampules of FSH on days three and four, until a satisfactory response was obtained.
3. Combo II: Two ampules of hMG were given beginning on day three of the menstrual cycle, supplemented by two ampules of FSH on days three and four, with further supplementation of FSH on subsequent days, until a satisfactory response was obtained.
4. 4-4 FSH: Two ampules of FSH were given beginning on day three of the menstrual cycle, supplemented by two ampules of FSH on days three and four, until a satisfactory response was obtained.

5. Other FSH: Two ampules of FSH were given beginning on day three of the menstural cycle, supplemented by other FSH beginning on day three, until a satisfactory response was obtained.

IS THERE A BEST BASIC STIMULATION REGIME?

Several previous studies from Norfolk have addressed this question.[4-6] For the purposes of this particular study, cases from series 1-27 (1 January 1981-30 June 1987) were analyzed according to the five stimulation protocols. Series 28-33 (1 July 1987-31 December 1988) were *not* included because subsequent to 1 July 1987, a heavy bias was introduced in the selection of stimulation protocols, according to a method to be described. Ideally, the cases for such a study should be randomized. This was not entirely possible. Some bias in case assignment to protocol arose because new patients were sometimes arbitrarily assigned to a particular protocol for trial, or repeat stimulations were sometimes shifted to an alternate protocol because of a poor previous response. However, to diminish bias as much as possible, groups with a high percentage of poor responders and couples with a male factor are stated but not analyzed by stimulation protocol because of the limited numbers. The suspected poor responders comprised two groups: those ≥ 40 years of age (n = 133) and those who failed to develop at least four follicles (n = 332). The males comprised a little more than 8% of the material (n = 189). This left 1,546 cycles for analysis (Table 1). There were no statistical differences among any of the protocols except that Combo I was significantly better than two hMG.

However, there was a clear trend for Combo I to be better than any of the other protocols. This trend was reinforced when multiple births were considered. The two hMG regime yielded 1.08 babies per birth, while Combo I yielded 1.35, suggesting that Combo I recruited more biologically competent eggs than the other protocols (Table 2).

VARIATIONS IN PATIENT RESPONSE

It was early recognized that there was a substantial difference from patient to patient in response to the same gonadotropin stimulation. Early on, patients were early classified into low, normal, and high responders,[7] and the pregnancy rate was enhanced with increase in response.[7] Shifting from one stimulation protocol to another in an attempt to improve response accounts for some of the bias in the material comprising Table 1. For poor responders it was often necessary to spend one stimulation cycle to realize this.

In an attempt to predict a patient's response pattern before stimulation, a number of tests designed to identify the patient's underlying hormonal milieu were carried out on 80 consecutive new patients in early 1987. All such patients were stimulated by Combo I.[8] While other tests were performed, the basal FSH and LH values on day three of the normal menstrual cycle proved as predictive as other more complicated tests. With the FSH and

Table 1. Norfolk Series 1–27 (1 January 1981–30 June 1987) ≥ 4 Follicles, No Males, Patients < 40 Years of Age

STIMU- LATION	CYCLES	TRANS- FERS	PREG- NANCIES	DELIVERIES + 20 WEEKS	TOTAL BABIES	% PREG NANT/ CYCLE	% PREG- NANT/ TRANS- FER	% DELIVERED/ TRANSFER	% BABIES/ TRANSFER
2 hMG	367	276	69	51	55	19 [a]	25 [b]	18.5 [b]	19.9
Combo I	537	483	156	94	127	29 [a]	32 [b]	19.5	26.3
Combo II	236	205	55	33	42	23	27	16.1	20.5
4-4 FSH	240	210	60	35	45	25	29	16.7	21.4
Other FSH	166	143	55	24	27	21	25	16.8	18.9

[a] $p = .0008$

[b] $p = .0508$

Table 2. Norfolk Series 1–27 (1 January 1981–30 June 1987) Multiple Babies at Delivery by Stimulation

STIMULATION	DELIVERIES	BABIES	% BABIES/DELIVERY
2 hMG	51	55	1.08
Combo I	94	127	1.35
Combo II	33	42	1.27
4-4 FSH	35	45	1.29
Other FSH	24	27	1.13

LH values, paired discriminant analysis identified seven distinct groups of patients with statistical differences among the means. Serum estradiol (E_2) response during stimulation, as well as the number of preovulatory oocytes aspirated and transferred, was highest in the groups with the lowest mean FSH/LH ratio, intermediate in the groups with FSH/LH ratios approximating unity, and lowest in the groups with a higher mean FSH/LH ratio. The pregnancy rates were greatly compromised in those groups with high FSH/LH ratios, and the continuing pregnancy rates were also compromised in the groups with very low FSH/LH ratios.

The preliminary study has been expanded for the purposes of this presentation. All new patients stimulated in 1987-1988 were subject to a similar but retrospective analysis, modified to include FSH/LH values which fell outside the limits of the seven groups identified by the original study. These outlying values were assigned to one of the designated original seven groups by a computer-generated table of probabilities relative to the ectopic values. This arbitrary assignment resulted in some expansion of the LH and FSH limits of each group compared to the original report. There were 498 new patient attempts at stimulation with Combo I, Combo II, 4-4 FSH, or other FSH. Of these, there were 232 with Combo I stimulation, the stimulation used in the 80 patients originally studied. The findings among patients stimulated by all protocols (n = 498) were essentially the same as among patients stimulated by Combo I (n = 232), as in the original study.

Considering the large group of 498 patients stimulated by all protocols, the pregnancy rates confirm, with minor changes, the results of the original study. Pregnancies were infrequent amont patients with high basal FSH values or high basal LH values. The best pregnancy rates seemed to occur when the FSH/LH ratio did not deviate too far from unity (Tables 3–5).

The pregnancy rates among patients with elevated basal FSH levels were quite unsatisfactory (groups 1 and 2), as they were among patients with rising LH levels (group 7). It is

Table 3. Norfolk Series 26–33 (1987–1988) (n = 498. Pregnancy Rates in the Seven Basal FSH/ LH Groups. New Patients Stimulated with Combo I, Combo II, 4-4 FSH, or Other FSH

GROUP	N	FSH	LH	FSH/LH	AGE
1	8	46.6 ± 7.5	26.0 ± 7.2	1.7923	37.3 ± 3.2
2	25	26.9 ± 6.8	17.3 ± 5.2	1.5549	36.7 ± 3.2
3	80	17.2 ± 2.9	12.8 ± 2.6	1.3438	35.3 ± 3.4
4	143	8.5 ± 2.2	8.8 ± 2.3	.9659	33.6 ± 3.7
5	66	13.3 ± 3.1	20.1 ± 3.5	.6617	33.5 ± 3.3
6	136	8.4 ± 2.5	14.4 ± 3.1	.5833	33.6 ± 3.9
7	10	11.3 ± 2.8	41.4 ± 18.9	.2729	33.0 ± 2.3

probably significant that there was over four years' difference in the mean age of patients as the basal FSH values rose. However, the level of basal LH seemed less related to age (Table 3).

The problems of stimulation were reflected in the cancellation rate, where there was a 63%, 32%, and 20% rate in groups 1, 2, and 7, respectively. This contrasted with a low of 5% in group 4 and an overall rate of 56/498 or 11.2% for the entire group (Table 4).

The egg data reflected a relative sparsity of eggs in the high FSH patients (groups 1 and 2). This probably accounted for the lower pregnancy rate in these groups. However, larger numbers of eggs were available in group 7, where the pregnancy rate was also low. This suggests either that the relatively large number of eggs harvested in the high LH group (group 7) were biologically inferior or that the luteal phase environment was inappropriate (Table 5.)

Considering the smaller group of 254 patients stimulated only by Combo I, the results are essentially the same except that the pregnancy rate in group 6 is also poor with this stimulation regime (Table 6).

The egg data, especially in group 6, again show adequate numbers of eggs. In the face of a relatively poor pregnancy rate, the egg quality and/or poor luteal environment becomes suspect as with the larger group of 498 stimulated by all protocols (Table 7).

Table 4. Norfolk Series 26–33 (1987–1988) (n = 498). Pregnancy Rates in the Seven Basal FSH/LH Groups. New Patients Stimulated with Combo I, Combo II, 4-4 FSH, or Other FSH.

GROUP	NO. OF ATTEMPTS	NO. OF CYCLES	NO. OF TRANSFERS	CANCEL. RATE %	NO. OF PREG.	DELIV. + >20 WKS	% PREG./ TRANSFER	% DELIV. + >20 WKS/TR
1	8	3	2	63	—	—	—	—
2	25	17	14	32	1	—	7.1	—
3	80	71	64	11	18	13	28.1	20.3
4	143	136	128	5	42	28	32.8	21.9
5	66	58	48	12	18	13	35.4	27.1
6	136	128	109	6	31	23	27.5	21.1
7	10	8	6	20	—	—	—	—

Table 5. Egg Data for 498 New Patients Stimulated by Combo I, Combo II, 4-4 FSH, or Other FSH

GROUP	EGGS RECOVERED PER CYCLE	EGGS INSEMINATED + FERTILIZED PER CYCLE	EGGS TRANSFERRED PER TRANSFER CYCLE
1	3.00	2.67	1.67
2	2.50	2.37	1.88
3	3.66	2.91	2.23
4	4.94	3.87	2.77
5	3.98	3.44	2.40
6	5.05	3.98	2.57
7	5.29	4.71	2.75

EFFECT OF GnRHa ON RESPONSE BY GROUP ACCORDING TO FSH/LH RATIO

In order to ascertain the effect of down regulation by GnRHa on response to various stimulation protocols by FSH/LH basal ratio, 157 attempts on 129 patients were analyzed. Leuprolide (Lupron; Tap Pharmaceuticals, Chicago, IL), 1 mg a day subcutaneously, was administered to selected patients beginning on day 20 to 23 of the previous cycle and continued until the gonadotropin stimulation was discontinued. The gonadotropin stimulation was started on day three as usual. In a few patients the expected menses did not appear, in which case the gonadotropin stimulation was started on theoretical day three, if the E_2 level was suppressed to < 30 mg/ml.

Many of the patients were assigned to leuprolide because they had previously been stimulated without pregnancy. However, several patients in groups 6 and 7 had been determined to have low FSH/LH ratios prior to any stimulation and therefore were placed on leuprolide during their first stimulation cycle. While the numbers are too few to draw any conclusions which are more than tentative, some trends on the value of leuprolide seem evident (Table 8).

In patients with very high FSH/LH basal ratios, leuprolide is not helpful (group 1). Its chief use may be in patients with relatively high LH values, those with a low FSH/LH ratio (group 6), where a pregnancy rate of 39.1% was obtained in contrast to 27.5% and 25.5% in the all-stimulation group and the combo group, respectively. However, the leuprolide group had a high miscarriage rate of 7/18 or 39%. Thus the term *delivery rate* for the leuprolide cases was 23.9%, compared with 21.1% for the all-stimulation group and 14.9% for the Combo I group for group 6 patients.

Table 6. Norfolk Series 26–33 (1987–1988) (n = 232). Pregnancy Rates in the Seven Basal FSH/LH Groups. New Patients Stimulated with Combo I begun on Day 3.

GROUP	NO. OF ATTEMPTS	NO. OF CYCLES	NO. OF TRANSFERS	CANCEL. RATE %	NO. OF PREG.	DELIV. + >20 WKS	% PREG./ TRANSFER	% DELIV. + >20 WKS
1	—	—	—	—	—	—	—	—
2	4	3	3	25	—	—	—	—
3	36	34	30	6	10	8	33.3	26.7
4	99	96	89	3	30	21	33.7	23.6
5	32	28	22	13	10	8	45.5	36.4
6	58	55	47	5	12	7	25.5	14.9
7	3	2	2	66	—	—	—	—

Table 7. Egg Data for 232 New Patients Stimulated by Combo I

GROUP	EGGS RECOVERED PER CYCLE	EGGS INSEMINATED + FERTILIZED PER CYCLE	EGGS TRANSFERRED PER TRANSFER CYCLE
1	--	--	--
2	4.00	4.00	3.67
3	4.03	3.00	2.50
4	4.88	3.81	2.91
5	4.16	3.40	3.23
6	5.10	4.12	3.10
7	1.00	1.00	--

Leuprolide, as used, did seem helpful if the FSH/LH ratio was extremely low, with very high LH values. For patients whose FSH/LH ratio approached unity, leuprolide did not seem to be helpful.

This point can be further studied by combining the data for groups 3-5 (more or less the old intermediate responder group).[7] For 85 leuprolide attempts (Table 8) the pregnancy rate and the term delivery rate were 30.3% and 21.1%, respectively, as compared with 35.5% and 26.2% for the Combo I group (Table 6) and 32.8% and 21.9% for the all-stimulation group (Tables 4 and 9).

Patients treated with leuprolide on average tend to have more pre-zygotes cryopreserved than non-treated patients. As indicated above, these data are not included. Thus the leuprolide patients will probably have more enhancement of their pregnancy rate than others.

There can be no doubt that stimulation after leuprolide requires a greater number of ampules of the gonadotropins (Table 10).

Furthermore, there seems to be a tendency for a slightly higher miscarriage rate in patients pretreated with leuprolide (Table 11).

The egg data seem to indicate that leuprolide-treated patients have more preovulatory eggs harvested and transferred than non-treated patients (Table 12). The finding that the term pregnancy rate does not reflect overall these large numbers of eggs again raises the question of the quality of these excessively recruited eggs or the quality of the luteal phase environment.

Table 8. Norfolk Series 26–33 (1987–1988) (n = 157). Pregnancy Rates in the Seven Basal FSH/LH Groups by All Stimulations Pretreated with Leuprolide

GROUP	NO. OF ATTEMPTS	NO. OF CYCLES	NO. OF TRANSFERS	CANCEL. RATE %	NO. OF PREG.	DELIV. + > 20 WKS	% PREG./ TRANSFER	% DELIV. + > 20 WKS
1	4	3	3	25	—	—	—	—
2	9	9	9	0	3	2	33.3	22.2
3	18	16	14	11	4	2	28.6	14.3
4	46	44	42	4	15	12	35.7	28.6
5	22	20	18	9	4	2	22.2	11.1
6	50	48	46	4	18	11	39.1	23.9
7	8	6	6	25	—	—	—	—

Table 9. Data on Intermediate Responders (Groups 3–5)

GROUP	NO. OF ATTEMPTS	NO. OF TRANSFERS	PREGNANCIES	DELIV. + > 20 WKS	% PREG./ TRANSFER	% DELIV. + > 20 WKS
All stim. (n=498)	289	240	42	28	32.8	21.9
Combo (n=232)	167	141	50	37	35.5	26.2
Leuprolide (n=157)	86	76	23	16	30.3	21.1

Table 10. Ampules Used with Various Stimulation Regimes

GROUP	ALL STIMULATIONS (n = 498)	COMBO I (n = 232)	LUPRON + ALL STIMULATIONS (n = 157)
1	20.3 ± 10.9	–	45.2 ± 23.2
2	23.8 ± 8.6	13.0 ± 3.4	39.5 ± 11.0
3	17.8 ± 5.6	14.7 ± 1.2	31.8 ± 13.8
4	15.5 ± 4.1	14.5 ± 1.7	25.9 ± 7.8
5	15.2 ± 4.6	13.7 ± 1.8	28.6 ± 10.1
6	14.9 ± 3.2	14.4 ± 1.5	24.3 ± 6.6
7	14.1 ± 3.4	13.3 ± 3.0	21.3 ± 7.1

Table 11. Abortion Rate by Stimulation Regime

GROUP	PREGNANCIES	ABORTIONS	% ABORTIONS/PREG
All stim. (n=498)	110	33	30.0
Combo (n=232)	62	18	29.0
Leuprolide (n=157)	44	15	34.1

EFFECT OF HIGH FSH EXOGENOUS THERAPY ON VERY LOW RESPONDERS

To determine whether very high doses of exogenous FSH could stimulate the ovary of patients who already had high basal FSH values, 85 patient cycles of groups 1–3 were treated with six ampules of FSH a day, beginning on day three, and continuing until it was judged that an optimum response had been obtained.

Table 12. Egg Data for 157 Patients Treated with Lupron Prior to Gonadotropin Stimulation

GROUP	EGGS RECOVERED PER CYCLE	EGGS INSEMINATED + FERTILIZED PER CYCLE	EGGS TRANSFERRED PER TRANSFER CYCLE
1	5.33	5.00	3.67
2	5.11	4.11	3.22
3	4.07	3.27	2.57
4	6.21	5.67	3.57
5	8.33	7.39	3.05
6	7.67	5.52	3.22
7	8.67	7.67	3.17

It may be that something can be accomplished by this approach (Table 13). The data are too sparse for more than a very tentative conclusion, but the trend is confirmed by the ability to recruit a few more eggs in the low response group (Table 14).

DISCUSSION

In view of the variation in response from patient to patient with the same gonadotropin stimulation, the search for a universal best stimulation is doomed to failure. On a physiological basis, this variation is not completely understood. However, advanced age and reductive ovarian surgery (data not shown) are two identifiable factors leading to low response. It is likely that other factors exist, as, for example, a predetermined limited supply of eggs or variation in the mechanism of the follicular apparatus for the acquisition of gonadotropin sensitivity, with low responders on one end of this spectrum and PCO-like patients on the other.

While the search should continue for improved ability to recruit more fertilizable eggs among patients with FSH/LH ratios which approximate unity, the real challenge is to provide a better outlook for those at each end of the FSH/LH ratio spectrum.

For low responders with a relatively high basal FSH, the use of even more exogenous FSH seems to be able to recruit an extra oocyte or two and thereby improve the pregnancy rate, which is quite unsatisfactory by "ordinary" stimuli. In our experience, leuprolide is clearly not helpful in this group of patients.

At the other end of the spectrum where the LH is relatively high, it seems to suggest that an occult PCO milieu is in place in spite of regular ovulation and menstruation. For

Table 13. Table 4. Norfolk Series 26–33 (1987–1988) (n = 85). Patients in Groups 1–3 Stimulated by Six Ampules of FSH per Day from Day 3

GROUP	NO. OF ATTEMPTS	NO. OF CYCLES	NO. OF TRANSFERS	CANCEL. RATE %	NO. OF PREG.	DELIV. + >20 WKS	% PREG./ TRANSFER	% DELIV. + >20 WKS
1	10	6	6	40.0	1	1	16.7	16.7
2	37	29	21	21.6	3	2	14.3	9.5
3	38	35	28	7.9	8	6	28.6	21.4

Table 14. Egg Data from 85 Patient Cycles Stimulated by six Ampules of FSH Beginning on Day 3

GROUP	EGGS RECOVERED PER CYCLE	EGGS INSEMINATED + FERTILIZED PER CYCLE	EGGS TRANSFERRED PER TRANSFER CYCLE
1	4.50	4.00	3.50
2	1.93	1.70	1.86
3	3.03	2.68	2.54

this type of patient, pre-stimulation treatment with leuprolide seems to be very helpful. However, if the FSH/LH ratio is excessively low—in the neighborhood of ± .27—due to a high LH, leuprolide, as used in this study, has not proved to be helpful.

For patients with an FSH/LH ratio approaching unity, our studies to date do not show that leuprolide is helpful over a standard Combo I stimulation.

The data presented in this paper, as well as studies by other groups, are consistent with the notion that there is no best universal stimulation protocol. Rather, best results are more likely if individual biological characteristics are considered in utilizing a stimulation protocol.

REFERENCES

1. G. S. Jones, The use of human menopausal gonadotropin for ovulation stimulation in patients for in vitro fertilization, *Infertility*, 6:11 (1983).
2. L. L. Veeck, Oocyte assessment and biological performance, *Ann. NY. Acad. Scie.*, 541:259 (1988).
3. H. W. Jones, Jr., A. A. Acosta, J. E. Garcia, B. A. Sandow, and L. Veeck, On the transfer of conceptuses from oocytes fertilized in vitro, *Fertil. Steril.*, 39:241 (1983).
4. H. W. Jones, Jr., Oocyte recruitment with human menopausal gonadotropin (hMG) and follicle stimulating hormone (FSH), *in*: "The Control of Follicle Development, Ovulation and Luteal Function: Lesions from In Vitro Fertilization," F. Naftolin and A. H. De Cherney, eds., Raven Press, New York (1987).
5. H. W. Jones, Jr., Ovarian stimulation regimes with hMG and FSH for oocyte recovery, *in*: "In Vitro Fertilization and Other Alternative Methods of Conception," S. S. Ratnam et al, eds., Parthenon, Park Ridge, N. J. (1987).
6. D. Navot and Z. Rosenwaks, The use of follicle-stimulating hormone for controlled ovarian hyperstimulation in in vitro fertilization, *J. In Vitro Fert. Embryo Trans.*, 5:3 (1988).
7. A. P. Ferraretti, J. E. Garcia, A. A. Acosta, and G. S. Jones, Serum luteinizing hormone during ovulation induction with human menopausal gonadotropin for in vitro fertilization in normally menstruating women, *Fertil. Steril.*, 40:742 (1983).
8. S. J. Muasher, S. Oehninger, S. Simonetti, J. Matta, L. M. Ellis, H.- C. Liu, G. S. Jones, and Z. Rosenwaks, The value of basal and/or stimulated serum gonadotropin levels in prediction of stimulation response and in vitro fertilization outcome, *Fertil. Steril.*, 50:298 (1988).

LIMITATIONS IN THE USE OF COMBINED
GONADOTROPIN RELEASING HORMONE ANALOG AND
HUMAN MENOPAUSAL GONADOTROPIN FOR IN VITRO FERTILIZATION

Z. Ben-Rafael,[a] Y. Menashe,[a] R. Mimon,[a] S. Lipitz,[a] D. Bider,[a]
M. Zolti,[a] U. Dan,[a] C. Pariente,[b] J. Shalev,[a] J. Dor,[a] D. Levran,[a] and S. Mashiach[a]

[a] Department of Obstetrics and Gynecology, [b] Institute of Endocrinology
The Chaim Sheba Medical Center and Sackler School of Medicine
Tel-Aviv University, Tel- Hashomer 52621 Israel

Ovarian stimulation is an important step in in vitro fertilization (IVF) therapy which has had a profound impact upon the success or failure of the whole procedure. It is now recognised that no single ovarian stimulation agent or dosage is optimal for all patients and that stimulation should be tailored and adjusted to best suit the patient's own hormonal status. However, as with every stimulation protocol, 20% to 40% of the patients are cancelled before ovum pick-up mainly due to premature Luteinizing Hormone (LH) surge.

Down regulation of pituitary function by gonadotropin releasing hormone analog (GnRH-a) before and during ovarian stimulation with human menopausal gonadotropin (hMG) can effectively decrease cancellation, hence it is not surprising that this drug has quickly gained wide popularity and is being used frequently for routine IVF. While there is no doubt that the combination of GnRH-a and hMG can significantly reduce cancellations before ovum pick-up through the prevention of untimely LH discharge (Fleming et al. 1985, Fleming et al. 1986), it is doubtful whether this combined regimen has other advantages. Treatment with GnRH-a is more expensive and it is time consuming (Mashiach et al. 1988) This treatment sometimes fails to improve ovarian response and is associated with complications, and pathological responses.

In this study we present our experience with the routine use of GnRH-a/hMG combination. Considering that most of the patients respond favourably to treatment with hMG only, we question the wisdom of routine use of GnRH-a for all the patients.

Advances in Assisted Reproductive Technologies, Edited by
S. Mashiach *et al.*, Plenum Press, New York, 1990

MATERIAL AND METHODS

Patients included in these studies were under 39 years of age. Couples with male factor were excluded. Two suppression protocols were used for all the patients. Long acting GnRH-a (Decapeptyl depot 3.2 C.R., Ferring, Sweden) was administered on either day 1–3 (follicular phase) or day 21–23 (luteal phase) of the menstrual cycle. Fifteen days after administration of GnRH-a stimulation with hMG/hCG was begun, independently of the day of menstruation. Follicular maturation was assessed by vaginal ultrasonography (El-cint, 5.0 MHz, Haifa, Israel) and serum 17 ß-estradiol (E_2) levels. Human chorionic gona-dotropin (hCG) was administered when two or more follicles >16 mm were visualized and serum E_2 level was >500 pg/ml. This paper does not deal with side effects like hot flushes, headaches, or rash. We divided the effect of GnRH-a according to the following phases of treatment.

Effects of GnRH-a:

> During the phase of suppression
> During the follicular phase (ovarian stimulation)
> During the luteal phase
> In the cycle after IVF/ET

During the Phase of Suppression

Induction of follicular maturation and ovarian cyst. The abnormality most frequently encounted during the phase of suppression was paradoxical stimulation of follicular growth and cyst formation. It is well documented that the administration of GnRH-a provokes an initial rise in gonadotropin with the so-called "flare up" effect, which is followed by down regulation of pituitary secretion of these hormones(Kenigsberg et al., 1986, Sandow 1983). This short-term rise in follicle stimulating hormone (FSH) and LH levels reaches a level that is capable of stimulating follicular growth. This follicular growth can be maintained and continued despite the subsequent down regulation and the decrease in peripheral FSH levels. We have previously noted that follicles that have grown as a result of GnRH-a stimulation are capable of luteinizing normally and probably ovulating after the administra-tion of an ovulatory dose of hCG (Bider et al. 1989).

Cyst formation is a common pathology of GnRH-a treatment that can be directly attrib-uted to the "flare up" effect of the analog. Cysts are formed regardless of whether GnRH-a is given during the follicular or luteal phase. Our experience indicated that ovarian stimula-tion in the presence of ovarian cysts can adversely affect follicular growth and outcome of IVF, while expectant management until the cysts regress spontaneously often results in prolongation of treatment for another two to three weeks, which requires the addition of a daily injection of GnRH-a.

Table 1. Number, Size and Hormonal Levels of Follicular and Luteal Cysts, Number of Days from Decapeptyl Injection to Aspiration

	Day 3 Follicular phase	Day 22 Luteal phase	p
No of patients	12/62 (19.3%)	8/36 (22.2%)	
Size of the cysts (mm)	23.8 ± 3	25.3 ± 2.2	NS
No of cysts	2.8 ± 0.5	1.8 ± 0.4	0.04
No of days to aspiration	11.2 ± 1.9	9.2 ± 2.2	NS
Serum E_2 before aspiration (pg)	431 ± 187	298 ± 143	NS
Serum P before aspiration (ng)	1.7 ± 0.5	4.9 ± 1 .6	0.06

In this study we prospectively followed up 98 patients who were treated with one of the suppression protocols, to study and compare the frequency of cyst formation during the luteal or follicular phase. Patients were instructed to have vaginal ultrasonographic examination and E_2 and progesterone (P) every three to four days during the suppression period.

Twelve of the 62 (19.3%) patients who received decapeptyl during the follicular phase and eight of the 36 (22.2%) patients who received decapeptyl during the luteal phase had developed cysts in response to GnRH-a. The lower limit was arbitrarily set as 16 mm and the size of these cysts varied up to 50 mm with a mean of around 25 mm. Our observation indicated that patients who receive GnRH-a on the follicular phase have a significantly higher number of cysts (2.8 ± 0.5) than patients who receive GnRH-a during the luteal phase (1.8 ± 0.4, p < 0.04) (Table 1). The cysts were aspirated 9–11 days after GnRH-a administration. During that period, patients in their luteal phase experienced bleeding while patients in the follicular phase did not bleed. Peripheral E_2 and P levels on the day of aspiration reached substantial levels; 431 ± 187 pg/ml and 1.7 ± 0.5 ng/ml, and 298 ± 143 pg/ml and 4.9 ± 1.6 in cases where GnRH-a was given during the follicular and luteal phase respectively. The differences in P levels were statistically significant (Table 1).

Cysts were evacuated as an ambulatory procedure without anesthesia with vaginal ultrasound guidance. The content of the cysts were usually clear, they contained few cells and no egg was identified in all these cases. Serum E_2 levels dropped quickly during the next 2–4 days to early follicular phase levels and two of the patients in the follicular phase menstruated.

Ovarian stimulation was started with hMG as scheduled - on day 16 after GnRH-a administration. The cysts usually remained in collapse and did not refill during stimulation. Serum E_2 levels on the day of hCG administration and the number of oocytes retrieved and fertilized was higher in patients receiving the analog on day 22 (Table 2). Two of eight patients in this group were pregnant while none of the 12 patients with cyst evacuation in the follicular phase group conceived.

The content of the so-called cyst was surprisingly similar to the content of normal fol-

Table 2. Hormonal Levels and Results of IVF Treatment in the Cycle Following Cyst Aspiration

	Day 3 Follicular phase	Day 22 Luteal phase	p
Serum E_2 after aspiration (pg)	31 ± 6.5	25 ± 4.4	NS
Serum E_2 on the day of hCG (pg)	885 ± 137	1346 ± 507	NS
Serum P on the day of hCG (ng)	0.55 ± 0.06	0.64 ± 0.8	NS
Number of oocytes	6 ± 0.9	11.2 ± 2.5	0.03
Number of fertilized eggs	3.2 ± 0.6	7 ± 2.1	0.04
E_2 in the cysts (pg /ml)	461 ± 56	401 ± 1 35	NS
P in the cysts (ng /ml)	7.6 ± 1.4	6.3 ± 1.9	NS

licles. E_2, P, and androgens were similar whether the cysts were collected during the luteal or follicular phase. However, FSH levels were surprisingly low.

This pathological condition of cyst formation might shed some light on an unresolved physiological issue. It is not completely clear whether the antrum of a growing follicle serves as a reservoir for the release of steroids, or is an excluded cavity as maintained by Short (1965). According to this observation, steroids are transferred from the site of production either to the circulation or to the follicular cavity,where they are excluded from further contribution to the circulation. Retention of steroids in follicular fluid might be enhanced by binding protein in the human follicular fluid (Ben-Rafael et al. 1986).

Our findings indicate that the cysts although devoid of cells when aspirated, have been functional and they continuously contribute some of their content into the circulation. This is supported by the fact that aspiration of these cysts caused a sharp drop in peripheral steroids. However, no direct correlation existed between steroids in the cysts and peripheral levels. The question of how these follicles grow to form cysts that can be luteinized but not ruptured is unresolved at this time.

Endometrium. Patients receiving GnRH-a during the luteal phase menstruated regardless of whether they developed cysts or not. In contrast, only two of the 62 patients who received GnRH-a during the follicular phase menstruated. As a result of the "flare-up" effect some of the patients had substantial elevation in E_2 and P levels during the suppression period without menstruation. Subsequently they were stimulated with hMG, hence the endometrium was again exposed to E_2 and P might have impaired the endometrial receptivity. This possibility is supported by the fact that no pregnancy occurred in the 12 patients who had aspiration of cysts during follicular phase. Until further studies are carried out to examine the effect of pathological responses such as follicular cysts on endometrial receptivity, we suggest that in cases where E_2 and P increased as a result of GnRH-a ovarian stimulation should be postponed until menstruation occurs. We also maintain that luteal phase suppression might be preferable.

Failed Suppression. Figure 1 illustrates some aspects of failure of GnRH-a to achieve the goal of suppression. The patient had mechanical infertility. It took her three months to reach pituitary suppression. The first monthly ampule of decapeptyl was not injected in full and the pituitary was free from suppression before the end of the month. As seen in the Figure, E_2 started to rise and follicular growth was detected (upper panel). The second injection caused "flare up" of the unsuppressed pituitary and follicular growth with presumed ovulation was noted. Only in the third month pituitary suppression was ascertained. The patient was then stimulated with hMG, oocytes were aspirated, and embryo transfer (ET) was achieved (lower panel). The main reason for failed suppression is failure to inject or absorb the whole content of the analog (Lemay et al. 1988).

Effects of GnRH-a during Ovarian Stimulation

Once suppression is accomplished safely and ovarian stimulation is begun several abnormalities were identified.

Lack of Improvement of Ovarian Stimulation. Clinically it was suspected that the response to GnRH-a / hMG combination is dependent on the patient's own basal response. Patients were divided into poor, normal and high responders according to the maximal E_2 levels reached on the hMG cycle which preceded the GnRH-a/hMG cycle.

In hMG cycle:

$E_2 <\ \ 400$ pg/ml   Poor responders
$E_2 < 1000$ pg/ml   Normal responders
$E_2 > 1000$ pg/ml   High responders

Patients who were treated with hMG only, and who had had ET and failed to conceive, were routinely offered another IVF cycle, which included down regulation of pituitary function with GnRH-a, and hMG stimulation. The last hMG-induced cycle was compared to the cycle with the combination of GnRH-a and hMG. We found that the effect of GnRH-a varied according to the patient's own response.

Pituitary down regulation with GnRH-a before hMG treatment in normal responders resulted in a prolonged phase of ovarian stimulation (follicular phase), and accordingly the number of ampules of hMG which were required to reach adequate ovarian stimulation more than doubled (Table 3). This observation is in accordance with our previous finding that every suppression protocol (Pill or GnRH-a) results in a prolonged latent phase and an increased number of hMG ampules (Mashiach et al., 1988). E_2 levels did not change markedly with or without the analog. However, the number of follicles observed and oocytes collected were significantly increased with the routine use of GnRH-a and hMG. The ratio of E_2/follicle observed was reduced significantly from 147 to 104 pg/ml (Table 3).

In contrast, in high responders the size of the cohort did not vary with the use of GnRH-a but E_2 levels were significantly lower and E_2 to follicle ratio decreased (Table 4). These findings indicate that the routine use of GnRH-a can substantially modify the number of follicles in the cohort and the number of oocytes in normal responders but not in high responders but the relative drop in peripheral E_2/follicular ratio might indicate that there is a lag period between the increase in size and E_2 production of the follicle.

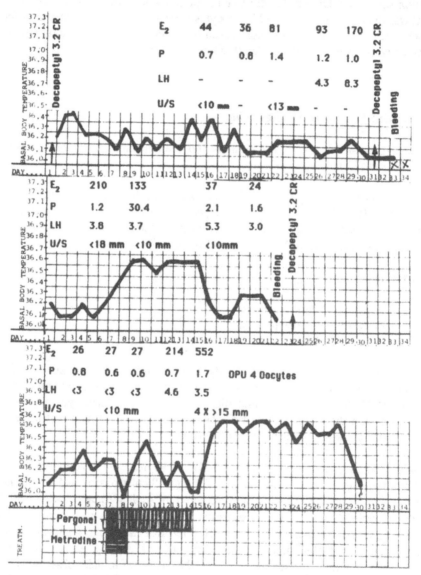

Fig. 1. Suppression with GnRH-a before stimulation with hMG. This patient required three-monthly injections before suppression was achieved.

Table 3. Comparison of the Results of Treatment in hMG Cycle and Combination of GnRH-a and hMG in Normal Responders

	hMG n = 23	GnRH-a/hMG n = 23	p
Days of hMG treatment	7.7 ± 0.3	11.3 ± 0.7	0.0001
No of ampules of hMG	17 ± 0.9	39.2 ± 0.7	0.0001
Peak E_2 levels	744 ± 53	764 ± 50	NS
Follicles ≥ 14 mm	5.9 ± 0.6	8.8 ± 0.9	0.002
No. of oocytes	3.9 ± 0.4	5.7 ± 0.7	0.005
E_2/Follicles ratio	147 ± 14	104 ± 10	0.01

Poor responders are characterized by:

Failure to develop more than 1–2 preovulatory follicles.
Failure to progress in E_2 levels despite increasing the dose of hMG.
Triggering of premature LH surge.

This can be considered a perimenopausal pattern of response. These patients often have elevated FSH and elevated FSH to LH ratio as well as high pretreatment serum E_2 levels. They represent about 3% - 4% of the treated population. In our experience in poor responders treated with GnRH-a, hMG, even in high doses often fails to elicit any ovarian response. The failure of this protocol to ameliorate the results of poor responders does not negate the possibility that other protocol employing GnRH-a for shorter periods or simultaneously with hMG might be found beneficial to these patients.

Relative Ovarian Insensitivity to hMG after GnRH-a Treatment. Administration of GnRH-a is associated with ovarian hyposensitivity to hMG treatment which is accentuated in poor responders. The latent phase, the time between initiation of stimulation with hMG and ovarian response, is significantly prolonged and the number of hMG ampules required to achieve adequate follicular growth is almost double compared with controls that are not suppressed by GnRH-a. Several theories were offered to explain this relative ovarian hyposensitivity; direct effect of the high dose of the analog through ovarian binding sites to GnRH; time required for a new cohort to grow due to possible disruption of the folliculogenesis; and down regulation of FSH/LH receptors in the ovary caused by the "flare up" effect of the analogue (Sheehan et al. 1982, Popkin et al., 1983).

Based on previous observations on the pharmacokinetics of hMG (Ben-Rafael et al., 1986) we hypothesized that the etiology for the ovarian hyposensitivity might be caused by suboptimal levels of circulating FSH. Hence, we evaluated serum FSH concentration during hMG stimulation in patients treated with GnRH-a / hMG and in controls treated with hMG only.

Table 4. Comparison of the Results of Treatment in hMG Cycle and Combination of GnRH-a and hMG in High Responders

	hMG (n = 25)	GnRH-a/hMG (n = 25)	p
Days of hMG treatment	7.6 ± 0.3	10.9 ± 0.7	0.0001
No of ampules of hMG	18.2 ± 1	36 ± 3.5	0.0001
Peak E_2 levels	1755 ±155	1348 ±120	0.005
No of follicles	10.6 ± 1	12.4 ± 1.1	NS
No of oocytes	8.2 ± 1	9.6 ±1.2	NS
E_2/Follicles	199 ± 21	122 ± 13	0.005

Thirty-four patients undergoing IVF and ET treatment were included in this study. The patients were divided into three groups. In group 1 and 2 GnRH-a was given on day 3 or 22 of the cycle, for 15 days prior to ovarian stimulation with hMG. Patients in the control group received a similar dose of hMG from day 3 of the cycle without GnRH-a.

The two study groups, as well as the control group, were comparable in terms of age, weight, number of years of infertility and number of previous IVF attempts. GnRH tests indicated complete pituitary suppression in the study group before initiation of hMG (Table 5).

The number of days required to reach adequate follicular growth (9.8 and 11 days) was significantly ($p < 0.01$) higher in the study than in the control group (6.8 days). This was due to the prolonged "latent phase", the period when E_2 levels are low and the ovary is not responding. However when the ovary started to respond, the "active phase" when E_2 levels increases daily was similar (between five and six days) in the study and the control group. As a consequence, the number of hMG ampules required to reach adequate ovarian response was significantly ($p < 0.001$) higher in group 1 (26.7 ± 2.1) and 2 (28 ± 3.2) than in the control (17.6 ± 1.3; Table 5).

In group 1 and 2 , basal FSH levels after 15 days of pituitary suppression and just be-fore initiation of hMG were significantly lower ($p < 0.05$) than in the control group (Fig-ure 2). In group 1 and 2, FSH levels increased from 9 mIU/ml to maximal level of 13 mIU/ml. In contrast, in the control group FSH levels rose from 14 mIU/ml before treatment to a maximal level of 17.6 mIU/ml on the last day of hMG stimulation. FSH levels were sig-nificantly higher in the controls than in the study group (Figure 2).

Peak serum E_2 levels and the outcome of IVF treatment were similar in the study and the control groups. The number of oocytes that were fertilized, cleaved and transferred and the pregnancy percentage did not differ significantly between the groups (Table 5).

Our findings indicate that low peripheral FSH levels which result from pituitary down reg-

Table 5. Comparison of IVF Outcome in Patients Treated with GnRH-a and Controls

	Decapeptyl 15 days	Decapeptyl 30 days	Control
Number of patients	12	12	10
Age	32 ± 0.9	33 ± 0.7	34 ± 0.5
Weight	55 ± 4	60 ± 6	56 ± 3.5
Years of infertility	6.2 ± 0.6	7.1 ± 0.7	6.6 ± 0.8
Previous IVF attempts	2 ± 0.4	1.5 ± 0.4	1.5 ± 0.3
Days of hMG stimulation	9.8 ± 0.9	11 ± 1.1	6.8 ± 0.9
Number of hMG ampules	26.7 ± 2.1	28 ± 3.2	17.6 ± 1.3
Oocytes recovered	6.5 ±1.6	6.3 ±1.1	5.3 ± 0.6
Fertilization	3.5 ± 1.4	2.8 ± 0.7	2.3 ± 0.2
Cleavage	3.1 ± 1.3	2.5 ± 0.6	1.8 ± 0.1
Transfer	2.2 ± 0.4	2 ± 0.4	1.8 ± 0.1
Pregnancies	2	1	1

ulation and from lack of endogenous contribution of FSH secretion might be the reason for the ovarian hyposensitivity in patients treated with the combination of GnRH-a and hMG.

Follicular response is highly dependent on serum levels of FSH. In both humans (Ben-Rafael et al. 1986) and monkeys (Zeleznik et al. 1986) the ovary starts to respond only when threshold levels of peripheral FSH are reached. In monkeys serum FSH levels of between 10-15 mIU/ml are insufficient and fail to initiate any follicular maturation, while levels above 15 mIU/ml initiated and maintained follicular growth.

The control group started with higher basal FSH levels and quickly reached FSH levels above threshold. In contrast, patients treated with GnRH-a started off with lower baseline levels of FSH and it took longer to reach these threshold levels. We suggest that these differences account for the longer latent phase and the so-called ovarian hyposensitivity of patients treated with GnRH-a. Once threshold levels of serum FSH are reached the response is similar in all groups. Furthermore, increasing the dose of hMG, which increases serum FSH concentration can partially reverse the effects of hyposensitivity and shorten the latent phase (Unpublished data).

To summarize, we suggest that the delay in ovarian response in patients treated with a combination of GnRH-a and hMG, is due to the lack of endogenous contribution of FSH which results in low circulating levels of FSH. A higher initial dose of hMG is required to reverse this effect.

Early Luteinization and Drop in E_2 Levels During Stimulation. Theoretically suppression with GnRH-a should prevent early luteinization by virtue of prevention of LH secretion. In our last series of GnRH-a treatment cancellation occurred in eight of 106 cycles. In four cases it was due to a rise in P levels or a drop in E_2 levels, or both, despite the suppression.

These two phenomena are probably related. A possible explanation for this apparently paradoxical response comes from studies on rats where it is maintained that granulosa cells are programmed to luteinize after certain cycles of cell division also without LH secretion. It is also possible that follicles can be luteinized by the small amount of LH contained in the hMG preparation.

Effects of GnRH-a During the Luteal Phase

Ovarian hyperstimulation syndrome (OHS). Early expectation that IVF will prevent or minimize the incidence of OHS by virtue of follicular aspiration did not materialize and the rate of OHS in IVF-treated patients does not seem to be reduced (Wildt 1986, Stucky 1987). It has been suggested that treatment of patients with polycystic ovarian disease (PCOD) with GnRH-a decreases OHS in IVF cycles (Radwanska et al. 1988). However this was not confirmed by others (Defazio et al. 1985). Our own findings show that the rate of OHS is not reduced and that almost all PCOD patients have at least grade 1 OHS.

The combined treatment of GnRH-a and hMG is capable of inducing a state of functional amenorrhea, however it cannot prevent abnormal responses that are of ovarian origin. We suggest that OHS and early luteinization are of ovarian origin and cannot be abolished by GnRH-a.

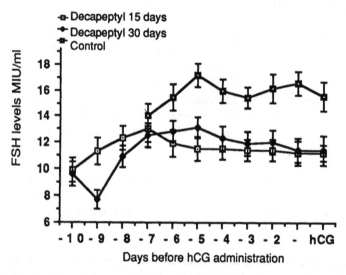

Fig. 2. Daily serum FSH levels in patients treated with GnRH-a and hMG and in controls treated with hMG only. Differences in FSH levels between the study and the control groups were statistically significant.

These abnormalities are not due to failure in pituitary suppression, but they occur despite the suppression. This suggestion is supported by the fact that such abnormal responses are also encountered in patients suffering from hypothalamic amenorrhea (Ben-Rafael 1983).

Corpus Luteum Insufficiency. The significance or existence of luteal phase defect, the diagnosis of corpus luteum insufficiency by hormonal parameters only, and the value of luteal phase support (LPS) in IVF treatments are controversial issues (Torode et al. 1987). It is known that normal corpus luteum function depends on normal pituitary function and LH pulses. Since the suppressive effect of long-acting GnRH-a can extend into the luteal phase, it is logical to assume that a luteal phase defect might accompany GnRH-a treatment, and that luteal support with P or hCG might be beneficial. We did not confirm this in a prospective randomized trial. Our result on a group of 58 patients treated with a combination of GnRH-a and hMG, indicated no differences in pregnancy rates between patients treated with three injections of 5,000 I.U. hCG during the luteal phase, patients treated with support of progestin suppositories (50 mg daily) or controls. Other studies have also failed to confirm statistically significant effects of luteal phase support (Torode et al., 1987). However, a recent study found an increase in pregnancy rate when hCG support was used (Smith et al. 1989).

Luteal phase support is probably justified in a small number of patients, perhaps those in whom a short luteal phase or low progesterone level is confirmed. It would be ideal to select the patients who might benefit from luteal phase support rather than treating all patients.

Effects of GnRH-a in the Cycle After IVF Treatment

Cycle Disturbances. Hormonal disturbances in the cycle following GnRH-a treatment have not yet been reported. Our findings indicate that the menstruation subsequent to the treatment was postponed in 15%–20% of the patients that were treated with long-acting GnRH-a and hMG. The period of amenorrhea lasted between 45–60 days. Menstruation resumed spontaneously without any treatment.

In summary, the introduction of GnRH-a to down regulate pituitary function before induction of follicular maturation is considered a major breakthrough in the ability to manipulate and modulate ovarian function. This treatment has the advantage of abolishing premature LH surge, thereby preventing cancellation of the cycle before ovum pick up. However it is questionable whether this combination has other advantages. GnRH-a/hMG use is more expensive, it prolongs the time required to complete a treatment cycle and it is associated with complications. Recently it was shown that embryos resulting from cycles suppressed with GnRH-a are more susceptable to freezing and thawing (Testart 1989). It therefore seems logical to use GnRH-a only under specific indications rather than routinely.

REFERENCES

Benadiva, C., Ben-Rafael, Z., Strauss, J. F.,Mastroianni, L., and Flickinger G.L., 1988, Ovarian response of individuals to different dose of human menopausal gonadotropin, *Fertil. Steril.*, 49:997.

Ben-Rafael, Z., Dor, J., Mashiach, S., Blankstein, J., Lunenfeld B., and Serr D.M., 1983, Abortion rate in pregnancies following ovulation induced by human menopausal gonadotropin/human chorionic gonadotropin, *Fertil Steril.*, 39: 157.

Ben-Rafael, Z., Mastroianni, L., Strauss, J. F., and Flickinger G.L., 1986, Differences in ovarian stimulation in human menopausal gonadotropin treated women may be related to follicle stmulating hormone accumulation., *Fertil Steril.* 46: 486.

Ben-Rafael, Z., Mastroianni, L., Meloni, F., Lee M.S., and Flickinger G.L., 1986, Total estradiol, free estradiol, sex hormone-binding globulin, and the fraction of estradiol bound to sex hormone binding globulin in human follicular fluid, *J. Clin. Endocrinol. Metab.*, 63: 1106.

Bider, D., Kokia, E., Lipitz, S., Blankstein, J., Mashiach, S, and Ben-Rafael, Z., Failure to improve ovarian response by combined GnRH agonist and gonadotropin therapy. (Submitted).

Defazio J., Meldrum, D.R., Lu, J.K.H., Vale W.W., Rivier J. E., Judd, H.L., and Chang R. J., 1985, Acute ovarian response to long acting agonist of gonadotropin releasing hormone in ovulatory women and women with polycystic ovarian disease, *Fertil. Steril.*, 44:453.

Fleming, R., Haxton, M. J., Hamileton, M.P.R., McCune, G.S. Black, W.B., Macnaughton, M.C., and Coutts, J.R.T., 1985, Successful treatment of infertile women with oligomenorrhea using a combination of LHRH agonist and exogenous gonadotropins, *Br. J. Obstet. Gynecol.*, 2: 368.

Fleming, R., and Coutts, J.R.T.,1986, Induction of multiple follicular growth in normally menstruating women with endogenous gonadotropins suppression, *Fertil. Steril.*, 45: 226.

Kenigsberg D., Littman, B. A., and Hodgen G.D.,1986, Induction of ovulation in primate models. *Endocrine reviews*, 7: 34.

Lemay, A., Sandow, J., Quisnel, G., Bergeron, J., and Merat, P., 1988 Escape from down regulation of pituitary ovarian axis following decreased infusion of luteinizing hormone releasing hormone agonist, *Fertil. Steril.*, 49: 802.

Mashiach, S., Dor, J., Goldenberg, M., Shalev, J., Blankstein, J., Rudak, E., Shoam, Z., Finelt, Z., Nebel, L., Goldman, B., and Ben-Rafael, Z., 1988, Protocols for induction of ovulation: The concept of programmed cycles. *in:* In vitro fertilization and other assisted reproduction, H.W. Jones, Jr. and C. Schrader., Eds., *Annals. of N.Y. Acad. Sci.*, 541: 37.

Popkin, R., Bramley, T.A., Currie, A., Shaw, R.W., Baird, D.T., Fraser, H.M. 1983, Specific binding of luitenising hormone releasing hormone to human luteal tissue. *Biochem. Biophys. Res. Commun.*, 114:750.

Radwanska, E., Rawlins, R.G., Turmmon, I., Maclin, V., Binor, Z., Damowski, W.P., 1988, Successful use of gonadotropin-releasing hormone agonist leuprolide for in vitro fertilization in a patient with polycystic ovarian disease and infertility unresponsive to standard treatment, *Fertil. Steril.*,49: 356.

Sandow, J., 1983, Clinical applications of LHRH and its analogues, *Clin. Endocrinol.*, 18: 571.

Sheehan, K.L, Casper, R.F, Yen, S.S.C., 1983, Induction of luteolysis by luteinizing hormone releasing factor (LRF) agonist: Sensitivity, reproducibility, and reversibility, *Fertil. Steril.*, 37:209.

Smith, E.M., Anthony, F.W., Gadd, S.C, Mosson., G.M., 1989, Trial of support treatment with human menopausal gonadotropin in the luteal phase after treatment with buserelin and human menopausal gonadotropin in women taking part in an in vitro fertilisation programme, *BMJ.*, 298:1483.

Stucky, B.G.A., Klogh E.G., Pullan P.T., Beilby J. A., Thompson, R. J., and Evans O.V., 1987, Continous gonadotropin-releasing hormone for ovulation induction in polycystic ovarian disease, *Fertil. Steril.*, 48:1055.

Testart, J., Forman, R., Belaish-Alart, J. C., Volante, M., Hazut, A., Strub, N., and Frydman, R., 1989, Embryo quality and uterine receptivity in in vitro fertilization cycles with or without agonists of gonadotropin releasing hormone, *Human. Reprod.*, 4:198.

Torode, H.W., Porter, P. N., Vaughan, J. I., Saunders D.M., 1987, Luteal phase support after in vitro fertilisation: A trial and rationale for selective use, *Clinc. Reprod. Fertil.*, 5:255.

Wildt, L., Diedrich, K., Van der Ven, H., Hasani, S. Al., Huber, H., and Klasen, R., 1986, Ovarian hyperstimulation for in-vitro fertilization controlled by GnRH agonist administered in combination with human menopausal gonadotropin, *Human. Reprod.*, 1:15.

Zeleznik, A.J., and Kubik C. J., 1986, Ovarian response in Macaques to pulsatile infusion of follicle-stimulating hormone (FSH) and luteinizing hormone: Increased sensitivity of maturing follicle to FSH, *Endocrinology*, 119:2025.

THE COMBINED PITUITARY SUPPRESSION/OVARIAN STIMULATION THERAPY: MYTHS AND REALITIES

V. Insler, E. Lunenfeld, G. Potashnik, and J. Levy

*Division of Obstetrics and Gynecology, Soroka Medical Center
and Faculty of Health Sciences
Ben-Gurion University of the Negev, Beer-Sheba, Israel*

Ovarian stimulation by human gonadotropins for induction of ovulation in anovulatory infertile patients has been in wide clinical use since the early 1970s. Some ten years later, gonadotropin therapy has been applied to obtain superovulation in spontaneous ovulators within the framework of IVF/ET programs. Many years ago[1] it was shown that ovarian stimulation by exogenous gonadotropins is much more effective and produces significantly more take home babies in amenorrheic women with low endogenous gonadotropins and estrogens (WHO Group I) than in anovulatory patients having some spontaneous, albeit derranged, pituitary and ovarian activity (WHO Group II). It has also become clear that the results and complications of gonadotropin stimulation (i.e. pregnancy rates, hyperstimulation incidence) are similar in anovulatory patients of Group II and in women subjected to IVF. The overall success rate in both types of treatment is 15%-20%,[2,3] while in amenorrheic women of WHO Group I pregnancy rates of over 80% have been steadily achieved.[4] It seems therefore that the presence of an active pituitary-ovarian axis represents a burden rather than an asset in the course of ovulation (or superovulation) induction.

The introduction of potent GnRH analogs capable of transient down regulation of the pituitary-ovarian axis seemed to carry the promise of significant improvement of the results of ovulation and superovulation induction. Indeed, within the last few years a number of reports appeared showing that the application of GnRH analogs is able to reduce the levels of gonadotropins,[5,6,7,8] to suppress the positive LH feedback mechanism,[9,10,11] to in-

Advances in Assisted Reproductive Technologies, Edited by
S. Mashiach *et al.*, Plenum Press, New York, 1990

crease the number of follicles developing in response to exogenous stimulation[12] and to produce a better synchronized follicular cohort.[14, 15, 16, 17] One most recent report[18] indicated also that GnRH analogs can induce final maturation of rat oocytes which subsequently developed into normal offspring.

The abovementioned benefits of combined pituitary suppression/ovarian stimulation therapy should be critically discussed on the basis of causes and reasons for failure in ovulation induction. This discussion should also try and separate the positive effects of each of the two components of the therapy.

Failure of ovarian stimulation with exogenous gonadotropins and, consequently, unacceptably low pregnancy rates may result from:

• lack of ovarian response
• inappropriate treatment or monitoring
• fertilization failure
• implantation failure
• premature luteinization of follicles
• dyssynchronized follicular growth

Lack of ovarian response: even to high doses of gonadotropins has been observed in some women with ovaries containing apparently normal follicles. According to recent studies,[19] the addition of growth hormone may increase and/or improve the ovarian response to exogenous gonadotropins. This might diminish the number of gonadotropin treatment failures. Clinical practice, however, indicates that this group of patients is rather small. Addition of growth hormone or other growth factors could nevertheless play an important role in reduction of the cost of therapy by significantly diminishing the dose requirement in most patients. On the other hand, there is no theoretical basis to presume that ovarian stimulation preceded by suppression of the pituitary/gonadal axis should be effective in women who do not respond to gonadotropins alone. Clinical observations also do not support this presumption.

Inappropriate treatment or monitoring: In recent years, inefficient treatment protocols or inadequate monitoring techniques have seldom been responsible for failure of ovulation (or superovulation) induction. This is particularly true when ovarian response to stimulation is assessed by both hormonal and sonographic methods. There is no doubt that the combined pituitary suppression/ovarian stimulation therapy is not easier to apply or to monitor than the treatment with gonadotropins alone.

Fertilization failure: Fertilization rate is primarily dependent on the quality of the male and female gametes. In IVF programs the laboratory standards and knowhow may also influence the results. On the other hand, there is no doubt that the number of ova available for fertilization (in vivo or in vitro) influences the outcome of therapy, i.e. increases the pregnancy rate.

Table 1. Clinical Parameters of Combined GnRH Analog/Gonadotropins Therapy and of Treatment with Gonadotropins Alone

Parameters	Gonadotropins		Gonadotr/Agonist	
	Mean	Range	Mean	Range
Duration of TTM (days)	16.6	(3–28)	11.1	(5–20)
Total dose (amp.)	26.2	(17–41)	16.5	(5–39)
Effective daily dose	1.7	(1–2.5)	1.7	(1–3)
Latent phase (days)	5.4	(2–9)	5.9	(3–9)
Active phase (days)	6.5	(5–10)	5.1	(3–7)

When discussing the results of combined hypophyseal suppression/ovarian stimulation therapy one should, however, clearly distinguish between the effects due directly to each of the two components. Such an analysis indicates that the increased number of ova produced by the combined therapy is the result of more intense ovarian stimulation. The combined treatment is longer and requires a much higher dosage of gonadotropins than ovulation induction by hMG/hCG alone (Table 1). It is certainly possible that ovaries stimulated by such doses of hMG, without the preceding suppression of the axis by GnRH analogs, would produce a similar oocytes harvest. A controlled crossover clinical study which could prove or negate this hypothesis has, to the best of our knowledge, not been performed. On the other hand, with regard to the quality of ova produced in response to the combined therapy, it is quite clear that the fertilization and cleavage rates are not signifi- cantly different from those obtained when using other treatment protocols. Thus, the possibility of a direct effect of GnRH analogs upon the follicle or the ovum, resulting in improved fertilization capacity is still questionable, at least in the human.

Implantation failure: This is obviously a consequence of the quality of zygotes entering (or introduced) into the uterus as well as the result of receptivity of the endometrium. The practical possibilities of assessing the quality of oocytes, both prior to and after fertilization are extremely limited. But it has been proven that the pregnancy rate in most IVF/ET programs increases significantly in proportion to the number of zygotes transferred.[20] A similar situation probably exists also in gonadotropin treatments followed by in vivo fertilization. It has been speculated[21] that the vast majority of gonadotropin-induced conceptional cycles are multifollicular with a consequent release and probably fertilization of a number of ova, of which only one is destined to produce a living fetus.

Again, analyzing the effects of the combined suppression/stimulation therapy, one must conclude that it is the stimulation component which is mainly responsible for the chain of events finally producing higher pregnancy rates. The more intense gonadotropin stimulation of the ovaries results in recruitment and maintenance of a larger follicular cohort. This generates a larger number of oocytes available for fertilization and, consequently, improves the implantation rate.

Untimely luteinization of follicles: It is generally accepted as one of the causes adversely affecting the results of ovulation induction in spontaneously menstruating (Group II) patients as well as the pregnancy rates of IVF programs.

The normal pituitary gland responds with an LH surge to the exponential rise of estrogen. The gonadotropin (or clomiphene) stimulated cycles are usually multifollicular and the estrogen levels represent the sum total of secretion of follicles being at different stages of development. In this situation, the pituitary may be forced to produce an LH surge despite the fact that the absolute levels of estradiol and the slope of its rise may not correlate with adequate maturity. This results in premature luteinization of follicles and adversely affects the quality of ova.[22] The incidence of premature luteinization differs in various reports.[12, 23, 24] In our IVF program the cancellation rate due to suspected premature luteinization has been approximately 20%.

It has been shown that the pituitary response to rising estrogen levels may be shut off by the application of potent GnRH analogs.[9, 12] The question as to what should serve as an accurate and clinically efficient tool for estimation of the analog's effect upon the pituitary response to estrogen levels is, however, still open. There seems to be no doubt that the estradiol benzoate (EB) test reflects accurately the hypophyseal capability to react to the slope of estradiol rise.

We performed 41 EB tests in 17 women. All initial and some of the repeated tests were positive, but in all cases the test became negative after application of the GnRH analogs for 14 to 21 days. The analysis of our material also indicated that there was a good time correlation between the pituitary response to EB and the levels of endogenous estradiol. In the vast majority of women having levels of E2 below 25 pg/ml the EB tests were negative, i.e. the pituitary positive feedback mechanism has been shut off. It might, therefore, be concluded that these levels of endogenous estrogen could serve as an indicator of the pituitary ability to respond to estrogen stimulation.

The capability of the combined suppression/stimulation therapy to avoid failures due to premature luteinization of follicles seems to be unequivocal. It is also obvious that this beneficial effect is due to the suppression component of the treatment.

Synchronization of follicular cohort: The advent of combined GnRH analog/gonadotropin therapy created great hopes for achieving a fully synchronized follicular cohort in response to application of pharmacological doses of hMG. The majority of reports describing the course and results of the combined therapy pointed out that, although the number of follicles was greater, their size was not uniform (Figure 1). The lack of synchronization of the follicular cohort produced by the combined suppression/stimulation therapy has also been indicated by the enormous individual variation of estrogen levels (Figure 2).

Experimental data regarding the physiology of follicular growth furnish a logical basis for understanding the reason why a relatively short application of GnRH analogs may not be able to suppress the function of both the pituitary and the ovaries to such an extent that subsequent stimulation with gonadotropins will result in the development of a fully syn-chronized follicular cohort.

In the mouse, the development of follicles from primordial to pre-ovulatory stage takes approximately five cycles.[25] This is also apparent in the rat (own unpublished data). Gou-geon[26] showed that, in the human, follicles require approximately 10 weeks to reach the preovulatory stage.

Our experiments in the rat indicated that FSH played an important role in follicular maturation during at least three out of five cycles required for structural and functional de-velopment of follicles. Elimination of the midcycle FSH surge in three consecutive cycles resulted in a significant reduction of the number and size of follicles (Table 2).

It seems that in order to be ready for the recruitment into the cohort which will finally produce the dominant follicle, a number of follicles must be "assigned", i.e. prepared by FSH during the preceding three cycles. In the combined pituitary suppression/ovarian stim-ulation therapy, the FSH levels are not fully suppressed and the GnRH analog is usually applied for a period of 14 to 21 days.

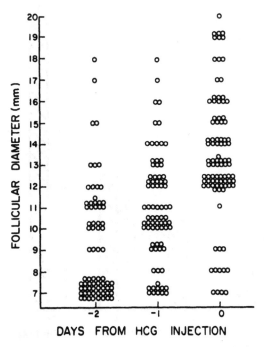

Fig. 1. Diameter of follicles developing during the combined pituitary suppression/ovarian stimulation therapy.

Table 2. The Influence of Midcycle FSH Surge on Follicular Development

No. of FSH surge eliminated	Number	Surface (SEM)	Follicles Diameter	(SEM)
0–1	10.29	(0.09)	1.17	(0.01)
2–3	7.65	(0.19)	0.76	(0.03)

It may therefore be presumed that the ovary still contains follicles at various stages of maturation with a different capacity to react to subsequent stimulation with pharmacological doses of gonadotropins. In such a situation, the development of a dyssynchronized follicular cohort is a rule rather than an exception. It is possible that in order to achieve the degree of ovarian suppression allowing a synchronized response to subsequent stimulation, the GnRH analogs should be applied for a few months rather than weeks. In order to ob-

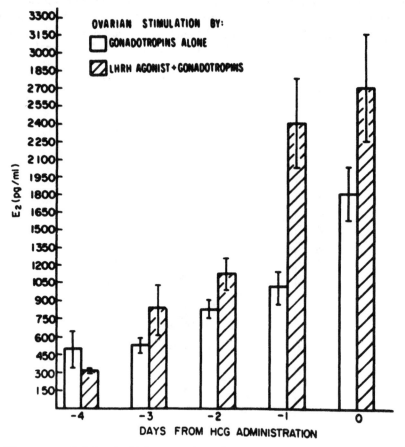

Fig. 2. Plasma estradiol levels during gonadotropin stimulation alone and during combined GnRH analog/gonadotropin therapy.

tain a better synchronized ovarian response to stimulation new treatment schemes based on prolonged use of presently available agonists or on application of potent synthetic GnRH antagonists will have to be developed. Basic studies concerning the presence and the extent of a direct effect of GnRH analogs upon the ovarian function would also help in better understanding and more efficient application of the combined pituitary suppression/ovarian stimulation therapy.

REFERENCES

1. V. Insler and B. Lunenfeld, Application of human gonadotropins for induction of ovulation, *in*: "Human Reproduction," A. Campos da paz, Y. Notake and M. Hayashi, eds., Igaku Shoin, Tokyo (1974).
2. American Fertility Society, In vitro fertilization — embryo transfer in the United States, 1985-1986: results from the National IVF-ET registry, *Fertil. Steril.*, 49:212 (1988).
3. B. Lunenfeld and V. Insler, "Diagnosis and Treatment of Functional Infertility," Grosse, Berlin (1979).
4. V. Insler, I. Eichenbrenner and B. Lunenfeld, Comparison of reproductive performance of various groups of anovulatory patients treated with human gonadotropins, *in*: "Clinical Application of Human Gonadotropins," G. Bettendorf and V. Insler, eds., George Thieme Verlag, Stuttgart (1970).
5. D. Meldrum, R. J. Lu, W. Vale, J. Rivier and H. L. Judd, Medical oophorectomy using a long-acting GnRH agonist - a possible new approach to treatment of endometriosis, *J. Clin. Endocrinol. Metab.*, 54:1081 (1986).
6. W. F. Growely, F. Comite, W. Vale, J. Rivier, D. L. Loriaux, and G. B. Cutler, Inhibition of serum androgen levels by chronic intranasal and subcutaneous administration of a potent luteinizing hormone-releasing hormone (LH-RH) agonist in adult men, *Fertil. Steril.*, 27: 1240 (1982).
7. S. S. Yen, Clinical application of gonadotropin-releasing hormone analogs, *Fertil. Steril.*, 39:257 (1983).
8. E. Lunenfeld, G. Potashnik and V. Insler, Combined GnRH agonist/gonadotropin therapy in women who failed to respond to other ovulation inducing therapy, *in*: Abstr. Israel Congress of Fertility and Sterility (1986).
9. L. Wildt, K. Diedrich, H. van der Ven, S. Al Jassani, H. Huebner and R. Klassen, Ovarian hyperstimulation for in vitro fertilization controlled by GnRH agonist administered in combination with menopausal gonadotropins, *Human Reproduction*, 1:15 (1986).
10. A. L. Shadmi, B. Lunenfeld, C. Bahari, E. Kokia, C. Pariente and J. Blankstein, Abolishment of the positive feedback mechanism: a criterion for temporary medical hypophysectomy by LH-RH agonist, *Gynecol. Endocrinol.*, 1:1 (1987).
11. V. Insler, G. Potashnik, E. Lunenfeld, I. Meizner and J. Levy, Ovulation induction with hMG following down regulation of the pituitary/ovarian axis by LH-RH analogs, *Gynecol. Endocrinol.*, (Suppl. 1) 2:67 (1988).
12. R. Fleming and J. R. Coutts, Induction of multiple follicular growth in normally menstruating women with endogenous gonadotropin suppression, *Fertil. Steril.* 45:226 (1986).
13. D. Feldberg, A. Yishaiahu, D. Ayalon, J. Voliovotsch, M. Schlaff and J. Goldman, The use of GnRH agonists in in vitro fertilization, Proc. Israeli Congress of Fertil. Steril. Abstr. 5., (1988).
14. S. Neveu, B. Hedon, J. Bringer, J. M. Chinchole, F. Arnal, C. Humeau, P. Cristol and J. L. Viala, Ovarian stimulation by a combination of a gonadotropin-releasing hormone agonist and gonadotropins for in vitro fertilization, *Fertil. Steril.*, 47:639 (1987).

15. G. Bettendorf, W. Braendle, V. Lichtenberg and Ch. Lindner, Pharmacologic hypogonadotropism - an advantage for gonadotropin stimulation, *Gynecol. Endocrinol.*, (Suppl. 1) 2:66 (1988).
16. B. Hedon, Use of GnRH agonists for in vitro fertilization, *Gynecol. Endocrinol.*, (Suppl. 1) 2:34 (1988).
17. R. Ron-el, E. Caspi, H. Nahum, A. Golan, A. Herman, Y. Soffer and Z. Weinraub, Improved pregnancy rate in IVF by combined long acting GnRH analog (D-TRP6-LH-RH) - gonadotropins and sequential embryo transfer, *Gynecol. Endocrinol.*, (Suppl. 1) 2:78 (1988).
18. R. Shalgi and N. Dekel, Embryonic development of fertilized oocytes induced to mature by GnRHa, Proc. Annual Meeting of Israel Endocrine Society, Abstr. 8 (1989).
19. R. Homburg, A. Eshel, I. Abdalla and H. S. Jacobs, Growth hormone facilitates ovulation induction by gonadotropins, *Clin. Endocrinol.*,29:113 (1988).
20. J. Webster, Embryo transfer, *in*: "In Vitro Fertilization — Past, Present, Future", S. Fishel and E. M. Symonds, eds., IRL Press, Oxford (1986).
21. V. Insler and G.Potashnik, Monitoring of follicular development in gonadotropin-stimulated cycles, *in*: "Fertilization of the Human Egg In Vitro", H. M. Beier and H. R. Lindner, eds, Springer Verlag, Berlin (1983).
22. A.O. Trounson, Factors controlling normal embryo development and implantation of human oocytes fertilized in vitro, *in*: "Fertilization of the Human Egg In Vitro", H. M. Beier and H. R. Lindner, eds., Springer Verlag, Berlin (1983).
23. J. E. Garcia, G. S. Jones, A. A. Acosta and G. Wright, Human menopausal gonadotropin-human chorionic gonadotropin in follicular maturation for oocyte aspiration. Phase I. *Fertil. Steril.*, 39:167 (1983).
24. J. Dor, A. Ellenbogen, Z. Ben-Raphael, S. Lipitz, D. Lev-Ran, A. Davidson, L. Nebel, D. M. Serr, E. Rudak and S. Mashiah, D-TRP6-GnRH analog for the IVF program, Proc. Israeli Congress of Fertil. Steril. Abstr. 5 (1988).
25. H. Peters and E. Levy, Cell dynamics of the ovarian cycle, *J. Reprod. Fert.*, 11:227 (1966).
26. A. Gougeon, Origin and growth of the preovulatory follicle(s) in spontaneous and stimulated cycles, *in*: Human In Vitro Fertilization; Inserm Symposium No. 24, J. Testart and R. Frydman, eds., Elsevier Science Publishers, (1985).

PAST PRESENT AND FUTURE OF GONADOTROPINS

Bruno Lunenfeld

*Bar-Ilan University,
and Institute of Endocrinology
Chaim Sheba Medical Center, Ramat Gan 52621, Israel*

The cornerstone to the conquest of infertility was laid in the beginning of this century. It took however nearly 80 years of work of many scientists from all over the globe to slowly unravel the puzzle of nature's most guarded secret, the control of the reproductive processes. Physiologists, biochemists, surgeons and physicians engaged in fundamental and applied research funded by international and national organizations, hand in hand with the pharmaceutical and diagnostic industries have slowly been able to reduce the often quoted figure of 12% infertility to a point that barely 2% of previously infertile women will not be able to experience motherhood.

Probably the most far-reaching discovery ever made in reproductive biology was the revelation by Crowe et al. (1966) in 1909 that the male and female reproductive systems are under the functional control of the anterior hypophysis by demonstrating that partial hypophysectomy in the adult dog provoked atrophy of the reproductive organs, and prevented sexual development in juvenile animals. However, only twenty years later, Zondek and Ascheim (1927) in Europe and Smith and Engel (1927) in the U.S.A. discovered the Gonadotropic hormones FSH, LH and hCG and obtained firm evidence that the male and female reproductive systems were under the functional control of gonadotropins secreted by the pituitary gland.

It took 30 more years to realize that the pituitary ovarian axis was controlled by the hypothalamus and 20 further years to discover Gonadotropin Releasing Hormone (GnRH) and their pulsatile release by the arcuate nucleus. However, only during the last decade, with

Advances in Assisted Reproductive Technologies, Edited by
S. Mashiach *et al.*, Plenum Press, New York, 1990

the discovery of non-steroidal gonadal factors such as inhibins, activins and growth factors, was the importance of intraovarian regulation in modulating ovarian response to gonadotropins realized. In the early 50's, the first attempts to use gonadotropins in ovulation induction in women were made. They were obtained from pregnant mares' serum(PMS) and animal pituitaries.

Gonadotropins of animal origin are no longer used for this purpose since humans rapidly produce antibodies to none-primate gonadotropins which neutralize their clinical effects. In 1954, Borth, de Watteville and myself (1954) were fortunate to demonstrate that kaolin extracts from pooled menopausal urine contained FSH and LH activity in comparable amounts. These extracts prevented Leydig cell atrophy and retained complete spermatogenesis in hypophysectomized male rats and were capable of inducing follicular growth and promotion of multiple corpora lutea in hypophysectomized female rats (Borth, Lunenfeld, Riotton, 1957).

Based on these observations, we predicted in 1954 that such extracts could open up interesting therapeutic possibilities (Borth, Lunenfeld, de Watteville, 1954). The recognition of the therapeutical potential of human gonadotropins stimulated the search for suitable sources for the extraction of these hormones. Most investigators were purifying gonadotropins from menopausal urine (hMG). However, the Stockholm group, lead by Karl Gemzell (Gemzell, Diczfalusy, Tillinger 1958), took the shorter route by processing human pituitaries (hPG). Because the potential dangers and the scarcity of post mortem pituitary glands (required for the production of hPG) eliminates the possibility of their wide scale use, attention was directed by pharmaceutical companies (Serono and Organon) to prepare purified extracts from menopausal urine for clinical use (Donini, Puzzuoli, Montezemolo 1964). Borth, Lunenfeld, and Menzi (1961) reported that this preparation was a potent ovarian stimulant in the human, capable of promoting multiple follicular development.

A survey of almost 22156 gonadotropin treatments published during the last several years indicates that this therapy for induction of ovulation has become universally accepted (Lunenfeld & Lunenfeld, 1988). The realization that FSH is capable of increasing the recruitment rate of new crops of small follicles, and is capable of stimulating multiple and normal follicular growth and development to the preovulatory stage with the yield of many fertilizable eggs, prompted the use of gonadotropins as the preferred treatment modality in most in vitro fertilization programs. With the use of gonadotropins for induction of super-ovulation in normally ovulating women, conceptual changes in our monitoring schemes had to be introduced. When attempting to induce ovulation in anovulatory women, the challenge was to imitate, as much as possible, the normal cycle and to aim at a single dominant follicle and prevent multiple follicular growth, multiple pregnancies and hyperstimulation.

In IVF or GIFT programs, in contrast, the conceptual idea is to use a super physiological dosage in order to obtain a large number of fertilizable eggs. For this purpose, many different protocols have emerged, each with its own merits and disadvantages and all using ultrasonograpic procedures to estimate both the number and size of the growing follicles and estradiol to estimate their functional integrity. However, enlarging recruitment rate of follicles, increasing asynchrony of follicular growth and deviating follicular development from luteinisation and atresia towards dominance will coincide with exaggerated estradiol levels.

The exaggerated production of estradiol will not always, due to the asynchrony of follicular development, coincide with a similar production of 'inhibins' or other intraovarian regulators. Thus, in ovulatory women with a normal pituitary ovarian axis, the exaggerated estradiol levels provoke, in about 15% of treatment cycles, an untimely spontaneous LH surge leading to cancellation of ovum pick up.

The possibility to desensitize the pituitary gland with superactive GnRH analogues and thus inhibit its capability to respond to estradiol with an LH surge (Shadmi, Lunenfeld, Bahari 1987), has prompted many investigators to use GnRH analogues prior to and in combination with hMG. This treatment regime has already proven its merits in making logistics easier and by significantly reducing cancellation rates in IVF programs and thus, increasing the overall success rate. However, the temporary functional gonadotropic specific hypophysectomy has introduced new challenges. It permits the design of stimulation protocols with a predominantly FSH environment during the recruitment phase without interference by endogenous gonadotropic secretion. It will eliminate an excessive endogenous LH environment which may interfere in normal follicular development and, by untimely inhibition of the meiosis inhibiting factors, provoke over-aging of eggs with a significant reduction in their fertilization or implantation capacity.

However, inhibition of endogenous gonadotropins following ovulation induction by hCG may, if not properly monitored, interfere with the luteal phase. The choice of the GnRH analogue to be used depending on its delivery system, its mode of administration and its biological half life, will dictate the optimal gonadotropic treatment protocol considering follicular recruitment and development, ovulation induction and the luteal phase. The choice of the GnRH analogue to be used depending on its delivery system, its mode of administration and its biological half life, will dictate the optimal gonadotropic treatment protocol considering follicular recruitment and development, ovulation induction and the luteal phase.

With the increased use of superactive GnRH analogues during the past three years, new information has accumulated which will help to design better protocols and improve our methodology in assisted conception. Even with the best protocols a number of patients will be poor responders. In some of them the response may be enhanced by purified FSH preparations. During the last few years, the importance of intraovarian regulation and the potentiating effect of growth hormone and growth factors on granulosa cell response to FSH has been demonstrated (Adashi, et al. 1988). It has been shown that in patients who need excessive amounts of gonadotropins, the concomitant administration of growth hormone and gonadotropins, can significantly reduce the amount of gonadotropins neccesary to stimulate follicular growth (Hamburg, et al. 1988 Blumenfeld, Lunenfeld 1989). It has also been shown by our group (Menashe, et al. 1989), that patients with a low growth hormone reserve (for example, patients who do not respond with elevation of growth hormone following clonidine administration (Fig. 1) need significantly more gonadotropins than those with a good reserve (Fig. 2). Although growth hormone deficient patients will respond to gonadotropins if a sufficient dose is used, it can be definitely stated that growth hormone and/or growth factors have a permissive modulating role on ovarian response to gonadotropins. Only well controlled clinical trials will show whether they have a definite role as an adjunctive therapy in the group of poor responders.

During the last two years, new generations of GnRH antagonists not producing exaggerated histamine release and capable of immediate inhibition of endogenous gonadotropins without prior stimulation have become available (Leal et al. 1988, Franchimont et al. 1989). Only the future will show whether they will play a significant role in IVF protocols.

In the past, the use of gonadotropins was confined as highly potential ovulation inducing drugs for first line therapy to amenorheic patients with hypopituitary hypoestrogenism and as a second line therapy, to anovulatory patients who failed to conceive following clomiphene citrate administration. Due to its potential in inducing superovulation, its use, since the early 80's has expended, exponentially, as the preferred drug in IVF, GIFT, and IUI programs.

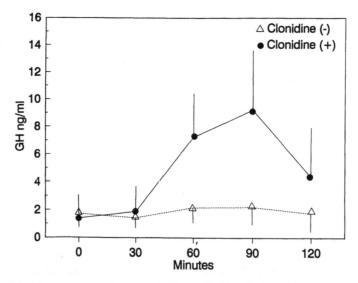

Fig. 1. Growth hormone response (mean ± SEM) to clonidine in 25 women prior to hMG administration for induction of ovulation.

Table 1. Women Aged 19–24 and Estimated No. of Infertile Women and Potential hMG Users

Country	women 19–34 × 1000	infertile women × 1000	potent. hMG users × 1000
West Europe	39.500	3.160	2.240
East Europe	45.200	3.620	2.570
U.S.A.	27.700	2.220	1.570
Canada	2.900	230	160
Japan	12.800	1.020	730
Australia	1.900	150	110

From figures on the size of the adolescent population published by the United Nation Population Division in 1988, I have estimated that the population size of women between the ages 19–34 in the developed world in 1990 will be about 130 million (Table 1). If we assume that at least 8% will be infertile, then the pool of the infertile population will be above 10 million, with about 700 000 new patients entering this pool every year between the years 1990 to 1995. If we assume, for the purpose of this intellectual exercise, that gonadotropins have a potential benefit for 5% of this infertile population as a first line therapy, for 70% of the clomiphene failures and for 45% of patients with infertility potentially treatable with IVF, GIFT or IUI (due to mechanical problems or male partner problems) then the potential Gonadotropin users could theoretically include nearly 7.5 million women with about half a million new candidates every year (Table 2). The future of infer-

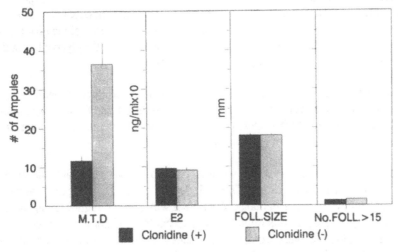

Fig. 2. Mean (± SEM) total dose of hMG (AMPS) and ovarian response as reflected by estradiol (E₂ ng × 10) follicular size (mm). No. of follicles 15 mm in clonidine positive and clonidine negative patients.

Table 2. New Women Entering 'Fertile Age Pool' Every Year 1990–1995

Country	new women × 1000	new infertile women × 1000	new potent. hMG users × 1000
West Europe	2.630	210	150
East Europe	3.010	240	160
U.S.A.	1.850	150	105
Canada	190	15	11
Japan	850	68	48
Australia	130	10	7

tility therapy clearly relies on the capacity to produce pharmaceutical grade human gonadotropins in sufficient quantities to meet this ever increasing world wide demand.

With recombinant DNA technology and highly refined cell culture techniques, Serono Laboratories, world wide and Integrated Genetics of Framingham, Massachusetts (U.S.A.) are actively producing recombinant DNA gonadotropins. What would currently require over 200 million liters of urine will ultimately be produced by genetically engineered cells in chemically defined culture medium comprising only a small fraction of that volume.

The initial challenge involved construction of appropriate vectors for inserting alpha and beta subunit DNA into well-characterized and stable cells. Unlike insulin and growth hormone, the gonadotropins are glycosylated heterodimeric peptides. The complex sugars are important for proper folding of the polypeptide backbone. The sites and extent of glycosylation determine tertiary structure, length of time for degradation (mostly by liver enzymes), the regions of the molecule exposed to target cell receptors, and exposure of the molecule to mechanisms that regulate metabolism in vivo. Whereas bacteria efficiently produce nonglycosylated peptides such as insulin, and yeast has been cloned for the production of certain vaccines, prokaryotic cells are incapable of correctly glycosylating the peptide subunits to produce biologically active gonadotropins.

Several mammalian cell lines have been constructed that produce authentic human LH, FSH and chorionic gonadotropins with activities indistinguishable from their natural counterparts. Thus recombinant DNA technology will provide, in the not too distant future, an almost endless supply of human gonadotropins.

It is, however, difficult to estimate when recombinant DNA gonadotropin will become available for therapeutic application. This will depend on quality and safety assurance by industry and the unpredictable demands and time it will take the regulatory agencies to approve a product produced by this relatively new technology. If our modern society accepts the challenge of allowing every woman who wishes to experience the joy of motherhood to obtain her goal, then industry will have to make the efforts to solve the production problems, society will have to bare the cost and it will remain for us to provide the technical services.

REFERENCES

Adashi, E. Y., Resnick, C. E., Svoboda, M. E., and Van Wyk, J. J., 1988, Somatomedin C enhances induction of LH receptors by FSH in cultured rat granulosa cells, *Endocrinol.*, 116:2369.

Blumenfeld, Z., and Lunenfeld, B., 1989, The potential effect of growth hormone on follicle stimulation with human menopausal gonadotropin in a panhypopituitary patient, *Fertil. Steril.*, 52:328.

Borth, R., Lunenfeld, B., and de Watteville, H., 1954, Activite gonadotrope d'un extrait d'urines de femmes en menopause, *Experientia*, 10:266.

Borth, R., Lunenfeld, B., and Riotton, G., et al., 1957, Activite gonadotrope d'un extrait des femmes en menopause (2e communication), *Experientia*, 13:115.

Borth, R., Lunenfeld, B., and Menzi, A., 1961, Pharmacologic and clinical effects of a gonadotropin preparation from human postmenopausal urine, *in:* "Human Pituitary Gonadotropins," A. Albert and M. C. Thomas, eds., Springfield.

Crowe, S. J., Cushing, H., and Homans, J., 1966, Cited by Lunenfeld, B. and Donini, P., Historic aspects of gonadotrophins, *in:* "Ovulation," R. B. Greenblatt, ed., J. B. Lipincott Co., Toronto.

Donini, P., Puzzuoli, D., and Montezemolo, R., 1964, Purification of gonadotropin from human menopausal urine, *Acta Endocr.* (Kbh), 45:329.

Franchimont, P., Almer, S., Mannaerts, B., Boen, P., and Kicivic, P. M., 1989, New GnRH antagonist ORG 30850: The first clinical experience, *Gynecol. Endocrinol.*, 3:(suppl. 1)13.

Gemzell, C. A., Diczfalusy, E., and Tillinger, G., 1958, Clinical effect of human pituitary follicle-stimulating hormone (FSH), *J. Clin. Endocrinol. Metabol.*, 18:1333.

Homberg, R., Eshel, A., Abdallah, H. I., and Jacobs, H. S., 1988, Growth hormone facilitates ovulation induction by gonadotropins, *Clinical Endocrinology* 29:113.

Leal, J. A., Williams, R. F., Danforth, D. R., Gordon, K., and Hodgen, G. D., 1988, Prolonged duration of Gonadotropin inhibition by a third generation GnRH antagonist, *J. Clin. Endocrinol. Metab.*, 87:1325.

Lunenfeld, E., and Lunenfeld, B., 1988, Modern approaches to the diagnosis and management of anovulation, *Int. J. Fert.*, 33:308.

Menashe, Y., Pariente, C., Lunenfeld, B., Dan, U., Frenkel, Y., and Mashiach, S., 1989, Does endogenous growth hormone reserve correlate to ovarian response to human menopausal gonadotropins, *Isr. J. Med. Sci.*, 25:296.

Shadmi, A. L., Lunenfeld, B., and Bahari, C., et al., 1987, Abolishment of the positive feedback mechanism: a criterion for temporary medical hypophysectomy by LH-RH agonist, *Gynecol. Endocrinol.*, 1:1.

Smith, P. E., and Engle, E. T., 1927, Experimental evidence regarding role of anterior pituitary in development and regulation of genital system, *Amer. J. Anatomy*, 40:159.

Zondek, B., and Ascheim, S., 1927, Das hormon des Hypohysenvorderlappens; Testobject zum Nachweis des Hormons. *Klinische Wochenscrift*, 6:248.

PHYSIOLOGY OF FOLLICLE DEVELOPMENT IN STIMULATED CYCLES

Torbjörn Hillensjö, Jan-Henrik Olsson, and Lars Hamberger

Department of Obstetrics and Gynecology, University of Göteborg
Sahlgrenska Hospital, S-413 45 Göteborg, Sweden

INTRODUCTION

The complicated interplay between the hypothalamo-pituitary-ovarian axis can obviously continue undisturbed for years in most women which is remarkable in the light of the difficulties facing the clinician who tries to improve upon nature in medically assisted reproduction. Follicle growth and maturation in the human is a prolonged process, the dynamics of which is not known in detail. Studies by Gougeon and Lefévre (1983) suggest that it takes about 85 days for one follicle to grow from the primary to the preovulatory stage i.e. almost three entire menstrual cycles. According to these authors the first growth phase requires about 60 days and appears to be independent of gonadotropins. The second growth phase requires about 14 days and the role of gonadotropin is uncertain. During the third and final growth phase the development of the follicle is clearly dependent on gonadotropin and, possibly, local growth factors. During its growth the follicle accumulates up to 50 million granulosa cells (McNatty 1978) and one important aspect of follicle growth is thus cell proliferation. Another important aspect of follicle development is differentiation, which is reflected by typical changes in follicular fluid levels of steroids - small antral follicles having an androgenic pattern, larger antral follicles having an estrogenic or even progestagenic pattern (McNatty 1978, Westergaard et al. 1986).

The key enzyme in follicular steroidogenesis is aromatase, the activity of which is regulated by FSH (Garzo and Dorrington 1984, Steinkampf et al. 1987). In the mid-follicular phase the leading follicle acquires a dominant role reflected by assymetric secretion of estradiol in ovarian venous plasma (Chikazawa et al. 1986), rising peripheral estrogens and sonographically recorded increase in follicular diameter (Leyendecker and Wildt 1983).

Advances in Assisted Reproductive Technologies, Edited by
S. Mashiach *et al.*, Plenum Press, New York, 1990

Table 1. Possible Causes of Disturbed Follicle Maturation in Stimulated Cycles

Excess FSH

- increased follicular regulatory protein
- increased inhibin subunits
- high estradiol

Direct effects of clomiphene isomers

- inhibition of progesterone synthesis
- inhibition of aromatase

Direct effects of GnRH analogs

- inhibition of FSH effects

The details of the regulation of proliferation and differentiation of the developing human follicle are as yet unknown. The role of classical growth factors as well as a number of postulated follicular fluid factors with local modulatory effects need further studies both in the natural and in the stimulated cycles.

In the stimulated cycle there may be, theoretically, a number of disturbances caused by excess gonadotropic stimulation or by direct interference of agents such as clomiphene citrate and GnRH analog at the ovarian level (Table 1).

We report here on some of our studies on human granulosa cells aiming at elucidating (i) the role of growth factors, particularly IGF-I in granulosa cells of normal and stimulated cycles and (ii) the possible direct effects of clomiphene isomers and GnRH on granulosa cell steroidogenesis.

MATERIAL AND METHODS

Granulosa cells were obtained (i) during the follicular phase of normally menstruating women scheduled for gynecological laparotomy for reasons unrelated to ovarian pathology, or from (ii) the follicular aspirates of women undergoing oocyte retrieval in the course of IVF/ET. These patients had been stimulated by a combination of clomiphene citrate (Clomivid®, Draco) hMG (Humegon®, Organon) and hCG (Pregnyl®, Organon or Profasi®, Serono). The cells were cultured as described elsewhere (Hillensjö et al. 1985) in medium 199 with or without fetal bovine serum as specified and in the presence of growth factors, gonadotropins, clomiphene isomers (supplied by Mr Jan Borg, Serono) or a GnRH analog as detailed below. As end-points the cellular uptake of ^{3}H- thymidine was analyzed as an indicator of DNA synthesis. The accumulation of progesterone and estradiol in the medium was analyzed by RIA.

Table 2. Effects of IGF-I on ^{3}H-thymidine Uptake in Human Granulosa Cells

	Natural cycle %	Stimulated cycle %
Control	100 (6)	100 (6)
IGF-I	194 ± 50 (6)*	303 ± 85 (6)**
LH	—	34 ± 10 (4)*

Granulosa cells were obtained from 6 follicles of normal cycles (cycle day 3–15) in women undergoing gynecological laparotomy, or from aspirates of 6 women during the course of IVF-treatment. After an initial 2-day culture in medium containing 1% fetal bovine serum, the media were changed to serum-free with and without IGF-I 10 ng/ml or LH 100 ng/ml. Cells were cultured for 18-20 h before ^{3}H-thymidine (0.5 µCi/ml) was added for the last 10–12 h of culture. Values shown are ^{3}H-thymidine uptake expressed as percent of control. Mean ± SE (n) indicates number of experiments.
* $p < 0.05$ vs control, ** $p < 0.01$.

RESULTS AND DISCUSSION

In our first series of experiments we found that IGF-I under serum-free conditions was able to significantly stimulate ^{3}H-thymidine uptake in granulosa cells from natural as well as stimulated cycles (Table 2). Under these conditions there was no effect on progesterone synthesis. However, in serum-containing medium we found that IGF-I potentiated the effect of FSH on progesterone synthesis during prolonged culture (4–6 days) of cells from natural cycle (data not shown). These biological effects in human granulosa cells are consistent with the demonstration of IGF-I receptors (Gates et al. 1987), secretion of IGF-binding protein in vitro (Suikkari et al. 1989) and possibly IGF-I (Adashi et al. 1985) in human granulosa cells and suggest that this growth factor may have a physiological role in human follicular development.

In our second series of experiments we examined the possible direct effects of clomiphene isomers on the granulosa cells. As we have previously shown (Olsson et al. 1987) the clomiphene isomers in vitro inhibited progesterone secretion (Table 3) and also estradiol secretion in the presence of androgen substrate (Table 3). Of the two isomers zu-clomiphene appeared somewhat more potent than en-clomiphene. Commercially available clomiphene citrate contains a mixture of the two isomers and we found that also the mixture of the two isomers (at 10^{-6} M) inhibited steroid secretion (data not shown). The follicular fluid levels of clomiphene during the actual treatment period in the early follicular phase is to our knowledge not known. Our findings therefore only shows a potential side effect of clomiphene during stimulated cycles, the clinical importance of which needs additional study.

In our third series of experiments we studied the possible direct effects of a GnRH analog on the FSH-response in granulosa cells of growing follicles. We found a consistent inhibitory effect of GnRH (10^{-6} M) in vitro on FSH-induced progesterone secretion (Table 4).

Table 3. Effect of Clomiphene Isomers on Steroid Secretion in Isolated Human Granulosa Cells

Treatment	(M)	Exp. 1 ng progesterone/well	Exp. II ng estradiol/well
Control	—	682 ± 59	86 ± 13
Enclomiphene	10^{-7}	—	68 ± 8
	10^{-6}	578 ± 21*	69 ± 8
	10^{-5}	170 ± 5**	29 ± 9**
Zuclomiphene	10^{-7}	—	59 ± 12
	10^{-6}	397 ± 33**	44 ± 4*
	10^{-5}	125 ± 11**	27 ± 6**

Granulosa cells obtained from follicle aspirates during oocyte retrieval in stimulated cycles were cultured for 2 days in modified medium 199 containing 10% fetal bovine serum and hLH 100 ng/ml, +/- clomiphene isomers. In ex. II testosterone 0.1 µg/ml was present in the medium. Accumulation of progesterone (exp I) or estradiol (exp II) was measured with RIA.
Mean ± SE of 3–6 determinations are shown.
* $p < =.05$ ** $p < 0.01$ vs control.

Table 4. Effects of GnRH Analog on Progesterone (pg/1000 cells x 48 h) Secretion in Human Granulosa Cells

	-FSH	+FSH
Control	541 ± 106	1649 ± 322**
GnRHa	475 ± 103	737 ± 169

Granulosa cells were isolated from 7 follicles (diameter 6–12 mm) on cycle day 6–10 of normally menstruating women during gynecological laparotomy for reasons unrelated to ovarian pathology. Cells were cultured for 2 days with and without hFSH (100 ng/ml) or D-ala^6-desgly 10-GnRH ethylamide (GnRHa, 10^{-6}M) Mean ± SE.
** $p < 0.01$ compared to absence of FSH.

This inhibitory response was only seen in granulosa cells derived from follicles (6–12 mm) in the early to midfollicular phases of the cycle. Granulosa cells obtained from larger follicles in the late follicular phase were challenged with LH in vitro and in this case there was no inhibition by the GnRH analog (data not shown). These results may indicate a potential effect of treatment with long-acting GnRH analog according to the long protocol in stimulation cycles for IVF/ET or GIFT. Whether the follicular fluid levels of GnRH actually reach the high concentrations necessary for inhibiting the FSH-response in vitro remains an open question. Data presented at this meeting suggest that the level of GnRH analog found in follicular fluid is only in the order of 10^{-10}M (Loumaye et al. 1989). Although high-affinity GnRH-receptors exist in rat granulosa cells (Hsueh and Jones 1981) only low-affinity GnRH-receptors have been reported in human ovarian tissue (Bramley et al. 1985). Our results may possibly relate to the interaction with endogenous GnRH-like peptides rather than to authentic GnRH (Aten et al. 1987).

CONCLUSIONS

Follicle development in the human is a prolonged process and the hormonal regulation of proliferation and differentiation has so far only been studied in the final growth phase. It is obvious that the gonadotropins play a major role in this regulation but increasing evidence suggests that growth factors such as IGF-I may be involved as local modulators or even mediators. In stimulated cycles excess gonadotropin may cause disturbances in follicle growth and maturation due to overproduction of local inhibitory factors and certain pharmacological agents used in conjunction with exogenous gonadotropins may, potentially, interfere with the hormonal response at the ovarian level.

ACKNOWLEDGEMENTS

This study was supported by grants from the Swedish MRC (5650, 2873, 5978). We thank Mrs Ann-Louise Dahl for typing the manuscript.

REFERENCES

Adashi, E.Y., Resnick, C.E., DÉrcole, A.J., Sveboda, M.E., and Van Wyk, J.J., 1985, Insulin-like growth factors as intraovarian regulators of granulosa cell growth and function, *Endocrine Reviews*, 6:400.

Aten, R.F., Polan, M.L., Bayless, R., and Behrman, H.R., 1987, A gonadotropin- releasing hormone (GnRH)-like protein in human ovaries: similarity to the GnRH-like ovarian protein of the rat, *J. Clin. Endocrinol. Metab.* 64:1288.

Bramely, T.A., Menzies, G.S., and Baird, D.T., 1985, Specific binding of gonadotropin-releasing hormone and an agonist to human corpus luteum homogenates: characterization, properties, and luteal phase levels, *J. Clin. Endocrinol. Metab.* 61:834.

Chikazawa, K., Araki, S., and Tamada, T., 1986, Morphological and endocrinological studies on follicular development during the human menstrual cycle, *J. Clin. Endocrinol. Metab.* 62:305.

Garzo, V.G., and Dorrington, J.H., 1984, Aromatase activity in human granulosa cells during follicular development and the modulation by follicle-stimulating hormone and insulin, *Am. J. Obstet. Gynecol.* 148:657.

Gates, G.S., Bayer, S., Seibel, M., Poretsky, L., Flier, J.S., and Moses, A.C., 1987, Characterization of insulin-like growth factor binding to human granulosa cells obtained during in vitro fertilization, *J.Recept. Res.* 7:885.

Gougeon, A., and Lefévre, B., 1983, Evolution of the diameters of the largest healthy and atretic follicles during the human menstrual cycle, *J. Reprod. Fertil.* 69:497.

Hillensjö, T., Sjögren, A., Strander, B., Nilsson, L., Wikland, M., Hamberger, L., and Roos, P., 1985, Effect of gonadotrophins on progesterone secretion by cultured granulosa cells obtained from human preovulatory follicles, *Acta Endocrinol.* 110:401.

Hsueh, A.J.W., and Jones, P.B.C., 1981, Extrapituitary actions of gonadotropin- releasing hormone, *Endocrine Rev.* 2:437.

Leyendecker, G., and Wildt, L., 1983, Control of gonadotropin secretion in women. *In:* Norman, R.L. ed: Neuroendocrine aspects of reproduction. AP Ny p. 295.

Loumaye, E., Coen, G., Pampfer, S., Van Kvicken, L., and Thomas, K., 1989, Administration of a GnRH agonist during ovarian hyperstimulation for IVF leads to significant concentrations of peptide in human follicular fluids. Abstract presented at the VI World Congress In vitro fertilization and alternate assisted reproduction, Jerusalem, Israel, April 2-7.

McNatty, K.P., 1978, Cyclic changes in antral fluid hormone concentrations in humans, *Clin. Endocrinol. Metab.* 7:577.

Olsson, J.H., Nilsson, L., and Hillensjö, T., 1987, Effect of clomiphene isomers on progestin synthesis in cultured human granulosa cells, *Human Reprod.* 2:463.

Steinkampf, M.P., Mendelson, C.R., and Simpson, E.R., 1987, Regulation by follicle-stimulating hormone of the synthesis of aromatase cytochrome P-450 in human granulosa cells, *Mol. Endocrinol.* 1:465.

Suikkari, A.M., Jalkanen, J., Koistinen, R., Bützow, R., Ritvos, O., Ranta, T., and Seppälä, M., 1989, Human granulosa cells synthesize low molecular weight insulin-like growth factor-binding protein. *Endocrinology* 124:1088.

Westergaard, L., Christensen, I. J., and McNatty, K.P., 1986, Steroid levels in ovarian follicular fluid related to follicle size and health status during the normal menstrual cycle, *Human Reprod.* 1:227.

SAFE GONADOTROPHINS

H. M. Vemer and E. H. Houwink

Scientific Development Group
Organon International BV, Oss, The Netherlands

Gonadotrophin therapy is a special kind of therapy in more than one way. Couples visiting an infertility clinic are not patients in the classical sense of the word: they are not really ill, their condition is not life-threatening and most of them can lead a normal life without treatment. That makes treatment highly elective in the sense that the benefit/risk ratio must be extremely high. Also the therapy itself has something special: we are not talking about chemical substances produced de novo in a laboratory, but about hormones, produced by the human body, excreted through body fluids and extracted from urine and purified in the factory.

Patients, doctors as well as manufacturers and authorities have always been alert to infections which might arise from these so-called biologicals. This alertness has of course been enlarged due to the occurrence of Creutzfeldt Jacob Disease virus in growth hormone extracted from human pituitaries. Especially Hepatitis-B and AIDS infections from blood products have been alarming. It is therefore of utmost importance to be sure that gonadotrophin preparations, derived from human urine, are free from contamination with AIDS as well as with other viruses. Hepatitis-B virus can be excreted via the urine (Karayiannis et al. 1985), also in symptomless carriers. Only one report appeared in the literature about a low virus contamination with HIV in the urine of a male homosexual (Levy et al. 1985).

During the process of extraction and production, most viruses are eliminated. Especially the routine treatment with high concentrations of ethanol of urines used for the production of gonadotrophins inactivates retroviruses like HIV and most other viruses.

Advances in Assisted Reproductive Technologies, Edited by
S. Mashiach *et al.,* Plenum Press, New York, 1990

53

To secure the absence of contaminating viruses, routine antigen testing for hepatitis-B and HIV are now being performed on all batches of bulk gonadotrophins of Diosynth, the exclusive supplier of Organon. The Hepanostika HBsAg is an in vitro diagnostic kit for the detection of hepatitis-B surface antigen, while the Vironostika HIV antigen is a kit for the detection of Human Immunodeficiency Virus antigens. These tests are enzyme-immuno-assays, based on a "sandwich"-principle. The solid phase contains the antibody to which the antigen binds. Then another incubation with labelled antibody follows, which in turn binds with the antigen (the sandwich). The label finally reacts with a substrate and the sub-sequent color can be read optically.

Up till now the absence of virus contamination in all batches of urine-derived bulk gon-adotrophins has confirmed the safety of our production procedures in this respect. Because virus testing is performed on concentrated bulk gonadotrophins, the safety of the final in-jection product is some orders of magnitude higher compared to the safety of released blood for transfusion: for Humegon and Pregnyl the sample size tested is more than 20% of a dai-ly dosage, while for whole blood this is less than 0.05% of a transfusion.

Due to newly developed test methods like polymerase chain reaction, having a very high sensitivity and specificity, it is now possible to detect even smaller amounts of virus than by ELISA antigen tests. The polymerase chain reaction, first described in 1985 (Saiki et al.), is based on a specific enzymatic amplification of target DNA in vitro. The 10^5 to 10^6 times amplified DNA fragments can be made visible by gel-electroforesis or by dot-spot/Southern blot hybridization with a specific probe. Using this test in a model system the amplification of one femtogram ($=10^{-15}$ gram) target DNA can be demonstrated by means of gel-electroforesis (Quint and Herbrink, 1988). In the near future we will be apply-ing routinely the polymerase chain reaction to bulk Humegon and Pregnyl to confirm the absence of viral contaminants with the most up to date methods.

It is concluded, that the preparations Humegon and Pregnyl can be regarded as safe for parenteral use due to the routine viral antigen testing with the available very sensitive and specific test methods on every single batch of bulk material.

REFERENCES

Karayiannis P, Novick DM, Lok ASF, Fowler MJF, Monjardino J, Thomas HC. Hepatitis-B virus DNA in saliva, urine and seminal fluid of carriers of hepatitis Bc antigen. *Brit Med J* 290: 1853-1855, 1985.

Levy JA, Kaminsky LS, Morrow WJW, Steiner K, Luciw P, Dina P, Hoxie D, Oshiro L. Infection by the retrovirus associated with acquired immunodeficiency syndrome: clinical, biological and molecular features. *Annals Internal Medicine*, 103: 694-699, 1985.

Quint WGV, Herbrink P. Monoklonale antilichamen en DNA/RNA-probes voor "snel-detectie" van micro-organismen bij infektieziekten. *Biotechnol. Ned.*: 125-129, 1988.

Saiki RU, Scharf S, Faloona F, Mullis KB, Horn GT, Erlich HA, Arnheim N. Enzymatic amplifi-cation of β-globin genomic sequences and restriction site analysis for diagnosis of sickle cell anemia. *Science* 230: 1350-1353, 1985.

IVF : LH SURGE VS HCG

Douglas Keeping

Queensland Fertility Group
225 Wickham Terrace, Brisbane 4000 Australia

INTRODUCTION

There is a prevailing attitude in the IVF literature that in stimulated cycles, an LH surge should not be allowed to occur, and that hCG must be given before such a surge occurs. Some examples are:

1. Navot and Rosenwaks (1988): "When the optimal stage of advanced follicular maturation is reached, the surrogate LH surge, in the form of hCG, is given. If hCG is inappropriately delayed, the elevated E2 may elicit, in a minority of patients, through positive feedback, spontaneous surges of serum LH. This LH rise:

   (1) May cause premature luteinisation
   (2) May confound the already complicated monitoring system of IVF patients
   (3) May require rescheduling of ovum pickups
   (4) LH surge may be attenuated
   (5) May adversely affect oocyte quality."

2. Talbert (1988): "When an LH surge occurs, most IVF programs recommend that the cycle be terminated. This recommendation is based on the observed lower pregnancy rates in such cycles."

   Lejeune et al. reported "no clinical pregnancies in 21 cycles with an LH surge, and 16 pregnancies out of 100 cycles when an LH surge did not occur."

Advances in Assisted Reproductive Technologies, Edited by
S. Mashiach *et al.*, Plenum Press, New York, 1990

3. Thatcher and De Cherney (1988): "In cycles scheduled for IVF, while similar medication regimes and cycle monitoring are used, it is imperative that the spontaneous surge does not occur... the goal of several new therapies... is to suppress the LH surge and ultimately ovulation."

The Queensland Fertility Group has sifted through this literature and it's considered opinion is that this is not absolutely factual.

The Queensland Fertility Group comprises six clinicians and six scientists, doing 35–40 IVF cycles per week. The hours of egg pickup are between 5.30am and 10pm, 365 days per year. To date the Group has between 600 and 700 babies. Before going further, it should be emphasised that these are babies, with nappies, hats, and going home in the back of somebody's car. They are not biochemical pregnancies, positive hCG's, or the plethora of other failures vaunted by IVF groups as a measure of success.

METHOD

This survey pertains to 5197 consecutive cycles which have been entered into the computer during the years 1985–1988, with an overall pregnancy rate of 16.4%.

Very briefly, the method of stimulation is similar to that used by most groups around the world. Clomid 100 mg is given from days 1–5 of the treatment cycle, gonadotrophins are used from day 5 with a daily dose starting at 75 I.U. for about two days and moving to 150 I.U. per day, and this is continued until the time of egg pickup. The exact dosage and regime varies, and is tailored to the individual to some extent. Over the past six years the majority of patients have been monitored with oestradiol, LH, and progesterone assays daily, escalating to two to three times per day around day 11 of the cycle. Over this time the vast majority of patients were monitored until they had an LH surge, and that is still the philosophy of the Group that we would rather the LH surge occurred. In a proportion of cases, hCG has been administered to trigger maturation and ovulation. When this is done 4000 units of hCG are given. There has been no attempt to perform a double blind controlled trial. Thus, there is no random selection into the groups of LH surge or hCG and this will be referred to later in the results. Egg pickup is timed to be 34 to 36 hours from the beginning of the surge or from the time of administration of hCG. The methods of egg pickup and embryo transfer are standard and are not referred to further in this presentation.

RESULTS

As stated earlier, these results pertain to 5197 consecutive cycles. Table 1 shows the proportion of women who were allowed to have an hCG surge and the proportion who had hCG administered prior to a surge.

Table 1. Basic Data (2)

	n	%
LH surge	4285	82.5
hCG	912	17.5
Total	5197	100

Table 2. Basic Data (3) Aetiology

	Total Number	% LH Surge
Tubal	2477	82.3
Oligospermia	733	84.7
Idiopathic	1343	83.6
Endometriosis	918	79.5
Antibodies	305	83.6
Total	5197	82.5

(N.B. Some multiple aetiology)

Table 3. Basic Data (4)   Egg Pickup: Method

	Total Number	% LH Surge
Laparoscopy IVF	3698	84.5
T/V U/S IVF	649	77.2
GIFT	850	77.8
Total	5197	82.5

Table 2 shows the proportion allowed to have an LH surge against etiology, to show that there is no great bias in selection, and that just over 80% of each aetiological group had an LH surge. Similarly Table 3 shows the proportion having a surge by method of egg pickup. Again there is no enormous difference between laparoscopic pickup, transvaginal ultrasound, or GIFT.

Table 4 shows the percentage of cases where eggs were not obtained at the egg pickup and also the proportion where there was failure to fertilise. It should be noted that the total cancellation rate prior to pickup is only 3%, which is considerably less than many groups around the world, who routinely cancel cycles if an LH surge occurs. We have little doubt which cancellation rates the patients prefer.

Table 4. Basic Data (5)

	LH Surge	hCG
Failure to get eggs	2.1%	1.0%
Nil fertilised	10.2%	6.3%
Cancelled prior to pickup (overall)		3%

Table 5. Basic Data (6)

	LH Surge	hCG
No. of cycles	4285	912
No. of eggs	16,257	4330
Eggs per pickup	4.39	5.58
Fertilisation rate	62.8%	64.1%
Pregnancy rate	16.5%	15.8%

Table 6. Pregnancy Rate (1)

		Total No.	No. Pregnant	% rate
Overall	: Surge	4285	708	16.5
	: hCG	912	144	15.8
Laparoscopy	: Surge	3123	418	13.4
	: hCG	575	74	12.9
Ultrasound	: Surge	501	64	12.8
	: hCG	148	20	13.5
GIFT	: Surge	661	226	34.2
	: hCG	189	50	26.5

Table 5 summarises the number of cycles, the total number of eggs, the eggs per pickup, the fertilisation rate, and the pregnancy rate, which is similar for both groups.

Table 6 shows the pregnancy rate for LH surge and hCG, overall, and for each of the three groups, laparoscopy, ultrasound, and GIFT. There is no major difference in any of the three between those having LH surge and those having hCG. There is a slightly higher success rate in GIFT in those having an LH surge but this is probably explained by the small numbers.

Table 7. Pregnancy Rate(2) Summary

	% Pregnant	
	LH Surge	hCG
Overall	16.5	15.8
Laparoscopy	13.4	12.9
T/V Ultrasound	12.8	13.5
GIFT	34.2	26.5

No statistics have been given. With the large numbers involved, very small differences attract a p value which is significant, but which is not helpful in distinguishing events of clinical importance.

Table 7 which is Pregnancy Rate (2) summarises in more simple terms the success rates shown in the last Table.

Moving from pregnancy rate to pregnancy outcome, one gets into an area of IVF which is notorious for its untruths. In Australia it is said that the following are the three lies most commonly told:

1. The cheque is in the mail.
2. I am from the Government and I am here to help you.
3. I promise I will still love you in the morning.

In IVF circles one can add a fourth great untruth as follows:

4. Our IVF success rate is x%. It does not need a great deal of cynicism to go along with such a philosophy. World IVF meetings tend to promote grandstanding, and the blurring of figures, mainly designed to show that a particular group has an almost 100% success rate, at a cost of about $1, at no inconvenience to the patient. We have learned over the last few years to be a little wary of these claims. The only objective way of reporting pregnancy rates that we have seen was suggested at another meeting by the Joneses from Norfolk Virginia. They suggested a sort of correlation matrix table, with the numbers of patients wanting IVF, accepted onto the program, starting a treatment cycle, reaching egg pickup stage, and so on, until the final one is live birth. These same categories are put on the vertical and horizontal axis and groups can then express their success in terms of a clearly defined numerator and denominator. So far this doesn't seem to have attracted great acceptance around the world, and part of its lack of success must be due to the fact that the results are then ruthlessly honest. Having said that, the following are the results of pregnancy outcome in our group.

Table 8 shows the miscarriage rate, ectopic rate, perinatal mortality, and take home

Table 8. Pregnancy Outcome

	LH Surge		hCG	
	No.	%	No.	%
Pregnant	523	100	97	100
Miscarriage	148	28.3	27	27.8
Ectopic	24	4.6	7	7.2
Perinatal Death	10	1.9	0	0
Take home baby	341	65.2	63	64.9
Take home baby/cycle		12.1		11.8

baby rate for both LH surge and hCG. These are expressed, using the denominator of all those patients who started a treatment cycle. There is no great difference between those having LH surge and those having hCG in terms of pregnancy outcome. The only apparent differences in perinatal mortality, are that the numbers here are too small to assume any clinical significance. The bottom line is the bottom line. That is, the number of babies that go home in a capsule on the backseat of a motor car, expressed as a proportion of those who started an IVF cycle.

The next three tables are tangential to the main issue of this paper. However it would appear that our Group is somewhat different in having such a large number of patients allowed to have an LH surge, that it gives us a fairly unique opportunity to look at some baseline data in terms of success. Our clinical judgement made us feel that high baseline LH levels, low and high oestrogen peaks, and those having an egg pickup at a time which appeared to be different from what one would expect in a natural cycle, might all have lower success rates.

Table 9 shows the success rate in terms of the baseline LH level. The baseline LH was divided up into those with a low baseline LH, which were too few to analyse, those plus or minus one standard deviation from the mean, which is the majority, those between plus one and plus two standard deviations, and those more than two standard deviations above the mean. It can be seen that the pregnancy rate up to plus two standard deviations from the mean is similar, but as expected once the baseline LH is more than plus two standard deviations the success rate drops, although there is still an achievable rate.

Table 10 shows the highest level of oestradiol reached during the cycle. The results here did not bear out our gut feelings, with a pregnancy rate which was fairly similar, for those with low and high oestrogen peaks, as well as those in the mid range.

Table 11 looked at the day of actual egg pickup, compared to the day we would have anticipated ovulation to have occurred based on her menstrual data. We accept that this is a

Table 9. Tangent (1)   Baseline LH Level

	Total Number	Number Pregnant	% Pregnant
Low (< - 1 S.D.)	too few		
Mean (± 1 S.D.)	1934	333	17.2
High (+ 2 S.D.)	189	30	15.9
Very high (> + 2 S.D.)	95	6	6.3

Table 10. Tangent (2)   Oestrogen Peak

	Total Number	Number Pregnant	% Pregnant
Low (< - 1 S.D.)	231	35	15.2
Mean (± 1 S.D.)	1709	285	16.7
High (+ 2 S.D.)	202	37	18.3
Very high (> + 2 S.D.)	82	11	13.4

Table 11. Tangent (3)   Day of Egg Pickup

	Total Number	Number Pregnant	% Pregnant
Early (< - 1 S.D.)	61	8	13.1
Mean (± 1 S.D.)	2002	343	17.1
Late (> + 1 S.D.)	155	16	10.3

very crude index, and it may be that the results do not help much. It would appear to show that whether an LH surge occurs prematurely, by our calculated criteria, or late, has no impact, or very little impact on the pregnancy rate.

SUMMARY

Table 12 shows the overall conclusion in terms of pregnancy rate and pregnancy outcome. It would appear that in our program, where 85% of patients have an LH surge, there is no distinguishable difference between those having a surge and those having hCG administered.

The advantages of either method are shown in Table 13. Certainly in our program there is a very low cancellation rate which has patient appeal. I think as well we have a better understanding of the timing of ovulation, and when hCG is given it may be given at a more optimal time than in those programs who are trying to pre-empt an LH surge. The disadvantages of an LH surge are logistical, in terms of long hours and perhaps more monitoring.

Table 12. Summary (1)

(1)	PREGNANCY RATE	:	SAME
(2)	PREGNANCY OUTCOME	:	SUBSTANTIALLY THE SAME

Table 13. Summary (2)

	ADVANTAGES OF LH SURGE		DISADVANTAGES OF LH SURGE
(1)	LOW CANCELLATION RATE	(1)	LONG HOURS
(2)	BETTER UNDERSTANDING OF TIMING OF OVULATION	(2)	MAYBE MORE MONITORING

Table 14. Summary (3)

	ADVANTAGES OF hCG		DISADVANTAGES OF hCG
(1)	SHORTER HOURS	(1)	TIMING - MAYBE TOO EARLY
(2)	PRE-EMPT POOR LH SURGES	(2)	HIGH CANCELLATION RATES
(3)	MAYBE LESS MONITORING		

The advantages of hCG are shown in Table 14. Again there are logistical advantages in terms of shorter hours and perhaps less monitoring. The other small group which may benefit, is that group which has a poor and seemingly inefficient LH surge and in these the administration of hCG may pre-empt such an event. The disadvantages of hCG administration are that the timing may be too early. The more one strives to avoid an LH surge, the more unphysiological will be the timing of administration. The high cancellation rates have been alluded to earlier.

Table 15 summarises our views on which method is better. The basic mixture of the two is better. Perhaps the most important thing is the timing of hCG administration. When we give hCG, in general terms it will be either coinciding with the natural LH surge, or just before, and we feel that such timing maybe very different from those groups who are trying to pre-empt an LH surge. With no figures whatsoever to back up such a statement, we would guess that the administration of hCG early enough to pre-empt most surges, may well be disadvantageous.

Table 16 shows the possible bias in interpreting results. In those programs that wish to prevent LH surge, any patients who do indeed have an LH surge may represent an abnormal group of premature surges, given that the surges occurred before early hCG may be given, and therefore analysis of such surges may appear to give a pessimistic result for the outcome of patients having a surge. Similarly programs that wish to await a surge may have

Table 15. Summary (4)

WHICH IS BEST?
NEITHER
PROBABLY A BALANCED MIXTURE OF THE TWO
MAY WELL DEPEND ON TIMING OF hCG

Table 16. Summary (5)

POSSIBLE BIAS

(1)   PROGRAMS THAT WISH TO hCG EVERYONE
THOSE WHO HAVE SURGE MAY REPRESENT AN ABNORMAL
GROUP OF PREMATURE SURGES.

(2)   PROGRAMS THAT WISH TO AWAIT A SURGE
THOSE WHO RECEIVED hCG MAY REPRESENT AN ABNORMAL
GROUP OF LATE SURGES, OR FAILURE TO SURGE.

Table 17. Summary (6)

THOSE WHO SAY LH SURGES DO NOT WORK IN IVF:

(1)   DON'T KNOW MUCH ABOUT OVARIES
(2)   DON'T KNOW ENOUGH ABOUT IVF

a bias in that patients receiving hCG may represent the extreme end of a spectrum, of late surges, or failure to surge, and who are finally given hCG as a last resort. These are just thrown in as a caution to the interpretation of results.

The final table, Table 17, summarises our view of those authors quoted at the beginning of this paper, and the countless dozens of others with similar quotes. We have approaching 700 babies, most of which have resulted from patients having an LH surge. We feel as clinicians that this gives us sufficient authority to state that those people who say LH surges do not work in IVF do not know much about ovaries, and don't know much about clinical IVF.

EXPERIENCE WITH GnRH-AGONISTS IN OVARIAN STIMULATION

**K. Diedrich, R. Schmutzler, Ch. Diedrich, H. van der Ven,
S. Al-Hasani, and D. Krebs**

*Department of Obstetrics and Gynecology, University of Bonn
Sigmund-Freud-Str. 25, D-5300 Bonn, F.R.G.*

Whereas Edwards and Steptoe[1] long considered the spontaneous cycle to be more advantageous for recovering a preovulatory oocyte for in vitro fertilization, it is now accepted that the stimulated cycle provides better results. The therapeutic principles used for ovarian stimulation in an in vitro fertilization program are the same as those used in treating normo- and hypogonadotropic ovarian insufficiency.

This includes clomiphene therapy with subsequent hCG administration and hMG/hCG therapy. These treatment schemes used by our working team produced good results, and they are still the most used therapeutic approaches.[2]

OVARIAN STIMULATION FOR IVF

First choice   Clomid
                                 Clomid/hMG
                                 hMG

Second choice   Pure FSH or FSH/hMG
                                 GnRH-agonist + hMG

Since sonography and additional estradiol and LH determination now provide a good assessment of the quality of the follicular maturation, we are now able to decide on a further course prior to puncture and determine whether the follicular puncture with oocyte retrieval

Advances in Assisted Reproductive Technologies, Edited by
S. Mashiach *et al.*, Plenum Press, New York, 1990

will lead to successful in vitro fertilization with subsequent embryo transfer. The result is that patients who have a disturbed follicular maturation are spared the frustration of follicular puncture. Eighteen percent of the cycles are cancelled prior to retrieval. The causes are as follows:

EFFICIENCY OF OVARIAN STIMULATION

only one dominant follicle	7%
low E_2 values	16%
E_2 decrease after administration (>30%)	8%
Endogenous LH surge (progesterone?)	13%

In these cases with a disturbed follicular maturation, we tried an ovarian stimulation with pure FSH[3] or GnRH-agonist[4] in a new treatment cycle.

Gonadotropin secretion is controlled by the pulsatile release of GnRH. Any change in this pulsatile secretion pattern and thus in continued GnRH-administration leads to a down regulation of the pituitary GnRH-receptors and results in a decreased gonadotrophin secretion. The outcome is a hypo-gonadotropic ovarian insufficiency which in addition to the decrease in FSH and LH, is indicated by a decrease in estradiol and progesterone. At this artificially induced stage of hypogonadotropic amenorrhea hMG stimulation is initiated.

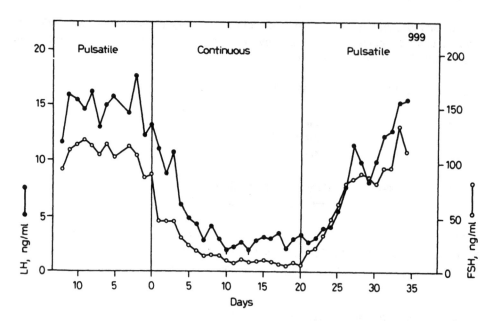

Fig. 1. Values of FSH and LH after treatment with GnRH-agonists.

The development of long-acting superactive GnRH-agonists has permitted a reversible blocking of the pituitary gonadotrophin secretion. Due to the molecular modification the synthetic GnRH-agonists are more resistant to enzymatic degradation at the pituitary cell membrane and there is an increased affinity for the GnRH-receptors. This results in a greater potency and duration of action of the GnRH-agonists as compared to native GnRH.

After an initial stimulatory effect with increased gonadotrophin levels cellular desensitization at the pituitary is achieved. This is caused by a GnRH receptor loss (down regulation) and an uncoupling of receptor from the secretory signal.

This combination of several active mechanisms resulted in an obvious reduction in LH and FSH secretion, thus suppressing temporarily and reversibly the steroid production and the ovarian response. This inhibitory effect of the GnRH-agonists on the gonadotrophic secretion is the rationale for adjuvant therapy with GnRH-agonists.[5, 6, 7, 8, 9]

Nevertheless, there is always therapeutic scope for GnRH-agonists when the therapeutic goal can be reached by suppressing ovarian estrogen secretion. The main scope for therapy with GnRH-agonists in gynecology is:

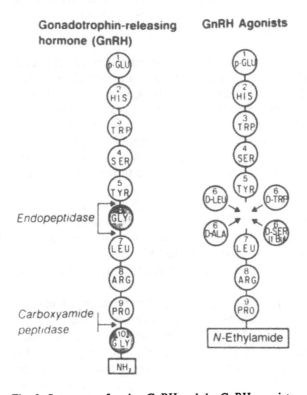

Fig. 2. Structures of native GnRH and the GnRH-agonists.

INDICATIONS FOR GnRH-AGONISTS

1. Treatment of infertility
2. Endometriosis
3. Uterine leiomyoma
4. Breast and ovarian (?) cancer
5. Precocious puberty

There are several applications for using GnRH-agonists, such as: nasal spray (4 × 300 µg/day), s.c. injection or even i.m. depot injection. The s.c. injections are started first with 500µg daily for seven to 10 days; then the treatment with GnRH-agonists is continued with 100 µg daily up to the time of the hCG administration, and lastly, intramuscular injection of 3.2 mg/4–6 weeks.

The fall in gonadotrophins FSH and LH, leads to a hypogonadotrophic amenorrhea, from which stimulation with hMG can afterwards be initiated. The protocols used so far can be classified in two types: the short and the long protocol. In the short protocol, the exogenous gonadotrophin stimulation already starts a few days after GnRH-agonist treatment. This protocol takes advantage of the initial stimulation under the agonist therapy. In the long protocol stimulation with exogenous gonadotrophins started after complete suppression of the pituitary - gonadal axis (Figure 3).[10]

The treatment with GnRH-agonists can be initiated at different times in the menstrual cycle. Starting the treatment in the second half of the luteal phase has proved to be a favourable time in the cycle for obtaining an ovarian suppression. In this phase, suppression period is short, since the ovaries cannot yet respond to the temporary endogenic gonadotrophin elevation induced by the GnRH-agonists. This also becomes evident in an absence of estrogen increase in these patients. Unlike the start of therapy with GnRH-agonists following commencement of the menstrual bleeding the flare-up of the gonadotrophins in this phase can be useful for this stimulation as can be shown in the short protocol (Figure 4).

In the long protocol ovarian stimulation with a combined therapy of GnRH-agonists and hMG, hMG-treatment is initiated when pituitary gonadotrophin secretion is sufficiently suppressed, which is indicated by low LH-values, basal estradiol and the occurrence of climacteric disturbances. The dosage for the hMG therapy is defined according to the usual protocols. Criteria for the hCG administration were also identical to other protocols.

In the following part of the lecture the results of ovarian stimulation using GNRH agonists and gonadotrophins obtained in our clinic or described in literature will be discussed. In one study, 119 cycles of 74 patients with primary or secondary infertility were treated. Th average age was 36 years. This was preceded by an average 2.6 cycles, in which a clomiphene- or hMG-stimulation was performed. The indication for in vitro fertilization was mainly tubal infertility frequently accompanied by other factors. Reasons for the GNRH-agonist treatment was a premature LH surge in 85.1%, hyperandogenemia in 16.2% and in-

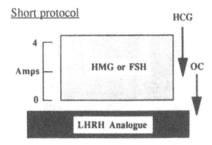

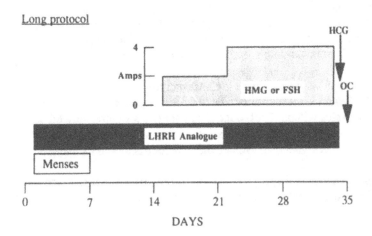

Fig. 3. Short and long protocol of ovarian stimulation with GnRH-agonists.

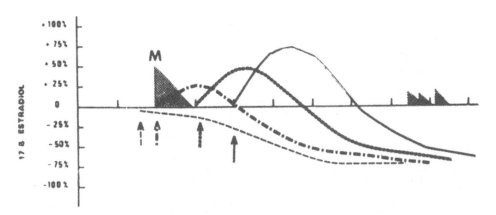

Fig. 4. Start of therapy of GnRH-agonists in the different phases of the cycles.[11]

cipient premature menopause in 6.8%. In some cases, a pretreatment with GnRH-agonists was employed due to endometriosis or leiomyoma of the uterus (10.8%). Decapeptyl was used as GnRH-agonists.[12]

Compared with the preceding cycles without GnRH-agonist treatment, the cancellation rate in these patients was reduced from 65% to 13%. These cancellations were mainly due to insufficient ovarian stimulation.

The combined GnRH-agonists and gonadotrophin treatment resulted generally in a better ovarian stimulation with higher estradiol values (1938 μg/ml versus 1390 μg/ml). More oocytes were gained with 3.9 as against 0.9 in the previous cycles. The number of embryos transferred per cycle increased to 1.8 from 0.5 in the previous cycles. The pregnancy rate in this unfavourable group of patients was 17%.

Two subgroups—hyperandrogenic and hypergonadotrophic patients—were looked at more closely. Seventeen cycles of twelve hyperandrogenic patients were treated. Compared with previous cycles without GnRH-agonist treatment, a significantly higher ovarian stimulation was achieved leading to an increase in the pregnancy rate from 0 to 25%. Although the cancellation rate was higher than average this subgroup achieved the highest pregnancy rate.

The hypergonadotrophic group consisted of only five patients. These patients were selected by elevated gonadotrophin levels of 20 - 30 U/ml and FSH/LH-ratio greater than one

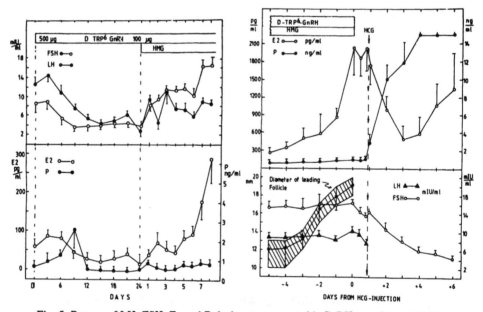

Fig. 5. Pattern of LH, FSH, E$_2$ and P during treatment with GnRH-agonist and hMG.

Table 1. Indications for the Therapy with GnRH-analogues (n = 74 patients)

	Number*	Percentage
Premature LH surge	**63**	**85.1**
CC- and HMG-resistant hypergonadotrophinaemia (FSH>LH)	**5**	**6.8**
Hyperandrogenaemia (LH>FSH)	**12**	**16.2**
GnRHa pretreatment (endometriosis,uterine myoma)	**8**	**10.8**

* The sum exceeds the total number of patients because of overlapping indications

Table 2. Comparison of hMG and GnRH-agonist/hMG for IVF

	HMG	GnRH-agonist/HMG
number of cycles	**195 (100%)**	**119 (100%)**
cancellation rate	**127 (65%)**	**16 (13,4%)**
number of oocytes/cycle	**0.9**	**3.9**
number of embryos/cycle	**0.5**	**1.8**
pregnancy rate/embryotransfer	**2 (3,5%)**	**14 (17,7%)**

GnRH-agonist cycles are compared to previous cycles without GnRH-agonist

indicating an incipient premature menopause. Although the pure ovarian stimulation could be slightly improved in terms of E_2 levels, number of oocytes and cancellation rate, no pregnancy could be established.

The treatment with GnRH-agonist was well accepted by patients and the temporary occurrence of hot flushes tolerated. Nevertheless, patient compliance for i.m. administration was substantially higher than for s.c. application.

In another study with 77 cycles the GnRH-agonists in combination with gonadotrophins were used as stimulation of first choice. The indication for in vitro fertilization corresponds to the distribution in the whole group of patients. With an embryo transfer rate of 84%, a pregnancy occurred in 15 cases producing a pregnancy rate of 23% per embryo transfer. Comparing these results with the general results of the University of Bonn during the period of 1981 to 1988, it becomes evident that the pregnancy rate per embryo transfer of 23% is exactly the same.

This shows that there is obviously no improvement in the quality of follicular maturation in normo-gonadotrophic cycles with normal ovarian function, or at least this is not revealed by the pregnancy rate. The conclusion we drew for our IVF program is that GnRH-agonists in combination with gonadotrophins should only be used as a therapy of second choice. As already mentioned, this group of patients does show a quite satisfactory pregnancy rate.

A study of the literature shows a pregnancy rate of between 20% to 36% when GnRH-agonists and gonadotrophins are used prior to in vitro fertilization. The difference in the pregnancy rates can be explained by the difference in the number of embryos transferred. In our group not more than three embryos were transferred. In the Rutherford-group from London 80% of the 30 pregnancies were achieved after the transfer of more than three embryos. Unfortunately this resulted in a high incidence (33%) of multiple pregnancies.

The other results correspond approximately to the success rates obtained in efficient IVF centers for normocyclic patients using gonadotrophins alone. Also a randomized study

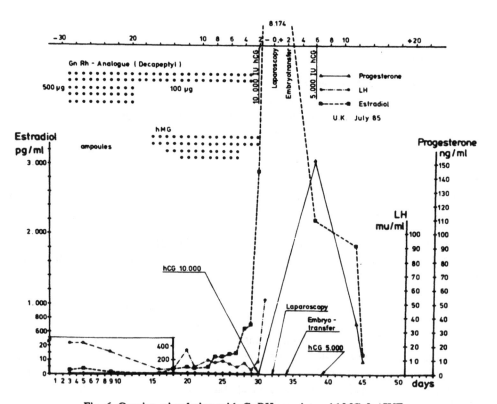

Fig. 6. Ovarian stimulation with GnRH-agonist and hMG for IVF.

Table 3. Ovarian Stimulation with GnRH-agonist/hMG (I. choice) and the Results for IVF at the University of Bonn

	GnRH-agonist/hMG	UFK Bonn 1981-1988
cycles	77	2389
embryotransferrate	84%	76%
pregnancies	15	430
pregnancy rate/cycle	19,5%	18%
pregnancy rate/embryotransfer	23%	23%

Table 4. Summary of Results with GnRH-agonists and hMG in Ovarian Stimulation

	cycles	cancellation rate	pregnancy rate/cycle
Rutherford et al. 1988	83	8%	36%
Lindner et al. 1987	74	-	23%
Smitz et al. 1987	84	8%	21%
Schmutzler et al. 1989	77	5%	19,5%

by Frydman[17] comparing long and short protocols of GnRH-agonists administered to patients who do not have major ovulatory dysfunction, showed a similar ongoing pregnancy rate per treatment cycle in both groups with 18% in the long protocol and 16% in the short protocol.

A crucial advantage of GnRH-agonists is the definitely low cancellation rate of between 5% to 8%. For patients this means much better effectiveness of this expensive infertility treatment.

The treatment with GnRH-agonists has a new and wide field in gynecology. Its effectiveness is achieved by suppressing ovarian function by a reversible medical hypophysectomy. Hence, this results in a favourable indication range for GnRH-agonists as an additional accompanying measure for ovarian stimulation therapy within the scope of in vitro fertilization, particularly in patients with a propensity for endogenic LH surge or with a PCO syndrome and endometriosis.

TREATMENT WITH GnRH-AGONISTS AND GONADOTROPHINS

Advantages:

reduced cancellation rate
increased success rate
improved synchronization of follicular development
timing of oocyte retrieval
new therapeutic approach for:
  endogenous LH surge
  PCO syndrome
  endometriosis

Disadvantages:

increased duration of stimulation
side effects of GnRH-agonists

This stimulation treatment produces a favourable synchronisation of follicular maturation and a reduction in the cancellation rate by avoiding the endogenic LH surge. In addition, it is easier to provide for follicle puncture on a workday.

A disadvantage of the therapy with GnRH-agonists is the longer time required for ovarian stimulation and the higher costs of hMG stimulation. In addition, temporary climacteric disturbances occur, which may be unpleasant.

REFERENCES

1. P. C. Steptoe, and R.G. Edwards, Birth after re-implantation of a human embryo. *Lancet* II:366, (1978).
2. K. Diedrich, H. van der Ven, S. Al-Hasani and D. Krebs, Ovarian stimulation for in-vitro-fertilization. *Human Reproduction*, 3:no.1, 39-44 (1988).
3. K. Diedrich, Ch. Diedrich, H. van der Ven, S. Al-Hasani, A. Werner and D. Krebs, Ovarian stimulation using pure FSH in an in-vitro-fertilization programme. *Human Reproduction*, 3: (Suppl.2), 23-38, (1988).
4. L. Wildt, K. Diedrich, H. van der Ven, S. Al-Hasani, H. Hübner, and R. Klasen, Ovarian hyperstimulation for in-vitro fertilization controlled by GnRH agonist administered in combination with human menopausal gonadotrophins, *Human Reproduction*, 1: no. 1, 15-19, (1986).
5. P. E. Belchetz, T.M. Plant, Y. Nakai, E.J. Keogh and E. Knobil, Hypophysial responses to continuous and intermittent delivery of hypothalamic gonadotrophin-releasing hormone, *Science*, 202: 631, (1978).
6. L. Wildt, A. Haüsler, G. Marshall, J.S. Hutchinson, T.M. Plant, P.E. Belchetz and E. Knobil, Frequency and amplitude of gonadotropin-releasing hormone stimulation and gonadotropin secretion in the rhesus monkey. *Endocrinology*, 109:376, (1981b).
7. J. Sandow, Clinical applications of LHRH and its analogues, *Clin. Endocrinol.*, 18:571 (1983).

8. S. S. C. Yen, Clinical applications of gonadotrophin-releasing hormone and gonadotrophin-releasing hormone analogues, *Fertil. Steril.*, 39:257 (1983).

9. D. H. Coy, J. A. Vilchez-Martinez, E. J. Coy, A. V. Schally, Analogues of luteinizing hormone releasing hormone with increased biological activity produced by D-amino acid substitutions in position 6. *J. Med. Chem.*, 19:423: (1976).

10. H. S. Jacobs, R. N. Porter, A. Eshel, and I. Craft, Profertility uses of LHRH analogues. *in:* LHRH and its analogues: Contraceptive and therapeutic Applications, Part 2, B. H. Vickery, and J. J. Nestror eds., MTP-Press, Lancaster, 1987.

11. B. H. Vickery and G. I. McRae, Effect of continuous treatment with super-agonist to LHRH when initiated at different times of the menstrual cycle in female baboons, *Int. J. Fertil.* 25:171 (1980).

12. R. K. Schmutzler, C. Reichert, K. Diedrich, L. Wildt, Ch. Diedrich, S. Al-Hasani, H. van der Ven and D. Krebs, Combined GnRH-agonist/gonadotrophin stimulation for in-vitro fertilization. *Human Reproduction*, 3: (Supp. 2),29: (1988).

13. A. J. Rutherford, R. J. Subek-Sharpe, K. J. Dawson, R. A. Margara, S. Franks and R. M. L. Winston, Improvement of in-vitro fertilization after treatment with Buserelin, an agonist of luteinising hormone releasing hormone, *Brit. Med. J.* 296:1765, (1988).

14. Ch. Lindner, W. Braendle, V. Lichtenberg, L. Bispink and G. Bettendorf, GnRH agonist/hMG-Stimulation für den intratubaren Gametentransfer (GIFT), Geburtsh. u. Frauenheilk. Sonderheft 1, 49. Jahrgang, S. 81, (1989).

15. J. Smitz, P. Devroey, M. Camus, I. Khan, C. Staessen, L. Van Waesberghe, A. Wisanto and A.C. Van Steirteghem, Addition of Buserelin to human menopausal gonadotrophins in patient with failed stimulations for IVF or GIFT. *Human Reproduction* 3: (Supp. 2)35, (1988).

16. R. K. Schmutzler, C. Reichert, K. Diedrich, L. Wildt, Ch. Diedrich, H. van der Ven, S. Al-Hasani and D. Krebs, Vergleich eines langwirksamen und kurzwirksamen GnRH-Analogs in Kombination mit Gonadotropinen zur In-vitro Fertilization unter verschiedenen Indikationsstellungen, Geburtsh. u. Frauenheilk. Sonderheft 1, 49. Jahrgang, S. 81, (1989).

17. R. Frydman, J. Belaisch-Allart, I. Parneix, R. Forman, A. Hazout and J. Testart: Comparison between flare up and down regulation effects of luteinizing hormone-releasing hormone agonists in an in-vitro fertilization program. *Fertil. Steril.*, 50: no.3, 471 (1988).

THE ASSOCIATION OF OVARIAN HYPERSTIMULATION
AND ZYGOTE INTRAFALLOPIAN TRANSFER

P. Devroey, C. Staessen, M. Camus, A. Wisanto, N. Bollen,
P. Henderix, and A. Van Steirteghem

Center for Reproductive Medicine, Academic Hospital and Medical Campus
Vrije Universiteit Brussel, Laarbeeklaan 101, 1090 Brussels, Belgium

INTRODUCTION

In 1986 we reported the first successful replacement of three zygotes in the fimbrial end of one healthy fallopian tube in a patient with female sperm antibodies.[1] More pregnancies have been established, especially in male and unexplained infertility.[2-5] The results of zygote intrafallopian transfer (ZIFT) in different indications i.e. unexplained, male infertility, endometriosis, polycystic ovarian disease (PCOD) and female sperm antibodies were analyzed.

MATERIALS AND METHODS

From January 1986 until December 1988, 219 ZIFT replacements were performed, for male infertility on 58 occasions, for unexplained infertility on 96, for PCOD on seven, for endometriosis on 12, for female antisperm antibodies on 17, after tubal surgery on three, for mixed indications on seven and in an oocyte donation program on 19 occasions. The pregnancies established after oocyte donation, tubal surgery and for mixed indications were excluded from this analysis.

The preliminary examinations included : repeated semen analyses, hysterosalpingogram/ hysteroscopy, laparoscopy, postcoital test and an endocrine evaluation.

Advances in Assisted Reproductive Technologies, Edited by
S. Mashiach *et al.*, Plenum Press, New York, 1990

Table 1. The Establishment of 92 Pregnancies after 219 ZIFT Replacements

N° of zygotes replaced	Replacements (n)	Pregnancies (n)	(%)
1	24	7	(29.20)
2	37	16	(43.20)
3	158	69	(43.70)
Total	219	92	(42.00)

Ovarian superovulation was induced by human menopausal gonadotrophins (hMG) (Humegon®, Organon, The Netherlands; Pergonal®, Serono, Belgium) alone or in association with either clomiphene citrate (cc, Pergotime®, Serono, Belgium) or gonadotropin releasing hormone analog (Suprefact®, Hoechst, Federal Republic of Germany). These stimulation schemes have been reported extensively by our group.[6—9]

Ten thousand I.U. human chorionic gonadotropins (hCG) (Pregnyl®, Organon, The Netherlands; Profasi®, Serono, Belgium) were injected and 36 h later a transvaginal ultrasound guided ovum pick-up was carried out.[10,11] The oocytes were cultured in Earle's medium as described elsewhere. Fresh or frozen semen was used. The sperm cells were selected by swim-up procedures and inseminated as reported previously.[12,13]

Approximately 18 h after insemination the presence of two pronuclei or more (PN) was assessed. If in one or more oocytes 2 PN were observed a ZIFT replacement was performed.

A video translaparascopic replacement was done. A loaded translucent Teflor catheter, 30 cm long, 0.75 mm diameter (K-KEI-2011, William A. Cook, Melbourne, Australia), containing up to three zygotes was introduced through a 10 cm long trocar (Labotect, Götingen, Federal Republic of Germany) in the left or right healthy fallopian tube for approximately three cm. The zygote(s) were injected and in the adjacent laboratory the catheter was checked to confirm that it was empty.

The supernumerary zygotes were either cryopreserved at the pronucleate stage with 1,2 propanediol (PROH) or left in culture and frozen at multicellular stage with dimethylsulphoxide (DMSO).[14—17]

Statistical analysis. The data were entered in a spread-sheet on an Apple Macintosh PC (Apple Computer, Inc., Cupertino, CA); a chi-square test was applied at the 5% level of significance.

RESULTS

As indicated in Table 1, 219 ZIFT replacements were performed, yielding 92 pregnancies.

Table 2. Pregnancy Outcome After the Replacement of One, Two or Three Zygotes

	One Zygote			Two Zygotes			Three Zygotes		
	T	P	(%)	T	P	(%)	T	P	(%)
Unexplained infertility	6	1		15	6		75	32	
Male infertility	13	3		9	4		36	16	
Endometriosis	0	0		2	1		10	7	
Antisperm antibodies	1	0		6	3		10	5	
PCOD	2	0		0	0		6	4	
	21	4	(19,5)	32	14	(43)	137	64	(46)

T = Transfers
P - Pregnancies

Table 3. Implantation Rate per Replaced Zygotes

N° of Zygotes/Replacement	N° of Zygotes Replaced	N° of Implanted Embryos	(%)
1	21	4	(19)
2	64	16	(23)
3	411	91	(22)

In Table 2 the percentage of pregnancies is reported if either one, two or three zygotes were replaced in unexplained infertility, male infertility, endometriosis, antisperm antibodies and PCOD. The pregnancy rate after respectively the replacement of one, two or three zygotes varied from 19.5%, 43% and 46%.

As demonstrated in Table 3 the implantation rate per replaced zygote was similar if one, two or three zygotes were replaced.

As expected the ongoing multiple pregnancy rate was higher when three zygotes were replaced (30.2%; 16/53) compared to two (9.1%; 1/11).

Out of 92 pregnancies 34 patients delivered 41 infants, including 5 sets of twins and one triplet. The three infants of the triplet died after premature labor. Thirty-five pregnancies were progressing normally and 23 ended in a miscarriage (25%). No ectopic pregnancy occurred.

Out of 730 supernumerary zygotes 439 (60%) were cryopreserved as zygotes (183) or as multicellular embryos (256) (Table 4).

Table 4. Cryopreservation of Supernumerary Zygotes/Embryos

Number of supernumerary zygotes	730	
Cryopreserved as zygotes		183
Number of zygote left in culture	547	
Cleaved embryos	426	
Cryopreserved as 4/8-celled embryos		247
Cryopreserved as blastocyst		9
Total N° of cryopreserved zygotes/embryos		439 (60%)

DISCUSSION

We have demonstrated that the overall pregnancy rate after ZIFT replacement was 42% for all the different indications: unexplained, male infertility, endometriosis, polycystic ovarian disease, female sperm antibodies, operated fallopian tubes, mixed cases and oocyte donation. Furthermore we observed a pregnancy rate of respectively 43% and 46% if two or three zygotes were replaced in unexplained infertility, male infertility, endometriosis, anti-sperm antibodies and polycystic ovarian disease. The implantation rate per replaced zygote was ≈20%, i.e. 19% for one zygote, 23% for two zygotes and 22% for three zygotes. The latter is doubled, compared to the implantation rate per replaced embryo in in vitro fertilization and embryo transfer (IVF/ET). In a retrospective analysis we calculated the implantation rate on 96 ZIFT replacements and 87 IVF/ET replacements for unexplained infertility. A significant difference was observed, 19 % after ZIFT and 11.1% after IVF/ET ($p < 0.001$).

Several arguments are acceptable to explain the higher implantation rate after ZIFT. The exposure to in vitro culture medium is reduced from ≈48 h to ≈18 h. For obvious reasons in vitro culture medium is replaced by in vivo "culture medium" of the fallopian tube. The early embryo cleavage occurs in the physiological tubal milieu. It has been demonstrated that some embryos are expelled out of the uterine cavity after traditional cervical replacements. The zygotes are probably not expelled from the fallopian tube and if they are they can be picked up by the fallopian tube out of the pouch of Douglas.[18] Furthermore the embryo reaches the uterine cavity at the appropriate time as in the natural condition.

The same phenomenon was observed if we compared the implantation rate after ZIFT (19.5%) versus GIFT (12.2%) in unexplained infertility respectively per replaced zygote and per replaced oocyte ($p < 0.05$). As indicated in Table 5 several differences exist between ZIFT and GIFT. The most important is that fertilization can be observed and that at the time of tubal replacement the oocytes are already fertilized. For obvious reasons we know that not all replaced oocytes by GIFT will fertilize. The advantage of ZIFT is that a laparascopy is avoided in the absence of fertilization. The important disadvantage of ZIFT is the two-step procedure spread over only two days.

Table 5. Comparison of ZIFT versus GIFT

	GIFT	ZIFT
Oocytes can be matured in vitro	-	+
Fertilization is observed	-	+
Detection of abnormal fertilization	-	+
Laparascopy is avoided if fertilization failure	-	+
Combination of ultrasound pick-up and laparascopic ZIFT	-	+

The most important side-effect of assisted procreation is the establishment of multiple pregnancies (twins/triplets) after the replacement of three zygotes. Since 16 out of 53 ongoing pregnancies (30.2%) after replacing three zygotes were multiple and only one out of 11 after replacing two zygotes, it seems reasonable to limit the number of replaced zygotes to two. A randomized prospective study would clarify this question, but this strategy is debatable from an ethical viewpoint.

The value of reducing the number of replaced zygotes to two was confirmed by a retrospective analysis. As shown in Table 2 the pregnancy rate after replacing two or three zygotes did not alter the pregnancy rate which varied, respectively, from 43% to 46%. Although these findings support the reduction to two zygotes a large number of attempts are needed to design a final strategy.

To avoid all risks of twin pregnancies the replacement of only one zygote could be proposed but this attitude will reduce the pregnancy rate. Furthermore, the usefulness of natural cycles where one zygote will be available could be investigated.

The abortion rate of 25% is acceptable. It is of great importance that no ectopic pregnancies were established.

These observations suggest that the ZIFT technique is of great value in longstanding non-tubal infertility. Since a maximum of three zygotes are replaced a cryopreservation program is needed to freeze the supernumerary zygotes/embryos. A 5% increase in pregnancies can be expected. Controlled studies are mandatory to confirm all the described aspects of the ZIFT technique.

ACKNOWLEDGMENTS

We thank the nursing and technical staff for their skillful assistance, and especially M. Van der Helst for typing the manuscript. This work was supported by a grant of the Belgian Fund for Scientific Medical Research (No. 3.0036.85).

REFERENCES

1. P. Devroey, P. Braeckmans, J. Smitz, L. Van Waesberghe, A. Wisanto and A.C. Van Steirteghem, Pregnancy after translaparascopic zygote intrafallopian transfer in a patient with sperm antibodies, *Lancet* i:1329 (1986).
2. P.L. Matson, D.G. Blackledge, P.A. Richardson, S.R. Turner, J.M. Yovich and J.L. Yovich, Pregnancies after pronuclear stage transfer, *Med. J. Aust.* 14:60 (1987).
3. J.L. Yovich, D.V. Blackledge, P.A. Richardson, P.L. Matson, S.R. Turner and R. Draper, Pregnancies following pronucleate stage tubal transfer, *Fertil. Steril.* 48:851 (1987).
4. M. Hamori, J.A. Stuckensen, D. Rumpf, T. Kniewald, A. Kniewald and M.A. Marquez, Zygote intrafallopian transfer (ZIFT): evaluation of 42 cases, *Fertil. Steril.* 50:519 (1988).
5. P. Devroey, C. Staessen, M. Camus, A. Wisanto, E. Degrauwe, G. Palermo, E. Van den Abbeel and A.C. Van Steirteghem, Zygote intrafallopian transfer: a new approach, *Gynec. Endocrinol (1990)*.
6. J. Smitz, P. Devroey, P. Braeckmans, M. Camus, I. Khan, C. Staessen, L. Van Waesberghe, A. Wisanto and A.C. Van Steirteghem, Management of failed cycles in an IVF/GIFT program with the combination of a GnRH analogue and hMG, *Hum. Reprod.* 2: 309 (1987).
7. P. Braeckmans, P. Devroey, M. Camus, I. Khan, C. Staessen, J. Smitz, L. Van Waesberghe, A. Wisanto and A.C. Van Steirteghem, Gamete intra-fallopian transfer: evaluation of 100 consecutive attempts, *Hum. Reprod.* 2: 201 (1987).
8. P. Devroey, A. Wisanto, J. Smitz, P. Braeckmans, L. Van Waesberghe and A.C. Van Steirteghem, Ovarian stimulation including in vitro fertilization, *Ann. Biol. Clin.* 45:346 (1987).
9. A.C. Van Steirteghem, L. Van Waesberghe, M. Camus, J. Deschacht, P. Devroey, J. Smitz and A. Wisanto, Hormonal monitoring for in vitro fertilization and related procedures, *Hum. Reprod.* 3 (suppl. 2): 1 (1988).
10. W. Feichtinger and P. Kemeter, Transvaginal sector scan sonography for needle guided transvaginal follicle aspiration and other applications in gynecologic routine and research, *Fertil. Steril.*, 45:722 (1986).
11. L. Hamberger, M. Wikland, L. Enk and L. Nilsson, Laparoscopy versus ultrasound guided puncture for oocyte retrieval, *Acta Europ. Fertil.* 17:195 (1986).
12. G. Palermo, P. Devroey, M. Camus, E. Degrauwe, I. Khan, C. Staessen, A. Wisanto and A.C. Van Steirteghem, Zygote intra-fallopian transfer as an alternative treatment for male infertility, *Hum. Reprod.*, 4:412 (1989).
13. I. Khan, M. Camus, C. Staessen, A. Wisanto, P. Devroey and A.C. Van Steirteghem, Success rate in gamete intra-Fallopian transfer using low and high concentration of washed spermatozoa, *Fertil. Steril.*, 50:19 (1988).
14. B. Lasalle, J. Testart and J.P. Renard, Human embryo features that influences the success of cryopreservation with the use of 1, 2 propanediol, *Fertil. Steril.*, 44:645 (1985).
15. A. Trounson and L. Mohr, Human pregnancy following cryopreservation, thawing and transfer of an eight-cell embryo, *Nature,* 307:707 (1985).
16. A.C. Van Steirteghem, E. Van den Abbeel, M. Camus, L. Van Waesberghe, P. Braeckmans, I. Khan, M. Nijs, C. Staessen, A. Wisanto and P. Devroey, Cryopreservation of human embryos obtained after gamete intra-fallopian transfer and/or in vitro fertilization, *Hum. Reprod.*, 2 593 (1987).
17. E. Van den Abbeel, J. Van der Elst, L. Van Waesberghe, M. Camus, P. Devroey, I. Khan, J. Smitz, C. Staessen, A. Wisanto and A.C. Van Steirteghem, Hyperstimulation: the need for cryopreservation of embryos, *Hum. Reprod.* 3 (suppl. 2):53 (1988).
18. J. D. Schulman, Delayed expulsion of transfer fluid after IVF/ET, *Lancet,* i:44 (1986).

CONTROL OF FOLLICULAR GROWTH
DURING THE PRIMATE MENSTRUAL CYCLE

Anthony J. Zeleznik

Departments of Physiology and Obstetrics and Gynecology
University of Pittsburgh School of Medicine
Pittsburgh PA 15213 USA

INTRODUCTION

The elucidation of the mechanisms responsible for the growth of follicles to the preovulatory stage is a challenge for both basic as well as clinical investigators. For the former, this problem provides a model system in which whole animal physiology must be combined with cellular and molecular biology to determine how one of the many follicles present in the ovary gains an advantage over all others such that it alone reaches maturity, ovulates and releases an oocyte for fertilization. At the level of clinical investigation and clinical practice, the understanding of the physiological and biochemical mechanisms responsible for folliculogenesis and follicle selection must provide the foundation for effective treatment of anovulation as well as ovarian stimulation for IVF and GIFT procedures. The goal of this chapter is to summarize studies performed in macaque monkeys that have provided the physiological explanation for the selection of the preovulatory follicle. Because of the remarkable similarity of the menstrual cycle of these animals with that of humans, it is hoped that information gained from these studies in monkeys will be directly applicable to humans.

OVERVIEW OF FOLLICULOGENESIS

Figure 1 illustrates a schematic representation of the stages of folliculogenesis. Primordial follicles are formed during embryonic development and perhaps for a short time after

Advances in Assisted Reproductive Technologies, Edited by
S. Mashiach *et al.*, Plenum Press, New York, 1990

birth. The initiation of folliculogenesis involves the transition of the primordial follicle to a primary follicle which is characterized by transition of the flattened epithelial cells of the primordial follicle to cuboidal granulosa cells. Although the primordial follicle can be maintained in a growth arrested state for as long as 50 years, once its transformation into a primary follicle occurs, the follicle will either develop to maturity or undergo death (atresia). Preantral follicular growth involves the cellular division of granulosa cells as well as growth of the oocyte. The largest preantral follicles present in the ovary typically have six to eight layers of granulosa cells. Based on studies performed largely in rats, large preantral follicles have very limited steroidogenic capacity and do not have LH receptors. However, granulosa cells from preantral follicles contain FSH receptors and are responsive to FSH in vivo and in vitro.[1-4] As discussed in the next section, because growth of preantral follicles occurs during both the follicular as well as the luteal phase of the menstrual cycle, it is unlikely that the hormonal changes responsible for the maturation of the preovulatory follicle play a major role in the growth of preantral follicles.

The maturation of the preantral follicle to the preovulatory stage involves major morphological and functional changes. Morphologically, the follicle acquires a fluid filled antral cavity. Functionally, the follicle develops the capacity to produce estradiol and progesterone via FSH-dependent induction of required enzymes. In addition, again through actions of FSH, the granulosa cells of the follicle acquire LH receptors. In the absence of FSH and LH, follicles are unable to mature beyond the preantral or early antral stages.[1-4]

The process of folliculogenesis culminating in the selection of a single follicle is usually described as a series of events beginning with recruitment of follicles during the early follicular phase of the menstrual cycle leading subsequently to the dominance of a single follicle at the time of ovulation. Subdividing folliculogenesis into distinct stages, although helpful in conceptualizing the overall process, implies that recruitment, growth and acquisition of dominance are independent phenomena. In reality, recruitment, growth and domi-

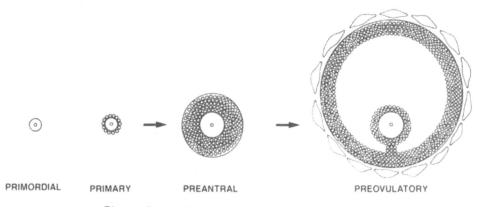

PRIMORDIAL   PRIMARY   PREANTRAL   PREOVULATORY

Fig. 1. Schematic representation of folliculogenesis.

nance are not independent events, but rather progressive stages of a continuing process. The remainder of this chapter deals with the process of folliculogenesis as it occurs throughout the entire menstrual cycle.

THE LUTEAL PHASE

Folliculogenesis is usually discussed within the confines of the follicular phase of the menstrual cycle because it is during this period that follicles mature to the preovulatory stage. However, because preovulatory follicular growth does not occur during the luteal phase of the cycle, an examination of follicular growth during this period should provide information regarding the mechanisms by which folliculogenesis is restrained during this period as well as providing information regarding the stimulus for the initiation of preovulatory folliculogenesis at the beginning of the next follicular phase.

Studies in humans in which the corpus luteum was removed either during the early, mid or late luteal phase demonstrated that regardless of the time of the luteal phase that luteectomy was performed, the length of time required for the development of a new preovulatory follicle was approximately fourteen days.[5] These results indicate that the inhibition of preo-

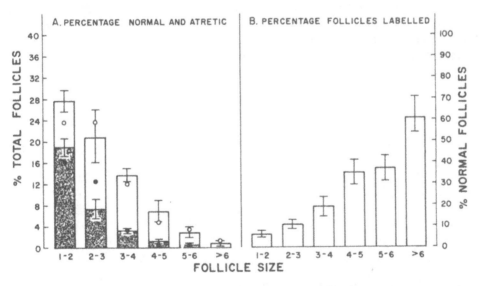

Fig. 2. Characteristics of preantral follicular growth during the luteal phase of the menstrual cycle. Panel A depicts the fractional distribution of healthy (□) and atretic (▦) follicles in relationship to follicle size in ovaries obtained during the mid luteal phase of the menstrual cycle. Panel B illustrates the percentage of follicles with labeled granulosa cells indicative of granulosa cell DNA synthesis. The open and closed circles represent data obtained from a single animal during the early follicular phase for comparison.

vulatory folliculogenesis during the luteal phase is readily reversed by the removal of the corpus luteum. The underlying question regarding the lack of preovulatory follicular growth during the luteal phase is whether the corpus luteum inhibits follicular growth directly or indirectly by suppressing gonadotropin secretion.

We initiated a study in monkeys designed to investigate preantral folliculogenesis during the luteal phase of the menstrual cycle.[6] For this purpose, animals were treated with radioactive thymidine during the luteal phase. Because thymidine is incorporated into DNA during cell division, the identification of labeled granulosa cells in a follicle would indicate that the follicle was growing during the exposure to the label. Ovaries were removed from the animals, serially sectioned and subjected to autoradiography. Figure 2 illustrates the results of the study. Panel A (right) shows the relative percentage of all normal (☐) and atretic (▦) preantral follicles in relationship to follicle size (number of granulosa cell layers). All stages of preantral follicles were observed in these ovaries obtained during the luteal phase. The results from one animal studied during the early follicular phase is superimposed in the open and closed circles. There was no apparent difference in the distribution of preantral follicles in the ovary from the follicular phase when compared to those obtained during the luteal phase. Panel B illustrates the incidence of follicles which contained labeled granulosa cells. A small number of 1–2 layered granulosa cells incorporated labeled thymidine. As follicular size increased, the percentage of labeled follicles also rose, with approximately 60% of the largest preantral follicles containing labeled granulosa cells. The superimposed values of the frequency of labeled follicles observed in the follicular phase ovary were nearly identical to those observed during the luteal phase. These data demonstrate that the hormonal environment of the luteal phase of the menstrual cycle does not interfere with the early stages of follicular growth.

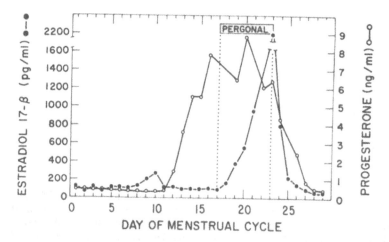

Fig. 3. Effect of administration of exogenous human menopausal gonadotropin during the luteal phase of the menstrual cycle. 75 I.U. hMG were given per day as shown by the shaded area.

The foregoing study demonstrated that preantral folliculogenesis is not impaired during the luteal phase of the menstrual cycle. To determine whether ovarian follicles retain their gonadotropin responsiveness at this stage, rhesus monkeys were treated with exogenous human menopausal gonadotropin during the mid luteal phase of the menstrual cycle. Figure 3 illustrates data from a representative animal. It can be seen that the administration of exogenous gonadotropin rapidly stimulated estrogen secretion and examination of ovaries via laparotomy revealed the presence of numerous large antral follicles. The ovarian responses to administration of exogenous gonadotropin during the luteal phase was indistinguishable to that seen when given during the early follicular phase.[7] The results of these studies demonstrate that the lack of follicular growth to the preovulatory stage during the luteal phase of the menstrual cycle is not due either to an impairment of the early stages of folliculogenesis or to alterations in gonadotropin responsiveness. Rather, the cessation of folliculogenesis during the luteal phase is most likely due to insufficient secretion of FSH and LH as a result of the feedback inhibition on the hypothalamic - pituitary axis imposed by the secretions of the corpus luteum.

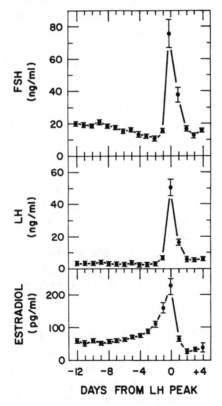

Fig. 4. Plasma concentrations of FSH, LH and estradiol throughout the follicular phase of the macaque menstrual cycle.

Figure 4 illustrates the temporal pattern of FSH, LH and estradiol during the follicular phase of the menstrual cycle of macaques. Like that seen in humans, the concentration of FSH during the early follicular phase is slightly elevated when compared with those seen during the mid to late follicular phase. In view of the absolute requirement for FSH in the process of preovulatory follicular growth, we investigated the hypothesis that the absolute concentrations of FSH throughout the follicular phase are responsible for both the initiation of preovulatory follicular growth as well as the selection of a single follicle.

To address this question we developed an experimental model in cynomolgus monkeys in which the absolute concentrations of FSH and LH could be controlled precisely by the intravenous infusion of highly purified FSH and LH.[8] To accomplish this, animals were equipped with jugular catheters which were connected to a remote sampling and infusion device. Animals were treated with a potent GnRH antagonist to interrupt spontaneous gonadotropin secretion and then received pulses of FSH and LH at the physiological pulse frequency of one pulse per hour. With this system the absolute concentrations of FSH and LH may be controlled independently by varying the amount of each gonadotropin delivered per pulse. Figure 5 illustrates the results of this study. The shaded areas represent data from control animals in which their plasma concentration of FSH was maintained at 8–10 mIU/ml and LH concentration was maintained at 12–15 mIU/ml for 14 days. As shown on the lower panel, there was no stimulation of estrogen secretion at this level of gonadotropic support. Histological examination of ovaries in these animals revealed the presence of many preantral and early antral follicles, but no large developing antral follicles. However, as shown by the solid lines, if FSH concentrations are elevated to 15–20 mIU/ml, estrogen secretion is stimulated. Moreover, once estrogen secretion becomes evident, the reduction of plasma FSH concentrations back to the 8–10 mIU/ml range is associated with continued estrogen secretion. Each of these animals contained large preovulatory follicles on the final day of FSH reduction. Moreover, the number of preovulatory follicles was related to the duration that FSH concentrations were maintained above the threshold value. Prolonging the duration of exposure of animals to elevated FSH concentrations resulted in progressive increases in the number of preovulatory follicles. Additional details of this study are provided in Reference 8. These data therefore demonstrate that there is a differential sensitivity of follicles to FSH that is dependent on the maturational state of the follicle. Concentrations of FSH in the 8–10 mIU/ml range are unable to initiate estrogen secretion and follicular growth. However, if follicular growth is first stimulated by FSH concentrations in the 15–20 mIU/ml range, subsequent reduction of FSH to the unstimulatory range does not result in atresia of the growing follicles.

Extrapolating the above data to the spontaneous menstrual cycle indicates that the initiation of follicular growth as well as the selection of the dominant follicle is accounted for directly by the absolute concentrations of FSH which in turn are governed by the feedback effects of ovarian steroids. During the luteal phase, plasma concentrations of FSH are maintained below 10 mIU/ml by the feedback actions of steroids secreted by the corpus luteum. This concentration of FSH is not sufficient to stimulate folliculogenesis beyond the preantral stage. Upon luteal regression, the feedback inhibition of gonadotropin secretion is relieved and plasma FSH concentrations rise to 15-20 mIU/ml which is effective in stimulating the final stages of follicular growth. As estrogen secretion is stimulated, plasma estrogen concentrations rise and suppress FSH secretion such that plasma FSH concentrations fall back to the 8-10 mIU/ml range.[9] Because this concentration of FSH is unable to

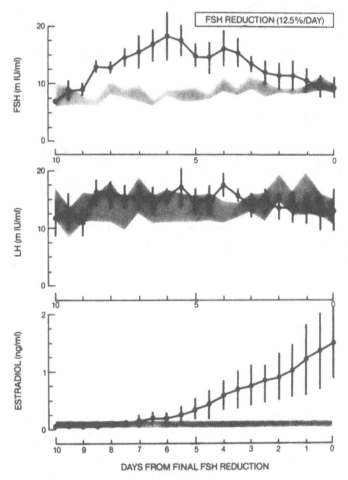

Fig. 5. Effect of pulsatile infusion of FSH and LH on ovarian folliculogenesis. All animals were treated with a GnRH antagonist to interrupt spontaneous gonadotropin secretion. The shaded areas represent data from control animals which received hourly pulses of hFSH and hLH in such a manner as to maintain plasma FSH and LH concentrations at 8–10 and 12–15 mIU/ml respectively. The solid lines show data from the experimental animals whole infusion of FSH was adjusted to provide a plasma FSH concentration of 15-20 mIU/ml until estrogen secretion became evident. Thereafter, the infusion of FSH was reduced by 12.5% per day over the next five days. Note that estrogen secretion continued in these animals in the presence of FSH concentrations that were unable to initiate follicular growth and estrogen secretion.

initiate antral folliculogenesis, new follicles do not enter the growing population. However, because the stimulated follicle gains increased sensitivity to FSH, it continues to mature until estrogen concentrations rise to such a level as to initiate the midcycle gonadotropin surge whereupon ovulation ensues.

The major unresolved issue regarding the selection process is the elucidation of the mechanisms responsible for the enhanced sensitivity of the maturing follicle that permits its continued growth in the presence of FSH concentrations that are unable to initiate the maturation of less mature follicles. Potential mechanisms that have been postulated include an increase in FSH receptors, changes in vascularity and possible influences of autocrine or paracrine factors.[10-11]

CLINICAL IMPLICATIONS

A frequent problem encountered during ovarian stimulation with exogenous gonadotropins is the lack of synchronous follicular growth. Based upon our studies in monkeys, the likely cause for this asynchrony is the continued maturation of less mature follicles during gonadotropin treatment. If the control of folliculogenesis in sub-human primates is similar to that in humans, a potential way to achieve greater synchrony of follicular growth would be to design an exogenous gonadotropin regimen that would initially elevate plasma FSH concentrations above the threshold for stimulation and thereafter progressively reduce FSH concentrations below the threshold value to restrict the entry of less mature follicles into the developing pool. Such a regimen, although potentially resulting in a fewer number of follicles, should provide a more synchronous population of maturing follicles which would reduce the difficulties in evaluating serum estrogen concentrations in the presence of numerous less mature follicles which contribute to the overall estrogen level, but most likely would not be of sufficient maturity to provide mature eggs.

REFERENCES

1. J. S. Richards, Maturation of ovarian follicles: actions and interactions of pituitary and ovarian hormones on follicular cell differentiation, *Physiological Reviews*, 60:51 (1980).
2. G. F. Erickson, and A. J. W. Hsueh, Stimulation of aromatase activity by follicle stimulating hormone in rat granulosa cells in vivo and in vitro, *Endocrinology*, 102:1275 (1978).
3. S. G. Hillier, A. J. Zeleznik, R. A. Knazek, and G. T. Ross, Hormonal regulation of preovulatory follicle maturation in the rat, *J. Reproduction and Fertility*, 60:219 (1980).
4. A. J. Zeleznik, and S. G. Hillier, The role of gonadotropins in the selection of the preovulatory follicle, *Clin. Obstet. Gynecol.*, 27:927 (1984).
5. D. T. Baird, T. Backstrom, A. S. McNeilly, K. S. Smith, and C. G. Wathen, Effects of enucleation of the corpus luteum at different stages of the luteal phase of the human menstrual cycle on subsequent follicular development, *J. Reproduction and Fertility*, 70:615 (1984).
6. A. J. Zeleznik, L. Wildt, and H. M. Schuler, Characterization of ovarian folliculogenesis during the luteal phase of the menstrual cycle in rhesus monkeys using thymidine autoradiography, *Endocrinology*, 107:982 (1980).

7. A. J. Zeleznik, and J. A. Resko, Progesterone does not inhibit gonadotropin induced follicular maturation in the female rhesus monkey, *Endocrinology*, 106:1820 (1980).
8. A. J. Zeleznik, and C. J. Kubik, Ovarian responses in macaques to pulsatile infusion of follicle stimulating hormone and luteinizing hormone: increased sensitivity of the maturing follicle to FSH, *Endocrinology*, 119:2025 (1986).
9. A. J. Zeleznik, Premature elevation of systemic estradiol reduces levels of follicle stimulating hormone and lengthens the follicular phase of the menstrual cycle in macaques, *Endocrinology*, 109:352 (1981).
10. A. J. W. Hsueh, E. Y. Adashi, P. B. C. Jones, and T. H. Welsh, Jr., Hormonal regulation of the differentiation of culture granulosa cells, *Endocrine Review*, 5:76 (1984).
11. D. W. Schomberg, Regulation of follicular development by gonadotropins and growth factors, *in*: "The Primate Ovary", R. L. Stouffer, ed., Plenum Press, New York (1987).

GROWTH FACTOR MODULATION OF GONADOTROPIN ACTION ON PORCINE GRANULOSA CELLS

Takahide Mori, Kenji Takakura, Masatsune Fukuoka,
Shunzo Taii, and Keiko Yasuda

Department of Gynecology and Obstetrics, Faculty of Medicine, Kyoto University
Sakyo-ku, Kyoto 606, Japan

INTRODUCTION

Ovarian functions are thought to be under the regulation of autocrine or paracrine control as well as endocrine control. Although the endocrine regulation of the ovary has been well described, little is known about the local intra-ovarian control. Considering that ovarian granulosa cells of developing follicles are confined to an isolated avascular environment and "cultured" in follicular fluid in vivo, substances concentrated in follicular fluid may play important roles in local intra-ovarian control.

Among the possible non-steroidal local ovarian regulators, several growth factors, including epidermal growth factor (EGF), fibroblast growth factor (FGF), platelet-derived growth factor (PDGF), insulin-like growth factor (IGF), angiogenic factors, and transforming growth factors,[1] have been reported to modulate gonadotropin actions on granulosa cells. In addition to these growth factors, we have recently reported that interleukin 1 (IL-1), an immunoregulator produced predominantly by activated macrophages, is one of the possible intra-ovarian regulators by demonstrating its remarkable inhibitory effect not only on progesterone production but also on the LH-induced morphological luteinization of cultured porcine granulosa cells.[2,3] Furthermore, we demonstrated interleukin 2 receptor/p55 (Tac) inducing activity (TIA) present in porcine follicular fluid, which is one of the major functions of IL-1. TIA in the follicular fluid was found to be derived from a non-steriodal

Advances in Assisted Reproductive Technologies, Edited by
S. Mashiach *et al.,* Plenum Press, New York, 1990

Table 1. Effects of IL-1 on LH-stimulated and Basal Progesterone Production

Days in culture	Progesterone Production (ng/10^5 cells/day)			
	Control	LH	LH+IL-1	IL-1
3	16.10 ± 1.3	394.5 ± 30.6	21.80 ± 1.2*	9.71 ± 0.20**
5	1.55 ± 0.05	1183.7 ± 96.5	3.04 ± 0.13*	1.06 ± 0.19**
7	0.87 ± 0.04	1734.0 ± 158.1	1.17 ± 0.09*	0.55 ± 0.02**
9	0.77 ± 0.08	1772.8 ± 180.1	0.95 ± 0.11*	0.58 ± 0.01

Values are mean ± SE of 3 determinations. Concentration of LH and IL-1 were 100 ng/ml and 25 ng/ml, respectively.

 * Significantly different from LH (P < 0.01).
** Significantly different from control (P < 0.01).

factor produced by granulosa cells and closely related to the follicular maturation. In the present communication, we discuss the possibility that IL-1 and TIA in the follicular fluid may contribute follicular development and luteinization by modulating gonadotropin actions on granulosa cells.

IL-1 AS A MODULATOR OF GONADOTROPIN ACTIONS ON GRANULOSA CELLS

Effects on LH-stimulated Morphological and Functional Luteinization

Cultured porcine granulosa cells obtained from medium-sized (3mm–5mm) follicles (M cells) have been shown to respond to LH with increased progesterone secretion and morphological transformation called luteinization.[4] These cells, in contrast, do not undergo such changes in the absence of gonadotropins or other stimulators. Using these cells we examined the effects of IL-1 on the luteinizing process of cultured granulosa cells in terms of morphology and progesterone secretion in response to LH.[2]

Table 1 shows the effect of IL-1 on LH-stimulated progesterone secretion by M cells cultured for nine days. In the absence of LH, progesterone secretion by these cells (basal progesterone secretion) was relatively low, whereas it responded to LH (100 ng/ml) with increased progesterone secretion by 10- to 1000-fold as compared with control. IL-1 (25 ng/ml) inhibited the LH-stimulated progesterone secretion as well as basal progesterone secretion in a dose-dependent manner. Significant effect of IL-1 was observed at as low as 50 pg/ml (1 unit/ml) and its effect was maximal at 5 ng/ml (100 unit/ml) (ID$_{50}$ = 0.5ng/ml).[2,3]

The morphology of cultured M cells, as observed by phase-contrast microscope, was fibroblastic in the absence of LH. When cultured in the presence of LH, in contrast, the-cells became epitheloid in shape and accumulated granules and lipid droplets in their cytoplasm. IL-1 (>1 ng/ml) blocked the morphological transformation of M cells induced by LH. The cells cultured in the presence of IL-1 together with LH remained fibroblastic during the whole culture period.[2]

Porcine granulosa cells obtained from large (> 5mm) follicles (L cells) appear to have more differentiated function as compared with M cells. L cells of the highest maturity, which have already been exposed to LH surge in vivo, are thought to have undergone functional luteinization and are able to luteinize in culture without exogenous addition of gonadotropins.[5] They show the morphological appearance characteristic of luteinized cells and secrete high amounts of progesterone even in the absence of LH in culture. We examined whether IL-1 affects progesterone secretion by these functionally luteinized granulosa cells using the cultures of L cells obtained from follicles larger than 7 mm in diameter (L8 cells). Basal progesterone secretion by these functionally luteinized granulosa cells was not affected by IL-1 as long as the highly differentiated features of these cells were maintained in culture.[3]

Attempts have been made to clarify the mechanism of the inhibitory effect of IL-1 on LH stimulated progesterone secretion by M cells.[6] The cells were pretreated with IL-1 for 12-60 h, after which they were examined for their ability to accumulate cAMP in response to LH. IL-1 reduced intra- and extra-cellular cAMP accumulation by M cells in time-and concentration-dependent manner. Significant inhibitory effect was observed at as low as 50-250pg/ml, and the effect was maximal at 100ng/ml (ID_{50} = 2 ng/ml). Pretreatment of M cells for 48 h with IL-1 (100 ng/ml) reduced cAMP accumulation in response to LH (100 ng/ml) by 73%–87%.

Experiments were then made to determine whether the inhibitory effect of IL-1 on LH-stimulated cAMP accumulation was due to its actions at LH receptor level or at the level of adenylate cyclase system. Pretreatment of M cells for 48 h with IL-1 (100 ng/ml) significantly reduced the binding of radioiodinated LH to its receptor by 36%. The same pretreatment also significantly reduced cAMP accumulation by the cells in response to forskolin by 28%–32%. Therefore it was suggested that the inhibitory effect of IL-1 on LH-induced cAMP accumulation is due to its action at the level of both LH receptor and adenylate cyclase system. Pretreatment for 48 h with IL-1 (25 ng/ml) reduced progesterone secretion by M cells in response to LH and forskolin to a similar extent as it reduced cAMP accumulation in response to these stimulators. Contrary to its inhibitory actions in the process of cAMP formation, IL-1 did not seem to affect post-cAMP steps of progesterone production, because IL-1 did not significantly inhibit progesterone secretion induced by a membrane-permeable cAMP analog, dibutyryl cAMP.

Effects on FSH-stimulated Functional Differentiation

IL-1 also affects FSH actions on granulosa cells. We studied the effects of IL-1 on FSH-induced differentiation in cultures of porcine granulosa cells obtained from small

Table 2. Effects of IL-1 on FSH-induced Formation of Functional LH Receptors in Cultured Porcine Granulosa Cells from Small Follicles

	cAMP (pmol/mg protein)
Control	313.8 ± 66.6
FSH	1046.8 ± 80.7
FSH + IL-1	203.7 ± 67.1*

Granulosa cells were cultured for 48 h in DME (2% FCS) in the presence and absence of FSH (500 ng/ml) and IL-1 (25 ng/ml). Media were then removed and replaced with serum-free media containing LH (500 ng/ml) and 3-isobutyl-1-methylxanthine (0.5 mM). The cells were incubated for 3 h, and cAMP content in culture media (extra-cellular cAMP) was assayed by radioimmunoassay. Data are mean ± SE.
* Significantly different from FSH (P < 0.01).

Table 3. Inhibitory Effect of IL-1 on FSH-induced Increase in Aromatase Activity in Cultured Porcine Granulosa Cells from Small Follicles

	Estrone	Estradiol
	(pg/10^5 cells/day)	
A	11.3 ± 0.9	108.5 ± 13.0
A + FSH	38.6 ± 3.0	224.2 ± 17.3
A + FSH + IL-1	11.3 ± 2.0*	71.4 ± 15.4*

Porcine granulosa cells from small follicles were cultured for 48 h in DME (2% FCS) in the presence and absence of androstenedione (A, 10^{-7} m), porcine FSH (500 ng/ml) and IL-1 (25 ng/ml). The spent culture media were assayed for estrone and estradiol by RIA. Data are mean ± SE.
* Significantly different from A + FSH (P < 0.01).

(1mm-2mm) follicles (S cells). These cells are relatively immature and are less responsive to LH stimulation[7,8] whereas they are already responsive to FSH stimulation.[9] Table 2 shows the inhibitory effect of IL-1 on FSH-induced functional LH receptor formation by S cells, as determined by the ability of the cells to accumulate cAMP in response to LH. IL-1 (25 ng/ml) completely blocked the FSH-induced increase in functional LH receptors. IL-1 also reduced basal levels of functional LH receptors, which is consistent with the inhibitory effect of IL-1 on LH receptor binding and adenylate cyclase activity observed in M cells. Table 3 shows the effect of IL-1 on estrogen secretion by S cells cultured in the presence of substrate, androstenedione. IL-1 (25 ng/ml), like its effect on functional LH receptor formation, almost completely blocked the FSH-induced increase in secretion of estrone and estradiol by S cells. These results suggest that IL-1 affects two major functions of FSH in porcine granulosa cell differentiation, induction of aromatase activity and formation of function LH receptors.

Effects of Proliferation of Granulosa Cells

In cultures of porcine granulosa cells in which LH- and FSH-induced differentation was examined, the cells were plated at relatively high cell density ($1 - 3 \times 10^5/cm^2$). The number of S and M cells counted at the end of cultures was consistently greater in the wells treated with IL-1, although it never exceeded the inoculated cell number in such high-density cultures.[3] IL-1 did not influence the number of L8 cells in high-density cultures. The increasing effects of IL-1 on the number of S and M cells were concentration-dependent and were inversely related with its effect on basal progesterone secretion by these cells. In order to ascertain the growth-promoting effect of IL-1 on granulosa cells, we examined its effects on the number of cultured granulosa cells plated at low density ($0.5 - 1.0 \times 10^4/cm^2$).[3] IL-1 (25 ng/ml), in the presence of 10% FCS, increased the number of S and M cells two-fold as compared with control after seven days in culture. IL-1 did not increase the number of L8 cells up to five days in culture, but significantly increased it after that. The proliferative response of L8 cells to IL-1 observed after five days in culture was coincident with decreased capacity of these cells to secrete progesterone, suggesting that the growth-promoting effect of IL-1 is inversely related with the capacity of the cells to secrete progesterone.

IL-2R/p55 (Tac) INDUCING SUBSTANCE IN FOLLICULAR FLUID

Assay of IL-2R/p55 (Tac) Inducing Activity in Follicular Fluid

YT cells, a human natural killer cell line,[10, 11] were used for the TIA assay of porcine follicular fluid (FF) or of conditioned medium (CM) of porcine granulosa cells. The cells, 5×10^5 cells/ml per well, were cultured in a 24-well plate for 24 h with or without the addition of samples or standard human recombinant IL-1α or IL-1β. The cells were then collected and stained with 20μ1 of an appropriately diluted fluoresceinated anti-human IL-2R/p55 (Tac) monoclonal antibody (FITC anti-Tac)[12] at 4°C for 30 min.

The fluorescent intensity of the cells was analysed by flow cytometry using Spectrum III (Ortho Pharmaceutical Co., NJ) on a linear fluorescence scale. TIA was expressed as the percentage of fluorescence positive cells. The flow cytometer was set so that approximately 10% of the YT cells without stimulation with samples or standard recombinant IL-1 were positive for Tac.

TIA in Porcine Follicular Fluid

FF from small follicles (sFF), medium-sized follicles (mFF) and large follicles (lFF) showed TIA in a dose-dependent manner and approximately 100 μl of sFF demonstrated an equivalent activity of 0.02 unit of human recombinant IL-1α. Among FF from the three stages of follicular maturation, sFF (n = 11) showed a significantly higher (P < 0.01) TIA than mFF or lFF (Table 4).

Table 4. IL-2R/p55 (Tac) Inducing Activity (TIA), Estradiol (E_2), Testosterone (T) and T/E_2 ratio of Porcine Follicular Fluid from Each Size of the Follicles

	TIA (%)	E_2 (ng/ml)	T (ng/ml)	T/E_2 ratio
sFF	30.3 ± 0.72*	4.06 ± 0.48	30.52 ± 3.04	8.05 ± 0.78
mFF	24.2 ± 0.46	26.78 ± 3.40	33.51 ± 4.21	1.29 ± 0.11
lFF	23.4 ± 0.65	58.27 ± 17.36	37.94 ± 7.39	0.96 ± 0.25

E_2 and T contents of porcine follicular fluid obtained from small (sFF), medium-sized (mFF) and large (lFF) follicles were determined by RIA. Data are mean ± SE.
* Significantly different from mFF and lFF (P < 0.01).

Relationship between Steroid Hormone Concentrations and TIA

Table 4 also shows the levels of estradiol (E_2), testosterone (T), and T/E_2 ratio of the follicular fluid from the various sizes of porcine follicles. As the follicular diameter increases, E_2 concentration increases and T/E_2 ratio decreases.

There was a significant negative correlation between TIA and follicular E_2 concentration (y [TIA] = 28.3 – 0.067 × [E_2], r = - 0.49, P < 0.05), while no significant correlation was found between TIA and follicular T concentration (y [TIA] = 27.4 - 0.023 × [T], r = -0.073, P > 0.05). Furthermore, a positive correlation was evident between T/E_2 ratio and TIA (y [TIA] = 23.7 + 0.72 × [T/E_2], r = 0.68, P < 0.01).

TIA in Conditioned Medium of Porcine Granulosa Cells

To elucidate whether or not granulosa cells produce the Tac inducing substance, the CM of porcine granulosa cells was assayed for TIA. The CM of granulosa cells from every stage of the maturation showed TIA distinctively, and exhibited a decreasing tendency of TIA in a graded manner with the maturation of the follicles; the intensity of TIA being in the order of small follicles, medium-sized follicles and large follicles. TIA in the CM of granulosa cells from each stage of the follicles rapidly decreased as the culture period was extended. Addition of LH or FSH in the cultures of granulosa cells did not affect TIA in the CM.

Effect of Sex Steroids on IL-2R/p55 (Tac) Induction

As FF contains high levels of sex steroids and the concentration of these steroids are closely related to the follicular development and maturation, the effect of these sex steroids on TIA was examined. The addition of E_2, progesterone, androstenedione and T at concentration of 10, 100 and 1000 ng/ml neither exhibited TIA by itself nor modulated IL-2R/p55 (Tac) induction by human recombinant IL-1 in the YT cells.

DISCUSSION

Ovarian FF contains a variety of bioactive substances, both steroidal and non-steroidal. It is the local microenvironment rather than the peripheral blood concentration of these substances which modulates the growth and maturation of follicles.[13, 14] Recently, an increasing number of non-steroidal factors including growth factors have been found to modulate follicular development possibly through autocrine or paracrine control.[1] To elucidate the mechanism of the follicular development and maturation, it is important to study the follicular microenvironment and bioactivity of the substances in FF.

In the present study, we reported that exogenously administered human recombinant IL-1 inhibited remarkably both progesterone production and the morphological luteinization of cultured porcine granulosa cells, and suggested that IL-1 might act as one of the putative luteinization inhibitors. IL-1 also modulates FSH actions on the cultured granulosa cells.

To elucidate the physiological role of IL-1 as a modulator of gonadotropin actions on granulosa cells, we examined IL-1 activity in the FF in terms of IL-2R/p55 (Tac) induction. Both the FF and the CM of granulosa cells showed TIA. Furthermore, it must be noted that the intensity of TIA in the FF and the CM decreased in association with the follicular development determined by follicular size and E_2 content. This inverse correlation between TIA and the follicular development suggests the possibility that the IL-2R/p55 (Tac) inducing substance produced by granulosa cells may play a certain regulatory role in the follicular maturation.

It is reported that the incidence of follicular atresia is relatively high when judged from strict criterion and that atretic follicles are characterized by low E_2 content with maintained androgen level in the follicle.[15, 16] Taking consideration of these reports, there is another possibility that this factor may take a part in follicular atresia because significant positive correlation was observed between TIA and T/E_2 ratio.

From an immunological viewpoint, it will be of interest to clarify if this activity is related to major defense mechanisms against infection and injury. Since the granulosa cells of developing follicles are confined to an isolated avascular environment and include oocytes that are unable to regenerate or proliferate, there might be a special local defense mechanism in follicles. The IL-2R inducing substance in the present study may contribute to this defense mechanism since the induced IL-2 receptor may mediate the IL-2 signals resulting in the proliferation and activation of lymphoid cells present in the perifollicular area.

Identification of this substance with porcine IL-1 is the next question to be answered. ALthough human[17] or murine[18] IL-1 has been cloned, and recombinant forms and antibodies to these are available, the characteristics of porcine IL-1 have not yet been precisely identified.[19] Therefore, the methods of identification for porcine IL-1 are limited. There is the possibility that this substance in porcine FF or the CM of porcine granulosa cells may not be IL-1α or β, since TIA in human FF obtained by the aspiration of follicles for in vitro fertilization could not be neutralized by anti-human recombinant IL-1α or β. In fact, it has been said that there may be another form of IL-1 other than IL-1α or β.[20, 21] Furthermore, the presence of IL-2R inducing activity is not specific evidence for IL-1 identification. Several other substances, including the ATL (adult T-cell leukemia)-derived factor,[10] tumor necrosis factor (TNF),[22] phorbol myristate acetate (PMA)[10] and forskolin,[23] can induce IL-2R. Though studies on the physiological role along with characterization of this substance are now in progress, our preliminary experiments suggest that it is non-steroidal, because it is heat labile, susceptible to trypsinization, concentrated by ultrafiltration (molecular cut at 30 kDa) and recovered in fractions between cytochrome C (mol wt 12,400) and albumin (mol wt 67,000) by gel filtration.

In Conclusion

1) IL-1 is a potent modulator of gonadotropin actions on cultured porcine granulosa cells.
2) TIA in porcine FF is a non-steroidal factor produced by granulosa cells. This factor may be one of the possible intra-ovarian regulators like some growth factors and may also participate in the local defense mechanism.

REFERENCES

1. A. Tsafriri, Local non-steroidal regulators of ovarian function, *in:* "The physiology of reproduction," vol IIA, The female reproductive system, E. Knobil and J. D. Neil, eds., Raven Press, New York (1987).
2. M. Fukuoka, T. Mori, S. Taii, and K. Yasuda, Interleukin-1 inhibits luteinization of porcine granulosa cells in culture, *Endocrinology*, 122:367 (1988).
3. M. Fukuoka, K. Yasuda, S. Taii, K. Takakura, and T. Mori, Interleukin-1 stimulates growth and inhibits progesterone secretion in cultures of porcine granulosa cells, *Endocrinology*, 124:884 (1989).
4. C.P Channing, Effect of stage of the estrous cycle and gonadotrophins upon luteinization of porcine granulosa cells in culture, *Endocrinology*, 87:156 (1970).
5. C. P. Channing, H. G. Brinkley, and E. P. Young, Relationship between serum luteinizing hormone levels and the ability of porcine granulosa cells to luteinize and respond to exogenous luteinizing hormone in culture, *Endocrinology*, 106:317 (1980).
6. M. Fukuoka, K. Yasuda, K. Takakura, S. Taii, and T. Mori, Inhibitory effects on interleukin-1 on luteinizing hormone-stimulated adenosine 3',5', - monophosphate accumulation by cultured porcine granulosa cells, *Endocrinology* (in press), (1989).
7. C. P. Channing, and S. K. Kammerman, Characteristics of gonadotropin receptors of porcine granulosar cells during follicle maturation, *Endocrinlogy*, 92:531 (1973).
8. C. P. Channing, F. Ledwitz-Rigby, Methods for assessing hormone-mediated differentiation of ovarian cells in culture and in short-term incubations, *in:* "Methods in Enzymology, J. G. Hardmann and B. W. O'Malley, eds., Academic Press, New York, vol 39:1835, (1975).
9. K. H. Thanki and C. P. Channing, Effects of follicle-stimulating hormone and estradiol upon progesterone secretion by porcine granulosa cells in tissue culture, *Endocrinology* 103:74, (1978).

10. J. Yodoi, K. Teshigawara, T. Nikaido, K. Fukui, T. Noma, T. Honjo, M. Takigawa, M. Sasaki, N. Minato, M. Tsudo, T. Uchiyama, and M. Maeda, TCGF (IL-2)-receptor inducing factor (s) 1. Regulation of IL-2 receptor on a natural killer-like cell line (YT cells), *J. Immunol.*, 134:1623 (1985).

11. M. Okada, M. Maeda, Y. Tagaya, Y. Taniguchi, K. Teshigawara, T. Yishiki, T. Diamantstein, K. A. Smith, T. Uchiyama, T. Honjo, J. Yodoi, TCGF (IL-2)-receptor inducing factor (s) II. Possible role of ATL-derived factor (ADF) on constitutive IL-2 receptor expression of HTLV-1(+) T cell lines., *J. Immunol.*, 135:3995 (1985).

12. T. Uchiyama, S. Broder, and T. A. Waldmann, A monoclonal antibody (anti-Tac) reactive with activated and functionally mature human T cells. I. Production of anti-Tac monoclonal antibody and distribution of Tac(+) cells., *J. Immunol.*, 126:1393 (1981).

13. K. P. McNatty, W. M. Hunter, A. S. McNeilly, and R. S. Sawers, Changes in concentration of pituitary and steroid hormones in the follicular fluid of human Graafian follicles throughout the menstrual cycle, *J. Endocrinol.*, 64:5553 (1975).

14. K. P. McNatty, and R. S. Sawers, Relationship between the endocrine environment within the Graafian follicle and the subsequent rate of progesterone secretion by human granulosa cells in vitro, *J. Endocrinol.*, 66:391, (1975).

15. W. S. Maxson, A. F. Haney, D. W. Schomberg, Steroidogenesis in porcine atretic follicles: Loss of aromatase activity in isolated granulosa and theca. *Biol. Reprod.*, 33:495 (1983).

16. C. G. Tsonis, R. S. Carson, J. K. Findlay, Relationship between aromatase activity, follicular fluid oestradiol-17β and testosterone concentrations, and diameter and atresia of individual ovine follicles, *J.Reprod. Fert.*, 72:153 (1984).

17. P. E. Auron, A. C. Webb, L. J. Rosenwasser, S. F. Mucci, A. Rich, S. M. Wolff, and C. A. Dinarello, Nucleotide sequence of human monocyte interleukin 1 precursor cDNA. *Proc. Natl. Acad. Sci.*, 81:7907, (1984).

18. P. T. Lomedico, U. Gubler, C. P. Hellman, M. Dukovich, J. G. Giri, Y. E. Pan, K. Collier, R. Semionow, A. O. Chua, and S. B. Mizel, Cloning and expression of murine interleukin-1 in Escherichia coli, *Nature*, 312:458 (1984).

19. J. Saklatvala, S. J. Sarsfield, and Y. Townsend, Pig interleukin-1. Purification of two immunologically different leukocyte proteins that cause cartilage resorption, lymphocyte activation, and fever, *J. Exp. Med.* 162:1208 (1985).

20. C. A. Dinarello, An update on human interleukin-1: From molecular biology to clinical relevance. *J. Clin. Immunol.*, 5:287 (1985).

21. J. J. Oppenheim, E. J. Kovacs, K. Matsushima, and S. K. Durum, There is more than one interleukin-1. *Immunology Today*, 7:45 (1986).

22. J. C. Lee, A. Truneh, M. F. Smith Jr, and K. Y. Tsang, Induction of interleukin-2 receptor (Tac) by tumor necrosis factor in YT cells, *J.Immunol.*, 139:1935 (1987).

23. S. Narumiya, M. Hirata, T. Nanba, T. Nikaido, Y. Taniguchi, Y. Tagaya, M. Okada, H. Mitsuya, and J. Yodoi, Activation of interleukin-2 receptor gene by forskolin and cyclic AMP analogues, *Biochem. Biophys. Res. Commun.*, 143:753 (1987).

FOLLICULAR RUPTURE DURING OVULATION: ACTIVATION OF COLLAGENOLYSIS

A. Tsafriri, D. Daphna-Iken, A.O. Abisogun, and R. Reich

Department of Hormone Research, The Weizmann Institute of Science Rehovot, P.O. Box 26, 76100, Israel

INTRODUCTION

The preovulatory surge of gonadotropins induces a series of orderly sequenced changes in the mature ovarian follicle. These include resumption of oocyte meiosis, luteinization of the follicle and finally the rupture of follicle wall and release of a fertilizable ovum. While initial studies on follicular response to gonadotropins dealt primarily with changes in steroidogenesis and resumption of meiosis, recently the biochemical changes associated with follicular rupture have been investigated.

Involvement of proteolytic processes in follicular rupture was first suggested by Schochet (1916). In view of the dense perifollicular meshwork of collagen fibers, it became obvious that collagenolysis would be a prerequisite for ovulation. Morphological studies have demonstrated a reduction in collagen fibers in the apical wall of rabbit follicles prior to ovulation (Espey, 1967). The regulation of collagenase activity in mammals is however a multifactorial process, involving both cellular and microenvironmental factors. These include cellular modulation of collagenase proenzyme release and regulation of its activation by local activators, inhibitors and changes in substrate susceptibility. Proteases have been implicated in activation of collagenase (Birkedal-Hansen et al., 1975; Horwitz et al., 1976). Here, we shall review some of our studies on preovulatory activation of ovarian collagenase and hence in follicular rupture at ovulation and the role of the plasmin-generating system, arachidonic acid (AA) metabolites and platelet activating factor (PAF) in this process.

Advances in Assisted Reproductive Technologies, Edited by
S. Mashiach *et al.,* Plenum Press, New York, 1990

OVARIAN COLLAGENOLYSIS AND FOLLICULAR RUPTURE

The extensive changes in follicular collagen observed prior to ovulation suggest that collagen degrading enzymes play a central role in ovulation. Indeed, collagenase like activity was demonstrated in ovarian tissues (Fukumoto et al., 1981; Morales et al., 1978, 1983). Nevertheless, until very recently, experiments to extract follicular collagenase and to demonstrate its ability to degrade follicular wall were not successful (Fukumoto et al., 1981), nor has a correlation between collagenase activity and ovulatory changes been demonstrated (Morales et al., 1978, 1983). This failure should probably be attributed to the fact that collagenase regulation in vivo is a multifactorial process involving local regulatory factors.

In order to circumvent the difficulties in demonstrating gonadotropin- induced changes in ovarian collagenase activity we have labeled collagen by local administration of [^{3}H]-proline ([^{3}H]-Pro). Some of the [^{3}H]-Pro incorporated into collagen molecules is enzymatically hydroxylated into hydroxyproline (Hyp). The presence of the latter is unique to collagen. Ovarian collagen content and collagenolysis were assessed by separating [^{3}H]-Hyp from acid hydrolysate of ovarian tissue by high performance liquid chromatography (Reich et al., 1985a).

Using this approach, significant degradation of labeled collagen was detected after the preovulatory surge of gonadotropins. Collagen degradation was blocked by Nembutal—a drug which prevents the secretion of gonadotropins, and hence ovulation. Moreover, upon administration of hCG on the morning of proestrus, the level of [3H]-Hyp in preovulatory follicles was reduced by 43% within 10 h (Reich et al., 1985a).

The role of collagenase in follicular rupture was corroborated by intrabursal administration of cysteine. It prevented ovarian collagenolysis as well as ovulation, and these two actions were dose-related (Reich et al., 1985a). Similarly, a microbial metallo-protease inhibitor, talopeptine, inhibited ovulation in vitro from hamster ovaries (Ichikawa et al., 1983) and a synthetic inhibitor, CI-1, inhibited ovulation in perfused rat ovary (Bränström et al., 1988) and in vivo (our unpublished observations).

ACTIVATION OF OVARIAN COLLAGENOLYSIS

The Plasmin Generating System

The demonstration of an increase in secretion of plasminogen activator (PA) by granulosa cells closely to ovulation (Beers et al., 1975) and the ability of plasmin, the product of PA action on plasminogen, to decrease the tensile strength of Graafian follicle wall (Beers, 1975) led to the suggestion that PA is involved in ovulation. It was hypothesized that plasmin activates latent ovarian collagenase, thus initiating the proteolytic processes culminating in the rupture of follicle wall at ovulation (Beers, 1975; Beers et al., 1975).

Rat granulosa cell PA activity was stimulated by LH, FSH, cAMP derivatives and prostaglandins E_1 and E_2 (Strickland and Beers, 1976; Beers and Strickland, 1978). In these studies, however, FSH was much more effective than LH in stimulating granulosa cell PA activity. This is in contrast to the physiological role of LH, and not FSH, as the ovulation-inducing hormone (Schwartz et al., 1975; Tsafriri et al., 1976). Recent studies seem to have resolved this discrepancy. It was found that FSH and LH were equally potent in enhancing rat granulosa cell PA activity provided that the cells were first properly stimulated to induce LH receptors by FSH or dbcAMP (Wang and Leung, 1983). Accordingly, in rat proestrous follicles LH and FSH were equally effective in stimulating PA activity (Reich et al., 1985b).

Comparison of in vitro accumulation of PA in cultures of rat granulosa cells, whole follicles and the residual tissue, mainly theca with some interstitial tissue, revealed that the granulosa cells contribute 80%–90% of total follicular PA activity (Reich et al., 1986). Furthermore, whereas most of granulosa PA activity is secreted into the culture medium, that of intact follicles or residual tissue is retained within the follicle and should be extracted prior to the assay from the tissue (Reich et al., 1985b; 1986). This finding suggests that the theca compartment serves as a specific barrier which prevents the secretion of PA into the extra-follicular compartment and, therefore, allows localized action of PA in follicular wall.

Two different molecular types of PA have been identified in mammals, the urokinase type (u-PA) and the tissue type (t-PA). Both molecular types of PA were identified in rat preovulatory follicles which, on the basis of electrophoretic mobility and neutralization with t-PA antiserum, could be characterized as t-PA and u-PA (Reich et al., 1986). The two molecular types were present both in the granulosa and in the theca compartments, the granulosa cells contributed 80% to 90% of the total follicular activity. Upon gonadotropin stimulation, a highly significant ($p < 0.001$) increase in t-PA was observed in whole follicles and in the granulosa cells. Likewise, rat granulosa cells and cumulus-oocyte complexes responded to gonadotropins by enhanced accumulation of t-PA (Canipari and Strickland, 1985; 1986; Liu et al., 1986; Ny et al., 1985). By measuring only the secreted PA, Canipari and Strickland (1985, 1986) reached the conclusion that the theca tissue (or the whole follicle) secretes u-PA and that this type is stimulated by the gonadotropin. However, our data indicating that only a minor fraction of follicular PA is secreted into the medium reconcile this apparent discrepancy. It appears, therefore, that t-PA is quantitatively the major PA stimulated by the preovulatory surge of gonadotropins in the rat and is hence, most likely, involved in ovulation.

Addition of steroids (1 µg/ml) into the medium in which unstimulated preovulatory rat follicles were cultured did not change follicular PA activity during a 6 h incubation period. By contrast, addition of estradiol-17β, progesterone or testosterone, but not 5α-dihydrotestosterone (all 1 µg/ml) to LH-stimulated follicles further enhanced PA activity. Furthermore, addition of inhibitors of steroidogenesis or of aromatase inhibited the LH-stimulated activity of PA ($p < 0.001$). LH-stimulated follicular PA activity could be restored in the presence of such inhibitors by the addition of progesterone, testosterone and estradiol-17β, but not the non-aromatizable 5α-dihydrotestosterone (Reich et al., 1985b).

The involvement of PA in follicular rupture was confirmed by a pharmacological approach. Several inhibitors of serine proteases, such as trans-amino-ethylcyclohexane blocked ovulation in hamster ovaries in vitro (Akazawa et al., 1983) and ε-amino caproic acid, benzamidine and bis(5-amidino-2-benzimidozylyl)methane (BABIM, kindly provided by Dr. J.D. Geratz of the University of North Carolina at Chapel Hill), blocked ovulation in rats (Reich et al., 1985b; Abisogun et al., 1988a) in vivo. Furthermore, antibodies to t-PA and a plasma derived plasmin inhibitor, α₂ antiplasmin, suppressed ovulation rate in rats (Tsafriri et al., 1989).

Delayed administration of these inhibitors (> 4 h after hCG) was ineffective in blocking ovulation (Reich et al., 1985b; Tsafriri et al., 1989), indicating that PA and plasmin are involved primarily in early changes leading to follicular rupture, i.e. collagenase activation. Collectively, these findings confirm the suggested role of the plasmin-activating system in ovulation.

Eicosanoids

The preovulatory surge of gonadotropins causes an increase in follicular prostaglandin synthesis (LeMaire et al., 1973; Armstrong and Zamecnik, 1975; Ainsworth et al., 1975; Bauminger and Lindner, 1975a). It was shown that this action of LH is not dependent upon steroid synthesis (Bauminger et al., 1975b), and that most follicular responses to LH, such as ovum maturation, activation of adenyl cyclase, steroidogenesis and luteinization proceed while prostaglandin synthesis is inhibited by drugs (reviewed by Tsafriri et al., 1986; Lipner, 1988). The exception is follicular rupture, which is prevented by administration of indomethacin (Tsafriri et al., 1972). The inhibitory action of indomethacin was overcome by administration of PGE_2 to rats (Tsafriri et al., 1972) or $PGF_{2\alpha}$ to rhesus monkeys (Wallach et al., 1975). Furthermore, in the perfused rabbit ovary $PGF_{2\alpha}$ was able to induce follicular rupture by itself or in hCG-indomethacin treated ovaries (Holmes et al., 1983). Collectively, these results suggest that prostaglandins have a physiological role in follicular rupture.

The recent elucidation of the lipoxygenase pathway of arachidonic acid cascade and availability of specific inhibitors of lipoxygenase led to examination of their role in ovulation. Unilateral administration of lipoxygenase inhibitors into the ovarian bursa of PMSG-stimulated immature or mature proestrous rats resulted in a dose-dependent suppression of ovulation, without affecting the contralateral ovary (Reich et al., 1983; 1987). Furthermore, ovarian lipoxygenase activity was demonstrated in homogenates of rat ovaries and follicles (Reich et al., 1985c; 1987) and of human granulosa cells (Feldman et al., 1986). In ovarian tissue of both the rat and the human, products of 5, 12 and 15 lipoxygenases were identified (Reich et al., 1987; Feldman et al., 1986) and their activity in the rat was stimulated considerably by hCG treatment. These studies suggest the involvement of lipoxygenase products in follicular rupture at ovulation. The scarcity of stable analogs has not yet allowed the identification of the metabolites involved in ovulation.

Collectively, eicosanoids of both cyclooxygenase and lipoxygenase pathways of arachidonic acid metabolism seem to be involved in follicular rupture. The finding that specific inhibition of any of these pathways blocks ovulation, further demonstrates the multiple regulation of the process. Arachidonic acid metabolites are involved in a multitude of physiological processes. Our studies implicate eicosanoids in the mediation of the gonadotropic activation of ovarian collagenolysis and ovarian vascular changes during the preovulatory period.

In several studies, doses of indomethacin which inhibited ovulation, did not affect LH/hCG-induced increase in PA activity (Shimada et al., 1983; Reich et al., 1985c; Espey et al., 1985). In spite of the fact that higher doses of indomethacin inhibited PA production in vivo (Espey et al., 1985), or in vitro (Canipari and Strickland, 1986), it seems that the inhibition of prostaglandin synthesis blocks ovulation through mechanism(s) which are not related to PA. Likewise, inhibitors of lipoxygenase which blocked ovulation did not affect follicular PA production (Reich et al., 1985c). This finding does not imply that PA is not involved in follicular rupture, but that increase in PA activity by itself is not a sufficient trigger of ovulation.

Addition of indomethacin (an inhibitor of cyclooxygenase) or nordihydroguiaretic acid (NDGA, an inhibitor of lipoxygenase) to ovarian extracts did not affect collagenolytic activity in vitro (Fig. 1). Likewise, indomethacin did not reduce collagenase activity extracted from PMSG-hCG-treated rat ovaries and activated in vitro by p-aminophenylmercuric acetate (Curry et al., 1986). These results seem to indicate that inhibitors of AA metabolism do not affect directly active ovarian collagenase. By contrast, when indomethacin or NDGA were administered to PMSG-treated immature rats concomitantly with hCG, ovarian collagenolysis in vitro was significantly reduced (Fig. 2). Thus eicosanoids of both, cycloxygenase and lipoxygenase pathways seem to play an essential role in gonadotropic activation of ovarian collagenolytic activity, required for follicular rupture at ovulation.

Periovulatory changes in ovarian and follicular microcirculation were described (Janson, 1975; Lee and Novy, 1978; Murdoch et al., 1983). These studies revealed initial increase in blood flow, which in the sheep was followed by a decrease (Murdoch et al., 1983). It is possible that eicosanoids are involved in ovulation by modulating ovarian vascular response to LH. In the ewe indomethacin prevented the second-phase reduction in follicular blood flow (Murdoch et al., 1983). In the rat, inhibitors of cyclooxygenase and of lipoxygenase did not affect LH-induced increase in ovarian or follicular blood-flow 3, 6 or 9 h after administration of gonadotropin. By contrast, these inhibitors prevented the hCG-induced increase in uptake of labeled iodinated bovine serum albumin (Abisogun et al., 1988b). While the precise mechanisms involved remain to be determined, all these results support the suggested role of eicosanoids in periovulatory changes in ovarian microcirculation and thereby also in ovulation.

Platelet activating factor (PAF)

PAF (1-Q-alkyl-2-acetyl-sn-glycero-3-phosphocholine) is a potent bioactive phospholip-id. In addition to triggering aggregation and degranulation of platelets, PAF induces margi-nation and activation of neutrophils. Furthermore, PAF is released upon activation of leu-kocytes. We have demonstrated a preovulatory increase in ovarian and follicular neutrophils, triggered by hCG, and this was blocked by inhibitors of eicosanoid synthesis (Abisogun et al., 1988c). In view of these changes common to ovulation and inflamma-tion, we have examined the possible involvement of PAF in ovulation. Unilateral injec-tion into the ovarian bursa of a specific PAF antagonist BN52021, isolated from the Chi-nese tree Ginkgo biloba (Braquet et al., 1985), resulted in a dose dependent inhibition of follicle rupture from the treated ovary. The inhibition was prevented when BN52021 and PAF were administered simultaneously. However, PAF was unable to induce ovulation on its own when administered to rats that were not stimulated by exogenous hCG and had their endogenous surge of LH blocked by Nembutal (Abisogun et al., 1989).

The PAF antagonist, in addition to inhibiting follicle rupture, suppressed the hCG-stimulated increase in ovarian collagen degradation and vascular permeability, similar to the

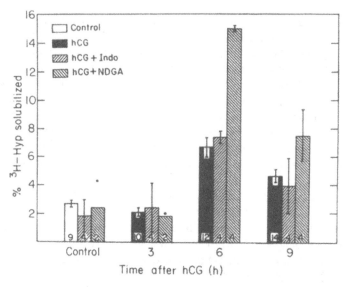

Fig. 1. Effect of eicosanoid synthesis inhibitors on ovarian collagenase activity in vitro. Indo-methacin (Indo), an inhibitor of cyclo-oxygenase ($10^{-5}M$) or nordihydroguiaretic acid (NDGA), an inhibitor of lipoxygenase ($10^{-4}M$) were added to ovarian extracts obtained at the indicated time after hCG administration. Endogenous collagenase activity was assayed in vitro during a 48 h incubation period of the 3H-proline-labeled ovarian collagen extract. The number of replicates is indicated on the columns; vertical brackets ± S.E.M. Data from Abisogun et al., in preparation.

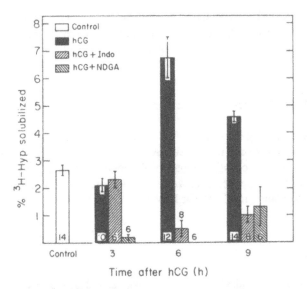

Fig. 2. Effect of eicosanoid synthesis inhibitors on ovarian collagenase activation by hCG. The inhibitors, indomethacin or NDGA were injected into the ovarian bursa concomitantly with hCG administration on the afternoon of the day of proestrus. Endogenous collagenase activity was assayed in vitro in ovarian collagen extracts obtained at the indicated time-intervals after hCG administration. Other details as in Fig. 1.

action of eicosanoid synthesis inhibitors. Administration of PAF partially reversed the inhibition of follicle rupture elicited by indomethacin or NDGA (Abisogun et al., 1989). These studies demonstrate the close interrelationship between eicosanoids and PAF in ovulation. The precise mechanism of PAF involvement in follicle rupture, as well as its cellular origin and targets of its action in the ovary, remain to be determined.

CONCLUDING REMARKS

The studies reviewed demonstrate the role of serine- and metallo-proteases in follicular rupture during ovulation. Furthermore, it seems that a protease cascade involving plasminogen activator, plasmin and collagenase is essential for ovulation. Use of pharmacological inhibitors of eicosanoid synthesis reveals additional mediators for LH/hCG-induced ovulation, products of cyclooxygenase and lipoxygenase pathways of AA metabolism. While eicosanoids do not affect directly collagenase activity, they are involved in the activation of ovarian collagenolysis by the gonadotropin. Likewise, the plasmin activating system seems to be essential for activation of ovarian collagenolysis, but increase in t-PA is not sufficient to ensure collagenolysis. Thus, in the presence of inhibitors of eicosanoid synthesis ovarian collagenolysis is prevented in spite of the increase in ovarian plasminogen activator. PAF appears to play an essential role in follicle rupture, and its action is closely related to eicosanoids. The involvement of protease cascade, eicosanoids and PAF in ovulation bears close resemblance to changes accompanying inflammation (Espey, 1980) and tumor cell invasion (Danø et al., 1985; Mignatti et al., 1986; Reich et al., 1988).

ACKNOWLEDGEMENTS

We thank Mrs. R. Slager and Mrs. A. Tsafriri for their skilful technical assistance and Mrs. M. Kopelowitz for secretarial help. A.O.A. is a recipient of the Macy Foundation Fellowship and this work constitutes part of the requirements for the Ph.D. degree of the Weizmann Institute of Science. This work was supported by a grant from the G.I.F., the German-Israel Foundation for Scientific Research and Development.

REFERENCES

Abisogun, A.O., Reich, R., Miskin, R., and Tsafriri, A., 1988a, Proteolytic processes at ovulation, *in:* "Development and Function of the Reproductive Organs", vol. II, M. Parvinen, I. Huhtaniemi and L.J. Pelliniemi, eds, Serono Symposia Review, Rome, 14:285.

Abisogun, A.O., Daphna-Iken, D., Reich, R., Kranzfelder, D., and Tsafriri, A., 1988b, Modulatory role of eicosanoids in vascular changes during the preovulatory period in the rat, *Biol. Reprod.*, 38:756.

Abisogun, A.O., Daphna-Iken, D., and Tsafriri, A., 1988c, Periovulatory changes in ovarian and follicular neutrophils: Modulation by eicosanoids, *Endocrinology*, 122:1083 (abstr.)

Abisogun, A.O., Braquet, P., and Tsafriri, A., 1989, The involvement of platelet activating factor in ovulation, *Science*, 243:381.

Ainsworth, L., Baker, R.D., and Armstrong, D.T., 1975, Preovulatory changes in follicular fluid prostaglandin F levels in swine, *Prostaglandins*, 9:915.

Akazawa, K., Matsuo, O., Kosugi, T., Mihara, H., and Mori, N., 1983, The role of plasminogen activator in ovulation, *Acta Physiol. Latino Amer.*, 33:105.

Armstrong, D.T., and Zamecnik, J., 1975, Preovulatory elevation of rat ovarian prostaglandin F and its blockade by indomethacin, *Mol. Cell. Endocr.*, 2:125.

Bauminger, S., and Lindner, H.R., 1975a, Periovulatory changes in ovarian prostaglandin formation and their hormonal control in the rat, *Prostaglandins*, 9:737.

Bauminger, S., Lieberman, M.E., and Lindner, H.R., 1975b, Steroid-independent effect of gonadotropins on prostaglandin synthesis in rat Graafian follicles in vitro, *Prostaglandins*, 9:753.

Beers, W.H., and Strickland, S., 1978, A cell culture assay for follicle-stimulating hormone, *J. Biol. Chem.*, 253:3877.

Beers, W.H., Strickland, S., and Reich, E., 1975, Ovarian plasminogen activator: Relationship to ovulation and hormonal regulation, *Cell*, 6:387.

Beers, W.H., 1975, Follicular plasminogen and plasminogen activator and the effect of plasmin on follicular wall, *Cell*, 6:379.

Birkedal-Hansen, H., Cobb, C.M., Taylor, R.E., and Fullmer, H.M., 1975, Trypsin activation of latent collagenase from several mammalian source, *Scand. J. Dent. Res.* 83:302.

Bränström, M., Woessner, J.F., Koos, R.D., Sear, C.H.J., and LeMaire, W.J., 1988, Inhibitors of mammalian tissue collagenase and metaloproteinases suppress ovulation in the perfused rat ovary, *Endocrinology*, 122:1715.

Braquet, P., Spinnewyn, B., Braquet, M., Bourgain, R.H., Taylor, J.E., Etienne, A., and Drien, K., 1985, BN52021 and related compounds: a new series of highly specific PAF-acether receptor antagonists isolated from Ginkgo biloba, *Blood Vessels*, 16:559.

Canipari, R., and Strickland, S., 1985, Plasminogen activator in the rat ovary. Production and gonadotropin regulation of the enzyme in granulosa thecal cells, *J. Biol. Chem.*, 260:5121.

Canipari, R., and Strickland, S., 1986, Studies on the hormonal regulation of plasminogen activator production in the rat ovary, *Endocrinology*, 118:1652.

Curry, T.E. Jr., Clark, M.R., Dean, D.D., Woessner, J.F. Jr., and LeMaire, W.J. Jr., 1986, The preovulatory increase in ovarian collagenase activity in the rat is independent of prostaglandin production, *Endocrinology*, 118:1823.

Danø, K., Andreasen, P.A., Grondahl-Hansen, J., Krigtensen, P., Nielsen, L.S., and Skriver, L., 1985, Plasminogen activators, tissue degradation and cancer, *Advances In Cancer Res.*, 44:139.

Espey, L.L., 1967, Ultrastructure of the apex of the rabbit's Graafian follicle during ovulatory process, *Endocrinology*, 81:267.

Espey, L.L., 1980, Ovulation as an inflammatory reaction - a hypothesis, *Biol. Reprod.*, 22:73.

Espey, L., Shimada, H., Okamura, H., and Mori, T., 1985, Effect of various agents on ovarian plasminogen activator activity during ovulation in PMSG-primed immature rats, *Biol. Reprod.*, 32:1087.

Feldman, E., Haberman, S., Abisogun, A.O., Reich, R., Levran, D., Maschiach, S., Zuckermann, H., Rudak, E., Dor, J., and Tsafriri, A., 1986, Arachidonic acid in human granulosa cells: Evidence for cyclooxygenase and lipoxygenase activity in vitro, *Human Reprod.* 1:353.

Fukumoto, M., Yajima, Y., Okamura, H., and Midoribawa, O., 1981, Collagenolytic enzyme activity in human ovary: An ovulatory enzyme system, *Fertil. Steril.*, 36:746.

Holmes, P.V., Janson, P.O., Sogn, J., Källfelt, B., LeMaire, W.J., Ahrén, K., Cajander, S., and Bjersing, L., 1983, Effects of $PGF_{2\alpha}$ and indomethacin on ovulation and steroid production in the isolated perfused rabbit ovary, *Acta Endocrinol.*, 104:233.

Horwitz, A.L., Kellman, J.A., and Crystal, R.E., 1976, Activation of alveolar macrophage collagenase by a natural protease secreted by the same cell, *Nature* (London), 264:772.

Ichikawa, S., Ohta, M., Morioka, H., and Murao, S., 1983, Blockage of ovulation in the explanted hamster ovary by a collagenase inhibitor, *J. Reprod. Fertil.*, 68:17.

Janson, P.O., 1975, Effects of luteinizing hormone on blood in the follicular rabbit ovary as measured by radioactive microspheres, *Acta Endocr.*, 79:122.

Lee, W., and Novy, M.J., 1978, Effects of luteinizing hormone and indomethacin on blood flow and steroidogenesis in the rabbit ovary, *Biol. Reprod.*, 18:799.

LeMaire, W.J., Yang, N.S.T., Behrman, H.R., and Marsh, J.M., 1973, Preovulatory changes in the concentration of prostaglandins in rabbit Graafian follicles, *Prostaglandins*, 3:367.

Lipner, H., 1988, Mechanism of mammalian ovulation, in: "The Physiology of Reproduction", vol. I, E. Knobil and J.D. Neill, eds., Raven Press, New York, p. 447.

Liu, Y-X., Ny, T., Sarkar, D., Loskutoff, D., and Hsueh, A.J.W., 1986, Identification and regulation of tissue plasminogen activator activity in rat cumulus-oocyte complexes, *Endocrinology*, 119:1578.

Mignatti, P., Robbins, E., and Rifkin, D.B., 1986, Tumor invasion through the human amniotic membrane: requirement for a proteinase cascade, *Cell*, 47:487.

Morales, T.I., Woessner, J.F., Howell, D.S., Marsh, J.M., and LeMaire, W., 1978, A microassay for the direct demonstration of collagenolytic activity in Graafian follicles of the rat, *Biochim. Biophys. Acta*, 524:428.

Morales, T.I., Woessner, J.F., Marsh, J.M., and LeMaire, W.J., 1983, Collagen, collagenase and collagenolytic activity in rat Graafian follicles during follicular growth and ovulation, *Biochim. Biophys. Acta*, 756:119.

Murdoch, W.J., Nix, K.J., and Dunn, T.G., 1983, Dynamics of ovarian blood supply to periovulatory follicles of the ewe, *Biol. Reprod.*, 28:1001.

Ny, T., Bjersing, L., Hsueh, A.J.W., and Loskutoff, D.J., 1985, Cultured granulosa cells produce two plasminogen activators and an antiactivator, each regulated differently by gonadotropins, *Endocrinology*, 116:1666.

Reich, R., Kohen, F., Naor, Z., and Tsafriri, A., 1983, Possible involvement of lipoxygenase products of arachidonic acid pathway in ovulation, *Prostaglandins*, 26:1011.

Reich, R., Tsafriri, A., and Mechanic, G.L., 1985a, The involvement of collagenolysis in ovulation in the rat, *Endocrinology*, 116:521.

Reich, R., Miskin, R., and Tsafriri, A., 1985b, Follicular plasminogen activator: Involvement in ovulation, *Endocrinology*, 116:516.

Reich, R., Kohen, F., Slager, R., and Tsafriri, A., 1985c, Ovarian lipoxygenase activity and its regulation by gonadotropin in the rat, *Prostaglandins*, 30:581.

Reich, R., Miskin, R., and Tsafriri, A., 1986, Intrafollicular distribution of plasminogen activators and their hormonal regulation in vitro, *Endocrinology*, 119:1588.

Reich, R., Haberman, S., Abisogun, A.O., Sofer, Y., Grossman, S., Adelmann-Grill, B.C., and Tsafriri, A., 1987, Follicular rupture at ovulation: Collagenase activity and metabolism of arachidonic acid, *in:* "The control of follicle development, ovarian and luteal function: Lessons from in vitro fertilization", F. Naftolin and A.H. deCherney, eds., Raven Press, p. 317.

Reich, R., Thompson, E.W., Iwamoto, Y., Martin, G.R., Deason, J.R., Fuller, G.C., and Miskin, R., 1988, Inhibition of plasminogen activator, serine proteinases and collagenase IV prevents the invasion of basement membrane by metestatic cells, *Cancer Res.*, 48:3307.

Schochet, S.S., 1916, A suggestion as to the process of ovulation and ovarian cyst formation, *Anatomical Record*, 10:447.

Schwartz, N.B., Cobbs, S.B., Talley, W.L., and Ely, C.A., 1975, Induction of ovulation by LH and FSH in the presence of antigonadotrophic sera, *Endocrinology*, 96:1171.

Shimada, H., Okamuka, H., Noda, Y., Suzuki, A., Tojo, S., and Takada, A., 1983, Plasminogen activator in rat ovary during the ovulatory process: independence of prostaglandin mediation, *J. Endocr.*, 97:201.

Strickland, S., and Beers, W.H., 1976, Studies on the role of plasminogen activation in ovulation, *J. Biol. Chem.*, 251:5694.

Tsafriri, A., Lieberman, M.E., Koch, Y., Bauminger, S., Chobsieng, P., Zor, U., and Lindner, H.R., 1976, Capacity of immunologically purified FSH to stimulate cyclic AMP accumulation and steroidogenesis in Graafian follicles and to induce ovum maturation and ovulation in the rat. *Endocrinology*, 98:655.

Tsafriri, A., Lindner, H.R., Zor, U., and Lamprecht, S.A., 1972, Physiological role of prostaglandins in the induction of ovulation, *Prostaglandins*, 2:1, 31:39.

Tsafriri, A., Braw, R.H., and Reich, R., 1986, Follicular development and the mechanism of ovulation, *in:* "Infertility", V. Insler and B. Lunenfeld, eds., Churchill Livingstone, Edinburgh, p. 73.

Tsafriri, A., Bicsak, T.A., Cajander, S.B., Ny, T., and Hsueh, A.J.W., 1989, Suppression of ovulation rate by antibodies to tissue type plasminogen activator and α_2-antiplasmin. *Endocrinology*, 124:415.

Wallach, E.E., Bronson, R., Hamada, Y., Wright, K.H., and Stersens, V.G., 1975, Effectiveness of $PGF_{2\alpha}$ in restoration of HMG-HCG induced ovulation in indomethacin-treated Rhesus monkeys, *Prostaglandins*, 10:129.

Wang, C., and Leung, A., 1983, Gonadotropin regulate plasminogen activator production by rat granulosa cells, *Endocrinology*, 112:1201.

INVOLVEMENT OF CALCIUM IN THE TRANSDUCTION
OF THE HORMONAL SIGNAL FOR INDUCTION OF OOCYTE MATURATION

Nava Dekel, Sari Goren, and Yoram Oron*

*Department of Hormone Research, The Weizmann Institute of Science
and
*Department of Physiology and Pharmacology
Sackler Faculty of Medicine, Tel Aviv, Israel*

The mammalian female is born with her ovaries containing the entire pool of oocytes, all of them arrested at the diplotene of the first meiotic prophase. Meiotic arrest in the oocytes persists throughout infancy. After the onset of puberty, at each cycle, one or more oocytes respond to the midcycle surge of luteinizing hormone (LH) and resume meiosis. Oocytes that resumed meiosis had generated an haploid nucleus and acquired the capacity to be fertilized. Resumption of meiosis is therefore referred to as oocyte maturation.

As mentioned above, oocyte maturation in vivo is stimulated by the pituitary gonadotropin LH (Lindner et al., 1974). This hormone also stimulates maturation in vitro of follicle-enclosed oocytes (Tsafriri et al., 1972). In the rat, gonadotropin releasing hormone (GnRH) can also induce oocyte maturation. The action of GnRH is mediated via specific receptors (Clayton et al., 1979; Jones et al., 1980; Dekel et al., 1988) and can be demonstrated both in vivo in hypophysectomized rats (Corbin and Bex, 1981; Ekholm et al., 1981; Dekel et al., 1985) and in vitro in follicle-enclosed oocytes (Hillensjö and LeMaire, 1980; Dekel et al., 1983). LH-induced oocyte maturation is associated with an increase in the follicular concentrations of cAMP (Lindner et al., 1974; Dekel et al., 1983). GnRH action on the other hand does not involve elevation of cAMP (Dekel et al., 1983). This observation suggests the existence of an alternate pathway for triggering of meiosis that is not mediated by the cAMP-second messenger system.

The results discussed in this paper suggest that the calcium-dependent protein kinase C (PKC) is a likely candidate for a second messenger system for hormonally-induced oocyte

Advances in Assisted Reproductive Technologies, Edited by
S. Mashiach *et al.*, Plenum Press, New York, 1990

Table 1: The Effect of Calcium on LH-Induced Oocyte Maturation

[CaCl$_2$] (M)	Oocyte maturation	
	Percentage ± SEM	Total number of oocytes
1.3	86 ± 2	56
0.6	31.5 ± 13.7	114
0.0	31 ± 13.8	84

Follicles were incubated with 1 μg/ml of ovine LH and the indicated concentrations of CaCl$_2$. After 6 h the oocytes were recovered and examined for maturation.

maturation. This hypothesis was initially based on the reports that GnRH-induced LH-release in the pituitary gonadotrophs involves activation of PKC (Hirota et al., 1985; Harris et al., 1985; Naor et al., 1985) and that PKC activators can induce oocyte maturation (Aberdam and Dekel, 1985). Since PKC is biochemically dependent on calcium, our studies were specifically directed towards investigation of the role of calcium in hormonally-induced oocyte maturation. Involvement of calcium in oocyte maturation has been studied in invertebrates and amphibians (Baulieu et al., 1978; Masui and Clarke, 1979; Morril and Kostellow, 1986) as well as in mammals (Tsafriri and Bar-Ami, 1978; Liebfried and First, 1979; Paleos and Powers, 1981; Jagiello et al., 1982; Maruska et al., 1984; Bae and Channing, 1985; Racowsky, 1986). In mammalian oocytes the data are often contradictory and no clear picture emerges of the role of calcium in this process.

Our experimental system is the isolated rat ovarian follicle. The oocytes enclosed by these follicles are maintained in vitro in meiotic arrest. These oocytes resume meiosis when incubated in the presence of hormones and other agents. At the end of the incubation period the follicles are incised, the cumulus-oocyte complexes recovered and the oocytes examined for maturation as indicated by the absence of the germinal vesicle.

If indeed calcium was involved in the induction of oocyte maturation we would assume that this cation will mimic hormone action stimulating the oocyte to resume meiosis. To test this hypothesis we incubated follicle-enclosed oocytes in the presence of increasing concentrations of calcium in the culture medium. A very poor effect, if any, was obtained even in the presence of high (20 mM) extracellular calcium. However, considering that influx of calcium into cell cytoplasm is very limited, we created culture conditions that will facilitate calcium entry by addition of calcium ionophores to the culture medium. In the presence of divalent ionophores, either ionomycin or A23187 (10^{-6} M), we did get induction of oocyte maturation which was directly dependent on the extracellular concentrations of calcium (Dekel et al., 1989). Analysis of the kinetics of calcium action reveals that induction of oocyte maturation is completed by 2 h. Calcium-induced oocyte maturation is therefore much faster than that of either GnRH or LH that require 4-5 h to complete their action (Dekel et al., 1983).

Table 2. The Effect of Voltage-Sensitive Calcium-Channel Modulators on LH-Induced Oocyte
Maturation

Agent (M)	[LH] (μg/ml)	Oocyte maturation	
		Percentage ± SEM	Total number of oocytes
--	0.1	83 ± 4	271
Verapamil (10^{-5})	0.1	89 ± 6	89
Nifedipine (10^{-5})	0.1	83 ± 4.5	252
Diltiazem (10^{-5})	0.1	87 ± 7.9	109
BAY-K-8644(10^{-6})	--	6 ± 4.6	132

Follicles were incubated with the indicated calcium-channel inhibitors or activator with or without the indicated concentrations of ovine LH. After 6 h the oocytes were recovered and examined for maturation.

Our results suggest that elevation of intracellular calcium is sufficient for induction of oocyte maturation. Is calcium essential for hormonally-induced oocyte maturation? A full response to LH is obtained in the presence of 1.3 mM calcium in the culture medium (Dekel et al., 1983). Decreasing extracellular calcium to 0.6 mM results in a 65% inhibition of hormone action. Total elimination of calcium by EDTA did not cause any further inhibition (Table 1). Taken together, the results of these experiments suggest that calcium is sufficient and essential for stimulation of oocyte maturation.

As mentioned above, lowering calcium concentrations in the culture medium resulted in a decreased response of the oocytes to LH. However, to our surprise, the increase in extracellular calcium did not potentiate but rather inhibited LH action (39% inhibition by 20 mM of $CaCl_2$). Inhibition of LH action by increasing calcium concentrations could result from a stimulated degradation cAMP, following activation of the cAMP phosphodiesterase by the calcium-calmodulin complex (Klee et al., 1980). Alternatively, inhibition of adenylate cyclase by high calcium concentrations, which has already been described for granulosa cells (Eckstein et al., 1986), could also account for the reduction in the response to LH.

As already mentioned the concentration of calcium in the incubation medium is 10^{-3} M. The intracellular concentration of calcium is 10^{-7} M. Calcium influx occurs according to the concentration gradient via specific channels present within the cell membrane. Two major types of calcium channels are described in the literature: a. voltage-gated calcium channels regulated by changes in the electrical potential of the cell membrane (Rauter, 1985), and b. ligand-gated calcium channels activated by binding of hormones to specific receptors (Neher, 1987). To test whether voltage gated channels are involved in the induction of oocyte maturation we have employed agents that either block or activate this specific type of calcium channel. The results of our experiments revealed that neither of the calcium channel blockers used, diltiazem, nifedipine or verapamil (Godfraind, 1984), inhibited hormone action (Table 2). Furthermore, the dihydropiridine BAY-K-8644, which is an activator of calcium channels (Garcia et al., 1984), failed to induce oocyte maturation (Table 2). The use of potassium chloride for membrane depolarization was also found to be inef-

Table 3. Maturation in Follicle-Enclosed Oocytes Incubated with KCl and Elevated Concentrations of Calcium

[KCl] (mM)	[Ca^{2+}] (mM)	Oocyte maturation	
		Percentage ± SEM	Total number of oocytes
50	1.3	11 ± 7	99
50	6.3	6.4 ± 3.5	93
50	11.3	10 ± 6.8	99

Follicles were incubated with 50 mM of KCl in the presence of the indicated concentrations of CaCl$_2$. After 6 h the oocytes were recovered and examined for maturation.

Table 4. The Effect of Fluphenazine on LH- and Calcium- Induced Oocyte Maturation

Fluphenazine (M)	LH (μg/ml)	CaCl$_2$ (M)	A23187 (M)	Oocyte maturation	
				Percentage ± SEM	Total number of oocytes
10^{-6}	--	1.3	--	1.5	65
--	0.1	1.3	--	81 ± 6.3	105
10^{-6}	0.1	1.3	--	88 ± 2.3	101
--	--	10	10^{-6}	33 ± 8	212
10^{-6}	--	10	10^{-6}	49 ± 11	213

Follicles were incubated in the presence or absence of either ovine LH or the calcium ionophore A23187 with or without flufenazine. After 6 h the oocytes were recovered and examined for maturation.

fective (Table 3). These results seem to suggest that ligand-gated calcium channels rather than voltage-gated channels may be involved in the hormonally-induced oocyte maturation.

Elevation of intracellular calcium concentration apparently turns on a cascade of biochemical events that finally lead to the biological response. One possible step in this biochemical pathway can be activation of calcium-calmodulin dependent enzymes. To test this possibility we have used fluphenazine, which interferes with the formation of the calcium-calmodulin complex. The lack of inhibitory effect of fluphenazine on either LH, or calcium induced oocyte maturation eliminates the possible involvement of such enzymes in this process (Table 4).

Activation of PKC is dependent on calcium. This enzyme can therefore serve as a second messenger system for calcium action. To test the involvement of PKC in hormonally-induced oocyte maturation we have employed the newly synthesized derivatives of isoquinolinesulfonamides, H-7 and H-8 that bind to protein kinases and inhibit their action (Hidaka et al., 1989). Addition of these agents clearly inhibits calcium action (from 52 ± 9% to 31 ± 5%, 113 and 165 oocytes, respectively; p = 0.05). Being not specific to PKC these kinase inhibitors have a clear disadvantage. However, since the dose dependency of their inhibition of LH-induced hormone action directly corresponded to their Ki towards

PKC (Dekel, unpublished), we suggest that it is their effect on PKC rather than that on PKA that is manifested in our experiments.

In conclusion, our results are compatible with the following sequence of events. Binding of the hormone to its receptors activates receptor-gated channels leading to an increase in intracellular calcium. This in turn leads to the activation of PKC that, by an as yet unknown mechanism, initiates oocyte maturation. This model actually suggests that for induction of oocyte maturation LH may utilize PKC rather than PKA as a second-messenger system. The idea that in addition to the cAMP-second messenger system, LH uses alternative pathways to elicit ovarian response, has already been raised by other investigators (Davis et al., 1986; Dimino et al., 1987; Allen et al., 1988). Our future experiments will seek further support for this intriguing possibility.

REFERENCES

Aberdam, E., and Dekel, N., 1985, Activators of protein kinase C stimulate meiotic maturation of rat oocytes. *Biochem Biophys. Res. Commun.*, 132:570.

Allen, R. B., Su, H. C., Snitzer, J., Dimino, M. J., 1988, Rapid decreases in phosphatidylinositol in isolated luteal plasma membranes after stimulation by luteinizing hormone. *Biol. Reprod.*, 38:79.

Bae, I., Channing, C. P., 1985, Effect of calcium ion on the maturation of cumulus-enclosed pig follicular oocytes isolated from medium-sized Graafian follicles. *Biol. Reprod.*, 33:79.

Baulieu, E. E., Godeau, F., Schorderet, M., and Schorderet-Slatkine, S., 1978, Steroid-induced meiotic division in Xenopus laevis oocytes: surface and calcium. *Nature*, 275:593.

Clayton, R. N., Harwood, J. P., and Catt, K.J., 1979, Gonadotropin-releasing hormone analogue binds to luteal cells and inhibits progesterone production. *Nature*, 282:90.

Corbin, A., and Bex, F. J., 1981, Luteinizing hormone-releasing hormone agonists induce ovulation in hypophysectomized proestrous rats: direct ovarian effect. *Life. Sci.*, 29:185.

Davis, J. S., Weakland, L. L., West, L. A., and Farese, R. V., 1986, Luteinizing hormone stimulates the formation of inositol triphosphate and cyclic AMP in rat granulosa cells. *Biochem. J.*, 238:597.

Dekel, N., Aberdam, E., Goren, S., Feldman, B., and Shalgi, R., 1989, Mechanism of action of GnRH-induced oocyte maturation. *J. Reprod. Fert. Suppl.*, 37:319.

Dekel, N., Lewysohn, O., Ayalon, D., and Hazum, E., 1988, Receptors for gonadotropin releasing hormone are present in rat oocytes. *Endocrinology*, 123:120.

Dekel, N., Sherizly, I., Phillips, D. M., Nimrod, A., Zilberstein, M., and Naor, Z., 1985, Characterization of the maturational changes induced by a GnRH analogue in the rat ovarian follicle. *J. Reprod. Fert.*, 75:461.

Dekel, N., Sherizly, I., Tsafriri, A., and Naor, Z., 1983, A comparative study of the mechanism of action of luteinizing hormone and gonadotropin-releasing hormone analog on the ovary. *Biol. Reprod.*, 28:161.

Dimino, M. J., Snitzer, J., and Brown K. M., 1987, Inositol phosphates accumulation in ovarian granulosa after stimulation by luteinizing hormone. *Biol. Reprod.*, 37:1129.

Eckstein, N., Eshel, A., Eli, Y., Ayalon, D., and Naor, Z. 1986, Calcium-dependent actions of gonadotropin-releasing hormone agonist and luteinizing hormone upon cyclic AMP and progesterone production in rat ovarian granulosa cells. *Mol. Cell. Endocr.* 47:91.

Ekholm, C., Hillensjö, T., and Isaksson, O. 1981, Gonadotropin-releasing hormone agonists stimulate oocyte meiosis and ovulation in hypophysectomized rats. *Endocrinology*, 108:2022.

Garcia, A. G., Sala, F., Reig, J. A., Viniegra, S., Frias, J., Fonteriz, R., and Gandia, L. 1984, Dihydropyridine BAY-K-8644 activates chromaffin cell calcium channels. *Nature*, 309:69.

Godfraind, T., 1984, Calcium modulators: Tools for cell biology. *in*: "Advances Cyclic Nucleotide and Protein Phosphorylation Research", Greengard, P., ed., Vol. 17, p 5.

Harris, C. E., Staley, D., and Conn, P.M., 1985, Diacylglycerol and protein kinase C: Potential amplifying mechanism for Ca^{2+}-mediated GnRH-stimulated LH release. *Mol. Pharmacol.*, 27:532.

Hidaka, H., Inagaki, M., Kawamoto, S., and Sasaki, Y. 1984, Isoquinolinesulfonamides, novel and potent inhibitors of cyclic nucleotide-dependent protein kinase and protein kinase C. *Biochemistry*, 23:5036.

Hillensjö, T., and LeMaire, W. J., 1980, Gonadotropin-releasing hormone agonists stimulate meiotic maturation of follicle-enclosed rat oocytes in vitro. *Nature*, London 287:145.

Hirota, K., Hirota, T., Aguilera, G., and Catt, K.J., 1985, Hormone-induced redistribution of calcium-activated phospholipid-dependent protein kinase in pituitary gonadotrophs. *J. Biol. Chem.* 260:3243.

Jagiello, G., Ducayen, M. B., Downey, R., and Jonassen, A., 1982, Alternations of mammalian oocyte meiosis I with divalent cations and calmodulin. *Cell. Calcium.*, 3:153.

Jones, P. B. C., Conn, P. M., Marian, J., and Hsueh, A.J.W., 1980, Binding of gonadotropin-releasing hormone agonist to rat ovarian granulosa cells. *Life. Sci.* 27:2125.

Klee, C. B., Crouch, T. H., and Richman, P.G. 1980, Calmodulin. *Ann. Rev. Biochem.*, 49:489.

Leibfried, L., and First, N. L., 1979, Effects of divalent cations on *in vitro* maturation of bovine oocytes. *J. Exp. Zool.* 210:575.

Lindner, H. R., Tsafriri, A., Lieberman, M.E., Zor, U., Koch, Y., Bauminger, S., and Barnea, A., 1974, Gonadotrophin action on cultured Graafian follicles: induction of maturation division of the mammalian oocyte and differentiation of the luteal cell. *Recent Prog. Horm. Res.*, 30:79.

Maruska, D. V., Leibfried, M. L., and First, N. L., 1984, Role of calcium and the calcium-calmodulin complex in resumption of meiosis, cumulus expansion, viability and hyaluronidase sensitivity of bovine cumulus-oocyte complexes. *Biol. Reprod.*, 31:1.

Masui, Y., and Clarke, H. J., 1979, Oocyte maturation. *Int. Rev. Cytol.* 7:185.

Morrill, G. A., and Kostellow, A. B., 1986, The role of calcium in meiosis. *Calcium and Cell Function*, 6:209.

Naor, Z., Zer, J., Zakut, H., and Hermon, J., 1985, Characterization of pituitary calcium-activated, phospholipid-dependent protein kinase: redistribution by gonadotropin-releasing hormone. *Proc. Natl. Acad. Sci. U.S.A.*, 82:8203.

Neher, E., 1987, Receptor-operated Ca channels. *Nature* 326:242.

Paleos, G. A., and Powers, R.D., 1981, The effect of calcium on the first meiotic division of the mammalian oocyte. *J. Exp. Zool.* 217:409.

Racowsky, C. 1986, The releasing action of calcium upon cyclic AMP-dependent meiotic arrest in hamster oocytes. *J. Exp. Zool.*, 239:263 .

Reuter, H., 1985, A variety of calcium channels. *Nature*, 316:391.

Tsafriri, A., Lindner, H. R., and Zor, U., 1972, In vitro induction of meiotic division in follicle-enclosed rat oocytes by LH, cAMP and PGE_2. *J. Reprod. Fert.* 31:39.

Tsafriri, A., and Bar-Ami, S., 1978, Role of divalent cations in the resumption of meiosis of rat oocytes. *J. Exp. Zool.* 205:293.

HYDRATION AS THE KEY EVENT IN MUCIFICATION OF THE RAT CUMULUS

Y. Barak, P. F. Kraicer, M. Kostiner, and Z. Nevo

Departments of Zoology and Chemical Pathology
Tel Aviv University, Ramat Aviv, Tel Aviv 69978, Israel

INTRODUCTION

The cumulus oophorus oocyte complex (OCC) is a unique structure in the mammalian female fertility system. The physiological role of the cumulus is to form the appropriate environment for the oocyte to develop, mature and prepare it for fertilization by the sperm. In the preovulatory stage, the cumulus exists as a compact-intact coat surrounding the oocyte. After maturation, the cumulus oophorus is expanded and transformed into a suspension of dissociated cells in a huge soft droplet of mucus, resulting in sperm-penetratable OCC.[1-3]

Maturational events in the cumulus oophorus are referred to collectively as mucification. Mucification has generally been assumed to entail stimulated biosynthesis of the extracellular matrix components, mainly proteoglycans and glycosaminoglycans (GAGs), induced by a surge of gonadotropic hormones. We found, however, in OCCs isolated from rats, that the major event in the mucification process is hydration of pre-existing hyaluronic acid (HA), not its de novo synthesis. Data have been accumulated indicating that the GAGs-HA in the preovulatory complexes are found as anhydrous-hydrophobic calcium salts, and removal of the calcium ions from the ECM facilitates the hydration-mucification process.

Advances in Assisted Reproductive Technologies, Edited by
S. Mashiach *et al.,* Plenum Press, New York, 1990

Material and Methods

Isolation of Oocyte-cumulus Complexes. Preovulatory OCCs were teased out, as previously described,[3] from large Graafian follicles of 25 day old rats, 44–48 hours after subcutaneous injection of 12 i.u. pregnant mare serum gonadotropin (PMSG, Sigma). Postovulatory complexes were isolated from distended oviducal ampullae of similar animals, treated additionally with 12 i.u. of human chorionic gonadotropin (hCG,Sigma) injected intraperitoneally 54-65 hours post PMSG injection, and sacrificed 16-18 hours later.

Culture Conditions. The isolated preovulatory OCCs were incubated in 37°C in an atmosphere of 5% CO_2 in air for about 20 hours. The culture medium was Eagle's basal medium with Hank's salts (Gibco) or Ca^{++}-free medium (Gibco-special order) supplemented with 15% of fetal calf serum. Incubation of the complexes was carried out in plastic petri dishes, 35mm diameter (Falcon, No. 3001) containing 1.5ml medium. All the reagents radioactive precursors, hormones and salts were added, if necessary, directly to the culture medium.

Assessment of mucification. The most common way to assess mucification is the utilization of testicular hyaluronidase. Mucification was therefore tested by the change in the susceptibility of the complexes to testicular hyaluronidase. OCCs that underwent both in vivo and in vitro mucification became vulnerable to five minutes exposure to 100U/ml of ovine testicular hyaluronidase (Sigma Type IV) at room temperature. Mucification was assessed when 2–3 peripheral layers of cells were dissociated from the cumulus mass, as described previously.[1,3]

Dispersal of cumulus cells from the complex was also tested in the presence of one of the following enzymes, Trypsin (Sigma) 0.25%; Streptomyces hyaluronidase (Callbiochem 50 I.U./ml; Paparin (Sigma) 1mg/ml.

Incorporation of [35] *S-sulphate and* [3]*H-glucosamine into GAGs molecules.* The synthesis of GAG molecules during mucification in vitro was monitored by measuring the incorporation of labelled precursors into isolated GAG.[4] ^{3}H-glucosamine and ^{35}S-sulfate were added to cultures of preovulatory isolated OCCs (pools of 100 OCCs each) in the presence or absence of agonists (LH, FSH, dbc-AMP). At termination the incubates were boiled in 0.1M acetate buffer, pH 5.4 and subjected to the action of 0.1% papain (Sigma) overnight at 65°C. The proteolytic products were removed by exhaustive dialysis against distilled water. Sodium chloride was added to the dialyzed solution to a final concentration of $0.03\underline{M}$. The GAGs were precipitated at 37°C by addition of a 10% w/v aqueous solution of cetylpyridinium chloride (CPC) to a final concentration of 0.5% w/v; a mixture of 1mg chondroitin-4 - and - 6-sulfates was added as carrier. The glycosaminoglycan-CPC complex was redissolved and dissociated by the addition of 1 ml of $2\underline{M}$ aqueous calcium chloride solution; 9ml of ethanolic ether (2:1) was added and the mixture cooled to -20°C overnight.[4] The precipitate of glycosaminoglycan was collected by centrifugation, redissolved in water, and diluted in Hydroluma (Lumac, The Netherlands) for radioactivity measurement.

Uronic acid determination. Determinations were performed on isolated GAGs from two separate pools of isolated preovulatory and postovulatory complexes, 1000 OCCs each. The GAG isolation procedure included as above boiling the samples for protein denaturation, papain digestion and exhaustive dialysis against distilled water to remove degraded products. The uronic acid determinations were run on lyophilized samples. Routine spectrophotometric determinations of uronic acid content were carried out by the modified Dische's carbazole reaction using glucuronic acid as standard.

Enzymatic digestion. Isolated GAGs obtained from 1000 OCCs each were incubated with 20 TRU (turbidity reducing units equals to international units) of Streptomyces hyaluronidase (E.C.4.2.2.2.; Calbiochem. Behring Corp., La Jolla, CA 92037, USA). The incubations were conducted in small volumes of hyaluronidase buffer (0.2 ml of 0.1 M sodium phosphate containing 0.15 M NaCl at pH 5.3) in a constant shaking water bath at 37°C overnight. The digests, including control samples containing the enzyme alone, were exhaustively dialyzed against distilled water to allow the low molecular weight GAG-degraded products to escape the dialysis tubing. Uronic acid content in the macromolecules was run on lyophilized samples.

Histochemical examinations. Groups of 20–30 preovulatory and postovulatory OCCs were immobilized on microscopic glass-slides coated with poly-L-lysine hydrochloride (Sigma) and fixed in either neutral buffer- formalin (NBF) solution for Alcian blue staining, or by Carnoy's fixative for Alizarin red staining. Alcian blue stains specifically all GAGs molecules in pH 2.6, while at pH 1.0 - this reagent stains only sulfated GAGs. The Alizarin red staining procedure was carried out exactly as described by Pearse.[7] The stain colours in an intensive red site of high concentrations of calcium ions.

RESULTS

Following the Biosynthesis of GAG Macromolecules

Preliminary in vitro studies were designed to compare the biosynthesis activities of pre- and postovulatory OCCs in regard to GAG molecules. The purpose of the experiment was to characterize and quantify the synthetic changes in composition and structure of the newly accumulated products of GAG substances in the ECM of the OCCs following mucification. The synthesis of GAG molecules during mucification in vitro was monitored by measuring the incorporation of labelled precursors into isolated GAG.[4] ^{3}H-glucosamine and ^{35}S-sulfate were added to cultures of preovulatory isolated OCCs (pools of 100 OCCs each) in the presence or absence of agonists[2,3] (LH, FSH, dbc-AMP). Both the total radioactive counts integrated into the GAG macromolecules of pre-and postovulatory OCCs and the actual increase (a factor of about 1.7) in the cultures undergoing mucification, were considerably lower than anticipated in light of the large amounts (many folds increase in volume) of ECM morphologically emerging between the postovulatory cumulus cells. The radioactivity detected in the isolated GAG macromolecules in many repetitious experiments in both pre- and postovulatory OCCs were within the range of $5\text{-}15 \times 10^3$ cpm per 100 OCCs similar to the results obtained by Eppig in mice OCCs.[5] This gap between the morphological events and the biosynthetic findings during mucification raised the possibility that the ECM in the postovulatory cumulus is mostly not a product of de novo synthesis. The other, more likely, alternative is a major physicochemical conformational change in the uronic acid-containing materials (uronic acid is a unique marker for GAG macromolecules) already present in the preovulatory complex.

Base Lines of Uronic Acid Content

To obtain the base line of the uronic acid content of pre- and post-ovulatory OCCs, routine spectrophotometric determinations of uronic acid content were then performed. The experiments were conducted on large batches of at least 1000 OCCs each, harvested either from proestrous Graafian follicles or oviducal ampullae. These amounts are necessary to obtain a substantial uronic acid content measurement, which falls in the proper sensitivity range of the modified Dische's carbazole reaction using glucuronic acid as standard.[6] An increase of only 70% in the uronic acid content was detected in the postovulatory complexes. A preovulatory complex contains approximately 20 ng of macromolecular uronic acid, whereas a postovulatory complex has 35 ng. These data show that during the mucification process, the increase in biosynthesis of GAG molecules is not in multifolds, just confirming the results obtained by following the biosynthesis of GAG molecules employing radiolabelled precursors seen in the previous paragraph.

These biochemical findings were further supported histochemically by Alcian blue staining of the various OCCs. At pH 2.6, which stains specifically all GAG molecules[7] both types of complexes are heavily stained (Fig. 1a&b). Further information was gained

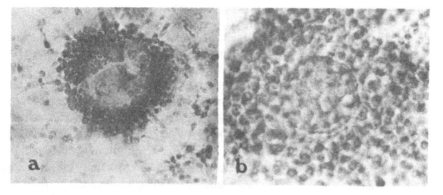

Fig. 1. Alcian blue staining of OCCs at pH 2.6[7]. (a) In the preovulatory complex, the blue staining is limited and concentrates in the pericellular regions of the cumulus cells. (b) In the postovulatory complex, the blue staining is distributed mainly in the ECM ($\times$ 400).

using Alcian blue staining at pH 1.0, which specifically stains sulfated GAGs. As seen in Fig. 2 a&b) of Alcian blue staining at pH 1.0 only postovulatory complexes have been stained intensively. These observations suggest that in addition to the minor alterations in the quantity of total GAGs, the other main changes involve both the localization of GAG molecules and the kind(s) of GAG populations.

This latest suggestion was further challenged biochemically by exposing isolated GAG macromolecules to a substrate (HA)-specific degrading enzyme, Streptomyces Hyaluronidase. The GAGs in the preovulatory OCCs were completely (100%) degraded, whereas only 38.3% of the isolated GAGs in the postovulatory OCCs have been eliminated. The calculated HA concentration is therefore 37 ng for a preovulatory complex, and about 20 ng for a postovulatory complex. Further analysis of the final degradation products di, tetra and hexasaccharides, with 4.5 unsaturated uronic acid at the nonreducing terminals are currently being investigated in our laboratory employing High Pressure Liquid Chromatography (HPLC).

The similar amounts of HA found in preovulatory and postovulatory complexes, and the restricted addition of newly synthesized non-HA GAG (i.e., GAG resistant to Streptomyces hyaluronidase), raised the question as to what changed and how, during mucification, to contribute to the huge changes in volume and mucoid appearance of the complexes?

In our experimentation with purified HA we found that increased concentrations of calcium ions cause changes in the viscosity and solubility of HA molecules. Very high concentrations of Ca^{++} (2-5 M $CaCl_2$) caused HA molecules to precipitate out of solution, whereas a similar ionic strength of sodium salts (e.g. NaCl) did not change HA solubility. Increased concentrations of Ca^{++} 0.5M, (but not of Na^+), specifically inhibited HA degrading enzymes (Streptomyces hyaluronidase, chondroitinase ABC and testicular hyaluronidase).

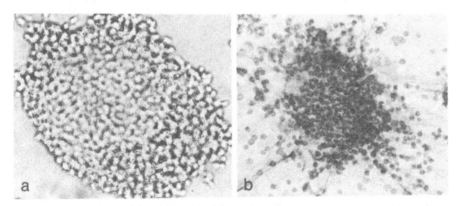

Fig. 2. Alcian blue staining of OCCs at pH 1.0[7]. (a) In the preovulatory complexes no staining is observed. (b) In the postovulatory complex, the blue staining is distributed in the ECM.

A survey examining the effect of various cationic atmospheres on mucification indicated that calcium ions play a regulatory role in mucification. Tables 1 & 2 illustrate the inhibition of both spontaneous and LH-induced mucification by increased concentrations of CA^{++} ions.

A complementary experiment employing a calcium chelator (EGTA 1.3 mM) induced expansion and swelling of preovulatory OCCs, i.e., spontaneous mucification.

Staining pre- and postovulatory OCCs with Alizarin, a calcium-specific histochemical procedure, resulted in an intensive red color of the ECM of preovulatory OCCs and a very pale red color of the postovulatory complexes, again indicating the absence of calcium from the cumulus of postovulatory OCCs (Fig.3 a&b).

DISCUSSION

The detailed sequential events leading to mucification is yet unknown. The only established fact, points at the involvement of the ovulating gonadotrophic hormones in the mucification process, in vivo as well as its stimulation in vitro.[1, 3] The biosynthesis of GAG molecules during mucification was considered as an outcome within the chain of events initiated by the gonadotrophins.[1, 3] The present study reconfirms the presence of a limited de novo synthesis of GAG molecules during mucification. However, the present findings direct the fucus at physicochemical changes occurring in pre-existing GAG molecules as the possible key event initiated during mucification. Furthermore, our analytical

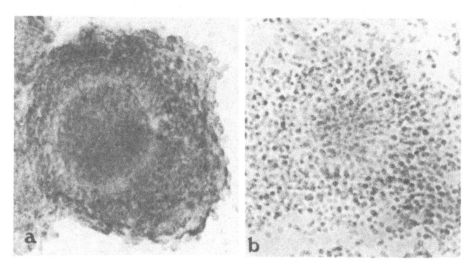

Fig. 3. Detection of calcium ions in OCCs stained by alizarin red method.[7] (a) In the preovulatory complexes, there is heavy red staining of the cumulus cells. (b) In the postovulatory complex, very mild red staining is noted (× 200).

Table 1. Inhibition of Mucification In Vitro, by Calcium Ions in the Presence of LH (10 mcg/ml)

Ca^{++} concern	-LH Control	0.4	1.3	2.6	3.9	8
No. of incubated OCCs	208	59	67	84	84	61
Proportion of mucified OCCs (%)	7.7	69.5	73.1	53.6	48.8	0

Table 2. Inhibition of Mucification In Vitro by Calcium Ions, w/o Agonists in the Medium

Ca^{++} concentration (mM)	0.4	0.6	0.9	1.3
No. of incubated OCCs	214	103	110	175
Proportion of mucified OCCs (%)	67.75	43.79	38.2	6.8

data regarding the GAG molecules show that the newly synthesized GAGs during mucification are rather sulfated GAG substances than HA ones. Other reports on the composition of the de novo synthesized GAGs during mucification, suggest synthesis of HA[5], or HA and sulfated GAGs.[14] This gap may stem either from the different animal system (rat vs. mouse), or differences in the interpretations of the results.

On the other hand, our findings suggest that the pre-existing GAG molecules (in the preovulatory OCCs) subjected to the so called conformational changes are hyaluronates (susceptible to Streptomyces hyaluronidase).

At this point one should try to associate the HA as the target molecule for the conformational change. HA in preovulatory OCCs must be, on one hand, stored in a compact-anhydrous hydrophobic form, and on the other hand, be available to change its properties and swell rapidly (by water) upon the appropriate signal(s).

Old scientific reports starting in the early fifties[11] have dealt with molecular stretching and contracting, gain and loss of water, changes in the dynamic shape and properties of polyelectrolytes gels, affected by alterations in their ionic atmosphere.

Calcium ions have emerged in our cations survey in vitro study, and confirmed by the Alizarin staining technique, as having a regulatory role on swelling-imbibition-mucification of the ECM of the cumulus.

It appears that a stechiometric interaction exists between HA and calcium ions based on neutralization of two free carboxylic groups of uronic acid residues by one calcium ion. Therefore, HA concentration in the system determines the effect of calcium ions, and not the absolute concentration of calcium.

Table 3. The Effect of Degrading Enzymes on the Disperse of Postovulatory Cumulus Cells

Enzyme	No. of examined OCCs	No. of completely dispersed OCCs after 5 mins	after 10 mins
Trypsin 0.25%	20	15	20
Papain 1mg/ml	30	30	-
Hyaluronidase testicular 1mg/ml	29	29	-
Hyaluronidase Streptomyces 50µ/ml	30	5	18

In the presence of high concentrations of calcium ions, the configuration of the HA molecule changes from a stretched-linear shape to a folded coil, unrecognizable by the degrading enzyme.[11] This suggestion is supported by the recent work of Heatley and Scott[12] who noted two distinct stable configurations for hyaluronan under water-rich and water-poor conditions, suggesting that this phenomenon might have biological implications. Thus, the transduction of the complexes from pre- to postovulatory stage involves a parallel shift of HA molecules from water-poor to water-rich conditions. However, it is not yet clear what causes the calcium mobilization (elimination) which enables the hydration of the cumulus ECM. We would like to postulate, that LH may lead to the release of a calcium binding substance, with an affinity to calcium greater than HA's. This assumption is based on our findings for the need of higher calcium concentrations to inhibit mucification in vitro in the presence of LH.

In summary, we would like to suggest the following working hypothesis: HA molecules do exist in preovulatory complexes as calcium salts. The molecules are anhydrous-hydrophobic in nature, densely packed, resistant to common degrading enzymes and have limited availability to Alcian blue staining. A signal, probably initiated by gonadotrophic hormones, induces the mobilization-removal of the calcium ions from the ECM of the OCC; however, the exact mechanism(s) is as yet unknown. This Ca^{++} elimination, changes the nature of the HA molecules, allowing their imbibition. This produces large cell-free spaces, permitting cellular dispersion and cell migration.[13]

Although the most commonly considered concept suggests GAG molecules as the adhesive substances, holding together the cumulus cells at the post ovulatory stage, an obscure possibility of other adhesive substances (e.g. proteins) still exist. This possibility is based on our preliminary observations: a) Guanidinium chloride which supposedly extracts and removes GAG molecules does not cause postovulatory cell dispersion. b) In addition to unpurified testicular hyaluronidase (containing proteases), pure proteases such as trypsin and papain do cause cell dispersion, whereas, a more purified hyaluronidase cause dispersion at a slower rate.

ACKNOWLEDGEMENT

We thank Zoharia Evron for technical assistance and Dr. Nava Dekel, Dept. of Hormone Res., The Weizmann Ins. of Science, Rehovoth, for critical reading of the manuscript.

REFERENCES

1. N. Dekel and P. F. Kraicer, Induction in vitro of mucification of rat cumulus oophorus by gonadoptrophins and adenosine 3',5'-monophosphate. *Endocrinology*, 102:1797-1802 (1978).
2. J. J. Eppig, Gonadotropin stimulation of the expansion of cumulus oophori isolated from mice: general conditions for expansion in vitro, *J. Exp. Zool.* 208:111-120 (1979).
3. Y.Barak, Kaplan, R., P. F. Kraicer & R. Shalgi, Influence of luteinizing hormone and progesterone on maturation and fertilizability of rat oocytes in vitro, *Gamete Res.* 5:257-262 (1982).
4. Z. Nevo, A. L. Howitz and A. Dorfman, Synthesis of chondromucuprotein by chondrocytes in suspension culture, *Dev. Biol.* 28:219-228 (1972).
5. J. J. Eppig, FSH stimulates hyaluronic acid synthesis by oocyte-cumulus cell complexes from mouse preovulatory follicles. *Nature* 281:483-484 (1979).
6. T. Bitter and H. M. Muir, A modified uronic acid carbozol reagent. *Annal. Biochem.* 4:330-334 (1962).
7. A. E. Pearse, Histochemistry: Theoretical & Applied, 3rd Ed., Little, Brown & Co. Boston, 1:672-763 (1968); 2:1404 (1972).
8. Z. Nevo, R. Gonzales, and D. Gospodarowicz, Extracellular matrix (ECM) proteoglycans produced by cultured bovine corneal endothelial cell, *Conn. Tiss. Res.*, 13:45-57 (1984).
9. I. Takazono and Y. J. Tanaka, Quantitive analyses of hyaluronic acid by high-performance liquid chromatography of streptomyces hyaluronidase digests, *Chromatogr.* 288:167-176 (1984).
10. G. J. Lee, D. W. Liu, J. W. Pav, and H. J. Tieckbelmann, Separation of reduced disaccharides derived from glycosaminoglycans by high-performance liquid chromatography, *J. Chromatogr.*, 212:65-73 (1981).
11. W. Kuhn, B. Hargitay, A. Katchalsky and H. Gisenberg, Reversible dilation and contraction by changing the state of ionization of high-polimer acid networks, *Nature* 165:514-516, (1950).
12. F. Heatley and E. Scott, A water molecule participates in the secondary structure of hyaluronan, *Biochem. J.*, 254:489-493 (1988).
13. B. P. Toole, Developmental role of hyaluronate, *Conn. Tiss. Res.*, 10:93-100 (1982).
14. A. Salustri, M. Yanagishita and V. C. Hascall, Synthesis and accumulation of hyaluronic acid and proteoglycans in the mouse cumulus cell-oocyte complex during FSH-induced mucification, *J. Biol. Chem.* (in press).

TUMOR NECROSIS FACTOR (TNFα)
AND THE ENDOGENOUS PROSTAGLANDIN INHIBITOR (PGIn)
IN BOVINE FOLLICLES

M. Shemesh,[a] R. Meiron,[a] M. Zolti,[a]
Z. Ben-Rafael,[b] and D. R.Wollach [c]

[a]Depts. of Horm. Res. & Immunology, Kimron Vet., Institute, Bet Dagan
[b]Dept. of Ob.-Gyn., Tel Hashomer, and
[c]Dept. of Virology, Weizmann Inst., Rechovot, Israel

INTRODUCTION

Ovulation in mammals is a complex course of events consisting of mainly ovum maturation, differentiation of granulosa cells (luteinization) and follicular rupture. Each of these mechanisms is regulated by specific inhibitors or inductors. Despite recent progress in our understanding of ovarian physiology and ovulation, the mechanism involved in follicular rupture remains relatively unknown.

Several studies have demonstrated an increase in follicular prostaglandin synthesis after gonadotrophic stimulation (LaMaire et al., 1973; Armstrong and Zamecnik, 1975). This synthesis and follicular rupture was inhibited by administration of indomethacin. The inhibitory action of indomethacin on follicular rupture was overcome by administration of PGE_2 in rats (Tsafriri et al.,1972); $PGF_{2\alpha}$ in rhesus monkeys (Wallach et al.,1975) and PGE_2 or $PGF_{2\alpha}$ in ewes (Murdoch et al.,1986). These findings led us to examine the role of a possible endogenous inhibitor of prostaglandin synthesis in bovine follicular fluid as well as changes in other regulatory factors. One of these regulatory factors is the tumor necrosis factor α (TNF), a protein produced by macrophages, which has been shown to participate in the regulation of cellular growth, macrophage prostaglandin synthesis and to alter follicular steroidogenesis in vitro.

Advances in Assisted Reproductive Technologies, Edited by
S. Mashiach *et al.*, Plenum Press, New York, 1990

Table 1. Effect of 5%-20% Bovine Follicular Fluid (bFF) upon Secretion of Progesterone (P_4) and Prostaglandin $F_{2\alpha}$ (pg/2 × 10^5 Cells in 24 h) by Cultured Bovine Granulosa Cells Derived from Large Follicles (1.5-2.0 cm diameter) at Mid-cycle During 24 hr of Incubation (means ±S.E.M.; n = 40–60)

	Mid-cycle				Preovulatory
	Control	5% bFF	10% bFF	20%bFF	10%bFF
P_4	1688 ± 83	860 ± 43*	1098 ± 80*	862 ± 86*	1720 ± 88
$F_{2\alpha}$	553 ± 41	191 ± 95*	233 ± 30*	254 ± 16*	575 ± 48

*$P < 0.01$; compared with appropriate control and preovulatory bFF (Tukey's honestly significant difference procedure.)

To examine this hypothesis bovine follicular fluid from different stages of the estrous cycle were analyzed for the presence of the putative inhibitor of PG synthesis. TNF production by granulosa cells (GC) and its presence in follicular fluid was also investigated.

MATERIALS AND METHODS

Follicular Fluid Collection and Culture of Granulosa and Theca Cells
Follicular fluid was obtained from preovulatory and mid- cycle follicles and extracted with charcoal. Theca and granulosa cells (GC) were prepared from large follicles obtained from Holstein-Friesian cows at the mid-point of the cycle. (Shemesh, 1979).

Measurement of Hormones and Prostaglandin Synthetase Activity
Media were extracted and the hormones analyzed by radioimmunoassay. Prostaglandin synthetase activity was measured using bovine seminal vesicle microsomal preparations and RIA as described previously (Shemesh 1979, Shemesh et al., 1984).

TNF Assay
The conditioned media from tissue incubations were assayed for the presence of TNF activity as described by Aderka et al., (1986). Briefly, TNF cytoxicity was quantitated using 9 mm microwells at 40,000 cells/well. Test samples were applied to the cells in serial dilutions in the presence of 50 µG/ml cycloheximaide. Twelve hours later, cell killing was quantitied by measuring the uptake of neutral red. One cytotoxic (CTX) unit was defined as the concentration at which 50% of the cells were killed.

RESULTS

The inhibitory effect of FF on PGF_{2a} production was examined in vitro by incubation of granulosa or theca cells in the presence or absence of FF in the incubation medium. Table 1 shows the effects of mid-cycle and preovulatory bFF on $PGF_{2\alpha}$ and progesterone

Table 2. Effect of 10% Mid-cycle Bovine Follicular Fluid (bFF) upon Secretion of Progesterone (P_4) and Prostaglandin $F_{2\alpha}$ (pg/mg tissue/24h) by Cultured Bovine Theca Tissue Obtained at Mid-cycle (mean±S.E.M.; n = 30)

	Control	10% bFF	Significance control vs 10% bFF*
P_4	3390 ± 460	855 ± 166	P < 0.009
$F_{2\alpha}$	742 ± 112	309 ± 77	P < 0.008

* Levels of significance were measured by Student's t-test.

Table 3. Effect of Mid-cycle Bovine Follicular Fluid (bFF) Combined with LH (1μg/ml) upon the Secretion of Progesterone and Prostaglandin $F_{2\alpha}$ (pg/2 × 10^5 cells/24 h) by Cultured Bovine Granulosa Cells derived from Large Follicles (Diameter 1.5–2.0 cm) Obtained at Mid-cycle During 24 h of Incubation (means ± S.E.M.; n = 40-60)

	Control	LH	20%bFF	LH + 20%bFF
P_4	1517 ± 48a	1780 ± 129b	813± 62c	1160 ± 100d
$F_{2\alpha}$	494 ± 28a'	706± 56b'	194 ± 38c'	402 ± 72d'

Non-matching superscripts (a,b,c,d,a',b',c',d') are significantly different; P < 0.05 by Tukey's honestly significant difference procedure.

secretion by cultured granulosa cells from large follicles. There was no significant difference in incubations to which charcoal-extracted calf serum or saline had been added. Data for controls were therefore pooled. Addition of 5,10 or 20% charcoal-extracted mid-cycle bFF to the culture medium caused a twofold decrease in the accumulation of both progesterone and $PGF_{2\alpha}$ by the cultured cells (P < 0.01). However, when preovulatory bFF obtained at estrus was used, the inhibitory effect on progesterone and $PGF_{2\alpha}$ secretion was not apparent.

Table 2 shows the effects of mid-cycle bFF on $PGF_{2\alpha}$ and progesterone secretion by cultured thecal tissue. The addition of 10% bFF to the culture medium caused a two- to four-fold decrease in the accumulation of both $PGF_{2\alpha}$ and progesterone respectively by cultured thecal tissue (P < 0.008).

The possibility that LH might overcome the inhibitory effect of mid-cycle bFF was also examined. The addition of 1 μg bovine LH/ml to the culture medium caused a significant (P< 0.05) increase in secretion of both progesterone and $PGF_{2\alpha}$ by the culture granulosa cells (Table 3). However, when 20% bFF and LH were both added, neither the inhibitory effect of the bFF nor the stimulatory effect of LH on progesterone and $PGF_{2\alpha}$ secretion was apparent.

The inhibitory activity of the FF was found in 20% ammonium sulfate precipitates after dialysis and lyophilyzation. This precipitate was then applied to Sephadex G-100 columns (Figure 1). The main peak of inhibitory activity was found in fractions 32–42 (40 KD). Aliquots of 20–24 µg protein/ml produced a 35% reduction in $PGF_{2\alpha}$ secretion by the granulosa cells. To further analyze the inhibitory action, ammonium sulfate precipitates derived from mid-cycle follicular fluid were studied for their effect on the conversion of arachidonic acid to prostaglandin derivatives by bovine seminal vesicle microsomal preparations. It was found that the crude extracts caused a dose dependent inhibition (1–6 mg) and that 6 mg of crude extract/ml caused a 100% reducton in conversion of arachidonic acid to its principle derivatives, PGE,PGF and thromboxane (Figure 2).

To examine the role of cytokines, specifically TNF, in regulation of PG synthesis, GC and theca cells derived from ovaries from different stages of the bovine estrous cycle were incubated in the presence of calcium ionophore (A23187) and TNF. As can be seen in Figure 3, TNF was a potent regulator of PG synthesis by both GC and theca cells. Basal pros-

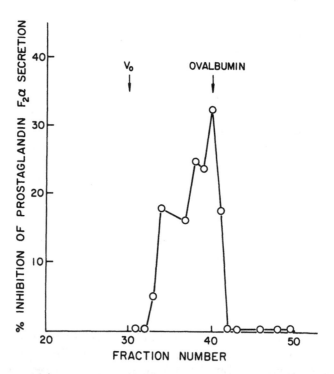

Fig. 1. Inhibition of granulosa cell prostaglandin production by fractions of an ammonium sulfate precipitate of bovine follicular fluid eluted from a Sephadex G-100 column. One tenth of each fraction was tested for activity. (Vo = void volume). Values are mean ± S.E.M.

taglandin synthesis was maximal prior to ovulation and remained high in the ovulated follicle for at least the initial 24 hr following ovulation. Basal $PGF_{2\alpha}$ production by theca cells from early cycle follicles was detectable but low. However the presence of TNF in the culture media resulted in a marked elevation of prostaglandin synthesis by the thecal cells regardless of the stage of the cycle. The highest elevation of PG synthesis in the presence of TNF was found in theca cells from preovulatory follicles.

To determine if GC can synthesize TNF as has been shown for macrophages, GC cells from various stages of the estrous cycle were incubated and the TNF secreted into the media was measured. As can be seen in Figure 4, this protein is synthesized by the GC. (FSH plays a role in the regulation of this TNF production as we have reported in this book (Zolti et al., 1989).

We next measured the content of the TNF in bovine FF from different stages of the estrous cycle. As can be seen in Figure 5, measurable amounts of TNF were found in FF from mid-cycle ovaries (eight to 12 days). The maximal activity was seen in follicular fluid obtained from follicles which had ovulated within the previous 24 hr. The activity remained high during 48 hr after ovulation.

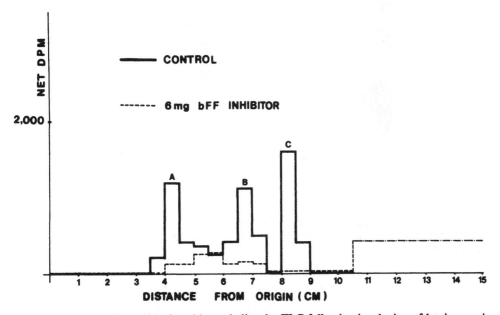

Fig. 2. Separation of arachidonic acid metabolites by TLC following incubation of bovine seminal vesicle microsomes with labeled arachidonic acid in the presence or absence of placental extract.

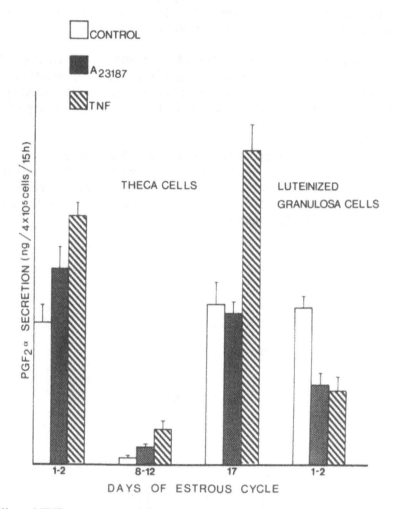

Fig. 3. Effect of TNF on prostaglandin synthesis by collagenase dispersed thecal cells from follicles of different stages of the estrous cycle. Data represents the mean ± S.E.M. of at least three experiments.

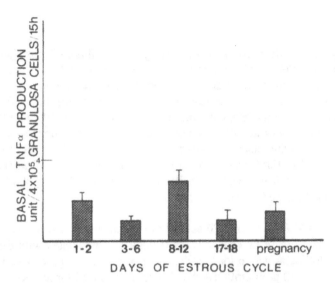

Fig. 4. TNF production by dispersed GC which were harvested from follicles at different stages of the cycle, incubated for 24 hr and the TNF production was measured as described in text.(means ± S.E.M.; N = 18).

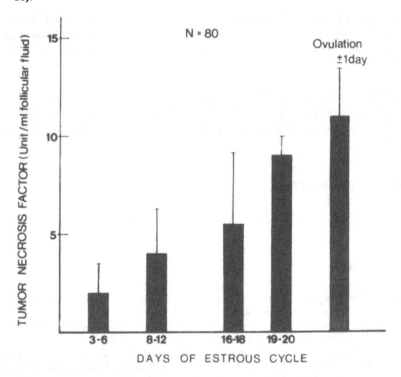

Fig.5. Concentration of TNF in bFF derived from early mid-cycle and preovulatory follicles (mean ± S.E.M.)

DISCUSSION

The data presented demonstrate that an endogenous inhibitor of prostaglandin synthesis is present in bovine FF of mid-cycle follicle. The level of this inhibitor becomes lower toward the time of ovulation. Similar observations for an endogenous prostaglandin inhibitor had been made for human FF by Carson and Trounson (1986). Interestingly, when human FF is treated with protease, the inhibitory activity is abolished and a substance with prostaglandin stimulatory activity becomes apparent. A stimulatory substance has also been found in the bovine placenta towards term, a time when the level of inhibitory substance is undetectable. Furthermore, this stimulatory factor of low mol wt can be seen even in mid-term placentae, if the extract is boiled for 5 min. (This boiling abolishes the PG inhibitory activity. (Shemesh et al., 1984).

The finding of TNF in bovine FF as well as the capacity of the GC to synthesize TNF suggests that the increase in TNF towards ovulation is associated with the degenerative changes in the follicular wall prior to follicular rupture. Furthermore, TNF is known as an essential factor for cell proliferation and differentiation as well for its necrotic activity. It is therefore possible that TNF is essential for granulosa cell transformation into luteal cells which occurs after the LH peak.

The finding that TNF stimulates PG synthesis in both ovarian granulosa and theca cells suggests that this factor plays a role in the rise in ovarian prostaglandin necessary for follicular rupture. This action may be autocrine, para- crine and/or involved in a number of intermediaries.

Finally, it is tempting to speculate that the decline in PG inhibitor as well as the rise in TNF production in the preovulatory follicle that occurs just prior to follicular rupture serves to insure the maximal production of prostaglandin that is essential for rupture of the follicle.

ACKNOWLEDGEMENTS

This work was sponsored by BARD grant IS-1257-87R.

REFERENCES

Aderka, D., Holtmann, H., Toker, L., Hahn, T. and Wallach, D., 1986, Tumor necrosis factor induction by sendai virus, *J. Immunol.*, 136:2938.

Armstrong, D. T. and Zamecnik, J., 1975, Preovulatory   elevation of rat ovarian prostaglandin F and itss blockade by indomethacin, *Mol. Cell. Endo.*, 2:125.

Carson, R., Trounson, A. and Mitchell, M., 1986, Regulation of prostaglandin biosynthesis by human ovarian follicular fluid: A mechanism for ovulation?, *Prostaglandins*, 32:49.

LaMaire, W. J. and Marsh, J. M., 1975, Interrelationships between prostaglandins, cyclic AMP and steroids in ovulation, *J. Reprod. Fertil., Suppl.* 22:53.

Murdoch, N. J., Peterson, T. A., Van Kirk, E. A., Vincent, D. L. and Inskeep, E. K., 1986, Interactive roles of progesterone, prostaglandins and collagenase in the ovulatory mechanism of the ewe, *Biol. Reprod.*, 28:1001.

Shemesh, M., 1979, Inhibitory action of follicular fluid on progesterone and prostaglandin synthesis in bovine follicles, *J. Endocr.*, 82:27.

Shemesh, M., Ailenberg, M., Lavi, S. and Mileguir, F., 1981, Regulation of prostaglandin biosynthesis by an endogenous inhibitor from bovine placenta, *in*: "Dynamics of Ovarian Function," N. B. Schwartz and M. Hunzicker-Dunn, eds., Raven Press, New York.

Shemesh, M., Hansel, W., Strauss, J. (III), Rafaeli, A., Lavi, S. and Mileguir, F., 1984, Control of prostanoid synthesis in bovine trophoblast and placentome, *J. Anim. Sci.*, 7:177.

Tsafriri, A., Lindner, H. R., Zor, U. and Lamprecht, S.A., 1972, Physiological role of prostaglandins in the induction of ovulation, *Prostaglandins* 2:1.

Wallach, E. E., Bronson, R., Hamada, Y., Wright, K. H. and Stersens, V. G., 1975, Effectiveness of PGF-2 in restoration of hMG-hCG induced ovulation in indomethacin-treated rhesus monkey, *Prostaglandins*, 10:129.

THE ROLE OF TUMOR NECROSIS FACTOR IN HUMAN GRANULOSA CELLS AND EARLY EMBRYO

M. Zolti,[1,2] R. Meirom,[1] M. Shemesh,[1] D. Wallach,[3]
S. Mashiach,[2] and Z. Ben-Rafael[2]

[1] Depts. of Hormone Research and Immunology, Kimron Vet. Inst.
[2] Dept. of OB/GYN, Sheba Med. Cent., Tel-Hashomer, and
[3] Dept. of Virology, Weizmann Inst., Israel

INTRODUCTION

Ovarian function is regulated by a complex system of endocrine, paracrine and autocrine controls. Recently there has been a growing interest in non-steroidal local ovarian regulators.[1-6] Specifically, immune cells[7-10] and related cytokines such as interleukin-1(IL-1) have been shown to affect ovarian function.[11-14] There is apparently also a reciprocal relationship between these factors and steroids, e.g. steroids can affect IL-1 production by macrophages.[15]

Tumor necrosis factor α (TNF), a 17 KD cytokine, is a product of activated macrophages[16] which was recently shown to be produced by rat and bovine granulosa cells (GC),[17] and also by macrophages in regressing rabbit corpus luteum.[18] Furthermore, TNF inhibits follicle stimulating hormone (FSH) or Tumor Growth Factor-β induced aromatase activity and stimulates progesterone synthesis in this tissue.[19-20]

TNF has a complex effect on endothelial cells. In vitro, TNF inhibits endothelial cell proliferation but in vivo it induces necrosis of tumor through vascular changes. However, if the vascular bed is normal TNF exhibits angiogenic activity.[16, 21, 22] The ability of TNF to stimulate different cells such as macrophages, fibroblast and smooth muscle to produce and secrete prostanoids[21, 23, 24] is another important action since prostanoids are involved in the process leading to follicular rupture.[25] TNF is also expressed in large amounts during

Advances in Assisted Reproductive Technologies, Edited by
S. Mashiach *et al.*, Plenum Press, New York, 1990

embryogenesis in mice,[26] but it is not known whether the early embryo can express TNF. In this study we therefore determined (a) if human oocyte GC and early embryo produce TNF; (b) if this production of TNF is regulated by gonadotropin and (c) what is the role of TNF in the production of steroids and prostanoids.

MATERIALS AND METHODS

Separation of GC from Follicular Fluid

GC were obtained from follicles stimulated by human menopousal gondotropin (hMG) and aspirated during vaginal ultrasound oocyte retrieval for in vitro fertilization (IVF) treatment. The treatment protocol for follicular stimulation consisted of hMG, three ampules/ day starting on day 3 of the cycle when estradiol concentrations reached levels above 400 pg/ml and ultrasonography demonstrated two or more follicles of 16 mm or more. Human chorionic gonadotropin (hCG) was administered 48 hr after the last dose of hMG.

Follicular fluid was centrifuged for 10 min at 400 g. The GC in the precipitate were combined and pooled. Separation of GC from RBC was achieved by centrifuging in sterile tubes containing Ficol for 30 min. at 600 g at room temperature.[27]

The GC formed a band at the interface where they were collected by aspiration, washed three times and resuspended in medium containing 0.5% bovine serum albumin. 3×10^5 cells per well containing 1 ml of media were used for incubations. The cultures were incubated overnight with or without FSH (25 ng/ml), hCG (10 ng/ml), prostaglandin F2α (PGF$_{2\alpha}$) (10 μM) or colony stimulating factor (CSF) (20% L929 supernate) alone or in combinations. Aliquots of medium were taken for analysis for TNF, progesterone (P), estradiol-17β (E$_2$) and PGF2α.

TNF Bioassay

The TNF cytoxicity was quantitated using 9 mm microwells containing 4×10^4 cells/ well. Test samples were applied to the cells in serial dilutions in the presence of 50 μg/ml cyclohexamide. Twelve hours later, cell killing was quantitated by measuring the uptake of neutral red. One cytotoxic unit was defined as the concentration at which 50% of the cells were killed. Each positive sample was neutralized by co-incubation for three hours at 4°C with monoclonal anti-human TNF.[28, 29]

Radioimmunoassays and Statistics

Progesterone, E$_2$-17β and PGF$_{2\alpha}$ were measured by radioimmunoassay. The intra- assay variations were 11%,10% and 12% and the interassay variations were 9%,7% and 9.0%, respectively.[30, 31] The student t-test was performed for all data comparisons data are presented as mean ± SEM.

RESULTS

Data presented in Fig. 1 demonstrated that human cultured GC secreted TNF (5–10 units/4×10^5/15 h incubation) as determined by bioassay and verified by specific neutralization by a monoclonal antibody to TNF.

To determine the role of the tropic hormones, CSF and $PGF_{2\alpha}$, in the regulation of TNF production, GC were incubated in the presence of FSH, hCG, CSF or $PGF_{2\alpha}$ alone or in combination. As can be seen in Fig. 1, FSH induced a significant enhancement (50%) in TNF production over the control. In contrast, this stimulatory effect was not apparent in the presence of hCG, $PGF_{2\alpha}$ or CSF. However, the combination of CSF and hCG induced a significant elevation (p < 0.003) in TNF production by cultured GC.

We next examine the effect of TNF on prostanoid production. As can be seen in Fig. 2, TNF (300 pM) induced a 3.4 fold increase in $PGF_{2\alpha}$ production by cultured GC. However in the presence of FSH this effect was not apparent. Furthermore, FSH alone was found to significantly (P < 0.001) inhibit $PGF_{2\alpha}$ production by the cultured GC.

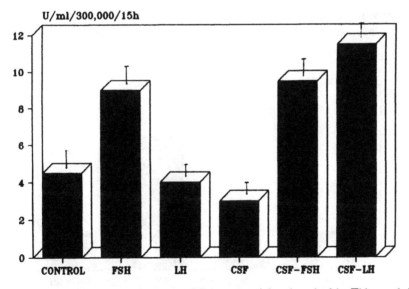

Fig. 1. Tumor necrosis factor production by GC (see material and methods). This graph is average of three independent experiments. FSH treatment or CSF + LH were significant over the control (p < 0.01).

To determine the possible role of TNF in steroidogenesis, P and E_2 production by the cultured GC were assayed. In the presence of TNF (Fig. 3), there was a significant elevation in P production, however E_2 production was not affected (data not shown). The increase in P production by TNF was of the same magnitude similar to that induced by $PGF_{2\alpha}$ (Fig.3).

TNF Secretion by the Early Embryo

Conditioned medium of oocytes cultured for 48 hr or more were assayed for TNF activity. Distinct TNF secretion was found in 14% of the 2- to 8-cell stage early-embryos. The cytotoxic activity of this TNF was neutralized by the specific monoclonal antibody. However, no TNF activity was detected in conditioned media from embryos which were beyond the 4-cell stage (including eight triploid and two normal embryos at the morulla stage) were not apparent for TNF activity.

To determine the source of TNF in the 2- 4-cell embryos we evaluated condition medium of sperm (n = 6) or non-fertilized oocytes (n = 20). Since sperm, oocytes and condition medium were devoid of TNF activity, we concluded that TNF is derived from the embryo or the surrounding cumulus.

DISCUSSION

Data presented demonstrates that culture human GC produce   TNF. This was proven by bioassay that could be neutrilized by specific monoclonal antibody to TNF. This cyto-

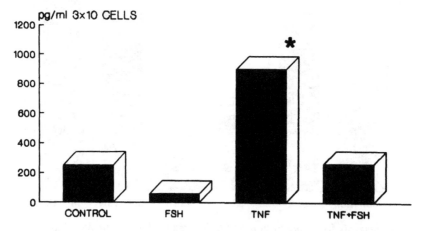

Fig. 2. $PF_{2\alpha}$ synthesis by GC (3×10^5) (see Material and Methods). TNF significantly elevated prostaglandin synthesis (p < 0.0001).

kine production was regulated by FSH and not hCG. However combination of hCG and CSF induced also significant elevation in TNF production by GC.

TNF is an important cytokine that has physiological and pathological actions (myeloid differentiation, tumor necrosis, sepsis, cahexia, angiogenesis, bone destruction).[16, 21, 32] TNF has been previously reported to be produced by bovine and rat GC.[17] In this study we demonstrated for the first time that it is also secreted by human GC. It is possible that TNF which has an important role in angiogenesis[21, 22] is produced by GC during follicular maturation as a result of FSH stimulation. Also, our finding that GC stimulated by TNF secrete prostaglandins which are essential for follicular rupture.[25]

Similarly, it has been recently reported that the immune cell has a role in follicular maturation and rupture.[33] It is possible that prostaglandin metabolites and TNF are responsible for the activation of the immune cells to exert their effect.

We have found that CSF-1, together with hCG, can stimulate TNF synthesis. Since CSF-1 is known to be produced by the endometrium around the time of ovulation,[34] it may function in the midcycle together with the luteinizing hormone to maintain TNF production during oocyte maturation and follicular rupture.

Mizuno has coined the expression of "ontogenic inflammation" to describe the inflammation around the area of cell division growth and differentiation which is necessary for maintenance of these processes.[26] He found that TNF is expressed during embryogenesis in a high concentration. We hypothesized that the embryo, which is a rapidly dividing and differentiating tissue, also produces TNF for maintenance of growth and differentiation.

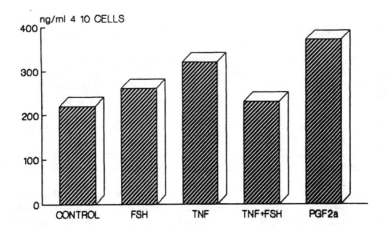

Fig. 3. Progesterone synthesis by GC (see material and methods). The increase in P by TNF was of the same magnitude as by PGF$_{2\alpha}$ (p < 0.01).

However, since we detected TNF in only 20% of the medium from 2–4 cell stage, it does not allow a clear conclusion as to the validity of this hypothesis. Likewise, it would also be desirable if it could be determined whether TNF is a positive marker for the ability to implant the early embryo. However, the clinical protocol used in an IVF program does not usually allow conclusions as to whether TNF is a positive or negative marker as several embryos were replaced in each recipient. Since TNF was undetected in oocyte and in morula stage embryos and as TNF is important for the expression of other cytokines that do appear later in development (Zolti, et al., submitted for publication), it might suggest that TNF is a positive signal for development.

The field of growth factor and cytokines and their effects on reproduction is a rapidly growing new area of investigation. In the present work we concentrated on the cytokines. However there is evidence that IL-1 acts on GC to stimulate proliferation, inhibits FSH induced hCG receptors and inhibits P synthesis. TNF appears to be an antagonist to IL-1 in this tissue. This is interesting as IL-1 and TNF are agonists in most other tissues studied.

Another cytokine, CSF-1, has been shown to be important in mice placental growth and inhibition of luteolysis.[34] Since there is an accumulating body of evidence that cytokines act on the hypothalmic-pituitary axis which in turn affect cytokine concentrations, examination of these reciprocal relationships would seem to be fruitful area for future research.[35,36]

In summary, the presence and the production of TNF by the GC and early embryo suggest that this cytokine has an important role in cell growth and differentiation, especially during the ovarian cycle.

REFERENCES

1. A. J. W. Hsueh, and E.Y. Adashi, Hormonal regulation of the differentiation of culture granulosa cell, *Endo. Rev.*, 5:76 (1984).
2. E. Y. Adashi, and C. E.Resnick, Antagonistic interactions of transforming growth factor in regulation of granulosa cell diffentiation, *Endocrinology*, 119:1879 (1986).
3. W. C. Dodson, and D. W.Schomberg, The effect of transforming growth factor β on follicle stimulating hormone induced differentiation of culture rat GC, *Endocrinology*, 12:512 (1987).
4. M. Knecht, P. Femy, and K. Catt, Bifunctional role of TGFβ during granulosa cell development, *Endocrinology*, 120:1243 (1987).
5. S. M. Bryant, J. A. Gale, D. L. Yanagihara, J. D. Campeau, and G. S. diZerega, Angiogenic mitogenic and chemotactic activity of human follicular fluid, *Am J. Ob. Gyn.*, 158:1807 (1988).
6. E. Y. Adashi, C. E. Resnick, M. E. Svoboda, and J. J. Van Wyk, A novel role of somatomedin C in the cytodifferentiation of granulosa cells, *Endocrinology*, 115:1227 (1984).
7. T. M. Kirsh, A. L. Friedman, R. L. Vogel, and G. L. Flickenger, Macrophages in corpora luteum of mice, Characterization and effect of steroid secretion, *Biol. Reprod.*, 25:629 (1983).
8. T. M. Kirsh, R. L. Vogel, and G. L. Flickenger, Macrophages, a source of luteotropic cybernins, *Endocrinology*, 113:1910 (1983).

9. W. G. Gorospe, and B. G. Kasson, Lymphokines from concannavalin stimulated lymphocytes regulates rat granulosa cell steroidogenesis in vitro, *Endocrinology*, 123:2462 (1988).

10. J. Halme, M. G. Hammond, C. H. Syrop, and L. M. Talbert, Peritoneal macrophages modulate human granulosa-luteal cell progesterone production, *J. Clin. Endo. Metab.*, 61:912 (1985).

11. P. E. Gottschall, A. Uehara, S. T. Hoffman, and A. Arimura, Interleukin-1 inhibits follicle stimulating hormone-induced differentiation in rat granulosa cells in vitro, *Bichem. Biophys. Res. Comm.*, 149:502 (1987).

12. P. E. Gottschall, G. Katsuura, S. T. Hoffman, and A. Arimura, Interleukin 1: an inhibitor of luteinizing hormone receptor formation in cultured rat granulosa cells, *FASEB J.*, 2:2492 (1988).

13. M. Fukuoka, T. Mori, S. Tah, and K. Yasuda, Interleukin-1 inhibits luteinization of porcine granulosa cells in culture, *Endocrinology*, 122:367 (1988).

14. M. Fukuoka, K. Yasuda, S. Taii, K. Takakura, and T. Mori, Interleukin-1 stimulates growth and inhibits progesterone secretion in cultures of porcine granulosa cells, *Endocrinology*, 124:884 (1989).

15. M. L. Polan, Gonadal steroids modulate human monocyte interleukin-1 (IL-1) activity, *Fertil. Steril.*, 49:964 (1988).

16. J. L. Old, Tumor necrosis factor, *Nature*, 330:602 (1988).

17. K. F. Roby, and P. F.Terranova, Localization of tumor necrosis factor (TNFα) in rat and bovine ovary using immunocytochemistry and cell blot: evidence for granulosa cell production. VII Ovarian Workshop: Paracrine Communication in the Ovary: Ontogenesis and Growth Factors. Serono Symposia, USA Press, (In press) (1988).

18. P. Bagavandoss, S. L. Kunkel, R. C. Wiggens, and P. L.Keyes, Tumor necrosis factor-α (TNF-α) production and localization of macrophages and T lymphocytes in the rabbit corpus luteum, *Endocrinology*, 122:1185 (1988).

19. N. Emoto, and A. Baird, The effect of tumor necrosis factor/cachectin on follicle stimulatory hormone induced aromatase activity in cultured rat granulosa cells. *Biochem. Biophys. Res. Comm.*, 153:792 (1988).

20. K. F. Roby, and P. F. Terranova, Tumor necrosis factor alpha alters follicular steroidogenesis in vitro, *Endocrinology*, 123:2952 (1988).

21. J. C. Stephen, and P. L. Libby, Human vascular smooth muscle cell, Target for and source of tumor necrosis factor, *J. Immunol.*, 142:100 (1989).

22. S. J. Leibovich, P. J. Polverini, H. M. Shepard, D. M. Wiseman, V. Shively, and N. Nuseir, Macrophage-induced angiogenesis is mediated by tumour necrosis factor-alpha, *Nature* 329:630 (1987).

23. M. Frater-Schroder, W. Risau, R. Hallmann, P. Gautschi, and P. Bohlen, Tumor necrosis factor type alpha, a potent inhibitor of endothelial cell growth in in vitro, is an angiogenic in vivo, *Proc. Natl. Acad. Sci.*, USA 84:5277 (1987).

24. P. Suffys, R. Begaert, F. Vanreg and W. Fiers, Reduced tumor necrosis factor induce cytotoxicity by inhibition of archidonic acid metabolism, *Biochem. Biophys. Res. Comm.*, 16:149(2):735-43 (1987).

25. D.T. Baird, "Reproduction in mammals:3. Hormonal control of reproduction," Cambridge University Press, Cambridge (1986).

26. D. Mizuno, and G. I. Soma, Endogenous and exogenous T.N.F. therapy (EET therapy): conceptural and experimental grounds. 22nd Forum in Immunology. pp. 282-285 (1988).

27. Z. Ben-Rafael, C. A. Benadiva, L. Mastroianni, Jr., C. J. Garcia, J. M. Minda, R. V. Iozzo, and G. L. Flickinger, Collagen matrix influences the morphologic features and steroid secretion of human granulosa cells, *Am. J. Obstet. Gynecol.*, 159:(6), 1570-1574 (1988).

28. D. Aderka, H. Holtmann, L. Toker, T. Hahn, and D. Wallach, Tumor necrosis factor induction by sendai virus, *J. Immunol.*, 136:2938 (1986).

29. A. Menger, H. Levny and J. Woollgy, Assays for tumor necrosis factor and related cytokines, *J. Immunol. Meth.*, 116:1, (1989).

30. M. Shemesh, W. Hansel, and J. F. Strauss (III), Modulation of bovine placental prostaglandins by endogenous inhibitor, *Endocrinology*, 115:1401, (1984).

31. M. Shemesh, W. Hansel, and J. F. Strauss (III), Calcium dependent cyclic nucleotide independent steroidogenesis in bovine placenta, *Proc. Natl. Acad. Sci.*, 81:6403, (1984).

32. M. Centrella, T. L. McCarthy and E. Ganalis, Tumor necrosis factor α inhibits collagen synthesis and alkaline phosphatase activity independently of its effect of deoxyribonucleus acid synthesis in osteoblast enriched bone cell cultures, *Endocrinology*,123:144 (1988).

33. A. O. Abisogun, P. Braouet, A. Tsafriri, The involvement of platelet activating factor in ovulation, *Science*, 243:381-3 (1989).

34. J. W. Pollard, R. J. Arceci, A. Bartoeci and E. R. Stanly, Colony stimulating factor. A growth factor for trophoblast. 3rd Conference on Reproductive Immunology (eds. T. Gill III and T. Keymann (In press).

35. A. Bateman, A. Singh, T. Kral, and S. Solmen, The immune hypothalamic pituitary adrenal axis. *Endo. Rev.*, 10:92 (1989).

36. E. W. Bernton, J. E. Beach, J. W. Holiday, R. C. Smallridge, and H. G. Fein, Release of multiple hormones by direct action of interleukin 1 on pituitary cells. *Science* 238:519 (1989).

THE OVARIAN RENIN-ANGIOTENSIN SYSTEM AND COMPARTMENTAL FUNCTION IN THE OVARY

R. Apa, T. Zreik, A. Lightman, A. Palumbo, T-K. Yoon,
P. Andrade-Gordon, A. DeCherney, and F. Naftolin

Department of Obstetrics and Gynecology
Yale University School of Medicine

The ovarian renin-angiotensin system (OVRAS) is present in all compartments of the ovary. Since it is gonadotrophin-sensitive and since the actions of its active principle, angiotensin II parallel gonadotrophin actions in the ovary, the ovarian renin-angiotensin system may be a mediator of gonadotrophin action in the ovary. The components of the renin-angiotensin system of the ovary including angiotensinogen, inactive and active renin, angiotensin I (AI), angiotensin II (AII) and perhaps angiotensin III (AIII), as well as angiotensin converting enzyme, have all been demonstrated through the use of immunohistochemistry, direct measurement, histochemistry and molecular biology. Obviously, since these factors respond to gonadotrophins, particularly LH, they become increasingly prominent as the gonadotrophin-sensitive compartments develop within the ovary. That is, the response of the ovarian compartments is marked by the increase of AII present. In an attempt to rationalize these compartmental developments we have formulated five stages of ovarian compartmental development (Figure 1). Initial follicle development is marked by thecal dominance, during which time the theca proliferates/differentiates from the stroma and androgens are the major steroid hormones. While these androgens may contribute to estrogens being formed by the developing granulosa layer, the androgens may also inhibit estrogen actions such as granulosa cell proliferation, thus mediating follicular atresia. Granulosa cell proliferation begins after the follicle has begun its development of theca cells. Initially, the granulosa cell-produced estrogen has actions on follicle contents, such as further granulosa proliferation and development of cell membrane receptors for gonadotrophin and prolactin; however, with granulosa cell dominance the developing follicles make sufficient quantities of estrogen to spill into the circulation where they prepare the reproductive tract for receiving the oocyte/embryo. In addition, at the peak of follicle development the abrupt rise of

Advances in Assisted Reproductive Technologies, Edited by
S. Mashiach *et al.*, Plenum Press, New York, 1990

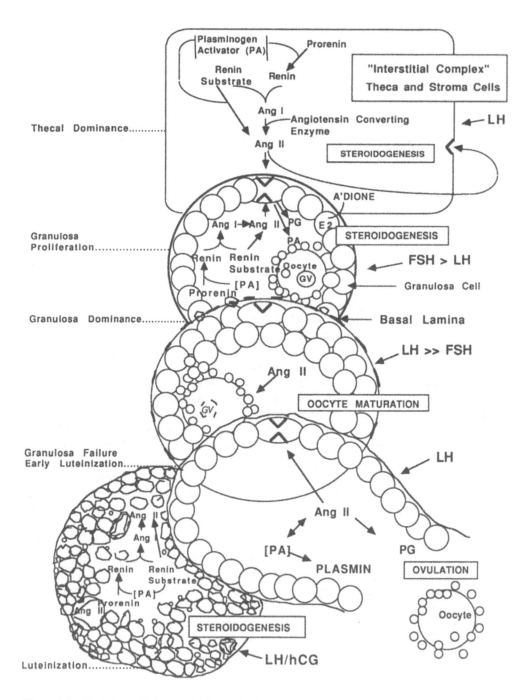

Fig. 1. Modified from: Palumbo A, Jones C, Lightman A, et al., Am J Obstet Gynecol 1989, 160:8–14.

estrogen causes the release of LH and FSH. Preovulatory granulosa cell failure then ensues as luteinization occurs; although the progesterone and testosterone may continue to rise as luteinization occurs,[1-3] the dominant effect of the gonadotrophin surge is a fall of estrogen which is accompanied by oocyte maturation and ovulation. Luteinization progresses further during corpus luteum formation, ultimately establishing sufficient levels of circulating progesterone to block estrogen induced actions such as endometrial proliferation.

Because we wish to know the importance of OVRAS in the above ovarian compartmental function, we haved asked several questions: (A) Can ovarian renin-angiotensin system blockers inhibit gonadotrophin actions? We found that angiotensin converting enzyme inhibitors are not completely effective in blocking ovarian angiotensin formation, which tends to rule them out as OVRAS blockers (unpublished data). While ACE inhibitors are not satisfactory pharmacological probes of OVRAS function, better results are obtained by AII receptor blockade with saralasin [Sar[1], Val[5], Ala[8]]. The germinal vesicle breakdown which is a marker of the preovulatory oocyte maturation and cumulus expansion during granulosa cell failure can be blocked with the use of saralasin and this blockade can be reversed by concurrent treatment with angiotensin II[4]. It has also been possible to show that saralasin blocks hCG-induced ovulation in PMS-primed immature female rats;[5] (B) Can angiotensin induce ovarian angiogenesis? Although angiotensin has been shown to induce experimental angiogenesis, this has not been specifically shown in the ovary. The practical problems for such a demonstration are considerable; however, the work of Fernandez and colleagues using angiotensin inserted into a corneal pocket in the rabbit eye to induce angiogenesis has kept our group interested in this possibility;[6] (C) Can OVRAS affect ovarian steroidogenesis? Paulson and colleagues have shown that administration of AII induces thecal cell production of androgens and progestins.[7] Thus, the early development of the thecal compartment has paradoxical effects. It furnishes the steroids which are both the precursor of estrogen to be made by the granulosa cells and which can arrest granulosa cell proliferation. It is apparently left to the state of the granulosa cell compartment to determine whether the amount of available aromatase will satisfactorily "detoxify" the thecal androgens before they can block granulosa cell proliferation resulting in atresia. Since Husain's group have demonstrated that AII causes increased estrogen secretion from quartered rat ovaries,[8] ostensibly from the granulosa cell compartment, we have attempted to delineate AII action on the pre-ovulatory granulosa cell compartment by incubating luteinized granulosa cells from women undergoing ovum harvest for in vitro fertilization.[9] We used luteinized granulosa cells, since they are the very same cells which during this period are bathed with levels of angiotensin in the range of 10^{-11}–10^{-9} M. When these cells are washed and then incubated for 24 hours with AII, there is a dose-dependent response to angiotensin II in the concentrations of progesterone, testosterone and estrogen in the culture medium. This increase can largely be obviated by adding saralasin to the culture at the same time as the angiotensin. It is particularly interesting that, while progesterone and testosterone in the medium continue to rise in a dose-responsive manner to added AII, estradiol does not. Therefore it appears that in these cells there is a biphasic response of aromatase, perhaps analogous with what is seen during preovulatory granulosa cell dominance followed by granulosa cell failure. To study this matter further we have measured aromatase activity using the tritium release method. We found that luteinizing granulosa cells which were kept for 24 hours in culture had a curve of aromatase activity in response to AII which was also biphasic, suggesting that increasing follicular AII, initially drives the granulosa cell dominance and, following the LH surge, contributes to granulosa cell failure.

In summary, we have shown that the ovarian renin-angiotensin system exists and responds in the compartments which perform ovarian function. Depending upon the stage of the compartmental development there is a more or less apparent role for angiotensin. For example: (a) during thecal dominance Paulson and colleagues have shown that angiotensin produces androgens and progestins. This appears to be under the control of LH; (b) during granulosa proliferation FSH is the primary driving hormone and angiotensin appears not to be very much in evidence, either in follicle fluid or in granulosa cells themselves; (c) during the early period of granulosa dominance, while the follicle grows and its cells proliferate the angiotensin in follicle fluid is not higher than plasma levels, nor are immunohistochemically high amounts of angiotensin present within the follicle; (d) finally, during the late follicular phase, with the development of the LH surge there is a rise in the angiotensin in the follicle fluid and this rise reaches levels which are physiologically important in driving granulosa lutein cell steroidogenesis with a biphasic curve for aromatase activity: High levels of follicular AII in the periovulatory period may be responsible for the differential granulosa lutein cell steroidogenesis seen during the period of granulosa cell failure when the estrogen falls.

ACKNOWLEDGEMENTS

MK421 (Enelapril) was a gift from Merk; anti AII was a gift from Dr. D. Ganten and Dr. A. Negro-Vilar; anti-renin was a gift from Dr. Inagami and Dr. Deschepper. R.A., A.L., and A.P. are Lalor Fellows supported by HD22970.

REFERENCES

1. K. P. McNatty, A. Makris, C. DeGrazia, R. Osathanondh and K. J. Ryan, The production of progesterone, androgens, and estrogens by granulosa cells, thecal tissue, and stromal tissue from human ovaries in vitro, *J. Clin. Endocrinol. Metab.*, 49:687 (1979).

2. K. P. McNatty, D. M. Smith, A. Makris, R. Osathanondh and K. J. Ryan, The microenvironment of the human antral follicle: Inter-relationship among the steroid levels in antral fluid, the population of granulosa cells, and the status of the oocyte in vivo and in vitro, *J. Clin. Endocrinol. Metab.*, 49:851 (1979).

3. H. L. Judd, and S. S. C. Yen, Serum androstenedione and testosterone levels during the menstrual cycle, *J. Clin. Endocrinol. Metab.*, 38:375 (1973).

4. A. Palumbo, A. Pellicer, A. H. DeCherney and F. Naftolin, Angiotensin action in oocyte maturation in rat. Abstract presented at the Society for Gynecological Investigation Meeting (1988).

5. A. Pellicer, A. Palumbo, A. H. DeCherney and F. Naftolin, Blockade of ovulation by an angiotensin antagonist. *Science* 240:1660-1661 (1988).

6. C. A. Fernandez, J. Twickler and A. Mead, Neovascularization produced by angiotensin II, *J. Lab. Clin. Med.*, 105:141-145 (1985).
7. R. J. Paulson, M. F. Hernandez, Y. S. Do and W. A. Hsuesh, Angiotensin II modulation of steroidogenesis by luteinized granulosa cells in vitro, Abstract presented at the Society for Gynecological Investigation Meeting (1989).
8. A. G. Pucell, F. M. Bumpus and A. Husain, Rat ovarian Angiotensin II receptors, Characterization and coupling to estrogen secretion, *J. Biol. Chem.*, 262:7076 (1987).
9. A. Palumbo, M. Alam, A. Lightman, N. MacLusky, A. H. DeCherney and F. Naftolin, Angiotensin II effects in vitro steroidogenesis by human granulosa lutein cells in culture, The Endocrine Society Meetings (Abs.) (1988).

THE PHYSIOLOGY OF HUMAN OVARIAN PRORENIN-ANGIOTENSIN SYSTEM

Joseph Itskovitz and Jean E. Sealey*

*Department of Obstetrics and Gynecology
Rambam Medical Center, Haifa 31096, Israel and
*Cardiovascular Center, The New York Hospital - Cornell University Medical College
New York, New York 10021, U.S.A.*

INTRODUCTION

Renin is classically considered to be an enzyme that is synthesized by the kidney and secreted into the circulation to affect a cascade of reactions which result in the formation of the biologically active octapeptide, angiotensin II. The latter plays a key role in the control of blood pressure and electrolyte homeostasis through its vasoconstrictive effect, and the stimulation of adrenocortical aldosterone biosynthesis.[1] In the kidney the mature active renin is processed from its biological inactive precursor, prorenin. A significant proportion of prorenin in the kidney is secreted as such by the kidney without being converted to active renin and it circulates in the blood at about 10 times the level of active renin. However, whereas plasma active renin appears to be entirely of renal origin, plasma prorenin is derived from the kidney and from extrarenal sources as well because it remains detectable in nephrectomized females and males.[2,3]

Renin gene expression and components of the renin-angiotensin system (RAS) have been reported to occur in several extrarenal tissues.[4] These include the brain, adrenal, heart, blood vessels, and reproductive organs. Although the presence of components of the RAS in these tissues suggests a possible locally functioning RAS, there is a question whether renin actually originates in all of these tissues or is sometimes derived from the plasma.

Advances in Assisted Reproductive Technologies, Edited by
S. Mashiach *et al.*, Plenum Press, New York, 1990

Furthermore, the physiologic significance of the putative local RAS in heart and vascular tissues is poorly defined. In contrast the evidence for a functional local RAS in reproductive organs is on a firmer basis. In this chapter we review the recent evidences for the existence of a local prorenin-angiotensin system in reproductive organs and particularly in the ovary. This ovarian RAS appears to operate differently and independently from renal renin and seems to be regulated by changes in prorenin concentration rather than renin.[5, 6]

BIOCHEMICAL PROPERTIES OF PRORENIN

Prorenin has higher molecular weight than renin   (56,000 vs. 48,000), and can be reversibly or irreversibly activated. It is irreversibly activated by serine proteases including trypsin; to measure prorenin trypsin, activation of prorenin is performed prior to an enzymatic or direct renin assay. The active renin that is formed after trypsin activation has characteristics similar to those of endogenous active renin with respect to pH optimum, reaction kinetics with homologous angiotensinogen, and inhibition by monospecific antirenin antibodies.[3, 7-9]

Prorenin can be reversibly activated by acid (pH 3.3). Studies employing purified prorenin from a human kidney suggest that acid activation involves conformational changes in the prorenin molecule, resulting in an unfolding of the prosegment away from the molecule and exposure of the active site. Prorenin can similarly be reversibly cryoactivated. It develops some intrinsic activity (about 30% to maximum) after three to four days of exposure to 0°C and becomes inactive again after 30-60 minutes exposure to 37°C. This cryoactivation seems to expose the peptide bond between the prosequence and the renin molecule because it can then be cleaved in chilled plasma by endogenous serine proteases.[10, 11] Therefore, it is not advisable to collect plasma samples into iced tubes, or to chill it before the incubation step of the enzymatic renin assay since some of the prorenin is irreversibly converted to active renin.

These observations suggest that under certain conditions the prosegment of the prorenin moves away from the active site resulting in a catalytically active prorenin molecule. Thus prorenin itself sometimes can duplicate the function of active renin.

Measurement of Prorenin and Active Renin

For the measurement of plasma prorenin, plasma active renin is measured first by enzymatic assay utilizing endogenous plasma renin substrate. The angiotensin I formed during the enzymatic assay is measured by RIA and is expressed as the hourly rate of angiotensin generation (ng/ml/hr). For the measurement of total plasma renin (prorenin + active renin), plasma prorenin is converted to active renin by limited proteolysis with sepharose-bound trypsin, and then the enzymatic assay active renin is carried out again. Prorenin is then calculated as the difference between the total renin and the active renin.[12]

THE OVARIAN PRORENIN-ANGIOTENSIN SYSTEM

The first clues for a possible role of the RAS in reproductive function came from the finding of a renin-like enzyme in the human placenta, amniotic fluid, and uterus.[13-15] Subsequently, prorenin was shown to be the major form of renin in these tissues.[16, 17] More recently, the ovary has been recognized to synthesize and secrete prorenin and appears to be the source of elevated plasma prorenin during the menstrual cycle at the time of LH surge, and during early pregnancy.[18-22]

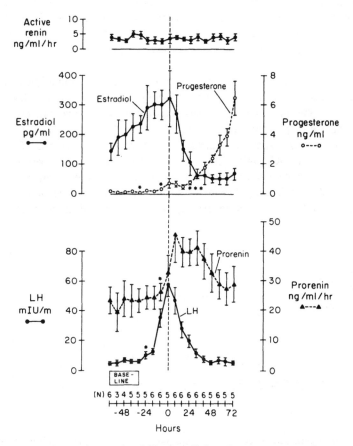

Fig. 1. Plasma prorenin, active renin, LH, estradiol and progesterone changes during midmenstrual cycle in six normal women (From Sealey et al.[19]).

Plasma Prorenin Throughout the Normal and Stimulated Menstrual Cycle

Studies in which plasma prorenin was measured throughout the normal menstrual cycle and in patients undergoing ovarian stimulation with gonadotropins suggest that the ovaries synthesize and secrete prorenin in response to LH/hCG stimulation. Prorenin increases two-fold during the mid-menstrual cycle at the time of LH surge[18, 19] (Figure 1). The rise in LH precedes the rise in prorenin by 8–16 hours and the surge of prorenin is sustained for about 40 hours. Prorenin often remains slightly above baseline during the luteal phase and then declines to baseline levels after the progesterone peak in the midluteal phase.[18] No consistent changes in plasma active renin are apparent throughout the menstrual cycle. The only significant increase in active renin is observed during the midluteal phase, concurrent with the progesterone peak. Two independent factors are possibly involved in the increase in plasma prorenin during the normal menstrual cycle. Whereas the elevated prorenin and active renin in the midluteal phase may be related to the renal secretion of prorenin and active renin secondary to the diuretic effect of progesterone, the prorenin rise concurrent with the LH surge is most likely due to prorenin secretion from the preovulatory follicle in response to LH. The finding of very high concentrations of prorenin in ovarian follicular fluid of the preovulatory follicle support this view[20] (see below).

The availability for study of patients undergoing ovarian stimulation with gonadotropins for the purpose of in vitro fertilization (IVF) gave us an opportunity to explore in greater depth a number of relationships between prorenin and ovarian functions. In contrast to the transient rise of prorenin in the normal menstrual cycle, hCG injection (due to its longer half-life) causes a sustained increase in prorenin[21, 22] (Figure 2). Plasma prorenin peaks on day 2–6 after hCG injection and returns to near baseline levels on days 12–14, when hCG is no longer detectable in the circulation. Plasma prorenin often rises to extremely high levels and the height of the plasma prorenin response is directly related to the number of preovulatory follicles on the day of follicular aspiration, suggesting that the

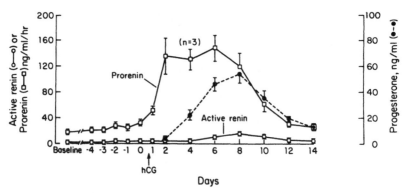

Fig. 2. Plasma prorenin, and progesterone levels (mean ± SEM) in five IVF patients treated with gonadotropins and hCG. Patients were treated with progesterone (25 mg/day) from day 3 throughout day 14. (From Istkovitz et al.[22]).

high plasma prorenin levels after hCG injection in ovarian-stimulated patients is not due to abnormally high prorenin secretion by one individual follicle, but rather the result of the presence of multiple preovulatory follicles[22] (Figure 3). Changes in plasma active renin are much less consistent than the changes in prorenin. Similar to the natural cycle active renin does not change with the initial rise in plasma prorenin and is significantly elevated at the midluteal phase, concurrent with the progesterone peak, and as discussed earlier, this rise in active renin may well be of renal origin. Thus in the ovary, prorenin is secreted independently of active renin.

To further characterize the dynamics of prorenin secretion from the ovary, we studied patients undergoing ovarian stimulation with gonadotropins for IVF in whom GnRH analog instead of hCG, was injected 36 hours before follicular aspiration to induce oocyte maturation.[23] In these patients plasma LH increased significantly within one hour and the surge was maintained for about 24 hours after the injection of GnRH analog. Consistent with the observations made in the periovulatory period of the normal menstrual cycle, plasma prorenin increased within eight to 12 hours and reached a peak 24 to 36 hours following the injection of GnRH analog. Due to the existence of multiple follicles in these ovarian-stimulated patients the peak of prorenin was much higher than observed in the normal menstrual cycle but was similar to the height of prorenin detected in ovarian-stimulated patients injected with hCG. However, in keeping with the prolonged luteotrophic effect of hCG, the prorenin surge was sustained for a longer period after hCG (Figure 4).

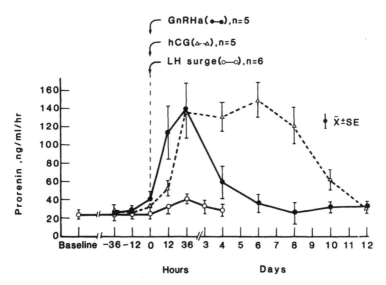

Fig. 3. Total serum renin levels (active renin comprised < 20% of the total renin) throughout the stimulated cycles of three representative patients with different numbers of ovarian follicles. Shown for comparison are the changes in total plasma renin throughout a natural cycle (From Itskovitz et al.[22]).

Taken together, these studies indicate that LH/hCG stimulate prorenin release from the preovulatory follicle and the corpus luteum. They also suggest that the ovarian renin system operates differently from the renal system because in the ovary, in contrast to the kidney, prorenin is secreted without concurrent secretion of active renin. It is also possible that more rapid changes occur in prorenin biosynthesis or secretion in the ovary than in the kidneys—being measured in terms of hours, rather than days.[3]

The ovaries do not appear to secrete substantial amounts of prorenin during the follicular phase, before the LH surge; the average plasma prorenin level during the follicular phase is similar to that found in postmenopausal women. However, the demonstration of an ovarian and peripheral venous gradient of prorenin during the follicular phase in six women undergoing elective hysterectomy for benign disease[24] suggests that some ovarian secretion of prorenin before the LH surge may exist, but is not reflected in the peripheral circulation or in the follicular fluids obtained at midfollicular phase (see below). Similarly, it is not clear whether the corpus luteum of the nonconception menstrual cycle synthesizes and secretes prorenin throughout the luteal phase. As discussed earlier the elevated levels of plasma prorenin during the early luteal phase appears to be of ovarian origin, whereas it is less clear whether the slightly elevated plasma prorenin during the midluteal phase is of ovarian or renal origin.[18] The detection of an ovarian and peripheral venous gradient of prorenin in two nonpregnant women during the luteal phase[24] and the demonstration of immunostaining for renin in luteal cells in ovaries from women with normal menstrual cycle[25] support the idea the corpus luteum synthesizes and secretes prorenin during nonconception menstrual cycles.

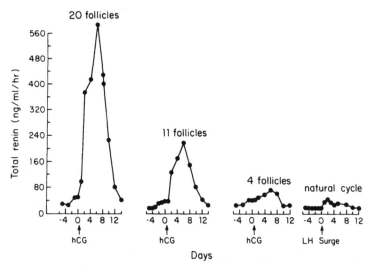

Fig. 4. Changes in plasma prorenin levels after the spontaneous LH surge in the normal menstrual cycle,[19] and after the injection of hCG[22] or GnRH analog (Buserelin 500 µg) in ovarian stimulated patients. Time 0 indicates the initiation of the spontaneous LH surge and the time of hCG or GnRH analog injection.

Components of the RAS in the Human Ovary

Further evidence of the existence of local RAS in the human ovary has been provided by direct measurements of the components of the RAS in ovarian follicular fluids,[20, 26-28] immunohistochemical localization of these components in ovarian tissues,[25, 29] and in vitro studies of cultured granulosa and theca cells.[29, 30]

Although human ovarian follicular fluid contains all components of the RAS, it is prorenin that appears to be the predominant form of ovarian renin detected in the follicular fluid in human,[20] baboon[31] and cow.[32] Prorenin is present in very high concentrations in follicular fluid of the hCG-stimulated preovulatory follicle.[20] Only 1% of the renin in ovarian follicular fluid is in the active form and it is most likely that this measurable active renin is artifactual because of inadvertant activation in vitro of prorenin. Activation of only 1% of the prorenin present in the follicular fluid would result in an active renin level that would be much higher than that in the concurrently collected plasma. So far, in primates only the kidney has been shown conclusively to secrete active renin, whereas other tissues (e.g. placenta, testis, adrenal) seem to secrete only prorenin.[16, 19, 24, 31, 33] It is unclear at precisely what time prorenin begins to be synthesized by the developing follicle but it seems to be either close to or at the time of the LH surge. While prorenin concentrations are about 100-fold higher in the fluids of the mature preovulatory follicles than in plasma, it is present in close to plasma levels in follicles aspirated at the midfollicular phase of the natural cycle. In follicular fluids aspirated 36 hours after hCG injection, prorenin levels are lower in fluids surrounding immature oocytes than in those surrounding mature oocytes (unpublished observations).

The theca cells seem to be the primary source of prorenin in the human ovary. In cultured human theca cells, prorenin production increased during the first 17 days of culture, peaking at around 10 days and then gradually decreased. Active renin levels were 10% or less of the prorenin levels. Prorenin was barely detectable in media form granulosa cells (obtained from either unstimulated or stimulated ovaries) cultured for 24 days.[29]

Immunohistochemical staining of human ovaries with anti-human renin antibody demonstrated the presence of renin primarily in theca cells and found little immunostaining in the granulosa cells of the developing follicles, thus confirming the culture results.[25, 29] Studies in cows[32] and rats[34] also suggest that the theca cells are the major source of prorenin. Staining for renin and angiotensin II of human luteinized granulosa cells obtained from ovarian stimulated IVF patients was also demonstrated,[25] but because luteinized granulosa cells do not seem to produce prorenin and renin in vitro[29] it is thus possible that prorenin and renin activity in this tissue arise from the neighboring theca cells or follicular fluid.

PRORENIN AND EARLY PREGNANCY

The prorenin level in plasma of normal pregnant women is about 10-fold higher than in nonpregnant women.[35-37] Prorenin is sythesized in the chorionic cells of the placental chorion laeve and is secreted into the amniotic fluid where it occurs in concentrations 10 times higher than in the circulation.[14–17] The decidua has also been suggested as a major source of prorenin which may contribute to the high levels of prorenin in the amniotic fluid or the plasma during pregnancy.[38]

The observations that maternal prorenin attains its maximal level within four weeks after conception when placental mass is small suggested that the source of elevated plasma prorenin is maternal.[37] We therefore measured prorenin levels in plasma of IVF patients who conceived after embryo transfer and have found that prorenin begins to rise on days 8 to 10 after embryo transfer, at the time when endogenous hCG from the implanting embryo is detectable in the circulation.[22, 39] (Figure 5). To further examine the ovarian contribution to the rising levels of prorenin after implantation we studied ovarian-failure patients who conceived after the transfer of donated eggs. In these patients no significant rise in plasma prorenin was detectable during the first four weeks after embryo transfer, indicating that the secretion of prorenin in early pregnancy is of ovarian origin[39] (Figure 6). The late (> 6 weeks of pregnancy) small rise of circulating prorenin in ovarian-failure patients suggests that the uteroplacental unit also contributes to the elevated levels of prorenin later in pregnancy.[40, 41]

Further support for the ovarian origin of prorenin in early pregnancy was provided by the observation that due to the existence of multiple corpora lutea in patients who conceived after ovarian stimulation, plasma prorenin is higher than in normal pregnancy[39]

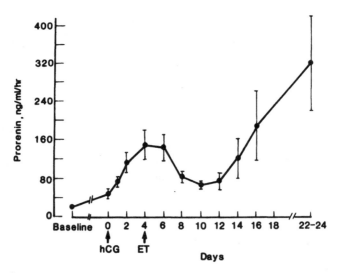

Fig. 5. Plasma prorenin (mean ± SEM) during early pregnancy in five ovarian-stimulated IVF patients. Baseline is day 3 of the menstrual cycle. ET = embryo transfer.

(Figure 6). Taken together our data are consistent with the view that the early rise in plasma prorenin after conception is the result of hCG stimulation of ovarian prorenin secretion, and that the ovary is the primary source of the rise in circulating prorenin during the first weeks of pregnancy. Ovarian prorenin appears to be one of the earliest nonsteroidal factor secreted by the corpus luteum of pregnancy. Its possible role in the maintenance of early pregnancy and maternal recognition of pregnancy is however not known.

REGULATION, MECHANISM OF ACTION AND PHYSIOLOGICAL ROLES OF THE OVARIAN PRORENIN-ANGIOTENSIN SYSTEM

Regulation of Ovarian Prorenin-Angiotensin System

Factors regulating ovarian renin gene expression, biosynthesis and secretion are poorly understood. Most of our current knowledge of factors regulating ovarian RAS is derived from studies in the rat. Renin mRNA in rat ovary and uterus has been detected with the aid of rat renin cDNA.[42] Renin mRNAs in the ovary, uterus and kidney of rat were indistinguishable by size according to their migration. The relative levels of renin mRNA in the

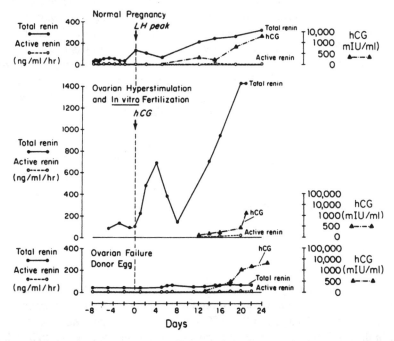

Fig. 6. Plasma total renin, active renin, and hCG levels during early pregnancy (upper panel), in ovarian-stimulated patient who conceived after IVF treatment (middle panel), and in patient with ovarian failure who conceived after induction of an artificial cycle and IVF of a donated oocyte (lower panel).

ovary and uterus was about 1/100th of that of kidney. Active renin was not present in measurable amounts in either ovary or uterus, but inactive renin was high.[42]

The identification of a putative estrogen regulator sequence in the 5' flanking region of the human renin gene[43] suggested that estrogen may be involved in the induction of ovarian renin gene expression. Kim et al[44] studied the effects of FSH injection on renin mRNA level, and on total renin content of the ovary of immature rats. FSH administration increased renin mRNA and total renin content after 36 hours. In hypophysectomized rats, the total renin content of the ovary was stimulated by estradiol as well as by FSH, suggesting that renin production may be increased primarily by estrogen produced in response to FSH, rather than by FSH itself. The apparent regulation of the ovarian renin level by estrogen supports the hypothesis that the induction of ovarian renin gene expression is a steroid-mediated process. Testosterone has been shown to influence renin mouse submandibular gland primarily by regulating the amount of renin mRNA available for translation.[45] Thus, it is possible that during the process of folliculogenesis ovarian renin gene is turned on by estrogen or its androgen precursors but the LH surge may be required before prorenin is to be secreted from the theca cells into the circulation or into the follicular fluid.

Down regulation of angiotensin II receptors by FSH has been demonstrated in cultured rat ovarian granulosa cells.[46] Angiotensin II on the other hand has been shown to increase testosterone secretion from rat ovarian slices.[47] These authors suggested that angiotensin II is involved in the initiation or maintenance of the events leading to follicular atresia. If this is correct, in the process of follicular development, the healthy and dominant follicles would have to be decoupled from angiotensinergic influences. High levels of prorenin are detectable in the bovine antral fluids of atretic follicles,[32] and autoradiographic studies provided evidence for selective expression of angiotensin II receptors on the rat atretic follicles, further supporting the supposition that the RAS regulates atresia.[48]

Although the human corpus luteum seems to secrete prorenin in response to hCG stimulation, and renin and angiotensin II were demonstrated immunohistochemically in human luteal cells,[25] and renin-like activity[49] and renin mRNA were detected in the rat luteal cells,[50] no information is yet available regarding factors controlling renin gene expression or renin biosynthesis and secretion from the corpus luteum.

Mechanism of Action

The results of the studies presented in this review suggest that in primates the extrarenal system in the ovary operates through prorenin, not active renin. Only prorenin increases in plasma during the LH surge or after hCG injection; active renin may increase too, but only later in the midluteal phase when the progesterone level rises enough to interfere with aldosterone action. In all organs of human and nonhuman primates studied (ovary, testis, adrenal), we could find no evidence of renin secretion, and so far, only the kidney has been shown conclusively to secrete active renin. Biochemical studies suggest that prorenin can be activated in vitro and once active renin is formed, angiotensin II is also formed in vitro. Thus the small amounts of renin detected in ovarian follicular fluid, amniotic fluid and in extrarenal tissue extracts, are likely the result of in vitro activation (conformational change in the prorenin molecule or proteolytic cleavage of the prosegment) of prorenin.

The mechanism for local activation of the prorenin-angiotensin system in the ovary and other extrarenal tissues is not known. We have hypothized that prorenin develops catalytic activity following binding of its prosegment to a prorenin receptor and the exposure of its active site. For such a locally active form of the RAS to function the putative prorenin receptors should be located near converting enzyme, angiotensinogen and angiotensin II receptors.[5, 6] Converting enzyme and angiotensinogen are widely distributed and angiotensin II receptors have been identified in the ovary and in other tissues where the renin gene is expressed.[51, 52] This form of RAS would enable local activation of prorenin which, through local generation of angiotensin II, could have paracrine or autocrine effects, without interfering with the normal function of the circulating renal RAS.

Physiological Roles of Ovarian Prorenin-Angiotensin System

Little is known about the physiological roles of the extrarenal RAS. The presence of prorenin in tissues that synthesize steroids, e.g. adrenal, testis, ovary and placenta, suggests that prorenin could have an effect on steroid biosynthesis.

Angiotensin II does not modify progesterone secretion or aromatase activity in cultured granulosa cells prepared from hypophysectomized, DES-treated immature rats.[46] Angiotensin II however, does increase testosterone secretion from rat ovarian slices,[47] suggesting as discussed earlier, a possible role for the RAS in follicular development and atresia. The high levels of prorenin detected in bovine[32] and human (unpublished data) follicular fluids aspirated from atretic follicles also support this view, but may merely reflect a common source of prorenin and testosterone from the hypertrophied theca cells.

Both stimulatory and inhibitory effects on steroid biosynthesis in human luteal cells or luteinized granulosa cells have been attributed to the RAS. Whereas angiotensin II had no effect on basal progesterone and estradiol production in long-term cultured luteinized granulosa cells obtained from IVF patients, it augmented hCG-stimulated production of progesterone. The angiotensin II inhibitor, saralasin, negated this effect but increased estradiol production.[53] In short-term (24 hours) culture of luteinized granulosa cells angiotensin II increased in a dose dependent manner progesterone and testosterone concentrations in the culture media, but estradiol decreased with increasing doses of angiotensin II.[54]

In bovine luteal cells, neither angiotensin II nor saralasin had any effect on basal progesterone production, but LH-stimulated progesterone production was reduced significantly by angiotensin II. Furthermore, angiotensin II reduced the level of LH-induced expression of mRNA encoding cholesterol side-chain cleavage (P450 scc) enzyme, which was reflected in reduced LH-stimulated progesterone production in the presence of angiotensin II in vitro.[55] Inhibitory influence of angiotensin II on adenylate cyclase system had been previously demonstrated in several tissues. The Ni unit of adenylate cyclase has been implicated in the inhibitory action of angiotensin II on adenylate cyclase and on cAMP pool and testosterone production of rat Leydig cells.[56]

The dual effect (inhibitory or stimulatory) of the RAS on ovarian steroidogenesis suggest that angiotensin II may operate through different kinds of angiotensin II receptor-dependent mechanisms (such as adenylate cyclase system or phospholipase C-mediated increase in intracellular calcium). Thus it is possible that the locally active ovarian RAS, depending on the type of cells and the stage of follicular or corpus luteum development, may exert both stimulatory and inhibitory effects on ovarian steroidogenesis.

Prorenin through local generation of angiotensin II could have other roles in ovarian function. Angiotensin II could participate in reproductive function as a direct vasoconstrictor agent, or even as a vasodilator agent through its ability to stimulate vasodilator prostaglandin production.[57] It may also induce neovascularization directly[58] or via stimulation of mRNA encoding basic fibroblast growth factor.[55] Other reports suggest that it is a growth factor.[59] The existence of high levels of prorenin in ovarian follicular fluid and plasma at the time of ovulation[19, 20] suggests a possible role for the RAS in the process of ovulation; blockage of ovulation by an angiotensin II antagonist in the rat has been recently reported.[60]

The association of prorenin with fluid accumulation and transfer in the kidney (glomerular filtration), chorion leave (amniotic fluid), ovary (follicular fluid), and the eye (vitreous fluid[61]), raises a possible role for ovarian prorenin in regulating fluid transfer across the blood-granulosa cells barrier (the basal lamina) and accumulation of fluid in the ovarian follicular antrum.

Although no clear role for ovarian RAS has yet been established in the human, and no reproductive dysfunction has been linked to abnormality in ovarian RAS, the observations that the human oocyte and embryo are bathed in fluids rich in prorenin suggest a possible role for prorenin in ovarian physiology and early embryonic development.

SUMMARY

All of the components of the renin-angiotensin system are present in the ovary. The striking difference between the ovarian renin system and the circulating renin system is that prorenin, not renin, is the final product of renin gene expression in the ovary. Ovarian prorenin concentrations are regulated, reaching very high levels at the time of ovulation and during early pregnancy when plasma LH and hCG are high, respectively. The ovarian sources of prorenin appear to be the thecal cells and the corpus luteum of pregancy. The function of the ovarian prorenin-angiotensin system has yet to be established. Preliminary reports suggest roles in the regulation of steroid biosynthesis, ovulation, atresia and/or neovascularization. The mechanism whereby ovarian prorenin becomes catalytically active is not understood, but may involve reversible exposure of the active site during binding to a specific site in the target organ. The next few years should see the characterization of the form, and the identification of the function of the ovarian prorenin- angiotensin system.

REFERENCES

1. J. H. Laragh, and J.E. Sealey, The renin-angiotensin-aldosterone hormonal system and regulation of sodium, potassium, and blood pressure homeostasis, *in:* "Handbook of Physiology: Renal Physiology", J. Orloff and R. W. Berliner, eds., Waverly Press, Baltimore (1973).
2. J. E. Sealey, R. P. White, J. H. Laragh, and A. L. Rubin, Plasma prorenin and renin in anephric patients, *Cir. Res.*, 41 (suppl.2):17 (1977).
3. J. E. Sealey, S. A. Atlas, and J. H. Laragh, Prorenin and other large molecular weight forms of renin, *Endocr. Rev.*, 1:365 (1980).
4. C. F. Deschepper, and W. F. Ganong, Renin and angiotensin in endocrine glands, *in:* "Frontiers in Neuroendocrinology", Vol. 10, L. Martini and W. F. Ganong, eds., Raven Press, New York 79,(1988).
5. J. E. Sealey, N. Glorioso, J. Itskovitz, and J. H. Laragh, Prorenin as a reproductive hormone: A new form of the renin system, *Am. J. Med.*, 81:1041 (1986).
6. J. E. Sealey, N. Glorioso, J. Itskovitz, S. A. Atlas, T. M. Pitarresi, J. J. Preibisz, C. Troffa, and J. H. Laragh, Ovarian prorenin, *Clin. Exp. Hypertens.*, A9 (8/9):1435 (1987).
7. S. A. Atlas, P. Christofalo, T. Hesson, J. E. Sealey, and L. C. Fritz, Immunological evidence that inactive renin is prorenin, *Biochem. Biophys.Res. Commun.*, 132:1038 (1985).
8. J. Bouhnik, J. A. Fehrentz, F. X. Galen, R. Seyer, G. Evin, B. Castro, J. Menard, and P. Corvol, Immunological identification of both plasma and human renal inactive renin as prorenin, *J. Clin. Endocrinol. Metab.*, 60:399 (1985).
9. S. J. Kim, S. Hirose, H. Miyazaki, N. Ueno, K. Higashimori, S. Morinaga, T. Kimura, S. Sakakibara, and K. Murakami, Identification of plasma inactive renin as prorenin with a site directed antibody, *Biochem. Biophys. Res. Commun.*, 126:641 (1985).
10. F. H. M. Derkx, B. N. Bouma, M. P. A. Schalekamp, and M. A. D. H. Schalekamp, An intrinsic factor XII-prekallikrein dependent pathway activates the human plasma renin-angiotensin system, *Nature*, 280:315 (1979).
11. J. E. Sealey, S. A. Atlas, J. H. Laragh, M. Silverberg, and A. P. Kaplan, Initiation of plasma prorenin activation by Hageman factor- dependent conversion of plasma prokallikrein to kallikrein, *Proc. Natl. Acad. Sci. USA*, 76:5914 (1979).
12. J. J. Preibisz, J. E. Sealey, R. M. Aceto, and J. H. Laragh, Plasma renin activity measurements: An update, *Cardiovasc. Rev. Rep.*, 3:787 (1982).
13. A. A. Hodari, R. Smeby, and F. M. Bumpus, A renin-like substance in the human placenta, *Obstet. Gynecol.*, 29:313 (1967).
14. S. L. Skinner, E. J. Cran, R. Gibson, R. Taylor, W. A. W. Walters, and K. J. Catt, Angiotensin I and II, active and inactive renin, renin substrate, renin activity, and angiotensinase in human liquor amnii and plasma. *Am. J. Obstet. Gynecol.*, 121:626 (1975).
15. E. M. Symonds, M. A. Stanley, and S. L. Skinner, Production of renin by in vitro cultures of human chorion and uterine muscle, *Nature* 217:1152 (1968).
16. G. M. Acker, F. X. Galen, C. Devaux, S. Foote, E. Papernik, A. Pesty, J. Menard, and P. Corvol, Human chorionic cells in primary culture: a model for renin biosynthesis, *J. Clin. Endocrinol. Metab.*, 55:902 (1982).

17. A. M. Poisner, G. W. Wood, R. Poisner, and T. Inagami, Localization of renin in trophoblasts in human chorion leave at term pregnancy, *Endocrinology*, 109:1150 (1981).

18. J. E. Sealey, S. A. Atlas, N. Glorioso, H. Manepat, and J. H. Laragh, Cycle secretion of prorenin during the menstrual cycle: synchronization with luteinizing hormone and progesterone, *Proc. Natl. Acad. Sci. USA*, 82:8705 (1985).

19. J. E. Sealey, I. Cholst, N. Glorioso, C. Troffa, I. D. Weintraub, G. James, and J. H. Laragh, Sequential changes in plasma luteinizing hormone and plasma prorenin during the menstrual cycle, *J. Clin. Endocrinol. Metab.*, 65:1 (1987).

20. N. Glorioso, S. A. Atlas, J. H. Laragh, R. Jewelewitz, and J. E. Sealey, Prorenin in high concentrations in human ovarian follicular fluid, *Science*, 233:1422 (1986).

21. J. E. Sealey, N. Glorioso, A. Toth, S. A. Atlas, and J. H. Laragh, Stimulation of plasma prorenin by gonadotropic hormones (Letter). *Am. J. Obstet. Gynecol.*, 153:596 (1985).

22. J. Itskovitz, J. E. Sealey, N. Glorioso, and Z. Rosenwaks, Plasma prorenin response to human chorionic gonadotropin in ovarian hyperstimulated women: correlation with the number of ovarian follicles and steroid hormone concentrations, *Proc. Natl. Acad. Sci. USA*, 84:7285 (1987).

23. J. Itskovitz, R. Boldes, A. Barlev, Y. Erlik, L. Kahana, and J. M. Brandes, The induction of LH surge and oocyte maturation by GnRH analogue (Buserelin) in women undergoing ovarian stimulation for in vitro fertilization (Abstr.), "International Symposium on GnRH Analogues in Cancer and Human Reproduction". Geneva, Switzerland, February 18-21 (1988).

24. R. J. Paulson, Y. S. Do, W. A. Hsueh, and R. A. Lobo, Gradients of prorenin and active renin in ovarian venous and peripheral venous blood samples obtained simultaneously, *Am. J. Obstet. Gynecol.*, 159:1575 (1988).

25. A. Palumbo, C. Jones, A. Lightman, M. L. Carcangiu, A. H. DeCherney, and F. Naftolin, Immunohistochemical localization of renin and angiotensin II in human ovaries, *Am. J. Obstet. Gynecol.*, 160:8 (1989).

26. L. A. Fernandez, B. C. Tarlatzis, P. J. Rzasa, V. J. Caride, N. Laufer, A. F. Negro-Vilar, D. E. DeCherney, and F. Naftolin, Renin-like activity in ovarian follicular fluid, *Fertil. Steril.* 44:219 (1985).

27. M. D. Culler, B. C. Tarlatzis, A. Lightman, L. A. Fernandez, A. H. DeCherney, A. Negro-Vilar, and F. Naftolin, Angiotensin II-like immunoreactivity in human ovarian follicular fluid, J. Clin. Endocrinol. Metab., 62:613 (1986).

28. F. H. M. Derkx, A. T. Alberda, G. H. Zeilmaker, and M. A. D. H. Schalekamp, High concentrations of immunoreactive renin, prorenin and enzymatically-active renin in human ovarian follicular fluid, *Br. J.Obstet. Gynaecol.*, 94:4 (1987).

29. Y. S. Do, A. Sherrod, R. A. Lobo, R. J. Paulson, T. Shinagawa, S. Chen, S. Kjos, and W.A. Hsueh, Human ovarian theca cells are a source of renin, *Proc. Natl. Acad. Sci. USA* 85:1957 (1988).

30. R. J. Paulson, Y. S. Do, W. A. Hsueh, P. Eggena, and R. A. Lobo, Ovarian renin production in vitro and in vivo: characterization and clinical correlation, *Fertil. Steril.*, 51:634 (1989).

31. J. E. Sealey, F. W. Quimby, N. Glorioso, and C. Troffa, Ovarian secretion of prorenin in the baboon (abstr.), "The Primate Ovary", Serono Symposia, Beaverton, Oregon, 16-17 May (1987).

32. D. Schultze, B. Brunswig, and A. K. Mukhopadhyay, Renin and prorenin- like activities in bovine ovarian follicles, *Endocrinology*, 124:1389 (1989).

33. J. E. Sealey, M. Goldstein, T. Pitarresi, T. T. Kudlak, N. Glorioso, S. A. Fiamengo, and J. H. Laragh, Prorenin secretion from human testis: No evidence for secretion of active renin or angiotensinogen, *J. Clin. Endocrinol. Metab.*, 66:974 (1988).

34. R. B. Howard, A. G. Pucell, F. M. Bumpus, and A. Husain, Rat ovarian renin: characterization and changes during the estrous cycle, *Endocrinology*, 123:2331 (1988).

35. W. A. Hsueh, J. A. Luetcher, E. J. Carlson, G. Grislis, E. Fraze, and A. McHarque, Changes in active and inactive renin throughout pregnancy. *J. Clin. Endocrinol. Metab.*, 54:1010 (1982).

36. J. E. Sealey, M. Wilson, A. A. Morganti, I. Zervoudakis, and J. H. Laragh, Changes in active and inactive renin throughout normal pregnancy, *Clin. Exper. Hypertens.*, 4(A):2373 (1982).

37. J. E. Sealey, D. McCord, P. A. Taufield, K. A. Ales, M. L. Druzin, S. A. Atlas, and J. H. Laragh, Plasma prorenin in first trimester pregnancy: Relationship to changes in human chorionic gonadotropin, *Am. J. Obstet. Gynecol.*, 153:514 (1985).

38. W. A. Hsueh, S. Kjos, P. W. Anderson, L. Macaulay, A. Sherrod, M. Koss, T. Shinagawa, and Y. S. Do, Human decidua is a major extrarenal source of prorenin (abstr.), *Hypertension*, 10:368A (1987).

39. J. E. Sealey, N. Glorioso, J. Itskovitz, C. Troffa, I. Cholst, and Z. Rosenwaks, Plasma prorenin during early pregnancy: Ovarian secretion under gonadotropin control? *J. Hypertns.*, 4(suppl 5):592 (1986).

40. F. H. M. Derkx, A. T. Alberda, F. H. De Jong, F. H. Zeilmaker, J. W. Makovitz, and M. A. D. H. Schalekamp, Source of plasma prorenin in early and late pregnancy: observations in a patient with primary ovarian failure, *J. Clin. Endocrinol. Metab.*, 65:349 (1987).

41. H. S. Brar, Y. S. Do, H. B. Tam, G. J. Valenzuela, R. D. Murray, L. D. Longo, M. L. Yonekura, and W. A. Hsueh, Uteroplacental unit as a source of elevated circulating prorenin levels in normal pregnancy, *Am. J. Obstet. Gynecol.*, 155:1223 (1986).

42. S. J. Kim, M. Shinjo, A. Fukamizu, H. Miyazaki, S. Usuki, and K. Murakami, Identification of renin and renin messenger RNA sequence in rat ovary and uterus, B*iochem. Biophys. Res. Commun.*, 142:169 (1987).

43. H. Miyazaki, A. Fukamizu, S. Hirose, T. Hayashi, H. Hori, H. Ohkubo, S. Nakanishi, and K. Murakami, Structure of the human renin gene, *Proc. Natl. Acad. Sci., USA* 81:5999 (1984).

44. S. J. Kim, M. Shinjo, M. Tada, S. Usuki, A. Fukamizu, H. Miyazaki, and K. Murakami, Ovarian renin gene expression is regulated by follicle-stimulating hormone, *Biochem. Biophys. Res. Commun.*, 146:989 (1987).

45. R. E. Pratt, V. J. Dzau, and A. J. Ouellette, Influence of androgen on translatable renin in RNA in the mouse submandibular gland, *Hypertension* 6:605 (1984).

46. A. G. Pucell, F. M. Bumpus, and A. Husain, Regulation of angiotensin II receptors in cultured rat ovarian granulosa cells by follicle-stimulating hormone and angiotensin II, *J. Biol. Chem.*, 263:11954 (1988).

47. F. M. Bumpus, A. G. Pucell, A. I. Daud, and A. Husain, Angiotensin II: An intraovarian regulatory peptide, *Am. J. Med. Sci.*, 295:406 (1988).

48. A. I. Daud, F. M. Bumpus, and A. Husain, Evidence for selective expression of angiotensin II receptors on atretic follicles in the rat ovary: an antoradiographic study, *Endocrinology* 122:2727 (1988).

49. R. R. Cabrera, D. C. Guardia, and E. De Vito, A renin-like enzyme in luteal tissue, *Mol. Cell. Endocrinol.*, 47:269 (1986).

50. A. Lightman, C. F. Deschepper, S. H. Mellon, W. F. Ganong, and N. Naftolin, In situ hybridization identifies renin mRNA in the rat corpus luteum, *Gynecol. Endocrinol.*, 1:227 (1987).

51. R. C. Speth, F. M. Bumpus, and A. Husain, Identification of angiotensin II receptors in the rat ovary, *Eur. J. Pharmacol.*, 130:351 (1986).

52. R. C. Speth, and A. Husain, Distribution of angiotensin-converting enzyme and angiotension II-receptor binding sites in the rat ovary, *Biol. Reprod.*, 38:695 (1988).

53. R. J. Paulson, M. F. Hernandez, Y. S. Do, W. A. Hsueh, and R. A. Lobo, Angiotensin II modulation of steroidogenesis by luteinized granulosa cells in vitro (abstr. 197), Society for Gynecologic Investigation, San Diego, California, March 15-18 (1989).

54. A. Palumbo, M. Alam, A. Lightman, A. H. DeCherney, and F. Naftolin, Angiotensin II affects in vitro steroidogenesis by human granulosa-lutein cells (abstr. 1075), The Endocrine Society, New Orleans, Louisiana, June 8-11 (1988).

55. D. Stirling, R. Stone, R. R. Magness, M. R. Waterman, and E. R. Simpson, Angiotensin II inhibits LH stimulated expression of mRNA encoding cholesterol side-chain cleavage (P450 scc), and stimulates expression of mRNA encoding basic fibroblastic growth factor in bovine luteal cells in primary culture. (abstr. 248), Society for Gynecologic Investigation, San Diego, California, March 15-18 (1989).

56. A. Khanum, and M. L. Dufau, Angiotensin II receptors and inhibitory actions in Leydig cells, *J. Biol. Chem.*, 263:5070 (1988).

57. J. C. McGiff, K. Crowshaw, N. A. Terragno, and A. J. Lonigro, Renal prostaglandins: Possible regulators of the renal actions of pressor hormones, *Nature*, 227:1255 (1970).

58. L. A. Fernandez, J. Twickler, and A. Mead, Neovascularization produced by angiotension II, *Lab. Clin. Med.*, 105:141 (1985).

59. T. R. Jackson, L. A. C. Blair, J. Marshall, M. Goedert, and M. R. Hanley, The mass oncogene encodes an angiotensin receptor, *Nature*, 335:437 (1988).

60. A. Pellicer, A. Palumbo, A. H. DeCherney, and F. Naftolin, Blockage of ovulation by an angiotensin antagonist, *Science*, 240:1660 (1988).

61. A. H. J. Danser, M. A. van den Dorpel, J. Deinum, F. H. M. Derkx, A. A. M. Franken, E. Peperkamp, P. T. V. M. de Jong, and M. A. D. H. Schalekamp, Renin, prorenin, and immunoreactive renin in vitreous fluid from eyes with and without diabetic retinopathy, *J. Clin. Endocrinol. Metab.*, 68:160 (1989).

PROLACTIN AS A POSSIBLE MODULATOR OF GONADOTROPIN ACTION ON THE OVARY

Yitzhak Koch and Nava Dekel

Department of Hormone Research
The Weizmann Institute of Science, Rehovot, 76100, Israel

In recent years it has been demonstrated by many investigators that various growth factors such as epidermal growth factor,[1,2] platelet-derived growth factor,[1] transforming growth factor ß[3] and especially insulin-like growth factor (IGF;[4,5]) can modify the action of the gonadotropic hormones on the differentiation and activity of the ovarian granulosa cells(GC). For reviews see references 6–8. In addition, recent evidence suggests that proteinacious substances that are produced in the ovary can modulate the follicular response to gonadotropins.[9]

IGFs constitute a family of homologous, low molecular weight, single chain polypeptide growth factors named for their distinct structural similarity (about 45%) to insulin. The IGFs were classified into two distinct groups: basic IGFs (IGF-I or somatomedin-C) and neutral IGFs (IGF-II). Whereas IGF-I gene expression is regulated by growth hormone, the regulation of IGF-II is largely yet unknown. IGFs circulate in the plasma bound to specific binding proteins.[10]

IGF-I was found to be mitogenic when added to bovine granulosa cells maintained in serum-free medium. IGF-I could replace insulin as the main mitogenic factor and it was effective at a concentration as low as 1 ng/ml.[11] IGF-I, on its own, does not stimulate progesterone secretion by GC. However, it has been recently demonstrated that when added together with FSH, IGF-I can enhance the FSH-mediated secretion of progesterone by cultured GC. Insulin was also effective but was several orders of magnitude less potent than IGF-I.[12] More recent data indicate that the thecal-interstitial cells of the ovary may also be a site of IGF-I reception and action.[13] IGF-I had no effect on its own on LH binding to GC but synergized with FSH in the induction of LH receptors.[14] The synthesis of receptors to IGF-I in GC is induced by FSH.[15, 16] Administration of FSH (10 µg/rat, twice daily) for three days induced a 2.5 fold increase in IGF-I receptor concentration in the ovary. Addition of growth hormone (GH, 100 µg/rat, twice daily) further augmented (by 50%) the FSH induced IGF-I receptor sites in GC, whereas co-administration of the FSH with a potent agonist of gonadotropin-releasing hormone (GnRH; 25 µg/rat twice daily) inhibited the FSH induction of IGF-I receptors. The in vivo maintenance of IGF-I receptors in GC could be achieved by administration of the gonadotropic hormones.[16]

Based on these results it was logical to try whether GH, which is known to regulate IGF-I synthesis, has an effect on the reproductive system. Indeed, it was found[17] that culturing rat GC in the presence of FSH (3 ng/ml) together with GH (100 or 300 ng/ml) enhanced the FSH-induced rise in LH receptor concentration and progesterone production. Using porcine GC it was demonstrated[18] that FSH (200 ng/ml) together with estradiol (1 µg/ml) stimulated the secretion of IGF-I to the medium. Likewise it was demonstrated[19] that administration of GH (100 µg) to rats resulted in a 50% increase of IGF-I concentration in the ovary as compared to serum levels. These results have prompted Homburg et al.[20] to combine treatment of human menopausal gonadotropin (hMG) together with GH for the induction of ovulation in seven anovulatory women. These investigators demonstrated that the amount, duration of treatment and the daily effective dose of hMG were all reduced due to the addition of GH.

The literature reviewed above demonstrates unequivocally the central role of IGFs in the modulation of the biological activity of FSH in the ovary. However, the notion that ovarian growth factors are regulated by growth hormone is still doubtful, since: a) The pattern of growth hormone secretion is unrelated to the different stages of the female reproductive cycle and does not follow the physiological variations in the release of the gonadotropic hormones. b) Administration of high doses of exogenous GH may induce a substantial increase in peripheral IGF levels which may then reach the ovary in an effective concentration. It is still unknown whether the physiological concentrations of GH are sufficient to induce the biological activities at the ovarian level. c) Some GH preparations possess lactogenic activities[21] and therefore may induce bioactivities that are not characteristic to GH.

We suggest, therefore, that ovarian IGF gene expression is mainly regulated by prolactin. This hypothesis is based on the following: a) GH, prolactin and human placental lactogen (hPL) probably evolved from a single molecule. Although they have subsequently diverged in both structure and function, they continue to share intrinsic lactogenic and growth-promoting activities. b) It has been already suggested that prolactin induces in the liver the production of somatomedin-like activity[22-24] as well as other growth factors which are still unidentified.[25, 26] c) It is well established that there are variations in prolactin secretion that are related to the different stages of the female's reproductive cycle.[27-29] For example, in the rat, a surge of prolactin secretion occurs concomitantly with the preovulatory surge of FSH and LH and a second surge of prolactin release occurs on the afternoon of the estrous stage, several hours after the estrous-morning surge of FSH.[27] In the human, it has been demonstrated that prolactin, like the gonadotropic hormones, is secreted in a pulsatile pattern.[28] Blood samples obtained from 10 women, at 15 min intervals for a period of eight hours, demonstrated that distinct pulsatile fluctuations of plasma prolactin levels were present in all subjects and that 96% of these pulses coincided with LH pulses. d) Receptors for prolactin are present in GC, and their synthesis is regulated by FSH.[30] Thus FSH is the main regulator of the synthesis of receptors for LH, prolactin and IGF-I in granulosa cells (GC).

It is well established that prolactin is an integral part of the reproductive system and its secretion is regulated in concert with the other hormones that are involved in reproduction. Receptors for prolactin are present in GC and part of the biological activity of prolactin is probably mediated by growth factors. Thus, we suggest that prolactin is the most likely candidate to serve as the physiological modulator of FSH action in the ovary by regulating the ovarian growth factor(s) gene expression.

Preliminary Results

Clinical studies: Preliminary studies were conducted at the Carmel Hospital, Haifa.[31] The results of these studies suggest that prolactin is capable of augmenting the ovarian response to exogenous gonadotropin administration. Four women suffering from severe polycystic ovary syndrome were treated by buserelin (Hoechst) 500 μg s.c. daily from the first day of the menstrual cycle. From the fourth day, the women were administered daily with hFSH (three days) followed by daily administration of hMG. Hormone treatment of these patients was discontinued on day 13 or 14 of the cycle since serum estradiol levels remained unchanged and since no follicular growth was observed by ultrasound. On subsequent cycles (six cycles) the same women have received similar hormonal treatment with one important variation: on days 3, 5 and 7 of the cycle they were treated in addition with a 5 mg dose of metoclopramide hydrochloride (Rafa), a potent adrenergic inhibitor. Such inhibitors are known to induce the release of prolactin from the pituitary gland. Indeed, plasma prolactin levels peaked within 60 min from 18 ± 3 ng/ml to 172 ± 11 ng/ml, and remained elevated for about four h. Serum levels of estradiol reached a peak of 443 ± 61 pg/ml on day 12–14 of the cycles and the number of leading follicles was 4.2 ± 0.6 follicles per cycle. Administration of hCG to these patients resulted in ovulation, as determined by serum progesterone levels. These results indicate that in contrast to chronic hyperprolactinemia which suppresses the reproductive functions, acute rises in serum prolactin levels may improve ovarian responsiveness to gonadotropin administration.[31]

Basic research studies: Mature cycling rats were administered with buserelin 2 μg/rat/day for three days. This treatment has previously been shown to down-regulate the gonadotrope cells of the pituitary gland and to suppress cyclicity. On the third day the rats were administered with pregnant mare serum gonadotropin (PMSG; 7.5 I.U./rat) with or without chloropromazine (1 mg/rat, s.c.). Chloropromazine is a potent stimulator of prolactin release. Forty-eight hours post-PMSG injection, all rats were administered with human chorionic gonadotropin (hCG; 4 I.U./rat) in order to induce ovulation. Ovulation was examined on the next morning; the oviductal ampulae were removed and the number of ovulated oocytes was recorded. A highly significant difference in ova shed between the two groups was demonstrated. Whereas the number of ova in the control group was 3.09 ± 0.70/rat (mean ± SEM) that of the chloropromazine treated rats was 8.30 ± 0.76/rat. Each group consisted of six rats and the experiment was repeated four times. Serum prolactin levels of the chlorpromazine-treated rats peaked within one hour after the drug administration and reached levels that were 35 fold higher than those of the control group. These results demonstrate that, in the rat, like in human, an acute rise in serum prolactin levels augments the bioactivity of the gonadotropin on the ovary.

The effect of prolactin on progesterone induced secretion by FSH was studied in cultured granulosa cells. GC were taken from hypophysectomized, diethylstilbesterol (DES)-treated rats and cultured for 72 h. In initial experiments we established that ovine FSH induced progesterone secretion in a dose-dependent pattern. We later chose an intermediate dose of FSH (20 ng/ml) which caused a 5.9 fold increase in progesterone secretion, as compared to the control cultures, and found that addition of increasing concentrations of ovine prolactin (10–1000 ng/ml) enhanced the FSH-induced release of progesterone secretion in a dose-related manner. Prolactin by itself does not induce progesterone synthesis in GC. The lowest dose of prolactin (10 ng/ml) that was tested in combination with FSH was already effective and doubled the FSH response. It is worthwhile to mention that the minimal effective dose of GH, in a similar system, was 100 ng/ml, whereas 30 ng/ml did not enhance the FSH-induced progesterone secretion.[17]

In conclusion, our working hypothesis suggests the ovarian growth factor(s) gene expression is mainly regulated by prolactin rather than GH, since (a) The pattern of GH secretion is unrelated to the physiological stages of the reproductive cycle, whereas prolactin is secreted in association with the cyclic variations of gonadotropin release. (b) GH and prolactin evolved from a single molecule and some GH preparations possess lactogenic activities. (c) Receptors for prolactin are present in the ovary. We present here supportive evidence for this new concept. Our preliminary experiments demonstrate that prolactin is indeed capable of promoting the FSH-induced progesterone secretion by cultured rat granulosa cells. In vivo, we have demonstrated, both in rats as well as in human, that acute increase in serum prolactin levels, during the follicular phase, augments the FSH-induced follicular development, resulting in increased ovulation rate.

REFERENCES

1. M. Knecht, and K.J. Catt, Modulation of cAMP-mediated differentiation in ovarian granulosa cells by epidermal growth factor and platelet-derived growth factor, *J. Biol. Chem.*, 258:2789 (1983).
2. M. K. Skinner, D. Lobb, and J. H. Dorrington, Ovarian thecal/interstitial cells produce an epidermal growth factor-like substance, *Endocrinology*, 121:1892 (1987).

3. P. Feng, K. J. Catt, and M. Knecht, Transforming growth factor β regulates the inhibitory actions of epidermal growth factor during granulosa cell differentiation, *J. Biol. Chem.*, 261:14167 (1986).

4. J. D. Veldhuis, and R. W. Furlanetto, Trophic actions of human somatomedin C/insulin-like growth factor I on ovarian cells: in vitro studies with swine granulosa cells. *Endocrinology*, 116:1235 (1985).

5. E. Y. Adashi, C. E. Resnick, M. E. Svoboda, and J. J. Van Wyk, A novel role for somatomedin-C in the cytodifferentiation of the ovarian granulosa cell, *Endocrinology*, 115:1227 (1984).

6. A. J. W. Hsueh, E. Y. Adashi, B. C. Jones, and T. H. Welsh, Hormonal regulation of the differentiation of cultured ovarian granulosa cells, *Endocr. Rev.*, 5:76 (1984).

7. E. Y. Adashi, C. E. Resnick, A. J. D'ercole, M. E. Svoboda, and J. J. Van Wyk, Insulin-like growth factors as intraovarian regulators of granulosa cell growth and function, *Endocr. Rev.*, 6:400 (1985).

8. K. J. Ryan, and A. Makris, Significance of angiogenic and growth factors in ovarian follicular development, *Adv. Exper. Med. Biol.*, 219:203 (1987).

9. S. A. Tonetta, D. L. Yanagihara, R. S. DeVinna, and G. S. diZerega, Secretion of follicle-regulatory protein by porcine granulosa cells, *Biol. Reprod.*, 38:1001 (1988).

10. G. L. Smith, Somatomedin carrier proteins, *Mol. Cell. Endocrinol.*, 34:83 (1984).

11. N. Savion, G. Lui, R. Laherty, and D. Gospodarowicz, Factors controlling proliferation and progesterone production by bovine granulosa cells in serum-free medium, *Endocrinology*, 109:409 (1981).

12. E. Y. Adashi, C. E. Resnick, M. E. Svoboda and J. J. Van Wyk, Somatomedin-C synergizes with follicle-stimulating hormone in the acquisition of progestin biosynthetic capacity by cultured rat granulosa cells, *Endocrinology*, 116:2135 (1985).

13. E. R. Hernandez, C. E. Resnick, M. E. Svoboda, J. J. Van Wyk, D. W. Payne, and E. Y. Adashi, Somatomedin-C/insulin-like growth factor I as an enhancer of androgen biosynthesis by cultured rat ovarian cells, *Endocrinology*, 122:1603 (1988).

14. E. Y. Adashi, C. E. Resnick, M.E. Svoboda, and J. J. Van Wyk, Somatomedin-C enhances induction of luteinizing hormone receptors by follicle-stimulating hormone in cultured rat granulosa cells, *Endocrinology*, 116:2369 (1985).

15. E. Y. Adashi, C. E. Resnick, M. E. Svoboda, and J. J. Van Wyk, Follicle-stimulating hormone enhances somatomedin C binding to cultured rat granulosa cells, evidence for cAMP dependence, *J. Biol. Chem.*, 261:3923 (1986).

16. E. Y. Adashi, C. E. Resnick, E. R. Hernandez, M. E. Svoboda, and J. J. Van Wyk, In vivo regulation of granulosa cell somatomedin-C/insulin-like growth factor I receptors, *Endocrinology*, 122:1383 (1988).

17. X. Jia, J. Kalmijn, and A.J.W. Hsueh, Growth hormone enhances follicle-stimulating hormone-induced differentiation of cultured rat granulosa cells, *Endocrinology*, 118:1401 (1986).

18. C. Hsu, and J. M. Hammond, Gonadotropins and estradiol stimulate immunoreactive insulin-like growth factor I production by porcine granulosa cells in vitro, *Endocrinology*, 120:198 (1987).

19. J. B. Davoren, and A. J. W. Hsueh, Growth hormone increases ovarian levels of immunoreactive somatomedin C/insulin-like growth factor I in vivo, *Endocrinology*, 118: 888 (1986).

20. R. Homburg, A. Eshel, H. I. Abdalla and H. S. Jacobs, Growth hormone facilitates ovulation induction by gonadotrophins, *Clin. Endocrinol.*, 29:113 (1988).

21. A. Gertler, A. Shamay, N. Cohen, A. Ashkenazi, H. G. Friesen, A. Levanon, M. Gorecki, H. Aviv, D. Hadary, and T. Vogel, Inhibition of lactogenic activities of ovine prolactin and human growth hormone (hGH) by a novel form of a modified recombinant hGH, *Endocrinology*, 118:720 (1986).

22. M. J. O. Francis, and D. H. Hill, Prolactin-stimulated production of somatomedin by rat liver, *Nature*, 255:167 (1975).

23. A. T. Holder, and M. Wallis, Actions of growth hormone, prolactin and thyroxine on serum somatomedin-like activity and growth in hypopituitary dwarf mice, *J. Endocrinol.*, 74:223 (1977).

24. R. M. Bala, H. G. Bohner, J. N. Carter, and H. G. Friesen, Effects of prolactin on serum somatomedin activity in hypophysectomized rats, *Clin. Res.*, 24:655 (1976).

25. C. C. W. Mick, and C. S. Nicoll, Prolactin directly stimulates the liver in vivo to secrete a factor (synlactin) which acts synergistically with the hormone, *Endocrinology*, 116:2049 (1985).

26. J. P. Hoeffler, and L. S. Frawley, Liver tissue produces a potent lactogen that partially mimics the actions of prolactin, *Endocrinology*, 120:1679 (1987).

27. Y. Koch, Y. F. Chow, and J. Meites, Metabolic clearance and secretion rates of prolactin in the rat, *Endocrinology*, 89:1303 (1971).

28. N. S. Cetel, and S. S. C. Yen, Concomitant pulsatile release of prolactin and luteinizing hormone in hypogonadal women, *J. Clin. Endocrinol. Metab.*, 56:1313 (1983).

29. L. Yogev, and J. Terkel, Timing of termination of nocturnal prolactin surges in pregnant rats as determined by the number of fetuses, *J. Endocr.*, 84:421 (1980).

30. C. Wang, A. J. W. Hsueh, and G. F. Erickson, Induction of functional prolactin receptors by follicle-stimulating hormone in rat granulosa cells in vivo and in vitro, *J. Biol. Chem.*, 254:11330 (1979).

31. A. Lissak, R. Auslender, M. Direnfeld, L. Kahana, J. Rosenfeld, Y. Sorokin, H. Abramovici, and Y. Koch, The effect of prolactin on ovarian responsiveness to gonadotropin administration in patients with severe polycystic ovary syndrome (PCO), *Israel J. Med. Sci.*, 25:290 (1989).

STUDIES ON PHARMACOKINETICS OF LHRH ANALOGUES AFTER DIFFERENT MODES OF ADMINISTRATION

J. Sandow, G. Jerabek-Sandow, and B. Krauss

Hoechst AG, D-6230 Frankfurt-80, Germany F.R.

INTRODUCTION

The LHRH agonist analogues have become valuable tools for endocrine investigations of reproductive functions. They are established drugs for therapy of reproductive disorders and assisted reproduction.[24, 25] LHRH agonists have a selective antigonadotropic action and suppress ovarian and testicular function in a reversible manner. In the control of gonadal steroid secretion, in reproductive disorders, and in oncology (prostate carcinoma and premenopausal mammary carcinoma), LHRH agonists have acquired an important therapeutic role. This is in part due to the fact that their action on gonadotropins and gonadal steroids is highly selective, with no side effects on the secretion of other hormones or on other organ systems.[24, 25, 31] To take full advantage of the clinical potential, different modes of administration are required for short-term use (injections), long-term use by self-medication (nasal spray formulations), and long-term use by injection of sustained release preparations. In gynaecology, nasal spray therapy has been investigated for contraception, endometriosis, leiomyoma uteri and other oestrogen-dependent conditions. More recently, long acting implants and microcapsules have been introduced into oncology, e.g. for adjuvant therapy of premenopausal mammary carcinoma. In this review, the pharmacokinetics of the different dosage forms will be discussed with reference to particular clinical indications, and the metabolism of LHRH agonists in experimental animals and in the human will be reported.

Peptide drugs are particularly well tolerated, because their metabolism by endogenous endo- and exopeptidases is rapid and leads to inactive metabolites. In this respect, peptide drugs differ markedly from conventional drugs. Biotransformation of LHRH agonists to

Advances in Assisted Reproductive Technologies, Edited by
S. Mashiach *et al.*, Plenum Press, New York, 1990

inactive metabolites (oligopeptides) is similar to endogenous LHRH (gonadorelin), because their structure is closely related. The rate of inactivation by enzymes is reduced by the chemical substitution, resulting in prolonged serum elimination. However, the capacity of peptidase enzymes is high, and even at doses used in oncology the rate of inactivation is similar to that found during lower doses used in reproductive disorders. During long-term treatment in experimental animals, and in clinical studies induction of drug-metabolizing enzymes has not been observed, whereas effects on testosterone-dependent male specific forms of cytochrome P-450 and on enzymes of steroid biosynthesis have been reported.[21]

The investigation of peptide pharmacokinetics and metabolism has made rapid progress due to the advances in analytical methodology. At the doses used in clinical studies, LHRH agonists can be measured conveniently by specific radioimmunoassay (RIA). This method is specific for the agonist structure, but limited in discriminating intact peptides and inactive metabolites. For a separation of metabolites from intact substance prior to RIA determination, high performance liquid chromatography (HPLC) has become the method of choice. We have studied the metabolism and pharmacokinetics of the LHRH agonist, [D-Ser (But) 6] LHRH (1-9) nonapeptide-ethylamide (buserelin) after different modes of administration in experimental animals, and in clinical studies.[25] In some investigations, serum concentrations and urinary excretion of other highly active agonists were also measured by RIA methods. Results on buserelin and on related nonapeptide or decapeptide agonists (nafarelin, goserelin, triptorelin) are discussed, and the clinical implications for suitable regimens in the different disorders are evaluated.

More recently, the clinical potential of LHRH antagonists has been confirmed. These peptides are substituted by multiple unnatural amino acids and enzyme resistance is greatly enhanced. The pharmacokinetics of LHRH antagonists[17] are quite different from those of the agonist analogues. The organ distribution studies with antagonists indicate long-lasting accumulation with a maximum after 12–24 h in liver, kidney and other tissues. In contrast to the agonists, tissue/plasma ratios indicative of accumulation are found not only in pituitary gland, liver, kidneys and ovaries, but also in the testis and adrenal glands. However, the endocrine effects on gonadal steroid secretion remain highly specific. Metabolic inactivation is greatly reduced, very potent LHRH antagonists have serum halflives of more than 24 hours. This new class of compounds with remarkable potential for immediate steroid suppression is closer to conventional drugs, due to their delayed inactivation and tendency to accumulate in tissues for a longer time period.[39]

ORGAN DISTRIBUTION

The first stage of investigation was to identify organs for preferential accumulation and inactivation in experimental animals. The organ distribution of 14-C, 3-H, or 125-I labelled LHRH (gonadorelin) and LHRH agonists (leuprolide, nafarelin, buserelin) has been investigated.[9, 27] Accumulation of drug equivalents is found in the liver, kidney, anterior pituitary and intestine (Fig. 1). In studies with 125-I-buserelin in rats, liver and kidney are the main inactivating organs.[28] There is a rapid transfer of drug from the plasma compartment into different tissues. High tissue/plasma ratios are reached in liver and kidney with a maximum about one h after i.v. injection.[27] In terms of the percent dose accumulated, the liver and kidneys contain the highest fraction of radioactive drug equivalents (intact agonists and metabolites). Metabolites are excreted in urine and in bile, the biliary metabolites are subsequently found in the intestine and in the feces. In the urine, a significant fraction of the dose is excreted, the time course of excretion of radiolabelled drug equivalents being proportional to the plasma concentration. The identification of urinary buserelin in the rat indicates that mainly inactive metabolites are excreted in this species.[33] Reports on the organ distribution of 125-I-buserelin,[27] and 3-H-LHRH or 14-C-nafarelin[9] reach similar conclusions on the organs which show accumulation.

Specific accumulation of a small fraction of the dose is found in the anterior pituitary gland (biological target organ), the time course of pituitary accumulation is proportional to receptor binding on gonadotrop cells, and indicative of the stimulation of LH-release. Only the anterior pituitary gland contains intact agonist (buserelin), 1 h after i.v. injection of 125-I-buserelin. In other tissues, no intact peptide is accumulated. After metabolic inactivation, the radiolabel 125-I may be found in the thyroid, salivary glands, and gastrointestinal mucosa. When 125-I labelled buserelin is administered to rats, there is a time-dependent increase of 125-I in the thyroid gland, resulting from deiodination of tyrosine, the site of labelling of buserelin.

There is also an increasing fraction of circulating 125-I in the plasma. In the organ distribution studies in rats, the fraction of intact peptide circulating in the serum or plasma

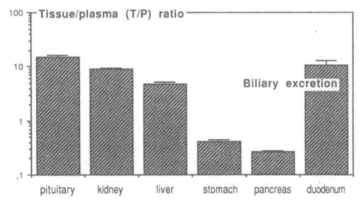

Fig. 1. Organ distribution of 125-I-buserelin in rats. The main inactivating organs are the liver and kidney, the physiological target organ is the anterior pituitary gland. Duodenal accumulation is due to biliary excretion of metabolites. A tissue/plasma ratio > 1 denotes accumulation of radioactive drug equivalents (metabolites). Mean and standard error of 4 rats, dose 2 µCi (10 ng peptide) per rat.

decreases with time after injection, as confirmed by bioassay and thin layer chromatography (TLC). The presence of 125-I (radioactive equivalents in the plasma) leads to erroneous half-life estimates unless removed by excretion before measuring the intact peptide. In rats treated with 125-I-buserelin i.v., the plasma obtained 1 h after injection contained mainly metabolites, when analyzed by thin layer chromatography (TLC) or by high performance liquid chromatography (HPLC). In studies with unlabelled buserelin (HPLC-separated plasma extracts of rats), 12.6 % intact buserelin are identified after one hour. The circulating 125-I is not retained by C18-silica extraction (SEP-PAK cartridges), an estimate of the circulating peptide fraction (agonists or antagonists) can be obtained by this separation. The fraction of intact buserelin in plasma is increased when the inactivating enzymes are saturated by a high dose of an LHRH antagonist administered prior to 125-I-buserelin.[28] With 14-C-nafarelin and 3-H-LHRH, it has also been observed that the fraction of intact peptide in the plasma decreases rapidly after treatment, whereas the total radioactivity circulating in the plasma remains high.[9] Significant differences exist with regard to protein binding. The hydrophobic agonist, nafarelin, is bound to plasma albumin to about 80%,[6] whereas LHRH nonapeptides (leuprolide, buserelin) have a protein binding of 15%–20%, similar to LHRH.[27,34]

METABOLISM

The metabolism of LHRH agonists has been studied with radiolabelled compounds, using HPLC and beta- or gamma-spectrometry. In later studies, we identified the metabolites of unlabelled buserelin in animals and in humans by HPLC and specific radioimmunoassay (HPLC/RIA). The biotransformation of buserelin to inactive metabolites was initially established in rats treated with 125-I-buserelin. The radioactive metabolites were isolated from

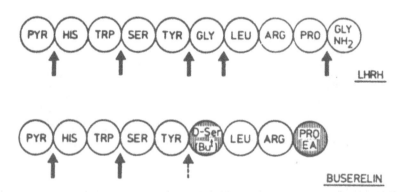

Fig. 2. Structure of LHRH and buserelin with enzyme cleavage sites. The longer duration of action of agonists is due to a reduced rate of inactivation. Pyroglutamyl peptidase, neutral endopeptidases and chymotrypsin inactivate buserelin. LHRH (gonadorelin) is also inactivated by the post-proline-cleaving enzyme.

liver, kidney, urine and plasma and compared with reference compounds. In subsequent studies, similar unlabelled metabolites were found in plasma, bile and urine of rats (Fig. 2). Results of different studies using metabolite identification by TLC and HPLC are in good agreement. In the pituitary gland of rats, intact buserelin and the inactive buserelin[2-9] octapeptide are found.[28] One hour after treatment, the liver of rats contains no intact buserelin, but a spectrum of smaller C-terminal metabolites. In the kidneys, intact buserelin and several C-terminal metabolites are present, one radioactive peak being identified as the buserelin[5-9] pentapeptide. This peak is identical with the main metabolite found in the urine of several species after unlabelled buserelin (using RIA for identification). This main urinary metabolite is found in all species investigated, including the human.[26]

Inactivation by Enzymes

In contrast to LHRH (gonadorelin), which is rapidly inactivated by several exo- and endopeptidases, the LHRH agonists are relatively resistant to enzyme degradation.[7, 15, 27] This is particularly evident in radioreceptor assays, when LHRH agonists are exposed to the enzymes contained in crude membrane preparations. Our studies with 125-I-LHRH and 125-I-buserelin have confirmed that similar enzymes are involved (pyroglutamyl-aminopeptidases, chymotrypsin-like enzymes), whereas the rate of degradation is strikingly different.[15, 27] During incubation of buserelin with pyroglutamyl-aminopeptidase, the (2–9) octapeptide is found as a result of cleavage of the N-terminal bond between pyroglutamic acid and histidine (Pyr1-His2). Intact buserelin, the (2–9) octapeptide, and other C-terminal metabolites, can be cleaved by chymotrypsin-like enzymes acting on the (Trp3-Ser4) and [Tyr5-D-Ser(But)6] bonds.[14] In the nonapeptide-ethylamide agonists, inactivation by the post-proline cleaving enzyme is blocked, whereas this enzyme can act on the decapeptide agonists like nafarelin and triptorelin. The AzaGly-substitution in goserelin[8] should afford same protection against the post-proline cleaving enzyme, but no reports on identification of metabolites of this compound have been published. Substitution by D-amino acids in position 6 does not completely prevent the inactivation by chymotrypsin, because of limited stereoselectivity. The buserelin (5-9)pentapeptide is gradually converted to the (6-9)tetrapeptide, and the nafarelin(5-10)hexapeptide is also converted to the nafarelin(6-10)pentapeptide by chymotrypsin.[7]

The LHRH antagonists are even more resistant to degradation by pyroglutamyl-aminopeptidase and chymotrypsin. It is difficult to assess the relative contribution of enhanced receptor affinity and/or enzyme resistance. In the hydrophobic decapeptide agonists, and in antagonists of related structure, an unprotected Pro9-glycinamide terminus is compatible with high potency, indicating that these peptides may be protected from receptor-associated enzymes after binding to LHRH receptors. In recent studies we found that the antagonistic potency of a group of LHRH antagonists[17] in the rats in vivo was closely correlated with their resistance to chymotrypsin in vitro, and with their receptor affinity for rat pituitary membranes.

Biliary Excretion

In organ distribution and autoradiography studies with labelled LHRH agonists, a significant fraction of the dose is excreted with the bile and found in the intestinal tract and in the feces.[7, 9] In rats treated with unlabelled buserelin, the main metabolite excreted in the bile was the buserelin (5-9) pentapeptide,[32] accounting for more than 95 % of the dose excreted in the bile. An active secretory mechanism concentrates the (5-9) pentapeptide in the bile. After injection of 1 mg/kg buserelin, there was a progressive time-dependent accumulation in the bile, and a bile/serum ratio of 140 was calculated for the major biliary metabolite 120 min after s.c. injection of buserelin. In the dog, the buserelin (5-9)pentapeptide is also found, but biliary excretion of buserelin is less than 1 %, indicating that renal elimination is quantitatively more important in this species. With 14-C- nafarelin, the biliary metabolites identified in the rat[7] were the (5-10)hexapeptide (27.7 %), and smaller amounts of the (5-7) tripeptide (5.2 %), the (1-7) heptapeptide (8.8 %), and the (1-9) nonapeptide (7.8%).

The biliary excretion of LHRH antagonist is less rapid than that of buserelin, but due to slow elimination, an increase in the intestinal accumulation due to biliary excretion with the feces is found for a time period of 32-48 h after i.v. injection. We have studied glycosylated LHRH antagonists such as [AcD-Nal1,D-Phe2,D-Phe3, His5,D-Ser(Rha)6,D-AlaNH2-10]LHRH, which have a good biology tolerance and cause only minimal histamine and mediator release at the site of injection.

Urinary Excretion

It is well established that immunoreactive LHRH (gonadorelin) is excreted in the urine of rats and humans in small quantities requiring concentration by extraction.[4] After administration of equal doses of LHRH and buserelin in rats by s.c. infusion for 14 days, the fraction of the dose of LHRH in the urine was 0.002 %, whereas 25%–30 % of the buserelin dose were excreted as immunoreactive (IR) buserelin.[33] The daily injection or infusion of 60 μg buserelin per rat induced a consistent urinary excretion during 14 days of treatment. The analytical separation of the urinary immunoreactive buserelin by high performance liquid chromatography (HPLC), and specific buserelin radioimmunoassay (RIA) as the detection method[33] confirmed that a significant fraction of the urinary buserelin excretion is intact peptide. The intact fraction may vary according to the rate of inactivation in different animal species. Several LHRH agonists are excreted in the urine, at daily amounts corresponding to 15%–30 % of the dose administered. We monitored the urinary excretion of LHRH agonists after microencapsulation and by s.c. infusion from minipumps in rats and found that the decapeptide triptorelin and the nonapeptide leuprolide are excreted to a significant extent in the urine of rats, providing a noninvasive parameter for studies on absorption and release (liberation) from sustained delivery systems.

A similar method is suitable for the pharmacokinetics of LHRH antagonists. The urinary excretion of LHRH antagonists depends on their lipophilicity, up to 20 % of the dose are excreted within 48 h after i.v. injection in rats. During this period, the antagonists are progressively metabolized as shown by the decreasing fraction of intact antagonists in the presence of more hydrophilic metabolites. Controlled release of antagonists will reduce their characteristic potential for histamine release. Serum concentrations and urinary excretion are well correlated, provided that a separation of intact antagonist and metabolites is performed.

METABOLITES IN ANIMALS AND IN THE HUMAN

The intact fraction of LHRH agonists in serum of several animal species was different, depending on the different rates of inactivation in each species. In the human, about 95 % of intact buserelin was found in the serum of patients with endometriosis,[16] and during infusions or implant treatment of patients with prostate carcinoma.[29] The spectrum of metabolites identified in the urine of the mouse, rat, dog, monkey and human was similar.[26] However, in each species different amounts of intact buserelin were present (Fig. 3). The fraction of intact buserelin in the urine in each species was well correlated with the dose requirement for suppression for gonadal steroid secretion. The mouse, being very insensitive to suppression requires an extremely high daily dose of 25 mg/kg (s.c.), and almost no intact buserelin was recovered from the urine, the peptide material consisting mainly of the (5–9) pentapeptide. In the rat, which also has a high dose requirement, 23% of intact buserelin were excreted. In the dog, which requires lower doses of buserelin for testosterone suppression than other species, a much higher fraction of intact buserelin was found (70%) than in the monkey and human (about 50%). Urinary metabolites of nafarelin identified in the monkey are the nafarelin (5–10) hexapeptide, (6–10) pentapeptide, the internal sequences nafarelin (5–7) tripeptide and (6–7) dipeptide are found by further degradation of the primary metabolites.

In the monkey and human, the fraction of intact urinary buserelin was 45-55%, and the buserelin (5–9) pentapeptide was the main urinary metabolite as in the other species. In clinical studies of treatment by buserelin injections, infusions, or by biodegradable implants the urinary excretion of immunoreactive buserelin was 16%–30% of the dose, depending on the experimental conditions. This immunoreactive peptide material consists of intact buserelin and smaller inactive metabolites, by HPLC/RIA about 50% intact buserelin were identified.

Excretion in Milk

Many peptides are excreted to a small extent in mother's milk, but generally their biological activity is low or absent due to rapid gastrointestinal inactivation (mainly by chymotrypsin), and by limited absorption from the gastrointestinal mucosa. Endogenous peptides (e.g. gonadorelin) may be secreted by the mammary gland, but do not crossreact with antisera raised against pro-ethylamide agonists. These peptide drugs are transferred to breast milk in proportion to their dosage and provide unequivocal analytical estimate of their distribution into mother's milk. Considerable interest has been directed at the contraceptive use of buserelin. In this context, the excretion of buserelin in the milk of nursing mothers was investigated in studies on post-partum contraception.[10] The dose administered was

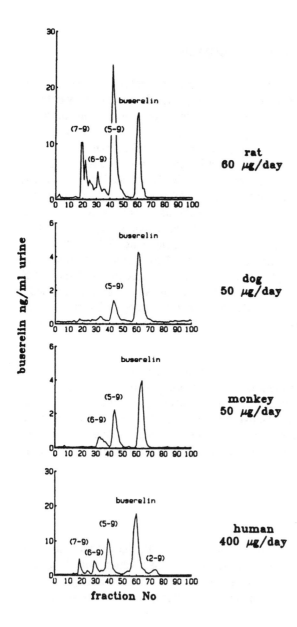

Fig. 3. Spectrum of intact buserelin and metabolites in the urine of several species. After treatment by injection or infusion of buserelin, the intact peptide and metabolites excreted in the urine were separated by HPLC and determined by specific RIA as the post-column detector. All metabolites have very low biological activity, and are inactive at concentrations resulting from human therapeutic doses. The buserelin (5-9)pentapeptide is the main metabolite in each species.

600 μg nasal spray, sufficient to maintain post partum contraception. The highest buserelin concentration found in one breast milk sample was 8.8 ng/ml, a maximum amount of 1–2 μg per feed was calculated as the dose likely to be ingested by infants. This dose is much below the orally effective dose of buserelin, which in adults is more than 500 μg/kg. Since chymotrypsin is present in the gastrointestinal tract of the neonate, a similar rate of inactivation may be assumed in nursing infants and in the adult. After nursing, there was no change in urinary LH excretion in the infants of mothers receiving the buserelin nasal spray. In this study, an oral dose of 600 μg buserelin had no effect on LH secretion of adult male test persons, and in previous studies up to 5 mg buserelin had no effect after oral administration. It is concluded that buserelin does not enter breast milk in biologically effective concentrations, and may safely be administered as a post-partum contraceptive by nasal spray.

PHARMACOKINETICS

A prerequisite for pulsatile stimulation of gonadotropin release is rapid dissociation and inactivation of LHRH, to avoid desensitization and restore receptor responsiveness after each LHRH pulse. Early studies with synthetic LHRH (gonadorelin) have shown, that after an initial rapid distribution phase, this decapeptide is eliminated rapidly with a half-life of 10-12 min.[13] In contrast, highly active LHRH agonists such as leuprolide, nafareline, goserelin and buserelin are eliminated much more slowly.[2, 8, 20, 38, 41] Their elimination half-life is of the order of 1–6 hours, depending on the compound and the method of determination. The peptides can be measured in the circulation by specific radioimmunoassays.[8, 14, 20, 22, 23, 41] Suitable antisera crossreacting with LHRH are available for the group of the LHRH decapeptide agonists with a C-terminal Pro-9-glycinamide. Other antisera detect nonapeptide agonists which contain a C-terminal Pro-9-ethylamide. These antisera do not crossreact with endogenous LHRH, and the practical limit of detection is therefore lower than for decapeptide antisera. A potential source of error that remains is crossreactivity with metabolites resulting from degradation of the LHRH agonist to be measured. Most antisera crossreact with the corresponding (2-9) octapeptides and (3-9) heptapeptides. We found that infusion of synthetic metabolites in animals is a valid method to check crossreactivity. Truly specific antisera should not detect metabolites under these conditions. Specific antisera should also detect only singular peak in HPLC-separated extracts of serum or urine samples. Identification of the intact peptide in serum is mandatory, the results of RIA measurements should be controlled by HPLC analysis of serum samples to separate bioactive intact peptide and inactive metabolites.[16] We have found that extraction of serum samples on C18-silica cartridges, high performance liquid chromatography (HPLC) for separation, and identification of intact agonist and metabolites by radioimmunoassay (RIA) with specific antisera is a highly selective and sensitive detection method.

In six endometriosis patients treated with an injection of 500 μg of buserelin (i.v.), the elimination half-life was 51-79 min.[16] The concentrations of immunoreactive (IR) buserelin measured in unextracted serum were 101 ng/ml (20 min after injection) and 1.12 ng/ml

(360 min after injection). In serum extracts analyzed by reversed-phase HPLC and RIA, intact buserelin was the main component (90% after 10 min, 74 % after 120 min and 52% after 360 min). The main serum metabolite was the inactive buserelin (5-9) pentapeptide (0.6% after 10 min, 19% after 120 min, and 12% after 360 min). In contrast to the slowly decreasing fraction of intact buserelin in the serum, the urinary excretion of intact buserelin remained constant during a 24 hour period after treatment. In the urine collected 0–1 h after treatment, 66% intact buserelin and 28% of the buserelin (5-9) pentapeptide were found. In the urine collected between 6-24 h, the ratio of intact buserelin to (5-9) pentapeptide remained similar. The urinary buserelin excretion in the first sample (0–1 h after treatment) was 345 µg/g creatinine, and in the last sample (6–24 h after treatment) the concentration was 4.7 µg/g creatinine. After s.c. injection of buserelin 1000 µg (Fig. 4), the serum half-life estimate was 80 min (between 1 and 8 h after injection).

The time course of buserelin elimination from the urine was parallel to that of serum elimination. The serum concentration and urinary buserelin/creatinine ratio were closely correlated,[36] a highly significant correlation of buserelin in serum and urinary excretion was also found in long-term studies with buserelin implants, and in experimental animals (rat, monkey). These results confirmed that serum elimination of buserelin is much slower than that of LHRH, the circulating agonist remains available to pituitary receptors for a long period and thus maintains desensitization and receptor down- regulation. High dose injections administered once per day s.c. maintain desensitization of pulsatile LHRH secretion for a 24 h period, as confirmed by the consistent testosterone suppression during daily injections of leuprolide 1 mg s.c. in prostate carcinoma.[37] For the initial treatment of prostate carcinoma, injections of 3 x 500 µg buserelin per day are used.[26]

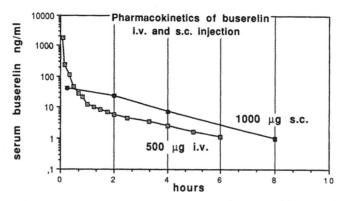

Fig. 4. Serum elimination of buserelin after i.v. and s.c. injection in patients with endometriosis. The elimination half-life after a dose of 500 µg i.v. or 1,000 µg s.c. is about 80 min. These doses are used for initial desensitization of gonadotropin release at the start of therapy.

Treatment by Nasal Spray

Nasal spray medication is a significant improvement for long-term therapy with peptides. The absorption rate of buserelin in humans corresponds to 2.5 or 3.3 % of the nasal spray dose, depending on the formulation.[34] Two bioequivalent formulations of buserelin have been investigated for their rate of absorption in comparison with s.c. injections (Fig. 5). The LH-releasing activity of the two nasal spray formulations is similar when administered at single doses of 400 µg (prostate carcinoma) or 300 µg (endometriosis). In patients with prostate carcinoma, an absorption rate of 2.5 % was found for the 400 µg dose of nasal spray administered three times per day. With the improved formulation used in endometriosis (single dose 300 µg three times per day), the rate of absorption was 3.3% (36). Single doses of 300 or 400 µg have a biological effect on LH-release comparable to s.c. injection of 10 µg buserelin. The serum concentrations after buserelin and nafarelin nasal spray have been monitored.[1, 14, 22, 23] Peak concentrations of immunoreactive buserelin are reached within 30-60 min. The time course of LH-release induced by the nasal spray is similar to that after an s.c. injection of buserelin. The urinary excretion of immunoreactive buserelin during treatment with nasal spray is dose-related, the total amount of peptide excreted in the urine during a 24 h treatment period increases proportionally to the dose increment (Fig. 5). In the urine, maximum concentrations of immunoreactive buserelin are reached one hour after nasal spray treatment. There is a time-dependent decrease of urinary buserelin excretion after nasal spray administration indicating a similar half-life ($t/2 = 80 - 90$ min) as that found in the serum.[36] The effective drug concentration during long-term treatment (fraction of the dose absorbed from the nasal mucosa) can be estimated by the dai-

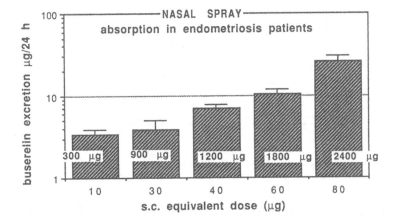

Fig. 5. Buserelin absorption from nasal spray in endometriosis. After nasal spray doses of 150-2,400 µg, a consistent absorption of 3.3% of the dose is found (weighted means 3.63, range 2.2-5.7 %). The s.c. equivalent doses (50-80 µg) corresponding to the nasal spray doses are indicated below each column. Means of 12-44 determinations per dose.

ly cumulative excretion (collecting the urine for 24 h during nasal administration of multiple doses), or by collection for an 8-h period after a single nasal spray dose.

In contraception, a single daily dose of buserelin nasal spray is administered, to reduce pituitary responsiveness and prevent the preovulatory LH-surge. In other gynaecological indications, 3-4 nasal spray doses of 300 μg per day are required to suppress oestrogen secretion and induce reversible amenorrhoea.[33] The urinary excretion of buserelin after nasal spray treatment (3 x 300 μg per day) indicates an absorption rate of 3.3 % of the dose, in comparison with the fraction of the dose found in the urine (Fig. 6) after buserelin injections (s.c.) 2 x 1,000 μg per day.[29,34] We found that over a dose range of 50-2,400 μg nasal spray, the absorption was consistent and reproducible in each patient (Fig. 5). In a study on nasal absorption of the daily maintenance dose of 1,200 μg nasal spray in prostate cancer patients, there was also a consistent absorption rate in each patient, buserelin excretion on two different test days was highly reproducible. The buserelin excretion calculated as μg/24 h or as μg/g creatinine for the same 24 h urine samples exhibited a highly significant correlation.

Treatment by Sustained Release

The initial studies on pharmacokinetics of buserelin administered by sustained release (implants or microcapsules) were performed with s.c. infusion systems (external mini-pump). Since LHRH (gonadorelin) and buserelin may be inactivated by enzymes in serum

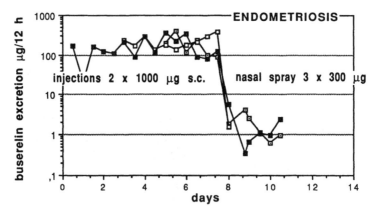

Fig. 6. Pharmacokinetics of buserelin by s.c. injection and nasal spray in 3 patients with endometriosis. There was a highly reproducible urinary excretion during daily s.c. injection of 2 x 1,000 μg buserelin per day. After changing to the nasal maintenance dose of 3 x 300 μg/24 h, a consistent average excretion of 3.83 μg/24 h was found in other patients monitored for 4-7 days. Measurement of urinary buserelin is the preferred non-invasive method for pharmacokinetic monitoring.

at room temperature, handling of the serum samples under careful refrigeration or the addition of bacitracin 10-3 M as an enzyme inhibitor is required to obtain reproducible and consistent results.[11] Buserelin in urine samples was stable when handled repeatedly during assay procedures, freezing and thawing did not affect the analytical result.[34] The serum concentrations and urinary buserelin excretion at a defined release (infusion) rate were determined (Fig. 7). Over a range of 50-300 µg buserelin/24 h, the serum concentrations and urinary excretion of buserelin were closely related to the rate of infusion.[18, 19, 34] A highly significant correlation was subsequently confirmed in several studies with buserelin implants in prostate carcinoma and endometriosis. In these studies, the urinary buserelin excretion required to maintain gonadal steroid suppression was established (critical limit of release). During the infusion studies, steady state conditions are reached after about 6 h, and the urinary excretion rate of immunoreactive (IR) buserelin (µg/g creatinine) remains constant. The monitoring of clinical studies is therefore considerably facilitated, urine samples may be obtained at predetermined intervals without having to perform a quantitative collection of urine for 24 h. After treatment with buserelin implants, the urinary buserelin excretion (µg/g creatinine) reaches a plateau on day 2-3 after treatment, with a gradual subsequent decline over a period of 4-12 weeks, depending on the dose and formulation.[30, 31] In patients with endometriosis, buserelin implants of 3.3 mg and 6.6 mg in polylactide/ glycolide 75:25 produced average urinary excretion rates of 10 and 20 µg/g creatinine during the plateau phase of release. In patients with prostate carcinoma, buserelin implants containing 6.6 mg peptide in a matrix of polylactic/ glycolic acid (75:25) produced urinary excretion rates of 10-15 µg buserelin/g creatinine, indicating a daily release rate of 50-75 µg buserelin when corrected for an average excretion of 1.5 µg creatinine/24 h, and a fractional excretion of 30 % of the dose in the urine.[3, 40] In several studies, the average serum concentrations during implant treatment with a dose of 3.3 mg or 6.6 mg (Fig. 8) were

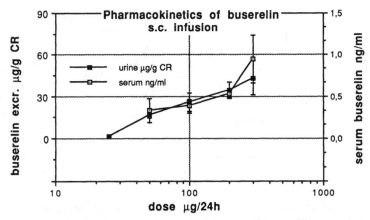

Fig. 7. Urinary buserelin excretion and serum concentration during prolonged s.c. infusion of 25-300 µg/day of buserelin from external minipumps, in women with endometriosis. There is a close correlation of both parameters over a wide dose range. Data of 3-6 patients per dose.

closely correlated with the urinary excretion rates (correlation coefficient 0.888-0.954 for 32-42 paired observations, respectively). The release rate in these studies was calculated from the buserelin/creatinine ratio in daily morning urine samples.

A very important consideration is the correlation of drug concentration and drug effect in clinical studies. During nasal spray therapy, the serum concentration is consistently decreasing with a half-life of 80 min. after each dose. It is therefore difficult to establish a critical lower limit of the serum concentration required for the maintenance of steroid suppression. An approximation of the daily dose required for suppression is found by calculating the total amount of urinary buserelin excreted during a 24 h period. The urinary excretion after a daily dose of 3 x 300 µg is equivalent to that after s.c. injection of 3 x 10 µg. During implant treatment (when the release rate is near linear during a 24 hours period), it is much easier to calculate the average fractional buserelin excretion (µg/g creatinine) and establish the critical lower limit of urinary excretion (and the corresponding release rate) compatible with clinical efficacy. Such studies have been performed in endometriosis (Fig. 9). After implant injection, ovarian activity is resumed at a critical lower limit of 0.2 - 0.4 µg/g creatinine as indicated by follicular maturation and a rise in serum estradiol, followed by ovulation and luteal progesterone concentrations.

Since the urinary excretion is closely correlated with the serum concentration, it can be calculated that a serum concentration of 100 - 200 pg/ml buserelin is required to maintain pituitary-ovarian suppression. In a study on the relationship of ovarian suppression and the associated buserelin release/urinary excretion rate in benign mastopathia, a similar limit was found.

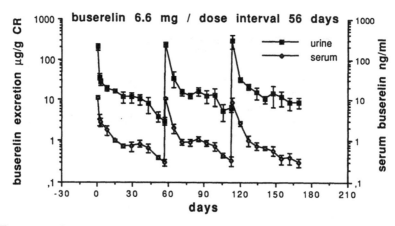

Fig. 8. Treatment of prostate carcinoma with buserelin implants. During repeated s.c. injection of implants (dose size 6.6 mg buserelin, dose interval 56 days), close correlation of the serum buserelin concentration and of the urinary buserelin excretion is found. Means and standard error of 8 patients treated for 6 months. Data of a collaborative study with Dr. Blom, Dept. of Urology, University of Rotterdam, Netherlands.[3]

Obviously, the oncology indications cannot be evaluated in the same manner, because dose renewal is mandatory before reaching the critical lower limit of suppression. However, ongoing studies in benign prostata hyperplasia (BPH) will elucidate the critical release rate for maintaining suppression in men.

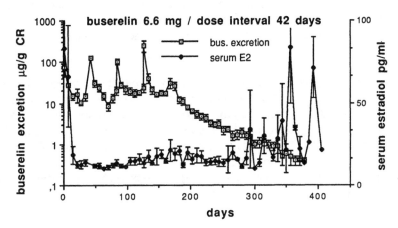

Fig. 9. Correlation of agonist release and clinical efficacy in patients with endometriosis. Four consecutive implant injections were administered at a dose interval of 6 weeks. After the last implant, the average buserelin excretion decreased to 0.2-0.4 µg/g creatinine, before ovarian activity (oestrogen secretion) was resumed. Means and standard error of 10 patients. Data of a collaborative study with Dr. H.M. Fraser, Edinburgh, Scotland.

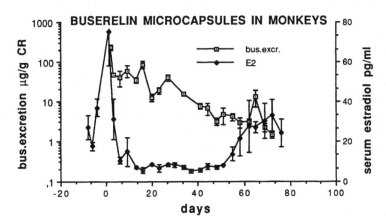

Fig. 10. Treatment of female rhesus monkeys with a micro-encapsulated buserelin formulation (polyhdroxibutyric acid/polylactid-glycolide 70:30). Buserelin excretion decreases to a critical lower limit (3 µg/g creatinine) before ovarian activity (oestrogen secretion followed by ovulation) can be measured. The critical lower limit is species-dependent, due to different distribution volumes and rates of metabolic inactivation.

Microencapsulation of LHRH Agonists

In gynaecology, microcapsule formulation of progestagens have been investigated for contraception, and similar injectable preparation can be prepared with LHRH agonists and antagonists. Their advantage is ease of administration and rapid biodegradation by erosion. We have monitored the release profile of experimental and clinically used microcapsule formulations of triptorelin (Decapeptyl™), leuprolide and buserelin. The release profiles were biphasic, consisting of an early phase by diffusion, and a secondary rise by polymer erosion.

In female monkeys, the duration of ovarian suppression after a buserelin dose of 6 mg in 60 mg microcapsules s.c. (Fig. 10) was 4-6 weeks, and similar critical limits of suppression were found (3 µg/g creatinine) for a different buserelin formulation which contains 3.6 mg buserelin in 40 mg microparticles of polylactide/glycolide.[12] The biodegradation of such microparticles of PLG 50:50 is rapid, 8 weeks after injection, the polymer material has disappeared from the injection site, and the tissue reaction is negligible. However, for long-lasting suppression, the use of implants appears to be preferable, because their duration of release can be extended to more than 3 months.[12, 33]

CONCLUDING REMARKS

A wide spectrum of clinical indications is covered by the LHRH agonists and in the future (for some indications) also by LHRH antagonists. Different dosage forms are required depending on the intended duration of action. In the oncology indications, long-lasting implants are preferable, because of the reduced frequency of treatment injections. In the gynaecological indications, the nasal spray treatment is preferable, when early reversibility is desired, particularly in the treatment of infertility. The prolonged suppression required in endometriosis and leiomyoma uteri before surgical intervention can be achieved both by injection of microencapsulated agonists, which suppress oestrogen secretion for 4-6 weeks, and by single implant injections which last for 4-12 weeks depending on the polymer composition. Implant treatment is recommended when prolonged reversible inhibition of ovarian activity is required, e.g. in severe endometriosis and leiomyoma uteri. In the development of such new dosage forms, the pharmacokinetic monitoring is essential to characterize the peptide concentrations needed to achieve a consistent clinical effect.

REFERENCES

1. Anik ST, McRae G, Nerenberg C, Worden A, Foreman J, Hwang JY, Kushinsky S, Jones RE, Vickery B (1984): Nasal absorption of nafarelin acetate, the decapeptide [D- Nal(2)6] LHRH, in rhesus monkeys I. *Journal of Pharmaceutical Sciences* 73, 684-885.
2. Barron JL, Millar RP, Di Searle (1982): Metabolic clearance and plasma half-disappearance

time of D-Trp6 and exogenous luteinizing hormone-releasing hormone. *J Clin Endocrinol Metab*, 54, 1169-1173.

3. Blom JHM, Hirdes WH, Schroeder FH, de Jong FH, Kwekkeboom DJ, van't Veen AJ, Sandow J, Krauss B (1989): Pharmacokinetics and endocrine effects of the LHRH analogue buserelin after subcutaneous implantation of a slow release preparation in prostatic cancer patients. *Urological Research* 17, 43-46.

4. Bourguignon JP, Hoyoux C, Reuter A, Franchimont P, Leinartz-Dourcy C, Vrindts-Gevaert Y (1979): Urinary excretion of immunoreactive luteinizing hormone-like material and gonadotropins at different stages of life. *J Clin Endocrinol Metab*, 48, 78.

5. Chan RL, Chaplin MD (1985): Identification of major urinary metabolites of nafarelin acetate, a potent agonist of luteinizing hormone-releasing hormone, in the rhesus monkey. *Drug Metab Dispos* 13, 566-571.

6. Chan RL, Chaplin MD (1985): Plasma binding of LHRH and nafarelin acetate, a highly potent LHRH agonist. *Biochem Biophys Res Commun* 127, 673-679.

7. Chan RL, Nerenberg CA (1987): Pharmacokinetics and Metabolism of LHRH analogs. *In:* LHRH and its Analogs, Part 2, (eds Vickery BH & JJ Nestor, MTP Press Ltd., Lancaster, pp. 577-593.

8. Clayton RN, Bailey LC, Cottam J, Arkell D, Perren TJ, Blackledge GRP (1985): A radioimmunoassay for GnRH agonist analogue in serum of patients with prostate cancer treated with D-Ser(tBu)-6-AzaGly-10-GnRH. *Clin Endocrinol* 22, 453-462

9. Chu NI, Chan RL, Hama KM, Chaplin MD (1985): Disposition of nafarelin acetate, a potent agonist of luteinizing hormone-releasing hormone, in rats and rhesus monkeys. *Drug Metab Dispos* 13, 560.

10. Dewart, PJ, McNeilly, AS, Smith, SK, Sandow, J, Hillier, SG and Fraser, HM (1987): LRH agonist buserelin as a post- partum contraceptive: lack of biological activity of buserelin in breast milk. *Acta Endocrinol* (Cph) 114, 185-192.

11. Fraser, HM, Yorkston, CE, Sandow, J, Seidel, H and von Rechenberg, W (1987): The biological evaluation of LHRH agonist (buserelin) implants in the female stumptailed macaque. *Br J Clin Pract* 41, 14-22.

12. Fraser, HM, Sandow, J, Seidel, H and Rechenberg, W v (1987): An implant of a gonadotropin releasing hormone agonist (buserelin) which suppresses ovarian function in the macaque for 3-5 months. *Acta Endocrinol* (Cbh) 115, 521- 527.

13. Handelsman DJ, Swerdloff RS (1986): Pharmacokinetics of gonadotropin-releasing hormone and its analogs. *Endocr Rev* 7, 95-105.

14. Holland FJ, Fishman L, Costigan DC, Luna L, Leeder S (1986): Pharmacokinetic characteristics of the gonadotropin-releasing hormone analog D-Ser(tBu)-6-EA-10 luteinizing hormone-releasing hormone (buserelin) after subcutaneous and intransal administration in children with central precocious puberty. *J Clin Endocrinol Metab* 63, 1065.

15. Horsthemke, B, Knisatschek, H, Rivier, J, Sandow, J & Bauer, K (1981): Degradation of luteinizing hormone-releasing hormone and analogs by adenohypophyseal peptidases. *Biochem Biophys Res Commun* 100, 753-759.

16. Kiesel, L, Sandow, J, Bertges, K, Jerabek-Sandow, G, Trabant, H and Runnebaum, B (1989): Serum concentration and urinary excretion of the LHRH agonist buserelin in patients with endometriosis. *J Clin Endocr Metab* 68 in press.

17. König W, Sandow J, Jerabek-Sandow G, Kolar C (1989): Glycosylated gonadoliberin antagonists. *In:* Peptides 1988: Proceedings of the 20th European Peptide Symposium (eds. G Jung, E Bayer), Walter de Gruyter Verl. Berlin, pp. 334-336.

18. Lemay A, Sandow J, Quesnel G, Bergeron J, Mérat P (1988): Transient escape from the down-regulation of the pituitary-ovarian axis during minipump subcutaneous infusion of luteinizing hormone-releasing hormone (LHRH). *Fertil Steril* 49, 802-812.

19. Lemay A, Sandow J, Bureau M, Maheux R, Fontaine J-Y, Mérat P (1988): Prevention of follicular maturation in endometriosis by s.c. infusion of luteinizing hormone- releasing hormone (LHRH) agonist started in the luteal phase. *Fertil Steril* 49, pp.410-417.

20. Nehrenberg C, Foreman J, Chu N, Chaplin MD, Kushinsky S (1984): Radioimmunoassay of nafarelin ([6-(3-(2-naphthyl)-D-alanine)]-luteinizing hormone-releasing hormone) in plasma or serum. *Anal Biochem* 141, 10.
21. Ohi H, Iwasaki M, Komori M, Miura T, Kitada M, Hayashi S and Kamataki T (1989): Effects of serum testosterone level with buserelin on the activities of drug and testosterone hydroxylase and on the content of a male-specific form of cytochrome P.450 in male rats. *Biochemical Pharmacology* 38, 535-538.
22. Reznik Y, Winiger BP, Aubert ML, Sizonenko PC (1987): Pharmacodynamics of [D-Ser (tBu)-6,desGly-10]GnRH-ethylamide (buserelin). *Acta Endocrinol* (Copenh) 115, 235-242.
23. Saito S, Saito H, Yamasaki R, Hosoi E, Komatsu M, Iwahana H, Maeda T (1985): Radioimmunoassay of an analog of luteinizing hormone-releasing hormone, [D-Ser(tBu)-6-desGly-10]ethylamide (buserelin). *J Immunol Methods* 79, 173-183.
24. Sandow, J (1987): Pharmacology of LHRH agonists. In: Pharmacology and Clinical Uses of Inhibitors of Hormone Secretion and Action (eds. B. Furr & A.E. Wakeling), Bailliere Tindall, London, pp. 365-384.
25. Sandow J (1989): Chemistry and metabolism of LHRH and its analogues. *In:* LHRH and its Analogues - their use in gynaecological practice (eds. R Shaw and JC Marshall),Butterworth & Co. (Publ.) Ltd., London, pp. 35-48.
26. Sandow, J and von Rechenberg, W (1987): Pharmacokinetics, metabolism and clinical studies with buserelin. *In:* LHRH and Its Analogs, Contraceptive and Therapeutic Applications, Part 2 (eds. Vickery, BH & Nestor Jr, JJ), MTP Press Ltd., Lancaster, pp. 363-382.
27. Sandow J, Clayton RN (1983): The disposition, metabolism, kinetics and receptor binding properties of LHRH and its analogues. Progress in Hormone Biochemistry and Pharmacology 2, 63-106.
28. Sandow, J, Jerabek-Sandow, G, Krauss, B & Stoll, W (1982): Metabolic and dispositional studies with LHRH analogs. *In:* LHRH Peptides as Female and Male Contraceptives, PARFR Series on Fertility Regulation (eds. Zatuchni, GI, Shelton, JD & Sciarra, JJ), Harper & Row, Philadelphia, pp. 338-354.
29. Sandow, J, Jerabek-Sandow, G, Krauss, B & von Rechenberg, W (1985): Pharmacokinetic and antigenicity studies with LHRH analogues. *In:* Future Aspects in Contraception, Part 2, Female Contraception (eds. Runnebaum, B, Rabe, T, Kiesel, L), MTP Press Ltd., Boston, pp. 129-147.
30. Sandow, J, Jerabek-Sandow, G, Schmidt-Gollwitzer, M (1985): Metabolism and pharmacokinetics of LHRH agonists. *In:* LHRH and its Analogues, Fertility and Antifertility Aspects (eds. Schmidt-Gollwitzer, M. & Schley, R.), Walter de Gruyter, Berlin, pp. 105-121.
31. Sandow, J, Fraser, HM & Geisthövel, F (1986): Pharmacology and experimental basis of therapy with LHRH agonists in women. *In:* Gonadotropin Down-Regulation in Gynecological Practice, Progress in Clinical and Biological Research, Vol. 225 (eds. R Rolland, DR Chadha and WNP Willemsen), Alan Liss, Inc., New York, pp. 1-27.
32. Sandow, J, von Rechenberg, W, Krauss, B and Jerabek Sandow, G (1987): Excretion of agonadorelin-agonist (buserelin) in the urine and bile of rats. *Naunyn Schmiedeberg's Arch Pharmacol* 335 (Suppl.) 23.
33. Sandow, J, Seidel, HR, Krauss, B and Jerabek-Sandow, G (1987): Pharmacokinetics of LHRH agonists in different delivery systems and the relation to endocrine function. *In:* Hormonal Manipulation of Cancer: Peptides, Growth Factors, and New(Anti) Steroidal Agents (eds. Klijn, JGM, Paridaens, R and Foekens, JA), Raven Press, New York, pp. 203-212.
34. Sandow, J, Fraser, HM, Seidel, H, Krauss, B, Jerabek-Sandow, G and von Rechenberg, W (1987): Buserelin: pharmacokinetics, metabolism and mechanisms of action. *Br J Clin Pract* 41, 6-13.
35. Sandow J, Fraser HM, Engelbart K, Seidel H, Donaubauer H and Rechenberg W v (1988): Preclinical studies of suppression of follicular maturation and oestrogen secretion in rats and monkeys. In: LHRH agonists in Oncology, ed K Hoeffken, Springer Verlag, Berlin, pp. 10-21.

36. Sandow, J (1988): Physiologische und pharmakologische Wirkungen von Buserelin: Präklinische Untersuchungen. [Physiological and pharmacological effects of buserelin: preclinical investigations], Symposium "Endometriose - das neue Therapieprinzip mit Supre-cur^R, Walter de Gruyter & Co. Verlag, Munich, in press.

37. Santen, RJ, Demers, LM, Max, DT, Smith, J Stein, BS and Glode, LM (1984): Long-term effects of administration of a gonadotropin-releasing hormone superagonist analog in men with prostate carcinoma. *J Clin Endocrinol Metab* 58, 397.

38. Sennello LT, Finley RA, Chu SY, Jagst C, Max D, Rollins DE, Tolman KG (1986): Single-dose pharmacokinetics of leuprolide in humans following intravenous and subcutaneous adminstration. *J Pharm Sci* 75, 158.

39. Vickery BH (1987): Pharmacology of LHRH antagonists. *In:* Pharmacology and clinical uses of inhibitors of hormone secretion and action (eds. BJA Furr and AE Wakeling), Baillière Tindall, London, pp. 385-408.

40. Waxman JH, Sandow J, Abel P, Farah N, O'Donoghue E, Fleming J, Cox J, Sikora K, Williams G (1989): Two monthly deport gonadotrophin releasing hormone agonist (buserelin) for prostatic cancer. *Acta Endocrinol* (Cph) 120, 315-318.

41. Yamazaki I, Okada H (1980): A radioimmunoassay for a highly active luteinizing hormone-releasing hormone analogue and relation between the serum level of the analogue and that of gonadotropin. *Endocrinol Japon* 27, 593-605.

OVULATION AND STRESS: PATHOPHYSIOLOGY

A.R. Genazzani, F. Petraglia, A. Monzani, G. Fabbri,
A.D. Genazzani, P. Di Domenica, and A. Volpe

Dept. of Obstetrics and Gynecology, Univ. of Modena
v. del Pozzo 71, 41100 Modena, Italy

The stress-induced impairment of the hypothalamus-pituitary-gonadal (HPG) axis has been shown in humans. Hypogonadism or secondary amenorrhea, in men and women respectively, are the most frequent reproductive disturbances caused by chronic stress. The common endocrine findings in human patients and in experimental animals exposed to stress are the low plasma concentrations of pituitary gonadotropins and gonadal steroid hormones. The mechanisms responsible for the stress-induced impairment of the HPG axis are still unknown. There is much evidence to suggest that the primary defect occurs in the hypothalamus. Stress activates neuropeptides and neurotransmitters which suppress gonadotropin releasing hormone (GnRH) activity. However, other authors suggest an impairment of the gonads as primary cause of HPG alteration.[1]

ROLE OF PROLACTIN

Prolactin (PRL) response to stress has been studied in animals and humans. It is well known that hyperprolactinemia is accompanied by hypogonadism and amenorrhea due to the inhibition of pulsatile GnRH release.

Different neuropeptides or neurotransmitters have been proposed to modulate the stress-induced increase of plasma PRL, some as possible prolactin-releasing factors (PRF): thyrotropin-releasing-hormone (TRH), peptide histide isoleucine (PHI), dynorphin, B-endorphin (B-EP), or as prolactin-inhibiting factors (PIF): dopamine (DA) and GnRH-associated peptide (GAP).[2]

Advances in Assisted Reproductive Technologies, Edited by
S. Mashiach *et al.*, Plenum Press, New York, 1990

The activation of PRF or the inhibition of PIF may cause the stress-induced increase of plasma PRL. Many results have shown that the blockade of DA activity is the major mechanism of action of several neurotransmitters and neuropeptides during stress. However, the PRL response to stress is still present after DA agonist administration, suggesting that other mechanisms regulate the PRL stress-induced increase.[3]

The pretreatment with propranolol, an antagonist of B-adrenergic receptors, blocks the stress-induced increase of plasma PRL levels, indicating a role of adrenergic receptors, probably blocking the DA release in the pituitary-portal vessel.[4]

A possible role of serotonine (SE), as putative PRF, increasing plasma PRL levels during stress has been shown.[5]

The endogenous opioid peptides stimulate the PRL secretion and the central administration of naloxone, an antagonist of opioid receptors, or of specific anti B-EP or dynorphin sera blocks the footshock induced PRL increase.[6] Therefore, opioids may be claimed as central mediators of the stress-induced PRL release. The opioid peptides stimulate the PRL secretion inhibiting DA and stimulating the serotononergic system.[7]

ROLE OF ENDOGENOUS OPIOID PEPTIDES

The opioid pathway is most probably very important in the hypothalamic control of the ovulatory mechanisms.

Naloxone, an opiate receptor antagonist, increases plasma LH levels in fertile women during the luteal phase, but not during the follicular phase of the menstrual cycle. The inhibitory activity of opioids on LH starts to be operative during the preovulatory days and plasma LH levels increase following naloxone injection. The naloxone induced-LH increase is related to the circulating estradiol levels.[8, 9] The low frequency of LH pulsatile during the luteal phase may also be ascribed to the increased activity of the opioids, probably related to the high progesterone levels.[10] In patients with hypothalamic secondary amenorrhea there is no increase of plasma LH following naloxone and this impairment of the opioid activity has been related to the hypogonadic state of these patients.[11, 12, 13] The same results have been shown in hypogonadic men.[14] The correlation with gonadal steroids is supported by the evidence that the treatment of induction of ovulation restoring a menstrual cycle also normalized the response of LH to naloxone.[15]

The increased opioid activity is therefore a possible causal factor of stress-related amenorrhea. Recent reports indicated that in three amenorrheic patients the treatment with naltrexone induced ovulatory cycles, supporting the possibility that inhibition of gonadotropin is an important factor in the genesis of the menstrual disorders.[16]

The correlation between gonadal steroids and the hypothalamic neurones regulating LH secretion are well known. Opioid peptides may represent a target of these peripheral hormones, modulating the messages on GnRH producing neurones. Because opioids have biological effects on several brain activities it is possible that the changes of LH secretion result from various conditions. The clinical findings showing an impairment of the HPG axis following psychological or physical stimuli, may be explained by the inhibitory effect of opioids on GnRH secretion. Indeed, the opioid activity increases following environmental changes, as an adaptive response. A classical psychoneuroendocrine disease, anorexia nervosa, is an example of hypogonadotropinemia and relates to high opioid activity. The behavioural changes leading to anorexia seem to be followed by the increased inhibition of opioids on plasma LH levels, as shown by the lack of effect of naloxone. The intracerebral administration of B-EP and dynorphin antiserum completely reversed the footshock-induced plasma LH decrease in rats.[17] This result employed the important role of opioids in the mechanisms leading from chronic stress to the inhibition of the HPG axis. Moreover, because the endogenous opioid peptides have many other biological activities in the brain, they may represent the link between the gonadotropin neuroendocrine regulation and hypothalamic functions. The opiatergic pathway may play the role of integrative system receiving multiple hormonal and neuronal afferences and contributing to the regulation of several brain physiological activities.

ROLE OF CORTICOTROPHIN-RELEASING-FACTOR AND GLUCOCORTICOIDS

Recently it has been shown that the injection of corticotropin-releasing-factor (CRF) into the the lateral ventricle of the brain markedly inhibits GnRH and LH secretion.[18] Indeed, the central injection of a CRF antagonist reverses the inhibitory action of stress on LH release.[19]

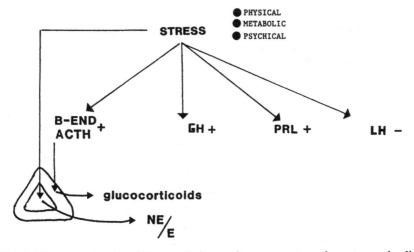

Fig. 1. Pituitary and adrenal hormonal changes in response to various stress stimuli.

At the present time, the exact mechanism through which CRF mediate the stress response in terms of LH release are still speculative. The finding that the central injection of B-EP antiserum reverses the CRF induced LH decrease, suggests an interaction between CRF and opioids.[17] Sirinathsinghji and others have reported that the anti-B-EP serum also prevents the CRF induced inhibition of lordosis behaviour in female rats, an action controlled by GnRH.[20] Additionally, immunoistochemical evidences show that CRF and B-EP neurons or nerve terminals are localized to GnRH terminals, both in the medial basal hypothalamus and in the media eminence,[21] thus supporting a functional interaction within the brain among these 3 neurohormones.

Moreover, glucocorticoids have a negative effect on pituitary release of LH, both in humans and in rats. Some reports indicate that cortisol might decrease LH secretion from cultured rat pituitary cells,[22, 23] inhibit plasma LH levels in monkeys and rats,[24, 25] and block ovulation.[25] Contrasting results are reported on the effect of glucocorticoids on plasma LH levels in humans, showing the decrease[26, 27, 28] or the ineffectiveness[29, 30, 31] of plasma LH after subchronic or chronic glucocorticoid treatments. The action of cortisol or corticosterone on the basal or GnRH-induced release of LH may be one of the stress-related responses leading to hypogonadism and/or amenorrhea. In fact, our recent data showed that hydrocortisone infusion inhibits the LH response to GnRH injection in amenorrheic patients. In healthy subjects the GnRH-induced LH increase was not effected by any drug infusion. Moreover, hydrocortisone infusion did not change the basal plasma LH levels either in patients or in controls.

The differences between the results reported here and those of previous studies might be related to the length of treatment, the type or dose of glucocorticoids used or the state of gonadal functions in the subjects. Plasma LH levels were studied following chronic administration of prednisolone in healthy women[26] and of prednisone in hirsute patients.[27]

Furthermore, the effect of subchronic treatment with dexamethasone (DXM) on plasma LH was studied in ovariectomized patients,[28] in patients with polycystic ovarian disease,[29] in healthy women[30] and in hypergonadotropinemic hypogonadal women.[31] While in these studies basal LH levels and the LH response to exogenous GnRH were inhibited by prednisolone[26] and prednisone,[27] DXM only reduced the postcastration gonadotropin rise,[28] without changing gonadotropin secretion.[29, 31]

In our study the short infusion of hydrocortisone inhibited the GnRH-induced response of LH in amenorrheic patients, while in fertile women it was not active. The major sensitivity of amenorrheic patients to the inhibitory effect of hydrocortisone on LH secretion might be explained by the low levels of plasma gonadal steroids. Indeed, Kamel and Kubajak[23] have shown that corticosterone interacts with gonadal steroids in modulating LH secretion in vitro. The above assumption might also explain the different results previously reported in humans between fertile and postmenopausal women.[26, 31] The evidence that hydrocortisone reduced the LH response to both GnRH injections suggests that in amenorrheic patients the drug inhibited the acute releasable pool and the functional reserve pool.

CONCLUSION

The inhibition of reproductive function induced by stress results from the interaction between the HPG axis and the other neuroendocrine systems. The increased secretion of PRL, opioids and of the hormones of the HPA axis after stress has been survived may block the activity of HPG axis.

The secretion of hypothalamic GnRH, pituitary gonadotropins and gonadal steroid hormones is impaired during the stress exposure by multiple neuroendocrine mechanisms and results in the lack of ovulation.

REFERENCES

1. R. Collu, W. Gibb and J. R. Ducharme, Effects of stress on gonadal function, *J. Endocrinol. Invest.* 7:529 (1984).
2. E. E. Muller, G. Nastico and U. Scapagnini, Neurotransmitters and Anterior Pituitary Function, Academic Press, New York (1977).
3. S. M. McCann, The role of brain peptides in the control of anterior pituitary hormone secretion, *in:* Neuroendocrines Perspective, E.E. Muller and R.M. MacLeod eds., vol.1, Elsevier Biomedical Press, Amsterdam (1972).
4. C. Denef and M. Baes, Beta-adrenergic stimulation of prolactin release from superfused pituitary cell aggregates, *Endocrinology*, 111:356 (1982).
5. R. Collu, P. Du Ruisseau and Y. Tache, Role of putative neurotransmitters in PRL, GH and LH response to acute immobilization stress in rats, *Neuroendocrinology*, 28:178 (1979).
6. F. Petraglia, W. Vale and C. Rivier, Beta-endorphin and dynorphin participate in the stress-induced release in the rat, *Neuro-endocrinology*, 28:178 (1987).
7. L. Grandison and A. Guidotti, regulation of prolactin release by endogenous opiates, *Nature*, 270:357 (1977).
8. M. E. Quigley and S. S. C. Yen, The role of endogenous opiates on LH secretion during the menstrual cycle, *J. Clin. Endocrinol.*, 51:179 (1980).
9. J. Blankstein, F. Y. Reyes, S. D. Winter and C. Faimen, Endorphins and the regulation of human menstrual cycle, *Clin. Endocrinol.*, 14:287 (1981).
10. M. Ferin, D. Van Vugt and S. Wardlaw, the hypothalamic control of menstrual cycle and the role of endogenous opioid peptides, *Recent. Prog. Horm. Res.*, 40:441 (1984).
11. M. E. Quigley, K. L. Sheehan, R. F. Casper and S. S. C. Yen, Evidence of an increased dopaminergic and opioid activity in patients with hypothalamic hypogonadotropic amenorrhea, *J. Clin. Endocrinol. Metab.*, 50:949 (1980).
12. S. E. Sauder, G. D. Case, N. J. Hopwood, R. P. Kelch and J. C. Marshall, The effects of opiate antagonism in gonadotropin secretion in children and in women with hypothalamic amenorrhea, *Pediatr. Res.*, 18:322 (1984).
13. F. Petraglia, G. D'Ambrogio, G. Comitini, F. Facchinetti, A. Volpe and A. R. Genazzani, Impairment of opioid control of luteinizing hormone secretion in menstrual disorders, *Fertil. Steril.*, 43:535 (1985).
14. J. D. Veldhuis, H. E. Kulin, A. Warner and S. J. Santner, Responsiveness of gonadotropin secretion to infusion of an opiate receptor antagonist in hypogonadotropic individuals, *J. Clin. Endocrinol. Metab.*, 55:649 (1982).
15. C. Nappi, F. Petraglia, G. Di Meo, M. Minutolo, A. R. Genazzani and U. Montemagno, Opi-

oid regulation of LH secretion in amenorrheic patients following therapis of induction of ovulation, *Fertil. Steril.*, 47:579 (1987).

16. L. Wildt and G. Leyendecker, Induction of ovulation by the chronic administration of naltrexone in hypothalamic amenorrehea, *J. Clin. Endocrinol. Metab.*, 64:1334 (1987).

17. F. Petraglia, W. Vale and C. Rivier, Opioids act centrally to modulate stress-induced decrease in LH secretion in the rat, *Endocrinology*, 119:2445 (1986).

18. F. Petraglia, S. Sutton, W. Vale and P. Plotsky, Corticotropin-releasing factor decreases plasma luteinizing hormone levels in female rats by inhibiting gonadotropin-releasing hormone release into hypophysial portal circulation, *Endocrinology*, 120:1083 (1987).

19. C. Rivier, J. Rivier and W. Vale, Stress-induced inhibition of reproductive function: role of endogenous corticotropin-releasing factor, *Science*, 231:607 (1986).

20. D. J. S. Sirinathsinghji, P. E. Whittington, A. Audsley and H.M. Fraser, Beta-endorphin regulates lordosis in female rats by modulating LH-RH release, *Nature*, 301:62 (1982).

21. M. A. Millan, D. M. Jacobovitz, L. R. Haugher and K. J. Catt, Distribution of corticotropin-releasing factor in primate brain, *Proc. Natl. Acad. Sci. USA*, 83:1921 (1986).

22. P. S. Li and W. C. Wagner, In vivo and in vitro studies on the effect of adrenocorticotropin hormone or cortisol on the pituitary response to gonadotropin-releasing hormone, *Biol. Reprod.*, 29:25 (1983).

23. D. E. Suter and N. B. Schwartz, Effects of glucocorticoids on secretion of luteinizing hormone and follicle-stimulating hormone by female rat pituitary cells in vitro, *Endocrinology*, 117:849 (1985).

24. A. K. Dubey and T. M. Plant, A suppression of gonadotropin secretion by cortisol in castrated male rhesus monkeys (Macaca mulatta) mediated by the interruption of hypothalamic gonadotropin-releasing hormone release, *Biol. Reprod.*, 33:423 (1985).

25. D. M. Baldwin and C. H. Sawyer, Effects of dexamethasone on LH release and ovulation in the cyclic rat, *Endocrinology*, 94:1397 (1974).

26. M. Sakakura, K. Takebe and S. Nakagawa, Inhibition of luteinizing hormone secretion induced by synthetic LRH by long-term administration with glucocorticoids in human subjects, *J. Clin. Endocrinol. Metab.*, 40:774 (1975).

27. A. E. Karpas, L. J. Rodriguez-Rigau, K. D. Smith and E. Steinberger, Effect of acute and chronic androgen suppression by glucocorticoids on gonadotropin levels in hirsute women, *J. Clin. Endocrinol. Metab.*, 59:780 (1984).

28. G. B. Melis, V. Mais, M. Gambacciani, A. M. Paoletti, D. Antonori and P. Fioretti, Dexamethasone reduces the postcastration gonadotropin rise in women, *J. Clin. Endocrinol. Metab.*, 2:237 (1987).

29. G. C. L. Lachelin, H. L. Judd, S. C. Swanson, M. E. Hauck, D. C. Parker and S. S. C. Yen, Long term effects of nightly dexamethasone administration in patients with polycystic ovarian disease, *J. Clin. Endocrinol. Metab.*, 55:768 (1982).

30. H. Vierhapper, W. Waldausl and P. Nowotny, Suppression of luteinizing hormone induced by adrenocorticotropin in healthy women, *J. Endocrinol.*, 91:399 (1981).

31. W. J. Georgitis, G. L. Treece and F. D. Hofeldt, Gonadotropin releasing hormone provokes prolactin release in hypergonadotropic hypogonadal women: a response not altered by dexamethasone, *Clin. Endocrinol.*, 19:319 (1983).

GnRH PEPTIDES AND CORPUS LUTEUM REGULATION

Tony Bramley

Department of Obstetrics and Gynecology
University of Edinburgh
37 Chalmers Street, Edinburgh EH3 9EW, Scotland, U.K.

INTRODUCTION

Agonist analogues of the hypothalamic decapeptide, gonadotrophin-releasing hormone (GnRH) have superior potency, resistance to degradation and longer duration of gonadotrophin release that GnRH. Repeated or continuous administration of agonist analogues to women induces a short period of gonadotrophin release, followed by down-regulation of pituitary GnRH-receptors, an ensuing decrease in gonadotrophin levels, and diminished secretion of ovarian steroids. A state of reversible hypogonadotrophic hypogonadism can be induced, with plasma oestrogen reduced to castrate levels ('medical gonadectomy'). GnRH agonist have had a profound impact on the treatment of a number of malignant and non-malignant gynecological diseases which are oestrogen-dependent (Vickery and Nestor, 1987; McLachan et al., 1986). Recently, GnRH agonists have been used increasingly in the induction of ovarian hyperstimulation in women undergoing in vitro fertilization and embryo transfer. The reversible hypogonadotrophic state induced by these analogues is followed by better follicular recruitment by exogenous gonadotrophins and suppression of premature luteinizing hormone (LH) surges which cause inappropriate early luteinization (Fleming and Coutts, 1986; Porter et al., 1984), a common cause of cancellation of the procedure (Kerin et al., 1984) and poor outcome of IVF (Bryski et al., 1988; Macnamee et al., 1988). GnRH agonist treatment allows the programming of oocyte recoveries (Zorn et al., 1987), and has been of particular benefit for women with polycystic ovarian disease (Filicori et al., 1988) and certain groups who respond poorly to other treatment regimens(Belaisch-Allart et al., 1989; Crosignani et al., 1988,1989).

Advances in Assisted Reproductive Technologies, Edited by
S. Mashiach *et al.,* Plenum Press, New York, 1990

However, greater amounts of gonadotrophins are required to induce adequate follicular development, significantly increasing treatment costs (Barnes et al., 1987). Moreover, certain protocols of slow-release GnRH agonist-gonadotrophin stimulation for IVF have been reported to depress luteal function, delay endometrial development (Smitz et al., 1988 a,b; Schmutzler et al., 1988; Van Steirteghem et al., 1988) and reduce embryo quality and subsequent implantation rates (Testart et al., 1989). Some of the reported effects on luteal function may be due to other factors, such as reduced pituitary LH release to support the corpus luteum during early implantation. However, direct effects of GnRH and its analogues have been reported on human ovarian cells, and on oocyte maturation (Dekel et al., 1983; Erickson et al., 1983) and uterine responsiveness (Pedroza et al., 1980) in the rat. Any effects of these analogues on the ovary, embryo, endometrium or developing trophoblast may have deleterious effects on the success of subsequent embryo transfer and implantation.

GnRH-like Peptides as Paracrine Regulators of Ovarian Function in the Rat

High-affinity GnRH-binding sites similar to pituitary GnRH receptors have been described in rat ovarian cells (Naor and Childs, 1985), and GnRH and its agonist analogues have direct effects on the gonads of hypophysectomized rats in vivo (Ying et al., 1981) and on gonadal cells in vitro (Hsueh and Jones, 1981). The magnitude of the GnRH effects observed depend on the end-point measured, the state of differentiation of the cells, the presence of gonadotrophins during exposure to GnRH and the time-scale over which the particular response was measured (Hsueh and Jones, 1981; Fraser et al., 1984; Ranta et al., 1984). Acute, stimulatory effects of GnRH on steroidogenesis in rat ovarian cells involve a constellation of changes including receptor clustering and internalization, calcium ion mobilization, enhanced turnover of phosphatidyl inositol and elevation of intracellular inositol phosphates, translocation and activation of protein kinase C, enhanced release of arachidonic acid, leukotrienes and prostaglandins (Bramley, 1989; Cooke and Sullivan, 1985; Dorrington et al., 1985), processes very similar to those involved in GnRH-stimulated release of gonadotrophins from the pituitary gland (Huckle and Conn, 1988; Conn et al., 1987). In contrast, inhibitory actions of GnRH on gonadal tissues are observed with longer-term GnRH treatment, and involve induction of a cyclic AMP phosphodiesterase, changes in the synthesis of gonadotrophin receptors, steroidogenic enzymes and cytoskeletal proteins, and the antagonism of gonadotrophin-dependent steroid secretion and cytodifferentiation (Hsueh and Jones, 1982; Cooke and Sullivan, 1985).

GnRH-like peptides can be isolated from rat gonadal tissues (Behrman et al., 1989), though such factors differ significantly from hypothalamic GnRH in size and/or elution from high performance liquid chromatography (HPLC), and often showed little or no cross-reaction with antibodies to mammalian GnRH (Hedger et al., 1987). Furthermore, GnRH is rapidly inactivated by rat gonadal cells. Finally, specific GnRH antagonists interfere directly with ovarian function, (Rivier and Vale, 1989; Birnbaumer et al., 1985), suggesting that GnRH-like peptides play an important paracrine role in the regulation of the rat ovary (Table 1).

Table 1. Criteria for establishment of a paracrine role for GnRH: (1) local production of GnRH in close proximity to target cells; (2) termination of GnRH action requires rapid local inactivation; (3) target tissue must possess specific GnRH receptors; (4) changes in hormone/receptor level must correlate with physiology; (5) GnRH must induce a biological response

Tissue	GnRH-like Factor ?	GnRH inactivation?	GnRH receptors?	Physiological Changes?	Response?
Ovary					
Rat	+	+	+	+	+
Pig	+	+	?	?	+
Cow	+	+	?	?	?
Sheep	+	+	?	?	?
Monkey	?	?	?	?	+
Man	+	+	+	+	±
Placenta	+	+	+	+	+

GnRH effects in the ovaries of other species

To date, fulfilment of all of the necessary criteria to establish unequivocally a paracrine role for GnRH-like peptides in the ovaries of other species has not been met (Table 1). GnRH-like peptides can be isolated from the ovaries of a number of species, including the pig and human (Behrman et al., 1989). Moreover, GnRH can affect steroid release by porcine granulosa cells in culture (Massicotte et al., 1980), though gonadal binding sites for GnRH could not be demonstrated in ovarian tissues of the sheep and pig (Brown and Reeves, 1983). GnRH binding sites were also reported to be absent from monkey (Asch et al., 1981) and human luteal tissues (Clayton and Huhtaneimi, 1982; Richardson et al., 1984; though see below). However, GnRH strongly suppresses the production of progesterone, oestradiol and cAMP by marmoset granulosa cells in culture. In the light of the altered potency of GnRH antagonists for human extrapituitary GnRH receptors (see below), it is interesting that a GnRH antagonist, rather than reversing the effects of agonist, potentiated inhibition further (Wickings, Hillier and Dixson, unpublished data).

GnRH Effects in the Human Ovary

Data on the effects of GnRH on human ovarian tissue are conflicting. Some found no effects of GnRH or its analogues on the response in culture of human granulosa cells (Casper et al., 1982; Dodson et al., 1988) or dispersed luteal cells (Tan and Biggs, 1983; Casper et al., 1984; Richardson et al., 1984). Others demonstrated inhibition (Tureck et al., 1982; Lamberts et al., 1982) or stimulation of steroid secretion in ovarian cells (Parinaud et al., 1988), whilst Ikoma et al., (1987), using slices of human corpus luteum, showed stimulation of basal secretion of progesterone and oestradiol in some corpora lutea, but inhibition of hCG-stimulated steroid secretion in other corpora lutea. (This latter study is of interest since we have shown previously that the concentrations of GnRH binding sites in human corpora lutea from different stages of the luteal phase vary markedly; Bramley et al., 1987).

Reconciling these different studies is not easy, since each differed with respect to the tissue used (slices, dispersed luteal cells, granulosa cells), the state of maturation of the cells used (long- versus short-term culture), the doses of GnRH, the agonist analogues used, the duration of incubation, and the presence of other hormones in the culture media.

Failure to detect GnRH binding sites in human luteal tissue (Clayton and Huhtaneimi, 1982; Richardson et al., 1984) was shown to be due largely to technical factors (Fraser et al., 1986). Moderate affinity binding sites for GnRH (Ka, 10^7 M^1) were described in human corpus luteum tissue which had high specificity for GnRH-like peptides and, in common with pituitary GnRH receptors, required both N- and C-termini of the molecule for binding (Bramley, 1989). However, unlike rat, sheep and human pituitary GnRH receptors (Wormald et al., 1985), these binding sites did not discriminate between GnRH and its superactive agonist analogues (Bramley, 1989). Furthermore, some GnRH antagonists with greatly enhanced potencies for pituitary receptors had reduced potencies towards human luteal GnRH binding sites (Fraser et al., 1986). These binding sites were associated largely with the luteal cell-surface membrane, though a small proportion were associated with intracellular membranes, suggestive of GnRH-receptor internalization (Bramley and Menzies, 1986). GnRH binding site concentrations varied throughout the luteal phase of the cycle and during early pregnancy, and correlated with a number of important indices of luteal function (Bramley, 1989). Little binding was detected in stromal, thecal and follicular tissue. Because of limitations on the availability of ovarian tissue, and because the affinities of the human corpus luteum and placental GnRH binding sites for a range of GnRH isoforms, agonist and antagonist analogues were very similar (unpublished observations) our recent studies on human extrapituitary receptors have used placental tissue, which is available readily and in large amounts.

GnRH Effects on the Human Placenta

A number of growth factors and steroid hormones can affect the release of human chorionic gonadotrophin from placental tissue explants and syncytiotrophoblast cells, among them GnRH. It is believed that cytotrophoblast cells produce a GnRH-like molecule (Petraglia et al., 1987) which acts on specific GnRH receptors in the syncytiotrophoblast to increase secretion of steroids, prostaglandins and chorionic gonadotrophin (but not placental lactogen) in vitro (Siler-Khodr, 1987) the response varying with gestational age. GnRH action in the placenta can be partially blocked by high levels of some but not all GnRH antagonists, and appears to involve interaction with steriospecific GnRH receptors, coupled to a GTP-binding protein (Escher et al 1988) which may modulate adenylate cyclase (Nulsen et al., 1988), though stimulation of hCG secretion by GnRH appears to be calcium-dependent (Mathialagan and Rao, 1989), and can be mimicked by cAMP or phorbol esters (Ritvos et al., 1988).

As mentioned above, human placental membranes contain moderate-affinity GnRH binding sites (Currie et al., 1981; Belisle et al., 1984; Iwashita et al., 1986) which, unlike pituitary GnRH receptors, do not discriminate between GnRH and its superactive agonists, but which show reduced potencies for a number of GnRH antagonist analogues. These binding sites are associated with human placental cell-surface membranes, and are not a reflection of the binding of degradation-resistant agonist tracers to GnRH-peptidases (Bramley, 1989).

The nature of the endogenous ligand which acts on the luteal and placental GnRH receptor is yet to be described. The GnRH gene was isolated and cloned from the human placenta (Seeburg and Adelman, 1984), hence GnRH mRNA must be expressed in this tissue. However, although the GnRH decapeptide was thought to be synthesized by human placental tissues (Tan and Rousseau, 1982), more recent studies have shown that most of this activity is non-identical to the hypothalamic decapeptide (Siler-Khodr, 1987; Mathialagan and Rao, 1986; Gautron et al., 1989). Placental GnRH (hCGnRH) may arise by differential processing of proGnRH mRNA. Indeed, placental proGnRH mRNA is known to differ from hypothalamic proGnRH mRNA in that the first intron is not spliced out in the placenta, and there is also evidence that hypothalamic and placental promoter regions may differ (Adelman et al., 1986; Seeburg et al., 1987). ProGnRH also codes for a protein, gonadotrophin-associated peptide (GAP), which stimulates the release of FSH and LH from the pituitary of the rat and human (Nikolics et al., 1985; Millar et al., 1986; Yu et al., 1988), and inhibits prolactin release in the rat, but not the sheep (Thomas et al., 1988) or human (Ishabashi et al., 1987). Furthermore, proGnRH can be processed post-translationally in different areas of the brain, giving rise to products other than GAP and GnRH (Acklands et al., 1988; Wetsel et al., 1988) and different physiological conditions can affect processing (Acklands et al., 1988). Thus, endogenous placental GnRH-like peptide may be proGnRH or a processed form of it.

Alternatively, other forms of GnRH have been isolated from hypothalami from other species which differ in aminoacid sequence from the mammalian hormone (King and Millar, 1987), and in some species more than one iso-form of GnRH may be expressed, with distinct anatomical distributions. Radiolabelled tracers of the different iso-forms of GnRH did not bind to rat or sheep pituitary membranes: however, salmon GnRH and chicken GnRH II bound to human placental membranes as well as GnRH agonist tracers (Bramley, 1989). GnRH tracer bound to a lesser degree. We have confirmed these findings (Fig. 1), and have established that the failure of the different iso-forms of GnRH to bind to rat (and sheep and rabbit) pituitary membranes, and of lamprey GnRH and chicken GnRH I tracers to bind to placental membranes, was not due to contamination of tracers by unlabelled peptide, to selection of the inactive radiolabelled peak, nor to the selective degradation of these tracers during the binding incubation (Fig. 1). However, studies of the degradation of GnRH tracers after exposure to sheep and pig luteal membranes, and to rat and rabbit placenta; indicated that these tissues, under the incubation conditions normally used to measure GnRH binding, almost completely inactivated every GnRH tracer tested (unpublished data). Thus, the inability to demonstrate GnRH receptors in pig, sheep and cow luteal tissue (Brown and Reeves, 1983) may be due, at least in part, to the presence of high levels of inactivating enzymes in these tissues. It will be of interest to look for GnRH binding sites in these tissues under conditions where peptide degradation is abolished.

In the light of these findings, we have up-dated the criteria which require to be established to demonstrate a paracrine role for GnRH in the regulation of the ovary and human placenta (Table 1). GnRH-like peptides may play a role as intragonadal regulators of ovarian function in a much wider range of species than was thought previously.

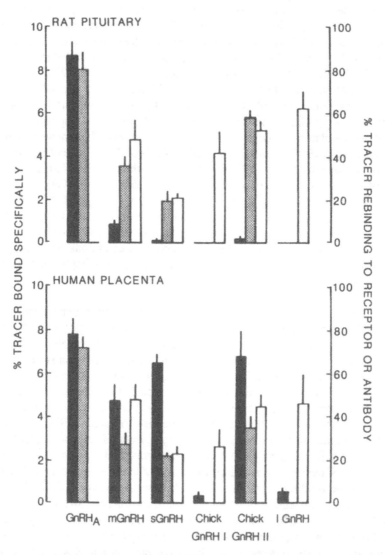

Fig. 1. Binding of radiolabelled GnRH and GnRH agonist to membranes of the rat pituitary and human placenta (solid bars), and the proportion of each tracer capable of rebinding to fresh rat pituitary and/or human placental membranes (hatched bars) or to a conformation-specific anti-GnRH antibody (open bars). Bars represent means of 2–8 separate experiments for each tracer.

ACKNOWLEDGEMENTS

I should like to thank Garry Menzies for his excellent technical assistance, members of the Department of Obstetrics and Gynecology for the provision of human tissues, Dr. J. King, MRC Regulatory Peptides Research Unit, University of Capetown, R.S.A., Dr. J. Sandow, Hoescht AG, Frankfurt, F.R.G. and Dr. R. Ellinwood, University of Illinois, U.S.A. for the provision of peptides and antiserum.

REFERENCES

Acklands, J.F., Nikolics, K., Seeburg, P.H. and Jackson, I.M. 1988, Molecular forms of gonadotropin-releasing hormone associated peptide (GAP): Changes within the rat hypothalamus and release from hypothalamic cells in vitro. *Neuroendocrinol.* 48: 376–386.

Adelman, J.P., Mason, A.J., Hayflick, J.S. and Seeburg, P.H. 1986, Isolation of the gene and hypothalamic cDNA for the common precursor of gonadotropin-releasing hormone and prolactin release-inhibiting factor in human and rat. *Proc. Natn. Acad. Sci.* U.S.A. 83:170–183.

Asch, R.H., van Sickle, M., Rettori, V., Balmaceda, J.P., Eddy, C.A., Coy D.H. and Schally, A.V. 1981, Absence of LH-RH binding sites in corpora lutea from rhesus monkeys Macaca mulatta,. *J. Clin. Endocr. Metab.* 53:215–217.

Barnes, R.B., Scommegna, A. and Schreiber, I.R. 1987, Decreased ovarian response to human menopausal gonadotropin caused by subcutaneous administration of gonadotropin releasing hormone agonist. *Fertil. Steril.* 47:512–515.

Behrman, H.R., Aten, R.F. and Ireland, J.J. 1989, Characteristics of an antigonadotropic GnRH-like protein in the ovaries of diverse mammals. *J. Reprod. Fertil. Suppl.* 37:189–194.

Belaisch-Allart, J., Testart, J. and Frydman, R. 1989, Utilization of GnRH agonist for poor responders in an IVF programme. *Hum. Reprod.* 4:33–34.

Belisle, S., Guevin, J-F., Bellabarba, D. and Lehoux, J-G. 1984, Luteinizing hormone-releasing hormone binds to enriched human placental membranes and stimulates in vitro the synthesis of bioactive human chorionic gonadotropin. *J. Clin. Endocr. Metab.* 59:119–126.

Birnbaumer, L., Shahabi, N., Rivier, J. and Vale, W. 1985, Evidence for a physiological role of gonadotropin-releasing hormone GnRH, or GnRH-like material in the ovary. *Endocrinology,* 116:1367.

Bramley, T.A. 1989, GnRH peptides and regulation of the corpus luteum. *J. Reprod. Fertil., Suppl.* 37:205–213.

Bramley, T.A. and Menzies, G.S. 1986, Subcellular fractionation of the human corpus luteum:distribution of GnRH agonist binding sites. *Molec. Cell. Endocr.* 45:27–36.

Bramley, T.A., Stirling, D., Swanston, I.A., Menzies, G.S., McNeilly, A.S. and Baird, D.T. 1987, Specific binding sites for gonadotrophin-releasing hormone, LH/chorionic gonadotrophin, low-density lipoprotein, prolactin and FSH in homogenates of human corpus luteum. II. Concentrations throughout the luteal phase of the menstrual cycle and early pregnancy. *J. Endocr.* 113:317–327.

Brown, J. L. and Reeves, J.J. 1983, Absence of specific luteinizing hormone-releasing hormone receptors in ovine and porcine ovaries. *Biol. Reprod.* 29:1179–1182.

Brzyski, R.G., Simonetti, S., Muasher, S.J., Seegar Jones, G., Droesch, K. and Rosenwaks, Z. 1988, Follicular atresia associated with concurrent initiation of gonadotropin releasing hormone agonist and follicle-stimulating hormone for oocyte recruitment. *Fertil. Steril.* 50:917–921.

Casper, R.F., Erickson, G.F., Rebar, R.W. and Yen, S.S.C. 1982, The effect of luteinizing hormone-releasing hormone and its agonist on cultured human granulosa cells. *Fertil. Steril.* 37:406–409.

Casper, R.F., Erickson, G.F. and Yen, S.S.C. 1984, Studies on the effect of gonadotropin-releasing hormone and its agonist on human luteal steroidogenesis in vitro. *Fertil. Steril.* 42:39–43.

Clayton, R.N. and Huhtaneimi, I.T. 1982, Absence of gonadotrophin-releasing hormone receptors in human gonadal tissue, *Nature*, Lond. 299:56–58.

Cooke, B.A. and Sullivan, M.H.F. 1985, The mechanisms of LHRH action in gonadal tissues. *Molec. Cell. Endocr.*, 41:115–122.

Conn, P.M., Huckle, W.R., Andrews, W.V. and McArdle, C.A. 1987, The molecular mechanism of action of gonadotropin releasing hormone GnRH, in the pituitary. *Recent Prog. Horm. Res.* 43:29–68.

Crosignani, P.G., Ragni, G., Lombroso, G.C., Scarduelli, C., de Lauretis, L., Caccano, A., Dalpra, L., Calvioni, V., Cristiani, C., Wyssling, H., Olivares, M.D. and Perotti, L. 1988, *Hum. Reprod.* 3. Suppl 2:29–33.

Crosignani, P.G., Ragni, G., Lombroso, G.C., Scarduelli, C., de Lauretis, L., Caccano, A., Dalpra, L., Calvioni, V., Cristiani, C., Wyssling, H., and Olivares, M.D. 1989, *J. Steroid Biochem.* 32:171–173.

Currie, A.J., Fraser, H.M. and Sharpe, R.M. 1981, Human placental receptors for luteinizing hormone releasing hormone. *Biochem. Biophys. Res. Commun.* 99:332–338.

Dekel, N., Sherizly, I., Tsafriri, A., and Naor, Z., 1983, A comparative study of the mechanism of action of luteinizing hormone and a gonadotropin releasing hormone analog on the ovary. *Biol. Reprod.* 28:161–166.

Dodson, W.C., Myers, T., Morton, P.C. and Conn, P.M. 1988, Leuprolide acetate:serum and follicular fluid concentrations and effects on human fertilization, embryo growth, and granulosa-lutein cell progesterone accumulation in vitro. *Fertil. Steril.* 50:612–617.

Dorrington, J.H., McKeracher, H.L., Chan, A. and Gore-Langton, R.E. 1985, Luteinizing hormone-releasing hormone independently stimulates cytodifferentiation of granulosa cells. *In:*"Hormonal control of the hypothalamo-pituitary-gonadal axis". pp 467–478. K. W. McKerns and Z. Naor, eds., Plenum Press, New York and London.

Erickson, G.F., Hofeditz, C. and Hsueh, A.J.W. 1983, GnRH stimulates meiotic maturation in preantral follicles of hypophysectomized rats. *In:*"Factors regulating ovarian functions". pp 257–261. G.S. Greenwald and P.F. Terranova, eds. Raven Press, New York.

Escher, E., Mackiewicz, Z., Lagace, G., Lehoux, J-G., Gallo-Payet, N., Bellabarba, D. and Belisle, S. 1988, Human placental LHRH receptor:Agonist and antagonist labelling produces differences in the size of the non-denatured, solubilized receptor. *J.Receptor Res.* 8:391–406.

Filicori, M., Campaniello, E., Michelacci, L., Pareschi, A.,Ferrari, P., Boletti, G. and Flamigni, C. 1988, Gonadotropin releasing hormone GnRH, analog surppression renders polycystic ovarian disease patients more susceptible to ovulation induction with pulsatile GnRH. *J. Clin. Endocrinol. Metab.*, 66:327–333.

Fleming, R. and Coutts, J.R.T. 1986, Induction of multiple follicular growth in normally menstruating women with endogenous gonadotropin suppression. *Fertil Steril.* 45:226–230.

Fraser, H.M., Sharpe, R.M. and Popkin, R.M. 1984, Direct stimulatory actions of LHRH on the ovary and testis. *In:*"LHRH and its analogs:a new class of contraceptive and therapeutic agents", pp 181–195. B.H. Vickery, J.J. Nestor and E.S.E. Hafez, eds., MTP Press, Lancaster.

Fraser, H.M., Bramley, T.A., Miller, W.R. and Sharpe, R.M. 1986, Extra pituitary actions of LHRH analogues in tissues of the human female and investigation of the existence and function of LHRH-like peptides. *In:* "Gonadotropin down-regulation in gynaecological practice". pp 29–54. R. Rolland, D.R. Chadha and W.M.P. Willemsen, eds. Alan R. Liss, Inc. New York.

Gautron, J.P., Pattou, E., Bauer, K., Rotten, D. and Kordon, C. 1989, LHRH-like immunoreactivity in the human placenta is not identical to LHRH. *Placenta*, 10:19–35.

Hedger, M.P., Robertson, D.M. and de Kretser, D.M. 1987, LHRH and "LHRH-like" factors in the male reproductive tract. *In:*"LHRH and its analogs:Contraceptive and therapeutic applications". pp. 141–160. B.H. Vickery and J.J. Nestor, eds. MTP Press, Lancaster.

Hsueh, A.J.W. and Jones, P.B.C. 1981, Extrapituitary actions of gonadotropin-releasing hormone. *Endocr. Rev.* 2:437–461.

Hsueh, A.J.W. and Jones, P.B.C. 1982, Regulation of ovarian granulosa and luteal cell functions by gonadotropin releasing hormone and its antagonist, *In:*"Intraovarian control mechanisms". pp. 223–262. C.P. Channing and S.J. Segal, eds. Plenum Press, New York.

Ikoma, H., Yamoto, M. and Nakano, R. 1987, The effect of luteinizing hormone releasing hormone LHRH, on steroidogenesis by the human luteal tissue. *Acta Obstet. Gynaecol. Scand.* 66:695–699.

Ishabashi, M., Yamaji, T., Takaku, F., Teramoto, A., Fukushima, T., Toyama, M. and Kamoi, K. 1987, Effect of GnRH-associated peptide on prolactin secretion from human lactotrope adenoma cells in culture. *Acta Endocr. Kbh,.* 116:81–84.

Iwashita M., Evans, M.I. and Catt, K.J. 1986, Characterization of a gonadotropin-releasing hormone receptor sites in term placenta and chorionic villi. *J. Clin. Endocr. Metab.* 62:127–133.

Kerin, J.F., Warnes, G.M., Quinn, P., Kirby, C., Godfrey, B. and Cox, L.W. 1984, Endocrinology of ovarian stimulation for in vitro fertilization. *Aust. N. Z. J. Obstet. Gynaecol.,*24:121–126.

King, J.A. and Millar, R.P. 1987, Phylogenetic diversity of LHRH. *In:*"LHRH and its analogs:Contraceptive and therapeutic applications, Part 2". pp. 53–73. B.H. Vickery and J.J. Nestor, eds. MTP Press, Lancaster.

Lamberts, S.W.J., Timmers, J.M., Osterom, R., Verleun, T., Rommerts, F.G. and de Jong, F.H. 1982, Testosterone secretion by cultured arrhenoblastoma cells:suppression by a luteinizing hormone-releasing hormone agonist. *J.Clin. Endocr. Metab.* 54:450–454.

Macnamee, M.C., Edwards, R.G. and Howles, C.M. 1988, The influence of stimulation regimes and luteal phase support on the outcome of IVF. *Hum. Reprod.* 3. Suppl. 2:43–52.

Massicotte, J., Veilleux, R., Lavoie, M. and Labrie, F. 1980, An LHRH agonist inhibits FSH-induced cyclic AMP accumulation and steroidogenesis in porcine granulosa cells in culture. *Biochem. Biophys. Res. Commun.* 94:1362–1366.

Mathialagan, N. and Rao, A.J. 1986, Gonadotropin releasing hormone in first trimester human placenta:isolation, partial characterization and in vivo biosynthesis. *J. Biosci.* 10:429–441.

Mathialagan, N. and Rao, A.J. 1989, A role for calcium in gonadotropin-releasing hormone GnRH, stimulated secretion of chorionic gonadotropin by first trimester human placental minces in vitro. *Placenta.* 10:61–70.

McLachan, R.I., Healy, D. and Burger, H.G. 1986, Clinical aspects of LHRH analogues in gynaecology:a review. *Br. J. Obstet. Gynaecol.* 93:431–454.

Millar, R.P., Wormald, P.J. and Milton, R.C. de L. 1986, Stimulation of gonadotropin release by a non-GnRH peptide sequence of the GnRH precursor. *Science, N.Y.* 323:68–70.

Naor, Z. and Childs, G.V. 1985, Binding and activation of gonadotropin-releasing hormone receptors in pituitary and gonadal cells. *Int. Rev. Cytol.* 103:147–187.

Nulsen, J.C., Woolkalis, M.J., Kopf, G.S. and Strauss, J.F. 1988, Adenylate cyclase in human cytotrophoblasts:characterization and its role in modulating human chorionic gonadotropin secretion. *J. Clin. Endocr. Metab.* 66:258–265.

Nikolics, K., Mason, A.J., Szonyi, E., Ramachandran, J. and Seeburg, P.H. 1985, A prolactin-inhibiting factor within the precursor for human gonadotropin-releasing hormone. *Nature, Lond.* 316:511–517.

Parinaud, J., Vieitez, G., Beaur, A., Pontonnier, G. and Bourreau, E. 1988, Effect of a luteinizing hormone-releasing hormone agonist Buserelin, on steroidogenesis of cultured human preovulatory granulosa cells. *Fertil. Steril.* 50:597–602.

Pedroza, E., Vilchez-Martinez, J.A., Coy, D.H., Arimura, A. and Schally, A.V. 1980, Reduction of LH RH pituitary and estradiol uterine binding sites by a superactive analog of luteinizing hormone-releasing hormone. *Biochem. Biophys. Res. Commun.*, 95:1056–1062.

Petraglia, F., Lim, A.T.W. and Vale, W. 1987, Adenosine 3',5'-monophosphate, prostaglandins, and epinephrine stimulate the secretion of immunoreactive gonadotropin-releasing hormone from cultured human placental cells. *J. Clin. Endocr. Metab.* 65:1020–1025.

Porter, R.N., Smith, W., Craft, I.L., Abdulwalid, N.A. and Jacobs, H.S. 1984, Induction of ovulation for in vitro fertilization using buserelin and gonadotropins. *Lancet*, 2:1284–1285.

Ranta, T., Knecht, M., Baukal, A.J., Korhonen, M. and Catt, K.J. 1984, GnRH agonist-induced inhibitory and stimulatory effects during ovarian follicular maturation. *Molec. Cell. Endocrinol.* 35:55–63.

Richardson, M.C., Hirji, M.R., Thompson, A.D. and Masson, G.M. 1984, Effect of a long-acting analogue of LH releasing hormone on human and rat corpora lutea. *J.Endocr.* 101:163–168.

Ritvos, O., Butzow, R., Jalkanen, J., Stenman, U-H., Huhtaneimi, I. and Ranta, T. 1988, Differential regulation of hCG and progesterone secretion by cholera toxin and phorbolester in human cytotrophoblasts. *Molec. Cell. Endocrinol.* 56:165–169.

Rivier, C. and Vale, W. 1989, Immunoreactive inhibin secretion by the hypophysectomized female rat:Demonstration of the modulating effect of gonadotropin-releasing hormone and estrogen through a direct site of action. *Endocrinology* 124:195–198.

Schmutzler, R.K., Reichert, C., Diedrich, K., Wildt, L., Diedrich, Ch., Al-Hassani, S., van der Ven, H. and Krebs, D. 1988, Combined GnRH-agonist/gonadotrophin stimulation for in vitro fertilization. *Hum. Reprod.* 3. Suppl.2:29–33.

Seeburg, P.H. and Adelman, J.P., 1984 Characterization of cDNA for precursor of human luteinizing hormone releasing hormone, *Nature*, Lond. 311:666–668.

Seeburg, P.H., Mason, A.J., Stewart, T.A. and Nikolics, K. 1987, The mammalian GnRH gene and its pivotal role in reproduction. *Recent Prog. Horm. Res.* 43:69–98.

Siler-Khodr, T.M. 1987, Placental LHRH-like activity, *In:*"LHRH and its analogs:Contraceptive and therapeutic applications, part 2". pp. 161–175. B.H. Vickery and J.J. Nestor, eds. MTP Press, Lancaster.

Smitz, J., Devroey, P., Camus, M., Khan, I., Staessen, C., Van Waesberghe, L., Wisanto, A. and Van Steirteghem, A.C. 1988a, Addition of Buserelin to human menopausal gonadotrophins in patients with failed stimulations for IVF or GIFT. *Hum. Reprod.* 3. Suppl. 2:35–38.

Smitz, J., Devroey, P., Camus, M., Deschacht, J., Khan, I., Staessen, C., Van Waesberghe, L., Wisanto, A. and Van Steirteghem, A.C. 1988b, The luteal phase and early pregnancy after combined GnRH-agonist/HMG treatment for superovulation in IVF or GIFT. *Hum. Reprod.* 3; Suppl. 2:585–590.

Tan, G.J.S. and Biggs, J.S.G. 1983, Absence of effect of LHRH on progesterone production by human luteal cells in vitro. *J. Reprod. Fertil.* 67:411–413.

Tan, L. and Rousseau, P. 1982, the chemical identity of the immuno-reactive LHRH-like peptide biosynthesized in the human placenta. *Biochem. Biophys. Res. Commun.* 109:1061–1071.

Testart, J., Forman, R., Belaisch-Allart, J., Volante, M., Hazout, A., Strubb, N. and Frydman, R. 1989, Embryo quality and uterine receptivity in in-vitro fertilization cycles with or without agonists of gonadotropin-releasing hormone. *Hum. Reprod.* 4:198–201.

Thomas, G.B., Cummins, J.T., Doughton, B.W., Griffin, N., Millar, R.P., Milton, R.C. de L. and Clake, I.J. 1988, Gonadotropin-releasing hormone associated peptide GAP, and putative processed GAP peptides do not release luteinizing hormone or follicle- stimulating hormone or inhibit prolactin secretion in the sheep. *Neuroendocrinol.* 48:342–350.

Tureck, R.W., Mastroianni, L., Blasco, L. and Strauss, J.F. 1982, Inhibition of human granulosa cell progesterone secretion by a gonadotropin-releasing hormone agonist. *J. Clin. Endocrin. Metab.* 54:1078–1080.

Van Steirteghem, A.C., Smitz, J., Camus, M., Van Waesberghe, L., Deschacht, J., Khan, I., Staessen, C., Wisanto, A., Bourgain, C. and Devroey, P. 1988, The luteal phase after in-vitro fertilization and related procedures. *Hum. Reprod.* 3; Suppl.2:161–164.

Vickery, B.H. and Nestor, J.J.Jr. 1987, Luteinizing hormone-releasing hormone analogs:Development and mechanism of action. *Semin. Reprod. Endocrinol.* 5:353–369.

Wormald, P.J., Eidne, K.A. and Millar, R.P. 1985, Gonadotropin-releasing hormone receptors in human pituitary:ligand structural requirements, molecular size, and cationic effects. *J. Clin. Endocr. Metab.* 61:1190–1194.

Wetsel, W.C., Culler, M.D., Johnston, C.A. and Negro-Vilar, A. 1988, Processing of the luteinizing hormone-releasing hormone precursor in the preoptic area and hypothalamus of the rat. *Molec. Endocr.* 2:22–31.

Ying, S-Y., Ling, N., Bohlen, P. and Guillemin, R. 1981, Gonadocrinins:peptides in ovarian follicular fluid stimulating the secretion of pituitary gonadotropins. *Endocrinology* 108:1206–1215.

Yu, W.H., Seeburg, P.H., Nikolics, K. and McCann, S.M. 1988, Gonadotropin-releasing hormone-associated peptide exerts a prolactin-inhibiting and weak gonadotropin-releasing activity in vivo. *Endocrinol.* 123:390–395.

Zorn, J.R., Boyer, P. and Guichard, A. 1987, Never on a Sunday:programming for IVF-ET and GIFT. *Lancet*, 1:385–386.

GnRH ANTAGONISTS:
NEW DEVELOPMENTS FOR INDICATIONS IN REPRODUCTIVE MEDICINE

Gary D. Hodgen[1], Daniel Kenigsberg[2], Claudio Chillik[3],
Douglas Danforth[1], Juan Leal[4], Keith Gordon[1], Richard Scott[1], and Robert Williams[1]

[1]*Professor and Scientific Director, The Jones Institute for Reproductive Medicine*
Director, Contraceptive Research and Development Program (CONRAD)
Department of Obstetrics and Gynecology
Eastern Virginia Medical School, Norfolk, Virginia 23510
[2]*Long Island In Vitro Fertilization, Port Jefferson, New York 11777*
[3]*Centro de Estudios Ginecologicos y Reproducion, Buenos Aires, Argentina*
[4]*The Women's Research Institute, Wichita, Kansas 67214-4716*

Here, it is our objective to summarize certain findings on the study of "first, second and third generation GnRH antagonists". A few points deserve emphasis:

1. GnRH antagonists achieve suppression of pituitary gonadotropin secretion by a mechanism(s) very different from the paradoxical stimulation/ inhibition which characterizes the response to GnRH agonists in achieving down-regulation.

2. Typically, at adequate doses, GnRH antagonists immediately "shut-off" FSH/LH secretion. After treatment on day 1, the serum concentrations of gonadotropins match the T 1/2 for clearance of LH (~25 min) and FSH (~2.5 hours).

3. Although the mechanism(s) by which GnRH antagonists inhibit FSH/LH secretion involves true receptor occupancy in competition with endogenous GnRH. As shown below, the data are not clear as to whether all GnRH antagonists work by similar entre.

4. Development of GnRH antagonists products for clinical indications has been delayed because some early analogs had excessive histamine release properties.

Advances in Assisted Reproductive Technologies, Edited by
S. Mashiach *et al.,* Plenum Press, New York, 1990

Table 1. Potential Therapeutic Applications of GnRH Agonists and Antagonists

a.	Precocious Puberty
b.	Prostatic Carcinoma
c.	Endometriosis
d.	Polycystic Ovarian Disease
e.	Uterine Fibroids
f.	Ovulation Induction, IVF or GIFT (Adjunctive)
g.	Chemotherapy/Irradiation (Prophylaxis)
h.	Contraception
i.	Menopausal Diagnosis (Osteoporosis Risk)

5. Recent evidence suggests that two or three of the "third generation" GnRH antagonists may be safe and effective through pre-clinical evaluation and initial phase I clinical trials. Surprisingly, a dose-dependent phenomenon has shown a couple of these newer analogs to possess long duration inhibition of FSH/LH secretion. One of these, Antide (also known as Nal-Lys), will be discussed in some detail.

A "FIRST GENERATION" GnRH ANTAGONIST

The hypothalamic-pituitary-ovarian axis can be "dissected" in a nonsurgical and reversible fashion by the administration of a potent gonadotropin-releasing hormone (GnRH) antagonist. We created a transient, functional lesion at the level of the pituitary gonadotrope by using a potent GnRH antagonist ([Ac pClPhe^1,pClDPhe^2,DTrp^3,DArg^6,DAla^{10}]-GnRH). In long-term castrate cynomolgus monkeys, doses of 0.05 to 2.0 mg/kg/day intramuscularly were administered for a total of 32 days. At doses up to 0.2 mg/kg/day, follicle-stimulating hormone (FSH) and luteinizing hormone (LH) in circulation were only moderately suppressed; these subjects responded to an estradiol challenge by manifesting an LH elevation or surge within 48 hours. At doses of 0.5 to 1.0 mg/kg/day, FSH and LH secretion was suppressed to or below the limits of assay detection within seven days, remaining in a severely hypogonadotropic state for the remainder of the treatment interval. Using 2 mg/kg/day, estradiol positive feedback for midcycle-like LH/FSH surges was fully inhibited. This suppression of gonadotropin secretion was rapidly reversible, in that circulating gonadotropin levels had returned to pretreatment castrate levels within 60 days after termination of GnRH antagonist treatments. These findings suggest that potent GnRH antagonists can effectively create a hypogonadotropic milieu without the initial enhancement of gonadotropin secretion that occurs during initiation of GnRH agonist therapy. "Medical hypophysectomy" through GnRH antagonist administration may permit a more direct and controlled approach to gonadal therapies such as ovulation induction (Kenigsberg et al., 1984).

In subsequent experiments, two important additional observations were made: 1) the GnRH antagonist blocked estrogen positive feedback for the LH/FSH in intact monkeys; 2) but an iv GnRH bolus led to a full LH secretory response similar to that of untreated control females (Kenigsberg and Hodgen, 1986).

A "SECOND GENERATION GnRH ANTAGONIST

Because of the excessive histamine release affect of the "First generation" analogs, next we tested the "Nal-Glu" GnRH antagonist (Schmidt, et al. 1984; Hahn, et al. 1985).

Pituitary sensitivity to a gonadotropin-releasing hormone (GnRH) challenge test before, during and after GnRH antagonist administration was compared in four ovariectomized female monkeys receiving GnRH antagonist intramuscularly (IM) at increasing doses of 0.3, 1.0, and 3.0 mg/kg/day over nine days. Three days before and three days after treatment, monkeys received vehicle alone. On experiment days 4, 7, 10, 13, and 16 100 µg of GnRH was administered intravenously (iv) and blood drawn at 0 and 30 minutes. Before treatment, tonic follicle-stimulating hormone (FSH) and luteinizing hormone (LH) levels were 248 ± 105 and 178 ± 31 ng/ml, respectively; after 0.3 mg/kg/day of GnRH antagonist, FSH and LH decreased to 30 ± 6 and 41 ± 4 ng/ml, respectively. After treatment with either 1 mg/kg/day or 3 mg/kg/day of GnRH antagonist, both gonadotropins were undetectable in serum. Monkeys with lower initial levels of gonadotropins were suppressed by 48 hours after

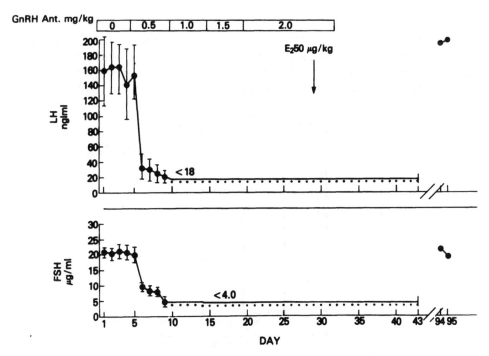

Fig. 1. GnRH antagonist suppresses FSH and LH levels in serum to below limits of assay detection. Long-term ovariectomized monkeys did not respond to an estrogen challenge test. Note the full recovery of gonadotropin secretion by 2 months after cessation of treatment. (Kenigsberg et al., 1984)

GnRH antagonist, while those with higher tonic gonadotropins were suppressed six days later (FSH: $r = 0.992$; LH: $r = 0.833$). The data show that initial physiologic status is predictive of the rapidity of the suppression response induced by a GnRH antagonist and that, after achieving pituitary suppression, responsivity to an iv GnRH challenge test may be restored before normal tonic FSH/LH secretion is regained (Chillik et al., 1987).

Using this Nal-Glu GnRH antagonist, early clinical trials are underway in men and women in several clinical research facilities (Pavlou et al., 1987; Bouchard, 1989).

A "THIRD GENERATION" GnRH ANTAGONIST

More recently, Folkers and Bowers (1987) synthesized a 5-D amino acid contained GnRH antagonist preliminarily called Nal-Lys, but which is now named Antide. We have begun studying Antide in depth because of two properties: 1) it's intrinsically very low allergic effect (same as native GnRH) and 2) its very long duration of LH/FSH inhibition when adequate doses were administered to long-term ovariectomized monkeys (Leal, et al., 1988).

These results demonstrate an unexpectedly prolonged duration of action of Nal-Lys-GnRHant at sc doses of 1.0 or 3.0 mg/kg. Since iv administration induced a similar lengthy duration of gonadotropin inhibition, a depot effect at the site of sc injection cannot explain these findings. Nal-Glu-GnRHant has no such prolonged effects. Possible explanations for these results include: 1) the Nal-Lys-GnRHant may be sequestered in peripheral

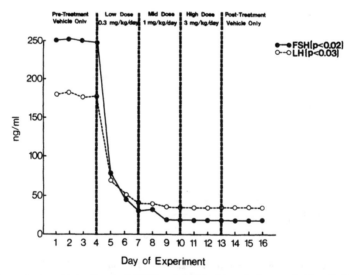

Fig. 2. Tonic serum concentrations of FSH and LH before, during and after GnRH antagonist treatment. Each point represents the mean of four animals. The decrease in serum FSH and LH throughout the study was significant ($p < 0.05$). (Chillik et al., 1987)

tissues and slowly released; 2) it may be highly resistant to enzymatic degradation and thereby recycled extensively in the circulation; and 3) this antagonist may have a noxious effect on anterior pituitary gonadotrophs or hypothalamic function. In fact, whether the mechanism of action involves hypothalamic or pituitary sites of action, or both, is unknown. However, it does not appear to act by altering the biological activity of the secreted LH.

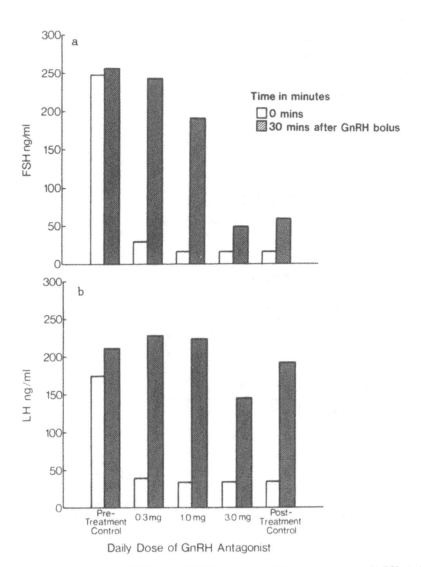

Fig. 3. Serum concentrations of FSH (a) and LH (b) before and 30 minutes after a GnRH challenge test (100 µg/IV). Each bar represents the mean gonadotropin level ± SEM, n = 4. (Chillik et al., 1987).

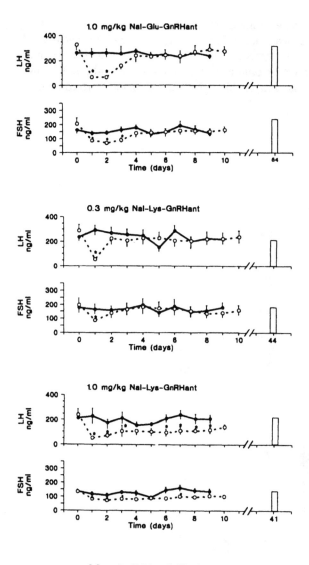

Fig. 4. Long-term effects of a sc single administration of GnRH antagonist on mean (±SE) serum LH and FSH concentrations in ovariectomized monkeys. Injection of vehicle (●), or GnRH antagonist (○) were made on day ○. The asterisks indicate significant differences from corresponding controls, (p < 0.05). For SI units, ng/ml - µg/L). (Leal et al., 1988)

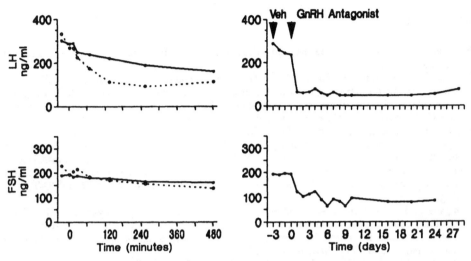

Fig. 5. Short-term (left) and long-term (right) effects of a single administration of vehicle (●) or Nal-Lys-GnRHant (3.0 mg/kg) (○) injected iv into ovariectomized monkeys. (For SI units, ng/ml -µg/L). (Leal, et al. 1988)

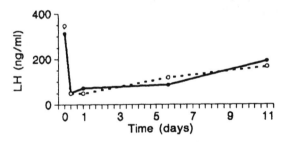

Fig. 6. Patterns of bioassayable (●) and immunoassayable (○) serum LH in a monkey given 3.0 mg/kg Nal-Lys-GnRHant, sc. (For SI units, ng/ml - µg/L). (Leal et al., 1988).

The combination of the low histamine release effect and prolonged duration of gonadotropin inhibition makes Nal-Lys-GnRHant of particular interest for clinical trials. However, the long duration of action and the mechanism(s) responsible deserve careful attention as a part of preclinical safety evaluations (Leal, et al. 1988).

REFERENCES

Bouchard, P., Maladies de la Reproduction, Hopital de Bicetre, Paris, France, Personal Communication.

Chillik, C.F., Itskovitz, J., Hahn, D., McGuire, J.L., Danforth, D.R., Hodgen, G.D., 1987, Characterizing pituitary response to a gonadotropin-releasing hormone (GnRH) antagonist in monkeys: Tonic follicle-stimulating hormone/luteinizing hormone secretion versus acute GnRH challenge tests before, during, and after treatment. *Fertil Steril* 48:480.

Folkers, K., Bowers, C., Tang, P., Feng, D., Okamoto, T., Zhang Y., Ljungqvist, A., 1987, Specificity of design to achieve antagonists of LHRH of increasing effectiveness in therapeutic activity. *In:* "LHRH and its Analogs," V. Vickery and J. Nestor (eds), MTP Press, Boston.

Hahn, D.W., McGuire, J.L., Vale, W.W., Rivier, J., 1985, Reproductive/ endocrine and anaphylactoid properties of an LHRH antagonist, ORF 18260 [Ac-D-Nal(2)[1], 4FDPhe[2], D-Trp[3]-D-Arg[6]]-GnRH. *Life* 37:505.

Kenigsberg, D., Littman, B.A., Hodgen, G.D., 1984, Medical hypophysectomy I: Dose-response using a GnRH antagonist. Fertil Steril 42:112. Kenigsberg, D., Hodgen, G.D., 1986, Ovulation-inhibition by administration of weekly gonadotropin-releasing hormone antagonist. *J Clin Endocrinol Metab* 62:734.

Leal, J.A., Williams, R.F., Danforth, D.R., Gordon, K., Hodgen, G.D., 1988, Prolonged duration of gonadotropin inhibition by a third generation GnRH antagonist. *J Clin Endocrinol Metab* 67:1325.

Pavlou, S.N., Wakefield, G.B., Island, D.P., Hoffman, P.G., LePage, M.E., Chan, R.L., Nerenberg, C.A., Kovacs, W.J., 1987, Suppressional pituitary gonadal function by a potent new luteinizing hormone- releasing hormone antagonist in normal men. *J Clin Endocrinol Metab* 64:931.

Schmidt, F., Sundaram, K., Thau, R.B., Bardin, C.W., 1984, [Ac-D-Nal(2)[1], 4FD-Phe[2],D-Trp[3],D-Arg[26]]-LHRH, a potent antagonist of LHRH, produces transient edema and behavioral changes in rats. *Contraception* 29:283.

IVF VERSUS GIFT

A. Albert Yuzpe

Professor of Obstetrics & Gynecology
University of Western Ontario, London, Ontario, Canada

The original concept of in vitro fertilization (IVF) and embryo transfer (ET) evolved from an attempt to circumvent the role of the fallopian tube in the reproductive process. The first successful IVF procedure resulting in the birth of Louise Brown in 1978 occurred in a woman with irreparable tubal disease. However, as is frequently the case with any new medical technology, the application of IVF rapidly expanded to include cases of infertility due to other causes, including endometriosis, male factor, idiopathic or unexplained infertility, or a combination of these.

Asch et al.,[1] considering the obvious importance of the fallopian tube in providing the "natural environment" where conception occurs, was able to achieve successful pregnancies through the technique of translaparoscopic gamete intrafallopian transfer (GIFT). This second monumental achievement in the growing list of new reproductive or assisted reproductive technologies, was based upon the foundations laid by IVF, i.e. controlled ovarian hyperstimulation and monitoring, as well as semen preparation.

For a short time the two procedures both involving laparoscopic-guided oocyte retrieval differed only in the fact that the GIFT technique required far less extensive oocyte manipulation, no oocyte laboratory, and allowed for return of the gametes to the fallopian tubes under the same anesthetic. However, with the addition of the vaginal ultrasound technology to oocyte retrieval, IVF suddenly became simplified and evolved into a procedure requiring no general anesthesia and only minimal analgesia. In addition, it was removed from a formal operating room setting. The elimination of the potential risk factors associated with operative laparoscopy under general anesthesia and the significant reduction in the basic costs associated with the technique suddenly made IVF more attractive than GIFT.

Advances in Assisted Reproductive Technologies, Edited by
S. Mashiach *et al.,* Plenum Press, New York, 1990

As transvaginal ultrasound-guided oocyte retrieval has gained in popularity however, it too has been shown to be associated with some degree of morbidity. Howe et al.[2] and Yuzpe et al.[3] each reported three cases of pelvic abscess complicating this technique.

The next step in the evolution of this technology has been to combine the best features of IVF/ET and GIFT. One of the advantages of IVF over GIFT has always been that fertilization and cleavage could be confirmed prior to embryo transfer and thus, the rate limiting factor in the success of IVF was felt to be in the "embryo replacement and or implantation process". At the same time, the development of PROST (pronuclear stage tubal transfer), ZIFT (zygote intrafallopian transfer) and TET (tubal embryo transfer) has combined the apparent benefits of both worlds, i.e. confirmation of fertilization in vitro, and replacement of the embryo or zygote into the more natural milieu of the fallopian tube. Oocytes are retrieved as for IVF using the ultrasound technique, and embryo or zygote replacement occurs generally via the laparoscopic route under general anesthesia.

A Comparison of IVF and GIFT

A comparison of IVF/ET and GIFT is a difficult, if not an impossible task. Each has its advantages and disadvantages as listed in Table 1. In addition, one must critically ask the question "Can or should the two procedures be compared?"

To answer this question one must consider patient populations and selection for each procedure. In order for a woman to become a candidate for GIFT or one of its modifications she must have at least one apparently normal, patent fallopian tube. The majority of IVF/ET subjects, on the other hand, suffer from irreparable tubal disease according to most published series and thus cannot truly be compared to a procedure, which for them would not be applicable, i.e. GIFT. Such comparison of outcomes would be analogous to a comparison of pregnancy rates between a group of women requiring fertility promoting surgery and one requiring ovulation induction.

A second area, where the two become difficult to compare, is in the male infertility (oligospermia) group. In many cases, bias is introduced during allocation of subjects to either one or the other treatment group. In such instances those males with poor semen quality are often assigned, along with their spouse, to the IVF/ET group in order to confirm the fact that this couple is in fact capable of achieving fertilization. Thus, even comparison of GIFT and IVF within the same diagnostic group is difficult without introducing bias.

In the case of the idiopathic infertility group, do either or any procedures really have a better outcome than no therapy, as advanced by Collins?[4] Curole et al.[5] reported a 44% cancellation rate among 355 GIFT cycles.When cycles were cancelled due to inadequate stimulation, or where intercourse did not follow discontinuation of treatment, no pregnancies ensued during that cycle. However, with natural intercourse there were three pregnancies (10%) and 12 pregnancies (24%) occurred when IUI was performed. Bellinge et al.[6] reported a sig-

Table 1. GIFT VS IVF - Advantages and Disadvantages

IVF
Advantages
    1. Minimal analgesic/local anesthesia
    2. Transvaginal ultrasound-guided oocyte retrieval
    3. Fertilization and cleavage can be confirmed
    4. Assessment of preembryo is possible
    5. Requires intact uterus/cervix only
    6. Permits cryopreservation of excess embryos

Disadvantages
    1. Expensive and elaborate technology
    2. Extracorporeal gamete manipulation and (?) implications
    3. Blind embryo replacement
    4. Outcome (?)

GIFT
Advantages
    1. Gametes are placed in their natural environment
    2. Fertilization occurs in normal site
    3. Avoids expensive and elaborate laboratory technology (see #5)
    4. Improved outcome compared to IVF (?)
    5. Permits in vitro fertilization and embryo cryopreservation

Disadvantages
    1. Operative technique requiring general anesthesia
    2. Some degree of surgical skill required
    3. Fertilization is not confirmed
    4. Requires normal fallopian tube(s)

---

nificant increase in the implantation rate (53%) in IVF cycles where high vaginal insemination was performed at the time of oocyte insemination compared to controls (IVF only - 23%). These findings were not substantiated, however, by Fishel et al.[7]

One might argue that those couples identified by Collins as conceiving spontaneously over time are those who never get to GIFT or IVF because they either conceive after prior treatment with timed intrauterine insemination combined with controlled ovarian hyperstimulation, or while on long waiting lists for IVF or GIFT. If such were true, then these new technologies might well be promoting pregnancies among those couples who would otherwise never conceive, perhaps due to some as yet unidentifiable defect in their reproductive process.

It is important on the basis of the aforementioned to critically evaluate the outcome of each procedure and to compare these results with subjects having similar diagnoses and who were treated in a similar manner. In the case of oligospermia, only randomized prospective studies between IVF and GIFT will provide the evidence necessary to formulate a conclusion regarding the preferred treatment.

What is GIFT really doing?

The suggested role of sperm capacitation and sperm mucus interaction in cases of normal fertility[8] has led some investigators to recommend or at least evaluate the possibility of improving pregnancy rates in GIFT cycles by promoting intercourse at or around the time of hCG administration. Marconi et al.[9] allocated alternate GIFT treatment cycles to either intercourse or abstinence groups where the latter avoided intercourse 48 hours before and after the GIFT procedure. In the GIFT/intercourse group, 13 of 16 patients conceived (81%) whereas five of 16 became pregnant in the abstinence group (31%). This significant difference in outcome requires further evaluation with larger numbers before any conclusion can be drawn.

Yovich et al.[10] compared results with IVF/ET, PROST, TET and GIFT. With the GIFT subjects, oocytes of the highest grading were selected for transfer. In the PROST group, selected pronuclear stage oocytes derived from higher grade oocytes were transferred. The pregnancy rate and also the number of pregnancy sacs per embryo transfer for IVF were significantly lower than for those other forms of assisted reproduction employing tubal transfer (12.5% per transfer for IVF versus 35.9% and 37.0% for GIFT and PROST respectively). Yovich postulates three reasons for improved pregnancy rates with the three tubal transfer techniques compared to IVF/ET: 1) the tubal milieu may be more favourable for preimplantation embryo development compared to the uterine cavity; 2) some unknown tubal factor may exist which improves embryo quality; 3) finally the uterine environment at 48 hours after oocyte retrieval may be less receptive or show hostility to the embryo when compared to the fallopian tube. This study concluded that PROST, TET and GIFT are preferable to IVF/ET in cases of nontubal infertility.

Khan et al.[11] evaluated the effect of total motile spermatozoa on pregnancy and abortion rates, since the number of sperm generally placed into the oviduct at the time of GIFT procedures is much greater than the expected number reaching the tube under normal conditions of intercourse. In addition the potential for polyspermy would then possibly be increased, leading eventually to abortion under such conditions. The study concluded that there was no significant difference in either the pregnancy rate per cycle or in the abortion rate when the number of sperm used per procedure varied from 100,000 to 2,500. A significantly greater pregnancy rate was noted in the idiopathic infertility group when compared to the other diagnostic categories, which included endometriosis, immunologic causes, andrologic causes, and a multiple factor group. This increase was noted at all sperm concentrations.

Result of IVF/ET and GIFT

A summary of the National IVF Registry results for 1987 were recently published[12] and are noted in Table 2. This information is based upon the reporting of 96 member clinics of SART (The Society for Assisted Reproductive Technologies). There is at least an equal number of nonreporting clinics in North America alone, and thus, the statistics may not re-

Table 2. National IVF/ET Registry 1987

	IVF/ET	GIFT	IVF+GIFT
Retrievals	8725	1968	199
Transfers	7561	—	—
Clinical Pregnancies (%)	1367 (16)* (18)**	492 (25)	55 (20)
Abortions(%)	344 (25)	116 (24)	—
Ectopics(%)	103 (7)	30 (6)	—
Deliveries	991	362	40
Babies	1260	489	49
Multiples(%)	235 (19.3)	103 (20.0)	9 (16.4)

* pregnancy rate per retrieval
** Pregnancy rate per transfer

flect the total picture. One would assume, however, that the largest clinics are reporting their results, and therefore that the majority of cases are still being included. when one considers all cases of infertility treated with either GIFT or IVF/ET after eliminating the tubal disease cases, the pregnancy rate with GIFT is higher in all diagnostic categories, whereas the abortion and ectopic pregnancy rates remain virtually the same.

SUMMARY

IVF/ET, GIFT, and its modifications PROST, ZIFT and TET, are all viable alternatives to assisted reproduction. Each has a specific role to play and each should be evaluated according to sound statistical and epidemiologic studies if reasonable comparison of methods is to be made.

It is amazing to believe that the evolution of these new technologies has progressed so rapidly from the "embryonic stage" in just ten short years. Embryo cryopreservation, zona drilling, oocyte and embryo donation, and other new technologies will certainly produce further rapid change in the next decade.

REFERENCES

1. R. H. Asch, L. R. Ellsworth, and J. P. Balmaceda, Pregnancy following translaparoscopic gamete intrafallopian transfer (Gift), *Lancet* 2:1034 (1984).
2. R. S. Howe, C. Wheeler, L. Mastroianni Jr., L. Blasco, and R. Tureck, Pelvic infection after transvaginal ultrasound-guided ovum retrieval, *Fertil. Steril.*, 49:726 (1988).
3. A. A. Yuzpe, S. E. Brown, R. F. Casper, J. A. Nisker, G. Graves, and L. Shatford, Transvaginal ultrasound guided oocyte retrieval for IVF. Accepted for publication February 9, (1989).
4. J. A. Collins, W. Wrixon, L. B. Janes, and E. H. Wilson, Treatment-independent pregnancy among infertile couples, *New. Eng. J. Med.*, 309:1201 (1983).

5. D. N. Curole, R. P. Dickey, S. N. Taylor, P. H. Rye, and T. T. Olar, Pregnancies in cancelled gamete intrafallopian transfer cycles, *Fertil. Steril.*, 51:363 (1989).
6. B. S. Bellinge, C. M. Copeland, T. D. Thomas, R. E. Mazzucchelli, G. O'Neill, and M. J. Cohen, The influence of patient insemination on the implantation rate in an in vitro fertilization and embryo transfer program. *Fertil. Steril.*, 46:252 (1986).
7. S. Fishel, J. Webster, P. Jackson, and B. Bahman, Evaluation of high vaginal insemination at oocyte recovery in patients undergoing in vitro fertilization, *Fertil. Steril.*, 51:135 (1989).
8. H. Lambert, J. W. Overstreet, P. Morales, F. W. Hanson, and R. Yanagimachi, Sperm capacitation in the human female reproductive tract, *Fertil. Steril.*, 43:325 (1985).
9. G. Marconi, L. Auge, R. Oses, R. Quintana, F. Raffo, and E. E. Young, Does sexual intercourse improve pregnancy rates in gamete intrafallopian transfer, *Fertil. Steril.*, 51:357 (1989).
10. J. L. Yovich, J. M. Yovich, and W. R. Edirisinghe, The relative chance of pregnancy following tubal or uterine transfer procedures. *Fertil. Steril.*, 49:858 (1988).
11. I. Khan, M. Camus, C. Staessen, A. Wisanto, P. Devroey and A. Van Steirteghem, Success rate in gamete intrafallopian transfer using low and high concentrations of washed spermatozoa, *Fertil. Steril.*, 50:922 (1988).
12. In vitro fertilization/embryo transfer in the United States: 1987 results from the National IVF-ET Registry. *Fertil. Steril.*, 51:13 (1989).

PATIENT SELECTION FOR ASSISTED PROCREATION

PierGiorgio Crosignani

III Department of Obstetrics and Gynecology
University of Milano
Via M. Melloni 52, 20129, Milano, Italy

Infertility is still difficult to define -

- It is a time-related phenomenon
- The indices used in diagnostic work differ in different places
- there is no agreement on the cut-off points for key indices
- the significance of some putative causes of infertility is not certain
- more than one case of infertility can coexist in a couple
- there is good evidence that all causes of infertility are not yet known.

With normal sexual practice and without contraception, pregnancy can be achieved within 12 months by 75% of couples.[1] Couples who do not conceive within this period should be regarded as possibly infertile.Only a few of these couples will turn out to be permanently sterile. For most other couples the causes of reduced fertility only prolong the time to achieving pregnancy, lowering the natural pregnancy rate per menstrual cycle.

A. Diagnosis of Infertility

In addition to being a chance-related event, infertility is a difficult issue because of the relative lack of concreteness of the indices used for its diagnosis. In fact, pregnancy is the only unquestionable proof of ovulation, tubal potency and sperm fertility. Other indices can give false indications. There is no general agreement about the diagnostic tests to be used to explain the infertility.

Advances in Assisted Reproductive Technologies, Edited by
S. Mashiach *et al.,* Plenum Press, New York, 1990

For instance, many studies do not include tests for the presence of sperm directed anti-bodies in the work up[2,3,4] while others do not use the postcoital test.[5,6]

The different tests used can give discordant results. A WHO multicentric trial recently reported 9% false positive and 45% false negative results obtained with hysterosalpingography, as compared with the results of pelviscopy for the same patients.[7] Another complication is that widely used key indices (plasma progesterone, semen analysis) are used with different cut offs. The meaning of abnormal findings such as varicocele, sperm-directed antibodies in the man or the woman and cervical mucus abnormalities is still questionable, and at best poorly defined. Last but not least, we do not know as yet all the causes of infertility.

All of this easily explains the very different rates of unexplained infertility reported by various groups. While some studies report a rate of unexplained infertility of less than 10% of the infertile couples studied,[8,9] others report values between 10% and 20%[10,11] and even higher rates of unexplained infertility have been reported.[12,13,14]

Since it is not the woman but the couple that is infertile, we quite often find causes for it in both,[15,16] as well as more than one cause in a single partner.

B. Patient Selection for Assisted Procreation

Assisted procreation has drastically reduced the area of hopeless sterility. The selection of the patients for the various strategies of assistance is primarily linked to the diagnostic definition.

1. *"No Choice" Couples.* When there is a diagnosis of absolute and permanent sterility, it is quite easy to indicate the method (if any) that should be able to overcome the cause of infertility (Table 1). These "no choice" couples should also have it explained in counselling that the rate of failure is high even after several attempts.

2. *Couples with Residual Chances of Spontaneous Pregnancy.* Patient selection is especially difficult when there is male subfertility or unexplained infertility, since no specific treatment can be suggested. It is therefore wise to carefully consider the following points:
    a) prior to indicating treatment, it is often mandatory to recheck old or ambiguous diagnostic results.
    b) All these couples have residual chances of spontaneous pregnancy (Table 2) and sometimes the pregnancy rate of "no treatment" for two to three years equals that achieved with several assisted attempts. Duration of infertility and an accurate diagnostic work up will indicate, with good approximation, the residual fertility and the expected time to pregnancy.
    c) The age of the woman is a limiting factor. At 23 she can wait two to three

Table 1. Assisted Procreation: Treatments for Couples Without any Chance of Natural Fertility

Cause of infertility	Methods of treatment
Absence of fallopian tubes	IVF
Untreatable azoospermia	donor insemination
Ovarian failure	ovum donation
Absence of the uterus	surrogacy

Table 2. Male Subfertility, Unexplained Infertility: % Chance of Conception in the Next One Year[2, 17]

Cause of infertility	Months of trying	
	< 36	> 36
Low sperm density (1×10^6)	25	18
Unexplained	40	13

Table 3. Methods Presented as Optional Choices for the Treatment of Male Subfertility or Unexplained Infertility

Method	Goals achieved with the method

Inseminations
  - intrauterine (IUI)
  - intraperitoneal (DIPI)
	- better sperm quality
	- ↑ n. of fertilizable oocytes (if associated with superovulation)

Gamete intrafallopian transfer (GIFT) Gamete intraperitoneal transfer (GIPT)	- ↑ n. of oocytes - better sperm quality - check the presence of both gametes - avoid their migration

In vitro fertilization and embryo transfer
  - in utero (IVF)
  - in tube (PROST - ZIFT)
	- ↑ n. of oocytes
	- better sperm quality
	- check the occurrence of fertilization
	- check the quality of the embryo

years for spontaneous pregnancy, at 38 it is almost impossible to suggest this strategy.

d) Several methods of assisted procreation are used to treat male subfertility and unexplained infertility. These methods try to increase the chance of fertilization "aspecifically" by improving the chances (Table 3).

Table 4. Assisted Procreation: Key Points in Decision Making

- Treat or wait?
- Results
- Invasiveness
- Quality of Pregnancy
- Availability of the Methods
- Costs

Table 5. Pregnancy after DIPI in Severe Male Infertility[18]

- Primary infertility for 6 years
  - ♀ 35 years
       regular menses, normal laparoscopy
  - ♂ 38 years
       from 1985, sperm analyses with severe asthenoteratospermia
       1986 - Houston - left varicocelectomy + clomiphene therapy
       1986 - Houston - 3 borderline-negative hamster test
       1987 - Houston - intrauterine insemination with washed sperm
       1987 - Texas   - IVF (5 oocytes, 1 embryo transfer)
       1988 - Norfolk - semen analysis ⟶ severe teratospermia
       1988 - Norfolk - IVF (3 oocytes, no embryo transfer)
       1988 - Milano - semen analysis ⟶ severe asthenoteratospermia
       1988 - Milano - DIPI combined with superovulation 6.3×10^6 spermatozoa inseminated
  - March 20, 1989 - trigeminal pregnancy (17th week)

All the strategies give results but controlled studies to compare them are still lacking. Therefore, when we suggest one of these methods, we have to take into account the pregnancy rate and invasiveness which is a critical parameter for its repetition. The quality of the pregnancies achieved is another critical issue, due to the expected high incidence of abortion and multiple pregnancies. Half of the pregnancies originated by IVF in women of about 40 will end in abortion.

With IVF or GIFT we can limit the number of embryos, and thus the potential number of twins, but this is not possible for the methods of inseminations associated with superovulation.

The diagnostic results that can be obtained with the treatment should also be especially considered in the severe cases of male subfertility. In vitro fertilization, for instance, is the only way to prove the ability of the sperm to fertilize the oocytes.

Last but not least important, is the "total" cost of the procedure, including the emotional cost of the failures. Whatever the method of assisted procreation we use, there are more failures than successes.

C. Conclusions

Table 4 summarizes the key points used for selecting patients, while Table 5 describes a case which shows how difficult it is to counsel a great proportion of the infertile couples.

REFERENCES

1. A. L. Southman, What to do with the 'normal' infertile couple, *Fertil. Steril.*, 11:543 (1960).
2. J. A. Collins, Y. So, E. H. Wilson, W. Wrixon and R. F. Casper, Clinical factors affecting pregnancy rates among infertile couples, *Can. Med. Assoc. J.*, 130:269 (1984).
3. E. R. Barnea, T. R. Holford and D. R. A. McInnes, Long-term prognosis of infertile couples with normal basic investigations: a life-table analysis, *Obstet. Gynecol.*, 66:24 (1985).
4. M. J. Haxton and W. P. Black, The aetiology of infertility in 1162 investigated couples, *Clin. Exp. Obst. Gyn.*, XIV:75 (1987).
5. P. R. Koninckx, M. Muyldermans and I. A. Brosens, Unexplained infertility: "Leuven" considerations, *Europ. J. Obstet. Gynecol. Reprod. Biol.*, 18:403 (1984).
6. R. J. Aitken, F. S. M. Best, P. Warner and A. Templeton, A prospective study of the relationship between semen quality and fertility in cases of unexplained infertility, *J. Androl.*, 5:297 (1984).
7. W.H.O. Task Force on the Diagnosis and Treatment of Infertility, Special Programme of Research, Development and Research Training in Human Reproduction, Comparative trial of tubal insufflation, hysterosalpingography, and laparoscopy with dye hydrotubation for assessment of tubal patency, *Fertil. Steril.*, 46:1101 (1986).
8. R. F. Harrison, Pregnancy successes in the infertile couple, Int. J. Fertil., 25:81 (1980).
9. A. K. Thomas and M. S. Forrest, A review of 291 infertile couples over eight years, *Fertil. Steril.*, 34:106 (1980).
10. B. S. Verkauf, The incidence and outcome of single-factor, multifactorial, and unexplained infertility, *Am. J. Obstet. Gynecol.*, 147:175 (1983).
11. J. A. Collins, C. A. Rand, E. H. Wilson and W. Wrixon, The better prognosis in secondary infertility is associated with a higher proportion of ovulation disorders, *Fertil. Steril.*, 45:611 (1986).
12. C. P. West, A. A. Templeton, M. M. Lees, The diagnostic classification and prognosis of 400 infertile couples, *Infertility*, 5:127 (1982).
13. B. E. Kliger, Evaluation, therapy and outcome in 493 infertile couples, *Fertil. Steril.*, 41:40 (1984).
14. M. G. R. Hull, C. M. A. Glazener, N. J. Kelly, D. I. Conway, P. A. Foster, R. A. Hinton, C. Coulson, P. A. Lambert, E. M. Watt and K. M. Desai, Population study of causes, treatment and outcome of infertility, *Br. Med. J.*, 291:1693 (1985).
15. J. Newton, S. Craig and D. Joyce, The changing pattern of a comprehensive infertility clinic, *J. Biosocial Science*, 6:477 (1974).
16. S. J. Behrman and R. W. Kistner, A rational approach to the evaluation of infertility, *in:* "Progress in Infertility", S. J. Behrman and R. W. Kistner, eds., Little Brown, Boston (1975).
17. T. B. Hargreave and R. A. Elton, Fecundability rates from an infertile male population, *Br. J. Urol.*, 31:71 (1981).
18. G. Ragni, G. C. Lombroso, L. De Lauretis, M. D. Olivares, C. Cristiani and P. G. Crosignani, Pregnancy after direct intraperitoneal insemination in severe male infertility resistant to other methods of assisted fertilization, *Human Reproduction* (in press).

ULTRASOUND-GUIDED CATHETERISATION OF THE FALLOPIAN TUBE FOR THE NON-OPERATIVE TRANSFER OF GAMETES AND EMBRYOS

John C. Anderson and Robert P. S. Jansen

Sydney IVF, 187 Macquarie Street, Sydney 2000, Australia
and
Department of Fertility, Royal Prince Alfred Hospital
Camperdown, Sydney 2050, Australia

INTRODUCTION

The normal site of fertilisation and early embryonic development in women is the ampullary - isthmic junction of the fallopian tube, just medial to its midpoint.[1] The fertilised ovum remains there for about 80 hours,[1] while it undergoes cleavage to reach the 16-cell stage, before compacting to a morula.After a relatively quick transit of the isthmus, the ovum lies free in the endometrial cavity as it forms a blastocele. As a blastocyst, it then hatches through the zona pellucida to implant itself in the endometrium.[3] In conventional IVF the embryo reaches the endometrial cavity after about 50 to 70 hours' culture, before compaction, when it has two to eight cells. There may be disadvantages in such early transfer to the uterus, but experience has not demonstrated any clinical benefit from longer culture in vitro. Experiments in mice and sheep have shown that tubal epithelial factors are important both in producing the best early embryonic growth rates in culture and, later, in enabling transferred embryos to survive beyond the stage of implantation.[4,5]

In principle, when the fallopian tubes are normal, it is sensible to make use of them whenever conception is to be assisted clinically, to provide the optimum environment for embryonic development. In practice, this use of the fallopian tubes has been possible only when laparoscopy under general anesthesia has been used, so that gametes[6] or early embryos[7,8] can be transferred to the outer, fimbriated ends of the tubes. IVF now rests on

Advances in Assisted Reproductive Technologies, Edited by
S. Mashiach *et al.*, Plenum Press, New York, 1990

transvaginal, ultrasound guided follicle aspiration for obtaining oocytes, often with local anesthesia alone. A non-operative method of transferring gametes or embryos would be attractive, as patients may benefit from an increased chance of conception without the need for laparoscopy.

CATHETERISATION OF THE FALLOPIAN TUBES FROM THE VAGINA

Anatomy

The extrauterine part of the tube has an average length of 11 cm, the medial 2–3 cm of which forms the tubal isthmus.[1] The medial isthmus is the narrowest part of the tube, with an average diameter of 0.4 mm (range 0.1–1.0 mm), mucosa that occupies 3–6 primary folds, and a firm thick muscular wall.[1] The interstitial or intramural portion of the tube links the isthmus to the funnel-shaped[9] angle of the endometrial cavity. The interstitial tube takes a convoluted course in most specimens dissected after fixation,[10, 11] but when intraluminal pressure or the flow of fluid is directed laterally from the uterine angle or funnel there are only minor deviations in its course.[12] The tubal isthmus at the time of ovulation produces mucus glycoproteins,[13, 14] which may be expected to lubricate and perhaps protect the mucosa. Lateral to the isthmus, the tubal ampulla is wide and thin-walled. To catheterise the tube from the vagina, a sufficiently fine catheter must be delivered to the uterotubal junction after negotiating an anteroposterior curve in the sagittal plane through the endocervix and a lateral curve in the coronal plane between the internal cervical os and the internal tubal ostium. As a basis for designing such a catheter we accepted that plastic catheters of approximately 0.6 mm have been used successfully to splint the isthmus during operative anastomosis of the fallopian tube.[15] Several procedures have been devised to image and to reach the tubes from the vagina with catheters, including radiography with opaque contrast medium,[16] hysteroscopy[17] and ultrasound.[18] Of these, ultrasound is the least likely to disturb the function of the mucosa in its support for the transport and development of the ovum.

TUBAL CATHETERISATION SET

Since the first system to catheterise the tube was described in detail[18] there have been changes to both the set and the technique.

Because damage to the tubal mucosa by the catheter was a possibility a platinum guide with a flexible tip was added. This is passed inside the catheter and ahead of it by one or two centimetres when the catheter reaches the uterotubal junction and must be withdrawn from the catheter before gametes or embryos are loaded. The design of the proximal part of the inner catheter was changed to accommodate a connecting embryo catheter, which is used to transfer embryos or gametes to the inner catheter once it is properly in place in the fallopian tube.

This transfer set (William Cook, Australia) comprises:

1. An opaque Teflon uterine cannula with an inherent curve for negotiating the endocervical canal and uterine cavity. The cannula has a single movable silicon O-ring which marks the distance the cannula has passed into the uterus (it will also make it obvious if later the catheter meets resistance and displaces the uterine cannula back towards the operator).

2. A malleable metal obturator, which fits inside the cannula to help it negotiate the endocervical canal.

3. A clear Teflon catheter drawn out to a 5 cm 2-French tip to enter the fallopian tube. The catheter fits inside the cannula after the obturator has been removed. There is a silicon O-ring marker to indicate how far it has passed into the tube. The catheter's proximal end is flared to accept the embryo catheter (No. 5 below).

4. A platinum-tipped wire guide, which seeks the lumen of the fallopian tube in case the course of the tubal lumen is tortuous. This guide is too flexible to be of use in seeking out the point at which the tubal ostium enters the lateral uterine angle: it is advanced beyond the end of the catheter only when the catheter has entered the proximal tube by 1cm. The wire guide has two movable "O" rings: one marks the position at which the tip is flush with the end of the catheter; the other marks how far into the tube it has passed.

5. A short, clear Teflon embryo catheter, which is used to transfer gametes or embryos into the fallopian catheter by fitting into the fallopian catheter's flared proximal end. The embryo catheter has a Luer-like fitting to accept a Hamilton syringe for precise displacement of its contents. Before the catheter system is used it is essential to make sure that the connection between the embryo catheter and the fallopian catheter is secure.

Catheterisation of the Tube with the Second Set

The set is used in association with a vaginal intracavity ultrasound probe. The following steps allow catheterisation of the tubes in most cases by operators skilled in transvaginal ultrasound imaging of the pelvic organs.

1. Establish the position and direction of flexion of the uterus.

2. Decide on which tube is to be catheterised. Take the cannula and hold it so that it points to the appropriate side. Take the obturator and hold it so that its curve lies forwards for an anteflexed uterus, backwards for a retroflexed uterus.

3. Load the obturator into the cannula so that the cannula's right or left-directed lateral

curve is temporarily overcome by the obturator's antero-posterior curve.

4. Pass the cannula and obturator through the cervical canal into the uterine cavity. Withdraw the obturator while twisting the cannula slightly in the direction of the right or left uterine angle, depending on the side to be catheterised (withdrawing the obturator allows the cannula to resume its laterally-directed curve).

5. Advance the cannula and wedge it into the lateral uterine angle against the internal ostium of the tube. Confirm that the cannula is in correct position by scanning the uterine angle in an oblique parasagittal plane.If the cannula is correctly placed its tip will be seen no more than 0.5 cm from the serosal surface of the uterus. Failure to reach this position will make passing the catheter difficult or impossible.

6. Load the wire guide into the catheter so that its tip is flush with the distal end. Set the leading silicon O-ring to mark the correct position of the guide. The second silicon O-ring is set 4 cm back to limit the wire guide's later advance into the tube.

7. Pass the combined catheter and wire guide through the cannula until resistance is met. Pass the catheter firmly for a distance of 0.5 cm into the first part of tube. Steady but generally slight resistance is felt at this stage, and the patient feels mild but steady discomfort. (A sharp or stabbing pain felt by the patient indicates that the catheter is twisting against resistance inside the uterine cavity.)

8. Advance the wire guide while holding the cannula and catheter still, until the two O-rings touch: the wire guide will then have been advanced 4 cm into the tube.

9. Advance the catheter while holding the cannula and wire guide still. Image the wire guide in the adnexa in an oblique parasagittal plane through the uterine angle (the most difficult part of the procedure, but necessary to confirm the correct position of the catheter).

10. Withdraw the wire guide. Injection of material opaque to ultrasound — namely moving air bubbles in a fluid medium — renders the catheter visible, within the real-time image of the myometrium, extending into the echoes of the adnexa and bowel immediately outside the myometrium.

This system means that air from the inner catheter (after the wire guide is withdrawn) must be expelled through the tube by medium before the gametes or embryos are transferred.

Use of the Set to Transfer Gametes and Embryos to the Tube

The following is a description of the use of the second catheter set, as manufactured by William Cook (Australia). It is not as simple to use as the first set but, because embryos or eggs are not preloaded, it may be more suitable for the less experienced operator. Before commencing, establish the internal volume of the particular catheter (it approximates 100-110 µl). The catheter is then passed into the tube as described above.

1. Connect the embryo catheter to a Hamilton syringe partly filled with culture medium. Flush the fallopian catheter in situ with this medium, to expel distally down the fallopian tube the air that was left in the catheter by removal of the wire guide.

2. Disconnect the embryo catheter + syringe from the fallopian catheter and draw up a volume of medium equal to the internal volume of the fallopian catheter, plus an additional 10 µl.

3. Draw up a tiny volume of air; draw up the gametes or embryos; draw up another small volume of air.

4. Reconnect the embryo catheter + syringe to the in situ fallopian catheter (a second operator makes this safer: it is crucial that there is no leakage at the connection).

5. Expel the contents of the syringe while scanning the adnexa. The gametes or embryos will be followed by a volume of medium equal to the dead space of the fallopian catheter plus an additional 10 µl for certain expulsion of catheter contents into the tubal lumen. Experienced operators may be able to image the air bubbles on each side of the gametes or embryos entering the adnexa.

6. Withdraw the catheter and check its contents microscopically. There ought not to be blood cells, gametes or embryos.

FALLOPIAN SPERM TRANSFER

By increasing the number of motile sperm in the immediate vicinity of the oocyte, direct transfer of sperm to the tube offers the prospect of increasing the chance of conception in a way comparable to IVF and GIFT while (a) avoiding the need for ovarian stimulation, (b) maintaining the correct follicular environment for the oocyte up to the time of ovulation, and (c) maintaining the tube's province of providing the correct milieu for fertilisation and early embryonic development.

Fallopian Insemination with Donor Sperm

In the first series of ultrasound-guided transfers using donor sperm, a five-fold improvement was seen in the chance of conception compared with intracervical insemination.[19]

Satisfactory catheterisations were accomplished in 46 of 50 cycles. Six of these cycles

Table 1. Fallopian Insemination—Donor

Cycles on which pregnancies occurred

AID	FID
27	1
40	1
30	3
8	1
12	2
4	1
15	4
15	3

were conceptional, representing an overall fecundability of 12%. This result compared with an expected fecundability for women with comparable durations of unsuccessful intracervical donor inseminations of 3.5% (the historical control group was not matched for clinical characteristics, similarity of donors, or intensity of monitoring). In the next 29 only two pregnancies occurred but many of these were refractory patients who had been treated in the first 50. Table 1 summarises the number of cycles undergone by FID patients. Four patients who conceived did so after more than one FID suggesting that no damage occurred to their tubes from the procedure.

Centrifugation and swim-up techniques as used generally for IVF and GIFT were used to prepare frozen-thawed donor spematozoa. A 30 µl sperm suspension, containing 50,000 to 75,000 motile spermatozoa, was injected into the fallopian tube isthmic lumen as the catheter was steadily withdrawn. Additional suspension was left in the uterine cavity and endocervix as the catheter was withdrawn from the genital tract. The best time for fallopian transfer of capacitated spermatozoa should theoretically be just after ovulation,but in 43 of 79 treatment cycles using donor semen the dominant follicle was still present at the time of fallopian sperm transfer; six of the eight pregnancies occurred in this subgroup. Because spermatozoa after centrifugation and swim-up show activated motility, it has been assumed that capacitation has occurred and that the spermatozoa are able upon transfer to fertilise an ovulated oocyte. This discrepancy between theory and practice in the timing of sperm transfer in relation to the time of ovulation needs to be resolved. The optimum concentration and volume of sperm suspension, and the best depth of transfer, also remain to be established.

Fallopian Insemination with Husband's Sperm - Unstimulated Cycles

Similar sperm preparative methods have been used with washed husband's semen in cases of mild to moderate oligospermia. No empirical evidence yet shows an advantage over intrauterine insemination or timed sexual intercourse (Table 2).

Table 2. Fallopian Insemination—Husband

Unstimulated cycles	
Patients	23
Cycles	38
Pregnancies	2 (5.3%)
Abortions	0
Ectopics	0

Table 3. Fallopian Insemination—Husband

Stimulated cycles

	Number	Pregnant	%
Clom	10	0	0
HMG	2	2	100
Clom/HMG	8	1	13
HMG +/-Clom	10	3	30

Fallopian Transfer Using Husband's Sperm-Stimulated Cycles

Benefits may also be derived from increasing oocyte number in association with fallopian sperm transfer by controlled ovarian stimulation. In 12 patients undergoing 20 controlled cycles there were three pregnancies with one abortion but no ectopics. These pregnancies all occurred in hMG stimulated patients. (Table 3)

FALLOPIAN OVUM TRANSFER

A pregnancy has been reported after fallopian catheterisation followed by transfer of sperm and ova.[20] There are several attractions of "ultrasound GIFT". Laparoscopic GIFT involves follicle aspiration and gamete transfer to the tubes within a short space of time - namely during the operation that makes the two parts of the GIFT possible. In cases where recovered oocytes are immature there is no opportunity to mature them by in vitro incubation, unless either GIFT is abandoned in favor of IVF or a second laparoscopy is performed within hours of the first. With ultrasound-guided transvaginal procedures, however, follicle aspiration using local anesthesia is distinct from catheterisation and gamete transfer: it technically does not matter whether the two procedures are separated by minutes, hours or days. The advantage is that oocyte maturity can be confirmed. As well as this it may be worth transferring ova still contained within cumulus. Should they be displaced to the distal reaches of the tube during transfer then there is the likelihood that cilial interaction with the cumulus would return the oocyte to the ampullary-isthmic junction, whereas embryos contained within naked zonas might be lost to the peritoneal cavity if propelled too far out into the ampulla.

Table 4. Ultrasound GIFT (3/87–10/87)

Patient	Age	Number of Eggs			
		Available	Returned		
			Right	Left	
1	38	1	0	1	*
2	31	6	2	2	
3	40	5	2	2	o
4	43	1	1	0	
5	26	3	0	3	
6	32	1	1	0	
7	38	4	2	2	
8	25	5	2	2	
9	40	1	0	1	
10	28	8	2	0	#

* Subsequent spontaneous pregnancy
o Chemical pregnancy
Difficult catheterisation - 3 eggs transferred but one returned

There were 10 patients in the first series of ultrasound GIFTs carried out between March 1987 and October 1987 (Table 4). One patient achieved a chemical pregnancy but none had an ongoing pregnancy. This was disappointing although from Table 4 it can be seen that four of the 10 had only one egg to return. This was because these patients were poor stimulators and had been cancelled from the laparoscopic GIFT programme.

FALLOPIAN EMBRYO TRANSFER

As pregnancies have been reported after the laparoscopic transfer of pronuclear pre-embryos[7] ultrasound GIFT was abandoned in favour of ultrasound embryo transfer (FET), removing the risk of transferring unfertilisable or polyspermic oocytes. In the first series (Table 5) seven patients were treated with FET (one of these was a 4-cell embryo). One of the seven patients conceived and has delivered at term; she had three pronuclear-stage embryos transferred to the right tube and one to the left tube.

Attempts to catheterise the second tube may be unsuccessful, thus leaving embryos or eggs untransferred and, worse, possibly pushing successfully transferred embryos out of the first tube during inadvertent injection into the endometrial lumen of the second catheter's contents. The practice therefore was changed to catheterise just one tube and transfer all the embryos or eggs to one side (Table 6). Experience was again disappointing with no pregnancies in 10 patients. It is possible that embryos were being expelled into the peritoneal cavity, as relatively large transfer volumes were in use. Laparoscopic and radiographic studies were carried out to establish optimal transfer volumes.

Table 5. Fallopian Embryo Transfer (8/87–4/88)

Patient	Age	Number of Eggs			
		Available	Returned		
			Right	Left	
1	39	2	2	0	
2	29	10	2	2	
3	30	6	3	1	Pregnant
4	37	8	1	3	*
5	37	15	1	0	4-Cell
6	41	8	1	3	*
7	28	4	2	2	

* Conceived in subsequent GIFT cycles

Table 6. Fallopian Embryo Transfer (4/88–10/88)

Patient	Age	Number of Eggs		
		Available	Returned	
			Right	Left
8	31	8	2	0
9	32	10	2	0
10	38	5	0	3
11	28	6	4	0
12	39	2	0	2
13	25	15	4	0
14	33	8	4	0
15	41	3	2	0
16	31	13	3	1

LAPAROSCOPIC AND RADIOGRAPHIC OBSERVATIONS

Laparoscopy and quantitative selective salpingography after catheterisation from the vagina revealed that 60 to 80 µl of dye or medium injected from a catheter situated near the ampullary-isthmic junction filled the tube and spilled in most women. Occasional patients did not spill until 200 µl had been injected. However, this means that the volume of medium injected carrying the gametes must be less than 60 µl. Selective salpingography injecting 30 µl only of a radio-opaque medium in one patient showed that the tube filled to the ampullary-isthmic junction immediately and spilled from the fimbrial end at 2 minutes. Some medium was still in the tube at 8 minutes but it had disappeared within 15 minutes (Fig.1). Limited observations of methylene-blue-stained cumulus masses in fixed and cleared hysterectomy-salpingectomy specimens after extracorporeal tubal catheterisation also showed that small volumes (e.g. 10 μl) may move oocytes or embryos several centimetres along the isthmus of the tube.

Transfer of gametes must therefore be carried out using the smallest possible volume and should be injected slowly, with the catheter inserted no more than 2 cm into the tube. Because no other pregnancies had occurred transferring embryos to the tube the original policy of transferring eggs was reinstituted, but now doing so slowly, using the smallest possible volume of medium (10 μl).The results are shown in Table 7.

Since this policy has been adopted three pregnancies have resulted from nine cycles. Two of these were ultrasound GIFT in whom oocytes-in-cumulus and sperm were transferred. The other resulted after oocytes were fertilised and three transferred at the pronuclear stage in a patient with male factor infertility in whom two of four previous IVF cycles had failed to produce fertilisation. One patient had only one oocyte, having been cancelled from the laparoscopic GIFT programme. Several difficulties with fallopian gamete or embryo transfer may remain. Myosalpingeal activity specific to the stage of the ovarian cycle, isthmic luminal resistance, and cilial activity in the endosalpinx all normally act together to position an ovum at the ampullary-isthmic junction, but catheterisation of the tube may nevertheless trigger propagated myosalpingeal contractions and could expel an embryo from the tube proximally or distally. Moreover, although it is usual with GIFT to transfer a small amount of air to the tubes along with the gametes, it is not known if this practice is

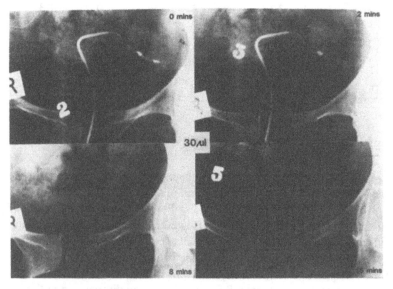

Fig. 1. When 30 μl of dye was injected into the proximal tube it filled the tube immediately but it had disappeared into the peritoneal cavity within 15 minutes.

Table 7. Ultrasound GIFT (11/88–1/89)

Patient	Age	Number of Eggs			
		Available	Returned		
			Right	Left	
11	31	13	0	4	
12	39	3	3	0	
13	38	4	4	0	
14	39	1	1	0	
15	37	4	0	4	Pregnant
16	36	6	3	0	#
17	31	5	3	0	Pregnant o
18	30	4	0	2	*
19	40	3	3	0	Pregnant

\# - Policy changed to maximum transfer of three

o - FET - Male factor : there was no fertilisation in the last two of four previous IVF cycle

* - one egg returned in the catheter and was not replaced .

harmful in association with embryos not shielded by the cumulus. The lumen of the tube has an unusual extracellular electrolyte composition (potassium ion concentrations may be higher than 20 meq per litre[21]),so more research is needed to determine the best chemical composition for the medium in which the embryos or gametes are transferred to the intimate environment of the endosalpinx.

In summary, no present practice in relation to fallopian embryo or gamete transfer should be considered definitive. Controlled studies are needed to determine optimum conditions.

COMPLICATIONS

All patients treated by fallopian catheterisation[18, 19, 22] understood that the procedure could be complicated by trauma to the fallopian tubes, infection or tubal pregnancy. These anticipated side effects have been uncommon. Among 172 catheterisations at Sydney IVF and at Royal Prince Alfred Hospital there have been two patients who developed unilateral pain and fever several days afterwards. In both cases endocervical cultures were negative, but antibiotics were administered, after which clinical resolution took place. In one, conception has since occurred after GIFT. In the other,hysterosalpingography has shown both tubes to be patent. Among 17 conceptions during the cycle of catheterisation (donor semen, n = 8; husband's semen in unstimulated cycles, n = 2; husband's semen in stimulated cycles, n = 3; semen and ova, n = 2; embryos, n = 2), there have been no tubal pregnancies. A further 11 pregnancies in patients previously catheterised have been viable and intrauterine. Observations of the first version of the catheter in women undergoing laparoscopic sterilization showed that extravasation could occur about 4 cm down the tube, where it turns away from the round ligament. The soft, flexible, self-seeking, platinum-tipped wire guide included in the second system is of the type used in selective venography to overcome the same complication with open-ended catheters in venous catheterisation. This system may be of use to inexperienced operators but experienced operators can pass the pre-loaded catheter without the wire guide. If it is passed no further than 2 cm this complication is avoided. Slow injection of the small volume results in the medium passing on into the tube without damage, and if eggs do pass beyond the fimbriated end, they have a chance of being picked up again by the tube. It is possible to recognise, by ultrasound, from the behaviour of the cannula, and the sensation of smooth passage of the catheter whether it is in the tube, and if it is not then the oocytes can be recovered in the catheter and incubated for in vitro fertilisation and later transfer to the uterus.

Pain during passage of the catheter is usually mild, but can occasionally be substantial and accompanied by vasovagal hypotension. Premedication with a narcotic analgesic and at-ropine should be offered. One patient on two occasions experienced bilateral cramps which lasted several days after bilateral catheterisations (a subsequent laparoscopy showed no sign of trauma, both tubes were patent, and the patient has since delivered a baby at term after laparoscopic GIFT).

The present catheter, the method for placing it and for transferring its contents to the tube may not be the best which can be devised. To date the pregnancy rates with fallopian gamete and embryo transfer have not been as high as after laparoscopic transfers. Further experience with small volumes, the position of the catheter in the tube, timing and sperm preparation will identify refinements which may improve pregnancy rates. Nevertheless nonoperative transfer of gametes or embryos to the fallopian tubes by ultrasound-guided catheterisation from the vagina is now an option in assisting conception. Experience with fallopian catheterisation will also improve our understanding of the normal physiology of reproduction.

REFERENCES

1. Woodruff D, Pauerstein CJ. The fallopian tube. Structure, function, management. Baltimore: Williams & Wilkins, 1969:46-66.
2. Croxatto HB, Ortiz ME, Diaz S, Hess R, Balmaceda J, Croxatto H-D. Studies on the duration of egg transport by the human oviduct. II. Ovum location at various intervals following lu-teinizing hormone peak. *Am J Obstet Gynecol* 1978: 132:629-34.
3. Hertig AT, Rock J, Adams EC, Menkin MC. Thirty four fertilized human ova, good, bad, in-different, recovered from 210 women of known fertility: a study of biologic wastage in early human pregnancy. *Pediatrics* 1959; 23:202-11.
4. Papaioannou VE, Ebert KM. Development of fertilized embryos transferred to the oviducts of immature mice. *J Reprod Fertil* 1986; 76:603-8.
5. Gandolfi F, Moor RM. Stimulation of early embryonic development in the sheep by co-culture with oviduct epithelial cells. *J Reprod Fertil* 1987; 81:23-8.

6. Asch RH, Balmaceda JP, Ellsworth LR, Wong PC. Gamete intra-fallopian transfer (GIFT): a new treatment for infertility. *Int J Fertil* 1985; 30:41-5.

7. Yovich JL, Matson PL, Blackledge DG, Turner SR, Richardson PA, Draper R. Pregnancies following pronuclear stage tubal transfer. *Fertil Steril* 1987; 48:851-57.

8. Balmaceda JP, Borrero C, Castaldi C, Ord T, Remohi J, Asch RH. Tubal embryo transfer as a treatment for infertility due to male factor. *Fertil Steril* 1988; 50:476-9.

9. Lisa JR, Gioia JD, Rubin IC. Observations on the interstitial portion of the fallopian tube. *Surg Gynecol Obstet* 1954; 99:159-69.

10. Sweeney WJ. The interstitial portion of the uterine tube - its gross anatomy, course, and length. *Obstet Gynecol* 1961; 19:3-8.

11. Rocker I. The anatomy of the utero-tubal junction area. *Proc Roy Soc Med* 1964; 57:707-9.

12. Rubin I.C. Observations on the intramural and isthmic portions of the fallopian tubes with special reference to so-called "isthmospasm". *Surg Gynecol Obstet* 1928; 4:87-94.

13. Jansen RPS. Endocrine response in the fallopian tube. *Endocr Rev* 1984; 5:525-51.

14. Jansen RPS. Cyclic changes in the human fallopian tube isthmus and their functional importance . *Am J Obstet Gynecol* 1980: 136:292-308.

15. Winston RML. Microsurgical tubocornual anastomosis for reversal of sterilization. *Lancet* 1977; 1: 284-5.

16. Platia MP, Krudy AG. Transvaginal fluoroscopic recanalization of a proximally occluded oviduct. *Fertil Steril* 1986; 44: 704-6.

17. Confino E, Friberg J, Gleicher N. Transcervical balloon tuboplasty. *Fertil Steril* 1986; 46:963-6.

18. Jansen RPS, Anderson JC. Catheterisation of the fallopian tubes from the vagina. *Lancet* 1987; 2:309-10.

19. Jansen RPS, Anderson JC, Radonic I, Smit J, Sutherland PD. Pregnancies after ultrasound-guided fallopian insemination with cryostored donor semen. *Fertil Steril* 1988; 49:920-2.

20. Bustillo M, Munabi AK, Schulman JD. Pregnancy after nonsurgical ultrasound-guided gamete intrafallopian transfer. *N Engl J Med* 1988; 319: 313.

21. Borland RM, Biggers JD, Lechene CP, Taymor ML. Elemental composition of fluid in the human fallopian tube. *J Reprod Fertil* 1980; 58: 479-82.

22. Jansen RPS, Anderson JC, Sutherland PD. Nonoperative embryo transfer to the fallopian tube. *N Eng J Med* 1988; 319: 288-91.

THE INTRATUBAL TRANSFER OF PRONUCLEUS
AND EARLY EMBRYONIC STAGE EMBRYOS

K. Diedrich, O. Bauer, H. v.d. Ven, S. Al-Hasani, and D. Krebs

Department of Obstetrics and Gynecology, University of Bonn
Sigmund-Freud-Str. 25, 5300 Bonn 1, F.R.G.

For 10 years now, in vitro fertilization (IVF) with subsequent embryo transfer (ET) has been successfully used in treated infertility. ET rates of between 70% and 80% testify to the progress obtained in IVF and embryo culture. Despite these advances made in recent years, the pregnancy rate after IVF is hardly higher than 20% per treatment cycle, even in the cases of experienced teams.[1,2] A comparison with the pregnancy rate per cycle under in vivo conditions, however, shows that the prospects per treatment cycle are not more favourable here either.

While IVF is indicated in cases of tubal infertility and can be regarded as the largest area of application, intratubal gamete transfer provides a new and promising infertility treatment wherever tubal function is normal. The advantage here is that both the fertilization and the early embryonic development take place under in vivo conditions. The pregnancy rates obtained are between 18% and 35%, depending on indications.[3,4]

Many working teams report a substantial fall in pregnancy rates after intratubal gamete transfer in cases of male disorders. Although the cause cannot always be established, the most likely reason would seem to be the lower fertilizing capacity of the poor spermatozoa. Therefore, the crucial question arises as to whether a combination of Gamete Intrafallopian Transfer (GIFT) and IVF in the form of intratubal embryo transfer is advantageous, not only for diagnostic but also for therapeutic reasons. IVF enables a few good quality spermatozoa, after being selected and improved from a poor quality ejaculate, to be brought together at an optimal point in time with a fertilizable oocyte in a tube-like medium,

Advances in Assisted Reproductive Technologies, Edited by
S. Mashiach *et al.,* Plenum Press, New York, 1990

Table 1. Pregnancy Rates per Cycle in Infertility Treatment

in vivo fertilization	**15-23%**	**Roberts et al., 1975** **Croxatto et al., 1978** **Buster et al., 1985**
in vitro fertilization	**14-25%**	**Edwards et al., 1986** **Cohen et al., 1988**
intratubal gametetransfer **("Gift")**	**18-35%**	**Asch et al., 1987** **Noss et al., 1987**

intratubal embryotransfer ?

which optimizes the chance of fertilization. Additional refinement of the IVF conditions, for example fertilization in a capillary tube, yields fertilizations even with low spermatozoa counts of between 300 and 500 sperm/ml.[5] Since IVF rates are now at a constant high level, this is a crucial opportunity for obtaining diagnostic information about the fertilizing capacity of the spermatozoa.

In this paper the indication, method and initial results of intratubal embryo transfer are reported. Compared with the intratubal gamete transfer, this treatment involves a possibility of using favourable and controlled (compared with intratubal gamete transfer) conditions of fertilization under in vitro conditions, which, on the one hand, are advantageous especially in the case of low quality sperm, while on the other hand, the embryo in its physiological tubal medium may have better development opportunities for subsequent implantation than in the case of intrauterine embryo transfer.

The indications for intratubal embryo transfer are the following:
Male Infertility
Immunological Infertility
Unexplained Infertility
Unsuccessful "GIFT"

1. *Male infertility:* After other male and conventional gynecological therapies, including several intrauterine inseminations have been exhausted, IVF with subsequent intratubal ET would seem to be appropriate in cases of male infertility for the following two reasons:
   a)   it is a diagnostic method enabling us to test the fertilizing ability of the spermatozoa;
   b)   it is a therapeutic method leading to a pregnancy and achieving the goal of treatment. If this therapeutic attempt leads to a successful fertilization with subsequent ET but without pregnancy, a new treatment by intrauterine insemination or

intratubal gamete transfer is certainly justified where the fertilizing ability of the sperms has been proved.

2. *Idiopathic infertility:* If, after many years, all the possibilities of infertility diagnostics and therapy have failed to produce a pregnancy, this new approach may be employed in cases of idiopathic infertility. Besides a prospective pregnancy, which would achieve the therapeutic goal, IVF in these patients also had the diagnostic value of detecting disorders in the gametes and of checking the fertilizing ability of the spermatozoa and the oocytes recovered.

3. *Unsuccessful intratubal gamete transfer or contra-indications:* If repeated attempts of intratubal gamete transfer do not produce a pregnancy despite sufficient spermatozoa and oocyte quality, it would seem reasonable to attempt an intratubal embryo transfer in order to establish whether fertilization with the gametes can be obtained at all. Even in the case of low sperm quality below the minimal criteria of intratubal gamete transfer (1 mill. progressively motile spermatozoa) it seems advisable to investigate first whether fertilization with this low-quality sperm is possible at all. Earlier work has shown that, even in cases of sperm quality well below these limits, pregnancy following IVF is nonetheless possible.[6]

4. *Immunological infertility:* Even in the rare cases of spermatozoa antibodies in the serum of the patient, it is useful to start with IVF in a medium free of serum and then to transfer the embryo into the tube.

Procedure of Intratubal ET

Ovarian stimulation is performed first for oocyte retrieval in preparation for the intratubal ET. The following therapeutic schemes are used:

1. Clomiphen/hMG/hCG
2. hMG/hCG
3. Pure FSH/hCG
4. GnRH-agonists/hMG/hCG

Oocyte retrieval was performed 36 to 38 hours after hCG administration. This group of patients underwent only a transvaginal follicle puncture monitored by ultrasound. Twenty hours after oocyte insemination, a check was made whether a fertilization with formation of the two pronuclei had taken place. Forty to 48 hours later the quality of the cultured embryos was assessed. Embryos with blastomeres of equal size and without fragments or other irregularities were judged to be good.

While IVF normally provides for the embryo to be transferred to the uterus in the 4- to 8-cell stage two days after insemination, we transferred the fertilized oocytes into a tube in the pronucleus or embryonal stage with tested tubal function. The transfer was performed pelviscopically using the same procedure as in intratubal gamete transfer. When the guiding catheter is in place, the pronucleus stage or the embryos to be transferred are aspirated by the transferring catheter and introduced approximately 2–3 cm inside the tubal lumen, where the content is then slowly injected. Two to three embryos are transferred.

This procedure was done in 53 patients per laparoscopy, whereas in 15 patients the embryos were transferred by the transvaginal and transuterine way as described by Bauer.[7]

Following pelviscopic or transvaginal intratubal pronucleus or embryo transfer, the patients rested in bed for 24 hours.

Intratubal Embryo Transfer: Procedures

Ovarian stimulation
Transvaginal oocyte retrieval
In vitro fertilization
Embryo culture

Intratubal embryotransfer — transvaginal
— per laparoscopy
— per hysteroscopy

The Australian team headed by Yovich[8] and the Brussels group[9] report good pregnancy rates after intratubal embryo and pronucleus transfer. While these groups reported pregnancy rates after IVF of 12% or 19% per treatment cycle, the results of intratubal pronucleus transfer (39% and 43%) were appreciably better. According to these reports, the method seems to yield better results than intratubal gamete transfer (35% and 27%). Nevertheless, the two groups used wider indications for this treatment, so that there was an idiopathic infertility in most patients. Only in a few cases was infertility due to male disorders (Table 2).

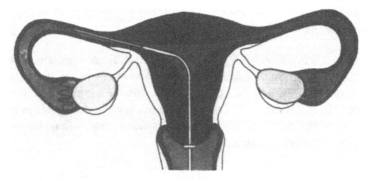

Fig. 1. Transvaginal transuterine ultrasonographically-guided embryo transfer.

Table 2. Intratubal Gamete versus Intratubal Embryo Transfer. Indications for Treatment Include: Unexplained Infertility, Male and Immunological Infertility

	Pregnancy rate	
	"Gift"	pronucleusstage
Yovich et al., 1988	35,9% (n=84)	39,7% (n=81)
Devroey et al., 1988	27% (n=63)	43% (n=57)

Table 3. Male Infertility and Intratubal Embryo Transfer (SEM and range)

	n=	sperm×10^6/ml	motility(%)	normal morphology(%)
Oligoasthenoteratospermia	38	11,3 ±3,8 (3,6-15,4)	22 ±6 (16-32)	26 ±7 (18-34)
Asthenospermia	9	48 ±17 (27-83)	23 ±7 (17-34)	48 ±8 (41-59)
Asthenoteratospermia	13	63 ±12 (42-95)	27 ±8 (21-34)	28 ±6 (19-33)
Oligozoospermia	8	7,5 ±4,3 (3,2-16,0)	43 ±7 (41-55)	46 ±5 (42-53)

At the University of Bonn since May 1987 an intratubal transfer in the pronucleus stage or in the 2– to 8-cell stage has been performed in 68 patients. The indication for this procedure was male infertility only. In 38 patients there was an oligoasthenoteratozoospermia, in nine patients an asthenozoospermia, in 13 patients asthenoteratozoospermia and in eight patients an oligozoospermia. The quality of the spermatozoa can be seen from Table 3.

In 53 patients the intratubal transfer was performed per laparoscopy and in 15 patients by the transvaginal way. The fertilization rate of the oocytes was 66% and 1–3 embryos were transferred into a tube. An average of 2.7 embryos were transferred per patient. Twenty-one pregnancies were achieved resulting in a pregnancy rate of 28% per laparoscopy and

Table 4. Transvaginal and Laparoscopic Intratubal Embryo Transfer (Male Infertility)

	n	per patient	pregnancies	
			n	%
pronucleus stage	25	2.8	7	28
embryonic stage	28	2.7	8	29

Table 5. Intratubal Transfer in the Pronucleus and Embryonic Stage

	cycles	number of embryos per patient	pregnancies	
			n=	%
per laparosopy	53	2.8	15	28
transvaginal (TV-test)	15	2.6	6	40%
total	68	2.7	21	31%

40% by the transvaginal way. The overall pregnancy rate was 31%. Only one abortion occurred but there was no ectopic pregnancy in this group of patients (Table 4).

In 25 patients the embryos were transferred in the pronucleus stage and in 28 patients in the 2–8 cell stage. An average of 2.7 embryos per patient were transferred into one of the functioning tubes. Among the embryos transferred, at least one embryo was assessed morphologically good (Table 5).

Hence, with a pregnancy rate of 31% in cases with male infertility in this admitted small group of patients, a good pregnancy rate was obtained that is rather higher than those reported following IVF or the GIFT procedure (Table 6).[10]

SUMMARY

Intratubal transfer at the pronucleus embryonic stages usefully combines the advantages of IVF and intratubal gamete transfer.

Table 6. Male Infertility and Pregnancy Rates: Intratubal Gamete versus Intratubal Embryotransfer

"Gift"	18%	Asch et al., 1987
intratubal embryotransfer	31% (n=68)	Diedrich et al., 1989
IVF and ET	17%	Jones et al., 1986

Following controlled and assured fertilization of the oocyte, the pronucleus stage or embryo is transferred from the unfavourable in vitro medium to the physiological tubal medium, where it can develop in its own time. In cases of male infertility, spermatozoa/oocyte contact is favourable under in vitro conditions and can be monitored. This method is particularly suitable where infertility is due to male disorders. In certain cases of idiopathic and immunological infertility too, and following unsuccessful attempts at intratubal gamete transfer, this new treatment is indicated and appears promising. Compared with IVF and intratubal gamete transfer, this method seems to provide a better pregnancy rate of infertility due solely to male disorders.

Whereas oocyte retrieval using transvaginal ultrasound puncture is made much easier, the pelviscopic method, which involves considerable outlays at present, can be facilitated by the employment of new techniques. This could be shown by our procedure of the intratubal transfer from the vagina monitored by ultrasound, which seems to be promising.

Clinical studies and comparisons between intratubal gamete transfer, IVF and the combination of the two techniques in the form of intratubal embryo transfer may soon enable us to define a detailed account of the advantages and indications of the various techniques.

The advantages of the intratubal embryo transfer are the following:

> Preincubation of the oocytes
> Optimal Spermatozoa-oocyte contact
> Controlled fertilization
> Physiological development of the embryo
> Reduced abortion rate (?)

REFERENCES

1. R. G. Edwards, Test tube babies, *Nature*, 293:253 (1981).
2. J. Cohen, R. Edwards, C. Fehilly, S. Fishel, J. Hewitt, J. Purdy, G. Rowland, P. Steptoe, and J. Webster, In vitro fertilization: a treatment for male infertility, *Fertil. Steril.*, 43:422 (1985).

3. R. Asch, Gamete intrafallopian transfer (GIFT). Results from multicentre studies. 12th World Congress on Fertility and Sterility, Singapore (1986).
4. Noss, Intratubarer Gametentransfer. 16. Jahrestagung der Deutschen Gesellschaft zum Studium der Fertilität und Sterilität, Bonn, August (1987).
5. H. H. Van der Ven, K. Hoebbel, S. Al-Hasani, K. Diedrich, D. Krebs, Befruchtung menschlicher Eizellen in Kapillarröhrchen mit sehr geringen Spermatozoenzahlen, *Begurtsh. u. Frauenheilk.* 47:630 (1987).
6. K. Diedrich, S. Al-Hasani, H. van der Ven, D. Krebs, Der intratubare Embryotransfer, *Geburtsh. u. Frauenheilk.*, 49:28 (1989).
7. O. Bauer, K. Diedrich, S. Al-Hasani, H. van der Ven, U. Gembruch, D. Krebs, Schwangerschaft nach transvaginalem intratubarem Embryotransfer, in press (1989).
8. J. L. Yovich, G. Blackledge, P. A. Richardson, P.-L. Matson, S. R. Turner, and R. Draper, Pregnancies following pronuclear stage tubal transfer, *Fertil. Steril.*, 48:851 (1987).
9. P. Devroey, P. Braeckmans, J. Smitz, L. van Waesberghe, A. Wisanto, A. van Steirteghem, L. Heytens, F. Camu, Pregnancy after translaparoscopic zygote intrafallopian transfer in a patient with sperm antibodies, *Lancet*, 1:1329 (1986).
10. H. W. Jones, G. S. Jones, M. C. Andrews, C. Bundren, J. Garcia, B. Sandrow, L. Veeck, Ch. Wilkens, J. Witmyer, J. A. Wortham, and G. Wright, The program for in vitro fertilization at Norfolk, *Fertil. Steril.*, 38:14 (1982).

TUBAL EMBRYO TRANSFER (TET) FOR THE TREATMENT OF COUPLES WITH OLIGO-ASTHENOSPERMIA

P. C. Wong, L. B. G. Hägglund, C. L. K. Chan, Y. C. Wong, C. Anandakumar,
V. H. H. Goh, S. C. Ng, A. Bongso, and S. S. Ratnam

INTRODUCTION

Initially GIFT was introduced as a treatment for unexplained infertility. However, it was soon applied to other causes of infertility when at least one healthy tube was present. The multicenter study on results after GIFT, in which we participated, showed an overall pregnancy rate of 29%, but only 15% in cases of male factor infertility (Table 1). Our own data at the National University of Singapore were similar with an overall pregnancy rate of 25% but only 17% in the male factor group (Table 2).

As the fertilization was felt to be a limiting step among couples with oligo-asthenospermia, a procedure including visualization of fertilization before replacement was proposed. Therefore the procedure of Tubal Embryo Transfer (TET) was introduced.[1,3]

MATERIAL AND METHODS

Fifty TET were performed in 50 consecutive couples with oligo-asthenospermia in this pilot study, starting late spring 1988. Most of these couples had previously failed one or more cycles of GIFT.

Induction of multiple follicular development was achieved by either clomiphene-hMG, pure FSH-hMG, or Buserelin-FSH stimulation. Daily estradiol analyses and ultrasound scans were performed to estimate the follicular response. When the follicles were found to be of appropriate size and number (> 3 follicles of > 15 mm diameter) and corresponding estradiol levels were detected (> 250 pg/ml per follicle) ovulation was induced with 10,000

Advances in Assisted Reproductive Technologies, Edited by
S. Mashiach *et al.,* Plenum Press, New York, 1990

Table 1. Pregnancy Rate in GIFT Cycles in the Multinational Cooperative Study, 1987

Aetiology	Cycles	Clinical Pregnancies	Success Rate %
Unexplained	796	247	31
Male	397	61	15
Endometriosis	413	132	32
Failed AID	160	66	41
Tubal Peritoneal	210	61	29
Cervical	68	19	28
Immunological	30	5	17
Premature ovifailure	18	10	56
Total	2092	601	29

Table 2. Pregnancy Rate in GIFT Cycles According to Aetiology in National University, Singapore

Aetiology	Cycles	Clinical Pregnancies	Success Rate %
Unexplained	51	16	31.4
Endometriosis	30	8	26.7
Oligospermia	35	6	17.1
Oligospermia + endometriosis	4	0	0
Others	18	4	22.2
Total	138	34	24.6

I.U. of hCG. A transvaginal ultrasound guided oocyte recovery was performed 36 hours after hCG administration and the oocytes were inseminated within six hours after oocyte recovery. Pronuclear stage was assessed approximately 18 hours after insemination and a video-laparoscopic tubular embryo replacement was performed 2–4 hours thereafter. A maximum of four embryos were replaced, two in each tube about 2–3 cm from fimbrial end. A METS 2020 transfer catheter was used and 5 µl of T-6 medium served as vehicle for the embryos in a total amount of maximal 15 µl of medium in the catheter. Luteal phase support was given.

RESULTS

In these fifty consecutive cycles, a clinical pregnancy was detected in 12 cycles (24%). Three of these pregnancies resulted in a clinical abortion (25%) and nine delivered a healthy baby. The pregnancy rate was found to vary with the number of embryos replaced and a much higher pregnancy rate was found if four embryos were replaced (37%) compared to the cycles where only two or three embryos were replaced (8.3% and 9.1% respectively) (Table 3).

Table 3. Pregnancy Rate in Tubal Embryo Transfer

No. of Embryos	No. of Cycles	No. of Pregnancies	%
2	12	1	8.3
3	11	1	9.1
4	27	10	37.0
Total	50	12	24.0

CONCLUSION

We found that TET had a much higher (24%) pregnancy rate in couples with oligo-asthenospermia than GIFT (17%). This was even more pronounced in cycles where four embryos could be replaced (37%).

These results have caused a change in our program, when it comes to type of treatment offered to infertile couples depending on their most likely cause of infertility. Couples with a tubal cause are still offered IVF as the treatment of choice and GIFT is offered to couples with an idiopathic infertility. Future couples with a male factor of infertility will be considered for TET and so will couples who have undergone repeated cycles of GIFT because of idiopathic infertility, but have failed to conceive.

REFERENCES

1. Devroey et al., 1986, Pregnancy after translaparoscopic zygote intrafallopian transfer in a patient with antibodies, Lancet, i: 329.
2. Blackledge et al., 1986, Pronuclear stage transfer and modified gameteintrafallopian transfer techniques for oligospermic cases, Med J Aust, 144: 444.
3. Yovich et al., 1987, Pregnancies following pronuclear stage fallopian transfer, Fertil Steril, 40: 851-857.

THE PRESENT PERSPECTIVES OF DIRECT
INTRAPERITONEAL INSEMINATION

Christiane Wittemer, Agnès Menard, Laurence Moreau, and Pierre Dellenbach

Service de Gynécologie-Obstétrique
Centre Médico-Chirurgical et Obstétrical
19 rue Louis Pasteur, 673023 Schilitigheim-cedex, France

INTRODUCTION

Direct intraperitoneal insemination (DIPI) as a treatment for certain cases of infertility has been used by our team since September 1985.[1] The present perspectives of this technique are being described in this paper. A brief recall of DIPI is first required.

DEFINITION

DIPI can be defined as being the insemination of in vitro capacitated sperm during ovulation in the pouch of Douglas.[2] The fallopian tubes and ovaries are bathed in the fluid of the peritoneal cavity. Capacitated sperm injected in peritoneal fluid reach the fertilization area through their own motility and the peritoneal liquid movements from Douglas' pouch towards the fallopian tubes.

CLINICAL AND BIOLOGICAL JUSTIFICATIONS OF DIPI

We suggest that the direct supply of capacitated sperm in the peritoneal fluid during ovulation could bypass certain obstacles which inhibit fertilization: in the case of cervical mucus insufficiency where spermatozoa cannot leave the hostile vaginal environment.

Advances in Assisted Reproductive Technologies, Edited by
S. Mashiach *et al.*, Plenum Press, New York, 1990

was also proposed in certain cases of oligoasthenospermia to supply the fertilization area with enough motile sperm. Lastly DIPI was also tried as a treatment in unexplained infertility. On the other hand, biological findings about peritoneal fluid physiopathology justify the use of DIPI.

The first experiment shows that peritoneal fluid volume fluctuates greatly throughout the menstrual cycle and reaches its maximum (about 20 ml) during ovulation (Table 1).[3]

The second experiment concerned sperm survival in peritoneal fluid. About 130 fluids were collected. Each of them was inseminated with husband's capacitated sperm and conserved for several days in the incubator. Motility was repeatedly observed with the use of the microscope after 1 hr, 6 hr, 24 hr, 48 hr, 72 hr and 96 hrs. In more than 60% of the cases, at least 10% motile sperm were found after 24 hrs. Sometimes a very good sperm survival was observed up to four days (Table 2).

Through DIPI, capacitated sperm are injected in peritoneal fluid at the same time oocytes are released, so that oocytes and sperm could meet in peritoneal fluid. It is for this reason that a study is being conducted to determine the capacity of human oocytes to be fertilized and then to cleave in peritoneal fluid in vitro. Our protocol includes normal ovulatory women during their IVF cycle. Just before oocyte puncture is performed, 5–10 ml of peritoneal fluid are collected by culdocentesis. If at least eight mature oocytes are retrieved, three of them are cultured in peritoneal fluid after insemination with 50,000 capacitated sperm.

Preliminary results show that in many cases embryos can be obtained in peritoneal fluid and three pregnancies have resulted.

RESULTS

Overall Results

Overall results obtained by our team with DIPI from September 1985 to March 1989 are presented. One hundred and sixty four couples were treated with one or several intraperitoneal inseminations. Thirty two pregnancies out of 300 DIPI cycles were initiated. Globally speaking there were 19.5% pregnancies per couple and 10.7% pregnancies per insemination cycle. These results are well worth analyzing in relation to the indications.

Results and Indications

Etiological characteristics of the 164 patients treated were: cervical, cervical + OATS, cervical + donor sperm, donor sperm, male subfertility, unexplained infertility. Out of 37 cases of unassociated cervical infertility, eight pregnancies were obtained: 21.6% pregnan-

Table 1. Volume of Peritoneal Fluid in Normal Ovulatory Women

Days of cycle	1-6	7-11	12-13	14-16	17-19	20-22	23-25	26-28
volume (ml)	4	6	9	22	17	16	13	7

Table 2. Sperm Survival in Peritoneal Fluid

Number of PF with motile sperm	110	124	80	49	15	5
Hours of incubation	1	6	24	48	72	96

cies per couple. In 45 cases associated cervical factor and male subfertility, 14 pregnancies were obtained: 31% pregnancies per couple. In 17 cases, donor sperm was used for DIPI because of poor quality of husband's semen. In 13 cases cervical infertility factor was associated. Only one pregnancy was obtained giving less than 6% pregnancies per patient. Twenty four couples where only a husband's slight oligoasthenospermia was detected, were included in DIPI protocol. Six pregnancies were obtained which is 25% pregnancies per patient. Lastly, in the remaining 41 cases, the couples' infertility was unexplained. Only three pregnancies were initiated; 7.3% pregnancies per couple.

These results tend to show that the best indication for intraperitoneal insemination is some form of cervical infertility alone or associated with male subfertility.

THE IMPORTANT ROLE OF THE MALE FACTOR

The role played in DIPI by the male factor proved to be important. This was especially assessed in a prospective series of cases treated in 1987. Firstly we studied sperm motility influence in fresh semen on DIPI results.

Initial sperm motility varies considerably from 10% to 90%. In the majority of cases initial motility was higher than or equal to 30%. If total motility was lower than 25% in fresh semen, the patient did not become pregnant. These results suggest that high initial motility factor might be instrumental in obtaining a successful outcome and, secondly, that severe asthenospermia with less than 20% motile sperm should not be treated by intraperitoneal insemination (Table 3).

In order to determine fresh semen qualities required for DIPI we also studied results obtained considering normally shaped sperm.

Teratospermia in fresh semen varies from 5% to 100% normal shaped spermatozoa. We can see that the best results are obtained with semen presenting at least 30% normally shaped sperm (Table 4).

Table 3. Sperm Motility and DIPI Results

Number of DIPI	0		3		9		19	32	22	11	11	3	0	
% motile sperm	0-		10-	20-		30	-40	-50	-60	-70	-80		-90	-100
Pregnancies	0		0		2		1	4	2	5	1	1	0	

Table 4. Normally Shaped Sperm and DIPI Results

Number of semen	11	22	51	25	10
% normally shaped	0	0 - 30	30 - 50	50 - 70	70 - 100
Pregnancies	1	2	15	5	2

Table 5. Number of Capacitated Sperm Injected

Number of DIPI	12	45	20	13	8	7	2	16	
Million spz injected	<1	1-	10	-20	-30	-40	-50	-60	>60
Pregnancies	0	5	6	2	1	0	1	2	

Principal information about the number of inseminated sperm during DIPI is provided in Table 5.

The average number of inseminated sperm is eight million. In 75% of the cases less than 30 million sperm were deposited in the pouch of Douglas. No pregnancy followed when less than one million sperm were inseminated. The optimal figure seemed to be around 10 million.

Sperm Quality Required for DIPI

Basing our conclusions on the three previous observations we have defined minimal semen qualities required for DIPI. Semen used for DIPI must have at least 30% motile and normally shaped spermatozoa.

Capacitation technique must provide at least three million motile sperm. This is why we systematically use discontinuous Percoll density gradient which increases motility and survival of capacitated sperm.

Exclusion of severe male infertility in DIPI protocols will be the conclusion of this part concerning DIPI indications.

OVULATION STIMULATION

The second part concerns another condition of success which is optimal stimulation with good ovarian response and correct timing of ovulation.

Intraperitoneal insemination has never been attempted during a spontaneous cycle and in all cases ovulation has been induced and timed using procedures perfected for in vitro fertilization (IVF).[4]

For normal ovulatory women traditional protocols with Clomid and/or human menopausal gonadotropin (hMG) and human chorionic gonadotropin(hCG) were used. Follicular growth and maturity were monitored daily by ultrasound and 17 beta estradiol. When follicular maturity was reached, ovulation was induced by using 5000 I.U. hCG.

THE SPERM PERITONEAL FLUID INTERACTION TEST

Because peritoneal fluid has been proven in certain cases to be toxic to sperm, we proposed the study of sperm-peritoneal fluid interactions in vitro and before each DIPI treatment. In this sperm-peritoneal fluid interaction test, the husband's sperm is capacitated in the same way as in the wife's peritoneal fluid and in Menezo B2 medium. Survival duration and motility quality are evaluated over 24 hr. The test is considered positive if at least 10% of the sperm is motile in the peritoneal fluid after 24 hr of incubation.

A correlation between the results obtained from this sperm-peritoneal fluid interaction test and the probability of pregnancy after DIPI can be made.[5] In a short series of 74 infertile couples, a systematic study of sperm-peritoneal fluid interactions was made. Fifty-five cases revealed a positive test; DIPI was performed and 13 pregnancies were obtained. Although the test was negative, DIPI was performed in 19 cases and only two pregnancies occurred.

At the present time we are trying to complete the sperm-peritoneal fluid interaction test with information about sperm penetration in oocytes zona pellucida in peritoneal fluid. Preliminary results tend to show that no penetration or weak sperm penetration in zona pellucida is a poor prognosis for further fertilization.

IMMUNOLOGICAL PROBLEMS

Our latest research has focused on immunological problems. The following question could indeed be asked: does DIPI create an immunological problem? More precisely, could repetitive injections of capacitated sperm induce immunization with negative effects on fertility at a later date?

The following protocol has been established:
- samples of peritoneal fluid and serum are collected from every woman during intraperitoneal insemination
- antisperm antibodies are searched using tray agglutination test[6]
- this search is first performed before the first insemination to detect previous immunization and then before each subsequent DIPI.

First results on this study concern 57 samples of peritoneal fluids:
- thirty-three samples were collected prior to a first insemination. Antisperm antibodies were found in peritoneal fluid and serum in only one case of unexplained infertility where immunological screenings were never done
- for six women, peritoneal fluid was collected before the first and second insemination. No antisperm antibodies were detected either before the first insemination, or prior to the second insemination
- nine peritoneal fluid samples were collected before a second DIPI. No antisperm antibodies were detected
- lastly three samples were collected from women having their third DIPI. Antisperm antibodies were not detected either

As far as this immunological problem is concerned, no antisperm antibodies attributable to DIPI can be detected at present. On the other hand, it must be pointed out that 75% of DIPI pregnancies occurred after the first cycle. Our present therapeutic strategy is to propose an IVF cycle after two or three unsuccessful DIPI cycles. It is therefore supposed that more inseminations are required to induce antisperm antibodies.

PLACE OF DIPI IN RELATION TO OTHER INFERTILITY TREATMENTS

In conclusion, what is the place of DIPI in relation to other infertility treatments? Elaborate infertility treatment techniques are numerous and the most frequent are: IVF, intratubal gamete transfer and intrauterine insemination (IUI).

In order to make the best choice we suggest that our strategies about infertility treatments be taken into consideration:
- in case of tubal disease an IVF protocol is performed
- in case of unexplained infertility, depending on patient age IVF or GIFT are proposed (< 37 years = IVF; > 37 years = GIFT)
- in case of severe male infertility IVF is recommended
- when a "slight" male subfertility exists, alone or associated with cervical infertility, as well as in unassociated cervical infertility, inseminations of DIPI or IUI are considered.

More precisely, we can see that in all infertility cases excluding tubal disease where IVF seems to be the best treatment, four techniques are now available: IVF, GIFT, DIPI, IUI.

On the one hand, IVF and GIFT are two sophisticated methods and on the other hand DIPI and intrauterine insemination are relatively simple and more innocuous treatments. No randomized study having been done, the ultimte choice of the team will depend on different factors: first, on its overall results obtained with each technique; second, on the cost of the technique versus its effectiveness; and third, very often on general preferences and experience of each team!

REFERENCES

1. A. Menard, Liquide péritonéal et fertilité, insémination intra-péritonéale par culdocentèse, ed Paris, thèse (1986).
2. A. Forrler-Menard, E. Badoc, L. Moreau, and I. Nisand, Direct intraperitoneal insemination. First results confirm, *Lancet* i:1468 (1986).
3. D. Bissel, Observation on the cyclical pelvic fluid in the female, *Am. J. Obstet. Gynecol.*, 24:271-273 (1932).
4. J. Garcia, A. Acosta, M. L. Andrews, H. R. Jones, T. Mantzavinos, J. Mayer, J. McDowell, B. Sandow, L. Veek, T. Whibley, C. Wilkes, and G. Wright, In vitro fertilization in Norfolk, Virginia 1980-82, *J. In Vitro Fertil. Embryo Transfer*, 1:24-28 (1984).
5. J. S. Pampiglione, M. C. Davies, C. Steer, C. Kingsland, B. A. Mason, and S. Campbell, Factors affecting direct intraperitoneal insemination, *Lancet* i:1336 (1988).
6. M. De Almeida, M. Herry, J. Testart, J. Belaisch-Allart, R. Frydman, and P. Jouannet, In vitro fertilization results from thirteen women with antisperm antibodies, *Human Reprod.*, 2:599-602 (1987).

LESSONS LEARNED FROM IVC (INTRA-VAGINAL CULTURE)

Claude Ranoux,[a] Machelle Seibel,[a] Herve Foulot,[b] Jean Bernard Dubuisson,[b] Francois Xavier Aubriot,[b] and Didier Rambaud[b]

[a]*Beth Israel Hospital, Harvard Medical School, 330 Brookline Avenue Boston, MA 02215, U.S.A.*
[b]*Port Royal University Clinic, Paris, France*

In vitro fertilization (IVF) represents a major technological advancement in fertility treatment. The steps include ovulation induction, oocyte retrieval, sperm preparation, fertilization and embryo transfer. In an attempt to simplify the fertilization process, we developed the intravaginal culture (IVC) at the Port Royal University Clinic in 1985.

This work is being continued and extended at Beth Israel Hospital. The IVC technique involves the fertilization of human oocytes in a tube without either air or CO_2. Many observations have evolved from IVC. This article will briefly describe the IVC technique and discuss the results and their implications.

TECHNIQUE

Following ovulation induction and oocyte retrieval, the oocytes are identified and consolidated in one petri dish. The 3 ml IVC tube is filled completely with culture medium that we use normally as growth medium and all oocytes (up to ten) are placed in this tube. Care is taken to avoid air bubbles in the cumulus mass as it will cause the oocytes to float. The oocytes will typically sink to the bottom of the tube. The spermatozoa are previously prepared by one of several standard techniques and a concentration of 30,000 to 60,000 motile spermatozoa are added per tube. As the volume of the tube is 3 ml, the concentration of motile spermatozoa is 10,000 to 20,000 per milliliter. The tube is hermetically closed without interposition of air or CO_2, and wrapped tightly in a cryoflex envelope

Advances in Assisted Reproductive Technologies, Edited by
S. Mashiach *et al.*, Plenum Press, New York, 1990

Table 1. Results of the Randomized Study

	100/IVF	100/IVC
Mean No. oocytes/retrieval	4.1	4.4
Mean No. oocytes fert/retrieval	3.7	3.9
Cleavage rate (%)	58.4	59.6
Transfer rate(%)	85	85
Pregnancy rate/retrieval (%)	23	22
No. ectopics	4	3
No. spon. abortions	4	4
No. live births	15	15

sealed at both ends to prevent any vaginal contamination. The tube is then placed into the vagina above a 6 or 6.5 cm diaphram where it remains for 44 to 50 hours. Following the intravaginal culture, the tube is removed, wiped of vaginal secretions, and the contents poured into a petri dish. The embryos are then transferred into the uterus as per usual. Excess embryos above four may be frozen at this time.

RESULTS

From October 15, 1985 to February 28, 1986, one hundred consecutive patients were treated with IVC.[1] The results were comparable to those obtained in our usual IVF program. The mean age of the patients was 33 years old, and the primary indication was tubal factor (66.3%). The majority of the stimulations were done by clomiphene citrate and human menopausal gonadotropin (hMG), and all oocyte retrievals were performed by laparoscopy. The cleavage, transfer and pregnancy rates per retrieval were successively 51.7%, 84% and 20%, with a live birth rate of 15 per retrieval.

To confirm these results, a second study was performed from June 1, 1986 to March 1, 1987 (Table 1). Two hundred patients were randomized to either IVF or IVC. Both patient population and indications were similar as were the ages (33.5 versus 33). More than half of the oocyte retrievals in each group were performed by ultrasound guidance. The results were not significantly different in the two groups with a birth rate of 15% per retrieval. Over 500 cases have now been performed with a 13.5% birth rate per retrieval.

LESSONS LEARNED

IVC Culture Tube

The design of the IVC tube has several special needs. It must be designed to prevent contamination from vaginal secretions. A cryoflex envelope was used for this purpose, but was not easy to prepare. A newly patented tube has been developed in which culture medium will be present and which necessitates no cryoflex envelope.

Fertilization in Microvolume

The realization that fertilization in vitro could be accomplished without the need of air or CO_2 opened the potential for various modifications. The question was raised; could fertilization be enhanced in a capillary tube which contains less volume, and therefore closer approximates the oocyte and spermatozoa? To this end, a 9 cm capillary tube (IMV, France) normally used for freezing of spermatozoa or embryos was employed.[2] The straw was filled with microvolume of 200 µl of pure B_2 as developed by Menezo (Api System, France). One to three oocytes were placed into the straw and 2,000 to 4,000 motile spermatozoa were added separately. The straw was incubated at 37˚c for 18 to 36 hours. The culture medium was emptied into a petri dish and fertilization was verified. Embryo transfer was performed in the usual fashion.

One hundred patients were included in this study. Criteria for inclusion was the retrieval of more than four oocytes. The first four oocytes were fertilized randomly by either IVC or IVF. Additional oocytes which numbered 322, were fertilized using the microvolume straw technique. One hundred sixty seven embryos were obtained for a cleavage rate of 51.8%. This was comparable to the results obtained from both IVF or IVC (52.2%). In three instances, ongoing pregnancy occurred following the transfer of embryos only resulting from the microvolume straw technique. These data unequivocally prove that normal fertilization can occur using a low sperm concentration and further verifies that it is unnecessary to change the culture medium if physiologic temperatures are maintained.

New Concepts in Fertilization

1) Maturation: The preincubation step is not required. Oocytes can be fertilized immediately after retrieval without any significant difference in either the cleavage rate or pregnancy rate.

2) Insemination: The concentration of motile sperm may be reduced. Only 10,000 to 20,000/ml motile spermatozoa are sufficient for acceptable fertilization and cleavage rates.

3) Embryo growth: The culture medium does not require change. Therefore, neither altering the serum concentration, removing the sperm, or providing air or CO_2 are required for normal embryo growth and development.

We have also observed a spontaneous cumulus dispersement in more than 95% of the embryos. This spontaneous cumulus dispersement is one proof of quality culture. Circumstances in which mechanical cumulus removal is required for all the embryos and oocytes of a same patient typically signifies a problem of culture (infection or toxicity before insemination).

Psychological Aspects

One interesting factor of IVC that was not anticipated initially was the psychological benefits of this technique. Many couples refused IVF for religious problems (fertilization out of the body) or because they were anxious about potential risks of laboratory error or congenital abnormality resulting from gamete manipulations. For these reasons, many couples were enthusiastic about IVC. They felt they actively participated in this important phase, the fusion between the two gametes. Once this technique was introduced, patients began to openly discuss their worries and fears which did not occur prior to the availability of IVC.

The Future

What is the next challenge? These new concepts provide the basis for envisioning the simplification of a highly technological procedure. If a biodegradable capsule could be developed in which gametes could be placed and subsequently inserted into the uterus, nature could once again prevail.

CONCLUSION

The IVC technique is less complicated and less expensive than conventional in vitro fertilization techniques. It permits a standardization of the biological steps. Because the complexity, cost, religious, ethical and legal problems of IVF are of great concern to many individuals, the advancement of the procedure is delayed. We think that IVC solves many of these concerns. For all of these reasons, IVC appears to have the potential to be the method of choice for the future.

REFERENCES

1. C. Ranoux, F.X. Aubriot, J. B. Dubuisson, V. Cardone, H. Foulot, C. Poirot, D. Chevallier, A New In Vitro Fertilization Technique: Intravaginal Culture, *Fertil. Steril.* 49:654(1988).
2. C. Ranoux, C. Poirot, H. Foulot, J.B. Dubuisson, F. X. Aubriot, O. Chevallier, V. Cardone, Human egg fertilization in capillary tube, *J. In Vitro Fert. Embryo Transfer.* 5:49 (1988).

THE BENEFITS OF TUBAL TRANSFER PROCEDURES

John L. Yovich, Rogan R. Draper, Simon R. Turner, and James M. Cummins

PIVET Medical Centre, Perth, Western Australia

INTRODUCTION

The PIVET IVF program was established following a successful pilot study in 1980/ 81 but despite increasing experience the annual results failed to improve after reaching a peak in 1984 when the overall pregnancy rate was around 17% and birth rate around 10% of cycles reaching the stage of oocyte retrieval. In fact, a number of independent reports from around the world published in the past year[1-4] indicated that the live birth rate from IVF procedures was only 9%–10% of retrieval cycles to 1986/7.

Therefore, following the first report by Asch et al.,[5] the technique of GIFT was introduced at the PIVET Medical Centre in late 1985 and immediately a concurrent study was conducted which was matched for the underlying infertility aetiology in order to compare the relative chance of pregnancy by IVF/ET and GIFT.[6] Figure 1 summarizes the findings from that study which compares the chance of pregnancy per transfer during an overlap period covering 1281 cycles of treatment, mostly between the years 1984–1986. That early study showed a significant benefit for the GIFT procedure where the pregnancy rate per transfer was 27% vs. 15% for IVF/ET (p < 0.001). Thereafter, all patients who had at least one patent fallopian tube were progressively converted into GIFT. However, we achieved no pregnancies in 17 cases of male factor infertility nor in five cases where the woman had antispermatozoal antibodies (ASABs) in her serum. The results for male factor infertility treated by GIFT were improved when the protocol was modified so that higher numbers of motile spermatozoa were transferred along with the oocytes.[7]

However, in many male factor cases we were unable to generate the minimum num-

Advances in Assisted Reproductive Technologies, Edited by
S. Mashiach *et al.,* Plenum Press, New York, 1990

ber of 250,000 motile spermatozoa required and hence developed the technique of pronuclear stage tubal transfer (PROST).[8] The procedure proved successful for male factor cases and was also seen to have advantages in the ovum donation program as donors (from GIFT and IVF/ET collections) and recipients could be hospitalized on different days, therefore maintaining the essential criterion of confidentiality between donor and recipient.[9] However, it was subsequently found that PROST also had its limitations, particularly in very severe male factor cases where the degree of oligo/asthenospermia was such that < 2 million progressively motile spermatozoa/ml were available in the semen. In such cases reinsemination, delayed fertilization, the use of enhancement adjuvants such as pentoxifylline[10] and the use of micromanipulation procedures such as zona splitting and zona drilling[11] required more prolonged culture so that the resulting cleavage stage embryos could be assessed prior to transfer. In addition, it was desired to transfer four-cell and eight-cell embryos which had previously undergone cryopreservation via the tubal route wherever possible. The early results using a laparoscopic technique of transfer showed a benefit over IVF/ET and subsequently transcervical tubal cannulation under ultrasound control as described by Jansen and Anderson[12] was employed to explore the possibility of reducing the need for general anesthesia and hospitalization. This modification has been termed transcervical TEST to differentiate from routine laparoscopic TEST.[13]

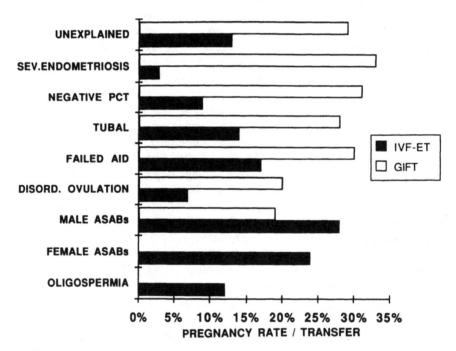

Fig. 1.   The observed pregnancy rates for patients matched by infertility category during the overlap years 1984-1986. GIFT; 109 pregnancies from 408 treatment cycles (27%), significantly higher (p < 0.001) than IVF/ET; 132 pregnancies from 873 treatment cycles (15%). Among subcategories, significance only demonstrable for severe endometriosis (p < 0.001).

Over the past three years a range of other developments have also been introduced in IVF, including the wider use of transvaginal ultrasound-directed oocyte recovery methods; the use of GnRH analogs in controlling ovarian stimulation; and the evolving use of luteal support regimens. These latter changes were applied universally in the IVF program at PIVET and hence the main aim of this data analysis was to evaluate the impact of tubal tranfer procedures against the traditional method of IVF/ET.

MATERIALS AND METHODS

The evaluation is based on all patients treated over a twelve-month period (January 1 to December 31 1988 inclusive) at PIVET Medical Centre. During that period 227 couples had a total of 415 treatment cycles which reached the stage of admission for oocyte retrieval. A further 18 women had a total of 26 transfer procedures in the ovum donation program. However, in the evaluation of pregnancy outcome, data is drawn from 505 pregnancies generated at PIVET to July 1988 and whose outcomes are fully known.

Ovarian Stimulation

The majority of patients had combined clomiphene citrate/hMG (Clomid; Merrell-Dow/ Pergonal; Serono) according to an established schedule.[14] An ovulation trigger of 10,000 I.U. hCG was given on the sixth day of successive estradiol (E_2) rise unless a premature surge occurred, which was then augmented with hCG. Approximately one-third of patients were given a GnRH analog (Leuprolide acetate; Lucrin; Abbott Australasia) using the long regimen[15] combined with hMG beginning on day 3 of the ensuing treatment cycle. The hCG trigger was then given on the seventh day of successive E_2 rise. In particular, Lucrin was given to all women who had raised basal LH including those with clinical polycystic ovary disease, those with raised androgens, those who had exhibited premature LH surges previously, those who were relatively insensitive to stimulation, requiring prolonged courses of hMG injections and, progressively, by the end of the year all women of 35 years and greater.

Procedures

1. **GIFT.** Applied the method of Asch 5 and modification of Matson et al.[7] where applicable. Included were cases of unexplained infertility, including negative sperm/ cervical mucus interaction, endometriosis and moderate oligospermia (> 5 but < 12 million progressively motile spermatozoa/ml of semen, and where at least 250,000 progressively motile spermatozoa could be harvested from the overlay).

2. **PROST.** Applied the methodology of Yovich et al.[8] and was preferred for cases of moderately severe male factor infertility including males with ASABs in the semi-

nal plasma (detected by the indirect immunobead test),[16, 17] synchronous ovum donations, cases where the female partner had circulating ASABs detected by the indirect immunobead test, cases of repetitive failure in GIFT and those cases within the GIFT program demonstrating only one or two follicles following ovarian stimulation.

3. **IVF/ET.** Applied standard methodology[18] and was reserved for cases of primarily tubal disorder, such that neither fallopian tube was patent or accessible for possible tubal cannulation. A number of such cases were included for laparoscopic TEST after previous tubal reconstruction but were converted to IVF/ET when the fallopian tubes proved not readily cannulable.

4. **TEST.** This was applied for cases of very severe male factor infertility (< 2 million motile spermatozoa/ml) where enhancement procedures were applied to improve fertilization or reinsemination was used and it was desired to assess the resulting cleavage stage embryos prior to transfer. It was also applied for cases of asynchronous embryo transfer, following cryopreservation of embryos, and was tentatively proposed for cases of uncertain tubal access (e.g. failure to conceive after reconstructive tubal microsurgery). It has also been applied more recently for synchronous and asynchronous noncommercial IVF surrogacy arrangements.[19] Generally the procedure was performed laparoscopically but on 18 occasions transcervical TEST was performed as part of a research study and included some cases with distal tubal disease preventing laparoscopic cannulation.[20]

5. **Ovum Donation.** Performed on 26 women by one of three procedures—ET (embryo transfer to uterus), PROST or TEST, usually with fresh pronuclear oocytes or cleaving embryos. In all cases hormone replacement therapy was applied using a previously described regimen based on that of Lutjen et al.[21] and more recently a modification proposed by Leeton,[22] using a fixed regimen of estradiol valerate 2 mg tds, beginning Proluton 50 mg imi bd any time after day 10 of the follicular phase (denoted as day 14), and adjusting to a transfer window between days 15 to 19 of the cycle.

Techniques

Husbands produce their semen samples two h prior to the GIFT procedure and 4 h after the IVF/ET, PROST and TEST procedures. In the latter cases, insemination is performed between six h and 16h after oocyte recovery. For PROST, 1–5×10^5 spermatozoa/ml are used to inseminate oocytes which have been recovered by the transvaginal ultrasound-directed technique using the disposable PIVET-COOK aspiration/flushing needle under intravenous Diprivan anesthesia, sometimes supplemented with light mask/airway general anesthesia. Up to 4 pronuclear stage (PN) oocytes are transferred (2×2 or 1×4) to the fallopian tubes the following day at laparoscopy. In normospermic GIFT, 100,000 spermatozoa (but at least 250,000 for oligospermics) are transferred with up to four oocytes, initially two to each fallopian tube or up to four in one tube if the other is not readily accessible, but latterly, transfer to one tube has become the preferred option for all cases. This is performed immediately after laparoscopic or transvaginal oocyte recovery, the latter also becoming the preferred option by the end of the year. For IVF/ET, oocyte recovery is performed by the transvaginal technique and oocyte insemination is precisely the same as for PROST. Two days later up to four embryos are transferred to the uterine cavity using a double catheter technique.[23] TEST is performed similarly but may include reinsemination of oocytes which have failed to display pronuclei 14h-18h postinsemination, the use of pentoxifylline and micromanipulation procedures where indicated. Oocytes are recovered by the transvaginal technique and up to four embryos may be transferred to the fallopian tubes —either by the laparoscopic or transcervical method, two or three days after insemination.

On 60 occasions (15%), a fifth and occasionally a sixth oocyte or embryo was transferred in patients who were thought to be of "low receptivity"—generally of advanced age and repeated failures on previous cycles. In such cases the additional oocytes or embryos were only transferred if no more than three of the total were graded > 3/4.[24]

Luteal Phase

Based on the advantages shown in previous reports from PIVET, IVF/ET cases are given Proluton 50 mg imi on days 0 to 4 inclusive and hCG 1000 I.U. on days 4, 7, 10, 13 of the luteal phase, dated from the oocyte recovery day.[25] Tubal transfer cases are not given Proluton but are given hCG in the aforementioned schedule.[26] Pregnancies are diagnosed by the detection of ß-hCG levels > 25 I.U. (Second International Reference Standard 61/6) and only where a subsequent rise can be shown over the ensuing week. A followup ultrasound is performed at week 7 and pregnancy outcomes are categorized into preclinical (wastage < 7 weeks), spontaneous miscarriage (after fetal heart is detected), blighted ovum (anembryonic gestational sac), ectopic gestation and advanced pregnancies (> 20 weeks) when subsequent deliveries are classified as births.

Pregnancy Support

Women who have vaginal bleeding in early pregnancy are offered hormone support therapy in a trial using medroxyprogesterone acetate 20 mg qid.[27] This is continued throughout the first trimester for those cases exhibiting positive fetal heart action at the ultrasound evaluation performed around the seventh week.

Statistics

Comparative evaluations of implantation and pregnancy rates used the X^2 test in appropriate contingency table. The chance of pregnancy and the risks of multiple pregnancy have been calculated from the binomial distribution using the following formulas:

the probability of an embryo implanting $= U^n C_r E^r (1-E)^{n-r}$
the probability of pregnancy $= U - U^r (1-E)^n$

where U = Uterine receptivity, E = Embryo viability, r = the number of embryos implanting, and n = the number of embryos transferred.[28] The calculations assume that U and E are independent and a maximum likelihood model was developed[29] to match the observed implantation and pregnancy rates.

The fertilization rate noted in the IVF program for normospermic cases (72.4%) was applied to estimate the number of embryos arising in GIFT procedures, to enable comparison of the relative effectiveness of the three tubal transfer procedures.

RESULTS

An overview of the results for 1988 is shown in Figure 2 which details 415 oocyte retrieval procedures with 371 transfers resulting in 141 pregnancies. It can be seen that 27% of cases were by the GIFT procedure, 33% had PROST, 25% had IVF/ET and 16% had one of the TEST procedures. The pregnancy rates per transfer ranged from a high of 43% for GIFT to a low of 25% for IVF/ET, the latter being significantly lower than for all tubal transfer procedures (p < 0.02). IVF/ET also had a lower pregnancy rate per retrieval procedure than the combined tubal group (p < 0.04). The likelihood of preterm delivery was not related to the conception procedure as the patterns were similar for both IVF/ET and tubal transfers but it was also noted that preterm delivery was almost twice as high as the general risk in Western Australia for singleton pregnancies (around 13% as opposed to 6.5%).

For most procedures there was minimal difference between the pregnancy rates per transfer or per retrieval except for PROST where the difference was 8% due to failed fertilizations related to a higher proportion of severe male factor cases. In Figure 3, the pregnancy rate is examined with respect to the implantation rate (number of gestational sacs arising [n = 173] per oocyte or embryo transferred [n = 1387]) and the birth rate per transfer procedure. There are two points of interest—firstly, the tubal transfer procedures all show significantly better implantation rates (p < 0.002), pregnancy rates (p < 0.02), and birth rates (p < 0.05). Secondly, although pregnancy rates are similar for the three tubal transfer procedures, implantation rates show a diminishing trend when one compares GIFT (E) with PROST and TEST (p < 0.08) which is inversely related to intralaboratory exposure.

Within the TEST program there was an apparent difference between the transcervical and laparoscopic techniques, being higher for the latter—pregnancy rates were 3/18 (16.7%) vs. 19/43 (44.2%) respectively, and implantation rates were 6/57 (10.5%) vs. 21/158 (13.3%) for laparoscopic TEST but the findings did not reach statistical significance (p < 0.08).

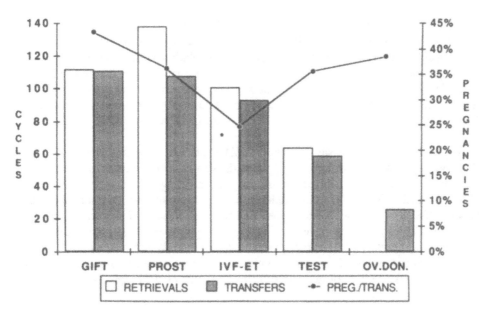

Fig. 2. Distribution of oocyte retrievals and transfer procedures conducted at PIVET during 1988 and associated pregnancy rates per transfer. Ovum donations were performed by ET and TEST (p < 0.02).

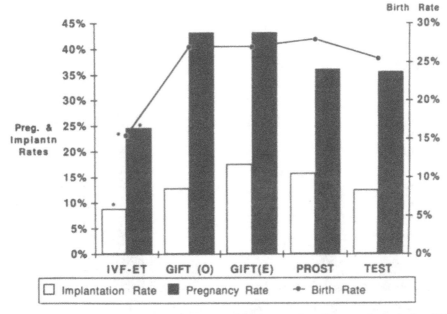

Fig. 3. Respective implantation, pregnancy and birth rates for all procedures during 1988. GIFT (E) refers to estimated implantation rate per embryo transferred, assuming 72.4% of oocytes become fertilized.

The pregnancy outcome from total pregnancies arising to July 1988 (Figure 4) shows that almost 35% (n=176) resulted in early pregnancy wastage, mostly from blighted ovum. Preclinical losses averaged around 7.5% being greater during the early years of IVF/ET and GIFT but lower in 1988 (4%). Ectopic pregnancies averaged around 8% and were highest in the TEST procedure, where all cases arose in transfers performed in patients with known tubal disease or a previous history of reconstructive tubal microsurgery. Just over 65% (n=329) of pregnancies proceeded beyond 20 weeks and were classified as births.

Eighty-five pregnancies were multiple (26%) and the distribution is shown in Figure 5. There was a much higher incidence of preterm delivery (Figure 6) but perinatal deaths were uncommon (Figure 7). 516 infants were delivered, providing an average of 1.3 infants per birth. This included one set of quadruplets following the transfer of five embryos at trans-cervical TEST and one set of quintuplets following the transfer of six pronuclear oocytes in PROST. The quadruplet and quintuplet pregnancies delivered in January 1989 at 33 and 31 weeks respectively; all infants required relatively little specialized care in the neonatal nursery and have thrived. There was a significantly better outcome of triplet and higher order pregnancies over that of twin pregnancies where the clinical protocol of management was less strict.

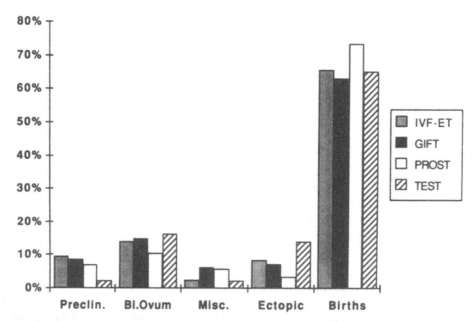

Fig. 4. Pregnancy outcome from 505 pregnancies arising from the PIVET IVF program to July 1988.

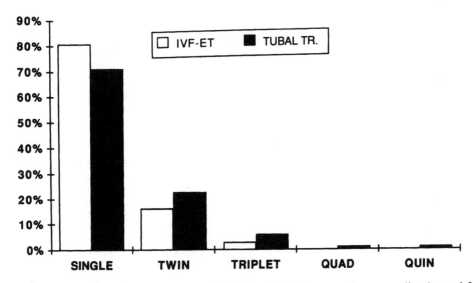

Fig. 5. Observed multiple pregnancy distribution of 397 pregnancies proceeding beyond 20 weeks.

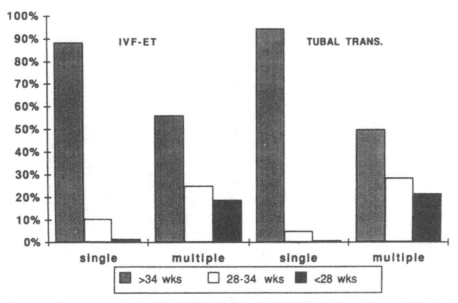

Fig. 6. Proportion of pregnancies which delivered preterm with respect to the conception procedure and the plurality; IVF/ET (n = 89) and tubal transfer (n = 167).

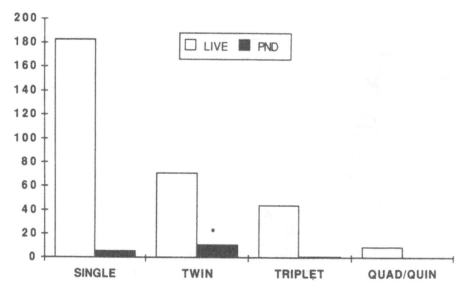

Fig. 7. Perinatal mortality among IVF-related deliveries with respect to the degree of plurality (p < 0.05).

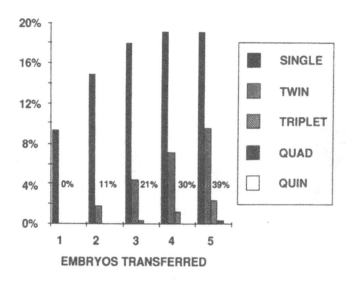

Fig. 8. The estimated chance of pregnancy and proportion of multiple pregnancies per embryo transferred in IVF/ET using a binomial distribution/maximum likelihood method based on 1988 data which provides an E factor of 0.20 and U factor of 0.47.

From the observed outcomes of the 1988 data, the maximum likelihood of probabilities for the E factor was similar for IVF/ET and tubal transfers (0.20 and 0.22 respectively) but significantly different for the U factor (0.47 for IVF/ET and 0.72 for tubal transfers). Applying the binomial estimates, the chance of pregnancy per oocyte or embryo transferred is 23% for IVF/ET (Figure 8) and 35% for tubal transfers (Figure 9) when three are transferred and is associated with a risk of multiple pregnancy around 20% for both (Figure 10). Higher pregnancy rates can be achieved by transferring greater numbers of oocytes or embryos but the multiple pregnancy rate rises accordingly (e.g. 40% when five are transferred).

DISCUSSION

Over the past three years both pregnancy rates and birth rates following IVF-related procedures have shown significant annual rises so that the chance of a live birth from procedures performed in 1988 was improved approximately 2.5-fold over the period. A number of factors have contributed to this improvement, including the greater utilization of tubal transfer procedures and the development of protocols designed to suit specific infertility subcategories. The procedure of IVF/ET, now reserved for all cases with tubal factor infertility, has also shown improving results over this period,[6, 8] supporting the multifactorial basis for improvement (e.g. the application of a GnRH analog for ovarian stimulation in poor prognosis cases, transvaginal ultrasound-directed oocyte recoveries, the use of a disposable aspiration/flushing needle, tighter clinical controls over procedures, and luteal sup-

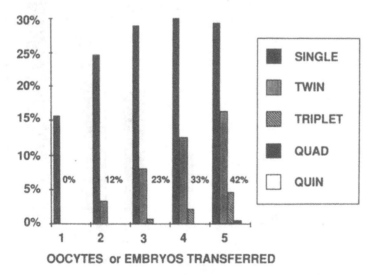

Fig. 9. The estimated chance of pregnancy and proportion of multiple pregnancies per embryo transferred in Tubal Transfer procedures using a binomial distribution/maximum likelihood method based on 1988 data which provides an E factor of 0.22 and U factor of 0.72.

port therapy). However, there remains a significant difference in the implantation and pregnancy rates between IVF/ET and the tubal transfer procedures. The findings of a similar E factor but a markedly higher U factor for tubal transfers indicate that the benefit of the latter relates somehow to the site at which oocytes or embryos are transferred. While several reasons have been proposed[8] the two favoured hypotheses are firstly, that there is a mechanical advantage in tubal transfer, with embryos being more likely to remain in situ; and secondly, that the early postovulatory uterine environment is unfavourable for many embryos and that a period of residence within the fallopian tubes is beneficial whilst the uterine cavity is undergoing appropriate developmental maturation to receive the embryo at the late morula/blastocyst stage.

On the basis of animal models[30] it was initially considered that improved results of GIFT over IVF/ET related somehow to adverse laboratory effects of in vitro fertilization and subsequent embryo culture. This is not borne out in the human situation as the pregnancy rates for TEST are not significantly lower than GIFT; although there may be some benefits in transferring oocytes or embryos at the earliest possible stage as there was a trend showing a reduction in the implantation rates for fertilized oocytes or embryos among GIFT, PROST and TEST progressively in descending order, noting that this is directly related to the period of laboratory exposure with each procedure adding an extra 24 hours re-

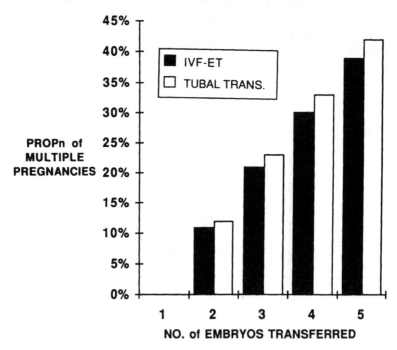

Fig. 10. Estimated risk of multiple pregnancy for IVF/ET and Tubal Transfer procedures using a binomial distribution/maximum likelihood method which provides an E factor of 0.20 and 0.22 respectively. The U factor is assumed to be 1.0 in conceiving women.

spectively. However, we are cautious of drawing firm conclusions from the data as patient selection criteria for the respective procedures may act to bias the results and any effects on implantation of the two-stage procedure required for PROST and TEST have not been specifically studied although none are apparent.

Evaluation of the pregnancy outcomes for the respective programs revealed that 65% of all pregnancies proceeded to births but there is still a very large early pregnancy wastage, mainly from the blighted ovum category. Applying the criteria of pregnancy diagnosis described, preclinical pregnancies have been a continuing feature among all the procedures although their incidence appears to be reducing over the past two years, possibly reflecting a general improvement in IVF experience. We have previously reported on a higher blighted ovum pregnancy rate with the GIFT procedure,[20] but this was not apparent during 1988 and may reflect the clinical preference of treating most male factor infertility by PROST or TEST. It may well be that the transfer of higher numbers of abnormal spermatozoa in the modified GIFT program for oligospermic cases[7] may have caused the development of a higher number of embryos with abnormalities, destined to have restricted development as blighted ovum pregnancies.

The ectopic pregnancy rate reached 12% in the TEST program and was entirely confined to those women having TEST procedures and in whom there was known underlying tubal disease, where the ectopic rate was > 30%.[20] Known underlying tubal disease or a history of tubal reconstruction is now a contraindication to any tubal transfer procedure at PIVET. However, ectopic pregnancy remains a significant problem in the IVF/ET program and the higher rate (9%) than that usually described elsewhere[31] may well reflect the progressive exclusion of all nontubal cases from the IVF/ET program. The most likely explanation for the high ectopic rate in IVF/ET is that embryos do travel into the fallopian tubes via the interstitial segment but may fail to return due to entrapment or tubal malfunction. Preliminary occlusion of the proximal fallopian tubes prior to inclusion in the IVF/ET program was proposed by the pioneers of IVF[32] and may well deserve consideration once again.

The main problem arising from IVF-related procedures is that of multiple pregnancies, with the attendant problem of increased risk of preterm delivery. This does impose major stresses upon neonatal nurseries but even higher order multiple pregnancies delivering preterm may have fully satisfactory outcomes for all infants and this is dependent upon factors within the obstetric management, e.g. at PIVET we have shown significant benefits of a protocol comprising bed rest from 20 weeks gestation, progestagen support and cervical suture at 10 weeks gestation.[33] Many IVF units limit oocyte or embryo transfers to a maximum of four and this is the upper limit proposed under guidelines from the Voluntary Licensing Authority[3] and the Reproductive Technology Accreditation Committee of the Fertility Society of Australia (recently circularized to Australian units). However, the multiple pregnancy rate is around 30% when four embryos are transferred (Figure 10), hence PIVET has now restricted the maximum number of oocytes or embryos transferred to three in order to keep multiple pregnancies below 20%.

Assisted reproduction by IVF remains a developing field of medical therapy and comprises a multidisciplined interaction, particularly from clinical operations and laboratory procedures involving gamete handling and embryo culture. We have shown that tubal transfer procedures in the described protocols have contributed to improving efficiency in assisted reproduction. The main aim now is to maintain the current effectiveness of the program for the alleviation of infertility, but further minimize the chance of triplet and higher order multiple pregnancies. In order to further reduce the numbers transferred but maintain pregnancy rates, improvements in embryo culture (to enable uterine transfer at a later stage) and a marker of embryo quality (for embryo preselection before transfer) are both required.

REFERENCES

1. Commonwealth Department of Community Services & Health. In Vitro Fertilisation in Australia. Canberra, *ACT*, April 1988.
2. Health Department of Western Australia, Epidemiology Branch. In Vitro Fertilisation & Related Procedures in Western Australia 1983-1987: a demographic, clinical and economic evaluation of participants and procedures. Perth, April 1988.
3. The Third Report of the Voluntary Licensing Authority for Human In Vitro Fertilisation and Embryology. London, April 1988.
4. U.S. Congress, Office of Technology Assessment. Infertility: Medical and Social Choices. Washington, D.C., 1988.
5. Asch RH, Ellsworth LR, Balmaceda JP. Pregnancy following translaparoscopic gamete intrafallopian transfer (GIFT). *Lancet* 1984;ii:1034.
6. Yovich JL, Matson PL. The influence of infertility aetiology on the outcome of in vitro fertilization (IVF/ET) and gamete intrafallopian transfer (GIFT) treatments. *Int J Fertil.* In Press.
7. Matson PL, Blackledge DG, Richardson PA, Turner SR, Yovich JM, Yovich JL: The role of gamete intrafallopian transfer (GIFT) in the treatment of oligospermic infertility. *Fertil Steril* 1987;48:608-612.
8. Yovich JL, Blackledge DG, Richardson PA, Matson PL, Turner SR, Draper R: Pregnancies following pronuclear stage tubal transfer (PROST). *Fertil Steril* 1987;47:851-857.
9. Yovich JL, Draper RR, Yovich JM, Edirisinghe WR, Cummins JM: Triplet pregnancy in a woman with primary ovarian failure following pronuclear stage tubal transfer (PROST). *Infertility.* In Press.
10. Yovich JM, Edirisinghe WR, Yovich JL: Preliminary results using pentoxifylline in a PROST program for severe male factor infertility.*Fertil Steril* 1988; 50: 179-181.
11. Odawara Y, Baker HG, Lopata A, Edirisinghe WR, Cummins JM, Wales RG, Yovich JL: Assisted fertilization of human oocytes using zona opening and low sperm concentrations. *In:* Proceedings of the Fertility Society of Australia, August 28-September 1, 1988.
12. Jansen RPS, Anderson JC: Catheterisation of the fallopian tubes from the vagina. *Lancet* 1987;ii:309-310.
13. Yovich JL, Draper RR, Turner SR: Trans-cervical TEST. *Fertil Steril* 1989 (submitted).
14. Yovich JL, Tuvik AI, Matson PL, Willcox DL: Ovarian stimulation for disordered ovulatory cycles. *Asia-Oceania J Obstet Gynaecol* 1987;13:457-463.
15. Yovich JM, Cummins JM, Yovich JL: Pituitary down-regulation with leuprolide confers significant benefits in IVF. Proceedings of the VI World Congress In Vitro Fertilization & Alternate Assisted Reproduction. Jerusalem, Israel, April 2-7, 1989.
16. Junk S, Matson PL, O'Halloran F, Yovich JL: Use of immunobeads to detect human antispermatozoal antibodies. *Clin Reprod Fert* 1986;4:199-206.

17. Matson PL, Junk S, Spittle JW, Yovich JL: The effect of antispermatozoal antibodies in semen upon spermatozoal function. *J Androl* 1988;11:101-106.
18. Yovich JL, Stanger JD, Tuvik AI, Yovich JM: In-vitro fertilization in Western Australia: a viable service program. *Med J Aust* 1984;140:645-649.
19. Yovich JL, Hoffman TD: Compassionate family surrogacy by IVF. *In:* In Vitro Fertilization and Alternate Assisted Reproduction, Edited by Z Ben-Rafael. New York, Plenum Publishing Corporation, 1989. In Press.
20. Yovich JL: Tubal transfers: PROST and TEST. *In:* Proceedings of the International Symposium on Gamete Physiology, California, November 6-10, 1988.
21. Lutjen P, Trounson A, Leeton J, Findlay J, Wood C, Renou P: The establishment and maintenance of pregnancy using in vitro fertilization and embryo donation in a patient with primary ovarian failure. *Nature* 1984;307:174-175.
22. Leeton J, Chan LK, Trounson A, Harman J: Pregnancy established in an infertile patient after transfer of an embryo fertilized in vitro where the oocyte was donated by the sister of the recipient. *J Vitro Fert Embryo Transfer* 1986;3:379-382.
23. Blackledge DG, Thomas WP, Yovich JL, Turner SR, Richardson PA, Matson PL: Transvaginal ultrasonically-guided oocyte pick-up in an in vitro fertilization program. *Med J Aust* 1986;145:300.
24. Marrs RP, Saito H, Yee B, Sato F, Brown J: Effect of variation of in vitro culture techniques upon oocyte fertilization and embryo development in in vitro fertilization. *Fertil Steril* 1984;41:519-523.
25. Yovich JL: Treatments to enhance implantation. *In:* Implantation: Biological and Clinical Aspects, Edited by M Chapman, G Grudzinskas, T Chard. Springer-Verlag. Berlin Heidelberg, 1988. pp.239-254.
26. Yovich JL: The diagnosis and therapy of implantation disorders. *In:* Proceedings of the XII World Congress of Gynaecology and Obstetrics, Rio de Janeiro, October 23-28, 1988.
27. Yovich JL: Medroxyprogesterone acetate therapy in early pregnancy has no apparent fetal effects. *Teratol* 1988;38:135-144.
28. Speirs AL, Lopata A, Gronow MJ, Kellow GN, Johnston, WIH: Analysis of the benefits and risks of multiple embryo transfer. *Fertil Steril* 1983;39:468-471.
29. Snedecor JW, Cochran WG: Statistical methods. Iowa: The Iowa State University Press, 1982. p.427.
30. Vanderhyden BL, Rouleau A, Walton EA, Armstrong DT: Increased mortality during early embryonic development after in vitro fertilization of rat oocytes. *J Reprod Fert* 1986; 77: 401-409.
31. National Perinatal Statistics Unit. Fertility Society of Australia: IVF and GIFT Pregnancies Australia and New Zealand 1987. Sydney, 1988.
32. Steptoe PC: Abnormal implantation, abortion and births after in vitro fertilization. *In:* Implantation of the Human Embryo, Edited RG Edwards, JM Purdy, PC Steptoe. London, Academic Press, 1985. pp.435-444.
33. Devine S, Yovich JL: Managing multiple pregnancies in an IVF program (Abstr). *In:* Proceedings of the International Conference for IVF Nurse Coordinators & Support Personnel. Jerusalem, Israel, March 30-April 2, 1989, p.83.

SELECTION CRITERIA FOR MICROSURGERY, GIFT AND IVF

Rainer Wiedemann, Thomas Strowitzki, Matthias Korell, and Hermann Hepp

Dept. of Obstetrics and Gynecology, University of Munich
Großhadern, Munich, Fed. Rep. of Germany

INTRODUCTION

In recent years, microsurgery, in vitro fertilization/embryo transfer (IVF/ET) and gamete intrafallopian transfer (GIFT) became well established methods for operative treatment of sterility. Definite concepts of patients' indications for the different treatment methods are now possible. Nevertheless, the choice of method often depends on the personal experience of the team and the equipment available in the hospital. In many cases no exactly defined indications are used in counselling the patient. Microsurgery, GIFT and IVF are often discussed competitively, although these methods were originally developed for different indications.

Modern sterility treatment should include individual counselling of the patient to obtain the best and adequate treatment modality for each individual couple. The main purpose of this paper is to verify clear indications for each operative sterility treatment and to discuss the results obtained under those preconditions.

It appears that it is somewhat difficult to compare microsurgery on the one hand, with GIFT and IVF on the other, because GIFT and IVF are treatments for one cycle only. After failure the entire procedure has to be repeated. Microsurgery in fact treats the tubal problem of the woman without limiting it to one cycle, which means microsurgery treats the patient, not the cycle. Apart from this, psychological and physical stress is primarily limited to the time of operation. These initial thoughts clearly show that a comparison of all three methods cannot offer entirely satisfactory results, but may allow a convincing approach to the patient's individual problem.

Advances in Assisted Reproductive Technologies, Edited by
S. Mashiach *et al.,* Plenum Press, New York, 1990

Table 1. Results Following Proximal Tubal Anastomosis in Relation to Histological Grading

	Stage I	Stage II	Stage III
Patients	27	24	18
Intrauterine pregnancies	11	2	0
Ectopic pregnancies	1	6	0
No pregnancy to date	11	16	18

In order to guarantee an appropriate treatment, the reproductive unit should offer all of the operative techniques which are available and established in reproductive medicine. Profound knowledge of the actual success rates of the different procedures in general and for the particular reproductive unit is an important pre-requisite as well as a clear concept of the wide indication spectrum.

MICROSURGERY

Before the introduction of IVF/ET, reconstructive surgery was the only treatment in tubal infertility. In spite of rising interest and improving success rates of IVF/ET, there still remain many absolute indications for microsurgery. These standard indications include peritubal and periovarian adhesions, proximal or distal tubal occlusions and the reversal of sterilization, if the remaining tubal length exceeds 4 cm. Acute inflammatory reaction and previous genital tuberculosis are accepted contraindications for microsurgery. Nevertheless, in many patients who had microsurgery based on standard indications, the results remained very poor. For a better estimation we have introduced a classification system for proximal and distal tubal occlusions. These scores were used in our patients undergoing microsurgical repair during operation for distal pathology and after histopathological grading in proximal pathology. Analysis was performed retrospectively in comparison to the achieved pregnancy rates.

Microsurgical proximal anastomosis was performed in 86 patients (Table 1). All of these patients had no additional tubal or other pathological findings. The follow-up of more than two years is now known in 69 women. Clinically, proximal tubal pathology can be subdivided into non-nodular lesions, mainly caused by inflammatory reaction and nodular lesions due to salpingitis isthmica nodosa or endometriosis. In our opinion this observation alone does not offer a satisfying prognostic factor. So an additional three-stage classification system was introduced by means of an histopathological grading.

Patients with a good prognosis were classified as Stage I. This includes findings of an intact tubal architecture, mild to moderate inflammatory reaction and all resected margins free of inflammation. In Stage II, expressing a poor prognosis, the epithelium is still intact. Connective tissue formation can be found, and an inflammatory reaction up to the muscularis. The resected margins are partially free of inflammation. In Stage III, the tubal

Table 2. Prognostic Score in Salpingostomy

Tubal wall	thin			thick
Adhesions	no	mild	severe	
Mucosa	> 75%	25% - 75%	< 25%	no
Score	0	1	2	3

Table 3. Prognostic Score in Salpingostomy (Staging) Score

Stage I	0– 5 pts
Stage II	6–10 pts
Stage III	11–16 pts

architecture is destroyed and the fibrosis of the muscularis is almost complete. There is inflammation in all layers and a complete obliteration of the lumen, giving an overall bad prognosis.

According to this histopathological grading, all 69 women with known follow-up were analyzed (Table 1). Nodular alterations were found in more than 40% of our patients (29 out of 69). Twenty-seven patients were classified as Stage I, 24 patients as Stage II, and 18 patients belonged to Stage III pathology. In Stage I patients, we had 11 intrauterine pregnancies and one ectopic pregnancy. The pregnancy rate decreased to two intrauterine pregnancies in Stage II patients, accompanied by six ectopic pregnancies, indicating a severely disturbed tubal function. In patients with Stage III disease, no pregnancy has occurred up to now.

Distal tubal pathology, as well as peritubal and periovarian adhesions, are classical indications for microsurgery. The aim of the treatment is to restore an almost normal ovum pick-up. There is a wide variation in results after fimbrioplasty with an overall pregnancy rate of 30% to 35%.[1] For a better estimation of prognosis, we have established a prognostic score for distal tubal pathology according to our histopathological grading in patients with proximal tubal occlusion. This score is based on intraoperative findings. Thickness of the tubal wall, extent of adhesions and alteration of the mucosa are analysed (Table 2). Each tube is graded separately, the total points of both sides are used for staging. If there is only one operated side, the points have to be doubled (Table 3).

Between 1978 and 1986 we operated on 178 patients with distal tubal occlusion. The follow-up of 135 women (75.8%) is documented. Only so-called pure cases of salpingostomy are included in the study. After salpingostomy we achieved a pregnancy rate of 32% regarding the patient (57 patients with 65 pregnancies). Thirty delivered a healthy baby, giving a baby-take-home rate of 16.9%. Seven clinical abortions and 20 tubal pregnancies occurred (11.2% of the patients). Fifty percent of the patients with pregnancies leading to a term delivery conceived more than 18 months after microsurgery.

Table 4. Salpingostomy—Pregnancies in Relation to Grading

	Stage I	Stage II	Stage III
Patients	20	72	33
Term pregnancies	8	9	3
	(40%)	(13%)	(9%)
Abortions	1	4	1
	(5%)	(6%)	(3%)
Ectopic pregnancies	4	13	0
	(20%)	(18%)	(0%)

One hundred and twenty-five patients could be analyzed retrospectively according to our score. In 20 Stage I patients, salpingostomy led to eight term deliveries (40%) and one abortion (5%). The rate of term pregnancies dropped to 13% in 72 Stage II patients and to 9% in 33 Stage III patients (Table 4). The rate of ectopic pregnancies was 20% in Stage I, and did not change significantly in Stage II with 18%. No ectopic pregnancy occurred in Stage III patients. The highest pregnancy rate was seen in patients with thin-walled hydrosalpinx without additional adhesions.

Retrospectively both gradings allow an excellent prognostic view of success after microsurgery. Nevertheless they are both based on intra- and post-operative examinations.

GIFT

The gamete intrafallopian transfer is a well established method since 1984.[2] At least one patent tube is an essential precondition. GIFT can be offered in three groups of classical indications: longstanding sterility, andrological subfertility and some forms of genital pathology, especially endometriosis.

Overall, from 1985 to 1988 we have treated 412 patients in 586 treatment cycles. We were able to achieve a pregnancy rate of 35.0%. These pregnancies include 50 clinical abortions (24.4%), 15 ectopic pregnancies (7.3%) and 34 multiple pregnancies (16.6%) (Table 5). So-called biochemical pregnancies are not listed.

Longstanding sterility, the first subgroup, covers a large variety of indications i.e. patients with so-called idiopathic sterility and/or women with ovarian insufficiency who have had previous hMG stimulation without consecutive pregnancy, as well as patients with immunological sterility or couples with moderate male factor. All patients passed through various forms of stimulation therapy or intrauterine inseminations in previous cycles.

One hundred and sixty-seven patients with longstanding sterility were treated in 212 treatment cycles. We could achieve the highest pregnancy rate of 42.6% (= 90 clinical pregnancies) in this subgroup. Additionally these patients showed the highest incidence of mul-

Table 5. GIFT - Results 1.7.85 - 31.12.88

	Total	Longstanding Sterility	Andrological Subfertility	Genital Pathology
Patients	412	167	137	163
Treatment cycles	586	211	195	248
Clinical pregnancies	205	90 (42.6%)	57 (29.2%)	81 (32.7%)
Spontaneous abortions	50	14 (15.6%)	16 (28.1%)	24 (29.6%)
Ectopic pregnancies	15	3 (3.3%)	4 (7%)	11 (13.6%)
Multiple pregnancies	34	26 (28.9%)	0	8 (9.9%)

tiple pregnancies including three quadruplets and four triplets. This unacceptably high rate of multiple pregnancies forced us to reduce the number of transferred oocytes to two or three since January 1987.

In patients with andrological subfertility, it is very difficult to establish a convincing definition of male factor. The WHO-definition explains andrological subfertility as a sperm count of less than 20 million/ml, a motility of less than 40% and more than 60% morphologically abnormal spermatozoa. According to Kerin,[3] we decided on a differentiation between moderate and severe male factor. A severe male factor is diagnosed when the sperm count shows less than 10 million/ml, and/or the motility is less than 30% and less than 30% normal forms can be found. Based on our clinical experience, patients with less than 800,000 motile spermatozoa/ml after swim-up were excluded from GIFT therapy.

In patients with male factor we had a pregnancy rate of 29.2% (137 patients treated). It is remarkable that no multiple pregnancy occurred in this subgroup, although always four oocytes were transferred. The rate of clinical abortions was 28.1% (16 patients) and four patients had an ectopic pregnancy (7.0%).

We have also treated 14 patients by GIFT with husband's isolated oligozoospermia. Nine pregnancies were achieved, indicating that isolated oligozoospermia cannot be considered as andrological subfertility.

Patients with various forms of endometriosis but patent tubes, alteration of the ovum pick-up caused by adhesions and patients with previous conservative surgery of the tubes formed the group of genital pathology. Overall we achieved a pregnancy rate of 32.7% in 163 patients and 248 treatment cycles (Table 6). Included is a high incidence of spontaneous abortions (29.6%), eight multiple pregnancies (9.9%) and 11 ectopic pregnancies (13.6%), giving the highest ectopic pregnancy rate of all.

In 55 of these patients GIFT was performed because of endometriosis. In 85 treatment cycles, a pregnancy rate of 42.4% (36 patients) could be achieved including 12 abortions and three multiple pregnancies (Table 6). It is of great interest to note that no ectopic pregnancy occurred in this subgroup.

Table 6. GIFT - Genital Pathology (1.7.85 - 31.12.88)

	Genital Pathology Total	Endometriosis	Pat. following tubal microsurgery
Patients	163	55	25
Treatment cycles	248	85	35
Clinical pregnancies	81 (32.7%)	36 (42.4%)	16
Spontaneous abortions	24 (29.6%)	12	5
Ectopic pregnancies	11 (13.6%)	0	6
Multiple pregnancies'	8 (9.9%)	6	0

Under the conditions of a controlled clinical study we have performed GIFT in 25 patients with patent tubes at least 24 months after microsurgery (Table 6). Patients were informed in writing of a possibly higher incidence of ectopic pregnancies. In 35 treatment cycles we induced 16 clinical pregnancies. Five pregnancies ended up as spontaneous abortions and six out of 16 patients developed an ectopic implantation. These data show an incidence for ectopic gestation of more than 30% compared to the pregnant patients.

Our results clearly show that GIFT is the method of choice in patients with longstanding sterility, male factor and in particular with endometriosis. GIFT in patients after microsurgery leads to a high incidence of ectopic pregnancies. In these patients recommendation of IVF should be discussed.

IVF/ET

The classical indications for IVF are bilateral salpingectomy, unsuccessful microsurgery or alterations of the inner genital tract not correctable by microsurgery respectively. Detailed indications will be discussed later on. The European results show a pregnancy rate of 16.5% per puncture in 1987.[4] The European-wide baby-take-home rate in 1987 was 11.0%. European pregnancy rate per transfer was 23% compared to 13.4% in the USA and 13.0% in Australia.

Our results are based on 384 treatment cycles up to now. Three hundred and forty-six transfers led to 60 clinical pregnancies, giving a pregnancy rate of 15.6% per puncture and 17.3% per transfer. Nineteen clinical abortions, three ectopic pregnancies and three twin pregnancies occurred.

CONCLUSIONS

Table 7 shows the operative treatment modalities in relation to the type of sterility. Patients after sterilization are still best treated by microsurgery. The reversal of sterilization offers a pregnancy rate per patient of more than 50%.[5] Non-nodular proximal tubal lesions

Table 7. Treatment of Sterility in Relation to Etiology

Type of Sterility	Treatment Methods
Tubal factor	Microsurgery, IVF, (GIFT)
Male factor	IUI, IPI, GIFT, IVF, TET
Longstanding sterility	Stimulation, IUI, GIFT, IVF, TET
Genital pathology (patent tubes)	Microsurgery, IVF, GIFT, TET

are also an absolute indication for microsurgical repair (Table 8). In general the pathology can be eliminated in total and the anastomosis can be done in healthy tissue. Pregnancy rates can reach 50% per patient.[6]

In a high percentage, nodular tubal pathology (s.i.n., endometriosis) cannot be totally eliminated. We observed a lower pregnancy rate in s.i.n. patients after microsurgery.[7] We still consider proximal nodular pathology as a relative indication for microsurgery. Depending on the stage of histological grading we recommend IVF in Stage III patients soon after microsurgery, because a realistic chance of a spontaneous conception after microsurgery cannot be offered to this group of patients.

In patients with Stage I and II we offer IVF after a period of at least 18 months after microsurgery depending on the age of the patient.

In sterility patients with patent tubes and endometriosis we prefer GIFT therapy. Pregnancy rates of 30% to 40% can be achieved.[8] A surgical approach is chosen in cases of extensive abdominal pain, additional tubal occlusion or a rapidly progressive disease.

Tubal phimosis and tubo-ovarian adhesions with disturbance of the ovum pick-up mechanism are still indications for microsurgery. Success rates of the surgical procedure are difficult to establish, because additional pathologies (e.g. chronic inflammatory reaction in the tubal wall) are often not mentioned in the literature. Results from literature are mainly based on small numbers of patients only. A pregnancy rate of 20% to 40% can be achieved.[1]

The thin-walled hydrosalpinx with and without additional adhesions is still a classical indication for microsurgery.

In general the operative findings correspond to Stage I as shown above with a pregnancy rate of up to 45%. Boer-Meisel et al. confirm these data in their work.[9] In counselling these patients the physician should take into account that 50% of the pregnancies occur 15 months or later after microsurgery. IVF should only be recommended in cases of tubal re-occlusion or in patients in whom a waiting period of up to 15 months is not acceptable because of advanced age.

From our findings the thick-walled hydrosalpinx causes a remarkably lower pregnancy

Table 8. Indications and Contraindications of Microsurgery

Absolute Indications	Absolute Contraindications
Reversal of Sterilization	Acute inflammation
Proximal tubal anastomosis	Previous genital TB
(non-nodular lesion)	
Distal tubal occlusion	Absence of the major part of the tubal ampulla
Tubal phimosis	Remaining tubal length < 3 - 4 cm
Adhesions	Additional infertility factors not amenable to treatment

Relative Indications	Relative Contraindications
Endometriosis (GIFT?)	Acute inflammation
Proximal tubal pathology (nodular)	Combined proximal and distal occlusion
(IVF?)	Reocclusion following previous microsurgery
Multiocular or thick-walled hydrosalpinx	thick-walled hydrosalpinx

rate and a higher incidence of tubal reocclusion. In these patients we perform a control lapa-roscopy half to one year after surgery and offer IVF in cases of reocclusion. In the occa-sional cases of multilocular (honey-comb) hydrosalpinx we prefer IVF primarily. We also recommend IVF in patients with combined proximal and distal occlusion and in women with reocclusion after microsurgery. We discuss the reoperation only in thin-walled hydro-salpinx with no additional adhesions. These recommendations are based on the findings of Winston,[10] who has demonstrated a pregnancy rate of less than 10% per patient in resalpingostomy.

Patients with known extensive adhesions are mostly classified to Stage III. The preg-nancy rate as mentioned above after microsurgery is poor. Extensive adhesions and frozen pelvis are considered as relative contraindications, because reformation of adhesions is very common.

In male factor sterility our data clearly show that GIFT offers a chance for a pregnancy of almost 30%. This rate neutralizes the disadvantage of failing knowledge about the fertili-zation rate. A maximum of six intrauterine inseminations should be performed prior to GIFT. Patients below our exclusion criteria of 800,000 motile sperm/ml still present a dif-ficult problem. Tubal embryo transfer after in vitro fertilization is a possible approach. We performed IVF for embryo transfer in 31 patients below our exclusion criteria. Only three oocytes out of 145 aspirated eggs could be fertilized. No pregnancy resulted from the trans-fers. These data show that this problem is not yet solved.[11]

Longstanding sterility offers excellent possibilities for GIFT after finishing noninva-sive treatment. All of our patients had different forms of stimulation and up to 15 IUI prior to GIFT. IUI allows a pregnancy rate of 10% per cycle and showed no difference in normal intercourse in couples with moderate male factor in Kerin's study.[3] In an unselected group of 345 patients we could achieve a pregnancy rate of only 3%. It is interesting to note that 36 of these couples were treated by GIFT afterwards with a pregnancy rate of 42%.

It should be the aim of modern counselling of patients to obtain an individual planning of each different sterility problem. The age of the patient has to be taken into account as well as the known fact that similar treatment methods result in completely different psychological and physical stress for the individual couple.

REFERENCES

1. Ch. Frantzen, Mikrochirurgische Tubenkorrektur, *Fertilität* 2:115 (1986).
2. R. H. Asch, L. R. Ellsworth, J. P. Balmaceda, and P. C. Wong, Pregnancy following translaparoscopic gamete intrafallopian transfer (GIFT), *Lancet* II:1034 (1984).
3. J. Kerin, and P. Quinn, Washed intrauterine insemination in the treatment of oligospermic infertility, Seminars in reproductive endocrinology 5(1):35 (1987).
4. J. Cohen, and J. de Mouzon, IVF results in Europe, VIth World Congress on In Vitro Fertilization and Alternate Assisted Reproduction, Jerusalem (1989).
5. K. Swolin, Tubal anastomosis, *Human Reprod.*, 3,2:177 (1988).
6. R. Wiedemann, P. Scheidel, H. Wiesinger, and H. Hepp, Die Pathologie des proximalen Tubenverschlusses -Morphologische auswertungen, *Geburtsh. u. Frauenheilk.*, 47:96 (1987).
7. R. Wiedemann, and H. Hepp, Zur differenzierten Indikationsstellung der operativen Techniken in der Reproduktionsmedizin Mikrochirurgie, IVF und ET, GIFT und TET, *Geburtsh. u. Frauenheilk,* 49:416 (1989).
8. R. H. Asch, Results of the multicentric international cooperative study of GIFT, 5th World Congress on In Vitro Fertilization and Embryo Transfer, Norfolk (1987).
9. M. E. Boer-Meisel, E. R. te Velde, J. D. F. Habbema, and J. W. P. F. Kardaun, Predicting the pregnancy outcome in patients treated for hydrosalpinx: a prospective study, *Fertil. Steril.*, 45,1:23 (1986).
10. R. M. L. Winston, Physiology and Pathophysiology of the fallopian tube, 3. Tagung der Arbeitsgemeinschaft für gyn. Mikrochirurgie, Zermatt, Schweiz (1988).
11. R. Wiedemann, U. Noss, and H. Hepp, Gamete intrafallopian transfer in male subfertility, *Human Reprod.*, 4,4:408 (1989).

TUBO-TUBAL ANASTOMOSIS VERSUS IN VITRO FERTILIZATION IN THE MANAGEMENT OF IATROGENIC INFERTILITY

Victor Gomel

Professor and Head, Department of Obstetrics and Gynaecology
University of British Columbia
Vancouver, Canada

In the last two decades, tubal sterilization has been used increasingly as a mode of contraception. This increase has resulted from changing social attitudes and the development of simpler tubal sterilization techniques, especially via laparoscopy. In North America, more than 800,000 tubal sterilizations are carried out each year.[1] This has led to a parallel increase in the demand for restoration of fertility. The most common cause for this request is change in marital union (67%).[2] This is not surprising in a mobile society with a high divorce rate such as ours; and considering further than sterilization procedures are frequently performed at a young reproductive age, in the midst of marital discord, during separation, or soon after divorce. The next two largest groups comprise women who desire more children and those who have experienced the loss of a child, usually during the first few months of life (Table 1).[3] In our study, women in these two groups were all in the same marital union, and were all sterilized during the puerperal period or soon after. These observations define a "risk population" with regard to tubal sterilization that include women in early reproductive years, those in an unstable marital union or in transitional or stress situtions such as separation, divorce and the puerperal or postabortal periods.[3-5]

TUBO-TUBAL ANASTOMOSIS

Microsurgery excels in tubo-tubal anastomosis. Due to the precision afforded by the technique, the result, especially in reversal of sterilization, is a normal tube albeit shortened.

Advances in Assisted Reproductive Technologies, Edited by
S. Mashiach *et al.*, Plenum Press, New York, 1990

Table 1. Reason for Requesting Reversal of Sterilization

Reason	Patients	
	No.	%
Change in marital status	201	67.0
Desire for more children (marital status same)	35	11.7
Crib death and death from other causes under 1 year of age	27	9.0
Tragedy	17	5.6
Psychological reason	20	6.7
Total	**300**	**100.0**

From V. Gomel, "Microsurgery in Female Infertility", Little, Brown and Company, Boston, 1983.

Operative Technique

After the induction of anesthesia, a Foley catheter is inserted into the bladder and connected to drainage. In addition, a pediatric Foley catheter is inserted into the uterine cavity to enable intra-operative chromopertubation. An extension tube connected on one end to the catheter and on the other to a syringe filled with dilute dye solution permits the syringe to be brought to the vicinity of the operative field.

A small Pfannenstiel incision is adequate for such procedures; however, if the patient has already got a midline or a paramedian scar, the same route of entry into the abdomen is used to avoid a second scar. Appropriate hemostasis of the abdominal wound is obtained. Prior to entering the peritoneal cavity, the gloves are thoroughly washed again to avoid the introduction of foreign material into the peritoneal cavity. Once the peritoneum is opened, a wound protector is inserted and a Kirschner retractor applied. The latter provides four-way retraction. The bowel is displaced into the upper abdomen and kept in place with a Kerlex pad soaked in the irrigation solution. Heparinized lactated Ringer's (5,000 I.U. of heparin per liter of lactated Ringer's) is used as an irrigation solution during the procedure to keep the exposed serosa moistened, and to expose bleeders in lieu of swabbing. A separate batch of the same solution is used to soak the abdominal pads that are used in the peritoneal cavity. The uterus and adnexa are elevated by loosely packing the pouch of Douglas with soaked Kerlex pads. The adnexa to be operated upon is similarly immobilized.[6] The microscope is brought into the operative field. Periadnexal adhesions that may be present are excised electrosurgically using 100 micron caliber insulated microelectrode (Martin Co., Tuttlingen, West Germany).

Anastomosis Technique

The proximal tubal segment is distended by transcervical chromopertubation. The tip of the occluded proximal stump is grasped and the tube is transected adjacent to the site of occlusion. Tubal transection is always effected sharply, usually with straight Iris scissors or an appropriate microsurgical blade. The occluded end is excised from the mesosalpinx electrosurgically. During the process of tubal transection and excision of the occluded end, it is mandatory to remain close to the tube to avoid damage to the mesosalpingeal vascular arcade. The patency of the proximal segment is confirmed by chromopertubation. Hemostasis is obtained by pinpoint coagulation of the more significant bleeders, which are located between the muscularis and the serosa. A special irrigator (Gomel irrigator, Martin Co., Tuttlingen, West Germany) permits the visualization of the bleeders under a fine stream of irrigation fluid. The bleeders are electrocoagulated with the use of the microelectrode. We must caution against division and electrocoagulation of the vascular arcade adjacent to the tube (in the mesosalpinx), excessive hemostasis and any electrocauterization of the mucosa since such actions frequently lead to failure. The cut surface is examined under the highest magnification of the microscope to ascertain normalcy. A normal tube is devoid of scarring, exhibits normal muscular and mucosal architecture and a pristine vascular pattern. If there is any abnormality, further excision is carried out until normal architecture is encountered. In the absence of significant luminal disparity, the occluded end of the distal segment is treated in a similar fashion.[7, 8]

End-to-end anastomosis of the two segments is performed in two layers using 8-0 Vicryl or PDS on a 130 micron caliber and 4 or 5 mm long tapercut needle. The muscularis is approximated first with four or more interrupted sutures. The number of sutures employed is dependent on the circumference of the tube at the anastomosis site. Whereas four interrupted sutures placed at cardinal points are usually sufficient to approximate the muscularis in an isthmic-isthmic anastomosis, many more are required in an ampullary-ampullary anastomosis to obtain satisfactory apposition. These sutures incorporate the muscularis and submucosa, avoiding the mucosa itself. The sutures are placed in a way to have the knots remain peripheral. The first suture is placed at the mesosalpingeal edge (six o'clock position) to align properly the two segments of tube. This initial suture is tied. A double throw on the first knot prevents slippage and permits application of optimal tension so as not to strangulate tissues. We usually place the additional sutures using a single strand of suture material. This approach offers better visibility, facilitates suture placement and prevents individual sutures from getting entangled. The loop between successive sutures is divided and thus each suture is tied individually. After the apposition of the muscularis, transcervical chromopertubation is carried out to ascertain patency at the anastomosis site. The serosa is joined next, usually with two continuous sutures, one of which is placed on the anterior and the other on the posterior aspect of the tube. The mesosalpinx is approximated in a similar manner.[8]

Tubo-tubal anastomosis may be (1) intramural-isthmic, (2) intramural-ampullary, (3) isthmic-isthmic, (4) isthmic-ampullary, (5) ampullary-ampullary, or (6) ampullary-infundibular. The type of anastomosis to be performed is dependent on the type of the prior sterilization procedure. Irrespective of the type of anastomosis required, the important factors that affect the anastomosis technique are (a) pre-existing tubal disease (usually in the proximal segment), (b) the relative luminal size of the two segments of tube, and (c) when significant luminal disparity exists, whether or not the ampullary stumps are free. In the presence of significant luminal disparity between the two tubal segments (as is frequently encountered in an isthmic-ampullary or intramural-ampullary anastomosis), accommodation can be accomplished with relative ease provided the ampullary stump is free. After the preparation of the proximal segment of tube as described previously, the ampulla is approached in the following manner. The ampullary stump is exposed, either by introducing a blunted flexible probe through the fimbriated end of the tube or by injecting a small amount of irrigation fluid through the fimbriated end and distending the ampullary segment. The serosa over the exposed ampullary stump is incised in a circular manner and excised from the tip of the stump. The tubal lumen is then entered and an opening is fashioned. Care is taken to make this window into the ampullary lumen only slightly larger than the lumen of the proximal segment. The anastomosis is then completed in two layers, as described previously. We recommend against the use of splints or tubal clamps, as these tend to hinder the surgeon and complicate the procedure.[8, 9]

Results

The published series on microsurgical tubo-tubal anastomosis for reversal of sterilization report viable pregnancy rates varying between 55% and 78% and tubal pregnancy rates between 1.5% and 6.5%. We have reported a viable pregnancy rate of over 75% and an ectopic pregnancy rate of under 2% among patients who had long-term followup.[9-16] Numerous factors affect the surgical outcome and include (1) technique of prior sterilization, (2) length of reconstructed tube, (3) type of anastomosis, (4) presence of a single versus two oviducts, (5) presence of pre-existing tubal disease (i.e. endometriosis, salpingitis isthmica nodosa), (6) surgical technique employed, and (7) status of the other fertility parameters.[8-10, 12, 14,16] The prior sterilization technique employed determines both the type of anastomosis required and the length of reconstructed tube obtained. The length of the reconstructed oviduct has significant influence upon the subsequent fertility. There is a significant drop in the subsequent fertility rate when the length of the reconstructed oviduct is 4 cm or less as measured on the antemesosalpingeal edge of the tube from the uterotual junction to the seromucous junction of the fimbriated end. In addition, the mean time interval between the reconstructive surgery and the occurrence of pregnancy is nearly three times as long in the group of patients who have tubes shorter than 4 cm.[8, 9, 13, 14] The presence of pre-existing tubal disease would, as expected, adversely affect the surgical outcome and increase the tubal pregnancy rate. Both the intra-uterine and tubal pregnancy rates would also be influenced by the nature of surgical technique employed.

Tubo-tubal anastomosis in the face of limited tubal destruction, especially in the isthmic segment, yields very high pregnancy rates. The highest pregnancy rates are achieved with the reversal of sterilization procedures effected with occlusive devices. Pregnancy rates subsequent to the microsurgical reversal of clip sterilization approach that of normal women.[8, 14, 16]

IN VITRO FERTILIZATION AND EMBRYO TRANSFER

In vitro fertilization (IVF) and embryo transfer offers an alternative treatment modality in infertility caused by tubal factors. The factors that influence the results of in vitro fertilization include (1) the age of the patient, (2) numbers of embryos transferred, (3) experience of the centre, (4) protocol for the induction of controlled superovulation, (5) availability of cryopreservation which permits replacement of supernumerary embryos at a subsequent cycle; this potentially increases the overall pregnancy rate from a single treatment cycle, (6) sperm parameters; this also influences the outcome of reversal of sterilization by tubo-tubal anastomosis.[17-20]

IVF Results

It is evident that the viable pregnancy rates obtained, per treatment cycle, varies among centers. The National IVF/ET Registry of the United States reported for the year 1986 a clinical pregnancy rate of 16.9% per transfer cycle. However, the viable pregnancy rate per transfer cycle was only 10.9%.[17] The reported rates per transfer cycle for 1987 were 16% for clinical and 12% for viable pregnancies.[18] The 1987 survey of the French IVF Registry reported a total (early) pregnancy rate of 17.2% per pickup cycle and 21.7% per transfer cycle among patients submitted to IVF because of tubal infertility.[19] These results are almost identical to those reported for 1986.[20] In our own in vitro fertilization program, the highest yield of pregnancies has been obtained in prior sterilization patients. The fact that the vast majority of these patients had proven prior fertility may explain this outcome. Among prior sterilization patients, IVF in our center yielded a clinical pregnancy rate of 21.7% and a viable pregnancy rate of 17.4% per pickup cycle.[21]

ANASTOMOSIS VERSUS IN VITRO FERTILIZATION

Because of the results yielded by microsurgical reconstruction, tubo-tubal anastomosis is our primary approach with previously sterilized women seeking restoration of fertility. In addition to the excellent viable pregnancy rates, this approach offers the following additional advantage: pregnancy occurs by natural means; more than one pregnancy can be achieved by the single intervention. It is, of course, mandatory to take into account the factors that affect the surgical outcome, which were described earlier. Therefore, proper counselling and investigation of the couple are necessary before recommending a reversal procedure.

Preliminary Investigation

The initial interview determines the reasons for which restoration of fertility is requested, the motivation on the part of the woman and her partner. The couple must be counselled regarding the preoperative investigation and the therapeutic alternatives. It is our practice to obtain the operative report and when available the pathology report of the sterilization prior to the initial interview. A comprehensive operative report, when available, is most useful in determining the potential for reversal and may also provide valuable information about the status of the pelvis and fallopian tubes at the time of sterilization. Sterilization techniques that destroy a small midtubal segment, especially isthmic, are technically easier to reverse, yield a longer reconstructed tube and better results. Clip sterilizations cause destruction of a short isthmic segment. Those performed by the application of Falope rings cause the destruction of approximately 2.5 to 3cm of tube. With other sterilization techniques, the damage is variable and largely dependent on the surgeon who performs the procedure.

The investigation should commence with the evaluation of the fertility status of the male partner and other fertility parameters of the woman. Preliminary hysterosalpingography, preferably using water soluble contrast material (Hypaque M60, Winthrop-Breon, New York, NY) permits the assessment of the uterine cavity, the cornual regions, and the length, internal patency and architecture of the proximal tubal segments. A subsequent laparoscopy permits the assessment of the length and status of the distal tubal segments including the fimbria. In addition, it enables the detection of pelvic and periadnexal adhesions or any additional lesions. Hysterosalpingography and laparoscopy are complementary in this regard and the information they yield is essential in counselling the patient about the prognosis. In some circumstances, when the sterilization has been performed with occlusive devices, since these are associated with predictable tubal destruction and when in the operative report the pelvic structures have been described in detail as being normal, especially with a woman in the same marital union, and if a great deal of time has not elapsed since the sterilization procedure, we may dispense with the preliminary laparoscopy.

SUMMARY

A detailed laparoscopic examination complemented by the hysterosalpingographic findings and the investigation of the other fertility parameters of the couple enable an accurate assessment of the prognosis and the selection of the appropriate therapeutic modality, either reversal of sterilization or in vitro fertilization.

In the absence of other infertility factors, our primary approach is a tubo-tubal anastomosis, especially if the reconstructed tube will attain a length of 4 cm or more. Tubal sterilizations are increasingly being performed via laparoscopy and also increasingly with the use of occlusive devices. Well-placed occlusive devices produce a measured and limited tubal destruction which renders subsequent anastomosis technically easier and associated with a higher pregnancy yield. Pregnancy rates subsequent to the microsurgical reversal of clip sterilization approach that of normal women. In addition to the excellent viable pregnancy rates yielded by tubo-tubal anastomosis, this approach offers the following advantages: Pregnancy occurs by natural means, more than one pregnancy can be achieved by a single intervention. On the other hand, the simplification of the various techniques associated with in vitro fertilization have changed this modality into an ambulatory procedure without affecting the outcome. In addition, IVF appears to yield higher pregnancy rates in this group of patients. With IVF success or failure occurs within the treatment cycle; in other words, the result is immediate. However, it must be emphasized that the viable pregnancy per treatment cycle, even in experienced centers, is markedly lower than that yielded by tubo-tubal anastomosis when performed by a well-trained surgeon.

In conclusion, tubo-tubal anastomosis and in vitro fertilization must not be viewed as competitive but rather as complementary therapeutic modalities since the failure of one will permit the use of the other, if applicable.

REFERENCES

1. H. B. Peterson, J. R. Greenspan, F. DeStephano, H. W. Ory, and P. M. Layde, The impact of laparoscopy on tubal sterilization in United States hospitals, 1970 and 1975 to 1978, *Am. J. Obstet. Gynecol.*, 140:811 (1981).
2. V. Gomel, Profile of women requesting reversal of sterilization, *Fertil. Steril.*, 30:39 (1978).
3. V. Gomel, "Microsurgery in Female Infertility", Little, Brown and Company, Boston, 184 (1983).
4. P. Thomson and A. Templeton, Characteristics of patients requesting reversal of sterilization, *Br. J. Obstet. Gynaecol.* 85:161 (1978).
5. G. S. Grubb, H. B. Peterson, P. M. Layde, and G. L. Rubin, Regret after decision to have a tubal sterilization, *Fertil. Steril.*, 44:248 (1985).
6. V. Gomel, Recent advances in surgical correction of tubal disease producing infertility, *Curr. Probl. Obstet. Gynecol.*, 1: No. 10, June (1978).
7. V. Gomel, Tubal reanastomosis by microsurgery, *Fertil. Steril.*, 28:59 (1977).
8. V. Gomel, "Microsurgery in Female Infertility", Little, Brown and Company, Boston, 187–199 (1983).
9. V. Gomel, Profile of women requesting reversal of sterilization: A reappraisal, *Fertil. Steril.*, 33:587 (1980).
10. J. A. Rock, C. A. Bergquist, H. A. Zacur, T. H. Parmley, D. S. Guzick, and H. W. Jones, Jr., Tubal anastomosis following unipolar cautery, *Fertil. Steril.*, 37:613 (1982).
11. A. H. DeCherney, H. C. Mezer, and F. Naftolin, Analysis of failure of microsurgical anastomosis after midsegment, non-coagulation tubal ligation, *Fertil. Steril.*, 39:618 (1983).
12. S. R. Henderson, The reversibility of female sterilization with the use of microsurgery: A report of 102 patients with more than one year of follow-up, *Am. J. Obstet. Gynecol.*, 149:57 (1984).
13. S. J. Silber, and R. Cohen, Microsurgical reversal of tubal sterilization: Factors affecting pregnancy rate, with long-term follow-up, *Obstet. Gynecol.*, 64:679 (1984).
14. P. J. Paterson, Factors influencing the success of microsurgical tuboplasty for sterilization reversal, *Clin. Reprod. Fertil.*, 3:57 (1985).
15. M. M. Spivak, C. L. Librach, and D. M. Rosenthal, Microsurgical reversal of sterilization: A six-year study, *Am. J. Obstet. Gynecol.*, 154:355 (1986).
16. J. A. Rock, D. S. Guzick, E. Katz, H. A. Zacur, and T. M. King, Tubal anastomosis: Pregnancy success following reversal of Falope ring or monopolar cautery sterilization, *Fertil. Steril.*, 48:13 (1987).
17. Medical Research International, The American Fertility Society Special Interest Group, In vitro fertilization/embryo transfer in the United States: 1985 and 1986 results from the Na-

tional IVF/ET Registry, *Fertil. Steril.*, 49:212 (1988).

18. Medical Research International and the Society of Assisted Reproductive Technology, The American Fertility Society, In vitro fertilization/embryo transfer in the United States: 1987 results from the National IVF-ET Registry, *Fertil. Steril.*, 51:13 (1989).

19. J. de Mouzon, J. Belaisch-Allart, J. Cohen, J. -B. Dubuisson, A. Guichard, J. Parinaud, A. Bachelot, and J. -J. Chalais, Dossier FIVNAT: Analyse des résultats 1987, Paris (1987).

20. J. de Mouzon, J. Belaisch-Allart, J. -B. Dubuisson, J. Montagut, J. Testart, A. Bachelot, C. Piette, Dossier FIVNAT: Analyse des résultats 1986 généralités, indications, stimulations, rang de le tentative, âge de la femme, Paris (1986).

21. University of British Columbia and University Hospital, Vancouver, B.C., Canada, IVF Program - unpublished data.

ECTOPIC PREGNANCY AND THE DECISION TO UNDERGO IVF/ET

Randle S. Corfman and Alan H. DeCherney

Department of Obstetrics and Gynecology
Yale University School of Medicine
New Haven, Connecticut

The formulation of a treatment plan for the patient with a history of one or more ectopic pregnancies is often a difficult one. Many times the decision for treatment must be made at the operating table, when the diagnosis is confirmed at laparoscopy or laparotomy. With the acceptance of conservative treatment of ectopic pregnancy, however, comes the opportunity for the physician to counsel the patient as to whether the patient's best chances of fertility are with expectant management or with the assisted reproductive technologies. A knowledge of the various treatment options and the potential of each for future fertility is necessary for an enlightened decision.

Several major advances have impacted upon the diagnosis of ectopic pregnancy, including laparoscopy, hCG testing and ultrasound. The discriminatory zone, which is the level of hCG at which an intrauterine sac is seen on ultrasound examination, has improved from 6,500 mIU or above when an abdominal transducer is employed[1] to 3,000 mIU when a vaginal transducer is used.[2] In addition, high-frequency vaginal ultrasound transducers and doppler flow studies hold great promise for detection of extrauterine pregnancies at an early gestation.

The transition from radical to conservative surgical approaches for the treatment of ectopic pregnancies has been made, permitting salvage of at least some modicum of reproductive function in the affected fallopian tube.[3] In fact, this conservatism has extended to even the means of surgical access to the ectopic pregnancy, as evidenced by the efficacy of laparoscopic treatment of the disease.[4] The application of high power density modalities, such as the laser and microelectrocautery, to both laparotomy and laparoscopic approaches prom-

Advances in Assisted Reproductive Technologies, Edited by
S. Mashiach *et al.*, Plenum Press, New York, 1990

Table 1. Outcome Following Conservative Surgery in Patients with Two Previous Ectopic Pregnancies

	No. pts	clinical preg No. %		intra-uterine No. %		ectopic preg No. %		term preg No. %	
DeCherney et al[13]	13	5	38	4	31	1	8	4	31
Tulandi[14]	16	11	69	8	50	3	19	5	31

ises to reduce tissue damage, perhaps improving the reproductive potential of the patient. Unfortunately, since the etiology of ectopic pregnancy in most instances is bilateral tubal damage secondary to infection, substantial improvement in the reproductive function of these fallopian tubes is unlikely.

The early diagnosis of ectopic pregnancy and the ability to perform conservative surgery in a relatively non-invasive manner, taken together, takes the treatment of ectopic pregnancy from the realm of an emergent to a non-emergent state, perhaps improving the reproductive potential of the patient with an ectopic pregnancy.

THE PATIENT WITH ONE PREVIOUS ECTOPIC PREGNANCY

The incidence of ectopic pregnancy is currently one in 100.[5] Whatever the etiology of the underlying abnormality it is known that after one ectopic pregnancy approximately 50% of the patients will be involuntarily infertile.[3, 6-8] Only one-third will carry a gestation to term.[3, 9, 10] The risk of having a second ectopic gestation ranges from 6% to 27%, thereby being increased several-fold when compared to the incidence of the first ectopic gestation.[3, 6, 7] It has also been observed that the incidence of a second ectopic pregnancy is higher in women who were infertile prior to the occurrence of the first ectopic pregnancy relative to fertile patients.[11]

In the patient who has undergone conservative surgical therapy for an ectopic pregnancy in her sole remaining tube a 61% intrauterine and 50% term pregnancy rate was observed.[12] The repeat ectopic pregnancy rate was 11%. The indication for the initial removal of the contralateral tube was often related to tubal pathology; the underlying pathology no doubt played a major role in subsequent tubal function affecting the results.

THE PATIENT WITH TWO PREVIOUS ECTOPICS

Two recent studies have examined the reproductive outcome after two tubal ectopic pregnancies. The first, by DeCherney et al.,[13] represents a retrospective chart review of 336 patients at Yale-New Haven Hospital with the diagnosis of ectopic pregnancies between 1976 and 1981. Thirty-two patients, or 9.5%, had two ectopic pregnancies, of which 23 were followed. As shown in Table 1, 13 of these patients attempted conception, five of

Table 2. Outcome Following Conservative Surgery in Patients with Two Previous Ectopic Pregnancies Who Conceived

	No. pts	intra-uterine preg		term preg		ectopic preg	
		No.	%	No.	%	No.	%
DeCherney et al[13]	5	4	80	4	80	1	20
Tulandi[14]	11	8	73	5	45	3	27

which (38%) were successful. Four of the 13 attempting conception (31%) achieved a term pregnancy, whereas 1 (8%) experienced a third ectopic. It follows that 60% of those attempting to conceive failed to do so.

Of the five patients who conceived, four (80%) realized a term pregnancy and one (20%) suffered a third ectopic gestation (Table 2). It should be noted that many physicians participated in the care of the patients so that use of microsurgical techniques, antibiotics and adhesion-prevention techniques were not controlled.

Tulandi[14] reported reproductive performance in twenty-four women with a recurrent tubal pregnancy and a surgically absent contralateral fallopian tube. A laparotomy and linear salpingostomy was performed by the same surgeon, using a high power density microelectrocautery. Eleven of sixteen (69%) patients attempting pregnancy were successful. Those unable to conceive totalled five (31%).

As shown in Table 1, five of 16 (31%) patients attempting pregnancy achieved term, with three of 16 (19%) suffering a repeat ectopic pregnancy. Of the 11 patients who conceived (Table 2) 45% (5/11) achieved term, with 27% (3/11) having an ectopic pregnancy.

Although differences noted between these studies are not statistically significant the trends observed are interesting. If one assumes that improvements in the diagnosis and management of ectopic pregnancies have occurred in the time interval between these studies it is possible that an improvement in reproductive outcome for these patients has actually been realized. This is supported by the improvement in percentage of patients attempting pregnancy who conceive after conservative treatment of an ectopic pregnancy (Table 1). The most obvious difference between the studies, in terms of pregnancy outcome (Table 2), is noted in the percentage achieving term gestation. This must reflect an increase in pregnancy wastage, although it is unlikely that the factors which contribute to this problem have significantly changed during the time between studies.

ASSISTED REPRODUCTIVE TECHNOLOGIES AND ECTOPIC PREGNANCY

The assisted reproductive technologies include in vitro fertilization-embryo transfer (IVF/ET), gamete intrafallopian transfer (GIFT) and tubal pre-embryo transfer (TPET).

Table 3.   IVF Treatment Outcome by Number of Retrieval Cycles Performed and by Number of Transfer Cycles Performed

	Total No.	clinical preg No. %	term preg No. %	ectopic preg No. %
By retrieval	8725	1367 16	991 11	103 7
By transfer	7561	1367 18	991 12	103 7

*adapted from the National IVF/ET Registry[15]

Currently, patients with a history of tubal disease, including those with previous ectopic pregnancies, are excluded from GIFT and TPET, both of which require at least one normal fallopian tube for transfer.

Although treatment outcome varies from institution to institution, results of the National IVF/ET Registry for 1987[15] report an overall clinical pregnancy rate of 16% and 18% by number of retrieval cycles or transfers performed, respectively (Table 3). The live birth rates were 11% and 12%, respectively.

It is interesting to note that the ectopic pregnancy rate, calculated either per retrieval or per transfer, was 7% (Table 3). It is also important to note that GIFT can result in ectopic pregnancy, with a rate of 6%.[15] The assisted reproductive technologies are not, therefore, a panacea devoid of risk for ectopic pregnancy.

Although Wood et al.,[16] and Guzick et al.,[17] found that successive cycles of treatment appeared to have equal chances of success in IVF/ET, others, including the Yale IVF Clinic,[18] find a plateauing of pregnancies after three to four cycles, yielding a cumulative pregnancy rate of about 35%. There is, unfortunately, no literature dealing with the cumulative ectopic pregnancy rate in IVF/ET. If one assumes that it also plateaus after three to four cycles an ectopic pregnancy rate after three to four cycles could approach 15–20%.

SUMMARY

For the purposes of this discussion it is assumed that the patient desires future fertility. Regardless of the patient's infertility history, the first ectopic pregnancy should be treated by conservative surgery.

After her second ectopic pregnancy, especially in the patient who was infertile prior to her ectopic pregnancy events, conservative surgery should be performed and the couple counseled to give strong consideration to IVF/ET. This is based on the fact that the couples' chance of attaining a term intrauterine pregnancy are virtually identical with either conservative surgery or expectant management and IVF/ET (30%-35%). The studies examined indicate that the risk of ectopic pregnancy with surgery is between 7%-19% and that it should be lower with IVF/ET, although a cumulative ectopic rate after several cycles of IVF/ET has not been established. It should be acknowledged, however, that ectopic pregnancies can and do result from the assisted reproductive technologies.

Interpretation of the present literature must be made with a clear understanding of the tremendous strides which have been made in recent years in the diagnosis and treatment of ectopic pregnancy. These advances permit the early diagnosis of this disease, permitting not only surgical intervention prior to tubal rupture but also allowing chemotherapeutic treatment options to be entertained. With this progress has come the ability to perform conservative surgery at laparoscopy, thereby decreasing the morbidity of the procedure. The potential for application of high power density modalities, such as the laser and microelectrocautery, to the surgical procedure should further impact upon improving reproductive potential following surgical intervention. Taken together we anticipate that future studies will reveal greater pregnancy rates following conservative surgery, necessitating reevaluation of our treatment protocols.

REFERENCES

1. R. Romero, N. Kadar, and P. Jeanty, et al, Diagnosis of ectopic pregnancy: value of the discriminatory human chorionic gonadotropin zone, *Obstet. Gynecol.*, 66:357 (1985).
2. B. S. Shapiro, M. Cullen, K. J. Taylor, and A. H. DeCherney, Transvaginal ultrasonography for the diagnosis of ectopic pregnancy, *Fertil. Steril.*, 50:425 (1988).
3. A. H. DeCherney, and N. Kase, The conservative management of unruptured ectopic pregnancy, *Obstet. Gynecol.*, 54:451, (1979).
4. J. L. Pouly, H. Mahnes, G. Mage, M. Canis, and M. A. Bruhat, Conservative laparoscopic treatment of 321 ectopic pregnancies, *Fertil. Steril.*, 46:1093 (1986).
5. Centers for Disease Control: Ectopic pregnancy-United States, 1981-1983. *MMWR* 35(18): 289-291 (1986).
6. E. W. Franklin, A. M. Zeiderman, and P. Laemmle, Tubal ectopic pregnancy: etiology and obstetric and gynecologic sequelae, *Am. J. Obstet. Gynecol.*, 117:220 (1973).
7. J. A. Schoen, and R. J. Nowak, Repeat ectopic pregnancy - a 16 year clinical survey. *Obstet. Gynecol.*, 45:542 (1975).
8. R. A. Bronson, Tubal pregnancy and infertility, *Fertil. Steril.*, 28:221 (1977).
9. J. G. Schenker, F. Exal, and W. Z. Polischuk, Fertility after tubal pregnancy, *Surg. Gynecol. Obstet.* 135:746 (1972).
10. S. Timonen, and U. Nieminen, Tubal pregnancy, choice of operative method of treatment, *Acta Obstet. Gynecol. Scand.*, 46:327 (1967).
11. M. A. Bruhat, personal communication.
12. A. H. DeCherney, R. Maheaux, and F. Naftolin, Salpingostomy for ectopic pregnancy in the sole patent oviduct: reproductive outcome, *Fertil. Steril.*, 37:619 (1982).
13. A. H. De Cherney, J. S. Silidker, H. C. Mezer, and B. C. Tarlatzis, Reproductive outcome following two ectopic pregnancies. *Fertil. Steril.* 43:82 (1985).
14. T. Tulandi, Reproductive performance of women after two tubal ectopic pregnancies. *Fertil. Steril.* 50:164 (1988).
15. Medical Research International and the Society of Assisted Reproductive Technology, The American Fertility Society: In vitro fertilization/embryo transfer in the United States: 1987 results from the National IVF-ET Registry, *Fertil. Steril.* 51:13 (1989).

16. C. Wood, R. McMaster, G. Rennie, A. Trounson, and J. Leeton, Factors influencing pregnancy rates following in vitro fertilization and embryo transfer, *Fertil.Steril.*, 43:245 (1985).
17. D. S. Guzick, C. Wilkes, and H. W. Jones, Cumulative pregnancy rates for in vitro fertilization, *Fertil. Steril.*, 46:663 (1986).
18. A. Hershlag, R. Loy, and A. H. DeCherney, Pregnancy rates in IVF-ET after consecutive cycles. In manuscript.

THE ENDOCRINOLOGY OF STIMULATED CYCLES
AND INFLUENCE ON OUTCOME

C.M.Howles [a] and M.C.MacNamee

Department of Endocrinology, Bourn Hall Clinic, Bourn, Cambridge, UK
[a] Serono Laboratories (UK) Ltd
99 Bridge Road East, Welwyn Garden City, Herts, AL7 1BR

INTRODUCTION

It is well established that the chances of pregnancy after carrying out assisted conception techniques (IVF; GIFT) are increased by adopting a superovulation strategy. A variety of superovulation regimens have been employed to promote the synchronous development of several preovulatory follicles which will yield a number of mature eggs. Further it has been clearly demonstrated that after IVF or GIFT, the incidence of implantation rises with an increase in the number of embryos or oocytes replaced.[1,2,3,4]

However, whilst being effective in initiating and maintaining multiple follicular development, the superovulation therapy should have no deleterious effect on follicular endocrinology or on the development of a receptive endometrium. This may not always be the case.

The purpose of this article therefore, is to highlight the importance of monitoring one particular hormone during superovulation for assisted conception.

The hormonal data in this article were obtained from patients attending for IVF at Bourn Hall Clinic, Cambridge, UK over a two year period (1986–1987). They were stimulated using clomiphene citrate (CC, SEROPHENE, Serono Laboratories (UK) Ltd, Welwyn Garden City) from days 2 to 6 of the cycle (100mg daily) and two ampoules of hMG (PERGONAL, LH 75 I.U:FSH 75 I.U; Serono) from day 5. Normally patients were admit-

Table 1. Incidence of the LH Surge in Stimulated Cycles (CC/hMG)

No. of surges	245	
Admissions	1730	(14%)
Oocyte recoveries	1363	(18%)
Replacements	1216	(20%)

ted on day 9 of the stimulated cycle when follicular development was assessed by ovarian ultrasonography and measuring blood levels of oestradiol, progesterone and LH.

In addition, urine samples were collected, approximately every three hours, from day 9 until hCG (PROFASI 5000 I.U. Serono UK) was administered. Normally, only three samples were assayed each day (0700, 1130, 1600). With these measurements the patients daily urinary LH profile could be assessed and if a rise above baseline occurred, any gaps were filled in, allowing the evolution of the surge to be closely followed.

The criteria used for giving hCG to induce ovulation were the presence of a leading follicle of 18 mm in diameter and oestradiol of 1.0 nmol/l per follicle (14 mm diameter).

The effect of superovulation on the LH surge and the influence of tonic LH levels on outcome will be reviewed.

The LH Surge in Superovulated Cycles

One of the major difficulties on the endocrinology of superovulated cycles is the occurrence of an endogenous LH surge. Prior to 1984 some 5% of patients attending Bourn Hall had ovulated prior to oocyte recovery due to poor detection of the surge. This can be a major problem as it has been reported that up to 20% of patients undergoing standard stimulation with CC/hMG will have had an endogenous LH surge prior to hCG (Table 1).[5,6]

Superovulation does not inhibit the surge but can cause it to be attenuated and aberrant and hence difficult to detect.[7,8] This is illustrated in Figure 1 which shows the mean plasma and urinary LH levels in 100 spontaneous and stimulated cycles.

Each urinary LH determination/per patient is the average of at least three separate samples representing no less than eight hours of urinary LH output. Note that tonic levels of LH during the late follicular phase are higher in stimulated cycles.

The presence of higher tonic levels of LH in stimulated cycles is probably a drug related effect.

For instance, Quigley et al.[9] have demonstrated a dose dependent increase in LH secretion after CC administration. Jeffcoate[10] reported that LH concentrations were significantly higher in the late follicular phase in IVF patients treated with CC/hMG compared to patients having hMG alone for routine ovulation induction.

Further, in a series of 22 patients treated on two separate occasions with CC/hMG and CC/FSH (METRODIN, Serono), urinary concentrations of LH were significantly lower (P= < 0.05) in the CC/FSH cycles (Figure 2).

Plasma LH concentrations followed a similar pattern but were only significantly lower in CC/FSH cycles on the day of hCG administration.

Figure 3 illustrates a typical attenuated LH surge during CC/hMG treatment. The reason for this attenuation is not clear but it may be due to the secretion of a factor which reduces the secretion of LH and FSH in response to oestradiol induced GnRH release.[11]

The majority of surges occurred in the morning (Figure 4) and at Bourn Hall all surge oocyte recoveries were carried out successfully between the hours of 0700 to 2000.

A 75% successful oocyte recovery rate has also been reported (for collections) between 0800 to 1800.[12]

Note that the onset of the LH surge after CC alone or in combination with hMG is similar to that observed in the spontaneous cycle (Table 2). Also there seems to be seasonal variations in the frequency of the LH surge (Figure 5).

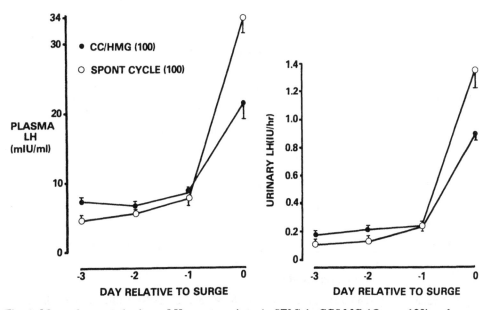

Fig. 1. Mean plasma and urinary LH concentrations (+ SEM) in CC/hMG (● ,n = 100) and spontaneous (○ , n = 100) during the periovulatory period.

Table 2. Onset of LH Surge (%) and Time of Day

Cycle	N	Time of Day (hour)		
		0300–1200	1230–1800	1830–0230
Spontaneous	100	67	15	18
CC alone	85	64	16	20
CC/hMG	245	65	23	12

Of note was the finding that in spontaneous cycles, an LH surge did not occur if the leading follicle was less than 16mm in diameter (Figure 6). Further, in CC/hMG cycles the minimum follicular diameter associated with recovery of mature oocytes after a surge was also found to be 16mm.

Premature Luteinization of Luteinized Unruptured Follicle Syndrome?

On the basis of these observations premature luteinisation was defined as a spontaneous LH surge occurring when the diameter of the leading follicle was less than 16mm in diameter and accompanied by a detectable rise in progesterone (> 6 nmol/litre). Using this definition rather than that of Fleming et al.[13], premature luteinisation was found not to be common, occurring in only six out of 245 patients. However, prior to the utilization of a rapid

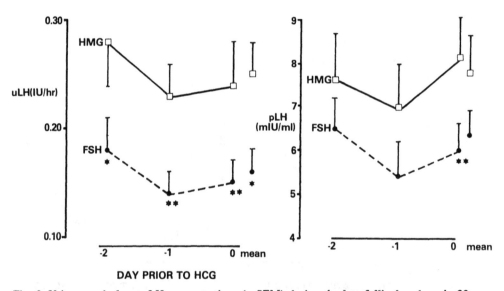

DAY PRIOR TO HCG

Fig. 2. Urinary and plasma LH concentrations (+ SEM) during the late follicular phase in 22 patients who were superovulated with CC/hMG (□) and in a successive cycle with CC/FSH (●). Overall mean for all estimates shown on far right. ** P < 0.01, * P < 0.05.

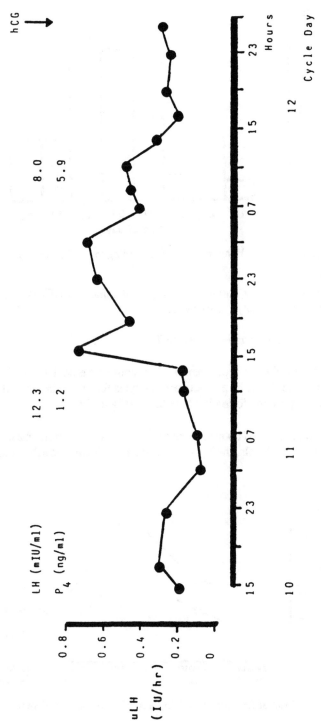

Fig. 3. Hormone profile of one patient who underwent an LH surge and ovulated prior to oocyte recovery.

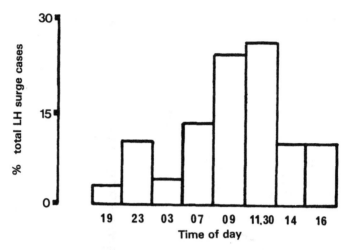

Fig. 4. Onset of LH surge (n = 245) according to the time of day.

and sensitive urinary LH assay, luteinised unruptured follicle (LUF) syndrome occurred more frequently in approximately 6% of cases.

An example of LUF is illustrated in Figure 7.

The rise in urinary LH to no more than 0.6 I.U./hour was associated with luteinisation but not follicular rupture. None of the recovered oocytes fertilized. In view of these data it seems plausible to propose the following hypothesis (Figure 8).

There are three ranges of LH concentrations associated with discrete phases of follicular development. Firstly tonic levels, during which normal follicular growth occurs. At higher

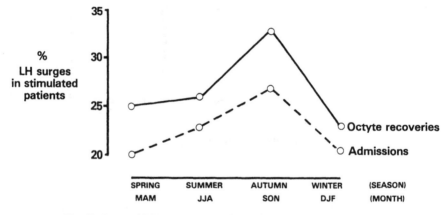

Fig. 5. Onset of LH surge (n = 245) according to the time of year.

concentrations follicular maturation ensues but not necessarily rupture. This leads to the recovery of post-mature oocytes with a low fertilisation potential (LUF). At higher concentrations of LH, both follicular maturation and rupture occur. Thus, the combination of higher tonic levels of LH and the attenuation of the LH surge during CC/hMG treatment make patient management very difficult for the IVF team.

However, in spite of all these difficulties, with adequate LH monitoring, the pregnancy rate of patients undergoing an LH surge was found to be similar to that in patients having hCG to time oocyte recovery (Table 3). This has been confirmed by other workers.[12]

Induction of the LH Surge by Progesterone Administration

Recent studies have shown that in those patients who do not have an endogenous LH surge prior to hCG, one can be elicited by administering progesterone.

For example, in two groups of patients who had 25mg progesterone (i.m.) six or four hours prior to hCG,[14, 15] 80% underwent an LH surge as a result of the injection (Table 4). This agrees with the concept that progesterone can facilitate the positive feedback effects of oestradiol in the induction of the midcycle gonadotrophin surge.[16]

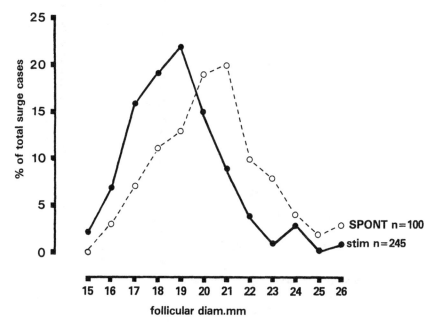

Fig. 6. Onset of LH surge in (CC/hMG, n = 245) and spontaneous (n = 100) cycles according to the diameter (mm) of the leading follicle.

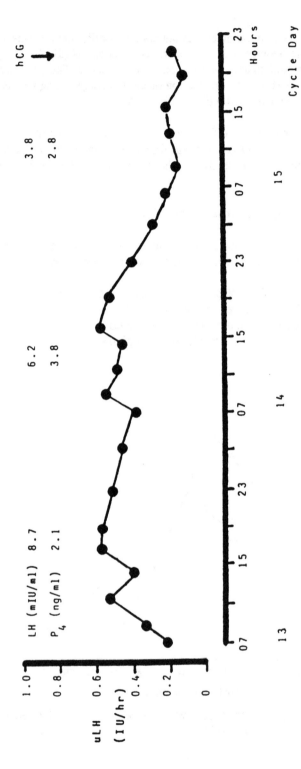

Fig. 7. Hormone profile of one patient who underwent luteinised unruptured follicle syndrome (LUFS).

Table 3. LH Surge V hCG Induced Oocyte Recovery

		LH Surge (245)	hCG Induced (245)
Oocytes/patient		6.1 ± 0.4	7.5 ± 0.3
Oocytes fertilised/cleaved		3.8 ± 0.3	4.8 ± 0.4
Embryo replaced (pregnant)	0	26	21
	1	26 (5)	18 (2)
	2	38 (7)	24 (4)
	3	152 (42)	177 (40)
	4	3 (1)	5 (2)
Pregnant/replacement (%)		55 (25%)	48 (21%)

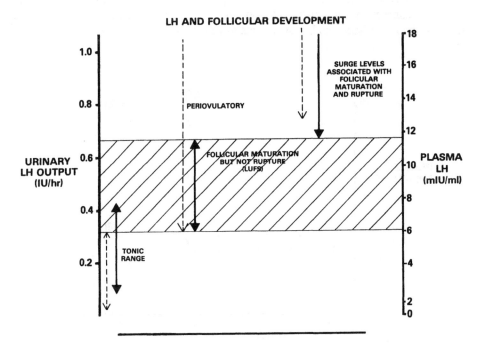

Fig. 8. Diagrammatic representation of urinary and plasma LH levels and stage of follicular development in CC/hMG (——) and spontaneous (-----) cycles. Through raising tonic LH levels and attenuating the LH surge, superovulation increases the incidence of LUFS (hatched area) which results in poor oocyte quality and lowered fertilisation potential.

Table 4. Progesterone Supplementation Prior to hCG and Outcome

	Progesterone Treated		Control	
Patients	59		59	
Surg pre hCG	47		4	
Preg/replacement (%)	16/57	(28)	10/51	(20)

(From refs. 14, 15)

The initial pregnancy rate was higher in the treated groups suggesting a possible advantage of eliciting a surge and/or rising progesterone which may result in the recovery of more mature oocytes and/or a more receptive endometrium.

Further studies on the measurement of LH by a highly specific monoclonal assay, have shown that plasma concentrations of LH rise to surge levels after an injection of hCG to induce ovulation (Figure 9).

This rise was seen again in 24/30 (80%) of patients studied. Interestingly all the patients who became pregnant, underwent a post-hCG LH surge (Table 5). However, further work is required on a large number of patients to validate this observation.

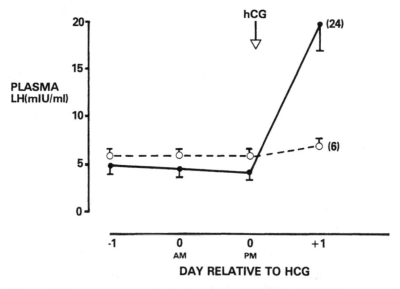

Fig. 9. Plasma LH levels prior to and after injection of 5000 I.U. PROFASI to induce final follicular maturation in 30 patients undergoing IVF. Twenty four patients (●) showed a rise in LH after injection whilst daily 6 (○) did not.

Table 5. Pregnancy Outcome and LH Pattern

LH Pattern	No. Pregnant
Rise after hCG	8/24
No rise after hCG	0/6

Tonic LH Levels and Effect on Outcome

There is now an increasing amount of evidence which suggests that high tonic levels of LH in the late follicular phase of IVF cycles may have a deleterious effect on fertilisation and embryo viability.[5, 17, 18, 19, 20] A similar relationship has also been suggested to exist in patients suffering from PCOD.[21, 22]

In a prospective study of over 400 IVF cycles these initial observations reported in 1819 have been confirmed (Figure 10).

Briefly, multivariate analysis of the data showed that tonic urinary LH levels in the late follicular phase were significantly lower in those patients who went on to conceive com-

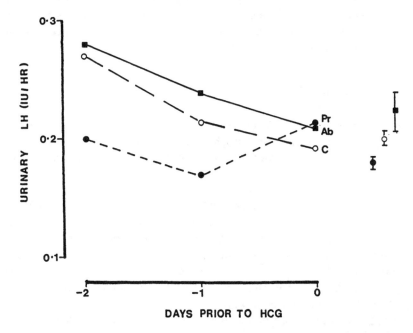

Fig. 10. Daily urinary LH output during the late follicular phase in ongoing pregnancies (● , n = 145), non-conception cycles (○ , n = 145) and in patients who aborted (■ , n = 48) prior to 12 weeks of gestation.

pared to those who did not. Further, in those patients who miscarried during early pregnancy, LH concentrations were similar to those who did not conceive but significantly higher than those whose pregnancy progressed.

What explanations can be offered for this apparent deleterious effect of high LH concentrations on embryo viability? Howard Jacobs and colleagues[23] have put forward the following hypothesis. Normally oocytes are held in the dictyotene stage of their first meiotic division until just before ovulation. Meiosis has been shown to be inhibited by oocyte maturation inhibitor (OMI), a low molecular weight peptide found in follicular fluid.[24] However, the midcycle LH surge inhibits the production and/or action of OMI, thus allowing maturation to occur at the appropriate time.[25] High concentration of LH prior to the ovulatory stimulus may prematurely initiate the resumption of meiosis, resulting in the release of post-mature oocytes. The most extreme examples of this effect are seen in LUF syndrome cases, where no fertili occurs. These oocytes if fertilised, develop into embryos with a low viability, which if they implant, result in an early pregnancy loss.

Some data from animals supports this hypothesis. In the sheep reduced fertilisation of oocytes and abnormal embryo development occurred in vivo after follicle growth in vitro had been carried out in the presence of elevated gonadotrophins.[26]

Secondly oocytes remain fertilizable for a longer period than they remain capable of giving rise to normal embryos. This is illustrated in the rat[27] and human[28] where increasing the interval between the known time of ovulation and insemination resulted in an increase in the rate of pregnancies which miscarried.

GnRH Analogue Agonists

How can late follicular phase levels of LH be reduced? Previously the effects of pharmaceutical agents on circulating gonadotrophin levels were briefly discussed. Treatment cycles which incorporate pure FSH were found to have lower levels of LH during this critical period.

A further reduction in endogenous LH can be achieved by employing a GnRH analogue agonist resulting in a transient state of hypogonadotrophic hypogonadism. A plethora of reports using these analogue agonists in combination with gonadotrophins have reported better pregnancy rates in patients with previously abnormal levels of LH or those who responded poorly to CC/hMG (eg. [29,30,31]) Many different types of protocol are used ranging from long to ultra-short. On the whole, better results are obtained from long protocols. This is probably because complete down regulation may not occur in short protocols prior to the critical LH sensitive phase in follicular growth. Secondly, progesterone may be inappropriately raised due to secretion from the rescued corpus luteum[32] which can result in a de-synchronization of endometrial development. In the ultra-short (SBG) protocol used at Bourn Hall, patients were screened by ultrasound scan and had P_4 measured prior to initiat-

Table 6. Outcome of Prospective Trial Comparing CC/hMG with a Short Buserelin Regimen (SBG)

		Treatment	
		CC/hMG	SBG
Patients		138	124
Discharged pre hCG		12	11
Oocytes/patients		7.6 ± 0.6	9.5 ± 1.0
Embryos replaced (preg)	0	5	2
	1	15 (1)	15 (2)
	2	14 (1)	15 (3)
	3	20 (6)	18 (9)
	4	72 (17)	63 (30)
Total Preg/replacement (%)		25/121 (21%)	44/111 (49%)

ing treatment. These precautionary measures resulted in a pregnancy rate equal to that reported for long protocols (Table 6).

In conclusion, the data presented suggest that superovulation can dramatically alter the inherent feedback mechanisms present in spontaneous cycles. High levels of LH in the late follicular phase of CC/hMG cycles seem to be associated with poor oocyte quality and lowered embryonic viability. By the judicious use of pharmaceutical preparations such as METRODIN and GnRH analogue agonists, tonic levels of LH can be reduced thus increasing the likelihood of a successful outcome.

ACKNOWLEDGEMENTS

The authors would like to express their gratitude to past and present colleagues at Bourn Hall for their contributions in this work, Ms Amanda Jemmett for typing the manuscript and Mrs C Clewlow for proof reading.

REFERENCES

1. R G Edwards, S B Fishel, J Cohen et al. Factors influencing the success of in vitro fertilization for alleviating human infertility. *J IVF & ET*, 1: 3, (1984).
2. P C Steptoe, R G Edwards, and D E Walters. Observations on 767 clinical pregnancies and 500 births after human in vitro fertilization. *Hum Reprod*, 1: 89, (1986).
3. J Testart, J Belaish-Allart, and R Frydman. Relationships between embryo transfer results and ovarian response and in vitro fertilization rate: analysis of the human pregnancies. *Fertil Steril*, 45: 237–243, (1986).

4. I Craft, T Al-Shawaf, P Lewis et al. Analysis of 1071 GIFT procedures—The case for a flexible approach to treatment. *The Lancet*, 1: 1094–1098, (1988).

5. R Punnonen, R Ashorn, P Vilja, et al. Spontaneous luteinizing hormone surge and cleavage of in vitro fertilized embryos. *Fertil Steril*, 49: 479–482, (1988).

6. L M Talbert. Endogenous luteinising hormone surge and superovulation. *Fertil Steril*, 49: 24–25, (1988).

7. I E Messinis, A Templeton, and D T Baird. Endogenous luteinising hormone surge during superovulation induction with sequential use of clomiphene citrate and pulsatile human menopausal gonadotrophin. *J Clin Endocrinol Metab*, 61: 1076–1080, (1985).

8. A Glasier, S S Thatcher, E J Wickings et al. Superovulation with exogenous gonadotrophins does not inhibit the luteinising hormone surge. *Fertil Steril*, 49: 81–85, (1988).

9. M M Quigley, A S Berkowitz, S A Gilbert et al. Clomiphene citrate in an in vitro fertilisation program: hormonal comparisons between 50- and 150-mg daily dosages. *Fertil Steril*, 41: 809–815, (1984).

10. S L Jeffcoate. The endocrine control of ovulation. *In*: IVF and Donor insemination. XIIth Study Group of the RCOG. RCOG Press, London. 1 (1985).

11. I E Messinis and A Templeton. Pituitary response to exogenous LHRH during superovulation induction with clomiphene is attenuated by a non-steroidal ovarian factor. *J Reprod Fert* Abstract Series No 2 (1988).

12. R F Caspar, H J Erskine, D T Armstrong et al. In vitro fertilization: diurnal and seasonal variation in luteinising hormone surge onset and pregnancy rates. *Fertil Steril*, 49: 644–648, (1988).

13. R Fleming, M J Haxton, M P R Hamilton et al. Successful treatment of infertile women with oligomenorrhea using a combination of an LHRH agonist and exogenous gonadotrophins. *Brit J Obstet Gynaecol*, 92: 369, (1985).

14. C M Howles, M C Macnamee and R G Edwards. The effect of progesterone supplementation prior to the induction of ovulation in women treated for IVF. *Hum Reprod*, 2: 91, (1987).

15. C M Howles, M C Macnamee and R G Edwards. Progesterone supplementation in the late follicular phase of an IVF Cycle: A 'natural' way to time oocyte recovery? *Hum Reprod*, 3: 409, (1988).

16. J H Liu and S S C Yen. Induction of midcycle gonadotrophin surge by ovarian steroids in women: a critical evaluation. *J Clin Endocrinol Metab*, 57: 797, (1983).

17. J D Stanger and J L Yovich. Reduced in vitro fertilization of human oocytes from patients with raised basal luteinising hormone levels during the follicular phase. *Brit J Obstet Gynaecol*, 92: 385, (1985).

18. C M Howles, M C Macnamee, R G Edwards et al. Effect of high tonic levels of luteinising hormone on outcome of IVF. *Lancet*, ii: 521, (1986).

19. C M Howles, M C Macnamee and R G Edwards. Follicular development and early luteal function of conception and non-conceptional cycles after human IVF: endocrine correlates. *Hum Reprod*, 2: 17, (1987).

20. M C Macnamee, C M Howles and R G Edwards. Pregnancies after IVF when high tonic LH is reduced by long-term treatment with GnRH agonists. *Hum Reprod*, 2: 569, (1987).

21. N A Abdulwahid, J Adams, Z M Van der Spuy et al. Gonadotrophin control of follicular development. *Clin Endocrinol*, 23: 613, (1985).

22. R Homburg, N A Armar, A Eshel et al. The influence of serum luteinising hormone concentrations on ovulation, conception and early pregnancy loss in patients with polycystic ovary syndrome. *Brit Med J*, 297: 1024, (1988).

23. H S Jacobs, R Porter, A Eshkol et al. Profertility uses of LHRH agonist analogues. *In:* "LHRH and its Analogues: Contraception and Therapeutic Application II". Eds: Vickery, B H and Nestor, J J, p 303. MTP Press, Lancaster, (1987).

24. S Winer-Sorgen, J Brown, T Ono et al. Oocyte maturation inhibitor activity in human follicular fluid: Quantitative determination in unstimulated and clomiphene citrate - and human menopausal gonadotrophin - stimulated ovarian cycles. *J IVF & ET*, 3: 218-223, (1986).

25. A Tsafriri and S H Pomerantz. Oocyte maturation inhibitor. *Clin Endocrinol Metab*, 63: 1284 (1986).

26. R M Moor and A O Trounson. Hormonal and follicular factors affecting maturation of sheep oocytes in vitro and their subsequent development capacity. *J Reprod Fertil*, 49: 101, (1977).

27. C R Austin. The Egg. *In:* Reproduction in mammals, Part 1 Germ Cells and Fertilization. 2nd Edn Cambridge Univ Press, Cambridge 58 (1982).

28. R J Guerrero and O I Rojas. Spontaneous abortion and ageing of human ova and spermatozoa. *New Eng J Medic*, 293, 573, (1975).

29. C M Howles, M C Macnamee and R G Edwards. Short term use of an LHRH agonist to treat poor responders entering an IVF programme. *Hum Reprod*, 2, 655, (1987).

30. A J Rutherford, R J Subak-Sharpe, K J Dawson et al. Improvement of in vitro fertilization after treatment with buserelin, an agonist of luteinising hormone releasing hormone. *Brit Med J*, 296, 1765, (1988).

31. H C Weise, K Fiedler and K Kato. Buserelin suppression of endogenous gonadotrophin secretion in infertile women with ovarian feedback disorders given human menopausal/human chorionic gonadotrophin treatment. *Fertil Steril*, 49, 399-403, (1988).

32. R Fleming, M E Jameson, M P R Hamilton et al. The use of GnRH analogues for combination with exogenous gonadotrophins in infertile women. *Acta Endocrinol* (Copenh), 19, Suppl 288: 77, (1987).

THE INFLUENCE OF AGE, UTERINE LENGTH, PRE-POST EGG RETRIEVAL, PLASMA ESTRADIOL DROP, AND THE NUMBER OF CLEAVED EMBRYOS TRANSFERRED TO THE UTERUS — ON OUTCOME FOLLOWING IVF/ET

G. Sher,   W.G. Vaught,   L.K. Vaught,   and V. K. Knutzen

Pacific Fertility Center
2100 Webster Street, Suite 220, San Francisco, California 94115

A study was undertaken to determine the effect of uterine depth, age of the female partner, the number of cleaved embryos transferred into the woman's uterus, and the periovulatory endocrine environment (as judged by the pre-post egg retrieval drop in plasma estradiol concentration), on clinical pregnancy rates following in vitro fertilization and embryo transfer (IVF/ET).

PATIENTS, METHODS AND MATERIALS

One hundred and sixteen patients under the age of 41 years were evaluated prospectively. In all cases, the cause of infertility was due to tubal occlusion, unrelated to endometriosis. None of the male or female partners had detectable amounts of antisperm antibodies as measured by the serum immunobead assay,[1] and all male partners had normal semen analyses, as assessed by sperm count, sperm motility, and sperm morphology. None of the female partners had been exposed to diethylstilbestrol (DES) in utero, nor did any have existing uterine pathology, congenital abnormalities of the uterus, or previous uterine surgery. At least one cleaved embryo was transferred in all cases.

All women underwent controlled ovarian hyperstimulation (COH) with human Menopausal Gonadotropin (hMG - Pergonal, Serono Laboratories) administered intramuscularly

Advances in Assisted Reproductive Technologies, Edited by
S. Mashiach *et al.,* Plenum Press, New York, 1990

from day 2 through day 8–10 of the cycle. The total dosage of hMG administered ranged from 1200 to 3000 IRP Units per cycle of treatment, and the duration of hMG administration was adjusted based upon serial plasma estradiol and follicular ultrasound evauations undertaken from cycle day 9. As soon as the plasma estradiol concentration reached or exceeded 550 pg/ml, and at least two ovarian follicles measured more than 18 millimeters, hMG therapy was discontinued for a period of 30 to 55 hours. Thereupon, hCG, 10,000 u, was administered intramuscularly. Blood was drawn for plasma estradiol measurement less than 12 hours after the hCG injection, and provided that the plasma estradiol concentration had shown a progressive rise, a laparoscopic oocyte retrieval was performed 33 to 35 hours following the hCG administration. All embryos were transferred within 48 to 55 hours following egg retrieval. The female partners all received 50 mg of Progesterone intramuscularly from the day of egg retrieval (i.e. luteal phase day 1 - LP1) through LP13. Quantitative serum Beta hCG assays were performed on LP11 and LP13 respectively. Evidence of a doubling effect in plasma hCG concentration was followed by discontinuation of Progesterone injections, and the intramuscular administration of hCG, 5,000 u, three times a week for an additional three weeks. Thereupon, a pelvic ultrasound examination was performed to look for one or more gestational sacs, and thereby diagnose a clinical pregnancy.

This study assessed the effect of four independent variables on pregnancy outcome following IVF/ET:

1.   Age: Women under the age of 41 years were categorized in two groups, those less than 35 years of age, and those between 35 and 41 years of age.

2.   Uterine depth: The length of the uterine cavity was measured from the ectocervic to the top of the uterine cavity through the transcervical introduction of a Hassen winged sound.

3.   Pre-post egg retrieval percentage drop in plasma estradiol concentration: Blood was drawn for plasma estradiol measurement less than 12 hours following hCG administration(A), and again 18–24 hours following egg retrieval (B). The percentage drop in plasma estradiol was calculated as follows:

$$\% \, E_2 \text{ drop} = \frac{A-B}{A} \times 100$$

4.   The number of embryos transferred to the uterus: Up to six, non-polyspermic embryos were transferred transcervically via a Jones catheter. Remaining embryos were cryopreserved, or discarded, based upon the patients' directive.

RESULTS

Figure 1 demonstrates the effect that age has on outcome following IVF/ET. It illustrates that women less than 35 years of age have an increased chance of achieving a clinical pregnancy following IVF/ET than do their older counterparts (P = 0.06).

Figure 2 clearly demonstrates that uterine depth influences outcome following IVF/ET. A sharp decline occurs in clinical pregnancy rates when uterine depth is less than 69 mm (P = 0.07).

Figure 3 demonstrates that transferring four to six embryos into the uterus improves the clinical pregnancy rate (P = 0.06).

Figure 4 demonstrates that when the pre-post egg retrieval percentage drop in plasma estradiol concentration exceeds 79 percent, the clinical pregnancy rate increases significantly (P = 0.03).

Table 1 presents the results of bilinear probit regression analysis of this data, and indicates that each variable has a compounding effect on the other as far as clinical pregnancy rates are concerned (P = 0.005). Figure 5 illustates this effect graphically.

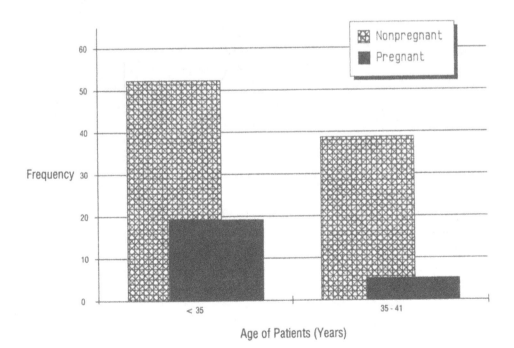

Fig. 1. Relationship between age of patient and pregnancy status.

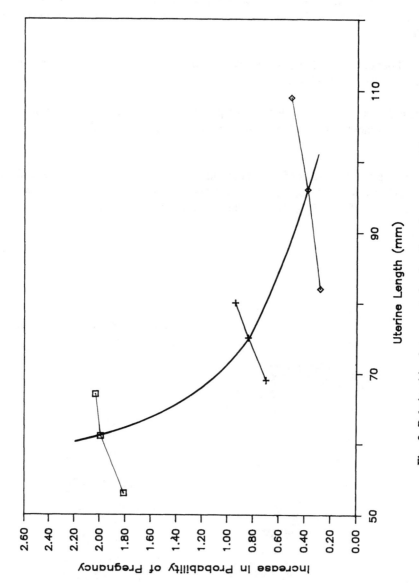

Fig. 2. Relationships between uterine length and pregnancy status.

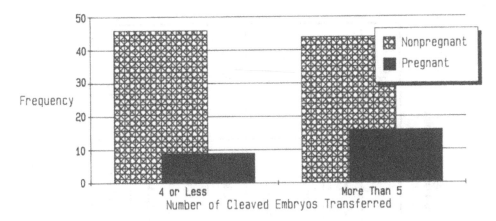

Fig. 3. Relationship between cleaved embryos transferred and pregnancy status.

DISCUSSION

The authors have previously reported on the influence of the female partner's age, the number of embryos transferred, uterine depth on clinical pregnancy rates following the performance of IVF/ET.[2] In addition, the results of this study clearly demonstrate that the percentage drop in plasma estradiol concentration from the immediate pre- to the immediate post-egg retrieval period influences success rates following IVF/ET. Furthermore, each of the abovementioned variables exerts both an independent and compounding effect on clinical pregnancy rates.

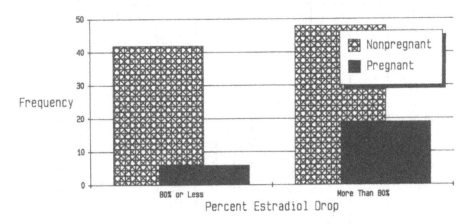

Fig. 4. Relationship between estradiol drop and pregnancy status.

Table 1. Probit Regression Analysis of Clinical Pregnancy Rate Using Three Explanatory Variables

VARIABLE	b_i COEFFICIENT	STD ERROR	Z	P VALUE
Ut Lt - 68mm or less	0.049417	0.033895	1.4579	0.07
Ut Lt - 68mm to 82mm	0.029639	0.027833	1.0649	0.14
Ut Lt - 83mm or above	0.019531	0.023108	0.84518	0.20
No of Embryos	0.45413	0.29607	1.5339	0.06
E_2 Drop (%)	0.55671	0.30724	1.8120	0.03
Age	-0.064327	0.040610	-1.5840	0.06
Constant	-1.4997	2.2730	-0.65978	0.25

Likelihood ratio test   LR = 19.9708   6 DF $p \leq 0.005$

It can be concluded that a woman aged less than 36 years, who has a pre-post egg retrieval plasma estradiol drop of greater than 79 percent, has a uterine length measuring more than 69 mm, and who subsequently has four to six embryos transferred to her uterus, has an optimal chance of achieving a clinical pregnancy following IVF/ET.

The manner in which the above-described variables affect pregnancy outcome is unclear. However, the following explanations are plausible.

1. The effect of age: It is possible that age impacts on the quality of oocytes, the potential of the uterine lining to be properly prepared for implantation (i.e. the normalcy of the follicular and luteal phases), and that pathological conditions such as uterine myomata, endometriosis or uterine polypi might be more likely to coexist in the older woman, thereby threatening successful implantation.

2. Uterine depth: It is postulated that smaller uteri have smaller volumes, and, accordingly, that embryos transferred to a small uterus would have a greater potential to reflux into the cervical canal, and vagina, or escape into or through the fallopian tubes following embryo transfer. This thesis is supported by the results of a separate presentation to this forum, which reveals this effect fluoroscopically.[3]

3. The number of embryos transferred to the uterus: It is probable that the more embryos transferred to the uterus, the more likely it is that at least one will be retained and will subsequently implant. This study demonstrates that the transfer of four to six embryos to a woman's uterus optimizes the chances of at least one embryo implanting successfully. The positive effect that the transfer of a large number of em-

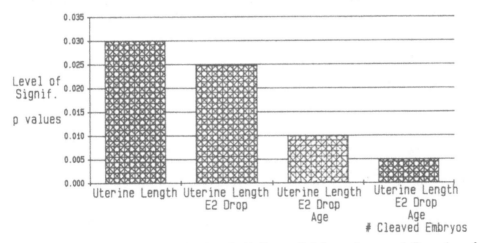

Fig. 5. Relationships between a) uterine length; b) % estradiol drop; c) age; and d) number of cleaved embryos on clinical pregnancies.

bryos to a woman's uterus has on clinical pregnancy rates has been clearly illustrated in a separate presentation to this forum. However, the same study demonstrates as significant increase in the incidence of multiple pregnancies.[4]

4.   Percentage drop in pre-post egg retrieval plasma estradiol concentration: This study clearly demonstrates that a pre-post egg retrieval drop in plasma estradiol concentration of greater than 79 percent increases the clinical pregnancy rate following IVF/ET. Explanations for this finding are at best speculative. However, it is possible that a hyperestrogenized endometrium in the immediate post-egg retrieval period might deter successful implantation in a similar manner to that which occurs when high doses of estrogen are administered to a woman following intercourse during the immediate post-ovulatory period of a natural cycle.

REFERENCES

1. Bronson, R., Cooper, G., Rosenfeld, D. L., Membrane-bound sperm-specific antibodies: their role in infertility, *in:* "Bioregulators in Reproduction", H. Vogel and G. Jaquello, eds., Academic Press, New York, 512 (1981).
2. Sher, G., Knutzen, V. K., Stratton, C. J., Chotiner, H. C., and Mayville, J., In vitro fertilization and embryo transfer - two years experience, *J. Obst. and Gyne.*, 67:3, 309-315 (1986).
3. Knutzen, V. K., Soto-Albors, C., Fuller, D., Sher, G., Shyrock, K., and Behr, B., Mock embryo transfer (MET) in early luteal phase, the cycle prior to in vitro fertilization and embryo transfer (IVF/ET): a preliminary report. VIth World Congress of In Vitro Fertilization and Alternate Assisted Reproduction, Jerusalem, Israel, April, 1989 (Abstract).
4. McNamee, P. I., Chun, B., Huang, T., Kosasa, T., Morton, C., and Terada, F., Experience with high ovarian stimulation, coupled with large numbers of embryos transferred per patient in an in vitro fertilization and embryo transfer IVF/ET program; VIth World Congress of In Vitro Fertilization and Alternate Assisted Reproduction, Jerusalem, Israel, April, 1989 (Abstract).

THE TIMING OF EMBRYO TRANSFER AND IMPLANTATION

Eliahu Caspi, Raphael Ron-El, Abraham Golan, Arie Herman,
Hana Nachum, Yigal Soffer, and Zvi Weinraub

*Department of Obstetrics and Gynecology, Assaf Harofe Medical Centre
Zerifin, Sackler School of Medicine, Tel-Aviv University, Israel*

The success rate in terms of clinical pregnancy after in vitro fertilization (IVF) and embryo transfer (ET) is mainly dependent on the number and quality of embryos transferred to the uterus. The optimal timing for transfer to the uterus of the developing embryos and the relationship between the cleavage stage at transfer and outcome are little explored.

We have previously investigated on a rather small number of cycles, the outcome of the transfer of four embryos on day 2 or day 3 after insemination or equally divided into two sequential transfers at day 2 and 3.[1] This prospective randomized trial showed a non-significant higher pregnancy rate at day 3.

The present results are those of a subsequent randomized study involving the transfer of four embryos on day 2 (44–48 hours) after insemination or delaying the transfer to day 3 (68–72 hours).

MATERIALS AND METHODS

The study includes 58 cycles in which at least four embryos were observed at 44–48 hours after insemination. They were randomly and equally allocated into the following two transfer protocols. In the first protocol of early transfer, the four embryos were transferred at 44–48 hours after insemination. In the second protocol the four embryos were transferred 68–72 hours after insemination. In both protocols the morphologically best embryos were

Advances in Assisted Reproductive Technologies, Edited by
S. Mashiach *et al.*, Plenum Press, New York, 1990

Table 1. Clinical Characteristics of the Patients

	Early transfer	Late transfer
No.of cases (cycles)	29	29
Women's age (mean ± SD)	32.9 ± 4.7	32.5 ± 3.9
Years of infertility (mean ± SD)	5.5 ± 2.8	5.5 ± 3.1
Tubal infertility	12	16
Unexplained infertility	14	7
Other categories	3	6

Table 2. Mode of Induction of Ovulation, Oocytes, Fertilization, Cleavage Rate and Embryos

	Early transfer	Late transfer
No. of cases (cycles)	29	29
D-TRP^6-/hMG cycles	21	23
hMG cycles	8	6
No. of oocytes	251	279
No. of oocytes/cycles	8.7	9.6
Fertilization rate	185/251 (74%)	198/279 (71%)
Cleavage rate	96%	94%
No. of embryos	178	187
No. of embryos frozen	42	35

transferred. At 44–48 hours and at the late transfer all embryos were observed for the number and morphology of the blastomeres which were graded according to Plachot et al.[2] The remaining good embryos of cycles with more than four were frozen. Table 1 shows the clinical characteristics of the patients. Table 2 shows the mode of induction of ovulation, the number of oocytes retrieved, the fertilization and cleavage rate and the number of embryos frozen in each study group. All these parameters were similar with no statistically significant difference.

RESULTS

Table 3 shows the results in terms of clinical pregnancies, proportion of embryos implanted and abortions which are very similar for the early and late transfers. The mean number of blastomeres of embryos transferred late was significantly (Table 4, $p < 0.001$) higher than those transferred early but no significant difference was found between the number of blastomeres and the proportion of good embryos (grade I-II) of pregnancy cycles as compared to no pregnancy cycles in all cases of both groups.

Table 5 shows the morphological quality of the embryos at early and late transfers. At 68–72 hours after insemination 19% of all the embryos (36/187) were of poor quality

Table 3. Results

	Early transfer	Late transfer
No. of cases (cycles)	29	29
No. of embryos transferred	116	116
No. of pregnancies	15 (52%)	16 (55%)
No. of fetuses	17	21
Proportion of embryos implanted	17/116 (15%)	21/116 (18%)
Multiple pregnancy rate	2/15 (13%)	5/16 (31%)
Abortions	3/15 (20%)	3/21 (14%)

Table 4. Number (Mean + SD) of Blastomeres and Morphology of Embryos in Early and Late Transfer, in Pregnancy and No Pregnancy Cycles

No. of blastomeres in:

Early transfer (range 2–6)	3.4 ± 1.1	$P < 0.001$
Late transfer (range 2–8)	5.5 ± 1.6	
Pregnancy cycles	5.1 ± 1.2	NS
No. pregnancy cycles	4.9 ± 1.4	

No. of blastomeres with good morphology (grade I–II)

Pregnancy cycles	85/124 (69%)	NS
No pregnancy cycles	69/108 (64%)	

Table 5. Morphology of Embryos at Early and Late Transfers

Morphological grade	Early transfer		Late transfer
I	105		93
II	53		58
III	15		14
	20 (11%) {	$p < 0.05$	} 36 (19%)
IV	5		22
Total	178		187

(grades III-IV) as compared to 11% (20/178) out of all the embryos at 44–48 hours after insemination, the difference is statistically significant (p < 0.05). Table 6 shows the morphological grade of all the embryos of the group with late transfer at 44–48 hours and 68–72 hours. The proportion of embryos with poor morphology (grade III-IV) rose from 14% (26/187) at 44–48 hours after insemination to 19% (36/187) after an additional incubation of 24 hours. This difference is not significant statistically.

Table 6. Morphology of Embryos of the Group with Late Transfer at 48 and 72 hours after Insemination

Morphological grade		48 hours		72 hours	
I		96		93	
II		65		58	
III		10		14	
	26 (14%)	{	NS	}	36 (19%)
IV		16		22	
Total		187		187	

Table 7. Pregnancy Rate of Two Controlled Studies in which Three or Four Embryos were Early or Late Transferred

	Early transfer		Late transfer	
	No. of transfers	Pregnant	No. of transfers	Pregnant
Edwards et al. (3 embryos/transfer)	24	8	19	9
Caspi et al. (4 embryos/transfer)	29	15	29	16
	53	23 (43%)	48	25 (52%)

DISCUSSION

Success of ET depends on many factors. The importance of some of them such as the number and quality of the embryos are quite known while the influence of other variables such as embryonal stage of development and endometrial condition at transfer are less clear. The present study shows that the transfer of four embryos, most of them of good morphology, at any time after insemination is associated with a high pregnancy rate and high multiple pregnancy rate. This is in accordance with other observations.[3, 4]

The incubation of the embryos for 24 hours more than usual is associated with a significantly more advanced stage of cleavage. The transfer of such embryos at 68–72 hours after insemination was not associated in this study with a significant improvement of pregnancy rate although the proportion of embryos implanted was higher, (18% versus 15% in early transfer) resulting in a higher multiple pregnancy rate of 31% versus 13% in the early transfer group. Two other studies have compared the pregnancy rate after early and late transfers similarly to the present study. The first one by Edwards et al.[5] was a controlled trial in which one, two or three embryos were transferred early or late and the pregnancy rate was 30% in late transfer and 24.7% in the early, the difference was not statistically significant.

The second study reported by Jones[6] was not a controlled one and the number of embryos transferred varied from one to six resulting in a pregnancy rate of 37% after late transfer compared to 26% after early transfer but in the group with late transfer there were a relatively large number of patients with multiple embryo transfers so that statistical comparison is impossible. If only the two controlled studies (Table 7) in which three embryos (Edwards et al.) or four embryos (the present study) are evaluated, then the late transfer was associated with a higher pregnancy rate of 52% per transfer compared with 43% after early transfer, the difference is statistically not significant. Thus it may be concluded that there is a slight trend of a beneficial pregnancy rate after late transfer and a further study involving a larger group of patients might show a statistically significant benefit in the transfer of more advanced developmental embryonal stage.

On the other hand an additional incubation time of 24 hours was associated with an increase (Table 6) in the number of embryos of poor morphology from 14% at 44–48 hours to 19% at 68–72 hours. This increase might be the result of a deleterious effect of laboratory conditions on the additional incubation time or simply a delayed manifestation of an initial inherent poor quality of these embryos. In any case this additional incubation time did not affect the implantation potential of the embryos transferred later, since the proportion of embryos implanted was higher at the late transfer.

This study indirectly indicated that the receptivity of the endometrium is probably similar for embryos transferred 44–48 hours or 68–72 hours after insemination.

REFERENCES

1. Caspi E, Ron-El R, Golan A, Herman A, Nachum H. Early, late and sequential embryo transfer in in vitro fertilization. *Fertil Steril* - in press.
2. Plachot M, Junca AM, Mandelbaum J, Cohen J, Salat-Baroux J, Da Lage C: Timing of in vitro fertilization of cumulus - free and cumulus - enclosed human oocytes. *Hum Reprod* 1: 237, 1986
3. Speirs AL, Lopata A, Gronow MJ, Kellow GN, Johnston WIH: Analysis of the benefits and risks of multiple embryo transfer. *Fertil Steril* 39:468, 1983
4. Yovich JL: Embryo quality and pregnancy rates in in-vitro fertilization. *Lancet* 1:283, 1985
5. Edwards RG, Fishel SB, Cohen J, Fehilly CB, Purdy JM, Slater JM, Steptoe PC, Webster JM. Factors influencing the success of in Vitro Fertilization for alleviating Human Infertility. *J In Vitro Fert Embryo Transfer* 1:2, 1984
6. Jones HW, Jr: Embryo transfer. *Ann NY Acad Sci* 442:375, 1985

OOCYTE QUALITY ACCORDING TO PROTOCOLS FOR CONTROLLED OVARIAN HYPERSTIMULATION AND PATIENTS' AGE

Ettore Cittadini and Roberto Palermo

Istituto Materno-Infantile, University of Palermo
Palermo, Italy

INTRODUCTION

In in vitro fertilization (IVF) and embryo transfer (ET), the induction of multiple follicular growth represents the first and necessary step for the efficient performance of the technique.

An ideal stimulation cycle is one in which steroidogenesis and folliculogenesis are adequate and multiple fertilizable oocytes are harvested with a high rate of transfer of multiple and rapidly cleaving embryos. Furthermore, an ideal cycle is one in which the embryonal stage and emdometrial maturation are synchronized for embryo implantation and growth.

Unfortunately, with any stimulation strategy, a wide range of ovarian responses may be expected and a relatively high proportion of these responses deviate from an ideal cycle.[1-3]

The central point concerning superovulation in the human female is the inherent asynchronism of pregonadotropin-dependent follicular cohort, a basic mechanism regulating the selection and dominance of a single ovulatory follicle.[4,5] As a consequence, with any regimen adopted for inducing multiple follicular growth, different nuclear status, functional viability and morphological appearance of the oocytes are the effect of the asynchronized growth.

Advances in Assisted Reproductive Technologies, Edited by
S. Mashiach *et al.,* Plenum Press, New York, 1990

OOCYTE QUALITY CLASSIFICATION

We are able to classify different categories of the maturational stage of the harvested oocytes evaluating the morphology of the oocyte-corona-cumulus complex (OCCC) and the nuclear status.

Mature oocytes (Figure 1a) represent the majority of the harvested oocytes (approximately 80%). Cumulus mass appears very expanded with a stretching ability of more than 5 cm. The cells of the corona layer have a radiant appearance and are expanded. The ooplasm is lightly colored and homogenous in granularity. An extruded polar body (PB) may be observed (this defines a Metaphase II oocyte). The lack of visualiziation of PB classifies a Metaphase I-mature oocyte.

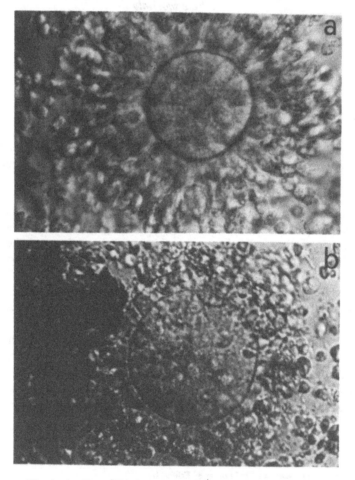

Fig. 1a–b. Example of a mature (a) and postmature (b) oocyte.

Postmature oocytes (Figure 1b) represent approximately 10% of total oocytes harvested. They usually show a scanty cumulus mass with granular appearance, clumped and slightly darkened cells. Stretching is still more than 5cm. The corona layer is expanded, separated from the zona pellucida and it shows slightly darkened cells. However, ooplasm appears granular and homogenous. The PB is usually observed.

"Luteinized oocytes" (Figure 1c) represent approximately 5% of total oocytes. A very darkened appearance of the corona cells (which is sometimes absent) and the cells of the cumulus mass, is the characteristic of this type of oocyte. The ooplasm is also darkened and may be vacuolized. The PB is often fragmented.

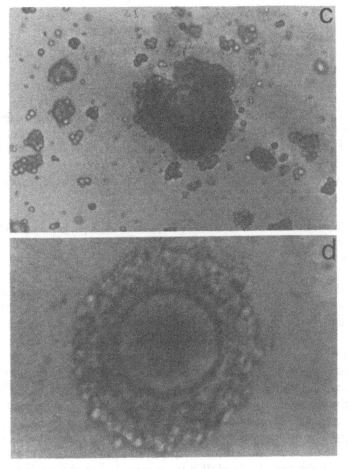

Fig. 1c–d. Example of a "luteinized" (c) and immature (d) oocyte.

Immature oocytes (Figure 1d) account for almost 5% of total oocytes harvested. This type of oocyte shows a non-stretchable, tightly packed cumulus mass. The corona is usually compact, and very close to the zona pellucida. Ooplasm is granular or smooth. The germinal vesicle is usually visible especially after the removal of the cumulus mass.

Atretic oocytes, very irregular in shape and with a degenerative appearance of the OCCC and ooplasm, are very rare. Usually these oocytes are not inseminated. This type of oocyte will not be taken into consideration for the analysis of the data.

This classification shows a significant correlation with the relative suitability of the oocytes within each category. Each class of oocyte shows, in the presence of adequate and fully capacitated sperms, a significant difference of the cleavage rates. Thus, mature, postmature, "luteinized" and immature oocytes differ significantly in cleavage rate in decreasing order.

PROTOCOLS FOR CONTROLLED OVARIAN HYPERSTIMULATION (COH)

Five protocols have been used for COH during a period of three years (1985-1988).

Protocol I was the sequential dosage of 300 I.U. of urinary follicle stimulating hormone (FSH), (hU-FSH, Urofollitropin, Metrodin, Serono Lab., Rome, Italy) on cycle day 3 and 4 and 150 I.U. of human menopausal gonadotropin (hMG) (Pergonal, Serono Lab., Rome, Italy) daily from day 5 until at least 600 pg/ml of plasma estradiol (E_2) were detected and at least two follicles more than or equal to 16 mm were observed at pelvic ultrasound.

Protocol II was the sequential dosage of 100 mg of Clomiphene Citrate (CC) (Clomid, Merrel, Rome, Italy) on cycle day 3 to 7 and 150 I.U. of hMG daily from day 6. In this protocol the same criteria were followed for menotropin discontinuation.

Protocol III was the concomitant administration of a gonadotropin-releasing hormone agonist (GnRHa) (Buserelin, Suprefact, Hoechst Pharmaceuticals, Frankfurt, West Germany) and menotropin. GnRHa was used to achieve pituitary desensitization and was begun on day 21 of the cycle preceding the stimulation with subcutaneous injection of 300 μg morning and evening. Menotropin stimulation was begun on day 3 to 5 provided that pituitary-ovarian function was adequately suppressed (E_2 less than 20 pg/ml). HMG and hU-FSH were used in combination beginning with 150 I.U. of hU-FSH plus 75 I.U. of hMG for three to five days with adjustments in the subsequent dosage according to ovarian response. Concomitant GnRHa-menotropin stimulation was continued until plasma E_2 was at least 600 pg/ml in the presence of at least two follicles of 17 mm at pelvic ultrasound.

Protocol IV differs from Protocol III in the intranasal route of administration of the agonist. Dosage was 200 μg every eight hours.

Table 1. Stimulation and Pregnancy Outcome Among COH Protocols

	PROT.I	PROT.II	PROT.III	PROT.IV	PROT.V
Stimulated cycles	340	188	135	89	54
No. of aspirations	200	199	98	60	34
(%)	58.8^a	$63.2^{a,b}$	72.6^b	67.4^b	62.9^b
Foll.> 15 mm/aspiration	7.9^c	5.9^d	$7.0^{c,d}$	$7.3^{c,d}$	$7.5^{c,d}$
Days of stimulation	9.1	9.5	12.1	11.7	13.1
Favourable E_2 pattern (%)	80.0	76.0	90.8	85.0	88.2
No. of transfers	166	95	81	49	27
No. of pregnacies	19	10	13	6	2
Pregnancy rate/ET	11.4	10.5	16.0	12.2	7.4
No. of miscarriages	5	5	3	4	2

(a) Protocol I vs Protocol II: $P < 0.001$; (b) Protocol II vs Protocols III, IV and V: $P < 0.05$; (c,d) Protocols I and II vs Protocols III, IV and V: $P < 0.001$.

Protocol V was the concomitant GnRHa-menotropin beginning the agonist on cycle day 1, and menotropin stimulation on cycle day 15 provided the adequate suppression. Dosage of the GnRHa and criteria for combined treatment discontinuation were the same followed for Protocols III and IV.

Ten thousand I.U. of human chorionic gonadotropin (hCG) (Profasi, Serono Lab., Rome Italy) were injected intramuscularly 24 to 36 hours after discontinuation of stimulation in all protocols. Oocyte retrieval was scheduled 34 to 36 hours after hCG injection and was carried out by a transvaginal or transurethral/transvesical ultrasound-guided approach.

Natural progesterone (P) in oil (Proluton, Schering, Milan, Italy),12.5 to 25.0 mg daily, was administered intramuscularly for luteal phase supplementation for two to three weeks, beginning on the day of ET.

STIMULATION OUTCOME AND OOCYTE CLASS PROPORTIONS AMONG PROTOCOLS

Comparison of stimulation and pregnancy outcome is shown in Table 1. Eight hundred and six cycles were analyzed. The cancellation rate was high in all protocols, expecially in Protocol I and II (only 58% and 63% had an oocyte retrieval). The cancelled cycles in protocols with pituitary desensitization were almost completely due to poor responses. A significant lower mean number of preovulatory follicles were stimulated using the sequence CC/hMG/hCG ($P < 0.05$). When concomitant pituitary desensitization was used a significantly longer period of treatment was required for adequate ovarian stimulation ($P < 0.001$). The

Table 2. Proportions of Oocyte Classes Among COH Protocols

OOCYE CLASS	PROT.I		PROT.II		PROT.III		PROT.IV		PROT.V	
	N.	(%)	N.	(%)	N.	(%)	N.	(%)	N.	(%)
Mature	920	(85.1)[a]	341	(73.3)[a]	362	(82.8)	213	(79.8)	143	(82.2)
Postmature	75	(6.9)[b]	70	(15.1)[b,c]	51	(14.1)	31	(11.6)	19	(10.9)[c]
"Luteinized"	50	(4.6)	25	(5.4)	17	(4.7)	12	(4.5)	8	(4.6)
Immature	36	(3.3)	29	(6.2)	7	(1.9)	11	(4.1)	4	(2.3)

(a) Protocol I vs Protocol II: $P < 0.05$; (b,c) Protocol II vs Protocols I and V: $P < 0.001$.

classification of E_2 patterns was simplified by differentiating only favourable and unfavourable patterns.[6] Any pattern in which no drop in E_2 values was observed between hCG day and the subsequent day was classified as favourable. No significant difference was observed in the proportions of E_2 patterns among protocols. No difference occurred in the pregnancy rate among protocols but only a trend of a better result when Protocol III was used.

Table 2 summarizes the proportions of oocyte classes among protocols. A lower proportion of mature oocytes was retrieved when Protocol II was used if compared with Protocol I ($P < 0.05$). The proportion of postmature oocytes was higher in Protocol II when compared with Protocol I and V. ($P < 0.001$). No differences occurred in the comparison of "luteinized" and immature oocytes among protocols. No clear evidence could be observed of a better synchronization of stimulated follicular growth with the use of concomitant pituitary desensitization and menotropin stimulation.

STIMULATION OUTCOME AND OOCYTE QUALITY
ACCORDING TO WOMEN'S AGE

The data concerning the 806 stimulation cycles were analyzed according to three age groups: ≤30 years, 31 to 35 years, and ≥36 years.

Almost 50% of the cycles (381 cycles) were carried out in the group ≥36 years of age. The women belonging to the intermediate age group underwent 266 cycles (32.9%), while only 160 cycles (19.9%) were carried out in the group ≤30 years of age.

Tables 3 and 5 summarize the stimulation and pregnancy outcome of the three age groups. The group of women ≥36 years of age had a trend of cancellation rate higher than younger women. In all age groups a significantly longer period of treatment was required for adequate follicular growth when GnRHa-menotropin stimulation was used. No differ-

Table 3. Stimulation and Pregnancy Outcome Among COH Protocols (≤30 years)

	PROT.I	PROT.II	PROT.III	PROT.IV	PROT.V
Stimulated cycles	64	40	31	14	11
No. of aspirations	44	27	27	10	8
(%)	68.8	67.5	87.1	71.4	72.7
Foll. >15 mm/aspiration	7.9[a]	6.2[b]	7.6[a,b]	7.6[a,b]	8.0[a,b]
Days of stimulation	9.1	9.5	12.3	12.1	13.5
Favourable E_2 pattern (%)	82.0	85.0	80.0	87.0	84.0
No. of transfer	32	22	17	7	5
No. of pregnancies	5	3	7	2	0
Pregnancy rate/ET	15.6	13.6	41.2	28.6	–
No. of miscarriages	2	1	0	1	–

(a,b) Protocols I and II vs Protocols III, IV and V: $P < 0.001$).

Table 4. Stimulation and Pregnancy Outcome Among COH Protocols (31–35 years)

	PROT.I	PROT.II	PROT.III	PROT.IV	PROT.IV
Stimulated cycles	117	73	37	23	15
No. of aspirations	81	51	27	16	10
(%)	69.2[a]	69.9[a,b]	73.0[b]	69.6[b]	66.7[b]
Foll. >15 mm/aspiration	7.0[c]	5.7[d]	7.0[c,d]	7.2[c,d]	7.2[c,d]
Days of stimulation	9.3	9.1	12.1	11.7	12.2
Favourable E_2 pattern (%)	80.0	70.0	85.0	85.0	82.0
No. of transfers	71	43	25	16	8
No. of pregnancies	8	6	1	0	1
Pregnancy rate/ET	11.3	14.0	4.0	–	12.5
No. of miscarriages	1	3	1	–	1

(a,b) Protocol II vs Protocols I, II, III, IV and V: $P < 0.001$;
(c,d) Protocol I and II vs Protocols III, IV and V: $P < 0.001$.

ences occurred in E_2 patterns among protocols in the age groups. A trend, but not of statistical significance, of a better pregnancy rate/ET was observed with the use of "desensitized" cycles (except for Protocol V) in the group of patients ≤30 years of age. A very low pregnancy rate (no viable pregnancies) occurred in the group 31 to 35 years of age with the use of concomitant GnRHa-menotropin stimulation, while a trend of a better pregnancy rate/ET was observed in "desensitized" cycles in the group of women ≥36 years of age. Consid-

Table 5. Stimulation and Pregnancy Outcome Among COH Protocols (≥36 years)

	PROT.I	PROT.II	PROT.III	PROT.IV	PROT.V
Stimulated cycles	144	83	72	55	27
No. of aspirations	80	42	44	30	14
(%)	55.6	50.6	61.1	54.5	51.9
Foll.> 15 mm/aspiration	5.2	5.3	6.3	5.7	6.3
Days of stimulation	9.1^a	10.2^b	$12.2^{a,b}$	$12.3^{a,b}$	$13.0^{a,b}$
Favourable E_2 pattern (%)	70.0	68.0	80.0	84.2	78.3
No. of transfers	63	30	39	26	14
No. of pregnancies	6	1	5	4	1
Pregnancy rate/ET	4.8	3.3	12.8	15.4	7.1
No. of miscarriages	2	1	2	3	1

(a,b) Protocol I and II vs Protocols III, IV and V: $P < 0.001$.

Table 6. Proportions of Oocyte Classes Among Protocols (≤30 years)

OOCYTE CLASS	PROT.I		PROT.II		PROT.III		PROT.IV		PROT.V	
	N.	(%)	N.	(%)	N.	(%)	N.	(%)	N.	(%)
Mature	216	(87.8)	103	$(78.0)^{ab}$	108	$(92.3)^b$	28	(84.4)	41	$(95.3)^a$
Postmature	5	$(2.0)^{c,d}$	9	$(6.8)^c$	9	$(7.7)^c$	6	$(18.1)^d$	2	(4.6)
"Luteinized"	16	(6.5)	10	(7.5)	0		0		0	
Immature	9	(3.7)	10	(7.5)	0		0		0	

(a) Protocol II vs Protocol V: $P < 0.01$; (b) Protocol II vs Protocol III: $P < 0.02$; (c) Protocol I vs Protocol II and III: $P < 0.01$; (d) Protocol I vs Protocol IV: $P < 0.001$.

ering all protocols together, the pregnancy rate/ET was significantly higher ($P < 0.05$) in the group of age ≤30 years (20.3%) than in the group of women of 31 to 35 years (9.8%) and in the group of women ≥36 years of age (9.5%).

Tables 6 through 8 give the comparison of the proportions of the oocyte classes among protocols and age groups.

Women ≤30 years old had a higher proportion of mature oocytes using Protocols III and IV if compared with Protocol IV ($P < 0.001$) and Protocol II and III ($P < 0.05$). Inter-

Table 7. Proportions of Oocyte Classes Among Protocols (31–35 years)

OOCYTE CLASS	PROT.I		PROT.II		PROT.III		PROT.IV		PROT.V	
	N.	(%)	N.	(%)	N.	(%)	N.	(%)	N.	(%)
Mature	378	(85.1)^{a,c}	141	(71.5)^{a,b}	95	(74.2)$_d$	76	(91.5)^b	29	(70.7)^c
Postmature	34	(7.6)^e	33	(16.7)$_d$	21	(16.4)$_f$	1	(1.2)^{d,e}	8	(19.5)^d
"Luteinized"	19	(4.2)	5	(2.5)$_h$	12	(9.3)^f	5	(6.0)$_{g,h}$	0	
Immature	13	(2.9)	18	(9.1)^h	0		1	(1.2)^{g,h}	4	(9.7)^g

(a,c) Protocol I vs Protocol II and V: P<0.001; (b) Protocol II vs Protocol IV: P<0.001; (d) Protocol IV vs Protocol II,III and V: P<0.001; (e) Protocol IV vs Protocol I: P<0.05; (f) Protocol II vs Protocol III: P<0.001; (g) Protocol IV vs Protocol V: P<0.05; (h) Protocol IV vs Protocol II: P<0.01.

Table 8. Proportions of Oocyte Classes Among Protocols (≥36 years)

OOCYTE CLASS	PROT.I		PROT.II		PROT.III		PROT.IV		PROT.V	
	N.	(%)	N.	(%)	N	(%)	N.	(%)	N.	(%)
Mature	326	(83.4)^a	97	(71.3)^{a,b}	160	(83.3)^b	112	(74.1)$_d$	71	(78.9)
Postmature	36	(9.2)$_{c,d}$	28	(20.5)^c	21	(10.9)$_{e,f}$	24	(15.8)^d	9	(10.0)
"Luteinized"	15	(3.8)^f	10	(7.4)^e	5	(2.6)^{e,f}	7	(4.6)^g	8	(8.8)
Immature	14	(3.5)	1	(0.7)^g	6	(3.1)	8	(5.3)^g	2	(2.2)

(a,b) Protocol II vs Protocol I nad III: P<0.05; (c) Protocol I vs Protocol II: P<0.001; (d) Protocol IV vs Protocol I: P<0.05; (e) Protocol III vs Protocol V and II: P<0.05; (f) Protocol III vs Protocol II: P<0.05; (g) Protocol IV vs Protocol II: P<0.05.

estingly enough, when concomitant GnRHa-menotropin stimulations were employed in this age group, no "luteinized" or immature oocytes were harvested.

In the age group of 31 to 35 years, the proportions of mature oocytes were higher in patients using Protocol I and IV when compared with Protocols II and V (P < 0.001). Lower rates of postmature oocytes occurred in Protocols IV while "luteinized" and immature oocytes were higher when Protocols IV and III were employed if compared with Protocols II and V (P < 0.01 and P < 0.05 respectively).

CONCLUSION

The evaluation of oocyte quality is critical to the embryology laboratory work up and success and for the evaluation of the adequacy of the COH protocols. Obtaining a synchronized follicular and oocyte maturation is the most important goal of any COH strategy. This retrospective study shows that by using different drugs and strategies for COH we could not observe a substantial difference indicating a synchronized follicular growth in the global population.

Interestingly, age may have a role in determining the synchrony of multiple follicular growth and this became evident when GnRHa-menotropin stimulation was applied. The inherent asynchrony of the oocytes of the cohort probably increases when the number of the oocytes within each cohort decreases. On the other hand, the relative proportions of follicles belonging to different maturation stages varies with the age[7] and this may reflect a different setting of the mechanisms regulating the follicular recruitment and growth. The wider asynchrony of oocytes of older ovaries could account for the high rate of poor responses in older women.

REFERENCES

1. J. F. Kerin, G. M. Warnes, Monitoring of ovarian response to stimulation in in-vitro fertilization cycles, *Clin. Obstet. Gynecol.*, 29:158 (1986).
2. J. M. Varyas, C. Morente, G. Shangold, R. P. Marrs, The effect of different methods of ovarian stimulation for human in in vitro fertilization and embryo replacement, *Fertil. Steril.*, 42:745 (1984).
3. H. W. Jones, A. A. Acosta, M. C. Andrews, J. E. Garcia, G. S. Jones, T. Mantzavinos, J. McDowell, B. Sandow, L. Veeck, T. Whibley, C. Wilkes, and G. Wright, The importance of follicular phase to success and failure in in vitro fertilization, *Fertil. Steril.*, 40:317 (1983).
4. A. Gougeon, Rate of follicular growth in the human ovary, in: "Follicular Maturation and Ovulation," R. Rolland, E. V. van Hall, S. G. Hillier, K. P. McNatty and J. Schoemaker, eds., Excerpta Medica International Congress Series 560, Amsterdam (1982).
5. G. D. Hodgen, D. Keningsberg, R. L. Collins, and R. S. Schenken, Selection and dominant ovarian follicle and hormonal enhancement of natural cycle, *Ann. N.Y. Acad. Sci.*, 442:23 (1985).
6. D. Navot, and Z. Rosenwaks, The use of follicle-stimulating hormone for controlled ovarian hyperstimulation in in vitro fertilization, *J. In Vitro Fert. Embryo Transfer*, 5:3 (1988).
7. A. Gougeon, Personal communication.

LUTEAL PHASE AND EMBRYONIC DEVELOPMENT IN GnRH-AGONIST/HMG-CYCLES FOR IN VITRO FERTILIZATION

R.K. Schmutzler, C. Reichert, K. Diedrich, Ch. Diedrich, H. van der Ven, S. Al-Hasani, and D. Krebs

University of Bonn, Department of Ob/Gyn, F.R.G.

INTRODUCTION

The combined GnRH-agonist hMG-stimulation has now become a worthwhile tool as a second line therapy in vitro fertilization (IVF) under special conditions such as premature LH-surge and hyperandrogenemia. Several authors even propose this treatment as a first line therapy with an increased overall pregnancy rate.

There are different agonistic compounds available with different suppressive potencies distinguished by their amino acid substitutions in positions 6 and 10 (Table 1). Besides these chemical differences the substances are also produced in different application forms. While nasal spray and subcutaneous injections are short acting forms that are given daily, intramuscular injections remain effective for at least one month by continuous release from the injected depot.

Pharmacological studies in rats showed that constant infusion of a GnRH-a, comparable to a single shot of the long lasting analogue induced a down regulation primarily on the pituitary level (Sandow 1983). An intermittent daily application, equivalent to subcutaneous treatment, however, was able to release attenuated LH-Surges even after two weeks. Therefore it was suggested that the subcutaneous application acts mainly by down regulation of ovarian receptors.

In order to investigate differences in the effect of short and long acting agonists we per-

Advances in Assisted Reproductive Technologies, Edited by
S. Mashiach *et al.,* Plenum Press, New York, 1990

Table 1. Aminoacid Sequences of GnRH and Some of its Agonists

	1	2	3	4	5	6	7	8	9	10
GnRH	pGlu-	His-	Trp-	Ser-	Tyr-	Gly-	Leu-	Arg-	Pro-	Gly NH$_2$
Leuprolid	1	2	3	4	5	D-Leu	7	8	9	Ethyl-amid
Buserelin	1	2	3	4	5	D-Ser | But	7	8	9	Ethyl-amid
Goserelin	1	2	3	4	5	D-Ser | But	7	8	9	AzGly
Nafarelin	1	2	3	4	5	D-Nal(2)	7	8	9	Gly-NH$_2$
Decapeptyl	1	2	3	4	5	D-Trp	7	8	9	Gly-NH$_2$

formed a study where the agonist D-6-TRP-LHRH (Decapeptyl, Ferring) was administered in different application forms.

MATERIAL AND METHODS

Seventy-four patients were treated in 119 cycles. The main cause of infertility was tubal blockage with 71.2%. The agonist was mainly given because of previously occurred premature LH-surges in 85.1%. The remaining cases include hyperandrogenaemia, insufficient ovarian stimulation and GnRH-pretreatment for other reasons.

In protocol A (Figure 1) we started in the early follicular phase on cycle day 1–4. The analog was given by daily subcutaneous (s.c.) injections. The initial dosage of 500µg/day was reduced to 100 µg/day after one week. Concomitant hMG-stimulation started after clinical and biochemical evidence of pituitary suppression, i.e. the occurrence of hot flushes and basal hormone level. This lasted an average of 23 days. In protocol B a single intramuscular (i.m.) injection of 3.2 mg Decapeptyl was given in the midluteal phase on cycle day 22. HMG-stimulation started here after a fixed interval of 14 days. Sufficient suppression was proven by a negative GnRH-test on day 14. In both protocols an initial hMG dosage of 150 I.U. per day was administered for three days and then increased to 225 I.U. When the dominant follicles reached a size of at least 18 mm in diameter 10.000 I.U. hCG were given and follicular puncture was performed 36 hours later. The luteal phase was supported with two hCG-injections of 5000 I.U. on day 2 and 7 after oocyte retrieval. In cases of severe hyperstimulation hCG was replaced by progesterone.

Hormone levels were measured by radio-immunoassay. Details are described elsewhere (Schmutzler, 1988).

RESULTS

When cycle parameters in both groups were compared (Table 2) a better ovarian stimulation was achieved in protocol A (s.c. cycles) indicated by higher estradiol levels although not statistically significant. A more pronounced difference was seen for progesterone levels and number of oocytes and embryos harvested resulting in a significantly higher pregnancy rate in protocol A (16.7%) compared to protocol B (5%). Pregnant cycles (Table 3) exhibited a stronger ovarian stimulation with significantly higher hormone level and oocyte and embryo numbers. A detailed analysis is given elsewhere (Schmutzler 1988).

As gamete numbers and luteal phase progesterone levels differed more than the follicular stimulation we investigated these parameters in more detail.

The quality of all recovered oocytes and embryos were compared. Concerning oocyte quality (Figure 2) there was no visible difference in the high quality class with about the

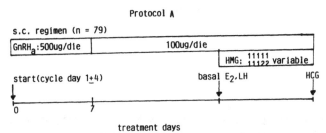

The suppression interval was varible and based on clinical and biochemical signs of pituitary suppression.

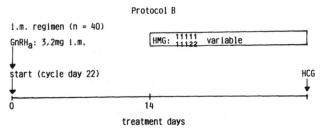

A fixed suppression interval of 14 days was chosen.

Fig. 1. Protocol A.

Table 2. Comparison of Cycle Parameters in Both Protocols

	s.c.-cycles (X ± SEM)	i.m.-cycles(X ± SEM)
E_2 the day of HCG (pg/ml)	2040 ± 169.6	1640 ± 191.5
Prog. the day after HCG (ng/ml)	4.0 ± 0.93	2.2 ± 0.3
No of oocytes/cycle	4.2	3.5
No of embryos/cycle	2.1	1.4
pregnancy rate / cycle	16.7%	5%

Table 3. Cycle Parameters in Pregnant Cycles (n = 14)

S.c. application	12
I.m. application	2
E_2 the day of HCG (pg/ml)	3010.8 ± 354[a]
P 1 day after HCG (ng/ml)	7.2 ± 3.0[a]
Oocytes/cycle	7.4
Embryos/cycle	3.7

[a]$\overline{X}$ ± SEM.

same percentage of mature preovulatory oocytes. However in the poorer quality classes the predominance of i.m. cycles was plainly recognizable. In particular, more degenerated eggs were found in protocol B.

Following 48 hours of culture this difference became more pronounced (Figure 3). A significantly higher percentage of healthy embryos developed in protocol A whereas in protocol B significantly more fragmented embryos were found. Sperm quality was checked to exclude an andrological influence on embryonic development. The results (Figure 4) revealed even a higher percentage of normozoospermia in protocol B.

The luteal phase was analyzed by measuring progesterone levels on day 1 and day 13 after hCG-application for ovulation induction. For patient convenience we did not draw blood in the midluteal phase. Figure 5 shows the progesterone levels on day 1 after hCG. Cycles are classified according to their progesterone levels in steps of 2 ng/ml. Cycles from protocol B (i.m.) are mainly represented in groups with low progesterone levels. Twelve days later (Figure 6) this disparity between both protocols increased. Groups are divided by steps of 10ng/ml. About 70% of cycles from protocol B revealed progesterone levels less than 10ng/ml compared to only 40% of cycles from protocol A.

DISCUSSION

The long acting agonist caused a stronger ovarian suppression than the short acting agonist although the total amount of GnRH-a released per day was about the same. The coincidence of a poorer ovarian stimulation and a lower pregnancy rate in protocol A compared to protocol B suggests a causal relationship between ovarian stimulation and pregnancy rate.

As the difference in pregnancy rates, however, is much stronger than in ovarian stimulation, it is suggested that the long acting agonist might exhibit a more pronounced ovarian suppression not only by hypophyseal down-regulation but also via direct ovarian effects. This is supported by differences seen in gamete quality and luteal phase progesterone levels that are unlikely to be due exclusively to differences seen in follicular development. Testart et al. 1989 also reported on a poorer embryo-quality in GnRH-a/hMG cycles compared to hMG-only cycles as he found a higher fragmentation and lower pregnancy rate for frozen-thawed oocytes coming from GnRH-a cycles. In rats it was shown in vitro that GnRH could induce atresia (Birnbaumer et al. 1985) and fragmentation (Hsueh et al. 1981) of oocytes.

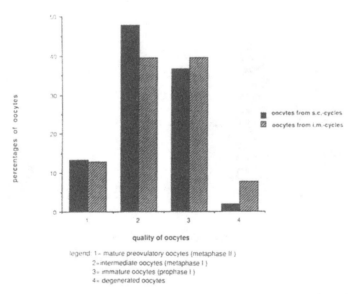

legend: 1 = mature preovulatory oocytes (metaphase II)
2 = intermediate oocytes (metaphase I)
3 = immature oocytes (prophase I)
4 = degenerated oocytes

Fig. 2. Percentage distribution of oocyte quality in I.M. versus S.C. cycles.

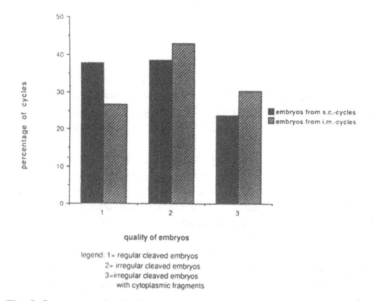

Fig. 3. Percentage distribution of embryo quality in I.M. versus S.C. cycles.

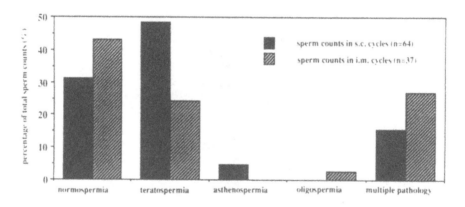

Fig. 4. Percentage distribution of sperm quality.

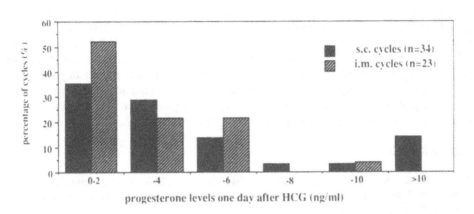

Fig. 5. Percentage distribution of progesterone levels one day after hCG in I.M. versus S.C. cycles.

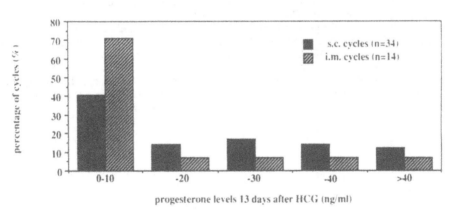

Fig. 6. Percentage distribution of progesterone levels 13 days after hCG in I.M. versus S.C. cycles.

The negative effect of GnRH on progesterone secretion of rat granulosa cells was intensively studied and proved by Hsueh et al. 1981. In vitro studies on human follicular cells are however contradictory. While Dodson et al. (1988) could not detect any effect of Leuprolide on luteal cells and embryonic development, Parinaud et al. (1988) described a progesterone decrease at high doses of concomitant Buserelin and hMG exposure to granulosa cells. In accordance with Parinaud et al. (1988) our results suggest that a stronger agonistic effect exerts a direct and negative influence on luteal phase progesterone secretion. In a successive study preliminary data indicate that a higher initial hMG dosage in protocol B is able to overcome the negative ovarian effects resulting in a similar pregnancy rate in both protocols (not yet published). This suggests a competitive interaction of GnRH-a and gonadotropins that might occur most likely in their intracellular pathways.

REFERENCES

Birnbaumer, L., Shahabi, N., Rivier, J., and Vale, W., 1985, Evidence for a physiological role of gonadotropin-releasing hormone (GnRH) or GnRH-like material in the ovary, *Endocrinology*, 116:1367.

Dodson, W.C., Myers, T., Morton, P.C., and Conn, P.M., 1988, Leuprolide acetate: serum and follicular fluid concentrations and effect on human fertilization, embryo growth and granulosa-lutein cell progesterone accumulation in vitro, *Fertil. Steril.*, 50:612.

Hsueh, A.J.W., and Jones, P.B.C., 1981, Extrapituitary actions of gonadotropin-releasing hormone, *Endocrine Reviews*, 2:437.

Parinaud, J., Beaur, A., Bourreau, E., Vieitez, G., and Pontonnier, G., 1988, Effect of a luteinizing hormone-releasing hormone agonist (Buserelin) on steroidogenesis of cultured human preovulatory granulosa cells, *Fertil. Steril.*, 50:597.

Sandow, J., 1983, Clinical applications of LHRH and its analogues, *Clin. Endocr.*, 18:571.

Schmutzler, R.K., Reichert, C., Diedrich, K., Wildt, L., Diedrich, Ch., Al-Hasani, S., van der Ven, H., and Krebs, D., 1988, Combined GnRH-agonist/gonadotropin stimulation for in vitro-fertilization, *Hum. Reprod.*, 3:29.

Testart, J., Forman, R., Belaisch-Allart, J., Volante, M., Hazout, A., Strubb, N., and Frydman, R., 1989, Embryo quality and uterine receptivity in in vitro fertilization cycles with or without agonists of gonadotropin-releasing hormone, *Human Reprod.*, 4:198.

FOLLICULAR PHASE GONADOTROPIN-RELEASING HORMONE AGONIST AND HUMAN GONADOTROPINS: A BETTER ALTERNATIVE FOR OVULATION INDUCTION IN IN VITRO FERTILIZATION

Jairo Garcia, Javad Bayati, Theodore Baramki, and Santiago Padilla

Department of Gynecology, Greater Baltimore Medical Center
Baltimore, Maryland, USA

INTRODUCTION

The efficacy of gonadotropin-releasing hormone agonist analog (GnRH-a) administered during the luteal phase of the preceding ovarian stimulation cycle has been demonstrated in in vitro fertilization (IVF).

Administration of GnRH-a during the early follicular phase stimulates both follicle stimulating hormone (FSH) and luteinizing hormone (LH) secretion for three days before down-regulation takes place, in two out of every three patients.

The purpose of this study is to compare luteal phase versus follicular phase GnRH-a as an adjuvant in ovulation induction with human menopausal gonadotropins (hMG) in IVF patients.

Materials and Methods

A total of 465 ovarian stimulated cycles for IVF were seen at the Women's Hospital Fertility Center at the Greater Baltimore Medical Center. Patients were prospectively allocated into three groups: group A (100 cycles) received the GnRH-a Leuprolide Acetate (Lupron, TAP Pharmaceuticals, North Chicago, IL) in the follicular phase (FPL); group B (95

Advances in Assisted Reproductive Technologies, Edited by
S. Mashiach *et al.*, Plenum Press, New York, 1990

cycles) received the GnRH-a in the luteal phase (LPL), and control group C (270 cycles) did not receive the GnRH-a.

The age of the patients ranged from 30–42 years. The indications for IVF were: Tubal factor 55%; endometriosis 20%; male factor 18%; diethylstilbestrol exposure (DES) 3%; normal infertile couple 3%, and others 1%.

Leuprolide Acetate was started on the third luteal phase day in the LPL group, and on the second menstrual cycle day in the FPL group. Patients received Leuprolide 0.75 or 1.0 mg/s.q./daily if body weight was less than or greater than 150 lbs., respectively. Leupro-lide was continued in both groups until the day of human chorionic gonadotropin (hCG) administration.

In the FPL group, 150 I.U. of pure FSH (Metrodin, Serono Laboratories, Randolph, MA), and 150 I.U. of hMG (Pergonal, Serono Laboratores, Randolph, MA) were adminis-tered daily, intramuscularly. FSH/hMG were started on the fifth cycle day, and given for at least two days. Thereafter, FSH was discontinued in a step-down fashion according to fol-licular development by ultrasound and serum E_2 levels. hMG was discontinued the day of hCG administration.

In the LPL group, 150 I.U. of FSH plus 150 I.U. of hMG were started after menses once the baseline serum E_2 levels were less than 20 pg/ml, and given for at least two days. FSH was tapered down in the same fashion as described for the FPL group, as well as the hMG.

In the control group, Leuprolide was not used. Patients were stimulated with 150 I.U. of pure FSH and 150 I.U. of hMG during the third and fourth cycle day, and with 150 I.U. of hMG daily thereafter.

All of the patients received 10,000 units of hCG administered intramuscularly when at least two follicles measured 16 mm in diameter by ultrasound in the FPL and LPL groups, or 15 mm in diameter in the control group.

Monitoring

1. Daily blood for E_2, LH and FSH.
2. Daily pelvic ultrasound was carried out once the E_2 levels were greater than 250 pg/ml.
3. Follicular aspiration was carried out by laparoscopy or transvaginal ultrasound 36 hours after hCG.

RESULTS

There was a total of 41 cancelled cycles: four, two and 35 in the FPL, LPL, and control

Table 1. Cancellation and Oocyte Retrievals in 465 IVF Cycles

	FPL	LPL	Control
Cycles	100	95	270
Cancellations	4	2	35
Oocyte retrievals	96	93	235

Table 2. FSH/hMG Regimens and Oocyte Retrieval Day in 424 IVF Cycles

	FPL	LPL	Control
FSH amps	7.5 ± 2	12.7 ± 7	4
HMG amps	15.6 ± 3	20.7 ± 3	10 ± 3
Retrieval day	13.8 ± 1.8	14.2 ± 2	11 ± 2

Table 3. Oocyte Quality

Ooctyes	FPL #	FPL (%)	LPL #	LPL (%)	Control #	Control (%)
Preovulatory	545	(64)[a,b]	501	(57)[a,b]	711	(48)[b]
Immature	178	(21)	194	(22)	327	(22)
Atretic	69	(8)[c, d]	85	(10)[c, d]	325	(22)[d]
Fractured zona	58	(7)	88	(10)	118	(8)

a: N.S.; b: $P < 0.005$; c: N.S.; d: $P < 0.05$

FSH/hMG groups, respectively. The cancellations occurred before oocyte retrieval due to poor response, with a single follicle detected by ultrasound (Table 1). Oocyte retrieval was carried out in 189 GnRH-a cycles with the following results:

The mean number of FSH and hMG ampules required per patient was smaller in the FPL group, 7.5 ± 1.8 and 15.6 ± 1.6 ampules, respectively, when compared with 12.7 ± 1.5 and 20.3 ± 1.5 ampules required by the LPL group of patients. There was no difference in the mean cycle day of oocyte retrieval, 13.8 ± 1.8 days for the FPL group, and 14.2 ± 2 days for the LPL group (Table 2).

There was a significant increase in the percentage of preovulatory oocytes in both FPL and LPL groups, 64% and 57%, respectively, when compared to 48% in the control FSH/hMG group. A significant decrease in the percentage of atretic oocytes was found in both FPL and LPL groups, 8% and 10% respectively, when compared to 22% in the FSH/hMG protocol. There was no change in the percentage of immature (21%, 22% and 22%) and fractured zona oocytes (7%, 10% and 8%), which remained constant in all three groups (Table 3).

Table 4. Pregnancy Outcome in 424 IVF Cycles

	FPL	LPL	Control
IVF cycles	96	93	235
Cycles w/o ET	10	3	31
Cycles with ET	86	90	204
Embryos/transfer	3.4	3.4	2.6
Pregnancy Cycle	35 (36.4%)[a,b]	25 (26.8%)[a,b]	56 (23.8%)[a]
Pregnancy/C/ET	35 (40.6%)[c,d]	25 (27.7%)[b]	56 (27.4%)[c]
Miscarriage	8 (23.8%)[d]	9 (36.0%)[d]	17 (28.0%)[d]

a: $P < 0.05$;   b: N.S.;   c: $P < 0.005$;   d: N.S.

Fertilization did not take place in 10 of the 95 FPL group cycles, in three of the 93 LPL group cycles and in 31 control group cycles. Lack of fertilization was due to a male factor in 28 cases (six in FPL, two in LPL and 20 controls), and to poor oocyte quality in 16 (four in FPL, one in LPL and 11 controls) (Table 4). There was no significant difference in the fertilization rate (per fertilizable oocyte) between the FPL and LPL groups, 71% and 75%, respectively.

The average number of embryos transferred was 3.4 and 3.3 in the FPL and LPL groups, respectively. The average of both FPL and LPL groups is higher than the concurrent 2.6 embryos per patient transferred in the control FSH/hMG protocol (Table 4).

The pregnancy rate was significantly higher in the FPL group, 36.4% per cycle, or 40.6% per cycle with embryo transfer than the 23.8% pregnancy rate per cycle and the 27.4% per cycle with embryo transfer in the control group. The FPL group's pregnancy rate was also higher than the 26.8% pregnancy rate per cycle, or 27.7% pregnancy rate per cycle with embryo transfer achieved in the LPL group, although the difference did not reach statistical significance.

The miscarriage rate was lower in the FPL group, 22.8% versus 36% observed in the LPL group, although the difference was not statistically significant (Table 4).

CONCLUSION

When comparing follicular phase Lupron with luteal phase Lupron protocols the following conclusions can be made:

1. There was no significant difference in follicular development, number of follicles, days of stimulation, number of fertilizable oocytes, fertilization and cleavage rate, or number of embryos transferred.

2. There was a significant difference between both groups since there was a smaller amount of pure FSH plus hMG requirement, a higher pregnancy rate, and a lower miscarriage rate in the FPL group.

3. Since follicular phase Lupron shows a greater pregnancy rate with a lower miscarriage rate and requires a smaller amount of hCG, the combination of all these factors makes the FPL protocol the ideal one to induce ovulation for IVF at the present time.

PROGESTERONE/ESTRADIOL RATIOS PREDICT OUTCOME IN IN VITRO FERTILIZATION

Charlotte J. Richards-Kustan,[a,b] G. John Garrisi,[a] Halina Wiczyk,[a] Lawrence Grunefeld,[a] John Mandeli,[a] Machelle Seibel,[b] Daniel Navot,[a] and Neri Laufer[a]

[a] Department of Ob/Gyn and Biostatistic, Mount Sinai Hospital
One Gustave L. Levy Place, New York, NY 10029
[b] Department of Ob/Gyn, Beth Israel Hospital
330 Brookline Place, Boston MA 02215

Despite increasing experience with in vitro fertilization (IVF), there are few endocrinological markers which can serve as a prospective predictor of IVF success. An initial investigation utilized periovulatory estradiol (E_2) levels after administration of low dose human menopausal gonadotropin followed by a coasting period before giving human chorionic gonadotropin (hCG). The levels of estradiol studied were lower than those typically seen in many IVF programs today.[1] Others have studied post-transfer progesterone/E_2 ratios and suggested a predictive value.[2] This report summarizes our efforts to further elucidate endocrinological markers that can be used to predict success or potential success in IVF.

METHODS

The seventy-nine patients chosen for this study had successfully undergone ovulation induction, ovum retrieval, IVF and embryo transfer during the time period of December 1986 to May 1987 at the Mount Sinai Hospital in New York City. Conception occurred in 21 of these patients. The remaining fifty-eight patients did not conceive. All women underwent ovulation induction with human menopausal gonadotropin 225 I.U. daily initiating day 2 or 3 of the menstrual cycle. Daily blood sample for E_2, progesterone (P), and daily pelvic ultrasound exams were obtained beginning on day 7 or 8 of the cycle. HCG, 10,000

Advances in Assisted Reproductive Technologies, Edited by
S. Mashiach *et al.,* Plenum Press, New York, 1990

units, was administered when the two largest follicles were greater than 15mm in average diameter and serum E_2 was >200 pg per follicle, with a total serum E_2 > 800 pg/ml.

Thirty six hours after hCG administration, ovum retrieval was performed either by laparoscopy or an ultrasound guided approach. Embryo transfers were performed in a routine manner utilizing either the supine or knee-chest position.

Progesterone in oil 50 mg IM was administered daily for 14 days following embryo transfer and weekly P levels obtained. If pregnancy was achieved, serum P levels were maintained above 20 ng/ml for 7–11 weeks.

The 79 patients were divided into five groups: 1) 58 who did not conceive (nonpregnant). 2) 13 who had normal conception and delivery (normally pregnant), which included seven singletons, four sets of twins and two sets of triplets. 3) Eight who conceived with pregnancies that did not go to term (abnormally pregnant pregnant). 4) All 21 who conceived (total pregnant). 5) The group of 66 patients who did not have a live birth (nonpregnant + abnormally pregnant pregnant) .

The day of hCG administration was designated day 0. Subsequently, serum E_2, P, and P/E Ratios were analyzed after converting progesterone levels to picogram quantities from the original nanogram quantities. Students t-testing was performed on the numbers and qualities of embryos in all five groups, after assigning the designation of 1-cell, 2-3 cell, and $\geq$ 4-cell embryos. Simple regression analysis was performed on embryo quality as compared to serum E_2, P levels and P/E ratios on days 0 and +1.

RESULTS

There was no significant difference between the groups in respect to the average day of initiation of therapy, the number of days of administration of hMG, the initial E_2, P levels or the initial P/E ratio.

The average age ot the abnormally pregnant pregnant patients was higher than that of the normally pregnant patients (36.4 vs 33.0, p < .05). Other age differences were insignificant. The number of high quality embryos transferred was greater in the abnormally pregnant pregnant vs nonpregnant (2.8 vs 1.5, p < .05). However, among all three conception groups (abnormally pregnant pregnant, normally pregnant, and total pregnant) there was no significant difference in numbers of embryos, or numbers of high quality embryos.

Examination of the regression analysis between the numbers of high quality embryos in each group and E_2, P and P/E ratios revealed that while there was no correlation between embryo quality and either E_2, P, there was a high correlation between higher embryo quality and P/E ratio in all the groups. However, the highest correlation in the nonpregnant

Table 1. Correlation Coefficients of Embryo Groups with P/E Ratios

	Day 0 P/E	Day +1 P/E
Nonpregnant		
≥ 4 cells	0.795*	0.313
2-3 cells	0.780*	0.256
1 cell	0.218	0.072
Normally pregnant		
≥ 4 cells	2.799	1.164*
2-3 cells	0.888*	0.373
1 cell	0.024	0.026
Abnormally pregnant		
≥ 4 cells	0.008	0.740*
2-3 cells	2.218	0.403
1 cell	0.000	0.000
Total pregnant		
≥ 4 cells	3.228	0.951*
2-3 cells	1.009*	0.388
1 cell	0.000	0.000
Nonpregnant and abnormally pregnant		
≥ 4 cells	0.853*	.345
2-3 cells	.791*	.275
1 cell	.217	.067

* Significant Correlation $p < .05$

group was seen on day 0. In the Pregnant groups, the highest correlation was seen on day +1 (See Table 1).

Among the 66 patients in group 5 who did not attain a live born, there was a weaker but significant correlation on day +1 between P/E and higher quality embryos. This heterogenous group may reflect lower quality embryos in the nonpregnant group, or inadequate preparation of the endometrium to receive high quality embryos (the abnormally pregnant pregnant group).

Examination of P/E ratios between all the groups on day 0 reveals that the P/E ratios were significantly lower among pregnant (total pregnant and abnormally pregnant pregnant) than the nonpregnant. In addition, there is a trend toward a significant difference between the normally pregnant and the nonpregnant, which would probably become apparent with larger numbers. A P/E ratio greater than 1.20 on day 0 predicts that pregnancy will not occur (See Figure 1).

Table 2. P/E Values Associated with Outcomes in In Vitro Fertilization

	Day 0	Day +1
Normally pregnant	.45 to 1.20	< 3.2
Pregnancy Loss	< .45	not predictive
Nonpregnant	> 1.20	> 3.40

In the abnormally pregnant pregnant group, the P/E ratio is lower than other groups on day 0. Since we have already demonstrated that embryo quality in this group was equal to the other pregnant groups, it may be assumed that the lower P/E ratio reflects inadequate endometrial development and ultimately failed implantation. In the normally pregnant group the P/E ratios on day +1 are lower than any other group. Furthermore, ninety-five percent confidence limits drawn around the three groups demonstrate that there is very little overlap (See Figure 2). Therefore, the P/E ratio on day +1 is predictive of outcome. Table 2 contains P/E ratios that are predictive of outcome in in vitro fertilization.

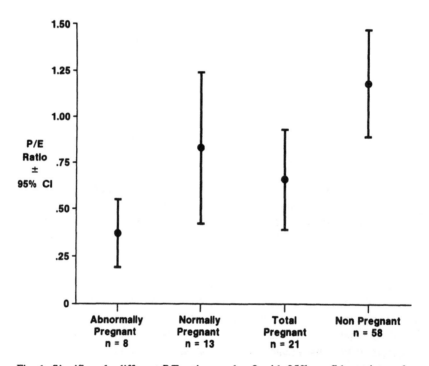

Fig. 1. Significantly different P/E ratios on day 0 with 95% confidence intervals.

Repeated measure analysis of variance (RPMANOVA) of the behavior of the P/E curves demonstrates that the behaviour of the abnormally pregnant is different from the normally pregnant group. In fact, there is a more rapid rise in the P/E ratio after hCG administration, indicating that despite production of equal numbers of equal quality embryos, there is something inherently different in steroid production between the two groups. Estradiol levels were not predictive of pregnancy in any of the groups. Neither did day 0 P predict pregnancy, although progesterone was higher in the normally pregnant than abnormally pregnant pregnant group (p < .001).

DISCUSSION

This study was undertaken to delineate endocrinological parameters that can be used to predict establishment of normal pregnancy in a group that has been selected for the highest success in IVF/ET based on their pattern of estradiol response using a high dose hMG pro-

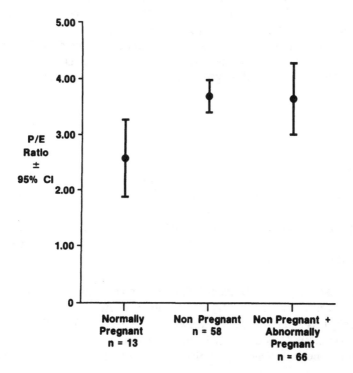

Fig. 2. Significantly different P/E ratios on day +1 with 95% confidence intervals.

tocol.[3] The parameters that have been used to predict success in in vitro fertilization in the past include rapid rise of E_2 before and after hCG administration, oocyte and sperm quality, embryo quality and number, and endometrial receptivity.[4] The obvious benefit of predictive parameters are avoidance of oocyte retrieval, which would result in tremendous savings in terms of emotional energy, physical pain and financial solvency.

The data suggests that the patients having a P/E ratio greater than 1.2 on day 0 have a reduced expectation of pregnancy. Although the fertilization rate in this group is identical to the other groups, the number of oocyte and high quality embryos available for transfer is reduced. In Group 1 (nonpregnant), Day 0 P/E ratios are correlated with the number of high quality embryos; similar correlation is seen in the pregnant groups, however it occurs on day +1. It would appear that either hCG was adminstered on the wrong day in Group 1, or that the cohort of follicles selected are inferior to those in the pregnant groups. Endometrial dyssynchrony on the basis of premature exposure to an unfavorable P/E ratio cannot be discounted as a contributing factor in the inability to conceive. Studies presently underway at Beth Israel Hospital confirm these results. P/E ratios on day +1 predict outcome of IVF/ET: those destined for normal pregnancy have a P/E ratio less than 3.2.

The abnormally pregnant pregnant group presents an opportunity to initiate therapeutic intervention on the day of oocyte retrieval based on a low P/E ratio on day 0. This group has the same number of high quality embryos as the normally pregnant group. However, there is a progesterone deficit that is apparent even before oocyte retrieval. The group is older than the normally pregnant group, perhaps accounting for altered steroidogenesis. Therefore, if the criterion for hCG administration has been met but there is an unfavorably low P/E ratio, early institution of progesterone therapy may be indicated.

CONCLUSIONS

Based on these data, several conclusions can be drawn. The number of high quality embryos is significantly increased in the groups that have an expectation for conception. However, those patients with an expectation for pregnancy loss constitute a unique group that have high quality embryos, but who have inadequate amounts of progesterone production on day 0. Based on these data the P/E ratio may provide a prospective therapeutic marker for early institution of progesterone.

The P/E ratio on day 0 can distinquish those with no expectation for conception (those with P/E ratios greater than 1.20). Day +1 P/E ratio is predictive of either no pregnancy or early pregnancy loss and thusly an unfavorable outcome after IVF/ET (those with P/E ratios greater than 3.40). This data may be used to cancel patients prior to oocyte retrieval with obvious cost benefits. Ongoing studies at Beth Israel Hospital confirm these results.[5]

REFERENCES

1. H.W. Jones, A. Acosta, M.C. Andrews, et al., The importance of the follicular phase to success and failure in in vitro fertilization. *Fertil. Steril.* 40:31, 1983.
2. A.A. Leidley-Baird, C. ONeill, M.J. Sinosich, et al., Failure of implantation in human in-vitro fertilization and embryo patients: The effects of altered progesterone/estrogen ratios in humans and mice. *Fertil Steril.* 45:69, 1986.
3. N. Laufer, A.H. Decherney, B.C. Tarlatzkis, F. Naftolin, The association between preovulatory serum 17 b estradiol pattern and conception in human menopausal gonadotropin-human chorionic gonadotropin stimulation, *Fertil. Steril.* 46:73, 1986.
4. P.M. Zartski, F.B. Kuzan, L. Dixon, M. Soules. Endocrine changes in the late follicullar and postovulatory intervals as determinants in in vitro fertilization pregnancy rate, *Fertil. Steril.* 47:137, 1987.
5. C.J. Richards-Kustan, G.J. Garrisi, H. Wiczyk, et al., P/E ratios predict outcome in in vitro fertilization, Submitted for Publication, June, 1989.

THE INFLUENCE OF ENDOMETRIAL PREPARATION PROTOCOLS
ON EMBRYO IMPLANTATION IN HUMAN
CRYOPRESERVATION TREATMENT

Jehoshua Dor, Edwina Rudak, Zion Ben-Rafael, David Levran*, Adrian Davidson,
Michal Kimchi, David M. Serr, Laslo Nebel*, and Shlomo Mashiach

*Department of Obstetrics and Gynecology and Department of Embryology**
Chaim Sheba Medical Center, Tel Hashomer
and Sackler School of Medicine, Tel Aviv University, Israel

INTRODUCTION

In promoting successful implantation of embryos following in vitro fertilization (IVF) and embryo transfer, there is always a dilemma as to what is more important, uterine environment[1] or qualities of the embryo itself.[2] Pregnancies following IVF treatment have been described previously in spontaneous cycles, stimulated cycles, and in egg donation program, using exogenous substitutional hormone therapy. In this study in order to evaluate the importance of the uterine environment, three protocols for uterine preparation prior to embryo replacement were compared while transferring frozen thawed embryos.

The three methods of uterine preparation that were studied prospectively were:

1. A spontaneous cycle
2. Gonadotropin stimulation treatment
3. Estrogen and progesterone substitutional therapy

We evaluated the daily 17 β estradiol and progesterone levels in three different patient groups in late follicular and early luteal phase and compared the pregnancy rates following replacement of frozen thawed embryos in the three protocols studied.

Advances in Assisted Reproductive Technologies, Edited by
S. Mashiach *et al.*, Plenum Press, New York, 1990

MATERIALS AND METHODS

The patients in the present study consisted of 109 women between the ages of 26 and 41 who underwent 124 replacements of thawed embryos. The patients had undergone IVF treatment because of tubal infertility, unexplained infertility, or male infertility suffered by their husbands.

Our protocol for IVF treatment has been described previously.[2] Ovum pick-up (OPU) procedures were performed either by laparoscopy or by transvaginal ultrasound (Elscint Ltd., Israel, sector vaginal probe 6.5 MHz). Oocytes and embryos were cultured in Earl's medium containing 8% human serum and oocyte culture and fertilization were performed as previously described.[3] Replacement of fresh embryos was carried out 42 to 72 hours after insemination.

Method of Freezing and Thawing

·A maximum of four embryos were replaced during the OPU cycle. Supernumerary embryos of regular appearance, minimal cytoplasmic fragmentation, and cleavage rates consistent with their developmental stage were selected for cryopreservation. Embryos were frozen at various cleavage stages from two pronuclei to early blastocytes, mainly up to the 8-cell stage. The two methods used for cryopreservation were:
1. Slow freezing and thawing, first described by Trounson et al.[4] using 1.5M dimethyl-sulphoxide (DMSO) cryoprotectant.
2. Rapid freezing and thawing, as described by Lassalle et al.[5] using 1.5M PROH with 0.1M sucrose as cryoprotectant.

Embryos were frozen in 0.2 ml of the cryoprotective medium in Nunc plastic cryotubes using a Planer R-204 biological freezer (Planer Products Ltd., Sunbury-on-Thames, UK). Because no difference in survival rates were found between these two methods of freezing embryos,[6] all frozen- thawed embryos were analyzed together.

Embryo Replacement

Three methods of cycle preparation were randomly used in patients undergoing replacement of thawed embryos:
(a) a spontaneous cycle, including daily monitoring of E_2, LH and P, and three times daily monitoring of these hormones when E_2 rose above 150 pg/ml and when a leading follicle of 17mm diameter was observed in ultrasonography. The day of embryo replacement was determined according to the progesterone rise and the stage of the embryo before freezing. The time equivalent to ovulation and "OPU" was considered to occur at around 24 hours after the first significant rise in plasma P levels. Thus, we could ensure that the embryo would arrive in a secretory endometrium.

(b) a stimulated cycle in which hMG was administered at a daily dose of 150 I.U. FSH + 150 I.U. LH from day 3 of the cycle, and hCG was administered using the same criteria as for a regular OPU cycle.[2]

(c) estrogen and progesterone replacement therapy, in which estradiol valerate (EV) was given from day 3 of the cycle at a dose of 1 mg/day, rising in 1 mg increments daily until 6 mg on day 8. From day 9, 1 mg EV/day is given, together with 100 mg progesterone/day from day 10, according to our egg donation protocol. The first day of progesterone administration is considered day of "OPU".[7]

RESULTS

To evaluate the endometrial receptivity in the transfer of frozen-thawed embryos, we analyzed and compared the progesterone (P), 17β estradiol (E_2) and the E_2 to P ratio in the three cycle preparation groups during the follicular and early luteal phases. The mean daily progesterone levels (Figure 1) were lowest in spontaneous cycles on the day of "OPU", OPU + 1; OPU + 2; OPU + 3; OPU + 4, while those patients receiving hMG + hCG stimulation had significantly higher progesterone levels ($p < .01$, ANOVA). This could be related to massive luteinization following hCG administration.

The mean daily E_2 levels are shown in Figure 2 and demonstrate a similar pattern, where the lowest levels occurred during spontaneous cycles and significantly higher levels

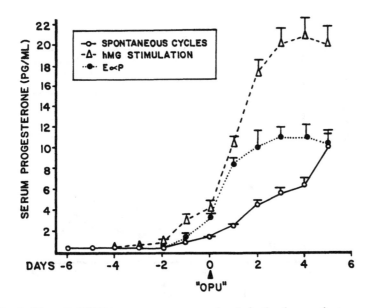

Fig. 1. Mean ($\pm$ S.E.M.) serum progesterone levels in the three patient groups.

Table 1. Mean E2/P Ratio in Early Luteal Phase in the Three Groups of Patients

	Spontaneous cycle	hMG+hCG cycle	E+P substitution
OPU	221.3	304.5	485.4
OPU + 1	82.5	116.9	54.0
OPU + 2	62.2	42.4	25.7
OPU + 3	88.7	30.4	30.4
OPU + 4	32.5	26.9	27.0

Not significantly different (Anova)

in the stimulated cycles ($p < .01$, ANOVA). Table 1 shows the E_2/P ratio in the early follicular phase from day of "OPU" to day of OPU + 4. There was no difference in the E_2/P ratio among the different protocols during the early luteal phase of the cycle.

To assess whether the hormonal levels might be different between conceptional and non-conceptional cycles as a factor which might affect pregnancy potential we compared mean daily progesterone and E_2 levels between pregnant and non-pregnant patients during the spontaneous cycles (Figures 3 and 4). There was no difference in the mean daily levels of P and E_2 between the two groups during follicular and luteal phase.

The method of cycle preparation did not significantly influence pregnancy rates, as shown in Table 2. Although the pregnancy rate was highest (16.1%) in spontaneous cycles

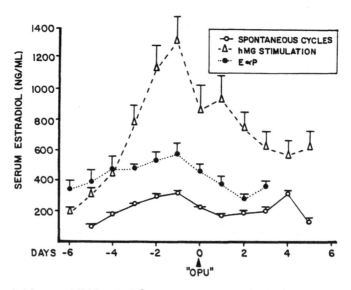

Fig. 2. Mean (± S.E.M.) of 17 β estradiol levels in the three patient groups.

Table 2. Cycle Preparation and Pregnancy Rate in Transfer of Frozen-Thawed Embryos

	Spontaneous cycle	hMG+hCG cycle	E+P substitution	Total
Number transferred	56	44	24	124
Pregnancies (%)	9 (16.1)	5 (11.4)	2 (8.3)	16 (12.9)
Delivered (%)	7 (12.5)	4 (9.1)	1 (4.1)	12 (9.6)

and the lowest in the substitutional protocol (8.3%), the differences between the three protocols did not differ significantly. Of the 16 pregnancies, 12 delivered, three terminated in a spontaneous abortion in the first trimester, and one was a tubal pregnancy in a spontaneous cycle. The spontaneous abortion rate was similar among the three protocols.

The embryonic implantation rate also did not differ significantly according to the method of cycle preparation (Table 3). The embryonic implantation rate was higher in spontaneous cycles (10.5%) than in either the hMG plus hCG cycles (6.5%) or E and P substitution cycles (4.5%), but the difference was not statistically significant.

DISCUSSION

In order to evaluate the importance of the endometrial environment for implantation of thawed embryos, we randomly compared three protocols of endometrial cycle preparation. The methods of cycle preparation compared in our study have been shown to have no sig-

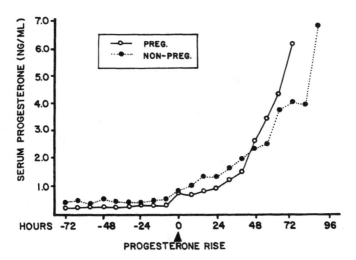

Fig. 3. Mean (± S.E.M.) of progesterone levels in the spontaneous cycles (pregnant vs. non-pregnant).

Table 3. Cycle Preparation and Implantation Rate of Frozen-Thawed Embryos

	Spontaneous cycle	hMG+hCG cycle	E+P substitution	Total
Embryos Replaced (number)	95	76	44	215
Mean Embryos Replaced	1.7	1.7	1.8	1.73
Embryos Implanted (%)	9 (10.5)	5 (6.5)	2 (4.5)	16 (7.4)

nificant effect on the implantation of pregnancy rates among 109 women who underwent 124 replacements of thawed embryos. Similar conclusion has been previously reported by Mandelbaum et al.[8] and Testart et al.,[9] when comparing spontaneous and stimulated cycles.

Although the absolute mean daily levels of progesterone and serum E_2 levels were highest in the stimulated cycles and lowest in the spontaneous cycles, the E_2/P ratio was virtually the same. This indicates that despite differences in the method of cycle preparation, the uterine environment in the luteal phase prior to embryo implantation does not differ significantly among the three groups. This was estimated by the E_2/P ratio rather than the absolute levels of these hormones. Stanger et al.[10] pointed out the importance of E2/P ratio in early luteal phase following OPU demonstrating that the hyper- and hypo-estrogenic response with low or high E_2/P ratio had significantly lower pregnancy rates than in cases with E_2/P ratio close to the mean.

The similarity between the uterine environment in spontaneous and stimulated cycles is

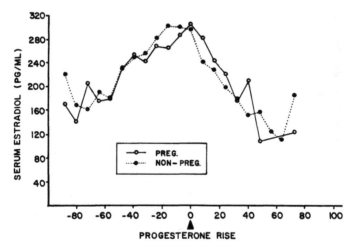

Fig. 4. Mean (± S.E.M.) of 17 β estradiol levels in the spontaneous cycles (pregnant vs. non-pregnant).

also supported by the findings of Frydman et al.[1] that in 24% of patients endometrial biopsy turned out to be dystrophic, independently of whether the cycle had been spontaneous or stimulated.

In spontaneous cycles, we found no difference in hormonal levels between pregnant and non-pregnant cycles, suggesting that the embryonic implantation potential is more important than endometrial environment as measured by hormonal levels during the luteal phase.

ACKNOWLEDGEMENT

We are indebted to Adrian Walter-Ginzburg, Ph.D. for her invaluable assistance in writing this chapter.

REFERENCES

1. R. Frydman, J. Testart, P. Giacomini, M. C. Imbert, E. Martin, and K.Nahoul, Hormonal and histological study of the luteal phase in women following aspiration of preovulatory follicle, *Fertil. Steril.*, 38:312 (1982).
2. J. Dor, E. Rudak, S. Mashiach, L. Nebel, D. M. Serr, and B. Goldman, Periovulatory 17 beta-estradiol changes and embryo morphologic features in conceptional and non-conceptional cycles after human in vitro fertilization, *Fertil. Steril.*, 45:63 (1986).
3. E. Rudak, J. Dor, S. Mashiach, L. Nebel, and B. Goldman, Chromosome analysis of multi-pronuclear human oocytes fertilized in vitro, *Fertil. Steril.*, 41:538 (1984).
4. A. Trounson, and L. Mohr, Human pregnancy following cryopreservation, thawing and transfer of an eight cell embryo, *Nature*, 305:707 (1983).
5. B. Lassalle, J. Testart, and J.P. Renard, Human embryo features that influence the success of cryopreservation with the use of 1,2 propanediol, *Fertil. Steril.*, 44:645 (1985).
6. J. Dor, E. Rudak, A. Davidson, Z. Ben-Rafael, D. Levran, M. Kimchi, D. M. Serr, L. Nebel, S. Mashiach, Endocrine and Biological factors contribution to the implantation of frozen-thawed embryos. Abstracts of the VI World Congress of In Vitro Fertilization and Alternate Assisted Reproduction, April 2-7, 1989, Jerusalem, Israel.
7. D. Levran, S. Mashiach, E. Rudak, J. Dor, J. Blankstein, D.M. Serr, B. Goldman, and L. Nebel, Evaluation of "pregnancy potential" of eggs and "endometrial receptivity" as determinators of the outcome of IVF and ET, *in:* "Abstracts of the Fifth World Congress on Human In Vitro Fertilization and Embryo Transfer," April 5-10, 1987, Norfolk, Virginia, USA, p. 108.
8. J. Mandelbaum, A.M. Junca, M. Plachot, M.D. Alnot, S. Alvarez, C. Debache, J. Salat Baroux, and J. Cohen, Human embryo cryopreservation, extrinsic and intrinsic parameters of success, *Hum. Reprod.*, 2:709 (1987).
9. J. Testart, B. Lassalle, R. Forman, A. Gazengel, J. Belaisch-Allart, A. Hazout, J.D. Rainhorn, and R. Frydman, Factors influencing the success rate of human embryo freezing in an in vitro fertilization and embryo transfer program, *Fertil. Steril.*, 48:107 (1987).
10. J.D. Stanger, L. Clark, M. Brinsmead, and M. Oliver, Day 3 luteal estradiol-progesterone ratios and pregnancy: analysis of factors, *in:* "Abstracts of the Fifth World Congress on Human In Vitro Fertilization and Embryo Transfer," April 5-10, 1987, Norfolk, Virginia, USA, p. 120.

THE REPLACEMENT OF FROZEN-THAWED EMBRYOS:
NATURAL OR ARTIFICIAL CYCLES?

B. Coroleu, A. Veiga, M. Moragas, F. Martinez, G. Calderón, and P.N. Barri,

Instituto Dexeus
Po Bonanova 67, 08017 Barcelona, Spain

INTRODUCTION

It is known that the success of in vitro fertilization (IVF) is directly related to embryo quality and endometrial receptivity.

Our freezing program started in May 1986, and the replacement of thawed embryos was performed initially in natural cycles according to Testart et al., (1986, 1987). The pregnancy rate using this protocol was 14% (Veiga, 1987). As the pregnancy rate obtained in IVF cycles and our oocyte donation program was much higher, we decided to start with stimulated cycles in the freezing program (Coroleu, 1988).

MATERIALS AND METHODS

The aim of this study was to compare the results obtained in six groups of patients (Figure 1).

Group A: Fourteen patients in whom embryo replacement was performed during spontaneous cycle;

Group B: Twenty eight patients in whom embryo replacement was performed in a natural cycle with administration of human chorionic gonadotropin (hCG) 10,000 I.U. in the luteal phase, 5,000 I.U. on the day prior to replacement, 2,500 I.U. 24 hours after replacement and 2,500 I.U. 72 hours after replacement;

Advances in Assisted Reproductive Technologies, Edited by
S. Mashiach *et al.*, Plenum Press, New York, 1990

Group C: Eighteen patients in whom embryo replacement was performed in a natural cycle with daily administration of 50 mg. of progesterone in oil in the luteal phase starting on the day prior to the replacement for 11 days.

Monitoring of ovulation of groups A, B and C was carried out with daily plasma (E_2 RIA), US and urinary LH (Hi-gonavis, Mochida) and plasma P_4 48 hours after LH-SIR.

Group D: Fourteen patients in whom embryo replacement was performed with administration of GnRH agonists (Buserelin, Hoechst Lab. s.c. 0.3 ml/12h.) to produce a medical hipophysectomy, and recreate the oocyte using exogenous estrogen and progesterone according to the protocol proposed by Navot et al., (Navot, 1986).
Buserelin was also administered during the luteal phase.

Group E: Eleven patients in whom embryo replcement was performed in the same way as group D, but GnRH administration was stopped when progesterone administration was started.

Monitoring of these two Groups (D and E) was limited to one plasma E_2 and P_4 analysis on the day prior to the replacement. The embryo replacement was performed three or four days after progesterone administration was initiated.

Group F: Ten patients in whom embryo replacement was performed with high stimulation of 150 I.U. of hMG on days three, five, seven and nine of the cycle, and 5000 I.U. of hCG administered when the E_2 reached 1000 pg/ml and the diameter of the biggest follicle was 20 mm.
The replacement took place 66 to 70 hours after hCG.
Luteal phase support was the same in Group B.
Monitoring was the same as in Groups A, B and C.

Twelve days after replacement plasma hCG was assayed to determine if pregnancy was established and ultrasound was performed at six weeks of amenorrhea.

In Groups D and E estrogen and progesterone treatment was continued until the 14th week of amenorrhea. In these Groups plasma E_2 and P_4 was assessed weekly to adapt the substituted treatment for each patient.

RESULTS

First of all, we wanted to be sure that there were no differences between the six groups concerning the age of the patients, the cause of infertility, the treatment with IVF, where the thawed embryos came from, and the number of embryos replaced per patient. The Table shows that the differences in such parameters are not statistically significant. (Table 1).

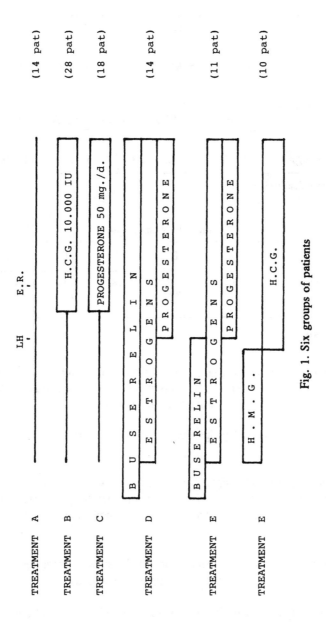

Fig. 1. Six groups of patients

Table 1. Age, Cause of Infertility and Previous IVF Treatment

	A	B	C	D	E	F
Age	32 ± 3.6	32 ± 3.6	33 ± 3.5	34 ± 3.8	34 ± 4.5	33 ± 2.8
Cause of infertility						
- Tubal factor	10 (71%)	18 (64%)	12 (66%)	9 (64%)	7 (63%)	7 (70%)
- Others	4	10	6	5	4	3
IVF previous cycle treatment						
- GnRH agonist	5 (36%)	14 (50%)	7 (40%)	7 (50%)	5 (46%)	4 (40%)
- Non agonist	9	14	11	7	6	6
No. embryos replaced/pt.	1.2 ± 0.4	1.5 ± 0.4	1.3 ± 0.6	1.3 ± 0.4	1.3 ± 0.4	1.3 ± 0.4

Differences not statistically significant

Table 2. Pregnancy Rate

	A	B	C	D	E	F
Pregnancy/patient	2/14*	8/28	5/18	0/14	6/11*	4/10
%	14.2	28.5	27.7	0	54.5	40
			28.3			

* Fisher exact test, p = 0.04

The next Table shows the pregnancy rate obtained in the six groups: the higher pregnancy rate was obtained in Group E, the group in which estrogen and progesterone, together with buserelin, until progesterone administration starts.

The differences are statistically significant between Groups A and E (Fisher Exact Test p = 0.04).

Even though no statistically significant differences were observed between Group A and all the other groups together, we wish to point out that the global pregnancy rate in the stimulated cycle was higher than that obtained in spontaneous cycles (28.3% versus 14.2%).

No pregnancy was obtained in Group D, in which estrogen/progesterone and buserelin was given if the agonist was administered during the luteal phase.

No statistically significant differences were observed between Group B (hCG luteal support) and Group C (progesterone luteal support) (pregnancy rate 28.5% versus 27.7%).

When the E_2 and P_4 levels in the six groups are compared, we can see that the differ-

Table 3. Comparison of Hormone Levels

	A	B	C	D	E	F
No of patients	14	28	18	14	11	10
E_2 (on day of E.R.) (pg/ml)	181 ± 63.7*	204 ± 66.6	165 ± 34	209 ± 78	299 ± 121*	209 ± 105
P_4 (on day of E.R.) (ng/ml)	7.3 ± 3.6**	12.9 ± 4.6	16.8 ± 7.2	22.2 ± 5.5	23.7 ± 6.1**	13.3 ± 5.6

* Mann-Whitney (U = 34; p < 0.02)
** Mann-Whitney (U = 1; p < 0.0001)

Table 4. Comparison of Pregnant and Nonpregnant

	Pregnant	Non pregnant
No of patients	25	70
IVF previous cycle treatment		
- GnRH agonist	9 (46%)	31 (44.2%)
- Non agonist	16	39
E_2 (on day of E.R.) (pg/ml)	262 ± 390	200.1 ± 184 N.S.
P_4 (on day of E.R.) (ng/ml)	14.6 ± 7.4	15.8 ± 12 N.S.
No embryos replaced/pt.	1.5 ± 0.5	1.3 ± 0.4 N.S.

ences in E_2 and P_4 between Group A and E are statistically significant (Estradiol levels Group A: 181 ± 63.7 pg/ml and Group E: 299 ± 121 pg/ml - Mann-Whitney test U = 34; p < 0.02. Progesterone levels Group A: 7.3 ± 3.6 ng/ml and Group E: 23.7 ± 6.1 ng/ml Mann-Whitney test U = 1, p < 0.0001) (Table 3).

If we classify our patients into two groups of pregnant and non-pregnant, we can see that no differences were observed in the previous IVF treatment cycle (agonist-non-agonist) in the E_2 and P_4 level on the day of embryo replacement or in the number of embryos replaced. (Table 4).

However, if we only consider the groups where no exogenous progesterone was administered (Groups A, B and F) we can see that the E_2 level is higher (244 versus 157) in the

Table 5. Results of Treatment when Progesterone was not Administered

	Pregnant	Non pregnant	
No. of patients	14	38	
E_2 (on day of E.R.) (pg/ml)	244.6 ± 239	157 ± 71.1	$p < 0.05$
P_4 (on day of E.R.) (ng/ml)	13.1 ± 3.8	7.4 ± 2.7	$p < 0.001$
No. embryos replaced/pt.	1.4 ± 0.5	1.3 ± 0.4	N.S.

pregnant group and P_4 levels are also higher in this group (13.1 versus 7.4). The number of embryos replaced is the same in both groups.

The differences were statistically significant, E_2 (Student test) $P < 0.05$ and P_4 $P < 0.001$. (Table 5).

CONCLUSIONS

In conclusion and according to the results obtained:
1. A higher pregnancy rate is obtained when the replacement is performed in a stimulated cycle.

2. The best results are obtained in the group where the replacement is carried out under GnRH agonist, estrogen and progesterone but stopping the agonist when progesterone administration is started.

3. Plasma progesterone and estradiol level is higher in the pregnant groups (A, B and F).

4. There are no differences between hCG or progesterone for luteal support.

5. It is also important to point out the possible negative effect of GnRH (s.c.) administration during the early luteal phase in embryo implantation.

REFERENCES

Coroleu, et al., 1988, Institute Dexeus oocyte donation programme, Human Reproduction, Abstracts from the Fourth Meeting of the European Society of Human Reproduction and Embryology, July 3–6 1988, Barcelona, Spain, 116.
Navot, et al., 1986, Artificially induced endometrial cycles and establishment of pregnancies in

the absence of ovaries, *The New England Journal of Medicine*, 314, 13:806-811.

Testart et al., 1986, High pregnancy rate after early human embryo freezing, *Fertil. Steril.*, 46, 2:268-272.

Testart et al., 1987, Factors influencing the success rate of human embryo freezing in an in vitro fertilization and embryo transfer program, *Fertil. Steril.*, 48, 1:107-112.

Veiga et al., 1987, Pregnancy after the replacement of a frozen-thawed embryo with < 50% intact blastomeres, *Human Reprod.*, 2, 4:321-323.

IMPROVEMENT OF OVARIAN RESPONSE TO INDUCTION OF SUPEROVULATION WITH COMBINED GROWTH HORMONE - GONADOTROPIN TREATMENT

George Coukos, Paolo G. Artini, Annibale Volpe,
Marta Silferi, Antonella Barreca*, Francesco Minuto*,
and Andrea R. Genazzani

*Department of Obstetrics and Gynecology, University of Modena, Modena, and
Chair of Endocrinology, University of Genova, Italy

INTRODUCTION

Recent in vitro data suggests that growth hormone (GH) and several peptide growth factors take part in the regulation of ovarian activity (Dorrington et al., 1987, Gospodarowicz et al., 1979; Jia et al., 1986; Knecht et al., 1983; Mondschein et al., 1984 and 1988; Schomberg et al., 1983). Among them, insulin-like growth factor-I (IGF-I) is the most studied in its role in modulating the growth and differentiation of the components of ovarian follicle. IGF-I has been shown to exert a strong stimulatory effect on the proliferation (Savion et al., 1981; Baranao et al., 1984) and the differentiation of granulosa cells (Baranao et al., 1984; Adashi et al., 1984 and 1986). Furthermore, the role of IGF-I in amplifying the action of gonadotropins on ovarian tissue, has been evidenced in the rat (Adashi et al., 1985). On the other hand, administration of GH in hypophysectomized estrogen-treated female rats increased the ovarian tissue IGF-I content (Jia et al., 1986). In vivo studies demonstrated that exogenous administration of GH is followed by an increase of the levels of plasma IGF-I in humans (Zapf et al., 1978).

On the basis of this evidence and the preliminary reports of Homburg et al. (1988a,b), the efficiency of combined gonadotropin-GH treatment for induction of superovulation was studied in some patients resistant to gonadotropins.

Advances in Assisted Reproductive Technologies, Edited by
S. Mashiach *et al.*, Plenum Press, New York, 1990

MATERIALS AND METHODS

A total number of 12 patients, aged 24–41 years, with spontaneous ovulatory cycles and infertility due to monolateral or bilateral tubal impatency, diagnosed by laparoscopy, were studied. Seven patients were 24–36 years old (plasma FSH: 4.7–11.6 mIU/ml, on 6th–8th day of spontaneous cycle); three of them were overweight with a body mass index (BMI) of 29–35. Five patients, aged 39–41 years (plasma FSH: 6.2–18.5 mIU/ml, on 6th - 8th day of spontaneous cycle) were also treated; one of them was overweight (BMI 30.4). All patients had previously proved resistant to high dosage gonadotropin therapy, i.e. 6– 17 ampules/cycle purified menofollitropin (Metrodin, Serono, Italy, 75 I.U. FSH/ampule), and 34–80 ampules/cycle human menopausal gonadotropin (Pergonal, Serono, 75 I.U. FSH + 75 I.U. LH/ampule). Cycles were cancelled within the 9th–12th day because of low plasma 17ß-estradiol (E_2) levels and/or poor follicular recruitment or growth.

In all patients pituitary GH response to exogenous growth hormone-releasing factor (GRF) was tested before administering the combined treatment. GRF (Groliberin, Kabi, Stockholm, Sweden) was administered at 1 mcg/Kg body weight (BW), bolus i.v.. Three patients—two obese and one of normal BW—from the group of younger women did not respond to the GRF-stimulatory test (GH < 5 ng/ml). From the other group, one patient out of five, who was obese, did not respond to the GRF-test (GH peak = 4.5 ng/ml).

During the combined treatment-cycle all patients were administered approximately the same dosage of gonadotropins as before, plus a standard dosage of GH, 0.1 I.U./Kg-BW/ day, s.c. (Genotropin—recombinant GH synthesized in E. coli K12—Kabi, Stockholm, Sweden). When at least one follicle reached 15 mm in diameter and plasma E_2 was approximately at 200 pg/ml per follicle, human chorionic gonadotropin (hCG) 10 000 I.U. (Profasi, Serono, Italy) was administered; follicular aspiration was performed by laparoscopy 32– 34 hours later.

Ovarian Response

The response to treatment was monitored through daily plasma E_2 measurement by radioimmunoassay (RIA) and transvaginal ultrasonic follicular evaluation, starting from day 7 of the stimulatory cycle.

Follicular Fluid GH, IGF-I and Steroid Evaluation

The follicular fluid (FF) content of GH, IGF-I, E_2 and progesterone (P) were measured in all patients treated with combined therapy who underwent laparoscopy. These patients were administered 6–10 ampules/cycle of menofollitropin and 30–65 ampules/cycle of hMG, plus the standard dosage of GH. The same hormones have been measured in the follicular fluids (FFs) from a control group of nine patients who normally responded to gonadotropin treatment. Induction of superovulation in these patients was performed with relatively high dosage of gonadotropins, i.e. 6–8 ampules/cycle of menofollitropin and 28–36 ampules/cycle of hMG.

Only clear-yellow FFs aspirated by laparoscopy were included in the study. A total number of 58 follicles were studied in the GH-treated group and 39 follicles in the control group. FFs were centrifuged at 2,000 rpm × 10 min and the supernatants were stored at -20°C until assay.

GH, IGF-I, E_2 and P Assay

IGF-I was measured by double-antibody RIA using materials provided by Nichols Institute (San Juan Capistrano, CA, USA) and a standard pure Met–IGF-I obtained by recombinant technology (a gift from Drs. Linda Fryklund and Anna Skottner, Kabi, Stockholm, Sweden). Sensitivity of the assay was 90 pg/ml; between assay coefficient of variation was 7.5%. In order to avoid interferences from binding proteins, the samples were treated with acid ethanol. GH was assayed by IRMA using a kit from Nichols Institute (San Juan Capistrano, CA, USA). E_2 and P were measured by RIA using kits from DBC (Los Angeles, CA). All samples were analyzed in a single assay to reduce experimental variability.

Statistical Analysis

Statistical analysis was performed using the Student's t-test and computing the correlation coefficient between groups.

RESULTS

The cycle with combined GH plus gonadotropins was cancelled in all patients aged 39–41 years, because of poor ovarian response. In these patients no significant differences existed between the sole gonadotropin- and the combined therapy-stimulated cycles, in terms either of follicular recruitment and growth (Fig. 1) or plasma E_2 increase. In all younger patients the combined GH-gonadotropin stimulation induced a satisfactory ovarian response (Fig. 1) with a better follicular recruitment (average 8, range 7–16 follicles/patient) and oocyte recovery (average 6.5 oocytes/patient). Laparoscopy was performed on the 11th–13th day. The average fertilization rate was three oocytes/patient. No pregnancy was obtained.

Follicular Fluid GH, IGF-I and Steroids

GH-treated patients showed FF levels of GH (4.5 ± 0.3 ng/ml, M ± SEM), E_2 (228.9 ± 8 ng/ml) and P (5.6 ± 0.2 mcg/ml) significantly higher than in controls (GH: 2.3 ± 0.2 ng/ml, p < 0.0001; E_2: 171.5 ± 8.1, p < 0.01; P: 4.3 ± 0.2 mcg/ml, p < 0.01), while no significant differences were observed in FF IGF-I levels (GH-treated: 158.6 ± 8 ng/ml; controls: 145.5 ± 11.6 ng/ml) (Fig. 2). A significant positive linear correlation was observed between FF GH and IGF-I levels in the GH-treated patients (r = 0.40, p < 0.05) (Fig. 2). No correlation was observed in the control group (data not shown).

DISCUSSION

These results show that the administration of GH increases the ovarian response to gonadotropins in women under 35 years of age, who previously displayed resistance to gonadotropin superovulation-induction, confirming the reports of Homburg et al. (1988 a,b).

The ovarian response to GH supplementation did not parallel the type of pituitary response to acute GRF administration. In fact, four out of seven younger patients showed normal pituitary response to GRF and normal plasma GH basal levels but gained a considerable benefit from GH supplementation, while no improvement of ovarian response to gonadotropins was achieved in one older patient showing low response to GRF. Rather than underlying altered pituitary GH secretion, age appears to be an important factor influencing the ovarian response to combined therapy. We cannot explain the different response to GH administration existing between the different age subgroups of patients. It is

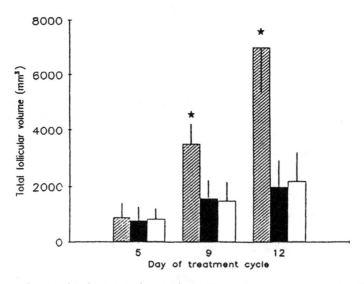

Fig. 1. GH supplementation improves the ovarian response in younger resistant patients (grey bars, p < 0.05) but not in older resistant patients (black bars). White bars represent all patients during the previous cycle of stimulation with sole gonadotropins.

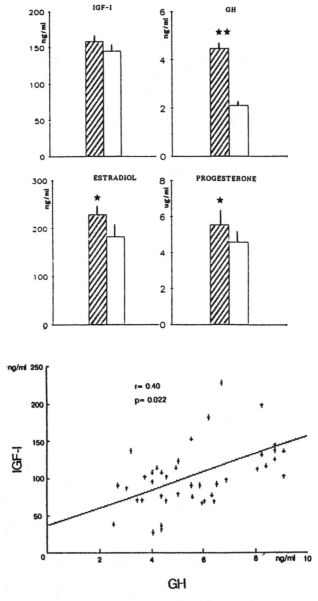

Fig. 2. Top: Follicular fluid levels of IGF-I, GH, E2 and P in GH-treated younger patients (grey bars) and in controls (white bars); * P < 0.01, ** p < 0.0001. Bottom: Linear correlation between follicular fluid GH and IGF-I in GH-treated younger patients.

known, however, that a decrease in fertility potential occurs after 35 years of age and a decrease of IVF treatment success has been reported in patients above 37–38 years of age (Edwards et al., 1983).

The improved ovarian response in GH-treated young patients is associated with significantly higher FF levels of GH, E_2 and P than in controls. Higher GH levels in the FFs of GH-treated patients might be due to a diffusion of plasma GH into the follicular compartment, while increased FF E_2 and P levels might reflect the effect of GH on ovarian steroidogenesis. Such an effect might be both direct (Jia et al., 1986) or locally mediated by IGF-I. The exogenous administration of GH in vivo induces an increase of ovarian IGF-I content (Davoren et al., 1985). The finding of no significant difference in FF IGF-I levels between GH-treated patients and controls, might apparently be contradictory to a GH-induced increase of ovarian IGF-I. However, since no low-responding patient underwent laparoscopy in the first stimulatory cycle, no data is available on FF IGF-I levels from resistant patients when stimulated with sole gonadotropins. It might be that GH administration increased FF IGF-I levels, bringing them up to levels similar to controls. The significant positive correlation existing between FF GH and IGF-I in GH-treated patients might sustain such a hypothesis.

In conclusion, the present data indicates that in stimulatory protocols for IVF/ET, GH associated with gonadotropins improves the ovarian response in young gonadotropin-poor-responsive patients. In these patients, intrafollicular levels of GH, E_2 and P are significantly higher compared to normally responsive patients stimulated with sole gonadotropins.

REFERENCES

Adashi, E. Y., Resnick, C. E., Svoboda, M. E., and Van Wyk, J. J., A novel role for somatomedin-C in the cytodifferentiation of the ovarian granulosa cell, *Endocrinology*, 115:1227, 1984.

Adashi, E. Y., Resnick, C. E., Svoboda, M. E., Van Wyk, J. J., Hashal, V. C., and Yanagishita, M.,Independent and synergistic actions of somatomedin-C in the stimulation of proteoglycan biosynthesis by cultured rat granulosa cells, *Endocrinology* 118:456, 1986.

Adashi, E. Y., Resnick, C. E., Brodie, A.M.H., Svoboda, M.E., and Van Wyk, J.J., Somatomedin-C mediated potentiation of follicle-stimulating hormone-induced aromatase activity of cultured rat granulosa cells, *Endocrinology* 117:2313, 1985.

Baranao, J. L.S., and Hammond, J. M., Comparative effects of insulin and insulin-like growth factors on DNA synthesis and differentiation of porcine granulosa cell, *Biochem. Bioph. Res. Comm.*, 124:484, 1984.

Davoren, J.B., Hsueh, A. J. W., Growth hormone increases ovarian levels of immunoreactive somatomedin C/insulin-like growth factor I in vivo, *Endocrinology* 118:888, 1986.

Dorrington, J. H., Bendell, J. J., and Lobb, D. K., Aromatase activity in granulosa cells: Regulation by growth factors, *Steroids* 50:411, 1987.

Edwards, R. G., and Steptoe, P. C., Current status of in vitro fertilization and implantation of human embryos, *Lancet* 2:1265, 1983.

Gospodarowicz, D., Bialecki, H., Fibroblast and epidermal growth factors are mitogenic agents for culture granulosa cells of rodent, porcine and human origin, *Endocrinology*, 104:757, 1979.

Homburg, R., Eshel, A., Abdalla, H. I., Jacobs, H. S., Growth hormone facilitates ovulation induction by gonadotrophins, *Clin. Endocrinol.*, 29:113, 1988a.

Homburg, R., West, C., and Jacobs, H. S., Controlled trial of co-treatment with growth hormone and gonadotrophins for induction of ovulation (Abstract). 178th Meeting of the Society for Endocrinology, *J. Endocrinol.*, 119 (suppl.):144, 1988b.

Jia, X. C., Kalman, J., and Hsueh, A. J. W., Growth hormone enhances follicle-stimulating hormone-induced differentiation of cultured rat granulosa cells, *Endocrinology*, 118:1401, 1986.

Knecht, M., and Catt, K. J., Modulation of cAMP-mediated differentiation in ovarian granulosa cells by epidermal growth factor and platelet-derived growth factor, *J. Biol. Chem.* 258:2789, 1983.

Mondschein, J. S., and Schomberg, D. W., Platelet-derived growth factor enhances granulosa cell luteinizing hormone receptor induction by FSH and serum, *Endocrinology*, 109:325, 1981.

Mondschein, J. S., Canning, S. F., and Hammond, J. M., Effects of transforming growth factor-beta on the production of immunoreactive insulin-like growth factor I and progesterone and on 3Hèthymidine incorporation in porcine granulosa cell cultures. *Endocrinology* 123: 1970, 1988.

Savion, N, Lui, G.M., Laherty R, Gospodarowicz D: Factors controlling proliferation and progesterone production by bovine granulosa cells in serum-free medium, *Endocrinology*, 109:409, 1981.

Schomberg, D. W., May, J. V., and Mondshein, J. S., Interactions between hormones and growth factors in the regulation of granulosa cell differentiation in vitro, *J. Steroid Biochem.* 19:291, 1983.

Zapf, J., Rinderknech, E., Humbel, R. E., and Froesch, E. R., Nonsuppressible insulin-like activity (NSILA) from human serum: reaccomplishments and their physiological implications, *Metabolism*, 27:1803, 1978.

THE PERMISSIVE ROLE OF GROWTH HORMONE IN OVULATION INDUCTION WITH HUMAN MENOPAUSAL GONADOTROPIN

Zeev Blumenfeld, Bruno Lunenfeld,* and Joseph M. Brandes

*Reproductive Endocrine Unit, Dept. Ob/Gyn, Rambam Med. Ctr., Technion, Haifa and
* Endocrine Institute, Tel-Hashomer, Israel*

INTRODUCTION

Although the central role of gonadotropins in the regulation of granulosa cell ontogeny is well established,[1,2] the variable fate of ovarian follicles subjected to comparable gonadotropic stimulation suggests the existence of additional intraovarian modulatory mechanisms.[1,3] In this connection, the role of several growth factors has been the subject of increasingly intense investigation.[1] Among potential modulators of granulosa cell ontogeny, the insulin-like growth factors (IGFs) appear unlikely suited to the task, combining replicative and in some instances cytodifferentiative properties.[1] On the other hand Somatomedin-C (Sm-C)/IGF-I seems to play an important role, not only in stimulating cell replication in tissues of diverse origin, but also in promoting the differentiation of different cell types.[1] Somatomedin-C (Sm-C)/IGF-I displays stringent growth hormone (GH) dependence[1,4,5] in vivo and in vitro.[1,4]

In 1972 Sheikholislam et al.[6] demonstrated that delayed puberty associated with isolated GH deficiency in human is readvanced by GH therapy. Recently Owen and colleagues[7] have succeeded in reducing the quantity of gonadotropins needed to achieve ovulation in infertile patients with hyposensitive ovaries undergoing ovulation induction by the addition of GH to hMG/hCG treatment. In light of their preliminary encouraging data we combined GH with gonadotropins in ovulation induction in a patient with panhypopituitarism who had demonstrated a poor response to gonadotropins in previous attempts of human menopausal gonadotropin (hMG)/human chorionic gonadotropin (hCG) ovulation induction, in another

Advances in Assisted Reproductive Technologies, Edited by
S. Mashiach *et al.*, Plenum Press, New York, 1990

patient with GH deficiency and in infertile patients with no response to Clomid, who were also poor responders to ovulation induction by gonadotropins.

MATERIALS AND METHODS

A 33 year old, white Jewish woman with primary amenorrhea was referred to our clinic due to primary infertility. The patient was diagnosed at the age of four years as suffering from panhypopituitarism and was treated in a pediatric endocrine clinic at another hospital. At the age of nine the thyroid function tests including radioactive I^{131} absorption were compatible with mild hypothyroidism, for which the patient was put on L-thyroxine replacement (0.1 mg/day, orally). During the four years previous to her referral, the patient underwent a gonadotropin-releasing hormone (GnRH) test (100 microgram I.V.). The serum luteinizing hormone (LH) and follicle stimulating hormone (FSH) concentrations were lower than 1 mIU/ml 15 minutes and immediately before and at 30, 60 and 90 minutes after the IV GnRH bolus. An intravenous (IV) injection of 0.25 mg corticotropin (ACTH) (Synacthen test), in 1986 caused the plasma cortisol to increase from 3.7 microgram/dl before injection to 14.8 microgram/dl 30 minutes afterwards and to 23.1 microgram/dl 60 minutes after the ACTH administration. However, repeat blood and free urine cortisol on admission were low; morning plasma cortisol—1.7 and 4.5 microgram/dl, and urinary free cortisol - less than 10 mg/24h. Serum GH concentrations were undetectable. Serum concentrations of glucose, creatinine, sodium, potassium, chloride, bicarbonate, bilirubin, calcium, phosphor, uric acid, cholesterol, albumin, globulin, alkaline-phosphatase, glutamic-oxaloacetic transaminase (GOT), glutamic-pyruvic transaminase (GPT) and blood osmolality - were all within normal limits. A hysterosalpingography revealed a hypoplastic uterus with bilateral patent tubes and normal spillage of contrast material in the peritoneal cavity. Synthetic progestatives (medroxy-progesterone-acetate 10 mg/d for five days) did not induce vaginal bleeding but sequential estrogen and progesterone administration brought about withdrawal bleeding. During the year before her referral the patient underwent four cycles of ovulation induction by hMG/hCG. In the first cycle the serum 17-ß estradiol (E_2) concentrations increased from less than 20 pg/ml to 500 pg/ml on the hCG administration day, after IM injection of > 90 ampoules of hMG (Pergonal, Teva Pharmaceutical Industries Ltd., Petah-Tiqva 49131, Israel) intramuscularly (IM) administered over 30 days. In the second hMG/hCG cycle the E_2 increased to 375 pg/ml on the hCG administration day, after consumption of 76 ampoules of hMG (Pergonal). In the third hMG/hCG cycle the E_2 concentration on the hCG administration day was 970 pg/ml, after consumption of 96 ampoules of hMG (Pergonal). The fourth hMG/hCG cycle was associated with E_2 increase to 3400 pg/ml (after seven days of seven ampoules of Pergonal/day), therefore hCG administration was withheld. The first three hMG/hCG cycles were associated with serum luteal progesterone (P) concentrations of 35-58 ng/ml.

On admission, the patient's physical examination was positive for undeveloped secondary sexual characters. Lack of pubic and axillary hair was noted. The breast were underdeveloped Tanner stage I-II. Pelvic examination revealed a normal cervix, an hypoplastic uterus with no palpable adnexal masses.

The patient's spouse semen analysis revealed sperm concentration $40 - 65 \times 10^6$ spermatozoa/ml, 40% motility, and 60% hamster eggs penetration (versus 93% penetration by normal control sperm). The husband's past history was remarkable for orchidectomy at the age of nine years due to tortion of one testicle. The remaining testicle was normal by size and consistency on examination, as well as the rest of his physical examination.

After induction of withdrawal bleeding with sequential estrogen and gestogen (Estradiol valerate 2 mg/day for 21 days and norgestrel 0.5 mg/day on days 12-21, Progyluton, Schering AG, Berlin/Bergkamen, FRG), the patient was started on Prednisone 15 mg/day, in addition to L-thyroxine 0.15 mg/day, and on day three of the cycle started receiving three ampoules of hMG (Pergonal, (75u FSH + 75u LH)/ampoule) daily, IM. On the 4th day of the cycle the patient started receiving 4u of GH IM (Nanormon, Nordisk Gentofte, Manufacturing Division of Nordisk Insulinlaboratorium, Denmark) three times/week. The hMG daily dose was increased to five ampoules/day on day 7 for two days, due to lack of increase in E_2 concentration for the first four days of hMG treatment, and subsequently decreased to three and later two ampoules/day. On the 14th day of the cycle, trans vaginal sonographic (TVS) visualization of the ovaries revealed two 17 and 18 mm sized follicles in the right ovary and four additional 14-16 mm sized follicles in each ovary. The hMG administration was witheld. Serum E_2 concentration was 819 pg/ml and 10,000u hCG were IM administered 70 hours after the last hMG injection. Due to negative post coital tests, intrauterine insemination (IUI) with concentrated husband's spermatozoa was performed, after a "swim-up" procedure of the washed sperm specimen 34 hours after hCG administration. The basal body temperature (BBT) chart was biphasic and mid-luteal progesterone—32.8 ng/ml. The hMG consumption/cycle was 36 ampoules of Pergonal. The patient received six injections of GH (4u each) during the 16 days of the follicular phase. The luteal phase was supplemented with three ampoules of hCG (2,500u each) three, six and nine days after the day of the first hCG (10,000u) administration. Menstrual bleeding occurred after a 15 days luteal phase.

After TVS revealed no ovarian follicles or persistent cysts, and a serum E_2 measurement of < 20 pg/ml, a second hMG/GH/hCG cycle was initiated on day 3 of the cycle. On the 12th day, E_2 was 786 pg/ml and P = 0.3 ng/ml. Administration of hMG was witheld and 10,000u hCG were IM administered 48 hours after the last hMG injection. The hMG consumption was 36 ampoules of Pergonal and the GH consumption was 16 units (four ampoules of 4u over 12 days). The mid-luteal P was > 40 ng/ml. Serum ß-hCG concentrations were 31, 45 and 497 mIU/ml on days 25, 27 and 32 of the cycle respectively, and on TVS an intrauterine single gestational sac was visualized on day 39 of the cycle. The patient continued on a normal pregnancy and was delivered of a 3200 gr male fetus by Caesarean section due to relative cephalo pelvic disproportion. The Apgar score was 8/9 at 1 and 5 minutes respectively.

Another 31 year old infertile patient who was treated by GH injections as a child due to GH deficiency was put on a combination of GH/hMG/hCG versus hMG/hCG in alternate cycles. The patient is the first out of 12 patients participating in a randomized, cross-over designed study, in collaboration with the clinical research department, Nordisk Gentofte A/S, Copenhagen, Denmark, as presented by the scheme in Figures 1 and 2.

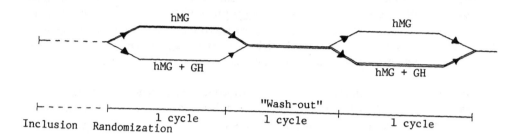

Fig. 1. Flow chart.

Cross-over design - 2 treatments
List of patients and sequence of treatment periods

1   BA

2   BA

3   AB

4   AB

5   AB

6   BA

7   BA

8   AB

9   AB

10   AB

11   BA

12   BA

A = Cycle with gonadotrophin alone

B = Cycle with Norditropin^R (B-hGH) + gonadotrophin

Fig. 2. Use of growth hormone as an adjunt to gonadotrophin in ovulation induction. A randomized cross-over study.

The second patient ovulated on both ovulation induction cycles (GH/hMG/hCG and hMH/hCG alone) without significant differences in 17β estradiol or Progesterone blood concentrations (Figure 3). On the GH/hMG/hCG cycle, 12u GH were administered intramuscularly 3 times/week until ovulation. The peripheral somatomedin levels were significantly higher during the follicular phase of the GH/hMG cycle than the levels during the "control" (hMG/hCG alone) cycle (Mean ± S.D., p < 0.01) (Figure 4).

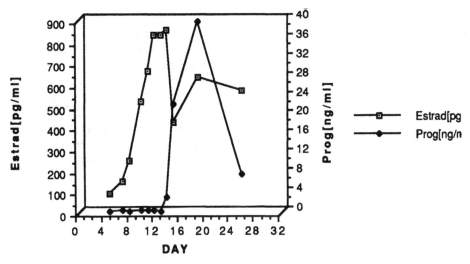

Fig. 3. Ovulation induction with GH/hMG/hCG and hMG/hCG alone.

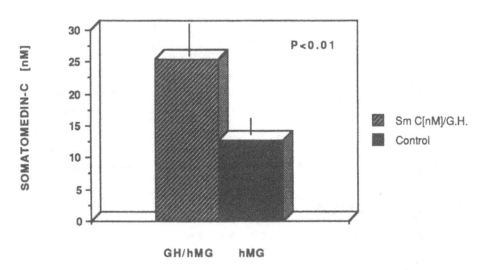

Fig. 4. Peripheral somatomadin-C levels in GH/hMG compared to hMG cycles.

The peripheral Sm-C levels during the follicular phase of the GH/hMG/hCG cycle were higher than the mid-luteal Sm-C level during the same cycle (25 ± 5) nanomoles/liter [nM] versus 15 [nM], respectively), as shown by Figure 5.

Three other infertile patients with poor response to hMG/hCG ovulation induction (necessitating 40–96 ampoules of hMG/cycle) have undergone Clonidine test, in which 150 microgram Clonidine hydrochloride (Normopressan, Rafa Laboratories Ltd., Jerusalem, Israel) were administered orally at 0800 hour after an overnight fast and peripheral blood was drawn immediately before and after 30, 60, 90, 120, 150 and 180 minutes thereafter for GH concentration measurements. Whereas a patient with normal response elicited a significant increase in GH concentration after Clonidine ingestion (B-E.M.) the other patients' GH concentrations did not increase after Clonidine administration (Figure 6).

To test the possibility that increased Sm-C/IGF-I levels may play a contributory role in conception, we measured Sm-C concentrations in follicular fluids of patients who conceived and in follicular fluids of those who did not conceive in GIFT or IVF/ET cycles. We did not find increased Sm-C concentrations in the follicular fluids retrieved from patients who conceived as compared to follicular fluids obtained from patients who did not conceive in those particular GIFT or IVF/ET cycles (Figure 7).

DISCUSSION

Recently, Sm-C/IGF-I and other growth factors have been found to have significant influences on gonadal physiology.[1] Although the pertinence of IGFs to granulosa cell physi-

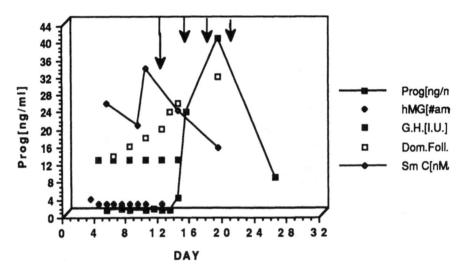

Fig. 5. Peripheral somatomadin-C levels in the follicular and luteal phase.

ology has received relatively limited attention, substantial experimental observations now support such possibility. It is beyond the purpose of this report to review exhaustedly the subject. Moreover, Adashi et al.[1] have recently reviewed the issue of IGFs as intraovarian regulators of granulosa cell growth and function. In brief, D'Ercole et al.[8] have observed that hypophysectomy of male rats, results in a 27.7% reduction in the total testicular content of immunoreactive (IR) - Sm-C/IGF-I, and that this effect can be reversed through systemic administration of oGH. Although hGH is known to activate both lactogenic and so-

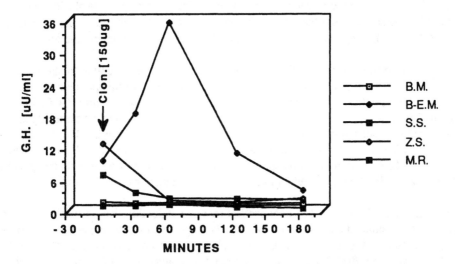

Fig. 6. GH concentration after clonidine ingestion.

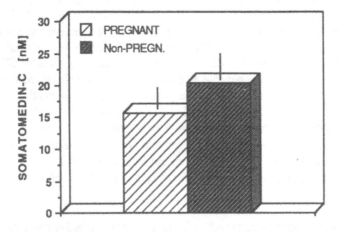

Fig. 7. Follicuclar fluid Somatomadin-C levels in conceptional and non-conceptional cycles.

matogenic receptors in the rat,[9] the ability of oGH (a somatogenic receptor ligand) to enhance testicular IR-IGF-I generation, suggests mediation via somatogenic, rather than lactogenic testicular receptors. These observations raise the possibility that the gonadal, like the hepatic content of Sm-C/IGF-I, may be growth hormone rather than gonadotropin-dependent.[1]

Recently, follicular fluid (FF) specimens from women undergoing in vitro fertilization (IVF) have been measured for IR-Sm-C/IGF-I[1] yielding a range of 0.3-0.81 standard somatomedin units/ml. Measured FF levels of insulin tend to be lower than those observed for IGFs and generally are too low to activate the high affinity granulosa/luteal cell insulin receptor.[1, 10-12] As such, these observations further underscore the possibility that IGFs rather than insulin may be physiologically relevant to the regulation of ovarian function.[1]

Although the pivotal role of FSH in the acquisition of granulosa cell aromatase activity is well accepted,[1, 2] the role of growth factors in the regulation of this process remains the subject of investigation.[1, 3] Concurrent treatment of granulosa cell culture (GCC) in vitro with Sm-C/IGF-I yielded a 2- to 3- fold enhancement of the FSH effect on proteoglycan biosynthesis.[1] Current treatment of GCC with Sm-C/IGF-I resulted in a 6.1-fold enhancement of the FSH effect on LH receptor induction.[1] The ability of Sm-C/IGF-I to synergize with FSH in the induction of LH receptors was dose- and time-dependent with an apparent ED_{50} of 6.2 ± (SE) 0.6 ng/ml and a minimal time requirement of ≤ 24 hours.[1]

The ability of Sm-C/IGF-I to enhance LH receptor induction was associated with increased hCG-stimulated progesterone biosynthesis suggesting that the newly acquired receptors are functional in nature.[1] These and other observations lend additional credence to the possibility that Sm-C/IGF-I may play an important regulatory role in the growth and differentiation of the granulosa cell *throughout* the life cycle of the ovarian follicle.[1]

The permissive action of GH has been initially suggested by observation in GH-deficient rats,[13, 14] mice[15-18] and humans (sporadic, hereditary or acquired GH deficiency).[19-24] Indeed, an association appears to exist between isolated GH deficiency and delayed puberty,[25] a process reversed by systemic GH replacement therapy of GH-deficient rats,[1, 13, 14] mice,[25, 26] and human subjects.[6, 27] Significantly, the mechanism(s) whereby GH therapy enhances pubertal development appears in all cases to involve an enhancement of gonadotropin dependent ovarian function.[1] Given the possibility that the gonadal, like the hepatic content of Sm-C/IGF-I may be GH-dependent,[8] it is tempting to speculate that the ability of GH to accelerate fetal pubertal maturation, may be due, at least in part, to the generation of intraovarian Sm-C/IGF-I and the consequent local potentiation of gonadotropin-dependent ovarian function.[1] Significantly, recent studies suggest a direct effect of GH on granulosa cell differentiation, an effect possibly mediated by endogenously produced IGFs.[28-30]

Our preliminary experience, as well as Owen et al.'s results are encouraging, suggesting a complementary or permissive role of GH to gonadotropins in ovulation induction. Addition of GH to hMG/hCG treatment significantly shortened the follicular phase and diminished the hMG consumption necessary to induce ovulation in Owen's patients. Our results, where hMG consumption decreased from 76–96 ampoules/cycle to 36 ampoules/cycle, when GH was added to hMG, commensurate with others' experience in vivo and with the in vitro data cited previously. Whereas Owen's group protocol used 20u GH three times weekly for 6 doses/cycle (120 units GH/cycle), our patient conceived while receiving only 4u GH × 3/week (16 units/cycle). The minimal effective dose of GH remains to be elucidated by future experience.

We believe the Clonidine test to have a discriminatory role in identifying the patients who may benefit from GH addition to ovulation induction. Our data do not support adding GH to hMG/hCG in patients with unexplained infertility who are not Clonidine non-responders. We conclude that the addition of GH to hMG/hCG ovulation induction may serve as a contributory adjunct in patients with GH deficiency and/or in poor responders to hMG/hCG ovulation induction.

ACKNOWLEDGMENT

The authors wish to thank Mrs. Bathia Navar for assistance in the preparation of this manuscript.

REFERENCES

1. Adashi, E.Y., Resnick, C.E., D'Ercole, J., Svoboda, M.E., and Van Wyk, J.J., 1985, Insulin-Like Growth Factors as Intraovarian Regulators of Granulosa Cell Growth and Function, *Endocr Rev*, 6:400.
2. Richards, J.S., 1979, Hormonal control of ovarian follicular development: A 1978 perspective, *Recent Prog Horm Res*, 35:343.
3. Hsueh, A.J.W., Adashi, E.Y., Jones, P.B.C., and Welsh, Jr. T.H., 1984, Hormonal regulation of the differentiation of cultured ovarian granulosa cells, *Endocr Rev*, 5:76.
4. Van Wyk, J.J., 1984, The somatomedins-biological actions and physiologic control mechanisms, *in:* "Hormonal Proteins and Peptides", Li C.H., ed., Vol. 12, Academic Press, Orlando, p 81.
5. Schoenle, E., Zapp, J., Humbel, R.E., and Froesch, E.R.. 1982, Insulin-like growth factor I stimulates growth in hypophysectomized rats, *Nature*, 296:252.
6. Sheikholislam, B.M., and Stempfel, Jr R.S., 1972, Hereditary isolated somatotropin deficiency: Effects of human growth hormone administration, *Pediatrics*, 49:362.
7. Owen, E., Homburg, R., Eshel, A., Abdulla, H., Mason, B., and Jacobs, H.S., Combined growth hormone and gonadotropin treatment for ovulation induction. Presented at the 44th annual meeting of The American Fertility Society, October 10-13, Atlanta, GA, 1988. Fertility and Sterility supplement, Abs. #003.
8. D'Ercole, A.J., Stiles, A.D., and Underwood, L.E., 1984, Tissue concentrations of somatomedin-C: Further evidence for multiple sites of synthesis and paracrine or autocrine mechanisms of action, *Proc Natl Acad Sci USA*, 81:935.

9. Ranke, M.B., Stanley, C.A., Tenore, A., Rodbard, D., Bangiovanni, A.M., and Parks, J.S., 1976, Characterization of somatogenic and lactogenic binding sites in isolated rat hepatocytes, *Endrocrinology*, 99:1033.

10. Rein, M.S., and Schomberg, D.W., 1982, Characterization of insulin receptors on porcine granulosa cells, *Biol Reprod*, 26 (Suppl 1):113.

11. Ladenheim, R.G., Tesone, M., and Charreau, E.H., 1984, Insulin action and characterization of insulin receptors in rat luteal cells. *Endocrinology*, 115:752.

12. Otani, T., Mauro, T., Yukimur, N., and Mochizuki, M., 1985, Effect of insulin on porcine granulosa cells: implications of a possible receptor mediated action, *Acta Endocrinol*, 108:104.

13. Advis, J.P., Ojeda, S.R., 1980, Activation of GH short-loop negative feedback delays puberty in the female rat, *Fed Proc*, 39:822.

14. Advis, J.P., Smith-White, S., and Ojeda, S.R., 1981, Activation of growth hormone short loop negative feedback delays puberty in the female rat, *Endocrinology*, 108:1343.

15. Snell, G.D., 1929, Dwarf, a new mendelian recessive character of the house mouse, *Proc Natl Acad Sci*, 15:733.

16. Smith, P.E., and MacDowell, E,C, 1930, Hereditary anterior pituitary deficiency in mouse, *Anat Rec*, 46:249.

17. Lewis, U.J., Cheever, E.V., and Vander Laan, W.P., 1965, Studies on growth hormone of normal and dwarf mice, *Endocrinology*, 76:210.

18. Cheng, T.C., Beamer, W.G., Phillips, III J.A., Bartke, A., Mollonee, R.L., and Dowling, C., 1983, Etiology of growth hormone deficiency in Little, Ames and Snell dwarf mice, *Endocrinology*, 113:1669.

19. Goodman, H.G., Grumbach, M.M., and Kaplan, S.L., 1968, Growth and growth hormone: II. A comparison of isolated growth hormone deficiency and multiple pituitary hormone deficiencies in 35 patients with idiopathic hypopituitary dwarfism, *N Engl J Med*, 278:57.

20. Rabinowitz, D., and Merimee, T.J., 1973, Isolated human growth hormone deficiency and related disorders, *Isr J Med Sci*, 9:1599.

21. Tanner, J.M., and Whitehouse, R.H., 1975, A note on the bone age at which patients with true isolated growth hormone deficiency enter puberty, *J Clin Endocrinol Metab*, 41:788.

22. Rivarola, M.A., Phillips, J.A. III, Migeon, C.J., Heinrich, J.J., and Hjelle, B.J., 1984, Phenotypic heterogeneity in familial isolated growth hormone deficiency type I-A, *J Clin Endocrinol Metab*, 59:34.

23. Magner, J.A., Rogol, A.D., and Gorden, P., 1984, Reversible growth hormone deficiency and delayed puberty triggered by a stressful experience in a young adult, *Am J Med*, 76:737.

24. Hammer, R.E., Palmiter, R.D., and Brinster, R.L., 1984, Partial correction of murine hereditary growth disorder by germ-line incorporation of a new gene, *Nature*, 311:65.

25. Bartke, A., 1964, Histology of the anterior hypophysis, thyroid and gonads of two types of dwarf mice, *Anat Rec*, 149:225.

26. Dereviers, M.M., Viguiermartinez, M.C., Mariana, J.C., 1984, FSH, LH and prolactin levels, ovarian follicular development and ovarian responsiveness to FSH in the Snell dwarf mouse, *Acta Endocrinol*, 106:121.

27. Kusano, S., Shiki, Y., Ichimura, K., Amemiya, S., Nozaki, Y., Ohyama, K., and Kato, K., 1984, The sexual development of idiopathic ˜rowth hormone deficiency - The problems of the absence of puberty or the acceleration ⊂   ʋne age on h GH therapy. Excerpta Medica International Congress Series 652:861 (Abstract 1202).

28. Jia, X.C., Kalmijn, J., and Hsueh, A.J.W., 1985, Growth hormone (GH) augments the FSH induction of LH receptors in cultured granulosa cells. Presented at the 32nd Annual Meeting of The Society for Gynecologic Investigation, Phoenix, AZ (Abstract 179P).

29. Jia, X.C., Kalmijn, J., and Hsueh, A.J.W., 1985, Growth hormone directly enhances FSH-induced differentiation of cultured rat granulosa cells. Presented at the 67th Annual Meeting of the Endocrine Society, Baltimore, MD (Abstract 819).

30. Moretti, C., Fabbri, A., Gnessi, L., Forni, L., Fraioli, F., and Frajese, G., GHRH stimulates follicular growth and amplifies FSH-induced ovarian folliculogenesis in women with anovulatory infertility. The Endocrine Society 70th Annual Meeting, New Orleans, LA, 1988, Abstract #195.

SYNCHRONIZATION OF DONOR AND RECIPIENT CYCLES IN AN EGG DONATION PROGRAM: FRESH VERSUS FROZEN/THAWED EMBRYO TRANSFERS

David Levran, Shlomo Mashiach, Edwina Rudak, Jehoshua Dor, Zion Ben-Rafael, and Laslo Nebel

Department of Obstetrics and Gynecology, The Chaim Sheba Medical Center
The Sackler School of Medicine, Tel-Aviv University

INTRODUCTION

A high pregnancy rate has been reported following egg donation in the range of 20%–40% per transfer procedure.[1,2,3] The procedure of egg donation is used mainly for non-cycling patients (ovarian failure). Besides preparing the endometrium in these patients by means of hormone replacement therapy[4,5] one should also synchronize the date of the endometrium with the age of the donated embryos. Synchronization can be achieved either by transfer of fresh embryos[1] or by freezing the embryos and replacing them on different timing.[2] In the present study we investigated the effect of synchronization by cryopreservation versus fresh embryo transfers on the results of treatment of egg donation.

MATERIAL AND METHODS

Cycles of egg donation were randomly allocated into one of the two protocols of synchronization; either by fresh embryo transfers or by freezing and thawing. In both groups the endometrium was prepared in a similar way and the embryos were transferred at the same stage of development. Two thirds of the retrieved eggs were left for the donors. The remaining third was distributed among one or more recipients. The donated eggs were then inseminated by the corresponding husband's semen. The resulting embryos were either fro-

Advances in Assisted Reproductive Technologies, Edited by
S. Mashiach *et al.,* Plenum Press, New York, 1990

Table 1. Indications for Egg Donation in the Recipient Patients

Diagnosis	No. Patients	No. Cycles of Treatment
Early menopause	36	55
Ovarian dysgenesis with normal karyotype	14	18
Chromosomal aberrations	6	10
Other (habitual abortions)	1	1
Total	57	84

zen or freshly transferred according to the protocol of synchronization that was planned for the given donation cycle.

Egg donors: According to the prevalent regulations in Israel, donation of oocytes is permitted only by patients who themselves require IVF treatment for infertility. Potential donors were asked to donate a third of the aspirated eggs. All donors signed an informed consent after being informed that donation of some of their eggs could reduce their chances to conceive at that cycle. The egg recipients remained anonymous for the donors.

Egg Recipients: Mainly patients suffering from primary ovarian failure (Table 1), with normal uterine cavity, normal sperm analysis and adequate response to our scheme of hormone replacement therapy (as judged from biopsies of the endometrium).

Hormone replacement therapy: The protocol of hormone replacement therapy consisted of estradiol valerate administered for six days in rising dosage, followed by a maintenance dose of 2mg/d and progesterone 100mg/d.[5]

Synchronization of cycles with fresh embryo transfers: The protocol for synchronization was based on the six days rule.[6] Replacement hormone therapy was started in the recipients when the first day of estradiol rise was noticed in the donors. Progesterone was started in the recipients on the day of ovum pick up (OPU).

Synchronization in cycles with cryopreserved embryos: All of the embryos resulting from cycles that were allocated for freezing were frozen regardless of their quality and if not totally damaged by the procedure were transferred into the recipient. The protocol of freezing and thawing was adopted from Lasalle et al.[7] Embryos at the stage of 2–4 cells were dehydrated using 1.5M, 1,2 propandiol, and 0.1M sucrose and then frozen in a programmed freezer (Planer Kryo-10). Rapid thawing was performed in the presence of 0.2M sucrose.

RESULTS

A total of 84 recipient cycles were studied. In 38 cycles fresh embryos were transferred, and in 46 cycles frozen/thawed embryos were transferred. Recipient patients in both groups

Table 2. Comparison between the Groups of Fresh and Frozen/Thawed Embryo Transfers

	Mean age of donors	No. Cycles	Total No. oocytes	Mean No. oocytes	Fertilization
Fresh embryo transfers	31.5 ± 4.0	38	94	2.5 ± 0.6	76%
Frozen/thawed embryo transfers	32.0 ± 3.9	46	118	2.6 ± 0.5	77%

Table 3. Comparison of Results of Egg Donation in the Groups Receiving Fresh or Frozen/Thawed Embryo Transfers

	Total No. embryos transferred	No. embryos implanted	Implantation of embryos	No. Cycles	No. Pregs	No. ongoing Pregs	Pregs/ cycle	% ongoing cycle
Fresh embryo transfers	71	17	24%	38	14	8	37%	21
Frozen/thawed embryo transfers	91	7	8%	46	7	3	15%	7
			P < 0.01				P < 0.05	

received the same number of eggs per cycle and had a similar percentage of fertilization (76%–77%), and no difference was found in the age of egg donors for the two groups (Table 2). Percent implantation of embryos in the fresh embryo transfer group was significantly higher (24% versus 8% in the frozen/thawed embryos. The higher implantation percent of fresh embryos is also reflected in a significantly better pregnancy rate per cycle of treatment in this group (37% vs 15% in the frozen/thawed embryo transfers group $P < 0.05$) (Table 3).

The pregnancy wastage in both groups was relatively high. Only eight out of 14 pregnancies in the fresh embryo transfers ended in normal delivery. In the frozen/thawed embryo group three out of seven patients delivered. The ongoing pregnancy rate for fresh embryo transfers in the recipients was 21% versus 7% in the frozen/thawed group (Table 3).

DISCUSSION

Cryopreservation of human embryos is being utilized routinely in many in vitro fertilization programs.[8, 9, 10, 11] Recently a correlation was reported between some morphologic parameters as seen on videocinematography recordings following thawing of frozen/thawed embryos and their "implanting capacity".[12] Very little is known today about the "implanting capacity" of embryos; it might be dependent on the "quality" of embryos as well as on the uterine environment. There is no available data as to what extent the process of freezing/thawing affects the pregnancy potential of embryos, if at all.

Cryopreservation could be used to facilitate the synchronization of cycles between the donors of eggs and egg recipients. The incorporation of cryopreservation into the egg donor program enables us to prepare the endometrium of the recipients without being dependent on the egg donor response. It is particularly advantageous in the cycling recipients but it also facilitates the process in patients with ovarian failure. Salat-Baroux et al. reported a 31% pregnancy rate when they replaced frozen-thawed embryos in recipients.[2] In our previous series we reported a 34% pregnancy rate in fresh embryo transfers in egg recipients,[1] which is very similar. It should be noticed that results of cryopreservation are most often presented per embryo transfer procedure. However there is always a part of embryos that would not be frozen because of their poor morphology or not transferred because of damages during the process of freezing and thawing. Therefore the cryopreserved embryos could be regarded as selected embryos. Since every single embryo in an egg donor program is precious, one should analyze the results of treatment based on the number of embryos originally donated including those lost in the process of freezing and thawing. Indeed, in the model used in this study, results were analyzed per embryos available for donation. Since the uteri for both groups were prepared by a similar protocol, the donated eggs were randomly allocated for donation, the egg donors were randomly enrolled into either fresh or frozen/thawed embryo donation, and all the embryos were treated the same way, the difference in the results between the two groups could be attributed mainly to the use of cryopreservation.

Results of the present study clearly indicate that per embryos available for donation, synchronization by fresh embryo transfer is preferable.

REFERENCES

1. Levran, D., Lopata, A., Rudak, E., Dor, J., Blankstein, J., Nebel, L.,and Mashiach, S., The pregnancy potential of oocyte cohorts & uterine receptivity in hyperstimulated egg donors and synchronized recipients, *Fertil. Steril.*, (in press, 1989).
2. Salat-Baroux, J., Comet, D., Alvarez, S., Antoine, Ju., Tibi,C., Mandelbaum, J., Plachat, M.. Pregnancies after replacement of frozen/thawed embryos in a donation program, *Fertil. Steril.*, 49:817, (1988).
3. Devroey, P., Braechmans, P., Camus, N., Khan, I., Sontz, J., Staesseus, C., Van Den Abbeel, E., Van Waesberghel Visanto, A., Van Steirteghem, A. C., Pregnancies after replacement of fresh and frozen thawed embryos in a donation program, *in:* "Future Aspects in Human In Vitro Fertilization, W.Feichtinger, and P. Kremeter, eds., Springer Verlag, Berlin, 133, (1986).
4. Lutjen, P., Trounson, A., Leeton, J., Findlay, J., Wood, C., and Remon, P. The establishment and maintenance of pregnancy using in vitro fertilization and embryo donation in a patient with primary ovarian failure, *Nature*, 307:174 (1984).
5. Levran, D., Mashiach, S., Rudak, E., Dor, J., Blankstein J, Serr, D, M., Goldman, B., and Nebel, L., Evaluation of the "Pregnancy Potential" of eggs and "Endometrial Receptivity" as determinators of the outcome of IVF and ET (Abst). Presented at the Fifth World Congress on

In Vitro Fertilization and Embryo Transfer, Norfolk, Virginia, April 5–10, 1987, p.108–O.C. 428.

6. Levran, D., Lopata, A., Nayudu, P., Martin, M., McBain, J., Bayly, C., Speirs, A., Johnston, I. W., Analysis of the outcome of in vitro fertilization in relation to the timing of hCG administration by the duration of estradiol rise in stimulated cycles, *Fertil. Steril.*, 44:335, (1985).

7. Lasalle, B., Testart, J., Renard . Human embryo features that influence the success of cryopreservation with the use of 1,2 propanediol. *Fertil. Steril.*, 44: 645 (1985).

8. Testart, J., Lassalle, B., Belaisch-Allart, J., Hazout, A., Forman, R., Rainhorn, J. F., Frydman, R., High pregnancy rate after early human embryo freezing, *Fertil. Steril.*, 46:168, 1986.

9. Marrs, R. P., Brown, J., Sato, F., Ogawa, T., Yee, B., Paneson, R., Serafini, P., and Vargyas, J. M., Successful pregnancies from cryopreserved human embryos produced by in vitro fertilization, *Am. J. Obstet. Gynecol.*, 156: 1503, (1987).

10. Cohen, J., Simons, F. S., Fehilly C, B., Edwards, R. G., Factors affecting survival and implantation of cryopreserved embryos, *J. In Vitro Fert. Embryo Transfer*, 3: 46, (1986).

11. Freeman, L., Trounson, A., Kirby, C., Cryopreservation of human embryos; Progress on the clinical use of the technique in human in vitro fertilization, *J. In Vitro Fert. Embryo Transfer*, 3:53 (1986).

12. Cohen, J., Weimer, K. E., and Graham, W., Prognostic value of morphologic characteristics of cryopreserved embryos. A study using videocinematography, *Fertil. Steril.*, 49:827(1988).

MOLECULAR MECHANISMS OF MAMMALIAN SPERM-ZONA PELLUCIDA INTERACTION: CROSSTALK BETWEEN SPERM AND EGG

Gregory S. Kopf[1], Yoshihiro Endo[1], Shigeaki Kurasawa[1], and Richard M. Schultz[2]

[1] Division of Reproductive Biology, Department of Obstetrics and Gynecology and
[2] Department of Biology, University of Pennsylvania, Philadelphia, PA 19104-6080

INTRODUCTION

Aside from its importance in understanding specific aspects of fertilization, studies on sperm - egg interaction provide an elegant general model to study intercellular communication and cellular activation. Sperm functions such as motility, metabolism, capacitation(in mammals) and the acrosome reaction are modulated by specific factors associated with the egg, its acellular or cellular investments, or fluids bathing both the male and female reproductive tracts. In some species the identity, structure and function of these factors are known, and their mode of action at both the physiological and biochemical levels are starting to be elucidated. Emerging from such studies is the idea that the interaction of some of these factors with sperm to bring about a physiological response occurs via processes analogous to ligand -receptor -intracellular second messenger signalling systems known to mediate somatic cell function in response to hormones, neurotransmitters, and growth factors.

A reciprocal situation occurs in the egg in response to the fertilizing sperm; sperm-egg fusion results in the awakening of a metabolically and synthetically quiescent egg, thereby initiating a series of events that ultimately allows the fertilized egg to embark on an active mitotic cell cycle with the subsequent development of a preimplantation embryo. Although the details of this sperm-induced egg activation are only starting to be understood it appears that this process may also occur through signal transduction pathways that mediate many ligand-induced somatic cell activation events.

Advances in Assisted Reproductive Technologies, Edited by
S. Mashiach *et al.,* Plenum Press, New York, 1990

This chapter will summarize experiments carried out in our laboratories designed to understand the biochemical basis of two specific fundamental cellular activation events that occur prior to and after sperm-egg fusion in the mouse. The first event is an egg-induced sperm activation — the zona pellucida-mediated acrosome reaction — that occurs subsequent to the species-specific sperm binding to the zona pellucida and is a prerequisite to successful fertilization. The second event is a sperm-induced egg activation — the establishment of the zona pellucida block to polyspermy — that prevents fertilization by supernumerary sperm. A molecular analysis of these two contact stimulated cellular activation events should prove invaluable to understanding the "crosstalk" between sperm and egg prior to and after fertilization, as well as to gaining insight into modes of intercellular communication and signal transduction between cells in general.

STRUCTURE AND FUNCTION OF THE MOUSE EGG ZONA PELLUCIDA

Species-specific sperm-egg recognition and interaction, sperm activation (i.e., the acrosome reaction), and an egg-induced block to polyspermy in the mouse all appear mediated by the extracellular matrix surrounding the egg called the zona pellucida (ZP). The ZP of the mouse egg is composed of three sulfated glycoproteins designated as ZP1, ZP2, and ZP3 (Bleil and Wassarman, 1980a; Shimizu et al., 1983; Wassarman, 1988), which are present in molar ratios of approximately 1:10:10, respectively. These glycoproteins are truly egg-associated products, as they are the major biosynthetic products of the growing oocyte that are secreted throughout the period of oocyte growth (Bleil and Wassarman, 1980b). ZP1 (M_r=200,000) (Bleil and Wassarman, 1980a) is a dimer connected by intermolecular disulfide bonds and may function to maintain the three dimensional structure of the ZP by cross linking filaments that are composed of repeating structures of ZP2/ZP3 heterodimers (Greve and Wassarman, 1985).

ZP2 (M_r=120,000 under nonreducing and reducing conditions; Bleil and Wassarman, 1980a), which is present in both oocytes and unfertilized eggs, may mediate the binding of acrosome-reacted sperm to the ZP (Bleil and Wassarman, 1986; Bleil et al., 1988). Upon fertilization, the egg effects a modification of ZP2 to a form called $ZP2_f$ (Bleil and Wassarman, 1981). This modification is brought about by the action of a protease (Moller and Wassarman, 1989), which is most likely secreted from the egg as a consequence of cortical granule exocytosis. $ZP2_f$ has a M_r=120,000 under nonreducing conditions, but under reducing conditions the M_r=90,000, suggesting that the proteolysis of the ZP2 molecule results in the generation of fragments that are held together by disulfide bonds. The biological consequence of the ZP2 to $ZP2_f$ shift is that $ZP2_f$ no longer will bind to acrosome reacted sperm (Bleil and Wassarman, 1986; Bleil et al., 1988).

ZP3 (M_r=83,000) accounts for both the sperm receptor and the acrosome reaction-inducing activities of the ZP of unfertilized eggs (Bleil and Wassarman, 1980c; 1983). The sperm receptor activity of ZP3 appears to be conferred by O-linked carbohydrate moieties and not by its polypeptide chain (Florman et al., 1984; Florman and Wassarman, 1985). Alpha-linked terminal galactose residues at the nonreducing termini of these O-linked oligo-saccharide chains may play a critical role in this binding activity (Bleil and Wassarman, 1988). On the other hand, the acrosome reaction-inducing activity of ZP3 may be conferred by both the carbohydrate and protein portions of the molecule (Florman et al., 1984; Wassarman et al., 1985), although the exact nature of the interaction between the protein and carbohydrate required for biological activity is not clear at this time. Associated with fertilization is a loss of both the sperm receptor and acrosome reaction-inducing activities of the ZP3 molecule (Bleil and Wassarman, 1980c, 1983). Since the electrophoretic mobility of ZP3 from fertilized eggs is similar to that of ZP3 from unfertilized eggs, it is thought the loss of these two important biological activities appear to be associated with a relatively minor biochemical modification of the ZP3 molecule.

SPERM-ZONA PELLUCIDA INTERACTIONS
PRIOR TO AND FOLLOWING SPERM-EGG FUSION

The properties of the ZP glycoproteins from unfertilized and fertilized eggs provide an elegant framework to build a model to explain the interaction of sperm with this specialized extracellular matrix. Since it has been previously demonstrated that only acrosome intact bind to the ZP of unfertilized eggs (Saling et al., 1979), the following sequence of events might explain sperm-ZP interaction both prior to and after sperm-egg fusion in the mouse. Species-specific sperm binding of acrosome intact sperm to the ZP is mediated via ZP3 (Saling et al., 1979; Florman and Storey, 1982; Bleil and Wassarman, 1980c), and once this binding has occurred ZP3 then mediates the acrosome reaction (Bleil and Wassarman, 1983). These initial interactions between acrosome intact sperm and ZP3 occur via sperm-associated ZP3 receptors presumably located in the plasma membrane overlying the head of the sperm (Bleil and Wassarman, 1986). The molecular nature of such a receptor(s) is unknown at this time. A sperm surface galactosyltransferase (Shur and Hall, 1982a,b; Lopez et al., 1985), as well as a trypsin inhibitor sensitive site (Saling,1981; Benau and Storey, 1987) may participate in this binding reaction. Secondary interactions of the acrosome reacted sperm with ZP2 then occur through the interaction of the exposed inner acrosomal membrane of the sperm (Bleil and Wassarman, 1986; Bleil et al., 1988); likewise, the identity of the presumed ZP2 "secondary" receptor(s) on the sperm inner acrosomal membrane is unknown. Once acrosome reacted, the sperm then penetrate the ZP, traverse the perivitelline space, and bind and fuse with the plasma membrane of the egg. Subsequent to sperm-egg fusion, the egg undergoes the cortical granule reaction (Szollsi, 1967; Barros and Yanagimachi, 1971,1972; Wolf and Hamada, 1977; Lee et al., 1988). Cortical granules, which reside within a few microns from the egg plasma membrane, are membrane-limited organelles that undergo exocytosis as a consequence of fertilization. This exocytotic event results in the release of cortical granule-associated enzymes into the perivitelline space, and converts ZP2 to $ZP2_f$ and modifies ZP3, which loses both its sperm receptor and acrosome reaction inducing activities. Although a proteinase that modifies ZP2 has been described in preparations of A-23187 activated mouse eggs (Moller and Wassarman, 1989), other proteases may also be involved in these egg-induced ZP modifications

(Gwatkin et al., 1973; Wolf and Hamada, 1977). As a consequence of these ZP modifications, acrosome-intact sperm no longer bind to the ZP, and sperm that are bound to the ZP and have acrosome-reacted can no longer interact and penetrate the ZP, since they are unable to establish secondary binding interactions with $ZP2_f$ (Bleil and Wassarman, 1986, 1988). Such egg-induced modifications of the ZP constitute the ZP block to polyspermy.

THE ZONA PELLUCIDA - MEDIATED ACROSOME REACTION AS A MODEL FOR SIGNAL TRANSDUCTION IN SPERM

ZP3 possesses a number of properties common to ligands that mediate cellular function in a receptor-mediated fashion. Although conserved at the genomic level in a variety of species (Ringuette et al., 1988), ZP3 subserves very specific functions as a component of the egg-associated extracellular matrix. (1) ZP3 is synthesized only by the growing oocyte (Bleil and Wassarman,1980b; Shimizu et al., 1983; Philpott et al., 1987). (2) There is little apparent amino acid sequence homology between ZP3 and any other known proteins or glycoproteins thus far examined (Ringuette et al., 1988). (3) ZP3 plays a critical role in sperm-egg interaction prior to fertilization (Wassarman, 1988); ZP3 mediates sperm binding and induction of the acrosome reaction (see above). (4) Both the receptor and acrosome reaction-inducing activities of ZP3 are observed in the nanomolar range (Florman and Wassarman, 1985; Bleil and Wassarman, 1986). (5) Mouse sperm appear to possess complementary binding sites for ZP3 that appear to display some degree of localization (over the acrosomal cap region) and are present in numbers (10,000–50,000 binding sites/cell) (Bleil and Wassarman, 1986); these numbers of sperm receptors are similar to that for receptor numbers in many hormonally-responsive cells. Thus, the ZP3-induced acrosome reaction has characteristics similar to receptor-mediated signal transduction processes observed in somatic cells.

Although poorly understood, some of the physiological responses of mammalian sperm to biological factors associated with the reproductive tract or the egg appear mediated through intracellular second messenger systems. Such responses include changes in ionic permeability (Babcock et al., 1979; Rufo et al, 1982; Lee and Storey, 1985; Thomas and Meizel, 1988) and cyclic nucleotide concentrations (Garbers and Kopf, 1980; Tash and Means, 1983). In addition, there are several ion requirements for the ZP-induced acrosome reaction. For example, binding of mouse sperm to the ZP (Saling et al., 1978) and the ZP-mediated acrosome reaction (Noland, Garbers & Kopf, unpublished) require extracellular Ca^{2+}, and bicarbonate is also required for the ZP-induced acrosome reaction (Lee and Storey, 1986). In addition, the ZP-induced acrosome reaction is accompanied by a loss of a transmembrane pH gradient (Lee and Storey, 1985). Little is known, however, about the relationships between ZP3-sperm interaction, changes in intracellular second messengers, changes in ion permeability, and the induction of the acrosome reaction.

The following section will outline studies from our laboratory that support the idea that a guanine nucleotide-binding regulatory protein (G-protein) in mouse sperm serves as a signal transduction element in coupling sperm-ZP interaction with the induction of the acrosome reaction.

THE IDENTITY OF GUANINE NUCLEOTIDE BINDING
REGULATORY PROTEINS IN SPERM AND THEIR ROLE
IN THE ZONA PELLUCIDA-MEDIATED ACROSOME REACTION

The guanine nucleotide-binding regulatory proteins (G proteins), which include G_s, G_i, G_o, and transducin, occupy critical roles as signal transducing elements in coupling many ligand-receptor interactions with the generation of intracellular second messengers (Casey and Gilman, 1988). These heterotrimeric plasma membrane-associated proteins are composed of distinct α-subunits, which contain a GTP binding domain, and more highly conserved β and γ-subunits. Activation of G proteins occurs upon GTP binding to the α-subunit, causing dissociation of α-GTP from the $\beta\gamma$ dimer. These dissociated subunits then exert their regulatory actions on respective targets, that ultimately result in ligand coupling to intracellular signalling systems. G proteins regulate the hormonally-responsive adenylate cyclase of somatic cells (Gilman, 1987), the cyclic GMP phosphodiesterase of retinal rod outer segments (Stryer, 1986), atrial K^+ channels (Mattera et al, 1989), and have been implicated in regulating phospholipase C (Cockcroft, 1987), phospholipase A_2 (Jelsema and Axelrod, 1987), and Ca^{2+} channels (Hescheler et al., 1987; Mattera et al., 1989). Since many of these intracellular signalling mechanisms appear to be important in the control of both invertebrate and mammalian sperm functions (Kopf and Garbers, 1980; Tash and Means, 1983), experiments were designed to determine whether sperm contain these signal transducing proteins (Kopf et al., 1986).

Sperm extracts were first assessed for the presence of the α-subunits of G-proteins by bacterial toxin-catalyzed ADP ribosylation. These toxins have been used as tools to identify the α-subunits of G-proteins and to investigate the functional roles of G-proteins in biological systems. Sperm extracts were examined for the presence of islet-activating protein (IAP; pertussis toxin) substrates. IAP catalyzes the ADP-ribosylation of the α-subunits of G_i, G_o, and transducin. Both invertebrate (abalone) and mammalian (human, mouse, guinea pig, bovine) sperm contain a single IAP substrate of $M_r = 41,000$. Peptide mapping of the mouse sperm $M_r = 41,000$ IAP substrate by limited proteolytic digestion demonstrates a similarity to α_i of mouse S-49 lymphoma cells. These data suggest that the mouse sperm $M_r = 41,000$ IAP substrate is α_i-like (Kopf et al., 1986). The presence of a G_i-like protein in mouse sperm was further substantiated using an antiserum directed against the β-subunit common to the different heterotrimeric G-proteins. Analysis of one-dimensional polyacrylamide gels of mouse sperm membrane extracts using Western blotting demonstrates the presence of a single immunoreactive band of $M_r = 35,000$, which comigrate with the β-subunit from somatic cell membranes.

Mouse sperm were also analyzed for the presence of specific substrates for cholera toxin-catalyzed ADP-ribosylation, indicative of the presence of G_s. No specific substrates were detected in any of the fractions assayed under a variety of conditions. The lack of our ability to detect G_s in sperm is consistent with the data of Hildebrandt et al. (1985), and are supported by the work of others who have examined the guanine nucleotide regulation of the sperm adenylate cyclases (Kopf and Garbers, 1980). It is possible, however, that G_s is present in sperm but that the current methods used to probe for its existence are inadequate at this time. The inability to reconstitute sperm adenylate cyclase activity with purified G_s from an exogenous source implies that this enzyme may have unique regulatory properties, and that G_s may not play a regulatory role in sperm function (Hildebrandt et al., 1985).

It can be concluded that mouse sperm, as well as sperm from a number of other species, contain G-proteins. In most species the G-protein appears to be G_i-like. The existence of these signal transducing proteins in sperm has been confirmed by other workers (Bentley et al., 1986). Additional biochemical, immunological, and molecular studies will be required to determine the relationship of the sperm G-proteins to somatic cell G-proteins, and to determine whether the sperm G-proteins are localized to different regions of the cell.

Function of G_i-like proteins in sperm was assessed by treating sperm with IAP. IAP can cross the plasma membrane and functionally inactivate G_i by ADP-ribosylating its α-subunit (Gilman, 1987; Ui et al., 1985). Capacitation of mouse sperm in the presence of IAP is not affected and these sperm bind to structurally-intact ZP to the same extent as control sperm incubated in the absence of the toxin. In addition, sperm motility and viability are not affected at any IAP concentration tested. Moreover, IAP can cross the sperm plasma membrane and ADP-ribosylate the α-subunit of the G_i-like protein under these incubation conditions, thus demonstrating that IAP is exerting its effect on the sperm G_i-like protein (Endo et al., 1987b).

Although the ability of the sperm to become capacitated is unaffected by IAP treatment, the ZP-induced acrosome reaction is inhibited in a concentration-dependent manner by IAP, with half-maximal effects at 0.1–1.0 ng/ml IAP. In these experiments the acrosomal status of mouse sperm was monitored using the fluorescent probe chlortetracycline (CTC) (Saling and Storey, 1979; Ward and Storey, 1984; Lee and Storey, 1985). CTC is fluorescent when present in a hydrophobic environment and complexed with calcium, and mouse sperm treated with this probe exhibit characteristic patterns of fluorescence that have been correlated with different stages leading to the completion of the acrosome reaction. Capacitated, acrosome-intact sperm give rise to a B pattern, which is characterized by a bright epifluorescence over the anterior portion of the head and midpiece; the post-acrosomal region displays no fluorescence. The S-pattern, which represents an intermediate stage that appears prior to the completion of the acrosome reaction, is characterized by bright fluorescence over the midpiece and punctate fluorescence over the anterior portion of the head. The appearance of this pattern correlates with the inability of the sperm to maintain a transmembrane pH gradient (Lee and Storey, 1985). The AR pattern is characteristic of sperm that have completed the acrosome reaction and is characterized by a lack of fluorescence over the anterior portion of the sperm head.

The inhibitory effect of IAP on the acrosome reaction is due to an effect on the ability of the sperm to progress from the B-pattern to the intermediate S-pattern; thus, the sperm remain in B. Inactivated IAP does not inhibit the ZP-induced acrosome reaction, and cholera toxin is also without effect. These inhibitory effects of IAP on the ZP-induced acrosome reaction were also confirmed at the electron microscopic level (Endo et al., 1988). These data suggest that functional inactivation of the mouse sperm G_i-like protein by IAP prevents the ZP-induced acrosome reaction and that the inhibition appears, at least, to be at the level of the B to S transition.

The specificity of the IAP effect on the ZP-induced sperm acrosome reaction was further defined in additional experiments since IAP has other properties in addition to its ability to ADP-ribosylate the α-subunit of various G-proteins (Ui et al., 1985). The effects of nonhydrolyzable GTP and GDP analogues on the IAP-mediated inhibition of the ZP-induced acrosome reaction were examined, since the β-subunit of the G_i $\alpha\beta\gamma$ heterotrimer appears to be required for the IAP-catalyzed ADP-ribosylation of the α_i-subunit (Neer et al., 1984). The rationale for this approach is as follows. Nonhydrolyzable analogues of GTP (e.g., guanosine-5'-O-3-thiotriphosphate; GTPγS) bind to the α_i-subunit with high affinity and dissociate the heterotrimer into GTPγS- α_i and $\beta\gamma$ subunits (Bokoch et al., 1984). It would be expected that if GTPγS penetrated intact sperm and bound to the α-subunit of the sperm G_i-like protein, the resultant subunit dissociation would prevent subsequent IAP-catalyzed ADP-ribosylation of the α-subunit and, as a result, prevent functional inactivation of the protein. There is precedent for limited cellular permeability of GTPγS (Minke and Stephenson, 1985). In contrast, incubation of intact sperm with a nonhydrolyzable GDP analogue(e.g., guanosine-5'-O-2-thiodiphosphate; GDPβS) would not be expected to inhibit the IAP-catalyzed ADP-ribosylation since this analogue would not cause subunit dissociation.

When intact sperm are capacitated in the presence of 100 μM GTPγS prior to the addition of 10 ng/ml IAP, sperm binding to the ZP and the subsequent acrosome reaction occurs to the same extent as control sperm incubated in the absence of both of these agents; this contrasts sharply with the inhibitory effects on the acrosome reaction observed when sperm are incubated with IAP alone. Capacitation of sperm in the presence of 100 μM GDPβS prior to the addition of IAP, on the other hand, does not abolish the IAP-mediated inhibition of the ZP-induced acrosome reaction. These data suggest that the IAP-sensitive component of mouse sperm is a G_i-like protein, which upon activation by GTPγS (but not GDPβS) loses its ability to become functionally inactivated by IAP-catalyzed ADP-ribosylation. Since the only IAP substrate found to be present in mouse sperm is the $M_r =$ 41,000 α-subunit of the G_i-like protein, it is likely that this protein is the IAP-sensitive component.

The experiments described thus far utilized mechanically-isolated, structurally intact ZP in which the acrosome reaction was monitored on sperm bound to the ZP. Although ZP3 has been demonstrated to account quantitatively for both the sperm receptor and acrosome inducing activities of the mouse ZP, it could be argued that IAP treatment of the sperm could affect the sperm surface such that a nonphysiological interaction between the sperm and the structurally intact ZP would occur; such a nonphysiological interaction could certainly be exacerbated by the steric constraints of an intact ZP. Such an interaction might ultimately result in a nonspecific inhibitory effect on the acrosome reaction. The locus of such an effect would most likely be downstream from the sperm binding site(s) since sperm binding to structurally intact ZP is not affected by IAP pretreatment of the sperm. The argument might also be made that when using such a sperm-ZP binding assay, the apparent inhibition of the acrosome reaction could result from a preferential binding to the ZP of a subset of IAP-treated sperm destined not to complete the acrosome reaction. In order to rule out these possibilities the effects of IAP treatment of sperm on the acrosome reaction induced by purified ZP3 was examined (Endo et al., 1988).

When compared to sperm capacitated in the absence of IAP, sperm capacitated in the presence of IAP are inhibited from undergoing the acrosome reaction in response to ZP3. As with structurally-intact ZP, this inhibition occurs at the B to S transition. In contrast to the inhibitory effects of IAP on the ZP3-induced acrosome reaction, the ability of sperm to undergo either a spontaneous acrosome reaction (i.e., in the absence of solubilized ZP3) or a nonphysiologically induced acrosome reaction (A-23187-induced) is not affected by the IAP treatment. Thus, the inhibitory effect of IAP on the ZP-induced acrosome reaction does not occur via a nonspecific inhibitory effect that is observed when monitoring acrosome reactions of sperm bound to structurally-intact ZP. More importantly, only acrosome reactions induced by the physiologically relevant ZP3 molecule are inhibited by IAP. These results are consistent with the idea that the IAP-sensitive site (G_i-like protein) in mouse sperm plays an important intermediary role in the acrosome reaction induced specifically by ZP3, the biologically relevant molecule present in the ZP.

Although the sperm-associated G_i-like protein appears to play a critical intermediary role in the ZP3-induced acrosome reaction, the intracellular signalling systems that this particular signal transducing protein are coupled to are not known at this time. Moreover, the biochemical responses of mouse sperm to ZP3 have not been examined to date. As stated above, G-proteins have been demonstrated to regulate a variety of intracellular second messenger systems (e.g., adenylate cyclase, cyclic GMP phosphodiesterase, phospholipases C and A_2), as well as ion channels (K^+; Ca^{2+}) in a variety of cell types (Casey and Gilman, 1988). Using the lipophilic fluorescent probe amino, chloro, methoxyacridine (ACMA) to monitor transmembrane pH gradients in mouse sperm, Lee and Storey demonstrated that the ZP-induced B to S transition, as monitored using CTC fluorescence, is accompanied by a parallel abolition of a transmembrane pH gradient as evidenced by the loss of ACMA fluorescence associated with the intact sperm (Lee and Storey, 1985). The loss of this transmembrane pH gradient is not a consequence of major permeability changes such as fusion of the plasma and outer acrosomal membranes. Presently, it is not clear whether the loss of this gradient is a cause, or a consequence, of the B to S transition. Nevertheless, it can be concluded that the ZP-induced B to S transition is associated with a modification of ion permeability in these cells.

Since IAP specifically inhibits the B to S transition the effects of this toxin on the loss of ACMA fluorescence of sperm bound to structurally intact ZP was examined. It was demonstrated that the IAP-induced inhibition of the B to S transition is accompanied by a parallel retention of ACMA fluorescence at all of the time points tested (Endo et al., 1988). These data demonstrate that the IAP-sensitive site is upstream from those event(s) associated with the loss of the transmembrane pH gradient normally accompanying the ZP-induced B to S transition, and suggest that the G_i-like protein might modulate such changes in ion permeability. Additional studies will be required to confirm whether such changes in ionic permeability are coupled to this protein, and whether other ionic changes might be coupled in a G-protein mediated fashion.

Studies from other laboratories have suggested that alterations in cyclic nucleotide metabolism and/or phospholipid metabolism may play important intermediary roles in the sperm acrosome reaction. Lee et al. have provided evidence for the potential role of poly-phosphoinositide turnover in regulating the ZP induced acrosome reaction in the mouse (Lee et al., 1987). Since these intracellular signalling systems are coupled to G-proteins in a receptor mediated fashion, it will be of interest to determine whether such second messenger systems are modulated in sperm by the ZP. Studies in our laboratory directed at answering such questions are currently in progress. A working model of the events comprising the ZP3-induced acrosome reaction is shown in Figure 1.

EGG ACTIVATION AS A MODEL OF SPERM-INDUCED SIGNAL TRANSDUCTION

The nature of the sperm-induced biochemical signal(s) that ultimately results in cortical granule exocytosis, the ZP and plasma membrane polyspermy blocks, and metabolic activation of the mammalian egg are presently not known. As observed with eggs of lower species, a fertilization-induced rise in intracellular Ca^{2+} concentration is at least one important component of early mammalian egg activation (Cuthbertson et al., 1981; Igusa and Miyazaki, 1986; Miyazaki et al., 1986). This sperm-induced intracellular Ca^{2+} rise appears to originate at the point of sperm-egg fusion and spreads over the cortex of the egg in a centripetal fashion (Miyazaki et al., 1986). As with eggs of nonmammalian species, this rise in free intracellular Ca^{2+} may be responsible either directly or indirectly for the cortical granule reaction thought to be involved in establishing the ZP polyspermy block. Those factors controlling the temporal and spatial changes in intracellular Ca^{2+}, as well as the mechanism by which this cation causes exocytosis, are only starting to be investigated.

Changes in egg polyphosphatidylinositide turnover may represent at least one response that may operate to control some of these aforementioned early events of egg activation in both nonmammalian and mammalian species. In eggs of lower species the fertilizing sperm is thought to activate a G-protein, which in turn may activate a phosphatidylinositol 4,5,-bisphosphate (PIP_2)-specific phospholipase C (Turner et al.,1984,1986, 1987; Kline and Jaffe, 1987). The activation of this enzyme results in the hydrolysis of PIP_2 to two intracellular second messengers, *sn*-1,2,diacylglycerol (DAG) and inositol 1,4,5-trisphosphate (IP$_3$) (Berridge,1987). The generation of these two second messengers initiates a bifurcating pathway of intracellular signalling in which DAG activates a Ca^{2+}, phospholipid-dependent protein kinase (PK-C) and IP$_3$ stimulates the release of Ca^{2+} from intracellular stores (Berridge, 1987). Increased diphosphoinositide and triphosphoinositide levels accompany fertilization in sea urchin eggs and these changes precede the cortical granule reaction (Turner et al., 1984). Furthermore, microinjection of sea urchin and amphibian eggs with IP$_3$ produces a cortical reaction similar to that observed during fertilization (Whitaker and Irvine, 1984; Busa et al. 1985). It has been postulated that the propagated fertilization wave of Ca^{2+} in sea urchin eggs may occur as a consequence of IP$_3$-mediated Ca^{2+} release with subsequent feedback release of additional IP$_3$ through Ca^{2+} activation of the phospholipase C (Swann and Whitaker, 1986; Ciapa and Whitaker, 1986). Since recent attention has focused on the role(s) of protein phosphorylations catalyzed by protein kinase C (PK-C) in the regulation of exocytosis (Kikkawa and Nishizuka, 1986), it is also possible that the DAG that

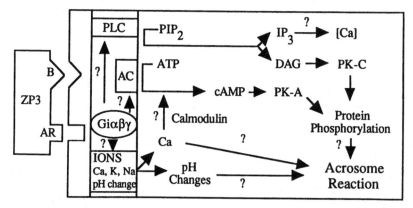

Fig.1. Hypothetical pathway by which ZP3-mediated signal transduction and induction of the acrosome reaction might occur in mouse sperm. The ZP3 molecule possesses distinct domains responsible for sperm receptor activity (**B**) and acrosome reaction-inducing activity (**AR**). The sperm-associated G$_i$-like protein functions as a signal transducing element that is activated upon ZP3- ZP3 receptor interaction, and couples receptor occupancy to a variety of potential intracellular second messenger cascade systems and/or changes in ion conductance. Such intracellular signalling systems include the polyphosphoinositide-specific phospholipase C (**PLC**), which hydrolyzes phosphatidylinositol 4,5,-bisphosphate (**PIP$_2$**) to form *sn* 1,2-diacylglycerol(**DAG**) and inositol 1,4,5-trisphosphate (**IP$_3$**). The liberated DAG activates a Ca^{2+} and phospholipid-dependent protein kinase (**PK-C**) that then phosphorylates proteins that ultimately lead to the induction of the acrosome reaction. An alternative/additional signalling pathway that may be modulated by the G$_i$-like protein is the adenylate cyclase (**AC**) system. Generation of cyclic AMP (**cAMP**), with resultant stimulation of a cAMP-dependent protein kinase (**PK-A**), could also lead to protein phosphorylation and acrosomal exocytosis. AC might be regulated by extracellular Ca^{2+} through the Ca^{2+}-binding protein, calmodulin. G$_i$-mediated effects on ion conductance (e.g., Ca^{2+}, K$^+$) might directly or indirectly lead to changes in intracellular pH. All of the above signalling systems may be integrated to bring about the induction of the acrosome reaction.

is generated upon phospholipase C activation plays a role in the cortical granule reaction. DAG induced PK-C activation may be responsible for the fertilization-induced activation of the sea urchin egg Na^+/H^+ exchanger, which brings about alkalinization of the intracellular pH required for DNA synthesis (Swann and Whitaker, 1985).

The fertilization-induced Ca^{2+} transients in hamster eggs can be generated when these eggs are microinjected with IP_3, and these effects are mimicked by the injection of GTP or GTPγS (Miyazaki, 1988). Both the sperm-induced and GTPγS-induced Ca^{2+} transients are inhibited by a prior injection of GDPβS. These data support the idea that sperm-induced Ca^{2+} transients in mammalian eggs might act through a similar G-protein- polyphosphoinositide signalling system. PK-C involvement in the generation of these Ca^{2+} transients has been suggested by experiments demonstrating that tumor promoting phorbol diesters induce transient oscillations in intracellular Ca^{2+} and parthenogenetic activation of ZP-free mouse eggs (Cuthbertson and Cobbold, 1985); these effects appear similar to those changes in Ca^{2+} concentrations observed during fertilization (Cuthbertson et al., 1981). The mechanism by which these effects occur are not clear since extremely high concentrations of phorbol diesters are required to bring about these effects.

Thus, it appears that common mechanisms of sperm-induced egg activation probably exist in both nonmammalian and mammalian eggs. In attempts to understand further mammalian sperm-ZP interactions prior to and after fertilization, we have investigated the role of polyphosphoinositide turnover in the mouse egg with regard to the egg-induced ZP polyspermy block. We have done this by examining each of the two arms of this bifurcating second messenger signal cascade system and demonstrating that activators of PK-C (either a diacylglycerol or phorbol diesters) and IP_3 cause egg-induced modifications in the ZP that, at least in one case, mimic only part of the egg-induced ZP modification that normally occurs after fertilization of mouse eggs. Mechanisms by which these agents effect these ZP modifications and the role of PK-C and IP_3 in the ZP polyspermy block will be discussed.

THE ROLE OF PROTEIN KINASE C IN THE EGG-INDUCED ZONA PELLUCIDA BLOCK TO POLYSPERMY

Treatment of either cumulus cell enclosed- or cumulus cell-free mouse eggs with biologically active phorbol diesters, such as 12-O-tetradecanoyl phorbol 13-acetate (TPA) or 4β phorbol 12,13 didecanoate (4β-PDD), results in a concentration- dependent inhibition of fertilization (Endo et al., 1987a). Similar inhibitory effects are seen with the more natural activator of PK-C, sn-1,2-dioctanoyl glycerol. These inhibitory effects are likely to be mediated by an activation of PK-C since fertilization is not inhibited when eggs are treated with 4α-PDD, which is a stereoisomer of 4β-PDD that does not activate PK-C. The inhibitory effect of these compounds on fertilization is not due to effects on sperm, since sperm treated with these compounds fertilize untreated eggs to the same extent as untreated sperm. Furthermore, the inhibitory effect of these compounds is not due to a direct effect on the

ZP, since mechanically isolated ZP treated with these agents bind and acrosome react sperm to the same extent as control, untreated ZP.

The inhibitory effects of these PK-C activators on fertilization appear to be due to an egg-induced modification of the ZP, for the following reasons. First, there is a strong correlation between the percentage of fertilization and sperm penetration through the ZP and entry into the perivitelline space at all of the concentrations of PK-C activators tested (Endo et al., 1987a). In other words, if the sperm are able to gain access to the perivitelline space, fertilization occurs. These observations suggested that the inhibitory effect of the PK-C activators on fertilization is at the level of the ZP and not at the level of the plasma membrane. This is supported by the fact that ZP-free eggs treated with these compounds are fertilized to the same extent as untreated ZP-free eggs (fertilization of ZP-free eggs most likely occurs by sperm that have undergone a spontaneous acrosome reaction). When the ZP from PK-C activator treated ZP-intact eggs were removed, radioiodinated, and analyzed two dimensional reduction gel electrophoresis , it was demonstrated that ZP2 was converted to $ZP2_f$, an alteration that has been shown to occur at the time of fertilization. Taken together, these data demonstrate that PK-C activators bring about an egg-induced modification of the ZP that could account for the observed inhibitory effects on fertilization.

When examining the effects of these PK-C activators on fertilization it was noted that the treated eggs bound similar numbers of sperm compared to the untreated or 4α-PDD-treated controls. Since the inhibitory effects of these agents appeared to be at the level of the ZP, it was apparent that a step distal to sperm binding had to be inhibited. The logical candidate was that sperm, once bound, failed to undergo an acrosome reaction. This was tested experimentally using the CTC assay previously described in this chapter.

As previously stated sperm bind to mechanically isolated, structurally-intact ZP from untreated eggs undergo time-dependent B to S and S to AR transitions, as assessed by the CTC assay. When capacitated sperm are incubated with structurally-intact ZP isolated from PK-C activator treated eggs, the B to S transition occurs with the same kinetics as sperm bound to ZP from untreated or 4α-PDD-treated eggs, but the S to AR transition is inhibited. Thus, although sperm bound to ZP from PK-C activator treated eggs can initiate an acrosome reaction (i.e., B to S transition) they fail to complete it, and accumulate in the S-pattern. These sperm are not irreversibly stuck in the S pattern, since further addition of either solubilized ZP from untreated eggs or A-23187 to S-pattern sperm will cause them to undergo an S to AR transition (Kligman et al., 1988).

The aforementioned data suggest that treatment of eggs with phorbol diesters or diC_8 results in an apparent dissociation in intact ZP of the sperm receptor activity from the acrosome reaction-inducing activity of the ZP (i.e., of ZP3). Normally, fertilization results in the loss of both of the biological activities of ZP3. It might be argued, however, that the apparent dissociation of these two biological activities of the ZP is due to primary effects to change the three dimensional structure of the ZP and not to changes in the intrinsic biological properties of the ZP3 molecule. This is not the case, since ZP3 isolated from either untreated eggs or TPA -treated eggs possessed similar amounts of sperm receptor activity (Endo et al., 1987c). ZP3 from 2-cell embryos, in contrast, did not possess sperm receptor activity; this is consistent with the inability of ZP from 2-cell embryos to bind sperm. Sperm receptor activity was assessed by a sperm binding competition assay, in which the inhibitory effect of solubilized ZP/ZP3 from the appropriate experimentally treated egg was measured on sperm binding to untreated ZP-intact eggs. Using the CTC assay, the acrosome reaction-inducing activity of the ZP3 from TPA treated eggs was also compared to that of control untreated eggs and 2-cell embryos. Capacitated sperm incubated with ZP3 from untreated eggs undergo the B to S and S to AR transitions. When ZP3 from TPA-treated eggs is incubated with sperm, the B to S transition occurs, but sperm are inhibited from completing the acrosome reaction as a consequence of their inability to undergo the S to AR transition. These results are identical to that observed with ZP isolated from TPA-treated eggs. In addition, ZP3 isolated from 2-cell embryos does not promote the B to S transition. It can be concluded that, 1) the altered biological activities of the ZP from PK-C activator-treated eggs is due to specific modifications of the ZP3 molecule, and 2) that PK-C activators cause an egg-induced dissociation of the sperm receptor activity from the acrosome reaction-inducing activity of ZP3. These observations are important for two reasons. First, since we now have a way to dissociate the sperm receptor activity from the acrosome reaction-inducing activity of the ZP3 molecule, we can use this tool to study the egg-induced biochemical modifications of the ZP3 molecule that are responsible only for the loss of acrosome reaction inducing activity; such studies may aid in understanding the structural and biochemical basis of the acrosome reaction inducing moiety of the ZP3 molecule. This was not possible before since ZP3 from unfertilized eggs possesses both biological activities and ZP3 from fertilized eggs has lost both biological activities. Second, since ZP3 from PK-C activator treated eggs inhibits the acrosome reaction by inhibiting the S to AR transition, it is now possible to use these ZP to study the biochemical correlates of the S pattern which were heretofore impossible to study since this pattern represents a true intermediate of the acrosome reaction and is transient in nature.

How might PK-C activators bring about this egg-induced dissociation of the sperm receptor activity from the acrosome reaction inducing activities of ZP3? Two models can account for these observations and both models are based on PK-C activators stimulating a partial cortical granule reaction. Results from preliminary experiments indicate that treatment of eggs with biologically active phorbol diesters, but not with inactive phorbol diesters, do in fact result in a partial reduction in the number of cortical granules (Ducibella, Kopf and Schultz, unpublished) detected with *lens culinaris* agglutinin (Cherr et al., 1988) when compared to either A-23187-treated or fertilized eggs (Ducibella et al., 1988).

The first model is based on an intrinsic heterogeneity in the cortical granule population and the selective PK-C stimulated release of one population of these granules that contain enzymes that modify only the moiety of ZP3 involved in inducing the acrosome reaction (Fig. 2A). Mouse eggs have been shown to contain a heterogeneous population of cortical granules, as demonstrated by differences in staining intensities in electron micrographs (Nicosia et al., 1977). Although this staining pattern may represent granules of identical content but at different stages of maturity, this pattern could as easily represent a true heterogeneous population of granules that have different protein compositions.

The second model is based on a homogeneous population of cortical granules and assumes two properties of the ZP3 molecule. The first assumption is that ZP3 has associated

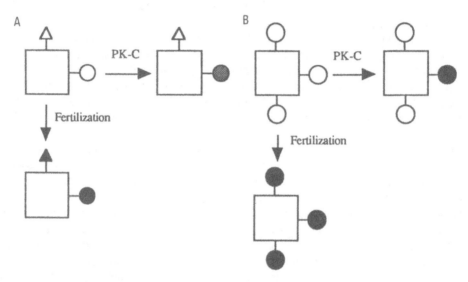

Fig.2. Model to account for the dissociation of sperm receptor and acrosome reaction-inducing activities in ZP3 isolated from eggs treated with activators of PK-C. A. One component (triangle) of ZP3 is responsible for sperm receptor activity and the other component (circle) is responsible for the acrosome reaction-inducing activity. Fertilization results in a modification of both of these activities (solid triangle and circle) and the loss of both of the biological activities of ZP3. Treatment of eggs with PK-C activators results in the selective release of a subpopulation of cortical granules that partially modifies the acrosome reaction-inducing component (stippled circle) but leaves intact the component responsible for sperm receptor activity. The consequence of this partial modification is that sperm initiate an acrosome reaction (B to S transition), but do not complete it (remain arrested at the S pattern). B. Multiple numbers of a single type of moiety (circles) are present in the ZP3 molecule. Successful sperm binding to ZP3 requires the interaction with only a single moiety, and this accounts for the concentration-dependence of ZP3 to inhibit sperm binding that is described by a hyperbolic function. Once this interaction has occurred additional interactions of the sperm with the other moieties of the ZP3 molecule are required to induce a complete acrosome reaction; this accounts for a concentration dependence of ZP3 to induce the acrosome reaction described by a sigmoidal function. Fertilization results in a modification (inactivation) of all the moieties on the ZP3 molecule(solid circles) and hence the loss of both biological activities. In contrast, the partial release of cortical granules that is stimulated by PK-C activators results in a submaximal modification of these moieties. Consequently ZP3 still possesses sperm receptor activity, but the suboptimal number of active moieties remaining are not sufficient to permit the cooperative interactions with the sperm necessary to induce a complete acrosome reaction. As a result sperm initiate (B to S transition), but do not complete the acrosome reaction because they remain arrested at the S pattern.

with it a single type of moiety that is present in multiple numbers; the interaction of these moieties with the sperm surface is responsible for both sperm receptor activity and the ability to induce a complete acrosome reaction (Fig. 2B). The second assumption in this model is that there are differences in the concentration-dependence of ZP3 to express sperm receptor activity and acrosome reaction-inducing activities. The interaction of acrosome intact sperm with the sperm receptor activity of the ZP3 molecule is described by a hyperbolic function (indicating single interactions between the sperm surface and ZP3 to mediate sperm binding), whereas the ability of ZP3 to induce a complete acrosome reaction (e.g., B to S to AR transitions) is described by a sigmoidal function (indicating cooperative interactions between the sperm surface and ZP3 to mediate the complete acrosome reaction). The reader is referred to the legend of Fig. 2 for additional clarification of these two models.

THE ROLE OF INOSITOL 1,4,5 - TRISPHOSPHATE IN THE EGG-INDUCED ZONA PELLUCIDA BLOCK TO POLYSPERMY

As reviewed earlier PIP_2 turnover, with the subsequent generation of DAG and IP_3 may represent an early fertilization response of both invertebrate and vertebrate eggs and may constitute at least one mechanism by which some of the early events of egg activation are mediated. Our previous experiments using PK-C activators would, in principle only activate one half of such a bifurcating second messenger cascade system, and the "partial" ZP modifications obtained under these experimental conditions is consistent with this idea. Therefore, we examined the effects of the other major second messenger in this signalling system — IP_3 — on egg activation, electrophoretic modifications of the ZP, and the sperm receptor and acrosome reaction inducing activities of the ZP (Kurasawa et al., 1989).

Mouse eggs were microinjected with various concentrations of IP_3 and the ZP from these individually injected eggs was examined for the ZP2 to $ZP2_f$ conversion. It was demonstrated that IP_3 induces a ZP2 to $ZP2_f$ conversion in a concentration dependent manner with an $EC_{50} \sim 5 - 10$ nM; a final intracellular concentration of 1 µM IP_3 results in approximately 80% of the eggs displaying the ZP2 conversion. The IP_3-induced conversion of ZP2 occurs in Ca^{2+}-depleted medium, but does not occur when eggs are microinjected with inositol phosphate analogues (1 µM final intracellular concentration) that have a reduced ability to release Ca^{2+} from intracellular stores (i.e., inositol 1,4,-bisphosphate; inositol 1,3,4-trisphosphate; inositol 2,4,5-trisphosphate). These data suggest that IP_3 microinjection results in the release of Ca^{2+} from intracellular stores, and that this Ca^{2+} release is somehow coupled to the cortical granule reaction and the resultant modification of ZP2 to $ZP2_f$.

Subsequently, experiments were designed to determine whether IP_3 microinjection also modifies the sperm receptor and/or acrosome reaction inducing activities of the ZP (those activities associated with ZP3). Sperm receptor activity was measured by examining the ability of sperm to bind to IP_3-microinjected eggs, and this binding activity was compared to both negative controls (untreated and vehicle-injected eggs) and positive controls (2-cell embryos). In order to design and interpret these experiments properly, however, it was critical to assay primary sperm binding (binding of acrosome intact sperm to ZP3) in the absence of secondary sperm binding (binding of acrosome reacted sperm to ZP2). The reason for this is that since IP_3 microinjection induces a ZP2 to $ZP2_f$ conversion, a reduction in sperm binding to IP_3 injected eggs could reflect either, 1) a reduction in the sperm receptor activity associated with ZP3, or 2) the inability of acrosome reacted sperm to establish a secondary binding via ZP2 since $ZP2_f$ will no longer bind acrosome reacted sperm. Therefore, reduced sperm binding could occur in ZP that retain the full complement of sperm receptor activity associated with ZP3 because sperm would acrosome react, could not establish secondary binding with $ZP2_f$, and subsequently fall off the ZP. The ideal way to circumvent these potential problems would be to incubate sperm under conditions that would permit normal sperm binding to the ZP but not allow the bound sperm to undergo the acrosome reaction.

Treatment of sperm with pertussis toxin is a convenient way to satisfy both of these conditions, since this toxin does not affect the ability of capacitated sperm to bind to the ZP but inhibits the sperm from undergoing the acrosome reaction (Endo et al., 1987b, 1988; see beginning of this chapter). Under these conditions a decrease in the extent of sperm binding to the IP_3-injected eggs should reflect a reduction in the sperm receptor activity of ZP3. When individual eggs microinjected with IP_3 (1 μM final concentration) are examined for binding activity of pertussis toxin-treated sperm, a reduction in sperm binding (at least 50%) occurs relative to vehicle-injected and untreated eggs. Analysis of the ZP2 to $ZP2_f$ conversion in the ZP from these same eggs reveals a remarkable 83% positive correlation between the reduction in the number of sperm bound to a given egg and the presence of a ZP2 to $ZP2_f$ conversion. These data suggest that IP_3 is causing a reduction in the sperm receptor activity of ZP3, a situation quite different from that seen with PK-C activators.

Fertilization normally results in the loss of both the sperm receptor and acrosome reaction inducing activities of ZP3. In order to determine whether IP_3 microinjection results in an egg-induced modification of the acrosome reaction activity of the ZP, ZP were isolated from these eggs, solubilized and monitored for this biological activity using the CTC assay. Results of such experiments demonstrate that, when compared to uninjected or vehicle-injected controls, the acrosome reaction activity of ZP from IP_3 injected eggs is greatly reduced.

These studies demonstrate that IP_3 microinjection results in a ZP2 to $ZP2_f$ conversion, as well as the loss of both sperm receptor and acrosome reaction-inducing activities of ZP3. This is in contrast to the effects of PK-C activators which bring about a ZP2 to $ZP2_f$ conversion, do not affect the sperm binding activity of ZP3, but cause the loss in the ability of ZP3 to induce a complete acrosome reaction. The modifications in the biological activity of the ZP induced by IP_3, therefore, more closely resemble those changes which normally occur after fertilization.

How might two second messengers in such a bifurcating signal transduction system bring about such different responses? The answer to this question may ultimately reside in a non-reciprocating interaction between the IP_3-and PK-C-modulated pathways. For example, IP_3 may activate PK-C indirectly by stimulating Ca^{2+} release from intracellular stores. This release of Ca^{2+} could, in principal, either directly activate PK-C or indirectly activate this enzyme by stimulating PIP_2 hydrolysis and subsequent DAG generation through the activation of phospholipase C. Thus, IP_3 microinjection ultimately results in the activation of both arms of this bifurcating pathway and the complete set of fertilization associated ZP responses is observed. On the other hand PK-C activation may not be able to feedback to stimulate IP_3 release, only one arm of the pathway is stimulated, and a partial ZP response occurs.

CONCLUSIONS AND FUTURE DIRECTIONS

Events in mammalian fertilization comprising sperm-egg recognition, egg-induced sperm activation, sperm-egg fusion, sperm-induced egg activation, and the ZP block to polyspermy are still not well defined at the present time. Our laboratories are focusing on the biochemical events involved in murine sperm-ZP interaction both prior to and after fertilization, as well as those events that comprise the block to polyspermy at the level of the ZP. An understanding of these events should provide a structural framework in understanding basic processes of cell-cell interaction and cellular activation, as well as providing insight into the biochemical mechanisms that underlie sperm-egg interaction and fertilization in the human.

Sperm from all species studied thus far contain G-proteins. The presence of such signal transducing proteins in these cells suggests that the regulation of sperm function might have control elements which are similar to ligand: receptor: G-protein: second messenger systems common to many somatic cells. This hypothesis is supported by experiments which demonstrate a potential intermediary role for the mouse sperm G_i-like protein in the acrosome reaction induced by ZP3. The specific function of this G_i-like protein in this important physiological event is not known at this time, although possible roles in regulating ionic movements, cyclic nucleotide metabolism, and polyphosphoinositide turnover are possible candidates. Studies directed at the localization and biochemical identity of the mouse sperm G_i-like protein, as well as the nature of the second messenger system(s) modulated by this protein, are in progress and should help to delineate the sequence of events involved in some of the early steps of sperm-ZP interaction. Based on studies in our laboratories of the biochemical mechanisms controlling the egg-induced polyspermy block, it appears as if the induction of the complete acrosome reaction may involve cooperative interactions between ZP3 moieties and complementary binding sites (receptors) on the sperm surface. The mechanism by which this occurs, as well as the role of signal transduction and intracellular second messenger pathways in coupling these interactions at the sperm surface with resultant physiological responses leading to acrosomal exocytosis, will be of great interest for future research.

Polyphosphoinositide turnover may play a key intermediary role in the early events of mammalian egg activation and the ZP block to polyspermy. Although it is likely that this process is mediated by a G-protein(s), the biochemical nature of the receptor for the sperm on the egg surface, the molecular identity of the G-proteins involved, and the coupling mechanisms of the putative egg-associated sperm receptor with this G-protein(s), as well as the consequences of this coupling, are still unknown. Biochemical and molecular approaches should enable us to answer questions such as whether sperm-egg fusion results in the generation of DAG and IP_3 through a G-protein modulated phospholipase C. The ultimate targets of DAG and IP_3 action within the egg are also of great interest. Does IP_3 serve to modulate intracellular Ca^{2+} in a manner similar to that seen in hamster eggs? Does PK-C play a role in cortical granule exocytosis and, if so, what is the role of protein phosphorylation in this important exocytotic event?

The role of cortical granules in mediating early events in mammalian egg activation has yet to be resolved. The biochemical basis for fertilization-induced cortical granule exocytosis is only starting to be determined in both invertebrate and mammalian species. Little is known about the contents of mammalian egg cortical granules and the role that specific granule components play in modifying the ZP. The answers to such questions should provide insight into the molecular mechanisms by which ZP modifications alter the receptivity of the egg for sperm. The question of heterogeneity in the cortical granule population is still to be addressed; the answers to such questions might provide insight into the apparent dissociations of ZP functions that we have observed experimentally. Work directed at isolating and characterizing mammalian cortical granules will help to resolve many of the above mentioned issues.

Finally, work directed towards defining the biochemical nature of the determinants in both ZP3 and ZP2, as well as the corresponding sperm-associated ZP3 and ZP2 binding site(s), involved in primary sperm binding, the induction of the acrosome reaction, and secondary sperm binding should aid in the development of a unifying hypothesis regarding sperm-egg recognition, sperm activation, and sperm-egg fusion.

ACKNOWLEDGMENT

Research performed by the authors was supported by grants from the National Institutes of Health (HD 19096 and 06274 to G.S.K., HD 18604 and 22681 to R.M.S., and HD 22732 to G.S.K. and R.M.S.), Mellon Foundation (G.S.K.), University Research Fund (R.M.S.), and Rockefeller Foundation (Y.E. and S.K.).

REFERENCES

Babcock, D.F., Singh, J.P. and Lardy, H.A. (1979) Alteration of membrane permeability to calcium ions during maturation of bovine spermatozoa. *Dev. Biol.* 69, 85-93.
Barros, C. and Yanagimachi, R. (1971) Induction of the *zona* reaction in golden hamster eggs by

cortical granule material. *Nature* (London) 233, 268-269.

Barros, C. and Yanagimachi, R. (1972) Polyspermy-preventing mechanisms in the golden hamster egg. *J. Exp. Zool.* 180, 251-266.

Benau, D.A. and Storey, B.T. (1987) Characterization of the mouse sperm plasma membrane zona-binding site sensitive to trypsin inhibitors. *Biol. Reprod.* 36, 282-292.

Bentley, J.K., Garbers, D.L., Domino, S.E., Noland, T.D. and VanDop, C. (1986) Spermatozoa contain a guanine nucleotide-binding protein ADP-ribosylated by pertussis toxin. *Biochem. Biophys. Res. Commun.* 138, 728-734.

Berridge, M.J. (1987) Inositol trisphosphate and diacylglycerol: Two interacting second messengers. *Ann. Rev. Biochem.* 56, 159-193.

Bleil, J.D. and Wassarman, P.M. (1980a) Structure and function of the zona pellucida: Identification and characterization of the proteins of the mouse oocyte's zona pellucida. *Dev. Biol.* 76, 185-202.

Bleil, J.D. and Wassarman, P.M. (1980b) Synthesis of zona pellucida proteins by denuded and follicle-enclosed mouse oocytes during culture in vitro. *Proc. Natl. Acad. Sci. USA* 77, 1029–1033.

Bleil, J.D. and Wassarman, P.M. (1986) Autoradiographic visualization of the mouse egg's sperm receptor bound to sperm. *J. Cell Biol.* 102, 1363-1371.

Bleil, J.D. and Wassarman, P.M. (1980c) Mammalian sperm-egg interaction: Identification of a glycoprotein in mouse egg zonae pellucidae possessing receptor activity for sperm. *Cell* 20, 873-882.

Bleil, J.D. and Wassarman, P.M. (1981) Mammalian sperm-egg interaction: Fertilization of mouse eggs triggers modification of the major zona pellucida glycoprotein, ZP2. *Dev. Biol.* 86, 189-197.

Bleil, J.D. and Wassarman, P.M. (1983) Sperm-egg interactions in the mouse: Sequence of events and induction of the acrosome reaction by a zona pellucida glycoprotein. *Dev. Biol.* 95, 317-324.

Bleil, J.D. and Wassarman, P.M. (1988) Galactose at the non reducing terminus of O-linked oligosaccharides of mouse egg zona pellucida glycoprotein ZP3 is essential for the glycoprotein's sperm receptor activity. *Proc. Nat'l. Acad. Sci., U.S.A.* 85, 6778-6782.

Bleil, J.D., Greve, J.M., and Wassarman, P.M. (1988) Identification of a secondary sperm receptor in the mouse egg zona pellucida: Role in maintenance of binding of acrosome- reacted sperm to eggs. *Dev. Biol.* 128, 376-385.

Bokoch, G.M., Katada, T., Northup, J.K., Ui, M., and Gilman, A.G. (1984) Purification and properties of the inhibitory guanine nucleotide binding regulatory component of adenylate cyclase. *J. Biol. Chem.* 259, 3560-3567.

Busa, W.B., Ferguson, J.E., Joseph, S.K., Williamson, J.R. and Nuccitelli, R. (1985) Activation of frog (Xenopus laevis) eggs by inositol trisphosphate. I. Characterization of Ca^{2+} release from intracellular stores. *J. Cell Biol.* 101, 677-682.

Casey, P.J. and Gilman, A.G. (1988) G protein involvement in receptor-effector coupling. *J. Biol. Chem.* 263, 2577–2580.

Cherr, G.N., Drobnis, E.Z., and Katz, D.F. (1988) Localization of cortical granule constituents before and after exocytosis in the hamster egg. *J. Exp. Zool.* 246, 81-93.

Ciapa, B. and Whitaker, M. (1986) Two phases of inositol polyphosphate and diacylglycerol production at fertilization. *Febs. Lett.* 195, 347-351.

Cockcroft, S. (1987) Polyphosphoinositide phosphodiesterase: regulation by a novel guanine nucleotide binding protein, G_p. *Trends Biochem. Sci.* 12, 75-78.

Cuthbertson, K.S.R., Whittingham, D.G. and Cobbold, P.H. (1981) Free Ca^{2+} increases in exponential phases during mouse oocyte activation. *Nature* (London) 294, 754-757.

Cuthbertson, K.S.R. and Cobbold, P.H. (1985) Phorbol ester and sperm activate mouse oocytes by inducing sustained oscillations in cell Ca^{2+}. *Nature* (London) 316, 541-542.

Ducibella, T., Anderson, E., Albertini, D.F., Aalberg, J. and Rangarajan, S. (1988) Quantitative studies of changes in cortical granule number and distribution in the mouse oocyte during meiotic maturation. *Dev. Biol.* 130, 184-197.

Endo, Y., Schultz, R.M. and Kopf, G.S. (1987a) Effects of phorbol esters and a diacylglycerol on mouse eggs: Inhibition of fertilization and modification of the zona pellucida. *Dev. Biol.* 119, 199-209.

Endo, Y., Lee, M.A. and Kopf, G.S. (1987b) Evidence for the role of a guanine nucleotide-binding regulatory protein in the zona pellucida-induced mouse sperm acrosome reaction. *Dev. Biol.* 119, 210-216.

Endo, Y., Mattei, P., Kopf, G.S. and Schultz, R.M. (1987c) Effects of a phorbol ester on mouse eggs: Dissociation of sperm receptor activity from acrosome reaction-inducing activity of the mouse zona pellucida protein, ZP3. *Dev. Biol.* 123, 574-577.

Endo, Y., Lee, M.A. and Kopf, G.S. (1988) Characterization of an islet activating protein-sensitive site in mouse sperm that is involved in the zona pellucida-induced acrosome reaction. *Dev. Biol.* 129, 12-24.

Florman, H.M. and Storey, B.T. (1982) Mouse gamete interactions: The zona pellucida is the site of the acrosome reaction leading to fertilization in vitro. *Dev. Biol.* 91, 121-130.

Florman, H.M. and Wassarman, P.M. (1985) O-linked oligosaccharides of mouse egg ZP3 account for its sperm receptor activity. *Cell* 41, 313-324.

Florman, H.M., Bechtol, K.B., and Wassarman, P.M. (1984) Enzymatic dissection of the functions of the mouse egg's receptor for sperm. *Dev. Biol.* 106, 243-255.

Garbers, D.L. and Kopf, G.S. (1980) The regulation of spermatozoa by calcium and cyclic nucleotides. *Adv. Cyclic Nuc. Res.* 13, 251-306.

Gilman, A.G. (1987) G proteins: Transducers of receptor-generated signals. *Ann Rev. Biochem.* 56, 615-649.

Greve, J.M. and Wassarman, P.M. (1985) Mouse egg extracellular coat is a matrix of interconnected filaments possessing a structural repeat. *J. Mol. Biol.* 181, 253-264.

Gwatkin, R.B.L., Williams, D.T., Hartmann, J.F., and Kniazuk, M. (1973) The zona reaction of hamster and mouse eggs: Production in vitro by a trypsin-like protease from cortical granules. *J. Reprod. Fert.* 32, 259-265.

Hescheler, J., Rosenthal,W., Trautwein, W., and Schultz, G. (1987) The GTP binding protein, G_o, regulates neuronal calcium channels. *Nature* (London) 325, 445-447

Hildebrandt, J.D., Codina, J., Tash, J.S., Kirchick, H.J., Lipschultz, L., Sekura, R.D. and Birnbaumer, L. (1985) The membrane-bound spermatozoal adenylyl cyclase system does not share coupling characteristics with somatic cell adenylyl cyclases. *Endocrinology* 116, 1357-1366.

Igusa, Y. and Miyazaki, S. (1986) Periodic increase of cytoplasmic free calcium in fertilized hamster eggs measured with calcium sensitive electrodes. *J. Physiol.* (London) 377, 193–205.

Jelsema, C.L. and Axelrod, J.(1987) Stimulation of phospholipase A_2 activity in bovine rod outer segments by the beta gamma subunits of transducin and its inhibition by the alpha subunit. Proc. *Nat'l. Acad. Sci.*, U.S.A. 84, 3623-3627.

Kikkawa, U. and Nishizuka, Y. (1986) The role of protein kinase C in transmembrane signalling. *Ann. Rev. Cell Biol.* 2, 149-178.

Kligman, I., Kopf, G.S. and Storey, B.T. (1988) Characterization of an intermediate stage of the zona pellucida-mediated acrosome reaction in mouse sperm. *Fertil. Steril.* 50 (Suppl.), 530–531.

Kline, D. and Jaffe, L.A. (1987) The fertilization potential of the Xenopus egg is blocked by injection of a calcium buffer and is mimicked by injection of a GTP analog. *Biophys. J.* 51, 398a.

Kopf, G.S., Woolkalis, M.J. and Gerton, G.L. (1986) Evidence for a guanine nucleotide-binding regulatory protein in invertebrate and mammalian sperm: Identification by islet-activating protein-catalyzed ADP-ribosylation and immunochemical methods. *J. Biol. Chem.* 261, 7327-7331.

Kurasawa, S., Schultz, R.M. and Kopf, G.S. (1989) Egg induced modifications of the zona pellucida of mouse eggs: Effects of microinjected inositol 1,4,5-trisphosphate. *Dev. Biol.*, in press.

Lee, M.A. and Storey, B.T. (1985) Evidence for plasma membrane impermeability to small ions in acrosome-intact mouse spermatozoa bound to mouse zonae pellucidae, using an aminoacridine fluorescent probe: time course of the zona-induced acrosome reaction monitored by both chlortetracycline and pH probe fluorescence. *Biol. Reprod.* 33, 235–246.

Lee, M.A. and Storey, B.T. (1986) Bicarbonate is essential for fertilization of mouse eggs: Mouse sperm require it to undergo the acrosome reaction. *Biol. Reprod.* 34, 349-356.

Lee, M.A., Kopf, G.S. and Storey, B.T. (1987) Effects of phorbol esters and a diacylglycerol on the mouse sperm acrosome reaction induced by the zona pellucida. *Biol. Reprod.* 36, 617-627.

Lee, S.H., Ahuja, K.K., Gilburt, D.J., and Whittingham, D.G. (1988) The appearance of glyco-conjugates associated with cortical granule release during mouse fertilization. *Development* 102, 595-604.

Lopez, L.C., Bayna, E.M., Litoff,D., Shaper, N.L.,, Shaper, J.H. and Shur, B.D. (1985) Receptor function of mouse sperm surface galactosyltransferase during fertilization. *J. Cell Biol.* 101, 1501-1510.

Mattera,R., Yatani,A., Kirsch,G.E. , Graf,R., Okabe,K., Olate,J., Codina,J.,Brown, A.K., and Birnbaumer,L. (1989) Recombinant α_i-3 subunit of G protein activates G_k-gated K channels. *J. Biol. Chem.* 264, 465-471.

Mattera, R., Graziano, M.P., Yatani, A., Zhou, Z., Graf, R., Codina, J., Birnbaumer, L., Gilman, A.G. and Brown, A.M. (1989) Splice variants of the α-subunit of the G protein G_s activate both adenylate cyclase and calcium channels. Science 243, 804-807.

Minke, B. and Stephenson, R.S. (1985) The characteristics of chemically induced noise in *Musca* photoreceptors. *J. Comp. Physiol. A.* 156, 339-356.

Miyazaki, S., Hashimoto, N., Yoshimoto, Y., Kishimoto, T., Igusa, Y. and Hiramoto, Y. (1986) Temporal and spatial dynamics of the periodic increase in intracellular free calcium at fertilization of golden hamster eggs. *Dev. Biol.* 118, 259-267.

Miyazaki, S. (1988) Inositol 1,4,5-trisphosphate-induced calcium release and guanine nucleotide-binding protein-mediated periodic calcium rises in golden hamster eggs. *J. Cell Biol.* 106, 345-353.

Moller, C.C. and Wassarman, P.M. (1989) Characterization of a proteinase that cleaves zona pellucida glycoprotein ZP2 following activation of mouse eggs. *Dev. Biol.* 132, 103–112.

Neer, E.J., Lok, J.M. and Wolf, L.G. (1984) Purification and properties of the inhibitory guanine nucleotide regulatory unit of brain adenylate cyclase. *J. Biol. Chem.* 259, 14222–14229.

Nicosia, S.V., Wolf, D.P. and Inoue, M. (1977) Cortical granule distribution and cell surface characteristics in mouse eggs. *Dev. Biol.* 57, 56-74.

Philpott, C.C., Ringuette, M.R. and Dean, J. (1987) Oocyte-specific expression and developmental regulation of ZP3, the sperm receptor of the mouse zona pellucida. *Dev. Biol.* 121, 568-575.

Ringuette, M.J., Chamberlin, M.E., Baur, A.W., Sobieski, D.A. and Dean, J. (1988) Molecular analysis of cDNA coding for ZP3, a sperm binding protein of the mouse zona pellucida. *Dev. Biol.* 127, 287-295.

Rufo, G.A., Singh, J.P., Babcock, D.F. and Lardy, H.A. (1982) Purification and characterization of a calcium transport inhibitor protein from bovine seminal plasma. *J. Biol. Chem.* 257, 4627-4632.

Saling, P.M. (1981) Involvement of trypsin-like activity in binding of mouse spermatozoa to zonae pellucidae. Proc. *Nat'l. Acad. Sci.*, U.S.A. 78, 62131–6235.

Saling, P.M. and Storey, B.T. (1979) Mouse gamete interactions during fertilization in vitro: Chlortetracycline as fluorescent probe for the mouse sperm acrosome reaction. J. *Cell Biol.* 83, 544–555.

Saling, P.M., Wolf, D.P. and Storey, B.T. (1978) Calcium-dependent binding of mouse epididymal spermatozoa to the zona pellucida. *Dev. Biol.* 65, 515-525.

Saling, P.M., Sowinski, J., and Storey, B.T. (1979) An ultrastructural study of epididymal mouse spermatozoa binding to zonae pellucidae in vitro: Sequential relationship to the acrosome reaction. *J. Exp. Zool.* 209, 229-238.

Shimizu, S., Tsuji, M. and Dean, J. (1983) In vitro biosynthesis of three sulfated glycoproteins of murine zonae pellucidae by oocytes grown in follicle culture. *J. Biol. Chem.* 258, 5858-5863.

Shur, B.D. and Hall, N.G. (1982a) Sperm surface galactosyltransferase activities during in vitro capacitation. *J. Cell Biol.* 95, 567-573.

Shur, B.D. and Hall, N.G. (1982b) A role for mouse sperm surface galactosyltransferase in sperm binding to the egg zona pellucida. *J. Cell Biol.* 95, 574-579.

Stryer, L. (1986) Cyclic GMP cascade of vision. *Ann. Rev. Neurosci.* 9, 87.

Swann, K. and Whitaker, M. (1985) Stimulation of the Na/H exchanger of sea urchin eggs by phorbol ester. *Nature* 314, 274-277.

Swann, K. and Whitaker, M. (1986) The part played by inositol trisphosphate and calcium in the propagation of the fertilization wave in sea urchin eggs. *J. Cell Biol.* 103, 2333-2342.

Szollsi, D. (1967) Development of cortical granules and the cortical reaction in rat and hamster eggs. *Anat. Rec.* 159, 431-446.

Tash, J.S. and Means, A.R. (1983) Cyclic adenosine 3',5'-monophosphate, calcium and protein phosphorylation in flagellar motility. *Biol. Reprod.* 28, 75-104.

Thomas, P. and Meizel, S. (1988) An influx of extracellular calcium is required for initiation of the human sperm acrosome reaction induced by human follicular fluid. *Gamete Res.* 20, 397-411.

Turner, P.R., Sheetz, M.P., and Jaffe, L.A. (1984) Fertilization increases the polyphosphoinositide content of sea urchin eggs. *Nature* (London) 310, 414-415.

Turner, P. R., Jaffe, L.A. and Fein, A. (1986) Regulation of cortical granule exocytosis in sea urchin eggs by inositol 1,4,5-trisphosphate and GTP-binding protein. *J. Cell Biol.* 102, 70-76.

Turner, P.R., Jaffe, L.A., and Primakoff, P. (1987) A cholera toxin-sensitive G-protein stimulates exocytosis in sea urchin eggs. *Dev. Biol.* 120, 577-583.

Ui, M.,Nogimori, K. and Tamura, M. (1985) Islet-activating protein, pertussis toxin: Subunit structure and mechanisms for its multiple biological actions. *In*: Pertussis Toxin, R.D. Sekura, J. Moss and M. Vaughan, eds., Academic Press, Inc., New York, p. 19-43.

Ward, C.R. and Storey, B.T. (1984) Determination of the time course of capacitation in mouse spermatozoa using a chlortetracycline fluorescence assay. *Dev. Biol.* 104, 287-296.

Wassarman, P.M., Bleil, J.D., Florman, H.M., Greve, J.M., and Roller, R.J. (1985) The mouse egg's receptor for sperm: What is it and how does it work? Cold Spring Harbor Symp. *Quant. Biol.* 50, 11-19.

Wassarman, P.M. (1988) Zona pellucida glycoproteins. *Ann. Rev. Biochem.* 57, 414-442.

Whitaker, M. and Irvine, R.F. (1984) Inositol 1,4,5-trisphosphate microinjection activates sea urchin eggs. *Nature* (London) 312, 636-639.

Wolf, D.P. and Hamada, M. (1977) Induction of zonal and oolemal blocks to sperm penetration in mouse eggs with cortical granule exudate. *Biol. Reprod.* 17, 350-354.

CARBOHYDRATED-MEDIATED SPERM-EGG INTERACTIONS IN MAMMALS

Ruth Shalgi*, Jeffrey D. Bleil, and Paul M. Wassarman

Department of Cell and Developmental Biology
Roche Institute of Molecular Biology, Roche Research Center
Nutley, New Jersey 07110, USA

FERTILIZATION PATHWAY IN MICE

The fertilization pathway in mice consists of several steps that must occur in a compulsory order prior to fusion of sperm and egg (reviewed in Wassarman, 1987a,b, 1988a). The steps include: (i) loose "attachment" and then tight species-specific "binding" of acrosome-intact sperm to the unfertilized egg's extracellular coat, or zona pellucida; (ii) completion of the acrosome reaction, an exocytotic event involving fusion of outer acrosomal and plasma membranes, by bound sperm; (iii) penetration of the zona pellucida by bound acrosome-reacted sperm (perhaps, by using a sperm proteinase, "acrosin"); (iv) fusion of acrosome-reacted sperm with the egg's plasma membrane. Within a few minutes of sperm-egg fusion, the zona pellucida is altered ("zona reaction") such that it becomes refractory to both binding of sperm and penetration by sperm; this constitutes a secondary block to polyspermy. These changes in the nature of the zona pellucida are thought to be brought about by cortical granule enzymes released into the extracellular coat as a consequence of the cortical reaction (fusion of cortical granule and plasma membranes). Thus, the steps in the fertilization pathway include species-specific cellular recognition, intracellular and intercellular membrane fusions, and enzyme-catalyzed modifications of cellular investments.

---

* Current address: Department of Embryology and Teratology, Sackler School of Medicine, Tel-Aviv University, Ramat Aviv, 69978 Tel-Aviv, Israel.

Advances in Assisted Reproductive Technologies, Edited by
S. Mashiach *et al.*, Plenum Press, New York, 1990

NATURE OF THE MOUSE SPERM RECEPTOR

The zona pellucida of mammalian eggs contains species-specific receptors to which sperm bind as a prelude to fertilization (Gwatkin, 1977; Wassarman, 1987a,b, 1988a; Yanagimachi, 1988). The mouse egg's zona pellucida is about 7 μm thick and contains about 3 ng of protein (reviewed in Wassarman, 1988b). In mice, egg ZP3, one of three different zona pellucida glycoproteins, has been shown to serve as a sperm receptor (reviewed in Wassarman, 1987a,b, 1988a; Bleil and Wassarman, 1980, 1986; Vazquez et al., 1989). ZP3 is present in more than a billion copies in the zona pellucida. Each acrosome-intact sperm binds, via plasma membrane overlying the anterior region of its head (the location of "egg-binding proteins"), to tens-of-thousands of copies of ZP3 at the surface of the zona pellucida. Bound sperm then undergo the acrosome reaction (membrane fusion), exposing the inner acrosomal membrane and, thereby, enabling them to penetrate the zona pellucida and fuse with the egg's plasma membrane (fertilization). Egg ZP3 is responsible for both sperm binding (sperm receptor) and induction of the acrosome reaction (acrosome reaction-inducer). Embryo ZP3 is ineffective as a sperm receptor and acrosome reaction-inducer. Thus, as expected, fertilization results in the inactivation of ZP3 (a consequence of the zona reaction).

ZP3 is synthesized by growing mouse oocytes as a 44,000 M_r polypeptide chain to which 3 or 4 asparagine-linked (N-linked), complex-type oligosaccharides and an undetermined number of serine/threonine-linked (O-linked) oligosaccharides are attached, giving rise to an 83,000 M_r mature glycoprotein (reviewed in Wassarman, 1988b; Salzmann et al., 1983). ZP3 is a single-copy gene that contains 8 exons (~8.5 kb transcription unit) encoding a 424 amino acid polypeptide chain (Kinloch et al., 1988; Kinloch and Wassarman, 1989). The amino-terminal 22 amino acids constitute a signal sequence that is cleaved from nascent ZP3, leaving a 402 amino acid polypeptide chain. Nascent ZP3 combines (via noncovalent interactions) with another zona pellucida glycoprotein, ZP2 (120,000 M_r) to form dimers that, in turn, polymerize into long filaments about 7 nm in width (Greve and Wassarman, 1985; Wassarman, 1988b; J.M. Greve and P.M. Wassarman, unpublished results). Filaments are interconnected (via noncovalent interactions) by a third zona pellucida glycoprotein, ZP1 (200,000 M_r), giving rise to a relatively thick extracellular matrix. ZP3 is located every 15 nm or so along each zona pellucida filament, resulting in the presence of tens-of-millions of copies of ZP3 at the zona pellucida surface of unfertilized eggs.

OLIGOSACCHARIDES AND SPERM RECEPTOR FUNCTION

Mouse sperm recognize and bind to a particular class of ZP3 O-linked oligosaccharides (about 3,900 M_r) which represents less than 10% of total O-linked oligosaccharides on ZP3 (Florman et al., 1984; Florman and Wassarman, 1985; Wassarman, 1989). The oligosaccharides are solely responsible for the glycoprotein's sperm receptor function; N-linked oligosaccharides and polypeptide chain play no direct role in sperm receptor function. The purified O-linked oligosaccharides, at nanomolar concentrations, prevent both binding of sperm to eggs and fertilization in vitro. A galactose, located in a-linkage at the nonreducing terminus of the oligosaccharides, serves as a determinant that is essential for binding of sperm to ZP3 (Bleil and Wassarman, 1988; Vazquez et al., 1989; Wassarman, 1989). Either removal of the oligosaccharide's terminal galactose by α-galactosidase or conversion of the sugar's C-6 alcohol to an aldehyde by treatment with galactose oxidase, inactivates both purified ZP3 and purified ZP3 oligosaccharides as a sperm receptor. In the latter case, sperm receptor activity can be restored by subsequent treatment of the substrates with sodium borohydride, thereby converting the C-6 aldehyde back to an alcohol.

It is of interest to note (in view of the section that follows), that treatment of purified ZP3 and ZP3 O-linked oligosaccharides with α-fucosidase results in destruction of their sperm receptor activity (Bleil and Wassarman, 1988; Wassarman, 1989). [Note: Several other exoglycosidases, including β-galactosidase, β-glucuronidase, β-N-acetylglucosaminidase (containing β-N-acetylgalactosaminidase), and neuraminidase had no effect on sperm receptor activity.] However, whereas release of galactose by α-galactosidase was detected by HPLC analysis of dansylated sugar derivatives, neither fucose nor any other sugar was detected using α-fucosidase. An explanation for the latter result remains to be determined in view of the effect of α-fucosidase on sperm receptor activity. Consequently, as yet, a strong case cannot be made for the involvement of fucose in sperm receptor activity.

IMMUNOLOGICAL ANALYSIS OF SPERM RECEPTOR OLIGOSACCHARIDES

The nature of ZP3 oligosaccharides was examined further by using a series of monoclonal antibodies directed specifically against O-linked carbohydrate epitopes. Immunoblots of zona pellucida glycoproteins were developed by using a secondary antibody conjugated to alkaline phosphatase, 5-bromo-4-chloro-3-indolyl phosphate (BCIP), and nitro blue tetrazolium (NBT). Of the monoclonal antibodies tested, one, designated LA4 (Dodd and Jessell, 1985; Jessell and Dodd, 1985), recognized ZP3, but not ZP1 or ZP2. The epitope recognized by this monoclonal antibody (IgM) contains a terminal galactose in α-linkage with a penultimate galactose (structure based on the type 2 [galactose (β1-4)-N-acetylglucosamine] lactoseries sequence) (Dodd and Jessell, 1985; Jessell and Dodd, 1985). Furthermore, the epitope for LA4 can include a fucose in α-linkage with the penultimate galactose residue. Consistent with the results of immunoblots, binding of monoclonal antibody LA4 to zonae pellucidae (isolated or surrounding ovulated eggs) was detected by immunofluorescence using a fluorescein-conjugated second antibody.

In view of results obtained with monoclonal antibody LA4, monoclonal antibodies directed against carbohydrate epitopes of blood group antigens A or B (types 1 and 2) (Chembiomed Ltd.) were tested with zona pellucida glycoproteins by immunoblotting and immunofluorescence (as above). Anti A (IgM) and anti-B (IgM) recognize the trisaccharides GalNAc-1,3-α-Gal-1,4 (or 1,3)-β-GlcNac-R and Gal-1,3-α-Gal-1,4 (or 1,3)-β-GlcNac-R, respectively (in each case, there is a fucose in α-1,2-linkage with the penultimate galactose residue). Consistent with results obtained with monoclonal antibody LA4 (see above), only anti-B recognized ZP3, albeit relatively weakly (anti-A did not react with any zona pellucida glycoprotein).

CONCLUDING REMARKS

Carbohydrates have been implicated as mediators of cell-cell interactions in several instances, including the case of sperm-egg interactions in invertebrates such as echinoderms and ascidians. Certainly the great diversity of known oligosaccharide structures is compatible with generation of species specificity for sperm-egg interactions. Here we have summarized several lines of evidence that suggest involvement of O-linked oligosaccharides in sperm-egg interactions in mammals. These oligosaccharides are present on ZP3, a zona pellucida glycoprotein that serves as a receptor for mouse sperm. Although the primary structure of the receptor oligosaccharides has not been determined as yet, certain features have been revealed by using exoglycosidases and monoclonal antibodies directed against carbohydrate epitopes. These features include a galactose residue at the nonreducing terminus in α-linkage with the galactose of an N-acetyllactosamine unit. The terminal galactose residue is essential for sperm recognition of and binding to ZP3. It is interesting to note that oligosaccharides with repeating N-acetyllactosamine units (lactosaminoglycans) are usually associated with developmentally regulated glycoconjugates (Pink, 1989; Feizi et al., 1981). In certain instances, such sequences are often terminated by a galactose in α-linkage with the galactose of the N-acetyllactosamine unit (Feizi et al., 1981; Dodd and Jessell, 1985; Jessell and Dodd, 1985). In fact, it has been reported that a porcine zona pellucida glycoprotein analogous to mouse ZP3 (i.e., the glycoprotein exhibits sperm receptor activity in vitro) has poly-N-acetyllactosamine glycans (Yurewicz et al., 1987). Hopefully, future experiments will lead to the identification of carbohydrate recognition and binding determinants on mammalian sperm receptors that restrict heterologous sperm-egg interactions. Finally, in view of results summarized here, it is likely that inactivation of ZP3, observed following either fertilization or artificial activation of mouse eggs, is attributable to modification of O-linked oligosaccharides by cortical granule enzymes.

ACKNOWLEDGEMENTS

We are extremely grateful to Drs. Thomas Jessell and Jane Dodd, College of Physicians and Surgeons of Columbia University, New York, NY for kindly providing the monoclonal antibodies directed against carbohydrate epitopes that were used in certain experiments summarized here.

REFERENCES

Bleil JD, Wassarman PM 1980 Mammalian sperm-egg interaction: Identification of a glycoprotein in mouse egg zonae pellucidae possessing receptor activity for sperm. *Cell* 20:873–82.

Bleil JD, Wassarman PM 1986 Autoradiographic visualization of the mouse egg's sperm receptor bound to sperm. *J Cell Biol* 102:1363–71.

Bleil JD, Wassarman PM 1988 Galactose at the nonreducing terminus of O-linked oligosaccharides of mouse egg zona pellucida glycoprotein ZP3 is essential for the glycoprotein's sperm receptor activity. *Proc Natl Acad Sci, USA* 85:6778–82.

Dodd J, Jessell TM 1985 Lactoseries carbohydrates specify subsets of dorsal root ganglion neurons projecting to the superficial dorsal horn of rat spinal cord. *J Neuroscience* 5:3278–94.

Feizi T, Kapadia A, Gooi HC, Evans MJ 1981 Human monoclonal autoantibodies detect changes in expression and polarization of the Ii antigens during cell differentiation in early mouse embryos and teratocarcinomas. *In:* Muramatsu T, Ikawa Y (eds) Teratocarcinoma and cell surface. North-Holland Biomedical Press, Amsterdam, p 167–81.

Florman HM, Bechtol KB, Wassarman, PM 1984 Enzymatic dissection of the functions of the mouse egg's receptor for sperm. *Dev Biol* 106:243–55.

Florman HM, Wassarman PM 1985 O-Linked oligosaccharides of mouse egg ZP3 account for its sperm receptor activity. *Cell* 41:313–24.

Greve JM, Wassarman PM 1985 Mouse egg extracellular coat is a matrix of interconnected filaments possessing a structural repeat. *J Mol Biol* 181:253–64.

Gwatkin RBL 1977 Fertilization mechanisms in man and mammals, Plenum Press, New York.

Jessell TM, Dodd J 1985 Structure and expression of differentiation antigens on functional subclasses of primary sensory neurons. *Phil Trans Roy Soc Lond, Biol* 308:271–81.

Kinloch RA, Roller RJ, Fimiani CM, Wassarman DA, Wassarman PM 1988 Primary structure of the mouse sperm receptor polypeptide determined by genomic cloning. *Proc Natl Acad Sci, USA* 85:6409–13.

Kinloch RA, Wassarman PM 1989 Nucleotide sequence of the gene encoding zona pellucida glycoprotein ZP3—the mouse sperm receptor. *Nucl Acids Res* 17:2861–63.

Pink JRL 1980 Changes in T-lymphocyte glycoprotein structures associated with differentiation. *Contemp Topics Mol Immunol* 9:89–113.

Salzmann GS, Greve JM, Roller RJ, Wassarman PM 1983 Biosynthesis of the sperm receptor during oogenesis in the mouse. *EMBO J* 2:1451–56.

Vazquez MH, Phillips DM, Wassarman PM 1989 Interaction of mouse sperm with purified sperm receptors covalently linked to silica beads. *J Cell Science* 92:713–23.

Wassarman PM 1987a The biology and chemistry of fertilization. *Science* 235:553–60.

Wassarman PM 1987b Early events in mammalian fertilization. *Annu Rev Cell Biol* 3:109–42.

Wassarman PM 1988a Fertilization in mammals. *Scientific American* 255 (December):78–84.

Wassarman PM 1988b Zona pellucida glycoproteins. *Annu Rev Biochem* 57:415–42.

Wassarman PM 1989 Role of carbohydrates in receptor-mediated fertilization in mammals. *In:* Bock G, Harnett S (eds) Carbohydrate recognition in cellular function. Ciba Foundation Symp, no. 45, John Wiley and Sons, Chichester, p 135–55.

Yanagimachi R 1988 Mammalian fertilization. In: Kobil E, Neill JD (eds) The physiology of reproduction. Raven Press, New York, vol 1:135–85.

Yurewicz EC, Sacco AG, Subramanian MG 1987 Structural characterization of the $M_r = 55,000$ antigen (ZP3) of porcine oocyte zona pellucida. *J Biol Chem* 262:564–71.

THE FUNCTION OF HUMAN SEMEN COAGULATION
AND LIQUEFACTION IN VIVO

Jack Cohen

Birmingham, England

INTRODUCTION

Microscopic examination of liquefied semen samples is the simplest, and must be the most common, investigation of infertility. The coagulation and liquefaction of human semen is that reproductive phenomenon most accessible to the reproductive biologist. Nevertheless we still lack a convincingly physiological explanation. Comparative studies[1] have regarded it as a not-very-effective semen retention mechanism, comparing it with the vaginal plug of some rodents. But these animals use the plug, with its included spermatozoa, as a "cork" behind a liquid ejaculate containing the effective sperms, which is forced up into the uterus by vaginal contraction.[2,3] This is very unlike the human situation but resembles that in the rabbit, where a minute jelly-blob lodges in the narrowest part of the 8-shaped vagina, enabling contraction of the anterior part to force some of the ejaculate into the higher reaches of the tract.[4] Attempts to homologise human with mouse/rat ejaculates have emphasized the high sperm content of the first fraction of human "split" ejaculates, obtained by masturbation; the relevance of this to ejaculation during copulation is very dubious. Suggestions[5,6] that these first sperms escape the inhospitable vaginal milieu, into a cervical reservoir, have been rendered less plausible by the discovery[7] that the spermatozoa do not experience vaginal acidity, because of the immediate buffering effect of the semen. Most of the work devoted to seminal coagulation/liquefaction[8,9,10] does not even attempt theories of its function in sperm presentation to the female tract.

An early puzzle in the search for mechanisms of sperm destruction and survival in female tracts was that IgG-coated sperms could be recovered from the female tract,[11] but that little if any IgG could be found in tract fluids, especially around ovulation-time in women.[12]

Advances in Assisted Reproductive Technologies, Edited by
S. Mashiach *et al.*, Plenum Press, New York, 1990

This puzzle was resolved by the discovery that IgG secretion, and leucocytosis, were elicited very promptly from the cervix of rabbits by spermatozoa.[13] This finding was incorporated into my sperm selection ideas,[14] and very similar "immune" responses were soon demonstrated in women:[15, 16] IgG appears within 5min of sperm presence at the cervix (additional to plasma transudate IgG[17] in the "lubricating fluid"); leucocytosis is well established by 15–20 min, although the maximum is not reached for up to four hours.[15] In the rabbit the first few (damaged-looking, "rapid transport") sperms can be found at the ovaries within a couple of minutes of mating, and Overstreet[18] has suggested that their function might be to prime the reproductive tract in anticipation of the fertilising sperms to come later. This suggestion corresponded with our view of the sperm-elicited IgG/leucocytosis, but we were unsure of the function; our prejudices inclined towards sperm selection, but we thought defence against vaginal infection at intromission was at least possible.[15]

Sequestration of most human sperms in the seminal clot would, of course, produce a similar physiological situation to that of the rabbit, but by very different means: the cervix would be stimulated by the few uninvolved sperms, so that the vaginal milieu would already have been leucocyte-and IgG-primed when most sperms were released into it. Many other mammals fitted this "double-sperm-presentation" pattern:[19] hamsters inseminated few sperms in early intromissions, many more 20 min later;[20] mice and rats had anterior vaginal presentation, followed by squeezing of sperms into the uterus 10–20 min later.[2] Pig and horse did not fit the pattern easily, but cattle, goats and sheep[21] showed leucocytosis and "rapid sperm transport" like rabbits. Feral cats also fitted, with a ~20 min interval within the intromission series.[22]

In this view, the "copulatory tie" shown by many rodents and, particularly, by dogs would be expected to show initial stimulation of the female response, then the male should wait to ejaculate the fertilising sperm-rich fraction until *after* the female has produced IgG and leucocytes. Alternatively, if the IgG/leucocytosis is primarily anti-bacterial then sperms might be produced early in the tie, to avoid phagocytosis (even though, in the rabbit, Taylor[23] demonstrated that a second buck ejaculating into a leucocyte-dominated vagina did *not* show reduction in relative fertility). However, dogs had been described as ejaculating nearly all of the sperms < 1 min after intromission into artificial vaginas (review in Christianson).[24] If this was also the case in vivo, it would favour anti-infection explanations of the female response, and would cast severe doubts on the "double-sperm-presentation" explanation of human semen coagulation/liquefaction. On the other hand, if events in the vagina of the bitch corresponded to the "double-presentation" pattern predicted by the theory (*despite* the AV evidence) this would be good evidence for a comparable explanation of the human situation.

MATERIALS AND METHODS

Twenty-three attempts were made to sample vaginal contents during the normal copulation of various collie-cross males with greyhound bitches, mostly first but some second matings, 48h later. I am indebted to David Hancock and family, lurcher breeders, of Little Aston, near Birmingham, for their co-operation and patience. A variety of different plastic tubes, of different compliance and diameters, with or without rigid outer guides, was tried. Five of the attempts were successful, in that the catheter was not displaced by the activities of the male, and he was not put off mating by its presence. Counts were made of squames, leucocytes, sperms, and erythrocytes (flushings by 5ml of warm Earle's MEM, Flow Laboratories, *via* a Craft pattern Embryo Replacement Catheter, inserted to at least 10cm; various volumes were recovered, and it was assumed that they sampled 5ml). Twelve of the bitches were sampled prior to intromission, and three of these gave subsequent copulatory samples. The last samples were recovered immediately after the penis was withdrawn. Tubing "dead space" was about 0.5ml, and the first 0.7 ml (or more) of fluid drawn into the syringe was discarded; no sample was less than 0.5ml (additional to this). Time post-insertion was written on each glass tissue-culture tube as the sample was being retrieved. Male "lubrication", dripping from the penis just prior to insertion, was sampled on many occasions.

Assessment was by phase-contrast microscopy (x140 and x560) of 3×10 μl wet preparations, under 18 mm² coverslips. The whole preparation was scanned briefly to locate any extreme heterogeneities, then 3 LP and/or 3 HP fields (area delimited by a field-iris-level rectangular mask) were counted, and the numbers multiplied to give numbers of cells/ml. Counts were begun within an hour of recovery of the fluids, except for series 5, when about 7h elapsed (because of a car accident). The counting took about 4h for each series, and re-sampling of initial samples towards the end of this period showed some changes in cell proportions: apparent sperm numbers in samples with many leucocytes were reduced by about 20%, but very variably. The three counts from each slide were usually (> 80%) within 20%, but the small sperm numbers, particularly, differed by as much as x6. Nevertheless, the overall picture obtained agrees with the more subjective assessment of sperm distribution gained by scanning the slides with an experienced eye. Those samples, midway through the ties, when *no* sperms were found in 3 LP fields were subsequently searched very thoroughly; there were never more than 27 sperms in all three 10μl samples (1×10^3/ml would have ~30 sperms in the samples).

RESULTS

Table 1 shows the results, and Figs 1-5 show each series graphically. The numbers of squames declined in successive samples (as did erythrocytes, data not shown), but numbers of sperms and of leucocytes showed some increments, especially in the post-tie samples. Spermatozoa were always present in the penile "drips" ($1–4.5 \times 10^5$/ml). None of the series corresponded to the "classical" description of nearly-all-sperms-first, in the first minute after intromission.[24] All five showed a large number of sperms in the last (post-tie) sample, and three of the five also showed some increment in sperm number in the penultimate sample, prior to withdrawal. Four showed some increase in leucocyte numbers during the tie (series 1 bitch, which did not, had been mated 48h earlier; the others had not), and four had

Table 1: Concentrations of Squames, Leucocytes and Spermatozoa in Fluid Recovered from the Vaginas of Copulating Bitches

	PRECOP M	PRECOP (M or F)	FIRST	SECOND	THIRD	FOURTH	FIFTH	POST-COP
DOGS 1								
VOLUME,ml recovered			1.5	5.5	5	6		3
TIME,min (0=insert)			3	8	16	20		23
SQUAMES/mm3			724	495	510	425		419
LEUCS/mm3			598	342	447	331		1257
SPERMS/mm3			71	171	6	0		1752
DOGS 2		F						
VOLUME,ml		4	0.5	1.5	6	3	3	4
TIME,min		-1	1	5	12	16	21	22
SQUAMES/mm3		842	431	225	431	168	56	842
LEUCS/mm3		237	150	655	861	2628	1600	3580
SPERMS/mm3 =		0	175	44	31	6	474	4075
DOGS 3		F						
VOLUME,ml recovered		2	3.5	0.5	2.1	6	3	1.6
TIME,min (0=insert)		-3	7	11	18	22	27	32
SQUAMES/mm3		125	237	256	56	37	144	212
LEUCS/mm3		47	0	0	0	162	144	187
SPERMS/mm3		0	876	1447	1066	187	1329	14244
DOGS 4		F						
VOLUME,ml collected		1	1	1	1	1	1	1
TIME,min (0=insert)		0	1	11	15	21	26	31
SQUAMES/mm3		125	237	193	56	37	143	0
LEUCS/mm3		0	0	0	0	187	225	842
SPERMS/mm3		0	0	1447	1066	187	1329	2334
DOGS 5	M	F						
VOLUME,ml	1.5	1	4	4	2	4	2	6
TIME,min	-1	-3	0	3	7	12	18	21
SQUAMES/mm3	44	449	143	106	25	131	144	480
LEUCS/mm3	81	125	50	100	174	44	94	225
SPERMS/mm3	454	0	755	406	268	125	75	399

PRE-COP M: pre-insertion sample from the dog's penile drips;
PRE-COP F: pre-insertion vaginal sample; FIRST, etc: numbers of cells/mm^3 in samples after injection of 5ml medium, at successive times (Min post-insertion) during the tie; POST-COP: numbers of cells in a sample taken after the dog withdrew.

most leucocytes, coincident with most sperms, in the last sample. Series 5, which seemed anomalous in other respects, also had a (relatively, but not absolutely) high squame count in the last sample too (the counting of series 5 was unfortunately delayed for some 7h, because of a car accident, and the tubes were not handled as carefully as the others).

Mr Hancock drew my attention to rhythmical movements of the peri-anal muscles of the dogs towards the end of each tie, after the bitches' strong abdominal contractions at 12-15 min; he had assumed this to be the dogs' "real" ejaculation, and was accustomed to using it as a signal that breaking of the tie was imminent. Series 5 probably did show this, perhaps more briefly than the others, and the tie was broken very promptly. All five of the bitches became pregnant (but all also had another mating to the same dog).

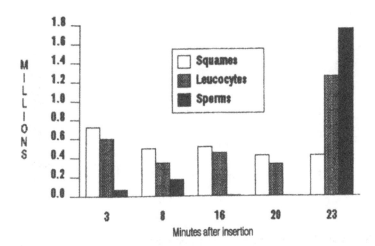

Fig. 1. This was the second mating, 48h after the first, of these two dogs. A few sperms were seen early, then *none* were found in the 16 and 20min samples; the last sample had many sperms and leucocytes. Squame numbers were approximately constant.

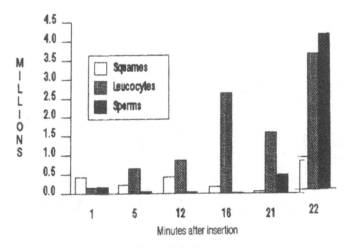

Fig. 2. This was a "first" mating that season, of a parous bitch. The vaginal milieu was sampled 1min prior to insertion of the penis. The tie was broken just prior to the last, 22min, sampling. A few spermatozoa in the first few minutes had disappeared at 16min, but more were found at 21min and a large number after the dog withdrew. Leucocytes increased dramatically during the tie.

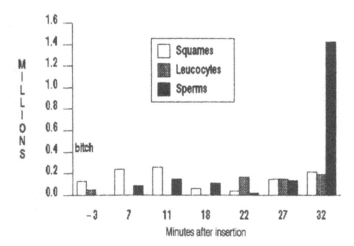

Fig. 3. This was another "first" mating. The first sample was taken at 7min into the tie, but a pre-tie sample had been taken from the bitch at -3min. A few sperms were found initially, and these declined; more were found at 27min, and many at 32min, after the dog withdrew at 31min.

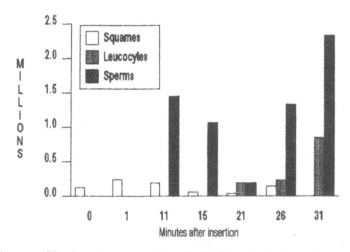

Fig. 4. This was a "first" mating too. Although the first sample was taken at intromission, when the dog produces many "drips" which contain a few sperms (see text), *no* sperms were found in this sample. Sperms found at 11min. declined in number, then more were found at the end. The last three samples showed some increase in leucocyte numbers.

DISCUSSION

All five of the successful samplings agreed that the sperms-first pattern was not shown by these copulating dogs; indeed, all showed some sperms early in the tie, and a large number later; the number dropped to almost none between these higher values. The pattern of numbers of squames gives good evidence that the sampling technique was adequate, and that the variation in sperm/leucocyte numbers did not simply reflect sampling differences. That only a small proportion of attempts was successful does not invalidate these results, because the failures were clearly irrelevant to the object of the experiment results were not discarded; there is no reason to believe that those dogs which succeeded despite the presence of the catheter were abnormal. So the successful samplings have demonstrated that the classical AV results (with most sperms ejaculated before the "leg-over" behaviour at about 1min after insertion)[24] are artefactual. All five of the series described here showed a pattern nearer to the expected "double-presentation" picture than to the AV-derived picture: a few sperms, probably from fluid like that which was sampled from the dogs before insertion, declined in numbers as the leucocyte population increased; then the dog ejaculated a large number of sperms after 20–30min; the tie was then promptly broken. Previous descriptions of dog ejaculations were a strong challenge to the double-sperm-presentation idea; however, the different picture obtained from monitoring the actual performance of copulating dogs supports the idea, and invalidates this challenge. So it is worth pursuing this interpretation of other mammalian copulatory tactics, including human semen coagulation and liquefaction.

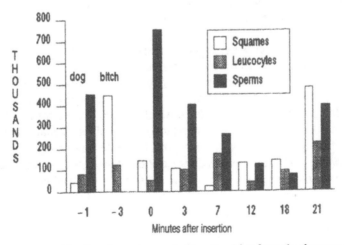

Fig 5. This, also, was a "first" mating. A sample from the drips from the dog was typical, with a few leucocytes (see text) and some squames. The counts for this series were made after a car accident, which introduced a 7h delay and some mistreatment of the tubes.

The classical view of mammalian sperm deposition tactics emphasises the problems faced by sperms in female tracts, in escaping the toxic vaginal fluids and phagocytosis whilst negotiating the several barriers deterring *any* sperm from reaching the eggs. In contrast, the view presented here does not see the female tract as inimical to *the fertilising kind* of sperm:[23] nor, in this view, are the cervical mucus utero-tubal junction and cumulus any barrier to the *right* sperms. However, in order to provide this discrimination among the sperms, which normally results in the prompt death of nearly all of them, the tract seems to need to switch into a special mode.

Such a switch, elicited by the sperms themselves, could be that normal prelude to pregnancy which is replaced in some clinical artificial reproduction programmes, resulting in improvement of pregnancy rates[25, 26] (or not).[27] However, it is certainly not mandatory; many pregnancies have been initiated without recent sperm input. It is, at first sight, much more puzzling that the complex mechanisms of sperm selection, which I suppose to be brought into play by this response, can be avoided in all IVF laboratories without apparent lack of fertility—or of ill results from using the "wrong" sperms. Good evidence that only a small proportion of sperms *can* fertilise[28] (contrary to my previous belief that all could, but were prevented from attaining the vicinity of the egg) has now been supported[19] by the elegant and surprising work of Talbot's group.[29] They showed that, although most homologous sperms fail to traverse the mammalian cumulus to the egg,[30, 31] heterologous (frog and sea-urchin) sperms *without* hyaluronidase could; even *Chlamydomona* could! So the "special" sperms are neutral objects[19] and I would suppose that, although normally discriminated by the successive mechanisms ("barriers") in the normal tract, they can be adequately filtered by the cumulus alone. However, if only a small proportion of sperms "should" (for whatever biological "reason") be permitted to fertilize, we should worry about artificial reproductive techniques (like partial zona dissection, PZD, but even removal of cumulus prior to insemination in vitro) which lessen such discrimination[32] in the interests of improving the chances of initiating a baby.

Less seriously, this new view of the female response to sperm deposition leads to suspicion of the usual sperm-cervical mucus interaction test (SCMIT), in which (clear, unreacted) human cervical mucus can be replaced by cow mucus or even chicken egg-albumen. Perhaps this test would be more informative in more physiological conditions, using the IgG-loaded, PMNL-containing fluid through which (a very few...) sperms attain the uterine "cavity"; unfortunately, the recovery and recognition of the tiny number of spermatozoa which are discriminated in this way is beyond the routine capability of most sperm laboratories. However, a method developed for isolation of the "good" sperms[32] can easily be adapted to give quantitative results; this uses lengths of polythene tubing (from Kwills).

REFERENCES

1. T. Mann, and C. Lutwak-Mann, "Male Reproductive Function and Semen," Springer-Verlag, N.Y., (1981).

2. M. Freund, Mechanisms and problems of sperm transport, *in:* "The Regulation of Mammalian Reproduction," S. J. Segal, R. Crozier, P. A. Corfman, and P G Conliffe, eds., Thomas., Springfield, Ill., 352-61, (1973).

3. N. T. Adler, On the mechanisms of sexual behaviour and their evolutionary constraints, *in:* "Biological Determinants of Sexual Behaviour," J. B.Hutchinson, ed., John Wiley and Sons, Chichester, 657-696 (1978).

4. I. J. Inkster, Sperm transport in the female rabbit (abstr). Proc. N. Z. Soc. Anim. Prod., 24:67-8, and unpublished film circa 1961 see *J.Reprod Fert.* 2:507-8 (1964).

5. J. M. Bedford, The rate of sperm passage into the cervix after coitus in the rabbit, *J. Reprod. Fert.* 25:211-8 (1971).

6. M. H. Johnson, and B. J. Everitt, "Essential Reproduction," Blackwell Scientific, Oxford (1984).

7. C. A. Fox, and B. Fox, A comparative study of coital physiology, with special reference to the sexual climax, *J. Reprod. Fert.*, 24:319-36 (1971).

8. R. S. McGee, and J. C. Herr, Human seminal vesicle-specific antigen is a substrate for prostate-specific antigen or P-30 , *Biol. Reprod.* 39:499-510 (1988).

9. H. Lilja, P. A. Abrahamsson, and A. Lundwall, Semenogelin, the predominant protein in human semen. Primary structure and identification of closely related proteins in the male accessory sex glands and on the spermatozoa, *J. Biol. Chem.*, 3 (264)1894-1900 (1989).

10. J. P. W. Vermeiden, R. E. Bernardus, C. S. ten Brug, C. H. Statema-Lohmeijer, J. Willemson, A. M. Brugma, and J. Schoemaker, Pregnancy rate is significantly higher in IVF procedure with spermatozoa isolated from non-liquefying semen in which liquefaction is induced with a-amylase, *Fertil. Steril.*, 51 149-52 (1989).

11. J. Cohen, and D. J. Werrett, Antibodies and sperm survival in the female tract of the mouse and rabbit, *J. Reprod. Fert.*, 42:301-310 (1975).

12. G. F. B. Schumacher, Soluble proteins of human cervical mucus. *In:* "Cervical Mucus in Human Reproduction", M. Elstein, K. S. Moghissi and R. Borth, eds., Scriptor, Copenhagen, 93-103 (1973).

13. A. Smallcombe, and K. R. Tyler, Semen-elicited accumulation of antibodies and leucocytes in the rabbit female tract, *Experientia*, 36:88-9 (1980).

14. J. Cohen, and K. R. Tyler, Sperm populations in the female genital tract of the rabbit, *J. Reprod. Fert.*, 60:213-221 (1980).

15. I. J. Pandya, and J. Cohen, The leucocytic reaction of the human cervix to spermatozoa, *Fertil. Steril.*, 43:417-21 (1985).

16. I. J. Pandya, Human cervix, secretions and spermatozoa, Ph.D. thesis, University of Birmingham, UK (1988).

17. G. Wagner, Vaginal transudation, *in:* "The Biology of the Fluids of the Female Genital Tract," F. K. Beller and G. F. B.Schumacher, eds., Elsevier, New York, 25-34 (1979).

18. J. W. Overstreet, Transport of gametes in the reproductive tract of the female mammal. *in:* "Mechanism and Control of Animal Fertilization," J. F. Hartmann, ed., Academic Press, New York, 499-543 (1983).

19. J. Cohen, and A. J-H., Adeghe, The other spermatozoa: fate and functions, *in:* "New Horizons in Sperm Cell Research", H. Mohri, ed., Tokyo, Japan Sci. Soc. Press., 125-134 (1986).

20. M. C. Chang, and D. Schaeffer, Number of spermatozoa ejaculated at copulation, transported into the female tract and present in the male tract of the golden hamster, *J. Hered.*, 48:197-209(1957).

21. P. E. Mattner, The distribution of spermatozoa and leucocytes in the female genital tract in goats and cattle, *J. Reprod. Fert.*, 17:253-261 (1968).

22. R. Tabor, Linnean Society meeting discussion (1985).

23. N. J. Taylor, Investigation of sperm-induced cervical leucocytosis by a double mating study in rabbits, *J. Reprod. Fert.*, 66:157-162 (1982).

24. Ib. J. Christianson, Reproduction of Dogs and Cats Bailliere- Tindall, 95 et seq.(1984).
25. B. S. Bellinge, C. Copeland, T. D. Thomas, R. E. Mazzucelli, G. O'Neill, and M. J. Cohen, Influence of insemination on implantation rate in human IVF, *Fertil., Steril.*, 46:252-6 (1986).
26. G. Marconi, L. Auge, R. Oses, R. Quintana, F. Raff, and E. Young, Does sexual intercourse improve pregnancy rates in GIFT? *Fertil. Steril.*,51:357-9 (1989).
27. S. Fishel, J. Webster, P. M. Jackson, and F. Bahman, Evaluation of high vaginal insemination at oocyte recovery in patients undergoing IVF, *Fertil. Steril.*, 51:135-8 (1989).
28. A. K. S. Siddiquey, and J. Cohen, In vitro fertilization in the mouse and the relevance of different sperm/egg concentrations and volumes, *J. Reprod. Fert.*, 66: 237-42 (1982).
29. P. Talbot, G. Carlantonio, P. Zao, J. Penkala, and L. T. Haimo, Motile cells lacking hyaluronidase can penetrate the hamster oocyte cumulus complex, *Devel. Biol.*, 108:387-398 (1985).
30. P. Talbot, Sperm penetration through oocyte investments in mammals, *Am. J. Anat.*, 174:331-46 (1983).
31. D. F. Katz, E. Z. Drobnis, and J. W. Overstreet, Factors regulating mammalian sperm migration through the female reproductive tract and oocyte vestments, *Gamete Res.*, 22:443-69 (1989).
32. T. Ahmad, J. C. Conover, M. M. Quigley, R. L. Collins, A. J. Thomas, Jnr., and R. B. L. Gwatkin, Failure of spermatozoa from T/t mice to fertilize in vitro is overcome by zona drilling. *Gamete Res.* 22 369-73 (1989).
33. J. Cohen, and S. H. Gregson, Antibodies and sperm survival in the female genital tract, *in:* "Spermatozoa, Antibodies, and Infertility," J.Cohen and W. H. Hendry, eds., Blackwell Scientific Publications, Oxford, 17-29 (1978).

THE ROLE OF IVF IN MALE INFERTILITY

Don P. Wolf

Division of Reproductive Biology and Behavior
Oregon Regional Primate Research Center, Beaverton, Oregon
and
Department of Obstetrics/Gynecology
Oregon Health Sciences University, Portland, Oregon

INTRODUCTION

In vitro fertilization was developed initially as a treatment modality for the hopelessly infertile patient with irreversibly damaged or absent fallopian tubes. With success and the development of other assisted reproductive technologies, however, indications for IVF have expanded to include endometriosis, immunologic, idiopathic, and male factor infertility. In evaluating the application of IVF to male factor infertility, attention must be drawn to definitions, for there is little consensus in the literature regarding the male factor. An abnormality in the individual parameters of the conventional semen analysis in the absence of a recognized female factor is often employed. Other definitions that have been used are based on the total number of sperm per ejaculate, the recovery of sperm in a swim-up fraction, the concentration of sperm showing various grades of motility and sperm morphology. Since it is clear that uniform and perhaps more rigorous definitions must evolve, I would like to suggest the use of the total number of morphologically normal, motile sperm per ejaculate since this fraction should contain the subpopulation of fertile cells. This approach could include a more rigorous evaluation of sperm morphology such as that described by Oehninger et al. (1988) as well as an objective method for describing motility including hyperactivation in washed populations (Robertson et al., 1988).

Advances in Assisted Reproductive Technologies, Edited by
S. Mashiach *et al.*, Plenum Press, New York, 1990

Table 1. Treatment Modalities for Male Infertility

None
Pooling ejaculates by cryopreservation
Immune suppression
Intrauterine insemination
Assisted reproductive technologies

AVAILABLE TREATMENT MODALITIES

The efficacy of IVF as a treatment modality for male factors must be considered in the context of available alternatives (Table 1). Based on several studies in the literature (Table 2), the treatment independent pregnancy rate for male factor is between 20% and 40% expressed as a 24- to 32-month cumulative conception rate. This translates into a monthly fecundity of 1%–2%. Other treatment modalities include the use of cryopreservation to increase the sperm numbers available for artificial insemination, immune suppression for antisperm antibody positive patients and intrauterine insemination. The ability to cryopreserve sperm for male factor patients, although possible in theory, has not proven efficacious. In fact, sperm from only one in four potential donors survive the conventional freezing procedures adequately to be used in sperm banking. Immune suppression with corticosteroids received attention in the 1970s as a treatment for infertile patients with high titers of antisperm antibodies. However, significant side effects and a general lack of success has since led to the abandonment of this approach. Theoretical advantages associated with the use of washed sperm in male infertility include the ability to isolate an elite population of highly motile cells for insemination, the ability to circumvent any infertility associated with sperm transport or a cervical factor, and delivery of sperm closer to the site of fertilization. A theoretical disadvantage is the poor colonization of cervical mucus that would be expected and hence the absence of a cervical reservoir of capacitating sperm. Success with intrauterine insemination (IUI) as a treatment modality for male infertility has varied over a wide range reflecting, in part, differences in patient selection and in the definition of the male factor. Conclusive support for this approach obtained by a rigorous prospective study is not yet available. A fecundity rate per treatment cycle of 7.4% was obtained in a recent collaborative European report (Sunde et al., 1988). In infertile patients with only modest abnormalities in the conventional semen analysis, several cycles of husband IUI with ovarian hyperstimulation seems justified before moving on to the more invasive and more expensive assisted reproductive technologies.

ASSISTED REPRODUCTIVE TECHNOLOGIES

Theoretical advantages associated with IVF or GIFT are related principally to the reduced concentration of sperm required to fertilize eggs. As few as 25,000 motile cells/ml can fertilize at maximum rates and since inseminations can be conducted in drops as small as 50 µl, the total number of motile cells required is small (Wolf et al., 1984). In general,

Table 2. Treatment Independent Conception Rates for Putative Male Factor Patients

Length of infertility		24- to 32-month cumulative conception rate (%)	Reference
--	<5 x 10^6 sperm/ml	23	Yates & de Kretser, 1987
--	5-20 x 10^6 sperm/ml	44	Yates & de Kretser, 1987
12-24	no female factor	42	Dunphy et al., 1989
24-48	no female factor	34	Dunphy et al., 1989
48-72	no female factor	22	Dunphy et al., 1989
72-96	no female factor	21	Dunphy et al., 1989
96-120	no female factor	23	Dunphy et al., 1989
>120	no female factor	22	Dunphy et al., 1989

despite differences in the definition of male infertility between reporting programs, fertilization rates in IVF are significantly subnormal when a male factor is present. Pregnancy rates, however, when expressed per embryo transfer are comparable to those observed with tubal infertility. As an example, Sharma et al. (1988) reported a fertilization rate of 71.3% for 1,736 cycles initiated for patients with tubal, idiopathic, or mild endometriosis as their etiology as opposed to 36.5% of 120 cycles initiated for male factor patients. The clinical pregnancy rates per cycle initiated and per embryo transfer in these two groups were 12.8% and 8.3% and 20.2% and 25.0%, respectively. This 8.3% value represents an increase in monthly fecundity of approximately fourfold over that reported above for the treatment independent rate. Awadalla et al. (1987) also concluded that fertilization levels are subnormal for male factor patients. These authors reviewed 25 male factor patients and compared fertilization outcome with matched non-male factor controls that were treated at the same time. In the control group, 70% of 62 oocytes fertilized as opposed to only 17% of 54 oocytes fertilized in the male factor group. Although many programs prefer to treat male factor patients by IVF as opposed to GIFT, results are available for the latter. Khan et al. (1988) reported a pregnancy rate of 36% per cycle initiated for idiopathic infertility versus only 17% for male factor or antisperm antibody positive patient categories.

ANTISPERM ANTIBODIES AND IVF

Although controversy exists concerning the significance of antisperm antibodies to IVF outcome, there is now little doubt that serum containing antisperm antibodies can inhibit human fertilization in vitro (Clarke et al., 1988) and there are strong suggestions that male factor patients showing high levels of bound antisperm antibodies are at risk for failed fertilization. Clarke et al., (1985) reported that control levels of fertilization (61%) were duplicated in patients with high levels of sperm-bound IgG antibody or lower titers (< 80%) of sperm with bound IgA antibodies. However, in a small group of patients with > 80% of sperm showing IgA-bound antisperm antibodies, fertilization levels were significantly reduced at 27% and no pregnancies were obtained. Furthermore, these data suggest that antisperm antibody positive patients appear with a frequency of 12%. This frequency is 3% in

Table 3. What is the Frequency of Failed IVF?

Etiology	Number	Failed fertilization (%)	Reference
All	943	8.9	Oehninger et al., 1988
All	135	8.8	Oregon Health Sciences University
All	56	17.0	Comhaire et al., 1988
No male factor	125	14.0	Englert et al., 1987
Male factor	16	37.0	Balmaceda et al., 1988
Male factor	111	18.0	Englert et al., 1987

our own program. It should be noted from these studies that, of the patients with antibodies, not all showed failed fertilization. Thus, not only do we have to diagnose patients with high titers of antisperm antibodies, but also that subset in which the antibodies will inhibit fertilization.

SPERM DYSFUNCTION OR "TRUE" MALE FACTOR

Failed IVF, provided egg quality is satisfactory, can be used as a measure of the frequency of "true" male factors, i.e., that which can be attributed to sperm dysfunction. For all patients in the Norfolk program, this value is around 9% which is similar to our own experience at Oregon Health Sciences University based upon 135 inseminations. The failure rate increases into the 20%–30% range when only male factor patients are considered, the latter defined by abnormalities in the conventional semen parameters (Table 3). This means that the false positive rate for diagnosing the male factor by semen analysis (abnormal semen parameters but fertile sperm) is very high at approximately 60% or twice the true rate! This high false positive rate undermines the use of the conventional parameters of the semen analysis as the principal diagnostic criterion for male factors.

Modalities that are suitable for the treatment of "true" male factor are few. Strategies might include endocrine therapy to increase sperm production in selected patients or exogenous treatment of available sperm. Methyl xanthines and adenosine derivatives have been used with various degrees of success for the latter purpose. At present, considerable attention has focused on the use of micromanipulation to assist fertilization in male factor cases. This can be done by directly injecting sperm under the zona pellucida into the perivitelline space or by first disrupting the zona pellucida followed by insemination. Approaches have also been described for the microinjection of sperm directly into the ooplasm.

ACKNOWLEDGMENTS

The secretarial and editorial assistance of Patsy Kimzey is acknowledged. This study, Publication No. 1645 of the Oregon Regional Primate Research Center, was supported by NIH grant RR00163.

REFERENCES

Awadalla, S. G., Friedman, C. I., Schmidt, G., Chin, N., and Kim, M. H., 1987, In vitro fertilization and embryo transfer as a treatment for male factor infertility, *Fertil.Steril.*, 47:807.

Balmaceda, J. P., Gastaldi, C., Remohi, J., Borrero, C., Ord, T., and Asch, R. H., 1988, Tubal embryo transfer as a treatment for infertility due to male factor, *Fertil.Steril.*, 50:476.

Clarke, G. N., Hyne, R. V., du Plessis, Y., and Johnston, W. I. H., 1988, Sperm antibodies and human in vitro fertilization, *Fertil. Steril.*, 49:1018.

Clarke, G. N., Lopata, A., McBain, J. C., Baker, H. W. G., and Johnston, W. I. H., 1985, Effect of sperm antibodies in males on human in vitro fertilization (IVF), *Am. J. Reprod. Immunol. Microbiol.*, 8:62.

Comhaire, F. H., Vermeulen, L., Hinting, A., and Schoonjans, F., 1988, Accuracy of sperm characteristics in predicting the in vitro fertilizing capacity of semen, *J. In Vitro Fert. Embryo Transfer*, 5:326.

Dunphy, B. C., Neal, L. M., and Cooke, I. D., 1989, The clinical value of conventional semen analysis, *Fertil. Steril.*, 51:324.

Englert, Y., Vekemans, M., Lejeune, B., Van Rysselberge, M., Puissant, F., Degueldre, M., and Leroy, F., 1987, Higher pregnancy rates after in vitro fertilization and embryo transfer in cases with sperm defects, *Fertil. Steril.*, 48:254.

Khan, I., Camus, M., Staessen, C., Wisanto, A., Devroey, P., and Van Steirteghem, A. C., 1988, Success rate in gamete intrafallopian transfer using low and high concentrations of washed spermatozoa, *Fertil. Steril.*, 50:922.

Oehninger, S., Acosta, A. A., Kruger, T., Veeck, L. L., Flood, J., and Jones, H. W., Jr., 1988, Failure of fertilization in in vitro fertilization: The "occult" male factor, *J. In Vitro Fert. Embryo Transfer*, 5:181.

Robertson, L., Wolf, D. P., and Tash, J. S., 1988, Temporal changes in motility parameters related to acrosomal status: Identification and characterization of populations of hyperactivated human sperm, *Biol. Reprod.*, 39:797.

Sharma, V., Riddle, A., Mason, B. A., Pampiglione, J., and Campbell, S., 1988, An analysis of factors influencing the establishment of a clinical pregnancy in an ultrasound-based ambulatory in vitro fertilization program, *Fertil. Steril.*, 49:468.

Sunde, A., Kahn, J. A., and Molne, K., 1988, Intrauterine insemination: a European collaborative report, *Hum. Reprod.*, 3(Suppl. 2):69.

Wolf, D. P., Byrd, W., Dandekar, P., and Quigley, M. M., 1984, Sperm concentration and the fertilization of human eggs in vitro, *Biol. Reprod.*, 31:837.

Yates, C. A., and de Kretser, D. M., 1987, Male-factor infertility and in vitro fertilization, *J. In Vitro Fert. Embryo Transfer*, 4:141.

IVF AND IMMUNOLOGICAL INFERTILITY

Norbert Gleicher

Department of Obstetrics and Gynecology, Mount Sinai Hospital Medical Center and Rush Medical College, Chicago, Illinois

INTRODUCTION

Does immunology come into play when in vitro fertilization (IVF) and other forms of assisted reproduction are considered? This is the basic question addressed by this review. And the answer has to be affirmative.

The fetus is an auto- as well as an allograft with half of its genome representing maternal "self" antigens and half paternal "allo" antigens. Reproductive immunologists have been primarily concerned with the paternal half because, based on immunologic transplantation principles, allogeneic tissue undergoes rejection by an immunologically competent recipient. The fetal allograft, however, does not. Reproductive immunology was less concerned with the fact that the fetal autograft may also elicit immune responses, similar to autoantigenic stimulation in other clinical situations. The interest in such autoimmune responses has, however, dramatically increased in recent years as it was recognized that autoimmune responses may interfere with normal reproduction.[1]

Reproductive immunology, as a consequence, encompasses today the investigation of the fetus as both an auto- and an allograft. Both of these characteristics of the products of conception appear to represent essential components for a normal reproductive process while, at the same time, both can be cause for reproductive failure if inappropriately stimulated.

Advances in Assisted Reproductive Technologies, Edited by
S. Mashiach *et al.*, Plenum Press, New York, 1990

459

The utilization of donor semen in human infertility established long ago that paternal antigen specificity was not required in order to achieve successful pregnancy. While some rather limited studies have suggested that seminal "desensitization" may be important for the achievement of normal pregnancy[2] this issue has always remained controversial and lacks experimental support. Experience with donor semen would, in fact, suggest that, if paternal antigens carry any level of responsibility for the success of pregnancy, prior exposure to those antigens is unnecessary. Whether the presence of the maternal "self" genome was equally unnecessary for successful pregnancy, was for a long time unknown in the human. Animal experience suggested that embryos could be successfully transferred within a species from mother to mother.[3] In the human, this experiment became possible only once IVF had been established as a clinical tool. Today we know that successful IVF can be achieved with both donor oocytes and donor sperm, resulting in a genetically defined *complete* allograft in contrast to the usual *semi*-allograft.[4] Whether reproductive efficiency varies in complete allografts from that in semi-allografts has not been established in the human. Pregnancies, appear, however, to progress normally, though numbers have remained too small to permit final conclusions.

The Fetus as an Allograft

A considerable volume of animal data has accumulated which suggests that the "allo" component of the fetus is essential to the establishment of normal pregnancy. Existing work by Clark,[5] Chaoat[6] and other investigators supports the concept that the allogeneic stimulation of the maternal immune system is required in order to achieve activation of a variety of immunological mechanisms which, in turn, allow for implantation to occur. Most of these mechanisms appear concentrated at the local uterine level and especially within the decidua.[3,5] Failure of appropriate allogenic stimulation results in failure of development of those local mechanisms and subsequent pregnancy wastage.[3]

The translation of this animal experience has led to some rather controversial human experimentations, in which lack of assumed allogeneic stimulation of the mother is taken as the reason for stimulation protocols, involving a variety of stimulating antigens.[7-10] Whether any of these protocols (often called immunization protocols), in fact improve upon outcome in comparison to untreated control couples has remained controversial. A number of controlled studies are underway and should resolve this issue. Successful attempts to improve IVF outcome through such stimulation protocols have not been reported so far.

Even if presently ongoing immunization studies suggest a benefit from such treatment, the best suited stimulant remains to be established. So far, husband cells,[7] donor (or pooled) leukocytes,[8] semen preparations[9] and trophoblast membrane preparations[10] have been suggested as antigenic stimulants. One wonders whether a more nonspecific stimulation of the female's immune system may not achieve the same effect at less risk and less cost.

The Fetus as an Autograft

In parallel to alloantigen, autoantigen has to enter the maternal circulation. Though clearly "auto" in its maternal genetic origin it represents a "foreign" antigen in the strictest sense of the definition since it is not derived from the mother but its parasitic fetus.[11]

Previously noted human IVF experience demonstrates that this foreign (auto) antigenic component does not appear essential to normal pregnancy since human pregnancies with donor oocytes have been successful.[4] A specific function for a possible auto-antibody response in the normal female is still tempting to postulate. Normal females have significantly higher normal autoantibody levels than normal males.[12] This is what probably predisposes female to the significantly increased risk towards autoimmune diseases in comparison to males. One can speculate that this predisposition would teleologically not have survived if it was not an essential component of an important biological process. We have suggested that this process is reproduction.[1]

If increased autoantibody levels are required in the female to neutralize the unusually large exposure to fetally derived "self" antigen, then one can postulate that if such exposure does not occur, an increased level of autoantibodies will remain in the maternal circulation. Alternatively, further antigenic stimulation would further increase autoantibody levels. Under both circumstances, unusually high levels of autoantibodies would be available in the maternal circulation.

Considerable evidence has accumulated which suggests that abnormally high levels of autoantibodies are associated with both infertility and pregnancy wastage.[11, 13] It is unclear at this point what the various autoantibody specificities are. It appears, however, that a preponderance lies with anti-phospholipid antibodies, especially of IgG isotype.

Anti-phospholipid antibodies undergo concentration in follicular fluid.[12] Since no such concentration occurs with other autoantibody groupings, one has to wonder whether this concentration of anti-phospholipid antibodies does not reflect local ovarian production. Such a hypothesis is further strengthened by the reported presence of anti-sperm antibodies in follicular fluid,[14] which generally correlates very closely with the presence of anti-phospholipid antibodies.[15] Moreover, Alexander et al.[16] recently correlated oocyte quality to the presence and distribution of T lymphocytes in follicular fluid.

In the IVF process, the presence of anti-phospholipid and anti-sperm antibodies has been correlated with adverse pregnancy outcome.[13, 15, 17] Preliminary and so far nonrandomized experience in our own and in another IVF program (K. Honea, Birmingham, Alabama, personal communications) suggests that the suppression of anti-phospholipid antibodies prior to an IVF attempt improves outcome. Kemeter and Feichtinger[18] reported in a randomized study that patients who were treated with corticosteroids (to suppress adrenal hormone production) experienced significantly improved pregnancy rates. It is tempting to speculate that at least part of the beneficial effect of corticosteroid therapy in these patients was due to the immuno-suppressive effect of the drug upon autoantibody levels.

Abnormal anti-phospholid antibodies, which appear to interfere with the IVF process occur in up to two-thirds of patients with endometriosis,[19] in a similar proportion of patients with unexplained infertility[11] and in close to 100 percent of patients with agglutinating sperm antibodies.[15] We, consequently, recommend the investigation of anti-phospholipid antibodies in patients with the above listed diagnoses before any IVF/GIFT attempt. Recent data from a double blinded multicenter study suggest that such autoantibody abnormalities need to be suppressed before normal conception can occur.[20] If such suppression has to take place during a stimulation cycle for IVF, then low dose corticosteroid therapy appears, in our experience, successful. Long-term therapy, which is especially indicated when autoantibody abnormalities are very profound, should be attempted with danazol.[20]

The Immuno-Endocrine System

Nobody would raise the question whether endocrinology is relevant to successful conception. Obviously it is! Nevertheless, the relevance of immunology is still frequently questioned despite the fact that increasing evidence suggests that endocrine and immune systems act in such close collaboration that they have to be considered one and the same.

Little doubt exists that implantation is as much an immunological as an endocrinological event. Specific immuno-active cell population are required for successful implantation.[4-6] These cells are apparently under endocrinological control.[4] Growth factors are increasingly recognized as the link between the traditional endocrine and immune systems. The investigation of growth factors is uniting endocrinologists and immunologists in a common quest. As so may times before, the investigation of the most basic reproductive processes promises to enhance our understanding of some very basic biological facts, with applications that reach far and beyond the field of reproductive research.

REFERENCES

1. N. Gleicher, and A. El-Roeiy, The reproductive autoimmune failure syndrome (RAFS), *Am. J. Obstet. Gynecol.*, 159:223-227 (1988).
2. J. J. Marti, and V. Herman, Immunogestosis: A new etiologic concept of "essential" EPH-gestosis, with special consideration of the primigravid patient, *Am. J. Obstet. Gynecol.*, 128:489-493 (1977).
3. G. Chaouat, The fetal allograft: Animal models in reproductive immunology, *in:* "Reproductive Immunology, Immunology Allergy Clin. North America," W. B. Saunders Co. in press.
4. P. Lutjen, A. Trounson, J. Leeton, J. Findlay, C. Wood and P. Renov, The establishment and maintenance of pregnancy using in vitro fertilization and embryo donation in a patient with primary ovarian failure, *Nature*, 307:174-176 (1984).

5. D. E. Clark, R. M. Slapsys, and B. A. Croy, Immunoregulation of host versus graft responses in the uterus, *Immunol. Today*, 5:11-18 (1984).

6. G. Chaouat, D. A. Clark, and T. G. Wegman, Genetic aspects of the CBA/JXDBA/2J and B10xB10, A model for murine spontaneous abortions and prevention by leukocyte immunization, 18th R.C.O.G. study groups. Early pregnancy loss. Mechanisms and treatment. R.C.O.G. W. R. Allen, D. A. Clark, T. J. Gill III, J. F. Mowbray, and W. R. Robertson, eds., R.C.O.G. Press, 89 (1988).

7. J. Mowbray, C. Cribbings, and H. Liddell, Controlled trial of treatment of recurrent spontaneous abortion by immunization with paternal cells, *Lancet*, i:941-943 (1985).

8. A. E. Beer, and J. F. Quebbetian, ZHU, nonpaternal leukocyte immunization in women previously immunized with paternal leukocytes. Immune response and subsequent pregnancy outcome, *in:* "Reproductive Immunology," D. A. Clark, and B. A. Croy, eds., Elsevier Science Publishers, 261 (1986).

9. C. J. Thaler, and J. A. McIntyre, Fetal wastage and nonrecognition in human pregnancy, *in:* "Reproductive Immunology, Immunology Allergy Clin. North America," W. B. Saunders Co., in press.

10. P. M. Johnson, K. V. Chia, and C. A. Hart, Trophoblast membrane infusion for unexplained recurrent miscarriage, *Brit. J. Gynaecol. Obstet.* 95:341-346 (1988).

11. N. Gleicher, A. El-Roeiy, E. Confino, and J. Friberg, Reproductive failure due to autoantibodies: unexplained infertility and pregnancy wastage, *Am. J. Obstet. Gynecol.*, in press.

12. N. Gleicher, A. El-Roeiy, Autoantibody profiles in normal females and males, Submitted for publication.

13. A. El-Roeiy, N. Gleicher, J. Friberg, E. Confino, A. B. Dudkiewicz, Correlation between peripheral blood and follicular fluid autoantibodies and impact on in vitro fertilization, *Obstet. Gynecol.*, 70:163-170 (1987).

14. G. N. Clarke, C. Hsieh, S. H. Hoh, M. N. Cauclu, Sperm antibodies, immunoglobulins and complement in human follicular fluid, *Am. J. Reprod. Immunol.*, 5:179-181 (1984).

15. A. El-Roeiy, G. Valessini, J. Friberg, Y. Shoenfeld, R. C. Kennedy, A. Tincani, G. Balestriere, and N. Gleicher, Autoantibodies and common idiotypes in sperm antibody positive females and males, *Am. J. Obstet. Gynecol.*, 159:370-375 (1988).

16. K. Droesch, D. L. Fulgham, H. C. Liv, Z. Rosenwaks, and N. J. Alexander, Distribution of T cell subsets in follicular fluid, *Fertil. Steril.* 50:618-621 (1988).

17. J. L. Yovich, J. D. Stanger, and D. Kay, In vitro fertilization of oocytes from women with serum antisperm antibodies, *Lancet*, i:369-372 (1984).

18. P. Kemeter, and W. Feichtinger, Prednisolene supplementation to clomid and/or gonadotropin stimulation for in vitro fertilization - a prospective randomized trial, *Hum. Reprod.* 1:441-444 (1986).

19. N. Gleicher, A. El-Roeiy, E. Confino, and J. Friberg, Is endometriosis an autoimmune disease? *Obstet. Gynecol.* 70:115-122 (1987).

20. A. El-Roeiy, P. Dmowski, N. Gleicher, E. Radwanska, L. Harlow, Z. Binor, I. Tummon, and R. G. Rawlins, Danazol but not gonadotropin-releasing hormone agonists suppresses autoantibodies in endometriosis, *Fertil. Steril.*, 50:864-871 (1988).

MICROSURGERY, ANDROLOGY, AND ITS ROLE IN IVF

Sherman J. Silber

St. Luke's Hospital
224 South Woods Mill Road, Suite 730
St. Louis, Missouri, 63017

I. PREGNANCY AFTER VASOVASOSTOMY FOR VASECTOMY REVERSAL: FACTORS AFFECTING LONG TERM RETURN OF FERTILITY IN 282 PATIENTS FOLLOWED FOR TEN YEARS

INTRODUCTION

Vasectomy is the most popular method of birth control in the world today.[22] More than a half million vasectomies are performed in the United States each year. Because of fear of child death in the developing world, changing views about family life in the western world, and the increasing prevalence of divorce and remarriage, there is now a large number of men requesting reversal of vasectomy. For many years the pregnancy rate after surgical reanastomosis of the vas had been very low, and a variety of explanations had been offered for the relatively poor success in reversing vasectomy.[9, 10, 23, 25, 31] With the advent of microsurgical techniques pregnancy rates improved considerably, suggesting that purely micro-mechanical factors were associated with the low success rates, but long-term follow-up on large numbers of patients were not available and the matter remains somewhat controversial.[36, 38] Theories for the consistently poor results with vasectomy reversal have included development of sperm antibodies, damage to the deferential nerve, and testicular damage.[4, 8, 12, 19, 21, 24, 28, 50] Yet some investigators questioned any correlation between sperm antibodies in the serum and subsequent fertility after vasovasostomy,[52] and the effect, if any, of vasectomy on the testis in humans and animals has also been very controversial.[40] Segregating the various studies by species has not cleared up the confusion. If any testicular damage occurs, the generally agreed upon mechanisms would be either autoimmune, or pressure related.[1, 6, 7, 18, 30, 40]

Advances in Assisted Reproductive Technologies, Edited by
S. Mashiach *et al.*, Plenum Press, New York, 1990

Table 1. Overall Long-term Pregnancy Rates in Patients undergoing Vasovasostomy. Ten Years Follow-up (Sperm seen in Vas Fluid)

	Combined 1975 and 1976-1977 Series	Original 1975 Series
Total Patients	282 (100%)	42 (100%)
Total Pregnant	228 (81%)	32 (76%)
Azospermic	24 (9%)	5 (12%)

The pressure increase subsequent to vasectomy has been well established, as well as the effect of this pressure on epididymal dilatation, perforation and sperm inspissation in the epididymis, causing secondary epididymal obstruction.[38, 41, 42] We found that the incidence of this confounding secondary epididymal blockage increased with the duration of time since vasectomy, and never occurred if there was a sperm granuloma at the vasectomy site.[36, 37, 40] Despite the dismal finding of no sperm in the vas fluid in patients with secondary epididymal blockage, the testicle biopsy always appeared normal.[41, 47] This apparent effect of pressure increase after vasectomy led to a suggestion that the testicular end of the vas not be sealed at the time of vasectomy, so as to lessen the pressure build-up, and possibly increase the ease of reversibility (notwithstanding the potentially damaging immunological consequences.[2, 34, 37] We wished to determine with the present study what the fertility rate would be for this favorable group of patients who had no epididymal damage as evidenced by sperm being present in the vas fluid.

We have carefully studied for nine to ten years a large group of patients who have undergone microsurgical vasovasostomy with no evidence of pressure induced secondary epididymal blockage. We attempted to relate in these patients presence or absence of varicocoele, post-operative semen analyses, pre-operative serum sperm antibody titers and quantitative evaluation of testicular biopsy to the chance for pregnancy. In this study we are reviewing the results in patients who were thought to have no epididymal blockage. Patients with no sperm in the vas fluid, all of whom exhibited secondary epididymal obstruction, will be the subject of a subsequent paper.

CONCLUSIONS

We wished to determine the eventual fertility of those vasectomy reversal patients who have no pressure induced secondary epididymal blockage. These patients underwent simple vasovasostomy because at the time of the reversal surgery, there were sperm present in

Table 2. Pregnancy Rate according to Distribution of Motile Sperm Count in Men with Sperm Patency following Vasovasostomy (Ten Year Follow-up)

Total Motile Sperm Count (per ejaculate)	Total Patients	Pregnant	Not Pregnant
0-10,000,000	32 (12%)	25 (78%)	7
10-20,000,000	31 (12%)	27 (87%)	4
20-40,000,000	32 (12%)	30 (93%)	2
40-80,000,000	79 (31%)	68 (86%)	11
80,000,000	84 (33%)	78 (92%)	6
TOTALS	258 (100%)	228 (88%)	30

Table 3. Pregnancy Rate according to % Sperm Motility in Men with Sperm Patency following Vasovasostomy (Ten Year Follow-up)

Motility	Total Patients	Pregnant	Not Pregnant
0-20	24	18 (75%)	6
20-40	70	66 (94%)	4
40-60	82	71 (86%)	11
60-80	62	55 (88%)	7
80	20	18 (90%)	2
TOTALS	258 (100%)	228 (88%)	30 (100%)

large numbers in the vas fluid. If there were no sperm in the vas fluid, the patients were excluded from the vasovasostomy series, and instead underwent vasoepididymostomy (see next section). We were able to obtain long-term follow up on 282 patients with sperm in the vas fluid who underwent vasectomy reversal eight to ten years ago. These patients were studied for pregnancy rate, post-operative semen parameters, duration of time since vasectomy, pre-operative serum antisperm antibody titers, the influence of varicocoele, and quantitative evaluation of testicle biopsy. All of the 44 patients with no sperm in the vas fluid who underwent vasovasostomy ten years ago remained azoospermic. Of the 282 patients with sperm in the vas fluid 228 (81%) eventually impregnated their wives. Twenty-four patients with sperm in the vas fluid (9%) were azoospermic and did not impregnate their wives. Of the 258 patients who had sperm patency, the pregnancy rate was 88% (Tables 1-5).

The number of mature spermatids per tubule in the testis correlated closely with the post-operative sperm count in patent cases. Quantitative evaluation of the testicle biopsy revealed normal spermatogenesis even in patients with azoospermia or severe oligospermia post-operatively. Failures were thus found to be due to blockage, either at the vasovasostomy site, or epididymal blockage unrecognized at the time of vasovasostomy. Sperm count

Table 4. Lack of Effect of Varicocoele (not operated on) on Pregnancy Rate following Vasovasostomy

	Number Patients	Patients With Varicocoele	Patients Without Varicocoele
Pregnant	228 (80.9%)	33 (78.5%)	195 (81.2%)
Not Pregnant	54 (19.1%)	9 (21.4%)	45 (18.8%)
TOTALS	282 (100%)	42 (14.8%)	240 (85.2%)

Table 5. Relationship of Serum Sperm Antibody Titers to Pregnancy Rate following Vasovasostomy

	Total Studied	Immobilizing Titre (Isojima) 2	10	Agglutinating Titre (Kibrick) 0	20
Husband Not Azospermic:					
Wife Pregnant	75	29 (39%)	18 (24%)	42 (56%)	30 (40%)
Wife Not Pregnant	11	1 (36%)	2 (16%)	6 (54%)	6 (54%)
Husband Azospermic	12	5 (42%)	3 (25%)	7 (58%)	5 (42%)
Entire Group Studied	98	38 (39%)	23 (24%)	56 (57%)	41 (42%)

had a minimal impact on the likelihood of pregnancy so long as there was patency, and there was no discrepancy between sperm count and actual testicular sperm production as determined by testicle biopsy.[47] Pregnancy was not related to presence or absence of a varicocoele, pre-operative serum sperm antibody levels, or testicle biopsy findings.

II. RESULTS OF MICROSURGICAL VASOEPIDIDYMOSTOMY: ROLE OF EPIDIDYMIS IN SPERM MATURATION

One hundred and ninety early patients with obstructive azoospermia caused by bilateral epididymal blockage have been followed for six years or longer after undergoing "specific tubule" vasoepididymostomy. At that time, we always attempted to perform the epididymal anastomosis as distally as possible so as to allow the greatest amount of epididymal length for sperm maturation. Thus the cases of vasoepididymostomy to the caput were more se-

Table 6. Corpus Epididymis: Lack of Relation of Post-op Spermatozoa Count to Pregnancy Rate

Spermatozoa Count (per cc)	Pregnant	Not Pregnant
Azospermic	0 (0%)	30
0 to 1 x 10^6	2 (67%)	1
1 to 5 x 10^6	5 (63%)	3
5 to 10 x 10^6	11 (65%)	6
10 to 20 x 10^6	6 (50%)	6
20 to 40 x 10^6	17 (81%)	4
40 x 10^6	32 (74%)	11
Unknown	5	0
	78 (56%)	61

Summary Footnote:

Patency Rate	78%
Overall Pregnancy Rate	56%
Pregnancy Rate in "Patent" Cases 72%	

verely diseased and had a greater number of blockages than the more common case of vasoepididymostomy to the corpus (Tables 6-15).

When anastomosis was performed at the corpus epididymidis, the "patency" rate was 78%, and the overall pregnancy rate was 56%. The pregnancy rate for "patent" cases was 72%, indicating that a high fertility rate can be obtained with spermatozoa that have not transited the full length of corpus epididymidis. By contrast, with vasoepididymostomy to the caput epididymidis there was a 73% "patency" rate, but the overall pregnancy rate was only 31%. The pregnancy rate for "patent" cases was 43%. Spermatozoa from the corpus epididymidis had a higher rate of fertility than spermatozoa from the caput epididymidis, but spermatozoa from proximal areas of the corpus have no less fertility than spermatozoa from the distal corpus epididymidis. The most remarkable observation was that in almost half the cases of caput anastomosis, spermatozoa which have never journeyed beyond the caput epididymidis were capable of causing pregnancy.

We now routinely perform all anastomoses at the caput for reasons I will now explain. We do not yet have data on vasoepididymostomy performed routinely at the caput epididymis for *all* cases of epididymal obstruction, but we suspect there will be no difference in the fertilizing potential of sperm from anywhere along the epididymal tubule, if all other factors are equal. By performing all anastomoses at the caput we anticipate remarkable im-

Table 7. Corpus Epididymis: Relation of % Directional Spermatozoa Motility to Pregnancy Rate in "Patent" cases

% Directional Motility	Pregnant	Not Pregnant
0 - 19	13 (48%)	14
20 - 39	18 (75%)	6
40 - 60	19 (76%)	6
60	22 (81%)	5

Table 8. Corpus Epididymis: Lack of Relation of Age of Wife to Pregnancy Rate in "Patent" Cases

Age Of Wife	Pregnant	Not Pregnant
25	14 (78%)	4
25 - 29	30 (68%)	14
30 - 35	30 (75%)	10
35	4 (67%)	2

provement in success rate. This runs counter to current dogma, and requires some explanation, as follows.

Epididymal Physiology: Is the Epididymis Just a Long Stupid Tube?

Because of advances in microsurgical techniques, it is now possible to bypass most cases of epididymal obstruction with a high incidence of technical success.[38, 42, 43] The fertilizing capacity of spermatozoa which have not traversed all sections of the epididymis can ideally be studied with this human clinical model. In every animal that has been studied, spermatozoa from the caput epididymidis are only capable of weak circular motion at most, and are not able to fertilize.[27] Spermatozoa from the corpus epididymidis can occasionally fertilize but the pregnancy rate is low.

But few of these previous animal studies allowed the spermatozoa time to mature and thereby possibly develop fertilizing capacity. Spermatozoa were simply aspirated from specific regions of the epididymis, and then promptly inseminated. In most studies where the epididymis was ligated to determine if time alone could allow spermatozoa maturation, the obstructed environment was so pathological that no firm conclusions could be reached.[13, 15, 29]

Table 9. Corpus Epididymis: Lack of Reltion of level of Corpus Epididymal Anastomosis to Pregnancy Rate in "Patent" Cases

	Pregnant	Not Pregnant
Proximal Corpus	7 (88%)	1
Mid Corpus	17 (74%)	6
Distal Corpus	54 (71%)	22

Table 10. Corpus Epididymis: Percent Pregnant at Varying Intervals Post-op in Relation to Sperm Count in "Patent" Cases

Sperm Count $(\times 10^6$ per cc)	6 Months	12 Months	18 Months	24 Months	24 Months
1 - 5	56%	22%	22%		
5 - 10	63%	25%		13%	
10 - 20	17%	49%		17%	17%
20 - 40	44%	28%	11%	11%	6%
40	34%	28%	6%	9%	22%
100%	41%	29%	8%	10%	12%

In 1969, Orgebin-Crist[27] pointed out that we still did not know with certainty from any of these animal studies whether the factors governing the maturation process of spermatozoa are intrinsic to the spermatozoa themselves and just require time, or whether spermatozoa must transit through most of the epididymis in order to mature. It was entirely possible that aging alone might mature the spermatozoa, and that spermatozoa might not need to pass through all of the epididymis in order to develop the capacity to fertilize. Yet because of the animal studies alluded to, and poor results in humans using non-microsurgical techniques, it has always been assumed that epididymal blockage carries a poor prognosis.[3, 16, 17, 32]

As far back as 1931, however, Young's[53] experiments in guinea pigs with ligation at various levels of the epididymis indicated to the contrary: "that the time consumed by spermatozoa in passing through the epididymis is necessary for a completion of their development, that the changes undergone during this period represent a continuation of changes which start while the spermatozoa are still attached to the germinal epithelium, and are not conditioned by some specific epididymal secretion. In fact he observed the same "inversion" of regions of sperm motility and non-motility in the obstructed epididymis that we have

Table 11. Corpus Epididymis: Lack of Relation of Sperm Count to Mean Time till Pregnancy in "Patent" Cases

Sperm Count $X10^6$ per cc	Mean Time Till Pregnancy
1 - 5	6.7 months
5 - 10	6.0 months
10 - 20	10.5 months
20 - 40	4.3 months
40	6.4 months

Table 12. Head of Epididymis: Relation of Pregnancy to Sperm Motility in 37 Cases with Patency

% Sperm Motility	Pregnancy Rate	Pregnant	Not-Pregnant
0 - 20%	15%	2 (13%)	11 (52%)
20%	58%	14 (87%)	10 (48%)
TOTALS	43%	16 (100%)	21 (100%)

Table 13. Head of Epididymis: Relation of Age of Wife to Pregnancy Rate in Cases with Patency

Age of Wife	Pregnancy Rate	Pregnant	Not-Pregnant
25 - 30	67%	12 (75%)	6 (29%)
30	21%	4 (25%)	15 (71%)
TOTALS	43%	16 (100%)	21 (100%)

Table 14. Head of Epididymis: Lack of Relation of Level of Successful Vasoepididymostomy to Previous Failure

	PREGNANT	NOT PREGNANT
Redo After Previous Failure	6 (38%)	10
Virgin Case	10 (29%)	25

Table 15. Head of Epididymis: Relation of Post-op "Patency" to Area of Anastomosis at Head of Epididymis

	PROXIMAL CAPUT	MID CAPUT OR MIXED	DISTAL CAPUT
"Patent"	9 (53%)	10 (71%)	17 (89%)
Non-Patent (Azoospermia)	8 (47%)	4 (29%)	2 (11%)

noted in clinical obstructive azoospermia. The more distal regions have the poorest motility and the more proximal regions have the best motility. Young concluded that in an obstructed epididymis the more distal sperm are senescent, while the more proximal sperm have had time to mature despite having not traversed the epididymis.

Our clinical experience with specific tubule vasoepididymostomy supports Young's original thesis.

Surgical Technique

All vasoepididymostomies were performed with the "specific tubule" technique we have described, which involves an end-to-end anastomosis of the inner lumen of the vas to the epididymal tubule, mucosa-to-mucosa in a leakproof fashion.[39, 42, 43] Virtually all of the earlier literature on vasoepididymostomy involves a longitudinal cut through the epididymal tunic and into the epididymal tubule which resulted in a random cutting of the epididymal tubule in many of its convolutions, which gives the appearance of many tubules leaking spermatozoa. The vas is sutured to that outer epididymal tunic hoping that a fistula would form. Because of the high rate of technical failure with that methodology, reliable data on the fertility of spermatozoa from the epididymis in the past has been difficult to obtain.

With the "specific tubule" technique used in this series, the epididymis is transected proximally until the point is reached where many spermatozoa are found (Fig. 1). Fluid at every level is examined under a phase contrast microscope in the operating room for the presence of and quality of spermatozoa. The anastomosis of the vas to the epididymis is performed at the transition point from no spermatozoa to the point where there is an abundant amount of spermatozoa in the fluid coming from the epididymal tubule (Fig. 2).

The fact that when a technically successful anastomosis to anywhere along the corpus epididymis is achieved, almost 72% of the wives get pregnant, with a mean time to conception of six months, clarifies the issue that spermatozoa do not necessarily have to traverse the entire corpus or cauda epididymidis in the human to achieve fertilizing capacity.

The lower pregnancy rates in previous clinical series most probably relate to a number of factors. We performed a specific tubule anastomosis rather than create a fistula which could lead to lower "patency" rates, and even poorer spermatozoa motility in the cases that are "patent".[3, 32] Newer microsurgical techniques have thus clearly improved the quality of spermatozoa in the ejaculate post-operatively.

It is fascinating that the numerical spermatozoa count had no impact on pregnancy rate, but spermatozoa motility did. This goes along with many clinical studies which demonstrate low spermatozoa counts in a high percentage of normal fertile males.[35, 46, 49, 54]

If the oligospermia is caused by partial obstruction (or epididymal dilatation), poor motility would result, and then fertility may be compromised. But if a patient's oligospermia is simply a reflection of his low testicular sperm production, fertility may not be poor.

In 1969, Marie Claire Orgebin-Crist[27] asked whether factors governing the maturation process of spermatozoa are intrinsic to the spermatozoa, or whether they reside in the epididymis. Epididymal ligation experiments have not always been clear in answering this question because they cause dilatation and epithelial disruption which negatively affects the motility of spermatozoa so retained.[13, 14, 15, 29, 53] Yet Young was able to draw a tentative conclusion in 1931 that indeed sperm maturation may be completely independent of epididymal transport. Others have made similar speculations regarding the corpus and cauda epididymidis.[5, 33]

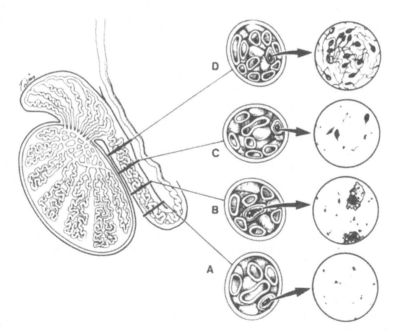

Fig. 1. Serial sectioning of epididymis until the site of proximal-most obstruction is bypassed.

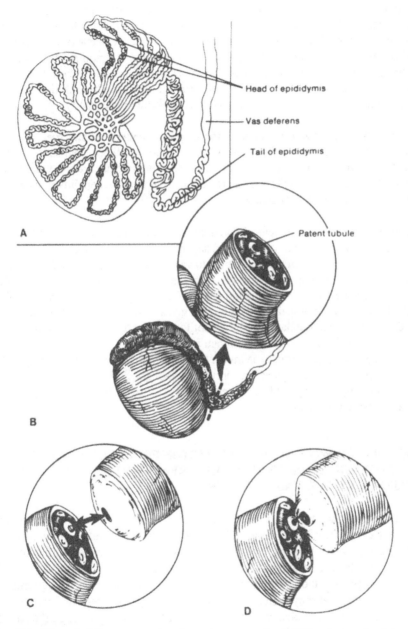

Fig. 2. "Specific tubule" anastomosis of vas lumen to the epididymis proximal to site of obstruction.

A vasoepididymostomy model such as ours, in which spermatozoa cannot traverse the full length of epididymis, would allow maturation to occur with time only in the vas deferens, and help clarify this issue. The fact that spermatozoa which could not have travelled through any portion of the corpus or cauda epididymidis were capable of fertilization, indicates that a full journey through the epididymis is certainly not required for maturation of spermatozoa sufficient to allow pregnancy. The fact that pregnancy occurred in almost half of the patent cases to the caput indicates that transit beyond the head of the epididymis is not an absolute requirement for spermatozoa to attain fertilizing capacity.

It should be emphasized that none of these patients underwent any special treatments such as in vitro fertilization or GIFT, and these pregnancies all occurred simply with natural intercourse. In the next several years we may find out whether with in vitro techniques more than 43% of these patients with spermatozoa from the caput epididymidis will or will not be able to accomplish fertilization.

Recent clinical cases have demonstrated that it is even possible in some circumstances for spermatozoa which have never traversed any length of epididymis to fertilize the human egg. In two cases reported of vasa efferentia to vas deferens anastomosis, the post-opertive ejaculate contained normally motile sperm, and the wives became pregnant.[44] In addition, pregnancy from aspiration of epididymal spermatozoa combined with in vitro fertilization and "ZIFT" in cases of unrepairable obstruction gives further evidence that transit through the epididymis is not a mandatory requirement for fertilization.[45,48] Finally, newer studies of epididymal sperm transport in the human indicate that the human epididymis is not a storage area, and indeed spermatozoa transit the entire human epididymis very quickly, in a mere two days; not eleven days as was previously thought.[20] Thus it is possible that in the human, the epididymis may not be as essential to spermatozoa development and fertility as it appears to be in most animals.

III. PREGNANCY WITH SPERM ASPIRATION FROM THE PROXIMAL HEAD OF THE EPIDIDYMIS: A NEW TREATMENT FOR CONGENITAL ABSENCE OF THE VAS DEFERENS

It has long been assumed that sperm must pass through a certain length of epididymis to mature, gain progressive motility, and become capable of fertilization. In every animal thus far studied, sperm from the non-obstructed proximal head of the epididymis exhibit only weak, circular swimming motions, and are incapable of progressive motility or fertilization of the egg. It is thought that only when sperm have traversed through most of the corpus epididymis that they mature sufficiently to become progressively motile, and are able to fertilize.[27] But our observations suggest that their journey through the epididymis may not be an absolute requirement and that sperm may only require a period of time to mature after leaving the testicle.

Congenital absence of the vas deferens accounts for 11% to 50% of cases of obstructive azoospermia, and heretofore has been considered basically untreatable.[11] This is a large and frustrating group of patients who have been shown on countless testicle biopsies to have normal spermatogenesis, and are theoretically making sperm quite capable of fertilizing an egg. Yet treatment up until the present time has been very dismal.[51]

Dr. Ricardo Asch and his team, along with our team have collaborated equally to develop a treatment protocol involving microsurgical aspiration of sperm from the proximal region of the epididymis, combined with in vitro fertilization (IVF) and zygote intrafallopian transfer (ZIFT), which now offers very good results in this previously frustrating group of couples.[35, 45]

We now have a method for microsurgical sperm aspiration from the proximal-most region of the head of the epididymis, combined with IVF, with the first documentation of fertilization and pregnancy utilizing this approach for the treatment of congenital absence of the vas deferens.

Induction of Follicular Development and Oocyte Retrieval

The female partners of men with azoospermia caused by congenital absence of the vas underwent induction of multiple follicular development with the following protocol: Leuprolide acetate (Lupron, TAP Pharmaceuticals, North Chicago, IL) 1 mg subcutaneously daily (0800 h) from day 1 of the menstrual cycle until the day of follicular aspiration. Patients received human follicle stimulating hormone (FSH) (Metrodin, Serono Laboratories, Inc., Randolph, MA) and human menopausal gonadotropins (hMG) (Pergonal, Serono) 150 I.U. intramuscularly (IM) daily (4:00 pm) from day 2 of the menstrual cycle until days 9 and 8 (patients 1 and 2), respectively. Human chorionic gonadotropin (Profasi, Serono, randolph, MA) 10,000 I.U. was administered IM (9:00 pm) on days 9 and 10, respectively.

Thirty-six hours after hCG administration, the patients underwent follicular aspiration in the operating room under intravenous sedation with titrating doses of 0.1 to 0.25 mg of Fentanyl (Sublimaze, Janssen Pharmaceutical, Inc., Piscataway, NJ) and 5 to 7 mg of midazolam HCl (versed, Roche Laboratories, Division of Hoffmann-La Roche, Nutley, NJ).

Follicular aspiration was performed using a transvaginal probe (GE H4222 TV) adapted to an ultrasound system (GE RT3,000, General Electric Company, Milwaukee, WI) with a needle set for ovum aspiration and follicle flushing (Labotect, Bovenden-Gotingen, FRG (#4060-2, length 300 mm, 1.4 mm outside diameter, 1.1 mm inside diameter) connected to a Craft Suction Unit (Rocket USA, Branford, CT) (#33-100) at a maximum vacuum pressure of 120 mm Hg.

Each case of follicular aspiration was performed without complications in less than 30 minutes and patients were discharged two hours after the outpatient procedure. The follicular fluids and follicular washings (with TALP-HEPES medium) were given immediately to the embryology laboratory adjacent to the operating room.

Epididymal Sperm Aspiration

At the same time, the husband undergoes scrotal exploration with the intention of aspirating sufficient numbers of motile spermatozoa to utilize for IVF of the aspirated eggs, with subsequent transfer into the wife's fallopian tube.

The surgical technique (Fig. 3) in the male was as follows: scrotal contents were extruded through a small incision, the tunica vaginalis was opened, and the epididymis was

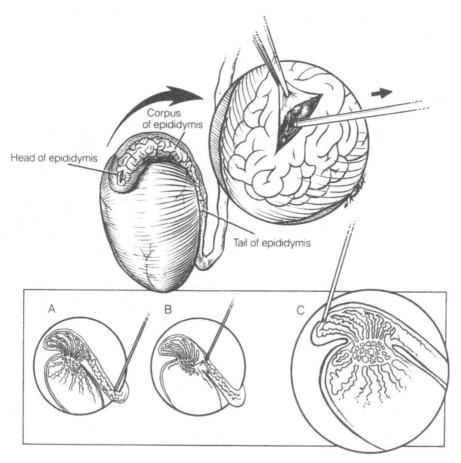

Fig. 3. Technique for epididymal sperm aspiration which begins in the distal corpus region of the epididymis, and moves proximally until motile sperm are recovered. In most cases, motility is observed only in the most proximal region of the epididymis.

exposed. Under 10-40X magnification with an operating microscope, a tiny incision was made with microscissors into the epididymal tunic to expose the tubules in the distal-most portion of the congenitally blind ending epididymis. Sperm were aspirated with a #22 medicut on a tuberculin syringe directly from the opening in the epididymal tubule. Great care was taken not to contaminate the specimen with blood, and careful hemostasis was achieved with microbipolar forceps. The epididymal fluid was immediately diluted in hepes buffered media, and a tiny portion examined for motility and quality of progression. If there was no motility or poor motility, another aspiration was made one-half centimeter more proximally. We thus obtained sperm from successively more and more proximal regions until progressive motility was found. In all cases, motile sperm were not obtained until we reached the proximal-most portion of the caput epididymis or vasa efferentia (Fig. 4).

Two days after insemination embryos are transferred to the fallopian tubes of each patient, via minilaparotomy using a technique similar to the one for gamete intrafallopian transfer (GIFT), via a Tomcat catheter (Monoject, St. Louis, MO) 2 1/2 cm inside the fimbrial ostium. The entire surgical procedure lasts approximately 30 minutes and the patients

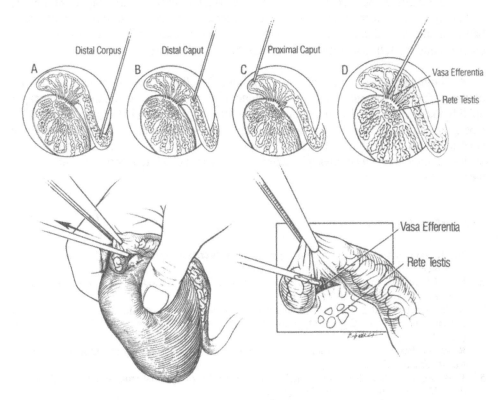

Fig. 4. Most motile sperm is found most proximally, usually in the vasa efferentia or rete testes fluid.

are discharged the next day and undergo an uneventful post-operative recovery. Patients receive progesterone in oil, 25 mg IM/day one day after the day of embryo transfer.

RESULTS

At present, of 32 cases, there have been ten pregnancies, two miscarriages, and eight ongoing or delivered live babies (25%).

Pregnancies which have occurred readily after vasoepididymostomy to the caput epididymis (and even in some cases to the vasa efferentia) suggest that immature sperm which have not had a chance to transit the epididymis might mature on their own during storage in the vas deferens.[44] If this theory were true, it might explain why we have been able to achieve success by aspirating more proximally, not being limited (because of theoretical considerations) to distal regions of the epididymis where the sperm are generally senescent and non-motile in the obstructed state.

Other factors in the success of this techique which may be equally important are: (1) Obtaining large numbers of oocytes in order to increase the odds for fertilization, (2) Incubation of sperm outside of the milieu of the obstructed epididymis, and (3) Transfer of the embryos into the fallopian tube (ZIFT) rather than into the uterus.

Although these results will have to be considered preliminary until greater numbers are obtained, for the moment it is safe to conclude: (1) Sperm from the proximal-most caput epididymis are capable of fertilization of the human egg in vitro, (2) Passage of time after emergence from the testicle may be adequate for sperm maturation without the absolute need for transit through the rest of the epididymis and (3) We now have an approach for achieving pregnancy in couples with a heretofore dismal condition, congenital absence of the vas deferens.

REFERENCES

1. N. J. Alexander, Autoimmune hypospermatogenesis in vasectomized guinea pig, *Contraception* 8:147-64 (1973).
2. N. J. Alexander, and S. S. Schmidt, Incidents of antisperm antibody levels in granulomas of men, *Fertil. Steril.*, 28:655 (1977).
3. R. D. Amelar, and L. Dubin, Commentary on epididymal vasostomy, vasovasostomy and testicular biopsy, *in:* "Current Operative Urology," Harper and Row, New York, 1181-1185 (1975).
4. R. Ansbacher, Sperm agglutinating and sperm immobilizing antibody in vasectomized men, *Fertil. Steril.*, 22:629 (1977).
5. J. M. Bedford, Development of the fertilizing ability of spermatozoa in the epididymis of the rabbit, *J. Exp. Zool.*, 162:319 (1966).

6. J. M. Bedford, Adaptations of the male reproductive tract and the fate of spermatozoa following vasectomy in the rabbit, Rhesus monkey, hamster and rat, *Biol. Reprod.*, 14:118-142 (1976).

7. P. E. Bigazzi, L. L. Kosuda, K. C. Hsu, and G. A. Andres, Immune complex orchitis in vasectomized rabbits, *J. Exp. Med.*, 143:380-404 (1976).

8. D. Brickel, J. Bolduan, R. Farah, The effect of vasectomy- vasovasostomy on normal physiologic function of the vas deferens, *Fertil. Steril.*, 37:807-810 (1982).

9. F. C. Derrick, Jr., W. Yarbrough, and J. D'Agostino, Vasovasostomy: Results of questionnaire of members of the American Urological Association, *J. Urol.* 110:556 (1973).

10. J. W. Dorsey, Anastomosis of the vas deferens to correct postvasectomy sterility, *J. Urol.* 70:515 (1953).

11. A. A. El-Itreby, and S. M. Girgis, Congenital absence of vas deferens in male sterility, *Inter. J. Fertil.* 6:409 (1961).

12. J. E. Fowler, Jr., and M. Mariano, Immunoglobin in seminal fluid of fertile, infertile, vasectomy and vasectomy reversal patients, *Urol.* 129:869-872 (1983).

13. P. Gaddum and T. D. Glover, Some reactions of rabbit spermatozoa to ligation of the epididymis, *J. Reprod. Fertil.*, 9:119 (1965).

14. P. Gaddum, Sperm maturation in the male reproductive tract: development of motility, *Anat. Rec.* 161:471 (1969).

15. T. D. Glover, Some aspects of function in the epididymis: experimental occlusion of the epididymis in the rabbit, *Int. J. Fert.*, 14:215 (1969).

16. H. C. Hanley, The surgery of male subfertility, *Ann. R. Coll. Surg.*, 17:159 (1955).

17. R. S. Hotchkiss, Surgical treatment of infertility in the male, *in:* "Urology", 3rd edition, N. F. Campbell and H. H. Harrison, eds., 671, (1970).

18. S. S. Howards, and A. L. Johnson, Affects of vasectomy on intratubular hydrostatic pressure in the testis and epididymis, *in:* "Vasectomy: Immunologic and Pathophysiological Effects in Animals and Man", I. H. Lepow, and R. Corzier, eds., Academic Press, New York, 55-56 (1979).

19. J. P. Jarow, R. E. Budin, M. Dym, B. R. Zirkin, S. Noren, and F. F. Marshall, Quantitative pathologic changes in the human testis after vasectomy: a controlled study, *N. Eng. J. Med.*, 313:1252-1260 (1985).

20. L. Johnson, and D. Varner, in press, Effect of age and daily spermatozoa production on transit time of spermatozoa through the human epididymis, *Biol. Reprod.*, (1988).

21. L. Linnet, and T. Hgort, Sperm agglutinins in seminal plasma and serum after vasectomy: correlation between immunological and clinical findings, *Clin. Exp. Immunol.*, 30:413 (1977).

22. L. Liskin, J. M. Pile, and W. F. Quillin, Vasectomy - safe and simple, *Popul Rep.*, 4:63-100 (1983).

23. R. G. Middleton, and D. Henderson, Vas deferens reanastomosis without splints and without magnification, *J. Urol.*, 119:763-764 (1978).

24. R. G. Middleton, and R. L. Urry, Vasovasostomy in semen quality, *J. Urol.*, 123:518 (1980).

25. V. J. O'Connor, Anastomosis of the vas deferens after purposeful division for sterility, *JAMA*, 136:162 (1948).

26. M. C. Orgebin-Crist, Sperm maturation in rabbit epididymis, *Nature*, 216:816 (1967).

27. M. C. Orgebin-Crist, Studies of the function of the epididymis, *Biol. Reprod.*, 1:155 (1969).

28. R. Pabst, O. Martin, and H. Lippert, Is the low fertility rate after vasovasostomy caused by nerve resection during vasectomy?, *Fertil. Steril.*, 31:316-320 (1979).

29. S. K. Paufler, and R. H. Foote, Morphology, motility and fertility in spermatozoa recovered from different areas of ligated rabbit epididymis, *J. Reprod. Fertil.*, 17:125 (1968).

30. S. K. Paufler, and R. H. Foote, Spermatogenesis in the rabbit following ligation of the epididymis at different levels, *Anat. Rec.* 164:339-348 (1969).
31. G. M. Phadke, and A. G. Phadke, Experiences in the reanastomosis of the vas deferens, *J. Urol.* 97:888 (1967).
32. R. Schoysman, and J. M. Drouart, Progres Recents Dans la Chirurgie de a sterilite masculine et feminine, *Acta Blic. Belg.*, 71:261 (1972).
33. R. Schoysman, and M. Bedford, The role of the human epididymis in sperm maturation and sperm storage as reflected in the consequences of epididymovasostomy, *Fertil. Steril.*, 46:293-299 (1986).
34. E. I. Shapiro, and S. J. Silber, Open-ended vasectomy, sperm granuloma and post-vasectomy orchalgia, *Fertil. Steril.*, 32:546-550 (1979).
35. R. Sherins, State of the art lecture, American Fertility Society, Toronto, (1986).
36. S. J. Silber, Microscopic vasectomy reversal, *Fertil. Steril.*, 28:1191-1202 (1977a).
37. S. J. Silber, Sperm granuloma and reversibility of vasectomy, *Lancet*, 2:588-589 (1977b).
38. S. J. Silber, Microscopic vasoepididymostomy; specific microanastomosis to the epididymal tubule, *Fertil. Steril.*, 30:565:576 (1978a).
39. S. J. Silber, Vasectomy and its microsurgical reversal, Urol. Clin. North Am., 5:573-584 (1978b).
40. S. J. Silber, Vasectomy and vasectomy reversal, *Fertil. Steril.*, 29:125-140 (1978c).
41. S. J. Silber, Epididymal extravasation following vasectomy as a cause for failure of vasectomy reversal, *Fertil. Steril.*, 31:309-315, (1979).
42. S. J. Silber, Microsurgery for vasectomy reversal and vasoepididymostomy, *Urol.*, 5:505-524 (1984).
43. S. J. Silber, Diagnosis and treatment of obstructive azoospermia, *in:* "Male Reproductive Dysfunction", R. J. Santen, and R. S. Swerdloff, eds., Marcel Dekker, New York, 479-517 (1986).
44. S. J. Silber, Pregnancy caused by sperm from vasa efferentia, *Fertil. Steril.*, 49:373-375 (1988a).
45. S. J. Silber, Pregnancy with sperm aspiration from the proximal head of epididymis: A new treatment for congenital absence of the vas deferens, *Fertil. Steril.*, 50, (1988b).
46. S. J. Silber, In press, Pregnancy after vasovasostomy for vasectomy reversal: a study of factors affecting long term return of fertility in 282 patients followed for ten years, *Human Reprod.*
47. S. J. Silber, and L. J. Rodriguez-Rigau, Quantitative analysis of testicle biopsy: determination of partial obstruction and prediction of sperm count after surgery for obstruction, *Fertil. Steril.*, 36:480-485 (1981).
48. S. J. Silber, T. Ord, C. Borrero, J. Balmaceda, and R. Asch, New treatment for infertility due to congenital absence of vas deferens, *Lancet*, 2:850, (1987).
49. R. Z. Sokol, and R. Sparkes, Demonstrated paternity in spite of oligospermia, *Fertil. Steril.*, 47:356 (1987).
50. N. J. Sullivan, and G. E. Howe, A correlation of circulating antisperm antibodies to functional success of vasovasostomy, *J. Urol.*, 117:189-191 (1977).
51. P. D. Temple-Smith, G. J. Southwick, C. A. Yates, A. O. Trounson, and D. M. De Kretser, Human pregnancy by in vitro fertilization (IVF) using sperm aspirated from the epididymis, *J. In Vitro Fert. Embryo Transfer*, 2:119 (1985).
52. A. J. Thomas et al, Microsurgical vasovasostomy: immunological consequences in subsequent fertility, *Fertil. Steril.*, 35:447 (1981).
53. W. C. Young, A study of the function of the epididymis, III. Functional changes undergone by spermatozoa during their passage through the epididymis and vas deferens on the guinea pig, *J. Exp. Biol.*, 8:151- 162 (1931).
54. Z. Zukerman, L. V. Rodriguez-Rigau, K. Smith, and E. Steinberger, Frequency distribution of sperm counts in fertile and infertile males, *Fertil. Steril.*, 28:1310 (1977).

NEW POSSIBLE METHODS FOR EVALUATION OF SPERM QUALITY

G. Paz,[1] S. Friedler,[1] L. Yogev,[1] A. Shullman,[1] H. Yavetz [1]
Z.T. Homonnai,[1] J. Lessing,[2] A. Amit,[2] I. Barak,[2] and I. Yovel [2]

[1] Institute for the Study of Fertility, and
[2] IVF Unit, Serlin Maternity Hospital, Tel Aviv Medical Center and
Sackler Medical School, Tel Aviv University, Tel Aviv, Israel

INTRODUCTION

Semen analysis is usually used for assessing male fertility based on the rationale and experience that quality of semen and sperm cells predict its capacity for fertilization. This is true in most cases, and is well-documented in the literature. Nevertheless, the crude criteria used such as sperm count, morphology and motility do not always predict the sperm cell fertilizing capacity.[1] Many other factors can be involved such as ATP contained,[2] fine configuration of the sperm cell,[3] existence of sperm antibodies,[4] etc. In these cases sophisticated methods for sperm quality evaluation are needed.

After ejaculation the sperm cell is challenged by physiological factors along the female tract. On its passage towards the site of fertilization in the oviduct, many selection factors filter the best motile and normal form cells.[5] It starts with the migration of spermatozoa from the vagina which infiltrate the cervical canal's mucus, traversing the uterus up to the fallopian tubes.[6] The sperm cell undergoes a process of anatomical and physiological changes and maturation defined as the capacitation of the cell.[7] When completed the sperm cell is judged by its capacity to be able to penetrate through the egg vestments such as the cumulus oophorus, corona radiata and zona pellucide. The latter will happen only after the acrosome reaction. The next in vivo physiological test of the sperm cell will occur at the fertilization site; the penetration into the ovum during the process of fusion of the gametes chromosomes, after decondensation of the sperm DNA.[8] Some of these events can also be tested in vitro by various laboratory techniques, such as: in vitro sperm penetration into cervical mu-

Advances in Assisted Reproductive Technologies, Edited by
S. Mashiach *et al.,* Plenum Press, New York, 1990

cus either by human or artificial cervical mucus.[9-11] The in vitro penetration test into zona free hamster oocytes is a valuable test for assessing the acrosome reaction[12] while the in vitro penetration into human ova is the best known for assessing fertilizing capacity of spermatozoa.

The in vitro zona free hamster ova penetration test for human washed spermatozoa is considered a good and reliable test to evaluate sperm capacity to fertilize, since this test assesses the events that take place just after penetration of the zona pellucida, such as sperm acrosome reaction, as well as the post-fusion events—nuclear sperm chromatin decondensation, formation of male and female pronuclei and release of the ova cortical granules.[8] One should also take into consideration the limitations of the test as cautioned by many authors in the past, especially concerning it's false negative predictive value.[1]

Normal passage of the sperm cell through the stages needed to achieve fertilization, depends on the integrity of the sperm cell membrane. This can also be evaluated by in vitro tests in addition to the zona free hamster test. Two tests are available: the hypo-osmotic swelling test[13] and the temperature shock test including the freezing test.

Spermatozoa with functional membranes will increase in volume when exposed to hypo-osmotic conditions, detectable by the curling of the fibers within the tail membrane. The results of the test correlate with the zona free hamster oocyte penetration test. This is true for normal fertile semen specimens although the limitations mentioned before are still valid for the hypo-osmotic swelling test as well.[1] In the Hamster test Chan et al[14] showed that the percentage of swollen sperm after hypo-osmotic treatment was not related to the hamster egg penetration results as determined by linear correlation and multiple regression analyses, and did not give additional information about the in vitro fertilizing capacity of sperm as evaluated by multivariate discriminant analysis. This was recently reported also by Avery et al.[15] On the other hand, Check et al[16] concluded that the hypo-osmotic test might be of value in predicting which couples have a better chance to bear a child. These inconclusive results are not in favor of the hypo-osmotic test as a routine predictive test for prognosis of fertilizing capacity of a semen specimen.

Human spermatozoa washed in IVF culture media can preserve the sperm cell viability and motility when stored at room temperature or refrigerated at 4°C for many days.[17] This method can be used for repeated artificial inseminations by using the same semen sample and avoiding expensive and, in some instances, even impossible freezing procedures.[18] Thus cooling of semen in appropriate conditions can be a useful clinical tool. A more stressful temperature shock is the freezing test, since during the process of freezing and thawing the integrity of the sperm cell membrane is challenged due to changes in ice crystallization and movement of water molecules crossing the sperm cell membrane.[19]

Cryopreservation of human spermatozoa has been attempted for many years.[20] Its methodology has been correlated with efficiency of sperm banking and preservation. Freezing and thawing has been proposed for testing the membrane stability and integrity.

The purpose of the present study was to evaluate the freezing test value in the prediction of the fertilizing capacity of semen samples derived from semen donors and husbands of women participating in the IVF program. In addition, the value of longevity of sperm motility was also tested by evaluating the value of four hr motility in the same system tested. In the case of these tests proving positive, they will be suggested as additional tools for evaluation of sperm quality.

METHODS AND MATERIALS

The study was carried out on 11 donors and 39 husbands of women treated for IVF. The protocol and technique of the Hakiriya Maternity Hospital's IVF program has been described elsewhere.[21] On the morning of oocyte harvest, husbands of patients were asked to deliver semen samples by masturbation (after three days of sexual abstinence). Following liquification, a volume of 0.5 ml was removed for a complete semen analysis and for freezing and thawing tests. The rest was used for in vitro fertilization. Fertilization of ova retrieved was carried out and recorded for each semen sample.

Eleven semen donors were screened for their sperm quality and freezability by the same protocol as the husbands. It should be stressed that semen was received, examined and tested about 2 hrs after its delivery to the hospital in the husbands group (masturbation was carried out at patients' homes), while the donor semen was checked and evaluated during the first hour after its delivery.

The protocol of semen cryopreservation is described in detail elsewhere.[22] In short the semen was diluted with a 7.5% glycerol based cryoprotectant mixture, equilibrized at 32°C for 10 minutes, then cooled in an electric cooler (Nicool LM10, France) at a rate of -1.7°C/min to -6.0°C followed by -5.0°C/min to -100°C prior to the immersion of the cryovials into liquid nitrogen. Thawing was carried out rapidly, in a 32°C incubator. The semen was examined for % post-thaw motility, defined as the ratio between the difference in motility % of the sample prior to and following cryopreservation to the initial motility.

The results of initial sperm concentration, sperm morphology and motility, % motility at four hours post delivery, % loss of motility after four hours and following cryopreservation and in vitro fertilization rate of ova were evaluated. Statistical analysis was performed by analysis of variance for repeated measurements using a statistical software (StatView TM512+, Brain Power, Inc., CA).

RESULTS

Table 1 gives the results of sperm quality and motility tests grouped according to donors and husbands of women belonging to the IVF program. The results clearly show that the donor sperm is superior in the crucial parameters of sperm quality: morphology, motility

Table 1. Sperm Quality; Concentration; Morphology and Motility of 3 Groups of Men. Numbers are mean ± SE, n = Number of Men, P = Significance of Differences

group	n	concentration million/ml	normal forms (%)	motility (%)
a Donors	11	122 ± 20.5	56 ± 1.9	59 ± 1.2
b Husband fertile	32	195 ± 23.4	50 ± 1.9	47 ± 2.6
c Husband non-fertile	7	143 ± 31	56 ± 4.2	49 ± 6.2
P; a vs. b		< 0.05	< 0.01	< 0.001

Table 2. Sperm Motility Tests; 4 Hours Motility, Post-thaw Motility and Loss of Motility (%). Numbers are mean ± SE, n = Number of Men, P = Significance of Differences

	group	n	Motility %			% loss of motility	
			Initial	After 4 hrs	Post-thaw	After 4 hrs	Post-thaw
a	Donor	11	59 ± 1.2	53 ± 1.7	47 ± 1.1	9 ± 2.5	18 ± 2.9
b	Husband fertile	32	47 ± 2.6	35 ± 2.4	24 ± 2.6	25 ± 3.8	53 ± 4.4
c	Husband non-fertile	7	49 ± 6.2	35 ± 4.9	17 ± 6.9	29 ± 4	68 ± 10.7
P; a vs. b			< 0.001	< 0.0001	< 0.0001	< 0.0001	< 0.0001

(P<0.01 and P<0.001 respectively). The concentration of spermatozoa was lower in the donor group (P<0.05) but this is not of any practical value. This difference can be explained on the basis of abstaining from sex for longer than three days in the husband groups.

Figure 1 depicts the relationship of the different motility values in the three groups tested and Table 2 summarizes the results of the sperm motil: ty tests: four hr motility and post-thaw motility and the calculated % loss of motility. It is clear from the results that the husband sperm motility is inferior in all variables measured. The percent initial motility, four hours motility and after thawing was significantly inferior to the donor sperm. An important parameter of sperm motility is the motility at four hrs post delivery of semen, compared to motility at the first hr (initial motility). This parameter is an excellent criteria for the longevity of spermatozoa. After four hours a mild loss of 9 ± 2.5% (mean ± SE) was recorded among the donors while the husbands' fertile sperm showed a marked decay of 25 ± 3.8% in sperm motility (P < 0.0001).

The correlation between initial motility and the four hr motility is given in Figure 2a. Our results indicate that the higher the initial motility, the better the four hr motility. This is clearly seen among the donor semen.

Freezing and thawing are supposed to test the integrity and temperature resistence of the sperm membrane. The post-thaw state is best assessed by the percent loss of sperm motility. Donor sperm lost about 18% of the initial motility when challenged by this test. In contrast, the husband sperm cells lost $53 \pm 4.4\%$ and $68 \pm 10.7\%$ of their initial motility in the fertilized and non-fertilized groups, respectively. Significantly, this was highly inferior ($P < 0.0001$) to the donor sperm performance. The results are depicted in Figure 2b, which illustrates the correlation between sperm initial motility and the post-thaw motility among the three groups of men.

Table 3 describes the ratio of ova fertilized to those inseminated (fertilization rate) by the sperm derived from the men of the three groups in the in vitro fertilization system. The same rate of fertilization was found between the donors and the fertile husband sperm.

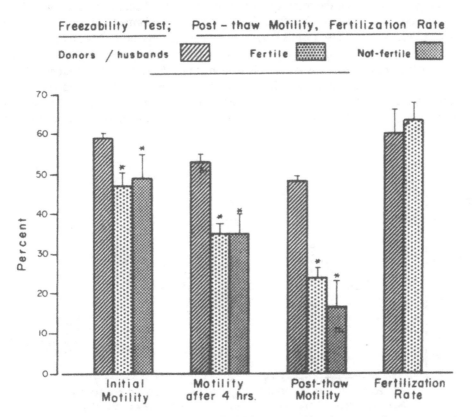

Fig. 1. Sperm motility and motility tests and fertilizing rates of the 3 groups of men; semen donors (11 men), husband fertile sperm (32 men) and husband non-fertile (7 men). Numbers are mean ± SE. Significant differences from donor by P < 0.001.

DISCUSSION

We introduced the possibility of using the freezing test as an additional test for evaluation of the integrity of the sperm cell membrane. This was based on the assumption that exposure of spermatozoa to a process of freezing and thawing, which was previously demonstrated to be useful in semen cryopreservation, might differentiate between viable fertile normal sperm and sperm cells with defects of the cell membrane's integrity.

To evaluate the test's predictive value on the sample's in vitro fertilization potential, we examined whether a correlation existed between percent loss of motility following freezing-thawing and results of the IVF procedure using the same, unfrozen semen. Based on the results of this study the test may have an important value, since the donors semen showed excellent results while the husband semen, although having initial good quality, was inferior in the tests such as freezability and loss of motility after four hrs. A preliminary study was carried out where we have shown that freezing semen two hours after its delivery does not change the quality of the motility tests. The same rate of loss of motility was found among donors and husbands when freezability tests were performed one and two hrs after delivery of ejaculates. These results may explain that the inferior quality and freezability outcome cannot be blamed on the period of two hrs that passed between semen delivery and the freezability testing. It seems reasonable to deduct that the results and differences found between the donors and husbands with fertile sperm are due to the appropriate selection of donors. On the other hand, we could not find any differences in sperm quality between the two husband groups, the one which fertilized at least one ova and the other which failed to do so.

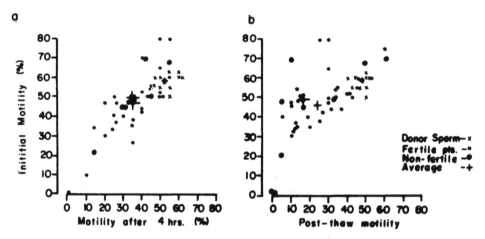

Fig. 2. Correlation between initial sperm motility and motility after four hrs (a) and post-thaw motility (b). The three groups of male donors, husband fertile and husband non-fertile are given by individual points shown in the figure.

Table 3. Fertilization Rate of the Sperm Derived from the Donors and Husbands with Fertile Sperm Numbers are mean ± SE, n = Number of Men

	group	n	Total Ova Fertilized/Inseminated	Mean Fertilization Rate (%)
a	Donors	11	243:472	60 ± 6.3
b	Husband fertile	32	122:219	63 ± 5.3

In addition to the efforts devoted to carefully selecting donors of high-quality semen, it is possible to use the freezability test for further selection of the donor semen. In a preliminary study we found that among 52 candidates for semen donations only 19 were found to meet the criteria of semen suitable for donations. The 33 semen samples failed the test due to low rate of motility following the freezing test (49.9 ± 0.8% and 22.6 ± 1.4% initial motility and post-thaw motility respectively). These values are similar to the values found in the husband non-fertile group of this study. Our findings also correlate well with prior observations[23] of great variation of donor semen freezability, indicating that sometimes semen of proven fertile males poorly withstands cryopreservation.

Cryopreservation done with the present technology obtained survival of motility of spermatozoa in these semen samples of the husbands in the range of 50% from the initial value. These results correlate well with the published results of semen cryopreservation, performed by similar methodology, as reported in a literature review.[24] They also correlate well with the hypo-osmotic swelling test demonstrating membrane defects in up to 40% of spermatozoa in a normal semen sample.[13] The post-thaw motility is considered to be a valuable parameter for sperm quality. A loss of 20% to 70% was reported following cryopreservation.[25-28] Freezing not only decreased the initial motility but also modified the velocity profile in the sperm population.[27,28] The cells that were mostly affected were the low motility cells, whereas rapid and progressive spermatozoa were significantly more resistant.[27] These results are in accordance with the results of the present study. The donor sperm cells initially having a superior quality were affected relatively mildly compared to the husband's sperm.

Studies were carried out on the value of the freezability test of sperm in predicting fertilization. The total concentration of motile spermatozoa after thawing was found to correlate well with the probability of achieving pregnancies by artificial inseminations.[29] A post-thaw concentration of less than 20 million motile spermatozoa correlated with up to only 20% of the total pregnancies obtained.[29] However, increasing post-thaw motility by an addition of agents such as caffeine did not improve pregnancy rates.[30] The conditions for fertilization in an in vitro system are quite different compared to in vivo and fertilization potential is examined in a more direct manner. Under these circumstances the freezability test is not reliable for predicting prognosis of fertilizing capacity. This was proven carefully in the present study. The great variability in the percent loss of motility of the spermatozoa, in addition to the dramatic loss of motility among the fertile husbands' specimens, hints at the low value of this test as being a predictive test for fertilization capacity. It has been shown that even a low degree of motility preserved the capacity to fertilize ova in an in vitro fertilization system. Furthermore, the husband fertile group was able to reach a rate of fertilization similar to the donor sperm, although the quality of sperm was significantly inferior; percent loss of post-thaw motility in the fertile husband was 53% while in the donor group it was only 18%. Moreover, no differences were found between the two husband groups in terms of sperm quality and motility tests.

No explanation other than intrinsic differences in the spermatozoal population characteristics can be offered for the observation that a high rate of motility loss, post-thaw, does not prevent the semen specimen from being fertile; on the other hand, intrinsic differences in the ova quality may be responsible for low rate of fertilization in the other specimens with a lower rate of motility loss.

Thus, the motility tests for membrane integrity are not sensitive enough for prediction of fertilization capacity of a semen sample.

SUMMARY

Fertilizing capacity was always correlated with sperm quality. Nevertheless, in vitro fertilization studies have shown that even low-quality semen treated properly can fertilize ova. Thus, additional tests for evaluation of sperm quality and fertilizing capacity were suggested. Among other factors, the sperm membrane integrity was challenged. The use of the hypo-osmotic test was not proven to be of great value, which leaves temperature shock tests as a possible tool for it's evaluation.

The purpose of the present study was to assess the possible use of two motility tests for evaluating sperm fertilizing capacity; the four hr post-delivery motility test and the post-thaw motility test.

The study was carried out on three groups of men. A donor group (11 men) and 39 husbands of couples participating in the IVF program subgrouped into those with fertile semen (32 men) and non-fertile semen (7 men).

Sperm quality of the donor samples were significantly superior in all motility parameters (initial $59 \pm 1.2\%$, four hours $53 \pm 1.7\%$ and post-thaw $47.0\pm1.1\%$), with a fertilization rate of $60.0 \pm 6.3\%$ (mean $\pm$ SE). Two hundred and forty-three ova were fertilized out of 472 inseminated. The husband fertile group had significantly lower motility values, and a significantly greater percent loss of motility. The four hr motility test showed a loss of $25.0 \pm 3.8\%$ compared to $9.1 \pm 2.5\%$ in the donor group and the post-thaw motility showed a loss of $53.0 \pm 4.4\%$ compared to $18.0 \pm 2.9\%$ in the donor group. All these results were statistically significant. The fertilization rate of this group was $63.0 \pm 5.3\%$

(122 ova out of 219), which is not different from the donor group. In the motility tests the non-fertile husband group performed similarly to the husband fertile group. These results show that the motility decay tests suggested here may be used for selection of donors but cannot be used as sensitive tests for prediction of sperm fertilization capacity.

ACKNOWLEDGEMENT

The authors express their gratitude to the technical help of Mrs. A. Gottreich (M.Sc.), Mrs. R. Rotem (M.Sc.), Mrs. T. Chavlin (B.Sc.), Mrs. A. Carmon (B.Sc.) and Mrs. F. Feldhun.

REFERENCES

1. J. A. Collins, Diagnostic assessment of the infertile male partner, *Curr. Probl. Obstet. Gynecol. Fertil.*, 10(5):173-224, (1987).
2. D. S.Irvine and J. R. Aitken, The value of adenosine triphosphate (ATP) measurements in assessing the fertilizing ability of human spermatozoa, *Fertil. Steril.*, 44:806 (1985).
3. E. Bostofte, J. Serup and H.Rebbe, The clinical value of morphological rating of human spermatozoa. *Int. J. Fertil.* 30:31 (1984).
4. W. R. Jones, Immunological infertility and its diagnosis, in: "Infertility: Male and Female," V. Insler and B. Lunenfeld eds., Churchill Livingstone Publishers, London (1986).
5. D. Mortimer and A. A. Templeton, Sperm transport in the human female reproductive tract in relation to semen analysis characteristics and time of ovulation, *J. Reprod. Fertil.*, 64:401 (1982).
6. J. Gonzales and F. Jezequel, Influence of the quality of the cervical mucus on sperm penetration: comparison of the morphologic features of spermatozoa in 101 postcoital tests with those in the semen of the husband, *Fertil. Steril.*, 44:796 (1985).
7. B. Lunenfeld, Diagnosis of male infertility, in: " Infertility: Male and Female," V. Insler and B. Lunenfeld,eds., Churchill Livingstone Publishers, London (1986).
8. C. Barros, A. Jedlicki and P. Vigil, The gamete membrane fusion test to assay the fertilizing ability of human spermatozoa. *Human Reproduction*, 3:637 (1988).
9. C. D. Matthews, A. E. Makin and L. W. Cox,Experience with in vitro penetration testing in infertile and fertile couples, *Fertil. Steril.*, 33:187 (1980).
10. A. Bergman, A. Amit, M. P. David, T. Z. Homonnai and G. Paz, Penetration of human ejaculated spermatozoa into human and bovine cervical mucus. 1. Correlation between penetration values, *Fertil. Steril.*, 36:363-367 (1981).
11. R. L. Urry, R. G. Middleton and D. Mayo, A comparison of the penetration of human sperm into bovine and artificial cervical mucus. *Fertil. Steril.*, 45:135 (1986).
12. R. Yanagimachi, H. Yanagimachi and B. J. Rogers, The use of zona-free animal ova as a test system for the assessment of the fertilizing capacity of human spermatozoa, *Biol. Reprod.*, 471:15 (1976).
13. R. S. Yeyendran, H. H. Van der Ven, M. Perez-Pelaez, B. G. Crabo and L. J. D. Zanneveld, Development of an assay to assess the functional integrity of the human sperm membrane and it's relationship to other semen characteristics, *J. Reprod. Fertil.*, 70:219 (1984).
14. S. V. W. Chan, E. J. Fox, M. M. C. Chan, W. L. Tsai,C. Wang, L. C. H. Tang, G. W. K. Tang and P. C. Ho, The relationship between the human sperm hypoosmotic swelling test, routine semen analysis and the human sperm zona free hamster ovum penetration assay, *Fertil. Steril.*, 44:668 (1985).

15. S. Avery, V. Bolton, B. A. Mason and C. Mills, The use of hypoosmotic sperm swelling test as an indicator for IVF outcome, *Human Reproduction* 3, (suppl.1), 94, (1988) (Absract 301).
16. J. H. Check, K. Nowroozi, C. H. Wu, A. Bollendorf, Correlation of semen analysis and hypoosmotic swelling test with subsequent pregnancies, *Arch. of Andrology*, 20:257-260 (1988).
17. J. Cohen, C. B. Fehilly and D. E.Walters, Prolonged storage of human spermatozoa at room temperature or in a refrigerator, *Fertil. Steril.*, 44:254 1985.
18. E. Kesseru and C. Carrere, Duration of vitality and migration ability of human spermatozoa cryopreserved at +4°C, *Andrologia* 16(5):429-433 (1984).
19. P. Mazus, Freezing of living cells: mechanisms and implications, *Am. J. Physiol.*, 247 (*Cell. Physiol.* 16:C125-C142) (1989).
20. J. K. Sherman, Synopsis of the use of frozen human semen since 1964: state of the art of human semen banking, *Fertil. Steril.*, 24:397 (1973).
21. J. B. Lessing, A. Amit, Y. Barak, A. Kogosowski, A. Gruber, I. Yovel, M. P. David and M. R. Peyser, The performance of primary and secondary unexplained infertility in an in vitro fertilization embryo transfer program, *Fertil. Steril.*, 50:903 (1988).
22. G. Paz, A. Gottreich, R. Rotem, H. Yavetz, L. Yogev and Z. T. Homonnai, The use of an electrical freezer in human sperm banking (submitted for publication).
23. A. B. Glassman, C. E. Benett, Semen analysis: Criteria for cryopreservation of human spermatozoa, *Fertil. Steril.*, 34:66 (1980).
24. J. K. Sherman, Cryopreservation of spermatozoa, *in:* " Infertility: Male and Female," V. Insler and B. Lunenfeld, eds., Churchill Livingstone Publishers, London (1986).
25. J. Cohen, P. Felten and G. H. Zeilmaker, In vitro fertilizing capacity of fresh and cryopreserved human spermatozoa: a comparative study of freezing and thawing procedures, *Fertil. Steril.*, 36:365 (1981).
26. P. Serafini and R. P. Marrs, Computerized stage freezing technique improves sperm survival and preserves penetration of zona free hamster ova, *Fertil. Steril.*, 45:854 (1986).
27. J. Auger and J. P.Dadoune, Computerized sperm motility and application of sperm cryopreservation, *Arch. of Androl.*, 20:103-112 (1988).
28. B. A. Keel, B. W. Webster and D. K.Roberts, Effects of cryopreservation on the motility characteristics of human spermatozoa. *J. Reprod. Fertil.*, 81:213-230 (1987).
29. N. C. Nielsen, J. Risum, K. Brogaard Hansen and U. Nissen, Obtained pregnancies by AID using frozen semen in relation to specific qualities of the semen, *Gynecol. Obstet. Invest.* 18:147 (1984).
30. H. Schneiden, The evaluation of cryoprotective media for human semen, *in:* " Frozen Human Semen, Developments in Obstetrics and Gynecology," Vol 4, D. W. Richardson and J. D. Symonds, eds., Martinus Nijhoff Publishers, The Hague (1980).

ULTRAMORPHOLOGIC CHARACTERISTICS OF HUMAN SPERM CELLS

B. Bartoov, F. Eltes, Y. Soffer*, R. Ron-El*, J. Langsam, H. Lederman,
D. Har-Even, and P. Kedem

*Department of Life Sciences, Bar-Ilan University, Ramat-Gan 52900, and
OB/GYN Department, Assaf Harofeh Medical Center, Zerifin, Israel

INTRODUCTION

Mammalian spermatozoa possess a unique topography which clearly defines the location of the various organelles responsible for the diverse functions the sperm cell must fulfill, before and during fertilization. A technique of in depth evaluation of sperm morphology might afford a wide spectrum of information, ranging from data concerning the integrity of the spermatogenetic process of the testes to the functional properties of the spermatozoa.

However, in humans, implementation of such an evaluative technique is confronted by two major obstacles: a) the absence of systematic classification of the structural pathology of human sperm. This classification is difficult to compile because of the well-documented pleomorphism of the cell, a phenomenon known as anisozoospermia. The pleomorphism hinders morphologists in their task of delineating the boundaries of normality of human spermatozoa and b) the minute dimensions of most sperm organelles which frequently place them beyond the limits of resolution of the optic microscope. At present, routine semen analyses employing a light microscope evaluate the morphological features of the sperm mainly from the overall appearance of the head and tail. The finer abnormalities which frequently subtend the patient's infertility, remain undetected (Zamboni, 1987).

We have defined and classified the ultramorphology of the normal human spermatozoon as well as 65 possible surface morphological malformations, by scanning electron microscope (SEM) and 42 internal morphological malformations, viewed through the transmission electron microscope (TEM) (Bartoov, et al., 1980, 1981, 1982; Glezerman and Bartoov, 1986).

Advances in Assisted Reproductive Technologies, Edited by
S. Mashiach *et al.*, Plenum Press, New York, 1990

A quantitative ultramorphological method directed at evaluating the potential of the human testis to produce fertile semen was developed. Instead of relating to each morphological characteristic separately, a single score, which reflects the weighted sum of the various malformation measurements and which can be used to differentiate between fertile and suspected infertile men, was computed, using a statistical discriminant analysis method (Nie et al., 1975). This quantitative ultramorphological (QUM) technique was elaborated further in order to reach two objectives: a) to understand sperm heterogeneity better and thus, to attain a dynamic morphological profile which might be a sensitive tool for evaluating and perhaps, monitoring testicular stress situations such as infection, trauma, hormonal inbalance, or response to drugs and b) to enable morphologists to use electron microscopy as a quantitative diagnostic tool for evaluating the functional potentials of the spermatozoa in the growing field of assisted reproduction. The present study describes the progress of our research towards the goals we have set.

METHODS

Transmission Electron Microscopy (TEM)

Sperm cells were separated from seminal plasma by centrifugation at $1650xg_{max}$ for 15 minutes at room temperature, prewashed twice by 0.1M Phosphate buffer pH 7.4 and fixed in 2% formaldehyde, 2% glutaraldehyde in 0.1 M phosphate buffer solution pH 7.4 for two hours at room temperature. The cells were postfixed in osmium tetraoxide. Block staining was performed by 0.5% uranyl acetate in veronal buffer (Ryter and Kellenber, 1958). Dehydration in graded alcohol solutions and embedding in Spurr (Ladd comp.) were done according to the method described by Glavert (Glavert 1980). Thin sections were cut with an LKB ultratome and examined with a JEOL 100 C TEM as described earlier (Bartoov, et al., 1980) at a magnification of about x 12,000. The TEM morphogram was obtained by examining randomly 50 cells, especially their internal organelles: nucleus, acrosome, post acrosomal lamina, axonema and outer dense fibers.

In addition to the 31 ultramorphological characters described previously (Glezerman and Bartoov, 1986) new ones were observed: acrosome equatorial region disorder, lack of the acrosome equatorial region, outer dense fibers disorder, tubular internal organization disorder, epithelial cells, white blood cells, spermatophages, immature germinal cells, red blood cells and bacteria. Ultramorphological characteristics such as partial acrosome with normal nuclear shape, empty mitochondria and breakage of Axonema were considered as artifacts and omitted from morphograms.

Scanning Electron Microscopy (SEM)

Sperm cells were separated from plasma and fixed as for the TEM preparation. The samples were dehydrated with graded alcohol and freon solutions and critical point-dried with

CO_2 and coated with gold (Cohen, 1974). The cells were examined with a JEOL JSM 35 SEM as described earlier (Bartoov, et al., 1980).

In addition to the 42 ultramorphological characters described previously (Glezerman and Bartoov, 1986) new characters were observed: acrosome disorder, basesball nuclear shape, partial posterior lamina, lack of posterior lamina, abaxial neck, large cytoplasmic droplet, exposed mitochondria, partial mitochondria, narrow mitochondria, long mitochondria, short tail, narrow tail, lack of tail, tail kinked toward the end piece, axonema disorder, epithelial cells, white blood cells, spermatophages, immature germinal cells, brush like cells, degenerative cells, red blood cells and bacteria. Characteristics as agenesis and lack of acrosome were considered as artifacts and omitted from morphograms.

Routine Semen Analysis

The routine analysis was performed as described previously (Glezerman and Bartoov, 1986).

STATISTICAL METHODS

The computer program used for all statistical analyses is SPSS-X (Norusis,1985). T-test for paired population and Pearson correlation test were conducted.

Factor Analysis of Sperm Ultramorphological Characteristics

Factor analysis is a statistical technique used to identify a relatively small number of factors that can be used to represent relationship among sets of many interrelated variables. The basic assumption of factor analysis is that underlying dimensions or factors can be used to explain complex phenomena. Our assumption was that the underlying dimensions of the sperm morphological heterogeneity will be related to spermatogenesis. In other words, we expected that the emerging factors will reflect the various stages of testicular maturation. This assumption led us to analyse the sperm characteristics according to their biological structure-function relationship measured both internally by TEM and externally by SEM.

The analysis was done in two stages. In the first stage a separate factor analysis was executed for specific characteristics of each of the cell organelles. Thus we were able to derive underlying maturation dimensions for each organelle. In the second stage a new factor analysis was executed, based on the scores created in the first stage. Thus the overall underlying maturation dimensions of the spermatogenesis process were revealed.

The 107 sperm cell ultramorphological characteristics observed by transmission and scanning electron microscopes were divided according to the six cell organelles: acrosome, head, nucleus, neck, mitochondria and tail. Six separate principal components factor analysis with varimax rotation was executed on each organelle. For each of the 16 emerged factors a separate score was constructed by averaging those characteristics that contributed to the factor (factor loading 0.5 and above). To overcome the problem of different ranges, all row data of characteristics were standardized by transformation to z score,before computing the factor scores.

The second degree factor analysis was executed on the 16 derived scores in order to establish the overall pattern of ultramorphological pathology of the sperm. The emerged factors from this analysis were then the basis for eight new comprehensive factor scores.

Discriminant Analysis

The purpose of the analysis is to find a single score which is a weighted sum of the original measurements and which most completely distinguishes between a test group and control group. The weights of the ultramorphological characteristics in the score, which is the discriminant function, are computed by taking into account the differences between their means, standard deviations and the correlations among the two compared groups. To the weighted means is added a constant which is also a function of data. After all weights and constant are computed, the mean scores are computed separately for each of the compared groups. The midpoint between these two mean scores is taken to be the discriminating value which was 0 in this study.

The different morphological sperm functional scores were based only on those morphological characteristics which were significantly correlated with the specific function, ($r > 0.2$, $P < 0.05$).

MORPHOLOGICAL SPERM FUNCTION POTENTIALS

Motility Potential

For the estimation of the morphological motility potential TEM ultramorphogram was used. Semen of 50 teratozoospermic men with motility higher than 10% and sperm count higher than 20×10^6 were examined. Spermatozoa were fractionated into a highly motile fraction (the upper one) and a submotile fraction (the lower one) using swim-up technique (Eltes et al. 1981). The TEM motility score (TMS) was derived using stepwise discriminant analysis (Norusis, 1985) between the sperm TEM characteristics of the highly motile and the submotile fractions.

Hamster Penetration Potential

For the estimation of the Hamster penetration potential sperm samples of patients who have undergone the cold Test-Yolk buffer Hamster penetration test were examined by TEM

SEM and routine analysis. The time period between the two tests was less than six months. One hundred and sixty-eight patients who had no evidence of antisperm antibodies present in their semen were divided into two groups: a. Men with penetration rate equal or higher than 35% (n=136, 'positive' group). b. Men who achieved less than 35% of sperm penetration into Hamster ova after two consecutive trials (n = 32, 'negative' group). The Hamster Penetration Score (HPS) was derived using stepwise discriminant analysis (Norusis,1985) between the sperm ultramorphological characteristics of the "negative" and "positive" groups.

Swim-up IUI Conception Potential

For the estimation of the conception success of swim-up IUI manipulation sperm samples of patients who have undergone the swim-up IUI treatment were examined by TEM, SEM and routine analysis. The time period between the treatment and the sperm analysis was less than six months. Fifty-two couples with no evidence of antisperm antibodies present and no female factor were divided into two groups: a. couples who achieved pregnancy (n = 13, 'positive' group), b. couples in which pregnancy did not result after three consecutive trials (n = 39 'negative' group). The swim-up IUI Pregnancy Potential Score (SPS) was derived using stepwise discriminant analysis (Norusis, 1985) between the sperm ultramorphological characteristics of the "negative" and "positive" groups.

In Vitro Fertilization Potential

a) The fertilization potential of sperm samples of patients who have undergone the IVF procedure was examined by TEM, SEM and routine analysis. The time period between the treatment and the sperm analysis was less than six months. Only cases with three or more retrieved eggs were included in the study. Eighty-six couples with no evidence of antisperm antibodies present and only mechanical female factor were divided into two groups: a. couples who achieved fertilization of at least one egg (n = 65, 'positive' group), b. couples with no fertilization resulted after two consecutive trials (n = 21). Cases in which the IVF resulted only in fragmented embryos were excluded from the study. The IVF Fertilization Potential Score (IVFS) was derived using stepwise discriminant analysis (Norusis,1985) between the sperm ultramorphological characteristics of the "negative" and "positive" groups.

b) For the estimation of the fertilization intensity potential of successful IVF cases, the 65 IVF successful cases ('positive' group) were subdivided into two groups: a. cases in which the fertilization of the retrieved ova was equal to or higher than 50% (n = 38, 'positive' subgroup), b. cases in which the fertilization was less than 50% (n = 27, 'negative' subgroup). The Fertilization Intensity Score (IIVFS) was derived using stepwise discriminant analysis (Norusis, 1985) between the sperm ultramorphological characteristics of the "negative" and "positive" subgroups.

Table 1. Morphological Motility Potential (TMS)

Significant morphological characteristics (n = 50, P < 0.0001)	Average frequency difference ± S.D.	Unstandardized function coefficients
Normal sperm cells	12.0 ± 5.8	+0.021
Immature germinal cells	16.9 ± 13.3	−0.051
Irregular acrosome	4.5 ± 6.2	−
Agenesis of acrosome	13.6 ± 11.3	−
Lack of acrosome	4.7 ± 3.8	−0.041
Lack of Post acrosomal lamina	11.0 ± 3.1	−0.032
Round nucleus	7.4 ± 3.2	−
Amorphous nucleus	9.3 ± 6.8	−0.046
Chromatin degradation	2.7 ± 1.2	−0.110
Chromatin subcondensation	12.2 ± 9.6	−
Mitochondrial aggregation	8.1 ± 6.4	−
9 + 2 disorder	4.3 ± 4.6	−
Tail kinked toward head	13.2 ± 11.9	−0.045

$$TMS = (\sum_{i=1}^{8} Bi \times Frequency\ i) + 2.312$$

RESULTS

Motility Potential

Using the swim-up technique and discriminant analysis we were able to discriminate 98% of high and low motile sperm fractions after swim-up. False positive = 1%, False negative = 1%, (can. cor. = 0.98, P < 0.0001) (Table 1). A positive score indicates that the sperm, from a morphological point of view, has a high motility potential and a negative score, indicates that the sperm cells have a low motility potential.

Hamster Penetration Potential

The human-hamster penetration assay with cold TEST-yolk buffer sperm incubation was performed on 168 patients. Based on the results of these analyses it was possible to divide the patients into two groups, a Positive group (n = 136) in which the penetration rate was equal or higher than 35% of the ova and a negative group (n = 32) in which the penetration rate was less than 35% of the ova. Using discriminate analysis we were able to create a Hamster Penetration Score (HPS) based on six SEM and TEM morphological characteristics (Table 2). This score correctly discriminates 74% of the high and low penetration groups. False Positive = 5%, False negative = 2.1% (can. cor. = 0.41, chi squared = 28.8, P < 0.0001).

Table 2. Human-Hamster Ova Penetration Potential (HPS)

morphological significant characteristics (T = TEM, S = SEM)	r^1 =	Ai^a	Bi^b
T-Agenesis of acrosome	−0.20	−	−
S-Acrosome disorder	−0.20	−	−
S-Round head	−0.17	-0.21	-0.44
S-Amorphous head	−0.17	+0.54	+1.34
S-Pin head	−0.21	−	−
S-Multinucleated head	−0.21	-0.35	-0.84
T-Sub acrosomal space	−0.20	+0.54	+0.68
S-Mitochondrial absence	−0.20	-0.34	-0.30
T-Tail kinked toward head	−0.20	-0.26	-0.27
R-% Motility^2	+0.23	−	−

[1] Correlations between ultramorphological characteristics and Human Hamster ova penetration capability expressed by high rate ($\geq$ 35%) and low rate (< 35% n = 168).
[2] From the routine semen analysis.
[a] Standardized canonical discriminant function coefficients.
[b] Unstandardized canonical discriminant function coefficients.

$$HPS = (\sum_{i=1}^{6} Bi \times Frequency\ i) + 0.80$$

Swim-up IUI Conception Potential

The QUM analysis was performed on 52 cases which underwent this technique. Using the discriminate analysis it was possible to achieve a swim-up pregnancy score (SPS) which was based on five morphological characteristics. This score correctly discriminated 92% of the conceptual and non- conceptual cases of swim-up IUI attempts (Table 3). False positive = 0%, False negative = 8% (can. cor. = 0.67, chi squared = 25.2, P < 0.0001).

IVF Fertilization Potentials

a. Ova Fertilization Potential

The IVF ova fertilization success score (IVFS) was achieved using the QUM and the following stepwise discriminate analysis. Using only four morphological characteristics: Vesiculated acrosome (TEM), Pin Head (SEM), Egg shape (SEM) and Postacrosomal lamina (SEM), we were able to correctly discriminate 80% of the cases (Table 4). False positive = 5%, False negative = 15% (can. cor. = 0.55 chi squared =29.7, P < 0.001).

From the 86 IVF cases which underwent the IVF study, 63 performed also the Human-Hamster penetration assay (HPA). When discriminate analysis was performed on these males taking into account the results of the QUM analysis and the percent of penetrated Hamster ova a new IVF score was established (IVFHS) which correctly discriminated 98%

Table 3. Swim-up-IUI Conception Potential (SPS)

morphological significant characteristics (T = TEM, S = SEM)	$r^1 =$	Ai[a]	Bi[b]
T-Acrosomal equatorial segment	−0.36	−0.33	−0.37
T-Partial acrosome	−0.25	+0.53	+0.62
S-Pin head	−0.28	−0.46	−0.43
S-Short tail	−0.33	–	–
T-9+2 disorder	−0.36	−0.38	−0.43
R-% normal[2]	+0.22	–	–
S-Epithelial cells	−0.28	−0.46	−0.68
S-Cytoplasmic droplet	+0.23	–	–
S-acrosomal region disorder	−0.31	–	–
S-Equatorial region disorder	−0.29	–	–

[1] Correlations between ultramorphological characteristics of and post-swim-up fertility success expressed by conception and nonconception (n = 52).
[2] From routine semen analysis.
[a] Standardized canonical discriminant function coefficients.
[b] Unstandardized canonical discriminant function coefficients.

$$SPS = (\sum_{i=1}^{5} Bi \times Frequency\ i) -0.90$$

Table 4. In Vitro Fertilization Success Potentials without Hamster Penetration Assay (HPA) IVFS, with HPA = IVFHS

morphological significant characteristics (T = TEM, S = SEM)	$r^1 =$	Ai[a]		Bi[b]	
		−HPA	+HPA	−HPA	+HPA
T-Acrosome agenesis	−0.40	–	–	–	–
T-Vesiculated acrosome	−0.21	−0.50	−0.56	−0.53	−0.67
T-Subcondensed chromatin	−0.21	–	–	–	–
S-Pin head	−0.35	−0.69	–	−0.78	–
S-egg-shaped head	−0.32	−0.52	–	−0.35	–
T-Amorphous head	−0.21	–	–	–	–
T-Postacrosomal lamina disorder	−0.25	–	–	–	–
S-Postacrosomal lamina disorder	−0.23	−0.48	−0.41	−0.55	−0.45
S-Mitochondrial absence	−0.29	–	–	–	–
S-Tail kinked toward head	−0.29	–	–	–	–
Percent of penetrated Hamster ova(HPA)	+0.42	–	+0.96	–	+0.05

[1] Correlations between ultramorphological charcteristics and IVF fertilization capability expressed by successful and unsuccessful cases (n = 86).
[a] Standardized canonical discriminant function coefficients.
[b] Unstandardized canonical discriminant function coefficients.

$$IVFS = (\sum_{i=1}^{4} Bi \times Frequency\ i) + 0.28$$

$$IVFHS = (\sum_{i=1}^{3} Bi \times Frequency\ i) -1.73$$

Table 5. IVF Fertilization Intensity Potential (IIVFS)

morphological significant characteristics (T = TEM, S = SEM)	$r^1 =$	Ai^a	Bi^b
T-Acrosome Agenesis	−0.40	−0.75	−0.39
T-Vesiculated acrosome	−0.21	−	−
T-Subcondensed chromatin	−0.21	−	−
S-Pin head	−0.35	−0.27	−0.31
S-Egg Shape head	−0.32	−	−
T-Amorphous head	−0.22	−	−
T-Post acrosomal Lamina	−0.25	−	−
S-Post acrosomal Lamina	−0.24	−0.60	−0.67
S-Absence of Mitochondria	−0.29	−	−
S-Tail Kinked Midpiece	−0.39	−0.67	−0.58
R-Speed (μ/sec)^2	+0.39	+0.32	+0.51

[1] Correlation between ultramorphological characteristics and the fertilization intensity of IVF expressed by high rate ≥ 50% and low rate (< 50%, n = 65).
[2] From routine semen analysis.
[a] Standardized canonical discriminant function coefficients
[b] Unstandardized canonical discriminant function coefficients

$$IIVFS = (\sum_{i=1}^{5} Bi \times Frequency\ i) + 2.30$$

of the successful and the unsuccessful cases of IVF attempts. False positive = 0%, False negative = 8% (can. cor. = 0.84, chi squared = 70.3, P < 0.0001).

Only two morphological characteristics: Vesiculated acrosome (TEM) and post-acrosomal lamina (SEM) and the percent of penetrated hamster ova were needed for the evaluation of this score (Table 4).

b. Fertilization intensity potential

The discriminant analysis was also performed on the routine and QUM analysis of the 65 IVF successful cases for the prediction of the high fertilization (≥ 50%) and low fertilization (< 50%) rates. The IVF intensity score (IIVFS) was based on four morphological characteristics: agenesis of acrosome (TEM), pin head (SEM), post acrosomal lamina (SEM), tail kinked toward midpiece (SEM) and the average sperm speed (μm/sec). This score correctly discriminated 79% of the cases (Table 5). False positive = 11%, False negative = 10% (can. cor. = 0.66, chi squared = 28.8, P < 0.0001).

A summary of the predictive values of the different IVF scores based on the routine semen analysis, QUM analysis and hamster penetration ova, their false positive and negative, is presented in Table 6. It appears that routine semen analysis is not sufficiently comprehensive to permit an accurate forecast of the success of IVF. The highest percentage of correct prediction was attained when the techniques of QUM analysis and penetration of Hamster ova were combined.

Table 6. Prediction of the IVF Success

Method of Analysis	No. of cases			Correct (%)	False %	
	Total	(+)	(−)	Prediction	(+)	(−)
Routine semen analysis	86	65	21	not predictable		
QUM analysis	86	65	21	80	5	15
QUM and routine analyses	86	65	21	76	6	17
QUM and % penetrating Hamster ova	63	51	12	98	2	0

Table 7. Prediction of the Different Functional Potentials of Human Spermatozoa by QUM Analysis

Sperm function	Routine semen analysis (LM)						EM QUM analysis						EM + LM analysis					
	*T	+	−	%	(+)	(-)	T	+	−	%	(+)	(−)	T	+	−	%	(+)	(−)
Motility	not done						100; 50; 50;		*98*		1	1	not done					
Human Hamster Penetration Assay	144; 118; 26			not predic.			168;136;32;		*74*		5	21	163; 131; 32			72	4	24
Conception Success in Swim-up IUI	44; 10; 34			*61*	32	7	52; 13; 39;		*92*		0	8	43; 11; 34			*91*	2	7
Fertilization Success IVF	86; 65; 21			not predic.			86; 65; 21;		*80*		5	15	86; 65; 21			76	6	17

* T = Total no. of cases; (+) = successful cases; (-) = not successful cases; % = correct prediction; (+) = percent false positive; (−) percent false negative.

The following sperm functions: motility, Human Hamster Ova penetration, conception success in swim-up IUI and fertilization success in IVF, the percent of their correct prediction; false positive and negative values achieved in the routine semen analysis, QUM analysis and a combination of both are summarized and presented in Table 7. It seems that the QUM analysis has advantage for the prediction of this function over the routine semen analysis.

Morphological Profiles

The study of morphological profiles was based on 1374 SEM and TEM ultramarphograms using the statistical method of factor analysis. The analysis was done in two stages: a) On the organellar level, b) overall cellular level. Eight secondary overall morphological profiles and their andrological definitions are presented in Table 8 and Figures 1-8.

Table 8. The Overall Ultramorphological Secondary Profiles

No.	Factor grouping[a]/item content[b]	Loading Factor
I	GROSS SPERMIOGENETIC FAILURE	
	Amorphous head	0.82
	Decondensed chromatin	0.76
	Malformation of postacrosomal lamina	0.67
	Agenesis of acrosome	0.70
	Tail kinked toward head	0.67
II	NUCLEAR ELONGATION ARREST	
	Round head	0.70
III	STRETCHED NUCLEAR SPERMIOGENESIS	
	Tapering head	0.76
	Stretched nucleus	0.68
IV	GROSS NECK SPERMIOGENETIC FAILURE	
	Mitochondria disorder	0.61
	Neck disorder	0.50
V	GROSS TAIL SPERMIOGENETIC FAILURE	
	Compressed tail	0.52
	Internal axonemal disorder	0.56
	External axonemal disorder	0.50
	Missing mitochondria	0.50
VI	INCOMPLETE CYTOKINESIS	
	Multihead	0.81
	Tail kinked toward midpiece, multitail	0.61
VII	INCOMPLETE EPIDIDYMAL MATURATION	
	Cytoplasmic droplet	0.68
	Nuclear membranes	0.60
	Acrosomal detachment	0.61
VIII	MINOR SPERM MALFORMATIONS	
	Acrosomal hypoplasia	0.57
	Compressed head	0.55
	Coiled Tail	0.54
	Narrow mitochondria	0.53

[a] Andrological definition
[b] Organellar morphological primary profile

Only the gross spermiogenetic failure profile was found to be significantly positively related to immature germinal cells appeared in the ejaculate ($r = 0.25$, $P < 0.001$). Other significant correlations between the difference of the overall ultramorphological profiles and different parameters of routine semen analysis are presented in Table 9.

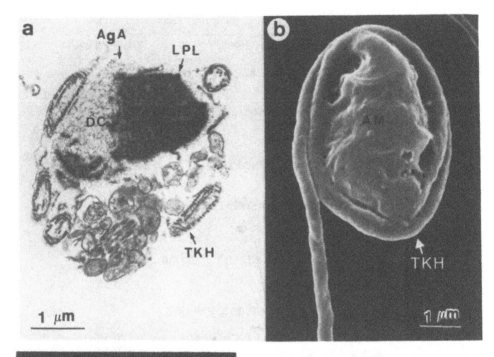

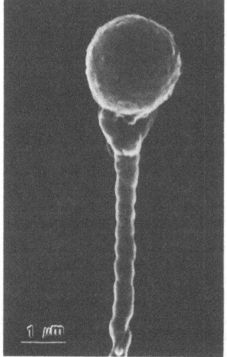

(Top) Fig. 1. Photomicrographs illustrating gross spermiogenetic failure (Profile I). (a) in TEM and (b) in SEM Dc = decondensed chromatin. LPL = Lack of Post acrosomal lamina, AgA = Agenesis of Acrosome, TKH = Tail kinked toward head, AM = Amorphous head shape.

(Left) Fig. 2. Photomicrograph illustrating Nuclear elongation arrest (Profile II). Round form head shape (SEM).

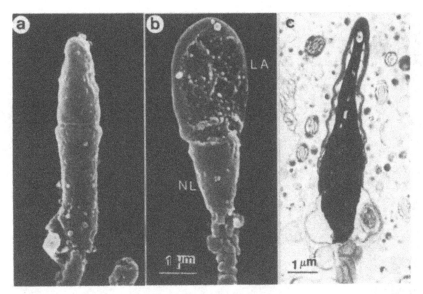

Fig. 3. Photomicrographs illustrating stretched nuclear spermiogenesis (Profile III). Tapering sperm head shape: (a) SEM, (c) TEM. Long acrosomal (LA) region, narrow long (NL) post-acrosomal region in SEM (b).

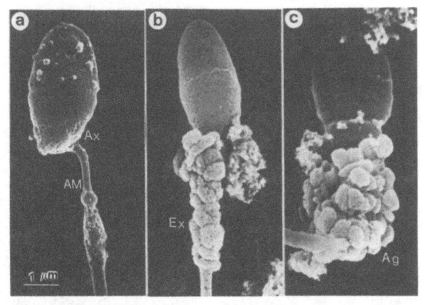

Fig. 4. Photomicrographs illustrating gross neck spermiogenetic failure in SEM (Profile IV). (a) Abaxial tail (AX), Absence of mitochondria (Am). (b) Exposed (EX) and (c) Aggregation (Ag) of mitochondria.

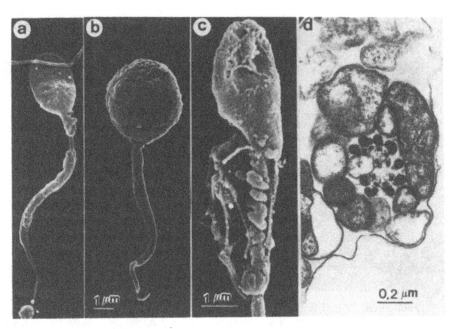

Fig. 5. Photomicrographs illustrating gross tail spermiogenetic failure (Profile V) (a,b) short and wide tail (same magnification), (c,d) axonemal disorder SEM and TEM respectively.

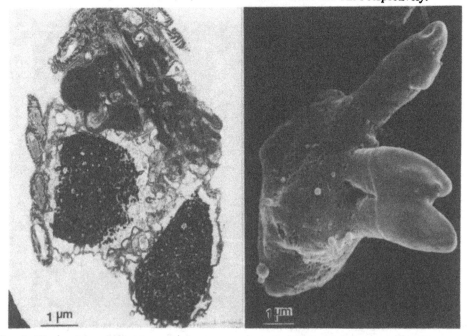

Fig. 6. Photomicrographs illustrating incomplete cytokinesis. (Profile VI). Multi-nucleated head and tail kinked toward midpiece. Left in TEM. Right in SEM.

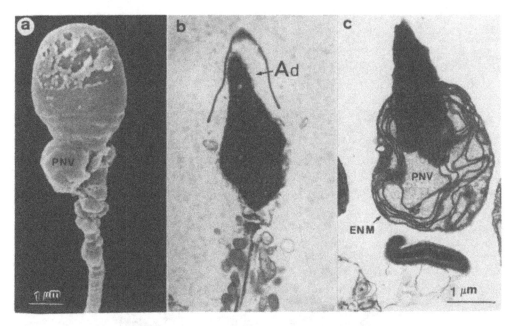

Fig. 7. Photomicrographs illustrating incomplete epididymal maturation (Profile VII). (a) Post nuclear vesicle (PNV, SEM), (b) Acrosomal detachment (Ad, TEM), (c) excess of nuclear membranes (ENM, TEM).

Table 9. Significant (p < 0.0001) Correlation between Ultramorphological Profiles and Different Parameters of Routine Semen Analysis

Ultramorphological profile	Viability % live (n = 1374)	Motility % mot., speed (n = 1374)		Seminal-plasma (n = 1374)
Gross spermiogenetic failure	−0.28	−0.36;	−0.28	−
Gross tail spermiogenetic failure	−0.28	−0.37;	−0.28	−
Nuclear elongation arrest	−	−0.27;	−	−
Incomplete epididymal maturation	−0.25	− ;	−	−
Minor sperm malformations	−0.25	−	−	−0.29*
				−0.37**

(−) no correlation
* Ca^{2+} level (mg %)
** Citrate level (mg %)

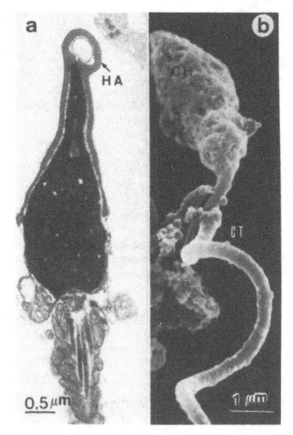

Fig. 8. Photomicrographs illustrating minor sperm malformations, (Profile VIII). (a) Hypoplasia of acrosome (HA, TEM). (b) Compressed head (CH) and coiled tail (CT, SEM).

DISCUSSION

Semen analysis contributes to the assessment of the male fertility potential by comparing test sperm parameters to those established for fertile males in the same laboratory or accepted by the WHO (1987). Although sperm morphology is considered to be a prominent and stable parameter of semen analysis of domestic animals, in humans, this parameter is hard to determine because of the heterogeneity of human sperm and the extremely small dimensions of most sperm organelles. QUM analysis was developed in order to overcome these difficulties. Clinical benefit from this analysis can be derived at different levels. At the basic level it permits estimation of sperm function potentials, using the organelle structure-function relationship. For example: tail, mitochondria and axonema are related to the motility potential, acrosome and post-acrosomal lamina to the penetration potential and the integrity of the nucleus to syngamy. Information at this level requires that the frequency of the malformations observed in each organelle be compared to those in normal fertile men.

A more sophisticated level of morphological information directed at assessing sperm function can be achieved from QUM analysis and the statistical method of discriminant analysis. This technique could be applied to samples of semen from men having either a positive or negative specific sperm potential. Thus, a score can be produced which can be used on the semen of men suspected of infertility to predict the existence of this specific sperm function. This level of information does not require a control group of fertile men.

The following morphological scores were established:

i. Motility Potential. Discriminant analysis of the TEM morphological characteristics of the highly motile fractions and those of submotile ones revealed a significant discriminant function. Thus, discriminant analysis permits that a morphological motility score based on eight discriminatory, weighed, morphological characteristics be attained. A positive score indicates that, from the morphological point of view, the sperm cells have a high motility potential and a negative score, that the sperm cells have a low motility potential. The morphological motility score enables the physician to compare this score with the actual motility obtained in routine semen analysis. When a discrepancy occurs between the two, i.e. positive score and asthenozoospermia, or what is called unutilized motility potential, one should be suspicious about the presence of antisperm antibodies or hostile seminal plasma. In both cases, sperm wash immediately after ejaculation is recommended.

ii. Hamster Ova Penetration Potential. Hamster ova penetration score was based on six morphological characteristics. A positive score indicates a high ova penetration potential of the human Hamster penetration assay and a negative score indicates a low penetration potential. Again a comparison between this score and the results of a patient's Hamster ova penetration assay might reveal the existence of unutilized penetration potential which may imply the presence of antisperm antibodies, possibly attached to the plasma membrane adjacent to the acrosome and post acrosomal lamina. In these cases, also, sperm wash immediately after ejaculation is recommended.

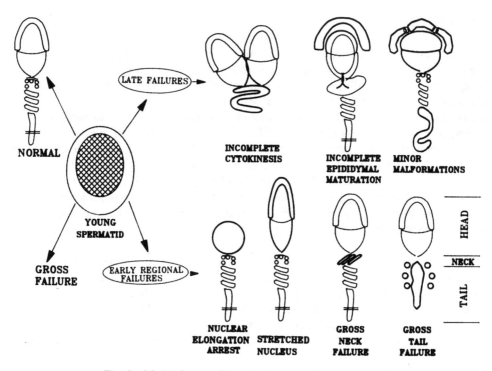

NORMAL

LATE FAILURES

INCOMPLETE
CYTOKINESIS

INCOMPLETE
EPIDIDYMAL
MATURATION

MINOR
MALFORMATIONS

YOUNG
SPERMATID

GROSS
FAILURE

EARLY REGIONAL
FAILURES

NUCLEAR
ELONGATION
ARREST

STRETCHED
NUCLEUS

GROSS
NECK
FAILURE

GROSS
TAIL
FAILURE

HEAD

NECK

TAIL

Fig. 9. Model for possible disturbances of spermiogenesis

iii. Swim-Up IUI Conception Potential. The morphological score of the conception success in the swim-up IUI technique was determined using only five morphological characteristics. The clinical importance of this score lies in its evaluation of the swim-up technique for infertile couples with a suspected male factor. A positive score should encourage the physician to pursue this technique. However, when this score is negative, the potential of success using this technique is low and the number of attempts in this direction should be limited.

iv. IVF Potential. The various parameters assayed in routine semen analysis, including sperm morphology by light microscope, were insufficient to predict success of IVF. This limitation of routine semen analysis was previously reported by Mahadevan and Trounson (1984), Cohen et al. (1985), Kruger et al. (1986) and Jeulin et al. (1986). However, using the QUM and discriminant analysis, it was possible to achieve IVF score using only four morphological characteristics. When the results of Hamster penetration assay were added to the QUM analysis, only two morphological characteristics: vesiculated acrosome viewed by TEM and post-acrosomal lamina viewed by SEM and the percentage of penetrated Hamster ova were sufficient to discriminate correctly 98% of the successful and unsuccessful cases of IVF attempts. A positive IVF score should be accompanied by the IVF intensity score. A positive intensity score indicates success in the fertilization of more than 50% of the retrieved ova; a negative score indicates less than 50% of the retrieved ova were fertilized. When both scores are positive, the physician should be encouraged to pursue this technique. When the IVF intensity score is negative, the advice of the IVF team is to choose a protocol allowing maximum egg retrieval. However, when the IVF score is negative, the number of the IVF trials should be limited. The IVF success score and the swim-up IUI conception success score which were calculated from the same ultramorphogram can assist the physician in his decision. When both scores are positive, the swim-up IUI technique should take preference since it is less expensive and less traumatic to the couple.

In this study we presented various function potentials of the human spermatozoa with prediction values ranging from 74% to 98%. We feel that QUM analysis when combined with the statistical method of discriminant analysis is a sensitive diagnostic and prognostic tool which can be utilized in the expanding field of assisted reproduction. A variety of scores can be produced when carefully preselected positive versus negative groups are presented. As for the scores already established, we feel that their sensitivity and specificity should be increased by enlarging the number of patients in each group. They should be verified on large independent, groups of patients.

In addition to the two levels of information achieved by QUM analysis, there is an additional level which may shed some light on the pathomorphology of human spermatozoa during spermiogenesis and epididymal maturation. To attain this, the statistical method of factor analysis following QUM analysis was used. From the eight morphological profiles presented in this study we would like to postulate the following possible disturbances of spermiogenesis and maturation of human spermatozoa. (Fig. 9).

It appears that testicular stress may cause gross spermiogenetic failure. If it takes place in the early stages, it is probably accompanied by serious malformations of most sperm organelles. This profile was the only one which correlated positively with the appearance of immature germinal cells in the ejaculate.

Other disturbances of spermiogenesis are also related to the early stages of development. However, these focussed on specific sperm regions, i.e. head, neck or tail. Four profiles followed this root: nuclear elongation arrest and stretched nucleus which are related to sperm head, gross neck failure and gross tail failure. The last three profiles are related to late failures: a) incomplete cytokinesis which most probably is related to the spermiation process of the seminiferous tubuli; b) incomplete epididymal maturation which include cytoplasmic droplets, access of nuclear membranes and detachment of the acrosome. Some of these malformations are perhaps due to insufficient formation of disulphide bridges during the epididymal transport; and finally, c) a profile which presents minor sperm malformations. This profile was the only one which was negatively correlated with the levels of the prostatic markers, Ca^{+2} and citrate found in the semen. It may be postulated that the malformations of this profile either occurred when sperm cells are exposed upon ejaculation to the seminal plasma or chronic prostatitis could cause a secondary effect on the late stages of sperm maturation.

In the future we hope to use these morphological profiles: (a) to discriminate the positive and negative groups of the various sperm function potentials described above. The integrated morphological component presented in the profile could produce more stable scores; (b) to evaluate different testicular stress situations such as infections, trauma, hormonal imbalance or response to drugs and irradiation treatments. Thus, QUM analysis supplemented with various statistical methods could serve as tools for monitoring the success of diverse treatments of male infertility.

ACKNOWLEDGEMENT

This work was supported in part by funds from the Health Sciences Research Center and the Lusinchi Research Fund, Department of Life Sciences, Bar-Ilan University .

REFERENCES

Bartoov, B., Eltes, F., Weissenberg, R. and Lunenfeld, B., 1980, Morphological characterization of abnormal human spermatozoa using TEM., *Arch. Androl.* 5:305-322.

Bartoov, N., Fisher, J., Eltes, F., Langsam, J. and Lunenfeld, B., 1981, A comparative morphological analysis of abnormal human spermatozoa. *in:* "Advances in Diagnosis and Treatment of Infertility," Insler, V., Bettendorf, G. eds, Elsevier North, Holland, 355-373.

Bartoov, B., Eltes, F., Langsam, J., Snyder, M. and Fisher, J., 1982, Ultrastructural studies in morphological assessment of human spermatozoa, *Int. J. Androl. Suppl.* 5:81-96.

Cohen,A. L. 1974, Critical point drying, *in:* "Principles and Techniques of Scanning Electron Microscopy," M.A. Hayat, ed., Reinhold Comp. pp. 44-105.

Cohen, J., Edwards, R., Pehilly, C., Fishel, S., Hewitt, J., Purdy, J., Rowland, G., Steptoe, P. and Webster, J., 1985, IVF: a treatment for male infertility, *Fertil. Steril.*, 43 No.3:422-432.

Eltes, F., Fisher, J. and Bartoov, B., 1981, Transmission electron microscope morphogram of progressive motile sperm cell fraction separated from teratozoospermic human Semen. *Arch. Andr.* 8:129-134.

Glavert,A. M, 1980, Fixation and embedding of biological specimens, *in:* "Practical Methods in Electron Microscopy," A. M. Glavert, ed, Elsevier-North, Holland.

Glezerman, M. and Bartoov, B, 1986, Semen analysis, *in:* "Infertility: Male and Female," Insler V, Lunenfeld, B., eds, Churchill, Livingstone, Edinburgh, London, Melborne and New-York, 243-271.

Jeulin, C., Feneux, D., Serres, C., Jouannet, P., Guilletnossos, F., Belaisch-allart, J., Frydman, R. and Testart. J., 1986, Sperm factors related to human IVF, *J. Reprod. Fertil.*, 76:735-744.

Kruger, T. F., Menkveld, R., Stander, F. S. H., Lombard, C. J., van der Merwe, J. T., van Zyl, J. A. and Smith, K., 1986, Sperm morphologic features as a prognostic factor in IVF, *Fertil. Steril.*, 46:No.6,1118-1123.

Mahadeven, M. M. and Trounson, A. O., 1984, The influence of seminal characteristics on the success rate of human IVF, *Fertil. Steril.*, 42: No. 3,400-405.

Norusis,M.J., 1985, "SPSS-X Advanced Statistics Guide", McGraw Hill Book Company, Chicago.

Nie, N. H., Heldi Hull, C., Jenkins, J. G., Steinberger, K. and Bent, D. H., 1975, "Statistical Package for the Social Sciences," Chap. 23, McGraw Hill, New-York.

Ryter, A. and Kellenber E., 1958, Etude au microscopic electronique de plasm contenant de e'acid desoxyribonucleiqe, Z. *Naturf.*13, 597.

World Health Organization WHO, laboratory manual for the examination of human semen and semen-cervical mucus interaction 1987, Cambridge University Press.

Zamboni, L., 1987, The ultrastructural pathology of the spermatozoa as a cause of infertility: the role of electron microscopy in the evaluation of semen quality. *Fertil. Steril.* 48:No.5, 711-734.

HEMIZONA ASSAY: ASSESSMENT OF HUMAN SPERMATOZOA DYSFUNCTION AND ITS CORRELATION WITH IN VITRO FERTILIZATION RESULTS

Sergio Oehninger, Richard Scott, Suheil J. Muasher,
Charles C. Coddington, Anibal A. Acosta, and Gary D. Hodgen

Jones Institute for Reproductive Medicine, Department of Obstetrics and Gynecology
Eastern Virginia Medical School
Norfolk, Virginia

INTRODUCTION

The hemizona assay (HZA) has been introduced as a new diagnostic test for the binding of human spermatozoa to human zonae pellucidae to predict fertilization potential.[1] In the HZA the two matched zona hemispheres created by microbisection provide several advantages: (1) the two halves (hemizonae) are functionally (qualitatively) equal zona surfaces, allowing a controlled comparison of binding; (2) the limited number of available human oocytes is amplified, since an internally controlled test can be performed in a single oocyte; (3) ethical objections to possible inadvertent fertilization of a viable oocyte are eliminated by using nonliving human oocytes cut into halves.[1]

Previous studies examined the feasibility of sperm binding to the hemizonae as well as the kinetics of tight sperm binding, showing maximal binding after four to five hours of coincubation.[1,2] Normal spermatozoa show equivalent binding to dimethylsulphoxide (DMSO; 2.0 M solution)-frozen stored and salt (magnesium chloride, $MgCl_2$; 1.5 M solution)-stored human hemizonae. Yanagimachi et al.[3] demonstrated earlier that the zona pellucida retains sperm binding potential following salt storage. Salt storage offers a simple and inexpensive means of accumulating and transporting human zonae pellucidae while maintaining functional properties for estimating sperm binding potential in the HZA.[2]

Advances in Assisted Reproductive Technologies, Edited by
S. Mashiach *et al.*, Plenum Press, New York, 1990

Franken et al.,[4] using DMSO-treated hemizonae, reported higher binding for patients achieving fertilization during in vitro fertilization (IVF) than for those patients with failure of fertilization. Tight zona binding was significantly and positively correlated with the percentage of motile sperm, the percentage of normal morphology, and seminal sperm concentration. In this previous report, although patients used as controls in the HZA had sperm counts of $\geq 20 \times 10^6$/ml with motility of $\geq 30\%$, the majority were infertile, and some had a slight degree of teratozoospermia.

Here, a prospective, blinded study was designed for two purposes: (1) to further assess the relationship between tight binding in the HZA and IVF outcome, and (2) to determine the sensitivity, specificity, and predictive value of the HZA. For optimal results, two strict criteria were established: (1) only proven fertile donors with normal semen parameters were used as internal controls in each test and (2) only IVF patients with at least two mature preovulatory oocytes at retrieval were included in the study.

This study further differed from previous work in that only one source of oocytes and one storage condition was used (human oocytes recovered from surgically removed ovaries and stored in salt solution) and that evaluation of male factor patients included a thorough work-up, as described below.

MATERIALS AND METHODS

Patients

A total of 28 men participating in the Norfolk IVF program (series 30 and 31, February through June 1988) were allocated to the study group. All presented after two or more years of infertility and had had at least three previous semen analyses. Seventeen patients had sperm concentrations of $\geq 20 \times 10^6$/ml, motility of $\geq 50\%$, and normal morphology.[5, 6] There were 11 male factor patients: two had oligozoospermia (< 20 million/ml), two had asthenozoospermia (< 50% motility), four had teratozoospermia ($\leq 14\%$ normal forms), and three had asthenoteratozoospermia. All had tested negative for serum and semen antisperm antibodies and bacteriologic studies.[7] Five patients had been further evaluated by a hamster oocyte/human sperm penetration assay (SPA), according to published protocols using BWW as the incubation medium and 3% bovine serum albumin as the protein source in a short incubation (6 hours) protocol.[8, 9]

Controls

Four proven fertile donors, who had fathered a child in the last two years, provided semen used as an internal control in each test. All specimens used had a sperm concentration of $> 80 \times 10^6$, motility of $\geq 50\%$, and normal morphology of $\geq 14\%$.[5, 6]

Semen Analysis

Semen evaluation was performed using an automated computerized system (Cell Soft Semen Analyzer, Labsoft Division of Cryo Resources, Ltd., New York, NY), as previously described.[4] Briefly, sperm images were analyzed at a rate of 15 frames/second. Lower limits included (1) sperm velocity $\geq$ 10 μm/second, (2) $\geq$ 1 frame of data for determining cell motility, and (3) a sperm head size of 4 to 25 pixels. The pixel size was set at 0.688 microns. For specimens with very low sperm counts or decreased motility, the sperm concentration was calculated using a hemacytometer; percentage of motility was assessed by scoring 100 sperm/microscopic slide. After sperm liquefaction was completed, slides were carefully prepared for morphologic assessment. The slides were stained with Diff Quick (American Scientific Products, McGraw, IL), and morphology was evaluated according to the strict criteria presently used in the Norfolk program.[5, 6, 10]

IVF/Embryo Transfer Procedures

Ovarian stimulation, oocyte retrieval, classification of oocytes, insemination, and embryo transfer were performed according to previously published protocols.[11, 13] All women had at least two preovulatory oocytes (metaphase I and II) at retrieval. The inseminating concentration varied from 50,000 sperm/ml/oocyte (non male-factor patients) to 500,000 sperm/ml/oocyte (male factors).[14]

Oocyte Handling and Cutting

Oocytes were obtained from surgically recovered ovarian tissue, following protocols already described.[1, 2, 15] Specimens were obtained from women aged 25 to 45 years who were undergoing pelvic surgery for benign and malignant gynecologic conditions. Zona-intact, immature oocytes (prophase I) were denuded of granulosa cells and placed in small plastic vials containing 0.5 ml of 1.5 M $MgCl_2$ (Mallinkrodt Chemical Works, St. Louis, MO), 0.1% polyvinylpyrrolidone, and 40 mM of Hepes Buffer, and stored at room temperature (23°C) for 6 to 30 days.

Before the assay, the desired number of oocytes was removed from salt storage and rinsed for 15 minutes in Ham's F-10 culture medium (Gibco Co., New York, NY) containing 7.5% heat-inactivated human fetal cord serum.Narishige micromanipulators (Narishige, Tokyo, Japan) were mounted on a phase-contrast inverted microscope (Nikon Diaphot, Garden City, NY) and used for cutting the oocytes. A detailed description of the cutting procedure has been published.[1, 2, 4]

Sperm Handling for HZA

An aliquot of semen (0.5 ml) was mixed with 1 ml of Ham's F-10 culture medium, supplemented with 7.5% human fetal cord serum, then centrifuged for five minutes at $400 \times g$. After a second wash, the final sperm pellet was overlaid with 0.5 ml of Ham's F-10, then incubated at 37°C in 5% CO_2 in air to achieve a swim-up separation of vigorously motile spermatozoa. After one hour the supernatant was removed and the sperm were used for the HZA.

Experimental Design

Twenty-eight HZA tests were performed. In 17 assays, infertile men with normal semen parameters were evaluated, while in 11 assays male factor patients were studied. In each of the 28 experiments, one salt-stored oocyte cut into halves provided one hemizona which was coincubated with the patient's semen; the matching hemizona was coincubated with the sperm of the fertile control. All patients were asked to provide a second semen sample for the HZA at 48 hours after insemination of the oocytes for IVF. Personnel carrying out the HZA had no prior knowledge of the patients' clinical history or IVF outcome.

During the IVF attempt, successful fertilization occurred in 22 cases, while poor or failed fertilization was observed in six cases (all male factor patients). A threshold value of 65% fertilization rate of preovulatory oocytes in IVF was used to distinguish group 1 (successful fertilization) from group 2 (poor fertilization). This cut-off value represents the 95% confidence interval of 112 couples undergoing IVF simultaneously, with a diagnosis of tubal factor and normal semen parameters, and fertilization of at least two preovulatory oocytes (mean fertilization rate of 90.1%).

All assays were done with 100 µl of the processed semen with a motile concentration of 500,000 sperm/ml. All sperm drops were held under oil in 35×10 mm plastic dishes (Falcon Plastics, Oxnard, NJ). The hemizonae were removed after four hours of incubation, and each was rinsed by quickly pipetting five times to dislodge loosely associated spermatozoa. The number of tightly bound spermatozoa on the outer surface was then counted with a phase-contrast microscope (200 ×).[1,2,4]

Statistical Analysis

All results are expressed as mean ± standard error of the mean. The two-tailed student's t test was used to compare the mean number of spermatozoa bound to the hemizonae, the hemizona index[1] (number of spermatozoa bound for patients/controls × 100), and semen parameters for patients with successful versus poor fertilization. As discussed above, a threshold value of 65% fertilization rate of preovulatory oocytes in IVF was used to distinguish these two groups. Group 1 had successful fertilization of ≥ 65% of preovulatory oo-

Table 1. Sperm Parameters and HZA Results According to IVF Outcome

	Group 1 (Fertizilation rate of preovulatory oocytes $\geq$ 65%) n = 22		Group 2 (Fertilization rate of preovulatory oocytes <65%) n = 6
PART A			
Patients' IVF Samples	Group 1	P Value	Group 2
concentration	116.0 ± 16.6	0.1 > p > 0.05	49.0 ± 14.5
% motility	71.3 ± 3.3	p<0.01	31.5 ± 6.3
% normal morphology	9.7 ± 1.3	0.1 > p > 0.05	4.3 ± 2.1
Patients' HZA Samples			
concentration	93.3 ± 13.9	0.2 > p > 0.1	51.0 ± 12.3
% motility	54.2 ± 3.4	p<0.02	33.1 ± 1.4
HZA binding	62.1 ± 10.9	p<0.02	7.3 ± 1.4
HZA index	120.4 ± 18.3	p<0.02	29.5 ± 7.5
PART B			
Controls' HZA Samples			
concentration	150.5 ± 13.3	p > .5	142.8 ± 16.6
% motility	64.6 ± 1.7	p > .5	66.8 ± 3.5
HZA binding	59.5 ± 11.2	p > .5	40.6 ± 11.9 [a]

[a] p<0.02 (Controls' vs patients' HZA binding within group 2, paired T test)

cytes. Group 2 had poor fertilization of < 65% of preovulatory oocytes. In the Norfolk program, patients with tubal infertility,[16] endometriosis,[17] DES exposure,[18] and unexplained infertility[19] have a fertilization rate of > 65% of preovulatory oocytes.

Within each of the two groups, comparison of the mean number of spermatozoa bound to the hemizonae from patients versus each internal control (proven fertile donors) was made by a paired t test.

RESULTS

Table 1 presents sperm parameters (concentration, percentages of motility and morphology) and HZA results (number of tightly bound sperm and HZA index) in patients and controls according to IVF outcome (group 1: ≥ 65% fertilization rate; group 2: < 65% fertili-

Table 2. Ability of HZA to Predict IVF Outcome

HZA Index	Poor Fertilization	Good Fertilization	TOTAL
<36	5	1	6
>36	1	21	22
TOTAL	6	22	28

zation rate). In the group with poor fertilization in vitro (n = 6), four patients had failure of fertilization (0% fertilization rate). In two cases 50% of the preovulatory oocytes were fertilized. As we have noted in our laboratory, in these cases of failed fertilization, low numbers of spermatozoa were observed at the level of the zona pellucida, with no sperm cells in the perivitelline space. Of the classic sperm parameters, groups 1 and 2 differed significantly only in the percentage of motility (p < 0.01) (Table 1, Part A). The mean number of tightly bound sperm in the HZA and the HZA index were significantly lower in the patients with poor fertilization rates in IVF (p < 0.02). Part B of Table 1 shows sperm parameters and HZA binding for the fertile controls used in the HZA for groups 1 and 2. Paired analysis showed significantly lower HZA binding in infertile patients than in controls within the group with poor IVF results (p < 0.02).

Objective comparisons of the matching hemizonae data are possible by calculating the ratio of the HZA index of tightly bound sperm. The scattering of data provided a natural breakpoint at an HZA index value of 36, allowing us to evaluate the HZA as a predictor of IVF results. Table 2 depicts the sensitivity (83%), specificity (95%), and the positive and negative predictive values (83% and 95%, respectively) of the HZA based on the present results.

The HZA results according to sperm quality are shown in Table 3. Sperm concentration, percentages of motility[20] and morphology,[5,6] mean number of sperm bound in the HZA, and the HZA index differed significantly when infertile patients with normal semen values were compared with male factor patients. Six of the 11 male factor patients correspond to the group with poor IVF results (< 65% fertilization rate).

Table 4 compares HZA, SPA, and IVF results in five of the male factor patients and in one non male-factor infertile patient. In this small sample, the HZA results correlated better than the SPA with IVF outcome.

DISCUSSION

This prospective, blinded study provides further evidence of the ability of the HZA to predict IVF outcome. With a sensitivity of 83%, a specificity of 95%, and an incidence of

Table 3. HZA Results According to Sperm Quality in Patients with Normal Semen and in Male Factor Patients

	Normal Sperm Parameters [a]	Male Factor Patients
	(n = 17)	(n = 11)
concentration	126.0 ± 18.2	64.2 ± 17.6 [b]
% motility	75.5 ± 2.5	42.9 ± 6.5 [c]
% normal morphology	12.1 ± 1.2	3.1 ± 0.9 [c]
HZA binding	72.2 ± 17.6	16.7 ± 5.0 [d]
HZA index	124.5 ± 23.2	59.2 ± 13.0 [b]

[a] Sperm concentration > 20 million/ml, motility > 50%[20], and $\geq$ 14% normal morphology[5,6]
[b] $p < 0.05$
[c] $p < 0.001$
[d] $p < 0.005$

Table 4. Comparison of the Mean No of Tightly-bound Sperm in the HZA, HZA Index, SPA, and IVF Results

Patient	Diagnosis	HZA Binding	HZA Index	SPA Patient vs Control %	Fertilization Rate of Preovulatory Oocytes in IVF %
1	asthenozoospermia	4	14	7 vs 54	0
2	teratozoospermia	7	35	23 vs 43	0
3	oligozoospermia	11	64	6 vs 32	0
4	oligozoospermia	13	16	0 vs 65	50
5	normal sperm	68	340	0 vs 76	80

one false-positive and one false-negative in 28 experiments, the HZA discriminated by identifying patients at risk for poor or failed fertilization. More experiments are needed to establish the threshold HZA index value and the intra-assay variation. The threshold HZA index value selected here (36) was chosen arbitrarily, considering the natural scattering of data found here. In addition, to maximize application of the HZA and its power of discrimination, it may be necessary to establish a minimal number of tightly bound sperm (probably 20 to 30, according to our current information)[21] for the fertile control specimen, thus assuring an acceptable oocyte binding capacity.

Fertilization of human oocytes is a complex process. Binding of spermatozoa to the zona pellucida represents an early, critical stage requisite for oocyte fertilization. Our results show a positive correlation between HZA results and IVF outcome. However, it is also important to realize that other physiological steps follow the binding of sperm to the zona and are necessary to achieve fertilization and development: sperm passage through the zona, binding to the oolema, penetration into the oocyte, sperm decondensation, and finally pronuclei formation and cleavage. Thus the observation of false-positive results should be an expected finding. False-negative results could be the consequence of inter-egg variation (variation in the binding ability of the oocytes) or from variation in interassay handling. Establishment of a minimal binding threshold for the control specimen may overcome its occurrence.

The interaction of human sperm and zonae pellucidae, as evaluated by the HZA, can thus provide important information, together with the percentage of motility and normal forms of a semen sample, in anticipating IVF outcome. In a previous study[4] we showed a positive correlation between the percentage of normal forms and the number of tightly bound sperm during the HZA. Ongoing experiments are underway to establish the minimal threshold for the total motile fraction of a semen sample required to perform and validate the HZA and to make the test results reliable in severely oligospermic patients.

SPA tests for sperm capacitation and acrosome action, as well as for penetrating ability and decondensation in a heterologous system. The HZA, using a homologous system, gives additional valuable information on spermatozoa-zonae pellucidae interaction. It is interesting that in a small sample of our population, the HZA could better predict IVF results than the SPA. This may suggest that dysfunctional sperm-zona binding may be the underlying cause in cases of failed fertilization, both in IVF and in natural reproduction. In IVF, sperm abnormalities are associated with a high number of cases of failed fertilization.[22]

In the current context of modern techniques for assisted fertilization (microinjection of a single spermatozoon into an oocyte[23] or under the zona pellucida[24] and zona drilling[25]), the HZA may provide physiopathological support for their use. When a dysfunctional sperm-zona interaction is diagnosed, these techniques—if their success is finally validated—may constitute a logical therapeutic alternative to be offered to patients.

In conclusion, in the HZA, male factor patients and patients with poor IVF rates have lower binding ability than selected fertile controls. The HZA is a valuable tool for evaluating dysfunctional sperm-zona binding, with good predictive value for IVF results.

REFERENCES

1. L. J. Burkman, C. C. Coddington, D. R. Franken, T. F. Kruger, Z. Rosenwaks and G. D. Hodgen, The hemizona assay (HZA): development of a diagnostic test for the binding of human spermatozoa to the human hemizona pellucida to predict fertilization potential, *Fertil Steril.*, 49:688 (1988).
2. D. R. Franken, L. J. Burkman, S. Oehninger, L. L. Veeck, T. F. Kruger, Z. Rosenwaks and G. D. Hodgen, The hemizona assay using salt stored human oocytes: evaluation of zona pellucida capacity for binding human spermatozoa, *Gamete Res.*, 22:45 (1988).

3. R. Yanagimachi, A. Lopata, C. B. Odom, R. A. Bronson, C. A. Mahi, G. L. Nicolson, Retention of biologic characteristics of zona pellucida in highly concentrated salt solution: the use of salt-stored eggs for assessing the fertilizing capacity of spermatozoa, *Fertil. Steril.*, 31:562 (1979).

4. D. R. Franken, S. Oehninger, L. J. Burkman, C. C. Coddington, T. F. Kruger, Z. Rosenwaks, A. A. Acosta and G. D. Hodgen, The hemizona assay (HZA): a predictor of human sperm fertilizing potential in IVF treatment, *J. In Vitro Fert. Embryo Transfer*, 6:44 (1989).

5. T. F. Kruger, A. A. Acosta, K. F. Simmons, R. J. Swanson, J. F. Matta, L. L. Veeck, M. Morshedi and S. Brugo, A new method of evaluating sperm morphology with predictive value for in vitro fertilization, *Urology* 30:248 (1987).

6. T. F. Kruger, A. A. Acosta, K. F. Simmons, R. J. Swanson, J. F. Matta and S. Oehninger, Predictive value of abnormal sperm morphology in in vitro fertilization, *Fertil. Steril.*, 49:112 (1988).

7. J. F. H. M. Van Uem, A. A. Acosta, R. J. Swanson, J. Mayer, S. Ackerman, L. J. Burkman, L. Veeck, J. S. McDowell, R. E. Bernardus and H. W. Jones Jr., Male factor evaluation in in vitro fertilization: Norfolk experience, *Fertil. Steril.*, 44:375 (1985).

8. R. J. Swanson, J. F. Mayer, K. H. Jones, S. E. Lanzendorf and J. McDowell, Hamster ova/human sperm penetration assay: correlation with count, motility and morphology for in vitro fertilization, *Arch. Androl.*, 12:69 (1984).

9. T. F. Kruger, R. J. Swanson, M. Hamilton, K. F. Simmons, A. A. Acosta, J. F. Matta, S. Oehninger and M. Morshedi, Abnormal sperm morphology and other semen parameters related to the outcome of the hamster oocyte human sperm penetration assay, *Int. J. Androl.*, 11:107 (1988).

10. T. F. Kruger, S. B. Ackerman, K. F. Simmons, R. J. Swanson, S. S. Brugo and A. A. Acosta, A quick, reliable staining technique for sperm morphology, *Arch. Androl.*, 18:275 (1987).

11. A. A. Acosta, R. E. Bernardus, G. S. Jones, J. E. Garcia, Z. Rosenwaks, S. Simonetti, L. L. Veeck and D. Jones, The use of pure FSH alone or in combination for ovulation stimulation in in vitro fertilization, *Acta Eur. Fertil.*, 16:81 (1985).

12. H. W. Jones Jr., A. A. Acosta, M. C. Andrews, J. E. Garcia, G. S. Jones, J. Mayer, J. S. McDowell, Z. Rosenwaks, B. A. Sandow, L. L. Veeck and C. A. Wilkes, Three years of in vitro fertilization at Norfolk, *Fertil. Steril.*, 42:826 (1984).

13. L. L. Veeck and M. Maloney, Insemination and fertilization, *in:* "In Vitro Fertilization - Norfolk," H. W. Jones Jr., G. S. Jones, G. D. Hodgen, Z. Rosenwaks, eds., Williams and Wilkins, p. 168, Baltimore (1986).

14. S. Oehninger, A. A. Acosta, M. Morshedi, L. L. Veeck, R. J. Swanson, K. Simmons and Z. Rosenwaks, Corrective measures and pregnancy outcome in in vitro fertilization in patients with severe morphology abnormalities, *Fertil. Steril.*, 50:283 (1988).

15. J. W. Overstreet, R. Yanagimachi, D. F. Katz, K. Hayashi, F. W. Hanson, Penetration of human spermatozoa into the human zona pellucida and the zona-free hamster egg: a study of fertile donors and infertile patients, *Fertil. Steril.*, 33:534 (1980).

16. A. A. Acosta, J. van Uem, S. B. Ackerman, J. F. Mayer, J. F. Stecker, R. J. Swanson, P. Pleban, J. Yuan, C. Chillik and S. Brugo, Estimation of male fertility by examination and testing of spermatozoa, *in:* "In Vitro Fertilization - Norfolk," H. W. Jones Jr., G. S. Jones, G. D. Hodgen, Z. Rosenwaks, eds., Williams and Wilkins, p. 126, Baltimore (1986).

17. S. Oehninger, A. A. Acosta, D. Kreiner, S. J. Muasher, H. W. Jones Jr. and Z. Rosenwaks, In vitro fertilization and embryo transfer: an established and successful therapy for endometriosis, *J. In Vitro Fert. Embryo Transfer*, 5:249 (1988).

18. S. J. Muasher, J. E. Garcia, and H. W. Jones Jr., Experience with diethylstilbestrol-exposed infertile women in a program of in vitro fertilization, *Fertil. Steril.*, 42:20 (1984).

19. D. Navot, S. J. Muasher, S. Oehninger, H-C Liu, L. L. Veeck, D. Kreiner and Z. Rosenwaks, The value of in vitro fertilization for the treatment of unexplained infertility, *Fertil. Steril.*, 49:854 (1988).

20. World Health Organization: Laboratory Manual for the Examination of Human Semen and Semen-Cervical Mucus Interaction, Cambridge University Press, p. 27 (1987).
21. G. D. Hodgen, L. J. Burkman, C. C. Coddington, D. R. Franken, S. C. Oehninger, T. F. Kruger and Z. Rosenwaks, Hemizona assay (HZA): finding sperm that have the "right stuff," *J. In Vitro Fert. Embryo Transfer*, 5:311 (1988).
22. S. Oehninger, A. A. Acosta, T. F. Kruger, L. L. Veeck, J. Flood and H. W. Jones Jr., Failure of fertilization in IVF: the "occult" male factor, *J. In Vitro Fertil. Embryo Trans.*, 5:181 (1988).
23. S. Lanzendorf, M. K. Maloney, L. L. Veeck, J. Slusser, G. D. Hodgen and Z. Rosenwaks, A preclinical evaluation of pronuclear formation by microinjection of human spermatozoa into human oocytes, *Fertil. Steril.*, 49:835 (1988).
24. A. Laws-King, A. Trounson, H. Sathananthan, and I. Kola, Fertilization of human oocytes by microinjection of a single spermatozoon under the zona pellucida, *Fertil. Steril.*, 48:637 (1987).
25. J. W. Gordon, Use of micromanipulation for increasing the efficiency of mammalian fertilization in vitro, *Ann. N.Y. Acad. Sci.*, 541:601 (1988).

INTRODUCTION TO THE TECHNOLOGY
OF MAGNETIC RESONANCE IMAGING IN INFERTILITY

Rudolf Klimek

Ob/Gyn Institute
Copernicus University School of Medicine, Cracow, Poland

INTRODUCTION

For more than one hundred years, investigators all over the world have been trying to develop techniques by which visualization of the body's interior, firstly only cavities, would be possible. The pioneer of this field, Phillip Bozzini,[1] wrote in 1807 "As it is impossible to obtain from chemistry a substance capable of illuminating the interior of the body cavities, and thus render these accessible to our eye, by means of a straight tube inserted therein, we have to conduct light from outside through the inserted tube."

From the time of his system built of mirrors and tubes with candlelight, through the "speculum utero-cystique" of P. S. Segelas (1821) to the first commercial cystoscopy of A. J. Desormeaux (1853), and, still later the first handbook of hysteroscopic technique published by S. Duplay and S. Clado (1898) and only recently leading to nuclear magnetic resonance (NMR) utilization, can we see both exterior and interior surfaces as well as whole atomic structures of the pelvic organs while examining a woman who is completely dressed.

What for P. Bozzini seemed impossible, illumination and visualization of the body cavities is now an everyday reality, without anesthesia or the need to insufflate or fill the body cavities. Furthermore, the natural fluids and secretions which originate in the highly complex reproductive organs may simultaneously be analyzed. Even the unquestionably useful contact microhysteroscopy, pioneered by K. Okhawa, will soon be replaced by NMR mi-

Advances in Assisted Reproductive Technologies, Edited by
S. Mashiach *et al.,* Plenum Press, New York, 1990

croscopic analysis. Photographic documentation of endoscopic examinations are also no longer difficult.

The introduction of NMR methods has changed many indications for other imaging techniques, e.g. hystero- and salpingoscopy, uterosalpingography, laparoscopy, "second look" operation, and diminishes the risk and complications of their therapeutic applications, e.g. cervical laceration, endometrial trauma, uterine perforation, and infections. These methods can even be used when there are contraindications to other procedures, such as infections or profuse bleeding.

Since the introduction of in vitro fertilization IVF) and embryo transfer (ET), many other techniques have been developed (intrauterine, intrafallopian and also intraperitoneal gametes transfer). Gamete intrafallopian transfer (GIFT) is recognized as highly successful, partially due to the fact that this method mimics some of the physiological mechanisms which permit fertilization at the natural site. All the above procedures also receive great support from the diagnostic possibility of the NMR technique which allows adequate simultaneous evaluation of developing embryo and endometrial environment. Finally, using NMR techniques and nonmagnetic instruments we can once again perform in vivo instead of in vitro fertilization. The exception to this are women with primary and premature ovarian failure.

The aim of this paper is to show that modern NMR technology not only decisively eases the diagnosis and treatment of infertility, but simultaneously indicates the necessity of limiting human reproductive experiments prior to adequately investigating those in vivo occurring natural processes which are currently available.

CLINICAL UNDERSTANDING OF NMR TECHNOLOGY

Magnetic resonance imaging and spectroscopy (NMR-S) are the methods of choice in contemporary diagnosis of infertility. In promoting their greater utilization we need to overcome certain barriers, among others, economical ones as well as achieving a better understanding of this newest of medical technologies.

Observed images and spectroscopic data in this technique are only the average results of processes which take place at the atomic level where matter (structure) and energy (function) are inseparable. Because of this, we can accurately evaluate them only from the scope of thermodynamics, although this requires changing many previous views and practices within medicine which for the past 150 years have relied on cellular theory.

NMR arises directly from radio-signals of atomic particles regardless of their site, whether they are present at the subcellular, cellular or tissue level as well as parts of body fluids or air in body cavities. As atoms, they react immediately to energetic and material changes in their environment, while the reorganization of subcellular and cellular structures need more time to react. Only then can they be recognized by USG and CT. For example, transient tissue ischemia is immediately localized by a three-dimensional MR image with the simultaneous spectroscopic possibility of detecting and evaluating metabolic changes while, for these same changes to be recognized by either USG or CT, a longer period of time elapses before being detectable. USG and CT scanners cannot register these rapid changes especially when they are reversible. The molecular events occur with awesome rapidity, often in less than 10^{-12}s, thus establishing the macroscopic picture. With a new technique (laser femtochemistry), it is now possible to observe chemical bond breaking, or bond formation on the femtosecond (10^{-15}s) time scale.[2] By using the NMR microscope, the observation of molecular motion of individual cells is a near clinical reality.

The main point of magnetic imaging is our ability to view only that part of the body which unites itself with two separate magnetic fields. Thus, observed phenomena made up of atoms are simultaneously structurally and energetically conditioned.

In medicine these considerations are essential for understanding the technology of development of NMR images. These images are conditioned, among other things, by magnetic field application, which constitutes a great energetic influence upon observed atoms. Thus, the appearance of observed body parts is dependent not only on their biological status, but also simultaneously on technological specifications of the scanner, values of the main magnetic field and the radio-frequency pulses as well as time elapsed between them. Furthermore, physicians can use special paramagnetic contrast agents to achieve better picture resolution and depth penetration.

Presently, real-time imaging at 0.5-2 T in a time of only 20ms could be obtained in any plane (axial, sagittal, and coronal) based on the gynecologist's preference and technical knowledge, which is of utmost importance while using specialized local receiver radio-frequency coils.

From the perspective of time, the initial determination of NMR tissue relaxation times reminds one of past early vivisections although they utilize the newest of technologies. There is nothing strange in the fact then that collection of samples for testing relaxation times inspired P.C. Lauterbur[3] to investigate these parameters in vivo. Therefore, thanks to spectroscopic investigations, a new field of imaging by NMR was originated which consequently changed contemporary medicine by permitting the examination of an organism's structure and function on an atomic level.

To overcome the many differences between relaxation times and imaging on the one hand, and existing medical views and practices on the other, required the unification of a singular concept of all heretofore separately treated phenomena on a biological, biochemical and biophysical level. By a beneficial coincidence, together with the progress in NMR knowledge, great leaps in the development of theoretical physics took place which led to the broadening of the 2nd Law of Thermodynamics through the description of self-organizing dissipative structures;[4] enabling an entirely new approach to come into existence, and allowing for a singular thermodynamic theory of human health and illness.[5]

NMR relaxation times were verified as nonspecific markers in the organism's state.[6] This halted the period of divergence of clinical estimation of relaxation times and obtained images, heretofore improperly interpreted by contemporary medical knowledge, which endeavoured to explain spatial-dynamic phenomena in a unilateral plane by either morphological, chemical or physical means.

New concepts are at times better introduced as examples rather than being strictly defined. Let us take the anatomical and histological examination of the uterus. Its division of peri-, myo-, and endometrium is an anatomical exercise whereas endometrial division into its two layers lies within the realm of histology or histopathology after their removal from the organism.

Currently, utilizing NMR in a fully dressed subject, we can view the abovementioned layers in appropriate enlargement on a monitor, both anatomically and histologically, and variations in their thickness, and even observe blood flow in real-time. Furthermore, in any part of this image not only can we designate relaxation times but we can also examine the concentrations of chosen chemical compounds during various stages of health or illness. Thus, NMR enables us to examine material without the necessity of removing tissues from the body and also allows more specific and precise testing in situ encompassing histological, biochemical, and physiological parameters.

In humans we can examine, in real-time and simultaneously, all processes from the atomic level up to and including their psycho-emotional changes seeing, quite simply, their internal state in a whole-body dynamic relationship.

NMR's diagnostic possibilities are well known though medicine's atomic era will truly begin at the moment when utilization of diagnostic and therapeutic methods within a magnetic field with nonmagnetic material is employed. This will be most readily observed and become apparent in a majority of in vivo and in vitro fertilization procedures.

NMR, which enables some women to fertilize again in vivo as opposed to their ability to conceive only through in vitro fertilization, has restored human dignity, and is not only limited to specific medical aspects.

INDICATION FOR IVF/ET IN TERMS OF NMR TECHNOLOGY

Similarly to other imaging modalities, we can take an ovum and place it in vitro to fertilize it, resulting in the creation of a human being, which is then placed back into its natural environment for further development. But two years ago, together with W. Froncisz and A. Fraczek[7] during an international NMR Symposium in Cracow, we showed that with the help of special NMR surface coils, the new possibility of in vivo fertilization exists. If we replace ultrasonographic techniques and add nonmagnetic instruments, NMR can be used to allow fertilization in vivo instead of removing ova from the body. Thanks to this new technique we eliminate the need for in vitro fertilization in women with ovaries.

The method described above reminds us of the now-forgotten gynecological procedure of implanting ovaries directly into the uterine wall enabling the egg close proximity with spermatozoa even with damaged or destroyed oviducts. In the past we transplanted the whole organ or simply attached its surface to the uterine cavity. This did not meet with any ethical or moral opposition, inasmuch as it was regarded as a purely medical activity, not relying on the removal of the ovum beyond the organism into an artificial, in vitro, environment. Presently we can do precisely that. Instead of translocating the entire organ we transfer a single ovum and, if need be, also its adjacent surroundings to the place of natural fertilization. Thus, in this way, theoretical and technical progress unravels a purely humanistic issue of contemporary man. It limits to a minimum the contact of gametes with an artificial environment as represented by the guided nonmagnetic needle used to transport them.

It is precisely this predictable and technically easy to perform procedure that not only resolves difficult technical issues but, simultaneously, prevents man from being threatened by an artificial environment during procreation, ensuring his natural reproduction and furthering survival of the species. Eliminating an artificial environment is important in the light of testing neoplastic cells in vitro. Purely mechanical, manipulation-induced damage to their membrane, in general, leads to a quickening of the progression of neoplastic growth. We should bear in mind that application of the IVF/ET method can lead to innumerable consequences resulting in mutation of individual zygotes left for even a short moment in a nonbiological environment. In terms of medical thermodynamics intentional micromechanical manipulation of zona pellucida or gamete surfaces or even application of a zona solvent is not to be tolerated. It is sufficient to be acquainted with reviews discussing the factors important in the incorporation or integration of biomaterials and devices by tissues. Biocompatibility of certain materials is constantly being improved upon through interdisciplinary cooperation between physicists, chemists, engineers, biomedical and clinical scientists.

We cannot forget that progress in physics led not only to the abovementioned possibility of in vivo fertilization where only recently in vitro fertilization was possible, but also to change our general understanding of results of these procedures.

The successful fertilization of an individual ovum by only one spermatozoon, chosen from amongst several millions, may be acknowledged. But does this success belong to the investigator or to the self-defense of life locked within the zygote, one of the biologically most vital cellular structures?

The possibilities of delivering a healthy child fertilized in vitro are less than those in vivo, inasmuch as the in vitro one dies more readily.[8] The negative influence of an artificial environment is greater than a real, natural one, as well as being more dangerous. It should not be forgotten that the process of evolution perfected the reproductive mechanisms over millions of years, whereas contemporary medicine wants to imitate the same, over a short period of experimentation covering only a few decades.

From a theoretical thermodynamic point of view, outlasting naturally-arising zygotes commands a huge selection in nature and is a result of cooperation between many factors. A 70% or 80% successful birth rate by naturally fertilized egg cells is a barrier which is continuously being changed, based on our ability to rapidly recognize the appearance of pregnancy.

Initially, biochemical methods, followed by biophysical ones, shifted our diagnosis of existing pregnancy closer towards the moment of fertilization. NMR microscopes will, in the near future, allow the inspection in reproductive organs of not only egg cells, but spermatozoa as well. This will enable us to confront the precise moment at which fertilization takes place and will again reduce the number of statistically successful pregnancies.

In accordance with thermodynamical laws singular fertilization is a process of simultaneous cooperation between the mother's organism, with her hundreds-of-thousands of eggs from both ovaries, and her maturing eggs just prior to ovulation, together with millions of sperm cells from one ejaculate. Just after fertilizing, zygotes in accordance with Nature's law die in far greater numbers as the moment of fertilization nears. In comparison with these natural phenomena, in vitro methods appear overly simplistic. (Fig.1)

An additional reason exists in order to conduct ourselves in accordance with ethico-moral issues. While we have the possibility of testing spectroscopically concentrations of various substances throughout the process of fertilization, we still need to perform massive testing in vivo before anyone will desire to produce different media for in vitro fertilization. Thanks to NMR we can already accurately calculate pH, concentrations of chemical compounds, blood flow, or followup nidation and organization of the embryo's immediate environment and the energetic state of its cells. We can act precisely in the same way as we took advantage of techniques in animal experiments in the past to give birth to the first child beyond the mother's body. Prior to developing NMR in vivo methods, such an action was justified. Nowadays this information directly from a human body is readily available.

We can justly say that NMR, through its natural union with medicine, changed it not

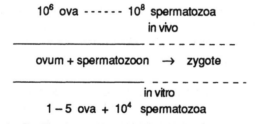

Fig. 1. Fertilization in terms of medical thermodynamics.

only in a biological way, but in a humanistic domain as well. By allowing women to choose in vivo fertilization in preference to and instead of in vitro methods, NMR is also returning a sense of human dignity to us.

CONCLUSIONS

The best illustration in concluding this paper is the indication for performing diagnostic testing in a magnet room with full anesthesia.

Specific modifications of anesthesia, nonmagnetic machines and carts, laryngoscopes, mercury sphygmomanometers, oximeters, and remote blood pressure devices are successfully being used to provide patient monitoring and anesthesia support in high-field magnetic resonance imaging installations. Thus, the ability to not only fertilize in vivo as mentioned above, but also to perform surgery under high-magnetic fields, has been achieved in practice.

On the other hand, most recent introductions of more economical NMR diagnostic techniques utilizing an ultra-low-magnetic field (0.02-0.06 Tesla) will also allow the use of ferromagnetic instruments and materials in the magnet room. Thus, in this brilliant way physics again opened up a new era in medicine. The only problem is that mankind must take advantage of these developments in a humanistic way and do so in a reasonable amount of time.

REFERENCES

1. P. Bozzini, Der Lichtleiter, Weimer Landes-Industrie-Comptoir (1807).
2. A. H. Zewall, Laser femtochemistry, *Science*, 242:1645 (1988).
3. P. C. Lauterbur, Image formation by induced local interactions, *Nature*, 71:97 (1973).
4. I. Prigogine, "From Being to Becoming", W. H. Freeman and Company, San Francisco (1980).
5. R. Klimek, Neurologic aspects of the dissipative structure of tumors, *Mat. Med. Pol.*, 42:91 (1980).
6. R. Klimek, P. C. Lauterbur, M. H. Mendonca-Dias, A discussion of nuclear magnetic resonance (NMR) relaxation time of tumors in terms of their interpretation as self-organizing dissipative structures, and of their study in vivo by NMR zeugmatographic imaging, *Gin. Pol.*, 53:493 (1981), *Mat. Med. Pol.*, 15:149 (1983).
7. R. Klimek, W. Froncisz, A. Fraczek, Possibilities in applying double surface coils in fertilizing in vivo as opposed to in vitro (Polish), NMR Seminary, p. 213, Cracow (1987).
8. H. W. Jones, Jr., and Ch. Schrader, "In Vitro Fertilization and Other Assisted Reproduction", *Ann. N.Y. Acad. Scien.*, 541 (1988).

DOPPLER ULTRASOUND IN INFERTILITY

Rajat K. Goswamy

The Fertility and IVF Centre
The Churchill Clinic, London SE1 7PW, and
Dept. of Obstetrics and Gynaecology
St. Thomas's Hospital, London SE1 7EH

In previous publications I have reported that uterine response to endogenous hormonal changes in the spontaneous ovarian cycle can be demonstrated with use of doppler ultrasound techniques (Goswamy and Steptoe, 1988a and Goswamy et al., 1988b). We demonstrated that uterine perfusion increases in response to rising estrogen levels during the follicular phase, decreases in the preovulatory phase in response to the preovulatory estrogen fall, and increases in the luteal phase in response to the combined effect of estrogen and progesterone. We also reported that in conception cycles uterine perfusion continues to increase in the late luteal phase in contrast with non-conception cycles where there is premenstrual decrease in perfusion as a result of falling progesterone levels, thus postulating that doppler ultrasound techniques could be used to diagnose pregnancy prior to the date of expected onset of menses.

In our second publication, we used the methodology, previously devised by us, to study uterine perfusion response in patients who had failed to conceive despite repeated multiple embryo replacement in in vitro fertilization (IVF) cycles. These patients were studied in spontaneous ovarian cycles and despite normal endocrine changes and ultrasound and basal body temperature changes in ovulatory cycles there was an inadequate uterine perfusion response in 50% of the patients recruited in the study. However, the numbers of patients reported in that paper were not large enough to prove by statistical analysis, that impaired perfusion is a cause of infertility.

Advances in Assisted Reproductive Technologies, Edited by
S. Mashiach *et al.,* Plenum Press, New York, 1990

This paper presents larger numbers of patients in this ongoing prospective study to confirm the hypothesis that decreased uterine perfusion response to endogenous hormone changes in spontaneous ovarian cycles is a cause of infertility, hitherto unconfirmed. It also describes a new quantitative method to measure uterine perfusion which will be shown to be superior to conventional formulae used in flow velocity wave form analyses in doppler ultrasound examinations.

MATERIAL AND METHODS

At the time of presentation of this paper, 254 patients had been recruited for this study. Recruitment was on the basis that three previous IVF attempts had been unsuccessful in achieving pregnancy, despite good gamete quality, in patients under the age of forty years; whereas patients over the age of forty years were included if two previous attempts had been similarly unsuccessful.

Our criteria for assessing gamete quality have been described previously (Howles et al., 1986; Steptoe et al., 1986). Briefly, if mean urinary levels were below 0.20 I.U./L and more than 60% of all oocytes recovered were fertilized by the sperm we presume that gamete quality was satisfactory.

In the initial two-and-a-half years of the study Doppler ultrasound studies were performed using a Philips SDD 600 spectrum analyser with a 3 MHz Doppler transducer which was offset on a 3.5 MHz imaging transducer attached to a Philips SDR 1550 imaging system (Philips Ultrasound Inc., CA, USA). Scans were performed abdominally with the patient's bladder full enough to visualise the uterine fundus in the longitudinal plane. An oblique longitudinal scan was performed to visualise the pulsations of the ascending branch of the uterine artery and the doppler gate was placed over this area to obtain the typical uterine artery wave form as previously described by Taylor et al. (1985). The uterine artery could be easily differentiated from the internal iliac artery by its shape and the auditory signal obtained was low in frequency and pitch as compared to the high pitched and harsh internal iliac artery sound. In the latter part of the study Doppler ultrasound studies were performed using a 7.5 MHz vaginal imaging transducer, attached to the Siemens SL2 ultrasound machine with built in 3 MHz Doppler crystals to obtain flow velocity wave forms from the ascending branch of the uterine artery in the oblique longitudinal plane.

The shape of the flow velocity wave forms was similar to those obtained with the abdominal route.

All studies were obtained during spontaneous ovarian cycles and the wave form classification and modified resistance index and modified S/D ratio previously described was noted.

A new computer programme using the Appleton Floscan system was used to devise another index, where the area under the systolic component of the flow velocity waveform

(*S) was divided by the area under the diastolic component (*D). We have called this ratio (*S/*D) the Perfusion Index (PeI).

However, until adequate numbers of patients were recruited onto the study to analyse the PeI, we continued to classify wave forms into Type O,A,B,C as previously described (Goswamy et al., 1988). Briefly, Type O and Type A waveforms obtained in the midsecretory phase were taken to indicate decreased perfusion of the uterine vascular bed, and Type B or Type C wave forms were assumed to indicate good uterine perfusion.

Patients with good uterine perfusion (Type B or C) were designated Group I patients and those with poor uterine perfusion were allocated Group II.

Group I patients were advised to go ahead with another IVF attempt with ovarian stimulation as in previous attempts. Group II patients were prescribed tablets Cycloprogynova 2 mg containing oestradiol valerate 2 mg and Norgesterel 0.5 mg (Schering Pharmaceutical Ltd., West Sussex), to be taken for twenty-one days each cycle starting on the fifth day of the period.

Repeat Doppler studies were then performed after the patients had been on this treatment for three months. Patients with midsecretory waveforms Type B or C were now considered to have satisfactory perfusion and advised to attempt IVF once again. In Group II patients ovarian stimulation was carried out as before and the patients continued to take oestradial valerate tablets 2 mg throughout the follicular and luteal phase.

Patients with Type O or A waveform despite cyclical were allocated to a third group and designated Group IIa. These patients had ovarian stimulation, oocyte recovery, IVF and subsequent cryopreservation of resulting embryos.

This was followed by further therapy using rising doses of oestradiol valerate until good Doppler wave forms were observed (Type B or C), and after progesterone supplementation as previously described thawed embryos were replaced in utero.

STATISTICS

Pregnancy rates of patients in Groups I, II, IIa and first attemptors were compared using Chi-square analyses, as were the pregnancy rates between Groups I, II, IIa and fourth attemptors.

RESULTS

A total of 254 women were recruited onto the study at the time of presentation of this paper, and 197 (77.5%) of these patients had completed hormone therapy and had another IVF attempt. Results of these 197 treatment cycles are presented here.

Table 1. Wave Form Classification Compared to Perfusion Index (PeI) in 187 Flow Velocity Wave Forms

	Mean ± S.D.
Type A (n = 65)	*,** 3.11 ± 0.87
Type B (n = 46)	* 1.98 ± 0.58
Type C (n = 76)	** 1.81 ± 0.50

* A B p + < 0.001; ** A C p = < 0.001; B C p = 0.12

Group I consisted of 107 (54.3%) women and Group II and IIa consisted of 90 (45.7%) patients. Nine (10%) out of the 90 patients were subsequently allocated to Group IIa and underwent high dose hormone therapy as decribed above.

The PeI was noted in 187 flow velocity wave forms and its correlation with the wave form classification is shown in Table 1.

Using Chi-square analysis, the mean PeI was highly significant between Type A and Type B wave forms (p = < 0.001) and between Type A and Type C wave forms (p = < 0.001). There was no significant difference between Type B and Type C wave forms (p = 0.12).

Therefore the PeI falls in the early follicular phase, rises in the pre-ovulatory phase, falls in the early luteal phase, continues to fall in conception cycles and rises prior to menses in non-conception cycles. This response to endogenous estrogen rise in the follicular phase, and a combination of estrogen and progesterone in the luteal phase is similar to that reflected by the wave form classifications previously described.

Patients in Group I were treated with IVF therapy using the same ovarian stimulation as in previous attempts. At the time of presentation of this paper 107 of these patients had completed therapy.

Patients in Group II were treated with cyclical hormone therapy, using tablets of Cyclo-progynova as previously described, and doppler studies were performed to assess uterine response in the mid-secretory phase. Of the 90 patients who had completed hormone therapy at the time of presentation of this paper, 58 (64%) achieved a Type C response, 22 (24%) achieved a Type B response. Ten (11%) women failed to achieve adequate uterine response with low dose therapy and were allocated to Group IIa. Patients in Group IIa were treated with rising doses of estrogen therapy in combination with progesterone therapy as in cases being treated for donor oocytes (Goswamy et al. 1988). The results of subsequent IVF therapy for patients in Group II and Group IIa are shown together because the number of patients in Group IIa is too small for meaningful statistical analysis.

Table 2.   Comparison of Subsequent IVF Therapy in Group I, II, and IIa Patients with First Attemptors

	Number of Patients	Number of Pregnancies	Pregnancy Rate (%)
First Attempt	1828	429	26.9
Group I	107	31	28.97
Group II & IIa	90	24	26.67

p = N.S.

Table 3.   Comparison of Subsequent IVF Therapy in Group I, II and IIa Patients with Women who have had 4 or more Attempts

	Number of Patients	Number of Pregnancies	Pregnancy Rate (%)
4 or more attempts	316	54	* 17.4**
Group I	107	31	* 28.97
Group II & IIa	90	24	- 26.67**

* p = < 0.001   ** p = < 0.021

Table 2 shows the results of subsequent IVF therapy of patients in Group I and Group II. The pregnancy rate in these groups is compared with first attemptors, using Chi-square analyses. There was no significant difference in the pregnancy rate between any of these groups.

Table 3 compares the outcome of subsequent IVF therapy in Groups I and II with patients who had four or more previous embryo replacements. There was a highly significant different between the pregnancy rate achieved by Group I women fourth attemptors (p = < 0.001) and a significant difference when the latter was compared to Group II (p = < 0.02).

Of the 10 patients in Group IIa, three (30%) achieved pregnancy with frozen-thawed embryos being replaced in cycles with rising doses of estrogen and progesterone therapy.

DISCUSSION

This paper has assessed uterine perfusion in women who have failed to achieve pregnancy despite repeated IVF attempts. In addition to confirming findings in previous publications, it also presents two additional methods to improve the use of this invetigation in the management of infertile women.

Firstly, by using vaginal ultrasound techniques, patients can be scanned with an empty bladder. This is not only more comfortable for patients, it also increases the accuracy of wave form analysis. I have described the effect of an excessively full bladder causing false positive diagnosis of increased uterine resistance, such that Type A wave forms may be converted to Type B or C wave forms by asking the patient to partially void the urinary bladder. This can be easily obviated with the use of vaginal ultrasound examinations.

Secondly, the use of the PeI indicates a significant difference between Type A and Type B wave forms. In a previous publication, there was no significant difference between the modified resistance index or the modified S/D radio, when these were used to differentiate between Type A and Type B waveforms (Goswamy et al., 1988b). By differentiating between Type B and Type A wave forms it is now possible to numerically separate the patients with poor perfusion from those with good uterine perfusion response.

Finally the hypothesis that decreased uterine perfusion is a cause of infertility is closer to being proved by the results from the ongoing study. If poor uterine perfusion is a cause of failed implantation in IVF patients, then the fall in pregnancy rates in fourth or subsequent attempts may be attributed to this. That is, if patients with good uterine perfusion were allowed to go ahead with another IVF attempt, their pregnancy rate should be the same as the first, second or third attemptors.

The converse would apply if patients with poor uterine perfusion were allowed to go ahead without any hormone therapy prior to another IVF attempt. However, giving hormone therapy does improve uterine perfusion in these patients and the pregnancy rate in subsequent attempts does improve significantly.

One could therefore conclude that decreased uterine perfusion is a cause of failed implantation. This would confirm that it is a cause of infertility hitherto unproven.

The reason why the uterus does not respond normally to endogenous hormone changes in spontaneous cycles is unknown. I would postulate that there seems to be some sort of desensitization of the receptors in the uterus to endogenous hormones which seem to wake these receptors to increase uterine perfusion, thereby increasing pregnancy rates.

If the mean pregnancy rate in the first three IVF attempts is 25% then in an arbitrary group of a hundred women 40 would not be pregnant after three attempts. Half of these would have poor uterine perfusion, i.e. 20% of all women undergoing IVF therapy! I end with a question. Should all patients have doppler studies prior to their first IVF attempt?

REFERENCES

Goswamy, R. K., and Steptoe, P. C., 1988, Doppler ultrasound studies of the uterine artery in spontaneous ovarian cycles. *Human Reprod.* 3, 6:721-726.

Goswamy, R. K., Williams, G., and Steptoe, P. C., 1988b, Decreased uterine perfusion - a cause of infertility, *Human Reprod.* 3, 8:955-959.

Howles, C. M., Macnamee, M., Edwards, R. G., Goswamy, R. K., and Steptoe, P. C., 1986, The effect of high tonic levels of luteinizing hormone on outcome of in vitro fertilization, *Lancet*, ii:521-522.

Steptoe, P. C., Edwards, R. G., and Walters, D. E., 1986, Observations 767 clinical pregnancies and 500 births after human in vitro fertilization, *Human Reprod.*, 1:89-94.

Taylor, K. J. W., Burns, P. N., Wells, P. N. T., Conway, D. I., and Hull, M. G. R., 1985, Ultrasound doppler flow studies of the ovarian and uterine arteries, *Brit. Med. Obstets. & Gynaecol.*, 92:240-246.

TRANSVAGINAL COLOR DOPPLER IN THE
STUDY OF UTERINE PERFUSION

Asim Kurjak and Ivica Zalud

*Ultrasonic Institute, Medical School University of Zagreb
Pavleka Miskine 64 41.000 Zagreb, Yugoslavia*

INTRODUCTION

The introduction of transvaginal Doppler sonography has markedly contributed to the refinement and accuracy of ultrasound diagnosis. In addition to numerous new morphological information, it offers new insights into the dynamic studies of blood flow within the female pelvis. This is particularly obvious when transvaginal color Doppler is used.[1] The most important advantage of this new and exciting diagnostic tool is the display of blood flow over the whole female pelvis, as compared with only one line of sight available with conventional pulsed Doppler.

We refer here to our own preliminary results obtained with transvaginal color Doppler technique in infertile patients.

SUBJECTS AND METHODS

Our study consisted of a group of 75 women. This was made up of 19 healthy fertile, eight postmenopausal and 48 infertile women. Transvaginal color Doppler studies were performed after B-mode ultrasound and classical gynecological examinations.

A 5 MHz transvaginal color Doppler probe was used for visualization of pelvic anatomy and blood flow studies. In addition to color Doppler, the machine (Aloka color Doppler

Advances in Assisted Reproductive Technologies, Edited by
S. Mashiach *et al.*, Plenum Press, New York, 1990

Table 1. Resistance Index (RI) in the Uterine Artery

Patients	N	Uterine artery (RI)	
		Left	Right
Fertile	19	0.86 ± 0.04	0.85 ± 0.07
Infertile	48	0.85 ± 0.06	0.86 ± 0.05
Postmenopausal	8	0.87 ± 0.08	0.89 ± 0.08
Total	75		

SSD 350, Aloka Co. Japan) is equipped with a conventional pulsed Doppler system which can be used simultaneously with B-mode imaging.

Color Doppler was used in an attempt to observe blood flow in both uterine arteries and to select an appropriate sample for spectral analysis. Peripheral impedance was then assessed using resistance index (RI) by pulsed Doppler.[2]

RESULTS

Both the left and right uterine arteries were easily seen in all subjects. The ascending branch of the uterine artery was seen just laterally to the cervix by performing an oblique longitudinal scan in all cases.

The brightness of color flow in postmenopausal women was generally not prominent and not easy to detect. No difference was detected in quality of color flow (i.e. brightness and pulsatility) in both uterine arteries between fertile and infertile women. More detailed analysis was performed using pulsed Doppler, where velocity with high pulsatility and very low diastolic flow was found. Some differences in peripheral impedance among subgroups of women were noticed (Table 1). However, a statistically significant difference could not be shown between the left and the right uterine artery. The RI was almost equal in the fertile group (left uterine artery: 0.86 ± 0.04; right uterine artery: 0.86 ± 0.05) and infertile group (left uterine artery: 0.85 ± 0.06; right uterine artery: 0.86 ± 0.05) and slightly higher in postmenopausal women (left uterine artery: 0.87 ± 0.09; right uterine artery: 0.89 ± 0.04). There were no statistically significant differences in RI. The diastolic flow was always present in all investigated subjects.

DISCUSSION

In the past five years the transvaginal approach for pelvic sonography has been suggested by many authors.[3-6] Blood flow studies were also performed using pulsed Doppler in order to assess a normal blood flow pattern and its physiological variations according to menstrual cycle.[7,8] We are the first group to study pelvic blood flow using transvaginal color Doppler. This technique is simple to perform and easy to interprete. Blood flow is presented by color and the information on location, direction and velocity of flow can be studied. In the majority of commercially available units, flow towards the transducer is coded in red, and flow away from the transducer is coded in blue. Turbulence is coded as the amount of green and yellow mixed with red or blue, resulting in a mosaic appearance. Flow velocity is directly proportional to the color brightness. To quantify blood flow pulsed Doppler focused on certain color area was used. The waveform patterns obtained allowed numerical analysis. Our results did not show a significant difference in quality of color flow in the observed groups of women.

Goswamy and Steptoe performed a transabdominal pulsed Doppler study of the uterine artery.[7] Their results indicated increasing uterine perfusion with rising levels of plasma oestradiol and progesterone and a direct correlation with falling oestrogen levels in the follicular phase. They concluded that Doppler study can be used effectively to monitor uterine response to the hormone environment.

Animal studies suggested that older animals are less fertile because of decreased uterine perfusion.[9] It may be hypothesized that the fertility of older animals may be enhanced by improving uterine perfusion, and that poor uterine perfusion may be responsible for failure of implantation of embryos. Goswamy et al. published their preliminary data which revealed a poor mid-secretory uterine response in 48% of the patients studied.[10] Our results did not show poor uterine perfusion in infertile women, and quality of color flow and RI were almost the same as in the cases of normal, fertile women.

To sum up, our study has shown that transvaginal color Doppler is a new, promising imaging modality for the assessment of uterine perfusion. However, further studies are needed before routine use is recommended.

REFERENCES

1. A. Kurjak, D. Jurkovic, Z. Alfirevic, and I. Zalud, Transvaginal color Doppler, *J. Clin. Ultrasound* 1989 (in press).
2. L. Porcelot, Applications cliniques de l'examen Doppler transcutane, *in:* INSERM, 7–11 Octobre, P. Personneau, ed., 213–240 (1974).
3. E. I. Timor-Tritsch, Y. Bar-Yam, S. Rotem, et al., The technique of transvaginal ultrasonography by the use of a 6.6 mHz transducer probe, *Am. J. Obstet. Gynecol.*, 158:1019–1024 (1988).
4. I. E. Timor-Tritsch, and S. Rotem, Transvaginal ultrasonographic study of the Fallopian tube, *Obstet. Gynecol.*, 70:424–428 (1987).
5. S. R. Schwimer, and J. Lebovic, Transvaginal pelvic ultrasonography, *J. Ultrasound Med.*, 3:381–384 (1984).
6. S. R. Schwimer, and J. Lebovic, Transvaginal pelvic ultrasonography: accuracy in follicle and cyst size determination, *J. Ultrasound Med.*, 4:61–66 (1985).
7. R. K. Goswamy, and P. C. Steptoe, Doppler utrasound studies of the uterine artery in sponta-

neous ovarian cycles, *Human Reproduction*, 3:721–726 (1988).

8. M. G. Long, J. E. Boultbee, M. E. Hanson, and R. H. J. Begent, Doppler time velocity wave-form studies of the uterine artery and uterus, *Br. J. Obstet. Gynaecol.*, 96:588–593 (1989).

9. C. E. Finch, and R. G. Gosden, Animal models for the human menopause, *in:* "Aging Repro-duction and the Climacteric", L. J. Mastroiani, and C. A. Paulsen, eds., Plenum, New York, 3–34 (1986).

10. R. K. Goswamy, G. Williams, and P. C. Steptoe, Decreased uterine perfusion - a cause of in-fertility, *Human Reproduction*, 3:955–958 (1988).

BLOOD FLOW MEASUREMENTS
IN THE UTERINE AND OVARIAN ARTERIES
IN PATIENTS UNDERGOING IVF/ET AND IN EARLY PREGNANCY,
PERFORMED WITH A TRANSVAGINAL DOPPLER DUPLEX SYSTEM

Israel Thaler, Joseph Itskovitz, Dorit Manor, and Joseph Brandes

Department of Obstetrics and Gynecology 'A,' Rambam Medical Center
Technion - Israel Institute of Technology, Haifa, Israel

INTRODUCTION

Recent technological advances in ultrasound imaging have contributed greatly to patient care in the fields of obstetrics and gynecology. Of particular importance was the recent development of the high frequency transvaginal sonography, which is emphasized by its extensive application in in vitro fertilization and embryo transfer (IVF/ET) programs. Such applications include monitoring of follicular development, follicle aspiration and scanning of the endometrium, the ovaries and other normal and abnormal pelvic structure. An extensive review on this subject is given elsewhere.[1]

A new and exciting development has emerged with the introduction of the transvaginal Doppler duplex systems. These are obtained by incorporating a pulsed, range-gated Doppler to the transvaginal real time scanner, transforming it into a high quality "flow probe" which offers a novel approach to the non-invasive hemodynamic evaluation of the female pelvis, both in the pregnant and non-pregnant situation. In this review the physical principles and the methodology of the transvaginal Doppler sonography will be described, followed by the results obtained in patients undergoing IVF/ET, both during the treatment cycle and in early pregnancy.

Advances in Assisted Reproductive Technologies, Edited by
S. Mashiach *et al.*, Plenum Press, New York, 1990

PHYSICAL PRINCIPLES

As is implied by its name, a duplex system consists both of a real-time scanning mode and a Doppler mode.

Imaging of pelvic structures

The female pelvis contains various soft tissue structures which have a similar acoustic impedance, and are therefore poor reflectors. The main advantage of using the transvaginal probe, is the short distance between it and the scanned organs. This proximity to the transducer makes it possible to increase its frequency, typically between 5–7 MHz, while attenuation is still acceptable at these distant ranges. As the vagina is rather an elastic organ, the probe can be manipulated to bring it as close to the scanned organ as possible, thereby placing it in the focal region of the transducer. The net result is a significant increase in image resolution, obtained by the high frequency of the transducer and by the application of a stronger focus. When a transvaginal duplex system is concerned, the application of a high frequency transducer which is placed near the vessel of interest generates high quality images, reflecting an axial resolution of 0.5 mm and a lateral resolution of 1.3 mm. Under such conditions, even as small a vessel as 1–2 mm can be picked up. Another important feature is the ability to magnify the image substantially ('zooming'), which, albeit at the cost of a somewhat decreased resolution, highlights small anatomical structures such as blood vessels. A more extensive review on the physical principle of transvaginal sonography is given elsewhere.[2]

Doppler Blood Flow Mmeasurements

The proximity of the probe to the pelvic vessels, also permits the application of a higher frequency Doppler transducer, thereby obtaining higher Doppler shifts at any given angle of insonation. Moreover, when high flow velocity is encountered, the limitations of the range-velocity product are largely overcome.[3] With the transvaginal approach, Doppler flow measurements of some pelvic vessels can be obtained at a smaller angle of insonation, thereby increasing the accuracy.[3] Transvaginal Doppler flow measurements are performed with an empty bladder, which is an important advantage over the transabdominal approach, where the bladder is fully distended. This causes ovarian and uterine displacement and can modify the impedance to flow in the vessels supplying these organs. All these properties turn the transvaginal Doppler duplex system into a versatile, accurate and indispensable tool for measuring blood flow characteristics in the female pelvis.

METHODOLOGY

All the measurements which will be described were taken with an Elscint ESI 2000

Doppler duplex system employing a 6.5 mechanical sector probe, especially designed for convenient intravaginal insertion and manipulation. A 5 MHz pulsed Doppler transducer is incorporated into the imaging probe. The transducer crystal is housed in a rounded spherical acoustic window, which enables good tisue contact.[1] During the examination the patient is lying in the supine position on a two level matress, so that the upper part of the body is on a higher level. With the knees bent and slightly elevated, the feet are resting on the lower level, some 30 cm below the upper one. This makes it possible for the examiner to manipulate the vaginal probe at different angles, applying the push-pull technique,[1] and to locate the vessels under study. A coupling gel is applied to the vaginal probe, which is then introduced into a rubber glove. The probe is then introduced into the vaginal fornix and a real-time image of the vessel is obtained. The line of insonation of the Doppler beam is adjusted (by manipulating the probe) so that it crosses the long axis of the vessel at as small an angle as possible. The sample volume is then placed to cover the entire cross-sectional area of the vessel. At this stage the system is switched to the dual mode operation, in which the two dimensional scanning and the range-gated pulsed Doppler operate in a quasi-simultaneous fashion. The flow velocity waveforms are displayed on the lower half of the screen after spectral analysis in real-time. Once a good quality signal is obtained, based on audio recognition, visual waveform recognition and maximum measured velocity, the image is frozen, including at least three waveform signals. A thump filter of 200 Hz is used to eliminate signals originating from vessel wall movements. At this point various calculations are performed:

1. The ratio between peak systolic and end diastolic flow velocity—commonly termed S/D ratio.
2. The difference between peak systolic and end diastolic flow velocity, divided by the maximum velocity of flow during systole. This parameter is also termed the resistance index - RI.
3. Volume flow rate - this measurement is based on the product of the mean velocity and the cross sectional area of the vessel. The latter is calculated from the vessel diameter (determined as the distance between two electronic calipers which are positioned over the vessel walls). The instantaneous flow rate is then integrated over time, to give the minute volume flow, expressed as milliliters per minute.

While the first two measurements are not angle dependent, the latter one is. This angle of insonation is determined by the user as the angle between the line of insonation (which is projected on the video screen) and the long axis of the vessel which is being measured. The calculations and analyses are displayed on an additional video display unit and are recorded on a video recorder. An immediate hard copy can also be obtained using a video printer.

ANATOMICAL CONSIDERATIONS

A schematic representation of the transvaginal probe in relation to the pelvic vessels is shown in Figure 1.

The main pelvic vessels which can be studied by the transvaginal Doppler duplex system are:

1. The uterine artery
2. The ovarian artery
3. The hypogastric artery
4. The arcuate artery
5. Fetal vessels

This review will focus mainly on the uterine and ovarian arteries.

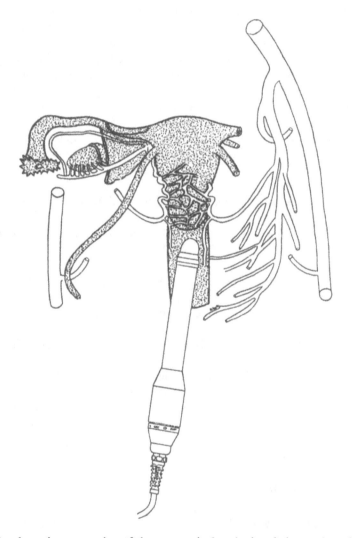

Fig. 1. A schematic presentation of the transvaginal probe in relation to the pelvic vessels.

The uterine artery: the ascending branch of the uterine artery is located in the parametrial region, at the level of the internal cervical os. The latter can be readily visualized, adjacent to the cervical canal. During diastole, relatively high flow rates are maintained, particularly in the luteal phase and in pregnancy. This 'attenuates' the pulsatility of the vessel, making it more difficult for the examiner to detect the arterial pattern which is so common in high resistance vessels (e.g. the hypogastric and the internal iliac artery). While the direction of flow is generally away from the transducer (negative velocity), it is possible in many instances to identify flow towards it, originating from the cervical branch of the uterine artery.

The ovarian artery: the ovarian artery originates from the aorta and reaches the ovary via the infundibulo-pelvic ligament. It can usually be detected lateral to the ovary in the inferior aspect (i.e. closer to the vaginal probe and in the direction of the pelvic wall). As the uterine artery forms anastomoses with the ovarian artery medial to the ovary and on the inferior border of the Fallopian tube, this approach eliminates erroneous measurements of signals originating from the distal branches of the uterine artery. Care should also be exercised so as to avoid sampling the hypogastric arteries, which are located adjacent to the lateral ovarian border. Fortunately, signals originating from these vessels demonstrate a distinct pattern which is different from those of the ovarian artery. Flow velocity signals can also be elicited from within the ovary itself, demonstrating a much lower impedance (reflected by the substantially higher flow velocity during diastole). However it is usually impossible to visualize these small intraovarian arterial branches, and to measure their diameter.

BLOOD FLOW IN THE UTERINE AND OVARIAN ARTERIES DURING CYCLE STIMULATION AND AFTER FOLLICLE PUNCTURE AND EMBRYO TRANSFER

Flow measurements of the uterine and ovarian arteries were obtained from 10 IVF/ET protocol patients who were treated with gonadotropins. The Doppler studies were performed every two to three days, starting from the third day of the cycle, prior to treatment. The day of hCG administration was marked as day 0 and all measurements were expressed in relation to this time (e.g. the first day of treatment was usually marked as day -9 to -7, and ET was performed on day +2). Daily measurements of serum E_2, progesterone and follicular growth were obtained in each patient. The Doppler measurements were obtained with an Elscint 2000 duplex scanner with a 6.5 MHz imaging probe to which a 5 MHz Doppler transducer was coupled. The details of the measurements were described in the previous section. Each vessel will be described seperately.

The Uterine Artery

The changes in the resistance index throughout the treatment cycle is shown in Figure 2. The values represent the left uterine artery. There was no difference between either side.

The values obtained initially before treatment were relatively low and demonstrated a gradual increase until the day of follicle aspiration. A rapid decline occurred thereafter throughout the luteal phase. Blood flow in the uterine artery during the treatment cycle is shown in Figure 3. Increased volume flow rates were measured prior to treatment. Blood flow then decreased gradually reaching the lowest levels at about day 0. Within 24 hours from the administration of hCG, there was a steep increase in flow rates in this vessel throughout the luteal phase.

The peri-ovulatory increase in RI and the decrease in volume flow rates coincided with the decrease in E_2 levels at midcycle. The subsequent increase in flow rates and the rapid decline in RI were associated with the exponential increase in progesterone levels in the luteal phase. The lower RI and the increased volume flow at the beginning of the cycle may express the situation in the late luteal phase of the preceding cycle (which was always untreated in this group of patients). The changes observed in the follicular phase reflect a 'readjustment' of uterine blood flow in the current treated cycle.

In spontaneous ovarian cycles, the resistance index in the uterine artery fell with a rise in estrogen level during the early follicular phase and increased in the periovulatory phase along with the drop in this hormone.[4] A decrease in uterine impedance was noted in the luteal phase along with rising progesterone levels. In this study a transabdominal off-set pulsed Doppler transducer (3 MHz) along with a mechanical sector imaging transducer (3 MHz) was used. While cases were observed in which no diastolic flow was present,[4] we did not find such a pattern in the uterine artery either before or following treatment i.e diastolic forward flow was consistently present at all phases of the cycle. However, such diastolic

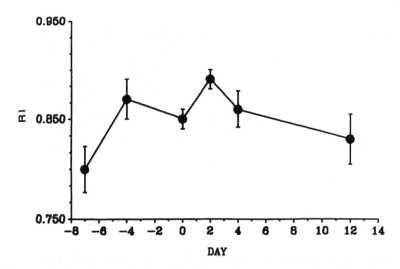

Fig. 2. Changes in the resistance index of the left uterine artery throughout the treatment cycle.

flow was not always continuous throughout diastole, particularly in the early follicular phase. In our experience, the transvaginal duplex system is superior to the transabdominal approach, as described above.

Ovarian steroids have been shown to alter α-1-adrenergic receptor numbers in periarterial sympathetic nerves and cause marked changes in uterine blood flow in gilts.[5] Injections of E_2 to non-pregnant sheep caused an increase in pelvic arterial blood flow.[6] Lowering of serum E_2 levels by GnRH agonists was associated with a decrease in diastolic blood flow in the uterine arteries, in patients with uterine fibroids.[7]

The Ovarian Artery

The changes in the resistance index in the left ovarian artery throughout the treatment cycle is shown in the upper part of Figure 4.

A midfollicular decline in the impedance to flow was observed, with a subsequent rise in day 0. A gradual steady decline in the RI occurred in the luteal phase. The flow profile was also obtained from within the ovarian substance itself as described above. The resistance index in this "luteal artery" is shown in the lower part of Figure 4. There were no significant changes in this vessel throughout the treatment cycle, although the values displayed an increased variability in the luteal phase. Volume flow in the ovarian artery increased from 40 ml/min prior to treatment to 70 ml/min in mid-cycle and 90–100 ml/min in the late luteal phase. Within the treatment cycle then, the ovarian flow more than doubled itself. No significant difference was found between the left and the right side for any of

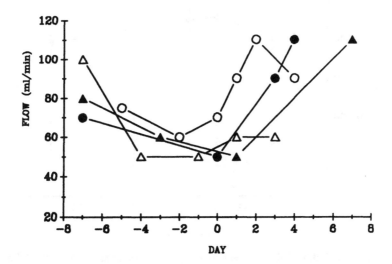

Fig. 3. Blood flow in the left uterine artery in four treated patients throughout the cycle.

the measured variables. The changes in vessel resistance and flow volume throughout the treatment cycle demonstrate, similar to the uterine artery, an association with estrogen and particularly with progesterone levels. Using a transabdominal duplex system with a distended bladder, it was found that the pulsatility index decreased throughout a normal ovulatory cycle, and was always lower in the ovary bearing the corpus luteum compared to the other side.[8] In our experience it is extremely difficult to locate the ovarian artery with certainty using the transabdominal route. Moreover, a full bladder is required in such measurements, which may cause ovarian displacement and modify the impedance to flow in the ovarian artery. We did not find any significant difference between the left and the right sides in stimulated cycles, which could be expected considering the fact that ovulation was being induced in both ovaries. In another study, transvaginal pulsed Doppler measurement of blood flow velocity in the ovarian artery was performed during cycle stimulation and after follicle puncture.[9] The pulsatility index was high four days before follicle puncture (i.e. day -2). It decreased markedly toward the day of embryo transfer. The impedance to flow was significantly lower in patients with high endocrine response compared to those with low endocrine response (i.e. E_2 levels below 200 pg/follicle > 15 mm on the day of ovulation induction). All patients in the latter study were treated with a long-acting gonadotropin-releasing hormone (GnRH) agonist prior to cycle stimulation. Doppler flow measurements were not performed following embryo replacement. Transvaginal pulsed Doppler assessment of blood flow to the corpus luteum was also performed in IVF patients following embryo transfer.[10] The flow velocity profiles were obtained from within the ovarian substance (the vascular supply to the corpus luteum) and from the supero-lateral margin of each ovary (representing the ovarian artery). There was a highly significant difference in the RI values

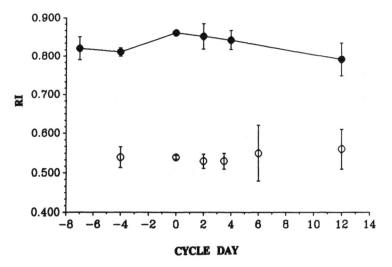

Fig. 4. The resistance index of the left ovarian artery during cycle stimulation and following follicle aspiration and ET. The upper trace represents values obtained from the main ovarian artery. The lower trace represents values obtained from the ovarian substance itself.

between patients who became pregnant and those who did not. No patient who became pregnant had a RI greater than 0.5 (although not all patients with a RI consistently below 0.5 became pregnant). In our study most of the RI values were above 0.5 (Figure 6). In fact, the RI in early pregnancy was in many cases above 0.5. By examining the data in the latter study,[10] it can be seen that the mean RI values in patients who did not become pregnant was 0.83 at three days and 0.9 at 10 days following ET. We have never observed such high values from signals obtained from within the ovary, and it is very likely that these values were derived from the ovarian artery itself as is demonstrated in Figure 4. Care should therefore be taken to differentiate between signals derived from the "luteal artery" as described above and from the main branch of the ovarian artery. The decline in vessel resistance as one moves from the ovarian artery to the luteal artery (about 40%) is more striking than when one moves from the ascending uterine artery to the radial or arcuate artery, and reflects the relatively high flow volume to the corpus luteum. In the rabbit, blood flow to the corpora lutea was seven-fold higher than to the ovarian stroma and follicles.[11] There was a positive correlation between blood flow to the corpus luteum and the rate of ovarian progesterone secretion. Lowering of blood pressure caused a marked decline in luteal blood flow and in ovarian venous progesterone concentration. This suggests that a high blood flow rate is necessary for an optimal secretion of progesterone from the fully developed corpus luteum. Studies in other species—sheep and guinea-pigs—have demonstrated that luteal tissue receives a very high blood flow with total ovarian flow rates being highest at times when functional corpora lutea are present.[12, 13] It should be recalled that part of the total ovarian blood flow is derived from the distal branches of the uterine artery. In the cow, a portion of the blood flowing to the ovary containing the fully functional corpus luteum is

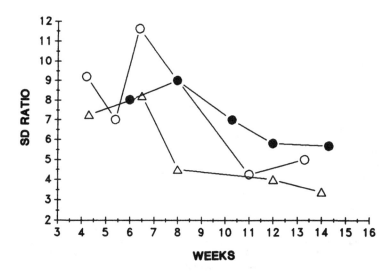

Fig. 5. Changes in systolic/diastolic flow velocity ratio in the uterine artery in three patients throughout the first trimester.

contributed by the ipsilateral uterine artery.[14] During peak ovarian blood flow, the uterine branch of the ovarian artery supplies 20%–40% of the total ovarian blood flow. Throughout the oestrus cycle, ovarian blood flow was positively correlated with systemic concentrations of progesterone and negatively correlated with systemic concentrations of E_2, while uterine blood flow demonstrated an inverse relationship to these hormones.[14]

UTERINE AND OVARIAN BLOOD FLOW IN EARLY PREGNANCY FOLLOWING IVF/ET

Blood flow to the uterus and ovary was measured in eight patients who became pregnant following IVF/ET. Measurements were taken as soon as pregnancy was confirmed by β-subunits of serum hCG and every two weeks thereafter. Changes in the SD ratio in the uterine artery in three different patients are shown in Figure 5.

A rapid and marked decline is noted, reflecting a substantial fall in the impedance to flow in the first 14 weeks of gestation. Changes in volume flow rates in the uterine artery are shown in Figure 6 for these patients.

A rapid increase is observed, reflecting the decrease in the impedance to flow as described above and, the increase in vessel diameter during this time. The resistance index in the ovarian artery also fell rapidly in the first trimester (Figure 7), while volume flow rates increased over 50 percent (Figure 8).

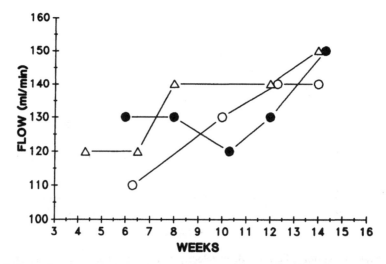

Fig. 6. Volume flow rates in the uterine artery in three patients throughout the first trimester.

The presence of high diastolic flow velocity in the uterine artery in the first trimester was described earlier.[8, 15] In a more recent study, a transvaginal duplex system was used to measure the pulsatility index in the main uterine artery from seven weeks of gestation.[16] A steep decline was observed until 24 weeks. We have observed a similar trend in the S/D ratio over the same period.[17] At the same time flow volume also increased secondary to an increase in vessel diameter and to a lower impedance.[17] This period of pregnancy is characterized by placental growth and development of new blood vessels, accompanied by trophoblastic invasion of the myometrial portion of the spiral arteries. The rate of rise of ovarian artery flow volume is higher during the first trimester compared to the uterine artery. Although the actual flow expressed in ml/min is similar in both arteries during this time period, the uterus is substantially heavier than the ovary, and therefore the ovary receives a very large blood supply per unit of weight (Figure 8). It is not completely understood why the ovary has such a rich blood supply. In the rabbit, a rapid increase in ovarian blood flow is observed until mid gestation.[18] The corpus luteum receives exceptionally high blood supply, similar to the carotid body.[18] A similar trend was observed in the rat ovary.[19] It was suggested that large ovarian flow is necessary for adequate steroidogenesis which is required for the maintenance of pregnancy.[18, 19] It is also possible that local concentrations of ovarian hormones result in vasodilation, a low vascular resistance and high flow rates.[18] This is also supported by the observation that the 'luteal' artery has a much lower impedance to flow compared to the main ovarian artery as was discussed earlier. The decline in vessel resistance is shown in Figure 7. No other account is published in the literature describing volume flow rates in the human uterine and ovarian arteries at such stages of gestation.

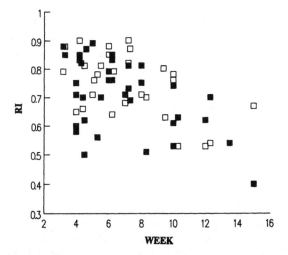

Fig. 7. Changes in the resistance index of the ovarian artery in both sides during the first trimester. The right side is shown by dark squares while the left side is shown by empty squares. No significant difference was found between either side.

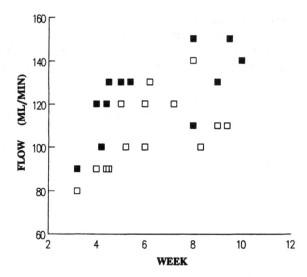

Fig. 8. Volume flow in the right (dark squares) and left (empty squares) ovarian arteries during the first trimester of pregnancy.

CONCLUSION

The high frequency transvaginal Doppler duplex system offers a novel approach for the investigation of perfusion characteristics in the uterine and ovarian circulation. It is more accurate and convenient than the transabdominal route, and is ideal for monitoring blood flow patterns in patients undergoing ovarian stimulation, infertility workup and various hormonal therapies. It also has an obvious potential for studying hemodynamic alterations in normal and abnormal pregnancies, from the very early stages of gestation.

ACKNOWLEDGEMENT

This study was supported by the Kerenyi Prenatal Research Fund.

REFERENCES

1.   I. E. Timor-Tritsch, I. Thaler, and S. Rottem, Review of transvaginal ultrasonography: a description with clinical application, *Ultrasound Quart.* 6:1 (1988).
2.   I. Thaler, A. Bruck, and Y. Bar-Yam, The vaginal probe—physical considerations, *in:* "Transvaginal Sonography," I. E. Timor-Tritsch, and S. Rottem, ed., Elsevier Science Publishing, Inc.,New-York (1988).
3.   S. H. Eik-Nes, K. Marsal, and K. Kristofferssen, Methodology and basic problems related to blood flow studies in the human fetus, *Ultrasound Med. Biol.* 10:329 (1984).

4. R. K. Goswamy, and P. C. Steptoe, Doppler ultrasound studies of the uterine artery in spontaneous ovarian cycles, *Hum. Reprod.* 3:721 (1988).
5. S. P. Ford, L. P. Reynolds, and D. B. Farley, Interaction of ovarian steroids and periarterial alpha-adrenergic receptors in altering uterine blood flow during the estrous cycle of gilts, *Am. J. Obstet. Gynecol.* 150:480 (1984).
6. N. J. Randall, R. W. Beard, I. A. Sutherland, J. P. Figueroa, C. J. Drost, and P. W. Nathanielsz, Validation of thermal techniques for measurement of pelvic organ blood flow in the nonpregnant sheep: comparison with transit-time ultrasonic and microsphere measurements of blood flow, *Am. J. Obstet. Gynecol.* 158:651 (1988).
7. W. H. M. Matta, I. Stabile, R. W. Shaw, and S. Campbell, Doppler assessment of uterine blood flow changes in patients with fibroids receiving the gonadotropin-releasing hormone agonist buserelin, *Fertil. Steril.* 49:1083 (1988).
8. K. J. W. Taylor, P. N. Burns, P. N. T. Wells, D. I. Conway, and M. R. G. Hull, Ultrasound Doppler flow studies of the ovarian and uterine arteries, *Br. J. Obstet. Gynecol.* 92:240 (1985).
9. J. Deutinger, A. Reinthaller, and G. Bernaschek, Transvaginal pulsed Doppler measurement of blood flow velocity in the ovarian arteries during cycle stimulation and after follicle puncture, *Fertil. Steril.* 51:466 (1989).
10. R. D. Baber, M. B. McSweeney, R. W. Gill, R. N. Porter, R. H. Picker, P. S. Warren, G. Kossof, and D. M. Saunders, Transvaginal pulsed Doppler ultrasound assessment of blood flow to the corpus luteum in IVF patients following embryo transfer, *Br. J. Obstet. Gynecol.* 95:1226 (1988).
11. P. O. Janson, J. E, Damber, and C. Axen, Luteal blood flow and progesterone secretion in pseudopregnant rabbits, *J. Reprod. Fert.* 63:491 (1981).
12. G. D. Niswender, M. A. Dickman, T. M. Nett, and A. M. Akbar, Relative blood flow to the ovaries of cycling and pregnant ewes, *Biol. Reprod.* 9:87 (1973).
13. P. O. Janson, and A. I. Albrecht, Methodological aspects of blood flow measurement in ovaries containing corpora lutea, *J. Apll. Physiol.* 38:288 (1975).
14. S. P. Ford, and J. R. Chenault, Blood flow to the corpus luteum-bearing ovary and ipsilateral uterine horn of cows during the oestrous cycle and early pregnancy, *J. Reprod. Fert.* 62:555 (1981).
15. H. Schulman, A. Fleischer, G. Farmakides, G. Bracero, B. Rochelson, and L. Grunfeld, Development of uterine artery compliance in pregnancy as detected by Doppler ultrasound, *Am. J. Obstet. Gynecol.* 155:1031 (1986).
16. J. Deutinger, R. Rudelstorfer, and G. Bernaschek, Vaginosonographic velocimetry of both main uterine arteries by visual vessel recognition and pulsed Doppler method during pregnancy, *Am. J. Obstet. Gynecol.* 159:1072 (1988).
17. I. Thaler, D. Manor, J. Itskovitz, S. Rottem, N. Levit, I. Timor-Tritsch, and J. M. Brandes, Changes in uterine blood flow during human pregnancy, *Am. J. Obstet. Gynecol.* In press (1989).
18. R. W. Abdul-Karim, and N. Bruce, Blood flow to the ovary and corpus luteum at different stages of gestation in the rabbit, *Fertil. Steril.* 24:44 (1973).
19. G. D. Niswender, T. J. Reimers, M. A. Dickman, and T. M. Nett, Blood flow: a mediator of ovarian function, *Biol. Reprod.* 14:64 (1976).

ULTRASONOGRAPHIC PREDICTION
OF OVARIAN HYPERSTIMULATION (OHS) AFTER IVF

Jacques Salat-Baroux, Charles Tibi, Sylvia Alvarez,
Ajuson Gomez, Jean Marie Antoine, and Dominique Cornet

Hôpital Tenon, Université Paris VI, France
4 Rue de la Chine, Paris 20è, France

The incidence of ovarian hyperstimulation syndrome (OHS) ranges from 3% to 6% for the moderate forms and from 0.3% to 4% the severe.[1] If it is easy to predict this complication in cases of Polycystic Ovarian Disease (PCO);[2] in all the other cases of stimulation, the reports concerning the correlations between the multiple variables (age, weight, type of stimulation, number of ampules of human Menopausal Gonadotropin (hMG) used) with OHS are conflicting. Daily monitoring of 17 B estradiol (E_2) and sequential pelvic ultrasonography proved to be beneficial, but they may obviate OHS only after six to eight days of stimulation and a variable number of ampules used. As the pathophysiology of this syndrome is unknown, it is of interest to indicate some predictive factors to avoid cancelling cycles or being faced with severe complications. In 1987, Blankstein et al.,[3] attempted to predict OHS by ultrasonographic determination of the number and the size of preovulatory ovarian follicles. Navot et al., in 1988,[4] using a stepwise logistic regression performed on 22 variables, identified a "high-risk group" for this syndrome: "young and lean patients, who after relatively few ampules of hMG, develop high estradiol levels and multiple small follicles".

The purpose of our study is to determine a predictive ultrasonographic feature, before using any stimulation, which might obviate OHS.

Advances in Assisted Reproductive Technologies, Edited by
S. Mashiach *et al.*, Plenum Press, New York, 1990

MATERIAL AND METHODS

We have carried out two studies: the first one was retrospective, in order to define the characteristics of the OHS and the ultrasonographic features of these patients and the second with the aim of illustrating the ultrasonographic prediction of OHS.

1) In the first retrospective study, 12 patients were selected out of 325 who had an ovarian stimulation for IVF, over a period of four months. For these twelve patients, the diagnosis of ovarian hyperstimulation was defined by: 1) Biological criteria: E_2 level ranging from 2500 to 4300 pg/ml (with a mean value of 2990 ± 690) on the day of hCG injection (DO); 2) Ultrasonographic criteria: polycystic ovarian response with more than 10 follicles of 14 mm on each ovary. All these patients were normally cycling patients without clinical or biological evidence of PCO, obesity or diabetes. On the fifth day of the cycle the plasmatic hormonal value for LH ranged from 1.5 to 4 mIU/ml, for FSH from 2.4 to 4.8 mIU/ml, for Prolactin from 4 to 22 ng/ml, and E_2 and androgen levels were also within the normal limits.

There was an ovarian stimulation of LHRH agonist (LHRH-a) in a long protocol in nine cases and hMG (three ampules a day for seven days) and FSH (in one case at the same dose). One patient had an association of LHRH-a in a short protocol plus hMG and the last patient had only hMG with an equivalent dose during the first seven days.

Six months after this biological and potentially clinical OHS, an ultrasonography study was carried out on these patients, on the fifth day of a spontaneous cycle: an ultrasound estimation of follicular number and size (especially 4mm to 8mm) was evaluated using an Edap Sector Scann, France, with a 5MHz probe. After determination of the ultrasonographic criteria a prospective study was carried out.

2) In the second "prospective" study, 109 patients with a mean age of 32.8 ± 3.2 years, enrolled in an IVF program, gave their consent for this study. They were all normally cycling women, with no PCO, and their spontaneous cycle duration was of 28.7 ± 1.5 days. The ovarian stimulation was performed using LHRH-a (DTRP 6) Ipsen Beaufour France in 52.3% of the cases and Buserilin (Hoechst Laboratory, France) in 47.7% of the cases. The duration of desensitization ranged from 15 to 26 days. This desensitization was followed by hMG (Humegon, Organon, France) at a standard dose of three ampules a day for eight to nine days.

The ultrasonographic criteria were indicated after complete desensitization using the same sector scan: uterus and ovarian diameter (with an error estimated of ±1 mm), number of follicles of 4–8 mm in diameter and their location in the ovaries. They were associated with plasma hormonal assay: E_2, $\Delta 4$ androstenedione, FSH, LH, and Prolactin evaluated on the same day as the ultrasonographic evaluation. LHRH and TRH tests were respectively performed when $\Delta 4$ was > 2 ng/ml or LH > 5 mIU/1 or LH/FSH > 2.5, and Prolactin > 20 ng/ml.

Table 1. Comparison between the Two Groups According to Basal Hormonal Values

	Δ4 ng/ml	Testo ng/ml	LH miu/ml	FSH miu/ml	PRL ng/ml
Group I n = 14 >10 F 4–8 mm	1.64 ±0.67	0.4 ±0.11	3.15 ±1.7	2.82 ±1.57	1.36 ±12
Group II n = 93 <F 4–8 mm	1.3 ±0.52	0.354 ±0.13	2.75 ±2.15	2.31 ±1.14	9.97 ±6.2
P	p = 0.032	NS	NS	NS	NS p = 0.08

We then tried, according to the ultrasonographic criteria of the first retrospective study, to define two groups of patients in the prospective study: a high risk of OHS group and a control group.

RESULTS

1) In the retrospective study, eight patients had an hCG injection: the day after hCG E_2 ranges form 2700 to 6300 pg/ml with a mean of 3515 ± 1804 pg/ml, the mean number of oocytes collected was 14.5 ± 2 and the majority of them had a moderate OHS syndrome defined by ovary enlarged to 8mm to 12mm, according to the classification of Rabau et al.[5] But, in all the 12 cases, the most striking feature was a mean number of more than 10 follicles of 4mm to 8mm in size, at least in one ovary, after ultrasound determination during a spontaneous cycle. This fact could constitute an ultrasonographic prediction of OHS. However, this retrospective study does not answer two questions: 1) How many of the 313 remaining patients without OHS in this series also had this criteria in a spontaneous cycle? 2) How many of them after stimulation, had a polycystic ovarian response without OHS? It was therefore necessary to carry out a prospective study.

2) In the prospective study, it was possible according to this ultrasound (US) criteria, to distinguish two groups of patients: Group I with a theoretical high risk for development of OHS (14 patients with at US examination, after desentization by LHRH-a, more than 10 follicles of at least 4mm–8mm on one ovary) and Group II with no risk of OHS (93 patients).

Table 1 shows that there is a statistically significant difference ($p < 0.03$) only for Δ4 androstenedione, when we compare the two groups according to the baseline hormonal values. But, after ovarian stimulation with the same protocol, we find a large difference con-

Table 2. Comparison between the Two Groups According to E_2 Values at Different Days of hMG Stimulation

	E_2D8	E_2D10	E_2D11	E_2 D of hCG	DhCG	Nb ampules
Group I n = 14 >10 F 4-8 mm	841.42 ±629.2 n = 7	1433.1 ±1073	1754 ±1219	1771 ±1267	10.07 ±1.54 n=13	28.14 ±4.3
Group II n = 93 <F 4-8 mm	398.78 ±337.3	823.6 ±636.2	1016.9 ±662.5	1052 ±678	11.13 ±1.6	31.8 ±7.7
P	p = 0.008	p = 0.004	p = 0.005	p = 0.005	p = 0.028	NS p = 0.08

Table 3. Frequency in the Two Groups of Polycystic Ovarian Response (POR) and Ovarian Hyperstimulation (OHS)

	POR right ovary	POR left ovary	OHS
>10 F 4-8 mm n = 14	75% 8/12	66.7% 8/12	42.8% 4/14
<F 4-8 mm n = 93	3.3.% 3/90	7.5% 7/92	5.3% 5/93
P	$p < 10^{-4}$	$p < 10^{-4}$	$p < 10^{-4}$

cerning E_2 response, on each day before and on the day of hCG injection, although the number of ampules injected was not different in the two groups (28.14 ± 4.3 in Group I versus 31.8 ± 7.7 in Group II) (Table 2). Furthermore, the day of hCG injection was statistically more precocious in Group I (10.07 ± 1.54 vs 11.13 ± 1.6 p < 0.02).

Table 3 represents, for the two groups, the rate of polycystic ovarian response and the rate of OHS defined by our criteria: 42.8% in Group I versus 5.3% in group II this difference is highly significant ($p < 10^{-4}$). If we try to establish a correlation between the baseline hormonal value and OHS in the 19 cases of polycystic ovarian response (more than 10 follicles of 14 mm on each ovary), we find a statistical difference only for Prolactin (8 ± 3.55 in the eight OHS patients without OHS versus 14.9 ± 13.1 in the 11 OHS positive patients ($p < 10^{-4}$)).

Table 4. Correlation between Basal Hormonal Values and OHS in 19 Cases of Polycystic Ovarian Response

	Δ4 ng/ml	Testo ng/ml	LH miu/ml	FSH miu/ml	PRL ng/ml	LH/FSH
OHS- n = 8	1.3 ±0.45	0.37 ±0.08	3.26 ±1.55	3.86 ±1.37	8 ±3.55	0.96 ±0.5
OHS+ n = 11	1.79 ±0.59	0.45 ±0.09	3.05 ±1.62	1.98 ±0.84	14.9 ±13.1	1.66 ±0.9
P	NS $p = 0.06$	NS $p = 0.06$	NS	$p = 0.002$	$p = 10^{-4}$	NS

Table 5. Hormonal Values According to the Ovarian Hyperstimulation Syndrome (OHS)

	Δ4 ng/ml	Testo ng/ml	LH miu/ml	FSH miu/ml	PROL ng/ml	LH/FSH	E_2D of hCG
OHS- n = 96	1.32 ±0.47	0.356 ±0.12	2.92 ±2.21	2.42 ±1.17	10.08 ±6.37	1.156 ±0.62	972 ±422
OHS+ n = 11	1.68 ±0.63	0.45 ±0.14	2.96 ±1.55	1.92 ±0.8	13.76 ±12.3	1.545 ±0.94	2468 ±465
P	0.013	0.014	NS	NS	NS	0.05	10^{-4}

If, we now compare the 11 OHS positive patients to the 96 OHS negative patients, we find, in Table 5, that there are statistically significant differences for Δ4, Testosterone and the ratio of LH/FSH but not for LH, between the two groups concerning the baseline hormonal levels.

Obviously, the level of E2 after ovarian stimulation is statistically higher in the group of OHS positive patients ($p < 10^{-4}$).

In Table 6, although the mean number of ampules is not different in the two groups, the day of hCG injection is more precocious, the age of patients lower in the OHS positive patients, and especially the number of small follicles (4–8mm at US in a previous spontaneous cycle) is significantly different in the Group of OHS positive patients (7.36 ± 3.05 versus 3.58 ± 3.16 in the group of OHS negative patients ($p < 10^{-4}$).

Table 7 summarizes the predictive value of this ultrasonographic test of OHS prediction: the positive predictive value was 42.8% and the negative was 93.4% with a sensitivity of 50% and a specificity of 91.4%.

Table 6. Comparison of Other Factors (Especially Follicles of 4–8mm) between the Two Groups (with or without OHS)

	Age	Foll 4–8 mm 10	Day of hCG Injection	Nb Ampules	Number of oocytes
OHS- n = 96	33.15 ±3.06	3.58 ±3.16	11.16 ±1.68	31.5 ±8.3	6
OHS+ n = 11	31.15 ±4.5	7.36 ±3.5	9.76 ±1.05	27 ±2.4	11.4
P	0.04	10^{-4}	0.004	0.06	10^{-4}

Table 7. Predictive Value of OHS by Ultrasound Criteria

Positive predictive value	42.8%
Negative predictive value	93.4%
Sensibility of the test	50%
Specificity of the test	91.4%

DISCUSSION

As ovarian hyperstimulation is a very serious complication, especially in Group II of WHO,[6] it was important to determine the high-risk factors in order to reduce its occurrence. While Haning et al.,[8–9] concluded that plasma E_2 is superior to ultrasonography in avoiding OHS, Blankstein et al.,[3] found that this syndrome can occur in spite of low or normal estrogen secretion. So they demonstrated that the risk becomes very high, in the presence of 11 or more preovulatory follicles, especially if more of them are small (>9 mm). Tal et al.,[9] report a positive correlation between small and secondary follicles and OHS. More recently, Navot et al.,[4] insisted on three predictive parameters (age, E_2 level on Day 0 and basal serum Prolactin) but also on the number of multiple small follicles at various maturational stages.

However, in all these series, the high risk factors were determined during ovulation induction, especially on Day O (of hCG injection). Our study originates in describing an ultrasonographic test for predicting OHS, before any ovarian stimulation is given (more than 10 follicles of 4mm to 8mm in at least one ovary). This test is more sensitive than baseline hormonal values, including LH and androgen levels. It enables the prediction of a higher response in E_2 after hMG induction, a polycystic ovarian response in 75% of the cases and an OHS in 42.8% of the cases. The differences with the control group are all statistically significant. And even if the positive prediction and sensitivity are only of respectively 42.8% and 50% the negative prediction of OHS and the specificity of the test are respectively 93.4% and 91.4%. We therefore confirm in this study the prominent role of small follicles in predicting OHS; moreover, this US test is of a predictive value in half of the cases before any ovarian stimulation. It could also provide reliable information on the number of ampules to be used before starting ovulation induction. However, these data are still to be verified on a larger series and it is of value to understand why these "small follicles", rather than the dominant follicles, could be such a valuable sign of a possible occurrence of OHS.

REFERENCES

1. B. Lunenfeld and V. Insler, Gonadotropins, in: "Diagnosis and treatment of function infertility," Grosse Verlag, Berlin 76-89 (1978).
2. J. G. Schenker and D. Weinstein, Ovarian hyperstimulation syndrome: a current survey, *Fertil. Steril.*, 30:255-268 (1978).
3. J. Blankstein, J. Shalev, T. Saadon, E. Kukia, J. Rabinovici, C. Pariente, B. Lunenfeld, D. Serr, and S. Mashiach, Ovarian hyperstimulation syndrome: prediction by number and size of preovulatory ovarian follicles, *Fertil. Steril.*, 47:597-602 (1987).
4. D. Navot, A. Relou, A. Birkenfeld, R. Rabinowitz, A. Brzezinski and E. Margolioth, Risk factors and prognostic variables in the ovarian hyperstimulation syndrome, *Am. J. Obstet. Gynecol.*, 159:210-215 (1988).
5. E. Rabau, A. David, D.M. Serr, S. Mashiach, and B. Lunenfeld, Human menopausal gonadotropins for anovulation and sterility, Am. J. Obstet. Gynecol., 98:92-98 (1967).
6. WHO Scientific Group, Agents stimulating gonadal function in the human, Geneva: World Health Organization - 1973 (Technical report series No 514).
7. R. V. Haning, C. W. Austin, I. H. Carlson, D. L. Kuzma, S. S. Shapiro, and W. J. Zweibel, Plasma estradiol is superior to ultrasound and urinary estriol glucoronide as a predictor of ovarian hyperstimulation during induction of ovulation with menotropins. *Fertil. Steril.*, 40:31-35 (1983).
8. R. V. Haning, I. M. Boehnlein, D. H. Carlson, D. L. Kuzman, and W.J. Zweibel, Diagnosis specific serum 17 beta-Estradiol (E2) upper limits for treatment with menotropins using a I direct E2 assay, *Fertil. Steril.*, 42:882-889 (1984).
9. J. Tal, B. Paz, I. Samberg, N. Lazarow, and M. Sharf, Ultrasonographic and clinical correlates of menotropin versus sequential clomiphene citrate: menotropin therapy for induction of ovulation, *Fertil. Steril.*, 44:342-349 (1985).

TEN YEARS EXPERIENCE WITH SIMPLIFICATIONS IN IVF

Wilfred Feichtinger and Peter Kemeter

Institute of Reproductive Endocrinology and In Vitro Fertilization
Trauttmansdorffgasse 3A, A-1130 Vienna, Austria

In January 1979, we initiated an in vitro fertilization (IVF) program in Vienna. The first successful treatments were achieved in 1982[1]. Since then it was the aim of our group to simplify the procedures. Care was taken not to diminish the success rate at each step of simplification.

STAGES OF SIMPLIFICATION

Stage I (1982). Simplifications during a research program at the university department: First, a routine protocol was established for the whole procedure. Laparoscopic oocyte recovery was performed in a theater with an adjacent laboratory during the routine operating theater program[2]. Monitoring of cycles and hormone assays were done in the outpatient sterility unit, ultrasound department or hormone laboratory, respectively. A simplified egg and embryo culture using a pre-prepared commercially available culture medium and a simpler gassing system were developed[1], and we started to reduce the bed rest time after embryo transfer (ET). The success rate of this system in terms of clinical pregnancies was 10% per treatment cycle and 17% per ET.

Stage II (1982-1983). Oocyte recovery, IVF and ET as outpatient procedures: The patients' cycles were monitored in the gynecological office which houses ultrasound facilities, hormone and IVF laboratories. Simple hemagglutination tests for urinary estrogen and LH-estimations were used in addition to ultrasound for monitoring Clomid and hMG/hCG controlled ovulation induction[3]. Oocytes were recovered by laparoscopy in a hospital under general anesthesia, but with the same-day discharge of patients. The recovered eggs were

Advances in Assisted Reproductive Technologies, Edited by
S. Mashiach *et al.*, Plenum Press, New York, 1990

transported to the laboratory in follicular fluid at 37°C in a thermos flask. ET was performed as an office procedure 24–48 hours after insemination, i.e. the day after laparoscopy. Bed rest after ET was reduced to a maximum of 30 minutes.

The success rate of this system in terms of clinical pregnancies was 10% per treatment cycle and 19% per ET in 1982.

I. Take a contraceptive pill (f.i. Marvelon) once a day
   commencing on the first day of menstruation.

II. Take the last pill on a TUESDAY, f.i. on: _____ (these pills must be
    taken not less than 18 days and no longer than 28 days).

III. The first day of the treatment is the SUNDAY following the tuesday on which you
     took the last pill. (The menstruation will start 2-7 days after the last pill, but this is
     not important for the treatment)

DAY OF TREATMENT:	1 Su	2 Mo	3 Tue	4 Wed	5 Thu	6 Fri	7 Sat	8 Su
Date:								
Prednisolone à 5 mg tbl.	1/2 tbl. in the morning, 1 tbl. in the evening daily throughout one month							first control at our institute in Vienna 8 o'clock in the morning
Clomiphene tbl.	2x1	2x1	2x1	2x1	2x1	---	---	see IV
HMG-Amp. à 75 U f.i. Humegon	2 Amp. HMG	---	2 Amp. HMG	---	2 Amp. HMG	---	2 Amp. HMG	

IV. You must report to our institute on the 8th day of the stimulation (= Sunday, at 8 o'clock
    in the morning) for the first sonography of the ovarian follicles and spermiogramm,
    if necessary.

Bring the following medications with you (but do not use them before you are told to):

1.) HCG-ampoules à 5000 IU I.OP., (f.i. Pregnyl)
2.) the remaining Humegon-ampoules
3.) tetracycline vag. suppositories for 5 days
4.) 2 spasmolytic rectal suppositories
5.) Duphaston tablets à 10 mg, after transfer 3 x 1 daily for two weeks

Fig. 1. Stimulation protocol by Kemeter/Feichtinger.

Stage III (1983-1985). Oocyte recovery, IVF and ET as office procedures: the patients' cycles were monitored as described above in the gynecological office. The stimulation protocol was changed to an individualized combined Clomid-hMG/hCG treatment. We introduced the technique of ultrasonically guided follicle aspiration under local anesthesia as described by Lenz et al[4] and Wikland et al[5] to make oocyte recovery, IVF and ET a completely independent office and outpatient procedure.[6]

Stage IV (1985 - update). Stimulation, embryo culture and transfer procedures were simplified during the following years.[7, 8, 9] Vaginal sonography was introduced as the method of choice for oocyte recovery,[10, 11, 12] it also proved to be important as the sole parameter for monitoring ovarian stimulation as it is also being used to improve the precision of the embryo transfer procedure. A simple pre-programmed stimulation protocol without hormone determinations is currently used with good success.[7] (Fig. 1)

The period and the start of stimulation is shifted by means of a contraceptive pill in such a way that stimulation is generally started on a Sunday. The patient takes Clomid 100 mg for five days, Prednisolone 7.5 mg for 30 days to suppress possible exaggerated adrenal androgens,[13] and she receives 150 I.U. hMG intramuscularly every other day from her doctor at home. From the eighth day of stimulation onward follicular growth is registered by daily ultrasound at the IVF center. 5000 I.U. hCG is given when the dominant follicle exceeds 18 mm diameter. Blood sampling for hormone estimations is not necessary and only one sample of urine is needed for the LH-estimation before hCG.

In comparison to Clomid only, Clomid plus hMG/FSH and hMG/FSH only, this protocol resulted in a significantly higher fertilization and pregnancy rate per follicular puncture (Table 1 and 2). The rate of abortions and extrauterine pregnancies, on the other hand, was decreased. When comparing repetitive IVF cycles with the first IVF cycle the pregnancy rate was decreased in the repeat cycles. The new protocol seems to be less stressful for the patients thus enabling more repetitions of IVF.

Advantages of programmed stimulation with pill—Clomid-hMG
- Programming (75% of IVF procedures on weekdays)
- Suppression of cysts, persisting corpora lutea, endometriosis, etc.
- Low incidence of LH-surge
- Low cancellation rate (13%)
- Pretreatment carried out by patient's home doctor
- Few injections (4-7)
- Reasonable number of good quality oocytes
- No overstimulations
- No hormonal tests necessary
- Low cost
- Low stress
- Frequently repeatable
- Good effectiveness

Table 1. Number of Follicles, Oocytes and Fertilizations with Different Stimulation Protocols (Mean and Standard Deviation)

		FOLLICLES	OOCYTES	FERTIL. OOCYTES
1.	CLOMIPHENE ALONE	3,3 ± 1,7	2,4 ± 1,5	1,4 ± 1,3
2.	CLOMIPHENE + HMG / FSH	* 4,8 ± 3,0	* 3,9 ± 2,7	* 2,2 ± 2,1
3.	HMG / FSH	* □ 6,1 ± 5,0	* □ 4,7 ± 3,4	* 2,2 ± 2,2
4.	KEMETER / FEICHTINGER	* □ 6,8 ± 3,3	* □ 4,5 ± 2,6	▨ 2,9 ± 2,3

    * SIGN. DIFFERENT TO CLOMIPHENE ALONE   ($p < 0,01$)

    □ SIGN. DIFFERENT TO CLOM.+ HMG / FSH   ($p < 0,01$)

    ▨ SIGN. DIFFERENT TO ALL OTHERS   ($p < 0,01$)

Table 2. Stimulation—Protocol and Pregnancy Rate

STIMULATION	NOT PREGNANT		PREGNANT		
	N	%	N	% PER PUNCT.	% PER ET
1. CLOMIPHENE ALONE	123	87,2	18	12,8	19,4
2. CLOMIPHENE + HMG/FSH	404	84,7	73	15,3	21,1
3. HMG/FSH ALONE	131	83,4	26	16,6	23,2
4. KEMETER / FEICHTINGER	218	78,7	59	21,3 *	25,7

    * SIGN. DIFFERENT TO 1. AND 2.

       $p < 0,05$

Recently, however, natural cycles have been used again for IVF in selected cases. A new approach towards "Natural IVF" is summarized as follows:

Natural IVF:
- Natural cycles - Ovulation induced with hCG
- No hormones
- Single oocyte collection by TVS - no anesthesia
- Intravaginal culture (Ranoux)[14]
- ET under sonographic control
- Office procedure

SIMPLE, LOW COST, FREQUENTLY REPEATABLE

DISCUSSION

These concepts are in contrast to many others who at the present are enthusiastic over LH-RH agonist pituitary suppression plus hMG stimulation. They claim an increased pregnancy rate would justify the higher cost and stress of such a treatment.[15, 16] This was not proved in a prospective randomized trial comparing LH-RH agonist + hMG to conventional stimulation.[17]

Even though there was a higher success rate, in our opinion this slightly higher pregnancy rate would not justify the considerable higher cost, the stressful treatment, the numerous injections, blood examinations, considerable risk of hyperstimulation cysts, and luteal phase deficiency, and therefore the low repeatability of this kind of treatment.

Now, with ten years of experience of IVF, it is high time to provide the patients with a simple, short, low-stress and low-cost treatment which is repeatable if necessary.

REFERENCES

1. W. Feichtinger, P. Kemeter, and S. Szalay, The Vienna program of in vitro fertilization and embryo transfer - a successful clinical treatment. *Europ. J. Obstet. Gynaec. Reprod. Biol.*, 15:205 (1983).
2. W. Feichtinger, S. Szalay, A. Beck, P. Kemeter, and H. Janisch, Results of laparoscopic recovery of preovulatory human oocytes from non-stimulated ovaries in an ongoing in vitro fertilization program, *Fertil. Steril.*, 36:707 (1981).
3. W. Feichtinger, P. Kemeter, C. Hochfellner, and I. Benko, Monitoring of Clomid-hMG-hCG controlled ovulation by ultrasound and a rapid (15 minutes) urinary estrogen assay, *Acta. Europ. Fertil.*, 14:107 (1983).
4. S. Lenz, J. G. Lauritsen, and M. Kjellow, Collection of human oocytes for in vitro fertilization by ultrasonically guided follicular puncture, *Lancet* II:63 (1981).
5. M. Wikland, L. Nilsson, L. Hansson, L. Hamberger, and P. O. Janson, Collection of oocytes by the use of sonography, *Fertil. Steril.*, 39:603 (1983).

6. W. Feichtinger, and P. Kemeter, In vitro fertilization and embryo transfer. An outpatient/ office procedure, *in:* "Recent Progress in Human In Vitro Fertilization," W. Feichtinger, and P. Kemeter, eds., Palermo, Cofese, 285, (1984).

7. P. Kemeter, and W. Feichtinger, Initial experience with a fixed stimulation schedule for in vitro fertilization without determination of hormone levels in the blood, *Fertilitaet* 5:14 (1989).

8. W. Feichtinger, and P. Kemeter, A simplified technique for fertilization and culture of human preimplantation embryos in vitro, *Acta. Europ. Fertil.*, 14:125 (1983).

9. W. Feichtinger, and P. Kemeter, Embryo transfer at an early stage - a possible reason for extremely reduced abortion rate, *in:* "Proceedings of the Vth Reinier de Graaf Symposium," Excerpta Medica, Amsterdam, 283, (1985).

10. W. Feichtinger, and P. Kemeter, Transvaginal sector scan sonography for needle guided transvaginal follicle aspiration and other applications in gynecologic routine and research, *Fertil. Steril.*, 45:722, (1986).

11. P. Kemeter, and W. Feichtinger, Transvaginal oocyte retrieval using a transvaginal sector scan probe combined with an automated puncture device, *Hum. Reprod.*, 1:24 (1986).

12. W. Feichtinger, and P. Kemeter, Ultrasound-guided aspiration of human ovarian follicles for in vitro fertilization, *in:* "Ultrasound Annual,"R. C. Sanders and M. Hill, eds., Raven Press, 25, (1986).

13. P. Kemeter and W. Feichtinger, Prednisolone supplementation to clomid and/or gonadotrophin stimulation for in vitro fertilization - a prospective randomized trial, *Hum. Reprod.* 1:441 (1986).

14. C. Ranoux, J. B. Dubisson, F. X. Aubriot, V. Cardone, H. Foulot, The first seven births using a new in vitro fertilization technique, *in:* "Abstracts of the Vth World Congress on In Vitro Fertilization and Embryo Transfer", April 5-10, Norfolk, Virginia, 90, (1987).

15. S. Neveu, B. Hedon, J. Bringer, J. M. Chincole, F. Arnal, C. Humeau, P. Cristol and J. L. Viala, Ovarian stimulation by a combination of gonadotropin-releasing hormone agonist and gonadotropins for in vitro fertilization, *Fertil. Steril.* 47: 689 (1987).

16. K. Droesch, S. J. Muasher, R. G. Brzyski, G. S. Jones, S. Simonetti, H. C. Liu, Z. Rosenwaks, Value of suppression with a gonadotropin-releasing hormone agonist prior to gonadotropin stimulation for in vitro fertilization, *Fertil. Steril.*, 51:292 (1989).

17. D. De Ziegler, C. Cornell, A. Hazout, H. Fernandez, A. Glissant, C. Baton, J. Bellaisch-Allart, R. Frydman, Randomised trial comparing clomiphene citrate-hMG, Abstracts of the VIth World Congress for In Vitro Fertilization and Alternate Assisted Reproduction, Jerusalem, April 2-7 p. 11 (1989).

COMPARISON OF CRYOPROTECTANTS AND STAGES OF HUMAN EMBRYO

J. Testart,[1,2] M. Volante,[2] B. Lassalle,[1] A. Gazengel,[2] J. Belaisch-Allart,[2]
A. Hazout,[2] D. de Ziegler,[2] and R. Frydman[2]

INSERM U 187[1] and Gynecologie-Obstetrique[2]
Hôpital Antoine BECLERE, 92 141, Clamart, France

Human embryos can be cryopreserved successfully by protocols using 1,2 propanediol (PROH), dimethylsulfoxyde (DMSO) or glycerol as cryoprotective agents. Each of these cryoprotectants give the optimal efficacy when used at a defined stage of embryo development, namely pronucleated or early cleavage (PROH), cleavage (DMSO) and blastocyst (glycerol). The aim of the present paper was firstly to review the success rates of each protocol from the data published by different teams. We wish also to present results obtained in Clamart using PROH for cryopreservation of 1- or 2-day embryos and to analyze the influence of several factors on these results.

SURVEY OF HUMAN EMBRYO CRYOPRESERVATION

Pregnancy rate depends on the number of embryos transferred simultaneously. Therefore we found it more objective to estimate the future of each frozen embryo at least until its implantation into the uterus.

There were 751 1-day and 1409 2-day frozen and thawed embryos from five published reports using PROH and sucrose (Table 1). Except for one report[2] the proportion of embryos suitable for transfer was 60% to 80% of those thawed and was not different between 1-day (57%) and 2-day (62%) embryos. Similarly pronucleated or cleaved frozen-thawed early embryos showed the same ability to implant after transfer (12.1% and 10.9% respectively). The implantation rate of frozen and thawed embryos was 6.%-6.9% (Table 1).

Advances in Assisted Reproductive Technologies, Edited by
S. Mashiach *et al.*, Plenum Press, New York, 1990

Table 1. Survey of Literature: Results of 1–2-day Embryo Freezing Using Propanediol (PROM 1.5 M) and Sucrose as Cryoprotectant

Group and ref	embryo stage	n thawed embryo	(% transferred)	n implanting embryos	(% of transferred)	(% of thawed)
CLAMART (present paper)	1-cell	222	(80)	19	(10.7)	(8.6)
	2-8-cell	780	(68)	48	(9.0)	(6.1)
NECKER (1)	2-10-cell	407	(61)	36	(14.5)	(8.8)
BRUSSELS (2)	1-cell	345	(38)	12	(9.2)	(3.5)
	2-4-cell	116	(28)	4	(12.5)	(3.4)
ATLANTA (3)	1-cell	102	(59)	16	(26.7)	(15.7)
	2-4-cell	106	(62)	8	(12.1)	(7.6)
RICHMOND (4)	1-cell	82	(76)	5	(8.1)	(6.1)
TOTAL	1-cell	751	(57)	52	(12.1)	(6.9)
	2-10-cell	1409	(62)	96	(10.9)	(6.8)

One thousand four hundred and forty-one cleaved embryos were frozen and thawed using DMSO (1.5 M) as reported by six different teams (Table 2). The proportion of embryos suitable for transfer (43%) was lower than in the case of either 1- or 2-day embryos frozen using PROH and sucrose (p < 0.001).

There was no difference between PROH and DMSO for the proportion of transferred embryos which implanted. Finally the rate of implantation of frozen and thawed embryos was lower when DMSO was used (4.2%) irrespective of the age of the embryo at the time of freezing (p < 0.01, Tables 1 and 2).

Freezing at the very early embryonic stages including pronucleated embryos has the advantage of limiting the time of in vitro culture before freezing. It is well known that increasing the culture time may lead to degeneration or cleavage blockage before the embryo reaches the expected stage for freezing. For example only one out of two or one out of four pronucleated embryos, remained suitable for freezing when cultured until the 8-10-cell or the expanded blastocyst stage, respectively.[7] Consequently success rates reported in Table 2 are overestimated compared with those reported in Table 1. From a clinical point of view those protocols using PROH could be two or three times more efficient than protocols using DMSO. In case of blastocyst cryopreservation using glycerol the high proportion of thawed embryos which implant (18.2%)[7] reflects only those developing in culture and the true success rate could be less than 5%.

Interesting work is actually in progress for ultrarapid freezing of animal or human embryos but although the survival rate of human frozen-thawed embryos is excellent,[10] questions arise for their developmental potential after transfer.

FACTORS AFFECTING EMBRYO CRYOPRESERVATION WITH PROH

According to the literature (see Table 1) all teams using PROH have similar protocols utilizing a final freezing solution containing PROH 1,5 M and sucrose and a rapid thawing rate.[11] Our team[12] and others[1, 2] have found a higher embryo survival after thawing of embryos demonstrating regular cell size and no cytoplasmic exudates at the time of freezing. Moreover the pregnancy rate did not decrease when embryos of a lower "quality" were transferred as fresh embryos.[13] This means that among embryos considered as normal according to cleavage rate and monospermy the lowest "quality" embryos may be those incapable of surviving stress of cryopreservation. From these considerations our attitude has been to freeze the best looking embryos.

We usually tried to freeze embryos at the pronucleated stage but this was complicated by the necessity to transfer cleaved embryos in the IVF cycle and by the inability to predict the developmental capacity of fertilized oocytes. Although we have proposed restraining the

Table 2. Survey of Literature: Results of 2–7-day Embryo Freezing Using Dimethylsulfoxyde or Glycerol as Cryoprotectant

Group and ref	cryoprotectant	embryo stage (âge)	n thawed embryos	(% transferred)	n implanting embryos	(% of transferred)	(% of thawed)
BRUSSELS (2)	DMSO 1.5 M	4-12-cell (2-3-day)	759	(37)	27	(9.6)	(3.6)
MONASH (5)	DMSO 1.5 M	(2-4-day)	396	(58)	18	(7.9)	(4.6)
ADELAIDE (6)	DMSO 1.5 M	2-cell to blastocyst	88	(11)	3	(30.0)	(3.4)
CAMBRIDGE (7)	DMSO 1.5 M	5-10-cell (2-3-day)	80	(35)	4	(14.3)	(5.0)
LOS ANGELES (8)	DMSO 1.5 M	(3-4-day)	63	(56)	2	(5.7)	(3.2)
TEL HASHOMER(9)	DMSO 1.5 M	2-cell to blastocyst	55	(76)	6	(14.6)	(10.3)
MONASH (10)	DMSO 3 M (ultrarapid)	3-8-cell (2-3-day)	12	(92)	0	(0.0)	(0.0)
CAMBRIDGE (7)	Glycerol	blastocyst	44	(52)	8	(34.8)	(18.2)
TOTAL	DMSO 1.5 M 2-cell to blastocyst		1441	(43)[a]	60	(9.6)	(4.2)[b]

difference with PROH (either 1-cell or 2-10-cell embryos, Table 1) : a : $p < 0.001$, b : < 0.01

Table 3. Embryo Survival at Thawing According to the Development Stage at Freezing

	Embryo stage (number of cells)						
	1	2	3	4	5	≥6	Total
number of frozen embryos	210	91	78	417	87	46	929
% intact at thawing	81.4	22.0	11.6	24.5	20.7	6.5	34.8
% transferred	81.4	56.0	44.9	76.0	66.7	76.1	71.8

Clamart, march'85 - march'89

number of freshly transferred embryos and increasing those of cryopreserved embryos[13, 14] such a policy was also dependant on ethical, psychologic, statistic and therapeutic considerations which varied since 1985, from the transfer of only one till three or four fresh embryos according to the period or to the patient. Moreover the use of ovarian stimulation under hypophysis suppression by a gonadotrophin hormone releasing factor agonist (GnRHa), which now concerns the great majority of IVF patients, was found unfavourable for the success of embryo cryopreservation.[15] These results were of importance to support our actual policy of transferring 3-4 embryos in the IVF cycle and freezing other embryos when available.

Embryo Stage at Freezing and Survival at Thawing

From 210 embryos frozen at the pronucleated stage 171 (81.4%, Table 3) were intact at thawing, a proportion 4-fold higher than that observed for cleaved embryos (20.7%, p < 0.001). Less than 25% of cleaved embryos were intact at thawing irrespective of the stage at freezing (Table 3). However 71.8% of all frozen-thawed embryos were judged suitable for transfer. This was because we have transferred not only those embryos keeping ≥ 50% of their initial cell number but all embryos keeping at least one cell after thawing. In fact these unlyzed embryos were shown to be capable of implantation and of giving birth to normal babies.[16, 17]

Survival at Thawing and Implantation

The implantation rates per single embryo transfer were 12.1%, 9.6% or 7.5% according to the proportion of surviving cells i.e. 100%, 50%-80% or 15%-40% respectively (not significant differences (Table 4). All three pregnancies obtained from one-fourth transferred embryos were ongoing pregnancies and two normal babies were born. These results rein-

Table 4. Implantation Rate per Single Embryo Transfer According to the Embryo Stage at Freezing and the Proportion of Cell Survivng Cryopreservation

per cent surviving cells at thawing	Embryo stage (number of cells) at freezing					
	1	2	3	4	>4	TOTAL
100	11.3(62)	0.0(9)	0.0(1)	17.8(45)	0.0(7)	12.1(124)
50-80		0.0(9)	16.7(6)	10.9(64)	6.7(15)	9.6(94)
15-40			0.0(2)	9.7(31)	0.0(7)	7.5(40)
TOTAL	11.3(62)	0.0(18)	11.1(9)	12.9(140)	3.4(29)	10.5(258)

Clamart, march'85-march'89

(   ) number of single embryo transfers

forced our policy to transfer all embryos which were not lyzed at thawing and demonstrate the ability of the human embryo to regulate its development in case of dramatic cell deficiency.

Most pronucleated embryos were transferred one day after thawing. This allowed the verification of the embryo's ability to cleave and permitted the unification of transfer on the day when most thawed embryos are transferred, i. e. two days after ovulation.

The in vitro culture after thawing of pronucleated embryos was of no consequence for their ability to implant. In addition to the 62 single transfers of embryos frozen at the pronucleated stage (Table 4) there were 77 pronucleated embryos transferred by a group of two or three simultaneously. Three out of 25 (12.0%) embryos which were not cultured, implanted. This compares with 13 out of 112 (11.6%) embryos that were cultured for 1-day in vitro. However certain pronucleated embryos cultured after thawing were judged unsuitable for transfer as they did not cleave in culture.

Implantation rates were similar after transfer of an embryo frozen at the pronucleated stage (11.3%) or the second day following in vitro insemination (10.2%). However no implantation was observed with 2-cell frozen embryos and only 3.4% of the embryos implanted when frozen at 5- to 8-cell stages (Table 4).

Patients' Stimulation Treatment in the IVF Cycle

We previously reported a detrimental effect of certain ovarian stimulation treatments in the IVF cycle on the survival and implantation of frozen embryos.[15] For this reason the retrospective analysis was performed in relation to patients' treatment.

Two groups of protocols of ovarian stimulation were recorded.

Non-GnRHa treatments were either clomiphene citrate + hMG[18] or programmed cycles.[19] Other stimulation treatments included the use of long acting or short acting GnRHa associated with exogenous gonadotrophin administration.[20] These protocols were described elsewhere.[21]

Table 5 shows a detrimental effect on embryo survival of protocols including GnRHa whatever the developmental stage of the frozen embryos. In average the proportion of intact embryos at thawing was 18.0% or 46.3% for patients treated with or without GnRHa and 59.8% or 80.0% of the embryos were suitable for transfer respectively.

Amongst the transferred embryos, 3% to 4% implanted when yielded in a GnRHa treated cycle, compared with 12% to 13% of those yielded in a non GnRHa treated cycle (p < 0.001, Table 6). Once the embryo survived after thawing (Table 5) implantation was also affected by the patients' GnRHa treatments in a similar proportion (3 to 4-fold) for pronucleated or cleaved embryos (Table 6). Although embryos originating from non-GnRHa cy-

Table 5. Embryo Survival at Thawing According to the Patient Stimulation Treatment in the IVF Cycle and the Embryo Stage at Freezing

	Embryo stage (number of cells)						
	1		2 or 4		3 or ≥ 5		Total	
patient treatment including GnRHa	yes	no	yes	no	yes	no	yes	no
number of frozen embryos	40	170	226	282	112	99	378	551
% intact at thawing	62.5[a]	85.9[a]	15.0[b]	31.2[b]	8.0[c]	21.2[c]	18.0[d]	46.3[d]
% transferred	62.5[e]	85.9[e]	61.9[f]	80.8[f]	54.5[g]	67.7[g]	59.8[h]	80.0[h]

Clamart, march '85-march '89

a, b, d, e, f, h : p < 0.001 ; c : p < 0.01 ; g : p = 0.05

Table 6. Implantation Rate According to the Patient Treatment in the IVF Cycle and the Embryo Stage at Freezing. Only transfers including embryos with comparable developmental stage (either 1-cell or > 1-cell) at freezing were recorded

| | Patient stimulation treatment in the IVF cycle | | | |
| | without GnRHa | | including GnRHa | |
embryo stage at freezing	1-cell	> 1-cell	1-cell	> 1-cell
number of transferred	115	287	33	201
number of implanting embryos	14	37	1	8
(%)	(12.2)	(12.9)	(3.0)	(4.0)

Clamart, march'85-march'89

Table 7. Pregnancy Rate According to the Number of Transferred Embryos and the Patient Stimulation Treatment in the IVF Cycle. In brackets, number of transfer cycles

| | patient stimulation treatment | |
number of transferred embryos	including GnRHa	without GnRHa
1	8.6 (58)	11.0 (200)
2	8.5 (47)	23.7 (76)
3	5.3 (19)	40.0 (20)
ALL	8.1[a] (124)	16.2[a] (296)

a : $p < 0,05$ Clamart, march'85-march'89

Table 8. Overall Success of Embryo Cryopreservation According to the Patient Stimulation Treatment in the IVF Cycle

number of embryos	patient stimulation treatment	
	without GnRHa	including GnRHa
frozen-thawed	551	378
transferred	416	221
implanting	54	10
% of frozen-thawed embryo	9.8[a]	2.6[a]
% of transferred embryo	13.0[b]	4.5[b]

Clamart,   march'85-march'89

a : p < 0.001 ; b : p < 0.01

cles were mostly transferred as single embryos (200/296 transfers, Table 7) pregnancy rate was twice that observed with embryos originating from GnRHa cycles (16.2% versus 8.1%, p < 0.05, Table 7). There was also a trend for a constant pregnancy rate when single (8.6%) double (8.5%) or triple (5.3%) transfers concerned embryos yielded in GnRHa cycles compared with embryos yielded in non GnRHa cycles (11.0%, 23.7% and 40.0% respectively, Table 7).

The overall success rate of embryo cryopreservation was dramatically affected by the protocol used for ovarian stimulation (Table 8). Accumulating the decreased embryo survival after thawing with the decreased implantation rate after transfer, only 2.6% of the embryos frozen in GnRHa treated IVF cycles implanted compared with 9.8% of those frozen in conventional IVF treatment (p < 0.001, Table 8).

Patient Treatment in the Transfer Cycle

There were 400 frozen-thawed embryos transferred in spontaneous cycles without any exogenous treatment.[22] One hundred embryos were transferred in cycles lightly stimulated with hMG (one ampoule on days 6, 8 and 10) and hCG (5000 I.U.) or following ovulation induction by hCG alone.[12] These 500 embryos were transferred at a time after ovulation corresponding to their developmental age (either one or two days).

There were 131 additional frozen-thawed embryos transferred in cycles in which ovulation was suppressed by GnRHa administration (Decapeptyl-Retard, 3.75 mg) followed by hormonal replacement beginning 10 days later. Hormonal replacement regimen was oral E_2

Table 9. Implantation Rate per Transferred Embryo According to Patient Treatment in both the IVF and the Transfer Cycles. In brackets, number of transferred embryos

Patient treatment in the transfer cycle	Ovarian stimulation in the IVF cycle	
	without GnRHa	including GnRHa
none (spontaneous cycle)	13.5[a] (296)	5.8[a] (104)
ovarian stimulation (gonadotrophin treatment)[1]	15.3 (85)	6.7 (15)
endometrium stimulation (steroïd treatment)[2]	3.2 (31)	3.0 (100)

Clamart, march'85-march'89

a : $p < 0.05$
(1) either follicle growth induction (hMG) or ovulation induction (hCG) or both
(2) exogenous E2 and P following GnRHa administration (see text)

valerate from day 1 to day 28 and vaginal micronized progesterone (P) from day 14 to day 28. In those cases embryo transfer occurred on the 3rd or 4th day of P administration.[23]

The implantation rate per embryo was higher with embryos coming from non-GnRHa than from GnRHa cycles for those transferred in spontaneous cycles (13.5% versus 5.8%, $p < 0.05$, Table 9). The same tendancy was evident for embryos transferred in gonadotrophin treated cycles (15.3% versus 6.7%, not significant, Table 9). In transfer cycles with steroid hormone replacement under GnRHa gonadotrophin inhibition the implantation rates were always low (3.0% to 3.2%) whatever the origin of the frozen embryos. These results confirm the deleterious effect on embryo viability of a combination of GnRHa and gonadotrophins. It is also doubtful whether the combination of the GnRHa and steroid hormones increase uterine receptivity. However we confirm our previous data[12] of comparable implantation rates for frozen-thawed embryos transferred either in spontaneous or in gonadotrophin lightly stimulated cycles.

Embryo Freezing in an Oocyte Donation Programme

Several differences are obvious between embryo freezing in conventional IVF or in IVF using donated oocytes and the results were presented separately. The oocyte donor has to be considered as a normal woman as its fertility was previously demonstrated. Moreover most of the recipient patients were devoid of ovarian function and the thawed embryo transfer cycle was managed in a different manner. No GnRHa was given in most cycles (26 out of 31) but steroid hormones (E_2 and P) were administered for endometrial preparation.[23]

Table 10. Additional Results from Embryo Frozen in an Oocyte Donation Programme

	ovarian stimulation in the IVF cycle		
	without GnRHa	including GnRHa	Total
number of thawed embryos	9	65	74
number transferred (%)	6 (66.7)	37 (56.9)	43 (58.1)
number implanting (%)	2 (33.3)	5 (13.5)[a]	7 (16.7)

Clamart, Jan'87-march'89

a : difference with IVF using couple's gametes (4.5 %, Table 8) : p < 0.05

Table 10 shows embryo survival and implantation in this programme. Most of the thawed embryos were from IVF cycles treated with GnRHa and gonadotrophins and in these cycles the proportion of transferred embryos (56.9%) was comparable to that observed in IVF using couple's gametes (59.8%, Table 5). There was a higher implantation rate per transferred embryo in the oocyte donation programme compared to conventional IVF (13.5% versus 4.5%, $p < 0.05$, Tables 8 and 10) and the same tendency appeared with embryos coming from non GnRHa treated cycles (33.3% versus 13.0%, not significant). Since embryos obtained with donor oocytes survive cryopreservation procedures in the same proportion as other embryos, it seems hazardous to explain their better implantation rate by the origin of the oocytes. In fact the implantation rate per frozen-thawed embryo coming from GnRHa treated cycles was similar in the oocyte donation programme (five out of 65 i.e. 7.7%, Table 10) and in conventional IVF (5.8% to 6.7%, Table 9) when the transfer cycle was devoid of GnRHa. Endometrium preparation using E_2 and P in the transfer cycle seems satisfactory only in cases of no suppression of hypophyseal function. We have already reported[15] a higher pregnancy rate following fresh embryo transfer in IVF cycles stimulated with GnRHa and gonadotrophins than with gonadotrophins alone. The present results support the hypothesis of an unfavourable uterine condition when the inhibition of endogenous gonadotrophins is not compensated by exogenous gonadotrophins administration. Endogenous or exogenous gonadotrophins seem necessary for embryo implantation even in the case of ovarian privation. Alternatively ovarian secretions may be modified under gonadotrophin privation with certain consequences at uterine level.

CONCLUDING REMARKS

At the present time human embryo cryopreservation provides the best results when early embryos are frozen in the presence of PROH and sucrose. Approximately 7% of 2160 frozen-thawed 1-or 2-day embryos implant compared with 4% of 1441 2-to 4-day embryos frozen in the presence of DMSO. Moreover freezing early embryos avoids the loss of certain embryos which occurs from lengthening in vitro culture duration. Using PROH, optimal implantation rates per frozen-thawed embryos are obtained with pronucleated and 4-cell frozen embryos. Ten per cent of the frozen 4-cell embryos which lose three cells when thawed are capable of initiating a normal pregnancy. Ovarian stimulation treatments under medical hypophysectomy induced by GnRHa appears responsible for lower survival rates after embryo thawing and a lower implantation rate after frozen-thawed embryo transfer. Frozen-thawed embryo transfer either in a spontaneous cycle or in a cycle lightly stimulated by gonadotrophins result in comparable pregnancy rates. Similar results occur with embryo originating from donated oocytes and transferred under steroid hormone replacement in patients suffering from ovarian failure. However when suppression of hypophyseal function by GnRHa is not balanced by exogenous gonadotrophin administration the implantation rate per transferred thawed embryo drops dramatically. This result underlines the need of FSH and/or LH for endometrium receptivity.

REFERENCES

1. Mandelbaum, J., Junca, A.M., Plachot, M., Alnot, M.0., Alvarez, S., Debache, C., Salat-Baroux, J., Cohen, J. Human embryo cryopreservation, extrinsic and intrinsic parameters of success. *Hum. Reprod.* 2:709 (1987).
2. Van den Abbeel, E., Van der Elst, J., Van Waesberghe, L., Camus, M., Devroey, P., Khan, I., Smitz, J., Staessen, C., Wisanto, A., Van Steirteghem, A. Hyperstimulation:the need for cryopreservation of embryos. *Hum. Reprod.* 3: 53 (1988).
3. Cohen, J., Inge, K.L., Wiker, S.R., Wright, G., Fehilly, C., Turner, T.G. Duration of storage of cryopreserved human embryos. *J. Vitro Fertiliz. Embryo Transf.* 5:301-303 (1988).
4. Fugger, E.F., Bustillo, M., Katz, L.P., Dorfmann, A.D., Bender, S.D., Schulman, J.D. Embryonic development and pregnancy from fresh and cryopreserved sibling pronucleated human zygotes. *Fertil. Steril.* 50: 273 (1988).
5. Freeman, L., Trounson, A., Kirby, C. Cryopreservation of human embryos: progress on the clinical use of the technique in human IVF. *J. Vitro Fertiliz. Embryo Transf.* 3: 53 (1985).
6. Quinn, P., Kerin, J. Experience with the cryopreservation of human embryos using the mouse as a model to establish successful techniques. *J. Vitro Fertiliz. Embr. Transfer* 3:40 (1986).
7. Cohen, J., Simons, R., Fehilly, C., Edwards, R. Factors affecting survival and implantation of cryopreserved human embryos. *J. Vitro Fertiliz. Embryo Transfer* 3: 46 (1986).
8. Marrs, R.P., Brown, J., Sato, F., Yee, B., Paulson, R., Vargyas, J.M., Ogawa, T. Clinical pregnancies after IVF and cryopreservation of human embryos. Am. Fertil. Soc., Canad. *Fertil. Androl. Soc.*, Abstr. 208 (1986).
9. Dor, J., Rudak, E., Levran, D., Kimchi, M., Nebel, L., Serr, D., Goldman, B., Mashiach, S. Pregnancies following cryopreservation of human embryos. Vth World Congr. IVF-ET, Abstr. PP-131, p. 77 (1987).
10. Trounson, A., Peura, A., Freemann, L., Kirby, C. Ultrarapid freezing of early cleavage stage human embryos and eight-cell mouse embryos. *Fertil. Steril.* 49:822 (1988).
11. Lassalle, B., Testart, J., Renard, J.P. Human embryo features which influence the success of cryopreservation with the use 1-2 Propanediol. *Fertil. Steril.* 44:645 (1985).
12. Testart, J., Lassalle, B., Forman, R.G., Gazengel, A., Belaisch-Allart, J., Hazout, A., Raihorn, J.D., Frydman R. Factors influencing the success rate of human embryo freezing in an in vitro fertilization and embryo transfer program. *Fertil. Steril.*, 48:107 (1987).
13. Testart, J., Lassalle, B., Belaisch-Allart, J.C., Forman, R., Hazout, A., Volante, M., Frydman, R. Human embryo viability related to freezing and thawing procedures. *Am. J. Obstet. Gynec.* 157:168 (1987).
14. Testart, J., Lassalle, B., Belaisch-Allart, J.C., Forman, R., Frydman, R. Cryopreservation does not affect future of human fertilized eggs. *Lancet* 6.9: 569 (1986).
15. Testart, J., Forman, R., Belaisch-Allart, J.C., Volante, M., Hazout, A., Strub, N., Frydman, R. Embryo quality and uterine receptivity in in vitro fertilization cycles with or without agonists of gonadotrophin releasing hormone. *Human Reprod.* 4:198 (1989).
16. Veiga, A., Calderon, G., Barri, R.N., Coroleu, B. Pregnancy after the replacement of a frozen-thawed embryo with < 50% intact blastomeres. *Human Reprod.* 2:321 (1987).
17. Testart, J., Lassalle, B., Belaisch-Allart, J., Forman, R., Hazout, A., Fries, N., Frydman, R. Human embryo freezing. *Ann. N.Y. Acad. Sci.* 541:532 (1988).
18. Belaisch-Allart, J., Frydman, R., Testart, J., Guillet-Rosso, F., Lassalle, B., Volante, M., Papiernik, E. In vitro fertilization and embryo transfer program in Clamart, France. *J. Vitro Fert. Embryo Transfer*, 1:51 (1984).
19. Frydman R., Forman, R., Rainhorn, J.D., Belaisch-Allart, J., Hazout, A., Testart, J. A new approach to follicular stimulation for in vitro fertilization : programmed oocyte retrieval. *Fertil. Steril.* 46:657 (1986).

20. Frydman, R., Belaisch-Allart, J., Parneix, I., Forman, R., Hazout, A., Testart J. Comparison between flare up and down regulation effects of LHRH agonists in an in vitro fertilization program. *Fertil. Steril.* 50:471 (1988).
21. Testart, J., Belaisch-Allart, J., Forman, R., Gazengel, A., Strubb, N., Hazout, A., Frydman, R. Influence of different stimulation treatments on oocyte characteristics and in vitro fertilizing ability. *Human Reprod.* 4:192 (1989).
22. Testart, J., Lassalle, B., Belaisch-Allart, J., Hazout, A., Forman, R., Rainhorn, J.D., Frydman, R. High pregnancy rate after early human embryo freezing. *Fertil. Steril.* 46:268 (1986).
23. Frydman, R., De Ziegler, D., Letur-Konirsch, H., Parneix, I., Volante, M., Bouchard, P., Testart, J. Different pregnancy rate after transfers of cryopreserved embryos in an oocyte donation program and in regular IVF. VI World Congr. IVF. and Alternate assisted reproduction, Abstr.

ULTRARAPID FREEZING

Alan Trounson and A. Henry Sathananthan

Monash and La Trobe Universities
Centre for Early Human Development
Monash Medical Centre, Clayton, VIC., 3168, Australia

INTRODUCTION

A very simple snap-freezing method has been developed for the cryopreservation of early embryos at the Monash Medical Centre in recent years (Trounson et al., 1987). The procedure does not utilize expensive biological freezers and is less time-consuming than conventional slow freezing methods, which have been in use since the first pregnancy was reported from a frozen embryo by Trounson and Mohr (1983). Ultrarapid freezing has proved to be very successful with mouse 2- and 8-cell embryos and has now been adopted for early human embryos in a number of centres in Italy, Belgium, Austria, Singapore and United States of America. Clinical trials are now underway and a few pregnancies have resulted after ultrarapid freezing. It is becoming increasingly necessary to cryopreserve excess human embryos generated from IVF programmes to be transferred to patients in subsequent menstrual cycles. Thus the technology of embryo freezing has become an integral part of IVF, as was reported in a recent survey by Van Steirteghem et al. (1989).

ULTRARAPID FREEZING METHOD

The protocol used for ultrarapid freezing is briefly as follows:

- Prepare 3.0 to 4.5M DMSO and 0.25M sucrose in phosphate buffered saline (PBS) or Hepes buffered M_2 medium containing bovine serum albumin (BSA).
- Draw about 40 µl of this solution into a clear plastic freezing straw.

Advances in Assisted Reproductive Technologies, Edited by
S. Mashiach *et al.*, Plenum Press, New York, 1990

Table 1. Two-Cell Embryos Transferred to Day 1 Pseudopregnant Mice

	Control	Ultrarapid Freezing
Implanted	92%	90%
Fetuses	80%	80%
Resorptions	12%	10%

Table 2. Survival after Ultrarapid Freezing of 8-Cell Mouse Embryos

Freezing Method	No. Embryos Frozen	No. Embryos Survived	No. Blastocysts Developed
Non-frozen	—	225	211 (94%)
Slow-frozen	148	98 (66%)	82 (84%)
Ultrarapid	214	200 (94%)	194 (97%)

- Pipette embryos directly into this freezing solution in the straw.
- Heat seal straw and allow embryos to equilibrate on a flat surface (3 min.).
- Plunge straw directly into liquid nitrogen.
- Thaw embryos in a 37°C waterbath.
- Expel embryos into 0.25M sucrose in PBS or M_2 (5–10 min.).
- Culture embryos in vitro or transfer to recipients.

Hints:
- Make freezing solution 24–72 hours before use (Trounson and Sjöblom, 1988).
- Do not irradiate freezing straws before use (Shaw et al., 1988).
- Do not agitate straws during equilibration or thawing (Trounson, 1989).
- Ensure that embryos are positioned on inner surface of straw (Trounson, 1989).
- Use protein concentrations of 4–32 mg/ml BSA (Shaw and Trounson, 1989).

ULTRARAPID FREEZING OF MOUSE EMBRYOS

Most of the basic research on the mouse model has been conducted by the Monash team over the past few years. In the original study, Trounson et al., (1987) showed that the viability of frozen-thawed 2-cell embryos in vitro and in vivo was equal to that obtained by slow cooling to -80°C in 1.5M DMSO. Their viability in vivo is shown in Table 1. Exposure of embryos to the freezing solution alone had no effect on embryo viability. These results were later confirmed with 8-cell embryos (Trounson et al., 1988) and it became evident that freezing in 3.5M DMSO was significantly better for embryo viability in vivo than freezing in 2.0M DMSO (Table 2). Further, fetal development of frozen 8-cell embryos in 3.5M DMSO was not significantly different to that of non-frozen controls. The lower viability of embryos frozen in 2.0M DMSO could be due to a reduction in cell numbers in blastocysts perhaps as a result of cryoinjury to cells undetected by phase-contrast microscopy and a time-lag in the rate of cell division during post-thaw culture. Subtle changes in cell structure (see below) may have also contributed to this reduction in viability. Blastocysts with fewer viable cells than controls may also account for the increased number of implanting embryos which do not continue normal development to fetuses when 2.0M DMSO is used (Trounson et al., 1988). Shaw and Trounson (1989) also showed that highly reproducible success rates with 3.0M–4.5M DMSO with 0.25M sucrose could be obtained with ultrarapid freezing if protein concentrations are maintained between 4–32 mg/ml BSA (Table 3). Other technical details recommended for ultrarapid freezing need some explanation. Irradiation of straws for sterilization may change their inner surface properties increasing gas bubble formation within embryos thereby reducing their survival and development (Shaw et al., 1988). During the 3 minute equilibration in DMSO and sucrose the straw should be kept still and horizontal on the bench to keep embryos on the inside surface rather than suspended in solution. This improves embryo viability for some unknown reason. When these simple precautions are taken embryo survival exceeds 90% and their viability in vitro and in vivo approaches that of non-frozen embryos.

ULTRARAPID FREEZING OF HUMAN EMBRYOS

Trounson and Sjöblom (1988) froze ultrarapidly excess human embryos from an IVF programme in Sweden to determine their capacity to survive and continue development in vitro after thaw. They used 3M DMSO and 0.25M sucrose in PBS with 4 mg/ml BSA. Of 20 embryos, ranging from pronuclear ova to 12-cell embryos, 85% survived and 88% of these continued cleavage after thawing. Two of the 17 embryos that survived developed into blastocysts. Clinical trials then began at Monash (Trounson et al., 1988) when 11

Table 3. Viability of 2-cell embryos after ultrarapid freezing in 1.5–4.5M DMSO as compared to controls

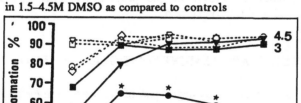

Frozen embryos (solid lines and symbols); Solution controls - dashed lines and hollow symbols; Diamonds = 0M DMSO. Statistically significant differences between frozen groups and their solution controls (P < 0.001) are starred (Reproduced from Shaw and Trounson, 1989).

Table 4. Ultrarapid Vs Slow Freezing of Human Embryos - Clinical Trial (1988)

	Slow Freezing (1.5M DMSO)	Ultrarapid Freezing (3M DMSO)
No. Embryos thawed (% survival)	97 (50%)	59 (73%)
No. Patients transferred embryos	35	29
No. Embryos transferred	47	44
No. Pregnancies (% rate)	6 (17%)*	2 (7%)**
No. Births	4	0

* 1 preclinical abortion, normal twins aborted (8 weeks)
** 1 ectopic (3 weeks), a normal fetus aborted (17 weeks)

cleavage stage embryos were frozen in 2M DMSO and 0.25-0.5M sucrose. Seven embryos survived completely intact whilst two had more than 50% of their original blastomeres intact, but no pregnancies were obtained after replacement in six patients. The latest results on ultrarapid freezing of human embryos are given in Table 4, where two pregnancies were obtained but miscarried. An ongoing clinical pregnancy is now progressing at the Epworth Hospital, Melbourne. Other groups in Italy, Belgium and Singapore are in the process of conducting clinical trials on ultrarapid freezing of human embryos and two pregnancies obtained were miscarried. The Belgian group (Gordts et al., 1989) have reported a pregnancy after transfer of eight cleaving embryos, which were frozen ultrarapidly, in six patients during subsequent spontaneous cycles. All 16 pronuclear stage embryos and 17/19 multipronuclear ova survived freeze-thaw, whereas only 4/18 cleaving embryos survived. This shows that the pronuclear stage is particularly suitable for ultrarapid freezing, as was also reported after slow freezing of human pronuclear ova with 1,2 propanediol (Testart et al., 1986; Van Steirteghem et al., 1987). A higher pregnancy rate per transfer of 23% (11/48 patients) was obtained by the second group.

ULTRASTRUCTURAL ASSESSMENT OF FROZEN EMBRYOS

Ultrastructural assessment is an integral part of our research in the development of IVF related technology. Normally embryo assessment is carried out by phase-contrast microscopy before and after freezing and during subsequent development in vitro. Although drastic changes in gross embryo morphology could be detected at this level, subtle changes within the embryo and in the structure of cellular organelles and of the cytoskeleton including spindle structure cannot be assessed by ordinary microscopy. Therefore critical morphological analysis of mouse and human embryos were routinely conducted during the development of various freezing protocols which included simple cooling to 0°C, freezing by slow and ultrarapid methods and by vitrification (Sathananthan et al., 1987; 1988a,b; Ng et al., 1988; Sathananthan and Trounson, 1989; Trounson and Sathananthan, 1989). Some ultrastructural studies have also been conducted on ultrarapid freezing of mouse and human embryos which have yielded exciting results (Sathananthan et al., 1988a; Ng et al., 1989). The preservation of fine structure of intact embryonic cells was remarkably good and was similar to that of embryos after slow freezing (Trounson and Sathananthan, 1989; Sathananthan et al., 1987; Ng et al., 1988) and of unfrozen controls (Sathananthan et al., 1986). Healthy human embryos ranging from pronuclear ova to morulae were also examined recently after ultrarapid freezing (Ng et al., 1989) and preservation of cell structure seems

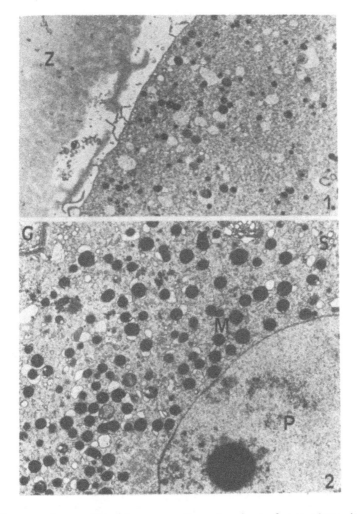

Figs. 1-5. Electron micrographs of human oocyte and embryos frozen ultrarapidly in 3.5M DMSO.

Figs. 1-2. Mature oocyte and two pronuclear ovum showing good preservation of structure. G = Golgi, M = mitochondria, P = pronucleus, S = smooth endoplasmic reticulum, Z = zona pellucida. x 7,000; x 12,000.

good, particularly in pronuclear ova. The preservation of cell structure after ultrarapid freezing is shown in Figures 1-12.

The major problems seem to be the total destruction of some of the blastomeres and formation of gas bubbles within cells (Fig. 3). Even if one blastomere survives in a 2 or 4 cell embryo it could be viable and result in a pregnancy (Ng et al., 1988). Other subtle ef-

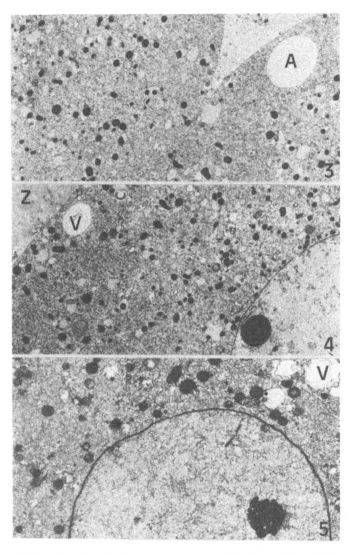

Figs. 3-5. Two-cell, 4-cell and 8-cell human blastomeres well preserved after freeze-thaw. Note air bubbles (A) in 2-cell embryo. All embryos had some fragments before freezing. V = vacuoles, Z = zona. x 7,000; x 7,000, x 11,900.

fects include disorganization of the cytosol to varying degrees in cells within the same embryo and formation of micronuclei in a few blastomeres after subsequent development in vitro (Sathananthan and Trounson, 1989). It appears that freezing of embryos in the mitotic phase should be avoided as the spindle is sensitive to cooling and this might result in

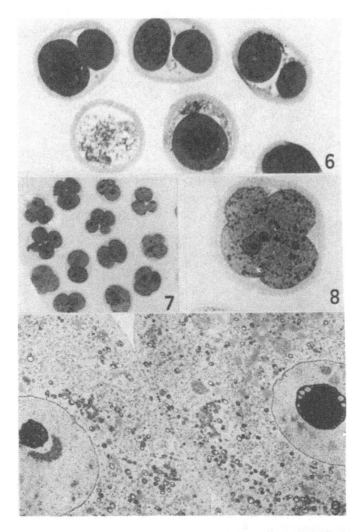

Figs. 6-12. Light and electron micrographs of mouse embryos frozen ultrarapidly in 3.5M DMSO.
Fig. 6. Two-cell embryos at thaw-note destruction of 1 or both blastomeres. x 200. Fig. 7. Embryos cultured after thaw of frozen 2-cell embryos x 100. Fig. 8. Micronucleus in a 4-cell embryo developed from a late 2-cell embryo. x 400. Fig. 9. Well preserved 4-cell embryo developed from a frozen oocyte. x 3,800. (Figs. 6, 8, 9 - reproduced from Sathananthan et al., 1988a).

chromosome scatter and consequent micronuclear formation (Figs. 8, 11, 12), as was demonstrated in mouse embryos (Sathananthan et al., 1988a). Nevertheless the mitotic phase is of short duration during cleavage and freezing this phase could be easily avoided. Zona damage was conspicuously minimal after ultrarapid freezing.

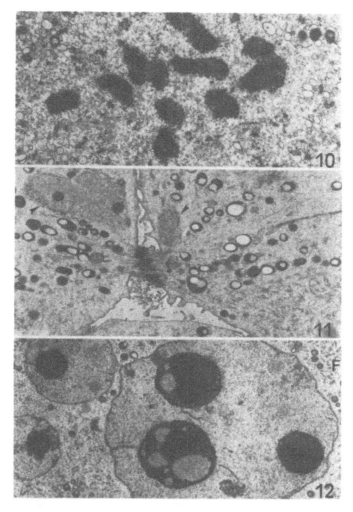

Fig. 10. Late 2-cell embryo fixed at 0°C after ultrarapid freezing - spindle microtubules are absent between chromosomes. x 11,900.

Fig. 11. Three-cell embryo developed from a later 2-cell frozen embryo. Micronuclei (arrows) are forming near equator of telophase spindle, while nuclei were forming at the poles (not shown). x 11,900.

Fig. 12. Four-cell embryo developed from a later 2-cell embryo. Two micronuclei are associated with the nucleus of a blastomere. x 7,400. F = fibrillar inclusions. (Figs. 11 and 12 - reproduced from Sathananthan et al., 1988a).

An ultrastructural investigation was undertaken to assess the effects of ultrarapid freezing on mouse oocytes and embryos using two cryoprotectants - 3.5M DMSO and 1,2 propanediol. The survival of frozen oocytes was low (33%–34%) when compared to that of 2-cell embryos (78%–79%) with either cryoprotectant. Postthaw development to blastocysts was 7%-15% for oocytes and 79%–80% for the embryos. Thus the oocytes were more vulnerable to ultrarapid freezing than were the 2-cell embryos. The oocytes were either preserved intact or were totally destroyed. The 2-cell embryos had either one or two blastomeres intact or both had disintegrated (Fig. 6). Those cells severely damaged by freeze-thawing showed disorganization of membrane components, cytosol and also nuclear components. Such changes were also induced by slow freezing or vitrification of human oocytes and embryos (Sathananthan et al., 1987; Ng et al., 1988). On the whole, ultrarapid freezing preserves cell structure quite well (Figs. 7-9). Spindle microtubules and fibrillar inclusions disappear on cooling when examined at 0°C after ultrarapid freezing (Fig. 9). These organelles reappear after post-thaw culture indicating progressive reorganization of cell structure on warming (Figs 9, 11). Chromosome analysis of 2-cell mouse embryos after ultrarapid freezing with these two cryoprotectants, however, showed no significant increase in aneuploidy but mitotic crossing over which was observed in 3.5% embryos after freezing with DMSO (Bongso et al., 1988). However, the stage of the cell cycle at which the embryos were frozen was not determined. Pronuclear ova and germinal vesicle oocytes are more resistant to cryopreservation and are evidently better preserved than mature oocytes. Embryos derived from frozen oocytes and 2-cell embryos generally appeared normal but 3/10 blastocysts examined had no inner cell masses.

CONCLUSION

The ultrarapid freezing technique is so simple, quick and cost-effective and produces reproducible results with early cleavage stage mouse embryos that it will eventually replace other methods of cryopreservation, particularly in smaller IVF programmes. Considering the size differences between pronuclear ova and corresponding blastomeres of early cleavage stages of mouse and human embryos and the differences that exist in their structural components particularly with respect to lipid, fibrillar inclusions and vesicular aggregates (present only in mouse cells), modification of this technique may be warranted to optimize survival of human embryos. Some major modifications are also required to cryopreserve blastocysts to ensure that high survival rates of late preimplantation embryos are maintained (Trounson, 1989). In general it seems advisable to ultrarapid freeze pronuclear ova and develop them to 2 or 4-cell stages in vitro before replacement in utero, as was also recommended for slow freezing methods (see review Trounson, 1989). Freezing of healthy 2-4 cell embryos should be preferred to later cleavage stages, as prolonged culture before freezing might induce subtle structural changes within blastomeres and promote partial fragmentation and nuclear abnormalities (Sathananthan and Trounson, 1989; Sathananthan et al., 1986). Poor quality frozen embryos lose their viability in vitro and result in increased rates of abortion after implantation (Lightman et al., 1989) and there is not much sense in freezing these ultrarapidly or by any other freezing method. Since the best embryos are transferred in the stimulated IVF cycle and invariably the excess poorer quality embryos are frozen for replacement in subsequent natural cycles, it should be feasible to randomly select the embryo for freezing at the pronuclear stage to minimize the bias in embryo selection and to make a valid comparison between frozen and non-frozen embryo viability in clinical trials.

REFERENCES

Bongso, A., Ng, S.C., Sathananthan, H., Mui-Nee, L., Mok, H., Wong, P.C. and Ratnam, S.S., 1988, Chromosome analysis of two-cell mouse embryos frozen by slow and ultrarapid methods using two different cryoprotectants, *Fertil. Steril.*, 49:908.

Gordts, S., Roziers, P., Campo, R. and Noto, V., 1989, Ultrarapid freezing of human embryos, Proc VI World Congress In Vitro Fertilization and Alternate Assisted Reproduction, Jerusalem, Israel, Abstract p.52.

Lightman, A., Itskovitz, J., Erlik, Y. and Brandes, J.M., 1989, Cryopreservation of unfavourable human embryos: survival and pregnancy rate of two freezing protocols, Proc VI World Congress In Vitro Fertilization and Alternate Assisted Reproduction, Jerusalem, Israel, Abstract p.53.

Ng, S.C., Sathananthan, A.H., Wong, P.C., Ratnam, S.S., Ho, J., Mok, H. and Lee, M.N., 1988, The fine structure of early human embryos frozen with 1,2 propanediol, *Gamete Res.*, 19:253.

Ng, S.C., Sathananthan, A.H., Bongso, A., Ratnam, S.S., Ho, J., Tok, V., Lee, M.N. and Mok, H., 1989, Ultrarapid freezing of human embryos (in preparation).

Sathananthan, A.H., and Trounson, A., 1989, Effects of culture and cryopreservation on human oocyte and embryo ultrastructure and function, *in:* "Ultrastructure of Human Gametogenesis and Early Embryogenesis", J. Van Blerkom and P. M. Motta, eds. Kluwer Academic, Massachusetts.

Sathananthan, A.H., Trounson, A. and Wood, C., 1986, "Atlas of Fine Structure of Human Sperm Penetration, Eggs and Embryos Cultured In Vitro," Praeger Scientific, Philadelphia.

Sathananthan, A.H., Trounson, A. and Freemann, L., 1987, Morphology and fertilizability of frozen human oocytes, *Gamete Res.*, 16:343.

Sathananthan, A.H., Ng, S.C., Trounson, A., Ratnam, S.S., Bongso, T.A., Ho, J., Mok, H. and Lee, M.N., 1988a, The effects of ultrarapid freezing on meiotic and mitotic spindles of mouse oocytes and embryos, *Gamete Res.*, 21:385.

Sathananthan, A.H., Trounson, A., Freemann, L. and Brady, T. 1988b, The effects of cooling human oocytes, *Human Reprod.*, 3:968.

Shaw, J.M. and Trounson, A., 1989, Effects of dimethyl sulphoxide and protein concentration on the viability of 2-cell mouse embryos frozen with a rapid freezing technique, *Cryobiology*, in press.

Shaw, J.M., Diotallevi, L. and Trounson, A., 1988, Ultrarapid freezing: effect of dissolved gas and pH of the freezing solutions and straw irradiation, *Human Reprod.*, 3:905.

Testart, J., Lassalle, B., Bellaisch-Allart, J., Hazout, A., Forman, R., Rainhorn, J.D. and Frydman, R., 1986, High pregnancy rate after early human embryo freezing, *Fertil. Steril.*, 46:268.

Trounson, A., 1989, Embryo Cryopreservation, *in:* "Clinical In Vitro Fertilization," C. Wood and A. Trounson, eds., Springer-Verlag, Berlin.

Trounson, A. and Mohr, L., 1983, Human pregnancy following cryopreservation, thawing and transfer of an eight-cell embryo, *Nature*, 305:707.

Trounson, A. and Sjöblom, P., 1988, Cleavage and development of human embryos in vitro after ultrarapid freezing and thawing, *Fertil. Steril.*, 50:373.

Trounson, A., and Sathananthan, A.H., 1989, Human oocyte and embryo freezing, *in:* "Opera Omnia of Marcello Malpighi: Ultrastructure of Reproduction," P.M. Motta, ed., Alan Liss, New York.

Trounson, A., Peura, A. and Kirby, C., 1987, Ultrarapid freezing: A new low-cost and effective method of embryo cryopreservation, *Fertil. Steril.*, 48:843.

Trounson, A., Peura, A., Freemann, L. and Kirby, C., 1988, Ultrarapid freezing of early cleavage stage embryos and eight-cell mouse embryos, *Fertil. Steril.*, 49:822.

Van Steirteghem, A.C. and Van den Abbeel, E., 1989, World results of human oocyte and embryo cryopreservation, Proc VI World Congress In Vitro Fertilization and Alternate Assisted Reproduction, Jerusalem, Israel, Abstract p.21.

Van Steirteghem, A.C., Van den Abbeel, E., Camus, M., Van Waesberghe, L., Braeckmans, P., Khan, I., Nijs, M., Smitz, J., Staessen, C., Wisanto, A. and Devroey, P., 1987, Cryopreservation of human embryos obtained after gamete intrafallopian transfer and/or in vitro fertilization, *Human Reprod.*, 2:593.

WORLD RESULTS OF HUMAN EMBRYO CRYOPRESERVATION

André C. Van Steirteghem and Etienne Van den Abbeel

*Center for Reproductive Medicine, Academic Hospital and Medical School
Vrije Universiteit Brussel, Brussels, Belgium*

INTRODUCTION

A survey on the practice of human embryo cryopreservation in the world was made up to December 31, 1988. It was based on the data provided by 132 centers in 18 different countries:

- **from Australia:** Queen Alexandra Hospital, University of Tasmania (J. Correy et al.); PIVET Medical Centre (J. Yovich et al.); The Royal Women's Hospital, Department of Reproductive Biology (Y. Du Plessis et al.); National Women's Hospital (Fertility Associates); Lingard Hospital; Lingard Fertility Centre (M. Brinsmead et al.); King Edward Memorial Hospital for Women, Concept Fertility Centre; Royal North Shore Hospital, Human Reproduction Unit (D. Saunders et al.); Integrated Fertility Services (G. Driscoll et al.); The Queen Elizabeth Hospital, The Reproductive Medicine Unit (G. Regan et al.); Infertility Medical Centre, Epworth Hospital (A. Trounson et al.);
- **from Austria:** Institute of Reproductive Endocrinology and IVF (W. Feichtinger et al.);
- **from Belgium:** Medical Centre for Fertility Diagnosis and IVF/ET (S. Gordts et al.); Institut de Morphologie Pathologique (J.M. Debry et al.); Fertiliteitskliniek, AZ-Antwerpen (L. Delbeke et al.); Centre de Procréation Médicalement Assistée, Université de Liège (A. Demoulin et al.); Centre for Reproductive Medicine, AZ, Vrije Universiteit Brussel (A.C. Van Steirteghem et al.);
- **from the Federal Republic of Germany:** Universitäts-Frauenklinik Erlangen (E. Siebzehnrübel et al.); Universitäts-Frauenklinik Bonn, Abteilung Gynäkologie/Geburtshilfe (K. Diedrich et al.); Medische Hochschule Hannover (A. Maass et al.); Universitäts

Advances in Assisted Reproductive Technologies, Edited by
S. Mashiach *et al.,* Plenum Press, New York, 1990

Frauenklinik Ulm (K. Sterzik et al.); Städt Krankenhaus Porz am Rein, Geburtshilfe-Gynäkologie Abteilung (K. Broer et al.); München 21 (H. Rjosk et al.); Karlsruher IVF-programm (E. Wetzel et al.); Städt. Klinik Darmstadt, Abteilung Gynäkologie & Geburtshilfe (G. Leyendecker et al.); Universitäts-Frauenklinik Göttingen (Kuhn et al.); Universitäts-Frauenklinik Tübingen (H. Tinneberg et al.); Frauenärtze/Gemeinschaftspraxis (G. Krüsmann, K. Fiedler et al.); Institute für Hormon-und Fortpfalzungsforschung (H. Bohnet et al.); Gemeinschaftspraxis Essen (T. Katzorke, D. Propping, L. Belkien); Universitäts-Frauenklinik Münster (J. Bordt et al.); Universitäts-Frauenklinik, Albert Schweitzer Straße 33 - 4400 Münster; Ludwig-Maximilians-Universität München, Frauenklinik/Klinik Großhadern (R. Wiedermann et al.); Christian-Aldrecht-Universität Kiel, Department of OB/GYN (L. Mettler et al.);

- **from France:** Centre Médico-Chirurgical et Obstétrical de la Sécurité Sociale, Strasbourg (P. Dellenbach et al.); Laboratoire de Biologie de la Reproduction, Faculté de Médicine de Grenoble (B. Sèle et al.); Hôpital Clinique Claude Bernard, Metz; Centre Hospitalier Metz, Service de Gyn-Obst. (Manini et al.); Hôpital Eduard Herriot, Centre de FIV CMU-Lyon (D. Boulieu et al.); Clinique Saint Jean, IFREARES, Nantes; Laboratoire de Fécondation in Vitro, CHU de Rennes (Le Lannou et al.); Unité INSERM 187, Département d'Obstét. & Gynécol. (J. Testart et al.); Laboratoire Albert Ier, Amiens (Bourdrel et al.); Clinique Jeanne d'Arc, Ile de la Réunion (Verrougstraete et al.); Centre FIV Paris Sud, Massy (Rivillon et al.); Laboratoire de Biologie de la Fertilité, Hotel Dieu, Paris (Vendrely et al.); Clinique St. Maurice, C.R.E.S. Lyon (Watrelot et al.); Groupe des Applicants Médical de la FIV, Faculté de Médicine/Service de Génétique, Lille (Deminatti et al.); Laboratoire de Cytologie, C.H.R.U. Morvan, Brest (Amice et al.); Centre Hospitalier Univ. de Nantes, Service D : Biologie de la Reproduction, Nantes (P. Barrière et al.); Hôpital Sud, C.H.U. de Grenoble (Racinet et al.); Grenoble University Medical School, Reproductive Biology Lab (Prof. Sele et al.); Hôpital Pellegrin, Maternité, Bordeaux (Janky et al.); Clinique Saint Sermin, Bordeaux (G. Discamps et al.); Hôpital Necker, Laboratoire de FIV-Unité 173 INSERM Paris (J. Mandelbaum et al.); Hôpital Bretonneau, Centre de Fécondation in Vitro, CHU Tours (Royere et al.); Hôpital Edouard Herriot, Laboratoire de Reproduction, Lyon (J.F. Guerin et al.); Centre FERTILY, Lyon (M. Cognat et al.); Polyclinique, Hotel Dieu, Unité de Fécondation in Vitro, Clermont-Ferrand (L. Janny et al.); Hôpital Tenon, Laboratoire FIV, Paris (J. Salat Baroux et al.); Centre Hospitalier Jean Rostand, Laboratoire de FIV, Paris (J. Cohen et al.); Clinique Marignan Paris (J. Cohen et al.); Clinique Mutualiste la Sagesse, Rennes (Arvis et al.); Association EPARP, Lille (J. Buvat); Centre Hospitalier Privé d'Aubervilliers (Allart et al.); Centre Hospitalier de CAEN, Maternité (Herlicoviez et al.); Hôpital de la Conception, Service de Gyn.-Obst., Marseille (Luciani et al.); Institut Médecine de la Reproduction, Marseille (Giorgetti et al.); Hôpital de la Maison Blanche, Reims (Harika et al.); Clinique Sainte Marie, Schoelcher (G. Audenay et al.); Hôpital International de l'Université de Paris, Fertility Centre Paris;

- **from the German Democratic Republic:** Klinik für Gynäkologie und Geburtshilfe der Karl Marx Universität (H. Alexander et al.); Wilhelm-Pieck-Universität Rostock, Department of Reproductive Medicine (R. Sudik et al.); Martin Luther Universität, Klinik und Poliklinik für Gynäkologie und Geburtshilfe, Halle;

- **from Greece:** Athens (H. Massouras et al.); "Blue Cross" Clinic, IVF & Infertility Centre (M. Tzafettas et al.); "Geniki Kliniki", Infertility and IVF Center Thessaloniki; University of Athens, Medical School (A. Comninos et al.);
- **from Hungary:** University Medical School, Department of Obstetrics and Gynaecology (G. Godo et al.); University Medical School of Debrecen, Department of Obstetrics and Gynaecology (B. Béla et al.); University Medical School of Pecs, Department of Obstetrics and Gynaecology;
- **from Israël:** Tel Aviv University, Assaf Harofe Medical Centre (R. Ron-El et al.); Rambam Medical Centre, Department of Obstetrics and Gynaecology (A. Lightman et al.); Hadassah University Hospital, IVF-Unit (N. Laufer et al.); Tel-Hashomer Hospital, Sheba Medical Centre (J. Dor et al.);
- **from Italy:** Instituto di Gynecologia e Ostetrica, Servizio FIV/ER, Torino (A. Di Gregorio et al.); Universita Cattolica del Sacro Cuore, Istituto di Clinica Obstet. e Ginecol., Roma (N. Garcea et al.); University of Milan, III Department of Obstetrics & Gynaecology (P. Crosignani et al.); Universita "La Sapienza", Servizio FIVET/GIFT, Roma (G. Micara et al.); IVF Group, CMR-Villa Regina, Bologna (L. Gianaroli et al.);
- **from The Netherlands:** Vrouwenkliniek, AZ-Utrecht (J. Van Kooij et al.); St. Annadal University Hospital, Department of Obstetrics and Gynaecology (J. Evers et al.); Vrije Universiteit Amsterdam, Medische Faculteit (J. Vermeyden et al.); Academisch Ziekenhuis Leiden (N. Naaktgeboren et al.); Erasmus Universiteit, Departement Endocrinologie, Groei en Voortplanting (G. Zeilmaker et al.);
- **from Scandinavia:** Regionsykehuset i Trondheim, Department of Obstetrics and Gynaecology (A. Sunde et al.);
- **from Singapore:** National University of Singapore, Department of Obstetricas and Gynaecology;
- **from South-Africa:** GrooteSchuur Hospital, Department of Infertility and IVF (K. Wiswedel et al.); Tygerberg Hospital, Reproductive Biology Unit;
- **from South-America:** Centro Colombiano de Fertilidad y Esterilidad, Bogota (E. Lucena et al.);
- **from Spain:** Instituto Dexeus, Departamento de Ostetricia y Ginecologia, Barcelona (P. Barri et al.); Hospital Clinico Universitario, Department of Obstetrics and Gynaecology, Valencia (A. Pellicer et al.); Hospital de Cruces-Cruces, Departamento de Ostetricia y Ginecologia, Bilbao (F. Rodriguez-Escudero et al.); Hopital Clinic i Provincial Barcelona, Servei Fecundacio Assistida; Instituto de Reproduccion CEFER, Barcelona (A. Verges et al.); Unidad de Reproduccion Humana, Departamento de Ostetricia y Ginecologia, Barcelona (P. Viscasillas); Clinica Mare Nostrum, Palma de Mallorca (B. Darder et al.);
- **from the United Kingdom:** The London Hospital, Department of Obstetrics and Gynaecology (T. Beedham et al.); University of Edinburgh, Department of Obstetrics and Gynaecology (D. Baird et al.); Ami Portland Hospital, Fertility Unit, London (M. Setchell et al.); John Radcliffe Hospital, Nuffield Department of Obstetrics and Gynaecology (D. Barlow et al.); Withington Hospital, Reproductive Medicine Unit, Manchester; Glasgow Royal Infirmary, University Department of Obstetrics and Gynaecology (Sir Malcolm MacNaughton et al.); BUPA Hospital, Norwich (R. Martin et al.); Jessop Hospital for Women, University Department of Obstetrics and Gynaecology (I. Cooke et al.); Rosie

Maternity Hospital, Cambridge (P. Braude et al.); University of York, Department of Biology (H. Leese et al.); Ninewells Hospital, Dundee (J. Mills et al.); King's College Hospital, Assisted Reproduction Unit, London (J. Parsons); The AMI Park Hospital, IVF Unit, Nottingham; The Princess Royal Hospital, Gavin Grown Clinic, Hull (A. Gordon et al.); Humana Hospital Wellington, Reproductive Studies (I. Craft et al.); BUPA Hospital, Manchester Fertility Services (S. Troup et al.); St. Mary's Hospital, IVF Unit, Manchester (P. Matson et al.); Southmead General Hospital, Bristol (A. McDermott et al.); The Glenn Hospital, Bristol (M. Hull et al.); Bristol Maternity Hospital, University of Bristol (A. McDermott et al.);

- **from the United States of America:** Sand Lake Hospital, Reproductive Biology Associates, Atlanta (J. Cohen et al.); Genetics & IVF Institute, Fairfax; University of Wisconsin-Madison; University of Rochester Medical Center, Infertility Unit (C. Graham et al.); Eastern Virginia Medical School, Jones Institute for Reproductive Medicine, Norfolk (V. Jones et al.).

At the occasion of the Fifth World Congress on In Vitro Fertilization and Embryo Transfer, which was held in April 1987 in Norfolk, Virginia, USA, a survey was made about the practice of human embryo and oocyte cryopreservation up to December 1986.[1]

MATERIALS AND METHODS

A simple questionnaire was sent to 483 centers in 18 different countries. The list of centers receiving the questionnaire in the country was appended,·and the respondents were asked to forward the questionnaire to groups that were not mentioned. Questions were asked about the practice of in vitro fertilization and embryo transfer (IVF/ET), gamete intrafallopian transfer (GIFT), zygote intrafallopian transfer (ZIFT) treatment, the fate of spare oocytes after GIFT, the fate of supernumerary zygotes and embryos after ZIFT and IVF/ET, the presence of the National Ethical Committee or legislation in the country, the experimental model, the supernumerary oocytes, zygotes or embryos that were cryopreserved, the protocols for freezing and thawing, the characteristics and monitoring of the replacement cycle, the outcome of thawing the zygotes and embryos as well as the outcome of replacement of frozen-thawed embryos.

The questionnaire dealt with the period until December 31, 1988. Table 1 lists per country the numbers of answers we received from the centers we contacted. As of March 1, 1989 we received 132 replies of 483 contacts i.e. 27.3% of the centers replied. A few answers arrived after this deadline could not be included in the survey.

RESULTS AND DISCUSSION

Legislation related to the cryopreservation of embryos was communicated by centers from some Australian states, German Democratic Republic, Israel, Norway and Spain.

Table 1. Number of Replies to the Questionnaire about Embryo Cryopreservation

Country	N° replies	N° contacts
AUSTRALIA	11	16
AUSTRIA	1	1
BELGIUM	5	16
FEDERAL REPUBLIC OF GERMANY	17	30
FRANCE	37	91
GERMAN DEMOCRATIC REPUBLIC	3	6
GREECE	4	5
HUNGARY	3	4
ISRAEL	4	16
ITALY	5	10
THE NETHERLANDS	5	8
SCANDINAVIA	1	2
SINGAPORE	1	2
SPAIN	7	10
SOUTH AFRICA	2	6
SOUTH AMERICA	1	2
UNITED STATES OF AMERICA	5	208
UNITED KINGDOM	20	50
Total	132	483

The number of oocytes replaced in the GIFT procedure was for 64 out of 87 centers limited to four; five groups mentioned replacing up to five oocytes and eight centers mentioned that they transferred up to six or more oocytes in the GIFT procedure (Fig. 1).

From the 31 centers performing ZIFT treatment, 26 replaced three or four zygotes in one or both fallopian tubes; only one group reported replacing 5 zygotes (Fig. 2).

One hundred and nineteen groups reported in the survey the number of embryos replaced during the IVF/ET treatment. Half of them (60) replaced up to four embryos, 34 groups a maximum of three embryos, while 15 and 9 centers transferred respectively five and six or more embryos (Fig. 3).

The destination of the supernumerary embryos obtained after GIFT combined with In Vitro Fertilization of the excess oocytes, ZIFT or IVF/ET was reported. As indicated in Fig. 4 most groups freeze the excess embryos; more rarely excess embryos are donated to patients requiring embryo donation, or discarded, or used for certain approved research projects; five centers mentioned that only a limited number of oocytes were inseminated.

In the survey cryopreservation of oocytes was reported by seven centers, while 81 groups mentioned that they were freezing the excess embryos. One of the questions inquired about the year the cryopreservation programme was started. Fig. 5 represents between 1982 and 1988 the cumulative number of programs where freezing and thawing of

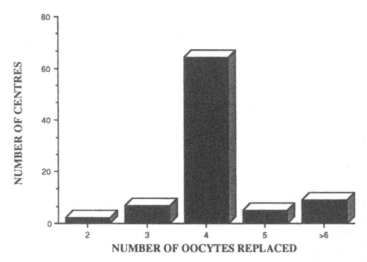

Fig. 1. Number of oocytes replaced during the GIFT treatment.

embryos was done. Especially in the years 1985, 1986 and 1987 a large number of groups started to freeze embryos.

A cryopreservation programme is useful in circumstances where the transfer procedure of fresh embryos is impossible. Although this occurs very seldom, 109 centers mentioned this event.

A major difficulty in an oocyte donation programme is the synchronization of the ovarian cycles of the donor and acceptor patients. The latter group can have functional ovaries

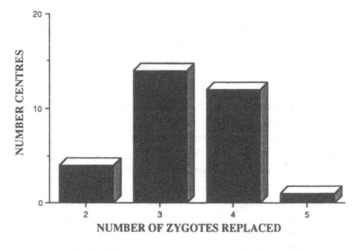

Fig. 2. Number of zygotes replaced in the course of the the the ZIFT procedure.

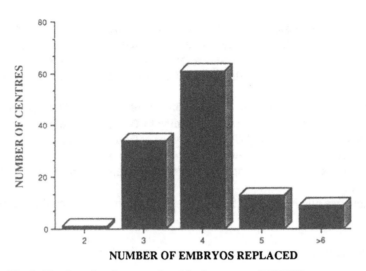

Fig. 3. Number of embryos replaced in the course of IVF-ET treatment.

or primary ovarian failure requiring steroid substitution therapy. The synchrony can be circumvented by a cryopreservation programme. Fourty-four centers reported that freezing and thawing embryos was used in the egg donation procedure.

Fourty-eight groups mentioned that they had an experimental model for their human programme. The animals used were the mouse (n = 35), the hamster (n = 8), the rat (n = 3), the rabbit (n = 1) and the monkey (n = 1).

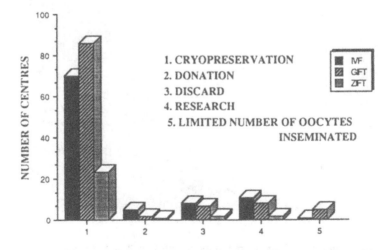

Fig. 4. Destination of excess embryos after GIFT, ZIFT, and IVF-ET.

Four different cryopreservation protocols were used. Some centers reported the use of more than one of these procedures, depending on the stage of development of the embryos at the time of freezing. The four procedures followed the protocols as reported in the literature, or introduced some minor modifications. The four freezing and thawing methods were:

1. Stepwise addition and removal of dimethyl sulphoxide (DMSO), seeding at -6°C or -7°C, and slow freezing and thawing according to Trounson and Mohr[2];
2. Addition of 1, 2 propanediol (PROH) with or without addition of sucrose, seeding at -6°C and rapid freezing and thawing, according to Testart et al.[3];
3. Addition of glycerol, seeding at -6°C, and rapid freezing and thawing according to Cohen et al.[2];
4. Ultrarapid freezing in dimethyl sulphoxide and rapid thawing, according to Trounson et al.[5]

Freezing was done either in glass cryules or in plastic straws.

The total number of embryos frozen up to December 31, 1988 was 30 850. Twenty-four centers used slow freezing and thawing with DMSO, 66 groups the PROH method, 13 centers the method with glycerol and three centers mentioned the use of the ultrarapid freezing method.

Most groups judged that the embryos were suitable for replacement after thawing if the zona pellucida was intact. Fifty-eight centers required that at least half of the pre-freezing number of blastomeres were intact, while seventeen groups mentioned that frozen-thawed embryos where only one blastomere had survived the freezing and thawing procedure,

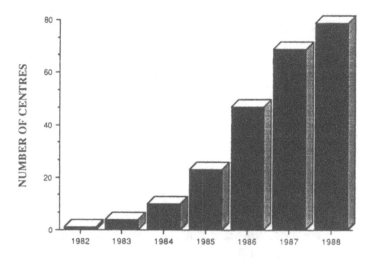

Fig. 5. Cumulative number of centres reporting cryopreservation of excess embryos after GIFT, ZIFT, ot IVF/ET.

Table 2. Outcome of Cryopreservation of Human Embryos

Embryos frozen and thawed			18322
Embryos suitable for replacement			10920 (56.6
Embryo replacements			6441
Pregnancies			738
Deliveries		302	
Babies born		329	
Abortions		216	
preclinical	96		
1st trimester	110		
2nd-3rd trimester	10		
Ongoing pregnancies		220	

would also been replaced. Embryos were kept in vitro for a variable period of hours between the end of the thawing procedure and the replacement. Certain groups mentioned that embryos frozen at the pronucleate stage were put in culture after thawing, usually for one day, and they were transferred only if they had developed to two- or four-celled embryos.

Cryopreserved embryos were replaced by 67 centers during a natural unstimulated cycle and/or by 49 groups in the course of a stimulated cycle. The replacement cycle was monitored in 70 centers by following serum and eventually urinary endocrine parameters and/or ultrasound monitoring was reported by 47 centers. Sixty-eight groups reported timing of the replacement in order to achieve a certain synchrony between embryonic stage and endometrial development; six groups mentioned asynchronous replacements. The maximum number of cryopreserved embryos that was replaced was limited to two for seven centers, to three embryos for 29 centers, to four embryos by 32 groups, while six centers mentioned replacing up to five or even more embryos.

The outcome of the cryopreservation practice is summarized in Table 2. In this summary all results were grouped together without any division in subgroups according to the origin of the frozen embryos, the cryopreservation protocol, the embryonic stage and quality, the transfer cycle, the number of transferred embryos, and so forth. From the 18 322 embryos that were frozen and thawed, 56.6% (10 920) were judged to be suitable for replacement; after 6 441 replacements where per transfer a mean of 1.7 embryos could be replaced, 738 pregnancies occurred. Overall this means that 4.1% of the 18 322 embryos implanted after replacement and that a pregnancy was achieved in 11.5 % of the replacements.

So far, 329 children have been born, including 27 sets of twins; an equal number of boys and girls were born. From these 738 pregnancies 216 (29.3%) were unsuccessful mostly due to preclinical (n = 96) or first trimester (n = 110) abortions. Two hundred and twenty pregnancies were still continuing at the time the survey was closed.

Four centers mentioned the occurrence of malformations. A therapeutic abortion was performed because of unilateral limb malformation diagnosed by ultrasound during pregnancy. One baby from twin boys had undescended testis, one baby from twin girls had pulmonary stenosis and one baby had a severe cardiac malformation, i.e. a double outlet of the right ventricle and a transposition of the great vessels.

ACKNOWLEDGMENTS

The skillful assistance of Mr. B. Van den Eede in collecting and analysing the data is greatly acknowledged. Mrs. M. Van der Helst typed the manuscript.

REFERENCES

1. A.C. Van Steirteghem, and E. Van den Abbeel, Survey on Cryopreservation, *Ann. N.Y. Acad. Sci* 541:571 (1988).
2. A. Trounson and L. Mohr, Human pregnancy following cryopreservation, thawing and transfer of an eight-cell embryo, *Nature* 305:707 (1983).
3. J. Testart, B. Lasalle, J. Bellaisch-Allart, A. Hazout, R. Forman, J.D. Rainhorn and R. Frydman, High pregnancy rate after early human embryo freezing, *Fertil. Steril.* 46:268 (1986).
4. J. Cohen, R.F. Simons, R.G. Edwards, C.B. Fehilly and S.B. Fishel, Pregnancies following the frozen storage of expanding human blastocysts, *J. In Vitro Fertil. Embryo Transf.* 2:59 (1985).
5. A. Trounson, A. Peura, L. Freemann and C. Kirby, Ultrarapid freezing of Early Cleavage stage human embryos and eight-cell mouse embryos, *Fertil. Steril.* 49:822 (1988).

INFLUENCE OF THE COOLING GRADIENT
AND THE CHOICE OF USED CRYOPROTECTANT
ON THE CRYOPRESERVATION OUTCOME OF HUMAN AND ANIMAL EGGS

S. J. Todorow, E. R. Siebzehnrübl, R. Koch, L. Wildt, and N. Lang

Unit of Reproductive Medicine
University Hospital of Obstetrics and Gynecology
Friedrich-Alexander University Erlangen, Universitätsstr, 21-23, D-8520 Erlangen

INTRODUCTION

The successful cryopreservation of human gametes and embryos conquered an indisputable although controversial place in modern reproductive medicine (Chen, 1986, Trounson et al., 1983). Until now, an impressive variety of methodological modifications have been used and the lack of consensus reached by different teams reflects the pioneering character of the work done. The present study was undertaken in an effort to investigate the following questions, the answers for which were incompletely, or not even given in the reviewed literature:

1. Direct comparison of the protective action of PROH and DMSO in a prospective randomized manner using different cryopreservation regimens.

2. Improvement of cryopreservation results by cryoprotectant (CP)-combination: potentiation of cryoprotective action and diminution of toxicity. This principal of pharmacology has been successfully applied by the vitrification of oocytes and embryos (Rall et al., 1985, Scheffen et al., 1986) but there has not been any report on its usage in conventional freezing systems.

3. Comparative evaluation of the role of low intermediary temperature (IT) and high IT freezing regimens on the cryopreservation of fertilized and non-fertilized eggs.

Advances in Assisted Reproductive Technologies, Edited by
S. Mashiach *et al.*, Plenum Press, New York, 1990

611

4. Assessment of the respective advantages or disadvantages of slow, "multiple step" addition and removal of the cryoprotectant compared to the "rapid" methods using sucrose in the equilibration and "washing out" phase.

The study was conducted on mouse, hamsters and human eggs and Pn- stages extending the validity of the conclusions to several species. The wellknown ethical controversy on the freezing of human embryos as well as the achievement of normal births in our clinic after cryopreservation of oocytes (Van Uem et al., 1987) and embryos (Siebzehnrübl et al., 1986) are the reason why we concentrated efforts in developing a better freezing technology for these particular developmental stages.

MATERIAL AND METHODS

Media: The fertilization of the mouse oocytes was conducted in M_2 medium supplemented with 30mg/ml Bovine Serum Albumine = BSA (Sigma). For the purpose of cultivation the protein supplementation was lowered (4mg/ml). The enzymatical denudation was carried out in Hepes buffered M_2-medium containing 150 I.U. hyaloronidase. The original recipes are given elsewhere (Quinn et al., 1982). Hamster eggs were handled in BSA supplemented Hams F10 media (4mg/ml). Human oocytes were recovered and cultured in modified Hams F10 supplemented with 10% (fertilization) or 20% (cultivation) patients preovulatory - or fetal cord serum. All cryopreservation experiments were conducted in modified PBS (pH = 7,4,osm $\cong$ 275 mosm) supplemented with 20% FCS (Sigma). The cryoprotectants DMSO (Merck) and PROH (Merck) were added separately or in combination till the 1.5 end equilibration was reached.

Ovulation Induction, Egg Recovery, IVF and Cultivation

Mouse: Infantile, four to six-week-old female mice (Cb6F1) were superovulated by successive i.p. application of 10 I.U. PMSG (Seragon- Organon) and 10 I.U. hCG (Primogonyl-Schering AG) 48 hrs apart. Oocyte retrieval was conducted by ampullary tube dissection, 14 hrs after the last injection. The removed cumulus masses were submitted to IVF or cryopreservation. In some groups the oocytes were enzymatically denuded from the surrounding granulous cells (150 I.U. hyaloronidase (Serva) for 10–15 minutes) before being processed further. For the purpose of IVF sperm was recovered from the cauda epididymis of 12–14 week-old adult male mice (Balb C). Pn-stages were collected under stereoscopic observation 22–24 hrs after the hCG-application. After cryopreservation and thawing the CP was completely removed by repeated washing (3 × 20% FCS in PBS). The cumulus masses and granulous free oocytes were given to the fertilization-microdrops, containing 3 - 5 × 10⁶/ml preincubated sperm (3 hrs). After 6 hrs the cells were transferred to the cultivation microdrops. Twelve hours later successful fertilization was checked by the presence of 2 cell stages (Whittingham et al., 1977).

Hamster: Female adult syrian gold hamsters were stimulated independently of the estrus cycle by 50 I.U. PMSG and 50 I.U. hCG i.p., 48 hrs apart. Oocyte recovery followed 16–18 hours later. Hamster Pn's were recovered 24–26 hours after hCG application and copulation.

Details on the fertilization and cultivation procedure have been published previously (Todorow et al., 1984).

Human: Human oocyte recovery was conducted in clomiphen, clomiphen hMG, hMG or FSH stimulated cycles by means of laparoscopy or ultrasound directed transvaginal follicular puncture. Details on the current IVF program and laboratory manual used were published elsewhere (Siebzehnrübl et al., 1986).

CRYOPRESERVATION

All animal experiments were carried out according to the protocols of a prospective randomized trial. Human oocytes were routinely frozen. Fertilization of those oocytes was conducted after thawing off in a successive cycle and sufficient sperm quality ($> 5 \times 10^6$ progressively moving sperm after "swim-up" of the ejaculate). Routinely up to four Pn's were cultivated for subsequent transfer in the same stimulated cycle. The excess diploid Pn's were frozen with the established DMSO-based cryopreservation technique (Trotnow et al., 1984). Polyploid Pn-stages were submitted to experimental protocols. In a few cases the morphological integrity of 48 hrs old unfertilized oocytes was evaluated after thawing.

Equilibration and Removal of the Cryoprotectant

The applied methods are schematically illustrated in Figure 1.

Rapid Method

The addition and removal of the cryoprotectant followed the procedure originally described by Renard et al. (1984, a.b.) and Testart et al. (1986, 1987). Dependent on the experimental group a change of CP was undertaken: 1,2 PROH, DMSO or combination of both. Before being given to the CP containing solution the cells were shortly equilibrated in 20% FCS/PBS (5 min.). The addition of sucrose at the end of the equilibration phase causes further dehydration of the cells and increment of the intracellular CP-concentration. Sucrose was given to the eggs during the depicted three steps of CP removal, in order to minimize the volumetric changes and the associated osmotic burden. The CP was completely washed out (20% FCS in PBS; 2×5 min) before processing the cells further for the purpose of IVF or cultivation.

Slow Multiple-step Method

The CP was added slowly in six steps of increasing molarity. The width of the steps was 0,25M and the duration of equilibration of the eggs in the different solutions is depicted in Figure 1. The CP removal was carried out analogously by transferring the cells to solutions of stepwise decreasing molarity.

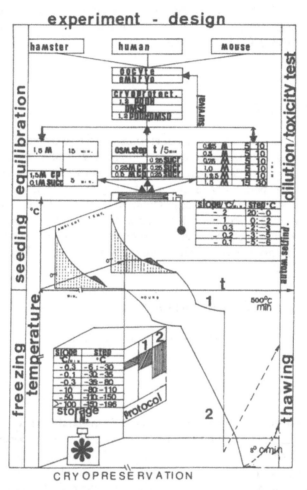

Fig. 1. Schematic illustration of the experiment design (Flow diagram). **Equilibrium phase:** Upper part: hamster, mouse and human eggs were equilibrated with the depicted CP-DMSO, PROH and combination of both. The two regimens are: "rapid method" (left side) and "slow method" (right side) of equilibration and washing out. **Seeding phase:** The cooling gradient which allows for the spontanous seeding to arise is depicted in the right upper corner. Freezing-Thawing-regimen: two different cooling and thawing gradients were used: Protocol 1 - high I.T. (35°C) and Protocol 2 - low I.T. (110°C). The corresponding thawing rates are: 500°C/min and 8°C/min.

Automatic Selfseeding

After equilibration, the eggs were transferred to the straws containing the desired CP mixture. The thermosensitive element of the cryo-equipment (Cryo Technik Erlangen 81) was immersed into a CP- containing straw (Figure 2). The straw carrier was moved down in a LN_2-vapour channel ("open vessel system") according to the preprogrammed temperature-time-gradient (Accuracy of ±0.01°C). The special shape of the straws allows the seeding spontaneously to arise in the end of the thin bent tail between 4.8°C and 7.4°C (Figure 2). The seeding progresses smoothly and extremely gently, reminding one of a climbing "spirale". The protectedly released "latent heat of crystallisation" is immediately set by the extremely sensitive movement of the carrier in the liquid nitrogen vapour channel, thus preventing the deleterious "overheating" of the probes (Trotnow et al., 1984). The program for seeding-induction is given in Figure 1. Automatic seeding combined with temperature measurement inside a container and the special control algorithm yield reproducible phase changes with an increase of temperature of less than 0.5°C.

FREEZE-THAWING REGIMENS

High Intermediary Temperature Method

After seeding induction the straws were cooled down to an intermediary temperature of -35°C. Cooling rate and timetable of the gradient is given in Figure 1. After the IT was

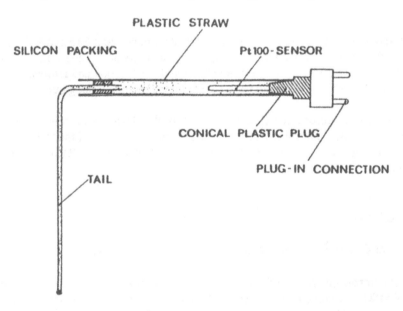

Fig. 2. Automatic and extremely gentle seeding is triggered by a crystal entering the plastic container via a thin plastic tube filled with cryoprotectant and reaching cooler areas of the vessel.

reached, the straw carrier was rapidly driven down the channel (-50°C/min.) and immersed into the liquid nitrogen. Subsequently the straws were moved to a liquid nitrogen container and after storage (24 hours to a month) thawed in a 37°C water bath (500°C/min.).

Low Intermediary Temperature Method

The straws were cooled down to an intermediary temperature of -80°C (-0.3°C/min) and then cooled further at a rate of -10°C to -110°C. The straws were then moved into the liquid nitrogen (-50°C/min). After the liquid nitrogen storage the cells were thawed at a rate of 8°C/min. The traditionally used DMSO based technology for freezing of human diploid embryos and oocytes differed to the point that the cells were cooled down to -70°C before being dropped into the liquid nitrogen container. Later there was no additional holding time at -30 to -35°C.

Statistics

Statistical analysis of data was performed using analysis of variance (Student-t-Test). Absolute numbers in different groups were compared by the X^2-2 Parameter test. Statistical package program - "Efistat".

Results

3622 mouse oocytes and Pn-stages, 2424 hamster oocytes and Pn's and 157 human oocytes and Pn's were submitted to four experimental protocols. (Tables 1,2,3, Figure 1): Protocol 1 and 2 - high IT (-35°C) by slow and rapid CP addition and removal; protocol 3 and 4 - low IT (-80°C) by slow and rapid CP equilibration and washing out.

The axiometric presentation (Figure 3) contains all relevant data from different group regimens and CP used and allows for the direct visual comparison of the following parameters: morphological survival rate (MSR), in vitro fertilization rate (FR), and cultivation or developmental rate (DR).

ANIMAL SYSTEMS

Influence of Different CPs on the Cryopreservation Success Rate

Direct comparison between three different CP equilibration regimens was undertaken: 1,2 PD,DMSO and combination of both. 825 mouse oocytes, 830 mouse Pn's, 597 hamster oocytes and 429 hamster Pn's participated in the 1,2 PD group. Comparative numbers were used in the DMSO group: 666, 644, 516 and 469. The achieved MSR, FR and DR

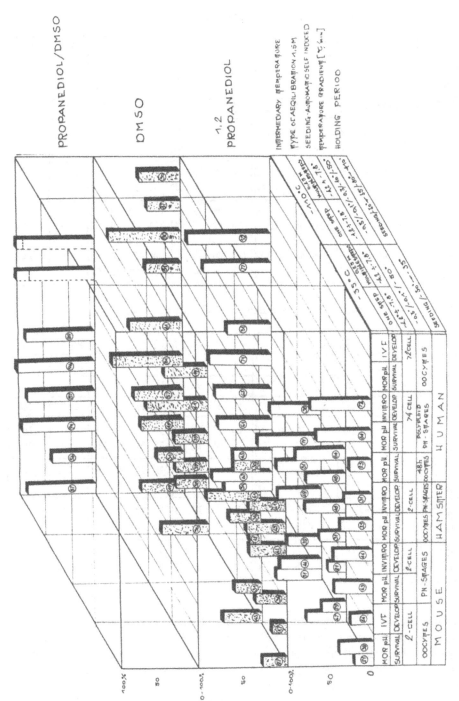

Fig. 3. Results of cryopreservation.

Table 1. Cryopreservation Results - Mouse System

C.P.	PROH								DMSO								PROH / DMSO	
eggs	oocyte				PNs				oocyte				PNs				oocyte	PNs
exp. prot.	1	2	3	4	1	2	3	4	1	2	3	4	1	2	3	4		
eggs thawed n	220	137	218	250	137	87	327	279	153	177	170	169	150	144	190	160	346	311
morph. surv n	71	54	109	158	59	27	160	171	41	76	95	83	54	17	62	79	301	292
%	32	39	50	63	43	31	49	61	27	43	56	49	36	12	39	49	87	94
IVF / D.R. n	19	11	22	68	26	13	82	108	7	24	36	26	22	7	24	22	147	239
%	27	21	44	43	44	48	51	63	17	31	38	31	41	36	39	28	49	82
zona damage n	64	26	11	12	29	29	6	16	46	35	3	10	25	59	8	3	7	9
%	29	19	5	5	21	34	2	6	30	20	2	6	17	41	4	2	2	3
degen. n	149	83	109	92	78	60	167	108	112	101	75	86	96	127	128	81	45	19
%	68	61	50	37	57	69	51	39	73	57	44	51	64	88	61	51	13	6

Table 2. Cryopreservation Results - Hamster System

C.P.	PROH								DMSO								PROH / DMSO	
eggs	oocyte				PNs				oocyte				PNs				oocyte	PNs
exp. prot.	1	2	3	4	1	2	3	4	1	2	3	4	1	2	3	4		
eggs thawed n	116	140	119	222	101	100	117	111	106	140	131	139	98	87	130	154	287	126
morph. surv n	42	67	85	158	47	72	81	93	49	85	93	124	51	34	71	109	275	120
%	36	48	71	71	47	72	69	84	46	61	71	89	52	39	55	71	96	95
IVF / D.R. n	-	-	-	-	15	46	47	66	-	-	-	-	21	10	35	66	-	112
%	-	-	-	-	31	64	58	71	-	-	-	-	41	30	49	61	-	89
zona damage n	24	36	-	4	21	16	7	3	17	29	12	1	22	12	3	8	8	2
%	21	26	-	2	21	16	6	3	16	21	9	1	23	14	2	5	3	2
degen. n	74	73	34	64	54	28	36	18	57	55	-	15	47	53	59	45	12	6
%	64	52	29	29	53	28	31	16	55	39	29	19	48	68	45	29	4	5

are given in Figure 3. There was not a statistically verified difference between the coefficients in the corresponding groups ($p > 0.05$). The same was proved for the pooled data from all four experimental protocols in both CP groups, although the success rates in the 1,2 PD group surpassed in most cases those of the DMSO group ($p > 0.05$). The results achieved in the "hamster system" were always better than those of the "mouse system" underlining the presence of species-specific susceptibility to the "cryo-damage". There were

Table 3. Cryopreservation - Human System

CRYOPROT.	PROH				DMSO		PROH/DMSO
EQUILIBR.	SLOW		RAPID		SLOW		
T° GRADIENT	LOW I.T. -110°C		HIGH I.T. -35°C		LOW I.T. -70°C		-110°C
EGGS	OOCYTES - AGED	PNs - POLYPL.	OOCYTES - AGED	PNs - POLYPL.	OOCYTES	PNs - DIPLOID	PNs - POLYPL.
n	12	31	8	17	38	39	12
157	43		25		77		12
MORPH. SURV. n/%	$3/25$	$24/77$	$4/50$	$11/64$	$14/37$	$20/51$	$10/86$
I.V.F. /DEVEL. n/%	-	$22/92$	-	$8/72$	$7/50$	$18/90$	$9/90$
% FROM ALL EGGS	-	71	-	47	18	46	75

346, 318, 311, 287 and 126 oocytes and Pn's used in the 1,2 PD/DMSO-combination group. All of them were submitted to the 4th experimental protocol. The comparison of the MSR (oocytes and Pn's) in the mouse 1,2 PD group (63%, 61%) and DMSO group (49%, 49%), with the combination group (87%, 94%) reveals the presence of statistically significant differences ($p < 0.05$). The same holds true for the MSR/DR of oocytes and Pn's in the hamster system: 1,2 PD (71%, 51%), DMSO (81%, 61%), combination group (96%, 89%). The mouse oocytes fertilization rate and DR improved considerably as well: 1,2 PD (43%, 63%), DMSO (31%, 57%) and the combination group (56%, 87%) ($p < 0.05$).

Influence of the Intermediary Temperature on the Cryopreservation Results

Six hundred and eighty seven mouse oocytes and 518 Pn's as well as 502 hamster oocytes and 386 Pn's were frozen in the high IT cryo-system (experiment 1,2). The numbers for the low IT cryo-system (experiment 3,4) are: 1153, 1267, 898, 638. The comparison of the mouse oocytes/Pn's MSR in experiment 1,2 (1,2 PD: 32%, 39%, 43%, 31%; DMSO 27%, 43%, 36%, 12%) and experiment 3,4 (1,2 PD:50%, 63%, 49%, 61%; DMSO: 49%, 49%, 56%, 39%) shows the better results achieved in the latter system ($p < 0.05$). The corresponding coefficients (MSR/DR) for hamster oocytes and Pn's underline the same conclusion: experiment 1,2 (1,2 PD: 36%, 31%, 48%, 64%: DMSO: 46%, 41%, 61%, 30%) experiment 3,4 (1,2 PD: 63%, 71%, 51%, 71%; DMSO 81%, 61%, 71%, 49%).

The combination group was not taken into account because there are data available only for the 4th protocol.

The zona pellucida breakdown rate of hamster oocytes and Pn's in experiments 1,2 (1,2 PD: 21%, 26%, 21%, 16%; DMSO: 16%, 21%, 23%, 14%) was significantly higher than in experiment 3,4 (1,2 PD: 0%, 2%, 6%, 3%; DMSO: 9%, 1%, 2%, 5%) (p < 0.01). The same holds true for the mouse oocytes and Pn's: experiment 1,2 (1,2 PD: 29%, 19%, 21%, 34%; DMSO: 30%, 20%, 17%, 41%) and experiment 3,4 (1,2 PD: 5%, 5%, 2%, 6%; DMSO: 2%, 6%, 4%, 2%) (p < 0.01).

Comparison Between the Slow, Multiple Step and the Rapid Method of Cryoprotectant Addition and Removal

Seven hundred and sixty one mouse oocytes and 804 Pn's were submitted to the rapid method of CP equilibration (experiments 1,3) compared to 1079 mouse oocytes and 981 Pn's with slow addition and CP removal (experiments 2,4). The corresponding numbers for the hamster system are: 457 oocytes, 446 Pn's (experiment 1,3) and 928 oocytes, 578 Pn's (experiment 2,4). Although there is a tendency of achieving better results in the latter compared to the former experiments, in most cases the differences were not significant (p > 0.05). There is no conclusive evidence that the slow stepwise cryoprotectant equilibration and washing out offers substantial advantages to the abovementioned rapid method.

Human System

The uneven distribution of the material used can be seen from Table 1. The DMSO-treated oocytes and diploid Pn's were transferred back to the patients, 24 hours after the successful cryopreservation (development to 2 or 4 cell stage). There were only triploid Pn's frozen in both experimental groups (1,2 PROH and PROH/DMSO). The MSR of the 48 hrs "aged" oocytes cannot be compared with the corresponding rate in the DMSO group. On the other hand the freezing regimes (DMSO group) applied to the diploid human Pn's and oocytes differed from that of the triploid Pn's frozen in both experimental groups (see Material and Methods). These circumstances restrict the validity of the conclusions made. The success rates are depicted in Figure 3. The MSR of 1,2 PD-treated Pn's (77%) and DMSO-treated Pn's (51%) differ significantly (p < 0.05). The DR in the former system is higher than in the latter (p > 0.1). On the whole 66% of the 1,2 PD-treated cells survive and develop further (low and high IT systems) compared to 46% in the DMSO group (p < 0.05). Although promising, the numbers used in the combination group (PD/DMSO) are still insufficient to produce any conclusive evidence on the potential benefit of CP-combination: MSR (86%) and DR (90%).

The 1,2 PD group allows a direct comparison of the rapid (experiment 1) to the slow method (experiment 4) of freeze-thawing and CP equilibration. The MSR and DR in the first group are lower (64%, 72%) than in the second (77%, 92%). The overall efficacy of the former (45%) compared to the latter system (71%) differs significantly (p < 0.05). The morphological survival rate of the "aged" oocytes is twice as high as in the "first" group (p < 0.05).

From 38 intact human oocytes frozen with DMSO 14 (37%) survived and seven embryos were produced after IVF (50%). The MSR of the Pn's in the DMSO system (51%) does not differ substantially from the corresponding rate for oocytes (37%) (p > 0.05). The DR of morphologically intact looking Pn's is much higher than the FR of oocytes which preserved their structural integrity after thawing, implying a higher susceptibility to cryo-damage of the latter.

DISCUSSION

Several cryoprotectants have been used to freeze human embryos and oocytes: Glycerol (Cohen, J. et al., 1985, Ashwood-Smith et al., 1986), PROH (Lassale, B. et al., 1985, Testart et al., 1986), DMSO (Chen, C. 1986, Al Hasani S. et al., 1986, Freeman et al., 1986). Numerous mechanisms account for the protective action and the substances used may express them in a different combination and degree: stabilization of the cytoplasmatic amorphous structure and inhibition of crystallisation, inhibition of deleterious membrane phase transitions, lowering of the eutectic point and minimization of the osmotic burden. The comparative information on different CP regarding their specific mechanisms of action is scarce. PROH inhibits crystallisation at a lower concentration than DMSO and Glycerol (Scheffen, B. et al., 1986). Methanol seems to be an "ideal" CP till -30°C but is almost ineffective below this temperature limit (Rall, W. F. et al., 1984). There are some data supporting the view that the optimal cryoprotective temperature intervals of DMSO and 1,2 PD differ: 1,2 propandiol down to -30°C and DMSO proved more effective at lower temperatures (Van der Abbel et al., 1987). In this study we looked for ways to boost the cryoprotective effect by CP combination and to decrease the associated toxicity by reducing the concentration of each substituent. This principle has been successfully used in the vitrification (Rall, W. F. et al., 1985, 1987). We report here for the first time on its application in conventional freezing systems. Through the combination of PROH/DMSO there is a significant improvement of the freeze-thawing success rates in the mouse and hamster experiments (see Results, Figure 3). Although promising, the results with human triploid eggs are still lacking the necessary numbers sufficient to prove a statistically verified difference to the other two systems.

In agreement with the observation of others (Bernard, A. et al., 1985) the prolonged 1.5 M PROH equilibration of intact unfertilized human oocytes very often leads to occurrence of "cytoplasmatic blebbing" in the perivitteline space and thus to structural damage of the egg. This would never happen in the 1.5 M PROH/DMSO system.

So far a randomized comparative study on the cryoprotective effect of PROH and DMSO has not been undertaken. In the mouse and hamster systems the 1,2 PD success rates in the different subgroups surpass, although not significantly, the corresponding coefficients reached in the DMSO subgroups. The difference attains significance only in the human system (p < 0.005). This supports the view of other authors on the benefits of 1,2 PROH over DMSO using systems (Van Steirteghem, A. C. et al., 1987). It should be noted that the accounted differences of the success rate could not only be due to the EP used but also to the differences of the freezing regimens used (see material and methods - low IT Regimen). This restricts the validity of the conclusion made.

Several authors undertook systemic investigations in order to compare low IT to high IT-DMSO using cryopreservation systems (Whittingham, D. G. et al., 1977, Critser, J. K., et al., 1986, Zeilmarker, G. H. et al., 1984). These studies complying with our results, underline the benefits of the former compared to the latter systems. We achieved pregnancies and births after "slow" freezing and thawing of DMSO treated human oocytes and Pn's (low IT regimen); (Siebzehnrübl, E. R. et al., 1986), Van Uem et al., 1987). There are authors supporting an opposing view on the freezing of mouse (Chen, C. 1986) and human oocytes (Trounson, A., 1986). Till now an analogous comparative study with PROH as a cryoprotectant has not been carried out. High IT-PROH based systems for freezing of animal and human early developmental stages are successfully established (Renard, J.P. et al., 1984, Lassale, B. et al., 1985). Trounson et al., (1986) report on the good results achieved by the freezing of six human oocytes in a low IT-PROH system. Our results with PROH prove that the complete dehydration of oocytes and Pn's in the "low IT" systems leads to higher MSR, FR and DR than in the "high IT" system. This holds true for all three species. Due to immediate transfer after thawing the high MSR of human Pn's and 2-cell stages (70%–80%) reported in the literature does not take into account the diminution of DR as a sequence of cryo-injury (Testart et al., 1986, Lassale et al., 1985). Recently published results on "rapid" freezing of human Pn's with 1,2 PD reveal that only 28% of 56 Pn's cleaved, although 71% survived morphologically immediately after thawing (Van Steirteghem, A. C. et al., 1986). In the "slow" freezing system we used, 71% of all initially frozen embryos cleaved 24 hours later. There are two modifications characterizing the denoted low IT system:

1. the introduction of an additional holding time (30 - 35°C, -0.1°C/min.) which allows for an additional dehydration to take place. This leads the intracellular CP concentration to rise by keeping down toxicity due to the low temperature;

2. slow cooling in additional steps from -80 to -110°C (-10°C/min.) and from -110°C to -150°C (-50°C/min.) conforms with the presumption of Trounson et al. (1986) that handling embryos below -80°C may have effects on embryo viability. Although the dehydration of the blastomeres below this temperature is complete (Rall, W. F. et al., 1984), the effects on the crystallisation structure of the zona pellucida are still unknown.

The analysis of the zona pellucida breakdown rate in the high and low IT systems reveals a three to four times lower rate in the first compared to the second system (Tables 1,2). The role of the intact zona pellucida for the successful embryo cryopreservation is indisputable.

The advantage of the automatic, selfinduced "smooth" seeding in an "open LN_2 system" has been discussed previously (Trotnow, S. et al., 1984, Siebzehnrübl, E. R. et al, 1986). There are studies pointing out the better diffusability of PROH and DMSO compared to glycerol, which makes them suitable for freezing of cells with large volume and low permeability - e.g. oocytes (Bernard, A. et al., 1983, 1985). In the adopted "rapid" system of equilibration, sucrose was given in order to achieve an additional dehydration and subsequent intracellular concentration increment of the CP before the freezing took place. The usage of sucrose in the "washing out" phase aims at reduction of osmotic burden for the cells (Szell, A. et al., 1985). The compared "slow" multiple step method of CP addition and removal gave equal or better results in all groups although the difference did not reach significance ($p > 0.05$). The prolonged equilibration of the CP did not show any additional toxic effect on the eggs. Volumetric changes during addition and removal of the CP are not observed. On the other hand, "one step" equilibration at room temperature can lead to shrinkage or reexpansion over 10% of the original volume with associated osmotic damage.

ACKNOWLEDGMENTS

This work was supported by a grant from the Sander-Foundation for Scientific Research, Munich. We gratefully appreciate this. We are indebted to Mrs. U. Woltering for her skillful technical assistance.

The protocols concerning research on human material were worked out according to the guidelines and with the approval of the Central Ethical Commission on Research on Human Embryos of the Federal Medical Council and the Ethical Commission of the Medical Faculty of the University.

REFERENCES

Al-Hasani, S., Tolksdorf, A., Diedrich, K., van der Ven, H., and Krebs, D., 1986, Successful in vitro fertilization of frozen-thawed rabbit oocyte, *Human Repr.*, 1:309-312.

Ashwood-Smith, M. J., 1986, The cryopreservation of human embryos, *Human Repr.*, 1:319-332.

Bernard, A., and Fuller, B. J., 1983, Cryopreservation of in vitro fertilized 2 cell mouse embryos using a low glycerol concentration and normothermic cryoprotectant equilibration: a comparison with in vitro fertilized ova, *Cryo-Letter*, 4:171-178.

Bernard, A., Imoedemke, D. A., and Shaw, R. W., 1985, Effects on cryoprotectants on human oocyte, *Lancet*, 1:632-633.

Chen, C., 1986, Pregnancy after human oocyte cryopreservation, *Lancet*, 19:884-886.

Cohen, J., Simons, R. F., Edwards, R. G., Fehilly, C. B., and Fishel, S. B., 1985, Pregnancies following the frozen storage of expanding human blastocysts, *J. of in vitro Fert. and Embr. Transf.*, 2:59-64.

Cohen, J., Simons, R. S., Fehilly, C. B., and Edwards, R. G., 1986, Cryopreservation of hamster oocytes: effects of vitrification or freezing on human sperm penetration of zone-free hamster oocytes., *Fertil. Steril.*, 46:277.

Fehilly, C. B., Cohen, J., Simons, R. F., Fishel, S. B., and Edwards, R. G., 1985, Cryopreservation of cleaving embryos and expanded blastocysts in the human: a comparative study, *Fertil. teril.*, 44:638-644.

Freeman, L., Trounson, R., and Kirby, C., 1986, Cryopreservation of human embryos: progress on clinical use of the technique in human in vitro fertilization, *J. of in vitro Fert. and Embr. Transf.*, 3:53-62.

Garcia, J., Maralles, A., and Garcia, A., 1986, Cryopreservation of mouse oocytes: influence of cooling rate, *Cryo-Letters*, 7:353-360.

Heyman, Y., Smorag, Katska, L., Vincent, C., 1986, Influence of carbohydrates, cooling and rapid freezing on viability of bovine nonmatured oocytes or 1-cell fertilized eggs, *Cryo-Letters*, 7:170-183.

Lassale, B., Testart, J., and Renard, J. P., 1985, Human embryo features that influence the success of cryoporeservation with the use of 1,2 propanediol, *Fertil. Steril.*, 44:645-651.

MacFarlane, P. R., Fragoulis, M., Ulherr, B., and Jay, S. D., 1986, Devitrification in aqueous solution glasses at high heating rates, *Cryo-Letters*, 7:138-145.

Michaelis, U., Burick, K., and Hahn, J. (1985), Kryokonservierung von Mäuseembryonen mit hohen Kühl- und Auftauraten unter Verwendung von DMSO, Glycerin und Propanediol als Gefrierschutzmittel, *Fertilatät*, 1:94-100.

Miyamoto, H., and Ishibashi, T., 1977, Survival of frozen-thawed mouse and rat embryos in the presence of ethylene glycol, *J. Repr. Fert.*, 50:373-375.

Plachot, M., Junca, A. M., Mandelbaum, J., Cohen, J., Baroux, J. S., and DaLage, C., 1986, Timing of in vitro fertilization of cumulus-free and cumulus-enclosed human oocytes, *Human Repr.*, 1:237-242.

Quinn, P., Barros, C., and Whittingham, D. G., 1982, Preservation of hamster oocytes to assay the fertilizing capacity of human spermatozoa, *J. Repr. Fert.* 66, 161-168.

Quinn, P., Stanger, J. D., and Whittingham, D. G., 1982, Effect of Albumine on Fertilization of mouse ova in vitro, *Gamete Research* 6:305-314.

Rall, W. F., Czlonkowska, M., Barton, S. C., and Polge, S. C., 1984, Cryoprotection of Day-4 mouse embryos by methanol, *J. Repr. Fert.*, 70:293-300.

Rall, W. F., and Fahy, G. M., 1985, Ice-free Cryopreservation of mouse embryos at -196°C by vitrification, *Nature*, 313:573-575.

Rall, W. F., Reid, D. S., and Polge, C., 1984, Analysis of slow-warming injury of mouse embryos by cryomicroscopical and physicochemical methods, *Cryobiology* 21:106-121.

Rall, W. F., Wood, M. J., Kisby, C., and Whittingham, D. G., 1987, Development of mouse embryos cryopreserved by vitrification, *J. Repr. Fert.*, 80:499-504.

Renard, I. P., and Babinet, C., 1984, High survival of mouse embryos after rapid freezing and thawing inside plastic straws with 1-2 propanediol as a cryoprotectant, *J. Exp. Zool.*, 230:443-449.

Renard, I. P., Nguyen, B. X., and Garnier, V., 1984, Two-step freezing of two cell rabbit embryos after partial dehydration at room temperature, *J. Repr. Fert.*, 71:573-580.

Scheffen, B., van der Zwalmen, P., and Massip, A., 1986, A simple and efficient procedure for preservation of mouse embryos by vitrification, *Cryo-Letters*, 7:260-269.

Siebzehnrübl, E. R., Trotnow, S., and Weigel, M., et al. (1986), Pregnancy after in vitro fertilization, cryopreservation, and embryo transfer, *J IVF ET* 3:261-263.

Siebzehnrübl, E. R., Schuh, B., van Uem, J., Koch, R., and Lang, N., 1987, Cryopreservation of rabbit oocytes as a model for the successful freezing of human ova, *Human Repr.*, (Abstr.) 2:63-64.

Testart, J., Lassalle, B., Allart, J. B., Forman, R., Hazout, A., Volante, M., and Frydman, R., 1987, Human embryo viability related to freezing and thawing procedures (1987), *Am. J. Obst. Gyn.*, 157:168-171.

Todorow, S., Todorow, I., Mirkoff K., and Kyutukchiev, B., 1985, Relation between human sperm kinesimetric data and penetration results in the zona-free-hamster egg test, *Amer. J. of Repr. Immun. and Microb.*, 7:85.

Trotnow, S., 1984, Cryopreservation of mammalian embryos, *in:* "Recent Progress in Human IVF", W. Feichtinger u. S. Kemeter, eds., 83.

Trounson, A., and Mohr, L., 1983, Human pregnancy following cryopreservation, thawing and transfer of eight-cell embryo, *Nature*, 305, 707-709.

Trounson, A., 1986, Preservation of human eggs and embryos, *Fertil. Steril.*, 46:1-12.

Trounson, A., and Freeman, L., 1985, The use of embryo cryopreservation in human IVF programmes, *Clin. Obstet. Gynecol.*, 12:825-831.

Van der Abbel, E., Van der Elst, J., and Van Steirteghem, A. C., 1987, Cryopreservation of 1-cell pronucleate stage mouse embryos with 1,2 propanediol or with dimethylsulphoide, *Human Repr.*, (Abstr.) 2:96.

Van Steirteghem, A. C., van den Abbel, E., Camus, M., and van Waesberghe, L., et al., 1987, Cryopreservation of human embryos obtained after gamete intra-Fallopian transfer and/or in vitro fertilization, *Human Repr.*, 2:593-597.

Van Uem, J. F. H. M., Siebzehnrübl, E. R., Schuh, B., Koch, R., Trotnow, S., and Lang, N., 1987, Birth after cryopreservation of unfertilized oocytes, *Lancet*, 3:752-753.

Whittingham, D. G., 1977, Fertilization in vitro and development to term of unfertilized mouse oocytes previously stored at -196°C, *J. Reprod. Fertil.*, 49:89-94.

Zeilmarker, G. H., Alberta, A. T., and Van Gent, I., et al. 1984, Two pregnancies following transfer of intact frozen-thawed embryos, *Fertil. Steril.*, 42:293-296.

BASIC CONCEPTS AND FUTURE TRENDS IN SPERM CRYOPRESERVATION

Don P. Wolf

Division of Reproductive Biology and Behavior
Oregon Regional Primate Research Center, Beaverton, Oregon
and
Department of Obstetrics/Gynecology
Oregon Health Sciences University, Portland, Oregon

INTRODUCTION

The "New Guidelines for the Use of Semen Donor Insemination: 1986" prepared and published by the American Fertility Society endorsed the use of fresh semen for therapeutic insemination (TI). The report reflected the bias that, if appropriate guidelines are followed, the risk of contracting an infectious disease with the use of fresh semen is low enough to be tolerable. The incentive for using fresh semen reflects the bias that pregnancy rates following the use of frozen semen are inferior. In 1988, the revised guidelines were published recommending the exclusive use of quarantined frozen semen for TI (The American Fertility Society, 1988). Moreover, a six- rather than a three-month quarantine period was specified. These recommendations, along with other concerns, have catalyzed interest in perfecting sperm cryopreservation protocols.

The freezing of human semen is an established procedure which has resulted in the birth of thousands of progeny worldwide with a very low incidence of birth defects (1%). Births have been reported following TI with sperm stored for up to 12 years and there is no reason to believe that this record won't be extended (Sherman, 1986). Currently, there are some 30 commercial- and university-based sperm banks in the United States and standards for cryobanking have been established by the American Association of Tissue Banks.

Despite the success realized with TI employing cryopreserved sperm, there seems little

Advances in Assisted Reproductive Technologies, Edited by
S. Mashiach *et al.,* Plenum Press, New York, 1990

room for argument concerning the viability of fresh as opposed to frozen sperm; the latter show significant reductions in motility and viability compared to their fresh counterparts and, once removed from the cryoprotectant, survival is limited to several hours. Freezing is detrimental for most cells with the plasma membrane as the principal target of cryodamage and intracellular ice crystal formation as the primary mechanism. Hence, the danger during freezing is not low temperature storage per se but rather transitions through the temperature zone where ice crystal formation is most likely to occur. Sperm from some mammals, but not man, are injured by exposure to low temperature alone, a phenomena called chilling injury or cold shock. Additionally, the mere exposure of sperm to cryoprotectants such as glycerol at ambient temperature can induce injury.

A number of different cryoprotective media and techniques have been described for the conventional cryopreservation of human sperm, based upon the use of glycerol as the principal, permeable cryoprotectant in the presence or absence of extenders such as citrated, TES-Tris or Hepes-fortified egg yolk. Because sperm contain relatively small amounts of cytoplasm and are relatively permeable to water, rapid but poorly controlled rates of freezing and thawing are tolerated.

EVALUATION OF CRYODAMAGE

An assessment of cryodamage with the objective of improving cryosurvival should include characterization of each individual step in the process, i.e., sample dilution, cryoprotectant addition, cooling to the freezing point and freezing. However, on a routine basis postthaw motility scores, in the presence of cryoprotectant, are most commonly used. A minimum standard is that 50% of the motile sperm originally present survive. Additionally, a stress test can be applied where sperm are first washed free of cryoprotectant and then incubated in an appropriate medium at 37° for up to 24 hours. Video-assisted digital imaging or comparable technology which provides quantitative assessments of sperm motility can be applied in monitoring sperm survival. Hyperactivation of human sperm, when expressed as a percentage of washed cells, is donor characteristic (Mack et al., in press). Therefore, this derived parameter may provide a convenient means for monitoring cryodamage as well. A summary of the quantitative changes observed in individual parameters of motility, including hyperactivation, during individual steps in the cryopreservation process are illustrated in Figure 1. While changes associated with cryoprotectant addition alone were minimal, substantial decreases were seen in all measured parameters except linearity following freeze-thawing.

With domestic animals, cryodamage is often monitored not only by motility but also by quantitative assessments of acrosomal status. The technology is now available (indirect immunofluorescence) for the rapid quantitation of acrosomal status in relatively large numbers of human sperm cells. Therefore, it seems appropriate to begin including acrosomal status quantitation as a measure of cryodamage. Some investigators are also utilizing the sperm penetration assay as a means of evaluating cryodamage (Critser et al., 1987). This

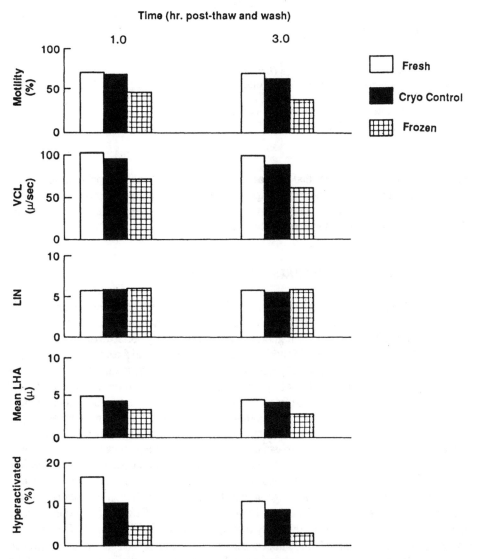

Fig. 1. Washed sperm motility parameters in single ejaculates exposed to individual steps in the conventional cryopreservation process. Semen was washed in human tubal fluid medium or Ham's F-10 medium containing maternal serum (7.5%) and analyzed by CellSoft (Cryo-resources, New York, NY) directly (fresh), after transient, ambient temperature exposure to the cryoprotectant (cryo control) or after freezing (frozen). For freezing, semen was diluted 1:1 with glycerated, TES, Tris-egg yolk freezing medium (Irving Scientific, Irving, CA) and cooled to -4° (30 min hold at 4° C) before exposure to liquid N_2 vapors and plunging into liquid N_2. VCL = curvilinear velocity at 30 frames/sec. LIN = linearity or VCL/VSL (straight line velocity). Mean LHA = mean lateral head amplitude. Hyperactivation as defined by Robertson et al. (1988).

Table 1. Techniques to Increase Banked Sperm Concentrations

---

1) Increase donor selection

2) Minimize dilution
    a) Add glycerol without extenders (Brown et al., 1988)
    b) Dilute 3 parts semen:1 part cryoprotectant (30% glycerol-egg yolk; Clarke et al., 1988)

3) Concentrate sperm post-thaw (Bordson et al., 1986)

---

approach is limited by the fact that only acrosome-reacted sperm are capable of fusing with the zona-free hamster egg and cryodamage includes acrosomal loss. Hence, false-positives may occur, i.e., fusion with zona-free eggs, which does not reflect the ability of capacitated, acrosome-intact cells to fertilize a zona-intact egg.

FRESH VERSUS FROZEN SPERM COMPARISONS

Current success rates for therapeutic insemination programs utilizing frozen semen are in the range of 65%–70% cumulative values over 12 insemination cycles. The corresponding levels for fresh semen, while higher, are not significantly so, at least when relatively small sample sizes are analyzed. Monthly fecundities for fresh and frozen semen are approximately 10%–14% and 5%–9%, respectively. In considering these differences, it must be recognized that the concentration of sperm used in the insemination unit is significantly higher for fresh semen TI since the frozen sample is diluted with cryoprotectant and the recovery of motile sperm postthaw is less than 100%. This difference may be as large as fivefold when the concentration or total number of motile sperm is contrasted. Based on a theoretical argument, a correlation exists between increasing pregnancy rates and increasing concentrations of motile sperm used in the insemination until, at some relatively high but undefined threshold level, further increases in the concentration of motile sperm transferred will have little or no influence on pregnancy or fecundity rates. In studies conducted at the University of Wisconsin (Richter et al., 1984; Brown et al., 1988), a sperm concentration dependency can be observed in pregnancy rates achieved with both fresh and frozen semen. Additionally, even at relatively high concentrations of cryopreserved sperm, while pregnancy rates are increased over lower cryopreserved sperm concentrations, a substantial difference still exists between the outcomes obtained for fresh versus frozen semen. Therefore, while the importance of utilizing higher concentrations of motile sperm in TI may be recognized, the threshold level for either fresh or frozen semen required for maximum pregnancy outcome has yet to be defined. Several strategies are available for increasing the motile sperm concentrations postthaw which include more rigorous donor selection (impractical as $\geq 80 \times 10^6$ sperm/ml is commonly required), protocols that minimize sample dilution during cryoprotected addition (Table 1) and, finally, approaches designed to stimulate motility in sperm postthaw.

CLINICAL PARAMETERS IN TI

Clinical parameters that may contribute to TI outcome include the timing of insemination, the number of inseminations, the route of insemination and the presence or absence of ovulation by the female partner. These parameters have been optimized adequately for fresh but not for cryopreserved semen. Timing for TI is usually based upon the results of home LH testing using commercially available kits. This approach is adequate. The importance of the number of inseminations to TI outcome with cryopreserved semen should be reevaluated despite findings which suggest that two inseminations per cycle produce fecundities similar to one (Iddenden et al., 1985). A single insemination with fresh semen provides a population of potentially fertile sperm for days, whereas cryopreserved samples probably do not survive for more than several hours. The route of insemination, cervical as opposed to uterine, also merits attention. We observed no difference in pregnancy rates for TI patients inseminated with fresh donor semen, or washed sperm, when a cervical versus a uterine route was compared (Patton et al., 1988). However, because of anticipated differences in sperm survival in the female reproductive tract, the route of insemination may emerge as an important parameter following TI with cryopreserved semen. The presence or absence of ovulation has obvious implications to TI outcome. This may impact 20% of TI cycles and may require the application of mild ovarian hyperstimulation protocols.

Fine tuning of the currently available conventional cryopreservation protocols including the use of controlled rate-freezing is not likely to produce marked improvements. The alternatives of increasing the quantity of sperm per insemination unit and increasing sperm motility postthaw by in vitro additives or treatments have already been mentioned. The evaluation of radically different approaches to cryopreservation is therefore appropriate. One such approach, vitrification, takes advantage of the ability of highly concentrated solutions of cryoprotectant to avoid crystallization when cooled to low temperatures. Instead of freezing, the cell suspension supercools and becomes so viscous that it forms a transparent solid called "glass." The sperm banking industry may benefit immensely by the application of vitrification or other similar radically different freezing protocols.

ACKNOWLEDGMENTS

The secretarial and editorial assistance of Patsy Kimzey is acknowledged. This study, Publication No. 1646 of the Oregon Regional Primate Research Center, was supported by NIH grant RR00163.

REFERENCES

Bordson, B. L., Ricci, E., Dickey, R. P., Dunaway, H., Taylor, S. N., and Curole, D. N., 1986, Comparison of fecundability with fresh and frozen semen in therapeutic donor insemination, *Fertil. Steril.*, 46:466.

Brown, C. A., Boone, W. R., and Shapiro, S. S., 1988, Improved cryopreserved semen fecundability in an alternating fresh-frozen artificial insemination program, *Fertil. Steril.*, 50:825.

Clarke, G. N., Hyne, R. V., du Plessis, Y., and Johnston, W. I. H., 1988, Sperm antibodies and human in vitro fertilization, *Fertil. Steril.*, 49:1018.

Critser, J. K., Arneson, B. W., Aaker, D. V., Huse-Benda, A. R., and Ball, G. D., 1987, Cryopreservation of human spermatozoa. II. Postthaw chronology of motility and of zona-free hamster ova penetration, *Fertil. Steril.*, 47:980.

Iddenden, D. A., Sallam, H. N., and Collins, W. P., 1985, A prospective randomized study comparing fresh semen and cryopreserved semen for artificial insemination by donor, *Int. J. Fertil.*, 30:54.

Mack, S. O., Tash, J. S., and Wolf, D. P., 1989, Effect of measurement conditions upon quantitation of hyperactivated human sperm subpopulations by digital image analysis, *Biol. Reprod.*, in press.

Patton, P. E., Burry, K., Wiley, M., and Wolf, D. P., 1988, A comparative study of intrauterine and intracervical insemination in a donor insemination program, *in:* Program Supplement of the American Fertility Society: 44th Annual Meeting (held in Atlanta, Georgia, October 10-13), p. S20 (abstract 057).

Richter, M. A., Haning, R. V., Jr., and Shapiro, S. S., 1984, Artificial donor insemination: fresh versus frozen semen; the patient as her own control, *Fertil Steril.*, 41:277.

Robertson, L., Wolf, D. P., and Tash, J. S., 1988, Temporal changes in motility parameters related to acrosomal status: Identification and characterization of populations of hyperactivated human sperm, *Biol. Reprod.*, 39:797.

Sherman, J. K., 1986, Current status of clinical cryobanking of human semen, *in:* "Andrology: Male Fertility and Sterility," J. D. Paulson, A. Negro-Vilar, E. Lucena, and L. Martini, eds., Academic Press, Orlando.

The American Fertility Society, 1988, Revised new guidelines for the use of semen-donor insemination, *Fertil. Steril.*, 49:211.

OOCYTE DONATION—STATE OF THE ART 1989

Richard T. Scott, Jr. [a] and Zev Rosenwaks[b]

[a] Division of Reproductive Endocrinology, Department of Obstetrics and Gynecology
Wilford Hall USAF Medical Center, Lackland AFB, TX 78236

[b] Division of Reproductive Endocrinology
Department of Obstetrics and Gynecology, New York Hospital
Cornell Medical School, 505 E. 70th Street, New York, NY 10021

In the past, infertility due to absent, inaccessible, or abnormal gametes in females was considered to be irreversible. Women who could not provide their own oocytes were forced to adopt if they wanted to have a family. However, the absence or inaccessibility of normal oocytes does not preclude implantation of an embryo derived from a donated oocyte or embryo. Although egg donation is analogous to sperm donation in couples where the male had abnormal or absent gametes, there are significant technical differences. While male gametes are available on demand in large numbers, and can be easily cryopreserved and stored, the same is not true for female gametes. Thus a number of approaches for egg and embryo donation have been proposed.

The first solution to this clinical problem was reported by Buster in 1983.[1] This technique involved intracervical insemination of a normal female volunteer with the infertile couple husband's sperm during the periovulatory period, uterine lavage with recovery of the embryo during the perinidatory period, and subsequent intrauterine embryo transfer to the infertile recipient in her early luteal phase. Although a number of successes were reported, concerns regarding transmission of infectious diseases, possible pregnancies in the volunteer, and relatively low pregnancy rates have limited the application of this technique.

Advances in Assisted Reproductive Technologies, Edited by
S. Mashiach *et al.*, Plenum Press, New York, 1990

As the technologies associated with in vitro fertilization (IVF) matured, another means of obtaining donor oocytes became available. Ooctyes could be obtained from donors using the same techniques used for women undergoing routine IVF. The oocytes could then be inseminated and cultured in vitro and transferred to the recipients' uterus in a fashion analogous to IVF patients, or they could be used for a tubal transfer procedure such as gamete intrafallopian transfer (GIFT), pronuclear stage transfer (PROST), or tubal embryo-stage transfer (TEST).[2-6]

Although oocyte donation (OD) is medically analogous to sperm donation, the procedure is technically more complex. Many technical and physiologic questions required solutions before successful egg donation became a reality. These include: 1) The possible sources of oocytes; 2) can endometrial receptivity be stimulated by exogenous hormonal replacement in women with ovarian failure? 3) what degree of embryonic endometrial dysynchrony can be tolerated to assure uterine receptivity and implantation? 4) What are the hormonal requirements for maintaining a pregnancy? The following report will address these issues.

OOCYTE DONATION

The oocyte donation program was established in 1983 as a means of providing oocytes to patients with ovarian failure, heritable genetic disorders, repetitively poor oocyte quality, or inaccessible ovaries. Since that time, a total of 73 transfers have been performed and a great deal has been learned about the physiology of human reproduction.

Stimulation of Endometrial Receptivity

The majority of the patients in the OD program had ovarian failure and were incapable of endogenous stimulation of endometrial receptivity. Therefore, a reproducible means of stimulating orderly endometrial development and receptivity was necessary if these women were going to conceive. Exogenous hormonal replacement regimens were developed which provided patients with near physiologic circulating estradiol (E_2) and progesterone (P) concentrations and which reliably produced orderly secretory development of the endometrium.[7,8]

Estrogen replacement was accomplished using a number of different preparations including oral micronized estradiol, transdermal estradiol patches, and estradiol impregnated vaginal ring.[8] By adjusting the dosage and timing of administration, all provided near physiologic circulating E_2 levels, although oral therapy provided supraphysiologic estrone concentrations.[9]

Progesterone replacement was accomplished with either vaginal suppositories or intramuscular injections. The most reliable serum concentrations were produced with the intramuscular injection of progesterone in oil and this is currently all that we are using. Serum

P levels in the normal range are consistently obtained with this regimen. Endometrial biopsies performed on cycle day 20 and 25 (in separate replacement cycles) confirm appropriate orderly secretory development.[8]

The routine hormonal replacement protocol provides 14 days of estrogen priming in which progressively increasing dosages of E_2 are provided followed by treatment with E_2 and P until a pregnancy is confirmed or excluded. Varying the length of the estrogen priming portion of the cycle from seven to 28 days had no detectable effect on the orderly secretory development which occurred once P supplementation was begun. Therefore, the development of secretory changes within the endometrium are dependent on the duration of P exposure and not the length of E_2 exposure or the total cycle length. This is not to say that some E_2 priming of the endometrium is not required, it simply states that the timing and degree of estrogen priming appears to be very flexible.

Pregnancies have been established in ovarian failure patients using these hormonal replacement protocols. This demonstrates that endometrial receptivity may be induced by exogenous E_2 and P in the absence of ovarian function.

Embryonic-Endometrial Maturation

Synchronization is maintained in natural conception because the hormonal milieu in which the oocyte develops is the same one that provides estrogen priming to the uterus. Similarly, the onset of progesterone production at the time of ovulation assures that embryonic age and the duration of progesterone induced secretory development are equivalent. In OD, there is no natural coordination of embryonic and endometrial and dysynchrony may occur. Embryonic-endometrial dysynchrony is defined formally as the difference in the developmental age of the embryo and the secretory development of the endometrium.

Gametes to be used in the OD transfers are handled identically to those of IVF patients.[10] This includes transfer of the embryos to the uterus of the recipient at approximately 36–42 hours following insemination. This means that these patients all receive embryos at the same approximate developmental stage and that differences in the degree of embryonic-endometrial dysynchrony reflect differences in the degree of endometrial maturation. Synchronization of endometrial and embryonic development is suggested when the embryos are transferred on the morning of the third day of P administration.

Before defining the optimal window of receptivity embryo transfers were performed between the second and tenth days of progesterone administration (Table 1). Pregnancies occurred in patients who had transfers on days 3–6 with ongoing pregnancies on days 3– 5.[11] The pregnancy rates were equivalent following transfers on each of days 3, 4, and 5. This indicates that the human can tolerate up to two days of embryonic endometrial dysynchrony induced by advanced endometrial maturation. Although the number of transfers is small, no pregnancies occurred when endometrial maturation was less than that of the embryo.

Table 1. Pregnancy Rates in Patients Attempting Pregnancy through Oocyte Donation in the Norfolk Program Based on the Cycle Day of Transfer. Pregnancy Rates were Equivalent on Days 3, 4, and 5

Days of Progesterone Treatment (n =)	Numbers of Transfers (n =)	Pregnancies	
		Total (n=; %)	Ongoing (n=; %)
2	5	0	0
3	32	10	9
4	14	6	5
5	13	6	4
6	5	1	0
7	2	0	0
10	2	0	0

It is worth emphasizing that dysynchrony is determined by the difference in embryonic age and secretory transformation. It is apparent that the duration of endometrial priming is very flexible as pregnancies occurred with as little as seven days or as much as 28 days of estrogen exposure prior to initiating P treatment.

Success Rates

The clinical success rates in patients attempting to conceive through OD provide insight into a number of clinically relevant questions. These include such issues as the effect of cryopreservation, the importance of embryo quality, and the limits of success rate using currently available techniques. The clinical pregnancy rates in OD patients are presented in Table 2.

Morphologic embryo grading correlates very well with success rates in OD patients (Table 2). Success rats for grade 1 and 2 embryos are higher than transfers of similar quality embryo delivery IVF cycles.[12] This emphasizes the impact which high quality embryos have on clinical success rates. Furthermore, since these pregnancy rates exceed those attained in strict tubal factor IVF patients, these data suggest that IVF patients may have some other factor such as altered endometrial development which decreases the probability of implantation.

Oocyte donation transfers provide an ideal system to evaluate possible effects of cryopreservation of embryos on clinical pregnancy rates. Whereas during IVF cycles, the histological status of the endometrium on the day of transfer is unknown, egg donation embryo transfers mandate precise advanced knowledge of endometrial development since they are hormonally manipulated and examined histologically. Thus we can easily assess the effect of cryopreservation on implantation. Indeed, if one controls the number of embryos trans-

Table 2. Pregnancy Rates in Oocyte Donation Patients Based on the Number of Embryos Transferred (2A), Embryo Quality (2B), and Cryopreservation (2C). Only Transfers within the Established Window of Transfer (Days 3 to 5 of Progesterone Therapy) are included

2A

Number of Embryos Transferred (n =)	Numbers of Transfers (n =)	Pregnancies Total (n=; %)	Pregnancies Ongoing (n=; %)
1	15	2 (13.3)	2 (13.3)
> 2	44	20 (45.4)	16 (36.4)

2B

Embryo Quality	Numbers of Transfers (n =)	Pregnancies Total (n=; %)	Pregnancies Ongoing (n=; %)
1	18	14 (77.8)	13 (72.2)[a]
2	20	7 (35.0)	5 (25.0)[b]
3	1	0 (0.0)	0 (0.0)
4	16	1 (6.3)	0 (0.0)
5	4	0 (0.0)	0 (0.0)

[a] > [b]; P < 0.01

2C

Type of Embryos	Number of Embryos Transferred (n =)	Numbers of Transfers (n =)	Pregnancies Total (n=; %)	Pregnancies Ongoing (n=; %)
Fresh	1	12	1 (8.3)	0 (0.0)
	≥ 2	35	18 (51.4)	14 (40.0)[a]
Cryopreserved	1	3	1 (33.3)	1 (33.3)
	≥ 2	9	2 (22.2)	2 (22.2)[b]

[a] > [b]; P > 0.01

ferred, any differences in pregnancy rates could reflect the impact of cryopreservation. In fact, the pregnancy rates are significantly lower following the transfer of cryopreserved embryos (Table 3) when compared to fresh embryo couples. This has direct clinical relevance since some authors have advocated the routine clinical application of cryopreservation in OD patients to aid with the logistics of synchronization.[13] Based on these data, it may be prudent to reserve the use of cryopreservation to those cases where fresh transfers are not possible.

Hormonal Replacement During Pregnancy

In ovarian failure patients, E_2 and P replacement continues to be necessary until placental steroidogenesis becomes adequate to maintain the pregnancy. Since no corpus lu-

Table 3. Different Sources of Donor Oocytes used by the Program Participants in the Oocytes Donation Program Survey

Type of Oocyte Donor	Percentage of Program
Relatives	55%
Excess oocytes from assisted reproduction patients	50%
Anonymous donors	41%
Nonanonymous donors	33%

teum is present, no significant endogenous source of E_2 or P exists. This raises the clinical question of what dosage and duration of E_2 and P replacement was necessary for these patients. Our clinical practice has been to maintain these patients on the same replacement dosages as they received prior to confirming the pregnancy. Dosages are not increased unless serum levels show a decline in circulating levels.

Deciding on the duration of hormonal replacement once pregnancy is established is more difficult because premature withdrawal could result in miscarriage. Since these patients are receiving constant dosages of E_2 and P, any significant increases in circulating E_2 and P levels must reflect the onset of placental steroid production. Significant increases in E_2 and P concentrations are detected approximately six and seven weeks after embryo transfer (eight and nine weeks of gestational age), respectively.[14] These data are clinically consistent with the work of Csapo which demonstrated that luteectomy after the eighth gestational week did not result in miscarriage.[15]

In clinical practice, we do not empirically discontinue these medications at this time. Serum E_2 and P levels are attained three times per week and when they rise above the 95% confidence interval of the concentrations during the first two weeks after embryo transfer, the patients' medications are decreased by half. If the patients' circulating concentrations continue to rise over the following week then all hormonal replacement is discontinued. This occurs on average approximately seven weeks post transfer which equates to a gestational age of nine weeks.

SURVEY OF OOCYTE DONATION RESULTS FROM MULTIPLE CENTERS

The Questionnaire

Several of the OD programs throughout the world were surveyed regarding multiple aspects of their programs. Information gathered included sources of oocyte donors, number and timing of transfers, pregnancy rates, route of transfer, and whether or not cryopreservation was being used.

Sources of Donor Oocytes

The single greatest obstacle to a successful OD program is the limited availability of donor oocytes. Possible sources include excess oocytes from IVF patients, women willing to undergo gonadotropin stimulation and donate oocytes at the time of tubal ligation, relatives, and anonymous and nonanonymous volunteers. The ethical and medicolegal issues involved are quite complex and beyond the scope of this discussion. The number of centers using each of the different donor sources are listed in Table 3. Obviously, each center may use one or more sources of donor oocytes.

Success Rates

The pregnancy rates which were reported were clearly effected by the number of embryos transferred, the timing of transfer, and whether or not the embryos had been cryopreserved. The success rates based on each of these factors are presented in Table 4.

Embryonic-Endometrial Dysynchrony

Intrauterine embryo transfers were performed on the first through the tenth day of progesterone therapy. Ongoing pregnancies occurred following transfers on days 2 through 5. This extends the window of transfer which we have found by one day, indicating that a two day old embryo can tolerate up to a two day advancement or a one day delay in endometrial maturation. Clearly, many of these centers used slightly different hormonal replacement regimens in their ovarian failure patients; however, the differences were predominantly in the duration and degree of estrogen priming.

Cryopreservation

The results from centers who performed both fresh and cryopreserved embryo transfers are presented in Table 4. These data are consistent with a decline in pregnancy rates when cryopreservation is used. In fact, this analysis does not take into account the number of embryos which do not survive the cryopreservation and thawing process but which would have been transferred and counted in the statistics of fresh transfers. Therefore, these data may actually underestimate the detrimental impact of the cryopreservation process.

Route of Transfer

The number of centers performing tubal transfer in OD patients was substantially lower than those using IVF. Pregnancies were reported using GIFT, ZIFT, and TEST. After controlling for the total number of oocytes or embryos transferred, the pregnancy rates were significantly higher in patients who underwent tubal transfers.

Table 4. Results of the Survey from 21 Centers who have Performed Oocyte Donation Transfers. Pregnancy Rates based on the Number of Embryos Transferred (4A), Whether or not the Embryos had Previously been Cryopreserved (4B), and Relative to the Length of Progesterone Exposure of the Endometrium (4C) are Presented

4A

Number of Embryos Transferred (n =)	Numbers of Transfers (n =)	Pregnancies	
		Total (n=; %)	Ongoing (n=; %)
1	288	42 (14.6)	26 (9.0)
2	325	75 (23.1)	54 (16.6)
3	201	52 (25.9)	39 (19.4)
4	127	51 (37.2)	37 (29.1)
5	40	20 (50.0)	15 (37.5)
6	28	10 (35.6)	8 (28.6)

4B

Type of Embryos	Number of Embryos Transferred (n =)	Numbers of Transfers (n =)	Pregnancies	
			Total (n=; %)	Ongoing (n=; %)
Fresh	1	182	25 (13.7)	14 (7.7)
	≥ 2	553	173 (31.2)	130 (23.5)[a]
Cryopreserved	1	106	17 (16.0)	12 (11.3)
	≥ 2	168	35 (20.8)	23 (13.7)[b]

4C

Days of Progesterone Treatment	Numbers of Transfers (n =)	Pregnancies	
		Total (n=; %)	Ongoing (n=; %)
2	264	55 (20.8)	40 (15.2)
2	505	106 (20.9)	97 (19.2)
4	98	36 (36.7)	32 (32.7)
5	22	9 (40.9)	6 (27.3)
6	13	2 (15.4)	0 (0.0)
7	3	0 (0.0)	0 (0.0)
8	0	-	-
9	0	-	-
10	2	0 (0.0)	0 (0.0)

While these data do suggest that the physiologic milieu provided by the fallopian tube may be of benefit, there are a number of factors which prohibit liberal extrapolation of these results. First, a number of individual IVF centers had success rates in OD patients which were equivalent to those achieved with tubal transfers. Thus, it is possible that the inter-program variation could explain the difference in pregnancy rates. Second, the source of the oocytes is not controlled for in this analysis. Finally, the potential bias of other confounding variables such as male factors were not considered in the analysis. These data do suggest that tubal transfers may be superior to IVF in this population; however, well controlled prospective studies will be required to resolve this question.

Table 5. Pregnancy Rates in Oocyte Donation Patients Undergoing either Tubal or Transcervical Transfer Procedures. Results are Presented per Number of Oocytes or Embryos Transferred since all of the Patients Undergoing Tubal Procedures had at Least Three Oocytes or Embryos Transferred

Number of embryos transferred	Route of transfer	Number of transfers (n =)	Pregnancies		Number of sacs per embryo (n =; %)
			Total (n=; %)	Ongoing (n=; %)	
3	transcervical	85	34 (40.0)	25 (29.4)	31 (12.2)
	tubal	14	8 (57.1)	6 (42.9)	18 (42.9)
4	transcervical	53	21 (39.6)	13 (24.5)	22 (10.4)
	tubal	27	17 (62.9)	14 (51.9)	36 (33.3)
5	transcervical	22	10 (45.5)	8 (36.4)	21 (19.1)
	tubal	7	6 (85.6)	6 (85.6)	8 (22.9)

CONCLUSIONS

Oocyte donation is now an established therapy for women with absent, abnormal, or inaccessible oocytes. Clinical pregnancy rates are well established and equal or exceed those seen with other assisted reproduction techniques.

Synchronization between embryonic and endometrial maturation is required and is based on the degree and duration of exposure of the endometrium to progesterone and is not directly related to the duration of estrogen priming. Pregnancies have occurred following the transfer of cryopreserved embryos; however, pregnancy rates are lower than with the transfer of fresh embryos. Therefore, the routine use of cryopreservation to make the logistics of embryonic-endometrial synchronization easier is not justified at the current time.

The optimal route of transfer for OD patients remains to be determined. Preliminary data does suggest that tubal transfers (GIFT, ZIFT, TEST) may be superior at current levels of technology.

A number of scientific and clinically relevant questions concerning OD remain to be answered: what is the optimal stimulation regimen; can the window of transfer be extended and can adequate safe and reliable sources of donor oocytes be made available?

REFERENCES

1. Buster, J. E., Bustillo M., Thorneycroft, I. H., Simon, J. A., Boyers, S. P., Marshall, J. R., Louw, J. A., Seed, R. W., and Seed R. G., 1983, Nonsurgical transfer of in vivo fertilized donated ova to five infertile women: report of two pregnancies, *Lancet* 2:23.

2. Lutjen, P., Trounson, A., Leeton, J., Findlay, J., Wood, C., and Renov, P., 1984, The establishment and maintenance of pregnancy using in vitro fertilization and embryo donation in a patient with primary ovarian failure, *Nature* 307:174.
3. Rosenwaks, Z., Veeck, L.L., and Liu H-C., 1986, Pregnancy following transfer of in vitro fertilized donated oocytes, *Fertil. Steril.*, 45:417.
4. Asch, R. H., Balmaceda, J. P., Ord, T., Borrero, C., Cefalu, E., Gastaldi, C., and Rojas, F., 1988, Oocyte donation and gamete intrafallopian transfer in premature ovarian failure, *Fertil. Steril.*, 49:263.
5. Formigli, L., Formigli, G., and Roccio, C., 1987, Donation of fertilized uterine ova to infertile women, *Fertil. Steril.*, 47:162.
6. Yovich, J. L., Blackledge, D. G., Richardson, P. A., Matson, P. L., Turner, S. R., and Draper, R., 1987, Pregnancies following pronuclear stage tubal transfer, *Fertil. Steril.*, 48:851.
7. Navot, D. M., Scott, R. T., Droesch, K. D., Kreiner, D. K., Liu H-C., and Rosenwaks, Z., 1989, Efficacy of human conception in vitro related to the window of implantation, *J. Clin. Endocr., Metab.*, 68:801.
8. Rosenwaks, Z., 1987, Donor eggs: Their application in modern reproductive technologies, *Fertil. Steril.* 47:895.
9. Droesch, K. W., Navot, D., Scott, R. T., Kreiner, D. K., Liu, H-C., and Rosenwaks, Z., 1988, Transdermal estrogen replacement in ovarian failure for ovum donation, *Fertil. Steril.*, 50:931.
10. Veeck, L. L., and Maloney, M., 1986, Insemination and Fertilization, *in:* "In Vitro Fertilization, Norfolk," H. W. Jones, G. S. Jones, G. D. Hodgen, and Z. Rosenwaks, eds., Williams and Wilkins, Baltimore, p. 168.
11. Navot, D., Droesch, K., Liu, H-C., Kreiner, D., Veeck, L., Steingold, K., Muasher, S. J., and Rosenwaks, Z., 1988, Efficiency of human conception in vitro related to the window of implantation. Presented at the Society for Gynecologic Investigation, Baltimore, March 17-20, Abstr. 419.
12. Veeck, L. L., 1988, Oocyte assessment and biological performance, *Ann. NYAS.*, 541:259.
13. Salat-Baroux, J., Cornet, D., Alvarez, S., Antoine, J. M., Tibi kC., Mandelbaum, J., and Plachot, M., 1988, Pregnancies after replacement of frozen-thawed embryos in a donation program, *Fertil. Steril.*, 49:817.
14. Scott, R., Navot, D., Droesch, K., Kreiner, D., Liu, H-C., and Rosenwaks, Z., A unique human in vivo model for studying early placental steroidogenesis and pregnancy maintenance, Presented at the Society for Gynecologic Investigation, Baltimore, March 17-20, 1988, Abstr. 420.
15. Csapo, A. I., Pulkinnen, K. O., and Wiest, W. G., 1973, Effects of luteectomy and progesterone replacement therapy in early pregnant patients, *Am. J. Obstet. Gynecol.*, 115:759.

SYNCHRONIZATION IN OVUM DONATION

P. Devroey, J. Smitz, M. Camus, L.Van Waesberghe, A. Wisanto,
and A. Van Steirteghem

Center for Reproductive Medicine, Academic Hospital and Medical Campus
Vrije Universiteit Brussel, Laarbeeklaan 101, 1090 Brussels, Belgium

INTRODUCTION

The establishment of the first pregnancy after embryo donation was reported in a patient with functional ovaries.[1] In 1984 the first birth after embryo donation was reported in a patient with absent ovaries.[2] Synchrony between the embryo stage and endometrial maturity is mandatory. Several possibilities exist to synchronize i.e. cryopreservation of the concepti and transfer after thawing at the appropriate moment in the recipient's cycle. Donor's and recipient's cycles can also be synchronized, using LHRH analogues. In this report we analyze the efficiency of LHRH analogues in a donor egg programme.

MATERIALS AND METHODS

From June 1, 1987 until December 1, 1988, 42 women agreed to donate their oocytes. After careful psychological and genetical evaluation they were accepted. All patients were treated with a combination of human menopausal gonadotrophins (hMG) (Humegon®, Organon, The Netherlands; Pergonal®, Serono, Belgium) and LHRH analogues (buserelin Suprefact®, Hoechst, Federal Republic of Germany).[3, 4] There exist two different ways to synchronize donor's and recipient's cycles by LHRH analogues:

1. Long protocol (n = 39). The donor started the administration of buserelin 6×100 μg/day, intranasally, from the first day of the cycle or from day 21. The administration of

Advances in Assisted Reproductive Technologies, Edited by
S. Mashiach *et al.*, Plenum Press, New York, 1990

buserelin was prolonged until the first day of the recipient's natural or artificial cycle start-ed. From the first day onwards 75 I.U. or 150 I.U. hMG were injected. The subsequent administration was individualized as described elsewhere.[5] Human chorionic gonadotrophin therapy (hCG) (Pregnyl®, Organon, The Netherlands; Profasi®, Serono, Belgium) was injected when a minimum of six follicles were present. Oocyte retrieval was done 36 hours after hCG injection either laparascopically in patients undergoing a tubal sterilization or vaginally under ultrasound guided puncture.[6,7]

2.   Short protocol (n = 3). The recipient received 1 mg oestradiol valerate until the onset of the recipient's menstrual period. The administration of buserelin was started the first day of the donor's cycle and the stimulation with hMG was started accordingly.

The retrieved oocytes were inseminated in vitro as described elsewhere.[8] If the recipient had healthy fallopian tubes zygotes are replaced in the fimbrial end of one healthy fallopian tube.[9] If the recipient had absent or damaged fallopian tubes 4–8-cell embryos were replaced via traditional clinical replacement.

A maximum of three zygotes or embryos were replaced, the supernumerary zygotes/embryos were cryopreserved for later use.[10-12]

Seventy-eight women were accepted in our programme. Sixty-three had ovarian failure and 15 had functional ovaries.

The recipients were treated with oestradiol valerate (E_2V) (Progynova®, Schering, Federal Republic of Germany) and micronized progesterone (Utrogestan®, Piette, Belgium), respectively. E_2V was administered on days 1–5 (1 mg), days 6–9 (2 mg), days 10–13 (6 mg), days 14–17 (2 mg), days 18–26 (4 mg) and days 27–28 (1 mg). Utrogestan (Piette, Belgium) was taken in a dose of 100 mg at 8 p.m. on day 14 and three times 100 mg at 8 a.m., 2 p.m. and 8 p.m. In cycles where an embryo replacement was performed, the oral ingestion of micronized progesterone was replaced by two different regimes:

(i)   on day 14 at 8 p.m. 50 mg progesterone I.M. and two times 50 mg at 8 a.m and 8 p.m. from days 15 to 26;

(ii) the intravaginal administration of micronized progesterone at the dose of three times 100 mg at 8 a.m., 2 p.m. and 8 p.m.[13]

The donation programme was approved by the Ethics Committee of the Academic Hospital.[14]

RESULTS

Three cycles were treated with a short protocol and 39 with the long protocol (Table 1).

Table 1. Synchronization of Donor's and Recipient's Cycle with LHRH Analogues

	Long Protocol	Short Protocol
Mean duration of desensitization (range)	20 (11-31)	
Mean length of hMG stimulation	12.5 (9-32)	11,13 and 15 days

Table 2. Mean Number of Oocytes Retrieved in the Long and Short Protocol

	Donors		IVF-Patients
	Long protocol	Short Protocol	
Cycles (n)	39	3	90
Oocytes (n)	539	41	143
Mean N° of oocyte/retrieval	13.8	13.6	7.2

Table 3. The Establishment of 19 Pregnancies after Synchronization

	N° of Donors	N° of Recipients	Pregnancies	(%)
Recipients				
Long protocol	39	63	16	(26)
Short protocol	3	15	3	(20)
Total	42	78	19	(24)

Table 4. Different Routes of Replacement

Mode of Replacement	Primary Ovarian Failure	Functional Ovaries
Uterine	11	3
Tubal	3	2

As indicated in Table 2 a mean of 13.8 oocytes were retrieved in the long and 13.6 in the short protocol.

Oocytes were donated by 40 volunteers and two sterilization patients.

As shown in Table 3, 19 pregnancies (24%) were obtained i.e. three in the short protocol and 16 in the long protocol.

The replacements were done either via the traditional cervical route or via the tubal route. (Table 4).

Table 5. Relation Between Day of hCG Administration and Outcome of Donation

Day of hCG administration		Pregnancy per Replacement
13	(n = 18)	5/23
< 13	(n = 17)	6/30
> 13	(n = 2)	3/10
Total	(n = 37)	14/63

Fourteen pregnancies were obtained in patients with ovarian failure. These pregnancies were established after a different duration of stimulation. (Table 5)

DISCUSSION

The synchronization of the recipient's cycle and the developmental stage of an 4–8-cell embryo after oocyte donation is mandatory. Two different procedures are available : cryopreservation or synchronization with LHRH analogues.

Cryopreservation is an acceptable procedure but in the course of freezing and thawing approximately 50% of the embryos do not survive. The latter indicates that the practice of cryopreservation has to be restricted to the supernumerary zygotes/embryos. As demonstrated, the use of LHRH analogues allows the synchronization between donor's and recipient's cycles (long protocol). Furthermore the duration of LHRH adminstration did not interfere with an adequate folliculogenesis.

In patients with ovarian failure the administration of 1 mg E_2V can be prolonged until the first day of the donor's menstrual period (short protocol). The latter is not feasable in patients with functional ovaries. Synchronization of donor's and recipient's cycle, if the recipient has functional ovaries, is only possible with a long protocol. Recipients with functional ovaries are impossible to synchronize with respect to the mid-cycle LH surge. In contrast, in patients with a substitution therapy, the artificial follicular phase can easily be corrected and adapted to the length of the donor's follicular phase by varying the dosage of oestradiol valerate.

We have demonstrated that pregnancies can be established on different days of hCG administration; apparently, although the numbers are too small, the length of the artificial follicular phase did not interfere with implantation rates.

In addition, the use of LHRH analogues have other advantages: it precludes the occurrence of premature and endogenous LH surges. Furthermore the number of retrieved oocytes were doubled compared to a group of in vitro fertilization patients. This finding is probably related to the fact that those patients have no endocrine pathology.

For these reasons the use of analogues is superior to other regimes of programmed cycles i.e. Norethisterone.[15,16]

In summary, the combination of LHRH analogues and hMG is the method of choice for synchronizing donor's and recipient's cycles in an egg donation programme. An elevated number of preovulatory oocytes were retrieved. A cryopreservation programme is mandatory to freeze the supernumerary zygotes/embryos and the embryos generated from failed synchronization in the case of functional ovaries.

ACKNOWLEDGMENTS

We thank all members of the Center for Reproductive Medicine of the Academic Hospital and Medical Campus of the Vrije Universiteit Brussel and Marleen Van der Helst for preparing the manuscript. A grant from the Fund for Medical Scientific Research is acknowledged and appreciated. (No. 3.0036.85).

REFERENCES

1. A. Trounson, J. Leeton, M. Besanko, C. Wood, and A. Conti, Pregnancy established in an infertile patient after transfer of a donated embryo fertilized in vitro, Br. Med., J. 286:835 (1983).
2. P. Lutjen, A. Trounson, J. Leeton, J. Findlay, C. Wood and P. Renou, The establishment and maintenance of pregnancy using in vitro fertilization and embryo donation in a patient with primary ovarian failure, *Nature,* 307:174 (1984).
3. J. Smitz, P. Devroey, P. Braeckmans, M. Camus, I. Khan, C. Staessen, L. Van Waesberghe, A. Wisanto, and A.C. Van Steirteghem, Management of failed cycles in an IVF/GIFT programme with the combination of GnRH analogues and hMG, *Hum. Reprod.,* 2:309 (1987).
4. T. Wildt, K. Diedrich, H. Van der Ven, S. Al Hasani, H. Hubner, and R. Klaser, Ovarian hyperstimulation for in vitro fertilization controlled by GnRH agonist administered in combination with human menopausal gonadotropins, *Hum. Reprod.,* 1:15 (1986).
5. P. Devroey, A. Wisanto, J. Smitz, P. Braeckmans, L. Van Waesberghe, and A.C. Van Steirteghem, Ovarian stimulation including in vitro fertilization, *Ann. Biol. Clin.,* 45:346 (1987).
6. W. Feichtinger, and P. Kemeter, Transvaginal sector scan sonography for needle guided transvaginal follicle aspiration and other applications in gynecologic routine and research, *Fertil. Steril.,* 45:722 (1986).
7. L. Hamberger, M. Wikland, L. Enk, and L. Nielsson, Laparoscopy versus ultrasound guided puncture for oocyte retrieval, *Acta Europ. Fertil.,* 17:195 (1986).
8. I. Khan, M. Camus, C. Staessen, A. Wisanto, P. Devroey and A.C. Van Steirteghem, Success rate in gamete intrafallopian transfer using low and high concentrations of washed spermatozoa, *Fertil. Steril.,* 50:922 (1988).
9. P. Devroey, P. Braeckmans, J. Smitz, L. Van Waesberghe, A. Wisanto, A.C. Van Steirteghem, L. Heytens, and F. Camu, Pregnancy after translaparoscopic zygote intrafallopian transfer in a patient with sperm antibodies, *Lancet,* i:1329 (1986).
10. A. Trounson, and L. Mohr, Human pregnancy following cryopreservation, thawing and transfer of an eight-cell embryo, *Nature,* 305:707 (1985).
11. B. Lasalle, J. Testart, and J.P. Renard, Human embryo features that influences the success of cryopreservation with the use of 1,2 propanediol, *Fertil. Steril.,* 44:645 (1985).

12. A.C. Van Steirteghem, E. Van den Abbeel, M. Camus, L. Van Waesberghe, P. Braeckmans, I. Khan, M. Nijs, J. Smitz, C. Staessen, A. Wisanto, and P. Devroey, Cryopreservation of human embryos obtained after gamete intra-fallopian transfer and/or in vitro fertilization, *Hum. Reprod.*, 2:593 (1987).
13. P. Devroey, G. Palermo, C. Bourgain, L. Van Waesberghe, J. Smitz, and A.C. Van Steirteghem, Progesterone administration in patients with absent ovaries, *Int. J. Fert.*, 34: 188 (1989).
14. M. Warnock, "A question of life", The Warnock Report on Human Fertilization and Embryology, Basil Blackwell, Oxford (1985).
15. A. Templeton, P. Van Look, M. A. Lumsden, R. Angell, J. Aitkin, and A.W. Duncan, The recovery of pre-ovulatory oocytes using a fixed schedule of ovulation induction and follicle aspiration, *Brit. J. Obstet. Gynecol.*, 91:148 (1984).
16. R. Frydman, J.D. Rainhorn, R. Forman, C.H. Belaisch, J. Allart, H. Fernandez, B. Lasalle, and J. Testart, Pregnancies following fixed schedule ovulation induction and embryo cryopreservation during diagnostic infertility laparoscopy, Future Aspects of Human In-Vitro Fertilization, W. Feichtinger and P. Kemeter, eds., Springer-Verlag, Berlin (1987).

IN VITRO FERTILIZATION - SURROGATE GESTATIONAL PREGNANCY

Wulf H. Utian, Robert Kiwi, James M. Goldfarb,
Leon A. Sheean, and Hanna Lisbona

*The Mt. Sinai Medical Center, Department of Ob/Gyn
1 Mt. Sinai Drive, Cleveland, Ohio 44106*

The purpose of this presentation is to review our early experience with in vitro fertilization, surrogate gestational pregnancy, a program that has resulted in a world medical and legal first.[1,2]

The idea for an attempt at IVF-surrogate gestational pregnancy was originally presented by one of us (Utian) in 1984 to the committee of the LIFE program (Laboratory for In Vitro Fertilization and Embryo Transfer) at the Mt. Sinai Medical Center of Cleveland. The LIFE program committee carefully considered the idea, and following agreement to proceed, the question arose as to the most appropriate direction to obtain institutional approval for this unique procedure. Informal legal and medical peer review suggested that this should best proceed initially through the Ethics Advisory Committee (EAC) of the institution, rather than through the Institution Research Review Committee (IRRC). Each step of an IVF program had already received appropriate approval and the major questions about this procedure were considered ethical rather than scientific.[9]

The matter was officially considered by the EAC, approved, and referred to the hospital Medical Executive Committee (MEC) where further deliberations resulted in a provisional approval for us to proceed. After the first successful birth, utilizing this technology, the entire outcome was reconsidered by the EAC, MEC and Board of Trustees. At this point the program was approved with certain stipulations.

This institution-wide discussion and decision-making process was considered extremely beneficial to the LIFE program[9] in that a large number of medical, ethical and legal issues that were raised and debated were incorporated into the final design of the approved pro-

Advances in Assisted Reproductive Technologies, Edited by
S. Mashiach *et al.,* Plenum Press, New York, 1990

gram. A part of these deliberations was made open to the public television team of NOVA (WGBH Boston) and was aired on national television in a program entitled "High Tech Babies".[10]

ETHICAL ISSUES

From the outset, it was recognized that our procedure could be strongly differentiated from so-called "traditional surrogacy". The latter results in a child representing the combined product of the husband of the infertile couple and a third party woman providing her egg and thus her genetic contribution. In this instance, the surrogate mother is therefore both the genetic and birth mother of the child. With paternity and maternity not in question, the custodial rights become more a legal issue than a medical one, and the resultant baby needs to be officially adopted.

In marked contra-distinction, IVF-surrogacy creates a genuine surrogate situation in which the third party carries the pregnancy for the infertile couple. Thus, the "birth mother" is separate from the "genetic mother". Ethical and legal questions of who is "mother" is raised, but there can be no doubt of the genetic question of parenthood.

The ethical issues for support of such a program therefore have to be examined from several perspectives. These would include the rights of the child-to-be, the infertile couple, the surrogate, her husband and children, the physicians and program offering the service and societal obligations in general.

The distinction between traditional surrogacy and IVF surrogacy was recognized in the deliberations of the Ethics Committee of the American Fertility Society in their report on ethical considerations of the new reproductive technologies.[3] In summation, this committee supported a role for IVF-surrogacy in reproductive medicine but made several recommendations for issues to be addressed in the research on IVF-surrogate mothers.

All of the above factors have been taken into account in the deliberations by the Ethics Advisory Committee, the Medical Executive Committee and other related bodies of the Mt. Sinai Medical Center of Cleveland. Specifically, some of the following issues became the most pertinent, and are listed below for completeness as discussion of these steps goes beyond the scope of this communication.

Surrogate
1. Payment - whether a surrogate should receive some form of reimbursement is highly debatable and distinction between reimbursement for time, risk and effort as opposed to payment for the baby is mandatory.
2. Ability to "out" - Safeguards have to be built into the program to allow the surrogate to withdraw from the situation prior to the onset of pregnancy.

3. Health risks - the surrogate needs to be appropriately informed of all the potential risks to her health and must accept these on an informed basis. Particularly, the risks of multiple pregnancies must be stressed because of the increased incidence with IVF.
4. Psychological risks - while these are not yet determined, it is important for the surrogate to be appropriately evaluated so that any undisclosed underlying psychological impediment to acting as surrogate can be excluded and protection provided against development of psychological harm.
5. Possible maternal bonding between IVF-surrogate and fetus in utero or after birth.
6. Change of mind during pregnancy - the rights of the surrogate under this heading raise numerous potential ethical conflicts.
7. Use of family as surrogate, e.g. mother, sister - the family inter-relationships under such circumstances are complex and raise numerous issues.
8. Visitation rights - this could be a contentious question.
9. Behavior during pregnancy, e.g. smoking, alcohol or drug abuse - the rights of the fetus as opposed to the rights of the surrogate mother come into question, and expectations by the infertile couple of the surrogate also are areas of conflict.
10. Obstetric risks, e.g. previous cesarean section - the need for operative delivery and the enhanced risks associated with this raise ethical questions as to how much risk a surrogate should take in assisting an infertile couple achieve their aim.
11. Health screening, e.g. AIDS screening - the rights of privacy concern all parties involved.
12. Excluding factors, e.g. age - apart from medical aspects, there are ethical questions regarding the selection of surrogate and who should be excluded.
13. Attitudes to amniocentesis and abortion - difference in attitudes between the infertile couple and the IVF-surrogate may bring up ethical issues.

Infertile Couple
1. Indications - while specific causes of infertility such as congenitally absent uterus may be regarded as a satisfactory indication, the ethical question of IVF-surrogacy for other non-medical reasons can be debated. For example, a professional woman not wishing to take off work, etc. could open an ethical debate.
2. Selection of surrogate - what factors should determine who should be the surrogate and what are the ethics of selection?
3. Age of the infertile couple - should an elderly couple employ the services of a young couple for the express purpose of achieving a desired infant?
4. Single parent - the question of donor sperm for an infertile single woman and use of surrogacy.
5. Homosexual couples - specific issues are generated with usage of donor gamete and surrogate.
6. Health screening, e.g. AIDS.
7. Attitudes to abnormal baby.
8. Attitudes to amniocentesis and abortion - can the couple legally or ethically force these procedures on the pregnant IVF-surrogate?

Child
1. Role of later relationship to surrogate gestational mother.
2. The effects of disclosing or not disclosing the use of a surrogate gestational mother or her identity to the child.

Surrogate's Husband
1. Right of refusal of wife's desire - raises major ethical issues of individual rights.
2. Limitation of conjugal rights - during the transfer cycle there should be limitation of conjugal rights.
3. AIDS screening - this raises the issue of privacy for the husband.

Surrogate's Child/Children
The question of harm to the surrogate's child/children is raised due to loss of "sibling" - the awareness of mother being pregnant but not bringing home baby.

Societal
1. Need to approve or disapprove.
2. Third party funding - is this a legitimate expense to the health care system?
3. Two-tier care - the issue of abuse of underprivileged women as surrogates for the more affluent.

The Provider Team
1. Rights of selective indications - the ethics of the team selecting the indications for care.
2. Rights of refusal - the issues can be extremely complex.
3. Provider of surrogates - the inter-relationship between the team as a provider of surro-

    gates and also of health care to the infertile couple can be complex and debatable.
4. Surrogate and infertile couple both as patients - numerous potential issues could be the eventuality of this complex situation. Combinations and permutations of the above issues lead to many potential debatable points of view. All of these issues need to be considered in planning a program.

LEGAL ISSUES

While there are still currently no federal statutes, regulations or requirements specifically directed at IVF-surrogacy, a plethora of archaic laws exist that could potentially cover one or other aspect of this procedure. These laws go beyond the scope of this communication. The subject of the new reproductive technologies and the law have been well reviewed elsewhere.[4,8] The rights of all parties concerned, including the potential unborn fetus, have been considered from numerous standpoints. Other than rights of the couple, the surrogate and husband and the unborn baby, legal issues including those of payment to the surrogate, the laws relating to child custody and adoption, and the laws of inheritance and of contracts should be considered. The major difference from traditional surrogacy and adoption is that the rearing couple will also be the genetic parents.[8]

CURRENT PROGRAM STRUCTURE

The program has been structured as follows taking legal, medical and ethical factors, as outlined, into account.

1. Indications: Ovulatory patients can be considered if they have:
    1. No uterus (previous hysterectomy or birth defect)
    2. Recurrent abortion (five or more losses)
    3. Severely abnormal uterus
    4. Post DES exposure with infertility or recurrent loss
    5. Severe medical contraindications to pregnancy but not to motherhood

2. Couples must recruit their own surrogates; there is no financial relationship between the surrogate and the provider of care

3. Criteria for selection of surrogate: the following are the minimal criteria for acceptance of a surrogate in this program. The surrogate:
    1. must be less than 35 years of age
    2. must have no known uterine abnormality
    3. must have regular menstrual cycles
    4. must understand the procedure and sign all required documents
    5. must meet independently with the Mt. Sinai Department of Psychiatry and the LIFE team for evaluation and to review the procedure
    6. should preferably be married and have her own children

The independent screening of the surrogate by the Department of Psychiatry adds a protective step for the welfare of the surrogate and also for the benefit of the program.

4. Contracts: the infertile couples are advised that before making formal application to the program they should determine if they can find an acceptable surrogate. They are strongly recommended to consult with their own attorneys to make certain that they understand their legal rights and obligations and the legal rights and obligations of others involved. They are also counselled to address such issues as:
    a. the standard of prenatal care to be followed by the surrogate
    b. what genetic testing, if any, will be done
    c. under what circumstances, if any, the pregnancy will be terminated
    d. Custody of the child
    e. inheritance rights of the child
    f. who will be listed on the birth certificate
    g. insurance, both life and health

Table 1. Results 1985 - 1987

ITEM	NUMBER
Applications received	50
Applications approved for surrogate screening	40
Couples/surrogates undergoing review and psychiatric evaluation	29
Couples rejected	2
Surrogates rejected	10
Couples/surrogate LIFE approved	
# couples with cycles initiated	15
# cycles initiated	20
# ovum donors to laparoscopy	16
Positive ovum recovery	15
Positive fertilization (1 or more oocytes)	14
Surrogates reaching embryo transfer	14
Pregnancies	4
Live births	2
Spontaneous abortion	1
On-going pregnancy	1 (twin pregnancy)

h. legitimacy of the child

i. payment of the surrogate's medical expenses

j. whether the husband and/or wife will have the right to determine the disposition of any frozen embryos in the event that they cannot reach agreement at a later date.

Any legal forms prepared by counsel for the couple and the surrogate have to be presented to the Mt. Sinai Medical Center and the LIFE team at the time the application is made to the program. Informed consent is undertaken as part of the signing of the appropriate contracts. Close contact is maintained between the program, the infertile couple and the pregnant surrogate to determine the pregnancy progress and outcome. We currently anticipate that all couples will also be re-evaluated one year following the birth of the baby.

CURRENT RESULTS

The initial results achieved between 1985 and 1987, a time of conservative evaluation of applications, are summarized in Table 1. The number of groups entered into active therapy represents 37.5% of applicants being accepted for evaluation to possibly undergo the procedure. The pregnancy rate for the 16 couples reaching ovulation aspiration is 25%. The number of cases is small but the success rate is promising. We anticipate that a pregnancy rate should generally be higher than that found for traditional IVF because of the embryo transfer to an individual (the surrogate) during a normal, unstimulated cycle with optimal hormonal and endometrial parameters.

The second baby was born on December 21, 1987 in the State of Virginia. There were no medical, psychosocial or ethical problems. Birth certificate registration is still pending decision by a Virginia State Court. One patient had a mid-trimester spontaneous abortion, and there is one twin pregnancy currently in progress.

CONCLUSION

We have been gratified by our early experience with this innovative approach to the treatment of human infertility. While we endorse further development of the technique, we emphasize the need for caution. Strict selection procedures of the infertile couples and potential surrogates are of paramount importance. Detailed counselling, appropriate informed consents and proper contracts are mandatory. Further longer-term followup from our program and any others that may develop are required before a final judgement can be made on the outcome of this procedure.

REFERENCES

1. W. H. Utian, L. Sheean, J. M. Goldfarb and R. Kiwi, Successful pregnancy after in vitro fertilization and embroy transfer from an infertile woman to a surrogate, *N. Eng. Jnl. Med.*, 313:1351-1352 (1985).
2. Wayne County Circuit Court, State of Michigan. C.A. Number 85-532014-DZ Honorable Marianne O. Battani. March 14, 1986.
3. Ethics Committee. American Fertility Society: Ethical considerations of the new reproductive technologies, *Fertil. Steril.*, 46:(Suppl.1) (1986).
4. M. I. Evans and A. O. Dixler, Human in vitro fertilization-some legal issues, *JAMA*, 245:2324-2327 (1981).
5. G. G. Blumberg, Legal issues in nonsurgical human ovum transfer, *JAMA*, 251:1178-1180 (1984).
6. M. M. Quigley and L. B. Andrews, Human in vitro fertilization and the law, *Fertil. Steril.*, 42:348-355 (1984).
7. J. C. Fanta, Legal issues raised by in vitro fertilization and embryo transfer in the United States, *Jnl. IVF and ET.*, 2:65-86 (1985).
8. J. A. Robertson, Embryos, families and procreative liberty: the legal structure of the new reproduction. *South Cal. Law Review*, 59:939-1041 (1986).
9. W. H. Utian, J. M. Goldfarb and L. A. Sheean, Implementation of an in vitro fertilization program, *Jnl. IVF*, 1:72-75, 1984.
10. NOVA - High tech babies. National public television. *WGBH*, Boston, November 4, 1986.

SUCCESS RATE OF SURROGATE GESTATIONAL PREGNANCIES USING IN VITRO FERTILIZATION DONOR OOCYTES

William Handel and Hilary Hanafin

Center for Surrogate Parenting, Inc.
Director/Psychologist
8383 Wilshire Blvd., Beverly Hills, CA

INTRODUCTION

This paper presents a review of a new procedure involving gestational surrogate mothers. The dramatic increase in in vitro clinics and procedures has logically extended to the transferring of embryos from an infertile couple to a gestational surrogate mother. The first such gestational surrogate birth was reported in 1987. Since that time there have been approximately 30 births worldwide. Due to the fact that the surrogates are implanted during unstimulated cycles and are young women with successful pregnancy histories, it was hypothesized that the success of in vitro fertilization will be greater when working with gestational surrogates.

The legal, psychological, and medical dynamics of such a solution to childlessness are indeed complex. There are few or no laws that address legal issues of working with a surrogate mother and there are few studies that outline the psychological risks and considerations. This paper will outline the procedures of one center which specializes in providing legal, psychological, and administrative services to all patients. It outlines issues that any medical team should consider prior to becoming involved in gestational surrogacy. Furthermore, the most current findings on the pregnancies using in vitro fertilization donor oocytes will be presented.*

---

*The medical procedures reported were conducted by Jirair Konialian, M.D., Century City Hospital, Los Angeles, CA, Jaroslav Marik, M.D., Beverly Hills Medical Center, Beverly Hills, CA, Richard Marrs, M.D., Hospital of the Good Samaritan, Los Angeles, CA.

Advances in Assisted Reproductive Technologies, Edited by
S. Mashiach *et al.*, Plenum Press, New York, 1990

SAMPLE

The sample consisted of 22 infertile couples who came to the "Center for Surrogate Parenting" in Los Angeles, California. These couples were of childbearing age, but unable to conceive due to hysterectomy, malformed uterus, extreme health risk of pregnancy, or unexplained infertility. Of the women, 17 ranged in age from 33 to 41 and four ranged in age from 42 through 47 years old. All were medically determined to still be ovulatory.

The gestational surrogate mothers consisted of 17 women who had met the psychological and medical requirements of the Center. They all had histories of uncomplicated pregnancies. Within this sample, 16 were under 32 years of age and one was 36. All were medically determined to be fertile, healthy and with regular cycles.

PROCEDURES

A. Psychological

Surrogate mother applicants and their spouses underwent an orientation and psychological screening process. Only women who had successful, uncomplicated pregnancies and who had children of their own were interviewed. Each candidate was told of the risks and demands of such a program. The orientation emphasized informed consent regarding legal, medical and psychological challenges. The psychological screening process involved assessing their intelligence, ability to keep commitments, social support systems, self-esteem, coping mechanisms, sensitivity towards others and stability. Furthermore, their motivations and their expectations were fully explored. Candidates were only accepted if they were motivated by factors other than money, and if they foresaw being a surrogate as a personally rewarding experience.

After initial interviews, selected candidates were given psychological testing. They were required to attend the mandatory support group meetings with other gestational surrogates. They were then medically evaluated for fertility and general health (including social disease testing).

The Center also provided a similar orientation for infertile couples, with each couple meeting with the staff psychologist to assess appropriateness for the program. Alternative solutions to childlessness, marital issues, expectations, informed consent and stability were discussed.

Accepted surrogate mother candidates and infertile couples were subsequently matched and introduced to each other under the guidance of a staff psychologist. Both sets of clients received psychological consultation throughout their participation.

B. Legal

The surrogate and the prospective parents have a legal relationship because she is carrying a child pursuant to a comprehensive and sophisticated contract that has been entered into. Prior to entering into this legal relationship, both the surrogate and the prospective parents must undergo a full and comprehensive legal consultation with independent counsel. This consultation includes a full description of the surrogate contract as well as the possible liabilities, duties and responsibilities of each party. The consultation includes an explanation of the ambiguous status of the contract in the United States. Very few jurisdictions have ever passed laws on the legality of surrogate parenting, and in particular, gestational surrogacy has never been addressed by any legislative or judicial body. Because there has never been a situation where a gestational surrogate attempted to renege on her contract or a couple was not willing to go forth under the terms of a gestational surrogacy agreement, the contract has never been tested.

Once the surrogate and the prospective parents had been legally informed of their rights and duties under the contract, a match was made and the parties once again went through the process of reading the contract together. This was done with the assistance of a video tape in which an experienced attorney again explained in full the legal aspects of the relationship.

Prior to the implantation, medical insurance was purchased on behalf of the surrogate and life insurance policies were set into place on behalf of the parties. The entire anticipated expense of the procedure, including medical expenses, payment to the surrogate, legal, administrative and psychological fees and all other anticipated miscellaneous expenses were placed in a trust account.

At approximately the sixth month of pregnancy, a legal petition was filed with the court of appropriate jurisdiction requesting a judgment for maternity and paternity on behalf of the biological parents. This legal action was initiated to establish the legal relationship between the prospective biological parents and the child. With the granting of the petition of maternity and paternity, the surrogate was deemed to have no parental rights to the child and the biological parents were legally deemed the natural parents of the child (while in utero). This petition eliminated the necessity of any adoption proceedings and legally recognized the intended relationship that the parties had created. The legal document contains the declaration of parentage and also orders the birth certificate be issued with the names of the biological parents as the natural parents of the child.*

---

*This legal procedure is quite different from the procedure involved when a surrogate is artificially inseminated. Under these circumstances, a suit for paternity is filed on behalf of the father and the infertile wife must subsequently adopt her husband's child. The surrogate relinquishes all parental rights to the adoptive mother, thus establishing a step-parent adoption procedure.

C. Medical

The doctor and patient decided whether the in vitro, ZIFT, or GIFT procedure was to be utilized. In the first embryo transfer attempt, estrogen and progesterone, along with hCG were used to synchronize the donor mother's cycle with the cycle of the specified surrogate mother. For all subsequent transfers, the donor mother took Lupron to help synchronize her cycle with her surrogate's cycle. Typically, up to three embryos were implanted with the ZIFT procedures and up to five embryos were implanted with the IVF procedure.

RESULTS

A. Medical

Of the 22 infertile couples, 21 of them underwent embryo transfer procedures. One couple was unable to produce the necessary oocytes. Thirty-one embryo transfers into surrogate mothers were conducted. There were a total of five dropped cycles due to problems with synchronization of the two women's cycles and four dropped cycles due to lack of proper oocyte development. Of the 31 transfers performed, 22 utilized the IVF procedure, five utilized ZIFT, and four utilized GIFT procedure. (The GIFT and ZIFT procedures have only been offered in the last year). Of the 31 transfer attempts, 10 resulted in pregnancy. Of these pregnancies, one resulted from GIFT, two resulted from ZIFT, and seven resulted from IVF. None of the five transfers conducted on behalf of donor mothers who were over 41 years have been successful to date. Of the pregnancies achieved, seven occurred on the first embryo transfer attempt, two were achieved on the patient's second trial and one occurred on the patient's fourth trial.

Currently, the status of the pregnancies is as follows: Four have delivered (one twin pregnancy); two are in the third trimester; one is in the second trimester; one is in the first trimester; and two of the pregnancies ended in miscarriage at nine weeks. The pregnancy rate for this sample group is 32% per transfer. The "take-home baby rate" per transfer is 25.8%. Overall, of the 22 couples attempting to end their childlessness, thus far eight of them will take home a baby, resulting in a 27.5% of the patients.

B. Legal

All of the gestational pregnancies in the program resulted in successful legal outcomes. These legal actions for maternity and paternity were filed in four different counties in California. The courts have been inclined to allow these as a matter of course. All have been granted with no opposition and extreme ease. Therefore, the biological parents' names were on the birth certificates immediately with the gestational surrogate having no parental rights.

C. Psychological

Of the four surrogate mothers who have delivered, all were able to relinquish the child without grief reactions or ambivalence. All of the pregnant surrogates appear to see themselves as clearly as a gestational mother and have not expressed a need or desire to keep the child. Furthermore the surrogates, pregnant or not, perceived their participation as a positive life experience and exhibited no regrets. Being able to have contact with the parents and with other gestational surrogates appear to be important valuables in their psychological resolution.

DISCUSSION

Although gestational surrogacy is in its infancy as an alternative to infertile couples, it appears that with the proper protections and safeguards it is an extremely successful method of creating a family. Obviously, many infertile couples prefer a biologically related child, and the possibility of gestational surrogacy allows the chance for these couples to fulfill this important need in their lives.

It is important to note that the success of gestational surrogacy is based on a comprehensive program of protections for all parties. These involve a very thorough legal grounding, as well as independent legal representation. It is critical that the attorneys who are involved are well versed in the area of reproductive technology law. The Center for Surrogate Parenting has been successful in its petitioning of the California Courts to recognize that gestational surrogacy needs to be viewed differently than other third party solutions to childlessness. The need for thorough psychological screening is essential for any successful program. It is impossible to underestimate the need for a strong psychological base and well qualified psychological experts in the field of surrogacy.

It is this involvement of experienced professionals that minimizes the medical team's responsibility and liability regarding the non-medical aspects of surrogacy. It is important that the medical community be protected by putting into place a comprehensive contract, appropriate insurance policies and a mutual understanding between the parties. With all the professionals in place (the legal, the psychological and the medical), the risks involved in gestational surrogacy are few and the potential is enormous.

It is clear that placing embryos in young women with unstimulated cycles results in a most favorable situation for pregnancy. The number of gestational surrogate procedures is increasing dramatically. With a conscientious team approach this alternative should be able to assist the infertile population as well as provide further medical understanding into the variables of in vitro, ZIFT and GIFT.

COMPASSIONATE FAMILY SURROGACY BY IVF

John Yovich[a] and David Hoffman[b]

[a] *PIVET Medical Centre, and*
[b] *T.D. Hoffman & Co, Perth, Western Australia*

INTRODUCTION

Surrogacy procedures are regarded as a highly controversial area of medicine with unresolved moral, ethical and legal issues. The same comments were applied to IVF only a decade earlier but at this stage many countries have established (by legal or voluntary measures) guidelines for the conduct of IVF procedures. Surrogacy has not developed to anywhere near that level and the subject continues to divide society and religions, as well as the legal and medical professions.

The concept of surrogacy is far older than that of IVF, including several biblical examples such as the son Hagar bore for Abraham on behalf of Sarah. In modern times, commercial surrogacy appears to have commenced in Japan (1963)[1] and flourished in the USA where the first advertisement for a surrogate appeared in 1975, formal arrangements being established soon after. It is interesting to consider that surrogacy contracts became prominent in the United States of America during the period when IVF was inhibited by a self-imposed moratorium consequential to funding restrictions imposed by NIH. By the end of 1988, it has been estimated that approximately 800 infants have been born through the assistance of commercial surrogacy agencies. By and large these have usually utilized the surrogate woman's egg but a small number of surrogacy arrangements have utilized embryos generated from the infertile couple by IVF, the first such case being reported in 1985.[2] Most surrogacy contracts have excluded access of the surrogate mother to the ensuing child and it would appear that approximately 1% of cases currently end up in legal contests as a consequence of the surrogate mother wishing not to relinquish the child. Several notable cases have been widely publicized in the media, e.g. Elizabeth Kane who was the first com-

Advances in Assisted Reproductive Technologies, Edited by
S. Mashiach *et al.*, Plenum Press, New York, 1990

mercial surrogate mother and who subsequently toured the United States promoting the concept but more recently has been touring the world, speaking against the concept, probably because she has no access to the child which she delivered in the late seventies. Baby Cotton was another notable case which was widely publicized in the press following a British court order to allow the American couple to take custody of the child following the surrogacy arrangement and hence return to the United States with that child. The problem causing most concern in such commercial surrogacy arrangements was highlighted by the Baby M case whereby the surrogate mother, Mary Beth Whitehead was reluctant to relinquish the child to the Stern couple who commissioned the arrangement and this was subsequently deliberated and decided upon by the courts, with custody being granted to the Sterns.

MATERIALS AND METHODS

At the PIVET Medical Centre, we are interested in assisting infertile couples, using established IVF techniques in a non-commercial concept denoted as "compassionate family surrogacy." Couples should meet the following criteria:

1. there must be an approved medical indication such as an absent or defective uterus;
2. there is to be no commercial fee;
3. both gametes must be derived from the infertile couple; where donor gametes are required, these must be provided by the usual anonymous process and must not be provided by the surrogate or her partner;
4. a relative (usually the woman's sister) or a very close friend should act as the surrogate.

These criteria should allow for an altruistic surrogacy arrangement within a family setting where assistance is being provided to an infertile couple but their own gametes are used. This should minimize the biological attachment between the surrogate mother and the child but also allow for long-term access between the child and the surrogate mother.

Ethical considerations for research projects undertaken by the first author have been provided by the Committee for Human Rights at the University of Western Australia. It was put to that Committee that surveys of the Australian community showed that surrogacy was not unacceptable[3] and that Justice Michael Kirby, previously of the New South Wales Law Reform Commission and currently Justice in the Australian High Court, indicated that fertility workers should pursue the area of surrogacy rather than wait for a legal framework to establish.[4] However, the UWA Committee for Human Rights found the matter to be "beyond (its) terms of reference" and thereafter an alternative ethics committee was established at the PIVET Medical Centre and approval given for the concept of compassionate family surrogacy.

Over the years a number of requests have been received from women seeking assistance by surrogacy and these are summarized in Table 1. Requests totalled approximately 90 cas-

Table 1. Reasons for Seeking Surrogacy

Infertility	:	Absent Uterus - cong./acquired
	:	Diseased Uterus, e.g. Asherman's
	:	Recurrent Pregnancy Wastage
	:	Repeated IVF Failure
Medical Conditions	:	Severe Renal Disease
	:	Unstable Diabetes
	:	Severe CVS or Resp. Disease
Genetic Diseases	:	Autosomal Dom., eg. Huntington's
	:	X-linked, e.g. Haemophilia A&B
	:	Autosomal Rec., eg. Thalassaemia
Social Reasons	:	To Avoid Pregnancy Discomforts
	:	Career, Aesthetic, Inconvenience

es, of which 38 were approved (Table 2); all cases having infertility related to known or suspected uterine factors. Although there was sympathy for the patients presenting with medical indications, none were approved as the ethics committee required a substantiated medical opinion that quality maternal survival was expected to be at least 18 years, mainly in the interests of the child. Also, genetic indications have not required the option of surrogacy as there have been adequate supplies of donor sperm, donor oocytes and embryos from within the infertility clinic. No cases were approved under the section categorized as Social Reasons.

Oocyte Retrievals

Couples were assessed according to the same protocol as others entering infertility programs.[5] Hormonal and ultrasound monitoring was utilized to detect ovarian function and ovarian stimulation was carried out by a clomiphene citrate/hMG schedule with hCG trigger (10,000 IU) given on the sixth day of consecutive E_2 rise; or a Lucrin/hMG schedule with hCG trigger given on the seventh day of E_2 rise. Oocyte recovery was performed 36 h later using the transvaginal ultrasound-directed technique in combination with the PIVET-COOK aspiration/flushing needle. The preparation of the surrogate was variable and is summarized in Table 3.

Where the surrogate was a local resident, the natural cycle was preferred but ovarian stimulation was applied if the natural cycle did not conform to normal parameters.[6] For those surrogate women travelling from interstate or overseas specifically for surrogacy, an artificial cycle was created using a hormone replacement schedule based on that of Lutjen et al.[7] and refined by J. Leeton (personal communication), with transfer being undertaken on day 15 to 19 of the artificial luteal phase.

Table 2. Approval for Surrogacy

INDICATION	NUMBER
INFERTILITY	
- Absent Uterus: Cong.	6
- Absent Uterus: Acquired	13
- Diseased Uterus	9
- Repeated IVF Failures	4
- Recurrent Pregnancy Loss	4
- Age > 40 years	2
MEDICAL INDICATIONS [Ethics Committee requires substantiated opinion that quality maternal survival expected to be > 18 years].	0
GENETIC INDICATIONS [Limited need due to availability of donor sperm, oocytes, embryos].	0
SOCIAL REASONS	0
Total Approved	38 cases

Table 3. Surrogate Preparation

Natural Cycle	: Nil
	: Norethisterone to synchronise cycle onset
	: Cryopreservation for luteal synchrony
Ovarian Stimulation	: CC/hCG trigger
	: CC/hMG/hCG trigger
Artificial Cycle	: Hormone Replacement Schedule
	33 day Leeton Schedule

Counselling and Consent

All parties to the surrogacy arrangement have a series of counselling sessions involving medical counselling, emotional and psychological assessment by infertility clinic counsellors, and legal counselling by the second author. Once the medical centre is satisfied that the surrogacy arrangement is appropriate and favourable to all parties, informed consent is provided in a contractual arrangement where the signatories are the surrogate mother and her partner, as well as the biological potential mother and father. These consent forms were modelled on recommendations by the Ontario Law Reform Commission.[8]

Table 4. IVF Surrogacy Management

INDICATION	SURROGATE	TECHNIQUES
*R-K-H syndrome	SISTER Inter-state	CC/HCG for Synch TEST - 5 embryos PREGNANT - Triplets
ASHERMAN'S syndrome	SR-IN-LAW Overseas	HRT TEST #1 - 2 embryos TEST #2 - 3 embryos
HYSTER. (Cancer)	SISTER Sterilization	CC/HCG IVF-ET - 4 embryos
HYSTER. (Rupt.Ut.)	SISTER Follicles Asp.	CC/HMG/HCG TEST - 4 embryos PREGNANT - twins
HYSTER. (Rupt.Ut.)	CLOSE FRIEND 9 years	Nat. Cycle/No Coitus IVF-ET - 2 embryos

Rokitansky-Kuster-Hauser

The biological mother provides consent to the ovarian stimulation and egg recovery procedures and both she and her husband consent to the procedures of IVF and embryo culture. They confirm their desire to adopt all children born from the surrogacy arrangement and accept responsibility for all defined expenses. They give an undertaking that they will not pay any fee to the surrogate mother but they do agree to pay a term life assurance contract for the surrogate woman in case of her demise during the pregnancy, to provide for her own family. The biological parents also give an undertaking to provide in their own will for all children born and accept all responsibility for those children. Furthermore, they agree to provide no information to the media or public concerning the arrangements (as this may prejudice their subsequent adoption of the children according to the Adoptions Act in Western Australia).

The potential surrogate mother agrees to receive the embryos and carry the pregnancy and both she and her husband sign a consent to allow the biological mother and father to adopt the child/children after delivery. They also agree to abide by all medical advice given for management of the pregnancy.

RESULTS

Five couples have had a total of seven treatment cycles as summarized in Table 4. Four treatment cycles involved tubal transfers with one pregnancy; two had embryo transfers to

Table 5. IVF Surrogacy Pregnancy

INDICATION	GAMETES	SURROGATE	METHOD	OUTCOME
*RKH Syndrome	H&W	Sister *sterilization* *M.,3 ch.,Vic.*	ET	Triplets *b. Oct.'88*
Hysterect.	H&W	Sister *M.,5 ch.,WA*	TEST	Twins *d. Apr.'89*

Rokitansky-Kuster-Hauser

the uterus (one pregnancy); and a further cycle did not reach the stage of transfer as a single cryopreserved embryo of Case #3 failed to survive thawing.

The two pregnancies which occurred are summarized in Table 5. The first involved a 26 year old woman with a congenital uterine absence due to Rokitansky-Kuster-Hauser syndrome. Gametes were collected from her husband and herself and subsequently five embryos (3 high-grade and 2 medium-grade)[9] were transferred to the uterine cavity of her sister who travelled interstate from Victoria for the procedure. The surrogate woman had three children, a stable marriage, and had previously undergone sterilization. Healthy triplet infants were born in October 1988 and the infertile couple (biological parents) were present at the delivery. Caesarean section under epidural was performed because of fulminating pre-eclamptic toxaemia at 33 weeks gestation. During the post-natal period a convivial atmosphere was apparent between the two families, including the husband of the surrogate woman, and her three children who appeared pleased to see the new arrivals and their subsequent transfer to the biological parents. After two weeks the surrogate woman and her family returned to Victoria and it is expected that the biological parents will be able to undergo a formal adoption procedure for their children.

The second case was a thirty-one year old woman who had a previous hysterectomy for a ruptured uterus. Embryos were generated from the gametes of herself and her husband and subsequently four cleaving embryos were transferred to the uterus of her sister who lives locally. The sister underwent ovarian stimulation as she suffered oligomenorrhoea and subsequently four embryos were transferred to one of her fallopian tubes. Her own follicles were aspirated and she kindly donated those oocytes to another infertile couple in the ovum donation program. Furthermore, she and her husband avoided coitus throughout the entire month. She is due to deliver a twin pregnancy at the end of April 1989 and again, the arrangement between the two families, including the husband and children of the surrogate woman, appears highly convivial.

DISCUSSION

The legal issues concerning surrogacy vary widely from country to country and in Australia there are State variations. For example, current legislation prevents surrogacy arrangements in Victoria and South Australia. In New South Wales, the Law Reform Commission has previously published information suggestive of a favourable attitude toward surrogacy,[10,11] but more recently it appears that anti-surrogacy legislation is being prepared. Similarly, the Governments of Queensland, Northern Territory, Tasmania, and now, Western Australia, appear to be preparing anti-surrogacy legislation.

The legal position in Australia is that the birth mother is the legal mother and this is therefore the case for surrogate mothers. The Adoption of Children Act (Western Australia) 1896-1985 allows for preferred adoption of the child by a relative or a married couple one of whom is a relative. However, the best possible position from which the child may be adopted, that is where the surrogate mother identifies the natural father in the child's birth certificate, may be denied by the Artificial Conception Act (Western Australia) 1985 which specifically states that a man is legally not recognized as the father if his sperm was used for inseminating the mother or for fertilizing an ovum transferred to her uterus. This section of the Act was designed to protect anonymous sperm donors but does create a hurdle within the adoptions system where embryo transfer is used; it remains to be tested if this will be interpreted as also applying to tubal transfer procedures, and whether the "fertilized ovum" definition in the Artificial Conception Act includes the embryo.

With respect to surrogacy legislation, it has already been stated that variable individual State legislation prevails. With respect to the impending Surrogacy Act in Western Australia, this is designed to achieve three aims: (1) discourage and deter all surrogacy; (2) uphold contracts arranged under adoption legislation; and (3) regard all commercial aspects as illegal. It would appear that this legislation is designed to legalize the existing surrogacy contracts and arrangements but deter any subsequent activities in this area.

Furthermore, the Fertility Society of Australia (FSA) through its Reproductive Technology Accreditation Committee, has recently (February 1989) criticized the structure of the PIVET Ethics Committee, perceiving it as being too supportive of the IVF unit. As this constitutes a threat to PIVET's accreditation by the FSA, which conducts a self-regulatory control over IVF units around Australia, the surrogacy program at PIVET has now been suspended, awaiting deliberation of the matter by the recently formed Cambridge Hospital Ethics Committee which has agreed to act as PIVET's IEC.

These various events appear to highlight the introductory comments to this paper, indicating that surrogacy remains a controversial area of medicine with a marked division of opinion within society, religious groups, legal, political and medical bodies. Nonetheless, from our limited experience with non-commercial, compassionate family surrogacy by IVF, we strongly believe that the technique as described, would create a favourable treatment mode for certain types of infertility which cannot be treated by any other method. It would appear that the non-commercial aspect and the ensured access of children arising from these arrangements with both their biological and surrogate mothers should be in the best interests of the child/children as well as the couples concerned. Whilst the antagonists of surrogacy would hold that unpredictable, complex and possibly adverse interactions may occur at some later stage, we believe that these would be no greater than normal intra-family interactions which have prevailed from the dawn of man.

ACKNOWLEDGEMENT

We gratefully acknowledge the expert laboratory and clinical skills of the PIVET IVF unit in managing the cases as well as providing facilities for the preparation of this manuscript.

REFERENCES

1. S. Downie, "Baby making: the technology and ethics," The Bodley Head, London (1988).
2. W. H. Utian, In vitro fertilization — surrogate pregnancy, *in:* "Proceedings of the VI World Congress in Vitro Fertilization and Alternate Assisted Reproduction, Jerusalem, Israel", April 2-7 (1989).
3. New South Wales. Law Reform Commission, "Artificial conception: Surrogate mother-hood: Australian public opinion." Sydney (1987).
4. M. Kirby, Surrogacy, *in:* "Proceedings of the Sixth Annual Meeting of the Fertility Society of Australia, Sydney", 11-14 November (1987).
5. J.L. Yovich, Infertility and Fertility: Female, *in:* "Treatment and prognosis: obstetrics and gynaecology," J.G. Grudzinskas, T. Beedham, eds., Heinemann Medical Books, London (1988).
6. J.L. Yovich, A.I. Tuvik, P.L. Matson and D.L. Willcox, Ovarian stimulation for disordered ovulatory cycles, *Asia-Oceania. J. Obstet. Gynaecol.* 13:457-463 (1987).
7. P. Lutjen, A. Trounson, J. Leeton, J. Findlay, C. Wood and P. Renou, The establishment and maintenance of pregnancy using in vitro fertilization and embryo donation in a patient with primary ovarian failure, *Nature,* 307:174 (1984).
8. Ontario Law Reform Commission Report on Human Artificial Reproduction and Related Matters (1985).
9. J.M. Cummins, L.M. Wilson, T.M. Breen and J.F. Hennessey, A formula for scoring human embryo growth rates in IVF: value in predicting pregnancy and in comparison with other es-timates of embryo quality, *J. Vitro Fert. Embryo Transfer.* 3:284 (1986)
10. New South Wales. Law Reform Commission, "Artificial conception. Discussion paper 3. Surrogate motherhood," Sydney (1988).
11. New South Wales. Law Reform Commission, "Artificial conception. Report 3. Surrogate motherhood," Sydney (1988).

SEMEN DONATION AND SEX SELECTION

Rhachi Iizuka

Department of Obstetrics and Gynecology
School of Medicine, Keio University
Tokyo 160, Japan

Follow-up studies of the children born from artificial insemination with donor's semen (AID)

In 1948, artificial insemination with donor's semen (AID) was performed for the first time in Japan at the Keio University hospital. Then the first successful cases of four pregnancies with donor's frozen semen were also reported by us. In Japan, since 1949, more than 7000 patients have become pregnant following AID with fresh or frozen semen, and almost all cases were treated at our clinic. In our cases, the donors are selected among the students of the medical school, Keio University. They are requested to undergo medical examinations to rule out hereditary disorders, hepatitis, and venereal diseases including AIDS. The information concerning the semen donors is closely guarded from the couples selected for AID therapy. They have to legitimize their child born as a result of AID.

Over the past decade, AID has been recognized as a practical therapy, and we therefore insisted on clinical investigations concerning their mental and physical developments.[1] However, the follow-up study of AID children was extremely difficult because of their special circumstances. Figure 1 represents our follow-up studies for the body weight and height in AID children up to five years old. The lines in the figure show the mean value of male or female children in Japan. The AID children indicated comparable or better development than the control children who were born naturally. Good physical development was also found in children between six to 17 years of age.

Advances in Assisted Reproductive Technologies, Edited by
S. Mashiach *et al.,* Plenum Press, New York, 1990

The mental development of AID children was estimated by use of the development quotient, DQ, for the infants under two and a half years old, and the intelligence quotient, IQ, for the older group. The DQ is distributed from 80 to 150, with a mean of 110.7, thus showing a rather high value compared with the control of 100. The IQ distribution was 84 to 150, with a mean of 111.7. The value is also above the average.

The school records of the AID children in elementary and junior high schools were also almost satisfactory or better. The emotional characteristics of AID children were also estimated. In both males and females all the characteristics investigated were good, and no abnormalities were found in their mental and emotional developments.

The mental and physical developments in AID children are comparable or higher than those born naturally. This might be due to the donor semen, which were provided by the medical students in our medical school. Another factor might be due to the long desire of the parents to have their own children resulting in more attention to their child's education.

Separation of human X- and Y-sperm by discontinuous Percoll density gradient.

Since 1980, we employed the discontinious Percoll density gradient centrifugation for washing and concentration of low quality semen for artificial insemination with husbands semen. Coincidental alteration of sex ratio was found in the delivered cases, with the females predominating the males in significant numbers.[2] This fact led us to the assumption that the Percoll density gradient might be available for the separation of human X- and Y-sperm.

The detection of human X- and Y-sperm was performed by the fluorescent staining with Quinacrine Mustard according to the method of Barlow and Vosa.[3] The sperm with a fluo-

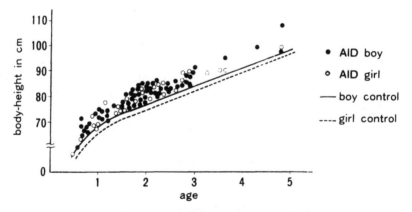

Fig. 1. Development of body height in AID children born from frozen semen.

rescent body at the center of the head were identified as Y-sperm and those without such a body as X-sperm. Rhode et al.[4] reported a successful separation of human X-and Y-sperm in a discontinuous density gradient centrifugation with sucrose; a higher percentage of X-sperm was found in the lower layer of the density gradient.

First, we adopted a discontinious Percoll density gradient of eight steps for the separation of human X- and Y-sperm. The quinacrine mustard staining suggested that the top layer was rich in Y-sperm, whereas it decreased along the density gradient, and the sediment was abundant in X-bearing sperm.[5] A series of subsequent experiments revealed the separation efficiency dependent on the number of steps in the density gradient, and centrifugal force and period.[6] It seemed difficult to prepare both types of cells simultaneously in one procedure. As the steps of density gradient increased, the purity of X-sperm recovered in the improved sediment. The improved isolation of human X-sperm was achieved by means of 12 steps of Percoll density gradient.[6] Figure 2 shows the profile of separation of X- and Y-bearing sperm in 12 steps Percoll density gradient, after centrifugation at 250 × g for 30 min. The purity of X-sperm was found to be 94% in the sediment with a mean recovery of 23% (Figure 3). Almost all the sperm in the sediment were found to be active and forward motile. X-chromosome-linked diseases usually affect males. Therefore, there is significance in ensuring the sex of newborn infants to avoid these diseases, and preselection of sex at the stage of fertilization, i.e. artificial insemination with the isolated X-bearing sperm, is most reliable for this purpose. The Japan Medical Association and the Japan Society of Obstetrics and Gynecology have issued a statement that a preconceived sex selection was necessary in terms of medical indication. At the present time, its clinical application should be limited to the incidence of severe sex-linked genetic diseases, especially for hemophilia.

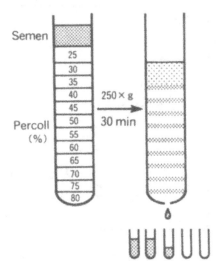

Fig. 2. Separation of human X- and Y-bearing sperm by means of Percoll density gradient of 12 steps.

At the preliminary stage for clinical application on the sex linked genetic diseases, the isolated X-bearing sperm fraction were inseminated in the infertile couples, who already had children, but visited Keio University Hospital complaining of secondary infertility.[7] To date, 10 couples have achieved pregnancy following AIH with the isolated X-bearing sperm (Table 1). Their prenatal courses were uneventful, and terminated in spontaneous deliveries. All the newborns were female and there were no abnormalities found.

Hysteroscopic insemination into tube (HIT)

Recently we examined the direct insemination into the fallopian tube under hysteroscopic guidance. Oligozoospermic semen was concentrated by means of the non-layer Percoll method with a slight modification, and the selection of progressively motile sperm was achieved by a swim down procedure into underlaid 80% Percoll. Each 0.5 ml of air, concentrated sperm, and 80% Percoll was introduced in a 2.5 ml disposable syringe, then it was laid down to extend the contact surface between sperm suspension and 80% Percoll layer, incubating at 37°C for 30 min. The progressively motile sperm penetrated into underlaid Percoll layer was recovered and adjusted to 4 - 5 × 10^6/ml.

Ovulation was estimated by ultrasonography; insemination was performed the next day after the diameter of the dominant follicle reached 19–20 mm. An aliquot (50–100 μl) of the isolated progressively motile sperm fraction was introduced into the fallopian tube under the hysteroscopic guidance.

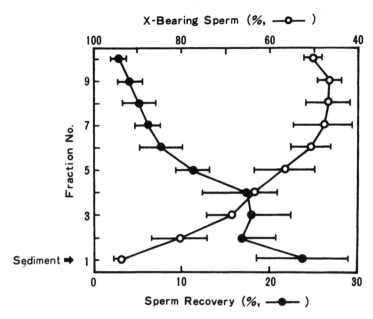

Fig. 3. Selective isolation of human X-bearing sperm in the discontinuous Percoll density gradient. Values are mean ± SD.

Table 1. Selective child bearing following AIH with the isolated X-bearing sperm

Case	Age	Steps of Percoll Gradient	No. of Male children	Baby Weight(g)	Baby Sex	Baby Abnormality
1	31	8	1	2810	♀	(−)
2	31	8	2	3280	♀	(−)
3	34	8	1	3250	♀	(−)
4	35	8	1	2680	♀	(−)
5	33	8	1	2860	♀	(−)
6	29	8	3	2590	♀	(−)
7	39	12	2	3010	♀	(−)
8	34	12	2	3890	♀	(−)
9	32	12	2	3200	♀	(−)
10	35	12	2	3100	♀	(−)

Thirty-nine patients with complaints of severe oligozoospermia, who had been treated with AIH more than 10 times without success, were treated with HIT, and resulted in three successful pregnancies. The first case terminated in spontaneous vaginal delivery. A normal male baby weighing 3008 g was born without abnormality.

The hysteroscopic insemination into the fallopian tube requires only 20×10^4 sperm to achieve pregnancy, and it would be a new approach for the treatment of severe oligozoospermia.

REFERENCES

1. F. Mochimaru, Artificial insemination with frozen donor semen: its current status and follow-up studies, *Keio J. Med.*, 28:333-348 (1979).
2. R. Iizuka, S. Kaneko, K. Kobanawa, and T. Kobayashi, Washing and concentration of human semen by Percoll density gradients and its application to AIH, *Arch. Androl.*, 20:117-124 (1987).
3. P. Barlow, and C. G. Vosa, The Y chromosome in human spermatozoa, *Nature*, 226:961-962 (1970).
4. W. Rgode, T. Porstmann, S. Prehen, G. Dorner, Gravitational pattern of the Y-bearing human spermatozoa, *J. Reprod., Fert.*, 42:587-591 (1975).
5. S. Kaneko, J. Yamaguchi, T. Kobayashi, and R. Iizuka, Separation of human X- and Y-bearing sperm using Percoll density gradient centrifugation, *Fertil. Steril.*, 40:661-665 (1983).

6. S. Kaneko, S. Oshio, T. Kobayashi, H. Mhori, and R. Iizuka, Selective isolation of human X- and Y-bearing sperm by differential velocity sedimentation in Percoll density gradients, *Biomed. Res.*, 5:187-194 (1984).
7. R. Iizuka, S. Kaneko, R. Aoki, and T. Kobayashi, Sexing of human sperm by discontinuous Percoll density gradient and its clinical application, *Human Reprod.*, 2:573-575 (1987).

RECENT PROGRESS IN BLASTOCYST IMPLANTATION RESEARCH

Koji Yoshinaga

Reproductive Sciences Branch, Center for Population Research
NICHD, NIH, Bethesda, Maryland 20892, USA

INTRODUCTION

Although there are many aspects to implantation research, this review will focus on the recent progress in the following three aspects: (1) embryonic signals for maternal recognition of pregnancy, (2) possible involvement of receptor-ligand reactions in the blastocyst-uterine interaction during implantation, and (3) in vitro model systems for implantation research.

EMBRYONIC SIGNALS FOR MATERNAL RECOGNITION OF PREGNANCY

Two Types of Trophoblast Cell Function in Luteal Maintenance

It is well established that progesterone plays an important role in preparation of the endometrium for implanting blastocyst and in subsequent maintenance of pregnancy. To secure an adequate supply of progesterone, blastocysts play active roles in regulating the luteal function. Two different types of blastocyst functions have been reported to maintain progesterone levels for successful establishment of pregnancy.

The first is the secretion of luteotropic hormone, namely chorionic gonadotropin, by trophoblast cells, which is commonly observed in the primates. Chorionic gonadotropin becomes detectable in peripheral blood soon after implantation. It is, however, possible that a blastocyst initiates chorionic gonadotropin secretion before implantation, because the

Advances in Assisted Reproductive Technologies, Edited by
S. Mashiach *et al.*, Plenum Press, New York, 1990

culture medium of a human pre-embryo at the blastocyst stage has been reported to contain a detectable amount of hCG (Fishel et al., 1984). The rescue of the corpus luteum from luteolysis by chorionic gonadotropin is also reported in non-human primates (Atkinson et al., 1975; Hodgen et al., 1975; Reyes et al., 1975; Jagannadha Rao et al., 1984; Stouffer et al., 1987; Webley et al., 1989). The progesterone supply by the corpus luteum is taken over by the placenta as pregnancy progresses.

The second type of the embryonic function to secure an adequate supply of progesterone is the secretion of trophoblastic protein, for example, of molecular weight of approximately 17,000 in the sheep. The protein is named oTP-1 in the sheep and bTP-1 in the cow. These proteins do not exert direct luteotropic effect on the corpus luteum, but they prevent the release from the endometrial epithelium of prostaglandin F2 alpha which is a potent luteolysin in these domestic animals.

Whether trophoblast cells of a blastocyst from any species of animals can secrete both chorionic gonadotropin and TP-1 is not known. It has however, recently been reported by Cross et al. (1989) that in the mouse constitutive antiviral activity is produced in the blastocyst after hatching from the zona pellucida and in the placenta at least until day 16 of pregnancy (day 0 is the sperm positive day), and that mRNA's encoding for interferon and IRF-1, (interferon regulatory factor-1, the protein that regulates expression of interferon beta gene) are abundant in the murine placentae and blastocyst. These results suggest that the rodent placentae secrete a protein similar to oTP-1 as well as a well documented prolactin-like luteotropin that takes over the luteotropic activity of the pituitary at mid-pregnancy.

Involvement of histamine, estrogen, prostaglandins, leukotrienes and platelet activating factor derived from the embryo in implantation and decidualization has recently been reviewed by Fleming et al. (1989) and Kennedy et al. (1989). The reader is also referred to other chapters in this volume.

A Possible Immunological Role of oTP-1

It has been demonstrated that both oTP-1 and bTP-1 belong to the interferons of alpha family (Imakawa and Roberts, 1989). The subsequent studies on the structure of mRNA for oTP-1 also show its similarity to those of mRNA's for the interferons of alpha family. In situ hybridization studies localized mRNA for oTP-1 in trophoblast cells of day 17 sheep embryo (Roberts et al., 1989).

In addition to the inhibitory effect on the release of prostaglandin F2 alpha from the endometrium, a role of oTP-1 as an immunological modulator has been suggested (Imakawa and Roberts, 1989). However, there are few reports on the immunological studies on oTP-1. Here, a possible immunological role of this protein will be postulated below.

Two types of cell surface molecules take part in immunological recognition. They are parts of the major histocompatibility complex. According to Feldman and Eisenbach (1988), normal cells have an appropriate ratio of these two types of molecules (H-2D and H-2K) which is recognized as foreign when transplanted into another individual. On the surface of metastatic tumor cells, one type (H-2D) of the molecules predominates because the corresponding gene (H-2D gene) is overly expressed. The tumor cells with predominating H-2D molecules on their surface are metastatic, because the host into whom the tumor cells are inoculated does not recognize them as foreign cells. When H-2K gene is transferred into the metastatic tumor cells, the transgenic cells produce more H-2K molecules and the H-2K/H-2D ratio falls within the range that the host can recognize them as foreign cells, and the degree of metastasis is markedly reduced. Feldman and Eisenbach also showed that the treatment with interferons alpha and beta in tandem of tumor cells elicited the expression of H-2D gene which, in turn made the tumor cells metastatic. On the other hand, treatment of the cells with interferon gamma elicited the expression of H-2K gene which, in turn, made the tumor cells non- metastatic.

Since oTP-1 is a protein of the interferon alpha family, it is conceivable that oTP-1 facilitates immunological tolerance of the maternal immune system by making the invading trophoblast cells appear as "not foreign" to the maternal organ, the uterus.

A possible mechanism postulated is that oTP-1 produced by trophoblast cells acts on themselves as an autocrine and stimulates production of cell surface molecules equivalent to the mouse H-2D through activation of the corresponding gene. The cell surface with predominant H-2D equivalent molecules makes the trophoblast cells which are about to interact with, or invade the endometrium, appear "not foreign" to the uterus leading to successful implantation.

Recently Mosinger and Forejt (1989) reported that transfection of LTM-5 EC (mouse embryonic carcinoma) cell line with H-2K gene resulted in rejection of EC graft in vivo without detectable changes in the cell surface molecules. They suggest that the expression level of the transfected gene may be too low to be detected and/or some soluble factor(s) is produced to influence the immunogenicity. Examination of the cell surface molecules of trophoblast cells may not, therefore, yield fruitful results. Clarification of the immunological roles of oTP-1 is eagerly awaited.

RECEPTOR-LIGAND REACTIONS IN IMPLANTATION

Blastocyst-Epithelial Cell Interactions

Enders and Schlafke (1967) described three stages of implantation process in the rat: apposition, adhesion and invasion. Although it is not clear what mechanisms are involved in the adhesion of blastocysts to the epithelial surface of the endometrium, appearance of stage specific glycoproteins (Anderson et al., 1986), reduction in the electrical negativity, and changes in the carbohydrate moiety of the glycocalyx have been implicated as possible mechanisms (Anderson, 1989). It has been demonstrated that a receptor-ligand binding reaction is involved in the process of fertilization; a glycoprotein in the zona pellucida, ZP-3, is a receptor for sperm acrosomal membrane components (Bleil and Wassarman, 1980). In the case of implantation, identification of the cell surface molecules pertinent to the receptive stage of the uterine sensitivity and of mature blastocysts as to their physical and/or biochemical characteristics is of utmost importance to determine whether these molecules are ligands or receptors. Possible involvement of ligand- receptor binding reactions during the process of implantation was discussed in detail elsewhere (Yoshinaga, 1989).

In guinea pigs, rhesus monkeys and humans, where implantation is intrusive, trophoblast cells penetrate the uterine epithelium between the cells. Since the epithelial cells are joined together by tight junctions (Zonulae occludentes), trophoblast cells must interact with these junctional complexes during the inter-epithelial cell penetration. Recently it was found that the tight junction contains specific proteins, ZO-1 (Stevenson et al., 1986; Anderson et al., 1988) and cingulin (Citi et al., 1988). It has been shown that another protein, uvomorulin (or E-cadherin), is present at the lateral-basal aspects of epithelial cells. Glasser et al. (1988) have recently succeeded in establishing a polarized uterine epithelial cell culture system, and have demonstrated that uvomorulin is indeed localized at the lateral-basal aspects of these cultured polarized epithelial cells. It is tempting to speculate that trophoblast cell membrane components may have specific affinity to these proteins, and they uncouple the junctional complexes as passing through between the epithelial cells. The trophoblast cells appear to be equipped with some components to interact with these proteins that serve as cell binders. As has been suggested for the tumor metastasis (Liotta, 1986), trophoblast cell surface may have either ligands or receptors to the uterine cell surface components. Coupling of some components of the trophoblast cell membrane with those of the epithelial cell membrane by ligand-receptor reactions and subsequent release of appropriate proteolytic enzymes to uncouple the existing junctions between the uterine epithelial cells may be considered as a possible means of trophoblast penetration through the uterine surface epithelium.

Penetration of the Basal Lamina and Stroma by Blastocyst

When trophoblast cells reach the basal lamina, they pause there before progressing into the stroma (Schlafke and Enders, 1975). Trophoblast cells contain receptors to laminin (Sutherland et al., 1988), and they are considered to anchor themselves to the basal lamina by means of a ligand-receptor reaction as laminin is one of the major components of the basal lamina. It is reasonable to consider that the anchored trophoblast cells then secrete proteolytic enzymes to disrupt the lamina, or stimulate neighboring cells which, in turn, secrete proteolytic enzymes. In the case of the rat implantation, the basal lamina is breached by decidual cells and not by trophoblast cells (Enders et al., 1981; Schlafke et al., 1985). This may be explained by secretion of proteolytic enzymes by decidual cells that are stimulated by some substance, for example, prostaglandin, from the trophoblast cells located on the other side of the basal lamina.

Decidualization of the stromal cells appears to favor the penetration by trophoblast cells by this mechanism, because the extracellular components found in the basal lamina are secreted by the decidualized stromal cells (Wewer et al., 1986).

IN VITRO MODEL SYSTEMS FOR IMPLANTATION RESEARCH

Why Are In Vitro Systems Needed?

In the rabbit and some domestic animals such as the pig, horse and sheep, blastocysts expand considerably prior to implantation to fill the center of the uterine lumen. The blastocysts that implant in this fashion, known as central implantation, are macroscopically visible and, with an appropriate surgical preparation, one can observe how the uterus grasps a blastocyst and how the latter adheres to the uterine wall (Boving, 1952). However, even in these species of mammals, the accessibility to blastocysts at the time and site of implantation is extremely limited and continuous observation of implantation process is not possible for an extended period of time. In other species of mammals, the size of blastocyst is minute and one cannot predict where in the uterus it is going to implant. It is, therefore, desirable if one can prepare a uterus, or even an endometrial tissue, which is hormone responsive in vitro, and make the uterine tissue receptive for implantation, and then place a blastocyst on the receptive uterine tissue to observe implantation process and to obtain samples for analysis. At present, not a single ideal model system for implantation research is available.

Use of ideal in vitro model systems for implantation research would provide the following advantages: one can obtain a large number of synchronized implantations with a known developmental stage and locations; thus one can better study stage-specific phenomena during the process of implantation; one can make analytical studies on the uterine sensitivity for implantation under precisely controlled conditions; and one can apply the information obtained in vitro to in vivo situations.

Blastocyst Development on Extracellular Matrix

Trophoblast outgrowth on the surface of a plastic dish is observed in a blastocyst culture system. This phenomenon is an expression of the capability of a blastocyst to grow and is considered of little use for the study of implantation because there is no interaction between the trophoblastic cells and uterine components. Studies on the behavior of a blastocyst on an extracellular matrix component, on the other hand, provide some useful information as to the affinity of trophoblast cells to certain substances which may reflect some, but limited, aspects of the blastocyst-uterine interaction during implantation.

The mechanism of tumor metastasis was hypothesized by Liotta (1986) that malignant tumor cells contain an increased amount of receptors for laminin, one of the major components of the basal lamina, by which the tumor cells anchor themselves to the basal lamina, and that they secrete, or stimulate other cells to secrete, proteolytic enzymes to disrupt the basal lamina for propagation of tumor cells. Armant et al. (1986 a,b) studied specific affinity of mouse blastocyst to extracellular matrix components in vitro. Sutherland et al. (1988) studied the behavior of mouse blastocysts cultured in vitro in serum-free medium; over the 24-36 hours posthatching, they became adhesive and attached on appropriate extracellula matrix component substrates, i.e. fibronectin, laminin, and collagen type IV in a ligand specific manner. When anti-extracellular matrix component antibodies are added to the culture medium, attachment and trophoblast outgrowth was inhibited. These in vitro studies suggest that the ligand-receptor binding mechanism is involved in the process if implantation. Recently Welsh and Enders (1989) examined the capacity to degrade a variety of extracellular matrices by rat trophoblast, mouse trophoblast, and rat uterine cells. They observed that rat uterine cells have a greater capacity to degrade a variety of complex extracellular matrices in vitro than either rat or mouse trophoblast cells. Based on their observations, these workers warned that cautious approaches be taken in in vitro studies of trophoblast invasiveness in these species.

Uterine Cell Culture Systems for Implantation Research

In their report on a mouse endometrial cell culture system, Sengupta et al. (1986) made a brief review on the so-called "in vitro models for implantation" citing a variety of substrates including strips of rabbit endometrium, whole uteri from immature mice maintained as organ culture, monolayers of "feeder cells" from endometrial and non-endometrial sources, and vesicles of uterine epithelium incubated in hanging drops. These workers then reported that dissociated mouse endometrial cells, when cultured on reconstituted collagen, formed tissue-like complexes that resemble endometrium in many ways: the stroma and epithelium had morphologic similarities to their counterparts in vivo. Earlier, Enders et al. (1981) reviewed "in vitro model systems" and found some model systems may be useful for examining some aspects of the blastocyst-endometrium interaction but are limited in one way or another as models for implantation. Improvement of a model system is, then, to expand the limitation in one way or another. The first improvement to be made is to establish endometrial tissue in vitro which responds to ovarian hormones as it does in vivo.

Sengupta et al. (1986) obtained endometrial-like tissues in vitro; multilayered stromal cells covered with epithelial cells which had microvilli and glycocalyx on the apical surface and lateral junctional complexes with adjacent cells. Glasser et al. (1988) successfully established a uterine epithelial monolayer culture system using immature rats. These epithelial cells cultured on EHS matrix-coated filters developed a cell polarity characterized by differences in morphology as well as in function with regard to the protein profiles of the apical and baso-lateral secretory compartments. These cells also maintained the responsiveness to estrogen validated by the secretion of two proteins identified as secretory markers of estrogen response in the intact uterus. Findlay et al. (1989) also reported successful establishment of a sheep uterine epithelial cell culture system using Matrigel-coated Millicell culture well inserts. The epithelial cells grown in this system retained their morphological characteristics for over two weeks. When the cells are obtained from ovariectomized ewes treated with ovarian hormones, the protein patterns of the culture media reflected the hormone treatment in vivo. Steroid treatments in vitro, however, were partially effective. These workers also demonstrated that oTP-1 or human alpha2 interferon added to the culture medium significantly attenuated the secretion of prostaglandins into the culture medium in a dose-dependent manner (Salamonsen et al., 1988; Findlay et al., 1989). The results indicate that this model system responded to the substance secreted by the sheep blastocyst and its analogue in the same fashion as observed in vivo.

Future Directions of In Vitro Models

The ultimate goal of developing in vitro model systems for blastocyst implantation is to understand clearly the mechanisms involved in human blastocyst implantation into the endometrium. Because one cannot experiment on human blastocysts from the ethical point of view, alternate approaches have been devised. One is the use of trophoblast cells taken from term placentae and the other is the use of non-human blastocysts. Kliman et al. (1986) demonstrated that the cytotrophoblasts purified from human term placentae differentiate in vitro into syncytiotrophoblasts and exhibited the endocrine function as they differentiate. Co-culture of cytotrophoblasts obtained from the first trimester or term placentae with human endometrial cells and tissues resulted in adhesion of trophoblast cells to the uterine cells and some degree of invasion into the uterine tissue (Kliman et al., 1989). Although this approach may provide some information as to the interactions between trophoblast cells and uterine cells and tissues, one major reservation to call this attachment "in vitro implantation" is that one does not know whether the uterine cells and tissues under these culture conditions are comparable to the uterine condition within the frame of the so-called implantation window when the uterus is receptive for blastocyst implantation. This criticism also applies to a culture system where marmoset blastocysts were co-cultured on a fibroblast monolayer (Hearn et al., 1988). Although this latter study was designed to examine the endocrine function of the developing embryo, attachment of a blastocyst to the fibroblast cells and development of embryo beyond the blastocyst stage may be compared to a type of ectopic pregnancy, rather than physiological implantation for establishment of pregnancy.

Development of blastocysts transplanted at ectopic sites such as the anterior chamber of the eye (Fawcett et al., 1947), under the kidney capsule (Fawcett, 1950; Kirby, 1960), and to the spleen (Kirby, 1963), and to the testis (Kirby 1963), provides some information as to their requirement for embryonic development, but it provides little information as to the actual requirements for implantation in utero. The uterine environment should be considered, at most of the time during the cycle, as an unfriendly one for a blastocyst to implant. It should be kept in mind that the uterus allows a blastocyst to implant only for a short period of time, probably four days out of the typical 28 day cycle in women (Mandelbaum et al., 1989), three days in the rhesus monkey (Hodgen, 1983), and less than 24 hours in the rat (Psychoyos, 1963). Without understanding this special uterine condition where the endometrium is rendered receptive for implanting blastocyst, attachment of a blastocyst to an extracellular matrix, to a monolayer of cultured cells or a tissue of foreign origin, or invasion into a mass of cells and tissues, should be regarded as model systems for ectopic pregnancy, not for blastocyst implantation. It is, therefore, very clear that the right direction to go is the study of the uterine receptivity in vitro. To achieve this goal, one has to establish ovarian hormone responsive endometrial tissues in vitro. It is conceivable that the uterine epithelium and stroma communicate with each other and paracrine mechanisms may be operating. Establishment of an ideal model system for implantation research is eagerly awaited.

REFERENCES

Anderson, J. M., Stevenson, B. R., Jesaitis, L. A. Goodenough, D. A., and Mooseker, M. S., 1988, Characterization of Z0-1, a protein component of the tight junction from mouse liver and Madin-Darby canine kidney cells, *J. Cell Biol.*, 106:1141-1149.

Anderson, T. L., Olson, G. E., and Hoffman, L. H., 1986, Stage-specific alterations in the apical membrane glycoproteins of endometrial epithelial cells related to implantation in rabbits, *Biol. Reprod.*, 34:701-720.

Anderson, T. L., 1989, Biomolecular markers for the window of uterine receptivity, *in:* "Blastocyst Implantation", K. Yoshinaga, ed., Adams Publishing Group, Ltd., Boston, 219-224.

Armant, D. R., Kaplan, H. A., and Lennarz, W. J., 1986a, Fibronectin and laminin promote in vitro attachment and outgrowth of mouse blastocyst, *Dev. Biol.*, 116:519-523.

Armant, D. R., Kaplan, H. A., Mover, H., and Lennarz, W. J., 1986b, The effect of hexapeptides on attachment of mouse embryos in vitro: Evidence for the involvement of the cell recognition tripeptide Arg-Gly-Asp, *Proc. Natl. Acad. Sci.*, USA, 83:6751-6755.

Atkinson, L. E., Hotchkiss, J., Fritz, G. R., Surve, A. H., Neill, J. D., and Knobil, E., 1975, Circulating levels of steroids and chorionic gonadotropin during pregnancy in the rhesus monkey, with special attention to the rescue of the corpus luteum in early pregnancy, *Biol. Reprod.*, 12:335-345.

Bleil, J. D., and Wassarman, P. M., 1980, Mammalian sperm-egg interaction: Identification of glycoprotein in mouse zonae pellucidae possessing receptor activity for sperm, *Cell*, 20:873-882.

Boving, B. G., 1952, Internal observation of rabbit uterus, Science, 116:211-214.

Citi, S., Sabanay, H., Jakes, R., Geiger, B., and Kendrick-Jones, J., 1988, Cingulin, a new peripheral component of tight junctions, *Nature*, 333:272-276.

Cross, J. C., Farin, C. E., Sharif, S. F. and Roberts, R. M., 1989, Interferon and interferon regultory factor-1 are present in placental tissue during pregnancy in mice. 71st Annual Meeting of Endocrine Society, Seattle, WA, Abstr. # 530.

Enders, A. C., and Schlafke, S., 1967, A morphological analysis of the early implantation stages in the rat, *Amer. J. Anat.*, 120:185-226.

Enders, A. C., Chavez, D. J., and Schlafke, S., 1981, Comparison of Implantation in utero and in vitro, *in:* "Cellular and Molecular Aspects of Implantation", S. R. Glasser and D. W. Bullock, ed., Plenum Press, New York, 365-382.

Fawcett, D. W., Wislocki, G. B., and Waldo, C. M., 1947, The development of mouse ova in the anterior chamber of the eye and in the abdominal cavity, *Am. J. Anat.*, 81:413-443.

Fawcett, D.W., 1950, Development of mouse ova under the capsule of the kidney. *Anat. Rec.* 108:71-92.

Feldman, M., and Eisenbach, L., 1988, What makes a tumor cell metastatic? *Scientific American*, November: 60-85.

Findlay, J. K., Salamonsen, L. A., and Cherny, R. A. 1989, Isolated endometrial cells as in vitro models to study the establishment of pregnancy, *in:* "Blastocyst Implantation", K. Yoshinaga, ed., Adams Publishing Group, Ltd., Boston, 75-81.

Fishel, S. B., Edwards, R. G., and Evans, C. J., 1984, Human chorionic gonadotropin secreted by preimplantation embryos cultured in vitro, *Science*, 223:816-818.

Fleming, S., O'Neill, C., Collier, M., Spinks, N. R., Ryan, J. P., and Ammit, A. J., 1989, The role of embryonic signals in the control of blastocyst implantation, *in:* "Blastocyst Implantation", K. Yoshinaga, ed., Adams Publishing Group, Ltd., Boston, 17-23.

Glasser, S. R., Julian, J. A., Decker, G. L., Tang, J.-P., and Carson, D. D., 1988, Development of morphological and fuctional polarity in primary cultures of immature rat uterine epithelial cells, *J. Cell Biol.*, 107:2409-2423.

Hearn, J. P., Hodges, J. K., and Gems, S., 1989, Early secretion of chorionic gonadotrophin by marmoset embryos in vivo and in vitro, *J. Endocr.*, 119:249-255.

Hodgen, G. D., Tullner, W. M., Vaitukaitis, J. L., Ward, D. N., and Ross, G. T., 1975, Specific radioimmunoassay of chorionic gonadotropin during implantation in rhesus monkey, *J. Clin. Endocrinol. Metab.*, 39:457-464.

Hodgen, G. D., 1983, Surrogate embryo transfer combined with estrogen-progesterone therapy in monkeys, Implantation, gestation, and delivery without ovaries, *JAMA*, 250:2167-2171.

Imakawa, K., and Roberts, R. M., 1989, Interferons and maternal recognition of pregnancy, *in:* "Development of Preimplantation Embryos and Their Environment", K. Yoshinaga, and T. Mori, eds., Alan R. Liss, Inc., New York, 347-358.

Jagannadha Rao, A., Kotagi, S. G., and Moudgal, N. R., 1984, Serum concentrations of chorionic gonadotrophin, oestradiol 17-beta and progesterone during early pregnancy in the south Indian bonnet monkeys (Macaca radiata), *J. Reprod. Fert.*, 70:449-455.

Kennedy, T. G., Squires, P. M., and Yee, G. M., 1989, Mediators involved in decidualization, *in:* "Blastocyst Implantation", K. Yoshinaga, ed., Adams Publishing Group, Ltd., Boston, 135-143.

Kirby, D. R. S., 1960, Development of mouse eggs beneath the kidney capsule, *Nature*, 187:707-708.

Kirby, D. R. S., 1963, Development of the mouse blastocyst transplanted to the spleen, *J. Reprod. Fert.*, 5:1-12.

Kirby, D. R. S., 1963, The development of mouse blastocysts transplanted to the scrotal and cryptorchid testis, *J. Anat.*, 97:119-130.

Kliman, H. J., Nestler, J. E., Sermasi, E., Sanger, J. M., and Strauss, J. F.III, 1986, Purification, characterization, and in vitro differentiation of cytotrophoblasts from human term placentae. *Endocrinology*, 188:1567-1582.

Kliman, H. J., Coutifaris, C., Feinberg, R. F., Strauss J. F. III, and Haimowitz, J. E., 1989, Implantation: In vitro models utilizing human tissues, *in:* "Blastocyst Implantation", K. Yoshinaga, ed., Adams Publishing Group, Ltd., Boston, 83-91.

Liotta, L. A., 1986, Tumor invasion and metastases - Role of the extracellular matrix: Rhoads Memorial Award Lecture, *Cancer Res.*, 46:1-7.

Mandelbaum, J., Junca, A.-M., Plachot, M., Cohen, J., and Alvalez, S., 1989, The implantation window in humans after fresh or frozen-thawed egg transfers. VI World Congress IVF and Altern. Assist. Reprod., Jerusalem, Abstr. p. 66.

Mosinger, B., and Forejt, J., 1989, Transfected H-2K gene as a cause of embryonal carcinoma cell rejection in vivo. *Immunogenetics*, 29:269-272.

Psychoyos, A., 1963, Precisions sur l'etat de "non-receptivite" de l'uterus, *C. r. hebd. Acad. Sci.*, 257:1153-1156.

Reyes, F. I., Winter, J. S., Faiman, C., and Hobson, W. C., 1975, Serial serum levels of gonadotropins, prolactin and sex steroids in the nonpregnant and pregnant chimpanzees, *Endocrinology*, 96:1447-1455.

Roberts, R. M., Farin, C. E., and Imakawa, K., 1989, Embryonic mediators of maternal recognition of pregnancy, *in:* "Blastocyst Implantation", K. Yoshinaga, ed., Adams Publishing Group, Ltd., Boston, 25-30.

Salamonsen, L. A., Stuchbery, S. J., O'Grady, C. M., Godkin, J. D., and Findlay, J. K., 1988, Interferon-alpha mimics effects of ovine trophoblast protein-1 on prostaglandin and protein secretion by ovine endometrial cells in vitro, *J. Endocr.*, 17:R1-R4.

Schlafke, S., Welsh, A. O., and Enders, A. C., 1985, Penetration of basal lamina of the uterine luminal epithelium during implantation in the rat, *Anat. Rec.*, 212:47-56.

Sengupta, J., Given, R. L., Carey, J. B., and Weitlauf, H. M., 1986, Primary culture of mouse endometrium on floating collagen gels: A potential in vitro model for implantation, *Ann. N.Y. Acad. Sci.*, 476:75-94.

Stevenson, B. R., Siliciano, J. D., Mooseker, M. S., and Goodenough, D. A., 1986, Identification of Z0-1: A high molecular weight polypeptide associated with tight junction (zonula occludens) in a variety of epithelia. *J. Cell Biol.*, 103:755-766.

Stouffer, R. L., Ottobre, J. S., and VandeVoort C. A., 1987, Regulation of the primate corpus luteum during early pregnancy, *in:* "The Primate Ovary", R. L. Stouffer, ed., Plenum Publishing Corp., New York, 207-220.

Sutherland, A. E., Calarco, P. G., and Damsky, C. H., 1988, Expression and function of cell surface extracellular matrix receptors in mouse blastocyst attachment and growth, *J. Cell Biol.*, 106:1331-1348.

Webley, G. E., Richardson, M. C., Given, A., and Hearn, J. P., 1989, The maternal recognition of embryo implantation in primates: Changing responsiveness of luteal cells of the marmoset monkey to luteotrophic and luteolytic agents during normal and conception cycles, *J. Reprod. Fert.*, (in press).

Welsh, A. O., and Enders, A. C., 1989, Comparisons of the ability of cells from rat and mouse blastocysts and rat uterus to alter complex extracellular matrix in vitro, *in:* "Blastocyst Implantation", K. Yoshinaga, ed., Adams Publishing Group, Ltd., Boston, 55-74.

Wewer, U. M., Damjanov, A., Weiss, J., Liotta, L. A., and Damjanov, I., 1986, Mouse endometrial stromal cells produce basement-membrane components, *Differentiation*, 32:49-58.

Yoshinaga, K., 1989, Receptor concepts in implantation research, *in:* "Development of Preimplantation Embryos and Their Environment", K. Yoshinaga, and T. Mori, Eds., Alan R. Liss, Inc., New York, 379-387.

PURIFIED HUMAN CYTOTROPHOBLASTS: SURROGATES FOR THE BLASTOCYST IN *IN VITRO* MODELS OF IMPLANTATION ?

C. Coutifaris, G. O. Babalola, R. F. Feinberg, L-C. Kao,
H. J. Kliman, and J. F. Strauss, III

Departments of Obstetrics & Gynecology and Pathology and Laboratory Medicine
University of Pennsylvania School of Medicine
Philadelphia, Pennsylvania, 19104, U.S.A.

INTRODUCTION

Although extensive studies on implantation have been performed with laboratory and domestic animals, it is difficult to extrapolate from these observations to the process of nidation in the human, because there are marked interspecies differences. Schlafke and Enders (1975), in an attempt to put some order in the diverse and oftentimes conflicting reports, proposed three possible modes for trophoblast implantation: *fusion* implantation, which is thought to take place in the rabbit, in which trophoblasts fuse with endometrial epithelial cells as the initial step in embryo-maternal interaction; *displacement* implantation which takes place in rodents, involves destruction of the uterine epithelium allowing the embryo to come into contact with the basal lamina prior to initiation of frank stromal invasion. The epithelial destruction is independent of the presence of the embryo and appears to be controlled by specific signals arising from the endometrial stoma. The third mode of implantation, *intrusive* implantation, appears to be operative in species with hemochoreal placentation, in which trophoblastic cells are markedly invasive. Intrusive implantation has been extensively examined in the ferret, where trophoblasts insinuate between endometrial epithelial cells prior to invading the basal lamina and endometrial stroma (Enders and Schlafke, 1972; Gulamhusein and Beck, 1973). Based on existing morphological observations, intrusive implantation has been proposed to be the mode of nidation in the human. However, lack of in vitro models for human embryo implantation has hindered our ability to study this process at the cellular and molecular level.

Advances in Assisted Reproductive Technologies, Edited by
S. Mashiach *et al.*, Plenum Press, New York, 1990

A method for isolation of cytotrophoblasts from human placenta was recently developed in our laboratories and this has allowed us to study these cells in vitro and further characterize their regulation (Kliman et al., 1986; Kliman et al., 1987). In the present paper, our current understanding of the cell biology and biochemistry of these cells will be reviewed. We believe that these cells, given their observed morphologic and functional differentiation, can be used to study some of the cellular and molecular events of human placentation in vitro. Development of in vitro models for trophoblast implantation can significantly impact on our understanding of the mechanisms involved in this basic reproductive process and can have potential applications in the areas of fertility regulation and human in vitro fertilization and embryo transfer.

THE TROPHOBLAST SYSTEM

It is well established that the mononuclear cytotrophoblast of the human placenta is the precursor cell to the terminally differentiated syncytial trophoblast (Pierce and Midgley, 1963; Boyd and Hamilton, 1970). Cytotrophoblasts isolated from human placentae and placed in monolayer cultures can be seen to initially aggregate and then fuse to form large multinucleated syncytia (Kliman et al., 1986). During this process of morphologic differentiation, the functional differentiation of these cells also occurs: while chorionic gonadotropin or placental lactogen are not elaborated by the mononuclear cytotrophoblasts, they are synthesized and secreted in substantial quantities by the differentiated syncytial structures (Feinman et al., 1986). Time-lapse cinematography has unequivocally documented the marked motility of these cells and also the sequence of events that lead to this morphologic differentiation. The first step in this process involves the development of multicellular aggregates which subsequently produce the syncytial structures through a process of cell fusion. This process is cell-specific since co-culture of cytotrophoblasts with fibroblasts or other cells of fetal origin does not result in fusion with non-trophoblastic cells. Thus, it is logical to assume that this homotypic (cytotrophoblast-cytotrophoblast) interaction requires specific cell surface components. The importance of cell adhesion molecules (CAM's) in the morphogenesis of tissues has been established and their characterization as transmembrane glycoproteins is well substantiated (Edelman, 1988; Horwitz et al., 1986). The rate-determining step in the formation of syncytia in vitro seems to be cell aggregation. Under two-dimensional monolayer culture conditions, aggregation of cytotrophoblasts occurs only if the culture media are supplemented with serum or if the culture surface has been precoated with extracellular matrix components such as fibronectin, laminin or types I, IV and V collagen (Feinman et al., 1986; Kao et al., 1988). It appears that the extracellular matrix proteins provide the latice for cellular motility, which then allows the cells to aggregate and ultimately fuse. If freshly isolated trophoblastic cells are placed in a shaking suspension culture, thus allowing random collision of the cells, aggregation occurs even under serum-free conditions (Babalola et al., 1989 and 1989a). Thus, the process of aggregation seems to depend on random motion of the trophoblastic cells with subsequent collision and adhesion rather than chemotactically mediated motion. If the freshly isolated cells are exposed to the inhibitor of protein synthesis, cycloheximide, cellular aggregation is inhibited. These findings suggest that specific plasma membrane proteins are involved in trophoblast cell aggregation, which are cleaved during enzymatic isolation of the cells prior to the initiation of the suspension culture. Protein synthesis and subsequent incorporation of these molecules into the plasma membrane is necessary prior to acquisition of the capacity to aggregate. We postulate, that a specific CAM is required for the trophoblastic aggregation process. Previous studies with CAM's have established that these plasma membrane glycoproteins are not solely anchor molecules but provide a mechanism for intercellular communication (Horwitz et al., 1986). Thus, it is conceivable that following aggregation, such intercellular signals may trigger the process of cellular fusion.

The process of morphologic differentiation from mononuclear cytotrophoblasts to syncytial trophoblasts is closely but not obligatorily coupled to endocrine differentiation of these cells (Kliman et al., 1986). The synthesis and secretion of placental protein hormones (i.e., hCG, hPL) markedly increases as syncytial formation proceeds. This is similar to the situation in vivo, in which cytotrophoblasts do not elaborate these hormones whereas syncytial trophoblasts do. Although the functional and morphologic differentiation of cytotrophoblasts in vitro are temporally linked, regulation of these processes can be separated: addition of the cyclic AMP analog, 8-bromo-cyclic AMP to the culture medium triggers the synthesis and secretion of chorionic gonadotropin by mononuclear cytotrophoblasts (Feinman et al., 1986). This occurs under conditions in which aggregation and fusion are inhibited (i.e.static culture in serum-free medium). The involvement of a cyclic nucleotide in this process suggests that, physiologically, functional differentiation with respect to hormone synthesis and secretion involves a signal initiated from a plasma membrane receptor coupled to adenylate cyclase. The ligand initiating this cascade is unknown at the present time. Since cell aggregation and fusion usually precede functional differentiation of these cells, it is attractive to speculate that CAM's or other closely related membrane components are involved in the trigger mechanism(s) for the functional differentiation of these cells.

TROPHOBLAST - ENDOMETRIAL INTERACTION

The initial steps in the process of human implantation are the *aposition* and subsequent *adhesion* of the trophoblast to the endometrial epithelial cells. It is postulated that this heterotypic (trophoblast-endometrial) interaction is mediated by specific cell surface components. Intrusion of the trophoblast in between epithelial cells is then achieved through processes, which are poorly understood at the present time. In order to study this initial interactive process, a two dimentional co-culture system was developed (Coutifaris et al., 1988; Coutifaris et al., 1989): human endometrial epithelial and glandular cells were isolated and a primary monolayer culture was established; these cells in culture form islands of flattened cells, which have been shown to be epithelial by immunostaining for cytokeratin and epithelial membrane antigen. Once the endometrial cultures were established (48–72 hours), purified cytotrophoblasts were added and their interactions were observed by time lapse cinematography and immunostaining for hCG at various times of co-incubation (24–

96 hours). Trophoblasts were seen to form aggregates surrounding the endometrial cell islands and at approximately 36–48 hours of coincubation were observed to penetrate in between the epithelial cells. As this intrusive process proceeded, a remodelling of the endometrial cells occurred behind the intruding trophoblasts, resulting in the complete encasement of these cells by the epithelium. Once the intruding trophoblast aggregates fused to form syncytia their motility ceased, they flattened out and formed large multinucleated cells, which were surrounded by the endometrial epithelial cells (figure 1). Interestingly, when malignant melanoma cells, instead of trophoblasts, were used in these coincubation experiments, the melanoma cells were observed to completely surround the epithelial cell islands but failed to penetrate into them in spite of their aggressive growth (Coutifaris et al., 1989). This is in marked contrast to the results from coincubation studies of either trophoblasts or melanoma cells with endometrial stromal cells: in these cultures both cell types intruded into the stromal cell islands. These observations suggest that trophoblasts possess the capacity to penetrate a confluent epithelium. This characteristic may be unique to the trophoblast since it was not exhibited by malignant melanoma cells. The cellular mechanism through which this intrusive process is achieved remains to be elucidated and the described two-dimentional in vitro system can now be used to explore it.

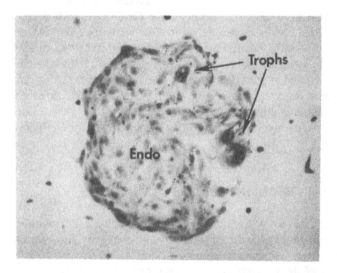

Fig. 1. Interactions of cytotrophoblasts with human endometrial epithelial cells in monolayer culture. Purified human endometrial epithelial cells were established in monolayer culture for 48 hours and subsequently coincubated with human cytotrophoblasts (500,000 cells) for 72 hours, fixed with Bouin's solution and immunocytochemically stained with antibodies against alpha-hCG using DAB as chromagen. Several darkly stained multinucleated syncytial trophoblasts (trophs) can be seen within the endometrial cell island (endo). Time lapse cinematography demonstrated that these syncytial structures formed by fusion of multiple mononuclear cytotrophoblasts, which at 0 time surrounded the epithelial cell island and subsequently proceeded to intrude into it. Once syncytial trophoblasts were formed their motility ceased and the endometrial cells remodelled themselves to eventually surround the trophoblasts.

INVASION

The final event of the process of nidation in the human is the "erosion" of the trophoblast through the basal lamina into the endometrial stroma and the establishment of hemochoreal placentation. Do cytotrophoblasts isolated from the placenta have invasive characteristics? Presently, it is believed that enzymes of trophoblast origin facilitate the invasion of the decidualized endometrial stroma. Recent evidence from our and other laboratories indicate that the ability to degrade extracellular matrix is an integral property of these cells (Fisher S., unpublished observations). One protease that has been shown to be elaborated by our cytotrophoblastic preparations is urokinase-type plasminogen activator (Queenan et al., 1987). Urokinase is produced by cytotrophoblasts in culture, and secretory activity of this enzyme is also up-regulated by 8-bromo-cyclic AMP (figure 2). Further, Northern blot analysis reveals increases in urokinase mRNA after cyclic AMP treatment (figure 2). The enzyme is presumably elaborated as an inactive proenzyme, which is activated by other proteases and appears to bind to specific receptors on the cell surface, localizing the proteolytic activity. Urokinase can be involved both in the direct digestion of extracellular matrix proteins (e.g. fibronectin, laminin) and in activation of other proteolytic enzymes (e.g. col-

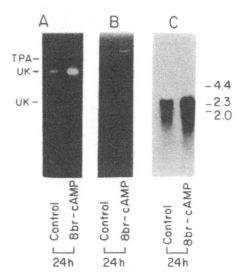

Fig. 2. A and B: Detection of urokinase-PA in conditioned media from cytotrophoblast cultures (1,000,000 cells/2mls) in the absence or presence of 8-bromo-cyclic AMP (1.5mM) for 24 hours. Media aliquots were electrophoresed in SDS-polyacrylamide gels containing gelatin and plasminogen (A) or gelatin alone (B). The locations of tissue-PA and high (~50K) and low (~30K) molecular weight u-PA activities are noted. The zymographs shown are representative of three separate experiments.
C: Hybridization blot analysis of u-PA mRNA extracted from cultured cytotrophoblasts. Total RNA (10 mcg) extracted from cells cultured in the absence or presence of 8-bromo-cyclic AMP for 24 hours was electrophoresed in agarose gels, transferred to nylon membrane and probed with specific 32P-labelled u-PA cDNA. Nucleic acid size markers are presented in Kb.

lagenase), which in turn degrade other extracellular matrix components. Urokinase activity is regulated in part by specific inhibitors which covalently bind to and inactivate the enzyme (Blasi et al., 1987). At least two types of plasminogen activator inhibitors, type I (PAI-1) and type II (PAI-2), are synthesized by human trophoblastic cells (Feinberg et al., 1988; 1989). The mRNAs encoding plasminogen activator inhibitor types I and II have been detected in primary cultures of human cytotrophoblasts and 8-bromo-cyclic AMP causes a reduction in the levels of the inhibitor mRNAs (figure 3). Thus, it appears that there is a reciprocal regulatory mechanism mediated by cyclic AMP in which expression of urokinase is increased with concomitant inhibition of genes encoding its inhibitors (figure 4). These observations suggest that the process of trophoblast invasion into the endometrium may be tightly regulated by coordinate control of genes encoding proteases and protease inhibitors.

CONCLUSIONS

In summary, purified mononuclear cytotrophoblasts isolated from the human placenta have been shown to: (i) possess the morphologic characteristics of in situ cytotrophoblasts; (ii) differentiate in vitro into syncytial trophoblasts through a process of aggregation and fusion, resembling the in vivo differentiation of these cells; (iii) during this pro-

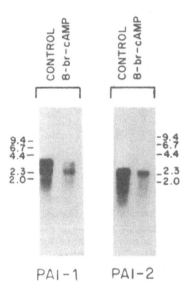

Fig. 3. Blot hybridization analysis of PAI-1 and PAI-2 mRNAs in cytotrophoblasts cultured in the absence (control) or presence of 1.5mM 8-bromo-cyclic AMP for 24 hours. Total RNA (10mcg) was probed with 32P-labelled nick-translated probes. Size markers (Hind III digest of lambda phage DNA) are presented in Kb.

cess of cytodifferentiation they also differentiate functionally, acquiring the capacity to synthesize and secrete the various protein and steroid hormones characteristic of the syncytial trophoblast; (iv) the mononuclear aggregates can be observed in vitro to intrude into endometrial epithelial cell islands mimicking the initial in vivo events of trophectoderm intrusion through the endometrial epithelium; (v) they elaborate proteolytic enzymes in vitro (i.e. urokinase), which have been previously implicated in the process of implantation and (vi) they have the capacity to regulate this process by the synthesis and secretion of specific protease inhibitors. Moreover, cyclic AMP controls some of these processes, offering us a tool to study the regulatory processes of these cells in vitro.

Based on morphologic criteria, the cellular mechanisms of human embryo implantation are unique to the human and there are marked differences compared to well studied laboratory and domestic animals. The elucidation of the cellular and molecular events of the initial trophoblast-endometrial aposition, adhesion and invasion requires the development of a **human** in vitro model of implantation. Since normal human blastocysts are not readily available for study, we propose that purified human cytotrophoblasts possess many of

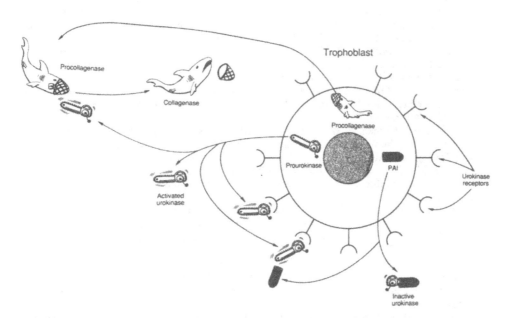

Fig. 4. Schematic representation of the sequential activation/inhibition of one of the trophoblast protease systems: the mononuclear cytotrophoblast (trophoblast) secretes the inactive pro-enzymes procollagenase and prourokinase. Upon secretion, the prourokinase is activated (urokinase) and, in turn, it activates the proteolytic enzyme collagenase. The urokinase also binds to its cellular membrane receptor and is inactivated by the plasminogen activator inhibitor (PAI), which is also secreted by the trophoblast. The level of the trophoblast enzymatic activity, and thus of its intrusive/invasive potential, depends upon the coordinate control of urokinase expression and that of its inhibitor.

the morphologic and functional characteristics of invading trophoblasts and can be used for the in vitro study of trophoblast - endometrial interactions. Basic knowledge acquired from such studies may have a dramatic impact on the further development and success of assisted reproduction.

ACKNOWLEDGEMENTS

Supported by USPHS grants HD-06274 and HD-00715 and grants from the Rockefeller, Lalor and Mellon foundations. Dr. Coutifaris is the recipient of the American Fertility Society-Ortho Distinguished Fellowship in Reproduction.

REFERENCES

Babalola, G. O., E. A. Soto, C. Coutifaris, H.J. Kliman, J. F. Strauss III 1989, Morphogenesis of the human placenta: characterization of the aggregation of cytotrophoblastic cells. 36th Annual Meeting of the Society for Gynecologic Investigation, San Diego, CA.

Babalola, G.O., C. Coutifaris, E.A. Soto, H.J. Kliman, H. Shuman, J.F. Strauss, III 1989a, Aggregation of dispersed human cytotrophoblastic cells: lessons relevant to the morphogenesis of the placenta, *Develop. Biol.* submitted for publication,

Boyd, J.D., W.J. Hamilton 1970, "The human placenta" pp. 167-174, MacMillan, London.

Blasi, F., J-D Vassalli, D. Keld 1987, Urokinase-type plasminogen activator: proenzyme, receptors, and inhibitors. *J. Cell. Biol.* 104: 801-804.

Coutifaris, C., H. J. Kliman and J. F. Strauss III 1988, Development of an in vitro model system for human embryo implantation.44th Annual Meeting of the American Fertility Society, Atlanta, GA. Abstract 039.

Coutifaris C., H.J. Kliman, P. Wu, J. F. Strauss III 1989, Specificity of trophoblast-endometrial interactions in a human in vitro implantation model system.36th Annual Meeting of the Society for Gynecologic Investigation, San Diego, CA. Abstract 489.

Edelman, G. M. 1988, Morphoregulatory molecules. Biochemistry 27: 3533-3543.

Enders. A.C., S. Schlafke 1972, Implantation in the ferret: Epithelial penetration. *Am. J. Anat.* 133: 291-316.

Feinberg, R.F., L-C Kao, G. Ringler, S. Murray, J. Queenan Lr., H.J. Kliman, D. Cines, T-C. Wun, J. F. Strauss, III 1988, Coordinate regulation of urokinase and plasminogen activator inhibitors in human cytotrophoblasts. 35th Annual Meeting of the Society for Gynecologic Investigation, Baltimore, Md. Abstract 344.

Feinberg, R.F., J.F. Strauss III, T-C Wun, H.J. Kliman 1989, Plasminogen activators PAs, and Plasminogen activator inhibitors PAIs, in human trophoblasts: markers of trophoblast invasion. 36th Annual Meeting of the Society for Gynecologic Investigation, San Diego, CA. Abstract 487.

Feinman, M.A., H. J. Kliman, S. Caltabiano and J. F. Strauss III 1986, 8-Bromo-3'5'-adenosine monophosphate simulates the endocrine activity of human cytotrophoblasts in culture.

Gulamhusein, A.P., F. Beck 1973, Light and electron microscopic observations at the pre- and early post-implantation stages in the ferret uterus. *J. Anat.* 115: 159-174.

Horwitz, A., K. Duggan, C. Buck, M. C. Beckerle, K. Burridge 1986, Interaction of plasma membrane fibronectin receptor with talin - a transmembrane linkage. *Nature* 320: 531-533.

Kao, L-C, S. Caltabiano, S. Wu, J. F. Strauss III and H. J. Kliman 1988, The human villous cytotrophoblast: interactions with extracellular matrix proteins, endocrine function and cytoplasmic differentiation in the absence of syncytium formation. *Dev. Biol.* 130: 693-702.

Kliman, H. J., J. E. Nestler, E. Sermasi, J. M. Sanger and J. F. Strauss III 1986, Purification, characterization and in vitro differentiation of cytotrophoblasts from human term placentae. *Endocrinology* 118: 1567-1582.

Kliman, H. J., M.A. Feinman, and J. F. Strauss III 1987, Differentiation of human cytotrophoblasts into syncytiotrophoblasts in culture. *In:* "Trophoblast Research" R. Miller and H. Thied, eds., Plenum Medical, New York, NY, Vol. 2, pp. 407-421.

Martin, O. and F. Arias-Stella 1982, Plasminogen activator production by trophoblast cells in vitro. Effect of steroid hormones and protein synthesis inhibitors. *Am. J. Obstet. Gynecol.* 142: 402-409.

Pierce, G.B. Jr. and A. R. Midgley Jr. 1963, The origin and function of human syncytiotrophoblastic giant cells. *Amer. J. Pathol.* 43: 153-173.

Queenan, J. T., Jr., L-C Kao, C. E. Arboleda, A. Ulloa-Aguirre, T. G. Golos, D. B. Cines, J. F. Strauss 1987, Regulation of urokinase type plasminogen activator production by culture human cytotrophoblasts. *J. Biol. Chem.* 262: 10903-10906.

Schlafke, S., A. C. Enders 1975, Cellular basis of interaction between trophoblast and uterus at implantation. *Biol. Reprod.* 12: 41.

A ROLE FOR PLATELET ACTIVATION FACTOR
IN EMBRYO IMPLANTATION

Chris O'Neill

Human Reproduction Unit
Royal North Shore Hospital of Sydney
St Leonards, NSW 2065, Australia

INTRODUCTION

The development in recent years of fertilization of oocytes and culture of early embryos in vitro has provided new insights into this phase of pregnancy. Evidence suggests that fertilization may be a relatively efficient process but that the development of embryos through to the blastocyst stage and implantation is very inefficient. Indeed, estimates in a number of species suggest that failure of preimplantation embryonic development and implantation may be as high as 35% in otherwise apparently normally fertile females. This represents an enormous loss of efficiency in reproduction.

This high incidence of reproductive failure suggests three obvious practical implications:

1. That a significant cause of idiopathic infertility is likely to be due to embryonic demise prior to implantation and that methods for detecting the presence of the embryo immediately following fertilization are necessary to diagnose this.

2. That even modest improvements in the implantation rate will have a significant influence on the treatment of human infertility and on agricultural production.

3. That further disturbance of implantation may offer the possibilities for new technologies of contraception.

Advances in Assisted Reproductive Technologies, Edited by
S. Mashiach *et al.*, Plenum Press, New York, 1990

Thus an understanding of the mechanism and signals involved in embryo-maternal communication may play a significant part in developing diagnostic and therapeutic tools for this stage of pregnancy.

Apart from the gonadotrophins that are secreted later in pregnancy, there has been proposed a variety of signals produced by the preimplantation embryo. These include: steroids, prostaglandins, leukotrienes, histamine, ovum factor, an immunosuppressive factor, stage specific proteins and platelet activating factor (PAF).

The aim of this review is to consider the evidence for the production of embryo-derived PAF and to consider the possible role(s) it may play in the establishment of pregnancy.

Platelet Activating Factor

Platelet activating factor, 1-0-alkyl-2-acetyl-sn-glycerol-3-phosphocholine (PAF) (Figure 1), is an ether phospholipid. The ether phospholipids are ubiquitous, although relatively minor, components of animal cells.[1] PAF was identified and synthesized following initial observations that platelets released histamine (platelet activation) following their interaction with activated neutrophils.[2] In 1972[3] it was shown that a soluble platelet activation factor was released from rabbit basophils sensitized to a specific IgE and challenged with antigen in vitro. This biologically active phospholipid was also discovered independently in another physiological system. It was apparent for some years that the renal medulla produced an endocrine-like, antihypertensive action which was associated with the polar lipid fraction. In 1979,[4] an alkyl ether analogue of phosphatidylcholine from choline plasmalogens was shown to have potent antihypertensive effects in the rat and was subsequently shown to be identical to PAF.[5] This hypotensive effect was shown to be independent of the platelet activating factor effects of PAF. The platelet activating action and the antihypertensive effects of this molecule were the first demonstrations of a biologically active phospholipid.

$$
\begin{array}{l}
\text{H}_2\text{C}-\text{O}-\text{CH}_2-(\text{CH}_2)_n-\text{CH}_3 \\
\quad | \\
\text{CH}_3-\text{C}-\text{O}-\text{CH} \qquad \text{O}^- \\
\quad \| \qquad\quad | \qquad\quad | \\
\quad \text{O} \qquad \text{H}_2\text{C}-\text{O}-\text{P}-\text{O}-\text{CH}_2-\text{CH}_2-\text{N}^-(\text{CH}_3)_3 \\
\qquad\qquad\qquad\qquad \| \\
\qquad\qquad\qquad\qquad \text{O}
\end{array}
$$

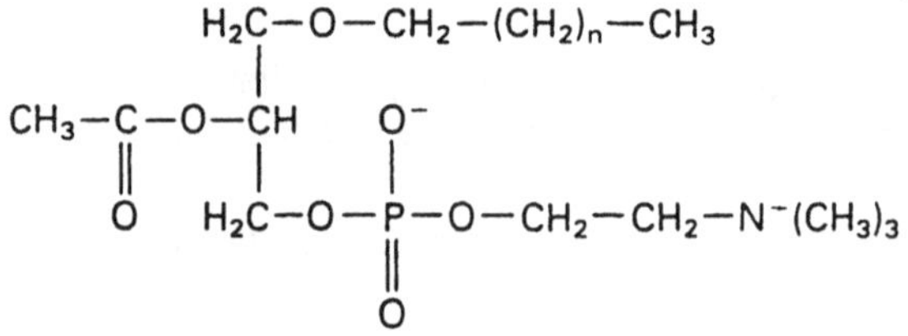

Fig. 1. The structure of synthetic PAF. Embryo-derived PAF from the 2-cell human (Collier et al.[19]) and mouse (O'Neill[20]) embryo has been shown to be homologous with this structure. In the biologically active molecules examined N = 12 or 14.

PAF has since been implicated as a mediator in an array of pathological and physiological responses (review).[6] This unique chemical can exert such a diversity of effects because it is active at extremely low concentrations (10^{-11}M), has a very short half life in blood, and because its effects are receptor mediated. Thus, PAF is a local autocoid, responses to which are dependent upon the type and location of the producing cells and on the target cells upon which it acts.

Biochemistry and Biosynthesis

PAF is classed as an ether phospholipid because instead of having a fatty acid moiety esterified to the glycerol backbone, a long chain alcohol undergoes condensation to produce an ether-linked hydrocarbon chain on the C_1 carbon of glycerol PAF is characterized by a short chain acyl group at position C_2 which is generally acetate, and this group is essential for the biological activity of the molecule. Hydrolysis of this C_2 acetate results in the lyso-PAF which is biologically inactive but which is cytotoxic at high doses. The native PAF has a phosphocholine group at its C_3 position that is also essential for its biological activity.

The biosynthetic pathways of PAF in some cell types are now becoming well defined. A recent review[7] provides a sound description of the pathways, their control and the enzymes involved. Current evidence indicates that there are two pathways of PAF production (Figure 2). First the conversion of 1-0-alkyl-2-acyl-sn-glyceryl-3-phosphocholine to the corresponding lysolipid by a deacylation reaction followed by its acetylation to create the active PAF (acetyltransferase pathway). Second, the alternative pathway involving the transfer of phosphocholine to 1-0-alkyl-2-sn-glycerol to produce PAF (cholinephosphotransferase pathway). The acetyltransferase pathway appears to require metabolic activation of cells while the cholinephosphotransferase pathway appears active at basal metabolic conditions in a number of cell types.

A precursor for both pathways is dihydroxyacetone phosphate, an intermediate of glycolysis. This is then converted to acyldihydroxyacetone phosphate, which also commonly forms the precursors of the esterified phospholipids. In the case of the ether phospholipids, including PAF, the acyl chain is replaced by a long chain fatty alcohol to form alkyldihydroxyacetone phosphate. This is reduced by an NADPH- dependent oxidoreductase to form 1-0-alkyl-2-lyso-sn-glyceryl-3-phosphate. This compound marks the bifurcation point of the two pathways. The pentose phosphate pathway, which is highly active in the early embryo,[8] is also a potential supply of dihydroxyacetone phosphate, while the NADPH produced is an important co-factor in PAF production.

Acetylation of the second carbon of the glyceryl backbone commits the compound to the phosphocholinetransferase pathway while acylation leads to the acetyltransferase pathway. PAF is produced via the cholinephosphotransferase pathway by dephosphorylation of 1-0-alkyl-2-acetyl-sn-glyceryl-3-phosphate to form 1-0-alkyl-2-acetyl-sn-glycerol. This is the substrate of CDP-choline cholinephosphotransferase resulting in the production of PAF. The acetyltransferase pathway is somewhat longer. 1-0-alkyl-2-_sn_-glyceryl-3-phosphate is also dephosphorylated followed by the addition of cholinephosphate. The 1-0-alkyl-2-acyl-_sn_-glyceryl-3-phosphocholine is deacylated by phospholipase A_2 to form 1-0-alkyl-2-lyso-_sn_-glyceryl-3-phosphocholine (Lyso-PAF). This is acetylated by an acetyl-CoA-dependent acetyltransferase to give PAF.

In principle, PAF can be produced de novo by either of these two pathways. It appears, however, that in reality de novo synthesis primarily occurs via the cholinephosphate pathway.[7] The acetyltransferase pathway generally appears to act via the utilisation of the stored precursor 1-0-alkyl-2-acyl-_sn_-glyceryl-3-phosphocholine. PAF is also readily converted back to the alkylacyl glycerophosphocholine (GPC) precursor providing a futile PAF metabolic cycle.

Following its production within cells, PAF is transferred to the plasma membrane from where it can be secreted. Extracellular protein is required for retrieval of PAF and albumin appears to serve this role well.[9] The hydrophobic regions of PAF have a high affinity for this protein. In the absence of extracellular protein, PAF secretion is reduced. The secretion of PAF means that this cell activator can potentially exert autocrine, paracrine or endocrine effects. The relative contribution to these physiological processes depends largely on the metabolic fate of PAF.

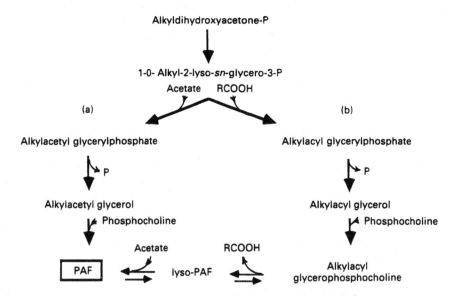

Fig. 2. Alternative pathways for PAF biosynthesis proposed by Sneider[7]: (a) the acetyltransferase pathway and (b) the phosphocholine transferase pathway. RCOOH is long-chain free fatty acids and P is phosphate.

The key intermediate of PAF catabolism is 1-0-alkyl-2-lyso-sn-glyceryl-3-phosphocholine (Lyso-PAF).[7] The enzyme responsible, acetylhydrolase (AH)[10] appears to be ubiquitous. It is found intracelullarly and in serum and plasma. Cell associated AH unlike serum AH is inhibited by phenylmethylsulfonylfuoride. This enzyme has a high affinity for PAF but not for the alkylacyl derivative. The specificity of the enzyme is demonstrated by its inability to hydrolyse the biologically active PAF analogue 1-0-alkyl-2-N-methylcarbamyl-sn-glyceryl-3-phosphocholine.[11]

The rapid degradation of PAF in the various pools limits its activity and hence its range of action to the immediate locality of its production. An endocrine role for PAF, as might be required for it to have luteotrophic actions, would be dependent on the presentation of PAF in a less labile form. Tissue culture experiments have shown that albumin is very effective in lifting PAF from the plasma membrane and acting as a carrier in the fluid phase. There have been a number of reports of PAF binding to other serum proteins. PAF binds with high affinity to alpha-1-acid glycoprotein.[12] Another serum protein of molecular weight 160-180K daltons was also shown to bind PAF.[13] It had a higher affinity for PAF than did albumin and offered protection from AH activity.[13]

The lyso-PAF produced by AH is potentially cytotoxic. Rapid further processing of lyso-PAF is required to avoid cell and tissue damage. The most well defined catabolic fate is that of its reacylation by an acyltransferase to form 1-0-alkyl-2-acyl-sn-glyceryl-3-phosphocholine (alkylacyl GPC). In most cells studied to date the fatty acid used in this esterification step is preferentially but not specifically arachidonic acid.[14] The alkylacyl GPC may then be either inserted into the membranes or undergo conversion to phosphatidylcholine.[14] The stored alkylacyl GPC may then serve as a substrate for the acetyltransferase pathway, thus creating a PAF cycle. It is important to note that PAF receptor antagonists[15] and inhibitors of cell activation[16] can prevent this cell associated catabolism of PAF.

Embryo-Derived PAF

It first became evident that the early embryo produces PAF by the observation[17] in mice, that splenic contraction and mild, but significant, thrombocytopenia occurred shortly after fertilization and throughout the preimplantation phase of pregnancy. Appropriate control experiments proved that this was due to platelet activation and was solely dependent upon the presence of the embryo.[17] It was further shown that the human[18, 19] and mouse[20] embryo at very early stages of development secreted a soluble platelet-activating factor. Purification and characterization showed that this agent in both the human and mouse was homologous with PAF.[19, 20]

The observation, however, that embryo-derived PAF induced significant maternal thrombocytopenia in a number of species suggested that monitoring maternal platelet count may be a means of diagnosing and monitoring early pregnancy.

Table 1. Production of PAF, as Assessed by the Reduction in Platelet Count in the Splenectomized Mouse Bioassay,[20] by 2-cell, 4- to 8-cell, Morula, Expanded Blastocyst, and Degenerate Mouse Embryos Cultured In Vitro for Varying Lengths of Time (mean ± SEM). (From Ryan et al., 1989) [22]

Experiment	Time in culture (hr)	Cell Stage		% original platelet count (n)
		Start	Finish	
1	1	2-cell	2-cell	94.5 ± 2.6 (8)
	4	2-cell	2-cell	91.8 ± 2.1 (12)
	6	2-cell	2-cell	81.5 ± 0.8 (4)
	24	2-cell	4- to 8-cell	85.5 ± 1.7 (29)
	24	Degenerate	Degenerate	100.3 ± 0.5 (11)
	-	Control Medium		100.6 ± 1.3 (5)
2	24	2-cell	4- to 8-cell	82.6 ± 1.4 (10)
	24[a]	4- to 8-cell	Morula	85.2 ± 3.4 (10)
	24[b]	Morula	Expanded blastocyst	94.0 ± 2.6 (10)
	24	Morula	Expanded blastocyst	81.9 ± 1.0 (6)
	24	Blastocyst	Expanded blastocyst	84.6 ± 0.6 (12)

[a] Total of 48 hr of culture
[b] Total of 72 hr of culture

A recent study[21] which attempted to validate such alterations as a response to early pregnancy showed their response to be quite variable so that, other than on a population basis, the peripheral platelet count has little predictive value.

Despite quite high rates of production of PAF by the early embryo,[18, 20] it could not be detected in the plasma or urine of women after successful embryo transfer. This is perhaps not surprising in view of the considerable dilution of embryo-derived PAF which would occur in blood, and also because of the well known lability of PAF in blood. Thus, direct measurement of PAF within body fluids may not be a useful means of monitoring pregnancy.

Production of PAF by the Early Embryo

Ryan et al.[22] undertook a study of the kinetics of secretion of PAF by embryos. 2-cell mouse embryos in culture produced increasing amounts of PAF for up to 6h and there was no further increase in PAF production after 24h. Degenerate embryos failed to produce PAF in culture (Table 1). After 72h in culture, 2-cell embryos developed to the blastocyst stage. However, a linear reduction in PAF activity was observed for culture media from three consecutive 24h periods during this period of culture in vitro. This appeared to be a factor of in vitro culture since morulae and blastocyst embryos (equivalent to a further 48h and 72h development, respectively), when freshly collected from the reproductive tract, produced the same level of PAF as two cells. Culture of 2-cell embryos in groups of 10-45 for 2-4h was carried out and then the amount of PAF determined in a bioassay compared with a standard curve produced by that of a synthetic PAF; 12.8 ± 1.0ng PAF/mouse embryo 24h was found to be produced.[22]

The level of PAF in culture medium is presumably a reflection of the balance between production and catabolism. The relative contributions of these to the loss of PAF activity are not clear at this time. Reduced secretion may be due to starvation of critical substrates or growth factors in vitro or possibly due to a negative feedback of PAF in the culture medium on further embryonic production. Degradation is known to occur by metabolism of PAF to its key intermediate,[7] lyso-PAF, by the enzyme acetylhydrolase, but its relative activity in the embryo is at this time not known.

It has been known for some time that culture of preimplantation embryos in vitro is detrimental to their subsequent pregnancy potential, and if PAF is an important embryonic product, then the adverse effect of culture on embryonic PAF production may well be related to the embryos reduced viability.

Similar observations have been made with human embryos produced by in vitro fertilization. Using a simple bioassay, it was found that embryos which resulted in pregnancy produced significantly more PAF in vitro than did embryos which failed to result in pregnancy following their transfer.[18] Of a further 85 embryos examined, 43% had a level of PAF that fell in the same range as the embryos that resulted in pregnancy. The production of PAF was related to the type of pharmacological treatment used to induce follicular development, the size and functional characteristics of the follicles from which they originated, the age of the culture medium in which they had developed, and the morphology and cleavage rate of the embryo itself.[18]

In a study using a highly and quantitative bioassay, similar results were observed.[19] The secretion of PAF by human embryos was found to be in the range 1.5–31.5 pmol/ml/24h and varied according to the morphology of the embryos. 4-cell embryos, considered morphologically normal at the time of transfer, produced more PAF than those which showed some defects (such as fragmentation or uneven blastomere size).

Studies currently under way with this bioassay show in a double-blind trial on nine separate occasions that there were no false-positive or -negative results. Under these conditions 228 media in which human 2- to 4-cell embryos had been cultured for 18h were tested for PAF. A wide range of PAF levels were observed from undetectable to 2700nM. Of these embryos, 121 (53%) gave a concentration significantly greater than the corresponding control culture tube 'PAF-positive' (Table 2). The average concentration of PAF in the media in which oocytes had failed to fertilize was 56.6 ± 8.0nM, not significantly different from the serum containing control culture tubes. The PAF-positive embryos were transferred to 30 patients who became pregnant and to 48 patients who failed to achieve pregnancy. Em-

Table 2. The Production of PAF by Embryos In Vitro as Assessed by In Vitro Platelet Aggregation Bio-assay,[19] Related to Whether or Not Pregnancy Occurred After Transfer

	Pregnant	Nonpregnant
Number of patients	30	48
Total number of embryos	94	134
Embryos transferred per patient (mean)	3.1	2.8
Number embryos which produce PAF (% of total number embryos)	55 (58.5%)	66 (49.2%)
PAF per embryo (nM) (mean ± s.e.m.)	295 ± 107*	76 ± 28
Number of patients with embryos transferred, none of which produced PAF (% of total in group)	4 (13%)	12 (26%)

* (p,0.03) Student t test

bryos transferred to women who achieved pregnancy produced more PAF (295nM) in vitro
than those transferred to women who failed to achieve pregnancy (74nM) (P < 0.03). The
proportion of PAF-positive embryos transferred to pregnant or to nonpregnant women was
not different (58.5% versus 49.2%). This suggests that while the production of PAF in itself was not an absolute marker for the capacity of embryos to produce PAF, enhanced production of PAF was associated with pregnancy. This might indicate that a certain minimal
production of PAF is required before pregnancy can result and this is consistent with the result of PAF supplementation data to be discussed later. If such a critical lower limit of
PAF production exists, its determination must await a more extensive investigation of the
correlation between the production of PAF and the subsequent success in cases where only
the embryo is transferred. This will take some time and require a multicentre study since
the number of cases in any one in vitro fertilization clinic where only one embryo is transferred is small.

The detection of PAF by bioassay might be compromised by a number of technical factors which include (i) enhanced metabolism of PAF prior to extraction, (ii) co-extraction
and purification of an inhibitor of the bioassay, and (iii) the binding of PAF to high-
affinity binding molecules in serum in some patients but not in others.

There was no evidence of enhanced activity of acetylhydrolase in any of the culture
tubes tested nor was this activity related to any of the endocrine parameters of the patients
undergoing treatment. Similarly, there was no evidence for a co-purifying inhibitor of the
bioassay, as has recently been reported in another experimental model.[23] High-affinity binding of PAF to serum molecules may occur. The organic extraction procedure utilized is designed primarily for recovery of PAF from its low-affinity binding to albumin. As noted
earlier a number of recent reports suggest specific binding to a number of serum proteins.
In particular, PAF and lyso-PAF were found to bind to lipoproteins.[24] It is possible that
the exaggerated steroid hormone production known to occur in women undergoing in vitro
fertilization might facilitate PAF binding to lipoproteins, since the sex steroids are known

to influence the class and type of lipoproteins in blood.[25] Furthermore, serum acetylhydrolase activity is positively correlated with the concentration of low-density lipoproteins and negatively correlated with high density lipoproteins.[26] Thus detailed analysis of the PAF carrier molecules may enhance the efficiency of extraction and purification.

At the 2- to 4-cell stage, some embryos may not produce PAF, but be still capable of resulting in viable pregnancies after transfer. Should this prove to be the case then clearly the use of PAF measurements in vitro as a measure of the pregnancy potential of the embryo is limited.

The results to date suggest that while PAF is clearly a marker for pregnancy potential and viability of early human embryos it cannot be said at this stage to be an absolute marker. Reduced or absent PAF production is associated with pregnancy failure but at this time there is not a 100% positive association between PAF production and pregnancy. An understanding of the role of PAF in pregnancy may help to define the likely diagnostic potential and limitations of embryo-derived PAF.

The Role of PAF in Early Pregnancy

PAF has a number of proinflammatory effects which implicates it as an attractive embryonic candidate for inducing the localized maternal response to implantation. It induces increased vascular permeability, causes vasodilatation and activates a number of inflammatory cells.[6] To investigate the role of PAF in early pregnancy, two inhibitors of PAF's activity were used to inhibit its actions during the preimplantation phase of pregnancy. Their effect on the establishment of pregnancy was monitored.

Iloprost [5-(E)(1S,5S,6R,7R)-7-hydroxy-6-(E)-(3S,4RS)-3-hydroxy-4-methyl-oct-1-en-6-yn-yl-bicylo3.3.0-octano-3-yliden-pentanoic acid) is a stable analogue of prostacyclin which stimulates denylcyclase activity leading to increased cellular cAMP and reduced cytoplasmic calcium. These effects inhibit the action of PAF in most cells. SRI 63-441 ([cis-(+/-)-1-[2-hydroxy[tetrahydro-5-[octadecylaminocarbonyl) oxy]methyl]-furan-2-yl]methoxy-phosphinyloxy]ethyl]-quinolinium hydroxide) is a specific PAF-receptor antagonist with potent pharmacological actions.[27] Both of these agents when administered to pregnant mice (on day 1–4 of pregnancy) significantly reduced the number of implantation sites.[28] They did not act by having gross toxic effects on the embryo either in vitro or in vivo. They appear to act by prevention of implantation rather than its subsequent disruption. The earliest sign of implantation is a local increase in the vascular permeability surrounding the site of implantation. This can be visualized by high molecular weight dyes. PAF inhibitors caused significantly fewer dye bands in the uterus on day 4 of pregnancy. There was no evidence of an effect of the antagonists on luteal endocrine profile.

In a further study,[29] experiments were performed to determine whether the PAF antagonist acted primarily at the maternal or embryonic level. The experiment used reciprocal em-

Table 3. The Percentage of Blastocysts that Implant after Transfer to Day 4 Pseudopregnant Recipients when either the Donor or the Recipient Females were Treated with the PAF-antagonist SRI 63-441). (From Spinks et al.)[29]

	Treatment of donors		Treatment of recipients	
	Control	SRI 63-441	Control	SRI 63-441
Implantation rate (%)	40.0	14.4*	42.3	36.4
No. of mice	11	9	11	11

* Logistic regression established a significant (p < 0.04) effect of treating the donor female with SRI 63-441.

bryo transfers, in which blastocysts from mice treated with PAF-antagonist (SRI 63-441) or saline (controls), from days 1–4 of pregnancy, were transferred to day 4 pseudopregnant recipients which were also treated with SRI 63-441 or saline on days 1–4 of pregnancy. The antagonist (40μg was administered at 16.00h on day 1 and at 09.00h on days 2–4 of pregnancy. The percentage of the transferred embryos which implanted was determined on day 8 of pregnancy. Treatment of the donor female had a dramatic effect on the implantation rate, resulting in a reduction of 64% (from 40% to 14.3%, P < 0.04), while treatment of the recipient female had no significant effect (Table 3). These results suggest that PAF-antagonists affected implantation at the embryonic level and did not adversely affect maternal physiology.

Such a conclusion was supported by the observations that pseudopregnant females treated with the antagonist SRI 63-441 were still responsive to an artificial decidual stimulus.[29]

The Effect of PAF on the Preimplantation Embryo

There are a number of lines of evidence to suggest that embryo-derived PAF may have its effects by acting as a direct autocrine growth factor for the preimplantation stage embryo.

The production of $^{14}CO_2$ from labelled lactate or glucose was measured to determine whether PAF had a direct effect on embryonic metabolism. Incubation of 2-cell embryos in medium supplemented with 1.0μg PAF/ml resulted in an 18% increase in CO_2 production from lactate. Higher doses resulted in a more variable response. The same dose (1μg/ml) of PAF was responsible for up to a 90% increase in the amount of CO_2 produced from glucose.[30] This was inhibited by the PAF antagonist SRI 63-441 (Ryan and O'Neill, unpublished results). Culture of embryos in medium supplemented with PAF did not significantly increase the proportion of embryos developing to the expanded blastocyst stage, however, there was a modest (P < 0.05) increase in the number of cells in each embryo (Table 4).

Table 4. Cell Number and Implantation, Foetal Development and Resorption Rates of Embryos Cultured in the Presence of 0, 0.1 or 1.0 µg PAF/ml (From Ryan et al.) [30]

Treat.	Cell No. ± sem (No. embryos)	Implan. Rate Day 8 (%)	Foet. Devel. Day 17 (%)	Res. Sites Day 17 (%)
Cont-A	*73.1 ± 2.1 (110)	82/200 (41)	45/132 (34)	21/66 (32)
0.1 PAF	78.2 ± 2.4 (97)	127/200 (63)[c]	59/132 (45)	24/83 (29)
Cont-B	*73.1 ± 2.1 (110)	86/160 (54)	53/132 (54)	9/62 (15)
1.0 PAF	81.9 ± 2.8 (79)[a]	107/160 (67)[b]	60/132 (45)	13/73 (18)

*same control group
values with superscripts differ significantly from controls:
[a] = $p < 0.05$;   [b] = $p < 0.025$;   [c] = $p < 0.001$

Cellular mechanisms by which PAF exerts this effect on carbohydrate metabolism have not yet been studied in detail. Its inhibition by the PAF antagonist SRI 63-441 suggests that PAF is acting via membrane receptors although these are yet to be identified on the embryo. In other cell types,[6] PAF binding to cell membranes induces cellular responses following transduction of the signal via membrane guanyl nucleotide regulatory proteins promoting generation of inositol triphosphate and diacylglycerol. Inositol triphosphates mobilize calcium and diacylglycerol promotes protein kinase C activity. Protein kinase C may have a role in enhanced carbohydrate metabolism. It effects glucose transfer in adipocytes and glycogenolysis of hepatocytes. In particular, phosphofructokinase may be a substrate for protein kinase C. Phosphofructokinase has been implicated as one of the key regulatory enzymes of early embryonic carbohydrate metabolism.

To assess the effect of increasing the available PAF on the pregnancy potential of cultured embryos, expanded blastocysts cultured from the 2-cell stage in medium with or without PAF were transferred to pseudopregnant recipients and implantation rates assessed on day 8 of pregnancy.[30] Supplementation of medium with 0.1µg PAF/ml resulted in a substantial increase in the implantation rate above control levels (Table 4). By taking into account the between animal variation, the observed effect could be considered to be due solely to enhanced viability of embryos rather than to any maternal effects following transfer.

This direct autocrine action was not restricted to the mouse. Supplementation of culture media with PAF also effected the metabolism of human polyploid embryos in vitro[31] (Table 5) as was the effect of PAF supplement of culture media on the pregnancy potential of human embryos produced by IVF[31] (Figure 3). Pre-embryo culture medium was supplemented with 0 (control), 0.186, 0.93, or 1.49 µmol/PAF. Pre-embryos were transferred to PAF containing medium 15–17h after insemination (i.e. just before syngamy) for 24h and then transferred to the uterus. For 185 women receiving control pre-embryos, the pregnancy rate (positive beta human chorionic gonadotrophin per oocyte retrieval) was 10.2%, while 166 women who received PAF treated pre-embryos (all concentrations combined)

Table 5. Oxidative Metabolism of 0.28mM D (U-^{14}C) Glucose and 1.12mM DL (1-^{14}C) Lactate by Polyploid Human Embryos in the Presence of Platelet Activating Factor. (From O'Neill et al.)[31]

Energy substrate	Dose PAF (μM)	No. embryos	pmoles CO_2/embryo/h (± s.e.m.)
Glucose	0	14	1.58 ± 0.37
	1.86	11	3.00 ± 0.57*
Lactate	0	6	2.00 ± 0.20
	1.86	4	2.66 ± 0.50

*significantly different from 0 μM PAF ($P < 0.05$) using Student's t test

achieved a pregnancy rate of 17.5%. This difference was significant. The pregnancy rates per pre-embryo transferred were 6.1% and 9.4% for the control and PAF groups, respectively. The percentage of positive pregnancy tests that resulted in a viable pregnancy (presence of fetal heart at eight weeks) was 79% in the controls and 76% in the PAF group. There was no difference in the average number of embryos transferred in either group.

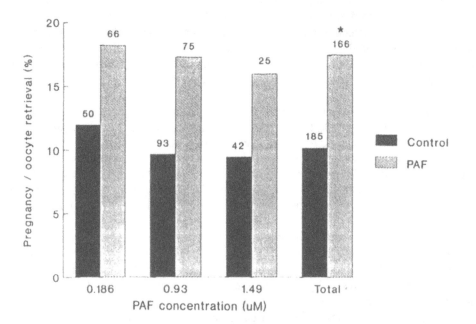

Fig. 3. The pregnancy rate for women after IVF who had pre-embryos transferred all of which had been cultured in the presence of 0 (control) 0.186, 0.93 or 1.9μM PAF. The numbers above each bar represents the number of women in each group * $p < 0.05$ by chi-square analysis (From O'Neill et al.).[31]

Although PAF promotes embryonic metabolism throughout the preimplantation phase, its actions appear to be particularly crucial at around the time of implantation in the mouse. Embryo transfer experiments with PAF-antagonist treated animals show that the antagonists exert their contragestational effect on the day of implantation.[29] The pre-implantation mouse embryo grows from the one-cell through to the blastocyst stage in defined culture medium without protein supplementation. Further development and implantation of the embryo requires metabolic activation. This leads to the outgrowth of the trophectoderm onto the substrate and differentiation of the inner cell mass. The metabolic activation can be induced by the provision of whole serum in the medium. Under these conditions the addition of a PAF-antagonist inhibited outgrowth[29] in a dose-dependent manner and at doses that did not affect development from the 1-cell to blastocyst stage.[28] This process is generally recognized to be analogous to implantation and its inhibition in vitro by PAF antagonists suggest that an essential role of embryo-derived PAF in the establishment of pregnancy may be as the autocrine trigger for blastocyst activation and the resulting differentiation of the trophectoderm into invasive trophoblast.

Maternal Actions of Embryo-Derived PAF

Apart from this apparent essential autocrine action of embryo-derived PAF, it also clearly exerts maternal effects. PAF induced platelet activation in maternal reproductive tract vasculature was reported for mice. A significant thrombocytopenia occurred in early pregnancy in mice, humans, the marmoset monkey[32] and cows.[33] Platelets are known to produce a range of potent growth factors and vasoactive agents. Indeed, blastocyst outgrowth in vitro was shown to require platelet dependent serum factors[34] and platelet factor IV supports blastocyst attachment and outgrowth.[35] Products of platelet activation also stimulates the secretion of progesterone by bovine luteal cells in vitro[36] while PAF directly stimulates P_4 production by human granulosa cells,[37] possibly suggesting a luteotrophic role for PAF.

PAF may also affect luteal function in a more indirect fashion. In vitro PAF caused a dose-dependent increase in the synthesis of PGE_2 by human endometrial granular tissue,[38] but had no effect on PGF_2 secretion. The resulting reduction of PGF_2/PGE_2 may be of benefit for luteal maintenance. In the bovine, both PAF and embryo conditioned culture medium significantly suppressed PGF_2 and increased PGE_2 secretion by bovine endometrial explants from nonpregnant animals. At the highest dose (5 μg PAF/ml) the prostaglandin secretion profile was the same as for explants from pregnant animals.[39] These effects were not caused by lyso-PAF and were inhibited by the PAF-antagonist WEB 2086. Endometrium from day 2 pregnant rabbits were shown to express high affinity binding sites for PAF.[40]

A similar inhibitory effect on PGF production was demonstrated in situ in sheep.[41] In ovariectomised sheep given an estrogen and progesterone regimen to mimic the normal luteal phase, oxytocin administered intravenously caused significant elevations in the circulating PGFM concentration. The intrauterine administration of PAF markedly suppressed such PGFM secretion. This suppression is consistent with an anti-luteolytic effect of PAF. Such an action was consistent with the observation that intrauterine instillation of PAF into the lumen of cycling sheep[41] and cows[33] caused a significant increase in the oestrous cycle length in such animals.

As well as these effects of exogenous PAF on uterine function, it has also been shown that the uterus of rats, rabbit and humans contain substantial concentrations of PAF. This PAF is apparently not secreted and a role for this uterine PAF is yet to be defined.

CONCLUSIONS

There is an accumulating weight of evidence implicating embryo-derived PAF in a variety of the important physiological adaptations in early pregnancy. To date its role as an autocrine growth factor for the preimplantation embryo appears to be essential for the establishment of pregnancy. However, it has a variety of maternal effects both by its direct action on a number of cells and indirectly by its activation of cells such as platelets. The range of its action and their control is still to be clearly defined (Figure 4).

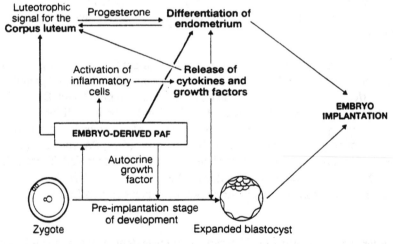

Fig. 4. This diagram shows the possible major sites of action of embryo-derived PAF, and the interaction of these functions in support of the establishment of pregnancy.

ACKNOWLEDGEMENTS

Some of the studies reported in this review were funded by the Australian NH and MRC and the Special Program of Research, Development and Research Training in Human Reproduction, World Health Organisation. I thank Alaina Ammit, Kristine Battye, Melanie Collier, John Ryan and Neil Spinks for their collaboration with many of these studies and Professor Douglas Saunders for his encouragement.

REFERENCES

1. H. K. Mangold, Ether lipids: Relic of the past and promise for the future, *La Rivista Delle Sostanze Grasse*, 65:349 (1988).
2. P. M. Henson, Release of vasocative amines from platelets induced by sensitized mononuclear leukocytes and antigen, *J. Exp. Med.* 131:287 (1970).
3. J. Benveniste, P. M. Henson, and C. G. Cochrane, Leukocyte-dependent histamine release from rabbit platelets. The role of IgE, basophils and a platelet-activating factor, *J. Exp. Med.*, 136:1356 (1972).
4. M. L. Blank, F. Snyder, L. W. Byers, B. Brooks, and E. E. Muirhead, Antihypertensive activity of an alkyl ether analog of phosphatidylcholine, *Biochem. Biophys. Res. Commun.*, 90:1194 (1979).
5. E. E. Muirhead, L. W. Byers, D. Desiderio, K. A. Smith, R. L. Prewitt, and B. Brooks, Alkyl ether analogs of phosphatidylcholine are orally active in hypertensive rabbits, *Hyperten.*, 3:107 (1981).
6. P. Braquet, L. Touqui, T. Y. Shen, and B. B. Vargaftig, Perspectives in platelet-activating factor in research, *Pharmacol. Reviews*, 39:97 (1987).
7. F. Snyder, The significance of dual pathways for the biosynthesis of platelet activating factor: 1-alkyl-2-lyso-sn-glycero-3-phosphate as a branch point. *In:* "New Horizons in Platelet Activating Factor Research," C. M. Winslow and M. Lee eds., John Wiley and Sons, Chichester (1987).
8. J. V. O'Fallon, and R. Wright Jr., Quantitative determination of the pentose phosphate pathway in preimplantation mouse embryos. *Biol. Reprod.*, 34:58 (1986).
9. J. C. Ludwig, C. L. Hoppens, L. M. McManus, G. E. Mott and R. N. Pinckard, Modulation of platelet activating factor synthesis and release from human polymorphonuclear leukocytes (PMN) - role of extracellular albumin. *Archs. Biochem. Biophys.* 241:337 (1985).
10. M. L. Blank, T. Lee, V. Fitzgerald, and F. Snyder, A specific acetylhydrolase for 1-alkyl-2-acetyl-sn-glycero-3-phosphocholine (a hypotensive and platelet activating lipid). *J. Biol. Chem.* 256:175 (1981).
11. J. T. O'Flaherty, J. F. Redman, J. D. Schmitt, J. M. Ellis, J. R. Surles, M. H. Marx, C. Piantadosi, and R. L. Wykle, 1-0-alkyl-2-N-methylcarbamyl-glycerophosphocholine: a biologically, potent, non-metabolizable analog of platelet activating factor. *Biochem. Biophys. Res. Commun.* 147:18 (1987).
12. P. H. McNamara, K. R. Brouwer, and M. N. Gillespie, Autacoid binding to serum proteins. Interactions of platelet activating factor (PAF) with human serum alpha-1-acid-glycoprotein. *Biochem. Pharmacol.* 35:621 (1986).
13. M. Matsumoto, and M. Miwa, Platelet activating factor-binding protein in human serum. *Adv. Prost. Thromb. Leuk. Res.*, 15:705 (1985).
14. R. Kumar, R. J. King, H. M. Martin, and D. J. Hanahan, Metabolism of platelet activating factor by type-II epithelial cells and fibroblasts from rat lung. *Biochim. Biophys. Acta.*, 917:33 (1987).

15. H. Lachachi, M. Plantavid, M. F. Simon, H. Chap, P. Braquet, and L. Douste-Blazy, Inhibition of transmembrane movement and metabolism of platelet activating factor (PAF-acether) by a specific antagonist, BN 52021, *Biochem. Biophys. Res. Commun.*, 132:460 (1985).

16. C. O'Neill, A. M. Ammit, R. Korth, and S. Flemming, Inhibitors of platelet activation as well as PAF receptor antagonists inhibited catabolism of PAF by washed rabbit platelets. *Lipids*, (Submitted).

17. C. O'Neill, Thrombocytopenia is an initial maternal response to fertilisation in mice. *J. Reprod. Fert.*, 73:567 (1985).

18. C. O'Neill, A. A. Gidley-Baird, I. L. Pike, and D. M. Saunders, A bioassay for embryo-derived platelet activating factor as a means of assessing quality and pregnancy potential of human embryos, *Fertil. Steril.*, 47:969 (1987).

19. M. Collier, C. O'Neill, A. J. Ammit, and D. M. Saunders, Biochemical and pharmacological characterisation of human embryo-derived platelet activating factor, *Human. Reprod.*, 3:993 (1988).

20. C. O'Neill, Partial characterisation of the embryo-derived platelet activating factor in mice, *J. Reprod. Fert.*, 75:375 (1985).

21. C. O'Neill, M. Collier, and D. M. Saunders, Embryo-derived PAF: Its diagnostic and therapeutic future, *Ann. N.Y. Acad. Sci.*, 541:398 (1988).

22. J. P. Ryan, N. R. Spinks, C. O'Neill, A. J. Ammit, and R. G. Wales, Platelet activating factor (PAF) production by mouse embryos in vitro and its effects on embryonic metabolism, *J. Cell. Biochem.*, 40:387 (1989).

23. M. Miwa, C. Hill, R. Kumar, J. Sugatani, M. S. Olson, and D. J. Hanahan, Occurrence of an endogenous inhibitor of platelet-activating factor in rat liver, *J. Biol. Chem.*, 262:527 (1987).

24. J. Benveniste, D. Nunez, P. Dunez, R. Korth, J. Bidault, and J. C. Fruchart, Performed PAF-acether and lyso PAF-acether are bound to blood lipoproteins, *FEBS Lett* 226:371 (1988).

25. M. J. Tikkanen, and E. A. Nikkila, Regulation of hepatic lipase and serum lipoproteins by sex steroids, *Am. Heart. J.*, 113:562 (1987).

26. D. M. Stafforini, T. M. McIntyre, M. E. Carter, and S. M. Prescott, Human plasma activating factor acetylhydrolase. Association with lipoprotein particles and role in the degradation of platelet-activating factor, *J. Biol. Chem.*, 262:4215 (1987).

27. D. A. Handley, J. C. Tomesch, and R. N. Saunders, Inhibition of PAF-induced systemic responses in the rat, guinea-pig, dog and primate by the receptor antagonist SRI 63-441, *Thromb. Haemost.*, 56:40 (1986).

28. N. R. Spinks, and C. O'Neill, Antagonists of embryo-derived platelet activating factor prevent implantation of mouse embryos, *J. Reprod. Fert.*, 84:89 (1988).

29. N. R. Spinks, J. P. Ryan, and C. O'Neill, Antagonists of embryo- derived platelet activating factor act by inhibiting the mouse embryos capacity for implantation, *J. Reprod. Fert.*, (In Press).

30. J. P. Ryan, N. R. Spinks, C. O'Neill, and R. G. Wales, Implantation potential and foetal viability of embryos cultured in the presence of platelet activating factor, *Proc. Aust. Soc. Reprod. Biol.*, 20:47 (1988).

31. C. O'Neill, J. P. Ryan, M. Collier, D. M. Saunders, A. J. Ammit, and I. L Pike, Supplementation of IVF culture media with platelet activating factor (PAF) increased the pregnancy rate following embryo transfer, *Lancet*, (In Press).

32. J. P. Hearn, G. E. Webley, A. A. Gidley-Baird, Embryo implantation in primates, *In:* "Implantation, Biological and Clinical Aspects," M. Chapman, G. Grudzinskas, and T. Chard, eds., Springer-Verlag, London (1988).

33. W. Hansel, Establishment of pregnancy: regulation of the corpos luteum, *Proc. 11th Int. Cong. Anim. Reprod. Artif. Insem.*, 5:61 (1988).

34. C. O'Neill, The role of blood platelets in the establishment of pregnancy, *In:* "Pregnancy Proteins in Animals," J Hau, ed., Walter de Gruyter, Berlin (1986).

35. M. C. Farach, J. P. Tang, G. L. Decker, and D. D. Carson, Heparin/Heparan sulfate is involved in attachment and spreading of mouse embryos in vitro, *Develop. Biol.*, 123:401(1987).
36. P. J. Battista, H. W. Alila, C. E. Rexroad, and H. Hansel, The effect of platelet-activating factor and platelet-derived compounds on bovine luteal cell progesterone production, *Biol. Reprod.*, 40:769 (1989).
37. C. O'Neill, Invited Review: Embryo derived platelet activating factor: A pre-implantation embryo mediator of maternal recognition of pregnancy, *Domest. Anim. Endocrin.*, 4:69(1987).
38. S. K. Smith, and R. W. Kelly, Effect of platelet-activating factor on the release of PGF-2a and PGE-2 by separated cells of human endometrium, *J. Reprod. Fert.*, 82:271 (1988).
39. T. S. Gross, W. W. Thatcher, C. O'Neill, and G. Danet-Desnoyers, Platelet activating factor alters the dynamics of prostaglandin and protein secretion by endometrial explants from pregnant and cyclic cows at day 17 post-estrus, *J. Reprod Fert.* (Submitted).
40. G. B. Kudolo, and M. J. K. Harper, Binding parameters of rabbit uterine PAF-receptors: Pilot study, *Biol. Reprod* (Suppl. 1):153 (1988).
41. K. M. Battye, G. Evans, and C. O'Neill, A role for platelet activating factor in maintenance of the ovine corpus luteum, *Proc. Aust. Soc. Reprod. Biol.* 21:32 (1989).

REGULATION OF TROPHOBLAST INVASION: THE REQUIREMENT FOR PROTEINASE CASCADE

S. Yagel[1,2], P.K. Lala[1], and D. Hochner-Celnikier[2]

[1] Department of Anatomy, University of Western Ontario, London, Canada, and
[2] Department of Obstetrics and Gynecology
Hadassah Mt Scopus, Hebrew University, Jerusalem

INTRODUCTION

Proteolytic degradation of extracellular matrix in tumor invasion has been widely documented.[1-3] A large number of reports have demonstrated increased amounts of proteinase activities in malignant tumor tissues as compared with control tissues. In some cases a direct correlation between the metastatic potential of tumor cells and the production of certain proteinases has been claimed.[1,4,5] Enzymes from the four main classes of proteinases, i.e. serine, cystein, acid and metalloproteinases have been associated with the invasive phenotype.

Activity or collagenase type IV and other metalloproteinases is tightly balanced in vivo and is inhibited by TIMP (tissue inhibitor of metalloproteinases). Down regulation of TIMP expression by antisense RNA to TIMP rendered the cell invasive, tumorigenic and metastatic in athymic mice.[6]

We used the amnion invasion assay,[7] in which amniotic membrane is denuded from its amniocytes and the examined cells are prelabelled with ^{125}I-iododeoxyuridine, to differentiate between the spontaneous and artificial metastatic potential, namely only cells that were able to complete the metastatic cascade were invasive in vitro.[8]

Degradation of extracellular matrix by trophoblast has been documented and suggested as an in vitro model for implantation[9,10] (Figure 1). High levels of plasminogen activator activity were detected in trophoblast-conditioned media[11,12] and urokinase type plasminogen activator production may be regulated by cAMP.[13]

Advances in Assisted Reproductive Technologies, Edited by
S. Mashiach *et al.*, Plenum Press, New York, 1990

However, it is not clear whether the invasive ability of the normal nonmetastatic trophoblast cells is an intrinsic property of these cells, independent of the microenvironment provided by the hormone-primed uterus and the accompanying decidual reaction, and if so, whether they share with tumor cells the same molecular mechanisms for invasion. In this paper, and in earlier studies[14] we address these questions by measuring the ability of in vitro grown first trimester human trophoblasts to invade human amniotic membrane under various conditions. The magnitude of this invasion was compared with that of several tumor lines: the JAR human choriocarcinoma line, the highly metastatic murine melanoma line B16F10, and the nonmetastatic murine mammary adenocarcinoma line C10. The results of these studies indicate that invasiveness is an intrinsic property of trophoblasts and that like metastatic tumor cells a proteinase cascade is required for basement membrane invasion.

MATERIALS AND METHODS

Cell Lines and Media

JAR human choriocarcinoma line was purchased from ATCC (Rockville, MD).

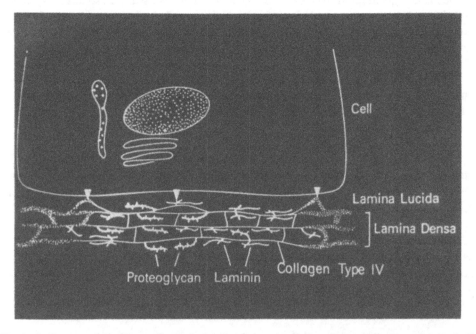

Fig. 1. A diagram of invasive cell attached to basement membrane. Proteolytic degradation of all components of the extracellular matrix is imperative for basement membrane penetration.

B16F10 murine melanoma cell line was provided by Dr. I.J. Fidler (Department of Cell Biology, M.D. Anderson Hospital, Houston, TX). This line has a high lung metastasizing ability. C10 murine mammary adenocarcinoma line was clonally derived in our laboratory and is incapable of metastasizing to the lung or any other organ from the primary transplant site. Cell lines were grown in the same medium as described later for trophoblasts.

The trophoblast cell lines TR-30-4 and TRJ were established from first trimester human placenta recovered from routine elective termination of pregnancy, as described elsewhere.[15,16] In brief, the placenta and the attached chorion, free from contaminating maternal decidua and blood clots, were rinsed repeatedly with phosphate buffered saline (PBS), and chorionic villi were minced into 1–2 mm pieces with scissors to isolate villus fragments. These fragments were cultured in RPMI 1640 medium (Grand Island Biological Co., NY) containing 7% FCS, streptomycin 20 mg/ml; penicillin 500 U/ml and amphotericin-B 25 μg/ml in closed culture flasks (Nucleon 25 cm^2, Nunc Denmark) for several days until nonadherent cells could be removed and discarded. Cultures were expanded for 1–2 weeks in these plastic tissue culture flasks in fresh medium. These cells grew as mixtures of mononuclear and multinuclear (syncytium), attached to the plastic with an approximate doubling time of 72 hr. The trophoblast lines were maintained by serial in vitro passages at 4–8 days intervals and these lines usually survived for eight to 13 passages. Their trophoblastic character was ascertained with structural and functional markers. With immunoperoxidase labelling methods, more than 85% of the cells stained positive for cytokeratin and all cells were negative for vimentin, indicating an epithelial origin. More than 75% of the cells reacted with a human anti-alpha chain chorionic gonadotropin (alpha-hCG) monoclonal antibody and 100% reacted with polyclonal anti-hCG antibody. The cells were negative for 63D3 antigen, which is a surface marker for human macrophages, but more than 90% of the cells expressed surface Trop-1 and Trop-2 markers characteristic of human trophoblast cells. RPMI 1640 supplemented as described above was used for the trophoblasts as well as for the tumor cell lines.

Invasion Assay

Subconfluent cultures of trophoblasts and tumor cells were incubated at 37°C for 24 hr in the presence of 0.3 μci/ml of ^{125}I-dUR (New England Nuclear Corp.; 2200 Ci/mole) in complete medium. The cultures were then washed, trypsinized and counted by hemocytometer. The specific labelling ranged between 0.96–1.11 dpm/cell under these conditions.

The amnion invasion assay was performed according to the method of Mignatti et al.[7] with the modifications as indicated below. Sterile silicone rubber rings were attached to the bottom of the culture wells (35mm diam; 6 wells per plate; Falcon, Jersey, NJ) with sterilized non-toxic silicone lubricant (Down Corning Corp. Midland, MI). RPMI medium was added to the wells to fill the inner diameter of the rings. The amnion membranes were fastened to the Teflon rings (invasion chambers) and then placed onto the silicone rubber supports in the culture plates with the stromal aspect facing down. They were then denuded of their epithelium by treatment with 0.25% of ammonium hydroxide followed by scraping and three washings with PBS. Two hundred microliters of RPMI medium plus 7% FCS, with or without enzyme inhibitors were added in the upper chambers, and the plates were incubated at room temperature for 60 min. ^{125}I-dUR-labelled. 1.5×10^5 cells suspended in 0.5 ml were added to the upper chamber of each well. A further 3ml of RPMI and 7% FCS were added, so that the level of fluid was the same inside and outside the Teflon rings (Figure 2).

The tissue culture plates were incubated at 37°C in 5% CO_2 atmosphere saturated with water. At the end of the 6-day incubation, the medium in the upper and lower chambers was removed. The viability of the cultured cells was determined by the trypan blue exclusion test. The layer of labelled cells growing on the epithelium-free amnion was washed twice with 1 ml of PBS and the washings were collected in a tube. A 4% Nadeoxycholate (DOC) solution in PBS was then added onto the upper chamber and the cell layer removed by gently scraping with a rubber policeman. The amnion was then washed three times with PBS, and the DOC and PBS washings were pooled in the same tube. The membrane was finally detached from the Teflon ring, washed again with PBS and put into a tube. The radioactivity associated with the amnion, the supernatant and the pellet of the upper and lower chamber medium, the PBS and DOC+PBS washings, were determined individually. The radioactivity associated with the amnion was expressed as percent of the total radioactivity. For some experiments, a 0.2 μm millipore filter was placed on the stromal aspect of the

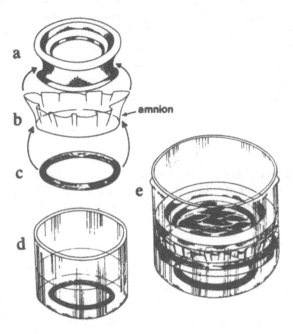

Fig. 2. The invasion and attachment chamber. The amnion is fastened to Teflon rings and then placed onto silicon rubber supports in culture flasks.

amnion, and the cells were recovered on the filter. This was done in order to examine the ability of the cells to cross the amniotic membrane.

Cell-Adhesion Assay

^{125}I-dUR-labelled cells were applied to amnion membranes in RMPI plus 0.1% FCS as described above. After 24 hr the non-attached cells were removed from the wells by washing and the remaining attached cells were harvested by trypsinization. Cells removed by trypsinization were expressed as a percentage of the total cells recovered. For cell adhesion to Matrigel, radiolabelled cells were applied in 1 ml of serum free medium to lymbro wells coated with 200 μl/well of Matrigel (Collaborative Research, Bethesda, MD).

Proteinase Inhibitors and Antibodies

Rabbit antimouse urokinase (uPA) antiserum was a gift from Dr. D. Belin, University Medical Center, Geneva, Switzerland, affinity-purified antimouse uPA from Dr. L. Ossowsky (Rockefeller University, New York, N.Y.); recombinant human collagenase inhibitor (HCI) from Dr. Carmichael (Synergen Inc., Boulder CO). Monoclonal human antiplasminogen antibody was purchased from ICN Biomedial Inc. (Lisle, IL); 1–10 phenanthroline, epsilon amniocaproic acid (EACA), mersalyl (salyrganic acid) and trasylol (apronitin) were purchased from Sigma Inc. (St. Louis, MO).

RNA Isolation, Electrophoresis, and Blot Hybridization

RNA was isolated from cells using the guanidine isothiocyanate-cesium chloride method. Equal amounts of RNAs (10 μg), quantified by absorbance at 260 nm, were denatured in MOPS/formaldehyde and electrophoresed in 0.8% agarose/formaldehyde gels and transferred to Gene plus paper (3M Scotch) by standard procedures.

Hybridizations with nick-translated cDNA probe were performed according to standard method. Dried filters were first moistened with $6 \times$ SSC (1X/SSC = 0,15 M NaCl and 0.015 M sodium citrate), and then placed in bags with 5–8 ml of prehybridization solution (50% formamide, 5XSSC, 0.1% each of Ficoll, polyvinyl-pyrollidone, bovine serum albumin, SDS, and 250 ng/ml denatured salmon sperm DNA) and incubated at 42°C for 4 h. This solution was replaced with fresh solution to which was added nick-translated cDNA (5 - 10×10^6 cpm/ml) and nick-translated lambda DNA. Hybridizations were continued for 18 h. Following hybridization, the filters were washed twice for 30 min at room temperature in 2 X SSC and 0.1% SDS, followed by two 45-min washes at 65°C in 1 X SSC and 0.1% SDS. After the washes the filters were blotted, wrapped in plastic, and placed with x-ray film (Kodak X-Omat) for autoradiography at -20°C. Autoradiograms were scanned with a Kontes densitometer (Vineland, NJ) to determine relative changes in mRNA levels.

A plasmid containing cDNA for human uPA was generously provided by the International Institute of Genetics and Biophysics Naples, Italy, and a plasmid containing cDNA for human collagenase type I, was a kind gift of Prof. P. Herrlich, the Institute of Genetics and Toxicology, Karlschuhe, W. Germany.

Scanning electron microscopy was performed on materials fixed with 2% glutarladehyde.

RESULTS

Attachment and Invasion

The basement membrane of the amnion provided an excellent substrate for the attachment and spreading of all cell types examined in this study. As shown in Figure 3, 48 hr after seeding of the trophoblast cells on the amnion basement membrane, the cells had covered the entire membrane. At 144 hr of incubation, some cells were seen in the stroma deeper to the basement membrane (Figure 4). When a millipore support was provided under the amnion trophoblast cells appeared on the filter after six days (Figure 5). The radioactivity associated with the amnion increased between 24 and 72 hr from 0.85 ± 0.1 to 3.1 ±

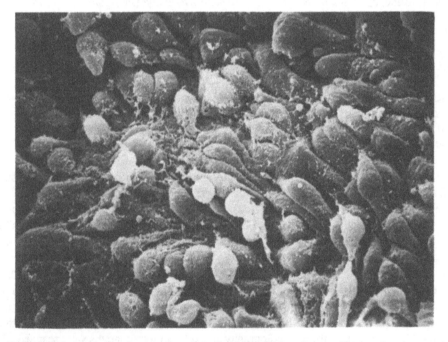

Fig. 3. Scanning electron micrographs of trophoblasts 72 hr after seeding on human amniotic membrane. The trophoblasts are seen to cover the entire membrane area. × 146.

0.12% for the JAR cells and from $0.7 \pm 0.01\%$ to $2.85 \pm 0.19\%$ for the B16F10 cells (positive controls), and it was not more than $0.75 \pm 0.05\%$ for the C10 cells (negative control). For the trophoblast cell line this increased between 24 and 144 hr from $0.2 \pm 0.05\%$ to a maximum of $3.9 \pm 0.43\%$. The variation from one trophoblast line to the other ranged 0–12% during the periods tested (data not shown).

Effect of Proteinase Inhibitors and Antibodies on Attachement and Invasion by Trophoblast, B16F10, and JAR Cell Lines

The results shown in Table 1 demonstrate the effects of the following inhibitors and antibodies on amnion invasion by trophoblast lines: HCI, a collagenase inhibitor; 1–10 phenanthroline, a metalloproteinase inhibitor; trasylol and EACA, serine proteinase inhibitors; and human antiplasminogen antibody. All these agents at the concentrations reported in Table 1 totally blocked amnion invasion by trophoblast cells. The radioactivity associated with the amnion in the presence of the inhibitors was not significantly different from that of the negative control, the non-invasive C10 line. The concentrations used, as described were not cytotoxic and did not affect attachment of cells to the basement membrane.

Results similar to those shown for trophoblast cells were obtained with JAR and

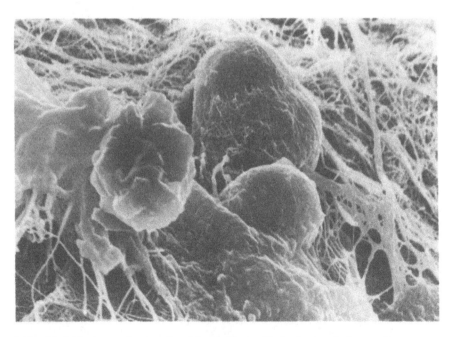

Fig. 4. Scanning electron micrographs of trophoblast cells in the amniotic membrane stroma after basement invasion. At 144 hr after seeding, trophoblast cells can be seen in the stroma at different stages of egress. × 1460.

Table 1. Effect of Metallo- and Serine-Proteinase Inhibitors on Invasion by Trophoblast. JAR and B16F10 Lines.

Inhibitor/Line	TR-J	TR-30-4	JAR	B16F10
Control	4.2 ± 0.24	3.8 ± 0.29	3.1 ± 0.16	2.8 ± 0.12
O-phenan* 10 µg/ml	0.75 ± 0.04	0.61 ± 0.02	0.6 ± 0.2	0.65 ± 0.27
HCl 5 µg/ml	0.3 ± 0.03	0.4 ± 0.009	0.6 ± 0.04	ND
Anti-uPA (purified)10 µg/ml	ND	ND	ND	0.25 ± 0.9
Anti-uPA (antiserum)10 µl/mg	ND	ND	ND	0.3 ± 0.01
Trasylol 200 U/ml	0.2 ± 0.0004	0.1 ± 0.03	0.3 ± 0.009	0.5 ± 0.02
EACA 500 µg/ml	0.1 ± 0.003	0.2 ± 0.007	0.4 ± 0.03	0.8 ± 0.05
Antiplasminogen antibody 2.2 µg/ml	0.1 ± 0.003	0.3 ± 0.003	0.1 ± 0.006	0.8 ± 0.02

Metallo-proteinase and serine-proteinase inhibitors were added to the invasion wells before the cells were plated on the basement membrane, as described under Materials and Methods. End point for assay was 144 hr for trophoblasts, 72 hr for JAR cells and B16F10 cells. For each inhibitor 3 to 6 samples were assayed. The results represent means ± SE.

*O-phenan = O- phenanthroline; ND = no determined

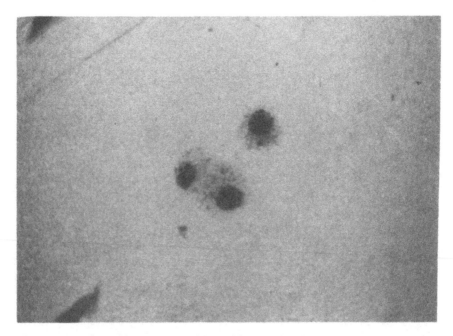

Fig. 5. Trophoblast cells recovered on a milipore filter underlying the amnion shown at 144 hr after culturing on the amnion basement membrane. The trophoblasts appeared on the filter in groups but single cells could also be seen. × 292.

Table 2. Effect of Mersalyl on Trophoblast and B16F10 Cell invasion through amniotic Membrane[1]

Inhibitor/Line Accelerator	TR-J	TR-30-4	B16F10
Control	4.2 ± 0.24	3.8 ± 0.19	2.8 ± 0.12
Mersalyl 0.2 mM	6.8 ± 0.23	5.7 ± 0.51	6.6 ± 0.27
Trasylol 200 U/ml	0.2 ± 0.004	0.1 ± 0.003	0.4 ± 0.01
Mersalyl 0.2 mM + Trasylol 200 U/ml	6.1 ± 0.19	5.5 ± 0.51	5.6 ± 0.21
Mersalyl 0.2 mM + O-phenan*, 10 µg/ml	0.65 ± 0.04	0.71 ± 0.06	0.55 ± 0.29

[1] Mersalyl, trasylol and O-phenantroline were added to the invasion chambers, as described under Material and Methods. Endpoint for assay was 144 hr for trophoblasts, 72 hr for B16F10 cells. For each assay three to six samples were measured.
The results represent mean ± standard errors.
* O-phenan = O-phenanthroline.

B16F10 cells (Table 1). For the B16F10 cells, we also studied the effects of rabbit anti-mouse uPA antiserum and an affinity purified antimouse uPA antibody (antiplasminogen activator antibodies). These could not be used for trophoblast and JAR cells because of a lack of cross-reactivity with the human enzyme. All these inhibitors and antibodies abrogated completely the invasion of the JAR and B16F10 cells into the amnion, and again, the radioactivity associated with the amnion membrane in the presence of the inhibitors was similar to the negative control C10 line.

Effect of Mersalyl on Amnion Invasion by Trophoblast and B16F10 Lines

Mersalyl, a mercurial compound known to activate procollagenase was tested for its effects on amnion invasion by trophoblast and B16F10 cells. Table 2 shows that mersalyl at 0.2mM concentration enhanced the invasion by about 1.7-fold for the trophoblast and by 2.1-fold for B16F10 cells. This effect was reversed by the metalloproteinase inhibitor 1–10 phenanthroline, but not by the plasmin inhibitor trasylol.

Effect of Metalloproteinase Inhibitor on Cell Adhesion

The attachment ability of the trophoblasts to the amnion or to the Matrigel was similar to the adhesion capability shown for the B16F10 melanoma cells (Tables 3,4). O-phenanthroline exerted no effect on the trophoblast ability to attach to both types of extra-cellular matrix, indicating for the proteinase inhibitor effect on the extracellular matrix degradation stage of the invasion process.

Table 3. Effect of O-phenanthroline on the Attachment Ability (Percent) of Trophoblast Cells to Amniotic Membrane

Cell line	Control	O-phenanthroline 10 μg/ml
TR-J	37.1 ± 3,4	34.3 ± 4.1
TR-30-4	35.4 ± 5.4	36.6 ± 3.7
B16F10	41.1 ± 8.4	38.3 ± 4.6

Cell attachment to amnion basement membranes. Radiolabelled cells were applied to amnion basement membranes and the percentage of attached cells was quantitated after 24 hr.

Table 4. Effect of O-phenanthroline on the Attachment Ability (Percent) of Trophoblast Cells to Matrigel

Cell lines	0'	5'	10'	15'	20'
TR-J	0	18.7 ± 1.7	28.2 ± 4.1	49.2 ± 6.7	68.8 ± 5.1
TR-J + O-phenanthroline					
10 μg/ml	0	19.1 ± 2.1	26.2 ± 2.1	54.8 ± 6.3	74.0 ± 8.3
B16F10	0	24.5 ± 3.1	32.8 ± 5.1	57.9 ±	72.5 ± 6.6

Time course of trophoblast cell attachment to Matrigel coated plastic. Trophoblasts were grown in the presence of 0.1% FCS. The results represent the mean ± S.D. of triplicate determinators.

Collagenase and TIMP Production and Collagenase and uPA Expression by the Trophoblast

Significant levels of collagenase and TIMP (Table 5) were detectable in the trophoblast-conditioned medium. Blot hybridization analyses of total RNA extracted from culture cytotrophoblast with specific uPA cDNA probe and specific collagenase type I probe revealed the presence of the 2.5 kilobase mRNA for the uPA, and the presence of the 3.7 kilobase mRNA for the collagenase. The level of both mRNAs was constant at 24 and 48 hr of culture. Gamma-actin mRNA (2.0 kilobase) was expressed in similar concentration at 24 and 48 hr of culture (Figure 6).

DISCUSSION

This study demonstrates that first-trimester human trophoblast cells growing in vitro have the ability to invade and to cross epithelium-free human amnion composed of basement membrane and connective tissue stroma. The degree of invasiveness of trophoblast lines measured with this assay was comparable to that of a highly metastatic murine melanoma line B16F10 or the human choriocarcinoma line JAR. Thus, the invasive ability of the human trophoblast appears to be genetically determined. This property is impor-

Table 5. Collagenase and TIMP Production by Trophoblasts in Culture

Cell line	Collagenase level ng/ml	TIMP level ng/ml
TR-J	182 ± 11.7	138 ± 15.1
TR-30-4	199 ± 21.3	127 ± 4.3

Fifteen thousand trophoblast cells were incubated for 24 hr and supernatants were assayed for collagenase and TIMP by highly sensitive radioimmunoassay.

tant for the process of implantation, when trophoblasts of the blastocyst penetrate the endometrial basement membrane, as well as for early placental development, when postimplantation trophoblasts invade uterine stroma, decidua, blood vessels and occasionally the myometrium.

In the present study, we have shown that trophoblasts share some of the molecular mechanisms of invasion with the metastatic cell lines JAR and B16F10. Plasmin generation was found to be an important step for the process of invasion by trophoblast, JAR and

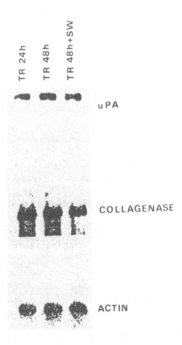

Fig. 6. Hybridization blot analyses of uPA, collagenase and gamma actin mRNAs extracted from cultured cytotrophoblasts. Total RNA (10 μg) extracted from cells at cultured period indicated, were electrophoresed in agarase gels, transferred to Gene plus and probed with specific [32]p-labelled cDNAs.

B16F10 cells. The invasion was blocked by antiplasminogen activator (uPA) antibodies, as well as by serine proteinase inhibitors, especially for plasmin. Trophoblasts have been reported to produce large amounts of plasminogen activators,[11,12] which may lead to the production of plasmin and eventual activation of procollagen. A direct role of collagenase in promoting invasion by these cell types is suggested by several data: (a) an activation of procollagenase (by mersalyl) enhanced the invasion, which was abrogated by the metalloproteinase inhibitor but not with plasmin inhibitor; (b) the invasion was blocked by collagenase inhibitor and metalloproteinase inhibitor; (c) these cells produced significant levels of collagenase, comparable to those reported earlier for invasive tumor cell lines.[17]

Thus it seems that trophoblast invasion ability is dependent on activation of the proteinases cascade: plasmin activator ---- plasminogen ---- plasmin ---- latent collagenase ---- collagenase, which is identical to the cascade reported for malignant tumor invasion.[7]

The properties described, which are important in the multistep process of basement membrane invasion, are shared by highly invasive and metastatic tumor cells as well as trophoblast cells that are invasive but nonmetastatic. Thus, genes coding for invasion-associated molecules in the embryonic trophoblast during ontogeny are possibly depressed during malignant transformation, and additional genetic or epigenetic changes must take place before the invasive cells acquires the metastatic phenotype.

REFERENCES

1. K. Tryggvason, M. Hoyhtya, and T. Salo, Proteolytic degradation of extracellular matrix in tumor invasion. *Biochim. Biophys. Acta*, 907:191-217, (1987).
2. L. A. Liotta, C. N. Rao, and S. H. Barsky, Tumor invasion and the extracellular matrix, *Lab. Invest.*, 49:636-649 (1983).
3. G. L. Nicolson, Cancer metastasis. Organ colonization and the cell-surface properties of malignant cells, *Biochim. Biophys. Acta.*,695:113-176, (1982).
4. L. A. Liotta, K. Tryggvason, S. Garbisa, I. R. Hart, C. M. Foltz, and S. Shafie, Metastatic potential correlates with enzymatic degradation of basement membrane collagen, *Nature*, 284:67-68, (1980).
5. B. F. Sloane, and K. V. Honn, Cysteine proteinases and metastasis. *Cancer Metastasis Rev.*, 3:249-263, (1984).
6. R. Khokha, P. Waterhouse, S. Yagel, P. K. Lala, C. M. Overall, C. Norton, and D. T. Denhardt, Antisense RNA-induced reduction in murine TIMP levels confers oncogenicity on Swiss 3T3 cells, *Science*, 243:947-950 (1989).
7. P. Mignatti, E. Robbins, and D. B. Rifkin, Tumor invasion through the human amniotic membrane: requirement for a proteinase cascade, *Cell*, 47:487-498 (1986).
8. S. Yagel, R. Khokha, D. T. Denhardt, R. S. Kerbel, R. S. Parhar, P. K. Lala, Mechanisms of cellular invasiveness: a comparison of amnion invasion in vitro and metastatic behaviour in vivo, *JNCI* 81:768-775, (1989).
9. R. H. Glass, J. Aggeler, A. Spindle, R. A. Pedersen, and Z. Werb, degradation of extracellular matrix by mouse trophoblast outgrowths: a model for implantation, *J. Cell Biol.*, 96:1108-1116 (1982).

10. S. J. Fisher, M. S. Leitch, M. S. Kantor, C. B. Basbaum, and R. H. Kramer, Degradation of extracellular matrix by the trophoblastic cells of first-trimester human placentas, *J. Cell Biochem.*, 27:31-41, (1985).

11. S. Strickland, E. Reich, and M. I. Sherman, Plasminogen activator in early embryogenesis: enzyme production by trophoblast and parietal endoderm, *Cell*, 9:231-240, (1976).

12. F. Blasi, J. D. Vassalli, and K. Dano, Urokinase-type plasminogen activator: proenzyme, receptor and inhibition, *J. Cell Biol.*, 104:801-804, (1987).

13. J. T. Queenan, L-C. Kao, C. E. Arboleda, A. Ulloa-Aguirre, T. C. Golos, D. B. Cines, and J. F. Strauss, Regulation of urokinase-type plasminogen activator production by cultured human cytotrophoblasts, *J. Biol. Chem.*, 262:10903-10906 (1987).

14. S. Yagel, R. S. Parhar, J. J. Jeffrey, and P. K. Lala, Normal nonmetastatic human trophoblast cells share in vitro invasive properties of malignant cells, *J. Cell Physiol.*, 136:455-463, (1988).

15. S. Yagel, R. F. Casper, W. Powell, R. S. Parhar, and P. K. Lala, Characterization of pure human first trimester cytotrophoblast cells in long term culture: growth pattern markers and hormone production, *Am. J. Obstet. Gynecol.*, 160:938-945, (1989).

16. S. Yagel, R. S. Parhar, and P. K. Lala, Trophic effects of first trimester human trophoblasts and hCG on lymphocyte proliferation, *Am. J. Obstet. Gynecol.*, 160:946-953 (1989).

17. L. Eisenbach, S. Segal, and M. Feldman, Proteolytic enzymes in tumor metastasis. II. Collagenase type IV activity in subcellular fractions of cloned tumor cell populations. *JNCI*, 74: 87-93, (1985).

THE IMPLANTATION WINDOW IN HUMANS AFTER FRESH
OR FROZEN-THAWED EMBRYO TRANSFERS

J. Mandelbaum*, A.M. Junca, M. Plachot, J. Cohen, S. Alvarez,
D. Cornet, M.O. Alnot, and J. Salat-Baroux

U 173 INSERM, Hôpital Necker
149 rue de Sèvres 75743 Paris, Cedex 15, France

INTRODUCTION

In mammals, the optimum time for embryo transfer greatly depends upon the species. In rodents, a high synchrony between the donor and the recipient is required.[1,2] In sheep[3] and cattle[4] too, better results are obtained with synchronous transfers. In other species, such as pigs[5], mares[6] and monkeys[7], an asynchrony up to one or even two days is possible.

Again in humans, one question arises from in vitro fertilization (IVF) programs: what is the optimum time for embryo transfer?

In an attempt to answer this question we first analyzed the results of frozen-thawed embryo transfers where the embryo age may be dissociated from the maternal chronology and then looked for the role of the maternal steroid environment in the receptivity to embryo implantation.

FROZEN-THAWED EMBRYO TRANSFERS: AN OPPORTUNITY
TO DISSOCIATE EMBRYO AND UTERINE AGES

The Freezing Program

Since 1986, embryo freezing was routinely used in our group for excess embryos resulting from IVF.

Advances in Assisted Reproductive Technologies, Edited by
S. Mashiach *et al.*, Plenum Press, New York, 1990

Table 1. Pregnancies After Transfer of Frozen-Thawed Human Embryos (1986, 1987, 1988)

Cycles	Transfers	Clinical Pregnancies (%)
Spontaneous	378	58 (15%)
Stimulated	141	22 (16%)
Artificial*	61	13 (22%)
Total	580	93 (16%)

X^2 TEST: NS (NECKER - TENON - MARIGNAN - SEVRES)
* Oocyte donation

The protocol, widely described[8,9], comprised Propanediol and sucrose[10] as cryoprotectants. In 1986, 1987, and 1988, 580 transfers of frozen-thawed (FT) embryos were performed either in spontaneous or lightly stimulated cycles using Clomifene Citrate (Clomid, Merrell) or human menopausal gonadotropins (hMG). Artificial cycles were applied to women without endogenous ovarian function in an oocyte donation program.

The transfer of FT embryos keeping at least 50% of their blastomeres intact (ie: 60% of thawed embryos) led to 16% clinical pregnancies but the differences between spontaneous, stimulated or artificial cycles were not significant (Table 1).

The rates of abortion (23%) and ectopic pregnancies (3%) did not differ from observations in classical IVF. Of the 69 evolutive pregnancies,15% (10/69) were twin implantations. At the present time, 61 deliveries have led to the birth of 70 normal healthy children. The single anomaly was a stillborn normal girl.

The Chronological Desynchronization in Spontaneous Cycles—(Table 2)

To determine the time of transfer in spontaneous cycles, patients were monitored from day 10 by morning daily plasma LH, estradiol and progesterone assays and sometimes the help of follicular ultrasound.

Transfers were considered to be in synchrony (S) when 2-day old embryos were replaced on the fourth day following the LH surge initiating rise (LH-SIR). Asynchronous transfers could be done in a uterus one day younger than the embryo (S - 1) or one or two days older than the embryo (S + 1; S + 2).

When comparing 249 synchronous and 117 asynchronous transfers where the timing of LH-SIR was accurate, it was obvious that for an identical mean number of embryos replaced in utero (2.2), transfers in synchrony led to 20.5% of clinical pregnancies instead of 11% for transfers out of synchrony, the difference being statistically significant (p < 0.05).

Table 2. Synchronous and Asynchronous Transfers of Frozen-Thawed Embryos in Spontaneous Cycles

Embryo-endometrium		Transfers		Clinical pregnancies (%)	
In synchrony (s)		249		51 (20.5%) a	
Out of synchrony	S − 1	117	30	13 (11%) b	2 (7%)
	S + 1		79		11 (14%) c
	S + 2		8		----

X^2 test: a - b: $p < 0.05$
            a - c: $p < 0.05$

S - 1 = endometrium one day younger
S + 1 = endometrium one day older
S + 2 = endometrium two days older

A one-day asynchrony in humans did not abolish the ability for the embryo to implant. However, transfers in a uterus one day older than the embryo appeared significantly less successful (14% of clinical pregnancies) than synchronous transfers (20.5% - $p < 0.05$). In the same way, transfers in a uterus one day younger only led to 7% of clinical pregnancies without however any statistical significance on this short series.

IS PROGESTERONE THE MAIN REGULATOR IN OVO-ENDOMETRIUM TOLERANCE ?

Frozen-Thawed Embryo Transfers

In 104 spontaneous cycles with FT embryo transfer, progesterone evaluation was performed until transfer. When looking at the correlation between the implantation rate, the day of transfer and the day of plasma progesterone rise above one ng per ml, no significant difference emerged (Table 3). There was, however, a trend towards a sharp decrease in the pregnancy rates (8% vs 18% to 28%) when transfers were carried out on the fourth day of elevated progesterone levels despite an average of 2.7 transferred embryos in this group.

Pharmacological Desynchronization

Animal experiments widely stated that endometrium receptivity to nidation depends on steroid environment and especially the timing of the sequence of estradiol and progesterone levels. By exogenous administration of progesterone, it is possible to create a pharmacological asynchrony between a young embryo and an overmature treated endometrium.

Table 3. Correlation Between Plasma Progesterone Levels before Transfer and FT Embryo Implantation in Spontaneous Cycles

Progesterone rise*	Embryos replaced	Transfers	Clinical pregnancies (%)
T-3	2.7	13	1 (8%)
T-2	2.0	39	7 (18%)
T-1	2.0	39	11 (28%)
T 0	2.8	13	3 (23%)

Results were not statistically significant.
X^2 TEST: NS
PROGESTERONE*: ≥ 1 ng/ml   (NECKER - TENON)
T O = Day of transfer

Table 4. Pharmacological Asynchronous Fresh Embryo Transfers in Humans

Progesterone treatment* starting from	Fair embryos**/ recovered oocytes (%)	Clinical pregnancies/ transfers (%)
hCG injection	899/1790 (50%) a	24/203 (12%) d
Oocyte pick-up	349/ 814 (43%) b	19/72 (26%) e
Embroy transfer	806/1623 (50%) c	44/206 (21%) f

SEVRES - NECKER

X^2 test:
- a-b: $p < 0.001$
- a-c: NS
- b-c: $p < 0.001$
- d-e: $p < 0.01$
- d-f: $p < 0.01$
- e-f: NS

*   Utrogestan: 300 mg/day
** Fair embryos: embryos either replaced or frozen

This was done in rodents: Schacht and Foote reported that in the rabbit the pre-ovulatory administration of progesterone decreased to 5% the implantation rate of embryos transferred to the uterus of treated recipients[11] and in the rat Dickmann also obtained a dramatic decrease in the percentage of transferred morulae developing in fetuses when the administration of progesterone began in the preovulatory period.[12] Such a pharmacological desynchronization was unwillingly performed by an early administration of progesterone, to women involved in the IVF program (Table 4).

The progesterone supply of the luteal phase usually started from oocyte pick-up or embryo transfer (micronized vaginal progesterone-Utrogestan, Besins-Iscovesco-at 300 mg per day). During a four month period, following a wrong prescription, it started from hCG injection, i.e. the preovulatory period.

Table 5. Morphological Aspect of Tripronucleated Human Embryos Cultured 40 Hours in Human Luteal Uterine Fluid

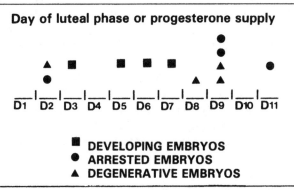

Day of luteal phase or progesterone supply

■ DEVELOPING EMBRYOS
● ARRESTED EMBRYOS
▲ DEGENERATIVE EMBRYOS

No difference was observed in the mean number of embryo replaced and the embryo morphological quality was not impaired by the preovulatory progesterone supply. However, the pregnancy rate per transfer significantly decreased during this period from 26 or 21 to 12% suggesting a deleterious effect on uterine receptivity to the conceptus.

Such results are consistent with the observations of Rosenwaks.[13] When embryos obtained after IVF of donated oocytes were transferred in patients without ovarian endogenous function, no pregnancies arose from transfers performed after the fifth day of progesterone substitution. So, an overmature endometrium may be refractory to implantation and if embryos are transferred out of synchrony and reach their own maturity during the uterine refractory period, they will be prevented from implanting.

THE UTERINE MILIEU AND ITS ROLE IN EMBRYO SURVIVAL

Some authors hypothesized that in rodents an embryo toxicity of the uterine fluid might be one of the factors involved in uterine refractoriness and consequently in the failure of asynchronous transfers: thus, uterine flushings of the refractory period were found to be blastocytotoxic by Weitlauf[14] and Psychoyos.[15] In humans, Aitken[16] only found an adverse effect of uterine flushings of the menstrual period on the hatching of mouse blastocyst in vitro.

We decided to investigate the action of human uterine fluid on the embryonic growth using as model the human tripronucleated egg, checked at Day 1 post-insemination, discarded from transfer but able to further cleave in vitro in 82% of the cases, although 15% only reach the blastocyst stage.[17]

Such triploid human embryos at the 2- to 6-cell stage 40 hours after insemination were cultured in straws, in aspirated uterine fluid of the luteal phase either pure or diluted with B_2

medium (Api-System, Montalieu-Vercieu, France). As the amount of uterine fluid decreased toward the late luteal phase, the greatest dilution (0.1–1 ml) was necessary at this period. Blood contaminated aspirates were discarded. Luteal uterine fluids were recovered either in spontaneous cycles or in artificial cycles. The aspect of the triploid embryos was evaluated after 40 hours of culture in uterine fluid and classified as cleaved, arrested or degenerative.

Twelve fluids were so tested (Table 5): only four (30%) allowed a further cleavage of the experimental embryos and were issued from the third to the seventh day of the luteal phase or the progesterone supply. On the other hand, six embryos cultured in fluids originating from the late luteal phase (days 8, 9 and 11) were either arrested or degenerative. The series is nevertheless too short to conclude and more data have to be collected to corroborate these findings.

In conclusion, the window of transfer in humans looks wider than in other mammals as the endometrium receptivity to the embryo certainly extends by several days. We absolutely need to better evaluate the length and control of this receptive period also called "implantation window" either in spontaneous or in superovulated IVF cycles. The difficulty to clarify this point lies in the continuous interaction between the various protagonists: the endometrium, target organ for ovarian steroids responds by modifications in its morphology and secretion to any alteration in the usual programme while in response, embryo development can be accelerated or decelerated.

Nevertheless, even if implantation could occur after asynchronous transfers, the success rate appears to be lower.

REFERENCES

1. Z. Dickmann, R.W. Noyes, The fate of ova transferred into the uterus of the rat, *J. Reprod. Fert.*,1, 197-212 (1960).
2. L.L. Doyle, A.H. Gates, R.W. Noyes, Asynchronous transfer of mouse ova, *Fertil. Steril.*, 14, 215-225 (1963).
3. N.W. Moore, J.N. Shelton, Egg transfer in sheep. Effect of degree of synchronization between donor and recipient, age of egg and site of transfer on the survival of transferred eggs, *J. Reprod. Fert.*, 7, 145-152 (1964).
4. L.E.A. Rowson, R.A.S. Rowson, R.M. Moor, A.A. Baker, Egg transfer in cow; synchronization requirements, *J. Reprod. Fert.*, 28, 427(1972).
5. C. Polge, Embryo transplantation in the pig. *In:* "Embryo transfer in mammals", Eds Merieux C., H. et Bonneau M., pp. 45-52 (1982).
6. N. Oguri, Y. Tsutsumi, Non surgical transfer of equine embryos, *In:* E.S.E. Hafez, K. Semm, "In vitro fertilization and embryo transfer". MTP Press Ltd Lancaster 287-295 (1982).
7. G. D. Hodgen, Surrogate embryo transfer combined with estrogen-progesterone therapy in monkeys, *J. Am. Med. Assoc.*, 250, 2167-2171 (1983).
8. J. Mandelbaum, A.M. Junca, M. Plachot, M.O. Alnot, S. Alvarez, C. Debache, J. Salat-Baroux, J. Cohen, Human embryo cryopreservation, extrinsic and intrinsic parameters of success, *Human Reprod.*, 2, 709-715 (1987).

9. J. Mandelbaum, A.M. Junca, M. Plachot, M.O. Alnot, J. Salat-Baroux, S. Alvarez, C. Tibi, J. Cohen, C. Debache, L. Tesquier, Cryopreservation of human embryos and oocytes, *Human Reprod.*,3, 117-119 (1988).

10. B. Lassalle, J. Testart, J.P. Renard, Human embryo features that influence the success of cryopreservation with the use of 1,2 propanediol, *Fertil. Steril*, 44, 645-651 (1985).

11. C.J. Schacht, R.H. Foote, Progesterone induced asynchrony and embryo mortality in rabbits, *Biol. Reprod.*, 19, 534-539 (1978).

12. Z. Dickmann, Effects of progesterone on the development of the rat morula, *Fertil. Steril.*, 21, 541-548 (1970).

13. Z. Rozenwaks, Donor eggs : their application in modern reproductive technologies, *Fertil. Steril.*, 47, 895-909 (1987).

14. H.M. Weitlauf, Factors in mouse uterine fluid that inhibit the incorporation of ^{3}H-uridine by blastocysts in vitro, *J. Reprod. Fertil.*, 52, 321-325 (1978).

15. A. Psychoyos, V. Casimiri, Uterine blastotoxic factors, 327-334. In: S. R. Glasser, D. W. Bullock," Cellular and molecular aspects of implantation", Plenum Publ (1981).

16. R.J. Aitken, J.B. Maathuis, Effects of human uterine flushings collected at various states of the menstrual cycle on mouse blastocysts in vitro, *J. Reprod. Fertil.*, 53, 137 (1978).

17. M. Plachot, A.M. Junca, J. Mandelbaum, J. Cohen, J. Salat-Baroux Embryonic development failures, *Reprod. Nutr. Develop.*, 28 (6B) : 1781-1790 (1988).

NON-INVASIVE BIOCHEMICAL METHODS FOR ASSESSING HUMAN EMBRYO QUALITY

H. J. Leese,[1] D. K. Gardner,[1] A. L. Gott,[1]
A. H. Handyside,[2] K. Hardy,[2] M. A. K. Hooper,[3]
A. J. Rutherford,[2] and R. M. L. Winston[2]

[1]Department of Biology, University of York, Heslington,York YO1 5DD, UK
[2]Institute of Obstetrics and Gynaecology, Royal Postgraduate Medical School
Hammersmith Hospital, Du Cane Road, London W12 OHS
[3]Sheffield Fertility Centre, 26 Glen Road, Sheffield S71RA

While human in vitro fertilisation (IVF) and Embryo Transfer techniques have helped overcome the problems of infertility in some thousands of couples, success rates remain disappointingly low. The major problem remains the embryo transfer stage where only a comparatively small proportion of embryos implant and are carried to term successfully. Embryos for transfer are assessed on the basis of their morphology and extent of development in culture. These methods are notoriously imprecise and there is a need for quantitative non-invasive assays of human embryo quality.

The criteria which these assays should fulfill have been summarised by Leese (1987) and Leese and Gardner (1989). Some factors whose uptake or release into the incubation medium might form the basis of non-invasive tests are shown in Figure 1. These factors represent physiological processes and it is hoped that their measurement will be a reflection of embryo quality.

The most intensively studied of these factors are those which relate to embryo energy metabolism. There are two reasons for this. Firstly, a good deal is known about the energy metabolism of the embryos of other species, particularly the mouse, from which comparisons may be made. Secondly, it is likely that the changes which the embryo brings about in the composition of its incubation medium during the first 48 hours following fer-

Advances in Assisted Reproductive Technologies, Edited by
S. Mashiach *et al.*, Plenum Press, New York, 1990

tilisation (the time for which human embryos are cultured in many IVF clinics) are greatest for the uptake of nutrients and formation of metabolic products concerned with energy metabolism. The initiation of high rates of nucleic acid and protein synthesis, which follow the activation of the embryonic genome and might be reflected in high rates of amino acid uptake, are not thought to begin in human embryos until the 4–8-cell stage (Tesarik, 1987; Braude et al., 1988) by which time the embryos will have been removed from culture. Similarly, the appearance of factors which signal the presence of the embryo to the endometrium would be expected to be released during later stages of development.

We have therefore concentrated on energy substrate utilisation, and measured the uptake of pyruvate and glucose by single human oocytes and preimplantation embryos. Pyruvate and glucose were chosen because they are consumed by preimplantation mouse embryos in culture; pyruvate is consumed preferentially during the early cleavage stages before glucose becomes the predominant substrate in the blastocyst (Leese and Barton, 1984; Gardner and Leese 1986). Furthermore, pyruvate and glucose have traditionally been added to a wide range of media used to culture human embryos.

The measurement of glucose and pyruvate uptake by single embryos requires the miniaturisation of conventional analytical methods. Such microscale assays as applied to human embryos were first reported by Leese et al. (1986) and have been described in detail by Leese and Gardner (1989). They are summarised in Figures 2 and 3.

Briefly, single embryos are incubated for 24 h in 4 or 5 µl of a modified T6 (HLT6) medium under paraffin oil at 37°C with a gas phase of 5% CO_2. The medium contains 0.47

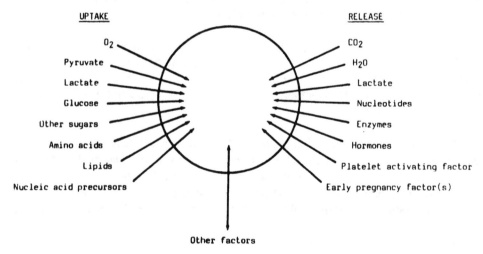

Fig. 1. Factors whose uptake or release into the incubation medium might form the basis of non-invasive tests.

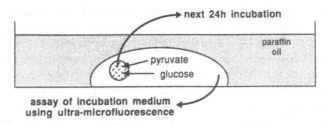

Fig. 2. Non-invasive measurement of substrate depletion by single human embryos.

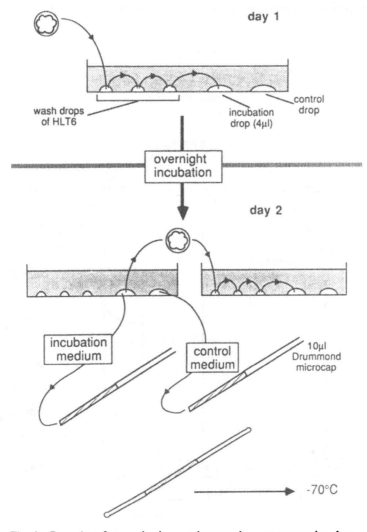

Fig. 3. Procedure for monitoring nutrient uptake over successive days.

mM pyruvate, 1 mM glucose and 5mM lactate, which reflect the concentrations of these nutrients in rabbit and mouse oviduct fluid. At daily intervals the oocyte or embryo is removed and transferred to a fresh 4 µl droplet for the next incubation period. 1 µl sample of the remaining medium and of medium from a control droplet were stored at -70°C prior to analysis for pyruvate and glucose using an ultramicrofluorometric technique (Leese et al., 1984; Leese and Barton, 1984). The analyses are carried out in 10 µl microdrops of reaction mixture under paraffin oil in glass wells attached to siliconized microscope slides. The reaction products (NADH and NADPH) are quantified using a fluorescence microscope (Leitz Fluovert) with photomultiplier and microphotometer attachments.

The unfertilised oocytes and spare embryos were obtained from patients attending the

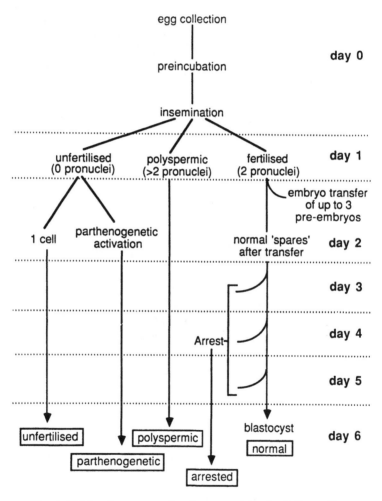

Fig. 4. Origin of oocytes and embryos used in metabolic studies.

IVF Clinic at Hammersmith Hospital, London as described in Figure 4. Full ethical permission for the work was obtained from the Ethics Committees of the collaborating institutions, the Voluntary Licensing Authority for Human In Vitro Fertilisation and Embryology of the Medical Research Council and the Royal College of Obstetricians and Gynaecologists, and the patients concerned.

Those fertilised spare embryos which developed to the blastocyst stage represented 58% of the total embryos used in the study. They were observed to have the highest pyruvate and glucose uptakes and were the only embryos to exhibit a pronounced increase in glucose uptake at the blastocyst stage (day 5 post-insemination) (Figures 5 and 6).

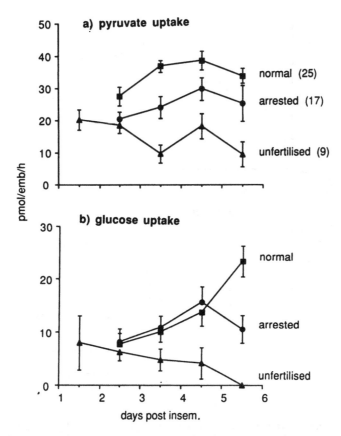

Fig. 5. Pyruvate (upper Panel) and glucose (lower panel) uptake by fertilised embryos which developed to the blastocyst stage by Day 6 (■); embryos which arrested during cleavage (●); and unfertilised oocytes (▲). Taken from Hardy et al., 1989, non-invasive measurement of glucose and pyruvate uptake by individual human oocytes and preimplantation embryos, Hum. Reprod., 4:188 (reproduced with permission).

Qualitatively, the nutrient uptake pattern exhibited by these "normal" human embryos is similar to those of the mouse (Leese & Barton, 1984; Gardner & Leese, 1986). There are however two important differences.

Firstly, the total nutrient uptake is considerably greater for human than mouse embryos, confirming the earlier results of Leese et al. (1986). This is the case even when the difference in size of the two types of embryo is taken into account. If the diameter of the

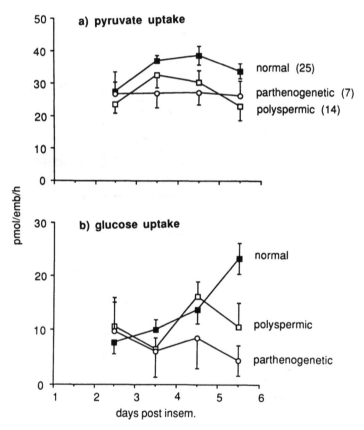

Figure 6. Pyruvate (upper panel) and glucose (lower panel) uptake by polyspermic embryos (□) and parthenogenetic embryos (unfertilised oocytes undergoing regular cleavage) (○). Uptake by fertilised embryos developing to the blastocyst stage by Day 6 has been included for comparison (■). Taken from Hardy et al., 1989, non-invasive measurement of glucose and pyruvate uptake by individual human oocytes and preimplantation embryos, Hum. Reprod., 4:188 (reproduced with permission).

mouse embryo is taken to be 80μ and that of the human 120μ, the respective volumes assuming each one to be a sphere, are 214 and 724 pl, a ratio of 1:3.4. However the average pyruvate uptake by single human embryos is about 35 pmol/embryo/h compared with 1–3 pmol/embryo/h in the mouse (Leese and Barton, 1984; Gardner and Leese, 1986; 1988), a greater than 10-fold difference. We have no explanation at the present time for the greater metabolic activity per unit volume of the human as opposed to the mouse embryo.

The second difference is that the pronounced decline in pyruvate uptake as glucose becomes the preferred substrate at the late morula stage is not observed in the human. Within the embryo, glucose may be fully oxidised to CO_2 and H_2O, incorporated into glycogen and other macromolecules, or oxidised to lactate. Mouse embryos form increasing amounts of lactate as they develop from the 2-cell stage to the blastocyst, even in the presence of air (Gardner and Leese, 1988). Wales et al., (1987), using a microscale radiochemical method, showed that single human embryos also form increasing amounts of lactate from glucose at the morula and blastocyst stages. We have therefore examined this "aerobic glycolysis" in single human embryos during their development in culture from day 2 post-insemination to the late blastocyst. The results are striking in that very large quantities of lactate are released into the incubation medium. We are currently investigating the possibility that some of this lactate may be derived from endogenous sources.

Diagnostic Capability of Non-invasive Metabolic Assays

The pioneering study was carried out on cattle embryos by Renard et al. (1980) who reported that 69% of bovine blastocysts consuming > 5μg glucose/h implanted successfully, whereas only 14% of those consuming < 5μg/h did so. There were no further reports until Gardner and Leese (1987) showed that the glucose uptake of single mouse blastocysts which implanted successfully following transfer was about 30% greater than those which failed to give rise to live offspring. The total number of embryos whose metabolism was measured prior to transfer was 50. We have since repeated this work on a further 31 embryos and found very similar results (Gardner and Leese unpublished).

We are about to embark on a long-term clinical study, initially to screen the pyruvate uptake of all the embryos produced by a given couple following IVF, i.e. those subsequently transferred together with those not transferred. If the results look promising and an association emerges between the mean nutrient uptake of those embryos selected for transfer and the subsequent rates of pregnancy, we would consider using pyruvate uptake alone as the means of selecting which embryos to transfer. We are encouraged in this approach by the marked differences in nutrient uptake between "normal" spare embryos, and those which were arrested during development, were parthogenetic, polyspermic or unfertilised (Figures 5 and 6) and Hardy et al. (1989).

ACKNOWLEDGEMENT

The work of Dr. D. K. Gardner was supported by the Science and Engineering Research Council. A. L. Gott is supported by the Wellcome Trust.

REFERENCES

Braude, P., Bolton, V., and Moore, S., 1988, Human gene expression first occurs between the four- and eight-cell stges of preimplantation development, *Nature*, 332:459.

Gardner, D. K., and Leese, H. J., 1986, Non-invasive measurement of nutrient uptake by single cultured preimplantation mouse embryos, *Human Reprod.*, 1:25.

Gardner, D. K., and Leese, H. J., 1987, Assessment of embryo viability prior to transfer by the noninvasive measurement of glucose uptake, *J. Exp. Zool.*, 242:103.

Gardner, D. K., and Leese, H. J., 1988, The role of glucose and pyruvate transport in regulating nutrient utilization by preimplantation mouse embryos, *Development*, 104:423.

Hardy, K., Hooper, M. A. K., Handyside, A. H., Rutherford, A. J., Winston, R. M. L., and Leese, H. L., 1989, Non-invasive measurement of glucose and pyruvate uptake by individual human oocytes and preimplantation embryos, *Hum. Reprod.* 4:188.

Leese, H. J., 1987, Analysis of embryos by non-invasive methods, *Hum. Reprod.*, 2:37.

Leese, H. J., and Barton, A. M., 1984, Pyruvate and glucose uptake by mouse ova and preimplantation embryos, *J. Reprod. Fert.*, 72:9.

Leese, H. J., Biggers, J. D., Mroz, E. A., and Lechene, C., 1984, Nucleotides in a single mammalian ovum or preimplantation embryo, *Anal. Biochem.*, 140:443.

Leese, H. J., Hooper, M. A. K., Edwards, R. G., and Ashwood-Smith, M. J., 1986, Uptake of pyruvate by early human embryos determined by a non-invasive technique, *Hum. Reprod.*, 1:181.

Leese, H. J., and Gardner, D. K., 1989, Embryo metabolism and viability in vitro, *in:* "CRC Handbook of IVF and ET: Laboratory Manual," A. Trounson and J. Osborn, eds., CRC Press Inc., U.S.A.

Renard, J. P., Philipon, A., and Memezo, Y., In vitro uptake of glucose by bovine blastocysts, *J. Reprod. Fert.*, 58:161.

Tesarik, J., 1987, Gene activation in the human embryo developing in vitro, *in:* "Future Aspects in Human In Vitro Fertilization," W. Feichtinger, and P Kemeter, eds., Springer-Verlag, Berlin and Heidelberg.

Wales, R. G., Whittingham, R. G., Hardy, K., and Craft, I. L., Metabolism of glucose by human embryos, *J. Reprod. Fert.*, 79:289.

PREGNANCY RATE AND PREGNANCY OUTCOME ASSOCIATED WITH LABORATORY EVALUATION OF SPERMATOZOA, OOCYTES, AND PRE-EMBRYOS

Lucinda L. Veeck

Jones Institute for Reproductive Medicine, Eastern Virginia Medical School
Norfolk, Virginia U.S.A.

INTRODUCTION

There are very few non-invasive methods currently available for determining the quality of gametes and pre-implantation embryos. Because sperm collection is easier than oocyte collection, and because the numbers of available sperm are so far in excess of available oocytes, the opportunity to examine large populations of sperm is possible within an in vitro fertilization setting. It follows that invasive testing may be performed on sperm without fear of compromising a given patient's chances for establishing pregnancy. Unfortunately, embryologists must rely primarily on gross morphology and developmental rate in order to make comparable assessments of the oocyte and developing conceptus; both an unreliable and uninformative means, yet the only means possible when the desire is to reduce potential danger to the oocyte and pre-embryo. Despite these limitations, various associations may be noted between the factors of gross analysis and the ability to contribute to ongoing pregnancy. While these factors may differ between sperm and oocytes, the additive effect of each factor may build a base upon which to estimate the potential for success.

SPERMATOZOA

In vitro fertilization is commonly used as treatment for couples with male factor infertility and for couples with multifactorial infertility in combination with abnormal semen

Advances in Assisted Reproductive Technologies, Edited by
S. Mashiach *et al.*, Plenum Press, New York, 1990

Table 1. Volume and Viscosity versus Fertilization, Transfer, Pregnancy, and Pregnancy Outcome. Norfolk Series 1–30

			VOLUME			
Volume	No. Cycles	% Fertilization Per Oocyte	% Cycles With Transfer	% Pregnancy Per Transfer	% Ongoing Pregnancy Per Transfer	% Ongoing Pregnancy Per Cycle
0–1ml	329	83%	88%	27%	18%	16%
1.5–5.0ml	1902	86%	90%	28%	18%	16%
5.5–8.0ml	136	85%	94%	27%	21%	20%
8.5+ml	15	89%	100%	40%	20%	20%
			VISCOSITY			
Viscosity						
Viscous (moderate to severe)	731	87%	91%	30%	20%	19%
Not Viscous	1665	85%	90%	27%	17%	15%

parameters. Oftentimes, male infertility is difficult to overcome even with the advantages that IVF offers—fewer sperm are required for insemination as compared to in vivo fertilization, one has the ability to manipulate numbers of sperm in the areas surrounding the oocyte, and the opportunity is available to assess fertilization success by examination of pronuclear development. Despite these manipulative and diagnostic advantages, sperm from some males repeatedly fail to fertilize oocytes, even when numbers of motile sperm are adequate. There is a growing recognition that the routine semen analysis parameters of concentration, motility, and standard morphologic assessment are inadequate for assessing the functional ability of sperm to penetrate and activate an oocyte. As a consequence, attention has been focused on developing new methods and revising old methods to assess fertilization potential. Here, fertilization, pregnancy, and pregnancy outcome is examined.

Correlation of Individual Semen Factors with IVF Success

Which factors of the basic semen analysis give us an estimate of the fertilizing ability of sperm collected for IVF? What is the result in terms of pregnancy and pregnancy outcome with normal and male factor populations? Tables 1–4 demonstrate the Norfolk experience as regards various parameters of male partner assessment and the relationship to clinical success.

Table 2. Sperm Concentration per Milimiter in Original Ejaculate versus Fertilization, Transfer, Pregnancy, and Pregnancy Outcome. Norfolk Series 1–30

Concentration/ml	No. Cycles	% Fertilization Per Oocyte	% Cycles with Transfer	% Pregnancy Per Transfer	% Ongoing Pregnancy Per Transfer	% Ongoing Pregnancy Per Cycle
$<20 \times 10^6$	111	62%	69%	28%	15%	10%
$>20 \times 10^6$	2214	87%	92%	28%	18%	17%
$>400 \times 10^6$	11	87%	91%	10%	0%	0%
200–399×10^6	213	90%	95%	26%	19%	18%
100–199×10^6	697	90%	93%	27%	19%	18%
50–99×10^6	857	88%	92%	29%	17%	16%
20–49×10^6	436	80%	87%	29%	19%	16%
$<10 \times 10^6$	41	48%	56%	22%	9%	5%
$<5 \times 10^6$	25	32%	36%	22%	11%	4%
$<1 \times 10^6$	6	5%	17%	0%	0%	0%

Volume and Viscosity

Not surprisingly, neither semen volume nor viscosity demonstrate any particular relationship to fertilization, pregnancy, or ongoing pregnancy rates (Table 1). Semen volumes in the hypospermic (< 1.0 ml) and hyperspermic (> 8.0 ml) ranges fail to demonstrate any reduction of reproductive performance, even with volumes as low as 0.2 ml and as great as 11.0 ml. The presence of viscous seminal fluid, while sometimes technically difficult to deal with, poses no detriment to IVF treatment.

Sperm Concentration

Sperm concentration in the range of 20–400×10^6/ml shows very little variation in relationship to fertilization rates, transfer rates, pregnancy rates, or pregnancy outcome (Table 2).

Below the level of 20×10^6/ml, one begins to notice a reduction in fertilizing ability, and this observation is greatly confirmed below the 10×10^6/ml level. Transfer rates per

Table 3. Motility in the Original Sample versus Fertilization, Transfer, Pregnancy, and Pregnancy Outcome. Norfolk Series 1–30

Motility	No. Cycles	% Fertilization Per Oocyte	% Cycles With Transfer	% Pregnancy Per Transfer	% Ongoing Pregnancy Per Transfer	% Ongoing Pregnancy Per Cycle
<30%	140	69%	81%	33%	21%	17%
>30%	2255	87%	91%	27%	18%	16%
80–100%	278	90%	95%	26%	16%	15%
50–79%	1365	88%	91%	28%	18%	16%
<20%	63	62%	78%	33%	22%	18%
<10%	6	62%	67%	25%	25%	17%

patient drop as a result of this diminishment of fertility, but it appears that rates of pregnancy are not significantly lower if the patient does reach the transfer stage. Of more interest is the observation that there is a tendency to produce more miscarriages in the very low concentration groups, and that this miscarriage rate, in combination with a lowered transfer rate per cycle results in an extremely low ongoing pregnancy rate per treatment cycle. Nonetheless, taken as a single parameter, sperm number alone is a poor indicator of reproductive potential as some males with concentrations well below 10×10^6/ml will possess the ability to fertilize all oocytes, while others in this range will fail to fertilize even one. This may be due to the fact that severe oligospermia rarely stands alone as a single abnormal semen parameter—in most cases one finds oligospermia present in conjunction with reduced motility and/or a reduction in normal forms. In the occasional couple with only an impairment of sperm number (not morphology), the prognosis for successful fertilization and ongoing pregnancy may be favorable, unlike the couple with multifactorial male problems.

Motility

The percentage of motile sperm in the original ejaculate plays a role similar to that of sperm concentration in terms of fertilization (Table 3). A slight reduction in fertilization rate is noticed below the 30% level with a resulting decrease in transfer rate. However, pregnancy rates and ongoing pregnancy rates after transfer are not affected to the same extent and are comparable to the higher motility groups. It was suggested by Burkman in ear-

Table 4. Morphology in the Original Sample versus Fertilization, Transfer, Pregnancy, and Pregnancy Outcome. Norfolk Series 25–28 (teratozoospermic patients and controls) (all > 20 × 10^6/ml and > 30% motility)

Morphology Pattern	No. Cycles	Average % Normal Forms	% Fertilization per Oocyte	% Cycles with Transfer	% Pregnancy per Transfer	%Ongoing Pregnancy per Transfer	%Ongoing Pregnancy PerCycle
>14% Normal Forms (Control Group)	41	18.4 ±3.5	94%	100%	44%	32%	32%
4–14% Normal Forms ("G" Pattern)	144	7.3 ± 2.8	86%	76%	44%	25%	19%
<4% Normal Forms ("P" Pattern)	47	1.4 ± 1.2	45%	62%	14%	7%	4%

lier data that a motility greater than 35% in original and washed sperm samples is associated with fertilization success.[1] It is possible that the practice of increasing the number of sperm for insemination in cases of low motility has significantly reduced the risk of fertilization failure. More recently, Acosta has observed that a reduction in motility down to 10% does not significantly impair reproductive performance.[2] The treatment of male factor motility disorders may be helped to some degree with the administration of FSH to the male partner (Acosta and Oehninger, unpublished data).

Morphology

Table 4 indicates that percent normal morphology plays the most significant role, as an individual semen parameter, in IVF reproductive success. A strict method of morphologic determination has been used in Norfolk for patients included in sperm morphology studies: Spermatozoa are considered normal when the head has a smooth oval configuration with a well-defined acrosome involving 40%–70% of the sperm head and an absence of neck, midpiece, or tail defects (Figure 1). No cytoplasmic droplets of more than half the size of the sperm head should be present. The length of a normal sperm head should be 5–6 μm and the diameter 2.5–3.5 μm. A microscopic eyepiece micrometer is used to do routine measurements. In contrast to other methods, borderline forms are counted as abnormal. At least 200 cells are counted on each of two slides which are stained by a quick stain technique.[3]

The amorphous head group is divided into two categories: slightly amorphous and severely amorphous. Slightly amorphous forms are those sperm with a head diameter of 2.0–2.5 μm with slight abnormalities in the shape of the head, but with a normal acrosome.

Severely amorphous heads are defined as those with no acrosome at all or with an acrosome smaller than 30% or larger than 70% of the sperm head. Completely abnormal shapes are also put into this category.

Neck defects are also classified into slightly amorphous and severely amorphous categories. The slight neck defect exhibits debris in the neck area or a thickness of the neck, but with a normally shaped head. Severe neck defects demonstrate a bend in the neck or midpiece.

All other abnormal forms—small, large, round, tapered, double head, double tail, coiled tail, cytoplasmic droplets—are classified according to standards set down by the World Health Organization classification.[3,4]

Spermatozoa with very long tails may represent an additional abnormal population incapable of normal fertilization. Although few references have been made to this aspect of sperm morphology in the literature, overly long tails have been observed, on occasion, in conjunction with failed fertilization (Veeck, unpublished data). Appleton and Fishel report a significant difference in the length of sperm tails in fertile versus infertile males: 32.2μm vs. 45.0μm.[5]

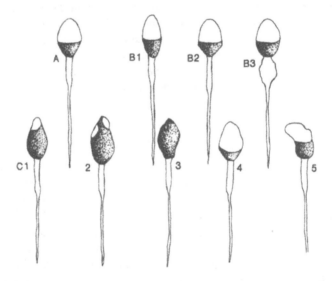

Fig. 1. Diagrammatic representation of spermatozoa stained with the Diff-Quik technique. (A) normal form (B1,B2) slightly amorphous heads (B3) slightly amorphous neck defect (C1-C5) severely amorphous forms by strict criteria.

In general, it can be observed that fertilization rates, pregnancy rates per transfer, ongoing pregnancy per transfer and per cycle are all reduced in patients with severe teratozoospermia. Additionally, an increase in fertilization rate can be achieved by increasing the concentration of sperm to oocyte in cases of teratozoospermia, but pregnancy outcome is still very poor.[3] This finding gives rise to some great concern about the potential of the newer micromanipulative techniques being applied in clinical situations, i.e. direct sperm microinjection into oocytes with sperm demonstrating inferior morphology. Animal data is lacking as far as live birth is concerned and success may be limited by the association with inferior sperm. Certainly, as many as 8% or more of the sperm in normal samples may possess gross chromosomal abnormalities,[6] and it is unknown at this time just how high that percentage may reach with sperm displaying high incidences of poor morphology.

Is Morphology Improved by the Standard "Swim-Up" Technique?

The standard "swim-up" technique employed by most IVF centers allows for the collection of a highly motile fraction of washed spermatozoa for oocyte insemination. While obviously selecting for sperm motility, this procedure also plays a role in the selection of normal forms. McDowell reported in 1986 that this technique provided a fraction of sperm with significant improvements in motility ($p < 0.001$) and improvements in morphology using old criteria ($p < 0.001$). [7]

More recent data generated in Norfolk utilized the methods of critical assessment to determine morphological improvements with the swim-up procedure. In a population of 73 consecutive IVF patients, the percent normal morphology improved 18.2% ($p < 0.05$) after the selection procedure.[8] Twenty-seven of these patients had a diagnosis of teratozoospermia (< 14% normal forms), and 21 demonstrated significant improvement in the total percent of normal forms after processing, six demonstrated only slight improvement or no improvement at all, and no samples showed a decrease in normal forms.[9]

Hamster Penetration Assay (HPA)

The hamster penetration assay, as performed in Norfolk, serves as an adjunct predictive tool to assess sperm function. As with many other fertility tests, this assay can show variability related to such factors as sperm concentration, preincubation time, albumin concentration, insemination time, and technique. Individual variability has also been noted. Less then 10% penetration is considered negative, 10% to 20% is considered borderline, and greater than 20% is strongly positive. Generally, correlation is good between normal HPA results (> 10% with proper controls) and the ability of sperm to achieve fertilization during an IVF attempt. However, a negative HPA (< 10% with normal controls of an acceptable range) is extremely variable in its prognostic value. If one examines the relationship between HPA results and sperm morphology, a better correlation exists. In males with normal count (> 20×10^6/ml), motility (> 30%), and morphology (> 14%), 86% will produce a positive HPA. In males with normal count, motility, but subnormal morphology (< 14%), 95% will produce a negative HPA result.[2]

Hemi-Zona Assay (HZA)

Franken et al., utilizing the hemizona assay (HZA), showed that semen samples classified as morphologically poor (< 4% NF) exhibited a significant impairment of sperm binding to the human zona pellucida.[10] Moreover, Oehninger et al. reported that IVF cases with poor fertilization rates (male factor patients) had lower sperm binding ability in the HZA as compared with patients with successful fertilization.[11]

It should be recognized that male subfertility likely involves a combination of sperm deficiencies which affect the functional ability of most or all of the spermatozoa within a sample. The ability of a single parameter to serve as a prognostic tool is limited by this multifactorial association. For instance, aberrations in sperm morphology may affect motility; systemic illness, fever and environmental toxins may produce damage to spermatocytes with resulting oligospermia, asthenospermia and/or teratozoospermia. Very low sperm numbers may or may not be found in conjunction with high percentages of sperm abnormalities. While it is sometimes difficult to weigh and assess clinical findings and relate these aspects to reproductive potential, a general prognostic value can be placed upon some specific parameters. It is our experience that the percentage of sperm with absolute normal morphology is one important factor related to the opportunity of establishing an ongoing pregnancy. Other factors such as initial concentration, concentration used for insemination, initial or processed motility, and results of biochemical, immunological or biological assays are more elusive. Functional variation may occur from patient to patient based upon the extent and severity of these findings.

OOCYTES

As embryologists working in clinical IVF settings, we are often prejudiced by the physical characteristics of the oocytes we see collected in the laboratory. More often than not, we tend to associate poor pregnancy outcomes with certain aspects of oocytes and early conceptuses; i.e. significant granularity of ooplasm, variations in size and shape of polar bodies, inclusion bodies within the ooplasm, and numerous other atypical observations. Which of these characteristics can actually be shown to interfere with pregnancy and ongoing pregnancy rates?

Ooplasmic Granularity

Homogeneous granularity of the ooplasm cannot be associated with significantly reduced fertilization rates. Of 8103 mature oocytes collected during the years of 1985 to

1988, 624 (7.7%) were noted as displaying significant granularity at the time of harvest. Fertilization rates were similar for "clear" and "granular" oocytes (91% vs 87%) and rates of triploidy were only marginally higher for those fertilized within the granular group (7% vs 10%). Pregnancy was more difficult to assess as many cycles had pre-embryos developing from both clear and granular oocytes transferred; it can be noted, however, that several term pregnancies were established solely from oocytes and/or pre-embryos exhibiting significant homogeneous granularity. Centrally dark granularity and patches of dark granules represent a separate observation and are probably associated with oocyte degeneration—when this situation is observed, fertilization is lower and a greatly reduced developmental potential follows even if pronuclear structures are noted. Additionally, it has been observed that oocytes and pre-zygotes that were considered granular at collection and at 12–18 hours post insemination often did not look granular after cleavage. Likewise, some oocytes which appeared clear at collection developed into pre-embryos with granular-appearing blastomeres.

Polar Body Size and Shape

Variation in the size of first polar bodies (overly large or small), while quite a subjective observation, may represent negative conditions for the oocyte: of 119 instances where these characteristics were noted, only 71% of the oocytes were observed to have two pronuclei on the day following insemination, and a full 15% displayed triploid fertilization. Oo-

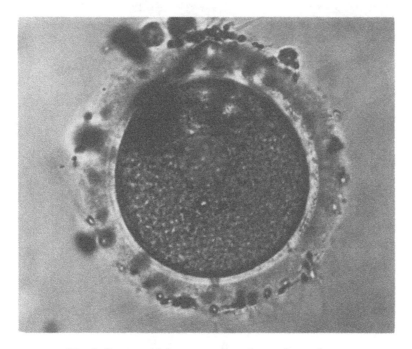

Fig. 2. Oocyte with large vacuole and granular ooplasm.

cytes with fragmented first polar bodies or with polar bodies which were completely detached from the oocyte proper fared even less well: a 17% triploidy rate was seen with fragmented polar bodies (13/76) and a 36% triploidy rate with detached ones, presumed to be postmature (8/22). These findings are similar to those reported earlier.[12]

Vacuoles and Inclusions

When cytoplasmic vacuoles were present at the time of oocyte harvest, a poor prognosis could be assumed for fertilization and pregnancy rates. In 124 oocytes with numerous and large cytoplasmic vacuoles, only 24 fertilized with two pronuclei (20%) and no pregnancies were established in five patients with transfer of only pre-embryos developed from vacuolated oocytes (Figure 2).

Inclusion bodies, such as the "refractile body" described in earlier publications, presented a poor prognosis for successful fertilization. This structure, 10 to 20 microns in diameter under brightfield microscopy, exhibits a highly refractile appearance and seems to be composed largely of lipid material and dense granules. It may be associated with early stages of oocyte degeneration. In an earlier study, only 2% of 47 oocytes displaying this condition were fertilized (Figure 3).[12]

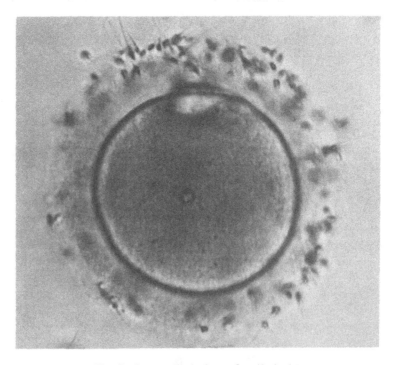

Fig. 3. Oocyte displaying refractile body.

Nuclear Maturation at Harvest

Oocytes collected at more advanced stages of maturation (metaphase II; MII) demonstrated the best chances for developing two pronuclei after insemination (Figure 4). Oocytes requiring between 5–15 hours in culture before attaining meiotic competence (metaphase I-mature; MI-M) demonstrated slightly lower rates of fertilization (Figure 5). Metaphase I oocytes that required longer than 15 hours in culture before reaching a metaphase II state (metaphase I-I; MI-I) demonstrated a very marked reduction in fertilizing potential, as did fully immature oocytes (prophase I; germinal vesical-bearing; PI) (Figure 6). A small group of oocytes, classified as "undetermined nuclear status at collection", but presumed to be mature or nearly mature by virtue of cumulus/corona expansion, fertilized in a fashion similar to metaphase II oocytes (Table 5). It should be noted that semen samples were routinely collected soon after oocytes were harvested, thus, spermatozoa were also kept in culture for the ensuing hours until oocytes were ready for insemination. Under these circumstances, it cannot be stated that lowered fertilization potential is correlated solely with oocyte maturity because reduction in sperm quality over time may also have been involved.

Pregnancy and ongoing pregnancy was evaluated in 2205 transfer cycles and correlated with original nuclear states of oocytes at harvest. Results are summarized in Table 6. Of

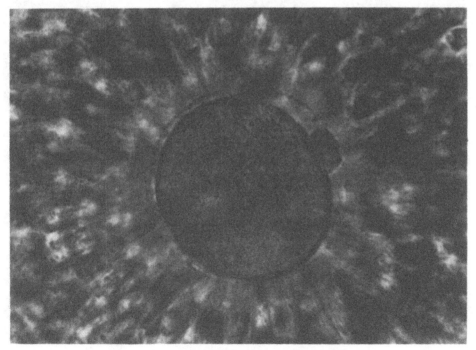

Fig. 4. Oocyte at metaphase II stage of maturation.

Table 5. Fertilization versus Maturation at Harvest. Norfolk Series 18–33 January 1985 to December 1988

	MII	Undeter.	MI-M	MI-I	PI
No. oocytes	4892	582	2629	424	1951
Fert./oocyte	94%	91%	86%	68%	49%
Abnormal fert.	8%	7%	7%	5%	3%

these transfers, 1225 can be placed into one of five "pure" transfer groups, groups denoting the transfer of one or more concepti of a single nuclear status at collection. For example, a "pure" MII group consists of transfer cycles in which only pre-embryos developed from metaphase II oocytes were transferred; a "pure" PI group had only the transfer of pre-embryos developing from prophase I oocytes. These pure groups would be opposed to "mixed" transfer groups where more than one conceptus was transferred, the concepti were of different maturational origins, and identification of the pre-embryo responsible for preg-

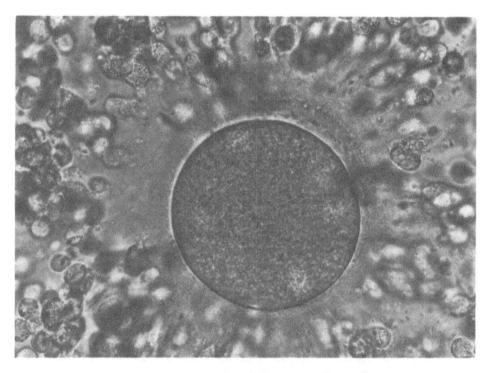

Fig. 5. Oocyte at metaphase I stage of maturation.

Table 6. Status of Nuclear Maturation and Association with Pregnancy. Norfolk Series 18–33 January 1985 to December 1988

Transfer Classification	No.	Preg/Transfer	Ongoing/Transfer
"Pure" MII	(769)	27%	17%
"Pure" MI-M	(272)	22%	15%
"Pure" Undetermined	(95)	20%	12%
"Pure" MI-I	(33)	15%	12%
"Pure" PI	(56)	5%	5%
"Mixed" MII + others	(922)	31%	21%
"Mixed" MI-M + others	(52)	25%	17%
"Mixed" MI-I + others	(6)	33%	17%

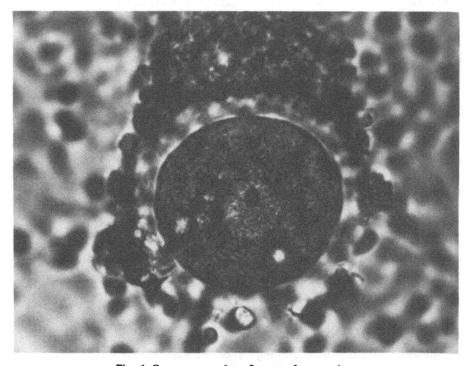

Fig. 6. Oocyte at prophase I stage of maturation.

Table 7. Pre-embryo Grading Blastomere Evaluation

Pre-embryo Grading blastomere evaluation
1 equal sizes, no fragments
2 equal sizes, minor fragments or blebs
3 distinctly unequal sizes, no fragments
4 equal or unequal sizes, significant fragmentation
5 few recognizable blastomeres, severe fragmentation

nancy was therefore impossible. Pure groups of metaphase II and metaphase I concepti demonstrated similar pregnancy and ongoing pregnancy rates although a tendency was noted for better results with advanced maturational stages. Pre-embryos developed from PI oocytes showed a significant reduction in the establishment of pregnancy, an observation which has been consistent in our laboratories over time. "Mixed" transfer cycles represented a higher overall pregnancy rate which can be attributed primarily to a larger number of pre-embryos transferred per cycle.

PRE-EMBRYOS

A grading system has been in effect in Norfolk since the beginning of 1987. It places a grade on the transfer cycle based upon the highest grade pre-embryo which is transferred (Table 7). Grades range from 1 (perfect morphology) to 5 (severe or complete fragmentation) (Figures 7–11). Pre-embryos are evaluated prior to transfer and the cycle classified accordingly. Pregnancy results for each morphologic grade are given in Table 8. Cycles with at least one pre-embryo of grade 1 morphology demonstrated significantly better success in establishing pregnancy and continuing pregnancy, although some degree of success was noted in each group.

Pre-Embryo Grading in Stimulated (IVF) versus Natural Cycles

Transfer cycles were evaluated for pregnancy based upon whether or not patients had been stimulated with gonadotropins during routine IVF trials or had been transferred in natural cycles as the result of oocyte donation or thawing of cryopreserved material. It was in-

Table 8. Pre-embryo Grading

Pre-Embryo Grading		
Grade	No.	Pregnancy/Transfer
1 perfect	359	46% (32)
2 minor frag.	443	28% (18)
3 irreg size	68	10% (7)
4 reg/irreg + mod/major frag.	315	19% (12)
5 severe frag.	115	5% (3)
Total	1300	28% (18)

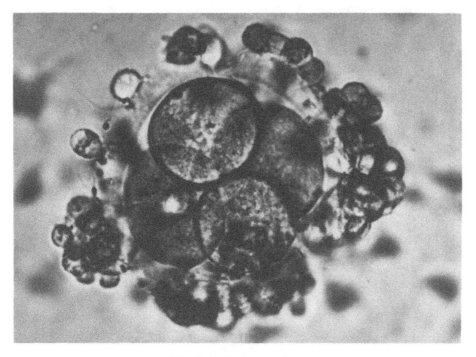

Fig. 7. Grade 1 pre-embryo.

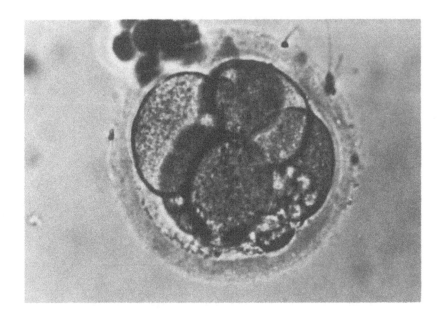

Fig. 8. Grade 2 pre-embryo.

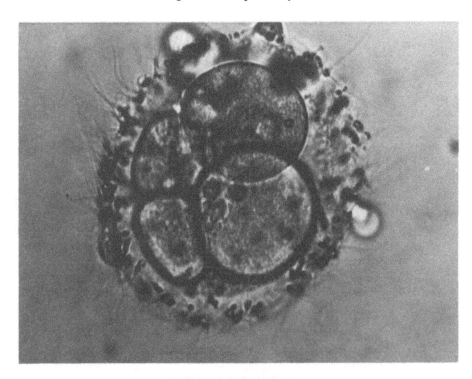

Fig. 9. Grade 3 pre-embryo.

teresting to note that of 120 transfers to patients in natural cycles, no pregnancy was established with grade 4 or grade 5 pre-embryos. In addition, an extremely high incidence of pregnancy was observed with grade 1 concepti (Table 9).

Growth Rates and Pre-Embryo Grading

Pre-embryos were additionally evaluated for their growth rate in culture prior to intrauterine transfer. Concepti were classified as either:

1. rapidly cleaving with eight or more blastomeres at 36 to 48 hours after insemination
2. cleaving at an average rate and displaying two blastomeres by 24 hours post insemination and four blastomeres by 40 hours post insemination, or
3. cleaving slowly and displaying only two blastomeres after 40 hours post insemination

When cleavage rates were combined with pre-embryo grading data, several interesting observations were noted. As a group, the slowly dividing pre-embryos scored low in their ability to establish pregnancies; nonetheless, if they were of grade 1 quality, their pregnancy potential equalled that of more rapidly developing concepti. Other authors have reported

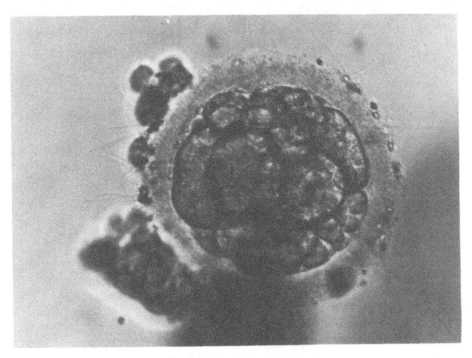

Fig. 10. Grade 4 pre-embryo.

Table 9. Pre-embryo Grading in Stimulated and Natural Cycles

	Stimulated Cycle		Natural Cycle	
Grade	No.	%Preg.	No.	%Preg.
1	325	45%	34	59%
2	400	28%	44	32%
3	64	9%	4	25%
4	285	21%	30	0%
5	106	6%	8	0%
Total	1180	28% (18) ongoing	20	29% (23) ongoing

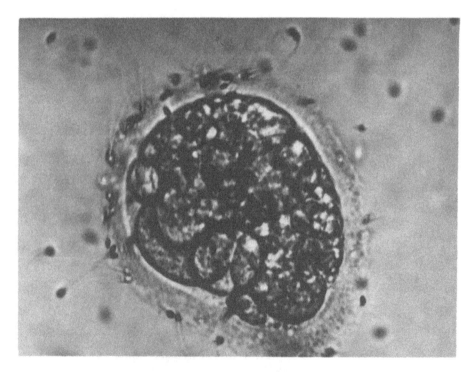

Fig. 11. Grade 5 pre-embryo.

Table 10. Growth Rates and Pre-embryo Grading. 1300 Cycles with Transfer

Grade	Rapid (77)	Average (1077)	Slow (146)	Total (1300)
1 (359)	39%	47%	50%	46% (32%)
2 (444)	14%	30%	11%	28% (18%)
3 (68)	0%	15%	0%	10% (7%)
4 (315)	7%	22%	12%	19% (12%)
5 (114)	0%	7%	2%	5% (3%)
Total (1300)	25% (16%) ongoing	31% (20%) ongoing	9% (5%) ongoing	28% (18%) ongoing

tendencies for enhanced survival with afragmented pre-embryos, particularly if cleavage rate was considered within average or better limits.[13-16] In addition, incidence of pregnancy was extremely low for all transfer grades except grade 1 in both rapid and slow cleaving groups (Table 10). Pre-embryos dividing at an average rate contributed to pregnancy at fair rates even when a grade of 4 was associated with the transfer.

DISCUSSION

Many numbers have been tabulated, combined with other numbers, and presented here in an effort to try and establish some sense of the long range potential of the individual

Table 11. Ideal versus Poor IVF Cycle. Norfolk Series 26–33

	Good Prognosis IVF n=226	Poor Prognosis IVF n=38
Sperm morphology	"N"	"P"
Conceptus morph.	1 or 2	3, 4, 5
Number transferred	4+	<3
Pregnancy/transfer	48%	8%
Ongoing/transfer	33%	5%

spermatozoon, oocyte, and early pre-embryo. One can further extrapolate the idealized IVF cycle from the data presented above and compare it with the cycle demonstrating the poorest prognosis (Table 11).

Little has been said about inherent genetic problems which may be present within any of these reproductive cells and which may in the end play a larger role in determining IVF success than any morphological estimation. Harvested oocytes may be chromosomally defective at rates as high as 30%–50%,[17-19] spermatozoa at rates reported to be 8%,[6] and pre-embryos at percentages of 26%–36%[20] or higher. Without genetically healthy gametes, success in establishing pregnancy is limited despite our laboratory quality control, efforts to enhance post-fertilization development, and dissection and tabulation of data. Ultimately, IVF success is really dependent on only two gross factors: pre-embryo quality and uterine receptivity. Until we fully understand all the mechanisms involved in producing the healthy gamete and receptive uterus, we can only try to establish relationships between that which we observe and that which constitutes the final outcome.

REFERENCES

1. L. J. Burkman, Experimental approaches to evaluation and enhancement of sperm function, *in:* "In Vitro Fertilization - Norfolk," H. W. Jones Jr., G. S. Jones, G. D. Hodgen, and Z. Rosenwaks, eds., Williams and Wilkins, Baltimore, Maryland, 201 (1986).
2. A. A. Acosta, T. Kruger, R. J. Swanson, K. F. Simmons, S. Oehninger, L. L. Veeck, P. Hague, P. Pleban, M. Morshedi, S. Ackerman, Role of IVF in male infertility, *Annals of the New York Academy of Sciences*, 541:297 (1988).
3. T. F. Kruger, A. A. Acosta, K. F. Simmons, R. J. Swanson, J. F. Matta, L. L. Veeck, M. Morshedi, and S. Brugo, A new method of evaluating sperm morphology with predictive value for IVF, *Urology*, 30:248 (1987).
4. S. Oehninger, A. A. Acosta, T. Kruger, L. L. Veeck, J. Flood, and H. W. Jones Jr., Failure of fertilization in in vitro fertilization: The "occult" male factor, *J. In Vitro Fert. Embryo Transfer*, 5:181 (1988).
5. T. C. Appleton, and S. B. Fishel, Morphology and x-ray microprobe analysis of spermatozoa from fertile and infertile men in in vitro fertilization, *J. In Vitro Fert. Embryo Transfer*, 1:188 (1984).
6. R. H. Martin, Comparison of chromosomal abnormalities in hamster egg and human sperm pronuclei, *Biol. Reprod.*, 31:819 (1984).
7. J. S. McDowell, Preparation of spermatozoa for insemination in vitro, *in:* "In Vitro Fertilization - Norfolk," H. W. Jones Jr., G. S. Jones, G. D. Hodgen, and Z. Rosenwaks, eds., Williams and Wilkins, Baltimore, Maryland, 162 (1986).
8. A. A. Acosta, C. F. Chillik, S. Brugo, S. Ackerman, R. J. Swanson, P. Pleban, J. Yuan, and D. Hague, In vitro fertilization and the male factor, *Urology*, 28:1 (1986).
9. R. T. Scott, S. Oehninger, R. Menkveld, L. L. Veeck, and A. A. Acosta, Critical assessment of sperm morphology before and after double swim-up preparation for in vitro fertilization (IVF), In press, *Arch. Androl.*
10. D. Franken, S. Oehninger, L. Burkman, C. C. Coddington, T. F. Kruger, Z. Rosenwaks, A. A. Acosta, G. D. Hodgen, The hemizona assay (HZA): A predictor of human sperm fertilizing potential in IVF treatment, *J. In Vitro Fert. Embryo Transfer*, 6:44 (1989).

11. S. Oehninger, C. C. Coddington, R. Scott, D. Franken, L. Burkman, A. A. Acosta, and G. D. Hodgen, Hemizona assay: Assessment of sperm dysfunction and predictor of in vitro fertilization outcome, *Fertil. Steril.*, 51 (1989).

12. L. J. Veeck, Oocyte assessment and biological performance, *Annals of the New York Academy of Sciences*, 541:259 (1988).

13. V. N. Bolton, S. M. Hawes, C. T.Taylor, J. H. Parsons, Development of spare human preimplantation embryos in vitro, An analysis of the correlations among gross morphology, cleavage rates, and development to the blastocyst, *J. In Vitro Fert. Embryo Transfer*, 6:30 (1989).

14. J. M. Cummins, T. M. Breen, K. L. Harrison, J. M. Shaw, L. M. Wilson, J. F. Hennessey, A formula for scoring human embryo growth rates in in vitro fertilization: It's value in predicting pregnancy and in comparison with visual estimates of embryo quality, *J. In Vitro Fert. Embryo Transfer*, 3:284 (1986).

15. F. Puissant, M. Rysselberge, P. Barlow, J. Deweze, and F. Leroy, Embryo scoring as a prognostic tool in IVF treatment, *Hum. Reprod.*, 2:705 (1987).

16. P. Claman, A. D. Randall, M. M. Seibel, T. Wang, S. P. Oskowitz, and M I. Taymoor, The impact of embryo quality and quantity on implantation and the establishment of viable pregnancies, *J. In Vitro Fert. Embryo Transfer*, 4:218 (1987).

17. M. Plachot, J. De Grouchy, A. M. Junca, J. Mandelbaum, J. Salat- Baroux, and J. Cohen, Chromosomal analysis of human oocytes and embryos in an in vitro fertilization program, *Annals of the New York Academy of Sciences*, 541:384 (1988).

18. R. H. Martin, M. Mahadevan, P. J. Taylor, K. Hildebrand, L. Long- Simpson, D. Peterson, J. Yamamoto, and J. Fleetham, Chromosomal analysis of unfertilized human oocytes, *J. Reprod. Fertil.*, 78:673 (1986).

19. H. Wramsby, K. Fredga, and P. Liedholm, Chromosomal analysis of human oocytes recovered from preovulatory follicles in stimulated cycles, *New Engl. J. Med.*, 316:121 (1987).

20. H. Wramsby, Chromosomal analysis of preovulatory human oocytes and oocytes failing to cleave following insemination in vitro. *Annal of the New York Academy of Sciences*, 541:228 (1988).

ASSESSMENT OF EMBRYO QUALITY AND VIABILITY THROUGH EMBRYO-DERIVED PLATELET ACTIVATING FACTOR

Usha Punjabi, Annie Vereecken, Luc Delbeke, Marlane Angle*,
Mieke Gielis, Christel Depestel, Chris Peeters, Jan Gerris,
John Johnston*, and Philippe Buytaert

Fertility Clinic, University Hospital of Antwerp, Belgium
* University of Texas, Dallas, U.S.A.

INTRODUCTION

Nearly 90% of apparently normal embryos fail to implant in the uterus following embryo transfer. This implantation failure might probably be due to a combination of embryonic and uterine factors.[1] O'Neill et al.,[2] reported that an increased vascular demand for blood platelets, resulting in mild thrombocytopenia, was an initial maternal response to pregnancy in mice[3] and humans.[2] In both these species, this effect was a direct consequence of a factor produced by the fertilized ova, the platelet activating factor (PAF).[4] This embryo-derived PAF showed similar chemical, biochemical and physiologic properties as the PAF, 1-0-alkyl-2-acetyl-sn-glyceryl-3-phosphocholine (PAF- acether).[5]

PAF appears during the pre-implantation period as an early signal of pregnancy. The production of this factor was shown to have a correlation with embryo quality and viability, therefore, this study was designed to monitor the production of PAF by the embryo in vitro by the release of 3H-serotonin from sensitized rabbit platelets, to evaluate its role in assessing embryo viability prior to transfer, and to establish correlations between PAF and clinical parameters.

Advances in Assisted Reproductive Technologies, Edited by
S. Mashiach *et al.*, Plenum Press, New York, 1990

Table 1. Grading of Embryos

Grade	Blastomeres	Cytoplasm
I	Equal size	No fragments, Nongranular
II	Equal size	Slight fragmentation, Granular
III	Unequal size	No fragments, Nongranular
IV	Unequal size	Slight fragmentation, Granular
V	Equal or Unequal size	Major fragmentation, Granular

Veeck, 1986.[6]

MATERIALS AND METHODS

Ovulation Induction

52 patients were selected at random for this study. They were given clomiphene citrate (clomid: Merrill) on days 3–7, supplemented with human menopausal gonadotropin (hMG; Serono, Italy) from day 6 onwards. Follicular development was monitored by daily peripheral blood Luteinizing hormone (LH), Oestrogens (E_2) and Progesterone (P4) assays together with ultrasonic measurements of follicular size. Ovulation was induced by administering human chorionic gonadotrophin (hCG Pregnyl, Organon). Laparoscopy or ultrasound for oocyte recovery was scheduled 35–37 hr post-hCG injection or 25–26 hr after an endogenous LH surge.

Fertilization and Embryonic Growth In Vitro

Ova were cultured in Earles balanced salt solutions (EBSS, Flow Laboratories) supplemented with 25 mM sodium bicarbonate, 1 mM sodium pyruvate, 100 IU/ml penicillin and 10% v/v heat-inactivated patient serum. Spermatozoa were added at a concentration of $50 – 150 \times 10^3$ depending on the quality of the semen sample. Fertilization was confirmed by the observation of pronuclei and the fertilized eggs were then transferred into 300 µl medium containing 15% v/v serum. After a further 24–30 hr one to four embryos were transferred into the uterine lumen in medium containing 75% v/v patient serum using a Wallace catheter (Wallace, England). Embryonic development was assessed using the criteria shown in Table 1.

Embryo-derived Platelet Activating Factor (PAF)

Following embryo replacement the 300 µl of culture medium was frozen at −20° C until assayed.

Extraction of PAF. The samples were adjusted to 0.8 ml with deionized water. Trace amounts (3500 cpm) of labelled PAF was added to each sample. The samples were first extracted with methanol : chloroform (2:1,v/v), sonicated for 5 min. and allowed to stand for another 20 min. at room temperature. The tubes were then centrifuged for 10 min. at 3000 g. The supernatant was separated and the residue re-extracted. The pooled supernatants were now extracted with water: chloroform (1:1, v/v) to effect phase separation. After two extractions the chloroform layers were pooled and evaporated to dryness under a stream of nitrogen in a water bath.

Thin - layer chromatography. A particularly effective means for the isolation and purification of PAF was through thin layer chromatography on Sicilia Gel G - 60 plates (Merck No. 5721). PAF (1-O-alkyl-2-acetyl-sn-glycero-3-phosphocholine) migrates (Rf = 0.21; Figure 1) between sphingomyeline and lecithin by utilizing a solvent system of chloroform - methanol - water - acetic acid (50:30:6:8, v/v). As standards phosphatidylcholine (2 mg/ml), sphingomyelin (15 mg/ml) and lyso-lecithin (1 mg/ml) were applied on the predried plates. The sample extracts were solved in chloroform-methanol (2:1, v/v) and subsequently applied. At the end of the run, the plates were removed and phospholipids visualised in a tank-filled with iodide-crystals at the bottom.

The lanes containing the PAF were sectioned and scraped into glass tubes. These samples containing PAF were then mixed with a solution of chloroform - methanol - water - acetic acid (1:2:0.7:0.1, v/v) and sonicated for 5 mins. After standing for 15 min. these tubes were then centrifuged for 20 mins at 3000 g. The clear solvent layer considered to be the PAF (lipid) extract was carefully removed from the residue. The residue was reextracted and the solvent layers pooled. To this chloroform—methanol (1:1, v/v) was added and mixed. The lower chloroform—rich layer was collected after centrifugation and evaporated under a stream of nitrogen. The tubes were rinsed with chloroform - methanol (2:1, v/v) and then analyzed for PAF activity.

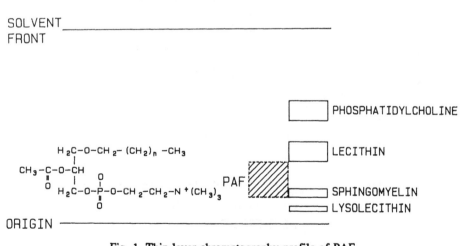

Fig. 1. Thin layer chromatography profile of PAF.

Preparation of washed 3H-serotonin labeled platelets. 3H- serotonin - labeled rabbit platelets were prepared by the procedure described by Pinckard et al.[7]

PAF assay. The dried extracts were dissolved in 100 μl bovine serum albumin solution. An aliquot was taken for recovery determination and another for the assay proper. Generally two different volumes were used for the assay. A 200 μl aliquot of platelet suspension was added to each sample followed by 20 μl of formalin solution. The samples were centrifuged and placed on ice.

PAF standard curve was treated in the same way. The percentage of serotonin secretion was determined by liquid scintillation spectroscopy relative to that released from the same volume of platelets after the addition of the scintillation fluid. The results were graphed linearly as the per cent serotonin secretion versus the PAF-concentration. Results of the unknown samples were read from this curve and corrected for recovery.

RESULTS

Out of the 52 patients who underwent embryo replacement of 1–4 cleaved embryos 18

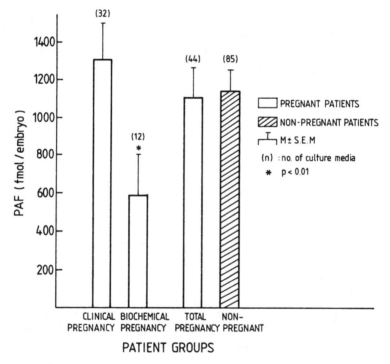

Fig. 2. Production of embryo-derived PAF in pregnant and nonpregnant patients.

(35 %) became pregnant: 13 clinical and 5 biochemical pregnancies. Embryos were considered PAF positive if the concentration was within the detection limit of the assay (>11 fmol/embryo).

A total of 120 culture media were analyzed for PAF. Our results showed 85 were PAF positive (71%), 26 PAF negative (22%) and 9 (8%) were difficult to analyze due to the oily nature of the samples. PAF concentration did not differ significantly between the patients who failed to become pregnant and those who became pregnant. But within the pregnant group the concentration was significantly higher ($p < 0.01$) in those patients who had an uneventful pregnancy (Figure 2). The activity of the 120 embryos was analyzed on the basis of the size of the follicles. There was no significant difference in the PAF concentration of embryos originating from follicles < 4 ml, 4 - 6 ml and > 6 ml in volume (Figure 3).

PAF concentration displayed a marginal correlation with the developmental stage of the embryo (Figure 4). Embryos produced significantly ($p < 0.02$) less PAF in the 5 or more cell stage. Morphological grading showed relatively no significant difference in the concentration of PAF produced (Figure 5). Culture conditions had an important effect on embryo quality. Embryos grown in media prepared more than two days earlier produced significantly more PAF than fresh medium (Figure 6).

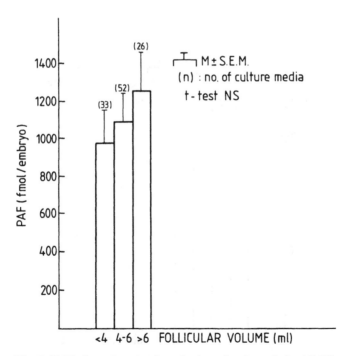

Fig. 3. Follicular volume and production of embryo-derived PAF.

The luteal phase endocrine profile were assessed in the pregnant and the non pregnant groups in an attempt to define the reasons for pregnancy failure. There was no difference in the P4 profile (Figure 8) for the two groups of women, but the E_2 profile for the women who failed to become pregnant was significantly (p < 0.004) elevated when compared with that of the pregnant group (Figure 7).

DISCUSSION

Recent reports describing the production of an embryo-derived PAF have suggested that the detection of this factor in embryo culture medium may provide a means of assessing human embryo viability.[3,4,8] The embryo-derived PAF has been detected by a bioassay that involves monitoring the relative change in the peripheral blood platelet count after intraperitoneal injection of embryo culture medium into splenectomized mice.[9] The sensitivity of the assay is less well defined, and the bioassay does not account for possible degradation of the activator by the animal after injection, nor does it take into consideration the presence of possible inhibitors.[4,9]

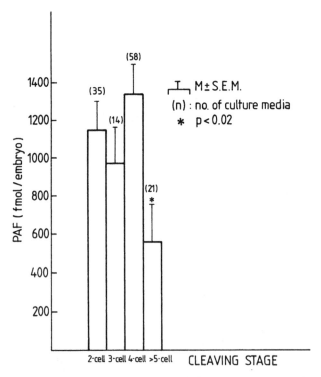

Fig. 4. Cleaving stage of the embryo and production of embryo-derived PAF.

Therefore, we adopted a technique whereby highly purified PAF was obtained by sequential methanol and chloroform extraction followed by thin layer chromatography. By these procedures PAF was separated from any of the known modulators (e.g., thrombin, ADP, collagen, prostaglandins or their intermediates) of platelet aggregation and/or secretion. The modified procedure for preparing washed platelets increased platelet yield with no contamination with leukocytes or erythrocytes and by consistent platelet responsiveness for aggregation to micromolar concentration of ADP.[7] PAF assay here is a functional one and a variety of factors influence the biologic activity of PAF to induce platelet secretion of serotonin, which forms the basis of the direct PAF assay. Platelet secretion is chosen rather than platelet aggregation to quantitate PAF because the extent of PAF platelet aggregation is dependent in part upon platelet release and reactivity to ADP while platelet secretion is not effected.[7] The concentration of PAF that is present in various biological samples may be regulated by the rate of its biosynthesis or degradation.

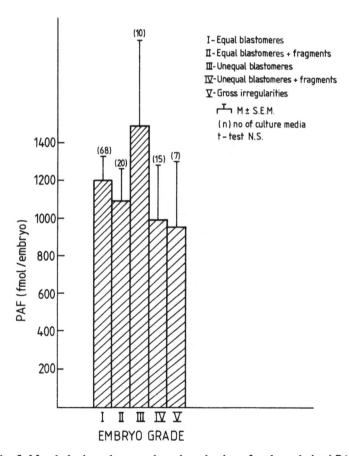

Fig. 5. Morphologic embryo grade and production of embryo-derived PAF.

PAF is inactivated by a specific PAF acetylhydrolase (1 - alkyl - 2 - acetyl - sn - glycero - 3 - phosphocholine acetylhydrolase ; EC 3.1.1.48) to form lyso PAF.[18] Both an intracellular and plasma form of the enzyme has been reported.[19] Since the embryo culture medium used in this study contained 15% (v/v) maternal serum, acetylhydrolase may have caused some loss of embryo-derived PAF activity and thus resulted in some false-negative results. But Farr et al.,[20, 21] who were the first to describe this enzyme in human plasma, showed that one of the characteristic features were that the enzyme could be inactivated by heating at 65° C for 30 mins. This finding was later confirmed by Maki et al.[26] Moreover, it was not possible to determine whether the large variability between embryos in the production of embryo-derived PAF was normal. It might perhaps represent some metabolic deficiency, or an artifact of fertilization and culture in vitro.

There was no significant difference in the PAF activity for various follicular volume ranges. This is in contrast to the findings of O'Neill et al.,[9] who obtained more PAF-positive embryos from moderately sized follicles (4 to 6 ml) compared with either larger or

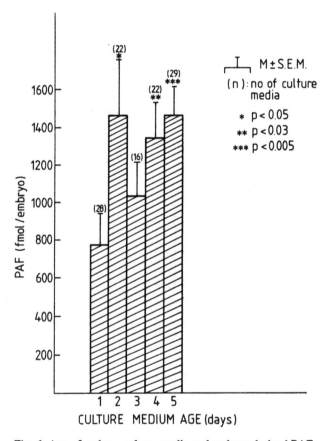

Fig. 6. Age of embryo culture media and embryo-derived PAF.

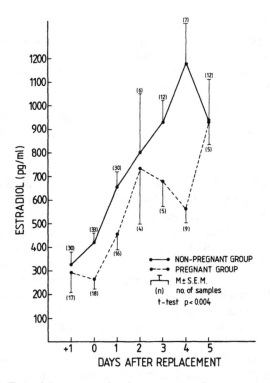

Fig. 7. Estradiol concentration in pregnant and nonpregnant group.

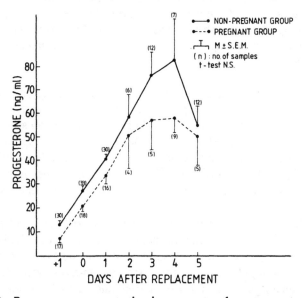

Fig. 8 . Progesterone concentration in pregnant and nonpregnant groups.

smaller follicles. The reason for this has not yet been defined. Establishing a correlation between transferred cenceptus and implanted embryo is often difficult because most transfers involve replacing more than one conceptus. There was a marginal correlation between the developmental stage of the embryo and PAF production. According to O'Neill[17] mouse embryos produce detectable levels of PAF within six hours of culture, and PAF production displays a correlation with the development of the embryo from 2-cell to blastocyst stages in protein free medium. Mouse embryos produce progessively less PAF in vitro with extended culture, compared to embryos of similar developmental stage collected freshly from the reproductive tract.

It was initially believed that blastomeres of normal concepti must be equal in size, regular in shape, free of extracytoplasmic fragments and blebs, and free of any cytoplasmic vacuolizaion or nuclear fragments. Such concepti are commonly observed, but perfect morphology is not required for implantation to occur. Often, concepti with less than desirable morphological aspects can be directly attributed to favourable in vitro ferilization results.[6] This was confirmed in our results, where a large proportion of morphologically normal embryos failed to display activity, and conversely, a number of embryos of poor appearance displayed activity. It is not known whether this variability in production of the factor(s) by apparently normal embryos was an artifact of induced ovulation, in vitro fertilization, and embryo culture procedures or whether this also occurs in natural conception.[8]

It has been previously shown[9, 12] that embryos grown in media that have been prepared more than three days previously resulted in significantly more pregnancies than fresh medium. It was shown here that significantly fewer embryos produced PAF when grown in medium that was three days old or younger. It seems reasonable to suggest that optimal embryo culture conditions should be defined as those which results in the production of PAF by the greatest proportion of embryos. The discovery of the factor provides, for the first time, a single assay for assessing the success of embryo culture and quality control of embryo culture procedures.

Media routinely used for culture of gametes usually contain glucose, lactate, and pyruvate,[13] the metabolism of which (via glycolytic and/or oxidative metabolic pathways) can produce energy for support of motility, etc. For most mammalian species evaluated,[14, 15] the requirement for a glycolyzable substrate at some point in capacitation suggests that glycolytic pathway is of considerable importance but that the degree of reliance on this pathway is species-specific. Hoshi et al.[16] observed that glucose is required by human sperm for optimal penetration of human zonae and zona-free hamster eggs.

Induced ovulation has previously been shown to cause ovarian hyperstimulation, resulting in elevated E_2 production in a significant proportion of women.[10] It is well established that high postovulatory levels of E_2 block implantation and O'Neill et al.,[8] showed that those women who displayed early pregnancy-associated thrombocytopenia, but failed to become pregnant, had a significant higher E_2 concentration than did pregnant women. Therefore, pregnancy was only established in those patients who displayed early pregnancy-associated thrombocytopenia and who had a satisfactory luteal endocrine profile prior to implantation. Appropriate endometrial preparation for implantation is dependent on the relative concentrations of the steroid hormones in the follicular and luteal phase of the cycle.[11] The hyperstimulation of E_2 production independently of P4 production, markedly alters the E_2/P4 ratio.

Our study shows that, there was a reduction in the implantation rate following ovarian stimulation when preovulatory serum E_2 values were excessive, even though a viable embryo was replaced. The adverse effect of excessive E_2 concentration may have operated at several levels, for e.g., by influencing oocyte and embryo quality [22, 23] or having a direct luteolytic action[24] or exerting a direct antinidatory effect on the endometrium.

Embryo-derived PAF seems to be strongly correlated with the viability of embryos and their pregnancy potential and may provide, for the first time, a quantitative, non-invasive means of determining which embryos are viable for transfer. It would also be a powerful tool for examing the effects of all aspects of the IVF and embryo transfer technique on the production of good-quality embryos.

The results reported here suggest that approximately 70% of embryos produced by IVF were viable. In IVF, in which multiple embryo transfers are routine this level of embryo viability should result in a much higher pregnancy rate. The current low pregnancy rate (18%) suggests that this figure of 70% is an overestimation of embryo quality or that other factors apart from embryo quality severely limit pregnancy potential. It seems likely that uterine receptivity for implantation or failure of the transfer procedure may be one such factor.

If viable embryos are obtained and a successful embryo cryopreservation program exits, we would propose that in case of excessive ovarian stimulation these embryos be frozen for replacement in later natural cycles.

REFERENCES

1. P. A. W. Rogers, B. J. Milne and A. O. Trounson, A model to show human uterine receptivity and embryo viability following ovarian stimulation for in vitro fertilization, *J. In Vitro Fert. Embryo Transfer*, 1:93 (1986).
2. C. O'Neill, A. A. Gridley-Baird, I. L. Pike, R.N. Porter, M.J. Sinosich and D. M. Saunders, Maternal blood platelet physiology and luteal phase endocrinology as a means of monitoring pre and post-implantation embryo viability following in vitro fertilization, *J. In Vitro Fert. Embryo Transfer*, 2:87 (1985).
3. C. O'Neill, Thrombocytopenia is an initial maternal response to fertilization in the mouse, *J. Reprod. Fert.*, 73:559 (1985).
4. C. O'Neill, Examination of the cause of early pregnancy - associated thrombocytopenia in mice, *J. Reprod. Fert.*, 73:567 (1985).
5. C. O'Neill, Partial characterization of the embryo - derived platelet activating factor in mice, *J. Reprod. Fertil.*, 75:375 (1985).

6. L. L. Veeck, Cleaved Human Concepti after IVF, *in* "Atlas of the human oocyte and early conceptus", C-L. Brown, ed., Williams and Wilkins, Baltimore 163 (1986).
7. R.N. Pinckard, R.S. Farr and D.J. Hanahan, Physicochemical and functional identity of rabbit platelet activating factor (PAF) released in vivo during IgE anaphylaxis with PAF released in vitro from IgE sensitized basophils, *J. Immuno.*, 123:1847 (1979).
8. C. O'Neill, I. L. Pike, R. N. Porter, A. A. Gridley-Baird, M. J. Sinosich, and D. M. Saunders, Maternal recognition of pregnancy prior to implantation methods for monitoring embryo viability in vitro and in vivo, *Ann. NY. Acad. Sci.*, 442:429 (1985).
9. C. O'Neill, A. A. Gridley-Baird, I. L. Pike, and D. M. Saunders Use of a bioassay for embryo-derived platelet activating factor as a means of assessing quality and pregnancy potential of human embryos, *Fertil. Steril.*, 47:969 (1987).
10. J. C. McBain, R. J. Pepperell, H. P. Robinson, M. A. Smith, and J. B. Brown, An unexpectedly high rate of ectopic pregnancy following induction of ovulation with human pituitary and chorionic gonadotrophin, *J. Obst. Gynaecol.*, 87:5 (1980).
11. J. E. O'Grady and S. C. Bell, The role of the endometrium in blastocyst implantation, *in*: "Development of Mammals", M. H.Johnson, ed., North Holland - Amsterdam, 165 (1977).
12. E. Amos, C. O'Neill, G. N. Kellow, I. L. Pike, and D. M. Saunders, Aged culture medium improves pregnancy rate in IVF and ET, *Proc. Fertil. Soc. Aust.*, 3:4 (1984).
13. B. J. Rogers, Mammalian sperm capacitation and fertilization in vitro: A critique of methodology, *Gamete Res.*, 1:165 (1978).
14. K. Niwa and A. Iritane, Effect of various hexoses on sperm capacitation and penetration of rat eggs in vitro, *J. Reprod. Fertil.* 53:267 (1978).
15. E. Dravland and S. Meizel, Stimulation of hamster sperm capacitation and acrosome reaction in vitro by glucose and lactate and inhibition by the glycolytic inhibitor - chlorohydrim, *Gamete Res.*, 4:515 (1981).
16. K. Hoshi, A. Saito, M. Suzuki, K. Hayashi and R. Yanagimachi, Effects of substrates on penetration of human spermatozoa into zona pellucida of human eggs and the zona-free hamster eggs, *Jpn J. Fertil. Steril.* 27, 439 (1982).
17. C. O'Neill, Embryo-derived platelet activating factor (Abs.) V World Congress on in vitro fertilization and embryo transfer, Norfolk, VA, PS-040, (1987).
18. M. L. Blank, T. C. Lee, V. Fitzgerald and F. Snyder , *J. Biol. Chem.*, 256:175 (1982).
19. M. L. Blank, M. N. Hall, E. A. Cress and F. Snyder, Inactivation of 1-alkyl-2-acetyl-sn-glycero-3 phoshocholine by plasma acetylhydrolase:Higher activities in hypertensive rats, *Biochem. Biophys Res. Commun.*, 113:666 (1983).
20. R. S. Farr, C. P. Cox, M. L. Wardlow and R. Jorgensen, Preliminary studies of an acid-labile factor (ALF) in human sera that inactivates platelet-activating factor (PAF), *Clin. Immunol. Immunopathol.*, 15:318 (1980).
21. R. S. Farr, M. L. Wardlow, C. P. Cox, K. E. Meng and D. E. Greene, Human serum acid-labile factor is an acetylhydrolase that inactivates platelet-activating factor, *Fed. Proc. Fed. Am. Soc. Exp. Biol.*, 42:3120 (1983).
22. H. W. Jones, A. Acosta, M. C. Andrews, J. E. Garcia, G. S. Jones, T. Mantzavinos, J. Mc Dowell, B. Sandow, L. Veeck, T. Whibley, C. Wilkes and G. Wright, The importance of the follicular phase to success and failure in in vitro fertilization, *Fertil. Steril.*, 40:317 (1983).
23. J. Dor, E. Rudak, S. Mashiach, L. Nebel, D. M. Serr, and B. Goldman: Periovulatory 17 β-estradiol changes and embryo morphological features in conception and nonconceptional cycles after human in vitro fertilization, 45:63 (1986).
24. J. N. Schoonmaker, K. S. Bergman, R. A. Steiner, and F. J. Karsch, Estradiol induced luteal regression in the rhesus monkey : evidence for an extra ovarian site of action, *Endocrinology* 110:1708 (1982).
25. J. M. Morris and G. Van Wageman, Interception:the use of post ovulatory estrogens to prevent inplantation, *Am. J. Obstet. Gynaecol.*, 115:101 (1973).

COCULTURE OF HUMAN ZYGOTES
ON FETAL BOVINE REPRODUCTIVE TRACT CELLS

Jacques Cohen, Sharon Wiker, Klaus Wiemer, Henry Malter, Carlene Elsner, Hilton Kort, Joe Massey, Andy Toledano, Dorothy Mitchell, and Robert Godke

Reproductive Biology Associates, Atlanta, USA
Department of Gynecology and Obstetrics, Emory University, Atlanta, USA
Department of Animal Science, Louisiana State University, Baton Rouge, USA

INTRODUCTION

When early cleaved human embryos are kept in culture, only one in four can be expected to develop into fully expanded blastocysts.[1] Alternatively, only 1% to 12% of them will implant and develop into full-term babies, when replaced into the uterus or fallopian tube before the third cleavage division commences.[2,3] Embryonic wastage following assisted reproduction can only in part be explained by an increased incidence of genetic abnormalities or loss at the time of replacement. Other more esoteric factors, like a reduced receptivity of the endometrium in stimulated menstrual cycles, probably play an important role as well.

Evidence for suboptimal culture conditions is only sparsely available. Embryonic morphology may change when cytotoxins are present in the culture system and likewise alterations in the pH or osmolarity may inhibit implantation due to metabolic shock. In a preliminary clinical trial we recently demonstrated that a monolayer of fetal bovine uterine fibroblasts could support human zygotes when coculture was maintained for one day prior to replacement.[4,5] The rate of cleavage seemed to enhance, and significantly more patients became pregnant following replacement of cocultured embryos as compared with those who had conventionally cultured embryos replaced. These findings support the assumption that alternate in vitro culture conditions may enhance embryonic development and implantation.

Advances in Assisted Reproductive Technologies, Edited by
S. Mashiach *et al.*, Plenum Press, New York, 1990

The results presented below give a short overview of the coculture procedures and the impact these helper cells may have on embryos and the general efficiency of assisted reproductive technology. A clinical trial in which patients whose embryos are either cultured conventionally or on fetal bovine uterine fibroblasts is still being conducted in our program at this time. The results of the first 159 patients allocated to this trial are presented below.

PREPARATION OF THE MONOLAYER AND COCULTURE

Fibroblasts were isolated from the endometrium of two healthy bovine fetuses, in the last trimester of gestation. The explant phase was maintained with Ham's F-10 culture medium supplemented with penicillin, streptomycin and amphotericin-B (HF-10). Monolayers were trypsinised with 0.05% trypsin in PBS. Cells were frozen at the seventh subpassage in DMSO. A sample of fibroblasts were screened for viruses.[4] Mono-layers for culture were prepared following at least one subpassage after thawing. Approximately 100,000 fibroblasts were pipetted into each well of four-well culture plates, 24 to 72 hours prior to their expected use as helper cells. Monolayered wells were checked for debris and percentage of cell coverage. Wells with 50% to 100% overgrown surface area were usually used for embryo culture. A lower percentage of monolayering was occasionally used and found to be supportive of embryonic growth. Multiple layers of fibroblasts were used on several occasions resulting in a decreased rate of embryonic growth, vacuolisation, and disappearance of cell boundaries.

Wells were washed once with protein-free HF-10 and twice with protein-free Earle's buffered saline solution (EBSS). Antimycotics were not used for well washing and coculture. Zygotes allocated to coculture were washed free of debris and spermatozoa and moved to wells, either containing fibroblasts and 15% human serum supplemented EBSS (coculture) or cell-free wells containing culture medium only (conventional culture). Culture was performed at 37° (under 5% CO_2, 5% O_2 and 90% N_2). Embryos remained 26 to 32 hours in either one of these two culture systems. Prior to replacement, embryos were washed three times and video-taped for subsequent morphologic analysis using a cell-free background. A maximum of three embryos were replaced at a time. Clinical pregnancy was defined when fetal heart activity was ultrasonically confirmed. Ongoing pregnancies were defined as either full term deliveries or beyond week 16 of gestation.

This study was performed on routine IVF patients during three clinical trials: (1) November 1987, (2) November 1988, and (3) February 1989. A total of 159 IVF patients with fertilization were included in this study. Every other patient was allocated to coculture, whereas embryos from in-between patients were cultured conventionally. Zygotes allocated to coculture were not selected on the basis of morphological differences. Videocinematography procedures and evaluation are extensively described elsewhere.[5,6] A total of 294 embryos were retrospectively analyzed by an observer who was not an integral part of the IVF team, and who did not collaborate on the coculture work.

Table 1. Clinical Pregnancies in the Coculture Trial

Type of culture	Proportion of clinically pregnant patients			
	Number of embryos replaced			All replacements
	1	2	3	
Conventional	1/15 (7%)	6/21 (29%)	12/45 (27%)	19/81 (24%)
Coculture	1/6 (17%)	8/18 (44%)	21/54 (39%)	30/78 (39%)

Table 2. Outcome of Pregnancies in the Coculture Trial

Proportion of Ongoing Pregnancies	Incidence of Multiple Pregnancies	Incidence of Implantation per Embryos Replaced
12/81 (15%)	2/19 (11%)	22/192 (12%)
27/78 (35%)	7/30 (23%)	37/204 (18%)

COCULTURE AND IMPLANTATION

A total of 396 embryos were replaced during this trial in 159 patients. The incidence of clinical pregnancy in 81 patients receiving conventionally cultured embryos was 24%, whereas 39% of patients who had cocultured embryos replaced became pregnant. The increase in implantation of cocultured embryos was independent from the number of embryos replaced (Table 1).

Seven out of 19 pregnant patients in the conventional culture group miscarried, whereas only 3/30 patients had a loss of pregnancy when cocultured embryos were used.

Although the proportion of ongoing pregnancies with coculture doubled per patient, a definite improvement in human IVF can only be confirmed when the incidence of implantation per single cocultured embryo is significantly increased. Although this variable increased considerably from 12% to 18% based on 192 control and 204 cocultured embryos, the trial will be continued in our program until appropriate statistical relevance can be shown.

THE MORPHOLOGY OF COCULTURED EMBRYOS

Significant differences in embryo morphology between conventionally and cocultured embryos were identified in four of 11 videocinematographic parameters. In general, cocultured embryos had more swollen blastomeres, less extracellular fragments, a higher blastomere-to-blastomere adherence, and their within-embryo zona pellucida thickness varied more frequently than conventionally cultured embryos (Table 3).

Table 3. Morphology of Conventionally and Cocultured Embryos

Morphologic Parameter	Percentage of Embryos		Significance
	Conventional	Coculture	
Low extracellular fragmentation	66%	80%	p = 0.02
Blastomere adherence	63%	84%	p = 0.003
Zona thickness variation increased	41%	60%	p = 0.0006
Swollen blastomeres	18%	51%	p < 0.0001

The total number of abnormal morphologic parameters per embryo was lower in cocultured than in conventionally cultured embryos. Only 30% of 117 conventionally cultured embryos were considered completely normal, whereas 52% of 104 cocultured embryos were without abnormalities.

THE FUTURE OF COCULTURE

EBSS is a relatively simple medium and does not provide the necessary nutrients for fibroblasts for an extended period of time. Nevertheless, tripronucleate zygotes developed well on fibroblasts when cultured for five or six days.[4] The coculture medium, which lacks fetal calf serum, does not promote cell-division, but the fibroblasts appear healthy and do not degenerate. Fetal calf serum was omitted because it is probably deleterious to human embryos. Hams F-10 (instead of EBSS) with bovine serum albumin (instead of fetal calf or human serum) may be a better modus for supporting human embryos and fetal calf cells simultaneously.

The nature of the benefits of a feeder-cell system is not well understood at present. "Conditioned" medium does not seem to support embryonic growth as well as direct embryo-cell contact. It is likely, considering the decrease in fibroblast cell growth using EBSS, that absorbence of ions played a major factor in this coculture system, rather than the secretion of growth factors. Other cell types (i.e. epithelial) or cells derived from different species (i.e. human) and different areas of the reproductive tract may be more supportive of embryonic growth. In addition, several types of cells can be combined in a synergistic effort to promote embryos.[7]

ACKNOWLEDGEMENTS

We would like to thank Mary Pat Mayer, Leigh Inge, Grace Amborski, and Robin Moseley for their valuable contributions.

REFERENCES

1. C. B. Fehilly, J. Cohen, R. F. Simons, S. B. Fishel, R. G. Edwards, Cryopreservation of cleaving embryos and expanded blastocysts in the human: a comparative study, *Fertil. Steril.*, 44:638 (1985).
2. J. L. Yovich, J. M. Yovich, W. R. Edirisinghe, The relative chance of pregnancy following tubal or uterine transfer procedures, *Fertil. Steril.*, 49:858 (1988).
3. The Society of Assisted Reproductive Technology, The American Fertility Society, In vitro fertilization/embryo transfer in the United States: 1987 Results from the National IVF-ET Registry, *Fertil. Steril.*, 51:13 (1989).
4. K. E. Wiemer, J. Cohen, G. F. Amborski, L. Munyakani, S. Wiker, R. A. Godke, In vitro development and implantation of human embryos following culture on fetal bovine uterine fibroblast cells, *Hum. Reprod.*, in press.
5. K. E. Wiemer, J. Cohen, S. R. Wiker, H. E. Malter, G. Wright, R. A. Godke, Coculture of human zygotes on fetal bovine uterine fibroblasts: Embryonic morphology and implantation. *Fertil. Steril.*, in press.
6. J. Cohen, K. L. Inge, M. Suzmann, S. R. Wiker, G. Wright, Videocine-matography of fresh and cryopreserved embryos: a retrospective analysis of embryonic morphology and implantation. *Fertil. Steril.*, in press.
7. K. E. Wiemer, R. A. Godke, unpublished results.

GAMETES- AND EMBRYO-OVIDUCT INTERACTIONS

S. Suzuki, Y. Oshiba, Y. Endo, S. Komatsu, H. Kitai, S. Omura,
M. Ohba, M. Ueno, and R. Iizuka

Keio University School of Medicine, Tokyo, Japan

INTRODUCTION

Although the time of ovulation in women can now be gauged with accuracy by means of echo sonography of the ovarian structure, there is as yet no comparable means of judging precisely when the embryos have passed from the fallopian tube to the uterus. It is worth reminding that what is occurring in the oviduct is still not clearly defined. The extent to which the post-ovulatory environment influences gametes maturation, fertilization and pre-implantation development is largely unknown. The oviductal environments, however, may have key roles in functional regulations of gametes and embryos (Figure 1).

Roles of the Calcium and Phospholipid Dependent
Protein Kinase in Sperm Acrosome Reaction

Much attention has focused on the role of protein phosphorylations catalyzed by the calcium, and phospholipid dependent protein kinase (PKC) in the control of the zona pellucida-induced mouse sperm acrosome reaction (Lee et al., 1987).

The sperm obtained from 10 - 12 week old DDY mice was suspended in the homogenization buffer (Kopf et al., 1986). Subcellular fractions (cytosolic, membrane and nuclear) were prepared using a method described previously (Cambier et al., 1987).

Advances in Assisted Reproductive Technologies, Edited by
S. Mashiach *et al.*, Plenum Press, New York, 1990

PKC activity was assayed in a reaction mixture (100 μl) containing 20 mM Tris-HCl pH 7.5, 10 mM $MgCl_2$, 0.4 mg/ml histone III-S, 0.2 mM $CaCl_2$, 25 μg/ml phosphatidyl-serine; PS, 4 μg/ml 12-0-tetradecanoylphorbol-13- acetate (TPA, Sigma), 39 μM [γ-^{32}P] ATP (300 mCi/mmol, Amersham) and 10 μl of sample. The basal level of histone phosphorylation was measured in the presence of 4 mM ethylene glycol bis (ß-aminoethyl ether) N,N,N',N',-tetra-acetic acid (EGTA) instead of $CaCl_2$, PS and TPA. The reaction mixture was incubated for 15 min at 30°C and stopped by immersion in an ice-water bath, and then transferred onto a P-81 filter (Whatman). Radioactivity was determined by Cherenkov counting.

It was indicated in Table 1 that PKC activity was present in the cytosolic fraction of mouse sperm and was translocated to the membrane after TPA treatment.

Sperm was incubated in phosphate-free modified Krebs Ringer (m-KRB) containing 3 mCi/ml [^{32}P] orthophosphate (ICN) for 1 hr. Radiolabeled sperm was then incubated for 15 min in the presence or absence of 20 μM of the Ca^{2+} ionophore A23187 followed by 5 min incubation in the presence or absence of 20 ng/ml TPA. After labeling, sperm was solubilized with extraction buffer (50 mM Tris-HCl pH 7.5, 1% Triton X-100, 1% sodium deoxycholate, 150 mM NaCl, 1 mM phenyl methyl-sulfonylfluoride; PMSF, 1 mM ethylenediaminetetraacetic acid disodium salt; EDTA, 5μM Na_3VO_4), and subjected to two dimensional gel electrophoresis (O'Farrell, 1975).

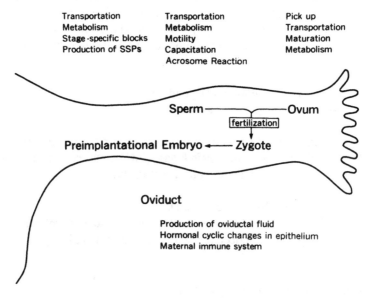

Fig. 1. Oviductal Environments.

Table 1. PKC Activities in Subcellular Fractions of Mouse Sperm

Treatment	Cytosol			Membrane			Nuclear		
	CaCl$_2$			CaCl$_2$			CaCl$_2$		
	EGTA	PS/TPA	Activity	EGTA	PS/TPA	Activity	EGTA	PS/TPA	Activity
Control	4443	5130	687	2048	2250	202	1372	1542	170
TPA	3830	4027	197	2048	2503	455	1378	1628	250

(pmol/min)

Protein phosphorylation of sperm treated with A23187 was similar to that of untreated sperm (Figure 2). However, when sperm was treated with A23187 prior to the addition of TPA, phosphorylation of an acidic 215 kDa protein and a basic 35 kDa protein was enhanced (Figure 2). This result suggests TPA and Ca^{2+} ionophore A23187 may have a synergistic effect on PKC-induced protein phosphorylation.

Our data indicate that PKC activities are in the sperm cytosol and are translocated to the membrane by TPA treatment. We also find that phosphorylation of a 215 kDa and a 35 kDa protein is enhanced by TPA treatment in vivo. The present data support the hypothesis that the activation of PKC and subsequent specific protein phosphorylation might be involved in the regulation of the zona pellucida-induced acrosome reaction.

Analysis of Cytoplasmic Factor in Developmental Cleavage of Mouse Embryo

One-cell embryos from certain mouse strains have been found incapable of developing beyond the two-cell stage in vitro (two-cell block). The partial pressure of oxygen in the gas phase of an embryo culture and ratio of pyruvate lactate concentrations in the medium are known to be important factors influencing embryonic development (Suzuki et al., 1988; Komatsu et al., 1989).

The present experiments were carried out to elucidate the mechanism of the two-cell block and gain some subsequent stages. On the basis of the results obtained, a reasonable explanation for this mechanism is presented.

When late two-cell stage embryos recovered from mated ICR mice were cultured in the medium under conditions, most were found to have undergone normal development as far as the eight-cell stage at 24 hr of incubation and to be compact. The blastocyst cavity was evident at 48 and by 72 hr, 32.3% of the embryos had shed their zonae pellucidae. In contrast, one-cell stage embryos recovered from mated mice of the same strain and under the same conditions showed arrested development at two-cell stage, as has also been noted previously. The addition of EDTA to the culture medium at a final concentration of 108 M, however, enabled the entire embryo population to develop to the morula stage within 72 hr. This effect of EDTA was also seen when it was injected into the cytoplasm of one-cell embryos. Ten picoliters of the culture medium containing 108 M EDTA induced cleavage through the four-cell stage in 24 embryos or through the morula stage in eight out of 35 injected embryos. This failed to occur by administration of the same amount of medium. None out of the same amount of medium, none out of 29 embryos exhibited developmental cleavage.

When pairs of blastomeres from different embryos were treated with phytohemagglutinin and then polyethylene glycol, they fused at 60%-87% efficiency, but development failed to progress beyond the two-cell embryos, 23 eight-cell stage within 72 hr. Thus, possibly a cytoplasmic factor(s) necessary for developmental cleavage is present in in vivo two-cell embryos but not in cultured two-cell embryos. Thus, a small amount of cytoplasm was transferred from cultured two-cell embryos, in vivo two-cell embryo or in vivo morulae

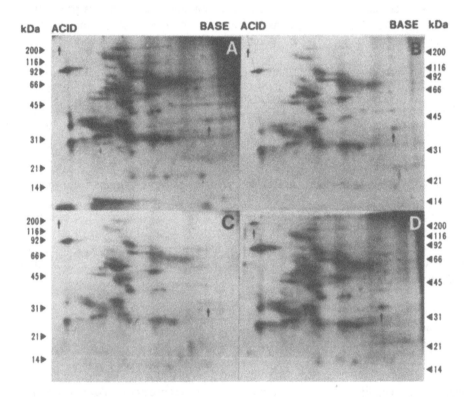

Fig. 2. Effects of TPA and Ca^{2+} ionophore on protein phosphorylation in intact mouse sperm. After sperm suspensions were radiolabeled with 3 mCi/ml 32p for 1 hr, the cells were treated follows: (A) no addition for 20 min, (B) no addition for 15 min and 20 ng/ml TPA for final 5 min, (C) 20 μM A23187 for 20 min, (D) 20 μM A23187 for 20 min and 20 ng/ml TPA for final 5 min. The reaction was terminated by the addition of lysis buffer, and the samples were subjected to two dimensional gel electrophoresis.

into one-cell embryos by microinjection. In a control experiment, the same from one-cell embryos. The cytoplasm from both in vivo two-cell embryos and morulae was effective for relieving the two-cell arrest, whereas the cytoplasm from one-cell embryos failed to have such an effect. These results support the possible involvement of cytoplasm in cleavage.

An examination was made to determine whether alteration in protein synthesis in embryos results, by means of exogenous EDTA administration, in obviating the two-cell block. One-cell embryos were collected as described into equal portions, one cultured for 24 hr in the presence of EDTA and the other in its absence for the same time. During that time, nearly all the embryos reached the two-cell stage. In vivo two-cell embryos were prepared as the control. Each group of embryos was labeled with ^{35}S methionine for 3 hr. None of the groups showed significant differences in the amount of ^{35}S methionine incorporated into the acid-insoluble materials per embryo, according to data of several independently conducted experiments. Eighty-seven proteins were identified from the position in the autoradiography by a short exposure and were common to all three groups. Two types of changes related to the two-cell block were observed in long exposed autoradiography: (1) a protein of ca. 100 kDa was present in both in vivo two-cell embryos and those cultured in EDTA, but not in embryos cultured without EDTA; (2) a protein of ca. 35 kDa was found in embryos cultured in the absence of EDTA, but not in those without it or in in vivo two-cell embryos. Both proteins were synthesized in cultured embryos whether EDTA was present or not, but not in in vivo two-cell embryos; the reason for this may possibly lie in the mode of cultivation. The phosphorylation of protein in embryos was then studied by labeling with ^{32}P orthophosphate. Phosphorylation in cultured two-cell embryos apparently has not been affected by the addition of EDTA to the culture medium; the patterns of ^{32}P labeled protein from two dimensional electrophoresis for both groups appear essentially the same. However, cultured and in vivo two-cell embryos clearly differed with respect to a protein of 25 kDa. In cultured embryos, this protein was more inclined to migrate toward an acidic area than in in vivo embryos. There is thus a possibility that this protein is phosphorylated in vitro to a greater extent than in vivo but the reasons for this are not understood. In fact, the intensity of the spot of this protein was stronger in cultured two-cell embryos than in in vivo two-cell embryos.

One-cell embryos from certain mouse strains were found incapable of developing beyond the two-cell stage in vitro, but a microinjection of EDTA effectively overcame this block. When two-cell arrested embryos were fused with embryos that had developed to the late two-cell stage in vivo, the fusants developed beyond the two-cell stage. Microinjection of cytoplasm of in vivo two-cell embryos into one-cell embryos also obviated the two-cell block. Analyses of ^{35}S labeled embryos by two dimensional polyacrylamide gel electrophoresis indicated changes in synthetic proteins possibly related to this block.

Biological Effects of Oviductal Environment on Early Embryonal Development

Progress in culturing embryos throughout their preimplantation stages at physiological rates has been hampered by species-specific blocks to development. There are blocks in various stages in in vitro culture. For example, the development of ICR mouse is arrested at the two-cell stage as previously described, that of pig at the four-cell stage and that of hamster at two- and four-cell stages (Leese 1988). In view of the continuing difficulties in culturing the embryos of certain species, the various blocks to development in culture, and the low rates of success in human in vitro fertilization procedure which to some extent are thought to be due to inadequate culture conditions, the co-culture of embryos and oviducts has become a valuable technique. Mouse embryos, which exhibit a two-cell block in vitro, have been successfully cultured to the blastocyst stage in mouse oviduct explants maintained in organ culture (Biggers et al., 1962). Whittingham achieved development of rat and mouse one-cell embryos to the blastocyst stage using inter- and intraspecific transfers to oviduct explants (Whittingham, 1986). Recently, Minami et al. showed that the explanted mouse oviduct can support the growth of some hamster early two-cell embryos to the morula and early blastocyst stages (Minami et al., 1988). Successful development of porcine embryos from the one- cell stage to the blastocyst stage has been accomplished using mouse oviducts in organ culture (Krisher et al., 1989).

We investigated the effect of human oviduct on early mouse embryonal development. Human oviducts were obtained from women of fertile age during elective gynecological laparotomy. The ampullary part was washed in phosphate buffered saline (PBS). It was cut into about 5 mm square pieces, settled on a stainless mesh in a culture dish with center well, immersed with minimal essential medium (MEM, Sigma), 10% human serum. Fertilized one-cell eggs were obtained 24 hr after human chorionic gonadotropin (hCG) injection from superovulated ICR mouse with human menopausal gonadotropin (PMSG). Seventy-two hr following co-culture, the eggs were observed to develop to eight-cell stage beyond two-cell block (Oshiba et al., 1988).

In order to know the oviductal contribution to the early embryonal development, we arranged the medium conditioned by the extracts from human oviductal epithelium dissociating and incubating overnight. The ampullary part was slit open and minced. The tissue was dissociated for 1 hr at 37°C in MEM, containing collagenase 1 mg/ml (180 U/mg solid), stirring gently inside the cell incubator. The supernatant was removed. PBS containing trypsin (0.25%) and EDTA (3mM) were added and the dissociation was repeated twice for ten minutes each. The supernatants were filtered, centrifuged at 300 g and the cell pellets washed twice in MEM containing 10% human serum. Cells were then plated in Falcon Petri dishes containing MEM supplemented with antibiotics, insulin (0.21 I.U./ml) and 10% human serum (Testas et al., 1986). On the following day, oviductal epithelial cells were well attached and they were subjected to electromicroscopic observations, whereas the medium was filtered for using as a conditioned medium involving the oviductal extracts.

The scanning electronmicroscope showed that the dissociated oviductal cell kept the microvilli, and that the ciliated cells also preserved them. (Figures 3,4). The transmission electronmicroscope indicated that the nuclei of cells were in the beginning of mitosis (Figure 5).

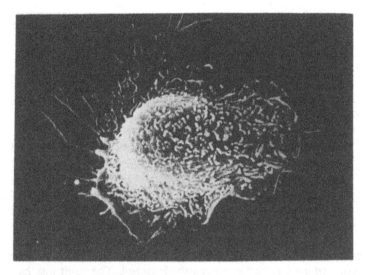

Fig. 3. SEM (x 12,000). Human oviductal cells were dissociated and attached to the culture dish after 24 hr incubation. This epithelial cell has microvilli.

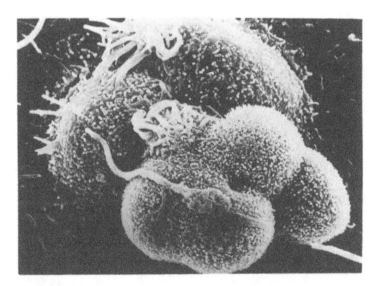

Fig. 4. SEM (x 10,000). The dissociated human oviductal epithelial cells preserve microvilli, and some have cilia.

Table 2. Cleavage of Mouse Embryos in Medium Conditioned by Epithelial Cells of Human Oviduct

	control MEM + Serum	Sample 1	control (MEM + Serum)		Sample 2	
	F_1	F_1	ICR	F_1	ICR	F_1
Total No.	19	55	24	19	51	27
2 cell	11 (57.9%)	47 (85.5%)	10 (41.7%)	15 (78.9%)	39 (76.5%)	26 (96.3%)
>2 cell	3 (15.8%)	36 (65.5%)	0 (0%)	4 (21.1%)	0 (0%)	15 (55.6%)

```
Sample 1: Oviduct collected from 34 y/o female on day 13 of cycle
Sample 2: Oviduct collected from 47 y/o female on day 13 of cycle
```

On the other hand, fertilized one-cell eggs of ICR and B6C3F1 mice were cultured in the conditioned medium obtained as previously described. The result was that the developmental rate beyond two-cell stage in conditioned medium was better than that in control. The two-cell block, however, was not overcome in this experiment (Table 2).

It is suggested that the milieu provided by co-culturing is not only non species-specific, but also sufficient enough for early embryonic development.

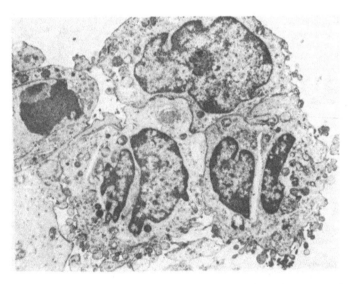

Fig. 5. TEM (x 15,000). The epithelial cells are in the beginning of mitosis. Mitochondria are around the nuclei.

REFERENCES

Biggers, J. D., Gwatkin, R. B. L., and Brinster, R. L., 1962, Development of mouse embryos in organ cultures of fallopian tubes on a chemically defined medium, *Nature*, 194:747-749.

Cambier, J. C., Newel, M. K., Justement, L. B., McGuire, J. C., Leach, K. L., and Chem, Z. Z., 1987, Ia binding ligands and cAMP stimulate nuclear translocation of PKC in B lymphocytes, *Nature*, 327:629-632.

Kopf, G. S., Woolkalis, M. J., and Gerton, G. L., 1986, Evidence for a guanine nucleotide-binding regulatory protein in invertebrate and mammalian sperm: Identification by islet-activating protein-catalyzed ADP-ribosylation and immunochemical methods, *J. Biol. Chem.*, 261:7327-7331.

Komatsu, S., Suzuki, S., Kitai, H., Endo, Y., Fukasawa, T., and Iizuka, R., 1989, Studies on in vitro 2-cell block mechanism in the mouse with special reference to cytoplasmic factors, Suppl. The International Symposium on Morphological Sciences, Motta, M., ed., Alan R. Liss, Inc. (in press).

Krisher, R. L., Petters, R. M., Johnson, B. H., and Bavister, B. D., 1989, Development of procine embryos from the one-cell stage to blastocyst in mouse oviducts maintained in organ culture, *J. Exp. Zool.*, 249: 235-239.

Lee, M. A., Kopf, G. S., and Storey, B. T., 1987, Effects of phorbol esters and a diacylglycerol on the mouse sperm acrosome reaction induced by the zona pellucida, *Biol. Reprod.*, 36:617-627.

Leese, J. H., 1988, The formation and function of oviduct fluid, J. Reprod. Fertil., 82:843-856.

Minami, N., Bavister, B. D., and Iritani, A., 1988, Development of hamster 2-cell embryos in the isolated mouse oviduct in organ cuture system, *Gamete Res.*, 19:235-240.

O'Farrell, P. H., 1975, High resolution two-dimensional electrophoresis of proteins, *J. Biol. Chem.*, 250:4007-4021.

Oshiba, Y., Ueno, M., Suzuki, S., and Iizuka, R., 1988, Studies on correlative relations between human cultured oviduct and mouse fertilized ova, presented at The 33rd Annual Meeting of Japanese Fertility and Sterility, Kyoto, Japan.

Suzuki, S., Komatsu, S., Kitai, H., Endo, Y., Iizuka, R., and Fukasawa, T., 1988, Analysis of cytoplasmic factor in developmental cleavage mouse embryo, *Cell Differentiation*, 24:133-138.

Testas, I. J., Garcia, T., and Baulieu, E. E., 1986, Steroid hormones induce cell proliferation and specific protein synthesis in primary chick oviduct cultures, *J. Steroid Biochem.*, 24(1):273-279.

Whittingham, D. G., 1968c, Intra- and inter-specific transfer of ova between explanted rat and mouse oviducts, *J. Reprod. Fertil.*, 17: 575-578.

EPITHELIAL CELLS FROM HUMAN FALLOPIAN TUBE IN CULTURE

T. Henriksen,[1] T. Tanbo, Th. Åbyholm,[2] B. R. Oppedal, and T. Hovig[3]

[1]Dept of Obstetrics and Gynecology, Aker Hospital, 0514 Oslo 5
and Institute for Surgical Research, University of Oslo,
Pilestredet, 0027 Oslo 1; Norway.
[2]Dept of Obstetrics and Gynecology, The National Hospital, 0027 , Oslo 1, Norway
[3]Institute of Pathology, The National Hospital, 0027 Oslo 1, Norway

INTRODUCTION

The fallopian tube is the site of gamete transport, fertilization and early embryo growth. The intraluminal milieu of the tubes represents expectedly optimal conditions for this part of the reproductive process. The fallopian luminal fluid is thought to be composed of a transudate from plasma as well as of components synthesized by the epithelial cells lining the inner surface of the tube (P.S.R. Jansen, 1984). The quality and quantity of the fallopian fluid is determined by the endocrine environment (M. Hammer, V. Larsson-Con, 1978; J. Lippes, R.G. Enders et al. 1972; J. Lippes, J. Krasner 1981). The fallopian epithelium is one of the hormonal targets since the cells show morphological alterations throughout the menstrual cycle indicating cyclicity in secretory activity (P.S.R. Jansen 1984).

A more specific study of the fallopian epithelial cells, especially with respect to which components these cells produce and how it is regulated requires isolated cells under controlable conditions. The present paper describes a method whereby epithelial cells from the fallopian tube can be cultured.

MATERIALS

Tissue culture dishes were obtained from Costar (cat. no. 3224). RPMI 1640 (cat no 04102400H), Penicillin-Streptomycin (cat. no. 043-5140) bovine calf serum (cat. no.

Advances in Assisted Reproductive Technologies, Edited by
S. Mashiach *et al.*, Plenum Press, New York, 1990

01662290) Dulbeccos PBS (cat. no. 041-4040), were all obtained from GIBCO. Transferrin (cat. no. A1887), human albumin (cat. no. A1887) insulin (cat. no. I 33505) were Sigma products. Trypsin-EDTA was provided by Flow Laboratories (cat. no. 1689149). Medium A consisted of RPMI 1640, Penicillin-Streptomycin (100 µg/ml) albumin 4 mg/ml, insulin (5 µg/ml) and transferrin (5 µg/ml). Medium A was sterilized by passage through a .22 µ Millipore filter. Medium B contained RPMI 1640, Penicillin-Streptomycin as above and 20% v/v bovine calf serum. Both media were kept at $-20°C$ until use.

Procedure for Isolation of Epithelial Cells from the Human Fallopian Mucosa

Three to five centimeters of the ampullary part of human fallopian tubes were obtained from healthy women undergoing either sterilization by minilaparotomy or hysterectomy. All the patients gave informed consent to donate parts of their fallopian tubes and the procedure was approved by the hospitals ethics committee.

The specimens were immediately placed in cold (4°C) Dulbeccos PBS and used within four hours. Under sterile conditions the fallopian tube was opened longitudinally and pieces of the mucosa were carefully cut out using a fine scissor. Care was taken to minimize contamination with submucosal connective tissue. The epithelial pieces were placed in 3 ml medium B in a 50 ml coned tube. The collected tissue was minced thoroughly with a scissor (20 minutes continuous chopping). Then another milliliter PBS was added and after mixing the cell suspension were left alone for 30 seconds to let the larger tissue pieces precipitate. The supernatant was transferred to a 15 ml tube and centrifuged at 500g for 10 minutes. The pellet consisting of tiny epithelial pieces (5-2000 cells each) were resuspended in medium B at a concentration of 10 cells per milliliter. 0.5 ml of the cell suspension were added per cm^2 culture dish. The cultures were kept at 37°C at 5% CO_2 in air for two days. Then the medium was removed, the cells were washed once with PBS and fresh medium A was added (0.3 ml per cm^2). Culture medium (medium A) was changed twice a week.

Culturing of Fibroblastic Cells from the Fallopian Wall

Small pieces of submucosal fallopian tissue were removed and immobilized at the bottom of the culture dishes by the weight of small glass pieces which were covered by medium B. The medium was changed twice a week. After about three weeks fibroblast-like cells grew out from the tissue fragments and the glass pieces were removed. These cells were either used as primary cultures or at first passage.

Morphologic Studies

The cells were evaluated by phase contrast microscopy twice a week especially with respect to possible overgrowth by fibroblasts.

Immunocytochemical Studies

For this part of the study the epithelial cells and the control fibroblastic cells were grown in Leighton tubes. The medium was removed, the cells washed twice with PBS and then fixed in 96 % ethanol for 30 seconds followed by 70 % ethanol. The slides were incubated with murine monoclonal antibodies to the cytoskeleton proteins (intermediate filament) cytokeratin PKK1 and vimentin (Labsystems, Helsinki, Finland) and as well as a polyclonal rabbit antibody to epidermal keratin (H.S.-Huitfeldt and P.Brandtzæg, 1985).

The PKK1 antibody has been shown to react with 56,48,45 and 41 kD keratin polypeptides. The antibody against the mesenchymal cytoskeleton protein vimentin reacts with a single 58 kD polypeptide (H. Holthøfer, A. Miettinen et al. 1983). The polyclonal rabbit antibody was raised against keratins that produce six major and two minor bands in SDS gel electrophoresis, all with molecular weight between 40 and 70 kD (H.S. Huitfeldt and P.Brandzaeg, 1985). Anti-PKK1 and anti-vimentin were applied at a 1:200 titer for 20 hours at room temperature. After washing antigen localization was accomplished by the alkaline phosphatase anti-alkaline phosphatase (APAAP) method (D.Y. Mason, 1985) as described elsewhere (P. Brandtzaeg, I. Dale, 1987). The slides were finally counterstained with hematoxylin and mounted in Apathys mounting medium (R.A. Lamb Laboratories, London UK). The mesenchymal cytoskeleton vimentin or cytokeratin PKKK1 were additionally combined with a polyclonal rabbit epidermal keratin R 505 in a three step paired immunofluorescence method including biotinylated horse antimouse IgG, fluorescein-labelled avidin and rhodamin-labelled swine anti-rabbit IgG (P. Brandzaeg and T.O. Rognum, 1983).

Electron Microscopic Studies

For electron microscopic studies the epithelial cultures were washed twice with Dulbeccos PBS and fixed with 2% glutaralehyde in 0.1 M cacodylate, pH 7.4 and thereafter in 1% buffered osmium tetroxide. For transmission electron microscopy the specimens were dehydrated in graded ethanols and embedded in Epon 812. The ultra thin sections were stained with lead citrate and uranyl acetate and examined in a Jeol 1200 EX electron microscope. For scanning electron microscopy the specimens were fixed as described above, critical points dried out and coated with gold-palladium. The specimens were examined in a Jeol 50 B scanning electron microscope.

Cell Proliferation

The cells were seeded evenly into eight tissue culture wells (2 cm^2). The cells were allowed to attach and spread for four days in medium B. Then all the floating cells were washed off by PBS. The cells in one duplicate culture were exposed to trypsin-EDTA for 10 minutes at 37°C . The bottom of the cultures were flushed repeatedly with a Pasteur-pipette to secure that all the cells had detached. The cell number in each well was determined twice in a hemocyometer. The remaining cultures were continued in fresh medium A. Duplicates were taken for cell counting at the indicated times using the same procedure.

RESULTS

The morphology of the epithelial cultures as compared to fibroblastic cells from the same organ is shown in Figure 1. The cells derived from the epithelial layer grew in sharp-

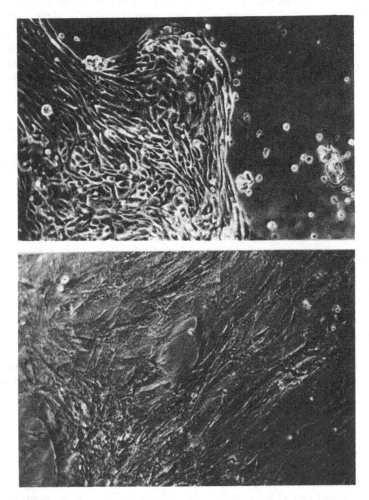

Fig. 1. Phase contrast microscopy (400 ×) of 18 days old cultures of epithelial cells from human fallopian tube (upper panel) and fibroblastic cells from submucosa of the fallopian tube (lower panel).

ly defined islands of epitheloid cells in contrast to the unrestricted growth of the stromal fibroblastic cells. The epitheloid growth pattern was retained for several weeks as demonstrated in Figure 2 where a three weeks old culture is shown.

Ultrastructurally the surface of the epithelial cells showed microvilli as demonstrated by scanning and transmission electron microscopy (Figures 3 and 4). Occasional cells showed ciliae (Figures 3 and 5). Desmosomes were found between neighbouring cells and the cytoplasm contained mitochondriae, numerous ribosomes, microtubules and intermediate filaments. Intracytoplasmatic vacuoles were also noted (Figure 6). The tubal epithelial cells stained positive for the cytokeratin antibody PKK1 and the polyclonal antibody to epidermal keratin R 505 but negative for vimentin. Using the double immunofluorescence method, however, we saw scattered vimentin positive cells, read as fibroblasts, in between the epithelial cells (Figure 7). The number of vimentin positive cells was less than 2% at 5 days in culture increasing to 5% after four weeks in vitro. A few of the cells in the older culture showed coexpression of cytokeratin and vimentin in double immune fluorescence. The tubal fibroblasts were all keratin negative and vimentin positive.

The replicative ability of the epithelial cells is demonstrated in Figure 8 where the increase in cell number per culture with time is presented. After approximately one week in culture no further increase in cell number was usually found. Thereafter the cultures stabilized with respect to cell number irrespective of whether the cells had reached monolayer or not. In the cases where monolayer was not present the cultures retained the morphology of well defined patches or islands of cells. Throughout the whole culture period there seemed

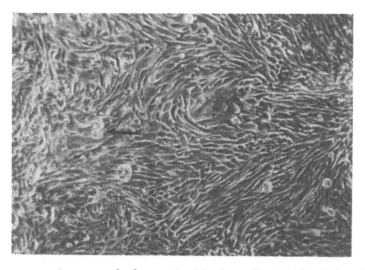

Fig. 2. Phase contrast microscopy of a four weeks old culture of epithelial cells from human fallopian tube. The epitheloid growth pattern is retained but there are signs of increasing cell detachment (arrow).

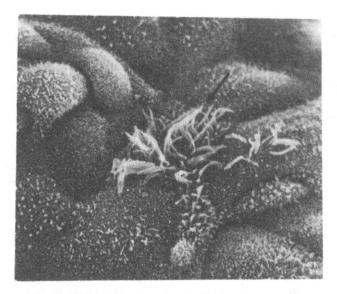

Fig. 3. Scanning electron microscopy of a 20 day old culture of human fallopian epithelial cells. The majority of the cells show microvilli . Occasional ciliae were are seen (arrow).

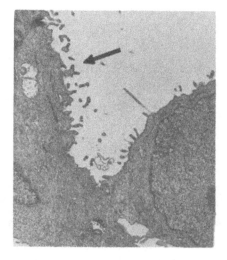

Fig. 4. Transmission electron microscopy of fallopian epithelial cells cultured for 20 days demonstrating microvilli.

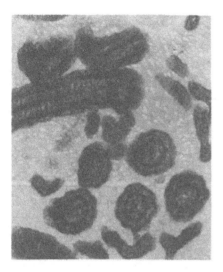

Fig. 5. Cross- and longitudinal sections of ciliae in a 10 day old culture.

Fig. 6. Two neighbouring cells connected by desmosomes. The cytoplasms contain free ribosomes, microtubules and filaments.

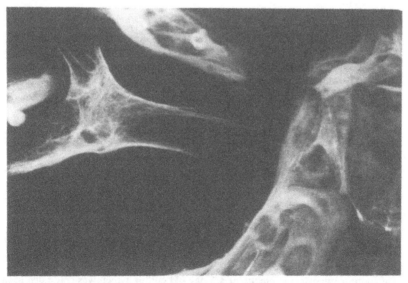

Fig. 7. Pattern of staining of cultured human fallopian epithelial cells incubated with fluorescenating antibodies to keratin and the mesenchymal vimentin. The majority of the cells stained positive for the epithelial filament protein keratin (orange) whereas about 2 % of the cells were vimentin positive (green).

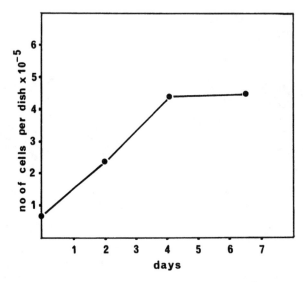

Fig. 8. Growth curve of human fallopian epithelial cells seeded in RPMI 1640 with 20 % bovine calf serum and then after two days continued in RPMI 1640 containing human albumin, transferrin and insulin.

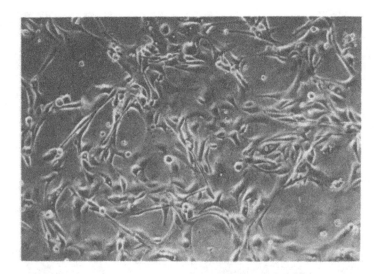

Fig. 9. Phase contrast microscopic picture of a 5 day old culture which was continued in presence of RPMI 1640 with 20 % bovine calf serum. The epitheloid growth pattern is disappearing in contrast to cells which received RPMI 1640 with albumin, transferrin and insulin from the second day in culture (Figure 1).

to be a slow loosening of cells which were replaced by regeneration among the remaining cells. However, after six to eight weeks a progressive net loss of cells was seen and the cultures vanished.

Attempts were made to use serum free medium (medium B) from day 0. Very few cells attached under these conditions. On the other hand if the epithelial cells were continued in serum containing medium (medium B) the epitheloid growth pattern was lost within a week (Figure 9).

DISCUSSION

The present work shows that epithelial cells from human fallopian tubes can be kept in culture as a monolayer for several weeks. The ultrastructural studies showed characteristic epithelial morphology of the cultured cells with microvilli, typical desmosomes as well as ciliae.The epithelial nature of the cells was confirmed by use of immunocytochemical methods demonstrating the presence of cytokeratin antigens in the majority of the cells (98%). Thus, vimentin positive cells were found only occasionally in young cultures (i.e. 10-14 days). Thereafter the percentage of vimentin positive cells increased up to approximately 5 %. We did not see overgrowth of fibroblasts as judged by phase contrast microscopy within the six to eight weeks period the epithelial cells could be held in culture. Nevertheless, the increasing number of vimentin positive cells do most probably represent an increasing number of fibroblastic cells since loss of cytokeratin expression combined with acquisition of vimentin antigens in the tubal epithelial cells is very unlikely within the present frame of time. In the older cultures, however, a few of the cells coexpressed cytokeratin and vimentin as demonstrated by double fluorescence.

The epithelial cells were found to proliferate well in culture as demonstrated by the growth studies. The increase in number of cells was not due to progressive overgrowth of fibroblastic cells because multiplication of the epithelial cells started before the percentage of vimentin positive cells started to increase. Furthermore, light as well as electron-microscopic studies of two to three weeks old cultures showed that the epithelial morphology was retained and that the number of vimentin positive cells was low.

Attempts to subculture the cells were not successful. We believe that the main cause was low plating efficiency which was observed when the cells were trypsinized and transferred to new dishes. We tried to passage the cells by loosening the cells mechanically without any improvement in the plating efficiency. Thus before effective passage of the cells can be achieved the plating efficiency must be considerably improved.

A very important part of the culture procedure was to change from serum containing medium to serum free two to three days after seeding. If the epithelial cells were allowed to continue in presence of serum their epithelial growth pattern progressively disappeared. Thus, the distinct islands of cells disintegrated and the individual cells aquired a more fibroblastic blastic appearance. A similar effect of serum has been described in studies of cultured epithelial cells from other organs (S.D. Chung, N. Alavi et al. 1982). On the other hand attachment of the fallopian epithelial cells seems to be dependent on some unknown serum factor(s) since attempts to seed the cells in absence of serum was unsuccessful. Fibronectin would be an obvious candidate as an attachment factor in serum but we were not able to establish cultures by combining serum free medium and dishes precoated with fibronectin.

We are aware of only one report on culture of human fallopian epithelial cells (P. McComb, L. Langley et al. 1986). This investigation was undertaken to study the ciliary activity in one patient with Kartageners syndrome. The cilia were found to be inactive. Confluent monolayers were reported to be obtained after six days. Unfortunately no data was given on the plating efficiency since we believe that the ciliary activity was the main cause for low attachment quotient in our study. In a recent report cultures of epithelial cells from ovine oviduct was established by using cells that were dislodged during flushing of the tubal lumen (F. Gandolfi and R.M. Moor 1987). The cultures were however, not further characterized with respect to morphology, immunoreactivity or retainment of epithelial characteristics as judged by electronmicroscopic studies.

In the present paper we have described a method to culture epithelial cells from human fallopian tubes. There is a need to improve the plating efficiency in order to obtain more cells from a given specimen and probably to be able to subculture the cells effectively. Nevertheless, cultures of epithelial cells from the human fallopian tube can be cultured in sufficient quantity to be used in studies.

ACKNOWLEDGEMENT

This investigation was supported by the Norwegian Research Council for The Sciences and the Humanities.

REFERENCES

Brandzaeg, P., Dale, I. and Fagerhol, M.K., 1987, Distribution of a formalin-resistant myelo-monocyte antigen (L1) in human tissues. I. Comparison with other leucocyte markers by paired immunofluorescence and immunoenzyme staining, *Am. J. Pathol.*, 87, 681-699.

Brandzaeg, P. and Rognum T.O., 1987, Evaluation of tissue preparation methods and paired immunofluorescence staining for immunocytochemistry of lymphomas, *Histochem J.*, 15, 655-689.

Chung, S.D.,Alavi, N., Livingston, D., Hiller, S. and Taub, M., 1982, Characterization of primary rabbit kidney cultures that express proximal tubule functions in a hormonally defined medium, *J. Cell Biol.*, 95, 118-126.

Gandolfi, F. and Moor. R.M., 1987, Stimulation of early embryonic development in the sheep by co-culture with oviduct epithelial cells, *J. Reprod. Fert.*, 81, 23-28.

Hammer, M. and Larsson-Con, V., 1978, Menstrual cycle enlargement of occlusive tubes after postmenopausal treatment with natural estrogen, *Acta. Obstet. Gynecol. Scand.*, 57, 189-196.

Holthofer, H., Miettinen, A., Letho, V.P., Lethonen, E. and Virtanen, I., 1984, Expression of vimentin and cytokeratin types of intermediate filament proteins in developing and adult human kidneys, *Lab. Invest.*, 50, 552-559.

Huitfeldt , H.S. and Brandzaeg, P., 1985, Various keratin antibodies produce immunohistochemical staining of human myocardium and myometrium, *Histochemistry* 83, 381-389.

Jansen, P.S.R., 1984, Endocrine response in the fallopian tube. *Endocrine Rev.* 5, 525-551.

Lippes, J., Enders, R.G., Pragay, D.A. and Bartilomew, W.R., 1972, The collection and analysis of human fallopian tubal fluid, Contraception 5, 85-92.

Lippes, J., Krasner, J., Alfonso, L.A., Dacalos, E.D. and Lucero, R., 1981, Human oviductal fluid proteins, *Fertil. Steril.*, 36, 623-629.

Mason, D. Y., 1985, Immunohistochemical labelling of monoclonal antibodies by the APAAP immunoalkaline phosphatase technique, *In:* "Immunohistochemistry," Bullock, G.R. and Petrisz, P., eds., Academic Press, New York, pp 25-42.

McComb. P., Langley, L., Villalon, M. and Verdugo, P., 1986, The oviductal cilia and Kartageners syndrome, *Fertil. Steril.*, 46, 412-416.

CYTOPLASMIC CONTROL OF THE TRANSFORMATION
OF SPERM NUCLEUS INTO MALE PRONUCLEUS

Andrzej K. Tarkowski

Department of Embryology, Institute of Zoology
University of Warsaw, Poland

INTRODUCTION

Studies on fertilization in mammals have emphasized the preparatory events of capacitation and acrosome reaction, and morphology and physiology of sperm penetration far more than the fate of the sperm nucleus in the egg cytoplasm. It is often assumed that fertilization is successfully accomplished once the spermatozoon has penetrated the oocyte. However, fertilization is not completed until karyogamy (which in mammals takes place shortly before the first cleavage division), when the maternal and paternal sets of chromosomes contribute to the common diploid metaphase plate. This event takes place several hours after sperm penetration.

Transformation of sperm nucleus into male pronucleus is a complex multistep process which consists of a series of molecular and morphological events, such as reduction of disulphide bonds in nuclear proteins, breakdown of the nuclear envelope, chromatin decondensation, exchange of protamines for histones, nuclear envelope reformation, pronuclear swelling and formation of nucleoli. DNA synthesis does not begin until both male and female pronuclei are morphologically fully developed, which occurs a few hours after sperm penetration.

Transformation of the sperm nucleus, like all nuclear events, is controlled by the cytoplasm. In order to understand this process, we need to know the answers to the following questions. First, are the abovementioned events controlled by a single or by separate cyto-

Advances in Assisted Reproductive Technologies, Edited by
S. Mashiach *et al.*, Plenum Press, New York, 1990

plasmic factors? Second, what are the dynamics of the factors involved? Third, what is their molecular nature? Experimental studies, some of which will be presented here, permit us to postulate the existence of cytoplasmic factors (and their dynamics), but cannot provide any information regarding their nature. In the case of mammals direct information on this point is scarce (and for technical reasons will probably remain scarce in future) and one has to extrapolate from studies which use other developmental systems, for instance amphibian eggs. However, such extrapolations should be made very cautiously.

Our experimental approach consists of subjecting sperm nuclei to the cytoplasm of immature or maturing oocytes or previously activated eggs, i.e. to a cytoplasmic environment different from that of the fully mature oocyte in metaphase II, at which stage fertilization normally occurs. This permits us to dissociate sperm penetration from egg activation, and to define when cytoplasmic activities responsible for various steps of male pronucleus formation start and cease to operate.

In this article I will confine myself to presenting our studies on the behaviour of the sperm nucleus in mouse oocytes which have not reached complete cytoplasmic maturity by the time of fertilization, and in oocytes which have been activated before sperm penetration. These topics seem to me relevant to in vitro fertilization (IVF) of human eggs for two reasons.

First, collection of human oocytes from preovulatory follicles rather than from the oviduct creates the risk of using eggs that have not yet attained cytoplasmic maturity. Cytoplasmic maturity need not be fully synchronous with nuclear maturity and unlike the latter cannot be determined by microscopic inspection. Second, development of microsurgical techniques which permit the spermatozoon to reach the surface of the vitellus without the need of penetrating the zona pellucida creates, or increases, the risk of fertilization of oocytes that earlier have undergone activation due to in vitro manipulations.

BEHAVIOUR OF SPERM NUCLEI IN M II OOCYTES
WHICH HAVE NOT COMPLETED CYTOPLASMIC MATURATION

As early as 1955 M.C.Chang formulated the notion that "although the nuclear maturation is a good indication of the maturity of the ova, the maturation of ooplasm as a whole should be considered as an important factor for fertilization" (Chang, 1955). Later studies have fully supported this idea. For instance, Iwamatsu and Chang (1972) have shown that mouse ovarian oocytes can be penetrated at any stage of maturation but sperm penetration will not induce activation unless the oocyte has completed (or is just about to complete) the first maturation division. However, acquisition of fertilizability (or generally speaking activability) is connected more closely with the acquisition of full cytoplasmic maturity (whatever it means in molecular terms) than with the exact stage of nuclear maturation. My collaborator Jacek Kubiak has recently shown in very precisely timed experiments that in F_1 (CBA × C57BL) mouse oocytes the capacity for activation develops gradually during metaphase II arrest. In other words, completion of the first meiotic division is not equivalent (at least in some oocytes) to the acquisition of fertilizability (Kubiak, 1989). On the other hand, it has been reported that in Lt/Sv strain of mice a high proportion (about 1/3) of ovulated eggs are arrested in metaphase of the first meiotic division and at this stage can be successfully fertilized or spontaneously or artificially activated (Kaufman and Howlett, 1986; O'Neill and Kaufman, 1987).

Maturation of oocytes in vitro appears to have a more adverse effect on cytoplasmic maturation than on nuclear maturation. In the past many workers have noticed that the developmental potential of oocytes which had attained metaphase II in vitro (often at a high rate) and at this stage were fertilized was poor, and definitely inferior to that of oocytes which matured in vivo and were naturally ovulated. Probably this was due to suboptimal culture conditions because in more recent studies the survival rate of mouse oocytes matured and fertilized in vitro was claimed to be normal (Schroeder and Eppig, 1984). A very interesting observation related to cytoplasmic maturation was made by Thibault and Gerard (1970, 1973) in the rabbit. They observed that in oocytes matured in vitro outside follicles (but not inside follicles) and subsequently fertilized, the transformation of the sperm nucleus was arrested at an early stage while the female pronucleus developed normally. This differential response of male and female chromatin to the same cytoplasmic environment is very puzzling and further studies on the hypothetical "male pronucleus growth factor" (MPGF) (Thibault and Gerard, loc. cit.) are needed.

In the abovementioned study Kubiak analysed the reaction of mouse zona-free oocytes to sperm penetration and parthenogenetic stimulation (8% ethanol) starting 11 hours after administration of hCG, i.e. virtually immediately after extrusion of the 1st polar body. He observed that a proportion of oocytes inseminated at 11 and 12 h either remain in metaphase II or extrude the 2nd polar body, but the haploid group of chromosomes left in the egg is rearranged on a new spindle. These oocytes, despite completing the second meiotic division, remain in the metaphase state (Kubiak has proposed calling this stage "metaphase III"). This is a very surprising observation because egg activation has often been considered as an "all or none" reaction. In both M II and M III oocytes that have been penetrated by a single spermatozoon, the sperm nucleus undergoes decondensation but subsequently recondenses and transforms into individual unichromatid chromosomes (Kubiak, 1989). Formation of G_1 chromosomes directly from sperm nuclei has previously been described in maturing mouse oocytes penetrated during prometaphase I or metaphase I (Clarke and Masui, 1986, 1987). There are some differences in the behaviour of sperm nuclei in these two types of oocytes (rate of chromosome formation, number of nuclei that can form chromosomes) which should be further explored.

Direct transformation of sperm nuclei into G_1 chromosomes has also been observed in penetrated human eggs that have not undergone activation (Schmiady et al., 1986); Plachot, personal communication). The oocytes examined by Schmiady and his colleagues did not resume second meiotic division, and all remained in metaphase II. I believe that the background of this aberrant reaction was the same as in mouse oocytes, namely cytoplasmic immaturity at the moment of sperm penetration.

In his study on the gradual development in mouse oocytes of the capacity for activation, Kubiak (loc. cit.) has observed also that sperm penetration is a much stronger activating stimulus than a parthenogenetic agent (ethanol): none of the oocytes penetrated later than 12 h post-hCG stayed in metaphase, while nearly 50% of the oocytes artificially activated at 13 h underwent second meiotic division and entered metaphase III. Parthenogenetic agents of all kinds are generally ineffective when applied to "early" oocytes and their efficiency increases with the age of oocytes (Tarkowski, 1976). It is believed that the state of metaphase is induced and maintained by meiosis-promoting factor (MPF) and that the transition from metaphase II arrest to interphase is brought about by inactivation of the chromosome condensation activity (reviewed by Masui, 1985). The above described observations could suggest that from a certain moment of oocyte maturation the activity of the chromosome condensation factor declines, and that egg activation becomes possible when this activity drops down to a certain level: higher in the case of fertilization and lower in the case of artificial activation. We have experimental evidence that in the egg cytoplasm this activity indeed decreases with post-ovulatory age (Czolowska et al., 1986). It is hard to believe, however, that the increasing susceptibility of oocytes to activation could be simply due to a decrease of chromosome condensation activity in the cytoplasm, because in the oocyte this activity is predominantly bound to chromosomes (Czolowska et al., loc. cit.). It can be stabilized by maintaining oocyte chromosomes in a condensed state with inhibitors of microtubule polymerization (colcemid or nocodazole) (Schatten et al., 1985; Maro et al., 1986; Borsuk and Mánka, 1988). As shown by Borsuk and Mánka (loc. cit.), in oocytes inseminated in the presence of colcemid the sperm nuclei undergo normal decondensation but subsequently re-condense and remain, together with metaphase II oocyte chromosomes, in a condensed state. However, contrary to untreated metaphase II oocytes which after sperm penetration are not activated and remain in metaphase (II or III), the colcemid-treated oocytes are activated: after removal of the drug they proceed to interphase, and the sperm nucleus then develops into a male pronucleus and the oocyte chromosomes into a female pronuclei or, more often, into several micronuclei. The re-condensation of the sperm nucleus under the influence of colcemid takes place only in the presence of oocyte chromosomes: in anucleate oocyte fragments sperm nuclei develop normally into male pronuclei. Thus colcemid does not act on sperm nuclei directly, but through maintaining chromosome condensation activity. This activity is somehow stabilized in the condensed chromosomes but can diffuse into the cytoplasm and reach the sperm nuclei.

BEHAVIOUR OF SPERM NUCLEI IN PREVIOUSLY ACTIVATED EGGS

Parthenogenic activation of mouse eggs does not produce an effective block against sperm penetration at the egg membrane level (Komar, 1982). This permits us to obtain sperm penetration into zona-free eggs at any time during the first cell cycle, and even during cleavage (Tarkowski and Tittenbrun, unpublished results; Borsuk and Tarkowski,

1989). The block against polyspermy at the level of zona pellucida ("zona reaction") does develop in parthenogenetic eggs but is slower and weaker than in fertilized eggs: sperm penetration is possible up to 3 h post-activation but after 1 1/2 h is very reduced (Komar, 1982; Komar and Kujawa, 1985). Thus with the exception of the earliest period after artificial activation, the zona pellucida constitutes an effective barrier for a spermatozoon on its way to the vitellus. However, because of the inefficient block against sperm penetration at the egg membrane level, zona-intact parthenogenetic eggs could probably be easily penetrated by sperm microsurgically injected under the zona pellucida.

The later sperm penetration occurs in relation to the moment of egg activation, the more limited the degree of transformation of the sperm nucleus. The capability of the egg cytoplasm to transform the sperm nucleus into a fully formed male pronucleus is lost sometime between three and five hours after artificial activation (Tarkowski and Tittenbrun, unpublished results; Borsuk and Tarkowski, 1989). Although in a certain proportion of eggs penetrated during the first hours after activation, sperm nuclei develop into pronuclei that appear normal in size and morphology, this normality may be only apparent. Although direct evidence is not yet available, it is very likely that asynchronous development of the female and male pronuclei may lead to developmental disturbances, probably as early as the first cleavage division.

In "older" parthenogenetic eggs sperm nuclei may start to decondense but are checked in further development. They stay in the egg cytoplasm unchanged until the first cleavage division, when they condense and are passively displaced to daughter blastomeres. Although their further fate was not followed in detail, it seems that they do not participate in further development.

Irrespective of the time of penetration, all sperm nuclei acquire the ability to be stained with Giemsa or toluidine blue which, according to Krzanowska (1982) and Miller and Masui (1982), is indicative of the reduction of disulphide bonds. This molecular event appears, therefore, not to be specific for the cytoplasm of metaphase II oocytes. However, the two subsequent events in sperm nucleus transformation—the breakdown of nuclear envelope and chromatin decondenstion—which in normal fertilization occur nearly simultaneously and very soon after sperm penetration, are stage-specific. The ability of the cytoplasm to break down the sperm nuclear envelope is lost earlier than its ability to decondense sperm chromatin. As a consequence there is a period during which sperm chromatin decondenses but within the original nuclear envelope (Borsuk and Tarkowski, 1989; Szöllösi, Borsuk and Szöllösi, personal communication). Such nuclei never develop into normal male pronuclei. The ability of the cytoplasm to decondense sperm chromatin is finally lost around 8 h post-activation. It reappears, however, during the first and later mitoses (Tarkowski, unpublished observations).

In the cytoplasm of parthenogenetic egg-fragments which have been enucleated, i.e. do not contain a female pronucleus, transformation of sperm nuclei is in general more successful than in nucleated egg-fragments. This difference may be connected with the different level of extrachromosomal constituents of the germinal vesicle karyoplasm required for pronuclear growth: high in the absence of a female pronucleus and low in its presence. Our earlier experiments have shown that extra-chromosomal constituents of the germinal vesicle are indeed required for the formation of pronuclei (Balakier and Tarkowski, 1980).

Although our understanding of the transformation of the sperm nucleus into the male pronucleus is still very limited, the experimental approach to this problem has proved very rewarding: there is suggestive evidence that several cytoplasmic factors are involved in this process and soon we should be able to determine precisely the periods of their activity.

REFERENCES

Balakier, H., and Tarkowski, A.K., 1980, The role of germinal vesicle karyoplasm in the development of male pronucleus in the mouse, *Exp. Cell Res.*, 128:79.

Borsuk, E., and Manka, R., 1988, Behavior of sperm nuclei in intact and bisected metaphase II mouse oocytes fertilized in the presence of colcemid, *Gamete Res.*, 20:365.

Borsuk, E., and Tarkowski, A.K., 1989, Transformation of sperm nuclei into male pronuclei in nucleate and anucleate fragments of parthenogenetic mouse eggs, *Gamete Res.*, 24.

Chang, M.C., 1955, The maturation of rabbit oocytes in culture and their maturation, fertilization and subsequent development in the fallopian tubes, *J. Exp. Zool.*, 128:379.

Clarke, H.J., and Masui, Y., 1986, Transformation of sperm nuclei to metaphase chromosomes in the cytoplasm of maturing oocytes of the mouse, *J. Cell Biol.*, 102:1039.

Clarke, H.J., and Masui, Y., 1987, Dose-dependent relationship between oocyte cytoplasmic volume and transformation of sperm nuclei to metaphase chromosomes, *J. Cell Biol.*, 104:831.

Czolowska, R., Waksmundzka, M., Kubiak, J.Z., and Tarkowski, A.K., 1986, Chromosome condensation activity in ovulated metaphase II mouse oocytes assayed by fusion with interphase blastomeres, *J. Cell Sci.*, 84:129.

Iwamatsu, T., and Chang, M.C., 1972, Sperm penetration in vitro of mouse oocytes at various times during maturation, *J. Reprod. Fert.*, 31:237.

Kaufman, M.H., and Howlett, S.K., 1986, The ovulation and activation of primary and secondary oocytes in Lt/Sv strain mice, *Gamete Res.*, 14:255.

Komar, A., 1982, Fertilization of parthenogenetically activated mouse eggs. I. Behaviour of sperm nuclei in the cytoplasm of parthenogenetically activated eggs, *Exp. Cell Res.*, 139:361.

Krzanowska, H., 1982, Toluidine blue staining reveals changes in chromatin stabilization of mouse spermatozoa during epididymal maturation and penetration of ova, *J. Reprod. Fert.*, 64:97.

Kubiak, J.Z., 1989, Mouse oocytes develop gradually the ability for activation during the metaphase II arrest, *Dev. Biol.*, 136, 537.

Maro, B., Johnson, M.H., Webb, M., and Flach, G., 1986, Mechanism of polar body formation in the mouse oocyte: An interaction between the chromosomes, the cytoskeleton and the plasma membrane, *J. Embryol. exp. Morph.*, 92:11.

Masui, Y., 1985, Problems of oocyte maturation and the control of chromosome cycles, *Develop. Growth and Differ.*, 27:295.

Miller, M.A., and Masui, Y., 1982, Changes in the stainability and sulphydryl level in the sperm nucleus during sperm-oocyte interaction in mice, *Gamete Res.*, 5:167.

O'Neill, G.T., and Kaufman, M.H., 1987, Ovulation and fertilization of primary and secondary oocytes in Lt/Sv strain mice, *Gamete Res.*, 18:27.

Schatten, G., Simerly, C., and Schatten, H., 1985, Microtubule configurations during fertilization, mitosis and early development in the mouse and the requirement for egg microtubule mediated motility during mammalian fertilization, *Proc. Natl. Acad. Sci. USA*, 82:4152.

Schmiady, H., Sperling, K., Kentenich, H., and Stauber, M., 1986, Prematurely condensed human sperm chromosomes after in vitro fertilization (IVF), *Hum. Genet.*, 74:441.

Schroeder, A.C., and Eppig, J.J., 1984, The developmental capacity of mouse oocytes that matured spontaneously in vitro is normal, *Devl. Biol.*, 102:493.

Tarkowski, A.K., 1975, Induced parthenogenesis in the mouse, *in:* "The Developmental Biology of Reproduction, 33rd Symposium of the Society for Developmental Biology", C.L. Markert and J. Papaconstatinou, eds.,pp. 107–129. Academic Press, New York.

Thibault, C., and Gerard, M., 1970, Facteur cytoplasmique nécessaire a la formation du pronucleus male dans l'ovocyte de Lapine, *C. R. Acad. Sc. Paris*, 270:2025.

Thibault, C., and Gerard, M., 1973, Cytoplasmic and nuclear maturation of rabbit oocytes in vitro, *Ann. Biol. Anim. Biochem. Biophys.*, 13:145.

PARTIAL ZONA DISSECTION FOR ENHANCEMENT OF SPERM PASSAGE THROUGH THE ZONA PELLUCIDA

Jacques Cohen, Henry Malter, Carlene Elsner,
Patricia Hunt, Hilton Kort, Joe Massey, Dorothy Mitchell,
Andy Toledo, Sharon Wiker, and Graham Wright

Reproductive Biology Associates, Atlanta, USA
Gamete and Embryo Research Laboratory, Gynecology and Obstetrics
and The Winship Cancer Center, Emory University
School of Medicine, Atlanta, USA

INTRODUCTION

Fertilization in mammals occurs in a number of intricate steps, culminating in the union of male and female genomes. Spermatozoa capacitate prior to sperm receptor binding on the zona pellucida (ZP), where the acrosome reaction is induced. Following ZP penetration membrane fusion occurs, triggering oocyte activation. This leads to the release of cortical granules and the zona reaction, causing a slow, but usually permanent block to polyspermy. Changes in the oolemma may initiate a fast, but often weak and temporary block, as well. Following decondensation and syngamy, the fertilization process is completed with the formation of a genetically new conceptus at the two-cell stage.[1] Pathological changes may occur during any of these steps, and fertilization may either be discontinued or result in a genetically abnormal embryo. Discontinued fertilization is common during human in vitro fertilization (IVF), especially when there is a sperm disorder. Although the methods for microsurgical fertilization proposed in recent years only alleviate abnormal ZP-binding, ZP-penetration, and/or membrane fusion, they may be welcome additions to IVF. Basically, three methods, each with its own advantages and disadvantages, have been proposed:

Advances in Assisted Reproductive Technologies, Edited by
S. Mashiach *et al.*, Plenum Press, New York, 1990

(i) Microinjection of a single intact spermatozoon into the ooplasm has been successfully applied to the rabbit.[2] Microinjection is often traumatic to the oocyte and requires artificial induction of the acrosome reaction and oocyte activation. Advantages are that it can be used with extreme low sperm counts and in species who lack a membrane block to polyspermy. (ii) Subzonal insertion of spermatozoa under the ZP has produced pregnancies in the mouse and human.[3, 5] Although it requires some type of sperm motility, it can probably be applied in extreme cases of male infertility. Artificial induction of the acrosome reaction and oocyte activation are not necessary. This method is presumable less traumatic than microinjection. (iii) The passage of spermatozoa through the ZP can be facilitated using acidic Tyrode's medium for zona dissolution (zona drilling)[4] or following mechanical opening using partial zona dissection (PZD).[6, 7] Both variations have yielded offspring in mice (zona drilling and PZD)[4, 8] and humans (PZD).[5] Although opening the ZP is nontraumatic, acidic Tyrode's may cause inhibition of further embryonic development in the human. Disadvantages of zona drilling and PZD are that it requires relatively large numbers of progressively motile spermatozoa and that it is inefficient in species lacking a fast block to polyspermy.[9, 10]

The use of PZD in human IVF and its possible consequences for fertilization, embryonic development, and implantation are discussed below with illustrations from studies performed in mice.

PZD IN OLIGO-ASTHENOSPERMIA

Twenty-five couples (29 cycles) with a history of male infertility were treated with PZD. Semen abnormalities in this program are defined according to WHO-standards, although more stringent cut-off limits are used for patients allocated to the PZD protocol. Oligospermia (8 cycles) was defined when there were less than 10×10^6/ml spermatozoa in the best fraction of a split ejaculate. Asthenospermia (10 cycles) was defined when there were less than 20% motile spermatozoa or when forward progression was consistently abnormal. Oligo-asthenospermia (11 cycles) was defined when there were less than 20×10^6/ml spermatozoa and less than 30% were motile. Methods for follicular stimulation, semen preparation, oocyte and embryo culture, microtool preparation, micromanipulation, and cryopreservation are described elsewhere.[6, 11–13]

Micromanipulation was performed in 5 µl droplets under oil in glass well slides using Nomarski optics. The PZD-protocol was applied stepwise: (i) Cumulus cells were removed with 0.1% hyaluronidase, three to nine hours following egg collection. (ii) Corona cells were partially removed with hypodermic needles, and (iii) the cytoplasm volume was reduced with 0.1 M sucrose. (iv) Approximately 50% of the oocytes were micromanipulated (*PZD-oocytes*), while the remainder was left in sucrose solution with intact ZP (*control oocytes*). (v) Sucrose was removed in three to four steps, and both groups of oocytes were separately inseminated 20 to 200 minutes after sucrose addition. The incision in the ZP was made with a microneedle stuck peripherally through the "one" and "11" o'clock posi-

Table 1. Results of PZD in 25 Infertile Men (29 cycles)

Patients	Fertilization		Polyspermy PZD	Cycles with pregnancy
	PZD	Control		
All	49/93 (53%)	25/85 (29%)	16%	6/29
Oligo (0)	16/20 (80%)	8/20 (40%)	0%	2/8
Astheno (A)	17/26 (65%)	10/27 (37%)	18%	3/10
OA	16/47 (34%)	7/38 (18%)	31%	1/11

Table 2. Number of PZD-Embryos Replaced and Implanted

No. of cycles	No. of PZD and control embryos	Total number of embryos	Number of fetal sacs	
			minimum PZD	maximum PZD
1	0 and 1	1	0	0
5	1 and 0	5	0	0
4	1 and 1	8	0	1
2	2 and 0	4	0	0
5	2 and 1	15	2	7
3	3 and 0	9	2	2
Total		42	4	10

tions of the ZP. The microneedle was rubbed with the holding pipet over the 12 o'clock position until the oocyte dropped off leaving an incision in the ZP. Maximally, three embryos were replaced at a time, and pregnancy was confirmed with observation of fetal heart activity.

Only 10/272 (4%) of oocytes (from male factor and immunological patients combined) were damaged. One oocyte was damaged during micromanipulation, and the remainder degenerated at corona dissection before (n = 5) or after (n = 4, pronuclei check) micromanipulation. The current status of the PZD-protocol in male factor patients is illustrated in Table 1. Fertilization following PZD occurred in 72% of patients whereas only 48% had fertilization in at least one of the ZP-intact control oocytes.

One of the patients who miscarried twins had Cushing's syndrome, and another singleton pregnancy miscarried in an older patient. Three patients delivered health infants. Two sets of non-identical twins (both having a boy and a girl) were delivered in November 1988, each 6 weeks premature. Both twins were obtained from replacements of two PZD and single control embryos (Table 2).

A third pregnancy delivered in February 1989. This baby resulted from replacement of three PZD-embryos only and originated as a twin pregnancy. A total of 42 embryos (32 PZD and 10 control) were replaced in 22 patients resulting in 10 fetal sacs of which four were twin pregnancies (24% per embryo, 27% per replacement, 21% per cycle).

Table 3.   PZD and Sperm Antibodies

| | In male partner | | In female partner | |
	PZD	control	PZD	control
Monospermy per patient	2/3	3/3	1/2	2/2
Polyspermy	3/3	0/3	1/2	1/2
Embryos replaced	1	8	1	4
Pregnancies (sacs)	3(5)		2(3)	

Table 4.   Reinsemination of Zona-intact and Dissected Oocytes

| Reinseminated oocyte | No. of patients (oocytes) | Incidence of fertilization | |
		monospermy	polyspermy
ZP intact	67 (161)	9%	3%
Zona drilled (PZD)	60 (215)	21%	22%

PZD AND SPERM ANTIBODIES

Semen analyses in five couples were repeatedly normal, but sperm directed antibodies existed in female (n = 2) or male partner (n = 3). The same PZD protocol as described above was applied to these immunological patients (Table 3). Fertilization occurred in all patients in all groups, but PZD zygotes were usually polyspermic. Antisperm antibodies in these patients may not interfere during membrane fusion but may interact prior to ZP-penetration. Subzonal insertion or zona drilling may therefore be contraindicated in these patients, as the oolemma is unable to restrict polyspermy when exposed to several fertilizing spermatozoa at the same time. Removal of cumulus and corona cells with hyaluronidase may effectively have alleviated immunological interactions causing the high success rate in this small series (Table 3).

PZD FOR REINSEMINATION

Assessment of human oocytes is often inaccurate causing difficulty in distinguishing between fertilization resulting from initial insemination and that from reinsemination. Reinseminated zona-intact oocytes from 60 patients were observed several times, and the incidence of reinsemination was only 12%.[6] Zona drilling has been performed routinely in one-day old unfertilized oocytes from consenting IVF patients since April 1988. Advantages are twofold: (i) it yields more monospermic embryos (Table 4); and, (ii) it provides a means for studying microsurgical techniques without jeopardizing the patient's other embryos. In our program, techniques are first developed in aged oocytes prior to application in young mature oocytes.

Table 5. Varying Sucrose-reinsemination Intervals and Reinsemination

Duration (min.)	Fertilized	Proportion of oocytes > 4 Pronuclei [a]	Parthenogenetic
4 - 20	19/34 (56%)	7/34 (21%)	0/34 (0%)
21 - 45	17/40 (43%)	5/40 (13%)	0/40 (0%)
46 - 72	16/41 (39%)	0/41 (0%)	4/41 (10%)

[a] Significant differences between groups

Zona dissection has increased the incidence of fertilization after reinsemination from 12% (19/161) to 44% (94/215). However, the frequency of polyspermy increased from 21% to 51%. Nevertheless, at least twice as many normal monospermic embryos can be obtained when PZD is applied to reinseminated oocytes. Zona dissection for reinsemination is successful irrespective of the original insemination results.[10] The frequency of polyspermy can be diminished by performing PZD 20 to 24 hours after oocyte collection, rather than 25 hours or later.

INHIBITION AND CORRECTION OF POLYSPERMY

Inhibition of Polyspermy after PZD

It has been postulated that the human oocyte, like the hamster oocyte, has no block to polyspermy at the level of the membrane.[9] Preliminary investigations following PZD of freshly collected oocytes indicated that polyspermy diminished, when the interval between sucrose addition and insemination in sucrose-free medium was lengthened.[13] In order to test the possibility of sucrose-induced oocyte activation, 115 one-day old oocytes were zona dissected prior to reinsemination using different sucrose addition-reinsemination intervals (Table 5). Sucrose was only used during micromanipulation for four to ten minutes.

A significant decrease in excessive polyspermy (more than four pronuclei) was found with increasing interval time. Additional evidence of sucrose triggered activation was derived from the presence of haploid parthenogenetic embryos when reinsemination was delayed. Sucrose or any other type of activation inducing mechanism may therefore be a useful tool in obtaining monospermy following PZD.

Correction of Polyspermy

A reliable method for enucleation of extra male pronuclei would be beneficial for human IVF especially since zona drilling increases sperm fusion.[14-16] Pronuclear extraction has been attemped in 66 polyspermic human zygotes in the absence of membrane relaxants. Sixteen of the zygotes had four or more pronuclei. The incidences of survival (36/66, 55%), syngamy (24/66, 36%), and normal cleavage (21/66, 32%) were independent from the number of extra male pronuclei present or extracted. Method variables were needle size, needle wall thickness, number of cytoplasmic insertions, the presence of an artificial hole in the ZP for needle piercing, and the use of a microprocessor controlled suction device rather than a manually controlled oil filled syringe. Extreme distortion of the oocyte shape (Figure 1A) was the most frequent cause of cell death, and this was caused by the inability of needles to pierce the ZP easily. Artificially made holes introduced next to the target pronucleus increased the survival rate dramatically (Figure 1B, 1C). Repeated insertion of the ooplasm was diminished with a microprocessor controlled suction apparatus. The use of acid Tyrode's medium for creation of a hole in the ZP (Figure 1C) adjacent to the target pronucleus appears to be the most reliable method. PZD prior to enucleation can still cause distortion of the oocyte shape (Figure 1C), when the PZD incision is not located exactly near the target pronucleus. Three of seven enucleated embryos developed to the morula stage, and one cavitated. Fourteen other embryos were fixed for chromosome analysis. Six chromosome counts could be performed: four embryos were diploid and two haploid. Human pronuclei often adhere firmly together, and disruption of their proximity during enucleation may cause incomplete syngamy. Paternal pronuclei could not be distinguished from maternal ones on the basis of size or sperm tail remnants. Pronuclei furthest away from the second polar body were therefore targeted for enucleation. Clinical application of this method still needs to be awaited, as it is currently unknown whether the relative position of the extra male pronucleus is a sufficiently reliable criterium for selection.

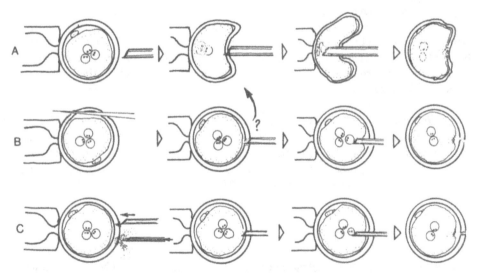

Fig. 1. Three methods for pronucleus extraction. (A) Large needle used for piercing the ZP and the membrane. (B) Small, thin-walled needle inserted through a PZD hole. (C) Thin needle inserted through a hole created by acidic Tyrode's medium.

Table 6. Corona Radiata Expansion and PZD in 27 Male Factor Cycles

Results per	Oocyte Treatment	Corona Expansion		
		Complete	Partial	Tight
Patient cycle	PZD	12/14 (86%)	8/10 (80%)	1/3 (33%)
	Control	11/14 (79%)	2/10 (20%)	0/3 (0%)
Oocyte	PZD	27/41 (66%)	19/43 (44%)	2/8 (25%)
	Control	17/40 (43%)	6/40 (15%)	0/9 (0%)

Table 7. Contraction of Aged Oocytes in Sucrose and PZD

	Parthenogenesis	Monospermy	Polyspermy
Spherical	3/29 (10%)	5/29 (17%)	0/29 (0%)
Pear-like	1/23 (4%)	8/23 (35%)	1/23 (4%)
Crenated	0/21 (0%)	3/21 (14%)	7/21 (33%)

OOCYTE FEATURES AFFECTING PZD

Corona radiata cells are not readily removed by hyaluronidase, and extensive manual dissection may damage the oocyte. Although the first polar body was extruded in occytes from 27 male factor patients treated with PZD, corona expansion differed from patient to patient. The expansion, assessed following cumulus removal, did not correlate well with nuclear maturation. The incidence of fertilization determined per ZP-intact oocyte or per patient enhanced when corona expansion increased (Table 6). However, following PZD, 44% of oocytes with only partially expanded corona were fertilized, whereas those with intact ZP fertilized in 15% of the cases.

Young mature human oocytes usually contract spherically when exposed to sucrose, whereas aged oocytes may contract in typically irregular fashions. Three types of contractions were observed: (i) spherical, (ii) pear-like, and (iii) crenated with cytoplasmic extrusions. The highest rates of polyspermy and fertilization were found when aged oocytes exposed to sucrose contracted in an irregular manner (Table 7).

The morphologic changes in sucrose may differentiate between oocytes which have only undergone nuclear changes, and those which also have mature cytoplasm.

EMBRYONIC DEVELOPMENT AND IMPLANTATION FOLLOWING PZD

Although introducing an incision in the ZP is a simple procedure, it may have considerable consequences for preimplantation development and implantation. It should therefore

Table 8. Corticosteroid/Antibiotic Trial in Patients who had PZD-embryos Replaced

Number of Embryos Replaced	Proportion of Pregnant Patients	
	(+)[a]	(-)
1	0/3	0/3
2	1/3	0/4
3	4/5	1/3

[a] (+) corticosteroids/antibiotics administered

be considered carefully despite an obvious lack of impairment in zona drilled mouse and human embryos. Non-pathogenic (even pathogenic!) micro-organisms may be present in the insemination suspension. Cytotoxins released from bacteria may impair embryogenesis. An intact ZP may delay or inhibit active transport of cytotoxic compounds. The possibility of viral infection should also be taken into account. Despite these possible hazards, zona dissection has yielded apparently normal embryos with a normal development rate and many of them implanted.[13] The protective role of the ZP in vitro may be marginal. However, the possibility of increased infection of zona-drilled embryos may present a problem in individual cases.

Blastomeres may be lost through large holes,[17] and this may affect blastocyst formation, possibly resulting in trophoblastic vesicles. Chimaeric development through aggregation of cells from embryos outside the ZP as a result of "over-drilling" has been postulated.

The presence of holes in the ZP can cause an immune response in the host uterus prior to embryonic compaction.[18] Hence, artificially twinned precompacted bovine embryos are embedded in agar and transferred to temporary recipients in order to avoid leucocyte invasion. The agar chips are removed after compaction. The mechanisms of possible leucocyte invasion in the human precompacted embryo are unknown and difficult to study. Theoretically, holes smaller than 20 µm in diameter may be traversed by "shape changing" polymorphonuclear leucocytes. Although PZD produces narrow crevices (less than 10 µm in diameter), they may be irregularly shaped and as long as 40 µm. Pressure on the ZP may cause leakage of embryonic cells as a result of embryo handling, interaction with the uterine wall, or following cleavage and blastomere expansion. In order to study these possibilities indirectly, we are conducting a prospective trial in patients who receive PZD embryos. Methylprednisolone (16 mg. daily) is administered for four days starting on the day following oocyte collection, in combination with tetracycline or ampicillin. Every other patient receives this treatment (Table 8). Thus far, patients who received corticosteroids and antibiotics, appear to have a higher chance of implantation. It should be noted, however, that the sample size of patients is still very small.

The presence of a hole in the ZP does not impair post-thaw survival rates of mouse and hamster embryos. Neither propanediol nor dimethylsulfoxide causes fracture of zona-drilled one- to 32-cell frozen embryos. Moreover, several frozen human PZD-embryos have survived the thawing process. One intact thawed PZD-zygote cleaved into two cells and was replaced with a partially (two of four blastomeres) intact non-drilled thawed control embryo; a singleton ongoing pregnancy was induced. It is likely that it was the zona-dissected thawed embryo which implanted, but absolute proof cannot be given. Thawed embryos with less than 3/4 of their blastomeres intact have not implanted in our program.

Both the timing and morphology of the hatching process of mouse and human PZD-blastocysts in vitro are altered. Thinning of the ZP in the expanding blastocyst does not occur in zona dissected embryos. Hatching usually occurs one day early as the blastocyst expands through the artificially made hole. One human PZD blastocyst divided into two halves during hatching, forming two twin blastocysts. Zona-dissected mouse embryos are frequently trapped during premature hatching resulting in the formation of "8-shaped" blastocysts. Such "trapped" blastocysts were never seen in ZP-intact controls. Although artificially made ZP holes may promote implantation, it is possible that it also increases the rates of blighted ova pregnancy and monozygotic twinning.

CONCLUSIONS

The incidences of fertilization and implantation increased in a small group of infertile men following microsurgical fertilization. Men with single sperm abnormalities, like oligospermia, benefitted especially from PZD. A drawback of PZD is that it is relatively inefficient, as it does not accurately control the number of spermatozoa crossing the ZP-incision. This is disconcerting since the block to polyspermy is weak on the level of the oocyte membrane. Even spermatozoa with severe motility disorders can cause multiple sperm fusion. Sucrose does not only assist the micro-manipulation process directly by shrinking the oocyte cytoplasm, enabling the investigator to pierce the perivitelline space non-traumatically, it also triggers activation, thereby reducing the possibility of polyspermy. Certain oocyte characteristics were found to inhibit fertilization even when a ZP-incision was present. The number of inseminated spermatozoa and the optimal duration of insemination needed for monospermic fertilization vary from patient to patient and oocyte to oocyte.

The presence of an artificial hole in the ZP may have considerable consequences for further embryonic development. Bacteria, cytotoxins, and viruses may jeopardize early embryos, and uterine leucocytes may attack the precompacted embryo after it is replaced. On the other hand, PZD may increase embryonic implantation by allowing hatching to occur more frequently and earlier during preimplantation development. The survival of human PZD embryos following cryopreservation is probably not impaired. Research into the mechanisms of human fertilization is needed to control monospermic fertilization, and the establishment of a reliable method for repairing polyspermic zygotes would promote the application of PZD even further.

ACKNOWLEDGEMENTS

We would like to thank Leigh Inge, Donna Reynolds, Beth Talansky, Jon Gordon, Carole Fehilly, Mary Pat Mayer, and Robin Moseley for their valuable contributions.

REFERENCES

1. F. J. Longo, "Fertilization," Chapman and Hall, London (1987).
2. Y. Hosoi, Development of rabbit oocytes after microinjection of spermatozoa, *in:* "Proc. of the 11th International Congress on Animal Reproduction and Artificial Insemination," Dublin (1988).
3. J. Mann, Full-term development of mouse eggs fertilized by a spermatozoon microinjected under the zona pellucida, *Biol Reprod.* 38:1077 (1988).
4. J. W. Gordon and B. E. Talansky, Assisted fertilization by zona drilling: a mouse model for correction of oligospermia, *J. Exp. Zool.*, 239:347 (1986).
5. S. C. Ng, A. Bongso, S. S. Ratnam, H. Sathananthan, C. L. K. Chan, P. C. Wong, L. Hagglund, C. Anandakumar, Y. C. Wong, and V. H. H. Goh, Pregnancy after transfer of sperm under zona, *Lancet*, 2:790 (1988).
6. H. E. Malter, J. Cohen, Partial zona dissection of the human oocyte: a nontraumatic method using micromanipulation to assist zona pellucida penetration, *Fertil. Steril.*, 51:139 (1989).
7. J. Cohen, H. Malter, C. Fehilly, G. Wright, C. Elsner, H. Kort, and J. Massey, Implantation of embryo after partial opening of the oocyte zona pellucida to facilitate sperm penetration, *Lancet*, 2:162 (1988).
8. H. E. Malter, unpublished results.
9. J. W. Gordon, L. Grunfeld, G. J. Garrisi, B. E. Talansky, C. Richards, and N. Laufer, Fertilization of human oocytes by sperm from infertile males after zona pellucida drilling, *Fertil. Steril.*, 50:68 (1988).
10. H. E. Malter, B. E. Talansky, J. W. Gordon, and J. Cohen, Monospermy and polyspermy after partial zona dissection of reinseminated human oocytes, *Gam. Res.*, in press.
11. J. Cohen, G. W. DeVane, C. W. Elsner, C. B. Fehilly, H. I. Kort, J. B. Massey, and T. G. Turner, Jr., Cryopreservation of zygote and early cleaved human embryo, *Fertil. Steril.*, 49:283 (1988).
12. J. Cohen, G. W. DeVane, C. W. Elsner, H. I. Kort, J. B. Massey and S. E. Norbury, Cryopreserved zygotes and embryos and endocrinologic factors in the replacement cycle, *Fertil. Steril.*, 50:61 (1988).
13. J. Cohen, H. Malter, G. Wright, H. Kort, J. Massey, and D. Mitchell, Partial zona dissection of human oocytes when failure of zona pellucida penetration is anticipated, *Hum. Reprod.*, in press.
14. R. G. Rawlins, Z. Binor, E. Radwanska, and W. P. Dmowski, Microsurgical enucleation of tripronuclear human zygotes, *Fertil. Steril.*, 50:266 (1988).
15. H. E. Malter and J. Cohen, Embryonic development following micro-surgical repair of polyspermic human zygotes, Submitted for publication.
16. J. W. Gordon, L. Grunfeld, G. C. Garrisi, D. Navot, and N. Laufer, Successful microsurgical diploidization of tripronuclear human zygotes, Submitted for publication.
17. B. E. Talansky and J. W. Gordon, Cleavage characteristics of mouse embryos inseminated and cultured after zona pellucida dripping, *Gam. Res.*, 21:277 (1988).

THE MICROINJECTION TECHNIQUE AND THE ROLE OF THE ACROSOME
REACTION IN MICROFERTILIZATION

A. Henry Sathananthan and Alan Trounson

La Trobe University and Monash Medical Centre
625 Swanston St., Carlton 3053 Australia

INTRODUCTION

An increasing number of men with male factor infertility seek assistance for conception by in vitro fertilization (IVF) and related techniques such as gamete intrafallopian tube transfer (GIFT). However, there is little prospect of success for men who have semen of severely reduced quality with low numbers of morphologically normal sperm which have forward progressive motility. Although sperm with only a single or double defect of low sperm numbers, impaired motility or high morphological abnormality, have a reduced chance of fertilizing eggs (Yovich and Stanger, 1984), pregnancy rates following IVF in these cases are not usually reduced (Yates et al., 1988). In such cases superovulation yields a number of eggs which can be fertilized and several early embryos can usually be replaced in utero. Pregnancy rates in GIFT for couples where the husband has reduced semen quality is disappointingly low (Jansen, 1988) and it is preferable to inseminate eggs in vitro and transfer pronuclear ova or cleaving embryos to the wife's oviduct or uterus. Three pregnancies have resulted in cases of severe oligoasthenozoospermia by the transfer of pronuclear ova into Fallopian tubes, after micro-insemination sperm transfer (MIST) under the zona (Ng et al., 1988, 1989c). The majority of men who have male factor infertility have multiple sperm defects so that fertilization cannot be achieved by conventional insemination methods in vitro or in vivo. There are also some men with apparently normal sperm who are incapable of fertilizing eggs in vitro because of some unknown problem in their fertilizing capacity or egg compatibility. This has led to research into ways of bypassing the zona pellucida which is the major barrier to sperm penetration of the oocyte.

Advances in Assisted Reproductive Technologies, Edited by
S. Mashiach *et al.*, Plenum Press, New York, 1990

OVERVIEW OF THE MICROINJECTION TECHNIQUE

This overview will examine the microinjection of sperm into human eggs to assist fertilization (microfertilization) with some reference to mouse work which is being carried out in our laboratory. Other methods of assisted fertilization are reported elsewhere in this book. Recent reviews of microinjection and micromanipulation have been written by Trounson et al., (1989) Ng et al., (1989a).

Microinjection specifically involves the transfer of sperm either into the perivitelline (PVS) of oocytes or directly into the ooplasm using micromanipulators. The egg is held firm by a suction pipette during the microinjection procedure. Two methods are in use to perforate the zona: (i) Mechanical perforation using the glass injecting needle (4–7 μm bore diameter) (Laws-King et al., 1987; Ng et al., 1988). (ii) Zona drilling where an acidified stream of medium is injected locally with a finely drawn glass pipette (Gordon et al., 1988; Ng et al., 1989b). Both techniques involve considerable micromanipulative skills and may cause lysis or damage to the egg if large bore glass needles are used. If excessive pressure is exerted by the suction pipette the egg cell may prolapse through larger perforations. Mechanical perforation seems to be more precise, the least harmful to the oocytes and the most successful to date. The size of the perforation can be better regulated by this method (Fig. 1), whereas with acid-drilling the aperture is much wider and variable in size as demonstrated by transmission electronmicroscopy (TEM) (Sathananthan et al., 1989a). Both single and multiple sperm injection have been attempted in the human and mouse.

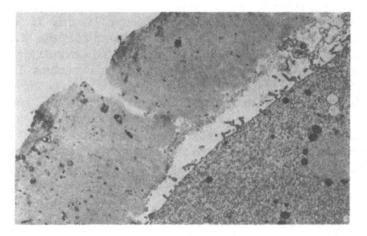

Fig.1. Narrow perforation in the zona of an unfertilized oocyte made by the sperm microinjection needle. The oolemma is not damaged. × 9100. (Reproduced from Sathananthan et al., 1989a).

SINGLE SPERM INJECTION

Single sperm are injected into the PVS of oocytes to ensure monospermic fertilization. Metka et al., (1985) were the first to report human fertilization using a micromanipulator. Of nine oocytes injected, one pronuclear ovum was obtained which cleaved to 4-cells. Our preliminary experiments (Laws-King et al., 1987; Sathananthan et al., 1989a) conclusively showed that single sperm injection of mature human eggs can result in high rates of fertilization, satisfying all criteria of normal fertilization namely evidence of two pronuclei and spermtail in the ooplasm (Fig. 2), absence of most of the cortical granules and abstriction of the second polar body. Seven fresh eggs microinjected produced five fertilized ova, two of which cleaved normally. Aged eggs (cultured 23–28 hours) before injection yielded lower rates of fertilization and some were parthenogenetically activated. Hence, it is important that fresh mature oocytes be used for microinjection. Normal fertilization and embryo development has also eventuated after single immotile sperm injection (Laws-King et al., 1987; Bongso et al., 1989a), which will be dealt with later. There are two major questions relating to sperm injection that need to be considered. Firstly, does the arbitrary selection of a single sperm for microinjection increase the chance of chromosomal abnormality in the resulting embryos? Secondly, are sperm of very poor quality more likely to be genetically abnormal and therefore increase the incidence of chromosomal defects in these embryos?

Our initial studies on microinjection of mouse eggs showed that 25% of these eggs could be fertilized and 81% of these developed into live fetuses or young, when transferred to foster mothers (Mann, 1988). This compares favourably to eggs fertilized in vivo (92%) or by routine IVF (90%). The mice that were born were morphologically normal and later bred normally. Further research in our laboratory showed that over 90% of mouse eggs fer-

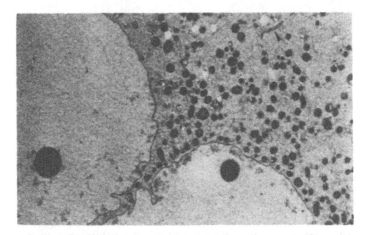

Fig.2. Normally fertilized ovum showing 2 pronuclei and a spermtail (arrow), 12 hours after single sperm injection under the zona, × 11,900. (Reproduced from Laws-King et al., 1987).

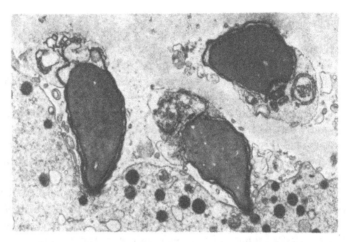

Fig.3. Abnormal sperm in the PVS of an unfertilized oocyte fixed 4 hours after multiple sperm injection under the zona. The acrosomes are intact and one sperm has sheared the inner zona. The egg is not activated as cortical granules are intact, × 19,600.

tilized by microinjection of a single sperm develop into blastocysts in vitro and also develop to fetuses in vivo at high rates (Lacham et al., 1989). These results indicate that single sperm selection for microinjection does not increase the incidence of embryonic abnormality. Whether this is applicable to the human remains to be investigated. Human sperm even from normal men, when examined by TEM, show a greater variety of sperm defects when compared to those of mice. Poor quality human sperm exhibit even greater morphological defects when examined in pellets or after microinjection into the PVS (Sathananthan et al.,

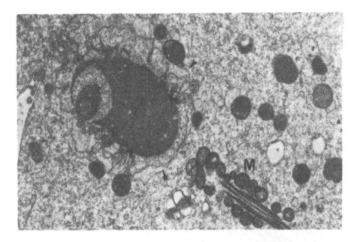

Fig.4. Abnormal sperm incorporated into the ooplasm of a polyspermic ovum, 24 hours after multiple sperm injection. Nuclear envelope has inflated (arrows) prior to chromatin decondensation. M = midpiece × 19,600. (Reproduced from Sathananthan et al., 1989a).

1989a) and abnormal sperm have also been incorporated into the ooplasm (Figs. 3,4). Martin et al., (1988) have reported a high rate of chromosomal abnormalities and an increased frequency of such abnormalities (30%) in microinjected hamster eggs, when compared to eggs inseminated by conventional methods (12%). Therefore, the prevalence of abnormal sperm, both morphological and genetical, raises some concern about the normality of embryos produced by microinjection. However, a normal pregnancy, screened by amniocentesis, is progressing after multiple sperm injection in Singapore (Ng et al., 1988). Obviously, genetic screening may be routinely advisable, when poor quality sperm are used for microinjection. Our view at Monash is that the chromosome normality of embryos produced by microinjection of sperm from men with severe sperm defects should be established before the technique is introduced into clinical practice.

MULTIPLE SPERM INJECTION

Fertilization rates could be appreciably increased by multiple sperm injection under the zona of human eggs (Lasalle et al., 1987; Ng et al., 1989a). Surprisingly, the rate of monospermic fertilization was quite high (19%) when compared to polyspermic fertilization (7%), after 7-10 sperm were injected into the PVS of 42 mature oocytes, (Ng et al., 1989c). Lasalle et al (1987) obtained 3 monospermic ova when 3-5 sperm were transferred into 7 eggs and polyspermy resulted when 10-12 sperm were injected. They also obtained high rates (55%) of polyspermy when 5-12 human sperm were injected into hamster eggs. We have obtained a dispermic ovum after injection of multiple sperm into the PVS (Fig. 5), confirm-

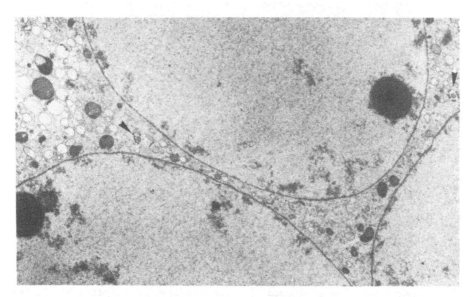

Fig.5. Tripronuclear ovum obtained 24 hours after multiple sperm injection into the PVS. Two sperm tail axonemes are evident (arrows), × 15,400. (Reproduced from Ng et al., 1989a).

ing that triploidy could result by simultaneous penetration of 2 sperm (Sathananthan et al., 1989a). So it appears that polyspermy may result when sperm numbers injected are too high. Monospermic fertilization is easily and non-invasively detectable in routine IVF (evidence of 2 pronuclei in the ooplasm) and the injection of multiple sperm clearly increases the chances of fertilization. Ultrastructural studies have also confirmed monospermic fertilization (Fig. 6) after multiple sperm injection (Sathananthan et al., 1989a). The ova appear to be morphologically normal and there is no reason to suppose that they are less viable than those fertilized in routine IVF. Three pregnancies have been reported after multiple sperm transfer though one ectopic has aborted (Ng et al., 1989c). The high incidence of monospermy in human oocytes strongly indicates that there exists an appreciable block to polyspermy at the level of the oolemma (vitelline block) in addition to the primary block in the inner zona (Sathananthan and Trounson, 1982). The oolemma is a mosaic membrane after fertilization, having incorporated both sperm plasma membrane and cortical granule membrane during sperm fusion and cortical granule exocytosis (Sathananthan et al., 1986a,b; Yanagimachi, 1988). Whether this vitelline block in mammals is a rapid electrical phenomenon has not been convincingly demonstrated nor has the role of cortical granule exocytosis in establishing the block been demonstrated equivocally. Sperm injection experiments will help us elucidate the nature of the vitelline block, since the zona is by-passed.

In our experiments in mice increasing the number of sperm injected under the zona has not dramatically increased the fertilization rate or polyspermy (Table 1). Over 40%–47% of the fertilized ova had two pronuclei, while only 4%–6% had multiple pronuclei, when 2–11

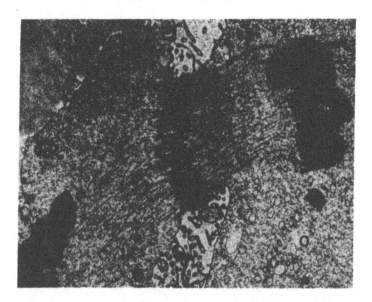

Fig.6.   Normally fertilized ovum obtained 4 hours after multiple sperm injection into the PVS. The second polar body (P) is being abstricted and discharged cortical granules are seen in the PVS (arrows). I = interbody, O = ooplasm, × 8,400.

Table 1. Fertilization and Polyspermy in Mouse Eggs Microinjected with Multiple Mouse Sperm under the Zona Pellucida

No. of sperm injected under the zona	No. of eggs injected	No. of eggs with 2 pronuclei (%)	No. of polyspermic eggs (%)
1 - 3	25	10 (40%)	0 (0%)
4 - 7	53	24 (45%)	2 (4%)
8 - 11	34	16 (47%)	2 (6%)

Reproduced from Trounson et al., (1989).

sperm were injected. Ng and Solter (1989) obtained 32 two-pronuclear ova and only two tripronuclear ova after injection of 3–5 sperm into the PVS of 101 mouse eggs. Significantly, a greater number of acrosome- reacted sperm were found in the PVS of mouse eggs in comparison to that of human eggs, after multiple sperm injection (Sathananthan et al., 1989b). This clearly shows that the mouse has a stronger vitelline block to polyspermy than that of the human. A vitelline block to sperm penetration has been reported in zona-free mouse eggs (Wolf, 1978). It is apparent that the human and mouse egg responds somewhat differently to the presence of multiple sperm.

Multiple sperm injection has a theoretical advantage in that some natural selection of sperm could occur at the level of the oolemma and both sperm and oocyte could have a key role to play in this process. This is particularly relevant to male infertility where there appears to be a higher percentage of structurally or genetically abnormal sperm. In our recent TEM study (Sathananthan et al., 1989a), we examined 100 sperm injected into the PVS of 16 oocytes after transfer of 5–10 sperm obtained from oligozoospermic men. Of these 76% had intact acrosomes and 33% of intact sperm were morphologically abnormal. All types of sperm defects including nuclear, acrosomal, midpiece and tail abnormalities were encountered. We are now evaluating sperm structure before and after injection to determine whether sperm abnormalities could arise by the sperm injection method per se (Martin et al., 1988). It is possible that some structural damage might occur when sperm are injected by fine-bore pipettes, especially in the case of mouse sperm which are considerably larger than human sperm. It is also possible that handling of sperm for microinjection may damage surface membranes involved in the acrosome reaction (AR) and sperm-egg membrane fusion. An intact plasma membrane in the midsegment of the spermhead is a prerequisite for gamete fusion (Sathananthan et al., 1986b).

There is evidence to show that injected sperm exit through the zona perforation in both human and mouse (Sathananthan et al., 1989 a,b). Both acrosome intact and reacted sperm were seen in the gap in the zona. Sperm move actively in the PVS many hours after microinjection and occasionally penetrate the inner zona. Sperm that exit penetrate the outer zona in the usual manner after completion of the AR (Fig. 7). Loss of sperm from the

PVS will lower the rates of fertilization especially if only one sperm is injected. A larger gap in the zona after acid-drilling will increase the chances of sperm loss from the PVS.

ZONA-DRILLING AND SPERM MICROINJECTION

Complete removal of the zona may result in polyspermy or reduced developmental potential since the preimplantation embryo is protected by this vestment during the first week of development. Localized zona-drilling combined with multiple sperm microinjection has been attempted by Ng et al., (1989b). The men had severe oligozoospermia and their spouses had previously failed IVF. Of 22 eggs microinjected four were damaged and the rest had not fertilized. Karyotyping of these eggs revealed a high incidence of possible arrest at anaphase II, after reinitiation of meiosis. In a parallel study where eggs were perforated mechanically in the presence of cytochalasin D, 15 unfertilized oocytes karyotyped did not show a similar arrest. Therefore zona-drilling with acid medium was not recommended by this group. It is possible that the acid in the medium lowered the pH in the vicinity of the egg, which interfered with the process of meiosis.

DIRECT SPERM INJECTION INTO THE OOPLASM

Sperm can be injected directly into the cytoplasm of mammalian eggs resulting in spermnuclear decondensation and pronuclear formation (Uehara & Yanagimachi, 1976; Thadani, 1980; Markert, 1983; Lazendorf et al., 1986). These pronuclear ova may develop to blastocysts, but oocyte mortality is quite high due to injury to the cell membrane. Since human sperm are relatively small, fine bore pipettes can be used to inject eggs. Lazendorf et al., (1988) reported that five (possibly eight) of 20 human oocytes degenerated, while six eggs developed pronuclei after microinjection. We have shown that multiple sperm could be injected with minimal injury to the ooplasm (Sathananthan et al., 1989a) and possible

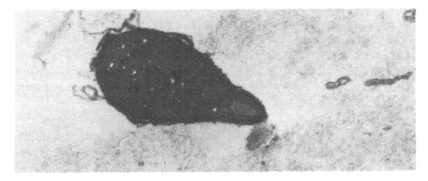

Fig.7. Acrosome-reacted sperm penetrating the outer zona having escaped from the PVS, fixed 24 hours after multiple sperm injection, × 27,300.

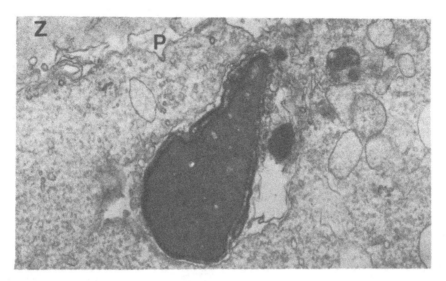

Fig.8. Spermhead, 24 hours after injection into the ooplasm of an intact egg. It is lying in a membrane-bound vacuole and its acrosome is swollen and vesiculating. P = PVS, Z = zona, × 35,700. (Reproduced from Sathananthan et al., 1989a).

sperm-oocyte membrane fusion could occur in membrane-bound vacuoles within the ooplasm resulting in pronuclear formation (Figs. 8,9). However, many of the eggs showed some evidence of localized degeneration when examined by TEM and spermheads, whether acrosome intact, reacting or reacted, remained unexpanded even 24 hours after injection. These also included abnormal sperm forms showing that normal sperm selection is not possible with the light microscope. If a single sperm is injected the chances of selecting an abnormal sperm is quite high particularly in cases of severe oligoasthenozoospermia. No pregnancies have eventuated by this method as yet.

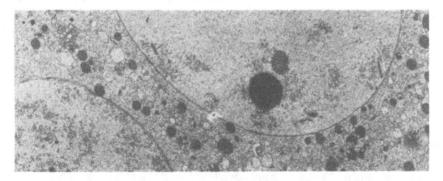

Fig.9. Two-pronuclear ovum developed from an egg 24 hours after multiple sperm injection into the ooplasm, × 7000.

Table 2. Fertilization of Mouse Eggs when Single Sperm from Different Segments of the Mouse Epididymis were Injected under the Zona

Site of sperm recovery in epididymis	No. of eggs injected with a sperm	No. of fertilized eggs
Caput	240	39 (16%)*
Corpus	175	63 (36%)
Cauda	177	60 (34%)

* $P < 0.001$.
Reproduced from Trounson et al., (1989)

MICROINJECTION OF IMMOTILE SPERM

Immotile human sperm injected under the zona will fertilize human eggs and activate embryonic development (Laws-King et al., 1987; Bongso et al., 1989). There is no direct correlation of genetic abnormality with decreased sperm motility, except for immotile cilia syndrome (Palmblad and Mossberg, 1984). Sperm motility is a prerequisite for zona penetration but is not essential for human sperm-oocyte membrane fusion (Sathananthan & Chen, 1986). Immotile sperm from patients with Kartagener's syndrome may be incorporated into hamster and human eggs provided they are allowed to make contact with the egg cell membrane in culture (Aiten et al., 1983; Ng et al., 1987). The egg seems to elicit a phagocytic response during sperm incorporation. Following positive evidence of fertilizability by immotile sperm, an attempt was made by Bongso et al., (1989a) to produce a pregnancy with sperm from the same patient by microinjection of single and multiple sperm into the PVS. Of five eggs, three were fertilized and developed into three to eight cell stages which were transferred into the uterus but did not implant. There are ethical concerns regarding the perpetuation of this syndrome, which is an autosomal recessive condition. The condition is compatible with normal life, provided the patients are treated for respiratory complications. There is a case report of a man with Kartagener's syndrome who has motile sperm and one son (Jonsson et al., 1982). For these reasons microinjection may be justifiable.

MICROINJECTION OF EPIDIDYMAL SPERM

Human eggs can be fertilized with epididymal sperm from men with obstructive azoospermia (Temple-Smith et al., 1985), but rates of fertilization are quite low (10%–12%) as with sperm with multiple defects (Yates et al., 1988). Obstructive azoospermia usually involves the cauda and much of the corpus epididymis, so that sperm could only be recovered from the caput epididymis for IVF. There are usually few progressively motile sperm in the caput which are unable to bind to or penetrate the zona of the egg (Trounson et al., 1989).

The ability of mouse sperm obtained from the three regions of the epididymis to fertilize eggs was examined after single sperm injection (Table 2). Fertilization rates were similar for cauda and corpus epididymis but that for caput sperm was approximately halved when compared to the other two segments. The reduced fertilizing ability of caput sperm may be attributed to the immaturity of sperm surface membranes which are not capable of fusion with the oolemma (Trounson et al., 1989). Capacitation and acrosomal reactivity of caput sperm clearly need to be investigated, to improve the chances of microfertilization in cases of obstructive azoospermia.

ROLE OF SPERM ACROSOME REACTION IN MICROFERTILIZATION

Sperm capacitation and the acrosome reaction (AR) are two important prerequisites for sperm-oocyte membrane fusion and subsequent sperm incorporation in the human (Sathananthan et al., 1986a,b). The AR is regarded as the morphological expression of sperm capacitation, which is a complex physiological process (Yanagimachi, 1988). Morphological changes have also been reported on the surface membranes of capacitated sperm. Only acrosome reacted sperm are capable of fusing with the human egg during monospermic or polyspermic fertilization (Sathananthan & Chen, 1986; Sathananthan et al., 1986b). It is therefore desirable to have a population of capacitated or acrosome reacted sperm for microinjection to ensure high rates of microinfertilization. This is particularly necessary if sperm are few or of poor quality and if single sperm injection is resorted to. The AR is usually induced in the culture media used in routine IVF during the washing, spinning and swim-up procedures. Acrosome-reacted sperm can be obtained one hour after washing with Whittingham's T_6 medium (Chen & Sathananthan, 1986; Ng et al., 1987) and sperm-oocyte membrane fusion can eventuate in two to three hours after insemination (Sathananthan & Chen, 1986; Sathananthan et al., 1986a,b). Routine sperm preparation methods have been used successfully for microinjection by Lasalle et al., (1987), Lazendorf et al., (1988) and Bongso et al., (1989a). However Laws-King et al., (1987) and Ng et al., (1988) have used other methods to synchronize sperm capacitation and possibly to harvest more acrosome reacted sperm. They used the Percoll and Ficoll methods of sperm separation (Mortimer et al., 1986; Bongso et al., 1989b). Laws-King et al., (1987) used the strontium substituted method which involves preparing sperm overnight. Obviously, the swim-up method cannot be used for immotile sperm and washing and incubating sperm with T_6 medium (four to eight hours) have produced reasonable fertilization rates (Ng et al., 1987; Bongso et al., 1989a).

In rodents, the AR is more uniform; after 30 minutes 8% of mouse sperm collected from the cauda epididymis in capacitating media are acrosome-reacted and this percentage increases to 25%–40% by two hours (Ward & Storey, 1984). Human sperm have only 10%–30% of acrosome reacted sperm after six hours incubation in a capacitating medium (Lee et al., 1987). A TEM study has shown that 10%–60% of sperm had reacted one to three hours after insemination at the surface of normally fertilized eggs of women who became pregnant by IVF (Chen & Sathananthan, 1986). In our recent TEM study of single or multiple sperm in microinjected eggs, we demonstrated both acrosome intact and acrosome

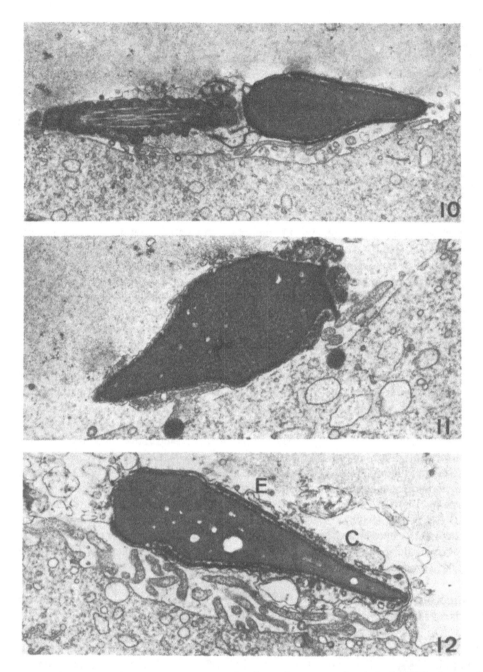

Figs. 10-12 Acrosome-intact, acrosome-reacted and reacting spermheads in the PVS of eggs after multiple sperm injection into the PVS. Note vesiculating acrosome cap (C) and intact equatorial segment (E) in fig. 12. × 21,000; × 27,300; × 27,300. (Reproduced from Ng et al., 1989a—Fig.10; and Sathananthan et al., 1989a—Figs.11 and 12).

reacted in the PVS of unfertilized oocytes (Figs. 10, 11). A few sperm were also in the process of undergoing the AR by vesiculation in the PVS (Fig. 12). Earlier studies demonstrated that such vesiculating sperm begin to fuse with the oocyte (Sathananthan et al,. 1986b) and the oocyte elicits a phagocytic response almost immediately, engulfing the sperm head. Hence the selection of acrosome reacted sperm for microinjection is not as critical as selecting a normal sperm, because the AR can be eventually completed in the PVS after injection. We have also quantitated the acrosomal status of 100 sperm in the PVS of 16 oocytes by TEM after multiple sperm injection and showed that 24% of the sperm were acrosome reacted or acrosome reacting while 76% were intact (Sathananthan et al., 1989a). The sperm were from oligozoospermic men and were separated by the Ficoll method (Bongso et al., 1989b) and eggs were examined three to 48 hours after injection. Hence the rates of capacitation and acrosome reactivity seem to vary considerably in different sperm populations and under different conditions. Human sperm seems more difficult to capacitate and acrosome react in vitro than mouse sperm after various treatments (Mortimer et al., 1988) and it may become necessary to develop better methods to increase fertilization rates and still preserve the integrity and fertilizing capacity of the gametes.

Another problem with human sperm is that it has a small acrosome and the assessment of the AR by light microscopy, is far more difficult when compared to that of the mouse. Rodents have a large swollen acrosomal cap in the anterior extremity of the spermhead and this region reacts before the rest of the acrosome (Sathananthan et al., 1989b). Assessment of the AR in the mouse is therefore much easier than that of human sperm, which is rather subjective with the staining methods now available. We have developed a more reliable method of assessing the human AR using fluorescein isothiocyanate-conjugated Concanavalin A (Holden et al., 1989). However, it is still difficult to visualize the earlier stages of the AR with the light microscope. The most reliable method is TEM, though this is very time-consuming and difficult. Hence it is desirable to compare the incidence of the AR obtained by staining or fluorescent methods and by TEM before one embarks in research on the microinjection of eggs.

REFERENCES

Aitken, R.J., Ross, A., and Lees, M., 1983, Analysis of sperm function in Kartagener's Syndrome, *Fertil. Steril.*, 40:696.

Bongso, T.A., Sathananthan, A.H., Wong, P.C., Ratnam, S.S., Ng, S.C., Anandakumar, C., and Ganatra, S., 1989a, Human fertilization by microinjection of immotile spermatozoa, *Hum. Reprod.*, (in press).

Bongso, T.A., Ng, S.C., Mok, H, Lim, M.N., Teo, H.L., Wong, P.C., and Ratnam, S.S., 1989b, Improved sperm density, motility and fertilization rates following Ficoll treatment of sperm in a human in vitro fertilization program, *Fertil. Steril.*, (in press).

Chen, C., and Sathananthan, A.H., 1986, Early penetration of human sperm through the vestments of human eggs in vitro, *Arch. Androl.*, 16:183.

Gordon, J.W., Talansky, B.E., Grunfeld, L., Richards, C., Garrisi, G.J., and Laufer, N., 1988, Fertilization of human oocytes by sperm from infertile males after zona pellucida drilling, *Fertil. Steril.*, 50: 68.

Holden, C.A., Hyne, R.W., Sathananthan, A.H., and Trounson, A.O., 1989, Development of an improved method for acrosomal status assessment as an indicator of human fertilizing ability, (submitted).

Jansen, R., 1988, Gamete intra-fallopian transfer, *in:* "Clinical In Vitro Fertilization", C. Wood and A. Trounson, eds., Springer-Verlag, Berlin.

Jonsson, M.S., McCormick, J.R., Gillies, C.J., and Gondos, B., 1982, Kartagener's syndrome with motile sperm. *New Engl. J. Med.*, 307:1131.

Lacham, O., Trounson, A., Holden, C., Mann, J., and Sathananthan, A.H., 1989, Fertilization and development of mouse eggs injected under the zona pellucida with single spermatozoa treated to induce the acrosome reaction, *Gamete Res.*, 23:1.

Lassalle, B., Courtot, A.M., and Testart, J., 1987, In vitro fertilization of hamster and human oocytes by microinjection of human sperm, *Gamete Res.*, 16:69.

Laws-King, A., Trounson, A., Sathananthan, A.H., and Kola, I., 1987, Fertilization of human oocytes by microinjection of a single spermatozoon under the zona pellucida, *Fertil. Steril.*, 48:637.

Lazendorf, S., Maloney, M., Ackerman, S., Acosta, A., and Hodgen, G.D., 1986, Fertilizing potential of acrosome-defective sperm following microsurgical injection into eggs, *Gamete Res.*, 19: 329.

Lazendorf, S.E., Slusser, M.S., Maloney, M.K., Hodgen, G.D., Veeck, L.L., and Rosenwaks, Z., 1988, A preclinical evaluation of pronuclear formation by microinjection of human spermatozoa into human oocytes, *Fertil. Steril.*, 49:835.

Lee, M.A., Trucco, G.S., Bechtol, K.B., Wummer, N., Kopf, G.S., Blasco, L., and Storey, B.T., 1987, Capacitation and acrosome reactions in human spermatozoa monitored by a chlortetracycline fluorescence assay. *Fertil. Steril.*, 48:649.

Mann, J.R., 1988, Full term development of mouse eggs fertilized by a spermatozoon microinjected under the zona pellucida, *Biol. Reprod.*, 38: 1077.

Markert, C.L., 1983, Fertilization of mammalian eggs by sperm injection, *J. Exp. Zool.*, 228: 195.

Martin, R.H., Ko, E., and Rademaker, A., 1988, Human sperm chromosome complements after microinjection of hamster eggs, *J. Reprod. Fert.*, 84: 179.

Metka, M., Haromy, T., Huber, J., and Schurz, B., 1985, Artificial insemination using a micromanipulator, *Fertilitat*, 1:41.

Mortimer, D., Curtis, E.F., and Dravland, J.E., 1986, The use of strontium-substituted media for capacitating human spermatozoa: An improved sperm preparation method for the zona-free hamster egg penetration test, *Fertil. Steril.*, 46:97.

Mortimer, D., Chorney, M.J., Curtis, E.F., and Trounson, A.O., 1988, Calcium dependence of human sperm fertilizing ability, *J. Exp. Zool.*, 246:194.

Ng, S.C., and Solter, D., 1989, Fusion of male germ cells (from male pronucleus to pachytene spermatocyte) with metaphase II oocytes in the mouse, (in preparation).

Ng, S.C., Sathananthan, A.H., Edirisinghe, W.R., Ho, K.C., Wong, P.C., Ratnam, S.S., and Ganatra, S., 1987, Fertilization of a human egg with sperm from a patient with immotile cilia syndrome: Case report, *in:*"Advances in Fertility and Sterility", S.S. Ratnam, E.S. Teoh and C. Anandakumar, eds., Parthenon, Lancaster.

Ng, S.C., Bongso, A., Sathananthan, A.H., Chan, C.L.K., Wong, P.C., Hagglund, L., Anandakumar, C., Wong, Y.C., and Goh, V.H.H., 1988, Pregnancy after transfer of multiple sperm under the zona, *Lancet*, 2:790.

Ng, S.C., Bongso, A., Sathananthan, A.H., and Ratnam, S.S., 1989a, Micromanipulation: its relevance to human IVF, *Fertil. Steril.*, (in press).

Ng, S.C., Bongso, T.A., Chang, S.I., Sathananthan, A.H., Ratnam, S.S., Chan, C.L.K., Wong, P.C., Hagglund, L., and Anandakumar, C., 1989b, Transfer of human sperm into the perivitelline space of human oocytes after zona-drilling or zona puncture (submitted).

Ng, S.C., Bongso, T.A., Ratnam, S.S., and Sathananthan, A.H., 1989c, Micro-insemination sperm transfer — a new technique for treatment of severe oligoasthenoteratozoospermia, (in preparation).

Palmbald, J., and Mossberg, B., 1984, Ultrastructural, cellular and clinical features of the immotile-cilia syndrome. *Ann. Rev. Med.*, 35: 481.

Sathananthan, A.H., and Trounson, A.O., 1982, Ultrastructure of cortical granule release and zona interaction in monospermic and polyspermic human ova fertilized in vitro, *Gamete Res.*, 6: 225.

Sathananthan, A.H., and Chen, C., 1986, Sperm-oocyte membrane fusion in the human during monospermic fertilization, *Gamete Res.*, 15: 117.

Sathananthan, A.H., Trounson, A.O., and Wood, C., 1986a, "Atlas of fine structure of human sperm penetration, eggs and embryos cultured in vitro", Praeger Scientific, Philadelphia.

Sathananthan, A.H., Ng, S.C., Edirisinghe, W.R., Ratnam, S.S., and Wong, P.C., 1986b, Human sperm-egg interaction in vitro, *Gamete Res.*, 15: 317.

Sathananthan, A.H., Ng, S.C., Trounson, A.O., Bongso, A., Laws-King, A., and Ratnam, S.S., 1989a, Human micro-insemination by injection of single or multiple sperm: ultrastructure, *Human Reprod.*, (in press).

Sathananthan, A.H., Trounson, A.O., Mann, J.R., Peura, A., and Lacham, O., 1989b, Mouse microfertilization by single or multiple sperm injection: ultrastructure, (in preparation).

Temple-Smith, P.D., Southwick, G.J., Yates, C.A., Trounson, A.O., and de Krester, D.M., 1985, Human pregnancy by in vitro fertilization (IVF) using sperm aspirated from the epididymis, *J. Vitro Fert. Embryo Transfer*, 2:119.

Thadani, V.M., 1980, A study of hetero-specific sperm-egg interactions in the rat, mouse and deer mouse using in vitro fertilization and sperm injection, *J. Exp. Zool.*, 212: 435.

Trounson, A., Peura, A., and Lacham, O., 1989, Fertilization of mouse and human eggs by microinjection of single sperm under the zona pellucida, *J. Reprod. Fert.*, (in press).

Uehara, T., and Yanagimachi, R., 1976, Microsurgical injection of spermatozoa into hamster eggs with subsequent transformation of sperm nuclei into male pronuclei, *Biol. Reprod.*, 15: 467.

Ward, C.R., and Storey, B.T., 1984. Determination of the time course of capacitation in mouse spermatozoa using a chlortetracycline fluorescence assay, *Dev. Biol.*, 104: 287.

Wolf, D.P., 1987, The block to sperm penetration in zona-free mouse eggs, Dev. Biol., 64:1.

Yanagimachi, R., 1988, Mammalian fertilization, *in:*"The Physiology of Reproduction", E. Knobil and J. Neill, eds., Raven Press, New York.

Yates, C.A., Thomas, C.J., Kovacs, G.T., and de Krester, D.M., 1988, Andrology, male factor infertility and IVF, *in:*"Clinical In Vitro Fertilization", C. Wood, and A. Trounson, eds., Springer-Verlag, Berlin.

Yovich, J.L., and Stanger, J.D., 1984, The limitations of in vitro fertilization from males with severe oligospermia and abnormal sperm morphology, *J. Vitro Fert. Embryo Transfer.*, 2:119.

MICROINSEMINATION OF HUMAN OOCYTES

Soon-Chye Ng, Ariff Bongso, Henry Sathananthan, and Shan S. Ratnam

Department of Obstetrics and Gynaecology
National University of Singapore
Lower Kent Ridge Road, Singapore 0511

INTRODUCTION

The severely oligozoospermic patient usually has a combination of multiple sperm defects[1] with very poor chances of spontaneous conception. In vitro fertilization (IVF) with embryo replacement (ER) has also not improved pregnancies in such cases. De Kretser and co-workers reported that when there is a combination of three or more defects in the semen analysis, fertilization in vitro diminishes to less than 8.0%.[2] Even tubal embryo transfer (TET),[3] otherwise known as pronuclear stage transfer (PROST) or tubal embryo stage transfer (TEST),[4] does not offer hope for such patients as their spouses' oocytes still need to be fertilized before the transfer of embryos into the fallopian tubes. Zona drilling to assist sperm penetration when first described[5] also offered hope for such patients but clinical trials using the technique did not produce any pregnancy.[6] Moreover, zona drilling with acid Tyrode's solution has been found to affect the meiotic process, probably by anaphase II arrest.[7] Recently, Cohen and co-workers described a new approach in which the zona of oocytes were partially dissected using force only via the micromanipulator (partial zona dissection, PZD).[8] Using this technique, they have reported three pregnancies,[9] which have since increased to six.

In PZD,[9] the manipulated oocytes were transferred into an insemination dish containing motile sperm densities of 5.0–20.0×10^6/ml to allow the motile spermatozoa to find their way through the dissected zona of the oocyte. However, for cases with severely low sperm densities, the spermatozoa need to be deposited directly into the oocyte ("micro-insemination", previously known as "micro-fertilization"), as opposed to "macro-

Advances in Assisted Reproductive Technologies, Edited by
S. Mashiach *et al.*, Plenum Press, New York, 1990

insemination" observed with zona-drilling and PZD (see review by Ng et al.).[10] Such micro-insemination can be done into the perivitelline space (Micro-Insemination Sperm Transfer, or MIST), or directly into the egg cytoplasm (Micro-Insemination Micro-Injection into Cytoplasm or MIMIC).[10] See also review by Gordon and Laufer.[11]

MATERIALS AND METHODS

This clinical trial was conducted with the informed consent of the patients after explanation of the risks involved to the resulting embryo. The study was also approved by the National University of Singapore, the National University Hospital and the Singapore Science Council.

Patients were included into the sperm transfer study only when their husbands' final sperm concentration was very poor ($< 1.0 \times 10^6$/ml) or poor concentration ($< 2.5 \times 10^6$/ml) and very sluggish motility. These patients were separated into two groups: (1) initial semen analysis with severe oligozoospermia ($< 5.0 \times 10^6$/ml); (2) initial semen analysis revealing abnormal sperm parameters not severe enough to consider sperm transfer, but these parameters were poor after 'swim-up' on the day of the oocyte recovery. The males had original semen analyses with motile sperm densities of between 0.07 and 2.01×10^6/ml (for group 1) and between 1.20 and 21.0×10^6/ml (for group 2); and total sperm densities of between 0.5 and 5.3×10^6/ml (for group 1) and between 12.9 and 76.0×10^6/ml (for group 2). Viable and normal forms were also low. Semen collected on the day of sperm transfer yielded very low motile sperm densities (Table 1).

For MIST, the entire semen sample was used and spermatozoa were prepared by the Ficoll entrapment procedure.[12] After separating the motile spermatozoa from the supernatant, the spermatozoa were concentrated by centrifugation in a micro-centrifuge (MSE Micro-Centуar, Landsborough, UK) at 6,500 rpm for 10 minutes, and the supernatant removed. Ficoll solution (5% in PBS, 0.1 ml) was added and the suspension kept at 4°C until ready for use. Recently, the method was modified by allowing a further 1–2 hours of incubation before micro-centrifugation to improve sperm acrosome reaction, and the final sperm solution was suspended in T6 medium with 20 mM HEPES and 10% heat-inactivated human serum (HS) instead of 5% Ficoll in PBS.

The sperm suspension was introduced into the sperm micropipette (Drummond, Pensylvania, USA) from the rear end and displaced into the tip with an Eppendorf Microinjector 5242 (Eppendorf, Hamburg, FRG). The sperm micro-pipette was prepared with the Narishighe micro-puller (#PB-7, Tokyo, Japan) and the Narishighe micro-grinder (#EG-4, Tokyo, Japan). The external diameter of the tip for MIST was 10–15 μm and the internal diameter was 8–12 μm. The angle of the bevel for the sperm micro-pipette was between 25° and 35°. The holding pipette was a Pasteur pipette drawn out on an ethanol flame. Both sperm and holding pipettes were bent over the ethanol flame to allow horizontal displacement over the microscope stage.

Table 1. Micro-Insemination Sperm Transfer: Sperm Parameters

	Severe Oligozoospermia	Final Oligozoospermia
Number of cases	15	14
Average age	31.5	34.0
Previous semen analysis:		
Density (X10^6/ml) ± SD	2.64 ± 1.31	37.76 ± 55.84
Motility (%) ± SD	23.00 ± 18.12	30.44 ± 16.45
Fresh semen analysis:		
Motile Density(X10^6/ml) ± SD	2.28 ± 3.79	3.84 ± 3.03
Total Density (X10^6/ml) ± SD	6.13 ± 8.88	15.90 ± 18.71
Post-washing:		
Motile Density (X10^6/ml) ± SD	0.34 ± 0.29	0.50 ± 0.65
Post-microcentrifugation:		
Motile Density (X10^6/ml)	1.18 ± 0.99	1.69 ± 1.98

The females for group 1 included 15 patients with a mean age of 31.5 (range 24 to 36 years), while those for group 2 included 14 patients with a mean age of 34.0 (range from 28 to 38 years). The women were stimulated with one of the folowing three regimes:

1. Clomiphene (Clomid, Merrell-Dow, Italy) 50 mg b.i.d. from day (D) 2 to 6 + human menopausal gonadotropin (hMG, Pergonal, Serono, Switzerland) 2 ampoules daily from D3 (1 patient);

2. Purified follicle-stimulating hormone (FSH, Metrodin, Serono) + hMG combination, with two and one ampoules respectively daily from D2 to D4, followed by two ampoules of hMG daily from D5 (20 patients);

3. Gonadotropin-releasing hormone agonist (Buserelin, Hoechst, Frankfurt, FRG) 400 μg eight hourly from D22 of prior menstrual cycle until hCG is given + Metrodin three ampoules daily from 20th day of Buserilin administration for four days, followed by two ampoules per day (eight patients).

They were monitored with daily serum estradiol (E_2) and ultrasound scanning. Human chorionic gonadotropin (hCG, Profasi, Serono) at 5,000 or 10,000 I.U. was administered when at least two follicles reached an average of 16 mm in the presence of satisfactory E_2 (> 300 pg/ml/follicle ≥ 14 mm). Two hundred and fifty-eight oocytes were collected transvaginally under ultrasonic guidance 34 hours from hCG administration. They were then in-

Table 2. Micro-Insemination Sperm Transfer: Fertilization

	Severe Oligozoospermia	Final Oligozoospermia
Metaphase II eggs	85	110
Sperm transferred: M ± SD	6.00 ± 3.38	6.81 ± 2.81
Accidental MIMIC	16	23
PN1 oocytes (%)	2 (2.4)	6 (5.5)
PN2 ova (%)	8 (9.4)	33 (30.0)
PN3+ ova (%)	2 (2.4)	0
Unfertilized (%)	65 (75.6)	63 (57.3)
Damaged (%)	8 (9.4)	8 (7.3)
Concomitant IVF		
Number of eggs	7	26
Fertilized (%)	0	5 (19.2)
MIST embryos transferred	15	29
Number of patients	9	12
Pregnancies (% per		
patient transferred)	2 (22.2)	1 (8.3)

cubated in a 5:5:90 gas mixture at 37°C in T6 medium with - 10% heat-inactivated human serum (HS).[13] Four to five hours after oocyte collection, the cumulus-corona complex was removed after exposure to 0.1% hyaluronidase in T6 medium. There were 20 oocytes at metaphase I, 10 at prophase I and 195 were at metaphase II, while 33 oocytes were subjected to IVF. Only the metaphase II oocytes were manipulated, within one hour. Each oocyte was manipulated and was transferred into T6 medium with 20 mM HEPES and 10% heat-inactivated human serum just before micro-manipulation (Zeiss motorized micro-manipulator MR with the Ziess IM35 inverted microscope, Carl Zeiss, Oberkochen, FRG). The zona was then directly punctured by the micropipette and between one and 10 motile (ideally about six) spermatozoa were introduced into the perivitelline space (PVS). An average of 6.00 ± 3.38 and 6.81 ± 2.81 respectively were transferred. After micro-insemination, the oocytes were washed three times in T6 medium before incubation overnight in 5% CO_2, 5% O_2 and 90% N_2 at 37°C. The oocytes were checked for pronuclei 14–18 hours later, and if fertilized they were transferred into either the patient's fallopian tubes (tubal embryo transfer, pronuclear stage) or uterine cavity (embryo replacement, 2–4 cell stage).

RESULTS

MIST was performed on 195 metaphase II oocytes and 16 (8.2%) were damaged (Table 2). Eight oocytes had two pronuclei (9.4%) in group 1, while 33 of 110 oocytes had two pronuclei in group 2 (30.0%). IVF was also used as a control only when the sperm concentration and motility was thought to be within the parameters for fertilization; in spite of that, for group 1 there was no fertilization in seven oocytes (two patients), and for group 2, five of 26 oocytes fertilized (five patients, 19.2%). Although the fertilization following MIST was better (30.0% versus 19.2%), this was not significant. In spite of multiple sperm transfer, polyspermy was seen only in two oocytes in group 1 (2.3%). Parthenogenetic activation was seen in two (2.3%) and six (5.5%) oocytes in groups 1 and 2 respectively. In group 1, a total of 15 embryos and zygotes were replaced in nine patients, with two pregnancies; (the numbers are more than the eight fertilized because seven embryos were healthy looking the next day). In group 2, 29 zygotes/embryos were replaced in 12 patients, with one resulting pregnancy.

The first pregnancy was reported elsewhere.[14] She has since progressed and a karyotyping of amniotic cells after amniocentesis showed 46XY. Her date of delivery is in early May 1989. In the second pregnancy the husband also had severe oligoasthenoteratozoospermia (3 ml of semen with a count of 1.1×10^6/ml very sluggishly motile sperm and 4×10^6/ml total sperm). After sperm preparation, a final count of 1×10^6/ml motile sperm in 0.20 ml of medium was obtained. Of 10 oocytes collected, six were subjected to MIST while the remaining four were inseminated in vitro. None of the latter fertilized, while two monospermic zygotes and one dispermic zygote were obtained after MIST. The monospermic zygotes were transferred into the left fallopian tube at pronuclear stage. She developed a left tubal ectopic pregnancy and an intra-uterine pregnancy which aborted. As the tubal pregnancy was treated conservatively with Methrotrexate, no tissue was available for karyotyping; the missed abortion revealed a 46 XX karyotype. The third pregnancy was from group 2. She was a 41-year-old female, with the husband being 48 years old and with oligoasthenoteratozoospermia. His previous sperm densities were between 15.6 and 17.8 $\times 10^6$/ml total spermatozoa, with motilities of 12%–16% and viability of only 26%. On the day of the oocyte collection, his sperm density was only 1.1×10^6/ml motile sperm, and 3.1×10^6/ml total sperm count. After Ficoll entrapment and micro-centrifugation, only 1.0×10^6/ml total sperm count was obtained. The wife was stimulated with a combination of FSH/hMG and had eight oocytes. After removing the corona and cumulus, seven were at metaphase II. Four zygotes with two pronuclei were obtained, and one other had one pronucleus. The four normo-spermic zygotes were replaced by TET. She became pregnant but the ultrasound at eight weeks amenorrhoea revealed a gestational sac without a fetal heart. Karyotyping of the chorionic villi again revealed normal karyotype (46XX).

DISCUSSION

Patients with such low sperm densities have not been reported to have pregnancies after IVF. Motility of sperm is also important and has direct correlation with fertilization of human oocytes in vitro.[15] The current method of sperm preparation described allows concentration of the few motile sperm available in semen for MIST. The disadvantages of MIST are the high rate of damage to the oocytes (8.2%), and the low rates of fertilization (9.4%

for group 1). However, the polyspermy and parthenogenetic rates for MIST are low (2.3% each). The technique now offers hope for such patients.

Fertilization has been reported after transfer of single sperm[16] and multiple sperm,[17] but with sperm from men with normal semen analysis. Laws-King et al.[16] reported that of seven human oocytes cultured 6–9 hours after collection which had single sperm transfers, five were fertilized. Three of these were cultured further in vitro and two underwent normal cleavage. The rest were examined by transmission electron microscopy (TEM) or were karyotyped.[16] Another 12 human oocytes were cultured for 23–28 hours before single sperm transfer; of these, three were fertilized but another three were parthenogenetically activated. TEM confirmed fertilization in both categories of these oocytes. Lassalle et al.[17] reported transfer of 3–5 sperm and 10–20 sperm into human oocytes. Normospermic rates were reported at 42.9% (3/7) and 0% (0/3), respectively.[17] Both oocytes that fertilized when 10–20 sperm were transferred were polyspermic (2/3).[17]

There is the possible risk of chromosomal abnormalities related to the higher incidence of morphological sperm abnormalities. Of 100 sperm examined by TEM in the PVS of 16 human oocytes after transfer of 5–10 sperm from oligozoospermic men, 76% had intact acrosomes and 33% of intact sperm were morphologically abnormal.[18] Nuclear, acrosomal, midpiece and tail defects were encountered in these abnormal sperm forms. In a survey of male patients attending a subfertility clinic, Chandley and co-workers[19] reported that 2% of them were chromosomally abnormal; this frequency rose to 6% among men with a mean sperm count of less than 20 million/ml, and was 15% among azoospermic men. Earlier studies[20] also demonstrated that the majority of chromosomally abnormal infertile men were oligozoospermic and azoospermic. Moreover, in a study on 1000 human male pronuclear chromosomes complements from 33 normal donors, Martin[21] reported an abnormality rate of 8.5% (5.2% aneuploidy and 3.3% structural abnormalities). There is also a high incidence of chromosomal abnormalities in human oocytes collected after ovarian stimulation.[22, 23] Such findings may explain the possible increased incidence of spina bifida and transposition of great vessels from babies born after IVF and GIFT in Australia and New Zealand.[24]

Recently, Martin and co-workers[25] reported a high incidence of chromosomal abnormalities in human spermatozoa haploid complements following micro-injection (MIMIC) into hamster oocytes. The spermatozoa were from normal donors, but pretreated with sonication or TEST-yolk buffer to reduce motility. Following sonication very few spermatozoa complements were found, and the karyotypes showed a very high degree of structural chromosomal abnormalities (91%), mainly multiple breaks and rearrangements.[25] TEST-yolk buffer preincubation of spermatozoa had a significantly lower frequency of structural chromosomal abnormalities (39%).[25] Spermatozoa from the same donors gave a 3%–11% chromosomal abnormalities following fertilization of zona-free hamster oocytes.[25] However, there is no definite evidence for a direct correlation of genetic abnormality with decreased sperm motility except for immotile cilia syndrome.[26]

It is significant that there has been a low incidence of polyspermy after MIST. There is no evidence for any form of oolemma "selection", though there is a report that 97% of human sperm nuclei from donors with poor penetration of zona-free hamster eggs (< 10%) underwent decondensation after direct cytoplasmic injection into hamster oocytes.[27] There is also increasing evidence for the existence of 'sperm receptors' resident on the oolemma. Earlier studies with zona-free eggs suggested that there was some form of block to excessive polyspermy.[28, 29, 30] In a study with murine gametes, S.C. Ng and D. Solter (unpublished data) transferred between 3–5 motile spermatozoa into oocytes before and after ethanol activation.[31] Of 101 oocytes in which there was no ethanol activation, 32 (31.7%) developed two pronuclei and a second polar body at eight hours. There was also a very low incidence of polyspermy, with only two oocytes with three pronuclei, and none with more than three pronuclei, at eight hours. The fertilization (monospermy and polyspermy) of oocytes was not significantly different when the eggs were exposed to ethanol 1–2 hours after the multiple sperm transfer. However, when the eggs were exposed to ethanol 1/2 to one hour before sperm transfer, only eight of 60 oocytes were fertilized. The data supports the hypothesis that sperm receptors or sperm blocks were present on the oolemma of zona-intact oocytes because the polyspermy rates were low in spite of transfer of 3–5 spermatozoa. Furthermore, ethanol activation prior to sperm transfer probably resulted in removal or inactivation of sperm receptors because of the low fertilization rate. The findings support the results of low polyspermy in this study (2.4% of 85 oocytes in the group with severe oligozoospermia, and 0% of 110 oocytes in the group with final oligozoospermia). The presence of sperm receptors on the oolemma may explain the low fertilization rates of inter-species fertilization in zona-free eggs, with the exception of golden hamster eggs.[32] In this species, it is possible that the sperm receptors are not so specific. Interestingly, it was reported recently that heteroantibodies found in human sera reacted against the hamster oolemma, though these oolemma antigens were distinct from antigens present on the surface of human spermatozoa.[33]

In conclusion, micro-insemination technique reported in this study has resulted in the world's first micro-insemination human pregnancy, with delivery due in early May 1989. This result is reproducible because we have three pregnancies, but unfortunately one was an ectopic while the other resulted in a missed abortion. It must be realised that in spite of low fertilization (10% for severe oligoasthenoteratozoospermia), the pregnancy rate is better than the second group where there is extremely poor sperm on the day of oocyte collection. This is probably due to the "better" fertility status of the wife in the former group. This technique of MIST should offer new hope for patients with severely poor sperm quality.

ACKNOWLEDGEMENTS

Funding for this study was by the National University of Singapore, and the Singapore Science Council. We also wish to acknowledge the contributions of our colleagues (Drs. Wong PC, Chang CLK, Hagglund L, Anadakumar C, Wong WC, and Goh VHH), and our technicians (Ms. Tok V, Chang PSH, Helen Mok, Lim MN, Ng PL, and Jean Ho), and Ms Shanti N for secretarial assistance.

REFERENCES

1. World Health Organization Task Force on the Diagnosis and Treatment of Infertility. Towards more objectivity in diagnosis and management of male infertility. (Prepared by F. H. Comhaire, D. de Kretser, T. M. M. Farley, and P. J. Rowe), *Int. J. Androl.*, Suppl.7:1 (1987).
2. D. M. De Kretser, C. A. Yates, J. McDonald, J. F. Leeton, G. Southwick, P. D. Temple-Smith, A O. Trounson, and E. C. Wood, The use of in vitro fertilization in the management of male infertility, *in:* "Gamete Quality and Fertility Regulation", R. Rolland, M. J. U. Heineman, S. G. Hillier, and H. Vemer, eds., Excerpta Medica, Amsterdam (1985).
3. P. C. Wong, A. Bongso, S. C. Ng, C. L. K. Chan, L. Hagglund, C. Anandakumar, Y. C. Wong, and S. S. Ratnam, Pregnancies after human tubal embryo transfer (TET): a new method of infertility treatment, *Singapore J. Obstet. Gynecol.*, 19:41 (1988).
4. J. L. Yovich, J. M. Yovich, and W. R. Edirisinghe, The relative chance of pregnancy following tubal or uterine transfer procedures, *Fertil. Steril.*, 49:858 (1988).
5. J. W. Gordon and B. E. Talansky, Assisted fertilization by zona drilling: a mouse model for correction of oligospermia, *J. Exp. Zool.*, 239:347 (1986).
6. J. N. Gordon, L. Grunfeld, G. J. Garrisi, B. E. Talansky, C. Richards, and N. Laufer, Fertilization of human oocytes by sperm from infertile males after zona pellucida drilling, *Fertil. Steril.*, 50:68 (1988).
7. S. C. Ng, T. A. Bongso, S. I. Chang, A. H. Sathananthan and S. A. Ratnam, Transfer of human sperm into the perivitelline space of human oocytes after zona-drilling or zona-puncture, *Fertil. Steril.*,52:73 (1989).
8. J. Cohen, H. Malter, C. Fehilly, G. Wright, C. Elsner, H. Kort, and J. Massey, Implantation of embryos after partial opening of oocyte zona pellucida to facilitate sperm penetration, *Lancet*, 2:162 (1988).
9. H. Malter, and J. Cohen, Partial zona dissection of the human oocyte: a non-traumatic method using micromanipulation to assist zona pellucida penetration, *Fertil. Steril.*, 51:139 (1989).
10. S. C. Ng, T. A. Bongso, A. H. Sathananthan, and S. S. Ratnam, Micro- manipulation: its relevance to human IVF, (Review), *Fertil. Steril.*,(Feb. 1990).
11. J. W. Gordon, and N. Laufer, Applications of micromanipulation to human in vitro fertilization, *J. In Vitro Fertilization and Embryo Transfer*, 5:57 (1988).
12. T. A. Bongso, S. C. Ng, H Mok, M. N. Lim, H. L. Teo, P. C. Wong, and S. S. Ratnam, Improved sperm density, motility and fertilization rates following Ficoll treatment of sperm in a human in vitro fertilization program, *Fertil. Steril.*, 51:850 (1989).
13. S. C. Ng, S. S. Ratnam, H. Y. Law, W. R. Edirisinghe, C. M. Chia, M. Rauff, P. C. Wong, S. C. Yeoh, C. Anandakumar, and H. H. V. Goh, Fertilization of the human egg and growth of the human zygote in vitro: the Singapore experience, *Asia-Oceania J. Obstet. Gynecol.*, 11:533 (1985).
14. S. C. Ng, T. A. Bongso, S. S. Ratnam, A. H. Sathananthan, C. L. K. Chan, P. C. Wong, L. Hagglund, C. Anandakumar Y. C. Wong, and H. H. V. Goh, Pregnancy after transfer of multiple sperm under the zona, *Lancet*, 2:790 (1988).
15. T. A. Bongso, S. C. Ng, H. Mok, M. N. Lim, H. L. Teo and S. S. Ratnam, The influence of sperm motility on human in vitro fertilization, *Arch. Androl.*, 22:49 (1989).
16. A. Laws-King, A. Trounson, H. Sathananthan, and I. Kola, Fertilization of human oocytes of micro-injection of a single spermatozoon under the zona pellucida, *Fertil. Steril.*, 48:637 (1987).

17. B. Lassalle, A. M. Courtot, and J. Testart, In vitro fertilization of hamster and human oocytes by microinjection of human oocytes by microinjection of human sperm, *Gamete Research*, 16:69 (1987).
18. A. H. Sathananthan, S. C. Ng, A. O. Trounson, T. A. Bongso, A. Laws- King, and S. S. Ratnam, Human micro-insemination by injection of single or multiple sperm: ultrastructure, *Human Reprod.*, In press (1989).
19. A. C. Chandley, P. Edmond, S. Christie, L. Gowans, J. Fletcher, A. Frankiewicz, and M. Newton, Cytogenetics and infertility in man, I. Karyotype and semen analysis, *Ann. Hum. Genet.*, 39:231 (1975).
20. B. Kjessler, Facteurs genetiques dans la subfertile male humaine, *in:* "Fecondite et Sterilite du male. Acquisitions recetes", Masson, Paris (1972).
21. R. H. Martin, The chromosome constitution of 1000 human spermatozoa, *Hum. Genet.*, 63:305 (1983).
22. H. Wramsby, K. Fredga, and P. Liedholm, Chromosome analysis of human oocytes recovered from preovulatory follicles in stimulation cycles, *New Engld. J. Med.*, 316:121 (1987).
23. T. A. Bongso, S. C. Ng, S. S. Ratnam, A. H. Sathananthan, and P. C. Wong, Chromosome anomalies in human oocytes failing to fertilize after insemination in vitro, *Human Reprod.*, 3:645 (1988).
24. P. A, L. Lancester, Congenital malformations after in vitro fertilization, *Lancet*, 2:1392 (1987).
25. R. H. Martin, E. Ko, and A. Rademaker, Human sperm chromosome complements after microinjection of hamster eggs, *J. Reprod. Fert.*, 84:179 (1988).
26. C. M. Rossman, J. B. Forrest, R. M. K. W. Less, A. F. Newhouse, and M. T. Newhouse, The dyskinetic cilia syndrome: abnormal ciliary motility in association with abnormal ciliary ultrastructure, *Chest*, 80:860 (1981).
27. S. E. Lanzendorf, J. F. Mayer, J. Swanson, A. Acosta, M. Hamilton, and G. D. Hodgen, The fertilizing potential of human spermatozoa following microsurgical injection into oocytes, 5th World Congress on IVF and ET, Program Supplement, AFS, 65 (1987).
28. A. Pavlok, and A. McLaren, The role of cumulus cells and the zona pellucida in fertilization of mouse eggs in vitro, *J. Reprod. Fertil.*, 29:91 (1972).
29. Y. Toyoda, and M. C. Chang, Sperm penetration of rat eggs in vitro after dissolution of zona pellucida by chymotrypsin, *Nature*, 22:589 (1968).
30. K. Niwa, and M. C. Chang, Requirement of capacitation for sperm penetration of zona-free rat eggs, *J. Reprod. Fertil.*, 44:305 (1975).
31. K. S. R. Cuthbertson, Parthenogenetic activation of mouse embryos in vitro with ethanol and benzyl alcohol, *J. Exp. Zool.*, 226:311 (1983).
32. R. Yanagimachi, Zona free hamster eggs; their use in assessing fertilizing capacity and examining chromosomes of human spermatozoa, *Gamete Research*, 10:187 (1984).
33. R. Bronson, and G. Cooper, Detection in human sera of antibodies directed against the hamster egg oolemma, *Fertil. Steril.*, 49:493 (1988).

FOOTNOTE/ADDENTUM:

The first patient delivered on 20th April, 1989. The baby was a healthy boy weighing 3.356 kg.

SPERM INJECTION INTO THE PERIVITELLINE SPACE OF THE OOCYTE

L. Mettler and K. Yamada

*Department of Obstetrics and Gynecology and Michaelis Midwifery School
University of Kiel Michaelisstrasse 16, D-2300 Kiel FRG*

INTRODUCTION

Human in vitro fertilization and embryo transfer (IVF/ET) (Steptoe and Edwards 1978) provides an approach to the treatment of sterility, designated as andrological, i.e. oligo- and asthenozoospermia or idiopathic causes (Cohen et al., 1985). However, the success rates of pregnancies in these patients generally remained relatively low (Riedel et al., 1985).

Microinjection of spermatozoa or sperm nuclei into oocytes has been attempted in animal models. Activation of oocytes and formation of male and female pronuclei (Uehara and Yanagimachi 1976, 1977), diploid chromosome sets of fertilized embryos (Thadani 1980) and blastocyst formation (Markert 1983), were observed.

With the development of techniques in human IVF/ET, the microinjection of spermatozoa into oocytes has theoretically become a tempting way to overcome the difficulties in treating infertility with unknown etiology or with andrological causes. Quite recently, microinjection of spermatozoa into the perivitelline space of oocytes has been tried in the mouse (Barg et al., 1986; Yamada et al., 1988) and in the human (Lassalle et al., 1987; Laws-King et al., 1987, Metka et al., 1985, 1987). As the direct injection of spermatozoa into the ooplasm may injure the gametes, this aspect warrants careful scrutiny.

Advances in Assisted Reproductive Technologies, Edited by
S. Mashiach *et al.*, Plenum Press, New York, 1990

In earlier experiments, the development of pronuclei, cell division and diploid chromosome sets were observed (Mettler et al., 1988; Yamada et al., 1988). The rates of fertilization were, however, relatively low (a maximum of 5%).

Capacitation and the acrosome reaction are believed to be inevitable prerequisites for spermatozoa to fertilize oocytes (Chang 1951; Yanagimachi 1983). In this study, capacitated and acrosome reacted spermatozoa were microinjected under the zona pellucida in order to evaluate their fertilizing capacity.

MATERIALS AND METHODS

Young mature CB6/F hybrid mice (Charles River/Wiga, Sulzfeld) were used in this study[1]. The animals were kept under clean, conventional conditions at room temperature of 22°C and a light/dark cycle, of 14 h of light and 10 h of darkness. Laboratory food and water were given ad libidum.

The culture medium consisted of a modified Ham's F-10 (Lopata et al., 1980). In short, Ham's F-10 (Flow Laboratories, Berlin) was supplemented with $NaHCO_3$ to give 25mM instead of the normal 14.3 mM $NaHCO_3$ 1.8mM L (+)-lactate (L 2000, Sigma, Munich), 0.75% bovine serum albumin (BSA Fraction V, Serva, Heidelberg), and 50 I.U./ml Penicillin G (Grünenthal, Stolberg).

Spermatozoa were obtained from the vas deferens after surgical removal. They were incubated for 2 h in culture medium for capacitation. Only upward swimming spermatozoa (C-spermatozoa) underwent induction of the acrosome reaction. These were put into the culture medium containing 12 mM of dbcGMP (N^2-2'-0-dibutyryl guanosine 3',5'-cyclic monophosphate, Sigma, Munich) and 10 mM imidazole (Sigma, Munich) in which almost all the spermatozoa were reported as having acrosome reacted (C & AR-spermatozoa) in the guinea pig (Santos-Sacchi and Gordon 1980) and in the mouse (Talansky et al., 1987). Acrosome reaction was confirmed on fixed specimens using the triple staining technique (Talbot and Chacon 1981) with minor modifications. After at least 20 minutes of treatment, the spermatozoa were put into a droplet of medium containing 1.5% BSA in a glass chamber just beside the droplets of oocytes for microinjection. These were covered by silicone oil as described in previous studies (Mettler et al., 1988; Yamada et al., 1988).

As control studies, both C- and C & AR-spermatozoa were put together with oocytes in concentrations ranging between 30,000 and 100,000/ml for normal IVF.

Oocytes were retrieved from the ampullae of superovulated females at 08.30 h. Superovulation was induced by s.c. injection of 5 I.U. of pregnant mare serum gonadotrophin (Seragon from Ferring, Kiel) followed by 5 I.U. of human chorionic gonadotrophin (Primogonyl from Schering, Berlin) 48 h later. Cumulus oophorus complexes from each ovary were collected separately.

Each ipsilateral pool was exposed to 10 I.U./ml hyaluronidase solution for 10–15 minutes to remove cumulus cells. Bovine testicular hyaluronidase (Serva, Heidelberg) was dissolved in culture medium at an initial concentration of 200 I.U./ml. A volume of 50 μl of this initial solution was pipetted into each of several culture tubes and stored at -20°C. Before use, 0.95 ml of culture medium was added per tube, yielding the required final concentration of 10 I.U./ml.

The contralateral cumulus oophorus complex masses were put individually into culture media at 09.00 h for the usual IVF-protocol. After removal of cumulus, the oocytes were washed and prepared for the microinjection. One or two oocytes were put into droplets of culture medium supplemented with 1.5% of BSA in a glass chamber just beside the droplets of spermatozoa.

Microinjection

Injecting pipettes were prepared by drawing out glatt capillary tubes using a horizontal pipette puller (Leitz, Hamburg). They were bevelled, using a rotating grinding wheel (Bachofer, Reutlingen) so that the inner diameter of the pipettes was 10 μm. Oocyte-holding suction pipettes were prepared by drawing out capillary glass tubes manually and smoothing the tips with the aid of a fine gas flame.

All the micromanipulatory procedures were performed under an inverted microscope (Olympus, Tokyo). Two micromanipulators (Leitz) were used. The air injection and suction systems were regulated by an electronic microinjector (Eppendorf, Hamburg), connected through polyethylene tubing.

Only actively moving spermatozoa were aspirated by suction into the injecting pipette and injected into the perivitelline space (Figure 1a–d) as previously reported (Yamada et al., 1988). If the injected spermatozoa were no longer motile in the perivitelline space, another active one was injected. The injection of a second sperm was, however, rarely necessary.

The mean time required per microinjection was three to five minutes. The maximal time that the oocytes were exposed to room conditions of temperature, etc. (during microinjection) was 10–15 min. The oocytes used for microinjection were retrieved 13.5 h after hCG treatment and were microinjected within 3.5 h.

After 8–12 h the injected oocytes were observed under the phase contrast microscope and estimated as to the presence of pronuclei and/or second polar bodies. Oocytes which had a second polar body and two pronuclei were deemed to be fertilized. After one and two days they were screened again. Embryos showing cell division after microinjection of spermatozoa were given colchicine to permit chromosomal analysis of air-dried specimens (Tarkowsky 1966). Male DNA was detected with an in situ hybridization technique. Statistical analysis was performed by the chi-square-test.

RESULTS

Acrosome Reaction of Spermatozoa

According to the triple staining technique, spermatozoa with intact acrosomes stain brown with Bismarck brown around the post-acrosomal region. Rose Bengal stains the acrosome pink. Dead spermatozoa are distinguished by their staining with Trypan Blue. According to these criteria 17% of the C-spermatozoa and 93% of the C & AR-spermatozoa had acrosome-reacted.

In Vitro Fertilization

The overall fertilization rate of the oocyte which were inseminated with C-spermatozoa was 79.3%, whereas that with C & AR-spermatozoa was 1.6% (Table 1). Of those fertilized oocytes, 77.7% reached the morula or blastocyst stage.

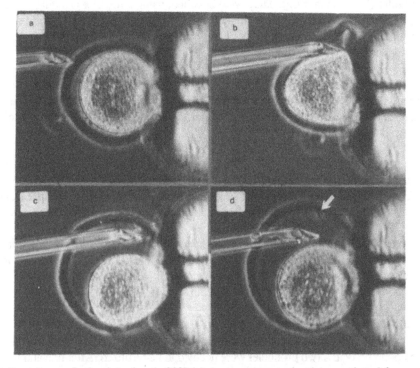

Fig. 1. Procedures of microinjection (× 300) (a) A spermatozoon has been aspirated from a droplet of spermatozoa into the injection pipette, which is brought into the droplet containing the oocyte. The oocyte is held by suction with the oocyte-holding suction pipette. (b) The injection pipette is tangentially introduced to the zona pellucida, and in an S-shaped way—(Yamada et al., 1988) (c)—into the perivitelline space. (d) Then the spermatozoon is injected (arrow).

Table 1. IVF with Capacitated (C) and Capacitated/Acrosome-reacted (C&AR)-Spermatozoa

	Experiments	Oocytes	Fertilized[a]		
	n	*n*	*n*	%	Mean fertilization rates ± SEM %
C-spermatozoa	16	858	680	79.3	81.29 ± 3.05
C&AR-spermatozoa	10	368	6	1.6	1.36 ± 0.65

[a]Oocyte with a second polar body and two pronuclei.

Table 2. Microinjection of Capacitated (C) and Capacitated/Acrosome-reacted (C&AR)-Spermatozoa into the Perivitelline Space

	Experiments	Oocytes	Fertilized[a]		
	n	*n*	*n*	%	Mean fertilization rates ± SEM %
C-spermatozoa	6	150	8	5.3	5.65 ± 1.49
C&AR-spermatozoa	10	209	41	19.6	18.87 ± 3.87

[a]Oocyte with a second polar body and two pronuclei.

Microinjection of C- and C & AR-Spermatozoa into Perivitelline Space

Oocytes with damaged ooplasm could be recognized during or just after micromanipulation, as they lost their integrity immediately after being damaged. The mean survival rate after macroinjection of spermatozoa was 85%. The number of oocytes in which motile spermatozoa were injected and those with a second polar body and two pronuclei formation are shown in Table 2. Figure 2 shows one of the fertilized oocytes with a second polar body in the perivitelline space and two pronuclei within the ooplasm. Consequently 19.6% of the injected oocytes were evaluated as being fertilized. All of these fertilized oocytes showed cleavage division, and 70% developed into four cell stages (Fig. 3) and morulae (Fig. 4). With in situ hybridization partly male chromosome sets were detected.

DISCUSSION

Although 2 h incubation in culture medium (in 5% CO2) is sufficient to capacitate spermatozoa, and is a procedure in routine use, it by no means guarantees that every spermatozoon becomes capacitated. This study further showed that acrosome-reacted spermatozoa could only rarely fertilize the oocytes in vitro. These findings are compatible with the results of other authors (Talansky et al., 1987; Yanagimachi 1983).

Injection of capacitated, acrosome-reacted, but immotile spermatozoa into the perivitelline space in the mouse resulted in no evidence of fertilization at all in a previous report (Barg et al.,1986). However, the injection of motile spermatozoa, without further treatment to induce acrosome reaction, into the perivitelline space, gave rise to fertilization in a previous report (Mettler et al., 1988; Yamada et al., 1987, 1988). In this study those findings were again confirmed and the rate of fertilization was 5%.

In the present study the acrosome reaction was induced in most of the living spermatozoa with the aid of a cyclic mononucleotide derivative without rendering them immotile. They were subsequently injected into the perivitelline space. The obtained fertilization rate was 19.6% and was higher than any previous reports concerning the microinjection of spermatozoa under the zona pellucida. Quite noteworthy is the fact that, as far as acrosome-reacted spermatozoa are concerned, the fertilization rate was distinctly higher per microinjec-

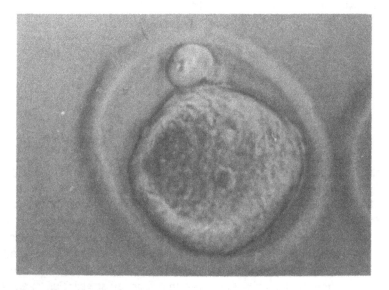

Fig. 2. A fertilized oocyte after microinjection of a C & AR-spermatozoon — a second polar boy and two pronuclei are visible (× 600).

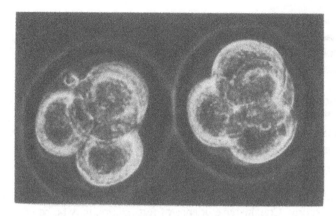

Fig. 3. Mouse 4-cell stage after microinjection (× 600).

tion than per IVF. Thus, it was convincingly demonstrated that the success of spermato-zoal microinjection is significantly promoted when the acrosome reaction is induced prior to microinjection.

Nevertheless, it must be said that this did not compare with the results of IVF. The discrepancy between the rate of success in inducing the acrosome reaction (90%) and the efficiency of such microinjected spermatozoa (20%) in bringing about fertilization can only be attributed to other experimental factors about which one may only speculate. The importance of capacitation of spermatozoa has been frequently discussed (Chang 1951), though

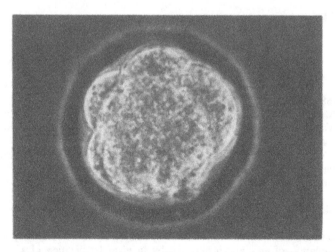

Fig. 4. Mouse morula stage after microinjection (× 600).

this whole mechanism has never been fully elucidated. Moreover, it cannot be said that every capacitated spermatozoon can fertilize an oocyte. Furthermore, the oocytes themselves must complete their activation process, including the release of cortical granules, to start the process of fertilization.

It is, therefore, necessary to establish methods by which capacitated spermatozoa can be distinguished from non-capacitated live murine spermatozoa. How the oocytes are activated during and after the microinjection has also to be ascertained. Only then will it be possible to state that the reduced fertilization rates per microinjection are due to failure of gamete activation.

Undoubtedly, the applicability of microinjection for human IVF/ET still requires further stringent scrutiny. Conclusive evidence of fertilization may be based on three criteria: (i) the presence of the second polar body and two pronuclei; (ii) the presence of diploid sets of chromosomes with Y-chromosomes or the proof of Y-specific DNA sequences per chromosomal DNA hybridization in situ, in male embryos; (iii) the birth of live normal offspring after embryo transfer over three generations.

In this study only criterion (i) was used as evidence for successful fertilization. Further work is in progress to prove the success of fertilization per microinjection based on criteria (ii) and (iii).

ACKNOWLEDGMENTS

We acknowledge the cooperation of U. Dieckmann, A. F. G. Stevenson and R. Schneider.

REFERENCES

Barg, P. E, Wahrman, M. Z., Talansky, B. E. and Gordon, J. W. 1986, Capacitated acrosome reacted but immotile sperm, when microinjected under the mouse zona pellucida, will not fertilize the oocyte, *J. Exp. Zool.*, 237:365-374.

Chang, M. C., 1951, Fertilizing capacity of spermatozoa deposited into the Fallopian tubes, *Nature.*, 168:697-698.

Cohen, J., Edwards, R. G., Fehilly, C., Fishel, S., Hewitt, J., Purdy, J., Rowland, G., Steptoe, P., and Webster, J., 1985, In vitro fertilization: a treatment for male infertility, *Fertil. Steril.*, 43:422-432.

Lassalle, B., Courtot, M., and Testart, J., (1987), Fertilization by microinjection of human sperm in hamster and human oocytes, *in*: "Future Aspects in Human In Vitro Fertilization," W. Feichtinger, and P. Kemeter, eds., Springer Verlag, Berlin, Heidelberg, 111-118.

Laws-King, A., Trounson, A., Sathananthan, H., and Kola, I., 1987, Fertilization of human oocytes by microinjection of a single spermatozoon under the zona pellucida, *Fertil. Steril.*, 48:642.

Lopata, A., Johnston, I. W. H., Hoult, I. J., and Speirs, A. I., 1980, Pregnancy following intrauterine implantation of an embryo obtained by in vitro fertilization of a preovulatory egg, *Fertil. Steril.*, 33:117-120.

Markert, C. L., 1983, Fertilization of mammalian eggs by sperm injection, *J. Exp. Zool.*, 228:195-201.

Metka, M., Haromy, T., Huber, J., and Schurz, B., 1985, Apparative Insemination mit Hilfe des Mikromanipulators, *Fertilität*, 1:41-44.

Metka, M., Haromy, T., Huber, J., and Schurz, B., 1987, Artificial insemination using a micromanipulator, *in*: "Future Aspects in Human In Vitro Fertilization," W. Feichtinger and P. Kemeter eds., Springer Verlag, Berlin, Heidelberg, 119-121.

Mettler, L., Yamada, K., and Kuranty, A., 1988, Mikroinjektion von Spermatozoen in die Eizelle, *Geburtsh.u.Frauenheilk.*, 48:625-629.

Riedel, H. H., Langenbucher, H., Buck S., and Mettler, L., 1985, IVF - eine Form der Therapie für die andrologisch bedingte Sterilitat, *Fertilität*, 1:67-75.

Santos-Sacchi, J., and Gordon, M., 1980, Induction of the acrosome reaction in guinea pig spermatozoa by cGMP analogues, *J. Cell Biol.*, 85:798-803.

Steptoe, P., and Edwards, R. G., 1978, Birth after the reimplantation of a human embryo, *Lancet*, ii:336.

Talansky, B. E., Barg P. E., and Gordon, J. W., 1987, Ion pump ATPase inhibitors block the fertilization of zona-free mouse oocytes by acrosome-reacted spermatozoa, *J. Reprod. Fertil.*, 79:447-455.

Talbot P., and Chacon, R. S., 1981, A triple stain technique for evaluating normal acrosome reaction of human sperm, *J. Exp. Zool.*, 215:201-218.

Tarkowski, A. K., 1966, An air drying method for chromosome preparations from mouse eggs, *Cytogenetics* 5:394-400.

Thadani, V. M., 1980, A study of hetero-specific sperm-egg interactions in the rat, mouse and deer mouse using in vitro fertilization and sperm injection, *J. Exp. Zool.*, 212:435-453.

Uehara T., and Yanagimachi, R., 1976, Microsurgical injection of spermatozoa into hamster eggs with subsequent transformation of sperm nuclei into male pronuclei, *Biol. Reprod.*, 15:467-470.

Uehara T., and Yanagimachi, R., 1977, Behaviour of nuclei of testicular, caput and cauda epididymal spermatozoa injected into hamster eggs, *Biol. Reprod.*, 16:315-321.

Yamada, K., Kuranty, A., Michelmann, H. W., Mettler, L., and Semm, K., 1988, Künstliche Befruchtung - Mikroinjektion von Spermatozoen in Mäusen-Oozyten, *Fertilität*, 4:133-136.

Yanagimachi, R., 1983, Fertilization, *in*: "In Vitro Fertilization and Embryo Transfer," Academic Press, London 65-80.

RESULTS OF IN VITRO FERTILIZATION
OF HUMAN OOCYTES AFTER ZONA DRILLING

R. Pijnenborg, R. Ongkowidjojo, C. Meuleman, F. Cornillie,
I. Van der Auwera, and P.R. Koninckx

Department of Obstetrics and Gynaecology
Catholic University of Leuven, Belgium

INTRODUCTION

In vitro fertilization has opened up new possibilities for treating male subfertility problems. With appropriate culture conditions it is possible to obtain oocyte fertilizations and pregnancies with semen samples of poor quality. Statistically, the percentage of fertilized oocytes is significantly lower with subfertile than with normal fertile semen (Cohen et al., 1985; Mahadevan et al., 1984). In the individual couple however, it is not yet possible to predict whether a particular semen sample of a subfertile man will be able to fertilize oocytes, although some sperm characteristics such as motility, morphology and concentration have some predictive value (Mahadevadan & Trounson, 1984; Jeyendran et al., 1986; Jeulin et al., 1986; Liu et al., 1988).

In order to increase the fertilization of oocytes with subfertile semen some of the potential barriers that prevent fertilization may be removed. One major barrier for fertilization is the zona pellucida, the acellular capsule that surrounds the oocyte. Possible approaches consist of drilling holes in the zona pellucida in order to facilitate the access to the oocytes, subzonal insertion of sperm cells or sperm microinjection into the ooplasm (Gordon & Laufer, 1988). The lack of adequate experimental work for the last two methods prompted us to develop a zona drilling programme. Previous experimental animal work had indeed demonstrated the safety of the method (Gordon & Talansky, 1986) and recently the first trials using human oocytes in 10 IVF patients were reported by Gordon et al. (1988).

Advances in Assisted Reproductive Technologies, Edited by
S. Mashiach *et al.*, Plenum Press, New York, 1990

MATERIALS AND METHODS

Patients

Twenty-six infertile couples with a male subfertility problem were selected to undergo the zona drilling procedure during IVF. Eight patients participated twice, and one patient three times in the programme, totalling 36 IVF cycles with zona drilling. Seventeen patients had participated previously in our IVF programme without micromanipulation: seven (11 cycles) with at least one oocyte fertilized in that cycle and 10 (13 cycles) with no fertilizations at the previous trials. In 12 patients the quality of the sperm looked so suspicious that it was decided that zona drilling had to be included during the first IVF trial. In all cases up to six randomly selected oocytes were micromanipulated while the rest of the oocyte-cumulus complexes were left intact as controls and were inseminated and handled further according to standard IVF procedures.

Stimulation protocol and oocyte retrieval

HMG (Humegon, Organon) stimulation and LHRH (Buserelin, Hoechst) desensitisation was started at the second day of the menstrual cycle. LHRH was given as a nasal spray at a daily dose of 600 μg until hCG (Pregnyl, Organon) injection. HMG was given intramuscularly at a dose of 3×75 I.U. per day during two days, continuing with 2×75 I.U. per day until the 5th or 6th day of the cycle. Further on the dose of HMG was individually adjusted following the blood oestradiol levels and the follicular diameter measured with transvaginal ultrasound. When at least two follicles had a diameter of 18 mm or more, and the blood oestradiol level increased exponentially during at least five days, 10,000 I.U. of hCG was given intramuscularly. A transvaginal ultrasonographic ovum pick-up was performed 34 to 35 hours after the hCG injection in which all visible follicles were punctured and aspirated.

Sperm processing and oocyte culture

Sperm samples were processed according to the washing and swim-up procedures (Mahadevan & Baker, 1984). Oocyte-cumulus complexes were cultured in 1 ml aliquots of modified Earle's medium (Purdy, 1982) supplemented with 8% cord serum in Falcon tubes in 5% CO_2 in air at 37°C. Insemination with 100,000 motile sperm cells was performed in these tubes or, if the sperm count was excessively low, sperm drops were placed under paraffin oil and oocytes were directly transferred to these drops. Oocytes were examined the next day, i.e. about 15–18 hours after insemination, for the presence of pronuclei as a sign of fertilization. After transfer of pronucleated oocytes to growth medium containing 15% cord serum cleavage was checked 24 hours later and embryo transfer to the patient's uterus was then performed.

Zona drilling procedure

Within each IVF cycle about half of the oocytes were randomly selected for the zona drilling procedure, while the remaining oocytes were left intact as controls. The drilling procedure used was analogous to the method described by Gordon et al. (1988). A blunt holding pipette and a fine drawn micropipette were fixed on Leitz micromanipulators. Prior to zona drilling oocytes were freed from their cumuli by a five minute exposure to a hyaluronidase solution (Hyason, Organon, 150 I.U./ml) followed by pipetting the oocytes in order to remove most of the adhering corona cells that might still be present after the enzymatic digestion. Oocytes to be drilled were transferred to a 0.2 ml drop of Earle's medium containing 10 mM Hepes.

According to the period and the actual drilling procedure used, we distinguish three groups of drilling cycles. Group I consists of 8 cycles and was chronologically our first period of zona drilling. In this initial period mechanical drilling was performed, i.e. piercing the zona pellucida with the micropipette. In the second Group (13 cycles) a chemical drilling method was used, using acid Tyrode (pH 2.3) as described by Gordon et al. (1988). The response of individual oocytes to the acid Tyrode treatment varied, particularly the ease with which the zona pellucida "burst" during the relatively short drilling time (up to two minutes) used in this group. Therefore during a third, chronologically the last, period (Group III, 15 cycles) Tyrode treatment was maintained until a bursting (explosion) of the zona pellucida was observed, an event that often occurred only after prolonged exposure, i.e. up to five minutes.

RESULTS

1. Observations on the oocytes during the drilling procedure

In the first period of zona drilling attempts were made to pierce a hole through the zona pellucida by mechanical thrusting with the micropipette. This procedure often required several minutes for one oocyte, and was continued until we had the visual impression that the tip of the micropipette was situated well within the perivitelline space. A temporary deformation of the oocyte was observed in all cases during the mechanical drilling procedure, but at the end of the manipulation the oocyte resumed its regular spherical shape.

In the second period focal solubilization of the zona with acid Tyrode was performed as described by Gordon et al. (1988). The different stages of chemical drilling are illustrated in Figure 1. Usually within 10 seconds after application of the salt solution part of the zona started to thin, presumably by dissolution of the outer layers. The remaining inner part resisted Tyrode treatment but occasionally it became elevated from the oocyte surface and was followed by bursting. Occasionally a tiny part of the oocyte surface was extruded out of the hole in the zona, and a complete hatching of the oocyte accidentally occurred. In this group manipulation was limited to about two minutes and the final "explosion" was only observed in a restricted number of cases.

In the third group of zona drilling the same approach was used as in group II, but here acid Tyrode application to each oocyte was maintained until bursting of the zona was observed. This phenomenon often required up to five minutes further treatment with Tyrode after the initial thinning of the zona. The explosion could sometimes be speeded up by applying a few gentle mechanical thrusts with the micropipette. Only in the latter case slight mechanical deformations were observed as in group I during the actual drilling procedure. In both groups of chemical drilling a slight contraction of the oocyte periphery near the drilling area was often observed. After finishing the actual drilling procedure an indentation of the oocyte in that area was often the result. At the time of fertilization control during the next day however, most of these oocytes had resumed their regular spherical shape.

2. Results of zona drilling on oocytes

The oocyte fertilization after zona drilling in randomly selected oocytes is summarized in Table 1. In the control groups, in which intact cumulus-oocyte complexes were inseminated as during a standard IVF procedure, 20%, 14% and 22% of the oocytes were fertilized respectively. Corresponding figures of the drilled oocytes were 9% (Group I, mechanical drilling), 10% (Group II, chemical drilling) and 10% (Group III, prolonged chemical drilling). The differences between control and drilled oocytes were not statistically significant in

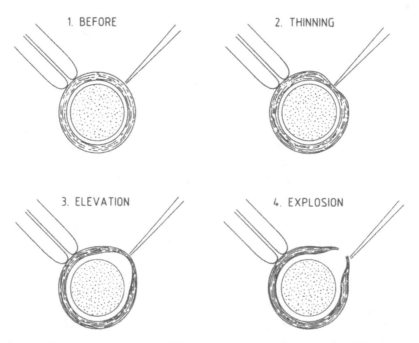

Fig. 1. Schematic representation of the different stages of chemical zona drilling. For explanation see text.

Table 1. Effect of Zona Drilling on Oocyte Fertilization

	No. of oocytes	No. of 2 PN (%)	No. of 3 PN (%)	No. of 1 PN (%)	No. of degen.(%)
MECHANICAL					
Control	30	6 (20)	0 (0)	0 (0)	6 (20)
Drilling	29	2 (9)	0 (0)	4 (14)	12 (41)
TYRODE A					
Control	58	8 (14)	0 (0)	0 (0)	7 (12)
Drilling	48	5 (10)	2 (4)	1 (2)	9 (19)
TYRODE B					
Control	102	22 (22)	7 (7)	0 (0)	15 (15)
Drilling	77	6 (10)	1 (2)	15 (24)	13 (21)

the separate groups. If the three groups were combined however, significantly more fertilizations were obtained in the control oocytes ($p < 0.01$).

Degenerated oocytes, evaluated by their dark and shrunken appearance, were seen at the time of fertilization control in the control groups (20%, 12%, 15%) as well as in the drilled groups (41%, 19%, 21% respectively). The differences were not significant, neither in the separate groups, nor in total.

Polyspermic fertilization (three or more pronuclei present at the day of fertilization control) was observed only in the control group III (7%) and the drilling groups II (4%) and III (2%). Again no significant differences could be found either combined, or in the separate groups.

Some of the oocytes in the drilled groups showed the presence of only one pronucleus which we classified as "parthenogenetic". The percentages were 14%, 2% and 24% in the three drilled groups, in contrast with 0% in all the control groups. The differences between control and drilled groups were not significant within each drilling group. If the three groups were combined however, significantly more parthenogenetic activations occurred in the drilled oocytes ($p < 0.00001$). If we compared the drilled oocytes of the three groups, we found a significant difference in parthenogenetic activation between groups II and III only ($p < 0.005$).

The further development of the fertilized oocytes after their transfer to growth medium was normal (Table 2). The majority of the pronucleated oocytes underwent regular cleavage and only two fertilized oocytes degenerated: one in control group III, and one in the mechanical drilling group I.

Table 2. Cleavage of Fertilized Ova after Zona Drilling

	No. of ova	No. of cleavages	No. of degenerations
MECHANICAL			
Control	6	5	0
Drilling	2	1	1
TYRODE A			
Control	8	8	0
Drilling	5	5	0
TYRODE B			
Control	13	10	1
Drilling	5	4	0

Table 3. Effect of Zona Drilling on Fertilization in Patients

	No. of patients	No. with positive IVF	Positive IVF Control only	Positive IVF Drilling only
Previous IVF				
+	11	6	3	0
−	13	6	2	4
No previous IVF	12	6	2	2
TOTAL	36	18	7	6

3. Results of zona drilling in patients

Of the 36 treated cycles 18 resulted in fertilization of at least one oocyte. In seven patients only control oocytes became fertilized, but in six patients fertilization was obtained only after zona drilling.

Table 3 shows the results of IVF and zona drilling in patients, taking into account the results of previous IVF attempts without zona drilling. Of 11 patients that had a positive IVF (i.e. at least one oocyte fertilized) during an earlier IVF trial, six patients had a successful IVF during the drilling cycle, but in this group fertilizations were obtained either in control oocytes only, or in both control and drilled oocytes. In other words in this group a successful IVF could not be ascribed to the use of the zona drilling procedure. On the other hand, in the 13 cycles of patients that had participated previously to IVF with negative results, six ended up with at least one oocyte fertilized, and in four of these in drilled oocytes exclusively.

4. Sperm characteristics in patients with zona drilling

Patients with a positive IVF during the drilling cycle, irrespective as to whether control or drilled oocytes were fertilized, had a significantly higher initial motility (Mean $39.4 \pm 17.5\%$ versus $16.0 \pm 17.4\%$; $p < 0.001$), higher motility after washing ($39.8 \pm 23.5\%$ versus $17.6 \pm 17.1\%$; $p < 0.005$) and higher sperm concentration after washing ($3.3 \times 10^6 \pm 2.8 \times 10^6$ versus $1.6 \times 10^6 \pm 1.9 \times 10^6$; $p < 0.05$). Differences between sperm concentrations in the initial semen sample ($16.1 \times 10^6 \pm 11.6 \times 10^6$ versus $8.0 \times 10^6 \pm 12.6 \times 10^6$) or percentage normal morphology (64.1 ± 9.8 versus $56.4 \pm 8.6\%$) were not statistically significant. Numbers of patients with fertilization in drilled oocytes only were too small for statistical evaluation of their sperm quality.

DISCUSSION

Micromanipulation techniques such as zona drilling offer an attractive approach to problems of oocyte fertilization with subfertile semen. However all manipulations are potentially harmful to gametes, and possible damaging effects of the zona drilling technique on oocytes must be considered. Gordon & Talansky (1986) emphasized the safety of the method, as based on the birth of normal young after transferring drilled fertilized mouse oocytes to pseudopregnant foster mothers. Also the oocyte degeneration rate in these experiments was very low (4.9%) in contrast with their first experiences with human oocytes, where only 31/47 (65%) survived the procedure (Gordon et al., 1988). Species differences in sensitivity and/or ease of drilling apparently exist between mouse and human.

Potential damaging factors leading to oocyte degeneration may be the mechanical deformation of the oocyte and, during chemical drilling, the effect of the very low pH on the oocyte surface where contraction and indentation was observed. Our data did not show an overall significant higher degeneration rate in the drilled oocytes.

A significantly higher number of drilled oocytes showed the presence of only one pronucleus at fertilization control, what we interpreted as parthenogenetic activation. Experimental parthenogenesis in laboratory animals can be induced by temperature shock, osmotic shock, cumulus removal with hyaluronidase or electrical stimulation (see Gwatkin, 1977 for a review). Of our three groups of zona drilling parthenogenesis was reduced in group II (chemical drilling, short period) with a significant difference between groups II and III. This would imply that mechanical manipulation and/or prolonged chemical treatment may be factors leading to activation.

We did not observe an increase in polyspermic fertilizations after zona drilling, and thus confirm Gordon's conclusion that in the human the block to polyspermia must be situated within the oocyte plasma membrane. Observations on oocytes with accidental zona fracture during aspiration however, led to the conclusion that zona pellucida damage may contribute to an increase in polyspermic fertilization with normal sperm (Lowe et al., 1988). On the other hand, in our series no fertilization was observed after inseminating accidentally hatched oocytes after the drilling procedure (personal observations).

We are concerned with the low rates in oocyte fertilization that we obtained after zona drilling. This is in contrast with the optimistic reports following mouse experimental work (Gordon & Talansky, 1986; Depypere et al., 1988). One reason for this discrepancy may be that diluted sperm as used in mouse experimental work may not represent an appropriate model for a bad sperm sample in human subfertility.

In our department no lower limit is imposed on sperm quality to allow participation to our IVF/zona drilling programme. Many sperm samples used were very bad indeed. Motility before and after washing and sperm concentrations after washing were higher for those IVF cycles that resulted in at least one fertilized oocyte. The logical conclusion is that partial removal of the sperm barrier is not enough for fertilization; sufficient motile spermatozoa should still be available to reach the oocyte. As the fertilization rates remain low in drilled oocytes, no real improvement of fertilization may be expected after zona drilling in patients that had a previous positive IVF. This is also supported by the data in Table 3. Only in patients with a previous negative IVF (or no IVF) a few cases had fertilizations of drilled oocytes only. There were too few cases like that to allow conclusions about the sperm quality within this category of patients.

We conclude that some minimal requirements of sperm concentration and motility should be fulfilled also for the application of zona drilling. More patients should be studied however before such criteria can be established. In addition further improvements of the micromanipulation techniques are necessary, including measures for avoiding oocyte damage or parthenogenetic activation.

SUMMARY

In order to evaluate the zona drilling technique to facilitate fertilization in IVF-couples with a male subfertility problem, zona drilling was applied randomly within each patient in 36 IVF-cycles (26 patients) using mechanical drilling (Group I: eight cycles, 59 oocytes), short chemical drilling with acid Tyrode (Group II: 13 cycles, 106 oocytes) and prolonged chemical drilling until zona bursting was observed (Group III: 15 cycles, 179 oocytes).

Fertilization rates of ocytes were comparable in control (C) and drilled (D) oocytes, i.e. between 10% and 20%. Oocyte degeneration rates were 20% (C) and 41% (D) (Group I), 12% (C) and 19% (D) (Group II), 15% (C) and 21% (D) in Group III. Parthenogenetic activation was observed in 0% (C) and 14% (D), 0% (C) and 2% (D), 0% (C) and 24% (D) of Groups I, II and III. Polyspermic fertilizations occurred in 0% (C) and 0% (D), 0% (C) and 4% (D), and 7% (C) and 2% (D) of Groups I, II and III respectively.

Considering the treatment cycles however, fertilization of only drilled oocytes occurred in six out of the 36 cycles. In patients that had a previous IVF cycle with at least one oocyte fertilized (11 cycles), zona drilling did not improve the results.

We therefore conclude that, although no improvement in the overall fertilization rate was observed, zona drilling can be beneficial in some patients.

REFERENCES

Cohen, J., Edwards, R., Fehilly C., Fishel, S., and Hewitt, J., 1985, In vitro fertilization: a treatment for male infertility, *Fertil. Steril.*, 43:422.

Depypere, H. T., McLaughlin, K. L., Seamark, R. F., Warnes,G. M., and Matthews, C. D., 1988, Comparison of zona cutting and zona drilling as techniques for assisted fertilization in the mouse, *J. Reprod. Fert.*, 84:205.

Gordon, J. W., Grunfeld, L., Garrisi, G. J., Talansky, B. E., Richards, C., and Laufer, N., 1988, Fertilization of human oocytes by sperm from infertile males after zona pellucida drilling, *Fertil. Steril.*, 50:68.

Gordon, J. W., and Laufer, N., 1988, Applications of micro-manipulations to human in vitro fertilization, *J. IVF. ET.*, 5:57.

Gordon, J. W., and Talansky, B. E., 1986, Assisted fertilization by zona drilling: a mouse model for correction of oligospermia, *J. exp. Zool.*, 239:347.

Gwatkin, R. B. L., 1977, "Fertilization mechanisms in man and mammals", Plenum Press, New York and London.

Jeulin, C., Feneux, D., Serres, C., Jouannet, P., Guillet-Rosso, F., Belaisch-Allart, J., Frydman, R., and Testart, J., 1986, Sperm factors related to failure of human in vitro fertilization, 1986, *J. Reprod. Fert.*, 76:735.

Jeyendran, R. S., Schrader, S. M., Van der Ven, H. H., Burg, J., Perez-Pelaez, M., Al-Hasani, S., and Zaneveld, L.J.D., 1986, Association of the in-vitro fertilizing capacity of human spermatozoa with sperm morphology as assessed by three classification systems, *Human Reproduction*, 1:305.

Liu, D. Y., Du Plessis, Y. P., Nayudu, P. L., Johnston, W. I. H., and Baker H.W.G., 1988, The use of in vitro fertilization to evaluate putative tests of human sperm function, *Fertil. Steril.*, 49:272.

Lowe, B., Osborn, J. C., Fothergill, D. J., and Lieberman, B.A. 1988, Factors associated with accidental fractures of the zona pellucida and multipronuclear human oocytes following in vitro fertilization, *Human Reproduction*, 3:901.

Mahadevan, M., and Baker, G., 1984, Assessment and preparation of semen for in vitro fertilization, *in:* "Clinical In Vitro Fertilization", C. Wood and A. Trounson, ed., Springer-Verlag, New York.

Mahadevan, M., Trounson, A. O., and Leeton, J. F., 1983, The relationship of tubal blockage, infertility of unknown cause, suspected male infertility, and endometriosis to success of in vitro fertilization and embryo transfer, *Fertil. Steril.*, 40:755.

Mahadevan, M., and Trounson, A. O., 1984, The influence of seminal characteristics on the success rate of human in vitro fertilization, *Fertil. Steril.*, 42:400.

Purdy, J. M., 1982, Methods for fertilization and embryo culture in vitro, *in:* "Human conception in vitro", R. G. Edwards and J. M. Purdy, ed., Academic Press, London and New York.

RAPID SEXING OF PREIMPLANTATION MOUSE EMBRYOS BY BLASTOMERE BIOPSY AND ENZYMATIC AMPLIFICATION

Jon W. Gordon, Luis M. Isola, and Michael W. Bradbury

Brookdale Center for Molecular Biology
Dept. of Ob/Gyn & Reproductive Science
Dept. of Medicine and Department of Geriatric and Adult Development
Mt. Sinai School of Medicine, 1 Gustave L. Levy Place, New York, NY 10029

INTRODUCTION

When infertile couples reach the point in their evaluation where in vitro fertilization (IVF) is considered as a remedy, they have usually exhausted all other modes of treatment. In the case of the female, previous testing and therapy may include diagnostic and/or therapeutic laparoscopies, hysterosalpingography, and prolonged cycle monitoring with or without hormone treatment. In the case of the males, various hormone treatment regimens, varicocoele surgery and tests such as the zona-free hamster penetration assay are typical. Given these circumstances, two important features characterize the typical IVF couple. First, their age is relatively advanced, and the female is often approaching the end of her reproductive life. Second, these couples have arrived at a juncture in their treatment course in which the likelihood of pregnancy is relatively low, and consequently, in which loss of a pregnancy, established after years of emotional stress, medical intervention and financial expense, is a disastrous event.

Given these circumstances, it is indeed unfortunate that IVF couples, by virtue of the females' advanced age, are at especially high risk for conceiving children with Down syndrome (trisomy 21), which could cause them to opt for pregnancy termination. In addition, these couples are at risk for myriad other genetic diseases which, though not necessarily increased in frequency in older individuals, pose an especially serious threat to the reproduc-

Advances in Assisted Reproductive Technologies, Edited by
S. Mashiach *et al.*, Plenum Press, New York, 1990

tive viability of these patients. For these reasons, a strong impetus exists to diagnose genetic disease in IVF embryos prior to the time of embryo transfer. Because current ovulation induction protocols often make many embryos available for transfer in a single IVF cycle, the ability to diagnose genetic disorders during cleavage could allow the culling of genetically diseased zygotes and selective transfer of the healthy embryos. In the situation where a surfeit of embryos allows for freezing, genetic diagnosis could rule out disease in embryos to be transferred in subsequent cycles, and could also provide a measure of confidence to couples requiring embryo donation.

There exist several complications to performing genetic diagnosis on the cleaving embryo, however. First, it must be possible to sample material from the embryo without incurring damage which would preclude subsequent implantation. Second, the diagnostic procedure must be effective even when very small amounts of starting material are available (1 –5 cells). Finally, the test must be rapid. If an embryo is sampled at the morula stage, 24– 30 hrs. are available before that embryo reaches the blastocyst, after which transfer to the uterus or freezing is required. If cells are sampled from the blastocyst, only a few hours may be available for completion of the assay.

In this chapter we will first review data which indicates that the embryo can be harmlessly sampled. We will then review efforts to remove blastomeres or trophoblast cells and perform genetic diagnosis. The current approaches to genetic evaluation consist of enzymatic assay and DNA analysis. The relative merits and deficiencies of each strategy will be delineated. This discussion will be presented in the context of our own efforts to perform genetic diagnosis by DNA analysis. In our work, determination of genetic sex has been the focus of effort.

PREIMPLANTATION EMBRYOS ARE AMENABLE TO BIOPSY

Preimplantation mammalian embryos exhibit two important characteristics which indicate that the opportunity does exist for safe biopsy. First, the cells of the embryo remain developmentally totipotent until at least the 8-cell stage. Tarkowski (1961) aggregated cells at the 8-cell stage and found that single composite embryos would develop. Mintz (1962) made similar observations, and further demonstrated that when such composite embryos were reimplanted into pseudopregnant females, adult genetic mosaics would result, and that both contributing genotypes could be found in all cell lineages (Mintz, 1965). Since the mice were not larger than normal, this observation demonstrates that cells at the 8-cell stage are able to undergo compensatory behavior in order to "make room" for cells of the other member of the aggregate pair. Further experiments show that developmental totipotency is retained by a significant portion of cells as far as the 8-cell stage. Tarkowski and Wroblewska (1967) dissociated embryos at the 4- or 8-cell stage, cultured and reimplanted the individual blastomeres, and obtained identical quadruplet embryos. However, when 8-cell embryos were dissociated, many blastomeres developed into "trophoblastic vesicles" without an inner cell mass (ICM) component. These experiments did not rule out totipotency for all cells at this stage, however, because very few cell divisions occurred between dissociation and blastulation. Gardner (1968) subsequently showed that ICM cells could be injected into the blastocyst cavity and become incorporated into the embryo proper. Thus, even at the blastocyst stage it is possible to insert cells into the ICM, either by adding to the size of the ICM or by displacing cells already occupying this component of the early embryo. This developmental plasticity has been exploited to produce chimeras from culture teratocarcinoma cells (Mintz and Illmensee, 1975; Papaiannou et al., 1975) or embryonal stem (ES) cells (Martin, 1981; Evans and Kaufmann, 1981).

These and related experiments also showed that the embryo contains more cells than are necessary to form an entirely normal term fetus. The observation that dissociated 4-cell embryos give rise to quadruplets indicates that at this stage, as many as four times the minimum number of cells required to produce a viable offspring are present. This number may actually fall as development progresses because of the commitment of cells to the trophoblast lineage. While it is not known how many cells form a normal ICM, use of integrated retrovirus proviral DNA as a tracer indicates that at most 8-cells contribute to all somatic lineages (Soriano and Jaenisch, 1986). However, it is strongly suggested by these and numerous similar experiments that cells can be safely removed from cleaving embryos without impairing development.

The ability to safely biopsy embryos thus creates the opportunity to perform genetic diagnosis during cleavage. Success of such diagnostic procedures depends on high sensitivity and speed. Two approaches can be taken. First, assays of gene expression can be conducted, and second, direct evaluation of DNA in the biopsied blastomere(s) can be made. The single most important of all genetic traits to determine is embryo sex. In humans, sexing is important because a variety of serious genetic disorders are sex linked, including Duchenne's muscular dystrophy, Lesch-Nyhan disease [hypoxanthine phosphoribosyltransferase (HPRT) deficiency], and hemophilia A (Factor VIII deficiency). In animals sexing is important because various agricultural enterprises (e.g. the dairy industry) could be economically more efficient if a preponderance of males or females could be born. Strategies for embryo biopsy include removal of single blastomeres from cleaving embryos or weverl trophoblast cells from the blastocyst (Figure 1).

DIAGNOSIS OF SEX USING H-Y ANTIGEN

Previous efforts to sex animal embryos have been based on an assay for a putative Y-chromosome—specific gene product, the histocompatability Y (H-Y) antigen. This antigen was identified by Eichwald and Silmser (1955) who noted that when reciprocal skin grafts were made between male and female mice of highly inbred mouse strains, the female graft was accepted by males while the male graft was rejected by females. A variety of studies subsequently were reported to indicate that the H-Y antigen was the male determining gene product (Wachtel et al., 1975). However, a prerequisite for the male determining gene product is that it be expressed prior to gonad determination. The report of Krco and Goldberg

(1976) that the antigen is expressed on 8-cell mouse embryos was consistent with this requirement, but also suggested that antibodies could be used to detect male embryos during cleavage.

Epstein and Travis (1980) used a cytotoxicity assay coupled with H-Y antibody exposure in an effort to kill XY embryos. When undamaged embryos were karyotyped, 92% (95/103) were XX. A small number of embryos with a few surviving cells were also karyotyped, and 16 of 17 were found to be XY. Shelton and Goldberg, used a similar assay, and reimplanted the embryos as an ultimate test of correct sexing. Of 17 embryos which survived the cytotoxicity assay and developed to term, 14 were XX (82%). Some success has been reported for use of H-Y antibody on other more economically valuable species. White et al. (1987) exposed porcine embryos to H-Y antibodies and then employed indirect immunofluorescence to identify XY embryos. Of 91 8-cell embryos tested, 46% fluoresced.

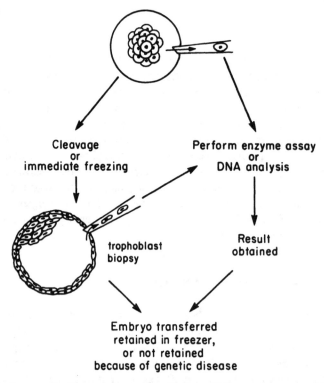

Fig. 1. Methods for genetic diagnosis by embryo biopsy. Cells can be removed from the morula (top) or several trophoblast cells can be sampled from the blastocyst (bottom left). Biopsy material is then subjected to enzymatic assay or DNA analysis for genetic diagnosis, and subsequent decisions regarding embryo transfer are made on the basis of the results.

When embryos were karyotyped to confirm the H-Y test, 48/59 were correctly sexed. White et al. (1987) performed a similar study on sheep embryos, and reported 88% accuracy (50/57 embryos tested) for sexing. The embryos were reimplanted with no apparent loss of viability resulting from exposure to reagents necessary for the sexing assay. Similar results have also been reported for bovine embryos (White et al., 1987) with 79% of fluorescing embryos proving to be XY and 89% of non- fluorescent embryos being XX. Eighty-two per cent accuracy has been reported for the assay on sheep embryos (Wood et al., 1988), and about 80% for bovine embryos when monoclonal anti-H-Y antibodies are used (Wachtel et al., 1988).

The difficulty with all of these reports is accuracy and reproducibility. In none of the previous H-Y assays was 100% accuracy achieved. A greater difficulty, however, is with preparations of H-Y antibody. These antibodies are usually of low titer, and consequently concentrated serum preparations must be added to embryos for adequate antibody binding. Moreover, H-Y antibody preparations are notoriously variable in performance, making reliability of the test lower.

DIAGNOSIS OF SEX BY HPRT ASSAY

Another approach to sexing on the basis of gene expression relies on assays for X-linked gene products in early embryos. The rationale for this approach is based on the phenomenon of X-chromosome inactivation. In mammals, X-inactivation takes place to compensate for the fact that XX cells have twice the dosage of X-linked genes as XY cells (Lyon, 1974). Heterochromatization of the supernumerary X corrects this disparity. However, X-inactivation does not occur in mammalian embryos until late preimplantation development (Gardner and Lyon, 1971). Thus, for the first few cleavage divisions, genes expressed from the X chromosome are expected to produce twice as much product in female (XX) zygotes as in males (XY).

The most successful effort to sex embryos by measuring differences in X- linked gene expression have been reported by Monk and her colleagues. This group first exploited the development of HPRT-deficient mice by manipulation of embryonal stem (ES) cells. ES cells are derived from blastocysts grown on feeder layers (Martin, 1981; Evans and Kaufman, 1981) and can frequently contribute to the germ cell population of chimeric mice after microinjection into the blastocyst cavity. Cells grown in the purine analog 6-thioguanine (6-TG) are killed if they express HPRT. Thus, ES cells deficient for this X- linked enzyme can be identified by culture in the presence of 6-TG. When this was done and ES cells were inserted into blastocysts, the ES cell-derived spermatozoa of a resultant chimeric mouse yielded pups with a complete deficiency for HPRT (Hooper et al., 1987). A similar result was obtained after retrovirus-mediated gene transfer was used to disrupt the HPRT gene (Kuehn et al., 1987). While HPRT deficient mice do not manifest any of the pathological features of human Lesch-Nyhan disease, they can serve as a model for detection of the deficiency by embryo biopsy.

Monk et al. (1987) used a microassay for HPRT on single blastomeres or on several cells removed from the trophoblast region of blastocysts to detect HPRT deficiency in mice. Results were confirmed by reimplanting the embryos and studying fetuses which developed. This result not only constituted a model approach for detecting Lesch-Nyhan disease in human embryos, it provided an avenue for embryo sexing by measurement of relative amounts of this X-linked gene product. Monk et al. (1988) subsequently reported success in detecting embryo sex by this means.

While these results are very exciting, they fall short of the "perfect" genetic diagnosis procedure for several reasons. First, the embryo biopsy must be done with sufficient care that integrity of the removed blastomere is retained. Second, the test relies on expression of HPRT during cleavage development. While this clearly takes place in mice, it is not apparent that this gene is turned on in other species prior to blastocyst formation. Finally, the test is limited to HPRT and sex diagnosis. The approach cannot be extended to testing for genes not expressed during early development. Ideally, a genetic testing strategy should be adaptable to diagnosis of autosomal genes which might not be expressed until late fetal or postnatal development, but mutations of which cause serious disease. Obvious examples of such disease genes are the sickle β-globin gene, the gene for Huntington's disease, and perhaps for the complex of genes which are triplicated in Down syndrome (trisomy 21). For these reasons, efforts to diagnose genetic traits by direct evaluation of DNA have also been under development.

DIAGNOSIS OF SEX BY CHROMOSOME ANALYSIS

The traditional method of DNA diagnosis is of course karyotyping. Because of the importance of sex determination and the obvious karyotypic difference between male and female embryos, sex karyotyping has received the most attention. Gardner and Edwards (1968) attempted sex diagnosis in rabbits. In this study 98 embryos were biopsied and 86 were successfully karyotyped. Of 33 sexed embryos transferred to the uterus, 20 reached term, and 18 were sexed to assess the accuracy of the original karyotyping procedure. Six females and 12 males were identified, and all corresponded correctly to the phenotype predicted.

While these results are impressive, examination of chromosomes for diagnosis of genetic disease has several drawbacks. First, it is difficult to be certain of obtaining an adequate metaphase preparation from very few cells. Thus, it may become necessary to expand a culture prior to karyotyping. In the process of culture, cells can undergo chromosomal rearrangements which are not present in the embryo from which the starting blastomere(s) was obtained. Second, expansion of the culture requires time, and thus, it is unlikely that a diagnosis can be obtained prior to the time that freezing or embryo transfer becomes necessary. Finally, karyotyping is not sufficiently sensitive to evaluate individual genes or even DNA rearrangements which might involve hundreds of thousands of base pairs; it can only be used to ascertain relatively gross features of the genetic endowment of the embryo such as sex or aneuploidy.

DIAGNOSIS OF SEX BY DNA HYBRIDIZATION

The potential to improve diagnostic strategies took a major leap forward with the advent of recombinant DNA technology, and in particular, its applications to the study of individual genes by Southern blot hybridization (Southern, 1975). The principle of nucleic acid hybridization has been applied to blastomeres removed from bovine embryos. A Y-chromosome—specific DNA probe was labelled with biotin and used on biopsied blastomeres in an effort to detect the Y chromosome in the absence of a metaphase chromosome spread (Leonard et al., 1987). However, the test was not specific enough for reliable use.

A. DNA analysis using the polymerase chain reaction. Recently, a new approach to studying small regions of DNA from very little starting material has been developed. This technique, termed the polymerase chain reaction (PCR) first involves the chemical synthesis of single stranded oligonucleotides which are situated at opposite ends of a small DNA segment (100-3000 bp). One oligonucleotide is identical to the coding strand of the DNA, while the other corresponds to the opposite, non-coding strand. These oligonucleotides are then added in great molar excess to a sample of total cellular DNA. After denaturation by boiling and reannealing, the oligonucleotides intercede for the endogenous DNA strand and form short heteroduplexes by base pairing. The oligonucleotides then serve as primers for strand extension by DNA polymerase. The polymerization step makes exact copies of the DNA between the priming sites, and also replicates the priming sites. Thus, after a second round of denaturation and reannealing, twice the number of annealing sites exist, and a subsequent round of polymerization exponentially doubles the number of copies of the DNA region to be studied. Because a single round of denaturation, annealing and polymerization requires only a few minutes, several hundred thousand copies of the target DNA segment can be produced in a few hours (Saiki et al., 1985). PCR methodology is illustrated in Figure 2.

Although PCR was first used on substantial amounts of purified DNA for diagnosis of point mutations such as sickle cell trait (Saiki et al., 1985), It soon became apparent that this technology could be used directly on cells without DNA purification. The technology was applied to prenatal diagnosis of sickle cell disease and of sex without prior DNA extraction (Kogan et al., 1987). In addition, momentum developed to refine the technology for diagnosis of genes in single cells. The discovery that genes transferred into cultured cells could occasionally integrate into the recipient genome by homologous recombination (Smithies et al., 1985), led investigators to attempt production of targeted mutations in ES cells. Since ES cells can contribute to the germ line of a chimeric mouse, success of such a venture would result in production of targeted mutations at any locus for which the gene or a part of the gene had been cloned. However, in order to detect homologous recombination events, which occur in only 0.1% of transformed cells, it was necessary to devise a sensitive assay. PCR presented one such opportunity (Kim and Smithies, 1988), and thus, diagnosis of traits in one or a few cells by PCR became an important goal.

In addition to its use for detection of homologous recombination events, however, PCR was considered for application to genetic diagnosis in gametes. Efforts to detect allelic differences in genes from single spermatozoa were at least partially successful (Li et al., 1988), though results could not be obtained in every case. Part of the difficulty with the PCR reaction, when it is applied to exceedingly small quantities of target DNA, is specificity of oligonucleotide annealing. Because oligonucleotides are present in enormous mo-

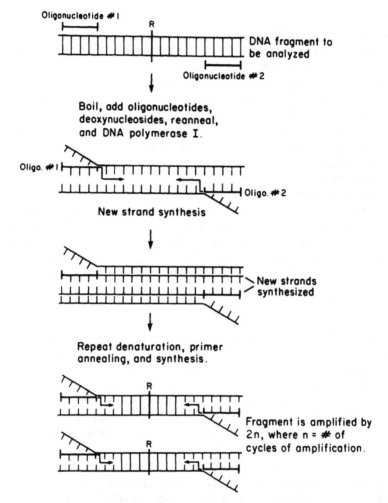

Fig. 2. The polymerase chain reaction (PCR, see text for details). R indicates a restriction endonuclease site within the amplified region which might be used to diagnose a mutation. Note that the procedure exponentially amplifies a selected region of DNA because the oligonucleotide priming sites are replicated in each PCR cycle (bottom).

lar excess, they can rarely anneal to genomic sites other than the desired target, even though the spurious annealing site might not have a perfect base match with the oligonucleotide. When large quantities of starting DNA are available for PCR such spurious annealing is not a problem because the large number of appropriate annealing targets assures a far more rapid annealing to the appropriate site and preferential amplification of the desired sequence. However, when the target for annealing is very low in concentration, and multiple rounds of PCR are required to obtain observable levels of amplification, extraneous sequences can become amplified (Saiki et al., 1988, Kim and Smithies, 1988). In order to reduce the obfuscation of results by nonspecific amplification, nucleic acid hybridization procedures can be conducted (Li et al., 1988). However, these techniques are intricate and time consuming.

B. Use of PCR in diagnosis of Embryo sex. In recognition of these limitations, both our laboratory and that of Handyside et al. (1989) sought to adapt PCR to amplification of a repeated DNA element which would serve as an indicator of an important genetic trait. The most convenient candidate trait for such an approach was also the most important— embryo sex. Sexing by PCR is made simple in principle by the fact that there exist highly repeated DNA elements unique to the Y-chromosome. Thus, when a single blastomere contains many copies of the annealing target for PCR, a single cell is, with regard to the initial annealing reaction, the equivalent of many cells. Use of a repetitive element as the target for PCR thus lessens the problem of nonspecific annealing.

Handyside et al. (1989) attempted to amplify a 149 bp repeated element from the human Y in biopsied human embryos. Both polyspermic and parthenogenetic embryos were included in the experimental group, and PCR results were compared with subsequent cytological analysis in the embryos which grew to the blastocyst. Of the 38 embryos originally harvested, 25 were available for analysis (66%). Of these 25, four males and 11 females were sexed by PCR and the results correlated with cytological analysis. This study had the limitations that several embryos were not successfully carried through the procedure, that 10 of the 25 tested embryos could not be confirmed by cytological analysis, and finally, that parthenogenetic embryos would be expected to be female and polyspermic embryos would be expected to be male. Finally, it was not possible to transfer the embryos for phenotypic confirmation of the sexing assay results. Nonetheless these results are highly encouraging that PCR could effectively determine embryo sex.

In simultaneous experiments in our laboratory we sought to identify sex by PCR using the mouse as a model. The advantages of this approach are the ready availability of experimental material and the fact that assayed embryos can be transferred to the uterus for phenotypic confirmation of the PCR results. The sequence we chose for synthesis of oligonucleotides was a Y-chromosomal element original cloned by Bishop et al. (1985) and designated pY353/B. This 1.5 kilobase (kb) *Eco*RI fragment is not as highly repeated on the mouse Y as is the human element used by Handyside et al. (1989) for sexing human embryos. Although this element was originally thought to be completely specific for the mouse Y chromosome, our preliminary Southern results analysis carried out prior to PCR suggested otherwise. Figure 3 shows such a Southern blot, which was performed to test both specificity of the cloned element for the Y-chromosome as well as the degree to which the element was repeated. As shown in Figure 3, the 1.5 kb fragment was represented more than 800 times on the Y chromosome of the DBA mouse, which is present in B6D2F1 males, the F₁ hybrid strain tested. However, some hybridization is also seen to DNA from the XX control.

Because no efficient method existed to identify the sequences in the pY353/B element which cross-hybridized to XX DNA. Therefore we cloned the fragment into the *EcoRI* site of pGEM3 (Promega), a plasmid adapted for sequencing, and arbitrarily sequenced several hundred bases at one end of the inserted by the dideoxy chain termination method (Sanger et al., 1977). The 102 bp region selected for amplification by PCR is shown in Figure 4, and the oligonucleotides chosen for priming the sequences are highlighted.

We next attempted PCR on a range of amounts of XX and XY mouse DNA to determine the degree of Y-chromasomal specificity of the arbitrarily chosen element. As can be seen in Figure 5, the 102 bp amplified fragment is present in both male and female DNA. However, clearly greater amounts are present in the XY samples, and when DNA equival-

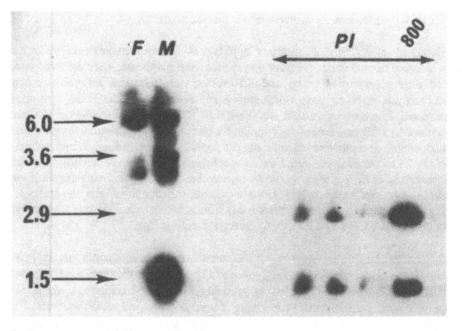

Fig. 3. Southern blot hybridization of *EcoRI* digested female F and male M DNA, loaded with increasing quantities of a recombinant plasmid containing pY353/B cloned into pGEM3. The lane labelled 800 represents 800 copies/cell of the plasmid. The 1.5 kb fragment of the plasmid is pY353/B, and the 2.9 kb fragment is pGEM3. Low levels of hybridization are seen in the 3.6 kb and 6.0 kb regions of the mouse genome, with lower homology in female than in male DNA.

```
5'------------->3'

GAA TTC ATA TAT ATG ACA GAG GCA ACA GCA TTG TGT CTG GTT TAA ACG GGC
CTT AAG TAT ATA TAC TGT CTC CGT TGT CGT AAC ACA GAC CAA ATT TGC CCG

AAG CAT CTT CAG TCC CAG GCA ATG AGC AAG TAT GAT ATT TGA AGG GAA TGG
TTC GTA GAA GTC AGG GTC CGT TAC TCG TTC ATA CTA TAA ACT TCC CTT ACC
```

Fig. 4. Sequence of pY353/B used for oligonucleotide synthesis and PCR. The two oligonucleotides are highlighted with a line above and below the respective sequences. The 5' end of the sequence is arbitrarily designated, and the 5'--->3' direction is shown at top.

ent in amount to that present in 2–3 cells is used for PCR, no detectable amplification was evident in the XX sample (Fig. 5). Given these results we proceeded to biopsy embryos and attempt a determination of genetic sex. Single blastomeres were removed from embryos in a medium consisting of modified Dulbecco's Phosphate buffered saline containing 106 mM NaCl, 2.7 mM KCl, 1.5 mM KH_2PO_4, 8.1 mM Na_2PO_4, 5.6 mM glucose, 25 mM sodium lactate, 0.33 mM sodium pyruvate, 3 mg/ml BSA and 2 mM EDTA. Embryos were kept in this chelating medium for 10 minutes at room temperature, then gently

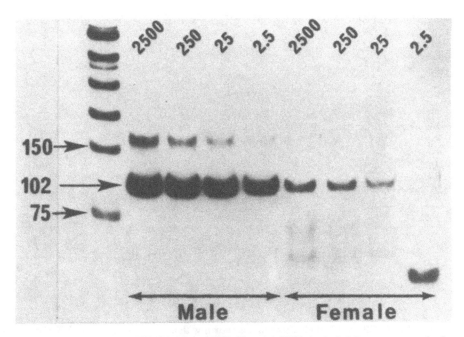

Fig. 5. Purified DNA from XY (Male) and XX (Female) DNA included in amounts equivalent to 2,500, 250, 25 and 2.5 cells, respectively, in the PCR reaction. The reactions were subjected to 45 rounds of amplification. Molecular weight markers (leftmost lane) consist of Hinf1 digested pBR322, end-labelled with T4 polymerase in the presence of 32P-ATP.

and repeatedly aspirated until a blastomere detached. The remainder of the embryo was transferred to a microdrop of M16 medium under oil. Tubes with blastomeres were analyzed using PCR.

In order to speed the analysis, radiolabelled dCTP was included in the PCR reaction for incorporation into the amplified segment. This allowed us to expose acrylamide gels directly to X-ray film to visualize the amplified band.

Figure 6 shows the results of such an effort. As demonstrated the degree of amplification falls into three categories: intense amplification comparable to the male control, weak amplification similar to the female control, and intermediate levels of amplification. Of XX embryos subjected to the assay over X experiments, 73% were considered to have decisively low or high levels of amplification, and were tentatively assigned the female or male sex, respectively.

Because we were working with a mouse model, we were able to reimplant the embryos and obtain phenotypic confirmation of the tentative sex determination made by PCR. Thus far, six embryos have developed after transfer, five of which were predicted to be males. In

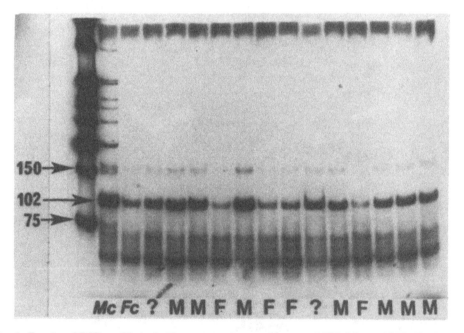

Fig. 6. Results of PCR on 13 single blastomeres shown with control XY (Mc) and XX (Fc) DNAs included in amounts equivalent to 2.5 cells. The large arrow shows the 102 bp band amplified in the reaction. The intensity of this band was used to score the embryos, with strong amplification considered to be indicative of an XY blastomere. The leftmost lane contains molecular weight markers. A small amount of marker DNA contaminates the Mc lane.

all six cases the sex indicated by PCR—five males and one female—was confirmed by direct examination of the gonads late in fetal development. In one experiment the embryos of indeterminate sex were transferred. Two developed, one of which was a male and the other a female. This test confirmed that in approximately 24% of assays the results were equivocal.

CONCLUSIONS AND PROSPECTS

The results presented here as well as those of Handyside et al. (1989) clearly show that it is possible to diagnose the genetic sex of embryos by PCR. These findings have significant implications for the future management of IVF, but the technique also requires refinement before it can reach its full clinical potential.

Two major improvements in the assay would make it an extremely powerful tool for improving the genetic health of IVF babies. First, the test must be successful in all embryos tested. While ambiguous results are tolerable as long as no errors are made for those embryos which are transferred, the increased pregnancy rate in IVF associated with multiple embryo transfer underscores the importance of obtaining a definitive diagnosis on every embryo. The other important refinement would be an increase in sensitivity which would allow single genes to be assessed for mutations. In this circumstance, disorders such as sickle cell disease and Tay Sachs disease (hexoseaminidase A deficiency) could be diagnosed.

A variety of conditions can be altered which change the character of the enzymatic amplification reaction. These include selection of the relative guanine:cytosine content of oligonucleotide primers, annealing temperatures, primer concentrations, supplementation of the reaction with additional DNA polymerase at one or more intervals after initiation of the procedure, etc. We are currently exploring all of these variables in an effort to improve the sexing assay, and we have confidence that the assay will be significantly refined. It is our view that not only will the sexing assay be perfected, but the ability to evaluate unique sequences in the genome from single biopsied blastomeres will soon be attained.

A profound additional implication of such an advance is that multiple genes could be simultaneously analyzed by inclusion of many sets of oligonucleotide primers, each specific for an individual locus. If the size of the amplified regions differed sufficiently to be resolved by gel electrophoresis, gel scanning devices could be used to examine many loci. This capability would not only allow diagnosis of multiple genetic diseases in a single assay, it could allow for selection of embryos with particularly beneficial genetic traits in cases of frozen embryo donation. For example, evaluation of histocompatability genes could allow couples receiving a frozen embryo to select a genotype which would allow for blood donation or even organ transplantation to or from close relatives. While such prospects appeared far fetched only a few years ago, the rapid development of PCR technology now makes them near realities.

When this time arrives, a discussion of the ethical aspects of prenatal diagnosis by blastomere biopsy must inevitably ensue. Will it be appropriate to refuse donor embryos because of their genotypic sex, or perhaps their hair color? While public acceptance of numerous birth control devices which block embryo implantation (e.g., the intrauterine device) indicates that cleaving embryos do not have the status of individuals in most societies, the discarding of such embryos on the basis of their genetic endowments (excluding instances where genetic disease is identified) is likely to create controversy. However this impending discussion is resolved, the development of PCR technology for ruling out genetic defects has the potential to make IVF embryos the healthiest of all, and to eliminate disastrous cases in which IVF pregnancies, established after arduous effort, undergo termination.

ACKNOWLEDGEMENTS

Work toward production of this manuscript was supported by National Institutes of Health Grants HD20484 and CA42103 and March of Dimes grant 1-1026 to JWG. This is manuscript No. 22 of the Brookdale Center for Molecular Biology, Mt. Sinai Medical Center.

REFERENCES

Bishop, C., Bousot, P., Baron, B., Bonhomme, F., and Hatat, D., 1985, Most *Mus musculus domesticus* laboratory mouse strains carry a *Mus musculus Y chromosome, Nature* 315:70-72.

Eichwald, E. J., and Silmser, C.R., 1955, Communication, *Trans Bull.* 2:154-155.

Epstein, C. J., Smith, S., and Travis, B., 1980, Expression of H-Y antigen on preimplantation mouse embryos. *Tissue Antigens* 15:63-67.

Evans, M. J., and Kaufman, M. H., 1981, Establishment in culture of pluripotential cells from mouse embryos, *Nature*, 292:154-156.

Gardner, R. L., Edwards, R. G., 1968, Control of sex ratio at full term in the rabbit by transferring sexed blastocysts. *Nature* 48:346-48.

Gardner, R. L., and Lyon, M.F., 1971, X chromosome inactivation studied by injection of a single cell into the mouse blastocyst. *Nature* 231:385-386.

Gordon, J. W., Isola, L. M., and Bradbury, M. W., 1989, Rapid sexing of preimplantation mouse embryos by blastomere biopsy and enzymatic amplification, Proc. Soc. Gynecologic Invest. #1, (abstr.).

Handyside, A. H., Penketh, R. J. A., Winston, R. M. L., Pattinson, J. K., Delhanty, J. D. A., and Tuddenham, E. G. D., 1989, Biopsy of human preimplantation embryos and sexing by DNA amplification, *Lancet*, 2:347- 349.

Hooper, M., Hardy, K., Handyside, A., Hunter, S., and Monk, M., 1987, HPRT-deficient (Lesch-Nyhan) mouse embryos derived from germline colonization by cultured cells, *Nature* 326:292-295.

Kim, H.-S., and Smithies, O., 1988, Recombinant fragment assay for gene targeting based on the polymerase chain-reaction. *Nucleic Acids Res.* 16, 8887-8903.

Kogan, S. C., Doherty, M., and Gitschier, J., 1987, An improved method for prenatal diagnoses of genetic diseases by analysis of amplified DNA sequences, *New Engl. J. Med.*, 317:985-990.

Krco, C. J., and Goldberg, E. H., 1976, H-Y (male) antigen: detection on eight-cell mouse embryos, *Science* 193:1134-1135.

Kuehn, M. R., Bradley, A., Robertson, E. J., and Evans, M. J., 1987, A potential animals model for Lesch-Nyhan syndrome through introduction of HPRT mutations into mice. *Nature* 326:295-298.

Leonard, M., Kirszenbaum, C., Cotinot, C., Chesné, P., Heyman, E., Stinnakre, M. G., Bishop, O., Delacroix, C., Vaiman, M., Fellous, M., 1987, Sexing bovine embryos using Y chromosome specific DNA probe, *Therio Genology*, 27:248 (Abstr.).

Li, H., Gyllensten, U. B., Cui, X., Saiki, R. K., Erlich, H. A., and Arnheim N., 1988, Amplification and analysis of DNA sequences in single human sperm and diploid cells, *Nature* 335:414-417.

Martin, G. R., 1981, Isolation of a pluripotent cell line from early mouse embryos cultured in medium conditioned by teratocarcinoma stem cells, *Proc. Natl. Acad. Sci. USA* 78:7634-7638.

Mintz, B., 1965, Genetic mosaicism in adult mice of quadriparental lineage, *Science* 148:1232-1233.

Mintz, B., and Illmensee, K., 1975, Normal genetically mosaic mice produced from malignant teratocarcinoma cells, *Proc. Natl. Acad. Sci. USA*, 72:3585-3589.

Monk, M., and Handyside, A. H., 1988, Sexing of preimplantation mouse embryos by measurement of X-linked gene dosage in a single blastomere. *J. Reprod. Fert.*, 82:365-368.

Monk, M., Handyside, A., Hardy, K., and Whittingham, D., 1987, Preimplantation diagnosis of deficiency of hypoxanthine phosphoribosyl transferase in a mouse model for Lesch-Nyhan syndrome. *Lancet* 2:423- 425.

Papaioannou, B. E., McBurney, M. W., Gardner, R. L., and Evans, M. J., 1975, Fate of teratocarcinoma cells injected into early mouse embryos, *Nature*, 258:69-73.

Saiki, R., Gelfand, D. H., Stoffel, S., Scharf, S. S., Higuchi, R., Horn, G. T, Mullis, K. B., and Erlich, H. A., 1988, Primer-directed enzymatic amplification of DNA with a thermostable DNA polymerase, *Science*, 239,:487-491.

Shelton, J. A., and Goldberg, E. H., 1984, Male-restricted expression of H-Y antigen on preimplantation mouse embryos. *Transplantation* 37:7-8.

Saiki, R. K., Scharf, S., Faloona, F., Mullis, K. B., Horn, G. T., Erlich, H. A., and Arnheim, N., 1985, Enzymatic amplification of β-globin genomic sequences and restriction site analysis for diagnosis of sickle cell anemia, *Science* 230:1350-1354.

Smithies, O., Grett, R. G., Boggs, S. S., Koralewski, M. A., and Kucherlapati R. S., 1985, Insertion of DNA sequences into the human chromosomal β-globin locus by homologous recombination. *Nature* 317:230-234.

Soriano, P., and Jaenisch, R., 1986, Retroviruses and probes for mammalian development: allocation of cells to the somatic and germ cell lineages, *Cell*, 46:19-29.

Southern, E. M., 1975, Detection of specific sequences among DNA fragments separated by gel electrophoresis, *J. Mol. Biol.*, 98:503-517.

Sanger, R., Nicklen, S., & Coulson, A.R., 1977, DNA sequencing with chain-terminating inhibitors, *Proc. Natl. Acad. Sci.*, 74:5463-5467.

Tarkowski, A. K., 1961, Mouse chimeras developed from fused eggs, Nature, 190:857-860.

Wachtel, S., Nakamura, D., Wachtel, G., Felton, W., Kent, M., and Jaswaney, V., 1988, Sex selection with monoclonal H-Y antibody. *Fertil. Steril.* 50:355-360.

Wachtel, S. S., Ohno, S., Koo, G., and Boyse, E., 1975, Possible role for H-Y antigen in the primary determination of sex. *Nature* 257:235-236.

West, J. D., Angell, R. R., Thatcher, S. S,, Gosden, J. R., Hastie, N. D., Glasier, A. F., and Baird D.T., 1987, Sexing the human Pre-embryo by DNA-DNA in situ hybridization, *Lancet*, 1:1345-1347.

White, K. L., Anderson, G. B., Berger, T. J,, BonDurant, R. H., and Pashen, R. L., 1987, Identification of a male-specific histocompatability protein on preimplantation porcine embryos. *Gamete Research* 17:107-113.

White K. L., Anderson, G. B., and BonDurant, R. H., 1987, Expression of a Male-specific factor on various stages of preimplantation bovine embryos, *Bio. Reprod.*, 37:867-873.

White, Kl., Anderson, G. B., Pashen, R. L., BonDurant, R. H., 1987, Detection of histocompatability-Y antigen: identification of sex of pre-implantation ovine embryos, *J. Reprod. Immunol.*,10:27-32.

Wood, T. C., White, K. L., Thompson, D. L. Jr., and Garza, F. Jr., 1988, Evaluation of the expression of male-specific antigen on cells of equine blastocysts, *J. Reprod. Immunol.* 14:1–8.

PROLIFERATION OF CELLS DERIVED FROM THE BIOPSY
OF PRE-IMPLANTATION EMBRYOS

A. L. Muggleton-Harris

MRC Experimental Embryology and Teratology Unit
St, George's Hospital Medical School
Cranmer Terrace, London, U. K.

Development of suitable methods for the culture of biopsied cells from the pre-implantation mammalian embryo is important if a single biopsied blastomere from a 2, 4, 8–16 cell embryo, or a small number of trophectoderm cells are removed from the blastocyst for genetic diagnosis. Our studies have centred on: developing techniques to biopsy and analyse embryonic and extra-embryonic cells and the maintenance and replication of biopsied cells in vitro. The techniques for in-situ hybridization, cytogenetics and many biochemical analyses depend on having sufficient numbers of cells to undertake these studies. Methods suitable for the enzyme analysis of trophectoderm biopsied cells, produced data from as little as 3–5 cells (Monk, Muggleton-Harris, Rawlings and Whittingham, 1988). A significant improvement has been made in the viability of biopsied pre-implantation embryos over those figures reported in our initial experiments, by maintaining a constant temperature during the micromanipulation and handling period. It is preferable to remove the manipulated embryo prior to removal of the biopsied cells. However, in the case of equivocal results, the original biopsy would have provided insufficient cells for reanalysis. The possibility of increasing the original biopsied cell number by culture or removing a greater number of cells must be considered. It is possible to remove at least 9–12 mural trophectoderm cells from the blastocyst which can reform an expanded blastocyst in vitro and produce live young in vivo (Muggleton-Harris, Rawlings and Findlay, unpublished results). Despite these results, it would be preferable to remove the minimum number of embryonic cells in the case of early cleavage stages, or extra-embryonic cells from the blastocyst and have them replicate in vitro to provide sufficient material should further analysis be required.

Advances in Assisted Reproductive Technologies, Edited by
S. Mashiach *et al.*, Plenum Press, New York, 1990

CULTURE MEDIA AND GROWTH FACTORS

The use of conditioned media from various cultured cell types has shown that the cells of the pre-implantation mouse blastocyst cells have a different response with regard to differentiation and growth patterns. Isolated mural trophectoderm cells readily attach to a gelatin or fibromectin substrate and remain in culture for a period (two to three days) before forming giant cells. Conditioned media from chorionic villus samples (CVS), kindly supplied by V. Bolton, Kings College Hospital, London, was used to enhance the viability of trophectoderm cells. Growth factors which are found in the conditioned media from Buffalo Rat Liver (BRL) cells repress the differentiation of embryonic stem (ES) cells derived from mouse inner cell masses (ICMs) (Smith and Hooper 1987). Therefore, BRL conditioned medium was used to encourage the proliferation of ICM or early embryonic cells. Growth or inhibitory factors present at embryogenesis have a direct role in development, a recent review by Heath and Smith (1989), discusses four sets of growth factors in detail and stresses that many growth factors are multifunctional. It is therefore advantageous when seeking to induce proliferation of cells, blastomeres or various subsets of cells from the blastocyst that a variety of growth factors and/or conditioned medium should be tried at different phases of their culture.

The culture of human pre-implantation embryos from 1-cell to the morula/blastocyst stage of development is not satisfactory at present. The success of various IVF laboratories ranges from 18–32%, therefore there is considerable room for improvement in the standard of culture conditions used. With the collaboration of V. Bolton, Kings College Hospital, London, two and four cell "spare" (which have been donated for research) human embryos have been cultured to the blastocyst stage using Hams F-12 media supplemented with L-glutamine plus 10% Fetal Bovine Serum (FBS). Due to the scarcity of human blastocysts, the ability to culture any "spare" 2–4 cell embryos through to the blastocyst stage provides an opportunity to further our studies on the biopsy and culture of extraembryonic material. In our laboratory twenty out of twenty-six (78%) have developed to the late morula/blastocyst stage using micro-drops of Hams F-12 media on gelatinised petri dishes, unfortunately very few of the blastocysts hatch from the zona (Muggleton-Harris, Findlay and Whittingham, 1990). To overcome this problem the zona has been "thinned" or removed with Acid-Tyrodes solution. In future studies the additional requirements for hatching of the blastocyst may be supplied by using MEM media. Lopata and Hay, (1989 a,b) reported that a significantly higher percentage of embryos develop to the blastocyst stage and hatched in the alpha-modification of minimum essential medium(MEM) in comparison with synthetic human tubal fluid (HTF). The presence of human cord serum did not significantly improve the number of embryos developing to blastocyst. The MEM will support development as effectively in the absence of a low concentration of serum as in its presence. However, the development to blastocyst was higher (26.8%) compared with their development in HTF (14.5%). Perhaps a combination of culturing the early embryos in Hams F-12, and then MEM to encourage hatching of the blastocyst may be more effective.

The viability of an embryo can be assessed by its ability to hatch and form an outgrowth in vitro, which is a standard routinely used to assess normal development and culture conditions for the mouse embryo. This criteria however, cannot be used for routine human embryo studies. By undertaking frequent morphological observations additional criteria such as cell proliferation, cleavage and growth patterns can monitor the progress of control and biopsied embryos providing the temperature and pH are rigorously maintained. A further test is to monitor the secretion of human chorionic gonadotrophin (hCG). Fischel and Edwards (1984) first measured the hCG secreted by human pre-implantation embryos, and a recent detailed study on hatched and intrazonal blastocysts detected hCG between day seven and eight after fertilization. Serum was shown to stimulate the output of hCG of both hatched and intrazonal blastocyst tissues (Lopata and Hay 1989b).

BIOPSY OF EMBRYONIC AND EXTRA-EMBRYONIC CELLS

Marilyn Monk's presentation has described our earlier work undertaken to diagnosis of HPRT-deficient male and carrier female mouse embryos by trophectoderm biopsy (Monk, Muggleton-Harris, Rawlings & Whittingham 1988). There are two obvious stages of development for which techniques need to be further developed in the mouse embryo, and then to apply them, if possible, to the human embryo. The 4–12 cell stage embryo provides the advantage that one (or more) blastomeres of a 4, 8 or 12 cell embryo can be removed and used to diagnose a deficiency of hypoxanthine phosphorobosyl transferase in a mouse model for Lesch-Nyan syndrome in a similar manner to that reported by Monk et al (1988) and Monk, Handyside, Muggleton-Harris and Whittingham (1989). However, there was no attempt to reanalyse the embryonic material before the embryo was transferred back into the in vivo environment. To assure confidence in the assessment for genetic defects one would require removal of at least double the amount of material for analysis and reanalysis if required, prior to further culture of the manipulated embryo to assess normal development. The embryo would then be transferred back into the in vivo environment at the morula or blastocyst stage. Even with the development of exciting new molecular techniques such as polymerase chain reaction (PCR) to amplify the DNA, there may be a requirement for reanalysis of the biopsied material. In the mouse we can put aside any "questionable embryo" but in the human this cannot be considered. A further advantage to using early cleavage stages for biopsy analysis is that one can transfer the manipulated embryo at an earlier developmental stage and achieve a greater success of live young born as opposed to that achieved at the blastocyst stage. This point is addressed with specific regard to the human embryo transfer by Drs. Edwards and Winston in their presentations. The advantages of biopsing trophectoderm (non embryonic tissue) of the mouse blastocyst has been discussed previously (Monk et al., 1988, 1989). One major advantage of biopsing trophectoderm from the blastocyst is that one is using extra-embryonic material and the embryonic ICM would remain undisturbed. Providing the biochemical assays or PCR techniques are accomplished within 8–24 hours the blastocyst can be biopsied, diagnosed and then be transferred back following a suitable period in culture during which the blastocoele cavity reforms and the blastocyst appears unaffected by the micro-manipulation procedures.

CELL AMPLIFICATION

Recent advance in techniques for gene insertion, DNA amplification and the correction of gene defects are now available for analysis of embryonic cells (Thompson, Clarke, Pow, Hooper and Melton, 1989; Joyner, Skarnes and Rossant, 1989; Zimmer and Gruss, 1989). However, as discussed earlier there is a distinct need to amplify the signal from the DNA of the biopsied cells and to have a further sample of that DNA should the results be equivocal. Therefore, (a) the biopsied material needs to be at least double the quantity required for the initial analysis, to enable reanalysis of the cell(s), or (b) the original biopsied cells must replicate at least twice in vitro to permit further analysis to be undertaken if required.

Initial studies on the culture of isolated blastomeres of the 2–4 cell mouse embryo using M16 medium (Whittingham, 1971) has shown that it can support replication*, however this medium was not successful for culturing isolated human blastomeres. Studies in our laboratory have utilised a variety of medium, conditioned medium and substrates for the culture of human embryos and cells. When two blastomeres from a 4-cell human embryo are cultured in Hams F-12 with L-glutamine, or in amniotic fluid, using micro-drops of media on fibronectin or gelatin coated dishes, the blastomeres replicate at least two to three times.

The patterns in which they cleave follow examples previously described for isolated cultured mouse blastomeres (Graham and Lehtonen, 1979). A number of the blastomeres which have 3–4 cell contacts formed a "crescent" shaped pattern, others have only one replicating blastomere and form a small group of cells approximately 1/4 of the original blastomere size (Fig. 1). Isolated abembryonic trophectoderm cells from the human blastocyst can replicate at least twice (Fig. 2), single trophectoderm cells within the zona which are cultured in amniotic fluid also replicate and form clusters of cells with multiple cell contacts (Muggleton-Harris & Findlay, unpublished results). Substrates appear to be extremely important with regard to modifying cell growth, replication and behaviour (Bowersox or Sorgenti, 1987; Muggleton-Harris and Higbee, 1987). Embryonic stem (ES) cell lines, in many instances, are derived on feeder layers of cultured cells which provide both substrate and growth factors (Robertson, 1987). A combination of Hams F-12, increased fetal bovine serum, conditioned media and the use of irradiated feeder layers was found to be successful for cloning individual human embryonic fibroblasts which were reconstructed from isolated karyoplasts and cytoplasts (Muggleton-Harris and Hayflick, 1976). A variety of

---

* Since presenting this data, a recent report by Wilton and Trounson (1989), has shown that a single blastomere from a mouse embryo can replicate and form a monolayer of cells (approx 10–30 cells). The blastomere(s) were cultured in amniotic fluid on different extra-cellular matrix components, one of which was fibronectin. These results support our observations in this manuscript on the positive aspects of varying both substrate and culture medium to encourage maximum growth potential of biopsied cells. L. G. Wilton & A. O. Trounson (1989), Biopsy of preimplantation of mouse embryos: Development of micro-manipulated embryos and proliferation of single blastomeres in vitro. Biol. Reprod. 40, 145–152.

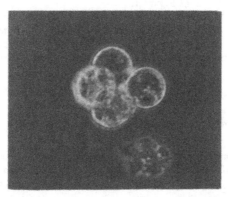

Fig. 1. Two blastomeres from a four-cell human embryo cultured in Hams F-12 medium, one blastomere has replicated twice forming four smaller cells.

conditions such as fibronectin, gelatin or agar-coated dishes have been used to encourage the growth of embryonic and extra-embryonic cells.

Growth and replication of subsets of cells dissected from the blastocyst have been obtained from specific experiments to study the replicative potential of the trophectoderm of the mouse pre-implantation embryo. Various dissections of the expanded blastocyst have been undertaken to examine whether the regulatory controls e.g. cell/cell interactions, blastocoele factors maintaining cell number and replication of both embryonic and extraembryonic cells could be manipulated. Previous work has shown that embryonic cells respond to factors and cell/cell interactions which exist at the blastocyst stage in the mouse embryo (Spindle, 1978; Ducibella, Albertini, Adriaansen and Biggers, 1975; Pedersen and Spindle, 1980). Regulation of cell growth and differentiation during early cleavage of embryos and at organogenesis can take place at the cellular level (Muggleton-Harris, Hardy and Higbee, 1987), and via cytoplasmic and or nuclear factors (Muggleton-Harris, 1988; Muggleton-Harris and Brown, 1988; Pratt and Muggleton-Harris, 1988; Muggleton-Harris,

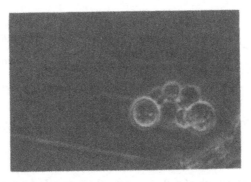

Fig. 2. Replicating isolated abembryonic trophectoderm cells in vitro.

Table 1. The effect of conditioned media on the manipulated components of a 3.5–4 day blasto-cyst. Dissection similar to that of Figure 1.

Medium	Degenerate ICM + Troph	Polar Troph TET cells	Reformed blastocyst or cell outgrowth
BRL	5 (9%)	8 (14%)	43 (77%)
BRL/CVS	2 (8%)	7 (28%)	16 (64%)
CVS	2 (7%)	13 (45%)	14 (48%)

This data is compiled from separate experiments using similar staged blastocysts and condi-tioned media.

BRL media is optimal for outgrowth. BRL/CVS media is optimal when both TET cells and out-growths are required. CVS media is optimal when TET cells are required.

Note that the level of degeneration following manipulation of the ICM and trophectoderm com-ponent are similar for each medium.

Whittingham and Wilson, 1982). To reduce or obviate possible factors effecting the growth potential of the trophectoderm and ICM cells, a varying number of the mural trophecto-derm cells were removed from the blastocyst by micromanipulative procedures (as shown in Fig. 3), this process also achieved an opening of the blastocoele environment thus dis-rupting the cell/cell interactions and factors within the blastocyst. The growth of the cells following the manipulative procedures were then examined in vitro. The relative benefits of various conditioned media on the proliferation of extraembryonic and embryonic cells fol-lowing the removal of mural trophectoderm from the blastocyst were also compared (Table 1); the results demonstrate that chorionic villus sample CVS conditioned media is optimal with regard to the proliferation potential of TET cells (45%). The BRL conditioned media appears to be important for blastocyst/ICM viability. Fifty percent of the manipulated blastocysts were able to reform expanded blastocysts following the removal of 1/4–1/2 of the mural trophectoderm. Whereas in CVS media only 27% were able to reform blasto-cysts. A combination of BRL/CVS conditioned media is optimal when both ICM and TET cells are required. The level of degeneration of isolated ICM/trophectoderm is similar for all

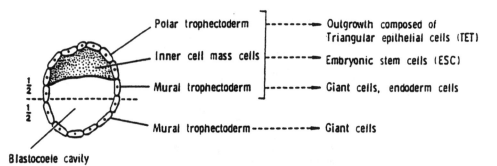

Fig. 3. Manipulation and growth of cells following the removal of mural trophectoderm from an expanded blastocyst.

three types of conditioned media. In order to identify and distinguish ICM and trophectoderm derived cell populations, some embryos were labelled with a viable fluorescent cell
marker. This procedure has been previously used by Fleming and George (1987) for cell
lineage studies during blastocyst formation in the mouse. After manipulation individual
blastocysts and/or the ICM/Trophectoderm components were cultured on gelatin coated 60
mm plastic petri dishes in small drops of medium under paraffin oil. The development,
morphology, replication and behaviour of the blastocysts ICM and trophectoderm were recorded daily. Viability and development of control non-operated zona-intact and zona-free
blastocysts exposed to similar experimental and culture conditions were also recorded. This
established a base-line for comparing the development and growth of blastocysts and manipulated components in each experiment.

The effect on cell growth and behaviour of the ICM/trophectoderm cells following the
removal of varying amounts of mural trophectoderm from the expanded blastocysts are
shown in Fig. 3. A small triangular-shaped epithelial trophectoderm (TET) cell was observed within 12–24 hours of culture following the removal of the majority of mural trophectoderm. These cells are not observed when a small amount of mural trophectoderm
(Fig. 4) is removed or when an intact blastocyst is cultured. The inner cell mass, in most
instances of manipulation and blastocyst culture can give rise to embryonic stem cell
(ESC) lines, and when the majority of mural trophectoderm is removed it has been noted
that the ICM/trophectoderm component attaches very quickly (12–24 hours) to the substrate. When the mural trophectoderm cells are present in large numbers attachment takes
24–48 hours and giant cells, endoderm and ICM cells are observed to form part of the outgrowth in vitro similar to that of non-manipulated blastocysts (Fig. 5). The dissected mural trophectoderm in all cases form giant cells and if there are sufficient numbers form a
vesicle prior to giant cell formation. Comparisons of the morphology and rapid growth
pattern of the TET cells identified in both FITC labelled and unlabelled experiments confirms that these cells are derived from the polar trophectoderm. By day five to six in culture giant cells have been observed on the periphery of TET cell colonies whereas mural
trophectoderm derived giant cells observed in blastocyst outgrowth form on the periphery

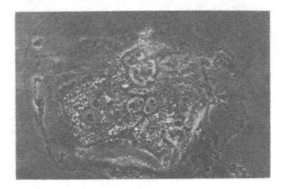

Fig. 4. Cultured mural trophectoderm forming giant cells.

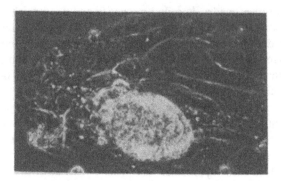

Fig. 5. Outgrowth from a cultured mouse blastocyst showing endoderm and ICM cells with peripheral giant cells.

of the ICM. The TET colony giant cells have little or no FITC label due to the rapid proliferation and growth of the original labelled trophectoderm cells. The TET cell population is in excess of 200 cells in Fig. 6 (Muggleton-Harris, 1989).

To establish that the manipulation and culture conditions used for these experiments did not present problems with regard to proliferation, growth and viability of the cells obtained in vitro, embryonic stem cells (ESC) derived from individual manipulated ICM/trophectoderm components were injected into suitable host blastocysts. The integration of the cells into the inner cell mass and contribution to the adult tissues of live young following embryo transfer to pseudo-pregnant recipients could be compared with established ESC used for previous studies (Hooper, Hardy, Handyside, Hunter & Monk 1987). The manipulated blastocysts ESC were found to contribute to the tissues of the adult mice (Muggleton-Harris, 1989).

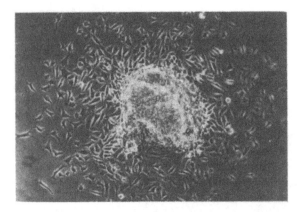

Fig. 6. Triangular eipthelial trophoblast (TET) cells derived from the polar trophectoderm.

Recent studies have indicated that polar trophectoderm cells can be successfully dissected from above the ICM of expanded blastocysts, thus FITC labelled polar trophectoderm can be injected in or co-cultured with various components of the blastocyst to study their response to the environment. Future studies on cultured mouse TET cells derived from the polar trophectoderm will establish whether there are similar morphological and cell surface markers occurring in vitro which can be correlated with those cells from which the ectoplacental cone (EPC) is derived in vivo e.g. electron microscope studies have described morphological characteristics of different trophectodermal cell types which give rise to the EPC (Mehrotra, 1988), and that cells of the cytotrophoblast and syntrophoblast have different transferrin receptors (Bierings, Anderson & Van Dijk, 1988), and proteolytic activities (Owers & Blandau 1971).

MANIPULATION OF THE HUMAN PRE-IMPLANTATION BLASTOCYST

Studies to encourage cell growth of embryonic and extra-embryonic cells of the human blastocyst have been undertaken in collaboration with C. Graham, Oxford University and V. Bolton, Kings College Hospital London, U.K. Should these studies prove successful it would establish a basis for diagnosing disorders in human pre-implantation embryos, and also establish human embryonic and extra-embryonic cell cultures for medical research. Expanded blastocysts from cultured 2–4 cell embryos and "spare" blastocysts (grade 2 or 3) have been micro-manipulated to isolate various components. Some blastocysts had the zona and/or approximately one half of the abembryonic trophectoderm removed, the remaining ICM plus trophectoderm cells was cultured on gelatin or fibronectin coated dishes in

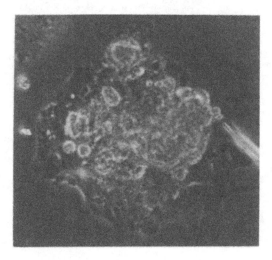

Fig. 7. Cultured manipulated ICM/trophectoderm component from a human blastocyst.

micro-drops of BRL/CVS media. The ICM/trophectoderm component attached to the substrate and formed an outgrowth (Fig. 7), there are far less giant cells when compared with those in Fig. 8, the ICM component appears to be at a suitable stage for disaggregation. A dezonaed non-manipulated human blastocyst outgrowth can be seen in Fig. 8, giant cells can be seen around the periphery of the ICM component. Other blastocysts were used to isolate five abembryonic trophectoderm cells, a similar number as that used for the mouse biopsy work, the manipulated blastocyst remained in the zona but degenerated after two to three days in vitro. However, if this procedure was repeated by thinning the zona with Acid-tyrodes prior to manipulation, the blastocyst hatched through the zona at the area of manipulation. When a blastocyst is manipulated within the zona to remove the ICM and most of the trophectoderm, individual trophectoderm cells which remain within the zona, can replicate at least three times and form small aggregates of cells with multiple contacts, when cultured in amniotic fluid. The ICM/trophectoderm component cultured in BRL/CVS medium quickly attaches (within 12 hrs) to the fibronectin surface, replicates and forms a flattened aggregate (Muggleton-Harris & Findlay, unpublished results). It is obvious from these initial studies that further work needs to be undertaken with a variety of culture conditions to enhance further cell replication. We have however, demonstrated that both extra-embryonic and embryonic cells of the human preimplantation embryo can be induced to proliferate in vitro. Therefore it is possible that one will obtain sufficient cells for re-analysis of biopsied material of genetic defects using cellular, cytogenic, biochemical and molecular techniques.

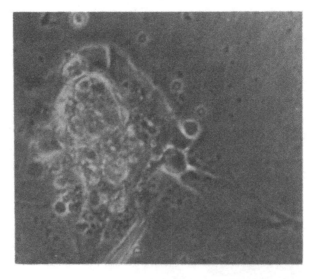

Fig. 8. A dezonaed non-manipulated human blastocyst outgrowth.

REFERENCES

1. M. B. Bierings, H. J. Adriaansen, and J. P. Van Dijk, The appearance of transferring receptors on cultured human cytotrophoblast and in vitro formed syncytiotrophoblast, *Placenta.*, 9:387-396 (1988).
2. J. C. Bowersox and N. Sorgente, Differential effects of soluble and immobilised fibronectins on aortic endothelial cell proliferation and attachment. *In vitro cell & develop., Biol.*, 23:759-764 (1987).
3. T. Ducibella, D. F. Albertini, E. Anderson, and J. D. Biggers, The preimplantation mammalian embryo; characterization of intercellular junctions and their appearance during development, *Dev. Biol.*, 45:231-250 (1975).
4. S. B. Fishel and R. G. Edwards, Human chorionic gonadotrophin secreted by preimplantation embryos in vitro, *Science.*, 223:816-818 (1984).
5. T. P. Fleming and M. A. George, Fluorescent latex microparticles: a non-invasive short term cell lineage marker suitable for use in the mouse early embryo, *Wilhelm Roux Arch. Develop. Biol.*, 196:1-11 (1987).
6. C. F. Graham and E. Lehtonen, Formation and consequences of cell patterns in preimplantation mouse development, *J. Embryol. & Exp. Morph.*, 49:277-294 (1979).
7. J. K. Heath and A. G. Smith, Growth factors in embryogenesis, Brit. Medical Bulletin, 45,2: 319-336 (1989).
8. A. L. Joyner, W. C. Skarnes and J. Rossant, Production of a mutation in mouse En-2 gene by homologous recombination in embryonic stem cells, *Nat.* 338:153-155 (1989).
9. M. Hooper, K. Hardy, A. Handyside, S. Hunter and M. Monk, HPRT-deficient (Lesch-Nyhan) mouse embryos derived from germ line colonization by cultured cells, *Nature*, 326:292-295 (1987).
10. A. Lopata and D. Hay, Human embryo and blastocyst culture, *Fertil. Steril.*, (in press) (1989a).
11. A. Lopata and D. Hay, The potential of surplus human embryos to form blastocysts, hatch from their zona and secrete human chorionic gonadotrophin in culture (Submitted to *Human Reproduction*) (1989b).
12. P. K. Mehrotra, Ultrastructure of mouse ectoplacental cone cells, *Biol. Struct. Morphog.*, 1:63-68 (1988).
13. M. Monk, A. Handyside, A. L. Muggleton-Harris, and D. G. Whittingham, Preimplantation sexing and diagnosis of HPRT deficiency in mice by biochemical assay, *Amer. Jour. of Med. Genet.* (in press) (1989).
14. M. Monk, A. L. Muggleton-Harris, E. Rawlings, and D. G. Whittingham, Preimplantation diagnosis of HPRT-deficient male and carrier female mouse embryos by trophectoderm biopsy, *Human Reprod.*, 3:377-381 (1988).
15. A. L. Muggleton-Harris, Cellular factors related to cessation of proliferation, and differentiation, *in*: "Growth Control during Cell Ageing," E. Wang and H. R. Warner, eds., CRC Press Florida USA (1988).
16. A.L. Muggleton-Harris, Proliferation and growth potential of polar trophectoderm cells following manipulation of the preimplantation mouse blastocyst (Submitted to development) (1989).
17. A. L. Muggleton-Harris and J. J. G. Brown, Cytoplasmic factors influence mitochondrial reorganisation and resumption of cleavage during culture of early mouse embryos, *Hum. Reprod.*, 3,8,1020-1-28 (1988).
18. A. L. Muggleton-Harris and L. Hayflick, Cellular ageing studied by the reconstruction of replicating cells from nuclei and cytoplasms isolated from normal human diploid cells, *Exp. Cell Res.*, 103, 321-330 (1976).
19. A. L. Muggleton-Harris and N. Higbee, Factors modulating mouse lens epithelial cell morphology with differentiation and development of a lentoid structure in vitro, *Develop.* 99:25-32 (1987).

20. A. L. Muggleton-Harris, K. Hardy, and N. Higbee, Rescue of developmental lens abnormalities in chimaeras on noncataractous and congenital cataractous mice, *Development*, 99:473-480 (1987).

21. A. L. Muggleton-Harris, D. G. Whittingham, and L. Wilson, Cytoplasmic control of preimplantation development in vitro in the mouse, *Nature.*, 299:460-462 (1982).

22. N. O. Owers and R. J. Blandau, Proteolytic activity of the rat and guinea pig blastocysts in vitro, *in:* "The Biology of the Blastocyst," R. J. Blandau, ed., University of Chicago Press USA, 207-221 (1971).

23. R.A. Pedersen and A. I. Spindle, The role of the blastocoele microenvironment in early mouse embryo differentiation, *Nature* (London), 284:550-552 (1980).

24. H. P. M. Pratt and A. L. Muggleton-Harris, Cycling cytoplasmic factors which promote mitosis in the cultured two-cell mouse embryo, *Development*, 104:115–120 (1988).

25. E. J. Robertson, Embryo-derived stem cell lines, *in:* "Teratocarcinomas and Embryonic Stem cells, a practical approach," E. J. Robertson, ed. IRL Press, Oxford, U. K. 71–112 (1987).

26. A. G. Smith and M. L. Hooper, Buffalo Rat liver cells produce a diffusible activity which inhibits the differentiation of murine embryonal carcinoma and embryonic stem cells, *Dev. Biol.*, 121:1-9 (1987).

27. A. I. Spindle, Trophoblast regeneration by inner cell masses isolated from cultured mouse embryos, *J. Exp. Zool.*, 203:483-489 (1978).

28. S. Thompson, A. R. Clarke, A. M. Pow, M. L. Hooper and D. W. Melton, Germ line transmission and expression of a corrected HPRT gene produced by gene targeting in embryonic stem cells, *Cell*, 56:313–321 (1989).

29. D. G. Whittingham, Culture of mouse ova, *J. Reprod. Fertil.*, (Suppl.), 14:7–21 (1971).

30. A. Zimmer and P. Gruss, Production of chimaeric mice containing embryonic stem (ES) cells carrying a homeobox Hox 1.1 allele mutated by homologous recombination, *Nat.* 338:150–153 (1989).

PREIMPLANTATION GENETIC DIAGNOSIS:
APPROACHES AND CURRENT STATUS

Yury Verlinsky, Eugene Pergament, and Charles M. Strom

Reproductive Genetic Institute, Reproductive and Medical Genetics
Department of Obstetrics and Gynecology
Illinois Masonic Medical Center, Chicago, Illinois

INTRODUCTION

The standard approaches to the prenatal diagnosis of genetic diseases are chorionic villus sampling, amniocentesis and, more recently, percutaneous blood sampling. Each of these procedures is associated with obstetrical risks to the mother and to the fetus. And, if a genetic abnormality is detected, the only alternative to continuing the pregnancy is an elective termination. For parents at high risk for a genetically abnormal conception, such as hemophilia, muscular dystrophy, or cystic fibrosis, the possibility as well as the experience of repeated pregnancy terminations inflict severe mental trauma. This would be avoidable if a genetic diagnosis could be made prior to implantation.

APPROACHES TO PREIMPLANTATION GENETIC DIAGNOSIS

A model for the preimplantation diagnosis of human genetic disorders has been previously proposed (Verlinsky et al., 1987). Patients will undergo ovarian stimulation to recruit several follicles which would then be transvaginally aspirated under ultrasound guidance and later inseminated. Each embryo would be cultured in vitro to the four-cell stage and then would undergo biopsy for the genetic analysis of the excised cells. If the biopsy is carried out at the four-cell stage and the genetic analysis could be completed within 24 hours, the unaffected embryos would be transferred to the patient in the same cycle. If the

Advances in Assisted Reproductive Technologies, Edited by
S. Mashiach *et al.*, Plenum Press, New York, 1990

Table 1. Summary of Experimental Results of Preimplantation Genetic Diagnosis at the Four-Cell (non-differentiated) Stage using Triploid Embryos

Total number of pre-embryos available:	50
Survival after microsurgery: Number of half-embryos successfully tissue cultured and available for genetic analysis	41
Number of half-embryos with development in-vitro	
Eight cell stage	2
Sixteen cell stage	3
Morula	35
Blastula	6
Minimum percent of embryos available for both genetic analysis and transfer	67%

biopsy is performed at a later time and/or the genetic analysis took more than 24 hours, the embryos would be cryopreserved until the genetic diagnosis was completed. Since preimplantation genetic diagnosis involves the transfer of only unaffected embryos, the possibility of initiating a genetically-abnormal pregnancy later requiring elective termination is avoided. As with any new approach to prenatal diagnosis, the two issues to be addressed are the safety of the biopsy procedure and the accuracy of the genetic analysis of the excised cells.

We wish to describe our experiences with preimplantation genetic diagnosis in humans. Two approaches have been employed: 1) biopsy of the pre-embryo at the four-cell or undifferentiated stage; and, 2) biopsy of the pre-embryo at the blastula or differentiated stage.

All of our experiments were conducted on human embryos rejected for transfer because of triploidy. Approximately 5% of all in vitro inseminated oocytes develop with three or more pronuclei, presumably because of polyspermy. The protocol to experiment on human triploid embryos rejected for transfer was evaluated and approved by the Institutional Review Board and the Ethics Committees of our institution.

EXPERIMENTAL RESULTS

The first set of experiments consisted of 50 triploid human embryos which underwent biopsy at the four-cell stage (Table 1). The biopsy instruments consisted of a holding pipette and an aspiration pipette, for the removal of the cells and their separation into two halves. The cells of each half-embryo were placed in an empty zona pellucida and cultured in vitro. Of 50 pre-embryos undergoing biopsy at the four-cell stage, 41 were successfully cultured in vitro as a monolayer and thereby were available for chromosomal, enzymatic

Table 2. Summary of Experimental Results of Preimplantation Genetic Diagnosis in the Blastula (differentiated) Stage using Triploid Human Embryos

Total number of blastula biopsies	18
Number established in culture:	12
Successful genetic analyses	
Karyotype	12
Hemoglobin gene analysis (PCR)	14
Survival in-vitro post-excision (> 2 days)	12

and DNA analyses. Of 50 pre-embryos individually placed in an empty zona pellucida, the short-term development of these triploid embryos was not impaired, as 41 reached the morula or blastula stages. Overall, 67% of the triploid embryos were technically available for both genetic analysis and transfer. Biopsy at the four-cell stage would allow transfer in the same cycle and would avoid the need for cryopreservation.

A second approach to the preimplantation diagnosis of genetic disease involves biopsy of the pre-embryo at the blastula stage. This approach can be viewed as an "early chorionic villous sampling" since trophectoderm tissue which develops into the placenta is excised. Biopsy of the trophectoderm tissue involves the following procedures: 1) the blastula is positioned by the holding pipette and glass needle prior to incision of the zona pellucida; 2) the glass needle is inserted into the zona pellucida; 3) herniation of trophectoderm tissue through incised zona pellucida occurs over a two hour period; and, 4) the glass scalpel or a microblade is used to excise the trophectoderm tissue. Following excision of the trophectoderm, there was some shrinkage of the entire embryonic and extraembryonic mass from the walls of the zona pellucida for a relatively short time and then recovery without any other morphological changes. The period of time required to complete the genetic analyses would require cryopreservation of the embryo and transfer in a subsequent cycle.

Table 2 summarizes the results of biopsies of trophectoderm tissues. Of 18 triploid embryos, 12 were established in tissue culture and thereby available for studies related to chromosomal aberrations and inborn errors of metabolism. In addition, DNA studies using the polymerase chain reaction were successfully conducted on excised tissue of 14 pre-embryos. Polymerase chain reaction or PCR is a technique for the amplification of specific sequences of DNA. One enormous advantage of PCR is the ability to analyze the DNA from minute quantities of DNA and we have developed techniques for the analysis of DNA from a single cell. The steps in the polymerase chain reaction involve 30 to 50 cycles of amplifying the DNA to a quantity whereby genetic analysis is direct and straightforward. For example, it is possible to isolate and amplify the DNA from a single embryonic cell to monitor for the presence of hemoglobin A and hemoglobin S. It must be emphasized that scrupulous attention must be given to avoid contamination of any reagent when analyzing a single cell.

CONCLUSION

In summary, we have performed a series of genetic analyses on human triploid pre-embryos which suggest that it will be possible to aspirate a cohort of follicles from women at high genetic risk, to biopsy and determine their individual genotypes, and then to transfer to these women only those embryos expected to produce a normal clinical outcome. This approach will be applicable for women at risk because of a familial chromosomal rearrangement or an inborn error of metabolism, and for those women at risk for a genetic disease amenable to analysis by DNA probes and recombinant DNA technology.

REFERENCE

Verlinsky, Y., Pergament, E., Binor, Z., and Rawlins,R., 1987, Genetic analysis of human embryos prior to implantation: future applications of in vitro fertilization in the treatment and prevention of human genetic diseases, in: "Future Aspects in Human In Vitro Fertilization," W. Feichtinger and P. Kemeter, eds., Springer-Verlag, Berlin.

THE FEASIBILITY OF PREIMPLANTATION-EMBRYO BIOPSY AND GENETIC SCREENING FOR PRENATAL DIAGNOSIS: STUDIES WITH MICE

Chris O'Neill,[a] Robert Lindeman,[b] Cynthia Roberts,[a] Ron Trent,[b] Urszula Kryzminska,[a] and Janice Lutjen[a]

[a] Human Reproduction Unit
Royal North Shore Hospital of Sydney, St. Leonards, 2065, NSW, and
[b] Clinical Immunology Research Centre
University of Sydney, Camperdown, NSW, 2006, Australia

INTRODUCTION

Prenatal diagnosis relies on the ability to biopsy sufficient material from the embryo or fetus to allow diagnosis of genetic disease without jeoparding the well-being of the embryo. Routinely, this is performed by amniocentesis in the second trimester of pregnancy or, more recently, by sampling the chorionic villus at 9–11 weeks of gestation. Using such techniques a broad range of inherited genetic diseases, aneuploidies and significant structural abnormalities of chromosomes can be diagnosed routinely.

In recent years, technical developments in the area of in vitro fertilization and molecular biology offer the real possibility of diagnosis of genetic disease in the preimplantation embryo. In this paper we describe our experience with the development of embryo biopsy techniques. These techniques have minimal detrimental effects on the viability of the embryo. The biopsy material itself may be used in classical cytogenetic analysis, as well as the new molecular biological approach of gene (DNA) amplification.

THE EFFECT OF EMBRYO BIOPSY ON EMBRYONIC VIABILITY

Two conditions required for embryo biopsy to be applicable for reliable genetic screen-

Advances in Assisted Reproductive Technologies, Edited by
S. Mashiach *et al.*, Plenum Press, New York, 1990

ing are: (1) the developmental potential of biopsied embryos is not impaired, and (2) sufficient cells are obtained to allow genetic diagnosis.

There have been a number of reports of successful embryo biopsy. Summers et al. (1988)[1] achieved pregnancies following biopsy of the marmoset monkey blastocyst. Mouse embryos were biopsied by cutting the mural trophectoderm of mouse blastocysts[2] or removal of blastomeres from the 8-cell stage embryo[3,4] to diagnose HPRT-deficiency. However, a relatively low pregnancy rate of biopsied embryos following transfer was achieved (30% for blastocyst and 40% for 8-cells). Biopsy of mouse embryos at the 4-cell and 8-cell stage, by removal of one blastomere resulted in good preimplantation development in vitro (94% of 4-cell biopsies and 80% of 8-cells developed to blastocysts)[5] and implantation following transfers to pseudopregnant recipients (53% for 4-cell biopsy).[6]

The advantage of biopsy at later developmental stages, such as the morula or blastocyst, is that relatively more cells (blastomeres) can be collected compared with biopsy at earlier developmental stages. It is therefore, essential to determine the developmental stage which allows maximum retrieval of cells at biopsy, with a minimum reduction in the pregnancy potential of the embryos following this procedure.

We compared the developmental potential, in vitro and in vivo, of mouse embryos biopsied at the 4-cell, 8-cell and morula stages.

Embryo Biopsy

Females of the Quackenbush mouse strain were superovulated. Embryos at the 4-cell, 8-cell and morulae stages were obtained and cultured in Quinn's medium with 3 mg BSA/ml. The 4-cell and 8-cell embryos were incubated in Ca^{++}/Mg^{++} free bicarbonate-buffered Quinn's medium with 3 mg BSA/ml for 2–3 h followed by short treatment (20 min) in Hepes-buffered Quinn's medium with 3 mg BSA/ml containing 5–7.5 μg cytochalasin B/ml and then biopsied. Morulae were prepared for biopsy by removing the zona pellucida by brief exposure (2–5 min) to 0.5% pronase (Calbiochem) or acidic Tyrode's solution of pH 2.5 for 30–50 sec. Biopsy was performed a short time after the zona pellucida was removed.

Micromanipulation was performed in 20 μl drops of Hepes-buffered Quinn's medium containing 3 mg BSA/ml under paraffin oil on a siliconised depression slide. At the 4- and 8-cell stage one (4-cell) or two (8-cell) blastomeres were removed. This procedure was performed using two types of micropipettes: (1) a holding pipette, which was used to hold the embryo in position, and (2) a biopsy pipette which, for 4- and 8-cell embryos had an internal diameter of approximately 10–20 μm and was shaped and ground to allow easy access through the zona pellucida. For morulae a micropipette was used as a knife and allowed approximately five cells to be sliced from the zona-free embryo.

Table 1. The Implantation Rate on Day 8 of Pregnancy

Developmental Stage	Implantation rate (% embryos implanted)		
	Biopsy (B)	Sham-control (S.C.)	Intact-control (I.C.)
4-cell	93/212 (49)[a]	35/61 (57)	135/231 (58)
8-cell	149/182 (82)	46/60 (77)	127/145 (88)
Morula	29/139 (21)[b]	72/212 (34)[c]	167/256 (65)

[a]: X^2 analysis, $P < 0.01$ (B vs I.C.)
[b,c]: X^2 analysis, $P < 0.001$ (b, B vs I.C.; c, S.C. vs I.C.)

For each experimental group there were two control groups: (1) sham-controls were of the same developmental stage and underwent the same treatments as biopsied embryos, including having the zona pellucida removed (morulae) or punctured (4-cells and 8-cells), but without removal of embryonic cells and (2) intact-controls were embryos collected at the same preimplantation stage as biopsied embryos and which underwent only undisturbed culture in vitro to the bastocyst stage, and transfer to recipient females.

Culture and Assessment of Viability of Biopsied Embryos

All embryos were placed in 10 μl drops of Quinn's medium and cultured to the blastocyst stage. The resulting expanded blastocysts were transferred to the uterus of Day 3 pseudopregnant recipients. Biopsied embryos were transferred to one uterine horn with either sham- or intact-controls placed in the contralateral horn. This was in a random manner to avoid bias. Both implantation rate and fetal viability were assessed by necroscopies on Days 8 or 17 of pregnancy respectively.

The implantation rate (Table 1) in recipients which achieved at least one implanting embryo is expressed as the number of implantation sites divided by the number of embryos transferred. There was no difference $(P > 0.05)$ in the implantation rates for biopsy, sham- and intact-controls for the 8-cell stage embryo group. The apparently lower implantation rate for sham-control was due to a smaller number of experiments using this control. For both 4-cell and morula stage embryo groups, biopsy impaired the implantation rate compared with intact controls (4-cell, $P < 0.01$). In the case of morulae, the implantation rate for both biopsy and sham- control groups was significantly lower $(P < 0.001)$ than intact-controls.

For both the 8-cell and morula stage embryos there was a nonsignificant trend $(P > 0.05)$ for higher resorption rates of biopsied embryos which had implanted compared to their controls. This trend resulted in significantly $(P < 0.05)$ fewer viable fetuses being present on Day 17 in both groups, compared to their controls (Table 2). Figure 1 shows

Table 2. The Implantation Rate on Day 17 of Pregnancy

Developmental Stage		Implantation rate (% implanting embryos)	Resorption rate (%)	Viability rate (% living fetuses)
8-cell	Biopsy (B)	64/98 (76)	31	44/98 (45)
	Sham-control (S.C.)	10/13 (77)	10	9/13 (68)
	Intact-control (I.C.)	45/66 (68)	18	37/66 (56)
Morula	Biopsy (B)	36/108 (33)	44	20/108 (19)
	Sham-control (S.C.)	18/78 (23)	22	14/78 (18)
	Intact-control (I.C.)	88/186 (47)	38	50/186 (27)

[a,b] ^{2}analysis, $P < 0.001$ (a, B vs I.C.: b, S.C. vs I.C.)
[c,d] ^{2}analysis, $P < 0.05$ (B vs I.C.)

Implantation (total number of implantation sites) and viability (excluding resorption sites) rates of 8-cell and morula embryos on day 17.

that while there was no difference in the fetal and placental weights of 8-cell biopsies versus controls, biopsy of morulae resulted in fetuses which were significantly smaller than their corresponding controls ($P = 0.02$).

The results of this study confirm that successful embryo biopsy can be performed at the 4-cell, 8-cell and morula stage in mice with subsequent successful pregnancy. It shows, however, that there were substantial differences in the developmental potential of embryos biopsied at different stages.

Besides the influence of the developmental stage at which the biopsy was performed, the experimental procedure may have resulted in a better overall development of 8-cells. In the case of 4-cell embryos, overnight culture from 2-cells was performed before the experiment was started, while 8-cells were collected fresh from the reproductive tract. This extended period in culture for the 4-cell group may have adversely affected their developmental potential.

The results suggest that the method used to perform the biopsy had an impact on the implantation rate. Puncture of the zona pellucida at the 4- and 8-cell stage by the pipette had only a slight effect on their potential for implantation (Table 1). However, removal of the zona at the morula stage had a marked effect on the implantation rate with a 48% reduction of sham-controls versus intact-controls. Thus a significant effect on the success of biopsy at the morula stage is the removal of the zona pellucida.

While the major adverse effects of biopsy were expressed as failure of implantation (4-cell and morula biopsy, Table 1 and 2), biopsied morulae also resulted in smaller fetuses (Figure 1). This may be due to a failure of these fetuses to compensate by this time of development (17th day of pregnancy) for the reduced cell number in the morulae. Totipotency of cells no longer occurs at the morula stage and differentiation into presumptive trophecto-derm and inner cell mass (ICM) has taken place, without visible morphological changes and differences between cells.[7] Therefore, it is impossible to judge whether presumptive embryonic or extraembryonic tissue is impaired by biopsy at this stage. Removal from the embryo or damage during biopsy of a few cells from the presumptive ICM may be directly reflected in poor implantation rate on Day 8 and decreased fetal weight on Day 17. The results suggest that the 8-cell stage embryos are most likely to be useful for biopsy before implantation. This stage is compatible with the logistics of the human IVF procedures. The 8-cell stage has the disadvantage of only providing two blastomeres for analysis. However, with the advent of gene amplification techniques this will be sufficient for the detection of gene defects.

CYTOGENETIC ANALYSIS OF BIOPSIES RESULTING FROM PREIMPLANTATION EMBRYOS

Cytogenetic analysis might be a potentially useful tool for the screening of embryo biopsies in cases of a parental chromosome rearrangement, for sex determination in cases where DNA studies are uninformative or in combination with in situ hybridization with specific DNA probes for chromosome indentification. Such techniques are entirely depen-

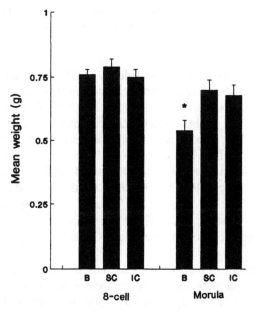

Fig.1. The weight (mean ± sem) of fetuses derived from 8-cell and morula stage embryo on day 17 of pregnancy (B - biopsy, S.C. - sham-control, I.C. - intact-control).

dent upon the production of embryo biopsy material which yields successful karyotypes with a high degree of efficiency. We therefore examined whether the biopsied material resulting from the studies described in the previous section were useful for karyotyping and in particular the influence of embryo developmental stage (and hence type of biopsy procedure) and various methods of preparation of biopsies on the success.

Blastomeres from 4- and 8-cell embryos were divided into one of three treatment groups.

A.   Culture (18 h) in Quinn's medium containing colchicine.
B.   Preculture (4–6 h) in Quinn's medium followed by exposure to colchicine overnight (16 h).
C.   Overnight culture (18 h) to Quinn's medium, followed by exposure to colchicine for 8 h.

Morula biopsy fragments were placed into one of the four treatment groups:

1.   Culture (18 h) in Quinn's medium containing colchicine.
2.   Short culture (4–6 h) in Quinn's medium containing colchicine.
3.   Preculture (2–4 h) in Quinn's medium, followed by exposure to colchicine overnight (16 h).
4.   Preculture (2–4 h) in Quinn's medium, followed by exposure to colchicine for 5–6 h.

Culture was conducted in 10 μl drops of Quinn's medium under oil with colchicine (Sigma Chemical Co., St. Louis, MO, USA) concentrations of 0.005, 0.01, 0.02, 0.05, 0.1, 0.25, 0.5 μg/ml.

Biopsies were treated with a hypotonic solution of 1% (w/v) sodium citrate for 90 seconds (1/4 and 2/8 biopsies) or five minutes (morula biopsies). These were then transferred with a siliconized micropipette to a 5 μl drop of 9:3:1 (1% sodium citrate:methanol:acetic acid) on a glass slide. The cytoplasm immediately darkened and the biopsy became loosely attached to the slide. A 5 μl drop of fixative (3:1, methanol:acetic acid) was gently dropped onto the slide at the edge of the first drop. As the biopsy flattened, a further 2–3 drops of fixative were added. Slides were air dried and stained for five minutes with 5% Giemsa (Gurrs R66 improved) or aged at 30°C for 1–2 days and G-banded using 2 × SSC at 60°C for 2 h, followed by 0.25% (w/v) trypsin at 4°C for 25–30s and stained as above. Biopsy preparations were viewed under 100× oil immersion and scored for the number of metaphase spreads. The chromosomes were counted and G-banded metaphases karyotyped.

A significant proportion of biopsies from the three developmental stages and treatment groups produced only interphase nuclei. The numbers of metaphases and interphase nuclei were compared between intact embryos and biopsies to determine whether the mitotic index was decreased after biopsy or the interphase nuclei observed were a result of the asynchronous division of blastomeres at the corresponding developmental stage (results not shown).

Table 3. Results of Cytogenetic Analysis of Biopsies With at Least One Metaphase

TREATMENT	Chromosome Count							Karyotype					
	n	$\bar{X}$	N	40^S	80	$<40^T$	39^A	$39/41^B$	40,XX	40,XY	60,XXY	80,XXXX	?
1/4 BIOPSIES	54	0	37	14	0	6	0	0	5	4	1	0	7
2/8 BIOPSIES	134	-	83	28^c	1	10	1	1	24	20	0	1	5
MORULA BIOPSIES	70	1.5	47	13	0	2	0	0	8	9	0	1	14

n = Total number of metaphases per biopsy
X = Average number of metaphase per biopsy
N = Number of biopsies with at least one metaphase
40^S = Analysis of solid stained metaphases
40^T = Hypoploidy probably due to technical loss
39^A = Hypoploidy confirmed in > 1 metaphase
$39/41^B$ = Misdivision in culture, confirmed by analysis of two metaphases
? = Overlapping of metaphases prevented accurate counting and analysis
c = One metaphase endoreduplicated

The total number of observed metaphases in biopsies exposed to colchicine for 18 h compared with the expected numbers based on intact embryos of the same developmental stage was significantly reduced. Thus the cleavage potential of blastomeres is reduced after biopsy. A short recovery in preculture did not raise the mitotic index but rather reduced it further due to the division of some blastomeres in that short period which did not reach the next mitotic cycle during the colchicine exposure.

Solid stained chromosomes were analysed during the early phase of the development of these techniques. Initial attempts at G-banding using the routine banding methods of this laboratory[8] were unsuccessful. Slides are normally aged at 60°C overnight but it was found that the chromosomes of these developmental stages were denatured, resulting in C- rather than G-banding. Aging of slides at 37°C greatly improved the quality of banding. 4- and 8-cell biopsies also exhibited a marked difference in chromosome morphology between treatments A and B. Immediate exposure to colchicine resulted in fuzzy chromosomes while those which were allowed a short recovery period in preculture resulted in chromosomes which could be karyotyped by G-banding.

The results of cytogenetic analyses of the three stages are listed in Table 3 (Figure 2). 4-cell and 8-cell biopsies were generally well spread with few chromosome overlaps. Overlapping was more often observed in morula biopsies and further complicated by overlapping cells caused by the compaction of these biopsies. Chromosome loss due to scattering as a result of cell breakage was greatest in 4-cell biopsies, which were particularly sensitive to hypotonic treatment.

The loss rate which was experienced varied markedly between experiments and did not appear to be related to the various culture regimens. Rather it may reflect embryo quality and/or operator skill on the day of biopsy or fixation.

The preculture treatment produced the best quality metaphases for cytogenetic analysis at all three development stages. The greatest success with karyotyping was achieved using 8-cell biopsies (Table 3). This was expected on the basis of the highest mitotic index for intact 8-cell embryos for the three developmental stages which were biopsied (results not shown). Although the mitotic index of precultured 4- and 8- cell biopsies was lower than that of biopsies placed directly into colchicine (probably due in part to some division during preculture), the number of successfully analysed biopsies was not significantly different (P > 0.05). As the mitotic index is less than 1.00 (8-cell embryos 0.85 ± 0.03), it is difficult to envisage a high success rate using these techniques. Longer colchicine exposure or synchronization may increase the metaphase yield but prolonged exposure to colchicine or other cell-cycle blocking agents may be detrimental.

The frequency of tetraploidy (3/167, 1.8%) (Table 3) observed in the biopsies is higher than seen in embryos of this species (unpublished results). One metaphase was obtained af-

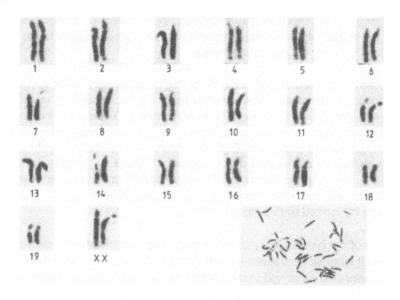

Fig. 2. A Karyotype from a biopsy of an 8-cell mouse embryo showing distinct G-banding. Insert shows metaphase spread.

ter direct exposure of the biopsy to colchicine however, the other two metaphases were from cultured biopsies and tetraploidy may have resulted from endoreduplication during culture, rather than being representative of the embryo karyotype. However, as only one metaphase was obtained in both cases the result could not be confirmed.

Cytogenetic studies of products of conception and the use of chorionic villus sampling for the prenatal diagnosis of chromosome errors have yielded a great deal of information on the incidence of mosaicism in both the trophoblast and embryo.[9, 10] Diagnosis at the preimplantation stage of embryonic development does not take into account errors in subsequent division. Further, it is preferable to base a cytogenetic diagnosis on the analysis of more than one metaphase, particularly with the chance of chromosome overlap or scatter.

If the success rate of this technique could be perfected, it might provide a good method of preimplantation diagnosis in high risk translocation carriers or sex-linked recessive traits without informative DNA studies. This will require the development of techniques which greatly enhance the proliferative capacity of biopsies in vitro.

For the majority of cases currently presenting for cytogenetic prenatal diagnosis (referred for advanced maternal age), this technique does not provide a viable alternative to chorionic villus sampling and amniocentesis since clinical testing requires a high degree of success. Even if a successful biopsy were made on every attempt, the mitotic index for these three developmental stages in mice suggests that metaphases will not be obtained for each sample. Further, the chromosomes may be overlapped or lost in preparation and so not allow a detailed analysis of the complete karyotype. However, if these technical problems are overcome, good quality G-banding can be achieved and may be a useful tool for research where some loss of embryos does not present a significant problem.

GENETIC ANALYSIS OF BIOPSIES BY GENE AMPLIFICATION

Conventional methods for DNA analysis include identification of modifications in DNA sequence by Southern blotting, using the hybridization of a DNA probe of known sequence with its complimentary copy in native DNA: In situ hybridization of such probes with the complimentary DNA within cells in preparations for microscopic examination will identify more extensive DNA deletions. All of these techniques require substantial amounts of tissue. Sufficient material can be obtained from sources such as fetal blood, amniotic fluid and the chorionic villus.

However, biopsies of the preimplantation embryo do not provide sufficient material. Culture of the biopsies is not efficient enough to offer any real hope that adequate material could be produced for conventional DNA analysis. The recent development of a technique for amplification of specific target sequences of DNA by the process known as the polymerase chain reaction (PCR) offers the possibility of reliable genetic screening with one or more cells.[11-18]

PCR is able to produce selective enrichment of specific DNA sequences by many orders of magnitude, greatly facilitating subsequent analysis. Knowledge of the base sequence at the boundaries of the segment to be amplified is a prerequisite for PCR. The reaction involves a three-step cycling process: (i) denaturation of double-stranded DNA, (ii) annealing of oligonucleotide primers, and (iii) primer extension. A cycle typically takes one to five minutes and is repeated 20 to 40 times. The PCR reaction contains a mixture of buffers, nucleotide "building blocks", primers, enzyme and nucleic acid template. Denaturation separates the complementary strands of DNA held together in the duplex by hydrogen bonds and is achieved by heating to 94°C–95°C. In the annealing process, primers are attached to the dissociated DNA strands—the oligonucleotide primers are designed to hybridise to opposite strands of template in a relative orientation such that their 5^1 to 3^1 extension products overlap. The primers are present in such vast molar excess that they are more likely to anneal to the dissociation strands than the strands are to reanneal to each other. Once annealing has occurred, the DNA polymerase catalyses the synthesis of new DNA strands by adding nucleotides complementary to those in the unpaired DNA strand to the annealed primer in a 5^1 to 3^1 direction. Since the extension products are capable of binding primers, each successive cycle doubles the amount of specific DNA synthesised (assuming 100% efficiency), resulting in an exponential increase in the number of copies of the segment bounded by the oligonucleotide primers. By the fortieth cycle, therefore, the theoretical number of copies is approximately 10^{11}, although amplification by factors of 10^5 to 10^7 is more commonly reported. The use of the thermostable DNA polymerase of *Thermus aquaticus* (Taq) has permitted automation of the procedure and increased product yield and specificity because of the higher annealing temperatures which can be utilised. Amplified sequences can be visualised directly in electrophoresis gels by ethidium bromide staining or by hybridisation to specific synthetic DNA probes. Sex determination of single cell biopsies from preimplation embryos by PCR amplification of high copy number Y chromosome specific sequences has been reported,[14] and a polymorphism closely linked to the cystic fibrosis defect has also been analysed by restriction enzyme digestion of material amplified from single oocytes.[15]

We have undertaken a preliminary study to investigate the possibility of identifying single gene defects by using mice with β-thalassaemia. Mice of known pedigree were bred to produce embryos of predictable thalassaemic status. Biiopsies were performed on morula stage embryos as described above and PCR was used to try to detect the normal β-globin gene. β-thalassaemia is a hemoglobinopathy that is characterised by reduced synthesis of normal β-globin chains. Human heterozygotes are asymptomatic, while homozygotes have profoundly ineffective erythropoeisis leading to severe anaemia, bone marrow hyperplasia, skeletal deformities and growth retardation. Repeated blood transfusion prevents death in early childhood but is associated with iron overload and its usually fatal consequences. β-thalassaemia in the Hbb[th-1] strain of mice (from the Jackson Laboratory, Bar Harbor) is due to a 3.7 kilobase deletion of the β-major globin gene[16] which is normally responsible for 80% of β-globin synthesis. Because the β-minor globin gene is intact, the disease is less severe in mice than in humans.

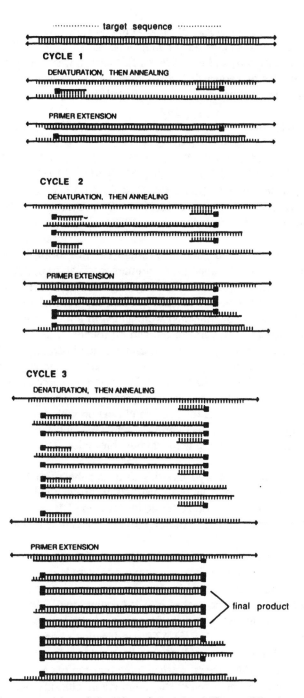

Fig. 3. A schematic representation of the primer induced specific amplification of target DNA sequences, to produce double-stranded DNA of defined sequence and molecular weight.

Genetic status of breeding stocks of mice were confirmed by gene mapping of tail tissue utilising a probe containing the β-major gene as a 7kb base Eco R1 fragment in PBR322 plasmid (donated by P. Leder, Harvard). Matings were set up between males and superovulating females of known thalassaemic status to produce embryos of predictable gene type: normal (Hbb/++) or homozygotes affected (Hbb/--).

Preliminary studies were undertaken to confirm the reliability of PCR in detecting thalassaemic status in DNA obtained from adult mice (results not shown). Oligonucleotide primers L1 (5'-CTACTCTTCCACATAAATGTC-3'), R1 (5'-GCTACTGAAGCTGTCCT AAGG-3'), L2(5'-CCAGCACTTCACAGTTCTC-3') AND R2 (5'-GCTTGGTAAACACA GAGGG-3'), and a detection probe P1 (5'-CTACCATTATTCTCTAAAC-3') were constructed using an Applied Biosystems Model 381A DNA synthesiser. PCR was performed according to method modified from Saiki et al.,1988.[17]

Reaction mixtures contained 50mM potassium chloride, 10mM Tris pH 8.3 at 25°C, 3mM magnesium chloride, the appropriate primers, each at 1µm, 300µM each of dATP, dCTP and dTTP, 100 µg/ml gelatin, and 500 ng of genomic DNA. Reaction mixtures were heated at 96°C for ten minutes to denature the DNA. 2.5 units of Taq polymerase or AmpliTaq (Perkin-Elmer Cetus Instruments) were added to a total reaction volume of 50 µl. The mixture was overlaid with paraffin oil, and 30 rounds of amplification performed in waterbaths or using a thermal cycler (Perkin-Elmer Cetus Instruments). The samples were placed at 56°C for 15 seconds for primer annealing. Extension took place at 72°C (30 seconds), and denaturation at 94°C (15 seconds). Extension was continued for seven minutes after the final cycle. 0.5µl of the amplification product using primers L1 and R1 was used as the DNA source for the second round of amplification with nested primers L2 and R2. The final reaction products were electrophoresed on a 6% polyacrylamide minigel, visualised and photographed after staining with ethidium bromide.

Amplifications from pre-embryo biopsies to detect the normal β-major globin gene were performed. Biopsy samples in 36 µl distilled water were boiled for twenty minutes. This mix then became the DNA source for 40 cycles using L1 and R1, followed by 40 cycles with nested primers L2 and R2. Control reactions containing no DNA were included.

Oligonucleotide probing was performed in 13 of the 36 embryos described in Table 4. Amplification products were transferred from polyacrylamide minigels to Zetaprobe (Bio-Rad) filters by electroblotting at 80 V for four hours in a tank system at 4°C in 0.5 × Tris-borate-EDTA buffer. Filters were soaked in 0.4M sodium hydroxide for ten minutes, neutralised in 2 × SSC, and prehybridised for four hours at 42°C in 5 × SSPE, 0.1% SDS, and 10 µg/ml salmon sperm DNA. A - ^{32}P-ATP end-labelled oligonucleotide probe P1 (2 × 10^6) cpm) was added to the prehybridisation mix, and hybridisation carried out overnight at the same temperature. Filters were washed three times in 6 × SSC at room temperature, followed by five minutes in a tetramethylammonium chloride washing solution (3M TMACl, 50mM Tris-HCl pH 8.0, 2mM EDTA, 0.1% SDS) and two 20 minute washes

Table 4. Detection of the Normal β-major Globin Gene by Amplification from Pre-embryo Biopsies of Predictable Thalassaemic Status

β major globin gene status	Result			
	Positive	Equivocal	Negative	Total
Present (Hbb/++)	11	5	4	20
Absent (Hbb/--)	2	0	14	16

in the same solution at 49°C. Autoradiography was performed overnight at -70°C using Kodak XAR-5 film and DuPont Quanta III intensifying screens.

Pre-embryo Biopsies and Ethidium Bromide Staining of PCR Products

Results of amplification of the normal genome from pre-embryo biopsies whose thalassaemic status could be predicted are summarised in Table 4. Normal b-major globin genes were detected in 11 out of 20 obligatory Hbb/++ embryos. In a further five biopsies a band could not be resolved with certainty, and in four cases it was absent. In 14 of 16 homozygous Hbb/-- biopsies, there was no amplification. However, in the remaining two a strong 86 bp band was present (Table 4).

Pre-embryo Biopsy and Oligonucleotide Probing of PCR Products

Probing with an end-labelled oligonucleotide confirmed the identity of the amplification product as normal β-major globin sequence in each of the eight cases where it was visible by ethidium bromide staining in the polyacrylamide gel, including the two false positive bands amplified from Hbb/-- embryos. One of two tested Hbb/-- biopsies which had been scored negative on the ethidium stained gell became positive after oligonucleotide probing. Three Hbb/-- embryos scored negative remained so following oligonucleotide probing.

The results obtained in this study showed that it is possible to detect a normal β-major globin gene from approximately five to ten copies of genomic DNA. Most normal (Hbb/++) pre-embryo biopsy samples produced visible bands on an ethidium-stained gel. In five cases, only indistinct bands were present, so that a confident diagnosis was not possible. However, rehybridising with an oligonucleotide probe increased the number of positives detected. If IVF is to be utilised, a rapid result is desirable, since early reimplantation optimises the likelihood of successful pregnancy. Results based on ethidium-stained gels of PCR products were rapid i.e. they were available within four hours of commencing amplification. On the other hand, the advantage of increased sensitivity with a radiolabelled oligonucleotide probe was offset by the additional day required for diagnosis.

False negative results using the PCR strategy would lead to these pre-embryos being passed over as candidates for reimplantation. The Taq polymerase has a low rate of base misincorporation, but it can be predicted[18] that the use of a very small number (1–2) of initial copies of DNA will increase the chance of sequence errors in a proportion of the final reaction product. In contrast, a false positive result may lead to a homozygous affected embryo being reimplanted. A number of precautions were taken to guard against the latter situation. Reagents were stored in aliquots in a separate laboratory from that in which PCR was performed. Separate pipettes were used to prepare the primer/dNTP/buffer amplification mixes from those used to transfer amplified material or genomic DNA. Reactions were prepared in a biosafety hood, and "no DNA" controls were included. Contamination with maternal DNA was minimised by washing by dilution prior to biopsy. The zona pellucida acts as a barrier which prevents any nonembryonic cells crossing, and the dissolution of the zona pellucida with acid-Tyrode's destroys any cells associated with the zona. Nevertheless, despite these precautions two false positives occurred among the obligatory Hbb/-- biopsies. False positives may result from carry over of previously amplified material,[19] since splashing of as little as 0.1 µl (from a PCR reaction which has generated 1012 copies) represents a contamination of up to 10^9 copies.[16]

CONCLUSION

This work has shown, using the mouse as a model for performing initial feasibility studies, that: (1) preimplantation stage embryos can be biopsied with little adverse effect on their development capacity, although the stage and method of biopsy greatly affects the viability of embryos; (2) full karyotypic analysis of biopsy material is possible but with a relatively low success rate due to technical difficulties and to the relatively low mitotic index of the biopsied cells. Thus cytogenetics and in situ hybridization is not currently a practical approach; and (3) single genes can be detected using DNA amplification (PCR). The latter is associated with some false-positives and -negatives but offers the option for detection of the normal genotype initially and thus exclusion of embryos with disease. In the future the continued use of alternative oligonucleotide primers will enable the direct detection of genetic defects in the model described above.

The marriage of embryo biopsy and gene amplification techniques offer an exciting practical new approach to prenatal diagnosis of genetic disease.

ACKNOWLEDGEMENTS

Some of these studies were supported by the IVF Advancement Foundation. R.L. is sponsored by an Australian National Health and Medical Research Council Postgraduate Medical Grant.

REFERENCES

1. P. M. Summers, J. M. Campbell, and M. W. Miller, Normal in vivo development of marmoset monkey embryos after trophectoderm biopsy, *Hum. Reprod.*, 3:389 (1988).
2. M. Monk, A. L. Muggleton-Harris, E. Rawlings, and D. G. Whittingham, Preimplantation diagnosis of HPRT-deficient male and carrier female mouse embryo by trophectoderm biopsy, *Hum. Reprod.*, 3:377 (1988).
3. M. Monk. A. H. Handyside, K. Hardy, and D. G. Whittingham, Preimplantation diagnosis of deficiency of hypoxanthing phosphoribosyl transferase in a mouse model for Lesch-Nyhan syndrome, *Lancet*, ii:423 (1987).
4. M. Monk, and A. H. Handyside, Sexing of preimplantation mouse embryos by measurement of X-linked gene dosage in a single blastomere, *J. Reprod. Fertil.*, 82:365 (1988).
5. L. J. Wilton, C. A. Kirby, and A. O. Trounson, Blastocyst and fetal development of mouse embryos biopsied at the 4-cell stage, *Proc. Aust. Soc. Reprod. Biol.*, 18:4 (1987).
6. L. J. Wilton, and A. O. Trounson, Biopsy of preimplantation mouse embryos development of micromanipulated embryos and proliferation of single blastomeres in vitro, *Biol. Reprod.* 40:145 (1989).
7. A. K. Tarkowski, and J. Wroblewska, Development of blastomeres of mouse eggs isolated at the 4- and 8-cell stage, *J. Embryol. Exp. Morphol.*, 18:155 (1967).
8. C. Roberts, and C. O'Neill, A simplified method for fixation of human and mouse preimplantation embryos which facilitates G-banding and karyotypic analysis, *Hum. Reprod.*, 3:990 (1988).
9. D. K. Kalousek, Mosaicism confined to chorionic tissue in human gestations, In: "First Trimester Fetal Diagnosis", M. Fraccaro, ed., Springer-Verlag, Berlin.
10. M. Verjaal, N. J. Leschot, H. Wolf, and P. E. Treffers, Karyotypic differences between cells from placenta and other fetal tissues, *Prenat. Diag.*, 7:343 (1987).
11. H. Li, U. B. Gyllensten, X. Cui, R. K. Saiki, H. A. Erlich, and N. Arnheim, Amplification and analysis of DNA sequences in single human sperm and diploid cells, *Nature*, 355:414 (1988).
12. A. J. Jeffreys, V. Wilson, R. Neumann, and J. Keyte, Amplification of human minisatellites by the polymerase chain reaction: towards DNA fingerprinting of single cells, *Nucleic Acids Res.*, 16:10953 (1988).
13. R. Higuchi, C. H. Von Beroldingen, G. F. Sensabaugh, and H. A. Erlich, DNA typing from single hairs, *Nature*, 332:543 (1988).
14. A. H. Handyside, J. K. Pattinson, R. J. A. Penketh, J. D. A. Delhanty, R. M. L. Winston, and C. G. D. Tuddenham, Biopsy of human preimplantation embryos and sexing by DNA amplification, *Lancet*, i:347-350 (1989).
15. C. Coutelle, C. Williams, A. H. Handyside, K. Hardy, R. M. L. Winston, and R. Williamson, Genetic analysis of DNA from single human oocytes: A model for preimplantation of cystic fibrosis, *Br. Med. J.*, 299:22-24 (1989).
16. L. C. Skow, B. A. Burkhart, F. M. Johnson, R. A. Popp, D. M. Popp, S. Z. Goldberg, W. F. Anderson, L. B. Barret, and S. E. Lewis, A mouse model for β-thalassaemia, *Cell*, 34:1043 (1983).
17. R. K. Saiki, D. H. Gelfand, S. Stoffel, S. V. Scharf, R. Higuchi, G. T. Horn, K. Mullis, and H. A. Erlich, Primer-directed enzymatic amplification of DNA with a thermostable DNA polymerase, *Science*, 239:487 (1988).
18. M. Krawczak, J. Reiss, J. Schmidtke, and U. Roesler, Polymerase chain reaction: Replication errors was reliability of gene diagnosis, *Nucl. Acid. Res.*, 17:2197-2201 (1989).
19. S. Kwok, and R. Higuchi, Avoiding false positives with PCR, *Nature*, 339:237 (1989).

GENETICS OF HUMAN PREIMPLANTATION DEVELOPMENT: IMPLICATION IN EMBRYO VIABILITY TESTING

Jan Tesarik

INSERM Unité 187, Service de Gynécologie-Obstétrique
Hôpital Antoine Béclère, 92141 Clamart, France

The ability of embryos produced by in vitro fertilization (IVF) to survive and continue their development after replacement into an appropriate environment in vivo is obviously a major factor contributing to the clinical efficacy of IVF and related techniques of assisted reproduction. Embryonic viability is influenced (1) by the quality of gametes giving rise to the embryo, (2) by the quality of in vitro conditions in which fertilization and early post-fertilization development occur, and (3) by fortuitous errors in biological mechanisms ensuring the normal course of the early developmental events. Although it seems that the quality of the gametes as well as that of the culture conditions can be further improved, there will always be a considerable proportion of non-viable embryos lacking any chance of establishing a normal pregnancy after transfer. If only viable embryos could be transferred, this would not only bring the chance of conception after transfer of a single embryo near the biological limit given by the coefficient of uterine receptivity (~0.43),[1] but this would also enable a substantial reduction of the cost of IVF and of the patients' discomfort associated with this procedure.[2]

A variety of methods for evaluating human embryonic viability have been proposed.[3] Even though many of them give, in statistical terms, a significant correlation with the establishment of pregnancy, none can determine unequivocally if a single embryo is viable and should be transferred or if it is non-viable and should be discarded. Yet it is just the possibility of this decision which would make the testing of embryo viability useful in the clinical practice.

It seems that the main drawback of the actually performed tests for embryo viability is

Advances in Assisted Reproductive Technologies, Edited by
S. Mashiach *et al.,* Plenum Press, New York, 1990

their largely empirical character. In general, some property of the embryo (morphology, speed of cleavage, uptake or secretion of different substances, etc.) is empirically taken as testing criterion and embryos are then classified into two or more categories according to this criterion. In fact, some embryos presenting bad results of these tests may subsequently recover, while those showing good results may later deteriorate. In order to construct more reliable and repeatable embryo viability tests, more information about the biological mechanisms controlling the early human development will be required.

One of the main biological events taking place in early embryos is the activation of the embryonic genome and its gradual engagement in the control of developmental events. In this paper I will summarize the available data about gene activity of human preimplantation embryos. In view of these data, the possibilities and limitations of testing human embryonic viability will be shown.

MANIFESTATIONS OF EARLY EMBRYONIC GENE ACTIVITY

Theoretical Background and Explanation of Terms

It is well known that development of all living cells is primarily guided by selective expression of their genomes. Early animal embryos represent a particular case. This particularity is given by the fact that the very early post-fertilization development is controlled, for some time, exclusively by genetic information originating from the oocyte (maternal genetic message) and stored in various forms, such as ribonucleic acid (RNA), enzymes, receptors, etc. At a certain point of development, which is specific for each species, the embryonic genome proper becomes active and assumes the developmental control. Embryos of different species may thus differ (1) as to the developmental stage at which the embryonic genome is activated, (2) as to the stage at which the activity of oocyte-derived message is lost, and (3) as to the developmental processes controlled, respectively, by products of the two genomic sources. It is evident that the knowledge of these three specific points concerning early embryonic gene activity of a particular mammalian species would be of great help to our understanding of vital activities of embryos of this species.

Manifestations of cellular gene activity in general may be studied at several different levels. An active gene must first be transcribed (*gene transcription*) from nuclear deoxyribonucleic acid (DNA) into messenger RNA (mRNA) which conveys the respective information from the nucleus to the cytoplasm. This type of gene activity is reflected by the appearance of newly synthesized RNA in the nucleus and some time later in the cytoplasm. This RNA can be demonstrated by labeling living cells with radioactive RNA precursors and detection of labeled RNA either, after extraction, by biochemical separation methods or in situ by autoradiography. When only limited amounts of cells are available, which is just the case of preimplantation embryos, only the latter detection technique is feasible. Alternatively, specific mRNAs (corresponding to specific known genes) can be detected in embryos by in situ hybridization with labeled nucleic acid probes (complementary DNA or

RNA sequences) and subsequent visualization of bound probes by autoradiographic or cyto-chemical methods.

Gene transcription is followed by *gene expression* when a specific protein is synthe-sized using the message mediated by mRNA. Newly appearing proteins can be detected by labeling with radioactive amino acids followed by extraction from cells, separation by elec-trophoresis on polyacrylamide gels and identification of labeled polypeptide bands by fluo-rography. Alternatively, specific proteins can also be detected in situ by immunocytochem-istry with specific antibodies.

Finally, the newly synthesized proteins assume their biological functions as structural and secretory proteins of cells, as enzymes modifying cellular metabolic pathways and gov-erning the production of other, non-proteinaceous structural and secretory components, or as regulatory factors involved in the responsiveness of cellular processes to different intra-cellular and extracellular signals. Only at this last level one can speak about the functional or phenotypical realisation of the corresponding genetic message and, more concretely, only at this level of expression embryonic genes can assume their function in developmen-tal control. The role of genetic message expressed at different stages of the preimplantation development in the control of individual developmental events can be studied by comparing the development of embryos with normal gene activity with those in which gene activity has been impaired either spontaneously or experimentally.

All the above steps in the phenotypical realization of genetic message are influenced by specific control mechanisms. Transcription is controlled by different transcription factors which act at the level of gene promoter regions and whose activity is modified by a variety of mechanisms such as phosphorylation, or by their interaction with other factors including regulatory proteins, hormones, metal ions, etc. *Posttranscriptional control* mechanisms regulate the rate of utilization of mRNA in the *translation process* (protein synthesis) and the rate of mRNA degradation. *Posttranslational control* mechanisms regulate the intracel-lular modifications of newly synthesized proteins (phosphorylation, glycosylation, integra-tion into supramolecular structures, etc.) which are necessary for their function.

Embryonic Gene Transcription

The transcriptional activity of human preimplantation embryos was first studied by au-toradiography using [³H]uridine as RNA precursor.[4,5] Using this methodology, the first em-bryonic RNA synthesis was detected at the 4-cell stage of human development. Moreover, there was a marked enhancement of RNA synthetic activity during the following cell cycle, in 6-8-cell embryos (Table 1). It is also the 6-8-cell stage of human embryogenesis when mRNA for a specific embryonic gene product ß-hCG first appears.[6]

However, when [³H]adenosine, another precursor of RNA synthesis, is used, slight in-corporation into pronuclei can be observed already in 1-cell zygotes.[7] Even though adeno-

Table 1. Survey of Data about the Beginning of Human Embryonic Gene Transcription

Stage	First manifestation of embryonic gene transcription	References
1-cell	Incorporation of adenosine into pronuclear RNA	7
2-cell	?	
4-cell	Slight incorporation of uridine into blastomere nuclei	4, 5, 10
6-8-cell	Appearance of ß-hCG mRNA	6
	Intense incorporation of uridine into blastomere nuclei	4, 5, 10

sine can be utilized by cells as precursor of both RNA and DNA syntheses, the early pronuclear adenosine incorporation precedes the beginning of pronuclear DNA synthetic phase, as demonstrated by parallel experiments using the specific DNA precursor thymidine.[7] The early pronuclear adenosine incorporation was thus interpreted as a very slight pronuclear RNA synthesis (Table 1). Moreover, it was shown that the normal development of pronuclear ultrastructure depended on this early gene activity.[7] It was suggested that this activity, whatever the exact type of synthesized RNA should be, makes part of chromatin preparatory changes rendering the nuclear template available for its future genetic function. It is actually not known if a similar kind of gene activity exists in the late pronuclear and 2-cell stages.

Embryonic Gene Expression

First biochemical expression of the human genome can be detected between the 4- and 8-cell stages of preimplantation development (Table 2) when α-amanitin-sensitive and stage-dependent changes in the pattern of newly synthesized polypeptides start to develop.[8] It is also the transition between the 4- and 8-cell stages when the first major phenotypical changes dependent on the embryonic genome occur at the level of blastomere ultrastructure (Table 2). These changes involve both the qualitative[9] and quantitative[10] modifications of cell organelles and their significance as markers of embryonic gene expression has been demonstrated.[9,11]

PERSISTENCE AND EXPRESSION OF OOCYTE-CODED MESSAGE DURING PREIMPLANTATION DEVELOPMENT

The beginning of embryonic gene expression does not imply automatically that the oocyte-derived developmental control mechanisms should be inactivated rapidly. This may

Table 2. Survey of Data about the Beginning of Human Embryonic Gene Expression

Stage	First signs of embryonic gene transcription	References
1-cell	—	
2-cell	—	
4-8-cell	α-Amanitin-sensitive synthesis of new proteins	8
4-8-cell	Qualitative and quantitative changes in cell organelles	9,10

be the case of the oocyte-coded (maternal) RNA which has been shown, in mouse embryos, to be degraded soon after the major outburst of embryonic gene activity,[12-14] but some oocyte-coded proteins show a remarkable longevity during the mouse preimplantation development.[15] In the absence of comparable data for human embryos we must suppose that there probably is a certain period in human preimplantation development during which both the oocyte-coded and embryo-coded message contribute to developmental control. This situation is graphically represented in Fig. 1. Components of the oocyte-coded message may be activated in a stage-dependent manner to guide the first three cell cycles after fertilization. Beginning with the activation of embryonic gene transcription between the 4- and 8-cell stages, the embryonic program probably assumes the main control over developmental processes, but some oocyte-coded systems may continue to be specifically activated at individual developmental stages and contribute to the regulation of cellular metabolism .

CONTRIBUTION OF THE OOCYTE- AND EMBRYO-CODED GENETIC MESSAGE TO EARLY DEVELOPMENTAL CONTROLS

When we summarize the available data about the human embryonic gene activity, the activation of the embryonic genome during the preimplantation development can be considered as a three-step process. The first step is characterized by the very early gene activity detectable in developing pronuclei of the zygote.[7] The period of the slight extranucleolar incorporation of uridine with the absence of nucleolar RNA synthesis[4,5] represents the second step, denoted as the early-cleavage pattern of gene activity.[10,11] Finally, the progressed-cleavage pattern of gene activity,[10,11] occurring in 8-cell and older embryos and characterized by intense labeling of both nucleolar and extranucleolar RNA with radioactive uridine, represents the third step of human embryonic gene activation. It is noteworthy that human embryos may often fail to progress normally from a given step of gene activation to the next one. Such abnormal embryos represent an excellent material for the study of the relationship between gene activity and early developmental events (Fig. 2). We have observed that, in the absence of the early pronuclear gene activity, pronuclei do not complete their development[7] and, consequently, embryogenesis is arrested at the 1-cell stage. On the other hand, the first two cleavage divisions can occur in the condition of inhibited mRNA synthesis[8] and major activation of the embryonic genome is not required for blastomere surface polarization and the development of intermediate junctions between blastomeres.[16] If embryos fail to change their gene activity from the early-cleavage to the progressed-cleavage pattern, they are still capable of forming morphologically normal compacted morulae[4] with distinct inner cell mass and trophectoderm cell types and with well developed gap junctions at intercellular contacts.[16] Nevertheless, the progressed-cleavage type of gene activity is necessary for the development of the blastocyst cavity and of the normal trophectoderm tissue.[11,16]

EMBRYONIC GENE ACTIVITY AND DEVELOPMENTAL BLOCKS

When mammalian embryos are cultured in vitro, their development is often blocked at a certain stage of preimplantation development. The block stage is specific for each species. Since the developmental arrest at these critical stages is mainly due to suboptimal culture conditions, it is more correct to speak about periods of increased embryonic vulnerability than about true developmental blocks. Such periods have been described in the mouse, pig, cow, sheep, and human (Table 3). Interestingly, in all these species the period of increased

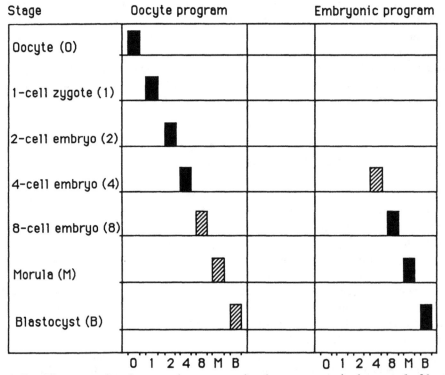

Fig. 1. Possible cooperations between the oocyte and embryo programs in the control of human preimplantation development. Both the oocyte-derived and embryonic developmental programs are activated in a stage-dependent manner (vertical bars). The leading program at each stage is represented by full black bars.

Table 3. Correlation between the Transition from Oocyte- to Embryo-Controlled Developmental Period and the Vulnerable Developmental Stages in some Mammalian Species (corresponding references in brackets)

Species	Transition period	Vulnerable period
Mouse	2-cell (17,18)	2-cell (19,20)
Pig	2-4-cell (21)	4-cell (22)
Cow	Late 8-cell (23-25)	8-16-cell (26,27)
Sheep	8-cell (28)	8-cell (28)
Human	4-8-cell (8, 10)	4-8-cell (8,11)

embryonic vulnerability corresponds to the period of major qualitative and quantitative changes in embryonic gene activity (Table 3). Accordingly, the critical period of human embryonic development is the transition between the 4- and 8-cell stages. However, even though the fate of most human embryos is decided at these stages, the result of this decision often becomes evident only some time later, during blastulation. On the other hand, the development of human embryos in vitro is arrested relatively frequently at the early cleavage stages.[29] This can be explained by the high proportion of developmentally incompetent oocytes in current hyperstimulation regimens so that the stored oocyte-coded message is not sufficient to ensure development throughout the relatively long oocyte-dependent period (cf. Fig. 1). In any case, the delayed gene activation in human embryos, as compared with mouse embryos, and the corresponding shift of the developmentally vulnerable period towards more advanced stages of the preimplantation development make any estimation of human embryonic viability before the 8-cell stage highly dubious.

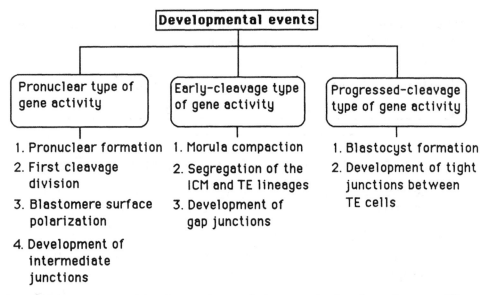

Fig. 2. The dependence of developmental events in human preimplantation embryos on different types of embryonic gene activity.

MARKERS OF EMBRYONIC GENE ACTIVITY:
A NOVEL APPROACH TO TESTING EMBRYO VIABILITY

We have shown previously that about 25% to 30% of blastomeres of human embryos cultured in vitro fail to undergo the early-cleavage–progressed-cleavage transition of gene activity.[4] This proportion is further increased when embryos are exposed to suboptimal culture conditions (bacterial growth in culture medium, instability of temperature and gas phase composition in the incubator). Polyploidy (mainly triploidy due to bispermic penetration) and blastomere multinucleation[30] also reduce the chances of embryos to activate normally their genomes (Fig. 3). Since the failure of the progressed-cleavage pattern of gene activity entails defective blastulation and abnormal development of the trophectoderm, the precursor of future trophoblast tissue, such embryos will obviously not implant normally. On the other hand, embryos that have passed through the vulnerable period and that have properly changed the pattern of gene activity have apparently a high chance of further normal development. It seems then that the reliability of embryo viability tests could be substantially improved if they were applied to 8-cell and older embryos. Many markers of the progressed-cleavage gene activity have been described;[6,8-11] unfortunately, all of them require invasive methods leading to the destruction of the tested cells. The search for non-invasive methods for the detection of the occurrence of the progressed-cleavage type of gene activity in human embryos is actually one of the most important tasks for human embryo research.

CONCLUDING REMARKS

The activation of the embryonic genome during the human preimplantation development occurs in several steps, the last of which, taking place between the 4- and 8-cell stag-

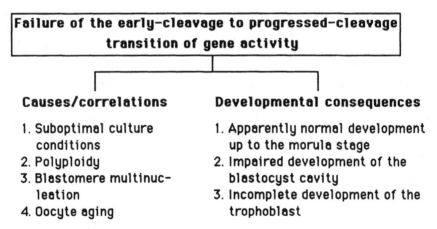

Fig. 3. Causes and developmental consequences of failed transition of embryonic gene activity from the early-cleavage to the progressed-cleavage pattern.

es is the most sensitive to environmental factors. Up to the 8-cell stage the developmental control is mainly exerted by oocyte-coded factors and many embryos originating from defective oocytes are eliminated during this period. Embryos that have properly changed their gene activity at the 8-cell stage have a great chance of further normal development. This implies that tests of embryo viability would be more reliable if applied to 8-cell or older embryos. These tests could possibly be based on the detection of specific embryonic products coded for by the embryonic genome and released to the culture medium. The current research aimed at the improvement of human embryonic growth in vitro to more advanced stages of the preimplantation development may also contribute to revision of the actual strategies of embryo transfer and to prolongation of culture with transfer of older embryos.

REFERENCES

1. Walters, D.E., Edwards, R.G., Meistrich, M.L. A statistical evaluation of implantation after replacing one or more human embryos. *J. Reprod. Fertil.* 74, 557-563 (1985).
2. Tesarik, J. Viability assessment of preimplantation concepti: a challenge for human embryo research. *Fertil. Steril.* in press (1989).
3. Leese, H.J. Analysis of embryos by non-invasive methods. *Hum. Reprod.* 2, 37-40 (1987).
4. Tesarik, J., Kopecny, V., Plachot, M., Mandelbaum, J. Activation of nucleolar and extranucleolar RNA synthesis and changes in the ribosomal content of human embryos developing in vitro. *J. Reprod. Fertil.* 78, 463-470 (1986).
5. Tesarik, J., Kopecny, V., Plachot, M., Mandelbaum, J., Da Lage, C., Fléchon, J.E. Nucleologenesis in the human embryo developing in vitro: ultrastructural and autoradiographic analysis. *Dev. Biol.* 115, 193-203 (1986).
6. Bonduelle, M.L., Dodd, R., Liebaers, I., Van Steirteghem, A., Williamson, R., Akhurst, R. Chorionic gonadotrophin-ß mRNA, a trophoblast marker, is expressed in human 8-cell embryos derived from tripronucleate zygotes. *Hum. Reprod.* 3, 909-914 (1988).
7. Tesarik, J., Kopecny, V. Nucleic acid synthesis and development of human male pronucleus. *J. Reprod. Fertil.* in press (1989).
8. Braude, P., Bolton, V., Moore, S. Human gene expression first occurs between the four- and eight-cell stages of preimplantation development. *Nature* (London) 332, 459-461 (1988).
9. Tesarik, J. Gene activation in the human embryo developing in vitro. In: Future Aspects in Human In Vitro Fertilization, W. Feichtinger and P. Kemeter (eds.), p. 251-261, Springer-Verlag: Berlin, Heidelberg (1987).
10. Tesarik, J., Kopecny, V., Plachot, M., Mandelbaum, J. Early morphological signs of embryonic genome expression in human preimplantation development as revealed by quantitative electron microscopy. *Dev. Biol.* 128, 15-20 (1988).
11. Tesarik, J. Developmental control of human preimplantation embryos: a comparative approach. *J. Vitro Fertiliz. Embryo Transf.* 5, 347-362 (1988).
12. Olds, P.J., Stern, S., Biggers, J.D. Chemical estimates of the RNA and DNA contents of the early mouse embryo. *J. Exp. Zool.* 186, 39-46 (1973).
13. Bachvarova, R., De Leon, V. Polyadenylated RNA of mouse ova and loss of maternal RNA in early development. *Dev. Biol.* 74, 1-8 (1980).
14. Piko, L., Clegg, K.B. Quantitative changes in total RNA, total poly (A), and ribosomes in early mouse embryos. *Dev. Biol.* 89, 362-378 (1982).
15. West, J.D., Green, J.F. The transition from oocyte-coded to embryo-coded glucose phosphate isomerase in the early mouse embryo. *J. Embryol. Exp. Morphol.* 78, 127-140 (1983).

16. Tesarik, J. Involvement of oocyte-coded message in cell differentiation control of early human embryos. *Development* 105, 317-322 (1989).
17. Flach, G., Johnson, M.H., Braude, P.R., Taylor, R.A.S., Bolton, V.N. The transition from maternal to embryonic control in the 2-cell mouse embryo. *EMBO J.* 1, 681-686 (1982).
18. Bolton, V.N., Oades, P.J., Johnson, M.H. The relationship between cleavage, DNA replication and gene expression in the mouse 2-cell embryo. *J. Embryol. Exp. Morphol.* 79, 139-163 (1984).
19. Whittingham, D.G., Biggers, J.D. Fallopian tube and early cleavage in the mouse. *Nature* (London) 213, 942-943 (1967).
20. Goddard, M.J., Pratt, H.P.M. Control of events during early cleavage of the mouse embryo: an analysis of the "2-cell block". *J. Embryol. Exp. Morphol.* 73, 111-153 (1983).
21. Tomanek, M., Kopecny, V., Kanka, J. Studies on RNA synthesis in early pig embryos [Abstr]. *Histochem. J.* 18, 138 (1986).
22. Hunter, R.H.F. Chronological and cytological details of fertilization and early embryonic development in the domestic pig, *Sus scrofa*. *Anat. Rec.* 178, 169-186 (1974).
23. Camous, S., Kopecny, V., Fléchon, J.E. Autoradiographic detection of the earliest stage of [^{3}H]-uridine incorporation into the cow embryo. *Biol. Cell* 58, 195-200 (1986).
24. King, W.A., Niar, A., Chartrain, I., Betteridge, K.J., Guay, P. Nucleolus organizer regions and nucleoli in preattachment bovine embryos. *J. Reprod. Fertil.* 82, 87-95 (1988).
25. Kopecny, V., Fléchon, J.E., Camous, S., Fulka, J.Jr. Nucleologenesis and the onset of transcription in the eight-cell bovine embryo: fine-structural autoradiographic study. *Mol. Reprod. Development* 1, 79-90 (1989).
26. Camous, S., Heyman, Y., Méziou, W., Ménézo, Y. Cleavage beyond the block stage and survival after transfer of early bovine embryos cultured with trophoblastic vesicles. *J. Reprod. Fertil.* 72, 479-485 (1984).
27. Eyestone, W.H., First, N.L. A study of the 8- to 16-cell developmental block in bovine embryos cultured in vitro [Abstr]. *Theriogenology* 25, 152 (1986).
28. Crosby, I.M., Gandolfi, F., Moor, R.M. Control of protein synthesis during early cleavage of sheep embryos. *J. Reprod. Fertil.* 82, 769-775 (1988).
29. Bolton, V.N., Hawes, S.M., Taylor, C.T., Parsons, J.H. Development of spare human preimplantation embryos in vitro: an analysis of the correlations among gross morphology, cleavage rates, and development to the blastocyst. *J. Vitro Fertiliz. Embryo Transf.* 6, 30-35 (1989).
30. Tesarik, J., Kopecny, V., Plachot, M., Mandelbaum, J. Ultrastructural and autoradiographic observations on multinucleated blastomeres of human cleaving embryos obtained by in-vitro fertilization. *Hum. Reprod.* 2, 127-136 (1987).

NUCLEAR TRANSFER

Randall S. Prather and Neal L. First

Department of Meat and Animal Science
University of Wisconsin-Madison
Madison, Wisconsin 53706 U.S.A.

INTRODUCTION

In the early part of this century scientists did not know if DNA (nucleoplasm) was equally inherited when the cells of an early embryo divided, since some proposed that it wasn't (Weismann, 1915). In 1938 Spemann proposed a *'fantastical experiment'* to determine if all of the nuclei of an early embryo are equivalent. He proposed to transfer nuclei from progressively more advanced cell stage embryos to enucleated oocytes. If there was indeed unequal inheritance, then at some point the nuclei would no longer be totipotent, or would be unable to direct development to all of the tissues of the adult. This would then signal the first inequivalence of the nuclei and provide information on the differentiation process. The necessary technology to complete the experiment was developed 14 years later by Briggs and King (1952). They (Briggs and King, 1952) showed that nuclei from Rana pipiens blastula stage embryos when transferred to activated, enucleated meiotic metaphase II oocytes redirected development from the 1-cell stage to the blastula stage. Interestingly they alluded to the fact that nuclei from more advanced cell stages seemed unable to redirect development at as high a rate as from the blastula stage. Subsequent experiments in Rana and Xenopus laevis illustrated that these nuclear transfer derived blastula stage embryos were capable of continued development through metamorphosis to adulthood (Fischberg et al., 1958; McKinnell, 1962). The techniques for nuclear transfer have been modified for application to mammals and have resulted in live offspring in sheep, cattle, rabbits and pigs but not mice or rats (reviewed by Prather and First, 1989). In introducing the concept of nuclear transfer we will first discuss factors affecting development after nuclear transfer in amphibia, then mammals and finally applications of this technology.

Advances in Assisted Reproductive Technologies, Edited by
S. Mashiach *et al.*, Plenum Press, New York, 1990

AMPHIBIA

Differentiation

During development from the 1-cell stage to the blastula stage the amphibian embryo relies on maternally derived and stored RNAs (Newport and Kirschner, 1982). Just prior to the blastula stage the amphibian embryo begins producing small amounts of its own RNA, this is followed by a major increase in RNA production at the midblastula stage, termed the midblastula transition (Newport and Kirschner, 1982; Nakakura et al., 1987). The decreased rate of development observed after transfer of nuclei beyond the blastula stage suggests that irreversible differentiation begins to occur in selected populations of nuclei within the blastula (Gurdon, 1964). However, the decreased rate of development observed from nuclei more advanced in development may not be entirely explained by differentiation alone. A major confounding factor is the change in the length of the cell cycle that occurs at the midblastula transition. Prior to the midblastula stage the length of the cell cycle is about 35 minutes with back to back M and S phases. This omission of G_1 and G_2 may explain the absence of transcription as RNA synthesis does not occur in M or S phases. As the embryo develops beyond the midblastula stage the cell cycle begins to progressively lengthen with the addition of both G_1 and G_2 phases while length of M and S phases remains unchanged; effectively lengthening the cell cycle as development proceeds. Nuclei that are transferred from cells that have long cell cycles would be required to divide earlier in their respective cell cycle after nuclear transfer as compared to nuclei from the blastula stage. Thus nuclei from more slowly dividing cells promote development at a higher rate if in G_2 when transferred (Von Beroldington, 1981), whereas the stage of the cell cycle is less important for nuclei from rapidly dividing cells (McAvoy et al., 1975; Ellinger, 1978).

Chromosomal Abnormalities

Although nuclei in G_1 undergo DNA synthesis after nuclear transfer to activated, enucleated metaphase II oocytes, in many instances the replication is not complete. This incomplete DNA replication results in chromosomal breakage and unequal inheritance among the daughter cells (Gurdon, 1964; DiBerardino and Hoffner, 1970). Later in development the chromosomal abnormalities are manifest as developmental restriction points or stages that the nuclear transfer embryo is unable to progress beyond. These restriction points are stably inherited as shown with retransfer; i.e., serial nuclear transfer (Briggs et al., 1964; DiBerardino and King, 1965).

Nuclear Reprogramming

As the embryo progresses beyond the midblastula stage, stage-specific RNAs are produced. Nuclear transfer results in a specific reprogramming of this RNA production as well as the accompanying morphological changes. For example, nucleoli are not discernable in pre-blastula stage embryos and rRNA production cannot be detected. At the blastula stage reticulated nucleoli appear coincidentally with the detection of rRNA production. When nuclei from embryos beyond the blastula stage are transferred to activated, enucleated oocytes, active rRNA synthesis ceases and nucleoli disappear. However, when the nuclear transfer embryo develops to the blastula stage reticulated nucleoli appear and rRNA production begins (Gurdon and Brown, 1965). Regarding mRNA production, in embryos undergoing normal development the 5 S^{ooc} gene undergoes transcription for a very short period of time at the gastrula stage. Nuclei derived from embryos beyond the gastrula stage (i.e., not producing the 5 S^{ooc} transcripts) that are transferred to activated oocytes direct the production of 5 S^{ooc} transcripts as the newly developing embryo passes through the gastrula stage (Wakefield and Gurdon, 1983). This specific genomic reprogramming results in reprogramming of the timing of specific morphological events, such as blastula formation. Therefore the reprogramming power of the oocyte cytoplasm appears to be highly specific.

The facilitation of the genomic reprogramming described above is thought to be mediated by proteins residing in the cytoplasm of the oocyte. Some of these cytoplasmic proteins migrate into nuclei of the normally developing embryo at specific cell stages (Stick and Dreyer, 1989). After nuclear transfer and activation of the oocyte these proteins migrate into the nucleus in a stage specific manner, whereas some nuclear proteins migrate out of the nucleus (Merriam, 1969; DiBerardino and Hoffner, 1975; Leonard et al., 1982). Together this exchange of protein between the nucleus and cytoplasm is thought to regulate genomic expression.

MAMMALS

Nuclear transfer in mammals has a much shorter history. In 1981 Illmensee and Hoppe reported the birth of three mice from nuclei derived from inner cell mass cells of a blastocyst stage embryo. They also reported that nuclei from trophectoderm could not direct similar development. These results corresponded with similar work reported by Modlinski (1978, 1981). However the results of Illmensee and Hoppe (1981) could not be repeated (McGrath and Solter, 1984a; Robl et al., 1986).[1] Undaunted by the discouraging results in mice, others more closely applied the amphibian nuclear transfer procedures to mammalian

---

[1] It should be noted that Illmensee and Hoppe (1981) used a surgical method of trnasfer; i.e., they penetrated the plasma membranes, whereas McGrath and Solter (1984a) and Robl et al. (1986) used a cell fusion method developed by McGrath and Solter (1983a) to transfer nuclei. In addition, contrary to the procedures used in amphibian nuclear transfer (Elsdale et al., 1960; Gurdon and Laskey, 1970) where a freshly activated, enucleated oocyte is used as a recipient, the procedures used by Illmensee and Hoppe (1981), McGrath and Solter (1983a, 1984a) and Robl et al. (1986) used a cell in interphase, either pronuclear or 2-cell stage. There are many confounding factors to consider when making comparisons among Illmensee and Hoppe (1981), McGrath and Solter (1983a, 1984a), Robl et al. (1986) and the amphibian results.

species other than the mouse. This has resulted in offspring from cleavage stage nuclei transferred to activated, enucleated oocytes in sheep (Willadsen, 1986, Smith and Wilmut, 1989), cattle (Prather et al., 1987), rabbits (Stice and Robl, 1988) and pigs (Prather et al., 1989a). An overriding conclusion is that the mouse may not be a good model species for mammalian nuclear transfer.

Nuclear Transfer in Mice and Rats

The methods for nuclear transfer in mammals using virus-mediated cell fusion was first described by McGrath and Solter (1983a). Embryos are first treated with cytochalasin B and colchicine, which disrupt microfilaments and microtubules, respectively. This treatment imparts an elasticity to the cells such that a small cytoplast or karyoplast can be aspirated into a small pipette and separated from the remainder of the cell without rupture of the membranes. The cytoplast or karyoplast can then be fused to a recipient cell with inactivated sendai virus. These procedures are highly efficient (reviewed by Robl and First, 1985).

The above procedures have been used to transfer pronuclei from one zygote to another and have shown both nuclear and cytoplasmic inheritance. For example, the hairpin-tail mutation in mice is inherited via the nucleus even though it is a maternally derived mutation (McGrath and Solter, 1984b), and stage specific embryonic antigen-3 is cytoplasmically inherited (McGrath and Solter, 1983b). Nuclear exchange at the 2-cell stage has shown that the 2-cell block to in vitro development in the mouse is dependent upon both the cytoplasm and nucleus, but that the nucleus alone, is responsible for strain differences in in vitro development beyond the 2-cell stage (Robl et al., 1988).

Nuclear transfer of cleavage stage nuclei in both mice and rats to enucleated pronuclear stage eggs rarely supports development to the blastocyst stage (McGrath and Solter, 1984a; Kono et al., 1988) as the nuclei are not fully reprogrammed (Barnes et al., 1987). When 8-cell stage nuclei are transferred to enucleated 2-cell stage blastomeres they can promote development to midgestation (Robl et al., 1986) and even to term (Tsunoda et al., 1987), but again they fail to be fully reprogrammed (Barnes et al., 1987).

Nuclear Transfer in Domestic Animals

The procedures for nuclear transfer in mammals other than the mouse have required a few modifications. In many species the cytoplasm is opaque due to inclusions within the egg. These inclusions can be centrifuged to one side of the cell to permit visualization of the pronuclei without adversely affecting development (Wall et al., 1985; Wall and Hawk, 1988). Nuclei or pronuclei can then be aspirated into micropipettes and transferred as for the mouse. Cell fusion can be mediated by sendai virus in the sheep (Willadsen, 1986), but not as effectively in the rabbit (Bromhall, 1975), rat (Kono et al., 1988) or cow (Robl et al., 1987). Thus an electrical cell fusion system (Berg, 1982) has been incorporated into nuclear transfer in all of the domestic species (Willadsen, 1986; Smith and Wilmut, 1989, Prather et al., 1987; Stice and Robl, 1988; Prather et al., 1989a). The electric pulse also conveniently activates the oocyte to resume meiosis and initiate development. Control pronuclear exchanges show that the procedures used for nuclear transfer (i.e., exposure to cytoskeletal inhibitors, in vitro culture, micromanipulation, cell fusion) are compatible with development to term (Robl et al., 1987; Prather et al., 1989a).

Nuclear transfer using multi-cell stage embryos has resulted in offspring from 8-cell stage (Willadsen, 1986) and inner cell mass cell sheep nuclei (Smith and Wilmut, 1989), 9- to 15-cell stage and 32-cell stage cattle nuclei (Prather et al., 1987; First, unpublished), 8-cell stage rabbit nuclei (Stice and Robl, 1988) and 4-cell stage pig nuclei (Prather et al., 1989a). Although the degree of reprogramming may not be large (except for the inner cell mass stage nuclei) the common denominator is that an oocyte in metaphase was used as a recipient cell. This is in contrast to the mouse and rat experiments where an enucleated interphase cell is used as a recipient. Development of mouse and rat nuclear transfer embryos is similar to cattle when an enucleated zygote is used as a recipient, it rarely results in cleavage (Robl et al., 1987). Thus the question arises 'What is responsible for the reprogramming?'.

As discussed earlier, protein association with the nucleus changes during development (Stick and Dreyer, 1989) and this association may bestow upon the nucleus the ability to differentiate. Thus at the zygote stage, cytoplasmic proteins that are important for gene expression are likely associated with the interphase chromatin. If the nucleus is removed during enucleation these essential proteins are also removed. A specific example of this type of protein dynamics is illustrated by the nuclear lamins. They reside on the inner nuclear envelope and are hypothesized to give 3- dimensional supramolecular organization to the nucleus. They are polymerized during interphase and depolymerized during mitosis. In the depolymerized state they are monomeric and dispersed throughout the cytoplasm (Gerace and Blobel, 1980). In mammals the lamins are composed of two major protein families that can be distinguished immunologically (A/C versus B). The lamin content of the nuclei of early mammalian embryos changes from the 1-cell stage to the 16-cell stage in mice (Schatten et al., 1985), cows and pigs (Prather et al., 1989b). If 16-cell stage nuclei are transferred to activated, enucleated oocytes the 16-cell stage nucleus acquires the lamin proteins from the cytoplasm of the 1-cell egg (Prather et al., 1989b). However, if the nuclei are transferred to either intact or enucleated zygotes the 16-cell stage nucleus acquires little, if any, of the lamin proteins. Thus the transferred nucleus is not comparable to a zygotic nucleus (Prather, First and Schatten, unpublished).

Differentiation involves stage specific expression of the embryonic genome. The first embryonic transcription as determined by α-amanitin inhibition of new zygotic transcription and protein synthesis is at the 4- cell stage in cattle (Barnes, 1988) and 8-cell stage in sheep (Crosby et al., 1988). Therefore, nuclei and cells from the 16- and 32-cell stage embryo discussed previously would be considered to have undergone initial differentiation.

Identicals

If two or more offspring resulted from a single donor embryo would they be identical? The answer is maybe! Extranuclear genetics would need to be identical for an identical individual to result. This includes mitochondria which have different genomes between strains of mice (Ferris et al., 1982), and likely centrosomes as well as other organelles that may contain their own DNA. With the present method of nuclear transfer many cytoplasmic components are transferred to the enucleated recipient cell along with the new nucleus. This would result in a mixing of the two cytoplasmic genotypes.

The source of donor nuclei would also need to be from a source where no genetic changes have occurred between the nuclei of the embryo. After the first cleavage division many genetic changes can occur. Translocations (King and Linares, 1983), diminution (Beerman, 1977), gene rearrangements (Alt et al., 1987), gene amplification (Tobler, 1975) and mutation can all occur to the nuclear genome. If this occurs in only one cell of a multi-celled embryo all of its descendants would have that genetic change and a mosaic embryo/animal would result. If this was the source of nuclei for nuclear transfer, then each nuclear transfer could be transferring a different genome. If these changes occurred after nuclear transfer in one nuclear transfer embryo and not all of them, then again they would not be identical.

For an identical individual to result cell migration patterns would also need to be similar. An example here is the melanoblast. Bisection of cattle embryos, a method that produces genomically identical offspring, both nuclear and cytoplasmic, has shown that the melanoblasts have different migration patterns. Two genetically identical calves may have a red patch of hair on their head. On one it may be above the right eye and on the other it may be below the right eye. This is the result of different uterine environments (Seidel, 1985). All other environmental factors starting with the uterine environment and continuing with environmental factors throughout development would need to be identical. This includes nutrition, disease exposure, injuries, etc., a feat almost impossible to achieve. Many of the factors that must be considered to conclude that two animals are identical are reviewed by Seidel (1983).

APPLICATIONS AND CONCLUSIONS

Genomically identical individuals would be valuable for a variety of purposes. Genomically identical individuals would be the perfect control for all types of experiments as all variation in response to a stimulus would be nongenetic. Thus experiments evaluating different diets, different environmental stresses, different drug regimes, etc., would require fewer numbers of animals. One would need to be careful that results are not extrapolated to all animals of a particular species, as results would only be valid for the genetics tested. Reciprocal nuclear transfer provides a method to study nuclear and cytoplasmic interactions, differentiation and dedifferentiation. Genomically identical individuals could offer many benefits to science.

The production of twin embryos by bisection at the morula or blastocyst stage has reached widespread commercial use with a frequency of completed pregnancies from transfer of half embryos often nearly equal to that from a whole embryo (Leibo, 1988).

Commercially, the application of nuclear transfer is also near. This would incorporate the production of large numbers of identicals by performing the first set of nuclear transfer with a single embryo, growing the resulting embryos to the morula stage and subjecting those to nuclear transfer. This process, serial nuclear transfer, could result in the production of large numbers of identical individuals. Some could be stored frozen while the others were transferred to surrogate dams and allowed to go to term. Once the genetic potential is evaluated the frozen store could be thawed and used as donors for repeated serial nuclear transfer. If a desirable embryo cell line is established, nutrition studies could evaluate the optimal diets for growth and a herd of identical animals could be sold with a guarantee to produce a certain level of production if maintained in a defined environment. The genetic gain possible with splitting embryos, thus producing two identicals, has been evaluated (Nicholas and Smith, 1983) and the selection pressure that could be applied with this technique could only result in greater genetic gain. Commercial aspects have been reviewed by Robl and Stice (1989).

REFERENCES

Alt, F.W., Blackwell, K. and Yancopoulos, G.D., 1987, Development of the primary antibody repertoire, *Science* 238:1079.

Barnes, F.L., 1988, Characterization of the onset of embryonic control and early development in the bovine embryo. Ph.D. dissertation, University of Wisconsin-Madison.

Barnes, F.L., Robl, J.M. and First, N.L., 1987, Nuclear transplantation in mouse embryos: assessment of nuclear function, *Biol. Reprod.* 36:1267.

Beerman, S., 1977, The diminution of heterochromatic chromosomal segments in Cyclops. *Chromosoma* 60:297.

Berg, H., 1982, Fusion of blastomeres and blastocysts of mouse embryos. *Bioelectricity Bioenergetics* 9:223.

Briggs, R. and King, T.J., 1952, Transplantation of living nuclei from blastula cells into enucleated frogs' eggs, *Proc. Natl. Acad. Sci.*, U.S.A. 38:455.

Briggs, R., Signoret, J. and Humphrey, R.R., 1964, Transplantation of nuclei of various cell types from neurulae of the Mexican axolotl (Ambrystoma mexicanum), *Dev. Biol.* 10:233.

Bromhall, J.D., 1975, Nuclear transplantation in the rabbit egg, *Nature* 258:719.

Crosby, I.M., Gandolfi, F. and Moor, R.M., 1988, Control of proteins synthesized during early cleavage of sheep embryos, *J. Reprod. Fert.* 82:769.

DiBerardino, M.A. and Hoffner, N., 1970, Origin of chromosomal abnormalities in nuclear transplants - a reevaluation of nuclear differentiation and nuclear equivalence in amphibians. *Dev. Biol.* 23:185.

DiBerardino, M.A. and Hoffner, N.J., 1975, Nucleocytoplasmic exchange of nonhistone proteins in amphibian embryos, *Exp. Cell Res.* 94:235.

DiBerardino, M.A. and King, T.J., 1965, Development and cellular differentiation of neural nuclear-transplant embryos of known karyotype. *Dev. Biol.* 15:102.

Ellinger, M.S., 1978, The cell cycle and transplantation of blastula nuclei in Bombiana orientalis, *Dev. Biol.* 65:81.

Elsdale, T.R., Gurdon, J.B. and Fischberg, M., 1960, A description of the techniques for nuclear transplantation in Xenopus laevis, *J. Embryo. Exp. Morph.* 8:437.

Ferris, S.D., Sage, R.D. and Wilson, A.C., 1982, Evidence from mtDNA sequences that common laboratory strains of inbred mice are descended from a single female, Nature 295:163

Fischberg, M., Gurdon, J.B. and Elsdale, T.R., 1958, Nuclear transplantation in Xenopus laevis. *Nature* 181:424.

Gerace, L. and Blobel, G., 1980, The nuclear envelope lamina is reversibly depolymerized during mitosis. *Cell* 19:277.

Gurdon, J.B., 1964, The transplantation of living cell nuclei. *Adv. Morpho.* 4:1.

Gurdon, J.B., 1986, Nuclear transplantation in eggs and oocytes, *J. Cell Sci.* 4 (Suppl):287.

Gurdon, J.B. and Brown, D.D., 1965, Cytoplasmic regulation of RNA synthesis and nucleolus formation in developing embryos of Xenopus laevis. *J. Mol. Biol.* 12:27.

Gurdon, J.B. and Laskey, R.A., 1970, Methods of transplanting nuclei from single cultured cells to unfertilized frogs' eggs. *J. Embryol. exp. Morph.* 24:227.

Illmensee, K. and Hoppe, P.C., 1981, Nuclear transplantation in Mus musculus: developmental potential of nuclei from preimplantation embryos, *Cell* 23:9.

King, W.A. and Linares, T., 1983, A cytogenetic study of repeat-breeder heifers and their embryos, *Can. Vet. J.* 24:112.

Kono, T., Shioda, Y., Tsunoda, Y., 1988, Nuclear transplantation of rat embryos, *J. Exp. Zool.* 248:303.

Leibo, S., 1988, Cryopreservation of ova and embryos. *In:* The American Fertility Society Regional Postgraduate Course. Hands-on IVF, Cryopreservation and Manipulation. April 25-29, Madison, WI.

Leonard, R.A., Hoffner, N.J. and DiBerardino, M.A., 1982, Induction of DNA synthesis in amphibian erythroid nuclei in Rana eggs following conditioning in meiotic oocytes, *Dev. Biol.* 92:343.

McAvoy, J.W., Dixon, K.E. and Marshall, J.A., 1975, Effects of differences in mitotic activity, stage of the cell cycle, and degree of specialization of donor cells on nuclear transplantation in Xenopus laevis, *Dev. Biol.* 45:330.

McGrath, J. and Solter, D., 1983a, Nuclear transplantation in the mouse embryo by microsurgery and cell fusion. *Science* 220:1300.

McGrath, J. and Solter, D., 1983b, Nuclear transplantation in mouse embryos, *J. Exp. Zool.* 228:355.

McGrath, J. and Solter, D., 1984a, Inability of mouse blastomere nuclei transferred to enucleated zygotes to support development in vitro, *Science* 226:1317.

McGrath, J. and Solter, D., 1984b, Maternal T^{hp} lethality in the mouse is a nuclear not cytoplasmic defect. *Nature* 308:550.

McKinnell, R.G., 1962, Intraspecific nuclear transplantation in frogs. *J. Heret.* 53:199.

Merriam, R.W., 1969, Movement of cytoplasmic proteins into nuclei induced to enlarge and initiate DNA or RNA synthesis, *J. Cell Sci.* 5:333.

Modlinski, J.A., 1978, Transfer of embryonic nuclei to fertilized mouse eggs and development of tetraploid blastocysts. *Nature* 273:466.

Modlinski, J.A., 1981, The fate of inner cell mass and trophectoderm nuclei transferred to fertilized mouse eggs. *Nature* 292:342.

Nakakura, N., Miura, T., Yamana, K., Ito, A. and Shiokawa, K., 1987, Synthesis of heterogenous mRNA-like RNA and low-molecular-weight RNA before the midblastula transition in embryos of Xenopus laevis, *Dev. Biol.* 123:421.

Newport, J. and Kirschner, M., 1982, A major transition in early Xenopus embryos: II. Control of the onset of transcription, *Cell* 30:687.

Nicholas, F.W. and Smith, C., 1983, Increased rates of genetic change in dairy cattle by embryo transfer and splitting. *Anim. Prod.* 36:341.

Prather, R.S., Barnes, F.L., Sims, M.L., Robl, J.M., Eyestone, W.H. and First, N.L., 1987, Nuclear transfer in the bovine embryo: assessment of donor nuclei and recipient oocyte. *Biol. Reprod.* 37:859.

Prather, R.S., Sims, M.L. and First, N.L., 1988, Nuclear transplantation in the porcine embryo, *Theriogenology* 29:290 Abstr.

Prather, R.S., 1989, Nuclear transfer in mammals and amphibians: nuclear equivalence, species specificity? *in:* "The Molecular Biology of Fertilization," H. Schatten and G. Schatten eds., Academic Press, New York.

Prather, R.S. and First, N.L., 1989, Nuclear transfer in mammalian embryos. *Inter. Rev. Cytol.* (in press).

Prather, R.S., Sims, M.M. and First, N.L., 1989a, Nuclear transfer in pig embryos. 3rd International Conference on Pig Reproduction, University of Nottingham, U.K. April 11–14 (in press) Abstr.

Prather, R.S., Sims, M.M., Maul, G.G., First, N.L. and Schatten, G., 1989b, Nuclear lamin antigens are developmentally regulated during porcine and bovine early embryogenesis. *Biol. Reprod.* (in press).

Robl, J.M. and First, N.L., 1985, Manipulation of gametes and embryos in the pig. *J. Reprod. Fert.* 33 (Suppl):101.

Robl, J.M., Gilligan, B., Critser, E.S. and First, N.L., 1986, Nuclear transplantation in mouse embryos: assessment of recipient cell stage. *Biol. Reprod.* 34:733.

Robl, J.M., Prather, R., Barnes, F., Eyestone, W., Northey, D., Gilligan, B. and First, N.L., 1987, Nuclear transplantation in bovine embryos. *J. Anim. Sci.* 64:642.

Robl, J.M., Lohse-Heideman, J.K. and First, N.L., 1988, Strain differences in early mouse embryo development in vitro: role of the nucleus. *J. Exp. Zool.* 247:251.

Robl, J.M. and Stice, S.L., 1989, Prospects for the commercial cloning of animals by nuclear transplantation, *Theriogenology* 31:75.

Schatten, G, Maul, G.G., Schatten, H., Chaly, N., Simerly, C., Balczon, R. and Brown, D.L., 1985, Nuclear lamins and peripheral nuclear antigens during fertilization and embryogenesis in mice and sea urchins, *Proc. Natl. Acad. Sci., U.S.A.* 82:4727.

Seidel, G.E., Jr., 1983, Production of genetically identical sets of mammals: cloning?, *J. Exp. Zool.* 228:347.

Seidel, G.E., Jr., 1985, Are identical twins produced from micromanipulation always identical? Proc. Ann. Conf. Artific, Insem. and Embryo Transfer in Beef Cattle (Denver), *Natl. Assoc. Anim. Breeders*, Columbia, MO, pp. 50-53.

Smith, L.C. and Wilmut, T., 1989, Influence of nuclear and cytoplasmic activity on the development in vivo of sheep embryos after nuclear transplantation , *Biol. Reprod.* 40:1027.

Spemann, H., 1938, Embryonic Development and Induction, Hafner Publishing Company, New York. p 401.

Stice, S.L. and Robl, J.M., 1988, Nuclear reprogramming in nuclear transplant rabbit embryos. *Biol. Reprod.* 39:657.

Stick, R. and Dreyer, C., 1989, Developmental control of nuclear proteins in amphibia, *in:* "The Molecular Biology of Fertilization," H. Schatten and G. Schatten eds., Academic Press, New York.

Tobler, H., 1975, The occurrence and developmental significance of gene amplification, *in:* "Biochemistry of Animal Development," R. Weber ed. Academic Press, New York.

Tsunoda, Y., Yasui, T., Shiods, Y., Nakamum, K., Uchida, T. and Sugie, T., 1987, Full term development of mouse blastomere nuclei transplanted into enucleated two-cell embryos, *J. Exp. Zool.* 242:147.

Von Beroldington, C.H., 1981, The developmental potential of synchronized amphibian cell nuclei. *Dev. Biol.* 81:115.

Wakefield, L. and Gurdon, J.B., 1983, Cytoplasmic regulation of 5 S RNA genes in nuclear transplant embryos, *EMBO J.* 2:1613.

Wall, R.J., Purssel, V.G., Hammer, R.E. and Brinster, R.L., 1985, Development of porcine ova that were centrifuged to permit visualization of pronuclei and nuclei. *Biol. Reprod.* 32:645.

Wall, R.J. and Hawk, H.W., 1988, Development of centrifuged cow zygotes in rabbit oviducts, *J. Reprod. Fert.* 82:637.

Weismann, A., 1915, Das keimplasma. Eine theorie de vererbung. Fischer 1892. English translation W.N.Parker and H. Ronfeld, Schribner's Sons, New York.

Willadsen, S.M., 1986, Nuclear transplantation in sheep embryos. *Nature* 320:63.

MMTV/N-ras TRANSGENIC MICE AS A MODEL FOR
ALTERED CAPACITATION MALE STERILITY AND TUMORIGENESIS

Ramon Mangues,[1] Irving Seidman,[1] Angel Pellicer,[1]
and Jon W. Gordon[2*]

*Department of Pathology and Kaplan Cancer Center
New York University Medical Center, New York, NY 10016 [1]
and Molecular Biology Center, Department of Geriatrics and Adult Development
and Department of Obstetrics and Gynecology and Reproductive Science
Mount Sinai School of Medicine, New York, NY 10029[2]*

INTRODUCTION

Infertility is generally defined as a failure to conceive following one year of unprotected intercourse; it affects about 10% to 15% of couples attempting a pregnancy for the first time (Romeny et al., 1975). Abnormalities in the male partner contribute to infertility in about 40% of these couples (Speroff et al., 1983).

A major problem that confronts physicians who treat the infertile couple is the inability to diagnose a particular cause of infertility in the male partner who has normal circulating levels of reproductive hormones (Smith et al., 1987). In most of the cases the cause of male infertility remains obscure, and the clearly defined causes are infrequent or rare (Sherins et al., 1986).

The established approaches to the study of male infertility are insufficient for unveiling its molecular basis. The classic semen analysis which defines sperm concentration, motility, forward progression and morphology does not reflect the functional competence of the spermatozoa. The occurrence of the conception cannot be predicted on the only basis of concentration and morphology of the spermatozoa in the ejaculate (Smith et al., 1977; van Zyl et al., 1975).

Advances in Assisted Reproductive Technologies, Edited by
S. Mashiach *et al.,* Plenum Press, New York, 1990

A major impact on medical practice has had the introduction of the sperm penetration assay because of the high correlation of its outcome with the fertilizing capacity in vivo and in vitro (Yanagimachi et al., 1976; Aitken et al., 1984; Aitken et al., 1985). It is known that the gamete membrane fusion occurs only if the spermatozoa have been capacitated and have undergone acrosome reaction (Moore et al., 1983). The zona-free hamster egg penetration assay apparently evaluates the ability of human sperm to capacitate, to undergo acrosome reaction, and to fuse with the vitelline membrane of the oocyte.

The availability of an in vitro test end point which represents the complete fertilizing capacity of the spermatozoa, should direct additional efforts to further dissect the mechanisms responsible for the impairment at each step of the multiple involved in the spermatogenic process.

Phenotypically identified, single gene mutations that have arisen spontaneously in inbred mice constitute important experimental approaches to the development of genetically defined mouse models of human diseases and they are being used to study human sterility.

Molecular genetics have provided valuable new tools for producing mutations injecting cloned DNA into the male pronucleus of a mouse zygote, which eventually will stably integrate into the genome and will be transmitted to the progeny (Gordon et al., 1980; Gordon et al., 1981). This approach provides potentially important tools for defining the requirements for successful male reproduction.

Using this technique we have generated a transgenic model which carries the N-ras activated oncogene under the control of the mouse mammary tumor virus (MMTV) promoter. The phenotypes seen in these animals include male sterility and neoplastic growth. The exploitation of the former one as a model for study of the infertility in humans is the main issue addressed in this paper.

Oncogenes are genes able to affect the normal cell growth and differentiation, resulting in cells with altered phenotype that can grow as tumors in animals. Proto-oncogenes are the unaltered forms of the oncogenes. They are highly conserved in evolution and, in contrast to their activated counterparts, appear to provide essential physiological functions in cell growth and differentiation in normal embryonic development as well as in the adult tissues. N-ras, first found in a human neuroblastoma cell line (Shimizu et al., 1983), belongs to the ras gene family which includes two more members, H-ras (DeFeo et al., 1981) and K-ras (Ellis et al., 1981) that were originally identified as the transforming principles of the Harvey and Kirstein strains of rat sarcoma virus, two acute transforming retrovirus generated by transduction of the respective cellular proto-oncogenes. All of them have a very similar exon-intron structure and encode strikingly homologous 21 Kd proteins (p21) Bos, 1989).

Some insight into the biological role of ras p21 has been gained from studying its structure and biochemical properties which has established a relation with the guanine-nucleotide-binding regulatory proteins (G proteins), although its specific function remains to be established.

We have analyzed the transgene expression in different organs and describe the molecular and functional studies directed to localize and decipher the mechanism by which the male sterility and neoplastic growth are induced. Using this model we hope to gain some insight into the molecular aspects of the spermatozoon differentiation and N-ras physiological role.

MATERIALS AND METHODS

Production of Transgenic Mice

Transgenic mice were produced as previously described (Gordon et al., 1980; Gordon et al., 1983). Briefly, immature B6D2F1 female mice were superovulated and mated to CD-1 males. The following morning the females were examined for the presence of vaginal plugs, and that afternoon the plugged females were sacrificed.

Fertilized ova were recovered and microinjected with a 9 Kb PstI-SmaI fragment (MMTV/N-ras oncogene) excised from the SP6-4 plasmid vector (Figure 1) and separated

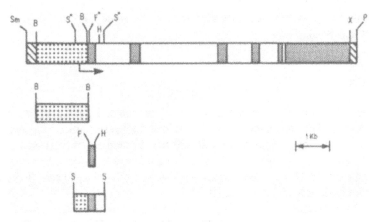

Fig. 1. Construction of the MMTV/N-ras fusion gene. The MMTV-LTR sequences contained within a 1.45 Kb BamHI fragment of pLTR2 plasmid were fused upstream to a Fnu4HI-XhoI fragment derived from a subclone of the N-ras oncogene (including 1st to part of the 6th exons) cloned from a NMU-induced mouse thymic lymphoma, as described (Guerrero et al., 1984). The low density stippled box represents the MMTV-LTR promoter sequences. The high density stippled boxes indicate the N-ras exons. The open boxes correspond to the intronic sequences, the hatched boxes represent the SP6-4 polylinker restriction sites. The arrow indicates the initiation of transcription. The code for enzymes is: X, XhoI; P, PstI; S, SacI; F, Fnu4HI; H, HaeIII; Sm, SmaI; B, BamHI. The asterisks indicate restriction sites not mapped throughout the gene.

by sucrose density gradient centrifugation. Approximately 2000 copies of the material were microinjected into each pronucleus.

Screening of founder animals was performed by extraction of DNA from spleen (Gordon, 1986), digestion with XbaI, and Southern hybridization using the MMTV/N-ras as a probe. Screening of offspring was performed by dot blot hybridization of 1 µg of tail DNA (Gordon, 1986).

DNA Analysis

Genomic DNA was extracted from tissues by sodium dodecyl sulfate lysis and proteinase K digestion, followed by phenol-chloroform-isoamyl alcohol extraction and ethanol precipitation (Ausubel et al., 1989). Restriction endonuclease digestions were carried out under the conditions recommended by the supplier. Twenty µg of DNA per sample were electrophoresed on 0.8% agarose gels and transferred to Biotrans nylon filters (ICN Biochemicals). The filters were then dried for one hour at 80°C and hybridized as previously described (Ausubel et al., 1989). Probes were labeled by random primed synthesis (Feinberg et al., 1984) or nick translation (Rigby et al., 1977). The filters were washed three times at 50°C in 0.1X SSC in 0.1% sodium dodecyl sulfate for 15 min, and autoradiographed.

The probe for DNA analysis was a 147 nt Fnu4HI-HaeIII fragment containing the first exon of N-ras oncogene previously isolated from a chemically induced thymic lymphoma (Guerrero et al., 1984).

RNA Analysis

RNA was isolated by guanidine thiocyanate extraction and CsCl gradient fractionation (Chirgwin et al., 1979). RNA yield was determined after resuspension, in diethylpyrocarbonate treated water, by UV absorption at 260 nm. Fifteen µg of total RNA per sample were electrophoresed on 1% formaldehyde gels transferred to nylon filters and hybridized with the first exon of N-ras probe labeled by random primed synthesis (Feinberg et al., 1984) or nick translation (Rigby et al., 1977).

The N-ras probe was the same as used for DNA analysis. The 0.24-9.5 Kb RNA ladder (Bethesda Research Laboratories) was used as RNA marker. Hybridizations and washes were carried out as described above. To ensure that equal amounts of RNA were loaded on the gels, the filters were hybridized to a b-actin probe (not shown).

RNase Protection

An 800 nt SacI fragment from the injected construct, which included the 3' most 374 nt from the MMTV LTR and 426 nt from N-ras, subcloned in Gem-3Z (Figure 1) was used to generate an antisense RNA probe with SP6 RNA polymerase.

The RNase protection assay was carried out using 10 μg of total RNA per sample, according to the following procedure (Ausubel et al., 1989): samples were precipitated in ethanol and redissolved in 30 μl hybridization buffer (40 mM PIPES pH 6.4, 400 mM NaCl, 1 mM EDTA, 80% formamide), containing 5×10^5 cpm of the probe RNA, denaturated at 85°C and incubated at 30°C for 10 hours. 350 μl ribonuclease digestion buffer (10 mM Tris-HCl, pH 7.5, 300 mM NaCl, 5mM EDTA) containing 40 μg/ml ribonuclease A and 2 μg/ml ribonuclease T1 were added to each hybridization reaction and incubated 30 min at 30°C. After adding 10 μl of 20% (wt/vol) SDS and 2.5 μl of 20 mg/ml proteinase K, the samples were incubated 15 min at 37°C. They were subsequently extracted with phenol/chloroform/isoamyl alcohol and precipitated in ethanol with carrier yeast tRNA. The pellets were redissolved in RNA loading buffer (80% v/v formamide, 1 mM EDTA, pH 8.0, 0.1% bromophenol blue, 0.1% xylene cyanol) and analyzed on denaturing 6% polyacrylamide/7M urea gel.

Histological Evaluation

Animals were sacrificed, tissues were taken, fixed in 10% formaldehyde and embedded in paraffin. Subsequently the blocks were sectioned at 4 μm, the preparations were routinely stained with hematoxylin and eosin and examined for pathological findings.

In Vitro Fertilization

In vitro fertilizations were performed as described by Gordon and Talansky (1986). Caudae epididymides and vasa deferentia were recovered from transgenic and control males and incubated in capacitation medium (Thadani, 1982) for two hours.

Unfertilized eggs were recovered by superovulation, 13.5 hours after administration of hCG. Oocytes were inseminated in 30 μl microdrops under mineral oil with 1×10^6 sperm/ml. After six to eight hours eggs were examined for the appearance of a second polar body and pronuclei. To confirm negative results, unfertilized eggs were incubated overnight and examined the following day for cleavage to the two-cell stage. In no case were eggs scored as unfertilized in initial examination found to cleave overnight.

When zona drilling was performed, the procedure was carried out exactly as previously described by Gordon and Talansky (1986). Zona-free inseminations were performed by incubating the oocytes in acid Tyrode's solution until the zonae were nearly completely dissolved, then transferring the eggs to M16 culture medium (Quinn et al., 1982) and gently pipetting the eggs until all remnants of the zonae were removed.

RESULTS

Establishment of Transgenic Lines

Six founder animals carrying between five and 50 copies/cell of the transgene were identified. Founder No. 5, a male, developed unilateral Harderian gland hypertrophy and was sterile. All other founders were able to breed, but three of them, one male and two females, did not transmit the transgene to progeny. The other two male founders, designated 4 and 7, transmitted the foreign DNA to progeny as a Mendelian trait at least once. The pedigrees were analyzed in detail for MMTV/N-ras expression and associated pathology.

In the line derived from founder No. 4, 14 out of twenty-four animals were transgenics, 10 of them resulting from crosses with wild type and four from crosses between F_1 transgenics.

In line 7, eleven out of fifteen animals had incorporated the transgene. This founder failed to sire additional progeny despite multiple matings prior to its death.

Expression of the Transgene

Once transgenic pedigrees were established it was of interest to determine the expression pattern of the oncogene in different organs in order to correlate it with subsequent pathological changes.

The expression of the fusion gene is due to the induction of the MMTV promoter as shown by Northern analysis (Figs. 2 and 3) and RNase protection studies (Figure 4). For Northern analyses, total RNA from tissues was hybridized with the first exon of N-ras as described in methods. The start site of transcription is located within the MMTV promoter as shown in the RNase protection assay in Figure 4. This allows highly specific detection of transgene expression. Transcripts initiated at the MMTV cap site protect a 438 nt fragment of labeled DNA, while RNA arising from the endogenous N-ras promoter protect a 120 nt fragment. Novel transcripts originating from the newly introduced construct, as shown in Figure 4, are absent in nontransgenic animals.

Higher levels of expression were observed in the organs with more significant abnormalities. Salivary gland, mammary gland and Harderian gland showed a consistent high level of expression in most of the animals. Intermediate levels of expression were seen in lung, prostate, testis and ovary.

Other tissues where we detected low level of expression include: thymus, brain, muscle, skin, bone marrow, kidney, gut and heart. We did not see any expression of the construct in pancreas or liver, although detection of the endogenous N-ras transcripts indicated adequate assay sensitivity.

RNA transcripts generated in representative samples of testis and lung, and ovary and submaxillary gland, in animals from both 4 and 7 transgenic lines are shown in Figures 2 and 3, respectively. The sizes of the transgene transcripts were different from those of the endogenous proto-oncogene, as expected from the different lengths of the MMTV promoter (Guerrero et al., 1988) and N-ras 5' untranslated sequences (Figure 1, Paciucci and Pellicer, unpublished).

The level of transgene expression can also be seen in representative samples of a different set of tissues in the RNase protection assay in Figure 4.

Pathological Alterations

Grossly, pathological changes exhibited distinct patterns that were sex-specific. Males showed a far greater incidence and severity of Harderian gland hypertrophy (Figure 5A), while females developed breast tumors (Figure 5B). The organs presenting proliferative alterations were Harderian gland, mammary gland and salivary gland. Harderian glands showed variable degrees of pathology: four benign growths in males and two hyperplasias in females were seen in the analyzed animals.

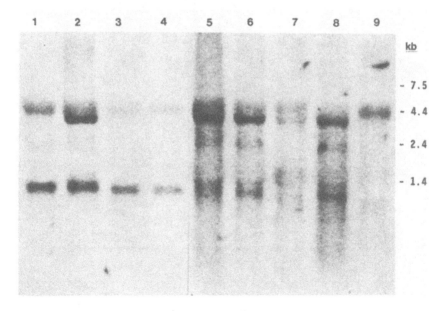

Fig. 2. Northern blot analysis of total RNA from testis and lung of line 4 and line 7 mice carrying the MMTV-N-ras transgene transcripts. Hybridization with the 1st exon N-ras (147 nt Fnu4HI-HaeIII fragment) probe. Time of exposure: 4 days. Lanes: 1, line 4 testis; 2 and 3, line 7 mice testis; 4, nontransgenic mouse testis; 5 and 6, line 4 female mice lung; 7 and 8, line 7 male mice lung; and 9, nontransgenic lung.

Both the incidence and the degree of the pathological findings in Harderian glands were higher in males than females. Six out of the eight males in line 7 presented Harderian gland hypertrophy, evidenced by bilateral exophthalmos and a change in the color of the gland from grey-brown to white-yellow. Although we saw some degree of hypertrophy (in most cases unilateral) in some of the female animals from the same line no exophthalmus was recorded. Two multifocal mammary adenocarcinomas were registered in female animals only. No benign hyperplastic growth has been seen so far in mammary tissue. In salivary gland, there was also variation in the severity of disease. We registered two instances of focal proliferation of the ducts, one in a male and one in a female, and an adenocarcinoma in a male animal. In lung, two metastatic carcinomas were recorded: one secondary to a primary mammary carcinoma and the second from an undetermined primary.

The latency period in Harderian gland tumors was shorter than in the other two malignancies. Figure 6 shows the histologic appearance of the mammary, submaxillary and Harderian tumors and lung metastases, occurring in mice from lines 4 and 7.

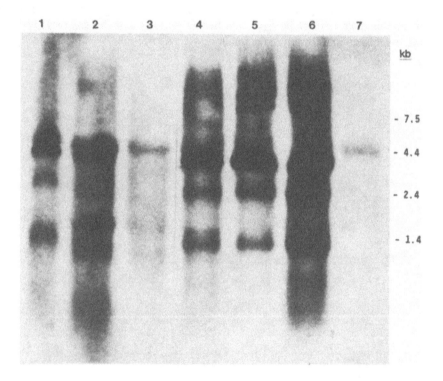

Fig. 3. Northern blot analysis of total RNA from ovary and submaxillary gland of line 4 and line 7 mice carrying the MMTV-N-ras transgene transcripts. Hybridization with the 1st exon N-ras (147 nt Fnu4HI-HaeIII fragment) probe. Time of exposure: three days. Lanes: 1 and 2, line 4 ovaries; 3, nontransgenic mouse ovary; 4, line 4 female mouse submaxillary gland; 5 and 6, line 7 male mouse submaxillary gland; and 7, nontransgenic submaxillary gland.

Male Infertility

A striking phenotypic feature of these transgenic mice was a severe compromise of male reproductive function. One male founder (No. 5) was never able to sire offspring despite an aggressive program of mating to outbred females. This animal died suddenly and was not available for necropsy. Founder No. 7 sired one litter of 15 progeny, 11 of which were transgenics. However, this animal subsequently became sterile. Moreover, all of its male offspring were totally sterile.

After several unsuccessful attempts to establish pregnancies from documented matings, F_1 males from the No. 7 line were tested for fertility by in vitro fertilization (IVF). One of

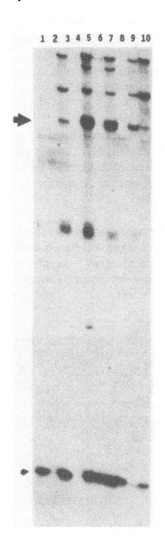

Fig. 4. RNase protection analysis. RNAs are transcripts corresponding to the MMTV/N-ras transgene in various tissues of transgenic lines 4 and 7. An 800 nt SacI fragment including the 3' most 374 nt from the MMTV-LTR and 426 nt from N-ras (Figure 1), subcloned in Gem-3Z, was used to generate the antisense RNA probe for this assay. A 120 nt protected fragment (indicated by the small arrow) is obtained for endogenous N-ras and a 438 nt protected fragment (indicated by the large arrow) is obtained only for transgenic animals. Lanes: 1, nontransgenic testis; 2, nontransgenic submaxillary gland; 3, line 4 mouse testis; 4, nontransgenic submaxillary gland; 5, line 7 mouse testis; 6, line 4 mouse testis; 7, line 7 mouse testis; 8, nontransgenic mouse thymus; 9, line 4 mammary gland; and 10, line 4 mouse brain.

the males was sacrificed at 12 months of age for this procedure. Secondary sex structures were grossly normal and sperm counts were normal. However, sperm were grossly deficient in motility, with only 10% of cells being motile. Moreover, motility characteristic of normal epididymal maturation (Yanagimachi et al., 1981) was never observed. When these sperm were used for IVF, no fertilization was observed, even when oocytes were subjected to zona drilling (Gordon et al., 1986) or complete removal of the zonae. This contrasted with results from a nontransgenic littermate, in which fertilization was 44% with zona intact eggs, and 100% with zona drilling or zona removal. A second F_1 male from line No. 7 was evaluated at 15 months. This animal had 0% sperm motility, and in most sperm, the heads were detached from the tails. Because of the complete absence of sperm motility, IVF was not attempted.

Line No. 4, while able to breed, also manifested reduced male fertility. After 12 months of age the founder and F_1 males were unable to sire progeny. A similar observation was made on one F_2 male at 12 months, though a transgenic littermate remained fertile. The infertile F_2 male was sacrificed for sperm examination. Only three of 167 cells counted were motile, and approximately 20% of sperm had detached heads. After IVF, 0/33 eggs were fertilized.

The only normally fertile transgenic male was founder No. 8. Southern blot hybridization suggested the presence of 1–5 copies/cell of the transgene. However, this animal failed to transmit the foreign gene to 30 offspring; therefore we conclude that this animal was a probable genetic mosaic, and was fertile because of the presence of the transgene in only a small subset of cells.

In summary, all pedigrees carrying the MMTV/N-ras construct manifested reduced male fertility which varied in its time of onset. In line No. 5 the founder was sterile, while in line 7 the founder was able to breed only once and all of its male progeny were infertile from birth. Line No. 4, the least severely affected, showed reduced fertility by 12 months, with several animals being completely infertile.

 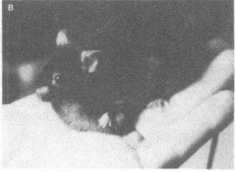

Fig.. 5A   Fig. 5B

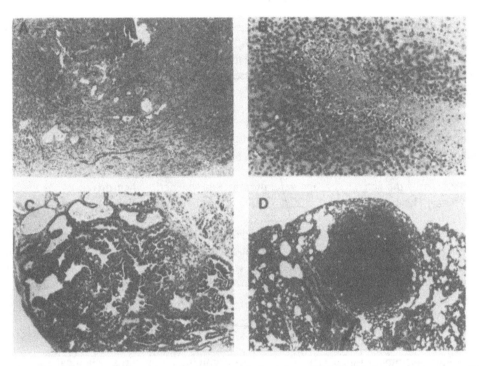

Fig. 6. Histologic appearance of tumors in tissues derived from MMTV/Nfiras transgenic mice. A) Carcinoma of mammary gland in a line 7 female (original magnification [OM] 4×). B) Submaxillary gland carcinoma in a male from line 4 [OM 10×]. C) Harderian gland tumor in a line 7 male. [Om 4×]. D) Metastatic nodule of mammary carcinoma in the lung of a line 4 female ([OM 4×].

DISCUSSION

The phenotypes described here as the consequences of the abnormally expressed N-ras oncogene help to approach the possible role of the N-ras protein in proliferation and differentiation in some mice tissues. These experiments establish, at the same time, a model for studying specific cases of male infertility.

The ability to transfer genetic information stably into the germ line of the laboratory mice has presented a powerful new tool with which to investigate the consequences of gene expression on specific cells in the context of the whole organism, and thereby to probe and perturb the complex processes involved in cell proliferation or/and differentiation.

In previous experiments, we repeatedly injected the N-ras oncogene protein coding information attached to its own promoter in the male pronucleus of a fertilized mouse egg and afterwards insert it into the oviduct of a pseudopregnant female, none of the newborn mice had integrated the transgene. We attributed this finding to the lethal effect of the high expression of the activated N-ras gene during embryogenesis when driven by its natural promoter. The expected output is 20%–30% of the newborn mice being transgenics.

Consequently, we turned to a hybrid construction for making the transgenics, using regulatory elements derived from another gene and aligned to transcribe the protein coding information from the activated N-ras. This approach presumes that tissue specificity resides in the 5' flanking region of a gene where the transcriptional promoter elements, as in most cases the enhancers, are located. We decided to recombine the mouse mammary tumor virus promoter (MMTV-LTR) with the N-ras oncogene in order to direct the synthesis of the oncoprotein in the cell types that MMTV-LTR specify.

Mouse mammary tumor virus is an RNA-containing tumor virus which causes mammary cancer in mice. Its transcription is regulated by steroid hormones; androgens have been demonstrated to induce the MMTV (Darbre et al., 1986) along with the progestagens and glucocorticoids, while the estrogens have not (Otten et al., 1988). It belongs to the group of retroviruses which do not carry an oncogene. As the other retroviruses, it replicates via a DNA intermediate. Unintegrated viral DNA as well as the integrated provirus are flanked by long terminal repeats (LTR), which arise during reverse transcription of viral RNA (Hsu et al., 1978). The LTRs play a role in viral DNA integration, RNA transcription, and the reverse transcription process (Hynes et al., 1984).

The MMTV most likely induces mammary tumors by insertional mutagenesis, through integration as a provirus next to a cellular proto-oncogene (int-1, int-2 or int-3) and activation of its expression (Varmus, 1984). It might be inferred that MMTV shows some cell-type specificity in its expression as well as its infectivity, though MMTV antigens have been detected in organs where it does not produce any alteration: seminal vesicle, prostate and coagulating gland (Imai et al., 1983).

The tissue specificity of MMTV has been repeatedly addressed in transgenic mice experiments where MMTV-LTR was combined with several oncogenes and it was concluded that in spite of the apparent tissue specificity of the virus the MMTV-LTR has a reasonable broad transcriptional specificity. In all cases studied MMTV-LTR invariably directed the expression to mammary, salivary and Harderian glands. In addition, and depending on the oncogene attached to this promoter a variety of other organs expressed the transgene, including lung, kidney, lymphoid tissues, ovary, testis. Different hybrid oncogenes produced distinct pathologies suggesting a cell specificity for oncoprotein action.

Most of the organs where we have detected expression of the MMTV/N-ras transgene have been identified as sites of expression for the other MMTV/oncogene constructs like c-myc (Stewart et al., 1984; Leder et al., 1986; Sinn et al., 1987), SV40 T and t antigens (Choi et al., 1987; Choi et al., 1988) c-neu (Muller et al., 1988; Bouchard et al., 1989) and v-H-ras (Sinn et al., 1987; Tremblay et al., 1989). The pathology we have registered could be attributed to the action of the construct as none of the nontransgenic littermates showed the alterations described in the results. Moreover, the RNase protection assay showed transgene transcripts initiated at the MMTV promoter in all pathological tissues.

Most of the phenotypes found for the MMTV/N-ras transgene are associated with proliferation (hypertrophy, hyperplasia or neoplasia). However, male sterility registered in these animals is an unusual finding in transgenic mice produced so far, including those that are carrying an oncogene and those that are not.

SV40 T antigen (Behringer et al., 1988), Py large T (Bautch et al., 1987) and Py middle T (Bautch et al., 1987) when attached to MMTV-LTR are expressed in testis without detectable consequence. On the contrary when c-neu was used for making the hybrid no expression in testis was detectable, although the expression of the transgene in epididymis was associated with bilateral hyperplasia of this structure and delayed infertility (Muller et al., 1988 and Brouchard et al., 1989). Enlargement of the seminal vesicle in these animals has also been reported (Brouchard et al., 1989). c-neu was the only oncogene which expression have been associated with pathological findings in the male reproductive system. No studies were done about the functional capability of the spermatozoa, thus no conclusion could be drawn about whether a specific interaction of the MMTV/c-neu in the process of sperm maturation or mechanical occlusion secondary to tissue hyperplasia was the cause of such pathology (Muller et al., 1988).

c-myc oncogene when driven by MMTV promoter is able to produce Sertoli cell neoplasm (Leder et al., 1986).

SV40 antigen is expressed in seminal vesicle, prostate and epididymis (Choi et al., 1987, Choi et al., 1988) and v-H-ras in testis and seminal vesicle (Tremblay et al., 1989) again without neoplastic growth or function alteration, when recombined with MMTV promoter.

Male sterility was directly associated with the reduced motility of the spermatozoa and its failure to penetrate even zona-free eggs, in the transgenic mice that we have produced, as shown in the in vitro fertilization studies and could be associated to an epidymal malfunction more than to a testicular cause.

The testis and epididymis are responsible for the production and maturation, respectively, of the spermatozoa. In the testis, the spermatozoon resulting from meiosis and differentiation has assumed its final shape and size, but it is still incapable of progressive motility and is unable to fertilize an egg. Sometime during its epididymal transit these two functions are acquired, however, the mechanism by which the epididymis exerts these changes on the transversing spermatozoon are largely unknown (Ewing et al., 1986). In most of the animals the motility and the ability to adhere zona pellucida and fertilize the egg are gradually acquired as the spermatozoa migrate into the distal region of the epididymis (Orbenin-Crist et al., 1969). Recent evidence suggests that the proteins produced and secreted to the lumen by the caput and corpus epididymis under androgenic stimulation may act as mediators of some of the epididymal influences on spermatozoa during maturation (Bardin et al., 1981; Orbenin-Crist et al., 1982; Tezon et al., 1987; Brooks, 1987; Holland et al., 1988a).

The association of N-ras oncogene expression with the impairment of the male reproductive system seen in the transgenic mice reported here sharply contrasts with the absence of any alteration in the system induced by H-ras, which encodes for a protein strikingly homologous to N-ras protein, when driven by the same promoter and expressed in the same tissues. This fact and the differential expression of the ras proto-oncogenes during the development (Leon et al., 1987) and spermatogenesis (Sorrentino et al., 1989) in mice strongly suggest a differential role for both proto-oncogenes in these processes.

Some proto-oncogenes are expressed and are likely to have a function in the spermatogenesis process in testis. raf-1, fos, and K-ras are predominately expressed before or during the meiosis (pachytene spermatocites), whereas mos, int-1, abl, pim-1 and N-ras have been detected in postmeiotic haploid spermatids (Sorrentino et al., 1989). H-ras is expressed in all stages of spermatogenesis. Moreover, there are testis specific transcripts of unusual size in early spermatids for pim-1, abl and mos that differ from those found in somatic cells(Propst et al., 1988a) and a differential temporal regulation in the expression of proto-oncogenes have been reported, making more evident their involvement in germ cell differentiation. Postmeiotic early spermatids express a pattern of genes different from the one seen in pachytene cells signalling that transcripts result from the novo transcription once meiosis is finished (Sorrentino et al., 1989).

Curtis Chubb (1989) has carried out a comprehensive study in inbred mice bearing single-gene mutations that induce infertility in males grouping the gene mutations according to the primary cause of infertility. The pretesticular or extratesticular endocrine disorders group includes mice homozygous for Ames dwarf (df/df) or dwarf (dw/dw) gene mutation and hypogonadal mice (hpg/hpg).

Grouped in primary testicular defects were atrichosis (at/at), sex reversed (X/X Sxr), Steel Dickie (Sl/Sl^d) and viable dominant spotting (W/W^v) as animal models for the Sertoli-cell only syndrome, and Blind-sterile (bs/bs), Flipper (fl/fl) oligotriche (olt/olt), quacking (qk/qk), testicular feminization (Tfm/Y) and hybrid sterile Hst-1^s/Hst-1^{ws}) mice.

Sterility was associated with posttesticular causes only in stubby mice (stb/stb), in which this pathology is due to impotence (Curtis, 1989). Holland and Orbenin-Crist (1988b) have recently described a mutant mouse model with normal sperm count, morphology and motility which sterility is attributed to the alteration of the process of sperm-egg interaction at the level of zona pellucida and vitelline membrane and which epididymis is apparently lacking a 18 Kd protein present in the wild type.

Proto-oncogene expression has been studied in some sterile mouse carrying spontaneously arisen or induced single gene mutations. In sex reversal (Sxr) and steel (Sl) mutants, which are lacking spermatids, none of the oncogenes described to be expressed in normal testis were present. Mutants such as testicular feminization (Tfm), hypogonadal (hpg) and mice carrying X-autosomal translocations, in which meiosis is blocked, lack transcripts in testicular tissue for mos and abl (Propst et al., 1988a) and int-1 (Shackleford et al., 1987) which expression is seen in round spermatids in normal testis. Quaking (qk) and some t-locus mutants, which germ cells are capable of completing spermatogenesis, but produce sperm incapable of fertilizing eggs, express testis specific mos and abl (Propst et al., 1988b) and int-1 transcripts (Shakleford et al., 1987).

Three transgenic models have been described to produce male sterility.

The IFN-b genes driven by mouse metallothionein I enhancer-promoter in which the expression of the transgene produce involuted testes containing few mature sperm and spermatocyte and spermatids degeneration (Iwakura et al., 1988). IFN-b is not expressed in normal testis and the high level of IFN seen in these transgenics in this tissue suggest a toxic effect on the spermatogenic process.

When major urinary protein (Mup) gene promoter was fused to the coding region of HSV thymidine kinase gene, testicular expression was the likely cause of male sterility observed (Al-Shawi et al., 1988). The same gene attached to MMTV promoter also causes male sterility (Ross et al., 1985), however when HSV TK is attached to its own promoter a low level of gene expression in testis is detected and no sterility produced. Thus, HSV tk gene is able to perturb spermatogenesis only when recombined to strong promoters that specify for testis expression.

None of the models described so far have been associated to specific cell malfunction, lacking studies about the mechanism by which the mutation or transgene affect spermatozoa function.

The tools to dissect the complex processes of spermatogenesis and epididymal maturation in a definite number of connected steps are now available and consequently it is possible to define the requirements for successful male reproduction. If any one of the necessary steps is damaged it will result in sterility. The model presented here will be useful in investigating the posttesticular causes of male infertility specifically the epididymal malfunction.

ACKNOWLEDGEMENTS

We thank Elizabeth W. Newcomb and Beth E. Talansky for generous assistance and helpful discussion, Isabel Guerrero for making the construction and Robert Lake for excellent technical assistance.

R.M. acknowledges support from Spanish Fondo de Investigaciones Sanitarias de la Seguridad Social (88/2797). These studies were conducted with support from CA 36327 and CA 16239 to A.P. and CA 42103, HD 20484 NIH Grants, and March of Dimes Grant 1-1026 to J.W.G. A.P. is a Leukemia Society Scholar.

REFERENCES

Aitken, R.J., Best, F. S. M., and Wagner, P., 1984, A prospective study of the relationship between semen quality and fertility in cases of unexplained infertility, *J. Androl.*, 5:297-303.

Aitken, R.J., 1985, Diagnostic value of the zona-free hamster oocyte penetration test and sperm movement characteristics in oligozoospermia, *Int. J. Androl.*, 8:348-356.

Al-Shawi, R., Burke, J., Jones, C. T., Simons, J. P., and Bishop, J. O., 1988, A Mup promoter-thymidine kinase reporter gene shows relaxed tissue-specific expression and confers male sterility upon transgenic mice, *Mol. Cell Biol.* 8(11):4821-4828.

Ausubel, F. M., Brent, R., Kingston, R. E., Moore, D. D., Seidman, J. G., Smith, J. A., and Struhl, J., (eds) 1989, Current Protocols in Molecular Biology, Greene Publishing Associates and Willey-Interscience, New York.

Bardin, C. W., Musto, N., Gunsalus, G., Kotite, N., Cheng, S.-L., Larrea, F., and Becker, R., 1981, Extracellular androgen binding proteins, *Ann. Rev. Physiol.*, 43:189-198.

Bautch, V. L., Toda, S., Hassell, J. A., and Hanahan, D., 1987, Endothelial cell tumors develop in transgenic mice carrying polyoma virus middle T oncogene, *Cell*, 51:529-538.

Behringer, R. R., Peschon, J. J., Messing, A., Gartside, C. L., and Hauschka, S. D., 1988, Heart and bone tumors in transgenic mice, *Proc. Natl. Acad. Sci., USA*, 85:2648-2652.

Bos, J. L., 1989, ras oncogenes in human cancer: a review, *Cancer Res.*, 49:4682-4689.

Bouchard, L., Lamarre, L., Tremblay, P. J., and Jolicoeur, P., 1989, Stochoastic appearance of mammary tumors in transgenic mice carrying the MMTV/c-neu oncogene, *Cell*, 57:931-936.

Brooks, D. E., 1987, Androgen-regulated epididymal secretory proteins associated with posttesticular sperm development, *In:* Orbegin-Crist M.-C., and Danzo, B. J., eds., *Ann. N.Y. Acad. Sci.*, 513:179-194.

Chirgwin, J. J., Przbyla, A. E., MacDonald, R. J., and Rutter, W. J., 1979. Isolation of biologically active ribonucleic acid from sources enriched in ribonuclease, *Biochemistry*, 18:5294-5299.

Choi, Y., Henrard, D., Lee, I., and Ross, S. R., 1987, The mouse mammary tumor virus long terminal repeat directs expression in epithelial and lymphoid cells of different tissues in transgenic mice, *J. Virol.*, 61:3013-3019.

Choi, Y., Lee, I., and Ross, S. R., 1988, Requirement for simian virus 40 small tumor antigen in tumorigenesis in transgenic mice, *Mol. Cell Biol.*, 8(8):3382-3390.

Darbre, P., Page, M., and King, R. J. B., 1986, Androgen regulation by the long terminal repeat of mouse mammary tumor virus, *Mol. Cell Biol.*, 6:2847-2854.

DeFeo, D., Gonda, M. A., Young, H. A., Chang, E. H., Lowy, D. R., Scolnick, E. M., and Ellis, R. W., 1981, Analysis of two divergent rat genomic clones homologous to the transforming gene of Harvey murine sarcoma virus. *Proc. Natl. Acad. Sci., USA*, 78:3328-3332.

Ellis, R. W., DeFeo, D., Shih, T. Y., Gonda, M. A., Young, H. A., Tsuchida, N., Lowy, D. R., and Scolnick, E., 1981, The p21 src genes of Harvey and Kirsten sarcoma viruses originate from divergent members of a family of normal vertebrate genes, *Nature*, 292:506-511.

Ewing, L. L., and T. S. K. Chang, 1986, The testis, epididymis, and ductus deferens, *in:* "Campbell's Urology," 5th ed. vol. 1, P. C. Walsh, R. F. Gittes, A. D. Perlmutter, and T. A. Stamey, eds., W. B. Saunders Company, Philadelphia, 200-233.

Feinberg, A. P., and Vogelstein, B., 1984, A technique for radiolabeling DNA restriction endonuclease fragments to high specific activity, *Anal. Biochem*, 137:266-267.

Gordon, J. W., 1986, A foreign dihydrofolate reductase gene in transgenic mice acts as a dominant mutation, *Mol. Cell Biol.*, 6:258-267.

Gordon, J. W., and Ruddle, F. H., 1981, Integration and stable germ line transmission of genes injected into mouse pronuclei, *Science*, 214: 1244-1246.

Gordon, J. W., and Ruddle, F. H., 1983, Gene transfer into mouse embryos: production of transgenic mice by pronuclear injection, *Meth. Enzymol.*, 101:411-433.

Gordon, J. W., Scangos, G. A., Plotkin, D. J., Barbosa, J. A., and Ruddle, F. H., 1980, Genetic transformation of mouse embryos by microinjection of purified DNA, *Proc. Natl. Acad. Sci. USA*, 77:7380-7384.

Gordon, J. W., and Talansky, B. E., 1986, Assisted fertilization by zona drilling: a mouse model for correction of oligospermia, *J. Exp. Zool.*, 239:347-354.

Guerrero, I., Pellicer, A., and Burstein, D. E., 1988, Dissociation of c-fos from ODC expression and neuronal differentiation in a PC12 subline stably transfected with a inducible N-ras oncogene, *Biochem. Biophys. Res. Com.*, 150:1185-1192.

Guerrero, I., Villasante, A., D'Eustachio, P., and Pellicer, A., 1984, Isolation, characterization, and chromosome assignment of mouse N-ras gene from carcinogen-induced thymic lymphoma, *Science*, 225 (4666):1041-1043.

Holland, M. K., and Orbegin-Crist, M.-C., 1988a, Characterization and hormonal regulation of protein synthesis by the murine epididymis, *Biol. Reprod.*, 38:487-496.

Holland, M. K., and M.-C. Orbegin-Crist, 1988b, Epididymal protein synthesis and secretion in strains of mice bearing single gene mutations which affect fertility, *Biol. Reprod.*, 38:497-510.

Hsu, T. W., Sabran, J. L., Mark, G. E., Guntaka, R. N., and Taylor, J. M., 1978, Analysis of unintegrated avian RNA tumor virus double-stranded DNA intermediate, *J. Virol.*, 28:810-818.

Hynes, N. E., Groner, B., and Michalides, R., 1984, Mouse mammary tumor virus: transcriptional control and involvement in tumorigenesis, *Adv. Cancer Res.*, 41:155-184.

Imai, S., Morimoto, J., Tsubura, Y., Iwai, Y., Okumoto, M., Takamori, Y., Tsubura, A., and Hilgers, J., 1983, Tissue and organ distribution of mammary tumor virus antigens in low and high mammary cancer strain mice, *Eur. J. Cancer Clin. Oncol.*, 19(7):1011-1019.

Iwakura, Y., Asano, M., Nishimune, Y., and Kawade, Y., 1988, Male sterility of transgenic mice carrying exogenous mouse interferon-b gene under the control of the metallothionein enhancer-promoter, *EMBO J.*, 7(12):3757-3762.

Leder, A., Pattengale, K., Kuo, A., Stewart, T. A., and Leder, P., 1986, Consequence of widespread deregulation of the c-myc gene in transgenic mice: multiple neoplasm and normal development, *Cell*, 45:485-495.

Leon, J., Guerrero, I., and Pellicer, A., 1987, Differential expression of the ras gene family in mice, *Mol. Cell Biol.*, 7:1535-1540.

Moore, H. D., and Bedford, J. M., 1983, The interaction of mammalian gametes in the female, *In:*

"Mechanism and Control of Animal Fertilization", Hartmann, J. F., ed., Academic Press, New York, 453.

Muller, W. J., Sinn, E., Pattengale, P. K., Wallace, R., and Leder, P., 1988, Single-step induction of mammary adenocarcinoma in transgenic mice bearing the activated c-neu oncogene, *Cell*, 54:105-115.

Orbegin-Crist, M.-C., 1969, Studies on the function of the epididymis, *Biol. Reprod.* (Suppl.), 1:155-160.

Orbegin-Crist, M.-C., and Fournier-Delpech, 1982, Sperm-egg interaction, Evidence for maturational changes during epididymal transit, *J. Androl.*, 3:429-433.

Otten, A. D., Sanders, M. M., and G. S. McKnight, 1988, The MMTV LTR promoter is induced by progesterone and dihydrotestosterone but not by estrogen, *Mol. Endocrinol.*, 2:143-147.

Propst, F., Rosenberg, M. P., and Van de Woude, G. F., 1988a, Proto-oncogene expression in germ cell development, *Trends in Genetics*, 4(7):183-187.

Propst, F., Rosenberg, M. P., Oskarsson, M. K., Russel, L. B., Nguyen-Huu, M. C., Nadeau, J., Jenkins, N. A., Copeland, N. G., and Van de Woude, G. F., 1988b, Genetic analysis and development regulation of testis-specific RNA expression of mos, abl, actin and Hox-1.4, *Oncogene*, 2:227-233.

Quinn, P., Barros, C., and Whittingham, D. H., 1982, Preservation of hamster oocytes to assay the fertilizing capacity of human spermatozoa, *J. Reprod. Fert.*, 66:161-168.

Rigby, P. W. J., Dieckmann, M., Rhodes, C., and Berg, P., 1977, Labeling deoxyribonucleic acid to high specific activity in vitro by nick translation with DNA Polymerase I, *J. Mol. Biol.*, 113:237-251.

Romeny, S., Gray, M. J., Little, B., 1975, The Health Care of Women in Gynecology and Obstetrics, McGraw-Hill, New York, 345.

Ross, S. R., and Solter, D., 1985, Glucocorticoid regulation of mouse mammary tumor virus sequences in transgenic mice, *Proc. Natl. Acad. Sci. USA*, 82:5880-5884.

Shackleford, G. M., and Varmus, H. E., 1987, Expression of the proto-oncogene int-1 is restricted to postmeiotic male germ cells and the neural tube of mid-gestational embryos, *Cell*, 50:89-95.

Sherins, R. J., and Howards, S. S., 1986, Male Sterility, In: "Campbell's Urology," 5th ed. vol. 1, Walsh, P. C., Gittes, R. F., Perlmutter, A. D., and Stamey, T. A., eds., Saunders Company, Philadelphia, 640-697.

Shimizu, K., Goldfarb, M., Perucho, M., and Wigler, M., 1983, Isolation and preliminary characterization of the transforming gene of a human neuroblastoma cell line, *Proc. Natl. Acad. Sci. USA*, 80:383-387.

Sinn, E., Muller, W., Pattengale, P., Tepler, I., Wallace, R., and Leder, P., 1987, coexpression of MMTV/v-Ha-ras and MMTV/c-myc genes in transgenic mice: synergistic action of oncogenes in vivo, *Cell*, 49:465-475.

Smith, D., Rodriguez-Rigau, L. J., and Steinberger, E., 1977, Relation between indices of semen analysis and pregnancy rate in infertile couples, *Fertil. Steril.*, 28:1314-1319.

Smith, R. G., Johnson, A., Lamb, D., and Lipshultz, L. L., 1987, Functional tests of spermatozoa. Sperm penetration assay. Urologic Clinics of North America, 14:451-458.

Sorrentino, V., McKinney, M. D., Giorgi, M., Geremia, R., and Fleissner, E., 1988, Expression of cellular protooncogenes in the mouse male germ line: A distinctive 2.4-kilobase pim-1 transcript is expressed in haploid postmeiotic cells, *Proc. Natl. Acad. Scie. USA*, 85:2191-2195.

Speroff, L., Glass, R. H., and Kase, N. G., 1983, Clinical Gynecologic Endocrinology and Infertility, Williams and Wilkins, Baltimore, 173.

Stewart, T. A., Pattengale, P. K., and Leder, P., 1984, Spontaneous mammary adenocarcinomas in transgenic mice that carry and express MTV/myc fusion genes, *Cell*, 38:627-637.

Tezon, J. G., Vazquez, M. H., De Larminat, M. A., Cameo, M. S., Pineiro, L., Piazza, A., Scorticati, C., and Blaquier, J., 1987, Further characterization of a model system for the study of human epididymal physiology and its relation to sperm maturation, In: Orbegin-Crist, M.-C., and B. J. Danzo, eds., *Ann. N.Y. Acad. Sci.*, 513:215-221.

Thadani, V. M. 1982, Mice produced from eggs fertilized in vitro at a very low sperm:egg ratio, *J. Exp. Zool.*, 219:277-283.

Tremblay, P. J., Pothier, F., Hoang, T., Tremblay, G., Brownstein, S., Liszauer, A., and Jolicoeur, P., 1989, Transgenic mice carrying the mouse mammary tumor virus ras fusion gene: distinct effects in various tissues, *Mol. Cell Biol.*, 9:854-859.

van Zyl, J. A., Menkveld, R., and van Kotze, W., 1975, Oligozoospermia, A seven-year survey of the incidence, chromosomal aberrations, treatment and pregnancy rate, *Int. J. Fertil.*, 20:129-132.

Varmus, H. E., 1984, The molecular genetics of cellular oncogenes, *Ann. Rev. Genet.*, 18:553-612.

Yanagimachi, R., Yanagimachi, H., and Rogers, B. J., 1976, The use of zona-free animal ova as a test system for the assessment of the fertilizing capacity of human spermatozoa, *Biol. Reprod.*, 15:471-478.

Yanagimachi, R., 1981, Mechanisms of fertilization in mammals, *In:* "Fertilization and Embryonic Development In Vitro," Mastroianni Jr., L., and Biggers, J. D., eds., Plenum Publishing Co., New York, 81-182.

HELPING PATIENTS END TREATMENT: THE IVF FOLLOW-UP CLINIC AS A TOOL FOR CONTINUING PSYCHOLOGICAL ASSESSMENT

Dorothy A. Greenfeld[1], Gad Lavy[1], David G. Greenfeld[2], Carole T. Holm[1], and Alan H. DeCherney[1]

[1] Department of Obstetrics and Gynecology of Yale University School of Medicine
New Haven, Connecticut
[2] Department of Psychiatry, Yale University School of Medicine

INTRODUCTION

Many couples enter an IVF/ET program without a clear idea of how many treatment cycles they intend to undergo. Others are determined to continue treatment as long as necessary to achieve a successful outcome. However, following review of patient performance the treatment team may conclude that continued attempts are not advisable. Reasons for recommending discontinuation of IVF/ET treatment include failure to conceive due to lack of response to ovarian stimulation, failed fertilization, and age factors.

Previous psychological studies in IVF/ET have established that, for the most part, participants are psychologically normal but emotionally pressured by the years of infertility and the demands of the treatment.[1,2] Other studies have included observations about decision making in IVF/ET treatment. For example, Mao and Wood[3] found that women who dropped out after one cycle often did so because of anxiety and cost. Leiblum et al[4] and Freeman et al[5] agreed that many drop out of the program because they find it too stressful and costly and many report that they want to pursue adoption. Callan et al.[6] reported that women who intended to stop treatment had older husbands, were more likely to be mothers, and were more likely to have had an IVF pregnancy. Those stopping and continuing did not differ in age, number of years of infertility, or number of IVF cycles. However, no previous studies have addressed the range of responses in women who are not responding to treatment and who are advised to consider stopping.

Advances in Assisted Reproductive Technologies, Edited by
S. Mashiach *et al.,* Plenum Press, New York, 1990

We report the responses of 25 couples who have been advised to end treatment. On the basis of these observations we make recommendations regarding the use of our new Follow-Up Clinic as a useful tool for psychological intervention and evaluation.

MATERIALS AND METHODS

Patient Population

Beginning in June 1988 the Yale IVF/ET Follow-Up Clinic assessed and evaluated 217 women who were undergoing IVF/ET treatment. Of those 217, 25 women were advised by the treatment team to discontinue IVF/ET. This advice was given by the physician, sometimes in conjunction with one or more staff members. Following the meeting of patient and physician, the clinical social worker met with these women (and with their husbands if they were present). The social worker conducted a supportive, semi-structured interview and rated the women's responses. The social worker subsequently followed up on these couples by telephone, generally within a few weeks.

The Treatment Team

The IVF/ET Follow-Up Clinic procedure begins with a team meeting where each case is reviewed. The team consists of all personnel involved in our IVF/ET program: physicians, nurses, lab personnel, research personnel, and a mental health professional (DAG). The case is presented by the physician who reviews all treatment cycles. The team evaluates the course of treatment to date and, on the basis of the evaluation may propose a protocol that would be more appropriate, may suggest additional tests for further evaluation, and may make an assessment of treatment prognosis. All of the team members contribute to the decision making process. However, the presenting physician's opinions about the termination of treatment are of particular importance, since the physicians' attitude has a direct bearing on the emphasis and tone with which closure of treatment is presented to the couple.

RESULTS

The women were classified based on their responses to the advice to terminate treatment.

Group 1: These couples welcomed the advice to end IVF/ET treatment and immediately began to explore other options such as adoption. Of the 25 subjects, 12 (48%) responded in this way and were assigned to Group 1. In fact, some of these subjects had actively initiated consultation in the Follow-Up Clinic as a means of obtaining assistance in achieving closure. The responses of this group in follow-up interview were consistent with their initial responses and give credence to the oft quoted phrase in IVF/ET psychological studies that couples often enter IVF/ET treatment as a means of seeking and finally achieving "closure" on treatment of infertility.[7]

Group 2: Couples in this group were highly ambivalent about continuing IVF/ET treatment, but they were not ready to stop at the time of evaluation and they insisted on at least one more treatment cycle to help gain their own sense of control over the situation. Eight of the twenty-five subjects (32%) were assigned to this group. In most instances only one additional cycle was necessary before couples were able to stop treatment. In follow-up interviews these couples reported that although they knew it was time to stop when so advised by the treatment team, they needed the time and an additional cycle in order to accept the inevitable. All of them were clearly grateful that they had the opportunity to make the decision themselves and in their own way.

Group 3: This small group of couples strenuously rejected the treatment team's advice and refused closure of any kind. Of the twenty-five subjects, five (20%) fell in this group. These couples reacted vehemently to the suggestion that they terminate treatment, making it clear that they would go elsewhere if they were not permitted to continue. For the most part, we attempted to work with these couples and to negotiate treatment planning with them, although two of the subjects in this group left our program and applied to other IVF/ET programs.

For the entire group of subjects the number of IVF/ET cycles (Mean 4.2, S.D. 2.3), number of years of infertility (Mean 7.4, S.D. 4.1), and age of subject (Mean 40.1, S.D. 4.1) were not significantly correlated with the subject's response to the advice to terminate treatment. However, subjects in Group 3 (couples who vehemently refused to accept the advice to end treatment) were more likely to be infertile as a result of male factor [r (23) = .46, P = .02].

DISCUSSION

It appears that an ongoing discussion of IVF/ET treatment is useful for most patients and necessary for many. Our experience in the IVF/ET Follow Up Clinic confirms what we have long suspected—that for many of our patients the resort to IVF/ET is a way of finally exhausting infertility treatment and coming to closure on the subject.

We have also learned that most patients greatly appreciate advice from the treatment team, but that many need to feel some degree of control over the situation and often need at least one additional cycle as a way of taking charge of the decision and "making it their own". Finally, there are some couples who cannot or will not stop treatment, even when advised that their treatment is not likely to succeed. It appears that these couples will continue to seek treatment regardless of the recommendation and if they are pressed they are likely to attempt to move to another center. The fact that couples in this group are more likely to be infertile as a result of male factor may suggest that the women are aware that they should be able to conceive and have more difficulty accepting the idea that they will not become pregnant. (This is particularly true for couples who are averse to the use of donor sperm.) In these cases the best strategy is usually to attempt to negotiate gently and carefully. It is helpful to be flexible and compassionate with these couples lest they go on a nearly endless round of IVF/ET clinics in search of a miraculous solution to their problems.

The varieties of couples' responses is illustrated in the following case vignettes:

Group 1: This is the group that accepts and is clearly relieved by the advice to terminate treatment. A typical response of patients in this group is that of Mrs. J., a 40 year old woman with a 12 year history of infertility. She said, "I really needed those four cycles of IVF/ET to make sure I was never going to be pregnant. Now that it's over I can get on with the process of adoption. I was struggling with the idea, but I couldn't go ahead until I knew for sure that pregnancy was out of the question".

Several months after she was advised to stop treatment, Mrs. J. appeared to be some-what distracted by the telephone call from the social worker. She said that she had been up all night with her eight week old adopted daughter and that "IVF/ET seems like a long time ago."

Group 2: Mrs. S. reacted to the advice that she end IVF/ET treatment with surprise and some distress, though she seemed willing to consider ending treatment. She said "Maybe it is time to stop, but my husband and I think it would be a mistake to give up too soon. The idea of ending treatment seems like a major decision for us. We need to be sure that IVF/ET isn't going to work." After an additional cycle the subject of termination of treat-ment was raised again with Mrs. S. She reacted with some sadness and grief, but she was able to accept her infertility and could begin to discuss alternate options.

Group 3: Mrs. C. reacted with great distress and vehement opposition when she was ad-vised to stop IVF/ET treatment. She said, "I'm much younger than most of your other pa-tients (she was 36 years old) and I feel I've been backed into a corner. My husband already has children from a previous marriage and so he won't agree to adoption. IVF is my only chance to have children, and I absolutely refuse to accept the idea that it won't work".

Mrs. D. was accompanied to the Follow-up Clinic by her husband. When they were ad-vised to stop treatment Mr. D. interrupted to ask about technical details of treatment and both partners acted as though they failed to hear the recommendation. Attempts to engage-both of them on the subject of their poor prognosis were fruitless. Both Mrs. C. and Mrs. D. have since gone to other IVF/ET programs.

SUMMARY AND RECOMMENDATIONS

Many couples enter IVF/ET treatment with little thought as to how many cycles of treatment they intend to undergo. The use of a IVF/ET Follow-up Clinic to monitor couples' progress in treatment gives the treatment team a formal opportunity to assess each couple and make specific treatment recommendations. Couples vary in their responses when they are advised that the team recommends that they end IVF/ET treatment. Some couples are relieved, and clearly welcome the end of their long and unsuccessful struggle to achieve pregnancy. Others acknowledge the wisdom of the advice but insist on at least one additional cycle in order to retain a sense of control and come to terms with their disappointment. A third group reacts with intense resistance to the advice, indicating their determination to continue treatment as long as necessary in order to achieve success.

After one or more failed cycles, each participant who has serious concerns about her participation can be seen in the Follow-up Clinic. Her previous medical treatment as well as each cycle of IVF/ET is reviewed and discussed by the treatment team. Although the original intent of the clinic was to provide a forum for a reevaluation and perhaps rethinking the patient's medical situation, we have found that this Follow-up Clinic also provides a useful tool for psychological assessment and intervention as well. Patients can participate to a degree in the decisions affecting their treatment. Certainly, it is an arena where patients can have their treatment histories reviewed, where they can feel that they are part of the discussion and decision making, and where they can be counseled and supported if they are advised to terminate treatment.

We recommend that couples be given ample opportunity to ask questions and to express their concerns about ending or continuing treatment. In order that this sometimes difficult process be managed smoothly and skillfully, we recommend that a mental health professional play an active role in the Follow-up Clinic.

REFERENCES

1. E. W. Freeman, A. S. Boxer, K. Rickels, R. Tureck, and L. Mastoianni, Psychological evaluation and support in a program of in vitro fertilization and embryo transfer. *Fertil. Steril.*, 43–48 (1985).
2. F. P. Haseltine, C. Mazure, W. DeL'Aune, D. Greenfeld, N. Laufer, B. Tarlatzis, M. L. Polan, E. E. Jones, R. Graebe, F. Nero, A. D. D'Luige, D. Fazio, J. Masters, and A. H. DeCherney, Psychological interviews in screening couples undergoing in vitro fertilization. *Annals N.Y. Acad. Sci.*, 422:504 (1985).
3. K. Mao, C. Wood, Barriers to treatment of infertility by in vitro fertilization and embryo transfer, *Med. J. Aust.*, 140:532–533 (1984).
4. S. R. Leiblum, E. Kemmann, D. Colburn, S. Pasquale, A. DeLisi, Unsuccessful in vitro fertilization: A follow-up study, *J. In Vitro Fertil. Embryo Trans.*, 4:46 (1987).
5. E. W. Freeman, K. Rickels, J. Tausig, A. Boxer, L. Mastrioianni, R. W. Tureck, Emotional and psychosocial factors in follow-up of women after IVF/ET treatment: A pilot investigation. *ACTA Obstet. Gynecol. Scand.*, 66:517 (1987).

6. V. J. Callan, B. Kloske, Y. Kashima, J. F. Hennessey, Toward understanding women's decisions to continue or stop in vitro fertilization: the role of social, psychological, and background factors. *J. of IVF/ET,* 5:363–369 (1988).
7. D. Greenfeld, C. Mazure, F. P. Haseltine, A. H. DeCherney, The role of the social worker in the in vitro fertilization program. *Soc. Work in Hlth. Care,* 10:71 (1984).

COUPLES' EXPERIENCES WITH IN VITRO FERTILIZATION: A PHENOMENOLOGICAL APPROACH

Judith Lorber[1] and Dorothy Greenfeld[2]

[1] Department of Sociology, Brooklyn College and Graduate School
City University of New York
[2] Department of Social Work, Yale-New Haven Hospital
New Haven, CT, USA

INTRODUCTION

This paper reports on couples' experiences with in vitro fertilization from a phenomenological perspective.[1] It discusses how the couples created the reality of the experience for themselves—how they shaped what they went through, and what the meaning of the experience was for them. It presents the results of separate telephone interviews with 20 husbands and wives who went through the IVF/ET Program, successfully and unsuccessfully, at Yale-New Haven Hospital in Connecticut in 1983-84. The questions addressed by this study were: First, what is the social meaning of IVF as a medical experience; and second, what is the social usefulness or latent function of IVF for couples who want a child?

For the most part, making the experience meaningful meant defining the situation so that they were agents of their experience while going through a difficult process. The latent function was to transform frustration into heroic effort, so no matter what the outcome, they had prevailed.

Advances in Assisted Reproductive Technologies, Edited by
S. Mashiach *et al.*, Plenum Press, New York, 1990

PREVIOUS RESEARCH

Despite admittedly low success rates in terms of live births,[2-4] most of the people who have gone through IVF programs do not seem to regret having done so,[5-8] which suggests that the experience is interpreted positively, even if there is no pregnancy or birth. Particular sources of stress from the procedure are anesthesia, surgical laparoscopy, embryo transfer, and the cost.[9] The "emotional roller coaster" that patients describe for all of infertility treatment is exacerbated by the possibilities for success and failure in in vitro fertilization,[10-14] with the most difficult time waiting at home after IVF to see if a pregnancy will actually ensue.[10] The results of this study show that IVF clients use a variety of coping mechanisms that allow them to feel in control.

RESEARCH DESIGN

Methods

In 1986, letters were sent to 75 couples who had entered the Yale-New Haven Hospital IVF/ET Program in 1983-84 by Dorothy Greenfeld, the social worker on the program, who did the initial assessment interviews.[15] In response to the first mailing, 19 agreed to be interviewed, 11 refused, and there was no response from 45. In a subsequent mailing, an additional five couples agreed to be interviewed, for a total of 24, a response rate of 32 percent. Because of time constraints, a total of 20 couples was interviewed, so the actual response rate was only 27 per cent. However, the range of outcomes was represented—four of the couples had a birth, two couples had a pregnancy but no birth, and 14 had no success.

Those who had a birth were asked about the pregnancy and delivery, and whether they agreed to have amniocentesis or other genetic tests. Those who became pregnant, but did not have a child, were asked whether they planned to try IVF again, or proceeded with other plans. Those who did not become pregnant at all were also asked if they planned to continue IVF, seek other treatment, try to adopt, or have adapted their lives to childlessness. All the husbands and wives were asked what effects going through the IVF program had on their labor force participation and the quality of their married life. (The Yale program did not admit unmarried couples at that time.) Finally, they were asked whether they told family members, close friends, colleagues, co-workers, and neighbors that they tried or are trying to have a baby through IVF.

Sample

The average age was 37.6 years for the women, with a range of 32 to 43 years old, and 37.8 years for the men, with a range of 27 to 43 years old. The majority of the respondents had completed four years of college; eight of the women and seven of the men had advanced degrees. Six of the women were professionals or managers and two owned a business. Four did white-collar work; one did paid work at home; five were homemakers. Nine of the men were professionals or managers, six owned a business, and three did white-collar work. There was no information on occupation for two of the couples.

Previous treatments for the women, most of whom had tubal problems, were surgery for 14; fertility drugs, two; exploratory laparoscopy, two; and fertility drugs and laparoscopy, two. There were no reports of treatment for the men. Four of the couples had had biological children.

These 20 couples started IVF treatment 56 times. Fourteen had 1–2 cycles, and five had 3-6 cycles. One couple had 16 attempts and 12 complete cycles in three years. From IVF, there were eight pregnancies, and four deliveries of five children.

FINDINGS

Trying "Everything"

In this study, for the couples who had one or two cycles, IVF was the "end of the line." They were ready to stop infertility treatments before going through IVF, but went through it because they were encouraged by their physicians. As one wife who had one cycle of IVF said, "We never intended to do it anyway but physicians talked us into it, but we had mixed feelings. If we'd been younger, we may have done it again." The husband of another couple said, "IVF was like one more step but not such a big thing. We had five years of treatment with the same doctor. IVF was the final treatment for us—we planned that before the cycle. Infertility treatment was too disruptive on our lives, our jobs. We were ready for a conclusion." (Ironically, this couple was later able to have two children without any treatment at all, after which the wife had her tubes tied).[16] For couples who have one or two cycles, the latent function of IVF was to have "tried everything."[17]

Becoming a Professional Patient

For those who go through multiple cycles, IVF becomes an illness career. One woman was treated at the Yale Infertility Clinic for five years before trying IVF. She does not ovulate without fertility drugs, which she had been on for two years. In the first IVF cycle, six eggs were produced and four embryos were transferred, but she did not get pregnant. In the second IVF cycle, 12 eggs were produced and nine embryos were transferred, again with no pregnancy. In the third cycle, 30 eggs were produced, four embryos were transferred, again, no pregnancy. She now has 18 embryos "on tap" which were frozen in five groups of four and three. She might not have continued if not for the freezing program, which lured her back. She had planned to come in for "thaw cycles," but the first month she didn't respond to Pergonal, and the second month she was too sick from asthma. She will soon have the "first batch" transferred. She calls Yale her "home away from home." In sociological terms, she has become a "professional patient."

Husband's Emotional Support

Another way of exerting control by the women was to involve their husbands emotionally in the process. One husband said, "It's a nightmare," because, despite an erratic schedule that included evening work, his wife insisted he be home every night for the "5 o'clock phone call." His wife needed him there with her when the phone call came because it is "usually bad news."

In another case where the husband was heavily involved, the couple were given positive pregnancy results but because of spotting, it immediately looked like the pregnancy was not viable. The husband said, "The problem with that part of the program was that you know about the conception when it happens." With another pregnancy that did not result in a viable fetus, the husband described the month from transfer to spontaneous abortion as the worst thing he had ever gone through emotionally.[18] In these cases, the knowledge of the process was emotionally devastating but the involvement of the husband brought the couple together emotionally.[10]

Eleven of the husbands and 13 of the wives felt the quality of the marriage improved during treatment, three of the husbands and five of the wives that it had deteriorated, and six of the husbands and two of the wives that it stayed the same. Where there had been a birth, three of the husbands and wives felt their relationship had improved, one husband that it had stayed the same, and one wife that it had deteriorated. For the two couples that had a pregnancy but no birth, neither husband nor wife felt there had been any change in their relationship.

The responses where IVF had failed to produce a child suggest that the latent effects are positive: eight of the husbands and 10 of wives felt the quality of their marriage had improved during IVF treatment. Three husbands felt it had stayed the same, and the 3 that it had deteriorated. Four of the wives felt their relationship with their husbands had deteriorated.

The open-ended responses indicate a great deal of mutual support and husband's empathy for the wives' pain and disappointment. One wife said that IVF treatment gave her husband a better understanding of what she was going through. During the seven previous years of treatment for infertility, she never really discussed her "bad days" with her husband. She protected him somewhat from all of the anxiety and depression of pregnancy failure. When they began treatment for IVF, the daily nine-hour commute they made together allowed them to discuss every aspect of the IVF ups and downs. The fact that her husband accompanied her made him much more active in her treatment than he had ever been before. As Seibel and Levin[14] say, "They often travel long distances and, indeed, have journeyed a long way together emotionally."

Using Alternative Medicine

One couple went through IVF four times before they had a child. The husband's theory of why cycle 4 worked and the others didn't illustrates in detail how a couple makes sense of their experience and exert control by using alternative medicine.

For cycle 1—they had no nutritional preparation and were staying in a strange place. One embryo was produced. For cycle 2, fifteen months later—the husband and wife meditated, they got involved in good nutrition, had emotional and physical preparation, and stayed in a nice place. The wife got pregnant but had a miscarriage at six weeks. In the husband's words, "a contributing factor may be that she stopped meditating." For cycle 3, nine months later—the couple prepared by seeing a midwife in a state different from where they lived and from where they had the IVF. They saw her twice a month for eight months. She helped them with, to quote, "relaxation development and gave wife acupuncture and herbal medicines to clear up scar tissue around tubes and help prepare her for emotional impact." Although the wife did not get pregnant, the husband credits this treatment with "more eggs and better sperm count." Six embryos were produced.

For cycle 4 — the husband believes the fourth time worked because they were more relaxed, had good nutrition, meditated, and weren't worried about money because he had a new job. They produced five embryos and a twin pregnancy. Unfortunately, one twin died in utero at five and a half months and that was emotionally devastating. Their means of control was to refuse amniocentesis because of the risk, and because they would not have aborted. Nonetheless, the patient felt too vulnerable and preoccupied at the beginning of the pregnancy to work, but then worked as a weaver in her home until the baby was born. The obstetrician, however, treated the pregnancy and birth as high risk, which is their way of exerting control over this new technology.

One of the couples who was successful in having a birth was not successful in being able to exert emotional control over the experience. The birth occurred after six cycles, a pregnancy that ended in a spontaneous abortion at three weeks, and a pregnancy that was terminated at 18 weeks after amniocentesis showed a genetic defect. The pregnancy that resulted in the birth was equally difficult emotionally. Placenta previa developed at 28 weeks, and the wife was hospitalized for 9 1/2 weeks.

Pursuing IVF and Adoption

IVF couples take control in another way—they pursue adoption while they are undergoing IVF, not after they have decided to stop treatment.[19] One couple put themselves on a waiting list for adoption during their first cycle of IVF treatment. They had two complete cycles, one year apart, with a chemical pregnancy in the first. They plan to continue IVF treatment, even though they are "just a phone call away" for a baby.

Another couple has a division of labor—she had Pergonal treatments, IVF, and has stocked up five batches of frozen embryos, which she plans to have transferred. He has been pursuing adoption, and the couple has learned from an agency that a baby will probably be theirs within a year.

Another woman never wanted IVF, but did it for her spouse—she wanted to adopt. She felt adoption was a much more "hands on experience" than IVF. She didn't like the lack of control she felt going through IVF. The adoption process gave her more control—she interviewed by telephone 19 women who answered ads the couple had placed in 25 cities around the country. She interviewed the final three women in person and flew to Texas to pick up the baby.

Disclosure as a Form of Control

While infertility may be experienced as a stigma, going through IVF can make a couple charismatic in that they are on the cutting edge of reproductive science. In addition, because of the emotional and physical trauma involved, the couples may be seen as heroes for having gone through so much to have a child.

Most couples, therefore, talked about their treatment. Even where there was no pregnancy, 79 percent told family, friends, co-workers and sometimes neighbors. The couples who had a pregnancy but no birth also communicated their situation openly.

All the couples who had children through IVF told "everyone" and plan to tell the children when they are old enough to understand. Although the mother of Yale's first twins preferred more privacy than her husband, she said she had made a scrapbook of the newspaper clippings, had kept a diary of the experience, and had even saved the petri dishes. She planned to show all these mementoes to the children when they were older.

CONCLUSIONS

This study of the phenomenon of the IVF experience from the patients' point of view found that the couples interviewed exerted control over a highly technical process by using alternative caregivers, giving each other emotional support, and stockpiling frozen embryos. IVF and adoption, although they are at opposite ends of the genetic kinship spectrum, did not seem to be mutually exclusive alternatives for these couples. In this study, because both took a long time and neither were under the couple's control, IVF and adoption were sometimes pursued concomitantly. Doing both at the same time seemed to compensate for the lack of control in each separate search for a child. In a larger sense, the use of IVF is an attempt to exert control over a malfunctioning body. It is an attempt to rectify the "disturbance in...ability to relate to and function in the world",[1] in this case, the world of family and kinship. Thus, even though IVF treatment is a technical failure for the majority of couples, it allows the couple to demonstrate to themselves and others that they are doing everything they can to combat their social illness—childlessness—and so, in social terms, it is interpreted as a successful phenomenon.

ACKNOWLEDGEMENTS

The research for this paper was supported by PSC-CUNY Grants 666-206 and 668-518. We would like to thank Alan DeCherney, M.D., for his support and cooperation.

REFERENCES

1. Baron RJ: An introduction to medical phenomenology: "I can't hear you while I'm listening." *Ann Int Med* 1985;103:606-611, quote at 609.
2. Blackwell RE, Carr BR, Chang RJ, DeCherney AH, Haney AF, Keye WR Jr, Rebar RW, Rock JA, Rosenwaks Z, Seibel MM, Soules MR: Are we exploiting the infertile couple? *Fertil Steril* 1987;48:735-39.
3. Paterson P, Clement C: What proportion of couples undergoing unrestricted in vitro fertilization treatments can expect to bear a child? *J Vitro Fertil Embryo Transfer* 1987;4:334-337.
4. Raymond CA: In vitro fertilization enters stormy adolescence as experts debate the odds. *JAMA* 1988;259:464-465, 469.
5. Alder E, Templeton AA: Patient reaction to IVF treatment. *Lancet* 1985;1 (Jan. 19):168.
6. Bainbridge I: With child in mind: the experiences of a potential IVF mother. *In:* Test-Tube Babies. WAW Walters and P Inger (eds). Melbourne: Oxford, 1982.
7. Crowe C: "Women want it": in vitro fertilization and women's motivations for participation. *Women's Studies Int Forum* 1985;8:57-62.
8. Leiblum SR, Kemmann E, Colburn D, Pasquale S, DiLisi AM: Unsuccessful in vitro fertilization: A follow-up study. *J Vitro Fert Embryo Transfer* 1987;4:46-50.
9. Nero FA, Greenfeld D, DeCherney AH: Sources of stress in an in vitro fertilization (IVF) program. Proc. Fifth Int Cong IVF and ET, Norfolk, VA 1987.
10. Callan VJ, Hennessey JF: Emotional aspects and support in in vitro fertilization and embryo transfer programs. *J Vitro Fert Embryo Transfer* 1988;5:290-295.
11. Garner CH: Psychological aspects of IVF and the infertile couple. *In:* Foundations of In Vitro Fertilization, CM Fredericks, JD Paulson, AH DeCherney (eds). Washington, DC: Hemisphere. 1987.
12. Mahlstedt PP, Macduff S, Bernstein J: Emotional factors and the in vitro fertilization and embryo transfer process. *J Vitro Fert Embryo Transfer* 1987;4:232-236.
13. Williams LS: "It's going to work for me." Responses to failures of IVF. *Birth* 1988;15:153-156.
14. Haseltine FP, Mazure C, De l'Aune W, Greenfeld D, Laufer N, Tarlatzis G, Polan ML, Jones EE, Graebe R, Nero F, D'lugi A, Fazio D, Masters J, DeCherney AH: Psychological interviews in screening couples undergoing in vitro fertilization. *Ann New York Acad Sci* 1985; 442:504-22.
15. Roh SI, Awadalla SG, Park JM, Dodds WG, Friedman, CI, Kim MH: In vitro fertilization and embryo transfer: treatment-dependent versus treatment-independent pregnancies. *Fertil Steril* 1987;48:982-986.
16. Callan VJ, Kloske, B, Kashima, Y, Hennessey, JF: Toward understanding women's decisions to continue or stop in vitro fertilization: the role of social, psychological, and background factors. *J Vitro Fert Embryo Transfer* 1988;5:363-369.
17. Greenfeld D, Diamond MP, DeCherney AH: 1987. Grief reactions following in vitro fertilization treatment. *J Psychosomatic Ob Gyn* 1987; 8:169-174.
18. Seibel MM, Levin S: A new era in reproductive technologies: the emotional stages of in vitro fertilization. *J Vitro Fert Embryo Transfer* 1987;4:135-140, quote at 136.
19. Callan VJ and Hennessey JF: IVF and adoption: the experiences of infertile couples. *Austral J Early Childhood* 1986;11:32-36.

PERCEIVED PSYCHO-SOCIAL FEATURES OF IVF COMPARED TO OTHER TREATMENTS FOR FERTILITY

S. Shiloh, S. Larom, S. Mashiach, and Z. Ben-Rafael

*The Chaim Sheba Medical Center and
Sackler School of Medicine, Tel Hashomer, Israel*

INTRODUCTION

Infertility has been described as a complex biopsychosocial crisis, which involved an interaction between physical conditions, medical interventions, social assumptions concerning parenthood, reactions of others and individual psychological characteristics (Taymor, 1978). In recent years, interest has increased in the psychological factor of infertility, accompanied by a change from clinical observations to more rigorous empirical studies. Topics which attracted the most attention were psychological causes of infertility (Christie, 1980; Mozley, 1976; Kaufman, 1984), and psychological and emotional impacts of infertility (Seibel & Taymor, 1982; Cook, 1987; Menning, 1984; Mahlstedt, 1985). Some investigators have concentrated on the "iatrogenic stress" caused by treatment of infertility (Taymor, 1978; Lalos et al, 1985; Daniluk, 1988). Some forms of stress are inherent in specific medical procedures. For example, embarrassment and sexual dysfunction as a result of the need to schedule sexual relations (Keye, 1984), pain and discomfort in physically intrusive procedures and ambivalence in artifical insemination by a donor (Berger, 1980).

Based on a cognitive approach, we believe that inner representation rather than "external reality" is what counts for human behavior and emotional states. Therefore, the aim of our study was to investigate how treatment for infertility is perceived, and what the content of the cognitive constructs is which individuals have regarding specific types of treatment.

A few studies have been conducted on specific beliefs and attitudes towards treatment of infertility. Wallace (1985) investigated cost-benefit perceptions of patients undergoing la-

Advances in Assisted Reproductive Technologies, Edited by
S. Mashiach *et al.*, Plenum Press, New York, 1990

peroscopic surgery, and concluded that the meaning of the event is important in determining psychological distress. Johnston et al. (1987) found that patients overestimate the likelihood of success in In Vitro Fertilization (IVF) procedures. A few studies have focused on attitudes towards treatment involving donors (Reading et al., 1982; Waltzer, 1982) and other studies have examined the general population's norms regarding acceptibility of new infertility treatment technologies (Matteson & Terranova, 1977; Walters & Singer, 1982; Daniels, 1986). To the best of our knowledge, no attempt has yet been made to capture the full range of cognitive elements that comprise the comprehensive concept of treatment for infertility.

The importance of the cognitive aspect in health-related behavior in general has been conceptualized in the Health Belief Model (Becker & Maiman, 1975). It led to many empirical studies, in which the model has been validated, and in which it was found that the most powerful predictor of health action is the perception of barriers and impediments to undergoing medical treatment (see review by Jans et al., 1984). Some evidence as to the importance of cognitions (perceptions and expectations) in treatments for infertility has recently been demonstrated by Callan et al. (1988). These authors have found that patients' beliefs regarding the effectiveness of the treatment, and their perception of social pressures to continue the treatment, could predict their decision to continue or halt IVF. These cognitive variables were better predictors of patients' decisions about IVF than the background characteristics and fertility histories of women and their husbands.

Based on the above considerations and findings, we believe that a better understanding of the concept "treatments for infertility", held by patients and by the general population, is an essential step in improving the understanding of the behavioral and emotional reactions of the patients to treatment. The purpose of the present study was, therefore:

1. to describe the cognitive elements which comprise the concept "treatment for infertility";
2. to compare patients' and non-patients' concepts of treatment for infertility;
3. to compare perceptions of IVF to perceptions of other treatments.

METHOD

Subjects

Participants in the study were divided into 2 groups, one of 50 women undergoing IVF treatment (patients), and 50 women who had at least 1 child and who had never suffered from infertility (non-patients). The mean age of the patients was 33.3 years (SD = 3.56), their mean education was 13.6 school years (SD = 2.23), none of them had children and their mean time of treatment for infertility was 6.3 years (SD = 3.28). The non-patient group was approximately the same age (mean = 34.26, SD = 5.61) and education (mean = 14.82, SD = 2.78). They had between 1 and 4 children (mean 1.92, SD = 0.85).

Instruments

A questionnaire was constructed in which 10 generic treatments for infertility had to be evaluated along 15 bipolar scales describing psychosocial features. The 10 treatments were: medical treatment (ovulation induction); pelvic reconstructive surgery; intrauterine manipulation (D & C, HSG); medical treatment for male; surgery for male; artificial insemination by husband semen; in vitro fertilization by husband semen; in vitro fertilization by donor semen and surrogate mother. Names of treatments were followed by 1–2 sentences explaining the procedure to those subjects who were unfamiliar with it.

The 15 description scales were derived from a pilot study in which another group of 10 IVF patients and 10 non-patients participated. These subjects were asked to openly comment on association, evaluations, concerns and beliefs regarding each of the 10 treatments. This procedure elicited 60 distinct answers. Two psychologists performed content-analysis on the data, and agreed upon 15 categories which covered the whole range of topics mentioned by subjects. After discussion, there was a 100% agreement between the two evaluators regarding item-category matching. Categories were defined as bipolar dimensions, to cover both positive and negative expressions of the same characteristic. The 15 scales defined in this way were:

1. Social norms—considered socially acceptable; considered socially unacceptable
2. Pain—causes pain and physical discomfort; does not cause pain and physical discomfort
3. Bonding to child—may cause difficulties in bonding to the baby and to parenthood functioning; can facilitate bonding to the baby and parenthood functioning
4. Couple's relationship—detrimental to the couple's emotional relationship; causes tightening and strengthening of couple's emotional relationship
5. Hope for success—elicits hope for success and optimism; does not elicit hope, expectation of failure
6. Physical threat—involves risk of physical damage or unhealthy side effects to the patient; does not involve any physical threat to the patient
7. Fetal health—elicits concern about fetal physical health; ensures fetal physical health
8. Humiliation—causes humiliation; does not humiliate
9. Medical sophistication—uses most sophisticated medical technology; simple and unsophisticated medicine
10. Sexual life—detrimental to normal sexual life; contributes to improvement in sexual life
11. Social exposure—much social exposure, cannot be hidden from others; enables total privacy and secrecy
12. Moral dilemma—elicits philosophical, moral or religious concerns; does not elicit any philosophical, moral or religious concerns
13. Artificialness—artificial and unnatural; does not interfere with the natural process
14. Self-esteem (F)—hurts woman's self-esteem; does not hurt woman's self-esteem
15. Self-esteem (M)—hurts man's self-esteem; does not hurt man's self-esteem

In the final version of the questionnaire, each dimension was presented as a 13-point Likert scale, in which both ends were defined, and the middle-point denoted equal agreement or disagreement with both ends. Subjects were requested to express their beliefs about each treatment on each dimension by circling numbers which represent their agreement with the given descriptions. For example;

causes pain   does not
and physical   6 5 4 3 2 1 0 1 2 3 4 5 6   cause pain
discomfort   and physical discomfort

At the end of the questionnaire, the same 10 treatments were presented again, this time to assess the subjects' preferences. Each treatment had to be evaluated on a 9-point scale ranging from 1 = "I would not consider using it even if this was the only treatment available", to 9 = "the most preferred treatment I would have chosen".

Procedure

Patients were recruited for the study during their visit to an IVF clinic in a large medical center. Non-patients were recruited from their place of work or at schools. The questionnaire and the purpose of the study was explained to the subjects by the researcher, who administered the questionnaire individually. The questionnaire was completed anonymously in 20–30 minutes.

RESULTS

The fifteen dimensions derived from the pilot study served as a basis for drawing a psychosocial profile for each of the treatments. A mean evaluation on each treatment dimension was computed for each group of subjects.

Multivariate analyses of variance (MANOVA) were performed to test inter-group differences in perceptual profiles of the treatments. No significant differences were obtained between patients and non-patients in perception of male treatments and treatments involving donors. However, some significant differences were found in the perception of female-centered treatments. In Figure 1, the perceptual profiles of patients and non-patients are presented for the IVF (husband's semen) treatment. Of the 15 inter-group comparisons, six significant differences were found. Patients perceived the "IVF husband" treatment as: less detrimental to the couple's relationship; less threatening to fetal health; less humiliating; less artificial; less damaging to self esteem and as involving more social exposure than non-patients.

In Figure 2 a comparison is presented between the "IVF husband" perceptual profile, and an aggregated profile of the other nine treatments, in the patient group.

Results indicate that patients regard IVF with husband's semen differently to other treatments on most dimensions (11/15). Two significant differences are on the negative side. "IVF husband" is evaluated as more painful and as involving more social exposure than other treatments. All other significant differences are in favour of "IVF husband" in comparison to other treatments. It is considered to be more socially acceptable, less harmful to bonding with child and to the couple's spiritual and sexual relationship, more successful and less humiliating, more sophisticated and less detrimental to the husband's self-esteem.

Mean preferences of patients and non-patients for the ten treatments are presented in Table 1.

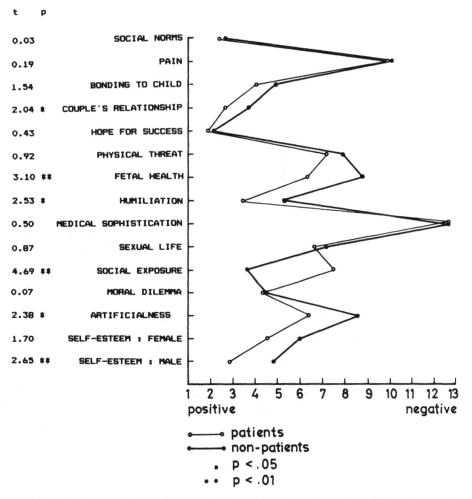

Fig. 1. Perceptual profiles of IVF (husband's semen) in patients (n = 50) and non-patients (n = 50).

The general rank order of treatment preferences in both groups is highly similar (Spearman rank-order correlation = 0.90). Both groups ranked donor treatments as the least preferred options. However, patients preferred "IVF-husband" to other treatments significantly more than non-patients (F = 4.03, p < 001). The treatment most favoured by non-patients was medical treatment for female, while "IVF-husband" was fourth in order of preference.

Finally, Spearman rank-order correlations were computed between mean preference scores and mean perceptual scores across the treatments. Results, presented in Table 2, indicate that the preferences of patients correlated significantly with the following dimensions: social norms; bonding to child; couple's relationship; hope for success; humiliation and

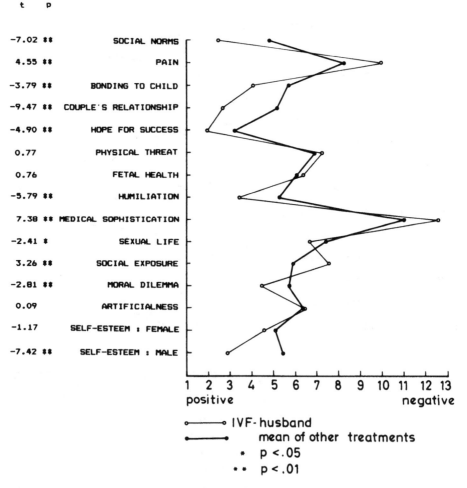

Fig. 2. Perceptual profile of IVF (husband's semen) versus a mean profile of other treatments (n = 50).

Table 1. Means and Standard Deviation of Treatment Preferences (1 = low, 9 = high)

Treatment	Patients (n = 50)		Non patients (n = 50)	
	Mean	SD	Mean	SD
Medical treatment -F	7.98	1.51	8.52	0.95
Pelvic reconstructive surgery	6.82	2.07	6.78	2.06
Intrauterine manipulation	6.94	2.30	7.24	1.95
Medical treatment -M	7.24	2.03	7.86	1.57
Surgery -M	6.76	2.32	6.24	2.04
Artificial insemination -H	8.52	0.86	7.98	1.47
Artificial insemination -D	3.10	2.54	2.86	1.85
IVF -husband	8.82	0.52	7.80	1.74
IVF -donor	3.04	2.44	3.04	2.06
Surrogate mother	2.28	1.86	1.62	1.30

moral considerations. The higher the preference, the less negative perception of those dimensions. The preferences of non-patients are generally correlated with the same dimensions, except for one significant difference: in comparison to patients, the non-patients' preferences are more correlated with perceived social exposure (-.85 and -.14 respectively; difference between correlations: $u = 2.11$, $p < .05$). Other differences between correlations did not reach statistical significance.

DISCUSSION

A central outcome of the present study has been the identification of attributes which are important in womens' judgment of treatments for infertility. Such a collection of perceptually salient features has been defined by cognitive scientists as a prototype or a frame (Smith, 1988; Minsky, 1975). These are structures around which our knowledge is organized, and which affect how we perceive, learn, remember, communicate and reason about information regarding the specific concept. In our study prototypes of treatment for infertility have been derived empirically from people's intuitions and judgment of treatment. The advantage of this technique is that it makes no prior assumptions about dimensions in use by subjects and therefore represents actual cognitions.

The role of lay prototypes and frames associated with medicine and health has recently been emphasized (Meyer et al., 1985; Bishop et al., 1987). Evans et al., (1986) conclude that "The fundamental starting point of any diagnosis and treatment is comprehension - of the patient by the physician, and of the physician by the patient... to be effective, the physician must discover not only the root causes of the patient's medical problem, but also something of the conceptual structure that supports both the patient's and the doctor's perceptions and understanding of the problem and the task at hand" (p. 1033).

Among other health-related prototypes, the "meaning of treatment" in relation to personal needs, to personal values and to personal conceptions of what constitutes quality of

Table 2. Spearman Rank Order Correlations Between Treatments' Mean Perceived Features and Preferences

Feature	Correlation with	Preference
	Patients	Non-patients
Social norms	-.83 **	-.78 **
Pain	.19	-.16
Bonding-child	-.90 **	-.62 **
Couple relationship	-.83 **	-.56
Hope-success	-.73 **	-.77 **
Physical threat	.04	.15
Fetal health	-.59	-.04
Humiliation	-.85 **	-.82 **
Medical sophistication	.27	.02
Sexual life	-.81 **	-.55
Social exposure[a]	-.14	-.85 **
Moral dilemma	-.63 *	-.80 **
Artificiality	-.36	-.66 *
Self-esteem -F	-.53	-.33
Self-esteem -M	-.47	-.33

* = p < .05
** = p < .01
a = the difference between correlations is significant at p < .05

life is becoming increasingly important in directing future research, as well as in reconsidering definitions of "compliance", "adherence" or "acceptance" of treatment (Gochman, 1988). Understanding the meaning of "treatment for infertility" is, therefore, of special importance, considering its association with, and impact on, some of the most personal and emotional aspects of life.

In our study, fifteen features have been identified as comprising the prototype of "treatment for infertility". Of these, the following were found especially important to patients (correlated most strongly with treatment preferences) - treatment's effects on bonding to the child; humiliation involved in the treatment; perceived social norms regarding the treatment; the treatment's impact on the couple's relationship and sexual life; the perceived success of the treatment and moral considerations. It is, therefore, recommended for health care providers to direct their attention to these topics when communicating with patients, since it is most likely that these constitute the patients' most salient concerns.

In Vitro Fertilization by husband's semen can be regarded as a specific instance of the "treatment for infertility" prototype. According to our results, IVF is generally perceived by patients as being more positive than other treatments on most scales. For example, IVF is considered to be more socially acceptable and less likely to lower the patients' self-esteem. This view corresponds to findings by Polit (1978) and Callan (1975), that couples, involved in IVF get special sympathy, and are judged more positively by the general population than other childless couples. Our subjects' perceptions are also compatible with findings of high self-esteem scores among IVF patients (Freeman et al., 1985). These positive perceptions may indicate the general halo of a sophisticated and prestigious treatment, which lacks the stigma associated with treatment involving donors. It can also be an expression of the patients' coping mechanism in resolving their cognitive dissonance regarding the treatment they are undergoing, by bolstering it in comparison to other treatments (Festinger, 1957).

Another feature of IVF which has been found significantly more positive than other treatments is its perceived success. Evidence of the high and unrealistic optimism attributed to IVF has been described by several authors (Johnston et al., 1987, Achmon et al., 1989). Among the explanations for these high expectations are the wide publicity of IVF successes through mass media and the fact that for most patients IVF is considered as their last chance for fulfilling their parenthood needs. Cognitive as well as motivational factors are therefore suggested as antecedents of the optimistic bias in the perception of IVF patients.

On only two scales has IVF been perceived less favorably than other treatments. It was perceived as more painful and as involving more social exposure. Pain and social exposure, however, are attributes that were found less significant for patients, and, therefore, despite these attributes, IVF by husband's semen received the highest preference scores by patients.

Comparisons between patients and non-patients yielded some additional insight into cognitive representations of treatments for infertility in general, and IVF in particular. The fact that significant inter-group differences were found only for female centered treatments is interesting. It appears that the experience of infertility did not affect the patients' conceptions regarding male treatment and treatment involving donors. This can be explained by the strong social stigma, and the general social taboos regarding male infertility and use of donor sperm, (Berger, 1980, Rowland, 1985, Woollett, 1985), which overrides the personal desire for parenthood needs and results in stereotypic perceptions of these treatments even by patients. Another explanation for the above finding is that social stereotypes influence mainly the perception of those treatments in which patients do not have personal experience (male and donor treatment). Therefore, for these treatments patients have the same perspective as non-patients. Only for female-centered treatment, for which patients have "episodic" versus "semantic" knowledge (Leventhal & Cameron, 1987), do their cognitions differ.

IVF by husband's semen was perceived by patients as generally more positive and preferred than non-patients. The above discussion on familiarity, high hopes, last chance and cognitive dissonance is relevant for this finding as well. However, it is noteworthy that non-patients' most preferred treatment was "medical treatment - female". It seems that non-patients prefer this therapeutic alternative due to its being familiar, non-invasive and therefore less threatening, whereas patients, after experiencing medical treatment for infertility (ovulation induction) perceive it as significantly more painful than non-patients.

Comparisons between the groups regarding the relative importance of the various features of treatment indicate a general tendency of patients to give more weight to factors within the family (couple's relationship, sexual life and bonding to the child). Non-patients give more weight to external and general factors, i.e. social exposure, artificialness and moral dilemma. These differences may indicate the cognitive process which patients have been through, changing from general concerns to more personal concerns regarding the treatment. Most interesting is the significant difference between groups regarding the weight of social exposure.

For non-patients, social exposure involved in treatment is highly correlated with its rejection. Patients gave less weight to this factor. It is reasonable that in early stages of their treatment, the issue of concealing the problem from others had been very important also for patients, while at their present stage, their problem had already been exposed to their environment, and therefore these considerations had become minor. This speculation needs confirmation in future research.

Our finding that preferences for treatments correlated with perceived features of treatments can lead to behavioral and emotional consequences. In the introduction, we described findings about relationships between perceived success of IVF and social pressures, and the decision to continue IVF attempts (Callan et al., 1988). Others have recently described relationships between overoptimistic beliefs in IVF and levels of anxiety (Achmon et al., 1989). A further investigation is still required to reveal more associations between patients' perceptions of various attributes of treatments and a list of emotional and behavioral outcome variables, e.g. coping with the demands of the treatment, general well being, adherence to regimens, persistence to continue attempts, choice of alternate solutions. Such a research effort can be made possible using the questionnaire we have developed, which gives a comprehensive picture of patients' beliefs. Uncovering of relationships beween specific cognitions and patients' reactions, may later serve practical purposes. The need for professional psychological support integrated in clinics for infertility has been stressed by several authors (e.g. Greenfeld et al., 1985, Freeman et al., 1985, Berger, 1977, Lalos, 1985). One of the tasks that professionals need to perform is the screening and detection of patients who are at special risk for psychological and behavioral difficulties (Hasteline et al., 1984, Johnston et al., 1984). It seems that the addition of a questionnaire such as ours, which covers the cognitive perspective, can make a significant contribution to directing efforts towards helping those who need special help, and pointing to the specific areas in which help is most needed for them.

ACKNOWLEDGMENTS

Our thanks go to Dr. A. Amit and his colleagues at the IVF Unit of Hakirya Hospital for their assistance at the pilot stage of this study.

REFERENCES

Achmon, Y., Tadir, Y., Fish, B., and Ovadya, Y., 1989, Attitudes and emotional characteristics in couples' reactions to in vitro fertilization, *Harefuah* — *J. Isr. Med. Assoc.*, 116:189-192.

Becker, M. H., and Maiman, L. A., 1975, Sociobehavioral determinants of compliance with health and medical care recommendations, *Med. Care*, 13:10-24.

Berger, D., 1977, The role of the psychiatrist in reproductive biology clinic, *Fertil. Steril.*, 28:141-145.

Berger, D. M., 1980, Couples' reactions to male infertility and donor insemination, *Am. J. Psychiatry*, 137:1047-1049.

Bishop, G. D., Briede, C., Cavazos, L., Grotzinger, R., and McMahon, S., 1987, Processing illness information: the role of disease prototypes, *Basic Appl. Soc. Psychol.*, 8:21-43.

Callan, V. J., 1985, Perceptions of parents, the voluntarily and involuntarily childless: a multidimensional scaling analysis, *Marriage & Fam.*, 47:1045-1050.

Callan, V. J., Kloske, B., Kashima, Y., and Hennessey, J. F., 1988, towards understanding women's decisions to continue or stop in vitro fertilization: the role of social, psychological, and background factors, *J. In Vitro Fert. Embryo Transfer*, 5:363-369.

Cook, E. P., 1987, Characteristics of the biopsychosocial crisis of infertility, *J. Counsel. & Develop.*, 65:465-470.

Christie, G. L., 1980, The psychological and social management of the infertile couple, *in:* "The Infertile Couple", R. J. Pepperell, B. Hudson, & C. Wood, eds., Churchill Livingstone, Edinburgh.

Daniels, K. R., 1986, New birth technologies: a social work approach to researching the psychosocial factors, *Soc. Work Health Care*, 2:49-60.

Daniluk, J. C. 1988, Infertility: intrapersonal and interpersonal impact, *Fertil. Steril.*, 49: 982-990.

Evans, D. A., Block, M. R., Steinberg, E. R., and Penrose, A. M., 1986, Frames and heuristics in doctor-patient discourse, *Soc. Sci. Med.*, 22:1027-1034.

Festinger, L. A., 1957, A theory of cognitive dissonance, Stanford University Press, Stanford.

Freeman, E. W., Boxer, A. S., Rickels, K., Tureck, R., and Mastroianni, L., 1985, Psychological evaluation and support in a program of in vitro fertilization and embryo transfer, *Fertil. Steril.*, 43:48-53.

Gochman, D. S., 1988, Health behavior research: present and future, *in:* "Health Behavior: Emerging Research Perspectives," D. S. Gochman, ed., Plenum Press, New York.

Greenfeld, D., Mazure, C., Haseltine, F. and Cherney, A., 1984, The role of social worker in the in vitro fertilization program, *Soc. Work Health Care*, 10:71-79.

Haseltine, F. P., Mazure, C., De l'Aune, W., Greenfield, D., Laufer, N., Tarlatzis, B., Polan, M. L., Jones, E. E., Graebe, R., Nero, F., D'Lugi, A., Fazio, D., Maters, J., and De Cherney, A. H., 1984, Psychological interviews in screening couples undergoing in vitro fertilization, *Ann. NY Acad. Sci.*, 442:504-522.

Janz, N., and Becker M., 1984, The health belief model: a decade later, Health Educ. Q., 11:1-47.

Johnston, M., Shaw, R., Bird, D., 1987, Test tube babies procedures: stress and judgements under uncertainty, *Psychol. and Health*, 1:24-38.

Johnston, W. I. H., Obe, K., Speirs, A., Clarke, G. A., McBain, J., Bayly, C., Hunt, J., and Clarke, G. N., 1984, Patient selection for in vitro fertilization: physical and psychological aspects, *Ann. NY Acad. Sci.*, 442:490-504.

Kaufman, S. A., 1984, Stress and infertility, *The Female Patient*, 9:107-114.

Keye, W. R., 1984, Psychosexual responses to infertility, *Clin. Obstet. Gynecol.*, 27:760-766.

Lalos, A., Lalos, O., Jacobsson, L. and Von Schoultz, B., 1985, Psychological reactions to the medical investigation and surgical treatment of infertility, *Gynecol. Obstet. Invest.*, 20:209-217.

Mahlstedt, P. P., 1985, The psychological component of infertility, *Fertil. Steril.*, 43:335-346.

Matteson, R. and Terranova, G., 1977, Social acceptance of new techniques of child conception, *J. Soc. Psychol.*, 101:225-229.

Menning, B. E., 1984, The psychology of infertility, *in:* "Infertility: Diagnosis and Management," J. Aiman, ed., Springer-Verlag, New York.

Meyer, D., Leventhal, H. and Gutmann, M., 1985, Commonsense models of illness: the example of hypertension, *Health Psychol.* 4:115-135.

Minsky, M., 1975, A framework for presenting knowledge, *in:* "The Physiology of Computer Vision," P. H. Winston, ed., McGraw-Hill, New York.

Mozley, P. D., 1976, Psychophysiologic infertility: an overview, *Clin. Obstet. Gynecol.*, 19:407-417.

Polit, D. F., 1978, Stereotypes related to family-size status, *J. Marriage and Fam.*, 40:105-114.

Reading, A. E., Sledmere, C. M., Cox, D. N., 1982, A survey of patient attitudes towards artificial insemination by donor, *J. Psychosom. Res.*, 26:429-433.

Rowland, R., 1985, The social and psychological consequences of secrecy in artificial insemination by donor (AID) programs, *Soc. Sci. Med.*, 21:391-396.

Seibel, M. M., and Taymor, M. L., 1982, Emotional aspects of infertility, *Fertil. Steril.*, 37:137-145.

Smith, E. E., 1988, Concepts and thought, *in:* "The Psychology of Human Thought", R. J. Sternberg and E. E. Smith, eds, Cambridge University Press, Cambridge.

Taymor, M. L., 1978, "Infertility," Grune and Stratton, New York.

Wallace, L. M., 1985, Psychological adjustment to and recovery from laparoscopic sterilization and infertility investigation, *J. Psychosom. Res.*, 29:507-518.

Walters, W. and Singer, P., 1982, "Test Tube Babies," Oxford University Press, Melbourne.

Waltzer, H., 1982, Psychological and legal aspects of artificial insemination (AID) - an overview, *Am. J. Psychother.*, 36:91-102.

Woollett, A., 1985, Childlessness; strategies for coping with infertility, *Int. J. Behav. Develop.*, 8:473-482.

PSYCHO-SOCIAL PROFILE OF IVF PATIENTS

Zvia Birman, A. Lewin, V. Soskolnie, D. Greenspan,
N. Laufer, and J. G. Schenker

*Department of Social Services and Department of Obstetrics
and Gynecology, Hadassah University Hospital, Jerusalem, Israel*

INTRODUCTION

Couples with difficulties in bearing children are described in the literature as couples who are in mourning. They mourn for the unborn child, for the unfulfilled dream, for a potential not realized, all these without the social support rendered in cases of other types of grief. The lack of support and the feeling of anger induce isolation of the couples, and with the passing of years, their psychological state deteriorates (Eck Menning 1980). Rosenfeld and Mitchell (1979) describe the response to problems of infertility as a pattern whose stages appear in the following order: shock, denial, anger, isolation, and in the end, acknowledgement (Williams 1977).

The common behavioral responses are seclusion, disorganization, fatigue and moodiness (Valentine 1986). Sexual functioning is also disturbed. It is difficult for the couple to endure inquiries into their situation. They avoid close involvement with their surroundings and sometimes also abstain from obtaining a complete assessment of their infertility problem so as not to be put in the position of receiving a "final" diagnosis. The last stage of the process is acknowledgement. That is to say, recognition of the difficulty of bearing children naturally. This last stage can lead to one of two resolutions: a decision to build a life style without children, or finding a way to fulfill the need to raise children by submitting themselves to treatment of the fertility problems, including surgery or in vitro fertilization (IVF), or by adopting a child. This decision can be assisted by psychological counselling. For this reason Rosenfeld and Mitchell (1979) see great importance in integrating psychological counselling into the treatment of infertility.

Advances in Assisted Reproductive Technologies, Edited by
S. Mashiach *et al.*, Plenum Press, New York, 1990

The treatment of infertility by IVF is a relatively new technique, since the pioneer work of Steptoe and Edwards in 1978. By extra corporeal cultivation and fertilization of human gametes, the techniques allow the achievement of pregnancy in couples who were previously considered infertile due to mechanical, male immunological or unexplained infertility. This innovative and remarkable treatment has produced an atmosphere of optimism amongst both patients and physicians, but this optimism is insufficient to diffuse the enormous stress accompanying the procedure. This is especially true in view of the relatively low success rate in becoming pregnant being in the range of 20% - 30%, with about 11% eventually giving birth.

The need for psychological counselling and support during the course of treatment was stressed upon in previous studies (Berger 1977, Whitmore 1979, Rosenfeld and Mitchell 1979, Eck Menning 1980, and Hertz 1982).

Further studies (Sahaj and Kent-Smith, 1988, Shatford and Hearn 1988) show that there are no significant differences between IVF patients and the average population in the psychological profile evaluation personality characteristics or life change of social support. The only remarkable factor noticed was the high level of anxiety during IVF treatment.

This study was undertaken in order to evaluate the psychological and social state of a randomly chosen group of patients undergoing IVF treatment in our department.

The main goal was to focus on points of possible conflict between treatment and patients' needs.

MATERIALS AND METHODS

Demographic, Socioeconomic Variables

Forty-six patients were randomly chosen from our IVF patients. All patients were married and their average age was 32. Their average education level was 12.5 years, which is above the average education level of Jewish women in Israel.

Seventy-nine percent of the study group worked before they were enrolled. Of the non-working women half stopped working because of the infertility treatment. The socioeconomic status: Most of the women were of the upper class, the remainder were middle class and no women were in the lower class. Husbands' status were similar (Table 1).

Variables Connected with IVF Treatment

The average length of marriage was 8.3 years. Twenty-six percent of them had one or two children and none of them had adopted children.

Table 1. Socio-Demographic Variables

No. of Patients:	46	
Age:	Mean = 32 years (± 3.7)	
Education:	Mean = 12.5 years (± 2.6)	
Employment:	Employed outside the home	= 79%
	Not employed	= 21%
Socioeconomic Level:	Middle to upper class	= 45%
	Middle class	= 34%
	Lower class	
	Not employed	

Table 2. Variables Connected with IVF Treatment

No. of years married	Mean = 8.32 years (± 3.8)	
No. of children	No children	= 70%
	1 child	= 20%
	2 children	= 4%
Duration of infertility	Mean = 7.26 years (± 3.4)	
Duration of medical treatment for fertility problems	Mean = 6.02 years (± 3.4)	
No. of IVF treatments		
1st cycle - 44%		
2nd cycle - 36%		
3rd cycle - 9%		
More than 3 cycles - 11%		

The mean duration of infertility was 7.2 years. Forty-four percent of the group were enrolled in their first IVF cycle (Table 2).

Types of Difficulties During Treatment

Upon meeting the women one received the impression that they were having a great deal of difficulties. We wanted to identify these and to see what could be done to decrease the weight of their problems. The main difficulty was day to day functioning. The second hardship was physical treatment (including pain). The third was lack of sufficient information. In spite of the difficulties, only one patient stated that she would withdraw from further treatment if the present treatment cycle failed. Eight-two percent answered that they would try again (Table 3).

Wish for Changes in IVF Program

As we maintained at the beginning we wanted to have a clear picture of points of possible conflicts between the patients' needs and IVF treatment. We asked them first about their difficulties and then we noted their wishes for changes.

Table 3. Types of Difficulties During Treatment

	Not difficult (or no problem) 1	Moderate 2	3	4	Very difficult (many problems) 5
Day-to-day functioning	31	8	8	25	28
Physical treatment	19	14	24	16	27
Insufficient information	32	16	24	11	13
Ultrasound	55	16	8	5	16
Blood tests	64	17	8	5	16
Team attitudes	66	17	9	3	6

Table 4. Wish for Changes in IVF Program

Shortening the waiting time (during treatment)	50%
More emotional support and attention from the team	44%
More information and instructions	39%
Shortening the bureaucratic procedures	33%
A separate department for IVF	18%

We fulfilled all the wishes with the exception of one. We could not, of course, help them with shortening of bureaucratic procedures (Table 4).

Dealing with wishes: We asked about the desire for a boy or girl. Twenty-six percent dreamed of twins; all the others (except two) were indifferent to the baby's gender.

We expected the women to manifest a high score of psychological distress. We attempted to obtain the description of this by using Derogatis and Melisaratos Brief Symptoms Inventory (The BSI, 1983). The BSI provides a profile of area of stress under nine headings. The profile shows the extent of distress.

The regular population's average score is around 50. We see women undergoing IVF treatment in the highest levels of distress. Very high scores of anxiety, hostility and depression (Figure 1) were noted. In these sub-scales the emotional distress is even more severe than distress of women suffering from chronic illnesses such as diabetes, terminal renal failure or even cancer (Soskolne and Kaplan De-Nour, 1987), which means that our patients are more anxious than women in life threatening conditions.

The literature and our experience shows that IVF patients are remarkably anxious and depressed and we have no doubt that by reducing the levels of distress we will achieve higher levels of success in IVF treatments. We found that women who became pregnant started the treatment in a better psychological state than women who did not succeed in conceiving. Unfortunately, the difference was not significant. We all have the feeling that improving the emotional state is vital, and we are all seeking ways of carrying this out.

We have no pretence of solving the psychological conflicts of our patients. We know that they are expected after long years of failure to become pregnant. They have to save their strength to overcome the demanding IVF treatment.

We found relaxation as a good way of fulfilling our wish to help the patients in a short time without the pains of conventional psychological therapy (Omer, 1986, Edmonsten, 1981).

We based our decision to use relaxation on the following reasons:

1. effective in reducing anxiety
2. demands for short professional intervention (± 1 hour per patient)
3. women can practise it by themselves whenever they feel the need
4. gives women control and self-confidence

We use relaxation techniques, with special guided imagination content which was constructed especially for this group. We use it after individual intake in which we choose the best way of helping the specific patient. For some of them it will be enough to practise relaxation. For others we will have supportive therapy in addition to the behavioral psychological help achieved by relaxation.

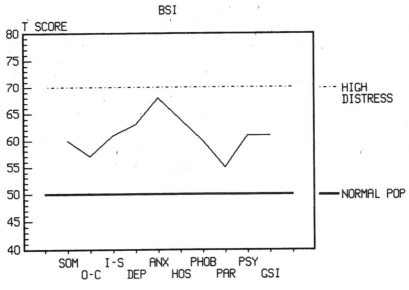

Fig. 1. Emotional distress.

The outcomes of using this technique cannot be proved as yet by statistics. Our experience leads us to believe that it works. Now we are going to validate our optimism by conducting a full scale study.

REFERENCES

Ballock, J. C., Iatrogenic impotence in an infertility clinic, Illustrative case, *Am. J. Obstet. Gynecol.* 124:476, 1974.

Berger, D., The role of the psychiatrist in a reproductive biology clinic. *Fertil. Steril.*, 28:141-145, 1977.

Bresnick, E., and Taymor, M., The role of counselling in infertility. *Fertil. Steril.*, 32:154-156, 1979.

Daniel, K.R., New birth technologies: A social work approach to researching the psychosocial factors., *Social Work in Health Care* 11: 49-60, 1986.

Derogatis, L.R., and Melisaratos N., The Brief Symptom Inventory: An introductory report, *Psychological Medicine* 13:595-605, 1983.

Eck Menning, B., The emotional needs of infertile couples, *Fertil. Steril.*, 34:313-319, 1980.

Edmonsten, W.E., Hypnosis and Relaxation: Modern Verification of an Old Equation, pp. 82-167, John Wiley and Sons, New York, 1981.

Elstein, M.: Effect of infertility on psychosexual function. *Br. Med. J.* 5:295, 1975.

Erickson, An experimental approach to psychogenic infertility, *in:* The collected papers of Milton Erickson Vol. II, Rossi E.L. Ed., New York, Irvington,1979.

Hertz, D., Infertility and the physician-patient relationship: A biopsychosocial challenge. *Gen. Hosp. Psychiatry* 4:95-101, 1982.

Omer H., Hypnotic-relaxation in the treatment of premature labor. *Psychosomatic Medicine* 1986.

Rosenfeld, D., and Mitchell M., Treating the emotional aspects of infertility: Counseling services in the infertility clinic. *Am. J. Obstet. Gynecol.*, 135:177-180, 1979.

Sahaj, D.A., and Kent-Smith, C., Psychosocial description of a select group of infertile couples, *J. Family Practice* 27: no. 4, 1988.

Shatford, L.A., and Hearn, M.T., Psychological correlates of differential infertility diagnosis in an in vitro fertilization program, *Amer. J. Obstet. Gynecol.* 158:1099-1107, 1988.

Soskolne, V., and Kaplan-DeNour Y., Emotional distress in various patient groups, 1987 (unpublished).

Statistical Abstracts of Israel. Published by the Central Bureau of Statistics Bul No. 7, 1986.

Valentine, D.P., Psychological impact of infertility: Identifying issues and needs, *Social Work in Health Care* 11:61-69, 1986.

Wallston, K.A., and Wallston, B.S., Development of the multidimensional health locus of control (MHLC) scales. *Health Education Monographs* 6:160-171, 1978.

Williams, L.S., and Power, P.W., The emotional impact of infertility in single women, *J. Am. Med. Wom. Assoc.*, 32:327, 1977.

EUROPEAN IVF RESULTS FOR 1987 (a) —
ATTEMPTS AND PREGNANCY OUTCOME

Jean Cohen [a] and Jacques De Mouzon [b]

[a] *Clinique Marignan, 3 rue de Marignan, 75008 Paris, and*
Hôpital de Sèvres, 141 Grande Rue, 92310 Sèvres, France
[b] *INSERM U 292, Centre Hospitalier de Bicêtre*
78 rue du Gl Leclerc, 94275 Le Kremlin-Bicêtre, Cedex, France

a) This work was done with the collaboration of the following centers:
1 - for Europe; 2 - for France

EUROPEAN TEAMS PARTICIPATING

AUSTRIA	GRAZ, WIEN-KEMETER, WIEN UNIV. FRAUENKLINIK
BELGIUM	CPMA, CRAF, DEVROEY, LEUVEN, LIEGE, TIENSEVE
DENMARK	BANGSUE
D.D.R.	BERLIN, LEIPZIG, ROSTOCK
FINLAND	HELSINKI, KUOPIO, OYKS, TAMPERE, TURKU
G.B.	HULL, MANCHESTER, PORTLAND, SHEFFIELD, SOUTHAMPTON
GREECE	BLUECROS TEHSSALONIKI, ENGONIKI, GENIKI THESSALONIKI MASSOURAS, METERA
ITALY	MILANO, BOLOGNA, SAPIENZA ROMA, TORINO
NETHERLANDS	ACA1, AZM, DYKSYT, ERASMUS
NORWAY	TROMSO, ULLEVAL, OSLO NATIONAL HOSPITAL
R.F.A.	BERLIN, BIELEFELD, ERLANGEN, FRAUENKLINIK BONN, GÖTTINGEN, HANNOVER, KARLSRUHER IVF PROGRAMM,

Advances in Assisted Reproductive Technologies, Edited by
S. Mashiach *et al.*, Plenum Press, New York, 1990

	KRUSMAN MÜNCHEN, MUNICH, TUBINGEN, ULM, WIEDEMAN MUNICH
SPAIN	BARCELONA HOSPITAL, BARRI BARCELONA, BILBAO, CE-FIVBA, GRANADA, OCTOBRE MADRID, VALENCIA, VISCASI-LAS BARCELONA
SWEDEN	HUDDINGE, KROLINSKA STOCKHOLM, LUND, MALMOE, LIN-KOPING, NILSSON, SOPHIA HEMMET STOCKHOLM
SWITZERLAND	CHUVLAUS LAUSANNE ULSTER QUEEN'S BELFAST
YOUGOSLAVIA	ZAGREB

FRENCH CENTERS FIVNAT 1987

PARIS ET REGION PARISIENNE: BAGNOLET CL. DHUYS, BOBIGNY CHU, BOULOGNE BELVEDERE, CLAMART BECLERE, NEUILLY CHEREST, NEUILLY HOP AMERICAIN, PARIS TENON, PARIS MARIGNAN, PARIS BAUDELOCQUE/ST VINCENT DE PAUL, PARIS PORT-ROYAL, PARIS BONSE COURS, PARIS LA PITIE/SALPETRIERE, PARIS DIACONESSES, PARIS HOTEL DIEU, POISSY CHR, SEVRES CHR, INSERM U 292 BICETRE.

PROVINCE: AMIENS CHU, AMIENS A.P.E.T.S., BAYONNE FIVETE COTE BASQUE, BESA CON CHU ST-JACQUES, BOIS GUILLAUME ST-ANTOINE, BORDEAUX HOP. SAINT-ANDRE, BORDEAUX IFREARES, BREST CHR, BRIVES CL. ST GERMAIN, CAEN CHU, CLERMONT FERRAND CHU, DEAUVILLE POLYCLINIQUE, DUNKERQUE MAT. DES BAZENNES, GRENOBLE CHU, GRENOBLE BELLEDONNE, LE MANS TETRE ROUGE, LE MANS CL. ST DAMIEN, LENS CL. BOIS BERNARD, LILLE CHR GYNEC SOCIALE, LILLE EPARP, LILLE CHR MATER PAV. OLIVIER, LOMME CTRE ST PHILIBERT, LORIENT CL. STE BRIGITTE, LYON CHU, LYON RHONE ALPIN, LYON C.R.E.S, LYON FERTILY, MARSEILLE HOP. CONCEPTION, MARSEILLE FIV PROVENCE, MARSEILLE CL. DU PRADO, MARSEILLE C.R.E.F., MARSEILLE BELLE DE MAI, METZ HOP. CL. BERNARD, MONTPELLIER CHU, NANCY MAT. PINARD, NANTES CHU, NICE A.N.E.R., NICE CHU, PERIGEUX ASS. FERTILITE PERIGOURDINE, REIMS CHU, REIMS COURLANCY, RENNES SAGESSE, ROUBAIX CHR A.R.A.M.P., STRASBOURG SHCILTIGEIM, ST. ETIENNE HOPITAL NORD, ST. ETIENNE CL. MICHELET, TOULOUSE CHU, TOULOUSE IFREARES, TOURS CHU, TOURS L.A.M., VALENCE GILLIOZ, VICHY LA PERGOLA.

OUTREMER: FORT DE FRANCE CL. ST. PAUL, POINTE A PITRE CL. ROSIERS.

ETRANGER: GENEVE BOURRIT.

MATERIAL AND METHODS

The present study reports the results of IVF attempts and pregnancy outcomes in Europe in 1987. The aim was to give an overview of the obstetric outcomes after IVF for the year 1987 (1/1 to 31/12). Most of the French IVF centers send their results to a central registry (FIVNAT = National IVF) and they were not asked to complete a form again. French centers fill in one form per attempt after the oocyte aspiration. One hundred and twenty European centers outside France were contacted, 69 agreed to participate and sent their forms before March 15, 1989. They filled in one form per unit. The following data were provided: 1) number of ovulation stimulation, oocyte recovery attempts, transfers and pregnancies (from fresh and thawed embryos); 2) attempt description: infertility diagnosis, stimulation regimen, oocyte recovery, transfer; 3) evolution of the pregnancy: abortion, ectopic, delivery; 4) pregnancy pathologies; 5) deliveries (term, mode, sex, birth weight, intensive care, outcome, birth defect) concerning one or more children.

Finally, the survey covered a total of 29,635 attempts (Table 1) with large disparities between the countries. Few centers of Great Britain answered because registration already exists there. We consider this study as a first attempt of a European survey of IVF.

Data were analyzed with a Vax 8530 computer from Centre de calcul de l'INSERM, Villejuif, France, using SAS software. The statistical methods used included student's test and chi square test.

RESULTS

Infertility origin is described in Table 2. Table 3 compares the indications for IVF between France and the rest of Europe. Tubal indications are only approximately 50% of the indications and French centers appear to be different. It is also the case for the stimulation regimen (Table 4 and 5). The analogs of LHRH are used in 60.3% of the French attempts versus 16.7% in the rest of Europe. One of the reasons is probably that this medication is totally reimbursed by the French Social Security, and not in the other countries.

Table 6 describes the attempts. One must remember that the forms filled in in France start only from the oocyte recovery.

Twenty-nine thousand, six hundred and forty-four attempts resulted in 4,866 pregnancies. The information about pregnancy outcome was unknown in 1,461 cases (1,282 in France and 179 in the rest of Europe). Table 7 shows the main results.

Among 3,405 pregnancies for which detailed data were obtained, 4.9% were ectopics, 0.6% were both intrauterine and ectopics, 26% resulted in abortion and 68.9% were delivered (Table 8).

Table 1. Sample Description

Country	Units	Attempts	Pregnancies
Austria	3	447	66
Belgium	6	3,193	548
Denmark	1	81	19
Finland	5	532	34
France	72	14,763	2,464
FRG	12	2,980	651
GB	5	496	37
GDR	3	381	34
Greece	5	694	72
Italy	4	1,020	134
Netherlands	4	1,917	362
Norway	3	654	
Spain	8	834	110
Sweden	7	1,238	160
Switzerland	1	123	10
Ulster	1	24	2
Yougoslavia	1	311	35

Table 2. Infertility Diagnosis

	n	%
Tubal pure	14,719	51.9
Male pure*	2,947	10.4
Azoospermia	872	3.1
Sperm Antibodies	293	1.0
Tubal + Male	3,597	12.7
Endometriosis	1,620	5.7
Unknown	2,119	7.5
Various and not precise	2,183	7.7
	28,350	100.0

*excluding azoospermia

The rate of ectopics was statistically different in France (8.1%) and in other countries (4.3%) (p < 0.001). This was not related to use of LHRH analogs. The rate of abortion was also different before 12 weeks, i.e. 18.8% in France versus 25.8% in other countries (p < 0.001). There is no explanation for these differences.

The outcome was very different in single and multiple pregnancies. Table 9 indicates that a large proportion of European centers transfer four or more embryos. This results in approximately 20% multiple pregnancies (Table 10).

Pathology during pregnancy (Table 11) was very high for multiple pregnancies. It re-

Table 3. Infertility Diagnosis

	Europe Except France (%)	FIVNAT (%)
Tubal pure	57.2	47.0
Male pure	7.6	13.0
Tubal + Male	10.1	15.1
Azoospermia	3.3	2.9
Sperm Antibodies	1.4	0.7
Endometriosis	4.4	6.9
Various/Unspecified	7.6	7.8
Unknown	8.4	6.6

$P < 0.001$

Table 4. Stimulation Regimen

		%
GnRH Analogs	11,187	39.0
Programmation	1,500	5.2
HMG	3,271	11.4
CC/hMG	10,443	36.4
FSH	1,204	4.2
Other/Unspecified	1,106	3.9

Table 5. Stimulation Regimen

	Europe Except France (%)	FIVNAT (%)
GnRH Analogs	16.7	60.3
Estroprogrestative/Danatrol	1.9	8.4
HMG + Clomiphene Citrate	18.4	5.0
HMG	53.1	20.9
FSH	5.8	2.8
Other/Unspecified	4.2	2.7

$P < 0.001$

sulted in a high incidence of cesarean sections (Table 12), premature births (Table 13) and transfer to pediatric unit (Table 14).

Single pregnancies appeared to be normal.

The rate of neonatal mortality is considerably higher in multiple pregnancies than in single ones (Table 15). The rate of congenital malformations is not different between single and multiple ones.

The congenital malformations were varied and did not show any specific pattern (Table 16).

Table 6. Attempts of IVF

Cancellation before oocyte puncture*	18.5%	
Spontaneous LH surge European ST	7.5%	
		$p < 10^3$
FIVNAT	3.1%	
Unsuccessful oocyte recovery**	2.5%	
Laparoscopic recovery Europe	34.0%	
		$p < 10^6$
FIVNAT	22.7%	

*   Europe except France ** France

Table 7. Main Results

	Oocyte Puncture	Transfer	Pregnancy	Delivery
Oocyte puncture	29,644			
Transfer	70.9%	21,029		
Pregnancy*	16.4%	23.1%	4,866	
			(3.405)**	
Delivery***	11.6%	16.4%	71.0%	2,347

  * Clinical pregnancies
 ** 3,307 pregnancies with information about outcome
*** Rate is estimated from pregnancies with known outcome

Table 8. Pregnancy Outcome

	n	%
Pregnancies	3,405	
Ectopic alone	168	4.9%
associated with IUP	19	0.6%
Spontaneous abortion	890	26.0%
< 12 weeks	802	23.6%
12 - 27 weeks	82	2.4%
Deliveries	2,347	68.9%

Table 9. Maximum Number of Transferred Embryos

Embryos	Units
2	1*
3	9
4	62
5	32
6	16
≥7	6

*Spontaneous cycles

Table 10. Multiple Pregnancies

	Europe (%)	FIVNAT (%)
Double	14.7	18.5
Triple	3.4	3.8
Quadruplets and more	1.2	0.3
Total	19.3	22.6

Table 11. Pregnancy Pathology

	Single (%)	Double (%)	Triple (%)
Toxemia	6.7	11.5	17.1
Threatening Premature Labor	10.7	37.8	46.1
IUGR	3.9	7.4	10.5
Hospitalization	11.2	28.1	42.1

No difference between FIVNAT and the rest of Europe.

Table 12. Mode of Delivery

	Single (%)	Double (%)	Triple (%)
Spontaneous	52.6	24.1	2.7
Instrumental	13.3	10.9	1.4
C. Section	34.2	65.0	96.0

Table 13. Time of Delivery

	Single (%)	Double (%)	Triple (%)
<27 Wk	0.6	3.1	1.6
27–32	1.6	7.3	30.1
33–37	14.4	54.2	60.3
38–41	79.3	34.7	7.9
≥42	4.1	0.8	0
Premature birth ≤36 weeks*	9.5	33.1	83.3

* FIVNAT

Table 14

	Single	Double	Triple
Sex Ratio	1.16	1.12	1.14
Weight < 2500	14.0%	46.4%	81.1%
Transfer to Pediatric Unit	9.0%	26.7%	30.8%
Intensive Care	3.2%	7.7%	31.3%

Table 15. Mortality-Malformations

Mortality	Single	Double	Triple
In utero	12.9%	16.9%	18.0%
Neonatal < day 8	5.0%	5.6%	45.0%
Neonatal day 8/day 28	1.4%	1.8%	9.0%
Total	19.3%	24.3%	72.0%
Malformations	2.3%	3.3%	2.3%

Table 16. Congenital Malformations

Chromosomal	5
Cardiopathy	8
Digestive, exomphalos	7
Genito-urinary	7
Cerebral	3
Extremities	4
Polymalformation	2
Cleft Palate	2
Angioma	2
Epstein	1
Broncho pulm. dysplasia	1
Pectoralis aplasia	1
DIV	2

CONCLUSION

The pregnancy rate in Europe seems good:
11.6% per cycle
16.5% per transfer

The delivery rate is 71%.

Ectopics are high (5.5%) and we need to understand more about the reason for ectopics, and perhaps arrive at some preventive decisions.

Twenty percent of pregnancies are multiple, 4% being triplets or more. The multiple pregnancies have more pathology, more prematurity, smaller for term babies, more mortality. We must see if they are related only to the number of embryos replaced or also to the indication, or the age of the mother, or the treatment and we must try to find a way to decrease or exclude triplets.

Finally, the type of malformations encountered in IVF are no different from those encountered in the normal population.

IVF RESULTS IN EASTERN EUROPE

László G. Lampé

Univ. Med. School Debrecen (Hungary)
H-4012 Debrecen

In the Eastern European countries, both research and clinical observations were delayed for several years after those of Edwards and Steptoe and other pioneers in this field. Despite increased efforts the magnitude of the gap was not reduced.

The first successful results of researchers and clinicians in the region of Brno, Czechoslovakia, deserve attention. Between 1984 and 1988, they were able to produce 24 pregnancies (with the exclusion of biochemical pregnancies) by means of in vitro fertilization (IVF). Of these, 14 were born alive, three were ectopic pregnancies, three were spontaneous abortions and three mothers are still pregnant.

In 1980, the same team developed a method similar to gamete intra-fallopian transfer (GIFT), and two years later they reported four pregnancies. Microsurgery for tubal reconstruction was performed at the time of ovulation and the removed ova were placed, together with the husband's sperm, into the freshly operated fallopian tubes. Their work was published in the Czech language, and since then several authors have quoted their publication.

In the German Democratic Republic, four University Departments (in Berlin, Halle, Leipzig and Rostock) have been involved in IVF and embroyo transfer (ET). They have already produced 12 babies, and six women were still pregnant at the time of questioning. The ET rate in Leipzig was 65%.

In Rostock, 89 GIFT procedures have been performed, resulting in 11 clinical pregnan-

Advances in Assisted Reproductive Technologies, Edited by
S. Mashiach *et al.*, Plenum Press, New York, 1990

cies and 10 further biochemical pregnancies. Four babies have been born, three women were still pregnant (one with a twin) at the time of questioning and four pregnancies ended in spontaneous abortions.

In the same departments, nine GIFT procedures were combined with simultaneous donor inseminations (AID). Of these, one baby was born, and two women are still pregnant.

In Hungary, all the University Departments of Obstetrics and Gynecology (Budapest, Debrecen, Pécs, Szeged) have been prepared and initiated IVF/ET and GIFT programs. As far as qualified personnel are concerned, the conditions are good. In contrast however, equipment for both surgical and laboratory procedures is rather deficient, which is probably responsible for the rather modest results. The first GIFT baby was born in 1988 in Pecs and further pregnancies have been achieved including two IVF/ET pregnancies (Csaba, 1989).

In the Soviet Union, Nikitin et al., (1987), from the Institute of Human Embryology, Leningrad, reported a series of 119 laparoscopic follicle punctures resulting in aspiration of 159 ova. A fertilization rate of 66% has been achieved. Embryo transfer was performed in 37 cases in the stage of 2–8 cell blastomeres. Implantation was confirmed in five cases. Three weeks later only one of them was alive, and this pregnancy was carried to term and a healthy baby weighing 3500 g was born.

In 1988 Leonov et al. reported good results in amenorrheic patients using GnRH treatment with the pulsatile microinfusion set of Ferring Co., supplemented with human menopausal gonadotropin (hMG) (Humegon, Organon) and human chorionic gonadotropin (hCG), 10,000 I.U. The follicle aspiration was made transvaginally.

By the end of 1988, in this center (All-Union Research Center for Mother and Child Health Care, Moscow), 16 healthy babies had been born after IVF procedure and a further 25 pregnancies were developing normally.

In Moscow too, at the Pirogov Medical Institute, an IVF and ET program was launched in 1987. The first baby was born in 1988, and further pregnancies were achieved later.

From Leningrad (Davidov et al., Postgraduate Medical Institute), and Kharkov (Grishenko et al.) several living pregnancies were reported by personal communication.

In the Soviet Union the initial treatment regimen for timing of ovulation consisted of clomiphene citrate with or without hCG. Recent protocols use hMG and hCG more widely.

In Yugoslavia, one of the leading institutes in IVF/ET is the Department of Obstetrics and Gynecology of the University of Ljubljana, Slovenia. They started the program in 1983, and achieved two pregnancies in 1984. In 1985, from an additional five pregnancies, seven babies were born, including two sets of twins. Since then, four births occurred and four other pregnancies were in the advanced stage at the time of data collection. In Zagreb

(Croatia), they started later, but by the date of reference they had 29 babies born due to the IVF/ET technique and over 30 other pregnancies were in the advanced stage.

Although precise assessment of the need for IVF/ET cannot be made because of the lack of available data, an increasing demand can be forecasted in the listed Eastern European countries.

Oocyte Aspiration

Oocyte aspiration in the institutions which started IVF/ET programs before 1984–1985, was performed by laparoscopy, while since 1987 the practice of ultrasonically-guided transvaginal follicle puncture and aspiration became more and more common. Nevertheless, to date the majority of the institutions still use laparoscopy.

Four to five years ago, hormonal assays were not always available in all laboratories. The lack of modern methods and a consequent mistiming of ovulation might be responsible for the initial very high failure rates of IVF and ET procedures in these institutions.

CONCLUSION

From this report it can be seen that in the Eastern European countries IVF and ET procedures have become acceptable in the treatment of selected infertility cases. In the early 1980s, only exceptional individual ambition and diligence made it possible to introduce this into certain institutions. The technical conditions were extremely deficient during that time; neither ultrasonographic methods, nor hormone assays were routinely available to predict the time of ovulation. Instrumental and biological conditions for culture and transfer of embryos was insufficient, and this explains the poor success rate of oocyte retrieval and embryo transfer in the early years.

In these countries, the need for the establishment of a concise ethical approach and legal regulation of the various methods of assisted reproduction is increasing. The most advanced countries are Czechoslovakia, Hungary, the German Democratic Republic and Yugoslavia.

At present, there is no uniformity in the established temporary regulation and ethical approach, while the method of preparation of proposals for legislation has been remarkably similar, with cooperation of medical doctors, lawyers, ethical experts and lay persons, including womens' organizations. The final regulation and legal codification will be made after a wide-scale investigation.

In general, although the success rates of IVF/ET procedures in the Eastern European countries are on the rise (Seppala et al. 1984) a definite gap still exists. The procedure is most successful among women under 30. Success rates are higher in secondary than primary infertility. The quality of the sperm is of high importance. The standard of equipment, the availability of appropriate laboratory facilities and a well trained, dedicated team are of decisive importance for the success of such programs. These latter conditions are to be improved in the Eastern European countries in order to make them capable of approaching the leading institutions of the world.

ACKNOWLEDGEMENTS

The reporter's thanks go to those making their results available for this survey. They are Professors J. Kobilkova, L. Pilka (Czechoslovakia), H. Bayer, K. Bilek, H. Wilken (GDR), C. N. Davidov, B. V. Leonov and A. I. Nikitin (Soviet Union), H. Meden-Vrtovec (Yugoslavia).

REFERENCES

Csaba, I., 1989, Personal communication.

Haake, K. W., Alexander, H., Baier, D., Glander, H. J., et al., 1988, Zur Problematik einer Mehrlingsschwangerschaft nach In-vitro-fertilisation und Embryotransfer, *Geburtsh. u. Frauenheilk*, 48:374–375.

Hoonakker, E. W., 1987, In vitro fertilization anno, 1987, *Organorama*, 4:11–12.

Hren-Vencelj, H., Meden-Vrtovec, H., Tomazevic, T., et al., 1985, The program of in vitro fertilization and embryo transfer at the gynecological clinic in Ljublana, *Zdrav. Vest.*, 54:249–252.

Leonov, B. V., Lukin, V. A., Moskalenko, N. V., et al., 1988, The use of GnRH agonist in the IVF program, *Summary of IVF/ET Scientific Meeting*, Moscow, 12–13.

Meden-Vrtovec, H., Tomazevic, T., Hren-Vencelj, H., et al., 1985, In vitro fertilization and embryo transfer, *Med. Razgl.* 24:557–565.

Nikitin, A. I., Kataev, E. M., Savitsky, G. A., et al., 1987, Extracorporal fertilization in the human with subsequent successful implantation of the embryo and birth of a child, *Archiv. Anatomii, Histologii, Embryologii*, 93:39–43.

Pilka, L., Ventruba, P., Soska, J., Cupr, Z., et al., 1987, In vitro fertilization and embryo transfer programme of the First Department of Obstetrics and Gynecology in Brno - Results in 1986, *Scripta Medica*, 60:355–362.

Pilka, L., Tesarik, J., Dvorak, M., Trávnik, P., 1982, Gravidity realized by transferring oocyte fertilized in vitro into ovarian tube, *Cs. Gynek.*, 46:564–570.

Ribik-Pucelj, M., Tomazevic, T., Meden-Vrtovec, H., et al., 1985, Laparoscopic oocyte retrieval in in vitro fertilization - embryo transfer procedure, *Jugosl. Ginekol. Perinatol.*, 25:9–12.

Semm, K., 1985, Derzeitiger Stand der in-vitro-fertilisation, Referat des 88. Deutschen Ärztetages in Lübeck-Travemunde, 15 Mai.

Seppälä, M., 1984, The world collaborative report on in vitro fertilization and embryo replacement: current state of the art in January 1984, *Annales of the New York Acad. Sci.*, XIV:558–563.

Tesarik, J., Pilka, L., Dvorak, M., Travnik, P., 1983, Ooctye recovery, in vitro insemination, and transfer into the oviduct after its microsurgical repair at a single laparotomy, *Fertility and Sterility*, 39:472–475.

Tomazevic, T., Hren-Vencelj, H., Meden-Vrtovec, H., et al., 1985, In vitro fertilization and embryo transfer programme in Ljubljana, *Acta Eur. Fertil.*, 16:133–137.

Wilken, H., 1987, Bericht über den 12. Weltkongress für Fertilität und Sterilität, *Zent. bl. Gynäkol.*, 109:608–613.
Zdanovskij, V. M., Kechivan, K. N., Anshina, M. B., Solomatina, E. V., 1988, First results of IVF/ET treatment in infertility, *Summary of IVF/ET Scientific Meeting*, Moscow, 15.

IVF RESULTS IN NORTH AMERICA

Martin M. Quigley

Society for Assisted Reproductive Technology
The American Fertility Society
Birmingham, Alabama

INTRODUCTION

The reporting of in vitro fertilization (IVF) pregnancy rates and related data have been an important part of each World Congress since the 1984 Third World Congress of In Vitro Fertilization and Embryo Transfer in Helsinki, Finland. The published report resulting from that Congress[1] detailed 10,028 stimulation cycles resulting in 600 live births throughout the world from the beginning of IVF through January 1984. As a point of comparison, in the United States alone, during 1987 there were 12,104 reported IVF stimulation cycles which resulted in 1075 live births. In this report I will be presenting the recent IVF and Gamete Intrafallopian Transfer (GIFT) results from the United States. This data obviously represents the collaboration of many individuals and is presented on behalf of the Society for Assisted Reproductive Technology (SART), an affiliated society of the American Fertility Society (AFS).

Society for Assisted Reproductive Technology

In the fall of 1985 SART (initially called the In Vitro Fertilization Special Interest Group) was incorporated as an affiliated society of AFS. Its purpose is to foster education and the collection of accurate data concerning the practice of IVF. The primary memberships in this Society belong to institutions or programs performing IVF. To be a member

Advances in Assisted Reproductive Technologies, Edited by
S. Mashiach *et al.*, Plenum Press, New York, 1990

of SART, a program must conform to the published minimum standards of AFS,[2] must be initiating at least 40 IVF stimulation cycles per year, and must have established success as defined by the birth of live infants to at least three women following IVF. Once accepted into membership, the program must participate in the IVF/ET Registry described below and must continue to be performing at least 40 IVF or GIFT stimulation cycles per year.

METHODS

The data which forms the basis of this report comes form two sources. The first is the National IVF/ET Registry. This registry was established in the fall of 1986 as a joint project of SART, AFS, and Medical Research International. The IVF/ET Registry has already produced two publications[3, 4] which reported the IVF results in the United States during 1985, 1986 and 1987. Some data from those prior publications, as well as previously unpublished data from the IVF/ET Registry, have been used in the preparation of this report.

The second source is from a U.S. Congressional Subcommittee's survey. In June of 1988, Representative Ron Wyden, Chairman of the U.S. House of Representatives' Subcommittee on Regulation and Business Opportunities, called for a survey of all United States IVF clinics. Working in collaboration with officers of SART and AFS, his Subcommittee Staff prepared a questionnaire which was distributed to all IVF Clinics in the United States in November 1988. The summary data from the 146 clinics performing IVF and/or GIFT which had been collated by March 9, 1989 have been used in this report.

The two sources of data show good correlation. For example, in 1987 the IVF/ET Registry reported on 8,725 oocyte retrieval procedures while the Wyden survey reported on 9,198 retrieval procedures. Similarly, the IVF/ET Registry's 1987 live birth per oocyte retrieval rate was 11.4% while the Wyden survey's live birth per oocyte retrieval procedure was 11.7%.

RESULTS

Retrievals

There has been a progressive increase in the number of IVF stimulation cycles in the United States over the last four years. Figure 1 summarizes the number of oocyte retrieval attempts reported to the IVF/ET Registry from 1985 through 1987 and reported to Representative Wyden's Subcommittee for 1987 and 1988. Using the Wyden survey data, in 1987 there were 9,198 IVF oocyte retrieval cycles which increased to 11,606 in 1988 (See Table 1). There has also been a progressive increase in the number of GIFT retrieval cycles performed per year. In 1987 there were 2,540 GIFT retrievals which increased to 3,182 in 1988 (See Table 2). Thus, in the United States at the present time, there are approximately one-fourth as many oocyte retrievals for GIFT as compared to retrievals for IVF.

In 1985 most oocyte retrievals for IVF were performed by means of laparoscopically guided follicular aspiration. As ultrasound techniques became more widely practised, by 1987 about two-thirds of the retrievals were performed using ultrasound directed aspiration. Almost all IVF retrievals are performed using ultrasound guidance at the present time.

Recent IVF Results

The data from the Wyden Subcommittee's Survey concerning IVF in 1987 and 1988 are summarized in Table 1. In 1987, there were 12,104 stimulation cycles, of which 24% were terminated prior to oocyte retrieval. Of the 9,198 oocyte retrieval procedures, 87.1% culminated in an embryo transfer. Additionally, 1,075 live births resulted, a rate of 8.9% per stimulation cycle, 11.7% per oocyte retrieval attempt, or 13.4% per embryo transfer.

In 1988, there were 14,619 reported stimulation cycles, of which 11,606 (79.4%) resulted in an attempted oocyte retrieval procedure. There were 10,135 embryo transfers (87.3% of the oocyte retrieval cycles). At the time the data were collected, 609 live births had occurred and 1,231 pregnancies were reported as continuing. Making the assumption that 70% of the reported continuing pregnancies would result in live births, the IVF live birth rate for 1988 is predicted to be 10.1% per stimulation cycle, 12.7% per oocyte retrieval procedure or 14.5% per embryo transfer.

GIFT

The United States experience with GIFT in 1987 and 1988 as reported to the Wyden Subcommittee is summarized in Table 2. In 1987 there were 3,488 GIFT stimulation cycles of which 27.2% were terminated prior to attempted oocyte retrieval. The 2,540 oocyte

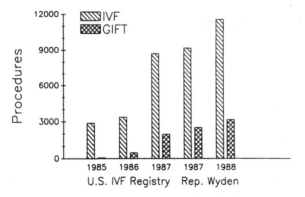

Fig. 1. Illustrated are the number of oocyte retrieval procedures performed by year for IVF and GIFT as reported to either the IVF/ET Registry or Representative Wyden's Subcommittee. Data for 1987 from both sources are illustrated and indicate the consistency of the data.

Table 1. Summary of Wyden survey IVF results

	1987	1988
Stimulation Cycles	12,104	14,619
Egg Recovery Procedures	9,198	11,606
Embryo Transfers	8,014	10,135
Live Births	1,075	609
Continuing Pregnancies	-	1,231

Percentages

	1987	1988
Scratched Cycles	24.0%	20.6%
Embryo Transfer (per aspiration)	87.1%	87.3%
Live Birth (per recovery procedure)	11.7%	-
Live Birth plus Continuing (per recovery procedure)	-	15.9%
Live Birth (per embryo transfer)	13.4%	-
Live Birth plus Continuing (per embryo transfer)	-	18.2%

This includes data from the 145 clinics performing IVF which were available for analysis on March 9, 1989.

Table 2. Summary of Wyden Survey GIFT Results

	1987	1988
Stimulation Cycles	3,488	4,193
Egg Recovery Procedures	2,540	3,182
Live Births	356	360
Continuing Pregnancies	-	534

Percentages

	1987	1988
Scratched Cycles	27.2%	24.1%
Live Birth (per recovery procedure)	14.0%	-
Live Birth plus Continuing (per recovery procedure)	-	28.1%

This includes data from the 128 clinics performing GIFT which were available for analysis on March 9, 1989.

retrieval procedures performed resulted in 356 live births, a-GIFT live birth rate of 14% per retrieval procedure.

In 1988, there were 4,193 stimulation cycles and 3,182 oocyte retrieval procedures. At the time the data was collated, there were 360 live births and 534 continuing pregnancies. Making the same assumption as above, i.e. that 70% of the continuing pregnancies would result in a live birth, the calculated 1988 GIFT live birth rate per attempted oocyte retrieval procedure is 23.1%.

Other Procedures

The Wyden survey only requested data on IVF and GIFT. Thus the most recent data available for frozen embryo transfers and donor oocyte pregnancies come from the IVF/ET Registry.[3,4] The data on clinical pregnancy rates for 1985 through 1987 for IVF, GIFT, frozen embryo transfers, and IVF with donor oocytes are illustrated in Figure 2.

Frozen Embryos

In 1986, there were seven clinical pregnancies from 112 frozen embryo replacement cycles (6.3%). In 1987 these rates improved with 50 clinical pregnancies resulting from 409 transfer procedures (10.2%).

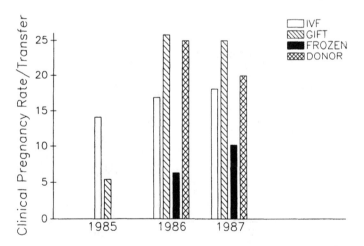

Fig. 2. The clinical pregnancy rates for IVF, GIFT, frozen embryo transfers, and IVF with donor oocytes as reported to the IVF/ET Registry for 1985, 1986, and 1987 are illustrated. The rates shown are for clinical pregnancies established per embryo transfer (or GIFT) procedure.

Donor Oocytes

The only clinic performing IVF using donor oocytes in 1985 and 1986 reported a 25% clinical pregnancy rate per transfer. By 1987, 17 clinics were performing IVF using donated oocytes. Twelve of 60 embryo transfers after IVF using donated oocytes resulted in a clinical pregnancy (20%).

Effects of Age on IVF and GIFT

According to the 1987 IVF/ET Registry data, most patients treated in the United States by IVF or GIFT are between the ages of 30 and 39, with an almost equal distribution between ages 30 through 34 and ages 35 through 39. Very few patients under age 25 are treated while a small, but significant, number of patients are over the age of 40. As shown in Figure 3, there is no apparent age effect on the percentage of patients who do not have an attempted oocyte retrieval or do not undergo embryo transfer. There is not a statistically significant difference in the rate of clinical pregnancy according to the wife's age, at least through age 43. As widely reported, the spontaneous abortion rate per clinical pregnancy progressively increased as women became older, with the exception that the second highest abortion rate was in the less than age 28 category. Overall, the higher abortion rate per clinical pregnancy resulted in the older group of women (ages 40-43) having a live birth rate per oocyte retrieval of 8.4%, compared to 12.5% in the 28-31 age group. However, even though there was a trend to lower live birth rate with increasing age, these differences are not statistically significant.

Somewhat surprisingly, the highest rate of clinical pregnancy establishment with GIFT

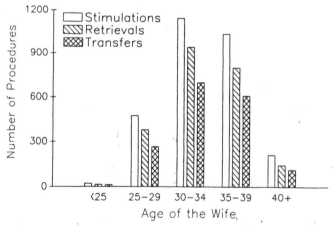

Fig. 3. Shown are the 1987 IVF/ET Registry data for the number of stimulation cycles, oocyte retrievals, and embryo transfers performed, analyzed by the age of the wife.

Table 3. IVF Anomalies Reported to the U.S. IVF/ET Registry

	1985	1986	1987
Chromosomal	1	3	8*
Other	2	9	11**
Liveborn Infants	257	311	1260

*4 induced abortions, 2 spontaneous abortions, 1 stillbirth .

**2 induced abortions, 2 neonatal deaths

was in women over the age of 39. There was also a much higher abortion rate in this group but the live birth rate per oocyte retrieval procedure was still highest in the over 39 group.

Congenital Abnormalities

The report compiled by Markku Seppala in 1984[1] provided the first firm evidence that IVF procedures did not result in increased rates of chromosomal and other anomalies. The published data from the IVF Registry[3,4] further substantiates that fact. As shown in Table 3, the number of IVF infants with chromosomal anomalies does not exceed the rate of six per thousand in live born infants resulting from natural conception. The rate of other abnormalities does not exceed the 1% to 5% rates usually quoted for naturally conceived, live born infants. Thus, the data continue to show that IVF does not result in an increase in congenital abnormalities.

Impact of Clinic Size on IVF Success

As expected, there is a direct relationship between the number of live births achieved and the number of oocyte recoveries performed, with the larger clinics producing the most live births. However, when the percentage of live births per oocyte retrieval procedure is compared to the number of oocyte recoveries (see Figure 4), the data are quite interesting. For the programs that performed at least 110 oocyte recoveries in 1987, the minimum live birth per oocyte retrieval rate was 7%. However, for the smaller programs (those doing fewer than 110 oocyte retrievals that year), there was a wide variation with live birth rates ranging from zero to almost 36%. The smallest programs, i.e. those doing less than 25 oocyte retrieval procedures per year were the ones most likely to have produced no live births.

One of the criteria for membership in SART is the performance of at least 40 IVF stimulation cycles per year. It was felt that, under normal circumstances, this was the minimum number of cycles per year that would allow a clinic to maintain the necessary expertise, especially in the embryo laboratory. When IVF data from the Wyden survey are examined, there were 72 clinics that initiated at least 40 IVF stimulation cycles in 1987. Seventy of these clinics (97.2%) reported at least one live birth in 1987. Of the 70 clinics with live births, the minimum live birth per oocyte collection procedure rate was 1.5% and the maximum rate was 25.3%. Three clinics reported rates over 20% and 16 reported rates over 15%. Conversely, nine programs had rates less than 5% and 20 had rates less than 8% (excluding the two clinics with no live births). The remaining 34 clinics had rates between 8% and 15%. These data are illustrated in Figure 5.

Live Birth Rates by Year

In Figure 6, the live births from IVF and GIFT in the United States reported to the IVF/ET Registry in 1985, 1986, and 1987, and reported to the Wyden Subcommittee for 1987 and 1988 are illustrated. As can be seen from the figure, there is excellent correlation of the data reported to both the Registry and Wyden Subcommittee for 1987. From 1985 through 1988, there has been a progressive, almost linear, increase in the number of live births for both IVF and GIFT. Similarly, Figure 7 shows the live birth rate for both IVF and GIFT per oocyte retrieval procedure. For IVF, there has been a progressive increase from 8.9% in 1985 to a projected 12.7% in 1988. Primarily due to the small number of procedures performed in 1985 and 1986, increasing success with the GIFT procedure is less clearly demonstrated.

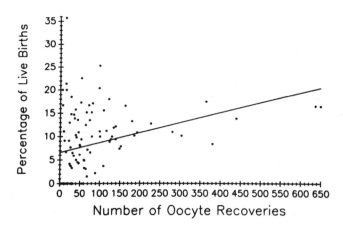

Fig. 4. Illustrated are the live birth percentages per oocyte retrieval, compared to the number of oocyte retrieval procedures performed, for each clinic reporting 1987 IVF data to Representative Wyden's Subcommittee.

DISCUSSION

In response to public concern about the proliferation of IVF programs and the reported difficulty for patients to obtain accurate data, a Subcommittee of the United States House of Representatives commissioned a survey in the fall of 1988 of all United States IVF and GIFT programs. These data were released on March 9, 1989 and provided for the first time a complete breakdown, on an *identified per clinic basis,* of the number of patients treated, stimulation cycles performed, live births, and continuing pregnancies, both for IVF and GIFT. Analysis of this data has shown that the programs that have not been successful to date are, in general, smaller, newer programs.

Overall, IVF and GIFT treatments are becoming more widely available and applied in the United States. This is illustrated by the increase in the number of clinics, stimulation cycles, and resultant live births each year (Figure 6). As the clinics have gained experience, there has been a progressive increase in the IVF success rate (as defined by live births per oocyte retrieval) over the last four years (Figure 7). Thus the increasing number of live births each year is a result of the combination of more cycles as well as a better per cycle success rate.

The available data from the IVF/ET Registry continues to support the hypothesis that IVF has not resulted in an increase in chromosomal or other congenital anomalies.

In summary, IVF and GIFT are now widely available in the United States for treatment of otherwise untreatable infertility. The success rates are progressively increasing with smaller, newer clinics less likely to have produced a live birth. The rate of chromosomal and other abnormalities tend to be within the range expected for natural pregnancies.

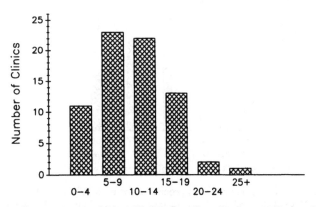

Fig. 5. The frequency distribution of live birth rates per IVF oocyte retrieval procedure for clinics that performed at least 40 IVF stimulation cycles in 1987 are shown. This data is from the Wyden survey.

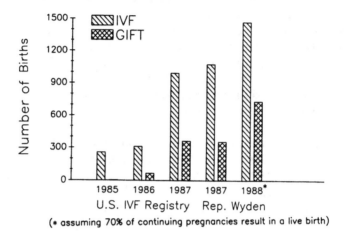

Fig. 6. Illustrated are the number of live births each year for IVF and GIFT, from data reported to the IVF/ET Registry for 1985, 1986, and 1987 as well as to Representative Wyden's Subcommittee for 1987 and 1988.

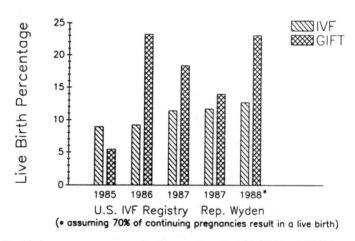

Fig. 7. The live birth rate per oocyte retrieval procedure for both IVF and GIFT as reported to the IVF/ET Registry for 1985, 1986 and 1987 and to Representative Wyden's Subcommittee for 1987 and 1988 are shown.

REFERENCES

1. M. Seppala, The world collaborative report on in vitro fertilization and embryo replacement: current state of the art in January 1984, *Ann. NY. Acad. Sci.*, 442:558 563, (1985).
2. American Fertility Society: Minimal standards for programs of in vitro fertilization, *Fertil. Steril.*, 41:13 (1984).
3. Medical Research International, The American Fertility Society Special Interest Group: In vitro fertilization/embryo transfer in the United States: 1985 and 1986 results from the National IVF/ET Registry, *Fertil. Steril.*, 49:212-215 (1988).
4. Medical Research International, Society of Assisted Reproductive Technology, The American Fertility Society: In vitro fertilization/embryo transfer in the United States: 1987 results from the National IVF-ET Registry, *Fertil. Steril.*, 51:13-19 (1989).

PREGNANCY RATES AND PERINATAL OUTCOME
IN AUSTRALIA AND NEW ZEALAND

Douglas M. Saunders[a] and Paul A.L. Lancaster[b]

[a] Human Reproduction Unit, Department of Obstetrics and Gynaecology
Royal North Shore Hospital of Sydney, St. Leonards, New South Wales 2065, Australia
[b] National Perinatal Statistics Unit, University of Sydney
Sydney, Australia

The National Perinatal Statistics Unit of the University of Sydney, in association with the Fertility Society of Australia, set up a natonal register of in vitro (IVF) pregnancies in 1983 when there was concern that IVF might increase the risk of congenital malformations or chromosomal abnormalities. The results were presented to the IV and V World Congress[1] and covered the first three reports[2-4] for pregnancies resulting from fertilizations between 1979-1985 (1,215 pregnancies). Gamete intra-fallopian transfer (GIFT) pregnancies were too few to be discussed at that time. The fifth and most recent report[5] published in December 1988 from 18 units in Australia and two units in New Zealand covered the 1987 fertilization cohort with deliveries up to September 1988. It included data on 920 IVF pregnancies in 1987 (a total of 3,247 in 1979-1987) and 548 GIFT pregnancies (a total of 816 in 1985-1987). In addition, the numbers and pregnancy rates for selected stages of IVF and GIFT for each unit were listed anonymously; the number of women treated by selected procedures such as embryo cryopreservation were also included. The IVF pregnancies after donor oocytes, sperm and embryos and frozen embryos were also reported.

One of the initial problems relating to pregnancy registers concerns the definition. Theterms "biochemical pregnancy" or "preclinical abortion" have been discussed previously.[6] Such pregnancies have been notified to the register since its inception, but reporting has declined in the last few years, probably because serum pregnancy tests are now not performed routinely in all units. A recent report of the Australian In Vitro Fertilization Colla-

Advances in Assisted Reproductive Technologies, Edited by
S. Mashiach et al., Plenum Press, New York, 1990

borative Group[7] showing live births as 69.0% of clinical pregnancies, but only 57.5% of all pregnancies including the preclinical abortions, stimulated considerable medical comment concerning high multiple pregnancy rates and consequent high perinatal mortality rates,[8] as well as adverse comment[9] about enormous resources being wasted for relatively low success rates. The report[10] of increased incidence of spina bifida and transposition of the great vessels also increased public concern but reinforced the need to obtain accurate data on relatively large numbers of IVF and GIFT births. The report of 103 U.S. and foreign clinics to the American Fertility Society Special Interest Group[11] calculated a "take home baby rate" for IVF in 1986 as 311 live births from 3,504 retrievals or 8.9%[12] and emphasized the importance of "truth in advertising" in order to avoid exploitation of the infertile couple.[13] At the V World Congress of IVF and ET, Howard Jones had commented that success rates appeared to be plateauing world-wide but the "take home baby" rates were still not available. The most recent report from the register of pregnancies in Australia and New Zealand[5] shows another increase in the number of IVF and GIFT pregnancies in 1987. For the first time it also contains data on pregnancy rates for all 20 units. This paper summarises some results of particular interest to referring obstetricians and gynaecologists and to the whole community.

MATERIALS AND METHODS

All pregnancies were reported to the Register on a form similar to the one described previously. As well, each unit was requested to submit data on the number of women treated and the number of cycles of treatment. If available, the numbers starting treatment and then reaching the stages of oocyte retrieval and embryo transfer were obtained. Combinations of IVF and GIFT in the same patient were included in the IVF data. From these summaries, it was then possible to derive pregnancy rates for IVF and GIFT for the various stages of treatment (Table 1). In addition, information was sought about the use of donated oocytes and frozen and thawed embryos (Table 2).

The information supplied was computed using a protected code for each Unit. Each Unit Director was made aware of the Unit's individual code letter, but not the letter of the other Units.

While it it possible to derive separate pregnancy rates for each successive stage of the treatment cycle, we prefer to relate these rates to the intermediate stage of oocyte retrieval when the women are committed to a significant procedure. From the perspective of infertile couples, it is more meaningful to report live birth rather than clinical pregnancies. In calculating live birth pregnancy rates, multiple live births were counted as one pregnancy.

RESULTS

In 1987, there were 6,796 IVF cycles reaching oocyte retrieval for IVF and 2,085 for

Table 1. IVF and GIFT Numbers and Pregnancy Rates for Selected Stages of Treatment, All Units, 1987

	IVF	GIFT	
Number of women starting a cycle of treatment	3612	1304	*
Number of women admitted Number of women admitted for oocyte retrieval:	3921	1240	*
Number of women with transferred embryos/gametes	3432	1223	*
Number of treatment cycles	7733	1954	
Number of treatment cycles			
Number of cycles in which embryos or gamete were transferred	5797	1739	
Number of clinical pregnancies	896	540	
Clinical pregnancies per 100 women admitted for oocyte retrieval	18.5	34.2	*
Clinical pregnancies per 100 oocyte retrieval cycles	13.2	25.9	
Number of pregnancies resulting in live births	643	366	
Live-birth pregnancies per 100 women admitted for occyte retrieval	13.3	23.8	*
Live-birth pregnancies per 100 oocyte retrieval cycles	9.5	17.6	

*   These data are incomplete.   Information from 3 units not available.

GIFT with total live-birth pregnancy rates per 100 retrieval cycles of 9.5 (range 3.0-15.0) for IVF and 17.6 (range 12.6-33.3) for GIFT. The total Australian data for IVF and GIFT is shown in Tables 1 and 2. The total number of certain selected IVF procedures and the numbers of embryos frozen, thawed or discarded is shown in Table 2.

Table 2. Type of Treatment or Procedure on Embryo, All Units, 1987

Women receiving donated oocytes	66
Women receiving donated embryos	4
Women having embryos frozen	903
Number of embryos frozen	2754
Women receiving thawed embryos	448
Number of thawed embryos transferred	1037
Number of embryos discarded	1787

Practical difficulties were experienced in obtaining information about "discarded embryos". Ten units stated that they had discarded embryos, seven did not report any discarded embryos, and three reported that their numbers were not available. At the moment there is some confusion about what is meant by "disposal of embryos" and whether to include, for example, polyspermic embryos, those embryos that are used for research, and surplus embryos that are not cryopreserved. A clearer definition is required for future reports.

Some key issues from the latest report[5] have been selected for special attention in this paper.

IVF data

The total number and outcomes for IVF pregnancies are shown in Figure 1 with an increasing trend in each successive year.

If the 101 preclinical abortions for 1987 were excluded, 69.7% of the 819 clinical pregnancies resulted in live births, similar to 68.9% for the years 1979-1987 combined.

The causes of infertility have not really changed over the years (Table 3), with tubal factors contributing the most patients (46.6%) for IVF.

Table 4 shows the outcome of IVF pregnancies related to the cause of infertility. The tubal ectopic pregnancy rate of 6.0% was significantly ($P < 0.05$ by Chi Square Analysis) higher than the rate of 2.6% for male factor infertility.

The trend towards the collection of more oocytes is shown in Table 5, and the increased number of embryos transferred in Table 6. More than three-quarters of IVF pregnancies in 1987 resulted from three or four embryo transfers.

Table 3. Causes of Infertility: IVF Pregnancies 1979-87 and GIFT Pregnancies 1985-87

	IVF Pregnancies %	GIFT Pregnancies %
Causes of infertility	1979-87	1985-87
Tubal	46.6	6.4
Male factor	7.2	12.5
Endometriosis	6.4	16.2
Other stated causes	2.8	8.1
Multiple causes	23.6	21.7
Unexplained infertility	13.6	35.1
Not stated	- -	- -
Total	100.0	100.0

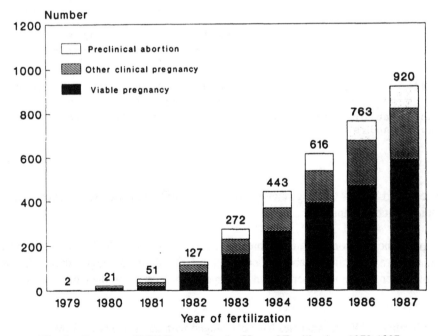

Fig. 1. Outcome of IVF Pregnancies by Year of Fertilization, 1979-1987.

Table 4. Outcome of IVF Pregnancy by Causes of Infertility, All Pregnancies, 1979-87

Outcome of Pregnancy	Causes of Infertility					
	Tubal	Male	Endome-triosis	Mult-iple	Unex-plained	Total
	%	%	%	%	%	%
Preclinical abortion	14.1	16.8	16.4	11.8	13.5	13.8
Termination	0.2	0.9	0.5	0.1	0.2	0.2
Ectopic Pregnancy	6.0*	2.6	4.3	4.6	2.8	4.8
Spontaneous abortion	20.6	19.4	18.8	19.5	19.0	19.7
Stillbirth	1.9	1.3	1.4	2.7	1.8	2.0
Live birth	57.2	59.1	58.5	61.3	62.6	59.3
TOTAL	100.0	100.0	100.0	100.0	100.0	100.0

*Tubal factor significant over male factor ($P<0.05$) by Chi Square Analysis

Tables 7 and 8 show the pregnancies and outcomes following the use of donor gametes or cryopreservation techniques.

The pregnancy outcome related to maternal age is shown in Table 9. For women aged 40 years or more, spontaneous abortion occurred in 40% of clinical pregnancies and, compared with other age groups, there was a corresponding decrease in the percentage of live births.

The plurality of IVF pregnancies has not changed with an 18.9% twin, 3.6% triplet and 0.2% quadruplet rate in 1979-87. The increase in preterm (< 37 weeks) and low birthweight

Table 5. Number of Oocytes Collected by Laparoscopy or Ultrasound Guidance, All IVF Pregnancies, 1979-86 and 1987

Number of oocytes collected	%	
	1979-86	1987
1	3.9	2.3
2	14.1	4.5
3	16.8	12.6
4	19.2	13.7
5	14.4	16.0
6	12.2	14.1
7	6.4	11.0
8 or more Not stated	13.1	25.9
Total	100.0	100.0

Table 6. Number of Embryos Transferred, All IVF Pregnancies

Number of embryos transferred	%		
	1979-85	1986	1987
1	13.0	6.9	6.5
2	27.2	18.9	15.1
3	35.1	33.1	32.0
4	21.3	37.2	44.7
5 or more	3.4	3.9	1.7
Not stated	-	-	-
Total	100.0	100.0	100.0

babies (< 2500G) observed for all pregnancies, including single pregnancies, and published previously[1] has shown little change at 17.2% and 15.9% respectively.

The preterm rate was higher in women aged less than 25 years or 40 years and over (Table 10) and was somewhat less for endometriosis as a cause (Table 11), but these differences were not significant.

Table 7. Number of IVF Pregnancies Following Donor Oocytes, Sperm or Embryos, and Frozen Embryos, 1979-87

Type of Pregnancy	1979-1986	1987	Total
Donor oocytes	12	10	22
Donor Sperm	164	57	221
Donor embryos	2	1	3
Frozen oocytes	4	-	4

Table 8. Use of Donor Gametes, Donor or Frozen Embryos, and Outcome of IVF Pregnancy, 1979-87

Pregnancy Outcome %	Donor Sperm n=221		Donor Oocytes n=22	Donor Embryos n=3	Frozen Embryos n=65
Preclinical abortion	32	14.5	9	-	9
Termination	-	-	-	-	2
Ectopic Pregnancy	4	1.8	5	-	5
Spontaneous abortion	44	19.9	18	33	17
Stillbirth	8	3.6	-	-	3
Live birth	133	60.2	68	67	65
Total	221	100.0	100	100	100

The obstetric outcome for IVF for single and multiple births is shown in Table 12, with increased perinatal mortality rates, as discussed previously.[1]

The incidence of major congenital malformations is shown in Table 13. Spina bifida and transposition of the great arteries occurred more frequently than expected.

GIFT Data

The outcomes of GIFT pregnancies for the period 1985-1987 are shown in Figure 2.

Table 9. Outcome of Clinical Pregnancies in Maternal Age Groups, IVF Pregnancies, 1979-87

Pregnancy Outcome%	Maternal age (years)					
	<25 n=86	25-29 n=887	30-34 n=1442	35-39 n=752	40+ n=70	Total* n=3247
Termination	-	0.2	0.3	0.2	2	0.3
Ectopic Pregnancy	10	5.6	5.2	5.9	7	5.6
Spontaneous Abortion	14	21.2	20.2	29.5	40	22.9
Stillbirth	3	2.6	2.7	1.2	-	2.4
Livebirth	73	70.4	71.6	63.3	52	68.9

* Total includes 10 pregnancies in which maternal age was not stated.

Table 10. Maternal Age and Duration of Single IVF Pregnancies of at least 20 Weeks Gestation, 1979-87

Gestational age - % in cohort weeks	Maternal age (years)					
	<25	25-29	30-34	35-39	40+	Total
20-31	2.6	3.1	3.9	3.2	7.7	3.6
32-36	23.1	12.0	13.4	13.3	30.8	13.6
37 or more	74.4	84.9	82.7	83.5	61.5	82.9
Total	100.0	100.0	100.0	100.0	100.0	100.0
Total <37	25.6	15.1	17.3	16.5	38.5	17.1

Table excludes 25 pregnancies with unstated gestational age and 3 pregnancies with unstated maternal age.

There were 760 clinical pregnancies after GIFT; 508 (66.8%) of these pregnancies resulted in live births. In comparing the causes of infertility for IVF and GIFT pregnancies (see Table 3), unexplained infertility was more common in the GIFT group. Among the small group of 52 pregnancies to women with tubal causes treated by GIFT, the incidence of ectopic pregnancy (19.2%) was significantly higher than for the other causal groups.

Spontaneous abortion rates of 23% for mothers less than 25 compared with 38% for those over 40. The preterm rate (< 37 weeks) was 13.5% for singleton pregnancies and 14.7% was the similar low birth weight (< 2500G) incidence.

Perinatal death rates after GIFT are shown in Table 15. The incidence of major congenital malformations was 3.1% (Table 13). There were five infants with a major urinary tract malformation.

DISCUSSION

With all the IVF and GIFT units in Australia and New Zealand contributing data to the register, it has been possible during the past five years to collect and analyse national results for a relatively large number of pregnancies. More recently, data from these units on numbers of women treated, and on numbers of treatment cycles, have enabled comparison of results in units of varying size and experience. Regular reports from the register have been widely publicised, ensuring that the benefits and risks of IVF and GIFT in treating infertility can be assessed not only by the medical and scientific groups but also by the general community. The spirit of cooperation that has arisen from the voluntary exchange of confidential data between the individual IVF units and the National Perinatal Statistics Unit

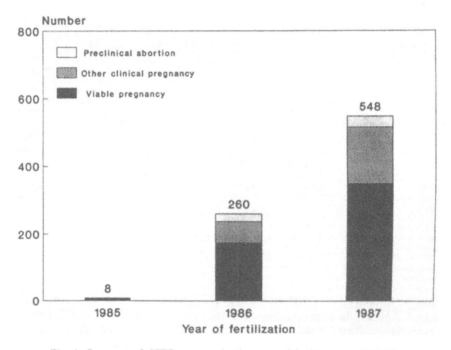

Fig. 2. Outcome of GIFT pregnancies by year of fertilization, 1985-87

Table 11. Causes of Infertility and Duration of Single IVF Pregnancies of at least 20 Weeks Gestation, 1979-87

Gestational age (weeks)		Cause of infertility					
		Tubal	Male	Endome-triosis	Mult-iple	Unex-plained	*Total
Per cent:	20-31	4.3	1.9	3.8	3.1	2.7	3.6
	32-36	13.0	16.2	7.5	15.9	13.1	13.6
	37 or more	82.7	81.9	88.8	80.9	84.2	82.8
	Total	100.0	100.0	100.0	100.0	100.0	100.0
	<37	17.3	18.1	11.3*	19.1	15.8	17.2

Total includes 51 pregnancies with 'other' or 'not stated' causes of infertility.
Table excludes 25 pregnancies with unstated gestational age.
* - Not significant.

Table 12. Outcome of Infants in Single and Multiple IVF Pregnancies of at least 20 Weeks Gestation, 1979-87

Outcome	Single	Multiple	Total
Total births	1545	991	2536
Stillbirth rate/1,000 total births	22.0	47.4	31.9
Neonatal death rate/1,000 live births	8.6	19.1	12.6
Perinatal mortality rate/ 1,000 total births	30.4	65.6	44.2

has had other beneficial effects. The Fertility Society of Australia and the units have developed guidelines for accrediting any group seeking to treat infertility by IVF and GIFT. A Reproductive Technology Accreditation Committee is now implementing those guidelines. Similar collaboration is required to achieve standard definitions for recording accurate information on selected procedures such as disposal of embryos.

Table 13. Major Congenital Malformations in Single and Multiple IVF Pregnancies of at least 20 Week Gestation, 1979-87

Outcome	IVF 1979-1987			GIFT 1985-1987		
	Single*	Mult-iple	Total	Single**	Mult-iple	Total
Total births	1552	991	2543	413	267	680
Congenital malform-ations						
- number	40	16	56	13	8	21
- rate %	2.6	1.6	2.2	3.1	3.0	3.1

```
*   Includes 7 induced abortions for fetal abnormality.
**  Includes 1 induced abortion for fetal abnormality.
```

Table 14. Outcome of Pregnancy by Causes of Infertility, GIFT Pregnancies 1985-87

Pregnancy Outcome %	Causes of Infertility					
	Tubal	Male	Endome-triosis	Mult-iple	Unex-plained	*Total
	n=52	n=101	n=131	n=176	n=284	n=816
Preclinical abortion	15.4	6.9	8.4	4.0	6.0	6.9
Termination	-	-	0.8	1.1	-	0.4
Ectopic Pregnancy	19.2**	1.0	3.1	3.4	4.9	4.8
Spontaneous Abortion	21.2	15.8	16.8	28.4	26.1	23.0
Stillbirth	1.9	2.0	4.6	2.3	2.5	2.7
Live birth	42.3	74.3	66.4	60.8	60.6	62.3
Total	100.0	100.0	100.0	100.0	100.0	100.0

```
*   Total includes 72 pregnancies with 'other' or 'not
    stated' causes of infertility.

**  Tubal factor significant over male factor (P<0.001,
    Fisher's Exact Test).
```

Table 15. Outcome of Infants in Single and Multiple GIFT Pregnancies of at least 20 Week Gestation, 1985-87

Outcome	Single	Multiple	Total
Total births	412	267	679
Stillbirth rate/1,000 total births	29.1	52.4	38.3
Neonatal death rate/1,000 live births	7.5	63.2	29.1
Perinatal mortality rate/ 1,000 total births	36.4	112.4	66.3

If the real "bench mark" of success can be defined as "live birth rates per 100 oocyte retrieval cycles" and agreed upon by all, then assessment of whether or not the rates are rising overall or whether or not they are acceptable when compared between units can begin.

An incidence of 6% for ectopic pregnancies where tubal factors were present, when compared with 2.6% in association with male factors and 0.9% in the community at large[15] suggests that hyperstimulation may have a role in the aetiology of ectopic pregnancy.

The high abortion rate for women over 40 probably justifies refusing women IVF over this age.

The increase in oocytes being obtained over the years for each operative procedure probably reflects the increased use of ultrasound. Although there appears to be more embryos being transferred without significant increases in multiple pregnancy rates, all attempts must be made to avoid grand multiple pregnancies by limiting the number of transfers.

Obstetricians still need to be made aware of the perinatal morbidity and mortality rates which remain unacceptably high. Ectopic pregnancy rates for GIFT pregnancies where tubal factors are present are still too high indicating that GIFT should not be attempted where tubal disease is suspected.

It is to be hoped that the work of the register does not create unwarranted anxiety or criticism of the new technology in the absence of accurate information on the obstetric outcomes and congenital abnormality rates in an identical patient group.

ACKNOWLEDGMENTS

We wish to acknowledge the support given by the 20 collaborating IVF and GIFT Units in Australia and New Zealand by sending their data to the Registry.

REFERENCES

1. D. M. Saunders, M. Mathews and P. A. L. Lancaster, The Australian register, current research and future role in: "In Vitro Fertilization and Other Assisted Reproduction". H. W. Jones and C. Schrader, eds., *Ann. N. Y. Acad. Sci.*, 541:7-21, New York (1988).
2. National Perinatal statistics Unit and Fertility Society of Australia 1984. In Vitro Fertilization Pregnancies, Australia 1980-1083. University of Sydney.
3. National Perinatal Statistics Unit and Fertility Society of Australia 1985. In Vitro Fertilization Pregnancies, Australia and New Zealand 1979–1984, University of Sydney.
4. National Perinatal Statistics Unit and Fertility Society of Australia 1987. In vitro Fertilization Pregnancies, Australia and New Zealand 1979- 1985, University of Sydney.
5. National Perinatal Statistics Unit and Fertility Society of Australia 1988. In Vitro Fertilization Pregnancies, Australia and New Zealand 1987, University of Sydney.
6. H. W. Jones, A. A. Acosta, M. C. Andrews, J. E. Garcia, G. S. Jones, T. Montzavinos, J. McDowell, B. A. Sandow, L. Veeck, T. W. Whibley, C. A. Wilkes and G. L. Wright, What is a pregnancy? A question for programs of in vitro fertilization, *Fertil. Steril.*, 40:728 (1983).
7. Australian In Vitro Fertilization Collaborative Group, In vitro fertilization pregnancies in Australia and New Zealand, 1979-1985, *Med. J. Aust.*, 148:429 (1988).
8. F. J. Stanley, In vitro fertilization - a gift for the infertile or a cycle of despair? Leading article, *Med. J. Aust.*, 148:425 (1988).
9. D. Bartels, High failure rates in in vitro fertilization treatments. Leading article, *Med. J. Aust.*, 147:474 (1987).
10. P. A. L. Lancaster, Congenital malformations after in vitro fertilization. (letter), *Lancet* 2:1392 (1987).
11. Medical Research International, The American Fertility Society Special Interest Group, In vitro fertilization/embryo transfer in the United States: 1985 and 1986 results from the National IVF/ET Registry, *Fertil. Steril.*, 49:212 (1988).
12. C. A. Raymond, IVF Registry notes more centres, more births, slightly improved odds, Medical News and Perspectives, *JAMA* 259:1920 (1988).
13. R. E. Blackwell, B. R. Carr, R. J. Chang, A. H. DeCherney, A. F. Haney, W. R. Keye, R. W. Rebar, J. A. Rock, Z. Rosenwaks, M. M. Seibel and M. R. Soules, Are we exploiting the infertile couple? *Fertil. Steril.*, 48:735 (1987).
14. H. W. Jones, Preface. in: "In Vitro Fertilization and Other Assisted Reproduction", H. W. Jones and C. Schrader, eds., *Ann. N.Y. Sci.* 541:xiii, New York (1988).
15. G. L. Rubin, H. B. Peterson, S. F. Dorfman, P. M. Layde, J. M. Maze, H. W. Ory, and W. Cates, Ectopic pregnancy in the United States, 1970 through 1978, *JAMA*, 249:1725 (1983).

SURVEY OF ASSISTED REPRODUCTIVE TECHNOLOGIES IN ASIA AND AFRICA (UP TO 31st JANUARY 1988)

S. S. Ratnam, S. C. Ng, and L. Hagglund

Department of Obstetrics and Gynaecology
National University of Singapore, Singapore

CONTRIBUTORS

T. Aono, T. A. Bongso, B. N. Chakravarty, D. Chan, S. P. Chang, Y. S. Chang, H. T. Chao, T. Chikasi, S. K. Desai, F. Le R. Fourie, N. Furuhashi, M. D. Hansotia, M. Hiroi, P. C. Ho, H. Hokanishi, K. Hoshi, H. Hoshiai, K. Ida, T. Ide, M. Igarashi, E. Iida, S. Isojima, T. Ito, O. Janizawa, K. Kamei, T. Kasegi, O. Kato, S. Kawakami, S. Kazeto, T. S. Kim, H. Kodama, H. Kubo, S. Y. Kuei, T. Kumasaka, M. S. Lee, T. Z. Lee, M. Leong, C. Leung, G. X. Lu, H. L. Lu, K. Mao, S. Matsuura, K. Mimami, H. Minaguchi, Y. Mio, T. Mori, H. Motoyama, K. Mzyazaki, Y. Nagata, H. T Ng, S. C. Ng, K. Niitsuma, O. Nishi, Y. Nishikawa, T. Ohno, H. Okamura, K. Ono, S. B. Pan, A. Rasad, Y. Sato, K. Sekiba, M. Seta, T. Shimizu, S. Soebijanto, H. Sugie, M. Suzuki, K. Takahashi, S. Takagi, M. Takayanagi, P. P. L. Tam, A. Tanaka, Y. Tomoda, C. R. Tzeng, I. Umera, E. Watanabe, K. L. Woh, S. Yajima, M. Yamada, M. Yamashita, T. Yoshida, L. Z. Zhang.

This is the first survey on Assisted Reproductive Technologies of Asian and African participants. There have been limited surveys of this nature in Asian and African countries, e.g. Suzuki et al. (1985). However, it must be realised that in vitro fertilization (IVF) in Asia (east of Israel) only began in 1982, with the first birth in May 1983 (Ng et al., 1984). Since then, there has been a slow increase in the number of centres in Asia involved in Assisted Reproductive Technologies (ART), with the largest numbers in Japan (see below).

Advances in Assisted Reproductive Technologies, Edited by
S. Mashiach *et al.*, Plenum Press, New York, 1990

1033

Table 1. Procedures Done by the Various Centers (by Countries)

IVF only	GIFT only	TET only	IVF and GIFT
Japan - 10 South Korea - 1	Japan - 14	Japan - 1	Hong Kong - 3 India - 2 Japan - 21 South Korea - 1 South Africa - 1 Taiwan - 4

IVF,GIFT and TET	GIFT+IVF
Indonesia - 1 Japan - 2 Singapore - 1	Taiwan - 2

Table 2. Indications for IVF, GIFT, TET and GIFT + IVF

Indications	IVF		GIFT		TET		GIFT+IVF	
pts %	pts	%	pts	%	pts	%	pts	%
Tubal	4913	61.76	147	7.01	1	8.30	9	7.96
Male problems	1565	19.67	625	29.80	5	41.70	14	12.39
Endometriosis	524	6.59	339	16.17	0	0.00	17	15.04
Idiopathic	641	8.06	916	43.68	3	25.00	68	60.18
Others	312	3.92	70	3.34	3	25.00	5	4.43
TOTAL	7955		2097		12		113	

Four questionnaires were sent out to known centres and to representative obstetrical and gynaecological centres in Asia and Africa. They were on "standard" IVF, Gamete Intra-Fallopian Transfer (GIFT), and Tubal Embryo Transfer (TET, otherwise known as Pronuclear Stage Transfer/PROST); cryopreservation of embryos; cryopreservation of oocytes; and donor embryos and oocytes. A total of 67 centres, mainly from Asia and only one from Africa (Republic of South Africa) replied with data. Two centres (from Taiwan) had a procedure in which GIFT is combined with IVF (presented here as "GIFT + IVF"). The data is

Table 3. Stimulation Regimes for IVF, GIFT, TET and GIFT + IVF

Stimulation	IVF cycles	%	GIFT cycles	%	TET cycles	%	GIFT+IVF cycles	%
Spontaneous	18	0.24	12	0.54	0	0.00	0	0.00
Clomiphene	208	2.77	5	0.23	0	0.00	0	0.00
hMG	69	0.92	27	1.22	0	5.00	7	5.52
FSH	1	0.01	0	0.00	0	0.00	0	0.00
Cld-hCG	108	1.44	11	0.50	0	0.00	0	0.00
Cld-hMG	309	4.12	22	0.99	0	10.71	15	11.81
Cld-hMG-hCG	4546	60.65	1133	51.20	7	53.85	35	27.56
hMG-hCG	1248	16.65	368	16.63	1	7.69	18	14.17
FSH-hCG	65	0.87	15	0.68	0	0.00	0	0.00
hMG-FSH-hCG	776	10.35	557	25.17	5	38.46	52	40.94
GRH/1-FSH-hCG	26	0.35	32	1.45	0	0.00	0	0.00
GRH/1-hMG-hCG	29	0.39	5	0.23	0	0.00	0	0.00
GRH/1-FSH-hMG-hCG	3	0.04	0	0.00	0	0.00	0	0.00
GRH/f-FSH-hCG	0	0.00	0	0.00	0	0.00	0	0.00
GRH/f-hMG-hCG	3	0.04	0	0.00	0	0.00	0	0.00
GRH/f-FSH-hMG-hCG	20	0.27	3	0.14	0	0.00	0	0.00
Others	67	0.89	23	1.04	0	0.00	0	0.00
TOTAL	7496		2213		13		127	

Key: GRH/1 = gonadotropin-releasing hormone agonist started in preceding luteal phase.

GRH/f = gonadotropin-releasing hormone agonist started in follicular phase with gonadotropins.

up to 31st January 1988, except for the Japanese centres (up to 31st December 1987). The number of centres practising the various procedures is listed in Table 1.

STANDARD IVF, GIFT & TET

110 data sets were available for analysis. The mean ages of the patients were from 27.7 to 35.0 years of age. There was a total of 10,177 patients (77.3% with primary subfertility). The indications are given in Table 2. The major indications for IVF were tubal (61.8%) for GIFT, idiopathic (43.7%) for TET, male problems (41.7%), and for GIFT + IVF, idiopathic (60.2%). Indications under "others" included immunological, polycystic

Table 4. Oocytes and Embryos Obtained

	IVF	GIFT	TET	GIFT+IVF
Oocyte recovery (cycles):				
Ultrasonic	2316	294	13	0
Laparoscopic	4543	4198	19	108
	----	----	---	---
Total OR	6859	4492	32	108
# oocytes collected	23998	10818	66	634
# embryos obtained	14142	694(extra)	36	194
# oocytes/embryos transf.	13276	8709	41	177(embryos)
Transfer cycles	5546	1971	11	108
Oocytes/recovery	3.5	2.4	2.1	5.9
Oocytes,Embryos/transfer	2.4	4.5	3.7	1.6

ovarian disease and combined. Interestingly, tubal indications were given for 7.0%, 8.3%, and 8.0% of patients who underwent GIFT, TET and GIFT+IVF respectively.

Stimulation regimes varied widely, as seen in Table 3. There was a total of 9849 cycles reported. For IVF, GIFT and TET, the most common regime was clomiphene with human menopausal gonadotropin (hMG) and human chorionic gonadotropin (hCG), at 60.7%, 51.2% and 53.9% respectively. For GIFT+IVF, the most common regimen was hMG combined with purified follicle-stimulating hormone (FSH) and hCG (40.9%). The use of gonadotrophin-hormone releasing hormone agonist (e.g. Buserelin) is fairly recent; they comprised only 1.1% of IVF cycles and 1.7% of GIFT cycles. "Others" include mainly clomiphene with FSH and hCG.

The number of oocytes and embryos obtained is tabulated in Table 4. The majority of oocytes were recovered by laparoscopy (77.2% of 11491 recoveries). Some GIFT cycles had ultrasonic recoveries (6.5%) and these are probably delayed replacements or transcervical replacements. A total of 11491 oocyte recoveries and 7636 replacements for all procedures were done. The average number of oocytes obtained per recovery was between 5.9 for GIFT+IVF and 2.1 for TET, while the average number of oocytes per replacement was 4.5 for GIFT and the average number of embryos per replacement was 3.6 for TET, 2.4 for IVF and 1.6 for GIFT+IVF. The reason why GIFT had a lower oocyte/recovery rate than

Table 5. Pregnancy Rates

	IVF	GIFT	TET	GIFT+IVF
Biochemical	315	139	0	11
1st trimester abortion	148	100	0	6
2nd trimester abortion	20	4	0	0
Ectopic	31	10	0	1
Vaginal delivery	140	96	0	11
LSCS delivery	175	102	0	7
Ongoing - 1st trim	34	21.	1	2
Ongoing - 2nd trim	36	31	0	7
Ongoing - 3rd trim	26	39	0	3
Clinical pregnancy/OR %	8.89	8.87	3.13	29.13
Clinical pregnancy/repl. %	11.00	20.45	9.10	34.26
Take-home preg/OR %	5.50	5.97	0.00	22.05
Take-home preg/repl. %	6.80	13.60	0.00	25.93

the oocyte/transfer rate is because of a much lower replacement rate (43.9% [1971/4492]) compared with IVF (80.9% [5546/6859]).

The pregnancy rates are shown in Table 5. The total pregnancy rates per recovery and per replacement were 13.52% and 20.35% respectively. The total clinical pregnancy rates per recovery and per replacement were 9.31% and 14.01%, respectively. The total take-home baby rates per recovery and per replacement were 5.87% and 8.83%, respectively. Table 5 breaks down the numbers into IVF, GIFT, TET and GIFT+IVF. As expected, GIFT+IVF resulted in the best pregnancy rates, including take-home baby rates, due to the higher total number of oocytes and embryos replaced per patient. The take-home pregnancy rate per oocyte recovery was 4.0 X better with GIFT+IVF than with IVF. Likewise, the take-home pregnancy rate per replacement was 3.8 X better with GIFT+IVF than with IVF. GIFT was 1.1 and 2.0 X better than IVF respectively. TET had the worst pregnancy rates without any deliveries (the only pregnancy obtained was from a centre in Japan); this is probably due to the early evolution of the technique at that time. There was a total of 531 deliveries from the four programmes.

There were nine reported abnormalities. They were:
a) 1st trimester - 47XX, +8; molar pregnancies (two cases);
b) 2nd trimester - anencephaly; hydrocephalus;
c) 3rd trimester - omphalocele; abnormality of ears; valgus; ventricular septal defect.

Table 6. Cryopreservation of Human Embryos

<u>Freezing cycles</u>:

PROH	153	(50.3%)
DMSO	129	(42.4%)
Ultra-Rapid	7	(2.3%)
Vitrification	0	
Others	15	(5.0%)
	304	

<u>Thaw cycles</u>:

Unstimulated	85	(54.9%)
Clomiphene	68	(43.9%)
hMG	0	
hMG & hCG	1	(0.7%)
hCG	0	
Others	1	(0.7%)
	155	

CRYOPRESERVATION OF EMBRYOS AND OOCYTES

A total of six centres in Asia and the only African centre to respond have a program of freezing embryos. Two hundred and forty-four patients had 304 cycles in which embryos were frozen. The stimulation and thaw cycles are detailed in Table 6. The most common method of freezing was with propanediol (PROH) - 50.3% with dimethylsulphoxide (DMSO) a close second - 42.4%. "Others" used are mainly with glycerol for blastocysts. Thaw cycles were mainly with the unstimulated cycle (54.9%), with clomiphene cycles comprising most of the rest(43.9%). There were 316 embryos thawed, but only 171 were replaced (54.1%). Moreover, there were only seven pregnancies, with four of them being biochemical pregnancies. There were only two deliveries (both from Singapore, with the first delivery in July 1987 [twins]). The remaining pregnancy was an ongoing pregnancy at 1st trimester from Hong Kong.

Oocyte freezing is only practised in one centre (in Taiwan), with only one patient donating five oocytes. The freezing method is with DMSO. Two oocytes were fertilized post-thaw, and both were replaced. However, no pregnancy resulted.

Table 7. Donor Oocytes and Embryos

	Oocytes	Embryos
Recipients:		
Ovarian Failure	6	0
Absence of Ovaries	0	0
Inaccessible Ovaries	0	1
Lack of Fertilization	2	4
Others	0	4
No.Donated Oocytes/Embryos	41	20
No.Fresh Embryos Transferred	16	20
No.Frozen-Thawed Embryos Transferred	10	0
No. Replacement Cycles (Fresh Embryos)	7	9
No. Replacement Cycles (Frozen-Thawed)	3	0
LSCS	1	1
Ongoing (2nd trimester)	1	0
Ongoing (3rd trimester)	0	1
Pregnancies (total)	2	2

DONOR EMBRYOS AND OOCYTES

There were oocyte and embryo donation programmes in three and two of the centres respectively (Table 7). There were eight recipients for donor oocytes (six from IVF/GIFT patients, and two from known donors), and nine recipients for donor embryos (all from IVF/GIFT patients). The recipients for donor oocytes were mainly ovarian failure (6/8), while donor embryos were mainly because of fertilization failure (4/9).

There were 41 donor oocytes; after fertilization with the husband's sperm there were 26 resulting embryos (63.4%). Sixteen embryos (16/26, 61.5%) were replaced fresh while the remaining 10 were frozen. The fresh embryos were replaced in seven cycles, while the frozen-thawed embryos were replaced in three cycles. There were two resulting pregnancies, with one delivery by LSCS (from Taiwan) and another ongoing in the second trimester (from Hong Kong) as at the end of January 1988.

There were a total of 20 donor embryos, and all were transferred fresh (in nine cycles). There were also two resulting pregnancies, again with one delivery by LSCS (from China) and another ongoing in the third trimester (from India) as at the end of January 1988.

ACKNOWLEDGEMENT

We are grateful to Mr. Piara Singh and Ms. Jamuna K for data analysis and to Ms. Shanti N for secretarial assistance.

REFERENCES

Ng, S. C. Ratnam, S. S, Law, H. Y., Rauff, M. Wong, P. C., Chia, C. M., Goh, H. H. V., Anandakumar, C., Leong, K. E., Yeoh, S. C., Pregnancy after in vitro fertilization and embryo transfer in Singapore, in: "In Vitro Fertilization, Embryo Transfer and Early Pregnancy," R. F. Harrison, J. Bonnar, W. Thompson, eds., MTP Press, Lancaster, 149-153 (1984).
Suzuki, M., Hoshiai, H., Yohkaichiya, T., Uehara, S., Mori, R., Nagai, H., In vitro fertilization and embryo transfer in Japan. 4th World Conference on IVF, Melbourne. Abstract 1.

ETHICAL, RELIGIOUS AND LEGAL DEBATE ON IVF AND ALTERNATE ASSISTED REPRODUCTION

J.G. Schenker

President of the Israeli Society of Obstetrics and Gynecology
Professor and Chairman of Obstetrics and Gynecology, Hebrew University, Hadassah
Jerusalem, Israel

Based on the views of:

Rabbi S. Goren   IVth Chief Israeli Rabbi - Jewish view
Prof. R. Klimek   Professor of Obstetrics and Gynecology, Cracow, Poland - Roman Catholic view
T. Kotaki   Buddhist priest of the Japanese Esoteric Buddhism - Buddhist view
Sheik Tahir al Tabari Chief Khadi, Haifa, Israel - Islamic view
Prof. P. Shifman   Faculty of Law, Hebrew University of Jerusalem, Israel - legal view

Infertility affects about 8% of the couples in the developed countries and up to 30% in some areas in the developing countries, representing 50 - 90 million people worldwide. Each year, two million infertile couples are diagnosed compared to the estimated 5.9 million new cancer cases or one hundred million new clinical malaria cases.

The question of etiology of infertility is of major importance if therapeutic or preventive measures are to be implemented. The regional distribution of major diagnostic categories of infertility show that in all the developing countries the rate of mechanical infertility is higher than in the developed countries. Sexually transmitted diseases play a major role in the couple's infertility, especially among male partners; 40% of African males and 29% of

Advances in Assisted Reproductive Technologies, Edited by
S. Mashiach *et al.*, Plenum Press, New York, 1990

South American males have a history of sexually transmitted diseases, compared to 13% of Asian males and 5% in males in the developed Western countries.

Professional, public, religious and personal opinions fuel ethical debates concerning the use of reproductive technology. In order to discuss these aspects, several basic issues should be mentioned, such as: the right to procreate, the moral status of the embryo, the parent-child binding, research on infertile patients, and the importance of truth telling.

Religion and medicine have been interrelated since the beginning of human history. The partnership of religion and medicine has not always been either holy or healthy, especially the field of reproduction and its control and therapy of infertility. A recent world-wide survey that was conducted by the Department of Obstetrics and Gynecology, Hadassah University Hospital, Ein Kerem, Jerusalem, Israel, shows that in many countries religion influences the policy and the application of the new techniques applied in infertility therapy.

What is the Attitude of the Various Religions to the Problem of Infertility and its Therapy?

Jewish Attitude: The Jewish attitude towards infertility can be learned from the fact that the first Commandment of God to Adam was "Be fruitful and multiply" (Gen. 1:28). This is expressed in a Talmudic saying from the second century which says: "Any man who had no children is considered as a dead man". This attitude arises from the Bible itself and refers to 1700 years before the Modern Era, from the words of Rachel, who was barren, to her husband Jacob: "Give me children or else I die" (Gen. 30:1).

The medical treatment for infertility is different for men and women. Any treatment for finding the cause of infertility and rectifying it in a woman is permissible. In the case of the male there are opinions limiting the search for infertility and its cure if it involves obtaining the male sperm by any method other than the natural one. On the other hand, there are many who permit such methods for curing a man of infertility. Even in the opinion of those who forbid such methods there are two alternatives for obtaining male sperm without direct ejaculation. It is strictly forbidden to destroy the sexual organs of human males.

Christian view: According to the Christian view there is no absolute right to parenthood. It is very important, but marriage does not have to have children for it to be a valid marriage. A Christian infertile couple should not indulge in behavior that undermines marriage or family, or that is unacceptable, in their efforts to become parents. Christians view parenting not as reproduction but as procreation. The difference relates to the purposes and act of parenting. The reason for procreation is to act as an agent of God - given creativity for the sake of God's purposes as discerned in Jesus Christ. Creating new life on God's behalf (procreation) means that the parents do not own their children as objects and children do not exist for their parents' needs.

The right to live is a fundamental one. It is the root of all rights and privilege a of human being. In the Roman Catholic Church there are principles which are very important and which guided the believers. The first principle is related to the protection of human life from its very beginning—from conception. The second is that procreation is inseparable from the psycho-emotional relation of parents. It means that procreation is not performed by the physician; the physician has to be in the position to help the parents achieve conception, but not the one who is a "baby maker". The third principle is related to the personal norm of human integrity and dignity and it should be taken into consideration in the medical decisions, especially in the field of infertility treatment.

Islamic view: According to the Islamic view, attempts to cure infertility are not only permissible, but even a duty. The Quran, as well as the Old Testament, gives a record of Abraham and Zakariya that to have progeny is a great blessing from God. The pursual of a remedy for infertility is therefore quite legitimate and should not be considered as rebellion against the fate decreed by God. Prophets of God who were childless incessantly asked their Lord to give them children. But the treatment of infertility should by no means trespass outside the legitimacy as ordered by God. To have offspring is considered a great blessing from God and fulfills an instinctive need.

The duty of the physicians is to help the barren couple to achieve successful fertilization, conception and delivery of a baby.

I, (Sheik Tabari) humbly believe that medical science and technology is creating a medical team which goes beyond the rules of God and Humanity. If this continues to be so, our generation will go down in history, should there be any history left, as a generation that rejected, disobeyed and revolted against God's rules as to the very sacred unit of the family. According to Islam, the unit of the family is Holy, composed of mother, father and child, without the intrusion of any element or body, whether it be uterus, semen or otherwise. Therefore, if we are to aid fertilization by helping parents who cannot, of their own physical composition, help or succeed in the fertilization of the specimen or the ova by certain medical means, anything beyond that is completely forbidden in Islam - or not acceptable. Also, any attempt to help these couples is to be done within the lifespan of the couple.

Buddhist view: (T. Kotaki presents his personal views based on the principles of Buddhism). Traditionally, Buddhism has imposed strict ethics on priests, while it has taken relatively lenient attitudes towards lay people. This means that Buddhist priests permit lay people to do whatever they want, as long as they do not harm others in a concrete way. This leads us to the thought that we do not have to accept infertility as it is. If medical treatment for infertility is available, we should make use of it. Generally speaking, having no children would be a greater threat to a marital relationship than the practice of modern technology for infertility. Historically, we see many societies where people do not respect women who cannot bear a child. Taking this into consideration, it is very difficult for any society to find a rational reason to reject a woman's desire to have a child. Therefore, any technology to achieve this goal, according to Buddhism, is morally acceptable. According to Buddhism treatment should be given to both unmarried as well as married couples.

Sophisticated new technologies for the treatment of infertility, such as IVF, GIFT and other methods, have been developed in the last decade. In addition to these new reproductive technologies, great progress has been made in the treatment of infertility with traditional medical and surgical approaches. A recent survey by the Department of Obstetrics and Gynecology, Hadassah Medical Center, revealed that IVF is practised in 704 centers in 52 countries in the five continents. The data from Asia, Australia, France, Israel, UK and USA, which included about 70% of all the centers practising these methods world-wide revealed the following results:

In 1987 the total number of treatment cycles was 51,360; the number of pregnancies achieved 6,130; the number of live births 3,908. The pregnancy rate per treatment is around 11%, while the live birth rate per treatment is about 8%.

What is the Attitude of the Different Religions to the Practice of These New Technologies?

Jewish view: Jewish law has a positive attitude to IVF and ET on condition that the egg and the sperm are from the married couple. In such a case it is permissible to take the husband's sperm in order to artificially inseminate the egg that was taken from his wife, and to reimplant it into her uterus. Children born from such fertilization are completely accepted according to Jewish Halachic law even for the Rabbinate. GIFT is also permitted, if the oocyte and the sperm are from the couple.

Christian view: The Vatican's statement on IVF is very clear. It does not accept IVF as a method for procreation, but accepts all babies born as a result of such a procedure. The arguments of the Roman Catholic church against the practice of IVF are:

    1. IVF involves disregard for human life
    2. IVF separates human procreation from human sexual intercourse

In 1959 Pope Pius XII declared that attempts of artificial human fecundation in vitro must be rejected as immoral and absolutely unlawful.

Islamic view: The procedure of IVF and ET is acceptable by Islam. It can be however be practised only if it solely involves husband and wife and if it is performed during the span of their marriage. According to the Islamic view the fusion of sperm and egg, a step further of the sexual act should take place only within the legal marriage contract.

Buddhist view: The Buddhist organizations in different countries do not have an authorized view on the new reproductive technology. IVF has been practised in Japan since 1982 and is also practised in other countries with a Buddhist population.

According to Buddhism there are three necessary factors for rebirth of a human being: the ovum, the sperm and Karma energy. This Karma energy is set forth by the dying individual at the moment of his or her death. The problem that arises is whether oocyte and sperm have to be in their natural environment or not. If the crucial factor in the formation of a human being is the Karmic energy, it is not required for sperm and ovum to be in their natural environment, and therefore IVF should be acceptable. On the other hand, if these three entities have to be in their natural environment, the new technology of reproduction becomes effectively unacceptable. The practice of IVF raises another dilemma from the Buddhistic point of view, since it is involved in procreation of more pre-embryos than the number that is implanted in the uterus.

Donation of Sperm, Gametes and Pre-embryos

Jewish view: According to Jewish Law it is forbidden to inseminate a married woman with sperm other than her husband's. Therefore insemination in vitro of a married woman's oocyte by donor's sperm is forbidden. The main problem is of incest, as the father of the child is an unknown sperm donor. In respect to children born as a result of donor insemination, even if the physician informs the couple of the identity of the sperm donor, there is still the fear that the same donor gave his sperm for insemination of other women who do not know his identity, and there is potential for incest, as is mentioned in the Bible "And the land became full of lewdness" (Lev. 19:29).

Referring to a man "who lay with many women and does not know their identity, or a woman who has lain with many men and does not know who is the father", the outcome of this might be a case of incest and the world will be filled with bastards. Some Jewish authorities have found a solution to this problem by using a non-Jewish donor, thereby preventing incest. With regard to the paternity of the child, whether artificial insemination is done in vitro or in vivo, the man from whom the sperm originated is considered the father of the child, the reason for this being that the owner of the sperm has no partner to his fatherhood or to the process of the egg's fertilization.

In the case of egg donation or embryo donation, the problem that arises is who should be considered the mother, the donor of the oocyte or the one in whose uterus the embryo develops, the one who gives birth. In case one of the women is Jewish and the other is not, the problem will rise about the status of the child, as to whether he is Jewish or not, since according to Jewish Law the religious status of the child is determined by his mother. This interesting subject has an apparent precedent in the literature. According to ancient tradition found in the Talmud and Midrashim, Dinah, the daughter of Leah and Jacob, was first conceived in Rachel's womb, and Joseph, the son of Jacob and Rachel was first conceived in Leah's womb, but in the end they were exchanged so that the male embryo which was in Leah's womb was removed and implanted in that of Rachel and the female embryo was removed from Rachel's womb and implanted in Leah's. The result was that Leah gave birth to a daughter called Dinah and Rachel gave birth to a son called Joseph. This description is based on the Talmud Tactate Berachot 60:A. Through the Bible Dinah is considered the daughter of Leah, and Joseph the son of Rachel. Contrary to the situation after artificial insemination where the father of the child is the sperm donor, with regard to motherhood in case of ovum or embryo donation, there is divisible partnership—ownership of the egg and the environment in which the embryo is conceived. Jewish law states that the child is related to the one who finished its formation—the one who gave birth. A judgment is found in the Bible that states that a person who starts an action but does not complete it and another person comes along and completes it, the one who completed the action is considered to have done all of it (Sota 13:8). But in spite of the sources mentioned above, it is very difficult to rely on them in giving a final determination on the maternity and Jewishness of the child.

Christian view: Through IVF, ET and AID human conception is achieved through the fusion of gametes of at least one donor other than the spouses who are united in marriage, therefore according to the Roman Catholic church and even other Christian churches, AID is forbidden. According to the Roman Catholic church AID is contrary to the unity of marriage, to the dignity of the spouses, to the vocation proper to parents and to the child's right to be conceived and be brought into the world in marriage and for marriage (according to the declaration of Pope Pius XII - 1949). Similar to the prohibition of sperm donation, oocyte donation and embryo donation are forbidden.

Islamic view: The practice of AID is strictly condemned by Islamic law. If the husband's infertility is beyond cure, the infertility should be accepted; according to Islamic law AID is considered adulterous. It enhances the chances of inadvertent brother-sister marriages in a community and it violates the legal system of inheritance. The procedure also entails the lie of registering the offspring of a man who is not the real father, and therefore leads to confusion of lines of geneology, the purity of which is of prime importance to Islam.

Ovum donation is similar to sperm donation in that it involves intervention of a third party other than the husband and wife, and it would not be permitted in Islam. Donation of embryos is also prohibited by Islamic law. It should be known that adoption of a child is prohibited by Islam.

Buddhist view: According to Buddhism donation of sperm and oocytes and even embryo is not prohibited, but it is suggested to refrain from these procedures as much as possible. The reasons for this are as follows: the parents in general may experience difficulties in taking care of a child who does not have their own genes, and especially when a malformed child is born by means of gamete or embryo donation; there is a danger that a donation of sperm or ovum from a third party would involve commercialization and create problems in our society; donation of gametes and pre-embryo would lead to eugenics, which may be reflected by strong social discrimination.

A Buddhist Sutra says that we must not seek what we do not have, meaning that we should exclude a third person's intervention in the process of reproduction, because once the donation is accepted, there will be many social problems to be settled regarding the process of reproduction. Unfortunately, we human beings do not have wisdom enough to overcome them. Therefore donation should be limited. However, a child who was already born by means of AID should be accepted as a legitimate child of the social father who has given consent to this practice. But they also have a right to know their genetic father when they reach maturity. And therefore the child's right to know his biological parents should be accepted.

Surrogate Motherhood

Surrogacy is practised in only six countries over the world. In several countries there is even a legislation and governmental regulation which prohibits the use of this technology as a cure for infertility.

Jewish view: The Jewish religion does not forbid the practice of surrogate motherhood. According to the Jewish law, if partial surrogacy is practised - a strange woman is inseminated by the sperm of a man and she completes the pregnancy by agreement, the child so born should be handed over to the owner of the sperm. In case of full surrogacy, when the embryo is transplanted to another woman, the question is not resolved, as was discussed in the case of ovum donation. From the religious point of view the child will belong to the father who gave the sperm, but to the mother who gave birth.

Christian view: The practise of surrogate motherhood is not accepted by the Christian religion, i.e. Roman Catholic, Protestant or Anglican. The objection is on the basis that surrogate motherhood is contrary to the unity of marriage and to the dignity of the procreation of the human person.

Islamic view: Surrogacy is not acceptable to Islam on the premise that pregnancy should be the fruit of a legitimate marriage. If surrogacy would be practised by the Moslems the child delivered would belong to the woman who carried it and gave birth to it, since in the Quran it is mentioned that our mothers are those women who provide the womb and give birth.

Buddhist view: There is no prohibition in Buddhism to the practise of surrogacy, but Mr. Kotaki is against this practise because of family ties, and legal and moral complications.

Cryopreservation of Pre-embryos

Cryopreservation of pre-embryos is at present nearly routinely practised in IVF programs. Even though pregnancy and live birth rates following this procedure are still very low, cryopreservation solves the problems of spare embryos and increases the total results of IVF and ET.

Jewish view: The freezing of the embryo raises the basic question of whether cryopreservation, which stops the development and growth of embryo, does not cancel all the rights of the pre-embryo's father. With regard to the mother, the problem is simplified since the embryo is transferred later into her uterus and will renew the mother-embryo relationship. If we declare that maternity is decided by completion of the development of the embryo into a fetus and neonate, the woman into whose uterus the embryo is transferred for implantation therefore has the right to motherhood. With regard to the relationship of the father whose main function is of fertilization of the oocyte in order to form the pre-embryo, the period of freezing may cause severing of the relationship between the child and his father. Freezing of sperm and pre-embryo can be permitted only when all the measures are taken to ensure that the father's identity will not be lost.

Roman Catholic view: "The freezing of embryos, even when carried out in order to preserve the life of an embryo, constitutes an offence against the respect due to human beings, by exposing them to great risks of death or harm to their physical integrity and depriving them, at least temporarily, of maternal shelter and gestation, thus placing them in a situation in which further offences and manipulations are possible". (Congregation for the Doctrine of Faith, Vatican, 1987).

Islamic view: According to Islamic law freezing of pre-embryo is acceptable.

Buddhist view: Buddhism accepts cryopreservation of pre-embryo due to the following reasons:

1. It saves some of the human embryos, which otherwise would be lost, for future use.
2. The patient's burden is reduced psychologically, financially, as well as physically.
3. The chances of pregnancy are improved by increasing the opportunity of the number of embryos transferred to the uterus per ovum collection.
4. The freezing of embryos should be accepted only during the period in which the ovum donor is in the reproductive age and should not be accepted after the death of one of the parents or the loss of the reproductive ability of the oocyte donor.

Experimentation and Research on Pre-embryos

Jewish view: According to Jewish law the embryo does not have full human standing. It is a person in fact but not in actuality. The embryo has no rights to purchase assets or property, and in the words of Maimonides "No man can transfer to whom has not come into the world". And the embryo is like someone who has not come into the world. In any case, it is forbidden to kill an embryo, and he who does so is judged almost as a murderer, as in the words of Maimonides "A son of Noah who kills a soul even an embryo in its mother's womb will be killed" (Kings 9:4). From this we learn that the embryo has rights of its own which cannot be wronged. Thus any action on the embryo which harms its development, growth and natural birth is forbidden according to Jewish law, except in the case where the embryo endangers the mother's life. In this case Jewish law states that the mother's life comes before that of an embryo whose head or most of its body has not come out into the light of the world.

Christian view: Christianity recognizes the pre-embryo as a human being from the stage of conception. No moral distinction is considered between zygotes, pre-embryos or fetuses. According to this statement any research on pre-embryos is not acceptable.

Islamic view: Islam places a very high value on ilm knowledge, but according to its view the ilm has boundaries. Some of the Islamic schools may accept performing research on excess embryos resulting from infertility therapy in order to increase our knowledge, which may be applied for the benefit of humanity. But creation of embryos for the purpose of research is unacceptable. To create a life in order to end it is to rival the actions that belong only to God.

Sheik Tabari: Islam places strong restrictions against any dealing with human life, starting from the first week, from the first feeling that life exists in the womb. Man has no right to affect this newly created life. Only in cases where a physician certifies that the mother's life is in danger, is it justified to put an end to the life of an embryo. Therefore, experimentation on human embryos may be permissible only in cases where the mother's life is in danger.

Buddhist view: The experimentation on pre-embryos has both negative and positive aspects. One of the negative aspects is that experimentation is in danger of undermining a sense of morality. On the other hand, the positive aspect of the research is that it has a potential possibility to contribute to our knowledge and improve the therapy of infertility and prevent genetic aberrations. Comparing the two aspects, it seems that the positive one outweighs the negative one. The experimentation should be performed on spare embryo; limited time 14 days.

There is an ongoing debate in society, especially among members of the medical profession as to the necessity of judicial regulations and public control of assisted reproductive technology. At present, most countries have not established legislation pertaining to the practice of this new technology, including IVF, ET, gamete and embryo donation, cryopreservation, surrogate motherhood and pre-embryo research. The legislation exists in only five countries; in some countries there is legislation regarding gamete donation, and in many countries proposals for legislation were presented to Parliament but were either not discussed or legislation not established. The reason for this is the fact that the law tends to lag behind social changes and scientific achievements. In many countries the new technology is practised according to regulations of the Ministry of Health or Social Affairs, in some countries according to statements of ethical committees. At least 100 committee statements have been prepared in different countries and by international bodies regarding the practise of the new technologies of reproduction. In some countries this practice requires licensing or authorization by governmental or professional authorities. Nevertheless, many units over the world are practising these new technologies on their own judgement without supervision of professional or governmental authorities.

Prof. Shifman: Normative regulations of reproductive techniques hold serious legal and moral issues. While medical needs may require that specific procedures are taken prior to the use of such techniques, such as proper screening of potential donors, the most difficult problem is the one which can be phrased as the problem of reproductive rights, namely, is it justified for society to put limitations on ones rights to be a parent by use of modern reproductive technologies?   In Israel there is no legislation but this technology is controlled by Public Health Regulation in In Vitro Fertilization, 1987. This regulation provides that an ovum donation should not be used in order to fertilize a single woman. IVF can be carried out with unmarried women only if they could use their own eggs. It should be noted, however, that the legal validity of these regulations is under serious doubt. Being a part of administrative legislations they may be treated as null and void if they exceed the competence of the legislator—ultra virus in the language of lawyers.

Thus the crucial issue is whether the law recognises the fundamental right to be a parent even by the use of IVF, AID or surrogacy, limitation of which is subject to strict scrutiny by the courts. This is not an easy question. As far as natural procreation is concerned, the tendency of many legal systems, generally speaking, is to refrain from intervention, at least direct intervention in which the individual's personal freedom and integrity of its body can be affected. In any case the right of privacy included the right of the individual, married or single, to be free from unwanted governmental intrusion into matters so fundamentally affecting a person and the decision whether to bear or beget a child. Natural procreation is totally different from adoption of children. In adoption, society is committed to the best interest of the child who has already been born, not to mention the simple economic facts of shortage of babies for adoption. In one sense, reproductive techniques are, in general, similar and close to natural procreation. Prima facie, ones right to privacy is affected by any limitation put by society on its reproductive rights. It is true, however, that modern reproductive techniques require involvement of third parties, namely physicians, in addition to the couple itself.

At first glance an argument could be made that if an individual expects positive support from society, it is only natural that society has a right to define the limits of the proper support it is ready to give the person. Yet, I am not sure that the use of IVF, for example, can be treated simply as a positive support given by society. In what does it differ from any other medical services which are given under the medical and financial usual terms? Is infertility something fundamentally different from any other medical defects which should be equally treated legally or medically? If surrogacy is requested only as a matter of convenience, for instance by a career woman who wants to avoid her own pregnancy, asking for a surrogate, one might perhaps claim that it is an abuse of medical services and in such a case, medical assistance should not be given.

But why deny other people who may remain infertile and childless if their access to IVF should be blocked ? It should be emphasised that the social price paid by society in reproductive techniques is not only a matter of access to modern technology. If an ovum or sperm are taken from a third party donor, the party's expectations are that parenthood should be based upon social relationships, not upon genetic facts. The social parents are expected to replace the biological parents. In other words, in order to satisfy the needs of infertile couples, society is supposed to deviate from the usual legal definition of parenthood which is rooted in biological terms.

It is therefore not surprising that religious systems are reluctant to approve the use of reproductive techniques if the use is made with genetic donation of a third party. Religious systems attach a great deal of importance to the natural kinships, a use of an anonymous genetic donation will prevent identification of the child's real biological parents. Such an identification is vital especially in order to avoid incest.

It should be noted that having full information on the child's genetic parents is not only a religious concern. For instance, the Swedish law of 1985 concerning the practice of AID. This law provides that the woman's spouse, who gave his consent to the performance of AID, and not the third party, is treated, by law, as the legal father to the child born after the use of AID. On the other hand, the law provides that the child, when it reaches the age of 18 years may trace his origins, and get information to the identity of his genetic father— namely the sperm donor. The new Swedish law has not resulted in a drastic decline of semen donors, since the law protects the potential donor releasing him from any parental responsibility and placing it upon the consenting spouse. The Swedish law is an example to a balance between two conflicting interests - the interest to satisfy the needs of a childless couple on the one hand, and the will to preserve the lineage of the child by keeping on record the identity of the genetic parents. A similar solution has been adopted by the Israeli legislator in the law of adoption. Yet, in the context of reproductive technologies, we witness the opposite phenomena, namely efforts made by physicians and others to extinguish any trace of donors' identity which stems from the fact that Israeli law lacks clear cut solutions as to the legal status of a child born as a result of the modern reproductive techniques.

The failure of the Israeli Parliament to legislate on these matters is a result of the hostile attitude taken by religious groups toward the use of genetic donations. While their objections did not change the reality of the situation in which reproductive techniques are freely performed, their negative attitudes contribute to the vague legal ruling on the matter.

It is extremely difficult to formulate a clear policy approved by law, and as a result, the administrative regulations made by the Ministry of Health place unjustified obstacles in the way of infertile couples, who could otherwise benefit from techniques such as IVF. It seems, however, that the legal validity of the above regulations lies in serious doubt, and any attempt to limit the use of IVF on moral or religious grounds, exceeds the competence of administrative legislation which could be based only upon medical considerations. Moreover, these limitations do not pay sufficient attention to the consideration of the prohibition of certain test-tube conceptions, for instance by the use of the surrogate mother, which denies the couple the opportunity to procreate, and might precipitate a serious challenge to marital harmony and to the peace between the spouses. Even from a religious point of view, one might claim that in the light of the Commandment "be fruitful and multiply", the use of a surrogate mother, at least if she is a single woman, should be looked upon as a religious duty and from a secular perspective, it seems very strange that one claims on the one hand that a woman has the right to control her own body and accordingly is entitled to terminate a pregnancy, and on the other hand denies her the right to be pregnant. My last point is that all legislation on these matters must begin with a general theory of parenthood. It seems that the starting point should remain the biological parenthood. We should deviate from the biological parenthood only in cases of an anonymous donor who does not actively participate in the child's conception. As a result, an ovum donor in IVF, like a sperm donor in AID should not be treated as a parent. On the other hand, in a conflict between the genetic mother and the gestational mother, when both actively participated in the child's conception, such as the surrogacy situation, it seems that priority should be given to the genetic mother. In this case the rule applies under which parenthood is usually defined in genetic terms, and an exception to the rule which is limited only to anonymous donation who did not take an active involvement in the conception itself.

PUBLISH OR PERISH—AN EDITOR'S PERSPECTIVE

Roger D. Kempers

Editor-in-Chief, FERTILITY AND STERILITY
Professor of Obstetrics and Gynecology
Mayo Clinic and Mayo Medical School, Rochester, MN 55905

INTRODUCTION

If there is one truth about authors, it is that every author wants to be published. However, the motivation for writing and publishing varies. Some authors are driven to write by vanity and some simply because they have something they want to tell. Others are driven by an intense career-dependent pressure. Academic promotion and access to competitive funding is often based on publications. Publication records often determine who will teach and who has access to research funds; and, consequently, research scientists find it necessary to publish early and often.

GOALS AND STANDARDS OF JOURNALS

A consequence of this need to publish is the generation each year of enormous quantities of scientific manuscripts submitted to journals for consideration for publication. It is little wonder that there is such a proliferation of biomedical publications, estimated today at nearly 20,000. Journals tend to focus on specific areas of interest, and their quality and prestige are the results of carefully determined and executed goals and standards.[1] These are clearly defined for *Fertility and Sterility*. It is ultimately the editor's responsibility to see that these are implemented and maintained. Among our goals are the following:

Advances in Assisted Reproductive Technologies, Edited by
S. Mashiach *et al.*, Plenum Press, New York, 1990

1. To publish definitive, timely, peer reviewed scientific and clinical studies in reproductive endocrinology.
2. To review and publish these studies promptly.
3. To produce a journal which is both educational and enjoyable to read.
4. To stimulate responsible debate on controversial issues.
5. To inform the readers on political, ethical, legal and socio-economic issues related to reproductive endocrinology.
6. To improve medical care by reporting the most recent advances in research and clinical experience.

EDITOR'S CONTROL OF DIRECTION OF THE JOURNAL

Journal editors exert considerable control over the direction of their journal and have much to do with content and ensuring editorial quality. Because of this, they have considerable control over the directions of research. Journal editors frequently are appointed because of their scientific credentials and not because they have received special training or experience in editing. Editing can be an extremely demanding occupation with a variety of responsibilities which include ultimate selection of acceptable manuscripts, supervising the financial aspects of the journal, assisting in establishing scientific policy, and in general influencing the education of their readers. In the midst of these responsibilities, editors have come to the realization that they also must play a central role in monitoring for and in dealing with fraud in science and scientific publishing.[2]

AUTHORS' EXPECTATIONS OF EDITORS

To maintain an efficient editorial office, editors should cultivate and maintain a smooth working relationship with authors. This, in turn, will enhance submission of the highest quality of manuscripts. Authors themselves have justifiable expectations of their editors, and the ability to satisfy these expectations has much to do with the continued submission of the best manuscripts and the success of a journal. What exactly do authors expect of editors? A partial list includes:[3]

1. Prompt notification of receipt of a manuscript or notification of delays in the review process.
2. Prompt decisions concerning acceptance, rejection, or need for revision.
3. Critiques which are thoughtful, sensible, and show evidence of a good understanding of the study.
4. Prompt publication following acceptance.
5. A final product with high-quality typography and reproduction of illustrations.

EDITORS' EXPECTATIONS OF AUTHORS

Journal editors, on the other hand, have very definite expectations of authors. Some of these are:

1. Careful observation of directions outlined for format in the instructions for authors.
2. Clarity and brevity in writing.
3. Clear statements of overall objectives of the research project, a well planned study design, careful statistical analysis, and sound conclusions avoiding speculation.
4. Accuracy of citations both in referencing within the text and in the recording of the bibliography.
5. High ethical standards including appropriate approval by Institutional Review Boards and limiting the number of authors only to those actively participating in writing the manuscript.
6. Honesty and integrity in avoidance of overlapping publications, duplicate publications, and plagiarism.

ROLE OF PEER REVIEW SYSTEM

Almost all authors hope to achieve acceptance of their manuscript by the most prestigious or visible journal in their field; however, acceptance is seldom guaranteed, and rejection is always a distinct possibility. We shall look at some of the factors which are relevant to acceptance and rejection later; but if a manuscript is rejected, what can be said about the care and treatment of the disappointed author? What are the editor's obligations to the author of a manuscript? I believe that when a manuscript is rejected, it should be accompanied by a courteous letter with sound editorial comments to support this decision. The philosophy behind rejection varies from one editor to another, and occasionally a rejection letter does not signify an irreversible decision. Opportunity may still be afforded for a continued sensitive dialogue between the editors, the reviewers, and the authors. I disagree with the view that the peer review system is only judgmental and believe that it also provides an educational experience. Editorial boards can make fallacious judgments, and it is not unusual to find that the ultimate disposition of the paper stimulates the authors to extensively revise their manuscript and again seek acceptance either in the original journal or another. Sensitive editors accept telephone calls and letters from authors asking for further identification of strengths and weaknesses of their manuscript and assist the authors in what can be a valuable learning experience culminating in a manuscript which merits acceptance. Authors should not be reluctant to communicate with the editor when they disagree with the reviewers' comments. Referees are not omniscient, and errors in judgment by referees can deny the publication of a potentially important report. Editors must be aware of the possibility of these serious errors; and by keeping open the lines of communication, afford an otherwise lost opportunity to recognize a significant contribution.

Although the pros and cons of the peer review system have been much debated, there is little doubt that peer review legitimizes an article. Few top-ranked journals have survived without a system of peer review. Because the drive to publish is so strong and because so many manuscripts are generated each year, it is only through the use of the peer review process that some control can be exerted over the proliferation of manuscripts and journals.

While the validity of the peer review system is taken for granted, few studies have addressed this subject. What is really known about bias, its political realities, its impact on research, the consequences of wrong decisions, and the philosophical weaknesses of the system? In May 1989, the American Medical Association is sponsoring the first International Congress on Peer Review in Biomedical Publications, and many of these issues will be debated. Perhaps some form of quality assurance of the peer review system will eventually evolve. Nevertheless, as the system now is utilized, I believe that the process assures that the best is published in the best place.

Reviewers provide guidance by giving either general or specific comments. Their fundamental analysis of a manuscript includes evaluation of its scientific validity by evaluating the study design, the analysis, the interpretation of the data, its originality, its relevance, and its importance to the field. Reviewers are selected for a variety of reasons such as research expertise, knowledge within a given field, dependability in providing unbiased reviews, promptness in generating the reviews, or because of an already demonstrated competence in providing thorough, constructive evaluation. A manuscript submitted to *Fertility and Sterility* is reviewed by a minimum of two referees; and often, a third reviewer must be utilized. Occasionally, even a fourth reviewer is sought. This occurs because occasionally disparate recommendations are generated. One reviewer may possess more competence than another in the same field and may be looking at different parts of the elephant. Reviewers may have different opinions regarding stylistic aspects of the manuscripts or its overall value or appropriateness for that journal. Disagreement is often legitimate, and often the third, usually a seasoned referee, having the benefit of the two previous reviews, can generate a final sound analysis. The ultimate decision, however, is made by the editor. Referees remain anonymous to the authors; however, it is our policy to provide referees of a particular article with the comments generated by the other reviewers. This allows the referees an opportunity to see another view of its strengths or weaknesses. This also serves as an educational experience for the reviewers. Most reviewers appreciate this courtesy by the editorial staff, and it creates interest and willingness in providing future reviews. Editors maintain files on referee performance noting their areas of expertise, depth of reviews, and promptness. The interface between the editors and the reviewers should be respectful, courteous, and efficient to ensure the vitality and success of the journal. The long hours spent in reviewing are usually a labor of love, and referees appreciate recognition or a special note of thanks or a telephone call. Reviewers' comments vary in depth and style. Experienced referees are usually diplomatic, attempt to be nonabrasive, provide positive as well as negative comments, and are cognizant of the sensitivity with which authors view their critiques. Editors often reword the comments or edit them extensively for appropriateness and for softening of the dialogue. Authors achieve secondary benefit from studying the reviews for they then have models to follow when they, in turn, are called upon to serve as referees.

ROLE OF VALID STATISTICAL ANALYSIS

Important in establishing the validity of an article is its ability to withstand the scrutiny of sound statistical analysis. The ultimate outcome of peer review often hinges on the biomedical researchers' understanding of statistical design inclusion methods. There are seven statistical questions which every biomedical research paper should address.[4]

"1. What is the objective of the research project?
2. What are the independent variables and dependent variables in the study?
3. What is the defined population of interest?
4. Was a repeated measure or block design used?
5. How was the sample chosen? Was it a true random sample? What are the biases present in the study?
6. What randomization procedures were used to assign subjects in their experimental treatment?
7. How was the sample size chosen?"

The first statistical question should be dealt with at the onset. Each research paper should clearly state its specific objectives. One of the most significant flaws and reasons why papers are rejected is that no research question has been asked. One of the authors has collected some data, something has been observed, but no one asked the question, "What are we trying to ask here?" An example of another common flaw is that of arriving at black and white conclusions drawn on the basis of the numerical value of the achieved level of statistical significance. A P-value of less than 0.05 is generally taken as a definitive conclusion of significance, yet interpretation of P-values either higher or lower than this figure may have important consequences.

More and more journals are establishing requirements for authors regarding appropriate study design and statistical analysis. Because of the importance of appropriate statistical evaluation, there is a growing involvement by statisticians in the scientific review process. Many journals now have salaried statistical consultants. There are not enough of them available to review every manuscript submitted to an editorial office; however, appropriate statistical review should be provided either by the reviewers or by an independent statistician. Many journals such as the *American Journal of Psychiatry*, the *New England Journal of Medicine*, the *Journal of Nutrition*, and the *Annals of Internal Medicine* require that nearly every manuscript published be reviewed for study design and statistical analysis by a statistician. *Fertility and Sterility* frequently requests that authors provide written documentation that a statistician has reviewed and approved the manuscript. In addition, we send many manuscripts to independent statistical consultants. Clearly, this contribution by statisticians is influencing research efforts and is elevating the level of credibility of published reports.

ETHICS IN WRITING AND MONITORING FOR FRAUD

A degree of trust is always associated with the review process. Editors and referees have neither the time nor the expertise to analyze all of the raw data. But, in the midst of our academic ranking system that has come to demand increasing quantities of publications for promotion and funding—publish or perish—editors have come to the realization that one of their responsibilities is monitoring for and responding to fraud in scientific publishing. Dr. Marcia Angell, Deputy Editor of the *New England Journal of Medicine* has had a special interest in this problem. She has described seven categories in the spectrum which ranges from simple deception to overt fraud.[5] While the more benign forms of deception are common, cases of overt fraud are rare. The types of fraud she identified were data fragmentation, loose authorship, duplication, selection of data bias, trimming of data, plagiarism, and fabrication. The common practice of fragmentation in publications consists of publishing a single study in fragments or parts in different journals or even in the same journal. Examples are separate papers publishing increments of a study as they accrue from a single clinical study, or taking a clinical study and publishing the epidemiology in one manuscript and the therapeutic aspects in another. A second benign, but common, form of deception is loose authorship. Often authorship of a scientific contribution is credited not only to the principal one or two writers but also to individuals who are only tangentially responsible such as coworkers in a department or the department chairman. Occasionally it is felt that having a coauthor with an established reputation will enhance the likelihood of acceptance of the publication. A recent study published in the *British Medical Journal* looked at the issue of the increasing number of authors per article.[6] A longitudinal study was made of representative articles during 1955 and 1985 published in *Lancet, Circulation,* and *Fertility and Sterility* and was subjected to careful statistical analysis. In all three journals the titles of the articles had become longer, and the number of authors and references had increased in the thirty year span. Interestingly, the number of words in the titles was significantly correlated with the number of authors. In other words, the longer the title, the more authors. The number of pages per article, however, had changed little. The number of authors had increased from one to two people to four to five people, and the length of the titles had increased by three to four words. In their conclusion, they humorously suggested that it would seem that the active contribution by some participants was limited to the title, which alone bears the thumbprint of all—publish or perish. Doctor Angell's third example of wrongdoing is the practice of duplicate publication. Occasionally, the identical manuscript is submitted simultaneously to journals in different specialties or to journals in different languages. While the title may be altered or the order of authors varied, the substance of the study and the data remain identical.

In the middle of the spectrum of fraud are a variety of maneuvers which some suspect investigators have designed to deal with inconvenient results. Selection of data and presenting only data which are not contradictory or have inexplicable findings creates this example of deception. In the haste to publish, contradictory findings requiring additional experimentation and investigation, which take time, may be buried within the manuscript. The first example here is called trimming. The deception occurs when data which do not conform to the hypothesis is actually altered by changing numbers and findings. Finally, at the extreme end of the spectrum of fraud are plagiarism and fabrication. Recent examples of these types of deception attributed to certain researchers at major medical centers in the United States and involving coauthors at senior levels who had not appropriately substantiated the data generated by their younger colleagues, have made the national press. Again, these examples reflect problems which can derive from the drive to publish at all costs.

What can be done to prevent this? Detection of the more benign forms of deception or fraud is the responsibility of the referees and ultimately the editor. Fragmentation can be minimized by insisting that copies of the authors' related manuscripts already published or in press be submitted along with the manuscript being reviewed. Similar or overlapping publications can then be appropriately scrutinized. Generally, the issue of fragmentation can be handled by asking the authors to combine the two or three manuscripts in question before further evaluation is carried out. Overlapping of publications presents a more difficult problem in diplomacy, but nontheless is obligatory for the editor. Similarly, the problem of loose authorship or too many authors must be dealt with by editors. Most journals limit the total number of authors on a manuscript to six or less. Obviously, exceptions to the rule may arise. It has been my experience that many junior authors, on being informed that their manuscript has too many coauthors, welcome the opportunity provided by outside pressure to drop some who had little to do with writing the report. They are then able to use the acknowledgement section to give appropriate credit.

Occasionally, authors write the editors expressing concern that the peer review system may permit competing investigators an opportunity to see new data being submitted to a journal and use that material unscrupulously. The reviewer has the advantage of seeing the pitfalls of the original study and thus allowing more rapid publication of his or her data. Again—publish or perish. Obviously, this is a method of plagiarism which is difficult for editors to control but can be minimized by insisting that reviews be returned within two to three weeks to expedite publishing of the original report. An editor's trust of his referees is vital to ethics of the publication process.

Editors respond to fraud in a variety of ways. Sanctions are instituted by *Fertility and Sterility* against authors who are guilty of duplicate publication. These authors are prevented from having their work considered for publication for a period of at least two years. Dealing with other types of fraud is often handled on a case-by-case situation. Institutions occasionally are the first to become aware of investigators' invalid studies and should assume some responsibility for alerting journal editors. Nevertheless, as Doctor Angell has stated,[5] "Given the intense pressure to publish, it is remarkable that science has done so well to maintain its ideals of dispassionate honesty and integrity." As editors we must continue to do our part in a cooperative effort to serve these ideals in order to preserve the future of scientific journal editing. The concept of thousands of editors, trained or in training, working in relative isolation is evolving toward more uniform practices. Organizations of editors are dealing with ethical issues such as fraud and duplicate publication and conducting workshops on scientific editing. A number of excellent textbooks on scientific editing are becoming available which are written for and about editors.

COMPUTERIZATION AND ELECTRONIC PUBLISHING

Editorial offices are rapidly turning to computerization, not only as a tool for more efficient managing of day-to-day manuscript processing, but also for journal publishing by telecommunicaton of manuscripts. Some editorial offices now permit manuscript transmission through modems using a central graphic system. While this system is evolving at the present time, most files cannot handle the mathematical notations, and hard copy must also be sent. A new area for publishers is the use of electronic publishing using both on-line and archival computer data bases. Storage and retrieval of information is now technically feasible. Computerized medical information services deliver, via an international telecommunications network, information to subscribers who use personal computers. Electronic publishing provides full text of various textbooks and journals to subscribers. The advantages of a full text data base include rapid access to published research and eliminates the possibility of overlooking relevant data which can occur because of the time consuming, arduous work of searching hard copy journals. Editors, on the other hand, can use communications and automated data bases to study patterns of publication in the literature which allow them to better evaluate submitted data and permit broader assessment of current research. Furthermore, it permits more rapid evaluation of papers and transmission of referees' comments, permits adaptation of more uniform styles of manuscript processing, and brings to the surface examples of ethical concerns such as duplicate publication.

Preliminary data indicates that subscribers to computerized medical information services continue to subscribe to hard copy journals at the same rate as before. Perhaps this is a reflection of the fact that total full impact of full text systems has not yet been felt, since no one data base system is well enough established to provide meaningful data on subscriptions and income to publishers.

My colleagues, we have considered an editor's thoughts on Publish or Perish. Scientific journal editing is an essential academic activity. It is an evolving and maturing field where the problem of ethics is forced on editors. "Editors are not naturally suspicious, and they believe in the integrity of their authors. They acknowledge definite responsibilities to authors as certainly authors have responsibilities to them. They are dedicated to the continuing challenge of publishing scientific research accurately, rapidly, and in the most highly polished form possible."[7]

REFERENCES

1. G. D. Lungberg, Opportunities and limits in medical publishing in the 1980s: "CBE Views", S. R. Geiger, ed., Council of Biology Editors, Bethesda (1985).
2. N. Grossblatt, Scientific fraud and the pressure to publish — the persistent shadow — is it

different today?: "CBE Views", S. R. Geiger, ed., Council of Biology Editors, Bethesda (1986).

3. R. G. E. Murray, What is an editor for?: "CBE Views", S. R. Geiger, ed., Council of Biology Editors, Bethesda (1983).

4. R. G. Marx, E. K. Dawson-Saunders, J. C. Bailar, B. B. Dan, and J. A. Verran, Interactions between statisticians and biomedical journal editors, *Stat. Med.* 7:1003–1011 (1988).

5. M. Angell, Editors and fraud.: "CBE Views", S. R. Geiger, ed., Council of Biology Editors, Bethesda (1983).

6. I. Ben-Shlomo, G. Goodman, A place in the sun? *Br. Med. J.* 297:1631 (1988).

7. R. D. Kempers, Benign deception: fragmentation, overlapping publications, and loose authorship, *Fertil. Steril.* 51:(May) (1989). In press.

INDEX

CPSIA information can be obtained
at www.ICGtesting.com
Printed in the USA
LVHW101549310520
657067LV00008B/711